D1573118

WITHDRAWN

COLD SPRING HARBOR SYMPOSIA
ON QUANTITATIVE BIOLOGY

VOLUME LXIX

www.cshl-symposium.org

Institutions that have purchased the hardcover edition of this book are entitled to online access to the companion Web site at www.cshl-symposium.org. Please contact your institution's library to gain access to the Web site. The site contains the full text articles from the 2004 Symposium and the Symposia held in 1998–2003 as well as archive photographs and selected papers from the 61-year history of the annual Symposium. Institutions requiring assistance with activating their online accounts should contact Kathy Cirone, CSHL Press Subscription Manager, at cironek@cshl.edu.

COLD SPRING HARBOR SYMPOSIA ON QUANTITATIVE BIOLOGY

VOLUME LXIX

Epigenetics

www.cshl-symposium.org

Meeting Organized by Bruce Stillman and David Stewart
COLD SPRING HARBOR LABORATORY PRESS
2004

COLD SPRING HARBOR SYMPOSIA ON QUANTITATIVE BIOLOGY VOLUME LXIX

©2004 by Cold Spring Harbor Laboratory Press
International Standard Book Number 0-87969-729-6 (cloth)
International Standard Book Number 0-87969-731-8 (paper)
International Standard Serial Number 0091-7451
Library of Congress Catalog Card Number 34-8174

Printed in the United States of America
All rights reserved

COLD SPRING HARBOR SYMPOSIA ON QUANTITATIVE BIOLOGY
Founded in 1933 by
REGINALD G. HARRIS
Director of the Biological Laboratory 1924 to 1936
Previous Symposia Volumes

I (1933) Surface Phenomena
II (1934) Aspects of Growth
III (1935) Photochemical Reactions
IV (1936) Excitation Phenomena
V (1937) Internal Secretions
VI (1938) Protein Chemistry
VII (1939) Biological Oxidations
VIII (1940) Permeability and the Nature of Cell Membranes
IX (1941) Genes and Chromosomes: Structure and Organization
X (1942) The Relation of Hormones to Development
XI (1946) Heredity and Variation in Microorganisms
XII (1947) Nucleic Acids and Nucleoproteins
XIII (1948) Biological Applications of Tracer Elements
XIV (1949) Amino Acids and Proteins
XV (1950) Origin and Evolution of Man
XVI (1951) Genes and Mutations
XVII (1952) The Neuron
XVIII (1953) Viruses
XIX (1954) The Mammalian Fetus: Physiological Aspects of Development
XX (1955) Population Genetics: The Nature and Causes of Genetic Variability in Population
XXI (1956) Genetic Mechanisms: Structure and Function
XXII (1957) Population Studies: Animal Ecology and Demography
XXIII (1958) Exchange of Genetic Material: Mechanism and Consequences
XXIV (1959) Genetics and Twentieth Century Darwinism
XXV (1960) Biological Clocks
XXVI (1961) Cellular Regulatory Mechanisms
XXVII (1962) Basic Mechanisms in Animal Virus Biology
XXVIII (1963) Synthesis and Structure of Macromolecules
XXIX (1964) Human Genetics
XXX (1965) Sensory Receptors
XXXI (1966) The Genetic Code
XXXII (1967) Antibodies
XXXIII (1968) Replication of DNA in Microorganisms
XXXIV (1969) The Mechanism of Protein Synthesis
XXXV (1970) Transcription of Genetic Material
XXXVI (1971) Structure and Function of Proteins at the Three-dimensional Level
XXXVII (1972) The Mechanism of Muscle Contraction
XXXVIII (1973) Chromosome Structure and Function
XXXIX (1974) Tumor Viruses
XL (1975) The Synapse
XLI (1976) Origins of Lymphocyte Diversity
XLII (1977) Chromatin
XLIII (1978) DNA: Replication and Recombination
XLIV (1979) Viral Oncogenes
XLV (1980) Movable Genetic Elements
XLVI (1981) Organization of the Cytoplasm
XLVII (1982) Structures of DNA
XLVIII (1983) Molecular Neurobiology
XLIX (1984) Recombination at the DNA Level
L (1985) Molecular Biology of Development
LI (1986) Molecular Biology of *Homo sapiens*
LII (1987) Evolution of Catalytic Function
LIII (1988) Molecular Biology of Signal Transduction
LIV (1989) Immunological Recognition
LV (1990) The Brain
LVI (1991) The Cell Cycle
LVII (1992) The Cell Surface
LVIII (1993) DNA and Chromosomes
LIX (1994) The Molecular Genetics of Cancer
LX (1995) Protein Kinesis: The Dynamics of Protein Trafficking and Stability
LXI (1996) Function & Dysfunction in the Nervous System
LXII (1997) Pattern Formation during Development
LXIII (1998) Mechanisms of Transcription
LXIV (1999) Signaling and Gene Expression in the Immune System
LXV (2000) Biological Responses to DNA Damage
LXVI (2001) The Ribosome
LXVII (2002) The Cardiovascular System
LXVIII (2003) The Genome of *Homo sapiens*

Front Cover (*Paperback*): Identical twins, New York City, 2003. ©Randy Harris.

Authorization to photocopy items for internal or personal use, or the internal or personal use of specific clients, is granted by Cold Spring Harbor Laboratory Press, provided that the appropriate fee is paid directly to the Copyright Clearance Center (CCC). Write or call CCC at 222 Rosewood Drive, Danvers, MA 01923 (508-750-8400) for information about fees and regulations. Prior to photocopying items for educational classroom use, contact CCC at the above address. Additional information on CCC can be obtained at CCC Online at http://www.copyright.com/

All Cold Spring Harbor Laboratory Press publications may be ordered directly from Cold Spring Harbor Laboratory Press, 500 Sunnyside Boulevard, Woodbury, NY 11797-2924. Phone: 1-800-843-4388 in Continental U.S. and Canada. All other locations: (516) 422-4100. FAX: (516) 422-4097. E-mail: cshpress@cshl.edu. For a complete catalog of all Cold Spring Harbor Laboratory Press publications, visit our World Wide Web Site http://www.cshlpress.com/

Web Site Access: Institutions that have purchased the hardcover edition of this book are entitled to online access to the companion Web site at www.cshl-symposium.org. For assistance with activation, please contact Kathy Cirone, CSHL Press Subscription Manager, at cironek@cshl.edu.

Symposium Participants

AAPOLA, ULLA, Institute of Medical Technology, University of Tampere, Tampere, Finland
ADAMS, CHRISTOPHER, Dept. of Genomics, Invitrogen Life Technologies, Carlsbad, California
AIZAWA, YASUNORI, Johns Hopkins University, Baltimore, Maryland
ALLIS, C. DAVID, Lab. of Chromatin Biology, Rockefeller University, New York, New York
ALLISON, ANNE, Dept. of Microbiology, University of Virginia, Charlottesville
ALLSHIRE, ROBIN, Wellcome Trust Centre for Cell Biology, University of Edinburgh, Edinburgh, Scotland, United Kingdom
ALMEIDA, RICARDO, Institute of Cellular and Molecular Biology, University of Edinburgh, Edinburgh, Scotland, United Kingdom
ALMOUZNI, GENEVIEVE, Institut Curie, Unité Mixte de Recherche, Centre National de la Recherche Scientifique, Paris, France
ALTSCHULER, STEVEN, Center for Genomic Research, Harvard University, Cambridge, Massachusetts
AMARIGLIO, NINETTE, Dept. of Pediatric Hematology and Oncology, Sheba Medical Center, Tel Hashomer, Israel
AN, WENFENG, Dept. of Molecular Biology and Genetics, Johns Hopkins University, Baltimore, Maryland
ANTAO, JOSE, Dept. of Molecular Biology, Harvard Medical School, Boston, Massachusetts
ARAI, YOSHIO, Dept. of Radiation Oncology, Veteran's Administration of Pittsburgh, University of Pittsburgh, Pittsburgh, Pennsylvania
ARROYAVE, RANDY, Dept. of Biological Sciences, City University of New York, Hunter College, New York, New York
AUGER, ANDREANNE, Centre de recherche de l'Hôtel-Dieu de Québec, Laval University, Québec, Canada
AVITAHL-CURTIS, NICOLE, Models of Disease, Novartis Institutes for Biomedical Research, Cambridge, Massachusetts
AVRAM, DORINA, Center for Cell Biology and Cancer Research, Albany Medical College, Albany, New York
BALCIUNAITE, EGLE, Dept. of Pathology, School of Medicine, New York University, New York, New York
BARLOW, DENISE, Center for Molecular Medicine, Institute of Microbiology and Genetics, University of Vienna, Vienna, Austria
BARTEL, DAVID, Dept. of Biology, Whitehead Institute for Biomedical Research, Massachusetts Institute of Technology, Cambridge, Massachusetts
BARTHOLOMEW, BLAINE, Dept. of Biochemistry, Southern Illinois University School of Medicine, Carbondale, Illinois
BARTOLOMEI, MARISA, Dept. of Cell and Developmental Biology, School of Medicine, University of Pennsylvania, Philadelphia
BATISTA, PEDRO J., Program in Molecular Medicine, Instituto Gulbenkian de Ciência-Portugal and School of Medicine, University of Massachusets, Worcester
BAUER, MATTHEW, Div. of Biological Sciences, University of Missouri, Columbia
BAULCOMBE, DAVID, John Innes Centre, Norwich, United Kingdom
BEAUDET, ARTHUR, Dept. of Molecular and Human Genetics, Baylor College of Medicine, Houston, Texas
BECKER, PETER, Adolf Butenandt Institute, University of Munich, Munich, Germany
BEIER, VERENA, Dept. of Functional Genome Analysis, German Cancer Research Center, Heidelberg, Germany
BELOZEROV, VLADIMIR, Dept. of Biochemistry and Molecular Biology, University of Georgia, Athens
BENDER, JUDITH, Dept. of Biochemistry and Molecular Biology, School of Public Health, Johns Hopkins University, Baltimore, Maryland
BENDER, LAUREL, Dept. of Biology, Indiana University, Bloomington, Indiana
BERGER, SHELLEY, Dept. of Molecular Genetics, Wistar Institute, Philadelpha, Pennsylvania
BERNARDS, RENÉ, Div. of Molecular Carcinogenesis, The Netherlands Cancer Institute, Amsterdam, The Netherlands
BESTOR, TIMOTHY, Dept. of Genetics and Development, Columbia University, New York, New York
BIBIKOVA, MARINA, Dept. of Molecular Biology, Illumina, Inc., San Diego, California
BICKMORE, WENDY, Dept. of Chromosome Biology, Human Genetics Unit, Medical Research Council, Edinburgh, Scotland, United Kingdom
BIRCHLER, JAMES, Div. of Biological Sciences, University of Missouri, Columbia
BIRD, ADRIAN, Wellcome Trust Centre for Cell Biology, University of Edinburgh, Edinburgh, Scotland, United Kingdom
BIRNBAUMER, LUTZ, National Institute of Environmental Health Sciences, National Institutes of Health, Research Triangle Park, North Carolina
BOCCUNI, PIERNICOLA, Dept. of Molecular Pharmacology and Therapeutics, Memorial Sloan-Kettering Cancer Institute, New York, New York

BOCKLANDT, SVEN, Dept. of Human Genetics, University of California, Los Angeles
BOND, DIANE, Research and Development, Electrophoresis Platforms, Applied Biosystems, Foster City, California
BONIFER, CONSTANZE, Dept. of Molecular Medicine, St. James University Hospital, University of Leeds, Leeds, United Kingdom
BOURC'HIS, DEBORAH, Dept. of Genetics and Development, Columbia University, New York, New York
BOYER, LAURIE, Dept. of Biology, Whitehead Institute for Biomedical Research, Massachusetts Institute of Technology, Cambridge, Massachusetts
BRENNAN, JENNIFER, Dept. of Genetics, University of North Carolina, Chapel Hill
BROMAN, KARL, Dept. of Biostatistics, Johns Hopkins University, Baltimore, Maryland
BROWN, CAROLYN, Dept. of Medical Genetics, University of British Columbia, Vancouver, British Columbia, Canada
BROWN, WILLIAM, Dept. of Genetics, Queens Medical Centre, University of Nottingham, Nottingham, United Kingdom
BUCHBERGER, JOHANNES, Dept. of Cell Biology, Harvard Medical School, Boston, Massachusetts
BUNGER, MAUREEN, Lab. of Molecular Carcinogenesis, National Institute of Environmental Health Sciences, National Institutes of Health, Research Triangle Park, North Carolina
BUSSEMAKER, HARMEN, Dept. of Biological Sciences, Columbia University, New York, New York
BYSTRICKY, KERSTIN, Dept. of Molecular Biology, University of Geneva, Geneva, Switzerland
CABELLO, OLGA, Dept. of Molecular and Cellular Biology, Baylor College of Medicine, Houston, Texas
CAI, HAINI, Dept. of Cellular Biology, University of Georgia, Athens
CAO, RU, Dept. of Biochemistry and Biophysics, Lineberger Comprehensive Cancer Center, University of North Carolina, Chapel Hill
CARBONE, ROBERTA, Dept. of Experimental Oncology, European Institute of Oncology, Milan, Italy
CARTER, NIGEL, The Wellcome Trust Sanger Institute, Hinxton, Cambridge, United Kingdom
CASLINI, CORRADO, Dept. of Medical Oncology, Fox Chase Cancer Center, Philadelphia, Pennsylvania
CASSEL, SALLIE, Invitrogen Life Technologies, Carlsbad, California
CEDAR, HOWARD, Dept. of Cellular Biochemistry, Hadassah Medical Center, Hebrew University, Jerusalem, Israel
CHADWICK, LISA, Institute for Genome Sciences and Policy, Duke University, Durham, North Carolina
CHAKRAVARTHY, SRINIVAS, Dept. of Biochemistry and Molecular Biology, Colorado State University, Fort Collins, Colorado
CHALKER, DOUGLAS, Dept. of Biology, Washington University, St. Louis, Missouri
CHAN, HUEI, Dept. of Pathology, University of Kentucky, Lexington
CHAN, SHIRLEY, Dolan DNA Learning Center, Cold Spring Harbor Laboratory, Cold Spring Harbor, New York
CHANDLER, VICKI, Dept. of Plant Sciences, University of Arizona, Tucson
CHEN, GENGXIN, Cold Spring Harbor Laboratory, Cold Spring Harbor, New York
CHEN, JACK, Cold Spring Harbor Laboratory, Cold Spring Harbor, New York
CHEN, XIN, Dept. of Developmental Biology, Stanford University School of Medicine, Stanford, California
CHENG, XIAODONG, Dept. of Biochemistry, Emory University School of Medicine, Atlanta, Georgia
CHERUKURI, SRUJANA, Dept. of Cell Biology, Cleveland Clinic Foundation, Cleveland, Ohio
CHESS, ANDREW, Dept. of Biology, Whitehead Institute for Biomedical Research, Massachusetts Institute of Technology, Cambridge, Massachusetts
CHONG, SUYINN, School of Molecular and Microbial Biosciences, University of Sydney, Sydney, New South Wales, Australia
CHOO, JUNGHA, Dept. of Biological Sciences, Korea Advanced Institute of Science and Technology, Daejeon, South Korea
CHUNG, JAE HOON, Dept. of Biological Sciences, Korea Advanced Institute of Science and Technology, Daejeon, South Korea
CIRO, MARCO, Dept. of Experimental Oncology, European Institute of Oncology, Milan, Italy
COCKERILL, PETER, Dept. of Molecular Medicine, St. James University Hospital, University of Leeds, Leeds, United Kingdom
COENRAADS, MONICA, Rett Syndrome Research Foundation, Cincinnati, Ohio
COHEN, AMIKAM, Dept. of Molecular Biology, Hebrew University Medical School, Jerusalem, Israel
COLOT, VINCENT, Unité de Recherche en Génomique Végétale, Centre National de la Recherche Scientifique, Evry, France
COMET, ITYS, Institut de Génétique Humaine, Centre National de la Recherche Scientifique, Montpellier, France
COONROD, SCOTT, Dept. of Genetic Medicine, Weill Medical College, Cornell University, New York, New York
COROMINAS, MONTSERRAT, Dept. of Genetics, Universitat de Barcelona, Barcelona, Spain
CURRIE, RICHARD, Center for Scientific Review, National Institutes of Health, Bethesda, Maryland
DAHARY, DVIR, Dept. of Cell Research and Immunology, Tel Aviv University, Tel Aviv, Israel
DAI, JUNBIAO, Dept. of Genetics, Development, and Cell Biology, Iowa State University, Ames, Iowa
DANNENBERG, JAN-HERMEN, Dept. of Medical Oncology, Dana-Farber Cancer Institute, Boston, Massachusetts
DAS, RAJDEEP, Cold Spring Harbor Laboratory, Cold Spring Harbor, New York
DAVID, GREGORY, Dept. of Medical Oncology, Dana-Farber Cancer Institute, Boston, Massachusetts
DAVIS, ERICA, Dept. of Genetics, Faculty of Veterinary Medicine, University of Liège, Liège, Belgium
DEAN, CAROLINE, Dept. of Cell and Developmental Biol-

ogy, John Innes Centre, Norwich, United Kingdom
DEFOSSEZ, PIERRE, Institut Curie, Unité Mixte de Recherche, Centre National de la Recherche Scientifique, Paris, France
DIAZ-PEREZ, SILVIA, Dept. of Human Genetics, University of California, Los Angeles
DJUPEDAL, INGELA, Dept. of Natural Sciences, Karolinska Institutet, University College Södertörn, Huddinge, Sweden
DORMAN, JANICE, Dept. of Epidemiology, University of Pittsburgh, Pittsburgh, Pennsylvania
DORMANN, HOLGER, Lab. of Chromatin Biology, Rockefeller University, New York, New York
DOSTATNI, NATHALIE, Institut Curie, Unité Mixte de Recherche, Centre National de la Recherche Scientifique, Paris, France
DRGON, TOMAS, Molecular Neurobiology Branch, Charles River Laboratories, National Institute on Drug Abuse, Baltimore, Maryland
DUAN, ZHIJUN, Dept. of Medical Genetics, University of Washington, Seattle
DUBOWITZ, VICTOR, Neuromuscular Unit, Dept. of Pediatrics, Imperial College London, London, United Kingdom
DUNCAN, ELIZABETH, Lab. of Chromatin Biology, Rockefeller University, New York, New York
DUNN, REBECCA, Dept. of Molecular Biology, Massachusetts General Hospital, Harvard Medical School, Boston, Massachusetts
EBERT, JOAN, Cold Spring Harbor Laboratory Press, Woodbury, New York
ECCLESTON, ALEX, *Nature*, Nature Publishing, San Francisco, California
ECKHARDT, FLORIAN, Dept. of Life Sciences, Epigenomics AG, Berlin, Germany
EKWALL, KARL, Dept. of Natural Sciences, Karolinska Institutet, University College Södertörn, Huddinge, Sweden
ELGIN, SARAH, Dept. of Biology, Washington University, St. Louis, Missouri
EMMONS, SCOTT, Dept. of Molecular Genetics, Albert Einstein College of Medicine, Bronx, New York
ESUMI, NORIKO, Dept. of Ophthalmology, School of Medicine, Johns Hopkins University, Baltimore, Maryland
EVANS, JASON, Cold Spring Harbor Laboratory, Cold Spring Harbor, New York
EZHKOVA, ELENA, Cold Spring Harbor Laboratory, Cold Spring Harbor, New York
FAN, YUHONG, Dept. of Cell Biology, Albert Einstein College of Medicine, Bronx, New York
FARMER, DEBORAH, Gene Regulation and Chromatin Unit, Hammersmith Hospital, Clinical Sciences Centre, Medical Research Council, Imperial College London, London, United Kingdom
FEIL, ROBERT, Institute of Molecular Genetics, Centre National de la Recherche Scientifique, Montpellier, France
FEINBERG, ANDREW, Epigenetics Unit, School of Medicine, Johns Hopkins University, Baltimore, Maryland

FELSENFELD, GARY, Lab. of Molecular Biology, National Institute of Diabetes and Digestive Kidney Diseases, National Institutes of Health, Bethesda, Maryland
FERRER, JORGE, Dept. of Endocrinology, IDIBAPS, Hospital Clinic de Barcelona, Barcelona, Spain
FIELDS, SCOTT, Dept. of Immunology, National Jewish Medical and Research Center, Denver, Colorado
FIRE, ANDREW, Dept. of Pathology and Genetics, Stanford University School of Medicine, Stanford, California
FISCHER, SYLVIA, Dept. of Molecular Biology, Massachusetts General Hospital, Harvard Medical School, Boston, Massachusetts
FISCHLE, WOLFGANG, Lab. of Chromatin Biology, Rockefeller University, New York, New York
FOUREL, GENEVIEVE, Unité Mixte de Recherche, Centre National de la Recherche Scientifique, Ecole Normale Supérieure, Lyon, France
FRAZER, KIMBLE, Div. of Pediatric Hematology and Oncology, Huntsman Cancer Institute, University of Utah, Salt Lake City
FRY, CHRISTOPHER, Program in Molecular Medicine, School of Medicine, University of Massachusetts, Worcester
GANN, ALEXANDER, Cold Spring Harbor Laboratory Press, Woodbury, New York
GASSER, SUSAN, Dept. of Molecular Biology, University of Geneva, Geneva, Switzerland
GASZNER, MIKLOS, Lab. of Molecular Biology, National Institute of Diabetes and Digestive Kidney Diseases, National Institutes of Health, Bethesda, Maryland
GENDREL, ANNE-VALERIE, Unité de Recherche en Génomique Végétale, Centre National de la Recherche Scientifique, Evry, France
GEORGES, MICHEL, Dept. of Genetics, Faculty of Veterinary Medicine, University of Liège, Liège, Belgium
GILES, KEITH, Dept. of Biology, Johns Hopkins University, Baltimore, Maryland
GOLDMAN, JOSEPH, Dept. of Molecular Biology, Harvard Medical School, Boston, Massachusetts
GOMEZ, JORGE, Dept. of Microbiology and Immunology, Emory University, Atlanta, Georgia
GOMOS-KLEIN, JANETTE, Dept. of Biological Sciences, City University of New York, Hunter College, New York, New York
GONZALO, SUSANA, Molecular Oncology Program, Spanish National Cancer Centre, Madrid, Spain
GOODMAN, LAURIE, *Journal of Clinical Investigation*, New York, New York
GOPINATHRAO, GOPAL, Cold Spring Harbor Laboratory, Cold Spring Harbor, New York
GOTO, DEREK, Cold Spring Harbor Laboratory, Cold Spring Harbor, New York
GOTTSCHLING, DANIEL, Fred Hutchinson Cancer Research Center, Seattle, Washington
GREWAL, SHIV, Lab. of Molecular Cell Biology, National Cancer Institute, National Intitutes of Health, Bethesda, Maryland
GROSSNIKLAUS, UELI, Institute of Plant Biology, University of Zurich, Zurich, Switzerland
GROTH, ANJA, Dept. of Cell Cycle and Cancer, Danish

Cancer Society, Copenhagen, Denmark
GRUNSTEIN, MICHAEL, Dept. of Biological Chemistry, University of California, Los Angeles
GUEGLER, KARL, Applied Biosystems, Foster City, California
GUENTHER, MATTHEW, Whitehead Institute for Biomedical Research, Massachusetts Institute of Technology, Cambridge, Massachusetts
HAIG, DAVID, Dept. of Organismic and Evolutionary Biology, Botanical Museum, Harvard University, Cambridge, Massachusetts
HAJKOVA, PETRA, Gurdon Institute, Wellcome Trust, Cancer Research U.K., University of Cambridge, Cambridge, United Kingdom
HALL, IRA, Cold Spring Harbor Laboratory, Cold Spring Harbor, New York
HAN, JEFFREY, Dept. of Molecular Biology and Genetics, School of Medicine, Johns Hopkins University, Baltimore, Maryland
HANNON, GREGORY, Cold Spring Harbor Laboratory, Cold Spring Harbor, New York
HARRIS, ABIGAIL, Abcam, Ltd., Cambridge, United Kingdom
HE, LIN, Cold Spring Harbor Laboratory, Cold Spring Harbor, New York
HE, YUEHUI, Dept. of Biochemistry, University of Wisconsin, Madison
HEARD, EDITH, Mammalian Developmental Epigenetics Group, Institut Curie, Unité Mixte de Recherche, Centre National de la Recherche Scientifique, Paris, France
HECHT, MERAV, Dept. of Cell Biochemistry and Human Genetics, Hebrew University, Jerusalem, Israel
HELD, WILLIAM, Dept. of Molecular and Cellular Biology, Roswell Park Cancer Institute, Buffalo, New York
HENIKOFF, STEVEN, Howard Hughes Medical Institute, Fred Hutchinson Cancer Research Center, Seattle, Washington
HIRATANI, ICHIRO, Dept. of Biochemistry and Molecular Biology, Upstate Medical University, State University of New York, Syracuse
HOEK, MAARTEN, Cold Spring Harbor Laboratory, Cold Spring Harbor, New York
HOHN, BARBARA, Friedrich Miescher Institute for Biomedical Research, Basel, Switzerland
HORIKOSHI, MASAMI, Lab. of Developmental Biology, Institute of Molecular and Cellular Biosciences, University of Tokyo, Tokyo, Japan
HOULARD, MARTIN, Département de Biologie Joliot-Curie, Service de Biochimie et de Génétique Moléculaire, CEA de Saclay, Gif-sur-Yvette, France
HSU, DUEN-WEI, Dept. of Biochemistry, Oxford University, Oxford, United Kingdom
HSU, MEI, Dept. of Microbiology and Immunology, Dartmouth Medical School, Lebanon, New Hampshire
HUA, SUJUN, Dept. of Genetics, Yale University School of Medicine, New Haven, Connecticut
INAMOTO, SUSUMU, Dept. of Biotechnology, Institute of Research and Innovation, Chiba, Japan
INGLIS, JOHN, Cold Spring Harbor Laboratory Press, Woodbury, New York
IOSHIKHES, ILYA, Dept. of Biomedical Informatics, Ohio State University, Columbus, Ohio
IRVINE, DANIELLE, Dept. of Chromosome Research, Murdoch Childrens Research Institute, Melbourne, Australia
ISHINO, FUMITOSHI, Medical Research Center, Tokyo Medical and Dental University, Tokyo, Japan
ISHOV, ALEXANDER, Dept. of Anatomy and Cell Biology, University of Florida, Gainesville
IWAMOTO, KAZUYA, Lab. for Molecular Dynamics of Mental Disorders, Brain Science Institute, RIKEN, Wako, Japan
JAENISCH, RUDOLF, Dept. of Biology, Whitehead Institute for Biomedical Research, Massachusetts Institute of Technology, Cambridge, Massachusetts
JAYARAMAN, LATA, Dept. of Oncology, Bristol-Myers Squibb, Princeton, New Jersey
JELINEK, TOM, Upstate Biotechnology, Inc., Lake Placid, New York
JENUWEIN, THOMAS, Research Institute of Molecular Pathology, Vienna Biocenter, University of Vienna, Vienna, Austria
JIA, SONGTAO, Lab. of Molecular Cell Biology, National Cancer Institute, National Institutes of Health, Bethesda, Maryland
JIMENEZ, JOSE, Dept. of Cellular and Molecular Physiology, Joslin Diabetes Center, Boston, Massachusetts
JOHNSTONE, KAREN, Dept. of Molecular Genetics and Microbiology, University of Florida, Gainesville
JORDAN, ALBERT, Dept. of Gene Regulation, Centre de Regulació Genòmica, Barcelona, Spain
JOSHI-TOPE, GEETA, Cold Spring Harbor Laboratory, Cold Spring Harbor, New York
KAJI, KEISUKE, Institute for Stem Cell Research, University of Edinburgh, Edinburgh, Scotland, United Kingdom
KAKKAD, REGHA, Center for Molecular Medicine, Institute of Microbiology and Genetics, Vienna Biocenter, University of Vienna, Vienna, Austria
KAKUTANI, TETSUJI, Dept. of Integrated Genetics, National Institute of Genetics, Shizuoka, Japan
KALANTRY, SUNDEEP, Dept. of Genetics, University of North Carolina, Chapel Hill
KAMBERE, MARIJO, Dept. of Molecular Biology and Biochemistry, Wesleyan University, Middletown, Connecticut
KAMPMANN, MARTIN, Lab. of Cell Biology, Rockefeller University, New York, New York
KANNANGANATTU, PRASANTH, Cold Spring Harbor Laboratory, Cold Spring Harbor, New York
KANNO, MASAMOTO, Dept. of Immunology, Hiroshima University, Hiroshima, Japan
KAPOOR, AVNISH, Dept. of Plant Sciences, University of Arizona, Tucson
KARPF, ADAM, Dept. of Pharmacology and Therapeutics, Roswell Park Cancer Institute, Buffalo, New York
KASSCHAU, KRISTIN, Center for Gene Research and Biotechnology, Oregon State University, Corvallis, Oregon
KATO, MASAOMI, Dept. of Integrated Genetics, National

Institute of Genetics, Mishima, Japan
KATO, TADAFUMI, Lab. for Molecular Dynamics of Mental Disorders, Brain Science Institute, RIKEN, Wako, Japan
KAUFMAN, PAUL, Dept. of Molecular and Cell Biology, Lawrence Berkeley National Laboratory, University of California, Berkeley
KE, QINGDONG, Dept. of Environmental Medicine, New York University, Tuxedo, New York
KELLY, WILLIAM, Dept. of Biology, Emory University, Atlanta, Georgia
KELSEY, GAVIN, Developmental Genetics Programme, The Babraham Institute, Cambridge, United Kingdom
KENNEDY, SCOTT, Dept. of Molecular Biology, Massachusetts General Hospital, Harvard Medical School, Boston, Massachusetts
KIDNER, CATHERINE, Cold Spring Harbor Laboratory, Cold Spring Harbor, New York
KIKYO, NOBUAKI, Stem Cell Institute, University of Minnesota, Minneapolis
KIM, JOHN, Dept. of Molecular Biology, Massachusetts General Hospital, Harvard Medical School, Boston, Massachusetts
KIM, JOOMYEONG, Div. of Genome Biology, Lawrence Livermore National Laboratory, Livermore, California
KIM, MI YOUNG, Dept. of Molecular Biology and Genetics, Cornell University, Ithaca, New York
KINGSTON, ROBERT, Dept. of Molecular Biology, Massachusetts General Hospital, Harvard Medical School, Boston, Massachusetts
KLAR, AMAR, Lab. of Gene Regulation and Chromosome Biology, National Cancer Institute, Frederick, Maryland
KLAUTKY, TRUDEE, Field Application Specialist, Invitrogen, Shrewsbury, Massachusetts
KLOSE, ROBERT, Wellcome Trust Centre for Cell Biology, University of Edinburgh, Edinburgh, Scotland, United Kingdom
KLYMENKO, TETYANA, Gene Expression Programme, European Molecular Biology Laboratory, Heidelberg, Germany
KOH, MINGSHI, Dept. of Biological Sciences, National University of Singapore, Singapore, China
KONFORTI, BOYANA, *Nature Structural & Molecular Biology*, New York, New York
KOUSKOUTI, ANTIGONE, Dept. of Molecular Genetics, Institute of Molecular Biology and Biotechnology, Foundation for Research and Technology, Herakleion, Crete, Greece
KOWENZ-LEUTZ, ELIZABETH, Dept. of Tumor Development and Differentiation, Max Delbrück Center for Molecular Medicine, Berlin, Germany
KRAUS, W. LEE, Dept. of Molecular Biology and Genetics, Cornell University, Ithaca, New York
KRIAUCIONIS, SKIRMANTAS, Wellcome Trust Centre for Cell Biology, University of Edinburgh, Edinburgh, Scotland, United Kingdom
KROGAN, NEVAN, Dept. of Medical Genetics, C.H. Best Institute, University of Toronto, Toronto, Ontario, Canada
KSANDER, BRUCE, Dept. of Ophthalmology, Harvard Medical School, Boston, Massachusetts
KUCEROVA, MARTINA, Dept. of Biological Sciences, Hunter College, City University of New York, New York, New York
KURATOMI, GO, Lab. for Molecular Dynamics of Mental Disorders, Brain Science Institute, RIKEN, Wako, Japan
KURODA, MITZI, Harvard-Partners Center for Genetics and Genomics, Harvard University, Boston, Massachusetts
KWON, HYOCKMAN, Dept. of Bioscience and Biotechnology, Hankuk University of Foreign Studies, Yongin, Seoul, South Korea
LABOURIER, EMMANUEL, Dept. of Research and Development, Ambion, Inc., Austin, Texas
LAI, ANGELA, Johnson and Johnson Research, Strawberry Hills, Australia
LAKEY, NATHAN, General and Administrative Offices, Orion Genomics, St. Louis, Missouri
LAKOWSKI, BERNARD, Dept. of Neuroscience, Nematode Genetics Group, Institut Pasteur, Paris, France
LANDE-DINER, LAURA, Dept. of Cellular Biochemistry, Hebrew University, Jerusalem, Israel
LEE, JEANNIE, Dept. of Molecular Biology, Massachusetts General Hospital, Harvard Medical School, Boston, Massachusetts
LI, MO, Dept. of Cellular Biology, University of Georgia, Athens
LI, YING, Dept. of Cellular and Developmental Biology, Fox Chase Cancer Center, Philadelphia, Pennsylvania
LIM, JAE-HWAN, Lab. of Metabolism, National Cancer Institute, National Institutes of Health, Bethesda, Maryland
LIPPMAN, ZACHARY, Cold Spring Harbor Laboratory, Cold Spring Harbor, New York
LIPSICK, JOSEPH, Dept. of Pathology, Stanford University, Stanford, California
LIU, LIANG, Dept. of Biology, University of Alabama, Birmingham
LOPES, ALEXANDRA, Dept. of Population Genetics, IPATIMUP, Instituto de Patologia e Immunologia Molecular da Universidade de Porto, Porto, Portugal
LOUIS, EDWARD, Dept. of Genetics, University of Leicester, Leicester, United Kingdom
LOWE, SCOTT, Cold Spring Harbor Laboratory, Cold Spring Harbor, New York
LU, XIANGDONG, Dept. of Biochemistry, University of North Carolina, Chapel Hill
LUBITZ, SANDRA, Dept. of Biotechnology, Dresden University of Technology, Dresden, Germany
LUSTIG, ARTHUR, Dept. of Biochemistry, Health Sciences Center, Tulane University, New Orleans, Louisiana
LYLE, ROBERT, Dept. of Genetic Medicine and Development, School of Medicine, University of Geneva, Geneva, Switzerland
MADHANI, HITEN, Dept. of Biochemistry and Biophysics, University of California, San Francisco
MAGGERT, KEITH, Dept. of Biology, University of Utah, Salt Lake City
MAJUMDER, SADHAN, Dept. of Molecular Genetics, M.D. Anderson Cancer Center, University of Texas, Houston

MAJUMDER, SARMILA, Dept. of Molecular and Cellular Biochemistry, Ohio State University, Columbus, Ohio
MANDIYAN, VALSAN, Dept. of Pharmacology, Yale University, New Haven, Connecticut
MANGONE, MARCO, Cold Spring Harbor Laboratory, Cold Spring Harbor, New York
MANIATAKI, ELISAVET, Dept. of Pathology, University of Pennsylvania, Philadelphia
MANN, CARL, Service de Biochimie et de Génétique Moléculaire, CEA de Saclay, Gif-sur-Yvette, France
MARIN BIVENS, CARRIE, Dept. of Research, Jackson Laboratory, Bar Harbor, Maine
MARTIENSSEN, ROBERT, Cold Spring Harbor Laboratory, Cold Spring Harbor, New York
MASUI, OSAMU, Dept. of Nuclear Dynamics and Genome Plasticity, Institut Curie, Unité Mixte de Recherche, Centre National de la Recherche Scientifique, Paris, France
MATOUK, CHARLES, Institute of Medical Sciences, University of Toronto, Toronto, Ontario, Canada
MCCANN, JENNIFER, Dept. of Molecular Medicine, Ottawa Health Research Institute, Ottawa, Ontario, Canada
MELLO, CRAIG, Program in Molecular Medicine, Cancer Center, School of Medicine, University of Massachusetts, Worcester
MELNICK, ARI, Albert Einstein College of Medicine, Bronx, New York
MEYER, BARBARA, Dept. of Molecular and Cellular Biology, Howard Hughes Medical Institute, University of California, Berkeley
MEYER, CLIFFORD, Dept. of Biostatistics, Dana-Farber Cancer Institute, Cambridge, Massachusetts
MEYER, ERIC, Laboratoire de Génétique Moléculaire, Unité Mixte de Recherche, Centre National de la Recherche Scientifique, Ecole Normale Supérieure, Paris, France
MIAO, FENG, Dept. of Diabetes, Beckman Research Institute, City of Hope, Duarte, California
MIN, JINRONG, Cold Spring Harbor Laboratory, Cold Spring Harbor, New York
MISHRA, NILAMADHAB, Dept. of Internal Medicine, Wake Forest University School of Medicine, Winston-Salem, North Carolina
MISHRA, RAKESH, Centre for Cellular and Molecular Biology, Hyderabad, India
MITO, YOSHIKO, Dept. of Basic Science, Fred Hutchinson Cancer Research Center, Seattle, Washington
MITTELSTEN SCHEID, ORTRUN, Gregor Mendel Institute of Molecular Plant Biology, Vienna, Austria
MIZUNO, YOSUKE, Center for Genomics and Bioinformatics, Karolinska Institutet, Stockholm, Sweden
MO, XIANMING, Dept. of Tumor Development and Differentiation, Max Delbrück Center for Molecular Medicine, Berlin, Germany
MOHAN, KOMMU, Centre for Human Genetics, Bangalore, India
MOHD-SARIP, ADONE, Dept. of Molecular and Cell Biology, Sylvius Laboratory, Centre for Biomedical Genetics, Leiden University Medical Centre, Leiden, The Netherlands
MONTGOMERY, NATHAN, Dept. of Genetics, University of North Carolina, Chapel Hill
MOONEY, MYESHA, Dept. of Microbiology, Emory University, Atlanta, Georgia
MORRIS, KEVIN, Dept. of Medicine, University of California at San Diego, La Jolla
MOTTAGUI-TABAR, SALIM, Center for Genomics and Bioinformatics, Karolinska Institutet, Stockholm, Sweden
MUCHARDT, CHRISTIAN, Expression Génétique Moleculaire, Centre National de la Recherche Scientifique, Institut Pasteur, Paris, France
MUNGALL, ANDREW, The Wellcome Trust Sanger Institute, Hinxton, Cambridge, United Kingdom
MURCHISON, ELIZABETH, Cold Spring Harbor Laboratory, Cold Spring Harbor, New York
MUTSKOV, VESCO, Lab. of Molecular Biology, National Institute of Diabetes and Digestive and Kidney Diseases, National Institutes of Health, Bethesda, Maryland
MYAKISHEV, MAX, Lab. of Metabolism, National Cancer Institute, National Institutes of Health, Bethesda, Maryland
NAGASE, HIROKI, Dept. of Cancer Genetics, Roswell Park Cancer Institute, Buffalo, New York
NAKATANI, YOSHIHIRO, Dept. of Cancer Biology, Dana-Farber Cancer Institute, Harvard Medical School, Boston, Massachusetts
NARAYAN, SANTOSH, Dept. of Pharmacology and Neuroscience, University of North Texas Health Science Center, Fort Worth, Texas
NISHIO, HITOMI, Dept. of Pediatrics, Mount Sinai School of Medicine, New York, New York
NOLAN, CATHERINE, Dept. of Zoology, University College of Dublin, Dublin, Ireland
NOWACKI, MARIUSZ, Laboratoire de Génétique Moléculaire, Unité Mixte de Recherche, Centre National de la Recherche Scientifique, Ecole Normale Supérieure, Paris, France
NUNEZ, SABRINA, Cold Spring Harbor Laboratory, Cold Spring Harbor, New York
OHTSUBO, HISAKO, Institute of Molecular and Cellular Biosciences, University of Tokyo, Tokyo, Japan
OKADA, YUKI, Dept. of Biochemistry and Biophysics, University of North Carolina, Chapel Hill
OMHOLT, STIG, Center for Integrative Genetics, Agricultural University of Norway, Aas, Norway
ONISHI, MEGUMI, Dept. of Cell Biology, Harvard Medical School, Boston, Massachusetts
OONO, KIYOHARU, Research Center for Environmental Genomics, Kobe University, Kobe, Japan
ORDWAY, JARED, Dept. of Biomarkers, Orion Genomics, St. Louis, Missouri
ORLANDO, VALERIO, Dept. of Epigenetics and Genome Reprogramming, Dulbecco Telethon Institute, Institute of Genetics and Biophysics, CNR, Naples, Italy
ORTIZ, BENJAMIN, Dept. of Biological Sciences, Hunter College, City University of New York, New York, New York

OZER, JOSEF, Dept. of Pharmacology, School of Medicine, Boston University, Boston, Massachusetts
PADDISON, PATRICK, Cold Spring Harbor Laboratory, Cold Spring Harbor, New York
PAGAN, JULIA, Dept. of Cancer and Cell Biology, Queensland Institute of Medical Research, Brisbane, Australia
PALLAORO, MICHELE, Dept. of Biochemistry, Institute for Research in Molecular Biology, Merck, Pomezia, Italy
PAN, QI, Dept. of Molecular Biology, Massachusetts General Hospital, Harvard Medical School, Boston, Massachusetts
PARO, RENATO, Zentrum für Molekulare Biology, University of Heidelberg, Heidelberg, Germany
PARRY, DEVIN, Dept. of Molecular Biology, Massachusetts General Hospital, Harvard Medical School, Boston, Massachusetts
PARTHUN, MARK, Dept. of Molecular and Cellular Biochemistry, Ohio State University, Columbus, Ohio
PARTRIDGE, JANET, Dept. of Biochemistry, St. Jude Children's Research Hospital, Memphis, Tennessee
PASZKOWSKI, JERZY, Dept. of Plant Biology, University of Geneva, Geneva, Switzerland
PELISSON, ALAIN, Institut de Génétique Humaine, Centre National de la Recherche Scientifique, Montpellier, France
PENNELL, ROGER, Ceres, Inc., Malibu, California
PETERSON, CRAIG, Program in Molecular Medicine, Medical Center, School of Medicine, University of Massachusetts, Worcester
PETRIE, VICTORIA, Dept. of Biochemistry, St. Jude Children's Research Hospital, Memphis, Tennessee
PFEFFER, PETER, Dept. of Reproduction, Agresearch Crown Research Institute, Hamilton, New Zealand
PHAM, ANH-DUNG, Dept. of Biochemistry, Howard Hughes Medical Institute, UMDNJ-Robert Wood Johnson Medical School, Piscataway, New Jersey
PIEDRAHITA, JORGE, Dept. of Molecular Biomedical Sciences, North Carolina State University, Raleigh, North Carolina
PIEN, STEPHANE, Dept. of Plant Systems Biology, VIB, University of Gent, Gent, Belgium
PILLUS, LORRAINE, Dept. of Biology, University of California at San Diego, La Jolla
PIRROTTA, VINCENZO, Dept. of Zoology, University of Geneva, Geneva, Switzerland
PLASTERK, RONALD, Hubrecht Laboratory, Netherlands Institute for Developmental Biology, Utrecht, The Netherlands
POGAČIĆ, VANDA, Friedrich Miescher Institute for Biomedical Research, Basel, Switzerland
POLLOCK, MILA, Cold Spring Harbor Laboratory Library, Cold Spring Harbor, New York
POOT, RAYMOND, Marie Curie Research Institute, Oxted, United Kingdom
POZZI, SILVIA, Dept. of Cancer Genetics, Roswell Park Cancer Institute, Buffalo, New York
PRASANTH, SUPRIYA, Cold Spring Harbor Laboratory, Cold Spring Harbor, New York
PRIOLEAU, MARIE-NOELLE, Laboratoire de Génétique Moléculaire, Unité Mixte de Recherche, Centre National de la Recherche Scientifique, Ecole Normale Supérieure, Paris, France
PROBST, ALINE, Dept. of Plant Biology, University of Geneva, Geneva, Switzerland
PRUITT, ROBERT, Dept. of Botany and Plant Pathology, Purdue University, West Lafayette, Indiana
PTASHNE, MARK, Dept. of Molecular Biology, Memorial Sloan-Kettering Cancer Institute, New York, New York
RAKYAN, VARDHMAN, Dept. of Epigenomics, The Wellcome Trust Sanger Institute, Hinxton, Cambridge, United Kingdom
RAMACHANDRAN, ILENG, Cold Spring Harbor Laboratory, Cold Spring Harbor, New York
RAMASWAMY, AMUTHA, Dept. of Biomedical Informatics, Ohio State University, Columbus, Ohio
RANDO, OLIVER, Bauer Center for Genomics Research, Harvard University, Cambridge, Massachusetts
RAUSCHER, FRANK, Dept. of Molecular Genetics, Wistar Institute, Philadelphia, Pennsylvania
RECHAVI, GIDEON, Sheba Cancer Research Center, Sheba Medical Center, Tel-Hashomer, Israel
REDDI, PRABHAKARA, Dept. of Cell Biology, University of Virginia, Charlottesville
REESE, KIMBERLY, Dept. of Cell and Developmental Biology, Howard Hughes Medical Institute, University of Pennsylvania, Philadelphia
REIK, WOLF, Developmental Genetics Programme, The Babraham Institute, Cambridge, United Kingdom
REINBERG, DANNY, Dept. of Biochemistry, Howard Hughes Medical Institute, UMDNJ-Robert Wood Johnson Medical School, Piscataway, New Jersey
RICE, JUDD, Dept. of Biochemistry and Molecular Biology, University of Southern California, Los Angeles, California
RICHARDS, ERIC, Dept. of Biology, Washington University, St. Louis, Missouri
RIDDIHOUGH, GUY, *Science*, American Association for the Advancement of Science, Washington, D.C.
RIDDLE, NICOLE, Div. of Biological Sciences, University of Missouri, Columbia
RINE, JASPER, Dept. of Molecular and Cellular Biology, University of California, Berkeley
RIVERA, ISABEL, Dept. of Biochemistry, Faculty of Pharmacy, University of Lisbon, Lisbon, Portugal
ROBERTS, CHARLES, Dept. of Pediatric Oncology, Dana-Farber Cancer Institute, Boston, Massachusetts
ROLLINS, ROBERT, Dept. of Genetics and Development, Columbia University, New York, New York
RONEMUS, MICHAEL, Cold Spring Harbor Laboratory, Cold Spring Harbor, New York
ROSEBROCK, ADAM, Dept. of Molecular Genetics and Microbiology, State University of New York, Stony Brook
ROSSMANN, MARLIES, Cold Spring Harbor Laboratory, Cold Spring Harbor, New York
ROSSON, GARY, Lineberger Comprehensive Cancer Center, University of North Carolina, Chapel Hill
ROZENBERG, JULIAN, Lab. of Metabolism, National Can-

cer Institute, National Institutes of Health, Bethesda, Maryland
RUGG-GUNN, PETER, Dept. of Surgery, University of Cambridge, Cambridge, United Kingdom
RUSCHE, LAURA, Dept. of Biochemistry, Duke University, Durham, North Carolina
RUSK, NICOLE, *Nature Methods*, New York, New York
RUSSANOVA, VALYA, National Institute of Child Health and Human Development, National Institutes of Health, Bethesda, Maryland
RUVKUN, GARY, Dept. of Molecular Biology, Massachusetts General Hospital, Harvard Medical School, Boston, Massachusetts
SAAVEDRA, RAUL, Scientific Review Branch, National Institute of Neurological Disorders and Stroke, National Institutes of Health, Bethesda, Maryland
SACCHI, NICOLETTA, Dept. of Cancer Genetics, Roswell Park Cancer Institute, Buffalo, New York
SANDERS, STEVEN, Gurdon Institute, Wellcome Trust, Cancer Research U.K., University of Cambridge, Cambridge, United Kingdom
SASAKI, HIROYUKI, Div. of Human Genetics, National Institute of Genetics, Mishima, Japan
SASSONE-CORSI, PAOLO, Institut de Génétique et de Biologie Moléculaire et Cellulaire, Illkirch, France
SATO, NORIKO, Tokyo Metropolitan Institute of Medical Science, Tokyo, Japan
SCHILDKRAUT, CARL, Dept. of Cell Biology, Albert Einstein College of Medicine, Bronx, New York
SCHLEGEL, ROBERT, Dept. of Oncology, Novartis Institutes for BioMedical Research, Cambridge, Massachusetts
SCHOENHERR, CHRISTOPHER, Dept. of Cell and Structural Biology, University of Illinois, Urbana
SCHOLZ, CHRISTIAN, Dept. of Hematology and Oncology, Charité Campus Virchow-Klinikum, Berlin, Germany
SCHONES, DUSTIN, Cold Spring Harbor Laboratory, Cold Spring Harbor, New York
SEISER, CHRISTIAN, Dept. of Medical Biochemistry, Vienna Biocenter, University of Vienna, Vienna, Austria
SELKER, ERIC, Institute of Molecular Biology, University of Oregon, Eugene
SHARP, JUDITH, Dept. of Biochemistry and Biophysics, University of California, San Francisco
SHARP, PHILLIP, Center for Cancer Research, Massachusetts Institute of Technology, Cambridge, Massachusetts
SHEN, CHANG-HUI, Dept. of Biology, College of Staten Island, City University of New York, Staten Island, New York
SHIBAHARA, KEI-ICHI, Dept. of Integrated Genetics, National Institute of Genetics, Mishima, Japan
SHINKAI, YOICHI, Dept. of Cell Biology, Institute for Virus Research, Kyoto University, Kyoto, Japan
SHOVLIN, TANYA, Dept. of Biomedical Sciences, University of Ulster, Coleraine, Londonderry, Northern Ireland, United Kingdom
SHUREIQI, IMAD, Dept. of Clinical Cancer Prevention, M.D. Anderson Cancer Center, University of Texas, Houston

SI, KAUSIK, Dept. of Neurobiology and Behavior, Columbia University, New York, New York
SIDDIQUI, KHALID, Cold Spring Harbor Laboratory, Cold Spring Harbor, New York
SIDHU, RAVINDER, Bone and Joint Research Unit, St. Bartholomew's Hospital, London, United Kingdom
SIGOVA, ALLA, Dept. of Biochemistry and Molecular Pharmacology, School of Medicine, University of Massachusetts, Worcester
SKIPPER, MAGDALENA, *Nature Reviews Genetics*, London, United Kingdom
SKOULTCHI, ARTHUR, Dept. of Cell Biology, Albert Einstein College of Medicine, Bronx, New York
SLAWSON, ELIZABETH, Dept. of Biology, Washington University, St. Louis, Missouri
SMITH, ANDREW, Cold Spring Harbor Laboratory, Cold Spring Harbor, New York
SMITS, GUILLAUME, IRIBHM, Université Libre de Bruxelles, Brussels, Belgium
SOLTER, DAVOR, Dept. of Developmental Biology, Max Planck Institute for Immunobiology, Freiburg, Germany
SOUTOGLOU, EVANTHIA, National Cancer Institute, National Institutes of Health, Bethesda, Maryland
SPADA, FABIO, Sars International Centre for Marine Molecular Biology, University of Bergen, Bergen, Norway
SPARMANN, ANKE, Dept. of Microbiology, State University of New York, Stony Brook
SPECTOR, DAVID, Cold Spring Harbor Laboratory, Cold Spring Harbor, New York
SPECTOR, MONA, Cold Spring Harbor Laboratory, Cold Spring Harbor, New York
SRIDHARAN, RUPA, Howard Hughes Medical Institute, University of California, Los Angeles
SRINIVASAN, LAKSHMI, Dept. of Animal Biology, School of Veterinary Medicine, University of Pennsylvania, Philadelphia
STANCHEVA, IRINA, Dept. of Biochemistry, University of Edinburgh, Edinburgh, Scotland, United Kingdom
STEBBINS, MICHAEL, *Nature Genetics*, New York, New York
STEWART, DAVID, Meetings and Courses Programs, Cold Spring Harbor Laboratory, Cold Spring Harbor, New York
STILLMAN, BRUCE, President and CEO, Cold Spring Harbor Laboratory, Cold Spring Harbor, New York
STRAHL, BRIAN, Dept. of Biochemistry and Biophysics, University of North Carolina, Chapel Hill
STRUNNIKOV, ALEXANDER, Lab. of Gene Regulation and Development, National Institute of Child Health and Human Development, National Institutes of Health, Bethesda, Maryland
SU, RUEY-CHYI, Dept. of Microbiology, Immunology, and Molecular Genetics, Howard Hughes Medical Institute, University of California, Los Angeles
SURANI, AZIM, Gurdon Institute, Wellcome Trust, Cancer Research U.K., University of Cambridge, Cambridge, United Kingdom
SUSSMAN, HILLARY, *Genome Research*, Cold Spring Har-

bor Laboratory Press, Woodbury, New York
SWIGUT, TOMASZ, Lab. of Molecular Vertebrate Embryology, Rockefeller University, New York, New York
SYMER, DAVID, Lab. of Immunobiology, National Cancer Institute, Frederick, Maryland
TACHIBANA, MAKOTO, Lab. of Mouse Models, Institute for Virus Research, Kyoto University, Kyoto, Japan
TAHILIANI, MAMTA, Dept. of Pathology, Center for Blood Research, Harvard Medical School, Boston, Massachusetts
TAIPALE, MIKKO, Gene Expression Programme, European Molecular Biology Laboratory, Heidelberg, Germany
TAKEDA, SHIN, Dept. of Plant Biology, University of Geneva, Geneva, Switzerland
TAM, PATRICK, Embryology Unit, Children's Medical Research Institute, Westmead, Sydney, Australia
TAMADA, HIROSHI, Stem Cell Institute, University of Minnesota, Minneapolis
THOMPSON, JEFFREY, Dept. of Biology, Denison University, Granville, Ohio
TIAN, YANAN, Dept. of Veterinary Physiology and Pharmacology, Texas A&M University, College Station, Texas
TIMMERMANS, MARJA, Cold Spring Harbor Laboratory, Cold Spring Harbor, New York
TOMPA, RACHEL, Dept. of Biochemistry and Biophysics, University of California, San Francisco
TORRES, HECTOR, National Institute of Environmental Health Sciences, National Institutes of Health, Research Triangle Park, North Carolina
TSIKITIS, MARY, Dept. of Pathology, School of Medicine, New York University, New York, New York
TSUCHIMOTO, SUGURU, Institute for Molecular and Cellular Biosciences, University of Tokyo, Tokyo, Japan
TUTEJA, JIGYASA, Dept. of Physiological and Molecular Plant Biology, University of Illinois, Urbana
TUZON, CREIGHTON, Dept. of Biochemistry and Molecular Genetics, Health Sciences Center, Universiy of Colorado, Denver
URA, KIYOE, Div. of Gene Therapy Science, Graduate School of Medicine, Osaka University, Osaka, Japan
VALER, MARC, Dept. of Research and Development and Marketing, Agilent Technologies, Waldbronn, Germany
VAN HOUDT, HELENA, Dept. of Plant Systems Biology, VIB, University of Gent, Gent, Belgium
VAN LOHUIZEN, MAARTEN, Div. of Moleular Genetics, The Netherlands Cancer Institute, Amsterdam, The Netherlands
VAN LOO, KAREN, Dept. of Molecular Animal Physiology, University of Nijmegen, Nijmegen, The Netherlands
VAN RIETSCHOTEN, JOHANNA, Dept. Molecular Cell Biology and Immunology, VU Medical Center, Amsterdam, The Netherlands
VAUGHN, MATTHEW, Cold Spring Harbor Laboratory, Cold Spring Harbor, New York
VAURY, CHANTAL, Université D'Auvergne, Clermont-Ferrand, France

VERDEL, ANDRE, Dept. of Cell Biology, Harvard Medical School, Boston, Massachusetts
VINCENZ, CLAUDIUS, Dept. of Biological Chemistry, Howard Hughes Medical Institute, School of Medicine, University of Michigan, Ann Arbor
VINSON, CHARLES, Lab. of Metabolism, National Institutes of Health, Bethesda, Maryland
WAGLE, MAHENDRA, Lab. of Developmental Neurobiology, Temasek Life Sciences Laboratory, Singapore, China
WALSH, MARTIN, Dept. of Pediatrics and Human Genetics, Mount Sinai School of Medicine, New York, New York
WANG, DUO, Dept. of Molecular Biology, Massachusetts General Hospital, Harvard Medical School, Boston, Massachusetts
WANG, SONG, Dept. of Biological Sciences, Stanford University, Stanford, California
WANG, YANMING, Lab. of Chromatin Biology, Rockefeller University, New York, New York
WARBURTON, PETER, Dept. of Human Genetics, Mount Sinai School of Medicine, New York, New York
WEBSTER, KYLIE, Dept. of Genetics and Bioinformatics, Walter and Eliza Hall Institute of Medical Research, Parkville, Victoria, Australia
WEST, ADAM, Div. of Cancer Sciences and Molecular Pathology, Western Infirmary, University of Glasgow, Glasgow, Scotland, United Kingdom
WEST, KATHERINE, Div. of Cancer Sciences and Molecular Pathology, Western Infirmary, University of Glasgow, Glasgow, Scotland, United Kingdom
WHITEHEAD, NEDRA, Applied Sciences Branch, Centers for Disease Control, Atlanta, Georgia
WICKNER, REED, Lab. of Biochemistry and Genetics, National Institute of Diabetes and Digestive and Kidney Diseases, National Institutes of Health, Bethesda, Maryland
WILHELM, BRIAN, The Wellcome Trust Sanger Institute, Hinxton, Cambridge, United Kingdom
WILLIAMSON, CHRIS, Mammalian Genetics Unit, Medical Research Council, Harwell, Oxfordshire, United Kingdom
WILSON, JON, Dept. of Protein Structure, National Institute for Medical Research, London, United Kingdom
WITKOWSKI, JAN, Banbury Center, Cold Spring Harbor Laboratory, Cold Spring Harbor, New York
WONG, LEE, Dept. of Chromosome Research, Murdoch Children's Research Institute, Royal Children's Hospital, Parkville, Victoria, Australia
WU, HAO, Dept. of Molecular and Medical Pharmacology, University of California, Los Angeles
WU, JIANG, Dept. of Pathology and Developmental Biology, Howard Hughes Medical Institute, Stanford University, Stanford, California
WYSOCKA, JOANNA, Lab. of Chromatin Biology, Rockefeller University, New York, New York
XIANG, TIANHAO (TIM), Dept. of Reproductive Technology, ViaGen, Inc., Austin, Texas
XU, GUOLIANG, Shanghai Institute for Biological Sciences, Chinese Academy of Sciences, Shanghai, China

XUAN, ZHENYU, Cold Spring Harbor Laboratory, Cold Spring Harbor, New York
YANG, SEUNG KYOUNG, Dept. of Biological Sciences, Korea Advanced Institute of Science and Technology, Daejeon, South Korea
YANG, XIANG-JIAO, Dept. of Medicine, McGill University, Montréal, Québec, Canada
YE, XIAOFEN, Dept. of Molecular Oncology, Fox Chase Cancer Center, Philadelphia, Pennsylvania
YILDIRIM, EDA, Lab. of Signal Transduction, National Institute of Environmental Health Sciences, National Institutes of Health, Research Triangle Park, North Carolina
YOKOYAMA, KAZUNARI, Gene Engineering Division, RIKEN, Tsukuba Science City, Ibaraki, Japan
YOUNG, DALLAN, Dept. of Biochemistry and Molecular Biology, University of Calgary, Calgary, Alberta, Canada
YUSA, KOSUKE, Dept. of Social and Environmental Medicine, Osaka University, Osaka, Japan
ZAMORE, PHILLIP, Dept. of Biochemistry and Molecular Biology, School of Medicine, University of Massachusetts, Worcester
ZHANG, FENG, Dept. of Genetics, Development, and Cell Biology, Iowa State University, Ames, Iowa
ZHANG, YI, Dept. of Biochemistry and Biophysics, Lineberger Comprehensive Cancer Center, University of North Carolina, Chapel Hill
ZHANG, ZHIGUO, Dept. of Biochemistry and Molecular Biology, Mayo Clinic, Rochester, Minnesota
ZHAO, KEJI, National Heart, Lung, and Blood Institute, National Institutes of Health, Bethesda, Maryland
ZHONG, GAN-YUAN, Dept. of Agriculture & Nutrition, Pioneer Hi-Bred International, Johnston, Iowa

First row: R. Plasterk; A. Auger, D. Young, H. Van Houdt; D. Allis, E. Bernstein
Second row: E. Selker; P. Sharp, H. Cedar; E. Louis, D. Gottschling, S. Henikoff; E. Heard
Third row: Y. Fan, G. Fourel; P. Sassone-Corsi, B. Stillman; B. Ortiz, G. Felsenfeld
Fourth row: A. Surani, P. Hajkova; V. Pirrotta, J. Paszkowski, M. Timmermans

First row: L. Goodman; S. Emmons, G. Ruvkun, B. Meyer; E. Richards, D. Baulcombe, I. Stancheva
Second row: M. Macari demonstrating the online Oral History Project; S. Gasser, C. Schildkraut
Third row: E. Yildirim, S. Cherukuri, B. Meyer; H. Sussman, M. Bartolomei, J. Lee
Fourth row: G. Kelsey, E. Davis; P. Overbeek, O. Cabello; I. Hall, J. Birchler

First row: S. Kennedy, E. Meyer; D. Haig; R. Pruitt; H. Sasaki; C. Peterson, S. Berger
Second row: C. Mann, P. Kaufman; G. Rechavi, A. Cohen; M. Grunstein, A. Bird
Third row: D. Spector, D. Stewart, P. Sassone-Corsi; P. Cockerill, C. Williamson; U. Grossniklaus, Y. Lazebnik
Fourth row: Airslee Wine and Cheese; O. Rando, S. Altschuler

First row: C. Tuzon, K. Ekwall; A. Pelisson, Z. Lippman
Second row: B. Hohn, G.-Y. Zhong; W. Bickmore; J. Partridge, L. Gann
Third row: Symposium Cocktails on Blackford Lawn

First row: Lunchtime on Blackford Lawn
Second row: M. Gaszner; G. Almouzni, Y. Nakatani; S. Lubitz, K. Webster
Third row: J. Watson, R. Wickner; K. Giles

First row: Brunch chez Blackford; K. West, M.-N. Prioleau
Second row: P. Sherwood, A. Klar, J. Hicks; V. Pogačić, V. Beier
Third row: J. Birchler, R. Martienssen; Grace closed circuit TV

First row: P. Kannanganattu, V. Ellison; D. Barlow, R. Lyle; V. Chandler, J. Bender
Second row: A Chess; Y. He, C. Dean; W. Kelly, V. Colot; S. Elgin
Third row: D. Gottschling, R. Jaenisch; P. Zamore, D. Bartel; A. Skoultchi, A. Harris
Fourth row: Summary (D. Gottschling)

First row: Y. Okada, H. Chan; T. Bestor, X. Cheng; B. Strahl
Second row: Y. Arai; A. Eccleston, R. Kingston, M. van Lohuizen; W. Reik; N. Sato
Third row: J. Brennan; M. Hoek, D. Goto; R. Rollins, N. Rusk, W. Held, G. Smits
Fourth row: A. Feinberg, M. Stebbins; D. Solter, J. Witkowski, A. Surani; G. Rosson

Foreword

Until recently, the general view of heredity has been seen through the lens of DNA. Indeed, the 2003 Symposium on "The Genome of *Homo sapiens*" contributed to that view by emphasizing the importance of DNA sequence and its origins. But increasingly, investigators are exploring a set of secondary phenomena that give rise to heritable changes in gene function that occur without a change in the underlying DNA sequence—epigenetic mechanisms such as DNA methylation, histone acetylation, imprinting, RNA interference, gene silencing, and paramutation. A growing body of evidence indicates that epigenetic changes are important contributors to the pathogenesis of disease in humans, animals, and plants and may lie at the heart of many important gene–environment interactions. And so it seemed timely to hold a Symposium explicitly devoted to "Epigenetics."

Previous Symposia that have in part examined the role of the macromolecular context in which the primary genetic information is found include the 1941 Symposium on "Genes and Chromosomes: Structure and Organization," which emphasized a biophysical approach to these structures; the two closely separated Symposia that examined "Chromosome Structure and Function" (1973) and "Chromatin" (1977), at which latter meeting the nature of the nucleosome was unveiled; and the 1993 Symposium on "DNA and Chromosomes," by which time the human genome project with its focus on the primary sequence was well underway. The 69th Symposium, however, was the first to fully explore the heritable aspects to these and related biochemical phenomena.

The field of epigenetics as we know it today was prominently introduced at the 1951 Symposium on "Genes and Mutations." There, Ed Lewis presented data on position effect variegation in *Drosophila*, a phenomenon that has played an important role in the history of the field. Equally importantly, Barbara McClintock presented her ideas about heterochromatin and movable genetic elements, the so-called Ac-Ds system in maize that opened up understanding of transposition and its links to gene silencing and formation of heterochromatin. Some 53 years later, the 69th Symposium witnessed a rather complete molecular description of her ideas, including links to RNAi.

The 69th Symposium ran for 5 days and included 460 participants with 68 oral presentations and 210 poster presentations. We wish to thank Dr. David Haig for his superb Dorcas Cummings Memorial public lecture presenting a theoretical approach to epigenetics, on the subject of The Divided Self—Brains, Brawn and the Superego. We also wish to thank the first night speakers, Drs. Davor Solter, Barbara Meyer, David Allis, and Andrew Fire for their superb overview presentations, and we particularly thank Dr. Daniel Gottschling for agreeing to summarize the Meeting.

We thank Jenna Williams and Mary Smith in the Meetings and Courses office for efficiently running the meeting and John Inglis and his staff at the Cold Spring Harbor Laboratory Press, particularly Joan Ebert, Kathleen Bubbeo, and Danny deBruin, for publishing both the on-line and printed versions of the meeting manuscripts. Essential funds to run this meeting were obtained from the National Cancer Institute, a branch of the National Institutes of Health. In addition, financial help from the corporate benefactors, sponsors, affiliates, and contributors of our meetings program is essential for these Symposia to remain a success and we are most grateful for their continued support.

Bruce Stillman
David Stewart
March 2005

Sponsors

This meeting was funded in part by the **National Cancer Institute** and the **National Institute of Environmental Health Sciences**, branches of the **National Institutes of Health.**

Contributions from the following companies provide core support for the Cold Spring Harbor meetings program.

Corporate Benefactors

Amgen, Inc.
Aventis Pharma AG
Bristol-Myers Squibb Company
Eli Lilly and Company

GlaxoSmithKline
Novartis Institutes for BioMedical Research
Pfizer Inc.

Corporate Sponsors

Applied Biosystems
AstraZeneca
BioVentures, Inc.
Cogene BioTech Ventures, Ltd.
Diagnostic Products Corporation
Forest Laboratories, Inc.
Genentech, Inc.
Hoffmann-La Roche, Inc.
Johnson & Johnson Pharmaceutical Research
 & Development, L.L.C.

Kyowa Hakko Kogyo Co., Ltd.
Lexicon Genetics, Inc.
Merck Research Laboratories
New England BioLabs, Inc.
OSI Pharmaceuticals, Inc.
Pall Corporation
Schering-Plough Research Institute
Wyeth Genetics Institute

Plant Corporate Associates

ArborGen

Monsanto Company

Corporate Affiliates

Affymetrix, Inc.

Agencourt Biosciences Corporation

Corporate Contributors

Axxora, L.L.C.
Biogen, Inc.
EMD Bioscience
Epicentre Technologies

Illumina
Integrated DNA Technologies
IRX Therapeutics, Inc.
KeyGene

Foundations

Albert B. Sabin Vaccine Institute, Inc.

Contents

Symposium Participants — v
Foreword — xxiii

Epigenetic Reprogramming and Genomic Imprinting

Mechanism of Mouse Germ Cell Specification: A Genetic Program Regulating Epigenetic Reprogramming *M.A. Surani, K. Ancelin, P. Hajkova, U.C. Lange, B. Payer, P. Western, and M. Saitou* — 1

Epigenetic Mechanisms in Early Mammalian Development *D. Solter, T. Hiiragi, A.V. Evsikov, J. Moyer, W.N. de Vries, A.E. Peaston, and B.B. Knowles* — 11

Nuclear Cloning, Epigenetic Reprogramming, and Cellular Differentiation *R. Jaenisch, K. Hochedlinger, R. Blelloch, Y. Yamada, K. Baldwin, and K. Eggan* — 19

Chromosome Loops, Insulators, and Histone Methylation: New Insights into Regulation of Imprinting in Clusters *W. Reik, A. Murrell, A. Lewis, K. Mitsuya, D. Umlauf, W. Dean, M. Higgins, and R. Feil* — 29

Genomic Imprinting: Antagonistic Mechanisms in the Germ Line and Early Embryo *A.M. Fedoriw, N.I. Engel, and M.S. Bartolomei* — 39

Drosophila Su(Hw) Regulates an Evolutionarily Conserved Silencer from the Mouse *H19* Imprinting Control Region *S. Schoenfelder and R. Paro* — 47

The *Air* Noncoding RNA: An Imprinted *cis*-silencing Transcript *G. Braidotti, T. Baubec, F. Pauler, C. Seidl, O. Smrzka, S. Stricker, I. Yotova, and D.P. Barlow* — 55

The (Dual) Origin of Epigenetics *D. Haig* — 67

Chromosome Inactivation

Sex and X-Chromosome-wide Repression in *Caenorhabditis elegans* *B.J. Meyer, P. McDonel, G. Csankovszki, and E. Ralston* — 71

Targeting Dosage Compensation to the X Chromosome of *Drosophila* Males *H. Oh, X. Bai, Y. Park, J.R. Bone, and M.I. Kuroda* — 81

Mammalian X-Chromosome Inactivation: An Epigenetics Paradigm *E. Heard, J. Chaumeil, O. Masui, and I. Okamoto* — 89

A Continuity of X-Chromosome Silence from Gamete to Zygote *K.D. Huynh and J.T. Lee* — 103

DNA Methylation

Reading the DNA Methylation Signal *A. Bird and D. Macleod* — 113

Genome Defense and DNA Methylation in *Neurospora* *E.U. Selker* — 119

Role of De Novo DNA Methyltransferases in Initiation of Genomic Imprinting and X-Chromosome Inactivation *M. Kaneda, T. Sado, K. Hata, M. Okano, N. Tsujimoto, E. Li, and H. Sasaki* — 125

Gene Repression Paradigms in Animal Cells *L. Lande-Diner, J. Zhang, T. Hashimshony, A. Goren, I. Keshet, and H. Cedar* — 131

Control of Development and Transposon Movement by DNA Methylation in *Arabidopsis thaliana* *T. Kakutani, M. Kato, T. Kinoshita, and A. Miura* — 139

DNA Methylation of the Endogenous PAI Genes in *Arabidopsis* *J. Bender* — 145

Induced and Natural Epigenetic Variation *H. Yi, N.C. Riddle, T.L. Stokes, H.-R. Woo, and E.J. Richards* — 155

Histone Modification and Nucleosome Structure

Linking Covalent Histone Modifications to Epigenetics: The Rigidity and Plasticity of the Marks *Y. Wang, J. Wysocka, J.R. Perlin, L. Leonelli, C.D. Allis, and S.A. Coonrod* — 161

Steps Toward Understanding the Inheritance of Repressive Methyl-Lysine Marks in Histones
  D. Reinberg, S. Chuikov, P. Farnham, D. Karachentsev, A. Kirmizis, A. Kuzmichev,
  R. Margueron, K. Nishioka, T.S. Preissner, K. Sarma, C. Abate-Shen, R. Steward, and
  A. Vaquero ... 171
Noncovalent Modification of Chromatin: Different Remodeled Products with Different
  ATPase Domains H.-Y. Fan, G.J. Narlikar, and R.E. Kingston ... 183
Acetylation of Yeast Histone H4 Lysine 16: A Switch for Protein Interactions in Heterochromatin and
  Euchromatin C.B. Millar, S.K. Kurdistani, and M. Grunstein ... 193
Histone Deposition Proteins: Links between the DNA Replication Machinery and Epigenetic
  Gene Silencing A.A. Franco and P.D. Kaufman ... 201
Trilogies of Histone Lysine Methylation as Epigenetic Landmarks of the Eukaryotic Genome
  M. Lachner, R. Sengupta, G. Schotta, and T. Jenuwein ... 209
Histone H3 Amino-Terminal Tail Phosphorylation and Acetylation: Synergistic or Independent
  Transcriptional Regulatory Marks? C.J. Fry, M.A. Shogren-Knaak, and C.L. Peterson ... 219
Structural Characterization of Histone H2A Variants S. Chakravarthy, Y. Bao, V.A. Roberts,
  D. Tremethick, and K. Luger ... 227

Chromatin Structure and Dynamics

Epigenetics, Histone H3 Variants, and the Inheritance of Chromatin States S. Henikoff,
  E. McKittrick, and K. Ahmad ... 235
Chromatin Boundaries and Chromatin Domains G. Felsenfeld, B. Burgess-Beusse, C. Farrell,
  M. Gaszner, R. Ghirlando, S. Huang, C. Jin, M. Litt, F. Magdinier, V. Mutskov,
  Y. Nakatani, H. Tagami, A. West, and T. Yusufzai ... 245
Do Higher-Order Chromatin Structure and Nuclear Reorganization Play a Role in
  Regulating Hox Gene Expression during Development? W.A. Bickmore,
  N.L. Mahy, and S. Chambeyron ... 251
SIR1 and the Origin of Epigenetic States in Saccharomyces cerevisiae L. Pillus and J. Rine ... 259
Analyzing Heterochromatin Formation Using Chromosome Four of Drosophila melanogaster
  K.A. Haynes, B.A. Leibovitch, S.H. Rangwala, C. Craig, and S.C.R. Elgin ... 267
Two Distinct Nucleosome Assembly Pathways: Dependent or Independent of DNA Synthesis
  Promoted by Histone H3.1 and H3.3 Complexes Y. Nakatani, D. Ray-Gallet, J.-P. Quivy,
  H. Tagami, and G. Almouzni ... 273
The Chromatin Accessibility Complex: Chromatin Dynamics through Nucleosome Sliding
  P.B. Becker ... 281
Histone H2B Ubiquitylation and Deubiquitylation in Genomic Regulation
  N.C.T. Emre and S.L. Berger ... 289

Polycomb and Related Silencing Mechanisms

Polycomb Silencing Mechanisms in Drosophila Y.B. Schwartz, T.G. Khan, G.I. Dellino,
  and V. Pirrotta ... 301
Mechanism of Polycomb Group Gene Silencing Y. Zhang, R. Cao, L. Wang, and R.S. Jones ... 309
Emerging Roles of Polycomb Silencing in X-Inactivation and Stem Cell Maintenance
  I. Muyrers-Chen, I. Hernández-Muñoz, A.H. Lund, M.E. Valk-Lingbeek, P. van der Stoop,
  E. Boutsma, B. Tolhuis, S.W.M. Bruggeman, P. Taghavi, E. Verhoeven, D. Hulsman,
  S. Noback, E. Tanger, H. Theunissen, and M. van Lohuizen ... 319

Nuclear Organization and Dynamics

The Function of Telomere Clustering in Yeast: The Circe Effect S.M. Gasser, F. Hediger,
  A. Taddei, F.R. Neumann, and M.R. Gartenberg ... 327
Genetic Instability in Aging Yeast: A Metastable Hyperrecombinational State
  M.A. McMurray and D.E. Gottschling ... 339
Restructuring the Genome in Response to Adaptive Challenge: McClintock's Bold
  Conjecture Revisited R.A. Jorgensen ... 349
Poetry of b1 Paramutation: cis- and trans-Chromatin Communication V.L. Chandler ... 355

Heterochromatin and Transposon Silencing

RNA Silencing Pathways in Plants *A.J. Herr and D.C. Baulcombe*	363
Transposons, Tandem Repeats, and the Silencing of Imprinted Genes *R. Martienssen, Z. Lippman, B. May, and M. Vaughn*	371
Transposon Silencing and Imprint Establishment in Mammalian Germ Cells *T.H. Bestor and D. Bourc'his*	381
RNA Interference, Heterochromatin, and Centromere Function *R.C. Allshire*	389
RNA Interference, Transposon Silencing, and Cosuppression in the *Caenorhabditis elegans* Germ Line: Similarities and Differences *V.J.P. Robert, N.L. Vastenhouw, and R.H.A. Plasterk*	397

RNA Interference and Related Mechanisms

Plant RNA Interference in Vitro *C. Matranga and P.D. Zamore*	403
A Conserved microRNA Signal Specifies Leaf Polarity *M.C.P. Timmermans, M.T. Juarez, and T.L. Phelps-Durr*	409
RNA Interference and Epigenetic Control of Heterochromatin Assembly in Fission Yeast *H. Cam and S.I.S. Grewal*	419
Regulation of *Caenorhabditis elegans* RNA Interference by the *daf*-2 Insulin Stress and Longevity Signaling Pathway *D. Wang and G. Ruvkun*	429
Interrelationship of RNA Interference and Transcriptional Gene Silencing in *Drosophila* *M. Pal-Bhadra, U. Bhadra, and J.A. Birchler*	433
Functional Identification of Cancer-relevant Genes through Large-Scale RNA Interference Screens in Mammalian Cells *T.R. Brummelkamp, K. Berns, E.M. Hijmans, J. Mullenders, A. Fabius, M. Heimerikx, A. Velds, R.M. Kerkhoven, M. Madiredjo, R. Bernards, and R.L. Beijersbergen*	439

Epigenetic Regulation of Phenotypes

The New Field of Epigenomics: Implications for Cancer and Other Common Disease Research *H.T. Bjornsson, H. Cui, D. Gius, M.D. Fallin, and A.P. Feinberg*	447
Epigenetic Regulation in the Control of Flowering *J. Mylne, T. Greb, C. Lister, and C. Dean*	457
Transposons and Tandem Repeats Are Not Involved in the Control of Genomic Imprinting at the MEDEA Locus in *Arabidopsis* *C. Spillane, C. Baroux, J.-M. Escobar-Restrepo, D.R. Page, S. Laoueille, and U. Grossniklaus*	465
Toward Molecular Understanding of Polar Overdominance at the Ovine Callipyge Locus *M. Georges, C. Charlier, M. Smit, E. Davis, T. Shay, X. Tordoir, H. Takeda, F. Caiment, and N. Cockett*	477
Dscam-mediated Self- versus Non-Self-Recognition by Individual Neurons *G. Neves and A. Chess*	485
Prions of Yeast Are Genes Made of Protein: Amyloids and Enzymes *R.B. Wickner, H.K. Edskes, E.D. Ross, M.M. Pierce, F. Shewmaker, U. Baxa, and A. Brachmann*	489
A Possible Epigenetic Mechanism for the Persistence of Memory *K. Si, S. Lindquist, and E. Kandel*	497
An Epigenetic Hypothesis for Human Brain Laterality, Handedness, and Psychosis Development *A.J.S. Klar*	499
Summary: Epigenetics—from Phenomenon to Field *D. Gottschling*	507
Author Index	521
Subject Index	523

Mechanism of Mouse Germ Cell Specification: A Genetic Program Regulating Epigenetic Reprogramming

M.A. SURANI, K. ANCELIN, P. HAJKOVA, U.C. LANGE, B. PAYER, P. WESTERN,[*] AND M. SAITOU[†]

Wellcome Trust Cancer Research UK Gurdon Institute of Cancer and Developmental Biology, University of Cambridge, Cambridge CB2 1QR, United Kingdom

Specification of germ cells and their segregation from somatic neighbors is one of the most decisive events of early development in animals. There are two key routes to the initiation of the germ cell lineage. One is through the inheritance of preformed germ cell determinants or germ plasm as observed in *Drosophila melanogaster* and *Caenorhabditis elegans* (Eddy 1975; Wylie 1999). The other route, which is sometimes referred to as the stem cell model, occurs in the mouse, where a group of pluripotent cells are first established with seemingly equivalent potential from which both germ cells and somatic cells are derived (McLaren 1999; Saitou et al. 2003). These two modes of germ cell specification are referred to as preformation and epigenesis, respectively (Fig. 1). Although preformation is observed in most model organisms, it is apparently a less dominant mode of germ cell specification, and epigenesis may be ancestral to the Metazoa (Extavour and Akam 2003).

Primordial germ cells (PGCs), the precursors of sperm and eggs, are the source of totipotency, which is a unique state generated by this lineage. Throughout the period of specification of PGCs and the extensive epigenetic reprogramming of the genome, germ cells require robust mechanisms to ensure that they do not undergo differentiation into somatic cells or revert to an overtly pluripotent state. Furthermore, as transmitters of genetic and epigenetic information to all subsequent generations, PGCs exhibit profound epigenetic reprogramming of the genome, which in mammals includes erasure and initiation of parental imprints that are responsible for the functional differences between the parental genomes during mammalian development (Surani 2001). Genetic mechanisms involved in PGC specification produce a lineage with the essential property of extensive epigenetic reprogramming of the genome that ensues after they reach developing gonads (Surani 2001; Hajkova et al. 2002; Lee et al. 2002). Thus, mechanisms that are involved in PGC specification, maintenance, and epigenetic reprogramming are of considerable general interest in the context of studies on embryonic and adult stem cells.

Although specification of the germ line is one of the most crucial events of early animal development, the mechanisms involved are surprisingly not evolutionarily conserved. This is apparently necessary because of the vital differences in early development of different species (Saitou et al. 2003; Schaner et al. 2003; Blackwell 2004). Compared to *D. melanogaster* and *C. elegans*, for example, specification of germ cells in mammals is considerably delayed (McLaren 1999). This is evident since the first requirement in mammals is development of a blastocyst, and specification of PGCs occurs later after blastocyst implantation. It is the pluripotent epiblast cells of the egg cylinder of the early postimplantation embryo, which is the source of both PGCs and somatic cells, that are specified in response to signaling molecules (Lawson and Hage 1994; Lawson et al. 1999). Incidentally, the early epiblast cells are also the source of pluripotent embryonic stem (ES) cells, which also have the potential to generate both somatic and germ cells in vitro and in vivo (Hubner et al. 2003; Toyooka et al. 2003; Geijsen et al. 2004; Surani 2004). Despite the essential differences in the mechanisms of germ cell specification observed in different organisms, there are nevertheless certain underlying principles, such as the repression of the somatic program in germ cells, which may be conserved among organisms, as are some of the molecules and mechanisms for the maintenance and propagation of the germ cell lineage.

REPRESSION OF SOMATIC PROGRAM: A COMMON THEME DURING PGC SPECIFICATION

The repression of somatic cell fate is a common feature observed in founder germ cells of all the animals studied in detail (Seydoux and Strome 1999; Saitou et al. 2003; Blackwell 2004). Disruption to germ line silencing is detrimental to the maintenance of the germ cell lineage. At the onset of germ cell specification in *D. melanogaster* and *C. elegans*, which occur very early during development, global repression of transcription prevents germ cells from acquiring a somatic fate (Leatherman and Jongens 2003). In mammals where PGC specification occurs at a later stage of development, there is targeted and active repression of genes that are specifically expressed in the neighboring somatic mesoderm cells, which share common developmental ancestry (see below). Until re-

---

[*]Present address: ARC Center for Development and Biotechnology, Murdoch Children's Hospital, Parkville, Victoria 3052, Australia.
[†]Present address: Laboratory for Mammalian Germ Cell Biology, Center for Developmental Biology, RIKEN Kobe Institute, Chuo-ku, Kobe, Hyogo 650-0047, Japan.

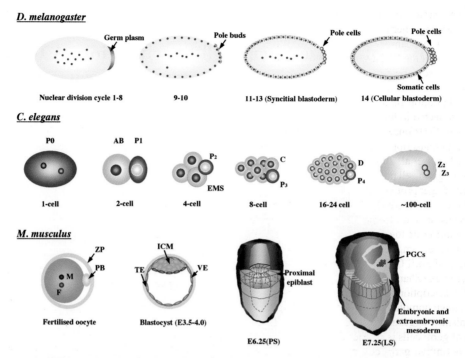

Figure 1. Specification of germ cells in different model organisms. In *Drosophila melanogaster*, those nuclei that enter the germ plasm induce the formation of pole cells, which give rise to germ cells. In *Caenorhabditis elegans*, the P0 blastomere divides to form the AB somatic blastomere and P1 germ line blastomere. The latter eventually gives rise to Z2 and Z3 cells, which develop into germ cells. In *Mus musculus*, germ cells originate from the pluripotent proximal epiblast cells of E6.25 embryos. By E7.25, a group of about 50 founder primordial germ cells (PGCs) are detected.

cently, it was not known that repression of the somatic program is likely to be a critical event during PGC specification in mice. We now have the first indications of a possible mechanism involved in this process, which may be a critical component of PGC specification.

Specification of germ cells in *Drosophila* represents one extreme, where pole cells, the precursors of germ cells, are set aside from the somatic lineage at the very outset even before the onset of zygotic transcription and cellularization. Specific mechanisms of transcriptional repression are used to escape from a somatic program adopted by the remaining cells. Notably, transcriptional repression in pole cells is achieved by the lack of phosphorylation of RNA polymerase II carboxy-terminal domain (CTD) repeat on serine at position 2 that is essential for transcriptional elongation and by the low levels of CTD phosphorylation on serine at position 5, a modification that is essential for transcription initiation (Seydoux and Dunn 1997; Van Doren et al. 1998; Zhang et al. 2003). Recent studies have shown that the regulation of transcriptional silencing in these putative germ cells requires *polar granule component* (*pgc*) gene, a noncoding RNA (Deshpande et al. 2004; Martinho et al. 2004). Absence of *pgc* results in the loss of repression, but this does not affect the formation of pole cells. The mutant pole cells display a precocious phosphorylation of ser2 of CTD. How precisely *pgc* works is currently unknown but it has been suggested that *pgc* might sequester critical components for PolII CTD phosphorylation for the transition from preinitiation complex to the elongation complex.

There are potentially other factors that are required in combination with *pgc* to induce transcriptional repression in pole cells. These include *germ cell–less* (*gcl*), since a loss of function of this gene also causes premature transcriptional activation in pole cells although the underlying mechanism is unclear (Leatherman et al. 2002). *nanos* and *pumilio* also have a role in transcriptional repression in germ cell precursors (Hayashi et al. 2004). Thus, there are overlapping and redundant mechanisms involved in regulating transcriptional repression in the *Drosophila* pole cells.

Global transcriptional quiescence is also seen during the formation of germ cell precursors in *C. elegans* (Seydoux et al. 1996; Seydoux and Dunn 1997). However, the mechanism differs from that in *Drosophila* because of key differences in their early development. In *C. elegans*, the single P blastomere divides asymmetrically to give rise to a germ cell (P1) and a somatic daughter (AB). Following symmetric division of the P4 blastomere, Z2 and Z3 blastomeres become restricted to the germ line fate (Seydoux and Strome 1999). However, as in *Drosophila*, the P blastomeres destined for the germ line fate are initially repressed for the RNA PolII transcription, while the contemporary somatic daughters (from AB somatic cells) are transcriptionally active (Seydoux and Strome 1999). In the absence of transcriptional repression, germ line blastomeres adopt a somatic fate (Fig. 1).

The early and rapid asymmetric division of a single blastomere generating a germ cell and a somatic cell in *C. elegans* requires a mechanism for transcriptional repression to be imposed in the P1 blastomere when the daughter cell acquires a somatic fate. However, initially, both the somatic and germ cell nuclei have an equivalent capacity for the initiation of transcription (Schaner et al. 2003). This is reflected in high levels of dimethylated lysine 4 of histone H3 (H3meK4), which is indicative of the potential for high transcriptional competence in germ cell blastomere (Schaner et al. 2003). However, transcription in the *C. elegans* germ cell precursors is repressed by PIE-1, a CCCH zinc finger protein that shows asymmetric segregation to the P blastomere (Mello et al. 1996; Seydoux et al. 1996; Seydoux and Strome 1999). PIE-1 competes with the recognition of the RNA polymerase II (PolII) CTD for the kinase complex, which is essential for the phosphorylation of PolII CTD and the release of the transcriptional complex (Zhang et al. 2003). PIE-1-mediated repression of transcription is unique to *C. elegans*, since there are no known homologs of the gene encoding PIE-1 in other organisms, suggesting that this mechanism for the specification of germ cells is restricted to *C. elegans*.

When the definitive germ cell precursors, Z2/Z3, are formed in *C. elegans*, PIE-1 is degraded but the transcriptional silencing still remains. This transcriptional silencing is evident in the loss of active chromatin markers such as H3K4 methylation. Interestingly, *Drosophila* pole cells also lack H3K4 methylation, but show high levels of H3K9 methylation. Altogether, germ cell precursors in both flies and worms carry hallmarks of transcriptional quiescence (Schaner et al. 2003). Loss of *pgc* in *Drosophila*, for example, results in the acquisition of H3K4 methylation and RNA PolII CTD serine 2 phosphorylation. These changes are associated with high transcriptional activity in pole cells similar to that in somatic cells (Deshpande et al. 2004; Martinho et al. 2004). Similar effects are seen in *C. elegans* following a loss of PIE-1. However, one of the differences in the specification of germ cells in flies and worms is that the early and rapid asymmetric division of a single blastomere generating a germ cell and a somatic cell in *C. elegans* does not allow for a chromatin-based repression in the germ cell precursors and is instead in a permissive state for transcription (Schaner et al. 2003). However, as soon as the fate is restricted in Z2/Z3, then the same chromatin remodeling is observed as in the fly pole cells. Interestingly, nanos is required for transcriptional silencing in pole cells of *Drosophila*, and the lack of H3K4 methylation in Z2/Z3 cells in worms may also require nanos homologs, nos-1 and nos-2 (Schaner et al. 2003). Understanding how nanos could link germ cell chromatin remodeling is of high interest.

These studies on *C. elegans* and *Drosophila* illustrate that germ cell specification in response to the inheritance of germ plasm has some common features although the key proteins involved in the transcriptional repression are not conserved. Germ cell specification in mice, however, differs more markedly since there is no involvement of preformed germ cell determinants, and PGC specification occurs relatively late during development (Lawson and Hage 1994; McLaren 1999). Studies on PGC specification and the role of intrinsic epigenetic mechanisms that regulate transcription are just beginning to reveal their critical role in establishing the germ cell lineage. The mechanisms of global transcriptional repression that underlie germ cell specification in *C. elegans* and *Drosophila* are not feasible in mammals, where transcriptionally active pluripotent epiblast cells need to respond to extrinsic signals for the specification of both PGCs and somatic cells (Lawson et al. 1999). Nevertheless, as described below, germ cell specification in mammals also requires mechanisms for repression of somatic specific genes that are active in the neighboring cells. This may indeed be the key event of germ cell specification in mice and probably in all mammals.

REPRESSION OF THE SOMATIC PROGRAM DURING PGC SPECIFICATION IN MICE

Mouse development involves the onset of embryonic transcription at the two-cell stage followed by differentiation of the trophectoderm and inner cell mass, which contribute to the placenta and the embryo proper, respectively (Fig. 1). Specification of PGCs in mice is an instructive process in which the pluripotent epiblast cells serve as the precursors of both fetal somatic and germ cells. Evidence shows that the proximal epiblast cells of the embryonic day (E) 6.5 egg cylinder that give rise to germ cells have to be primed by signaling molecules, including bone morphogenetic protein 4 (BMP4), before they acquire competence to give rise to germ cells (Lawson and Hage 1994; Lawson et al. 1999). Thus, specification of germ cells in mice is an instructive process. At about E7.5, approximately 50 nascent PGCs can be detected (Lawson and Hage 1994; McLaren 1999). However, not all cells in the proximal epiblast give rise to germ cells as the majority go on to form the somatic extraembryonic mesodermal lineages (Lawson and Hage 1994). Therefore, PGC specification may involve not only the induction of a germ cell–specific program, but also a need to repress genes that instruct a somatic fate on neighboring cells. Our recent studies suggest that repression of the somatic program may involve epigenetic mechanisms as an essential component of PGC specification.

The acquisition of germ cell competence in E6.25 proximal epiblast cells results in the up-regulation of *fragilis* (Saitou et al. 2002; Tanaka and Matsui 2002), a gene that encodes a member of a family of transmembrane protein (Fig. 2). These genes are the homologs of the interferon inducible human *Iftm* gene family, of which members are predicted to be involved in homotypic cell adhesion and cell cycle elongation (Friedman et al. 1984; Evans et al. 1990, 1993; Deblandre et al. 1995). During gastrulation, *fragilis* expression shifts to the posterior region and intensifies in a region where the founder PGCs are eventually detected at E7.5. Expression of *fragilis* is not detected in the BMP4 homozygous mutants, and it is substantially reduced in heterozygous embryos, which suggests that *fragilis* expression is BMP4 dose–de-

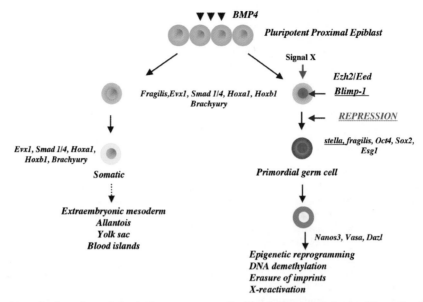

Figure 2. Specification and epigenetic regulation in the mouse germ cells. The pluripotent proximal epiblast cells of E6.25 embryos respond to BMP4 and other unknown signals (Signal X), which results in the initiation of *Blimp-1* expression in a small group of epiblast cells. There is repression of genes, including *Hox* genes that are up-regulated in the neighboring somatic cells. This is apparently a critical event during PGC specification in mice. Subsequently, there is extensive epigenetic reprogramming of the genome in the germ cell lineage, including erasure of parental imprints and X-reactivation, followed by initiation of parental imprints later during gametogenesis.

pendent (Saitou et al. 2002). At the end of the specification, PGCs exhibit expression of *stella,* a nuclear protein with a SAP-like domain, within the center of the *fragilis* positive cell cluster. These *stella*-positive cells appear to be the lineage-restricted germ cells (Saitou et al. 2002).

Analysis of the *stella*-positive founder PGCs at E7.5 (early bud [EB] stage) and their somatic neighbors indicates that the germ cells initially share the gene expression program of cells destined to form the extraembryonic mesoderm cells. In particular, *stella*-positive PGCs show repression of the region-specific homeobox genes, including *Hoxb1*, *Hoxa1*, and *Evx1*, which are all highly expressed in the neighboring somatic cells (Saitou et al. 2002). Since homeobox genes specify either the regional identity of cells along the body axis or induce differentiation of cells toward specific somatic cell lineages, our result suggests that the founder germ cells avoid somatic specification at least in part by preventing or suppressing homeobox gene expression (Fig. 2).

We have further evidence for active repression of the somatic program during PGC specification based on single-cell cDNA analysis at E7.25 (later streak [LS] stage) (Downs and Davies 1993). These cells are ~6 hours earlier than those described above, which express high amounts of *fragilis*, but not *stella* in significant amounts. Remarkably, they show high levels of *Evx1* expression, which is comparable to the levels in the neighboring somatic cells. *Evx1* is therefore clearly repressed in the founder PGCs, suggesting an active mechanism for the repression of this homeobox gene during PGC specification. Other genes, including *Smad1* and *Smad4* (U.C.

Lange, unpubl.), that are the transducers of BMP4 signaling are also apparently repressed, which suggests that germ cells shut off their response to BMP4 at the time of germ cell specification. These examples serve to illustrate that during PGC specification, there is probably extensive repression of genes essential for the specification of neighboring somatic mesodermal fate (Saitou et al. 2003). However, this proposed model requires further investigation.

To summarize the intrinsic events of PGC specification, we have proposed that the founder germ cells escape from a somatic cell fate through repression of the somatic program. At E7.0 (midstreak [MS] stage) shortly after gastrulation commences, cells located within the proximal-posterior region express genes such as *T* (*Brachyury*) that are associated with mesoderm formation. They are also strongly positive for the region-specific homeobox gene *Evx1*. At E7.25 (LS stage), there is a strong expression of *fragilis* in a subset of cells. Subsequently in cells that show expression of *stella,* there is a strong repression of *HOX* genes and of signal transducers *Smad1* and *Smad4*. These are the nascent PGCs detected at E7.5 (EB stage) with high expression of *fragilis* and *stella*, which eventually become lineage-restricted as germ cells. By contrast, in cells that do not up-regulate *fragilis* at the LS stage, expression of homeobox genes, such as *Hoxb1* and *Hoxa1*, begins and expression of *Evx1* and the *Smads* is maintained. Although *Oct4* as well as other genes, such as *Ezh2* and *Eed*, are expressed at high levels in both somatic cells and germ cells at the time of germ cell specification, they are subsequently maintained exclusively or at higher levels, respectively, only in germ cells. The repression of the somatic program during the critical period of PGC specification is clearly a key event (Fig. 2). Elucidation of the mechanism underlying this process is likely to lead to a significant advance in the understanding of PGC specification in mice (Saitou et al. 2002, 2003).

Blimp-1: A POTENTIAL REGULATOR OF PGC SPECIFICATION

To begin to address the mechanism of repression of the somatic program during PGC specification, we examined expression of candidate genes that could be involved in this process. Among these are two members of the polycomb group (PcG) proteins, Ezh2 and Eed, that are known to be associated with repression of *Hox* genes in *Drosophila* (Orlando 2003). These PcG group proteins are expressed during early development (O'Carroll et al. 2001; Erhardt et al. 2003) and to the same extent in both the founder PGCs and neighboring somatic cells. Despite this, it is possible that germ cell–specific factors may recruit these polycomb proteins to the *Hox* gene loci and repress their expression only in germ cells. Similarly, we also detected G9a expression in PGCs and somatic cells, which has the potential for methylation of histone H3 lysine 9 (H3meK9) within the euchromatin region (Tachibana et al. 2002). However, more significantly, we found that Blimp-1 was expressed exclusively in germ cells but not in the immediate neighboring somatic cells (Y. Ohinata et al., in prep.) (Fig. 2). Blimp-1, a transcriptional repressor, belongs to the RIZ family with a SET domain and five C2H2 type zinc fingers toward the carboxyl terminus (Turner et al. 1994; Kouzarides 2002; Lachner and Jenuwein 2002). Although, there is no evidence at present that this protein with a SET domain has methyltransferase activity (Kouzarides 2002; Lachner and Jenuwein 2002), there is evidence that Blimp-1 can form repressive chromatin complexes with other proteins (Makar and Wilson 2004). It is also worth noting that Blimp-1 may also have a function in transcriptional activation through its acidic carboxyl terminus (Sciammas and Davis 2004). Blimp-1 is widely expressed in many tissues during development and therefore it probably has distinct functions in a context-dependent manner (Chang et al. 2002).

In the context of PGC specification, Blimp-1 expression is first detected among a group of *fragilis*-positive cells, which precedes the critical period of repression of *Hox* genes in the putative germ cells, and of the expression of *stella* that marks the end of PGC specification (Y. Ohinata et al., in prep.). When *stella* starts to be detectable, the number of Blimp-1-positive cells exceeds that of *stella*-positive cells at the late streak stage. Preliminary studies also suggest that Blimp-1 is critical for PGC specification, because their development is affected in Blimp-1 null embryos. A potential role of Blimp-1 may be to repress genes, including *Hox* genes, and this process may be affected in the absence of Blimp-1. As a result, the Blimp-1 mutant cells may acquire an alternative fate or undergo apoptosis. Studies are in progress to elucidate the precise role of Blimp-1 in PGC specification.

There is a precedent for the proposed role of Blimp-1 as a regulator in cell fate specification during terminal differentiation of B cells into Ig-secreting plasma cells. This event is associated with both the global repression of the B cell genetic program (Turner et al. 1994; Shaffer et al. 2002; Sciammas and Davis 2004), as well as of transcriptional activation of certain genes associated with the differentiation of plasma cells (Sciammas and Davis 2004). Indeed, it was originally described as a "master regulator" of plasma cell differentiation (Turner et al. 1994), because overexpression of Blimp-1 in BCL1 cells is sufficient to cause plasma cell differentiation. More remarkably, Blimp-1 apparently causes plasma cell differentiation through overall suppression of mature B cell–associated gene transcription, including *cMyc, Id3, Pax5,* and others (Turner et al. 1994; Shaffer et al. 2002; Shapiro-Shelef et al. 2003; Sciammas and Davis 2004). The proposed repression of *Pax5*, detected in one study (Lin et al. 2002), may in turn lead to derepression of a number of other genes, leading to a major change in the transcriptome. Blimp-1-binding sites have been identified in these and other genes. These findings may serve as a paradigm for the specification of PGCs through suppression of somatic cell program.

A recent study suggests that a complex modular nature of Blimp-1 protein is capable of a variety of distinct functions, which is a hallmark of a significant and crucial regulator of transcription that could be critical for cell fate determination (Sciammas and Davis 2004). For example, there are several potential mechanisms by which Blimp-1 could cause transcriptional repression of target genes, although there is as yet no evidence that the SET domain at the amino terminus of Blimp-1 has the enzymatic property for histone tail modifications. It has long been known that *Blimp-1* is involved in the postinduction repression of human interferon-β (*IFNβ*) gene transcription in response to viral induction. This could be accomplished through a proline-rich region located toward the amino terminus of Blimp-1 that can recruit repressor proteins Groucho and HDAC2 (Ren et al. 1999). However, more recent evidence shows that Blimp-1 can also recruit G9a, a histone H3 methyltransferase for lysine 9 (H3-2mK9), which is known to be active during early mouse development (Gyory et al. 2004). G9a is widely expressed in early postimplantation embryos and in germ cells, and loss of function of G9a is early embryonic lethal (Tachibana et al. 2002). This provides a potential for a repressive chromatin to form that could repress the somatic program during specification of PGCs in mice. What is also important is that the C2H2 zinc fingers present in Blimp-1 have an essential role in inducing repression of *IFNβ* promoter (Keller and Maniatis 1992; Gyory et al. 2004). Thus, it is possible that Blimp-1 through its zinc fingers may target methyltransferase activity associated with G9a (and possibly other histone methylases in other contexts) to specific sites on a variety of genes. We also cannot entirely exclude a possibility that the SET domain of Blimp-1 itself may also have enzymatic activity capable of histone modifications. These studies provide potential models that are compatible with Blimp-1-mediated repression of specific sets of genes during germ cell specification (Fig. 3).

From all the studies carried out so far, it is evident that transcriptional repression of somatic cell fate is a key component of germ cell specification in model organisms, although the mechanisms differ significantly among them. Different mechanisms are adopted in *Drosophila* and *C. elegans*, where global transcriptional repression is rapidly induced in the cells destined for the germ cell lineage. In mice, germ cell–competent cells are transcriptionally active at the time of germ cell specification. It also appears that the PGC-competent cells are initially destined for a somatic cell fate but this program is repressed during the critical period of PGC specification.

PROPAGATION AND EPIGENETIC REPROGRAMMING OF THE GERM CELL LINEAGE

Specification of the germ cell lineage involves distinct mechanisms that are not evolutionarily conserved in different organisms, where genes such as *pgc* and *Pie1* have unique roles. Thus, distinct genetic pathways presumably generate specific chromatin states in founder germ cells. Some aspects of the propagation of the germ cell lineage may, however, rely, at least partially, on conserved homologous genes. Among these is Nanos, an RNA-binding protein and a translational repressor that is conserved among the germ cell lineage in different species (Schaner et al. 2003; Blackwell 2004). In *Drosophila*, *nanos* and *pumilio* are necessary for the maintenance of the germ cell lineage (Subramaniam and Seydoux 1999; Wang and Lin 2004). A recent study in *Drosophila* demonstrated that a lack of maternally inherited Nos leads to the loss of germ cell lineage by apoptosis, but when apoptosis was repressed, some of the Nos-negative germ cells adopted somatic fate and lost expression of the germ line–specific expression of Vasa (Hayashi et al. 2004). In *C. elegans*, *nos-1* and *nos-2* have overlapping roles and their loss of function results in sterile adults, which is associated with aberrant H3meK4 modifications in germ cells (Subramaniam and Seydoux 1999; Schaner et al. 2003). In mice as well, *nanos 2* has a role in maintaining spermatogenic stem cells, while loss of *nanos 3* results in the loss of both male and female germ cells earlier during their migration into the gonads (Tsuda et al. 2003). Thus *nanos3* does not affect PGC specification but it clearly has a role in the maintenance of the lineage (Fig. 2).

Also of particular note is the potential role of *Polycomb* (PcG) genes, *Ezh2* and *eed*, in the germ cell lineage. In mice, both these genes are expressed in embryonic cells at the time of germ cell specification, but their role is yet unclear. It is possible that this PcG complex plays an essential role through propagation of early cell fate decisions in germ cells and somatic cells. The role of Ezh2/Eed complex, which includes histone deacetylases in transcriptional repression, has been well documented (for review, see Cao and Zhang 2004). It has long been established that this complex can, for example, induce silencing of *Hox* genes. Ezh2 is a SET domain protein with the predominant activity for the methylation of histone H3 lysine 27 (H3meK27). The Ezh2/Eed complex is present throughout early preimplantation development in germ cells, oocytes, and in pluripotent stem cells in mice (O'Carroll et al. 2001; Erhardt et al. 2003). The role of these proteins in the mouse germ cell lineage is yet unclear.

In *C. elegans*, the role of PcG has been well established. Here MES-2 (counterpart of Ezh2), MES-3, and MES-6 (counterpart of Eed) function as a protein complex for transcriptional silencing (Seydoux and Strome 1999; Pirrotta 2002). As outlined above, PIE-1 is initially involved in transcriptional repression but this maternally inherited protein is eventually degraded at the 100-cell stage, while the expression of *nos* and *mes* genes continue. Recent studies reveal that the MES 2/3/6 complex is responsible for H3meK27 methylation, as is the case with Ezh2/Eed complex in mice and humans (see Cao and Zhang 2004). Mutations in *mes* genes showed a marked alteration in H3meK27 marks, which may explain the loss of transcriptional silencing in the germ cell lineage. This PcG complex is also detected in the mouse germ cell lineage from the time of specification and later when the germ cell lineage is established. The precise role of this complex in regulating the epigenetic status and maintenance of the germ cell lineage remains to be elucidated.

One significant property of the mouse germ cell lineage includes extensive epigenetic reprogramming, involving erasure and initiation of parental imprints (Surani 2001; Hajkova et al. 2002; Lee et al. 2002). In vivo, the erasure of imprints coincides with the entry of migrating PGCs into the developing gonads, which is a part of genome-wide DNA demethylation. Furthermore, in female embryos, the inactive X chromosome is also reactivated when germ cells reach the developing gonads. It is important to note that the erasure of imprints occurs in both male and female germ cells before the onset of sexual differentiation of developing gonads. By contrast, reinitiation of new parental imprints during gametogenesis is sex specific, as described elsewhere in this volume.

During maturation of the oocytes, there is also the maternal inheritance of epigenetic modifiers that are present in the mature oocyte, which have an essential role during early development (Erhardt et al. 2003). Among these are

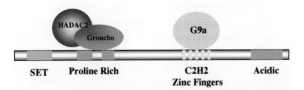

Figure 3. Blimp-1 protein and potential mechanisms of gene repression. Studies suggest that Blimp-1 could induce repression of target genes through a complex comprised of Groucho and HDAC2, or through recruitment of G9a histone methyltransferase. The five Kruppel-type zinc fingers may be essential for the recruitment of Blimp-1-dependent repressive chromatin to target sites. No function has yet been found for the SET domain that is present in Blimp-1. Blimp-1 can also induce transcription, perhaps through the acidic tail in the carboxyl terminus.

again the PcG proteins, Ezh2 and Eed. A detailed analysis of the role of these and other maternally inherited factors in the mammalian oocyte has yet to be carried out. Some of the factors present in the oocyte must also have a role in reprogramming of the somatic nucleus. It is of interest to consider if some of these factors and mechanisms are common to both the oocyte and PGCs at the time of the epigenetic reprogramming of the genome in the germ cell lineage.

GERM CELLS AND STEM CELLS: REVERSIBLE PHENOTYPES

Although PGCs are highly specialized cells with many unique properties, they retain an underlying pluripotency and show expression of pluripotency-associated genes such as *Oct4*. However, PGCs in vivo do not exhibit a tendency to undergo differentiation into diverse cell types, or to revert to an overtly pluripotent state. The maintenance of the phenotype could be attributed to the overlapping mechanisms that regulate transcriptional repression to ensure that PGCs do not undergo differentiation. PGCs are indeed highly refractory to differentiation even when transplanted to ectopic sites or into blastocysts. This is in contrast to the pluripotent epiblast or ES cells, which undergo differentiation into all the diverse cell types. This indicates that PGCs do not respond to the signals that induce differentiation of epiblast and ES cells.

There are, however, conditions in vitro that can erase epigenetic changes associated with the cell fate decision leading to PGC specification. When PGCs are cultured in vitro under specific conditions, which include exposure to FGF2, at least some of them revert to forming pluripotent stem cells called embryonic germ (EG) cells (Matsui et al. 1992; Donovan and de Miguel 2003). What intrinsic epigenetic mechanism leads to this phenotypic reversal is unknown. EG cells are very similar to ES cells that are derived from epiblast cells of blastocysts (Durcova-Hills et al. 2001). However, EG cells exhibit erased imprints like their precursor PGCs in vivo (Tada et al. 1998). Pluripotent ES (and to a large extent EG) cells can, in turn, give rise to all somatic cells and germ cells in vivo when introduced into blastocysts, as well as when allowed to undergo differentiation in vitro. These studies demonstrate that PGCs and pluripotent stem cell states are potentially reversible.

CONCLUSION

The specification of germ cell lineage in many organisms involves repression of the somatic program. The mechanisms during this critical event appear not to be evolutionarily conserved. At one extreme are the organisms in which the germ line cell fate is dependent on maternally inherited germ plasm. These components act on establishing a global transcriptional silencing, by blocking RNA polII but also inducing a general refractory chromatin state. In *Drosophila*, transcriptional silencing is established in pole cells prior to the onset of zygotic transcription. In *C. elegans*, germ cell and somatic blastomeres are created from a single cell by the first asymmetric cell division. Since normal transcription is essential in the contemporary somatic daughter cell, the chromatin initially is in an active state for transcription. Here, preferential segregation of PIE-1, a zinc finger protein, in germ cell has a critical role in transcriptional repression through a mechanism affecting RNA processing. In mice, specification of PGCs is an instructive process and it involves yet another distinct mechanism. Following implantation, pluripotent epiblast cells commence differentiation into somatic and germ cells in response to signaling molecules. Cells that give rise to PGCs are transcriptionally active and initially indistinguishable from the neighboring cells, which give rise to the extraembryonic mesoderm. Repression of the somatic program in germ cells is thus a critical event during PGC specification. Blimp-1 may be a potential regulator of this process, perhaps through recruitment of other proteins, such as Groucho, HDAC2, and/or G9a, to generate a repressive chromatin.

The maintenance of the germ cell lineage probably requires overlapping redundant repression mechanisms. Indeed, PGCs in mice are highly refractory to phenotypic changes, suggesting that robust mechanisms ensure their propagation. This may be essential to ensure that germ cells that retain an underlying pluripotency do not differentiate into somatic cells or revert to an explicit pluripotent phenotype. Among the likely mechanisms are PcG group proteins, Ezh2 and Eed in particular, which are known to have a critical role in transcriptional regulation of early embryos, germ cells, and pluripotent stem cells.

A particularly noteworthy property of mouse PGCs is the extensive epigenetic reprogramming of the genome, including erasure and reestablishment of parental imprints, genome-wide DNA demethylation, and reactivation of the inactive X chromosome. The discovery of the underlying mechanism of this process will advance understanding of the regulation of genome functions through erasure and reestablishment of new epigenetic states.

ACKNOWLEDGMENTS

We thank the Wellcome Trust and BBSRC for grants in support of this work and all of our colleagues for their comments during the course of this work.

REFERENCES

Blackwell T.K. 2004. Germ cells: Finding programs of mass repression. *Curr. Biol.* **14:** R229.
Cao R. and Zhang Y. 2004. The functions of E(Z)/EZH2-mediated methylation of lysine 27 in histone H3. *Curr. Opin. Genet. Dev.* **14:** 155.
Chang D.H., Cattoretti G., and Calame K.L. 2002. The dynamic expression pattern of B lymphocyte induced maturation protein-1 (Blimp-1) during mouse embryonic development. *Mech. Dev.* **117:** 305.
Deblandre G.A., Marinx O.P., Evans S.S., Majjaj S., Leo O., Caput D., Huez G.A., and Wathelet M.G. 1995. Expression cloning of an interferon-inducible 17-kDa membrane protein implicated in the control of cell growth. *J. Biol. Chem.* **270:** 23860.
Deshpande G., Calhoun G., and Schedl P. 2004. Overlapping mechanisms function to establish transcriptional quiescence in the embryonic *Drosophila* germline. *Development* **131:**

1247.

Donovan P.J. and de Miguel M.P. 2003. Turning germ cells into stem cells. *Curr. Opin. Genet. Dev.* **13:** 463.

Downs K.M. and Davies T. 1993. Staging of gastrulating mouse embryos by morphological landmarks in the dissecting microscope. *Development* **118:** 1255.

Durcova-Hills G., Ainscough J., and McLaren A. 2001. Pluripotential stem cells derived from migrating primordial germ cells. *Differentiation* **68:** 220.

Eddy E.M. 1975. Germ plasm and the differentiation of the germ cell line. *Int. Rev. Cytol.* **43:** 229.

Erhardt S., Su I.H., Schneider R., Barton S., Bannister A.J., Perez-Burgos L., Jenuwein T., Kouzarides T., Tarakhovsky A., and Surani M.A. 2003. Consequences of the depletion of zygotic and embryonic enhancer of zeste 2 during preimplantation mouse development. *Development* **130:** 4235.

Evans S.S., Collea R.P., Leasure J.A., and Lee D.B. 1993. IFN-alpha induces homotypic adhesion and Leu-13 expression in human B lymphoid cells. *J. Immunol.* **150:** 736.

Evans S.S., Lee D.B., Han T., Tomasi T.B., and Evans R.L. 1990. Monoclonal antibody to the interferon-inducible protein Leu-13 triggers aggregation and inhibits proliferation of leukemic B cells. *Blood* **76:** 2583.

Extavour C.G. and Akam M. 2003. Mechanisms of germ cell specification across the metazoans: Epigenesis and preformation. *Development* **130:** 5869.

Friedman R.L., Manly S.P., McMahon M., Kerr I.M., and Stark G.R. 1984. Transcriptional and posttranscriptional regulation of interferon-induced gene expression in human cells. *Cell* **38:** 745.

Geijsen N., Horoschak M., Kim K., Gribnau J., Eggan K., and Daley G.Q. 2004. Derivation of embryonic germ cells and male gametes from embryonic stem cells. *Nature* **427:** 148.

Gyory I., Wu J., Fejer G., Seto E., and Wright K.L. 2004. PRDI-BF1 recruits the histone H3 methyltransferase G9a in transcriptional silencing. *Nat. Immunol.* **5:** 299.

Hajkova P., Erhardt S., Lane N., Haaf T., El-Maarri O., Reik W., Walter J., and Surani M.A. 2002. Epigenetic reprogramming in mouse primordial germ cells. *Mech. Dev.* **117:** 15.

Hayashi Y., Hayashi M., and Kobayashi S. 2004. Nanos suppresses somatic cell fate in Drosophila germ line. *Proc. Natl. Acad. Sci.* **101:** 10338.

Hubner K., Fuhrmann G., Christenson L.K., Kehler J., Reinbold R., De La Fuente R., Wood J., Strauss J.F., III, Boiani M., and Scholer H.R. 2003. Derivation of oocytes from mouse embryonic stem cells. *Science* **300:** 1251.

Keller A.D. and Maniatis T. 1992. Only two of the five zinc fingers of the eukaryotic transcriptional repressor PRDI-BF1 are required for sequence-specific DNA binding. *Mol. Cell. Biol.* **12:** 1940.

Kouzarides T. 2002. Histone methylation in transcriptional control. *Curr. Opin. Genet. Dev.* **12:** 198.

Lachner M. and Jenuwein T. 2002. The many faces of histone lysine methylation. *Curr. Opin. Cell Biol.* **14:** 286.

Lawson K.A. and Hage W.J. 1994. Clonal analysis of the origin of primordial germ cells in the mouse. *Ciba Found. Symp.* **182:** 68.

Lawson K.A., Dunn N.R., Roelen B.A., Zeinstra L.M., Davis A.M., Wright C.V., Korving J.P., and Hogan B.L. 1999. Bmp4 is required for the generation of primordial germ cells in the mouse embryo. *Genes Dev.* **13:** 424.

Leatherman J.L. and Jongens T.A. 2003. Transcriptional silencing and translational control: Key features of early germline development. *Bioessays* **25:** 326.

Leatherman J.L., Levin L., Boero J., and Jongens T.A. 2002. Germ cell-less acts to repress transcription during the establishment of the Drosophila germ cell lineage. *Curr. Biol.* **12:** 1681.

Lee J., Inoue K., Ono R., Ogonuki N., Kohda T., Kaneko-Ishino T., Ogura A., and Ishino F. 2002. Erasing genomic imprinting memory in mouse clone embryos produced from day 11.5 primordial germ cells. *Development* **129:** 1807.

Lin K.I., Angelin-Duclos C., Kuo T.C., and Calame K. 2002. Blimp-1-dependent repression of Pax-5 is required for differentiation of B cells to immunoglobulin M-secreting plasma cells. *Mol. Cell. Biol.* **22:** 4771.

Makar K.W. and Wilson C.B. 2004. Sounds of a silent Blimp-1. *Nat. Immunol.* **5:** 241.

Martinho R.G., Kunwar P.S., Casanova J., and Lehmann R. 2004. A noncoding RNA is required for the repression of RNApolII-dependent transcription in primordial germ cells. *Curr. Biol.* **14:** 159.

Matsui Y., Zsebo K., and Hogan B.L. 1992. Derivation of pluripotential embryonic stem cells from murine primordial germ cells in culture. *Cell* **70:** 841.

McLaren A. 1999. Signaling for germ cells. *Genes Dev.* **13:** 373.

Mello C.C., Schubert C., Draper B., Zhang W., Lobel R., and Priess J.R. 1996. The PIE-1 protein and germline specification in C. elegans embryos. *Nature* **382:** 710.

O'Carroll D., Erhardt S., Pagani M., Barton S.C., Surani M.A., and Jenuwein T. 2001. The polycomb-group gene Ezh2 is required for early mouse development. *Mol. Cell. Biol.* **21:** 4330.

Orlando V. 2003. Polycomb, epigenomes, and control of cell identity. *Cell* **112:** 599.

Pirrotta V. 2002. Silence in the germ. *Cell* **110:** 661.

Ren B., Chee K.J., Kim T.H., and Maniatis T. 1999. PRDI-BF1/Blimp-1 repression is mediated by corepressors of the Groucho family of proteins. *Genes Dev.* **13:** 125.

Saitou M., Barton S.C., and Surani M.A. 2002. A molecular program for the specification of germ cell fate in mice. *Nature* **418:** 293.

Saitou M., Payer B., Lange U.C., Erhardt S., Barton S.C., and Surani M.A. 2003. Specification of germ cell fate in mice. *Philos. Trans. R. Soc. Lond. B Biol. Sci.* **358:** 1363.

Schaner C.E., Deshpande G., Schedl P.D., and Kelly W.G. 2003. A conserved chromatin architecture marks and maintains the restricted germ cell lineage in worms and flies. *Dev. Cell* **5:** 747.

Sciammas R. and Davis M.M. 2004. Modular nature of Blimp-1 in the regulation of gene expression during B cell maturation. *J. Immunol.* **172:** 5427.

Seydoux G. and Dunn M.A. 1997. Transcriptionally repressed germ cells lack a subpopulation of phosphorylated RNA polymerase II in early embryos of Caenorhabditis elegans and Drosophila melanogaster. *Development* **124:** 2191.

Seydoux G. and Strome S. 1999. Launching the germline in Caenorhabditis elegans: Regulation of gene expression in early germ cells. *Development* **126:** 3275.

Seydoux G., Mello C.C., Pettitt J., Wood W.B., Priess J.R., and Fire A. 1996. Repression of gene expression in the embryonic germ lineage of C. elegans. *Nature* **382:** 713.

Shaffer A.L., Lin K.I., Kuo T.C., Yu X., Hurt E.M., Rosenwald A., Giltnane J.M., Yang L., Zhao H., Calame K., and Staudt L.M. 2002. Blimp-1 orchestrates plasma cell differentiation by extinguishing the mature B cell gene expression program. *Immunity* **17:** 51.

Shapiro-Shelef M., Lin K.I., McHeyzer-Williams L.J., Liao J., McHeyzer-Williams M.G., and Calame K. 2003. Blimp-1 is required for the formation of immunoglobulin secreting plasma cells and pre-plasma memory B cells. *Immunity* **19:** 607.

Subramaniam K. and Seydoux G. 1999. nos-1 and nos-2, two genes related to Drosophila nanos, regulate primordial germ cell development and survival in Caenorhabditis elegans. *Development* **126:** 4861.

Surani M.A. 2001. Reprogramming of genome function through epigenetic inheritance. *Nature* **414:** 122.

———. 2004. Stem cells: How to make eggs and sperm. *Nature* **427:** 106.

Tachibana M., Sugimoto K., Nozaki M., Ueda J., Ohta T., Ohki M., Fukuda M., Takeda N., Niida H., Kato H., and Shinkai Y. 2002. G9a histone methyltransferase plays a dominant role in euchromatic histone H3 lysine 9 methylation and is essential for early embryogenesis. *Genes Dev.* **16:** 1779.

Tada T., Tada M., Hilton K., Barton S.C., Sado T., Takagi N., and Surani M.A. 1998. Epigenotype switching of imprintable loci in embryonic germ cells. *Dev. Genes Evol.* **207:** 551.

Tanaka S.S. and Matsui Y. 2002. Developmentally regulated expression of mil-1 and mil-2, mouse interferon-induced transmembrane protein like genes, during formation and differentiation of primordial germ cells. *Mech. Dev.* (suppl. 1) **119:** S261.

Toyooka Y., Tsunekawa N., Akasu R., and Noce T. 2003. Embryonic stem cells can form germ cells in vitro. *Proc. Natl. Acad. Sci.* **100:** 11457.

Tsuda M., Sasaoka Y., Kiso M., Abe K., Haraguchi S., Kobayashi S., and Saga Y. 2003. Conserved role of nanos proteins in germ cell development. *Science* **301:** 1239.

Turner C.A., Jr., Mack D.H., and Davis M.M. 1994. Blimp-1, a novel zinc finger-containing protein that can drive the maturation of B lymphocytes into immunoglobulin-secreting cells. *Cell* **77:** 297.

Van Doren M., Williamson A.L., and Lehmann R. 1998. Regulation of zygotic gene expression in *Drosophila* primordial germ cells. *Curr. Biol.* **8:** 243.

Wang Z. and Lin H. 2004. Nanos maintains germline stem cell self-renewal by preventing differentiation. *Science* **303:** 2016.

Wylie C. 1999. Germ cells. *Cell* **96:** 165.

Zhang F., Barboric M., Blackwell T.K., and Peterlin B.M. 2003. A model of repression: CTD analogs and PIE-1 inhibit transcriptional elongation by P-TEFb. *Genes Dev.* **17:** 748.

Epigenetic Mechanisms in Early Mammalian Development

D. SOLTER,* T. HIIRAGI,* A.V. EVSIKOV,† J. MOYER,† W.N. DE VRIES,†
A.E. PEASTON,† AND B.B. KNOWLES†

*Max-Planck Institute of Immunobiology, 79108 Freiburg, Germany; †The Jackson Laboratory,
Bar Harbor, Maine 04609

In one sense, development of multicellular organisms begins well before fertilization because the molecules and controlling mechanisms that direct early development are put in place during oogenesis. The informational content of a mammalian egg at fertilization is not only restricted to its DNA sequence, but also to various DNA and chromatin modifications, specific macromolecules (RNAs and proteins), and possibly the characteristic architecture of the cytoplasm and plasma membrane. These sources of stored, necessary information can be viewed as epigenetic controlling mechanisms (Fig. 1). In the ensuing text we will briefly touch on these subjects, delineating what is presently known about them and, more important, what still remains to be elucidated. The literature on these subjects is vast and space restrictions prevent us from citing each relevant paper, for which we apologize.

DNA, CHROMATIN, AND IMPRINTING

The egg and sperm genomes are transcriptionally inactive: The DNA is highly methylated, and the chromatin components, histones and protamines, are modified to ensure heterochromatization (Li 2002; Grewal and Moazed 2003; Reik et al. 2003). To secure the activation of the embryonic genome essential for further development, this must change. Upon fertilization the paternal genome undergoes rapid demethylation (Santos et al. 2002), followed by gradual demethylation of the maternal genome (Santos et al. 2002; Reik et al. 2003). The basis for this

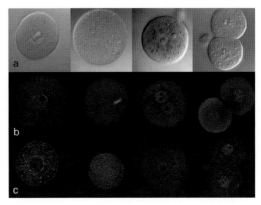

Figure 2. In the same development stages, (*a*) a differential interference contrast (DIC) image of fully grown oocyte, ovulated oocyte, zygote, and two-cell-stage embryo; (*b*) distribution of histone H3 monomethylated at lysine 9 as detected by immunofluorescence (Abcam, cat # ab9045, rabbit polyclonal antibody); and (*c*) localization of Polycomb group protein Suppressor of Zeste 12 (Upstate, cat # 07-379, rabbit immunoaffinity purified IgG).

differential DNA demethylation is not entirely clear, although it may be related to the observed differences in histone H3 methylation patterns.

Histone H3 when methylated at lysine 9 (H3/K9) marks heterochromatin and is present in high amounts in the germinal vesicle (Fig. 2b). In the fertilized egg, histone H3/K9 is prominent in the female pronucleus but less abundant or absent in the male pronucleus (Reik et al. 2003; Liu et al. 2004). The difference between male- and female-derived chromatin persists into the two-cell-stage embryo (Fig. 2b) (Liu et al. 2004), but is no longer detectable by the four-cell-stage embryo. Changes in methylation of histone H3 are regulated by histone methyltransferase activity and it is thus interesting to note that Suppressor of Zeste 12 (SUZ12) is detectable in pronuclei of the late, but not early, zygote and its abundance in the nucleus appears to increase in the two-cell stage (Fig. 2c). *Suz12* mRNA is an abundant transcript in the two-cell-stage mouse embryos (Evsikov et al. 2004), and human SUZ12 is part of a multiprotein complex together with the histone methyltransferase specific for lysine 9 (and lysine 27) of H3 histone (Kuzmichev et al. 2002). It is thus possible that de novo methylation of H3

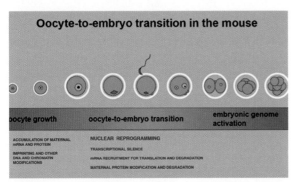

Figure 1. Oocyte-to-embryo transition in the mouse. Relevant events occurring during oocyte growth and during the oocyte-to-embryo transition, which are discussed in this paper.

histones in cleavage-stage embryos is mediated by a similar complex. Egg chromatin is also associated with a specific form of H1 linker histone, known as oocyte-specific H1FOO (Gao et al. 2004; Teranishi et al. 2004), which is slowly replaced by somatic H1 histone during cleavage. It is not clear how much this change contributes to the activation of the embryonic genome.

The observed differences in DNA and histone methylation between the paternal and maternal genomes (Reik et al. 2003) may have significant developmental consequences, or they could be developmentally irrelevant. Sperm chromatin is stripped of protamines soon after fertilization; therefore, its earlier DNA demethylation could simply reflect easier access to the denuded DNA. Examination of the timing of DNA demethylation following nuclear transfer of round spermatids or spermatocytes, both of which are devoid of protamines but compatible with normal development, would be informative in this respect.

Aside from the possible global differences between the paternal and maternal genome, imprinting is another epigenetic mechanism that leads to a long-term difference in expression of specific genes. During gametogenesis, imprinting imposes a mark on certain genes so that, following fertilization and into the adult, imprinted genes are expressed either from the maternal or paternal allele alone. The presence of numerous genes in the hemizygous state, in terms of expression, obviously poses certain genetic risks, which are presumably balanced by as yet unidentified gains. For our present discussion it is sufficient to say that interference with imprinting can lead to significant developmental disturbances (Reik and Walter 2001a; Kaneko-Ishino et al. 2003), and that imprinting is essential for normal development (McGrath and Solter 1984; Surani et al. 1984).

Despite extensive work, the exact molecular mechanism of imprinting is not known, though methylation of specific DNA sequences associated with imprinted genes clearly plays a role (Reik and Walter 2001a; Kaneko-Ishino et al. 2003). Not all imprinted genes are regulated in the same way, and several different and complex mechanisms have been described (Reik and Walter 2001a; Kaneko-Ishino et al. 2003). Methylation of a specific region as a functioning imprint is much more common in the maternal genome, regardless of whether the imprinted gene is expressed maternally or paternally. Because the paternal genome is subject to drastic early demethylation, it has been hypothesized that the preponderance of methylation imprints in the maternal genome evolved to protect imprinting marks from this early demethylation (Reik and Walter 2001b). Although imprinting plays an important role in mammalian development, and its dysfunction may be instrumental in several pathological states, its exact purpose and evolution is not clear and is the subject of much speculation (Wilkins and Haig 2003).

TRANSCRIPTION AND TRANSCRIPTIONAL CONTROL

Several mechanisms can account for the transcriptional silence of mature gametes. In mammals activation of the embryonic genome per se occurs in the cleavage stages but in the mouse some signs of transcription have been observed in the late zygote (Bouniol et al. 1995). Several approaches have been described to explore the changes in gene expression, which occur between the end of oocyte growth and completion of preimplantation development. All the methods described thus far rely on the presence, absence, increase, or decrease of mRNAs. One should bear in mind that, depending on the method used, these changes do not necessarily reflect a change in transcription.

We prepared and analyzed cDNA libraries representing different stages of preimplantation development (Rothstein et al. 1992). After sequencing randomly picked clones and clustering the expressed sequence tags, we were able to determine the nature and abundance of mRNAs present at a specific developmental stage (Evsikov et al. 2004). A similar approach has been used recently to compare the transcriptomes of preimplantation and postimplantation embryos and several stem cell lines (Sharov et al. 2003). Microarray analysis of global genome activity during preimplantation development has also been accomplished (Hamatani et al. 2004; Wang et al. 2004). Despite possible technical limitations and the problems of interpretation, these approaches do provide a first global insight into the molecular anatomy of early embryos. These studies will serve as a source of information to identify and analyze the expression, and eventually the function, of individual genes. Because significant increases and decreases in the abundance of individual mRNAs have been observed in this period when there is supposedly no transcription, and because these abundance levels change in the presence of inhibitors of RNA polymerase II (Hamatani et al. 2004; Wang et al. 2004), other mechanisms besides transcription of normal host genes are obviously at work. This brings us to the issue of mRNA metabolism, utilization, and controlled translation.

TRANSLATIONAL CONTROL

Transcriptional silence lasts about 48 hours in the mouse and in humans it is presumably longer. Although some mRNAs may be delivered by the sperm (Ostermeier et al. 2004), egg maturation, completion of meiosis, remodeling of sperm and egg genomes, and activation of embryonic genome are based on stored maternal mRNAs and proteins. Translation of stored mRNAs and their degradation can provide a rapid and timely response to the demand for a specific gene product without activating the transcriptional machinery. Indeed, controlled translation is a preferred molecular control mechanism in the oocyte-to-embryo transition and in the nervous system (Darnell 2002). In addition to controlled translation of stored maternal mRNA there is also the option for localization of maternal mRNAs in the egg, which has been extensively explored during amphibian and fruit fly development (Richter 1999; Solter and Knowles 1999; Stebbins-Boaz et al. 1999; van Eeden and St Johnston 1999; Darnell 2002).

The mechanism, which first masks and then activates maternal mRNA translation in the mouse egg, was initially suggested for tissue-type plasminogen activator and involves polyadenylation of the mRNA resulting in protein synthesis (Strickland et al. 1988). Subsequently, controlled translation of numerous genes during the period of transcriptional silence was described and may be of significant importance to the control of gene expression during the oocyte-to-embryo transition (Oh et al. 2000; Solter et al. 2002; Knowles et al. 2003). Very briefly, specific sequences found in the 3´UTR of certain mRNA species bind proteins, which protect them from degradation and translation. One of these 3´UTR sequences, the cytoplasmic polyadenylation element (CPE), binds a protein, CPEB, which in turn interacts with a protein complex containing maskin, localized at the 5´UTR end of the mRNA (Stebbins-Boaz et al. 1999). On initiation of oocyte maturation CPEB becomes phosphorylated, destabilizing the complex and allowing the mRNA to be polyadenylated, translated, and degraded (Hodgman et al. 2001). CPEB null-mutant mice are unable to make functional gametes because of the lack of formation of a synaptonemal complex (Tay and Richter 2001). Mice, bearing alleles of CPEB that can be conditionally mutated in the oocyte, should be employed to determine the role of CPEB in the later stages of gametogenesis, in oocyte maturation, and in progression through the zygote and cleavage stages. A northern blot showing expression of *Mphosph6*, using mRNAs isolated from various stages during the oocyte-to-embyro transition, illustrates the increased polyadenylation of this CPE-containing mRNA in the ovulated oocyte and zygote, and its deadenylation and degradation in the two-cell embryo (Fig. 3a). The cognate protein is synthesized normally in the ovulated oocyte and zygote (Fig. 3b) and is detected at different locations during the oocyte-to-embryo transition (Fig. 3c).

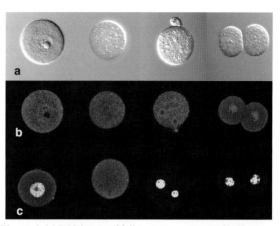

Figure 4. (*a*) DIC image of fully grown oocyte, ovulated oocyte, zygote, and two-cell-stage embryo; (*b*) localization of a 19S proteasome subunit PSMC4, as detected by immunofluorescence; and (*c*) localization of 20S proteasome in the same developmental stages. (Modified, with permission, from Evsikov et al. 2004 [©S. Karger AG, Basel].)

Transcripts for several components of this translational control system are found in mouse oocytes and early embryos. CPEB, as well as Tacc2 and Tacc3, the mouse equivalents of maskin (Hao et al. 2002; Evsikov et al. 2004), are present. Other proteins, which bind to 3´UTR motifs and are known to be involved in mRNA stabilization in other species, such as Elavl1 and Elavl2, are also abundantly expressed (Evsikov et al. 2004). While cytoplasmic polyadenylation is closely followed by or occurs in concert with translation, some fully polyadenylated mRNAs are not translated (Oh et al. 2000). This suggests the existence of still other mechanisms (Stutz et al. 1998) controlling translation. The picture of a versatile and complex temporal (and maybe spatial) control of mRNA utilization is slowly emerging.

Maternal mRNA storage and translation and modification of proteins made during oocyte growth are essential, temporally controlled initiation events. Just as these components are made available at the right time they should also be degraded in a temporal sequence. The RNA and protein degradation machinery and the molecules, which direct specific mRNAs and proteins to these degradation complexes, are present in the egg and early embryo (Evsikov et al. 2004). Both the 19S and 20S proteasome are present during the oocyte-to-embryo transition. The 20S proteasome appears to be localized to the nucleus (Palmer et al. 1994), while components of the 19S proteasome are found in both the nucleus and the cytoplasm (Fig. 4) (Evsikov et al. 2004). The regulatory 19S subunit may function in transcriptional activation (Gonzalez et al. 2002). The importance of a functional proteasome system is evidenced by an experiment we performed in which fertilized mouse eggs were exposed to the proteasome inhibitor, lactacystin. Incubation of zygotes with lactacystin arrests development, while exposure of embryos shortly before the first division takes place has little or no effect (Table 1). This suggests that protein and RNA degradation is essential for the transition toward activation of the embryonic genome.

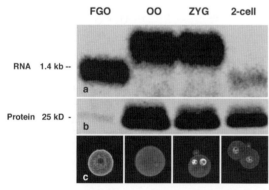

Figure 3. The maternal message of *Mphosph6* is under translational control. A northern blot, immunoprecipitation, and immunofluorescence show the relationship between the polyadenylation status of the transcript and translation of the protein. (*a*) The transcript is deadenylated in the full-grown oocyte (FGO), adenylated in the ovulated oocyte (OO) and zygote (Zyg), and undergoes degradation in the two-cell-stage embryo (2-cell). (*b*) Adenylation leads to translation of the protein in the OO, zygote, and two-cell-stage embryo. (*c*) The protein detected in the FGO may have been accumulated during oocyte growth. Immunofluorescence shows changing localization of MPHOSPH6. Anti-MPHOSPH6 guinea pig antiserum was a kind gift from Dr. Joanne M. Westendorf, Scripps Research Institute, La Jolla, California.

Table 1. Effect of Proteasome Inhibitor Lactacystin on Preimplantation Development of Mouse Zygotes Isolated at Different Time Points

Start of in vitro culture	Development in vitro					
	Control			10 μM lactacystin		
	No. of embryos	No. of two-cell (40 h)	No. of morulae (83 h)	No. of embryos	No. of two-cell (40 h)	No. of morulae (83 h)
18 h	19	19	16	24	0	0
20 h	20	19	19	27	3	0
22 h	10	10	9	21	13	0
24 h	42	41	40	166	166	145

All timing in hours post-hCG.

RETROTRANSPOSONS AND microRNA

A large part of the mammalian genome consists of interspersed repeats and transposable elements, whose role and effect on gene activity and function in genome evolution has been long debated (Britten 1997; Brosius 2003; van de Lagemaat et al. 2003; Kazazian 2004). Transcripts representing various classes of retrotransposons and repetitive sequences are abundantly represented (Fig. 5) in libraries of early mouse embryos (Evsikov et al. 2004; Peaston et al. 2004). The presence and transcription of various retrotransposons can have multiple, important effects. Activity of reverse transcriptase is essential in early embryos since development of one- to four-cell-stage embryos exposed to the reverse transcriptase inhibitor, neviparine, is arrested (Pittoggi et al. 2003).

In view of the high level of reverse transcriptase activity in the preimplantation period of development, particularly the late two-cell embryo (Evsikov et al. 2004), it is very likely that significant transposition of these viral genomes takes place. Retrotransposition of cellular gene transcripts may lead to integration and formation of pseudogenes in the two-cell-stage embryo. Indeed, two-cell-stage-embryo-expressed genes are more likely to have three or more pseudogenes than genes expressed exclusively in somatic cells (Evsikov et al. 2004; Zhang et al. 2004). Sequences derived from transposable elements can insert into cellular genes, creating mutations, or possibly novel regulatory elements, which could activate or silence targeted genes directly (Jordan et al. 2003; Han et al. 2004; Peaston et al. 2004; Pi et al. 2004), or through initiation of antisense transcripts (Kiyosawa et al. 2003; Han et al. 2004; Sémon and Duret 2004).

Transposable elements may also regulate gene expression and participate in chromatin remodeling via RNA interference (RNAi) mechanisms as shown in yeast (Schramke and Allshire 2003). While we do not know whether microRNAs are active in early embryos, embryonic stem cell–specific microRNAs have been described (Houbaviy et al. 2003). In addition, *Dicer1*, a gene essential for the RNAi pathway, is crucial for completion of early development; *Dicer1*-null-mutant embryos die soon after implantation (Bernstein et al. 2003). The absence of homozygous null-mutant females precludes analysis of the role of *Dicer1* during the oocyte-to-embryo transition. However, construction and conditional deletion of *Dicer* alleles in the oocyte will allow this important question to be addressed.

CYTOPLASMIC LOCALIZATION AND POLARITY

Accumulation of maternal mRNAs and proteins during oocyte growth is not sufficient for normal development; they have to be precisely localized in order to complete their function. The best and most detailed study of this phenomenon has been accomplished in *Drosophila* where mislocalization leads to abnormal development (van Eeden and St Johnston 1999). However, all metazoan organisms may require localized mRNA and protein to initiate development (for detailed reviews, see Etkin and Jeon 2001).

Does this principle apply to mammalian development? Because of the remarkable regulative capacity of the preimplantation mammalian embryo, which can develop normally after a substantial part of it has been removed or after the entire embryo has been duplicated by aggregation with another, it has been assumed that there is no cytoplasmic localization or polarity of developmental significance. In the last few years this view was challenged

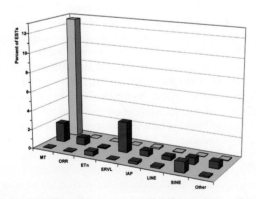

Figure 5. Differential expression of different classes of transposable elements in full-grown oocyte (*light blue*), two-cell-stage embryo (*dark blue*), and blastocyst CDNA library (*red*). MT, Mouse transcript; ORR, Origin Region Repeat; Etn, Early Transposon; ERLV, Endogenous Retrovirus L; IAP, Intracysternal A-type Particles; LINE, Long Interspersed Nucleotide Element; SINE, Short Interspersed Nucleotide Element.

and it was suggested that the first cleavage division in mouse embryo is determined by the position of the second polar body, which was represented as the animal pole, thus defining the animal–vegetal (A–V) axis (Gardner 2001; Zernicka-Goetz 2002), and also, though there was disagreement on this point, by the sperm entry point (Piotrowska and Zernicka-Goetz 2001; Davies and Gardner 2002). In addition, though somewhat controversially (Davies and Gardner 2002), it was proposed that following the first cleavage the two blastomeres possess different developmental fates, and that the one that contains the sperm entry point and divides first will contribute predominantly to the inner cell mass of the blastocyst (Piotrowska and Zernicka-Goetz 2001; Piotrowska et al. 2001; Zernicka-Goetz 2002).

The ideas that landmarks in the egg and zygote can be used to predict the topography of the blastocyst, and possibly beyond, and that the axes of egg and zygote bear a spatial relationship with those of later embryos are intellectually pleasing. Furthermore, if documented, this would bring mammalian development into the conceptual fold of the general developmental pattern of other metazoans. However, our recent results cast serious doubt on the notion that the first cleavage plane is predetermined and that the A–V axis of the mouse egg exists in any functional predictive sense (Hiiragi and Solter 2004). Using time-lapse photography we have shown that the second polar body, the putative marker of the animal pole, moves extensively before and after cleavage, and that it always ends up within the cleavage furrow. Thus, the plane of the first cleavage determines the final location of the second polar body and not the other way around. In addition, we documented that as the two pronuclei move from periphery of the zygote toward the center, they finally firmly appose before nuclear membrane breakdown. The plane of apposition coincides with the plane of the first cleavage and is independent of the position of the second polar body or the sperm entry point (Fig. 6).

In the majority of cases both pronuclei follow the most direct path toward the center with the result that the plane of apposition and the plane of the first cleavage bisect the shortest arc between the second polar body and the sperm entry point (Fig. 6a–d). However, in some cases the

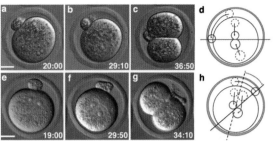

Figure 6. The first cleavage plane is specified by the topology of the two apposing pronuclei in the egg center. (*a–c, e–g*) Sequence of events for an embryo of each type. (*a,e*) Pronuclei formation. (*b,f*) The two pronuclei apposing in the center just before pronuclear membrane breakdown, thus defining the first cleavage plane. (*c,g*) Nuclei formed at the two-cell stage. In each frame, time is given in hours:minutes after hCG injection. Scale bars, 20 μm. (*d,h*) Schematic models of first cleavage plane specification in each type of embryo (corresponding to the types in *a–c* and in *e–g*, respectively). Note that in an embryo of the type shown in *e,f*, the first cleavage plane is specified by the two pronuclei rotating after their apposition in the egg center. *Broken lines*, the initial position of the second polar body and pronuclei prior to reaching the final position before cleavage (represented by a *solid line*); *broken blue line*, the presumed cleavage plane had the pronuclei not rotated; *red arrows*, the rotation of the pronuclei and resulting rotation of the cleavage plane.

pronuclei rotate once they reach the center of the egg. In those cases the plane of division is again determined by the final plane of apposition (Fig. 6e–h). It is important to note that, in the examples shown, the plane of division, instead of coinciding with the plane, which passes through the animal pole (Gardner 2001; Piotrowska and Zernicka-Goetz 2001; Piotrowska et al. 2001; Zernicka-Goetz 2002; Gardner and Davies 2003) and the sperm entry point (Piotrowska and Zernicka-Goetz 2001; Zernicka-Goetz 2002), is actually perpendicular to it. However, in all cases observed it is predicted by the apposition of the two pronuclei (Hiiragi and Solter 2004).

It seems that in this unique biological situation, when there are two nuclei within a round cell, their apposition signals the organization of the cytoskeleton and the mitotic spindle to insure the correct positioning and separation of chromatids, i.e., the plane of apposition corre-

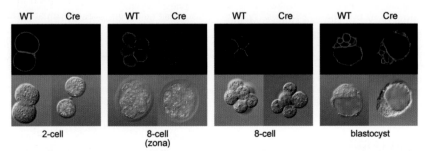

Figure 7. Embryos lacking maternal E-cadherin develop normally in spite of the lack of blastomere adhesion during early cleavages. E-cadherin, detected using a polyclonal antibody, is visible on the blastomere surface of a control (WT) two-cell-stage embryo, eight-cell-stage embryo, and blastocyst, but not on the blastomere surface of two-cell-stage and eight-cell-stage embryos lacking maternal E-cadherin (Cre, ZP3 cre removal of floxed E-cadherin during oocyte growth). Protein expressed from the paternal allele is visible at the blastocyst stage. 8-cell (zona) represents an eight-cell-stage embryo stained with the zona pellucida still surrounding it. *Bottom panel*, the DIC images.

sponds to the future metaphase plate (Hiiragi and Solter 2004). According to our results the first cleavage plane has no observable relationship with any morphological feature of the egg and zygote, thus the idea that the A–V axis of the egg relates to the future axes of the embryo cannot be supported. Whether the plane of the first cleavage is related to the plane of bilateral symmetry of the blastocyst, or its embryonic–abembryonic axis, remains to be determined. However, conditional deletion of E-cadherin in the oocyte (de Vries et al. 2000, 2004), which eliminates all intracellular contacts between blastomeres from the two-cell until the 16-cell stage, does not affect the development of these embryos (Fig. 7). Although the blastomeres are kept together within the zona, the individual cells are not fixed in place until compaction, in this case at the 16-cell stage, making a fixed relationship between the first cleavage plane and the position of cells in the blastocyst even less likely.

CONCLUSIONS

Early mammalian development is dependent on a wide range of stored epigenetic information. The DNA of sperm and egg is modified by methylation and particular genes are either paternally or maternally imprinted, which determines their subsequent expression pattern. The chromatin of the germ cells contains unique proteins, specially modified histones and protamines, which insure transcriptional silence, and these proteins must be modified or replaced for the embryonic genome to become active. Maternal mRNAs and proteins, made during oogenesis, are selectively utilized to insure the proper transition from the oocyte to the embryo. Various mechanisms regulate correct, timely translation and degradation of stored maternal mRNAs as well as posttranslational modifications and degradation of maternal proteins. The possible role of transposable elements and microRNAs during early development is just beginning to be explored. At this writing it appears highly unlikely that polarity exists in the mammalian egg and zygote or that the first cleavage plane determines bilateral symmetry in the blastocyst.

ACKNOWLEDGMENTS

This work was supported by CJ Martin Fellowship 007150 awarded by the NHMRC of Australia (A.E.P.), The Lalor Foundation (A.V.E.), Howard Hughes Memorial Institute (J.M.), The National Institutes of Health P20 RR018789 (W.N.dV.), HD37102, P30 CA 34196, The National Science Foundation EPS-0132384 (B.B.K.), and the Max Planck Society (D.S., T.H.). The authors are grateful to Dr. Joanne M. Westendorf for her gift of antibody to M Phase Phosphoprotein 6.

REFERENCES

Bernstein E., Kim S.Y., Carmell M.A., Murchison E.P., Alcorn H., Li M.Z., Mills A.A., Elledge S.J., Anderson K.V., and Hannon G.J. 2003. Dicer is essential for mouse development. *Nat. Genet.* **35:** 215.

Bouniol C., Nguyen E., and Debey P. 1995. Endogenous transcription occurs at the 1-cell stage in the mouse embryo. *Exp. Cell Res.* **218:** 57.

Britten R.J. 1997. Mobile elements inserted in the distant past have taken on important functions. *Gene* **205:** 177.

Brosius J. 2003. The contribution of RNAs and retroposition to evolutionary novelties. *Genetica* **118:** 99.

Darnell R.B. 2002. RNA logic in time and space. *Cell* **110:** 545.

Davies T.J. and Gardner R.L. 2002. The plane of first cleavage is not related to the distribution of sperm components in the mouse. *Hum. Reprod.* **17:** 2368.

de Vries W.N., Binns L.T., Fancher K.S,. Dean J., Moore R., Kemler R., and Knowles B.B. 2000. Expression of Cre-recombinase in mouse oocytes: A means to study maternal effect genes. *Genesis* **26:** 110.

de Vries W.N., Evsikov A.V., Haac B.E., Fancher K.S., Holbrook A.E., Kemler R., Solter D., and Knowles B.B. 2004. Maternal β-catenin and E-cadherin in mouse development. *Development* **131:** 4435.

Etkin L.D. and Jeon K.W., eds. 2001. *Cell lineage specification and patterning of the embryo*. Academic Press, San Diego, California.

Evsikov A.V., de Vries W.N., Peaston A.E., Radford E.E., Fancher K.S., Chen F.H., Blake J.A., Bult C.J., Latham K.E., Solter D., and Knowles B.B. 2004. Systems biology of the 2-cell mouse embryo. *Cytogenet. Genome Res.* **105:** 240.

Gao S., Chung Y.G., Parseghian M.H., King G.J., Adashi E.Y., and Latham K.E. 2004. Rapid H1 linker histone transitions following fertilization or somatic cell nuclear transfer: Evidence for a uniform developmental program in mice. *Dev. Biol.* **266:** 62.

Gardner R.L. 2001. The initial phase of embryonic patterning in mammals. In *Cell lineage specification and patterning of the embryo* (ed. L.D. Etkin and K.W. Jeon), p. 233. Academic Press, San Diego, California.

Gardner R.L. and Davies T.J. 2003. The basis and significance of pre-patterning in mammals. *Philos. Trans. R. Soc. Lond. B Biol. Sci.* **358:** 1331.

Gonzalez F., Delahodde A., Kodadek T., and Johnston S.A. 2002. Recruitment of a 19S proteasome subcomplex to an activated promoter. *Science* **296:** 548.

Grewal S.I.S. and Moazed D. 2003. Heterochromatin and epigenetic control of gene expression. *Science* **301:** 798.

Hamatani T., Carter M.G., Sharov A.A., and Ko M.S.H. 2004. Dynamics of global gene expression changes during mouse preimplantation development. *Dev. Cell* **6:** 117.

Han J.S., Szak S.T., and Boeke J.D. 2004. Transcriptional disruption by the L1 retrotransposon and implications for mammalian transcriptomes. *Nature* **429:** 268.

Hao Z., Stoler M.H., Sen B., Shore A., Westbrook A., Flickinger C.J., Herr J.C., and Coonrod S.A. 2002. TACC3 expression and localization in the murine egg and ovary. *Mol. Reprod. Dev.* **63:** 291.

Hiiragi T. and Solter D. 2004. First cleavage plane of the mouse egg is not predetermined but defined by the topology of the two apposing nuclei. *Nature* **429:** 1360.

Hodgman R., Tay J., Mendez R., and Richter J.D. 2001. CPEB phosphorylation and cytoplasmic polyadenylation are catalyzed by the kinase IAK1/Eg2 in maturing mouse oocytes. *Development* **128:** 2815.

Houbaviy H.B., Murray M.F., and Sharp P.A. 2003. Embryonic stem cell-specific microRNAs. *Dev. Cell* **5:** 351.

Jordan I.K., Rogozin I.B., Glazko G.V., and Koonin E.V. 2003. Origin of a substantial fraction of human regulatory sequences from transposable elements. *Trends Genet.* **19:** 68.

Kaneko-Ishino T., Kohda T., and Ishino F. 2003. The regulation and biological significance of genomic imprinting in mammals. *J. Biochem.* **133:** 699.

Kazazian H.H., Jr. 2004. Mobile elements: Drivers of genome evolution. *Science* **303:** 1626.

Kiyosawa H., Yamanaka I., Osato N., Kondo S., and Hayashizaki Y. 2003. Antisense transcripts with FANTOM2 clone set and their implications for gene regulation. *Genome Res.* **13:** 1324.

Knowles B.B., Evsikov A.V., de Vries W.N., Peaston A.E., and Solter D. 2003. Molecular control of the oocyte to embryo

transition. *Philos. Trans. R. Soc. Lond. B Biol. Sci.* **358:** 1381.
Kuzmichev A., Nishioka K., Erdjument-Bromage H., Tempst P., and Reinberg D. 2002. Histone methyltransferase activity associated with a human multiprotein complex containing the Enhancer of Zeste protein. *Genes Dev.* **16:** 2893.
Li E. 2002. Chromatin modification and epigenetic reprogramming in mammalian development. *Nat. Rev. Genet.* **3:** 662.
Liu H., Kim J.-M., and Aoki F. 2004. Regulation of histone H3 lysine 9 methylation in oocytes and early pre-implantation embryos. *Development* **131:** 2269.
McGrath J. and Solter D. 1984. Completion of mouse embryogenesis requires both the maternal and paternal genomes. *Cell* **37:** 179.
Oh B., Hwang S.-Y., McLaughlin J., Solter D., and Knowles B.B. 2000. Timely translation during the mouse oocyte-to-embryo transition. *Development* **127:** 3795.
Ostermeier G.C., Miller D., Huntriss J.D., Diamond M.P., and Krawetz S.A. 2004. Delivering spermatozoan RNA to the oocyte. *Nature* **429:** 154.
Palmer A., Mason G.G., Paramio J.M., Knecht E., and Rivett A.J. 1994. Changes in proteasome localization during the cell cycle. *Eur. J. Cell Biol.* **64:** 163.
Peaston A.E., Evsikov A.V., Graber J.H., de Vries W.M., Holbrook A.E., Solter D., and Knowles B.B. 2004. Retrotransposons regulate host genes in mouse oocytes and preimplantation embryos. *Dev. Cell* **7:** 597.
Pi W., Yang Z., Wang J., Ruan L., Yu X., Ling J., Krantz S., Isales C., Conway S.J., Lin S., and Tuan D. 2004. The LTR enhancer of ERV-9 human endogenous retrovirus is active in oocytes and progenitor cells in transgenic zebrafish and humans. *Proc. Natl. Acad. Sci.* **101:** 805.
Piotrowska K. and Zernicka-Goetz M. 2001. Role for sperm in spatial patterning of the early mouse embryo. *Nature* **409:** 517.
Piotrowska K., Wianny F., Pedersen R.A., and Zernicka-Goetz M. 2001. Blastomeres arising from the first cleavage division have distinguishable fates in normal mouse development. *Development* **128:** 3739.
Pittoggi C., Sciamanna I., Mattei E., Beraldi R., Lobascio A.M., Mai A., Quaglia M.G., Lorenzini R., and Spadafora C. 2003. Role of endogenous reverse transcriptase in murine early embryo development. *Mol. Reprod. Dev.* **66:** 225.
Reik W. and Walter J. 2001a. Genomic imprinting: Parental influence on the genome. *Nat. Rev. Genet.* **2:** 21.
———. 2001b. Evolution of imprinting mechanisms: The battle of the sexes begins in the zygote. *Nat. Genet.* **27:** 255.
Reik W., Santos F., Mitsuya K., Morgan H., and Dean W. 2003. Epigenetic asymmetry in the mammalian zygote and early embryo: Relationship to lineage commitment? *Philos. Trans. R. Soc. Lond. B Biol. Sci.* **358:** 1403.
Richter J.D. 1999. Cytoplasmic polyadenylation in development and beyond. *Microbiol. Mol. Biol. Rev.* **63:** 446.
Rothstein J.L., Johnson D., DeLoia J.A., Skowronski J., Solter D., and Knowles B.B. 1992. Gene expression during preimplantation mouse development. *Genes Dev.* **6:** 1190.
Santos F., Hendrich B., Reik W., and Dean W. 2002. Dynamic reprogramming of DNA methylation in the early mouse embryo. *Dev. Biol.* **241:** 172.
Schramke V. and Allshire R. 2003. Hairpin RNAs and retrotransposon LTRs effect RNAi and chromatin-based gene silencing. *Science* **301:** 1069.
Sémon M. and Duret L. 2004. Evidence that functional transcription units cover at least half of the human genome. *Trends Genet.* **20:** 229.
Sharov A.A., Piao Y., Matoba R., Dudekula D.B., Qian Y., Van-Buren V., Falco G., Martin P.R., Stagg C.A., Bassey U.C., Wang X., Carter M.G., Hamatani T., Aiba K., Akutsu H., Sharova L., Tanaka T.S., Kimber W.L., Yoshikawa T., Jaradat S.A., Pantano S., Nagaraja R., Boheler K.R., Taub D., Hodes R.J., Longo D.L., Schlessinger D., Keller J., Klotz E., Kelsoe G., Umezawa A., Vescovi A.L., Rossant J., Kunath T., Hogan B.L.M., Curci A., D'Urso M., Kelso J., Hide W., and Ko M.S.H. 2003. Transcriptome analysis of mouse stem cells and early embryos. *PLoS Biol.* **1:** 410.
Solter D. and Knowles B.B. 1999. Spatial and temporal control of maternal message utilization. In *Development: Genetics, epigenetics and environmental regulation* (ed. V.E.A. Russo et al.), p. 389. Springer, Berlin, Germany.
Solter D., de Vries W.N., Evsikov A.V., Peaston A.E., Chen F.H., and Knowles B.B. 2002. Fertilization and activation of the embryonic genome. In *Mouse development. Patterning, morphogenesis, and organogenesis* (ed. J. Rossant and P.P.L. Tam), p. 5. Academic Press, San Diego, California.
Stebbins-Boaz B., Cao Q., de Moor C.H., Mendez R., and Richter J.D. 1999. Maskin is a CPEB-associated factor that transiently interacts with eIF-4E. *Mol. Cell* **4:** 1017.
Strickland S., Huarte J., Belin D., Vassalli A., Rickles R., and Vassalli J.-D. 1988. Antisense RNA directed against the 3´ noncoding region prevents dormant mRNA activation in mouse oocytes. *Science* **241:** 680.
Stutz A., Conne B., Huarte J., Gubler P., Volkel V., Flandin P., and Vassalli J.-D. 1998. Masking, unmasking, and regulated polyadenylation cooperate in the translational control of a dormant mRNA in mouse oocytes. *Genes Dev.* **12:** 2535.
Surani M.A.H., Barton S.C., and Norris M.L. 1984. Development of reconstituted mouse eggs suggests imprinting of the genome during gametogenesis. *Nature* **308:** 548.
Tay J. and Richter J.D. 2001. Germ cell differentiation and synaptonemal complex formation are disrupted in CPEB knockout mice. *Dev. Cell* **1:** 201.
Teranishi T., Tanaka M., Kimoto S., Ono Y., Miyakoshi K., Kono T., and Yoshimura Y. 2004. Rapid replacement of somatic linker histones with the oocyte-specific linker histone H1foo in nuclear transfer. *Dev. Biol.* **266:** 76.
van de Lagemaat L.N., Landry J.-R., Mager D.L., and Medstrand P. 2003. Transposable elements in mammals promote regulatory variation and diversification of genes with specialized functions. *Trends Genet.* **19:** 530.
van Eeden F. and St Johnston D. 1999. The polarisation of the anterior-posterior and dorsal-ventral axes during *Drosophila* oogenesis. *Curr. Opin. Genet. Dev.* **9:** 396.
Wang Q.T., Piotrowska K., Ciemerych M.A., Milenkovic L., Scott M.P., Davis R.W., and Zernicka-Goetz M. 2004. A genome-wide study of gene activity reveals developmental signaling pathways in the preimplantation mouse embryo. *Dev. Cell* **6:** 133.
Wilkins J.F. and Haig D. 2003. What good is genomic imprinting: The function of parent-specific gene expression. *Nat. Rev. Genet.* **4:** 359.
Zernicka-Goetz M. 2002. Patterning of the embryo: The first spatial decisions in the life of a mouse. *Development* **129:** 815.
Zhang Z., Carriero N., and Gerstein M. 2004. Comparative analysis of processed pseudogenes in the mouse and human genomes. *Trends Genet.* **20:** 62.

Nuclear Cloning, Epigenetic Reprogramming, and Cellular Differentiation

R. Jaenisch,[*†] K. Hochedlinger,[*] R. Blelloch,[*] Y. Yamada,[*] K. Baldwin,[‡] and K. Eggan[*†]

[*]Whitehead Institute for Biomedical Research, Cambridge, Massachusetts 02142; [†]Department of Biology, Massachusetts Institute of Technology, Cambridge, Massachusetts 02139; [‡]Department of Biochemistry and Molecular Biophysics, Columbia University, New York, New York 10032

Almost half a century ago the technique of nuclear transplantation (NT) was pioneered in amphibians (Gurdon 1999; Di Berardino et al. 2003). These experiments demonstrated that nuclei of somatic cells are totipotent but that the ability to generate live animals decreases with the developmental age of the donor nucleus. The generation of Dolly from an adult mammary gland cell demonstrated that at least some cells within an adult organism retain totipotency and are able to direct development of a new animal (Wilmut et al. 1997). After Dolly, additional mammalian species were successfully cloned from somatic cells, albeit with a low efficiency, as in most cases only 0.5–10% of reconstructed oocytes develop into apparently healthy adults (Rideout et al. 2001).

Epigenetic regulation of gene expression is recognized to be one of the key mechanisms governing embryonic development and cellular differentiation and, when misdirected, can lead to disease. Therefore, it is of major interest to define the specific epigenetic states that distinguish between the genomes of embryonic, somatic, and diseased cells. In normal development the process of differentiation from embryonic to differentiated cells involves alterations in the epigenetic conformation of the genome, such as DNA methylation or chromatin modifications (Li 2002; Jaenisch and Bird 2003).

In disease, abnormal modifications to chromatin can contribute to the malignant transformation of cells. For example, epigenetic changes can promote cell proliferation, inhibit apoptosis, and induce angiogenesis during tumorigenesis by activating oncogenes and silencing tumor suppressor gene (Jones and Baylin 2002; Felsher 2003). Thus, we have to understand the epigenetic states that distinguish the genome of embryonic cells from that of different somatic cells and transformed cells (see Fig. 1). To alter the phenotype of cells in a rational way we first need to define the molecular parameters that distinguish these different cell types. Nuclear cloning represents an unbiased tool to begin unraveling these mechanisms as it provides a functional readout of the epigenetic changes induced by the egg cytoplasm. In this paper, we will summarize recent work from our laboratory that focuses on the relation between the state of differentiation and the efficiency of epigenetic reprogramming by nuclear transfer. This is relevant for the possible therapeutic applications of the nuclear transfer technology. Moreover, we will discuss the potential use of nuclear transfer for studying the genome of cancer cells. This review is largely based upon summaries of our work in a review commissioned by the President's Council on Bioethics (Jaenisch 2004) and in the Proceedings of a recent Novartis conference (Jaenisch et al. 2004).

EPIGENETIC REPROGRAMMING IN NORMAL DEVELOPMENT AND AFTER NUCLEAR TRANSFER

The majority of cloned mammals derived by nuclear transfer (NT) die during gestation, display neonatal phenotypes resembling "Large Offspring Syndrome" (Young et al. 1998), often with respiratory and metabolic abnormalities, and have enlarged and dysfunctional placentas (Rideout et al. 2001). In order for a donor nucleus to support development into a clone, it must be reprogrammed to a state compatible with embryonic development. The transferred nucleus must properly activate genes important for early embryonic development and also suppress differentiation-associated genes that had been transcribed in the original donor cell. Inadequate "reprogramming" of the donor nucleus is thought to be the principal reason for developmental failure of clones. Since few clones survive to birth, the question remains whether survivors are

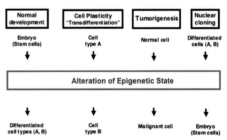

Figure 1. Importance of epigenetic changes during cellular differentiation, transformation, and nuclear cloning. Epigenetic alterations are essential during normal development when embryonic cells give rise to all differentiated cell types of the body. Epigenetic changes may also be involved in the "transdifferentiation" of one differentiated cell type into another differentiated cell type. In addition, epigenetics plays an important role in the malignant transformation of cells by activating oncogenes and silencing tumor suppressor genes. Conversely, nuclear transfer can globally reprogram the epigenetic state of a differentiated cell into that of an embryonic (stem) cell.

normal or merely the least severely affected animals, making it to adulthood despite harboring subtle abnormalities originating from faulty reprogramming (Rideout et al. 2001).

Evidence obtained over the last few years has given insights into molecular changes that are abnormal in cloned as compared to normal animals. Table 1 summarizes some of the epigenetic differences that distinguish cloned from normal animals as result of faulty reprogramming. For the following discussion it is useful to compare the different stages of development following nuclear transplantation. The stages of development that are depicted in Table 1 and that will be discussed in sequence are (i) gametogenesis, (ii) cleavage, (iii) postimplantation, and (iv) postnatal development.

(i) The most important epigenetic reprogramming in normal development occurs during gametogenesis, a process that renders both sperm and oocyte genomes "epigenetically competent" for subsequent fertilization and for faithful activation of the genes that are crucial for early development (Latham 1999; Rideout et al. 2001). In cloning, this process is shortcut and most problems affecting the "normalcy" of cloned animals may be due to the inadequate reprogramming of the somatic nucleus following transplantation into the egg. Since the placenta is derived from the trophectoderm lineage that constitutes the first differentiated cell type of the embryo, one might speculate that reprogramming and differentiation into this early lineage are compromised in most cloned animals. Indeed, results obtained in our laboratory and by others indicate that the fraction of abnormally expressed genes in cloned newborns is substantially higher in the placenta as compared to somatic tissues (Humpherys et al. 2002; Fulka et al. 2004). In contrast to epigenetic reprogramming that occurs prezygotically, it appears that postzygotic reprogramming such as X-chromosome inactivation (Eggan et al. 2000) and telomere length adjustment (Lanza et al. 2000; Tian et al. 2000; Wakayama et al. 2000; Betts et al. 2001) are faithfully accomplished after nuclear transfer and, therefore, would not be expected to impair survival of cloned animals.

(ii) During cleavage, a wave of genome-wide demethylation removes the epigenetic modification present in the zygote so that the DNA of the blastocyst is largely devoid of methylation. Between implantation and gastrulation, a wave of global de novo methylation reestablishes the overall methylation pattern, which is then maintained throughout life in the somatic cells of the animal (Monk et al. 1987; Jaenisch 1997; Reik et al. 2001). In cloned embryos abnormal methylation at repetitive sequences (Kang et al. 2001) and frequent failure to reactivate Fgf4, Fgf2r, and IL6 (Daniels et al. 2000) have been observed. To investigate gene expression, the activity of "pluripotency genes," such as Oct4, that are silent in somatic cells but active in embryonic cells was examined in cloned embryos. Strikingly, the reactivation of *Oct-4* (Boiani et al. 2002; Bortvin et al. 2003) and of *"Oct-4-*like" (Bortvin et al. 2003) genes was shown to be faulty and random in somatic clones. Because embryos lacking *Oct-4* arrest early in development (Nichols et al. 1998), incomplete reactivation of *Oct-4*-like genes in clones might be a cause of the frequent failure of the great majority of NT embryos to survive the postimplantation period. Also, a number of studies have detected abnormal DNA methylation in cloned embryos (Bourc'his et al. 2001; Kang et al. 2002; Santos et al. 2002; Mann et al. 2003). Though it is still an unresolved question to what extent the epigenetic modification of chromatin structure and DNA methylation, which occurs in normal development, needs to be mimicked for nuclear cloning to succeed, the available evidence is entirely consistent with faulty epigenetic reprogramming causing the abnormal gene expression in cloned animals.

(iii) The most extensive analysis of gene expression has been performed in newborn cloned mice. Expression profiling showed that 4–5% of the genome and between 30% and 50% of imprinted genes are abnormally expressed in placentas of newborn cloned mice (Humpherys et al. 2002). This argues that mammalian development is surprisingly tolerant to widespread gene dysregulation and that compensatory mechanisms assure survival of some clones to birth. However, the results suggest that even sur-

Table 1. Normal versus Cloned Embryos

Stage	Normal embryos	Cloned embryos	References
Gametogenesis	Genome "competent" for activation of "early" genes, establishment of imprints	None	
Cleavage	Global demethylation	Abnormal methylation	1
	Activation of embryonic ("Oct4-like") genes	Stochastic / faulty activation of "Oct4-like" genes	2
Postimplantation	Global de novo methylation, X-inactivation, telomere length adjustment (postzygotic events)		3
	Normal imprinting and gene expression	Abnormal imprinting, global gene dysregulation	4
Postnatal	Normal animal	Large offspring syndrome, premature death, etc.	5

References. 1. Monk et al. 1987; Bourc'his et al. 2001; Dean et al. 2001, 2003; Kang et al. 2002. 2. Boiani et al. 2002; Bortvin et al. 2003. 3. Eggan et al. 2000; Lanza et al. 2000; Tian et al. 2000; Wakayama et al. 2000; Betts et al. 2001. 4. Humpherys et al. 2002; Fulka et al. 2004. 5. Young et al. 1998; Ogonuki et al. 2002; Tamashiro et al. 2002; Hochedlinger and Jaenisch 2003.

viving clones may have subtle defects that, though not severe enough to jeopardize immediate survival, will cause an abnormal phenotype at a later age.

(iv) The generation of adult and seemingly healthy adult cloned animals has been taken as evidence that nuclear transfer can generate normal cloned animals, albeit with low efficiency. Indeed, a routine physical and clinical laboratory examination of 24 cloned cows of 1–4 years of age failed to reveal major abnormalities (Lanza et al. 2001). Cloned mice of a corresponding age as that of the cloned cows (2–6 mon in mice vs. 1–4 years in cows) also appear "normal" by superficial inspection. However, when cloned mice were aged, serious problems, not apparent at younger ages, became manifest. One study found that the great majority of cloned mice died significantly earlier than normal mice, succumbing with immune deficiency and serious pathological alterations in multiple organs (Ogonuki et al. 2002). Another study found that aged cloned mice became overweight with major metabolic disturbances (Tamashiro et al. 2002). Thus, serious abnormalities in cloned animals may often become manifest only when the animals age. This raises the question: Is it possible to produce truly "normal" clones?

"Normal" Clones: Do They Exist?

A key question in the public debate over cloning is whether it would ever be possible to produce a normal individual by nuclear cloning. The available evidence suggests that it may be difficult if not impossible to produce normal clones for the following reasons. (1) As summarized above, all analyzed clones at birth showed dysregulation of hundreds of genes. The development of clones to birth and beyond despite widespread epigenetic abnormalities suggests that mammalian development can tolerate dysregulation of many genes. (2) Some clones survive to adulthood by compensating for gene dysregulation. Though this "compensation" assures survival, it may not prevent maladies from becoming manifest at later ages. Therefore, most, if not all, clones are expected to have at least subtle abnormalities that may not be severe enough to result in an obvious phenotype at birth but will cause serious problems later, as seen in aged mice. Different clones may just differ in the extent of abnormal gene expression: If the key "Oct-4-like" genes are not activated (Bortvin et al. 2003), clones die immediately after implantation. If those genes are activated, the clone may survive to birth and beyond.

As schematically shown in Figure 2, the two stages when the majority of clones fail are immediately after implantation and at birth. These are two critical stages of development that may be particularly vulnerable to faulty gene expression. Once cloned newborns have progressed through the critical perinatal period, various compensatory mechanisms may counterbalance abnormal expression of other genes that are not essential for the subsequent postnatal survival. However, the stochastic occurrence of disease and other defects at later age in many or most adult clones implies that such compen-

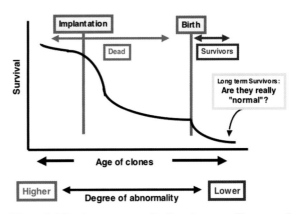

Figure 2. The phenotypes are distributed over a wide range of abnormalities. Most clones fail at two defined developmental stages, implantation and birth. More subtle gene expression abnormalities result in disease and death at later ages.

satory mechanisms do not guarantee "normalcy" of cloned animals. Rather, the phenotypes of surviving cloned animals may be distributed over a wide spectrum from abnormalities causing sudden demise at later postnatal age or more subtle abnormalities allowing survival to advanced age (Fig. 2). These considerations illustrate the complexity of defining subtle gene expression defects and emphasize the need for more sophisticated test criteria such as environmental stress or behavior tests.

Is It Possible to Overcome the Problems Inherent in Reproductive Cloning?

It is often argued that the "technical" problems in producing normal cloned mammals will be solved by scientific progress that will be made in the foreseeable future. The following considerations argue that this may not be so.

A principal biological barrier that prevents clones from being normal is the "epigenetic" difference between the chromosomes inherited from mother and from father (i.e., the difference between the "maternal" and the "paternal" genome of an individual). DNA methylation is an example of such an epigenetic modification that is known to be responsible for shutting down the expression of nearby genes. Parent-specific methylation marks are responsible for the expression of "imprinted genes" and cause only one copy of an imprinted gene, derived either from sperm or egg, to be active while the other allele is inactive (Ferguson-Smith and Surani 2001). When sperm and oocyte genomes are combined at fertilization, the parent-specific marks established during oogenesis and spermatogenesis persist in the genome of the zygote (Fig. 3A). Of interest for this discussion is that within hours after fertilization, most of the global methylation marks (with the exception of those on imprinted genes) are stripped from the sperm genome whereas the genome of the oocyte is resistant to this active demethylation process (Mayer et al. 2000; Oswald et al. 2000). This is because the oocyte genome is in a different "oocyte-appropriate" epigenetic state than the

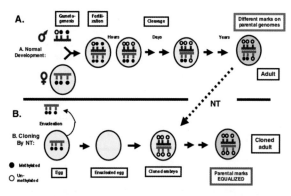

Figure 3. Parental epigenetic differences in normal and cloned animals. (*A*) The genomes of oocyte and sperm are differentially methylated during gametogenesis and are different in the zygote when combined at fertilization. Immediately after fertilization the paternal genome (derived from the sperm) is actively demethylated whereas the maternal genome is only partially demethylated during the next few days of cleavage. This is because the oocyte genome is in a different chromatin configuration and is resistant to the active demethylation process imposed on the sperm genome by the egg cytoplasm. Thus, the methylation of two parental genomes is different at the end of cleavage and in the adult. Methylated sequences are depicted as filled lollipops and unmethylated sequences as empty lollipops. (*B*) In cloning a somatic nucleus is transferred into the enucleated egg and *both* parental genomes are exposed to the active demethylating activity of the egg cytoplasm. Therefore, the parent specific epigenetic differences are equalized.

sperm genome. The oocyte genome becomes only partially demethylated within the next few days by a passive demethylation process. The result of these postfertilization changes is that the two parental genomes are epigenetically different (as defined by the patterns of DNA methylation) in the later-stage embryo and remain so in the adult in imprinted as well as nonimprinted sequences.

In cloning, the epigenetic differences established during gametogenesis may be erased because both parental genomes of the somatic donor cell are introduced into the egg from the outside and are thus exposed equally to the demethylation activity present in the egg cytoplasm (Fig. 3B). This predicts that imprinted genes should be particularly vulnerable to inappropriate methylation and associated dysregulation in cloned animals. The results summarized earlier are consistent with this prediction. For cloning to be made safe, the two parental genomes of a somatic donor cell would need to be physically separated and separately treated in an "oocyte-appropriate" and a "sperm-appropriate" way, respectively. At present, it seems that this is the only rational approach to guarantee the creation of the epigenetic differences that are normally established during gametogenesis. Such an approach is beyond our present abilities. These considerations imply that *serious biological barriers* exist that interfere with faithful reprogramming after nuclear transfer. It is a safe conclusion that these biological barriers represent a major stumbling block to efforts aimed at making nuclear cloning a safe reproductive procedure for the foreseeable future (Jaenisch 2004).

DERIVATION OF CLONED ANIMALS FROM TERMINALLY DIFFERENTIATED CELLS

A question already raised in the seminal cloning experiments with amphibians was whether terminal differentiation would diminish the potency of a nucleus to direct development after transfer into the oocyte (Di Berardino 1980; Gurdon 1999). The recent isolation of rare adult stem cells from somatic tissues and reports of their developmental plasticity raise an important question: Do viable clones predominantly or exclusively result from adult stem cells randomly selected from the donor cell population (Hochedlinger and Jaenisch 2002b)? These cells might be similar to embryonic stem (ES) cells that require less reprogramming and support postimplantation development with high efficiency. In order to better define the influence of the donor nucleus on the development of cloned animals, we have compared cells of different developmental stages for their potency to serve as nuclear donors.

Table 2 summarizes the potential of blastocysts derived from normal zygotes with that derived from ES cell and somatic donor cell nuclei after transfer into oocytes. The results show that ES NT embryos develop to term at a 10–20-fold higher efficiency than embryos from cumulus or fibroblast donor cells. The main conclusion that can be drawn from the observations summarized in Table 2 is that the nucleus of an undifferentiated embryonic cell is more amenable to, or requires less reprogramming than, the nucleus of a differentiated somatic cell. The epigenetic (or pluripotent) state of the genome in an ES cell may more closely resemble that of the early embryo, which enables ES cells to serve as more effective nuclear donors. This notion was directly tested by analyzing the expression of genes active in the different donor cells with the expression pattern seen in cloned blastocysts derived from the respective donor cells (see above; Bortvin et al. 2003).

Assuming that the epigenetic state of somatic stem cells resembles that of embryonic stem cells it appears possible that most surviving clones could be derived from nuclei of rare somatic stem cells present in the heterogeneous donor cell population rather than from nuclei of differentiated somatic cells as has been assumed (Liu 2001; Hochedlinger and Jaenisch 2002b; Oback and Wells 2002). Because in previous experiments no unambiguous marker had been used that would retrospectively identify the donor nucleus, this possibility could not be excluded.

Table 2. Higher Survival of Mice Cloned from ES Cell Donor Nuclei as Compared to Somatic Donor Cell Nuclei

Donor cells	Survival to adults (from cloned blastocysts) (%)	References
Cumulus	1–3	1
Fibroblasts	0.5–1	1
Sertoli cells	1–3	1
B, T cells, neurons	<0.001	2
ES cells	15–25	3

References. 1. Wakayama and Yanagimachi 1999, 2001. 2. Hochedlinger and Jaenisch 2002a; Eggan et al. 2004; Li et al. 2004. 3. Eggan et al. 2001.

Thus, it had not been resolved whether the genome of a truly terminally differentiated cell could be reprogrammed to an embryonic state. To address this issue, we used nuclei from mature B and T cells and from terminally differentiated neurons as donors to generate cloned mice.

Monoclonal Mice from Mature Immune Cells

The monoclonal mice were generated from nuclei of peripheral lymphocytes where the genetic rearrangements of the immunoglobulin and TCR genes could be used as stable markers revealing the identity and differentiation state of the donor nucleus of a given clone. Because previous attempts to generate monoclonal mice had been unsuccessful (Wakayama and Yanagimachi 2001), we used a two-step cloning procedure by producing, first, embryonic stem (ES) cells from cloned blastocysts and, in a second step, monoclonal mice by tetraploid-embryo complementation (Fig. 4) (Nagy et al. 1993; Eggan et al. 2001). Animals generated from a B or T cell donor nucleus were viable and carried fully rearranged immunoglobulin or T cell receptor genes in all tissues (Hochedlinger and Jaenisch 2002a). As expected, the immune cells of the monoclonal mice expressed only those alleles of the Ig and TCR locus that had been productively rearranged in the respective donor cells used for nuclear transfer and the rearrangements of other Ig or TCR genes was inhibited.

Our results allow two main conclusions. (i) They constitute the first unequivocal demonstration that nuclei from terminally differentiated donor cells can be reprogrammed to pluripotency by nuclear cloning. The frequency of directly deriving cloned embryos from mature B and T cells (instead of the two-step procedure used in our experiments), while difficult to estimate, is likely significantly lower than that of deriving clones from fibroblasts or cumulus cells (possibly less than 1 in 2000 operated embryos; Table 2). This is consistent with the notion that genomic reprogramming of a terminally differentiated cell may be extremely inefficient. (ii) Nuclear cloning allows the detection of even subtle genetic alterations within a single cell. This is of potential medical interest, as the cloning of somatic cells taken from a patient suffering of a complex disease, such as Parkinson's or diabetes, would generate ES cells that are isogenic to the patient. In vitro differentiation of these ES cells into dopaminergic neurons or beta cells, respectively, may provide an experimental in vitro system to study a complex human disease for which no animal model is available.

Cloned Mice from Mature Olfactory Neurons

Nuclei from postmitotic cells that have irreversibly exited the cell cycle as part of their program of differentiation have not been demonstrated to retain the capacity to direct embryogenesis in mammalian cloning experiments. These considerations have led to the suggestion that postmitotic cells might be refractory to epigenetic reprogramming or alternatively might have acquired changes in the DNA that could limit their developmental potential (Rehen et al. 2001). Consistent with this notion are previous experiments that failed to generate live mice from neurons (Yamazaki et al. 2001). This lead to the suggestion that the DNA of postmitotic neurons might undergo rearrangements to generate neural diversity and that these changes in DNA sequence may prevent these nuclei from reentering the cell cycle and directing embryogenesis (Vassar et al. 1993; Zhang and Firestein 2002). One particularly clear example of neuronal diversity is provided by the olfactory sensory epithelium. In the mouse, each of the two million cells in the olfactory epithelium expresses only 1 of ~1500 odorant receptor (OR) genes such that the functional identity of a neuron is defined by the nature of the receptor it expresses. The pattern of receptor expression is apparently random within one of four zones in the epithelium suggesting that the choice of receptor gene may be stochastic. One mechanism to permit the stochastic choice of a single receptor could involve DNA rearrangements (Chun and Schatz 1999).

We have generated fertile adult mouse clones by transferring the nuclei of postmitotic olfactory neurons into enucleated oocytes (Eggan et al. 2004). In a similar approach as used for the generation of the monoclonal mice (see Fig. 4), in a first step ES cells were derived from the cloned blastocyst and cloned mice were derived subsequently by tetraploid complementation. As summarized in Table 2, the efficiency of deriving cloned ES cells from olfactory neurons was in the same range as that for nuclei from immune cells. These observations indicate that a postmitotic neuronal nucleus can reenter the cell cycle and can be reprogrammed to pluripotency. A more recent study confirmed these results (Li et al. 2004).

The generation of mice cloned from a mature olfactory neuron allowed investigating whether OR choice involves irreversible DNA rearrangements. Mice were cloned from an olfactory neuron that expressed one allele of the P2 gene (which is one of the 1500 OR genes in the repertoire). The appropriate neurons for nuclear transfer were picked under the fluorescent microscope because the donor mice carried a green fluorescent protein (GFP)

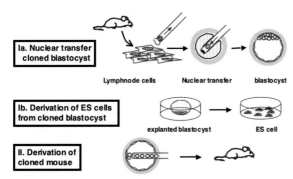

Figure 4. Two-step procedure for the derivation of monoclonal mice from mature lymphoid donor cells. (*I*) Nuclei from peripheral lymph node cells were transferred into enucleated eggs and cloned blastocysts were derived. The blastocysts were explanted in vitro and cloned ES cells were derived. (*II*) In a second step, monoclonal mice were derived by tetraploid complementation (Eggan et al. 2001; Hochedlinger and Jaenisch 2002a).

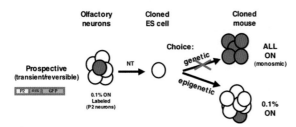

Figure 5. Nuclear cloning of mature olfactory neurons.

marker inserted into the P2 gene rendering the P2 neurons fluorescent. If OR choice involved DNA rearrangements, the prediction would be, in analogy to the monoclonal mice described above, that a mouse cloned from a P2-expressing neuron would express this receptor in all olfactory neurons and the repertoire of receptor expression might be altered (Fig. 5). Alternatively, if OR choice involved a reversible epigenetic mechanism the cloned animals should have an identical P2 expression pattern to the donor mouse and a normal repertoire of receptor expression. An exhaustive analysis of OR expression showed that the OR repertoire in the cloned mice was indistinguishable from that of wt mice. In addition, the DNA of mice derived from sensory neurons revealed no evidence for rearrangements of the expressed P2 olfactory receptor gene. These results indicate that the mechanism of OR choice is fully reversible and does not involve genetic alterations as seen in the maturation of B and T cells.

TOTIPOTENCY OF NEURONAL NUCLEI

The two-step cloning procedure used to produce mice from neuronal nuclei generates mice in which the neuronal-derived ES cells give rise to all embryonic tissues while cells from the tetraploid host blastocyst contribute the embryonic trophectoderm (Eggan et al. 2001). Thus, cloning of lymphocytes or neurons via an ES cell intermediate did not reveal totipotency of a nucleus from a terminally differentiated cell (Rossant 2002). To demonstrate totipotency of mature olfactory sensory neuron (OSN) nuclei, we transplanted nuclei from the cloned ES cells into enucleated oocytes to generate recloned mice (Eggan et al. 2004). The cloned pups had enlarged placentas but displayed no overt anatomic or behavioral abnormalities and survived to fertile adults, consistent with previous cloning experiments (Eggan et al. 2001). These observations demonstrate that nuclei of terminally differentiated olfactory neurons can be reprogrammed to totipotency directing development of both embryonic and extraembryonic lineages.

REPROGRAMMING OF CANCER NUCLEI BY NUCLEAR TRANSPLANTATION

The cloning of mice from terminally differentiated lymphocytes and postmitotic neurons has demonstrated that nuclear transfer provides a tool to selectively reprogram the epigenetic state of a cellular genome without altering its genetic constitution (Hochedlinger and Jaenisch 2002a; Eggan et al. 2004). Thus, nuclear transfer allows one to globally analyze the impact of epigenetics on the malignant state of a cancer cell. Historic experiments in frogs have demonstrated that kidney carcinoma nuclei can be reprogrammed to support early development to the tadpole stage (McKinnell et al. 1969). A similar result was recently obtained in mice, where nuclei from a medulloblastoma cell line were able to direct early development, albeit with low efficiency, resulting in arrested embryos (Li et al. 2003). However, these experiments did not unequivocally demonstrate that the clones were derived from cancer cells as opposed to contaminating non-transformed cells (Carlson et al. 1994). Moreover, the experimental setup did not allow the distinction between abnormalities caused by the nuclear transfer procedure versus abnormalities caused by the donor nucleus.

We have used nuclear transfer as a functional assay to determine whether the genomes of different cancer cells can be reprogrammed by the oocyte environment into a pluripotent embryonic state. Following nuclear transfer of different tumor cells, clones were allowed to develop to blastocysts and then explanted in tissue culture to derive ES cells. The resulting ES cells (hereafter denoted as NT ES cells) were then analyzed to confirm the tumor cell origin and tested in multiple assays for their developmental and tumorigenic potential. This modified cloning procedure (i) circumvents abnormalities associated with nuclear transfer (Hochedlinger and Jaenisch 2003) and (ii) permits a detailed analysis of the developmental (Hochedlinger and Jaenisch 2002a; Rideout et al. 2002; Eggan et al. 2004) and tumorigenic potential of the reprogrammed nucleus.

We have shown that the nuclei of leukemia, lymphoma, breast cancer, and melanoma cells were able to support preimplantation development into normal-appearing blastocysts and hence differentiation into the first two cell lineages of the embryo, the epiblast and trophectoderm, without signs of abnormal proliferation. Therefore, the malignant phenotype of these tumor types can be suppressed by the oocyte environment and permit apparently normal early development. However, none of the blastocysts cloned from four different hematopoietic tumors (1095 nuclear transfers) and a breast cancer cell line (189 nuclear transfers) produced NT ES cell lines after explantation in culture. Only the genome from a doxycycline-inducible RAS melanoma model gave rise to two NT ES cell lines from a total of 590 nuclear transfer experiments (Hochedlinger et al. 2004). These ES cells were able to differentiate into most, if not all, somatic cell lineages in teratomas and chimeras including fibroblasts, lymphocytes, and melanocytes. This occurred despite severe chromosomal changes documented by comparative genomic hybridization. These data suggest that the secondary chromosomal changes associated with malignancy do not necessarily interfere with preimplantation development, ES cell derivation, and a broad nuclear differentiation potential. However, chimeras produced from these embryonic stem cells developed cancer with higher penetrance, shorter latency, and an expanded tumor spectrum when compared with the donor mouse model. That is, all the chimeras developed melanomas shortly after induction of RAS. Interestingly, they also developed other malignant tumors including rhabdomyosarcomas and a malignant peripheral nerve sheath tumor. These later tumors had constitutively activated RAS and showed an identical comparative genomic hybridization profile as the melanomas suggesting shared genetic pathways among these very different tumor types.

In contrast to the melanoma model, we found that whereas the transfer of nuclei from embryonal carcinoma cells resulted in morphologically normal blastocysts from which ES cell lines could be produced with high efficiency, the resulting ES cells had the same developmental and tumorigenic potential as the parental EC cell lines (Blelloch et al. 2004). Strikingly, the nuclei from the three EC cell lines used in the study (F9, P19, METT-1) conferred their distinctive developmental and tumorigenic potentials to the resulting NT ES cells. This suggested that genetic alterations within the embryonic carcinoma nuclei were responsible for limiting the developmental potential of the NT ES lines. Consistent with this, comparative genomic hybridization showed shared and unique genetic alterations between the different EC cell lines. These findings support the notion that cancer results from the deregulation of stem cells and further suggest that the genetics of embryonic carcinomas will reveal genes involved in stem cell self-renewal and pluripotency.

Together, our findings demonstrate the general use of NT as a functional assay for characterizing commonalities among different types of cancer. Similarly, the nuclear transfer approach should be useful for the analysis of complex genetic disorders, such as diabetes and heart disease, in order to characterize and manipulate the multiple alleles affected in these diseases. At present, no other method has the power of amplifying the genome of a single cell with complex genetic alterations into a population of pluripotent ES cells.

OUTLOOK: NUCLEAR REPROGRAMMING, CELLULAR PLASTICITY, AND THE PROSPECTIVE FOR CELL THERAPY

Immune rejection is a frequent complication of allogeneic organ transplantation due to immunological incompatibility. To treat the "host versus graft" disease, immunosuppressive drugs are routinely given to transplant recipients, a treatment that has serious side effects. ES cells derived by nuclear transplantation are genetically identical to the patient's cells, thus eliminating the risk of immune rejection and the requirement for immunosuppression (Hochedlinger and Jaenisch 2003). Moreover, ES cells provide a renewable source of replacement tissue allowing for repeated therapy whenever needed. Indeed, we have recently demonstrated for the first time that NT could be combined with gene therapy to treat a genetic disorder (Fig. 6) (Rideout et al. 2002).

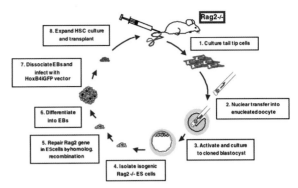

Figure 6. Scheme for therapeutic cloning combined with gene and cell therapy. A piece of tail from a mouse homozygous for the recombination activating gene 2 (Rag2) mutation was removed and cultured. After fibroblast-like cells grew out, they were used as donors for nuclear transfer by direct injection into enucleated MII oocytes using a piezoelectric-driven micromanipulator. Embryonic stem (ES) cells isolated from the NT-derived blastocysts were genetically repaired by homologous recombination. After repair, the ntES cells were differentiated in vitro into embryoid bodies (EBs), infected with the HoxB4iGFP retrovirus, expanded, and injected into the tail vein of irradiated, Rag2-deficient mice (see Rideout et al. 2002).

A key issue of transplantation medicine is the availability of isogenic functional cells that are of sufficiently high quality and can be obtained in large quantities. Therapeutic cloning would, in principle, solve this problem. Indeed, the recent generation of human embryonic stem cells by nuclear transfer into human eggs has provided evidence that this approach is a technically feasible strategy for the treatment of human disease (Hwang et al. 2004). Yet, serious obstacles such as the availability of human eggs and ethical considerations impede the application of therapeutic cloning for the treatment of patients suffering from disorders such as Parkinson's or diabetes. As an alternative approach adult stem cells have been proposed because numerous studies have claimed that these cells can give rise to many or all cells of the adult by a process designated as "transdifferentiation." However, the process of transdifferentiation is remarkably inefficient; some of the experimental claims have been difficult to reproduce or have alternative explanations (Terada et al. 2002; Ying et al. 2002; Alvarez-Dolado et al. 2003; Wang et al. 2003).

As depicted in Figure 7, the nuclear transplantation from a differentiated donor cell into the egg can alter the epigenetic state of the donor nucleus so that it is able to direct development of a new animal. This is accomplished by the action of reprogramming factors that are present in the egg's cytoplasm and that induce the genome to assume an epigenetic conformation appropriate for an embryonic state. It will be a major goal of future work to use the nuclear transfer technology as an experimental tool for defining the nature of the egg's reprogramming factors and the mechanisms of their action. Indeed, a recent report has demonstrated that Oct4 is reactivated in mammalian somatic nuclei after having been transplanted into *Xeno-*

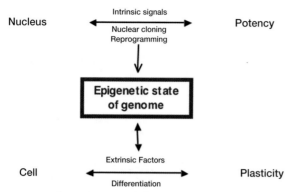

Figure 7. Epigenetic state of a genome determines nuclear potency and cellular plasticity. Nuclear transfer experiments have shown that factors within the oocyte can reprogram the epigenetic state of a differentiated cell into that of an embryonic cell, thus increasing the developmental potency of its nucleus. A better understanding of the factors that direct epigenetic reprogramming may allow us to directly alter the epigenetic state of differentiated cells and turn them into embryonic stem cells for use in cell replacement therapy.

pus oocytes (Byrne et al. 2003). Thus, it may be possible to understand the molecular basis of epigenetic reprogramming and to establish rational strategies for altering the potential of somatic cells (compare Fig. 1). The hope would be that such an approach eventually would allow the reprogramming of a patient's cells into different embryonic stem cells that could be used for cell replacement therapy without the need for nuclear transplantation.

ACKNOWLEDGMENTS

The work summarized in this article has been supported by grants from the National Institutes of Health.

REFERENCES

Alvarez-Dolado M., Pardal R., Garcia-Verdugo J.M., Fike J.R., Lee H.O., Pfeffer K., Lois C., Morrison S.J., and Alvarez-Buylla A. 2003. Fusion of bone-marrow-derived cells with Purkinje neurons, cardiomyocytes and hepatocytes. *Nature* **425**: 968.

Betts D., Bordignon V., Hill J., Winger Q., Westhusin M., Smith L., and King W. 2001. Reprogramming of telomerase activity and rebuilding of telomere length in cloned cattle. *Proc. Natl. Acad. Sci.* **98**: 1077.

Blelloch R., Hochedlinger K., Yamado Y., Brennan C., Kim M., Mintz B., Chin L., and Jaenisch R. 2004. Nuclear cloning of embryonal carcinoma cells. *Proc. Natl. Acad. Sci.* **101**: 14305.

Boiani M., Eckardt S., Scholer H.R., and McLaughlin K.J. 2002. Oct4 distribution and level in mouse clones: Consequences for pluripotency. *Genes Dev.* **16**: 1209.

Bortvin A., Eggan K., Skaletsky H., Akutsu H., Berry D.L., Yanagimachi R., Page D.C., and Jaenisch R. 2003. Incomplete reactivation of Oct4-related genes in mouse embryos cloned from somatic nuclei. *Development* **130**: 1673.

Bourc'his D., Le Bourhis D., Patin D., Niveleau A., Comizzoli P., Renard J.P., and Viegas-Pequignot E. 2001. Delayed and incomplete reprogramming of chromosome methylation patterns in bovine cloned embryos. *Curr. Biol.* **11**: 1542.

Byrne J.A., Simonsson S., Western P.S., and Gurdon J.B. 2003. Nuclei of adult mammalian somatic cells are directly reprogrammed to oct-4 stem cell gene expression by amphibian oocytes. *Curr. Biol.* **13**: 1206.

Carlson D.L., Sauerbier W., Rollins-Smith L.A., and McKinnell R.G. 1994. Fate of herpesvirus DNA in embryos and tadpoles cloned from Lucké renal carcinoma nuclei. *J. Comp. Pathol.* **111**: 197.

Chun J. and Schatz D.G. 1999. Rearranging views on neurogenesis: Neuronal death in the absence of DNA end-joining proteins. *Neuron* **22**: 7.

Daniels R., Hall Y., and Trounson A. 2000. Analysis of gene transcription in bovine nuclear transfer embryos reconstructed with granulosa cell nuclei. *Biol. Reprod.* **63**: 1034.

Dean W., Santos F., and Reik W. 2003. Epigenetic reprogramming in early mammalian development and following somatic nuclear transfer. *Semin. Cell Dev. Biol.* **14**: 93.

Dean W., Santos F., Stojkovic M., Zakhartchenko V., Walter J., Wolf E., and Reik W. 2001. Conservation of methylation reprogramming in mammalian development: Aberrant reprogramming in cloned embryos. *Proc. Natl. Acad. Sci.* **98**: 13734.

Di Berardino M.A. 1980. Genetic stability and modulation of metazoan nuclei transplanted into eggs and oocytes. *Differentiation* **17**: 17.

Di Berardino M.A., McKinnell R.G., and Wolf D.P. 2003. The golden anniversary of cloning: A celebratory essay. *Differentiation* **71**: 398.

Eggan K., Akutsu H., Hochedlinger K., Rideout W., Yanagimachi R., and Jaenisch R. 2000. X-chromosome inactivation in cloned mouse embryos. *Science* **290**: 1578.

Eggan K., Akutsu H., Loring J., Jackson-Grusby L., Klemm M., Rideout W.M., III, Yanagimachi R., and Jaenisch R. 2001. Hybrid vigor, fetal overgrowth, and viability of mice derived by nuclear cloning and tetraploid embryo complementation. *Proc. Natl. Acad. Sci.* **98**: 6209.

Eggan K., Baldwin K., Tackett M., Osborne J., Gogos J., Chess A., Axel R., and Jaenisch R. 2004. Mice cloned from olfactory sensory neurons. *Nature* **428**: 44.

Felsher D.W. 2003. Cancer revoked: Oncogenes as therapeutic targets. *Nat. Rev. Cancer* **3**: 375.

Ferguson-Smith A.C. and Surani M.A. 2001. Imprinting and the epigenetic asymmetry between parental genomes. *Science* **293**: 1086.

Fulka J., Jr., Miyashita N., Nagai T., and Ogura A. 2004. Do cloned mammals skip a reprogramming step? *Nat. Biotechnol.* **22**: 25.

Gurdon J.B. 1999. Genetic reprogramming following nuclear transplantation in Amphibia. *Semin. Cell Dev. Biol.* **10**: 239.

Hochedlinger K. and Jaenisch R. 2002a. Monoclonal mice generated by nuclear transfer from mature B and T donor cells. *Nature* **415**: 1035.

———. 2002b. Nuclear transplantation: Lessons from frogs and mice. *Curr. Opin. Cell Biol.* **14**: 741.

———. 2003. Nuclear transplantation, embryonic stem cells, and the potential for cell therapy. *N. Engl. J. Med.* **349**: 275.

Hochedlinger K., Blelloch R., Brennan C., Yamada Y., Kim M., Chin L., and Jaenisch R. 2004. Reprogramming of a melanoma genome by nuclear transplantation. *Genes Dev.* **18**: 1875.

Humpherys D., Eggan K., Akutsu H., Friedman A., Hochedlinger K., Yanagimachi R., Lander E., Golub T.R., and Jaenisch R. 2002. Abnormal gene expression in cloned mice derived from ES cell and cumulus cell nuclei. *Proc. Natl. Acad. Sci.* **99**: 12889.

Hwang W.S., Ryu Y.J., Park J.H., Park E.S., Lee E.G., Koo J.M., Jeon H., Lee B.C., Kang S.K., Kim S.J., Ahn C., Hwang J.H., Park K.Y., Cibelli J.B., and Moon S.Y. 2004. Evidence of a pluripotent human embryonic stem cell line derived from a cloned blastocyst. *Science* **303**: 1669.

Jaenisch R. 1997. DNA methylation and imprinting: Why bother? *Trends Genet.* **13**: 323.

———. 2004. The biology of nuclear cloning and the potential of embryonic stem cells for transplantation therapy. *The President's Council on Bioethics:* http://www.bioethics.gov/reports/stemcell/appendix_n.html.

Jaenisch R. and Bird A. 2003. Epigenetic regulation of gene ex-

pression: How the genome integrates intrinsic and environmental signals. *Nat. Genet.* (suppl.) **33:** 245.
Jaenisch R., Hochedlinger K., and Eggan K. 2004. Stem cells, nuclear cloning and the epigenetic state of the genome. In *Stem cells: Nuclear reprogramming and therapeutic applications* (ed. J. Gearhart). Novartis Foundation Symposium. (In press.)
Jones P.A. and Baylin S.B. 2002. The fundamental role of epigenetic events in cancer. *Nat. Rev. Genet.* **3:** 415.
Kang Y., Koo J., Park J., Choi Y., Chung A., Lee K., and Han Y. 2001. Aberrant methylation of donor genome in cloned bovine embryos. *Nat. Genet.* **28:** 173.
Kang Y.K., Park J.S., Koo D.B., Choi Y.H., Kim S.U., Lee K.K., and Han Y.M. 2002. Limited demethylation leaves mosaic-type methylation states in cloned bovine pre-implantation embryos. *EMBO J.* **21:** 1092.
Lanza R.P., Cibelli J.B., Faber D., Sweeney R.W., Henderson B., Nevala W., West M.D., and Wettstein P.J. 2001. Cloned cattle can be healthy and normal. *Science* **294:** 1893.
Lanza R.P., Cibelli J.B., Blackwell C., Cristofalo V.J., Francis M.K., Baerlocher G.M., Mak J., Schertzer M., Chavez E.A., Sawyer N., Lansdorp P.M., and West M.D. 2000. Extension of cell life-span and telomere length in animals cloned from senescent somatic cells. *Science* **288:** 665.
Latham K.E. 1999. Mechanisms and control of embryonic genome activation in mammalian embryos. *Int. Rev. Cytol.* **193:** 71.
Li E. 2002. Chromatin modification and epigenetic reprogramming in mammalian development. *Nat. Rev. Genet.* **3:** 662.
Li J., Ishii T., Feinstein P., and Mombaerts P. 2004. Odorant receptor gene choice is reset by nuclear transfer from mouse olfactory sensory neurons. *Nature* **428:** 393.
Li L., Connelly M.C., Wetmore C., Curran T., and Morgan J.I. 2003. Mouse embryos cloned from brain tumors. *Cancer Res.* **63:** 2733.
Liu L. 2001. Cloning efficiency and differentiation. *Nat. Biotechnol.* **19:** 406.
Mann M.R., Chung Y.G., Nolen L.D., Verona R.I., Latham K.E., and Bartolomei M.S. 2003. Disruption of imprinted gene methylation and expression in cloned preimplantation stage mouse embryos. *Biol. Reprod.* **69:** 902.
Mayer W., Niveleau A., Walter J., Fundele R., and Haaf T. 2000. Demethylation of the zygotic paternal genome. *Nature* **403:** 501.
McKinnell R.G., Deggins B.A., and Labat D.D. 1969. Transplantation of pluripotential nuclei from triploid frog tumors. *Science* **165:** 394.
Monk M., Boubelik M., and Lehnert S. 1987. Temporal and regional changes in DNA methylation in the embryonic, extraembryonic and germ cell lineages during mouse embryo development. *Development* **99:** 371.
Nagy A., Rossant J., Nagy R., Abramow-Newerly W., and Roder J.C. 1993. Derivation of completely cell culture-derived mice from early-passage embryonic stem cells. *Proc. Natl. Acad. Sci.* **90:** 8424.
Nichols J., Zevnik B., Anastassiadis K., Niwa H., Klewe-Nebenius D., Chambers I., Scholer H., and Smith A. 1998. Formation of pluripotent stem cells in the mammalian embryo depends on the POU transcription factor Oct4. *Cell* **95:** 379.
Oback B. and Wells D. 2002. Donor cells for cloning: Many are called but few are chosen. *Cloning Stem Cells* **4:** 147.
Ogonuki N., Inoue K., Yamamoto Y., Noguchi Y., Tanemura K., Suzuki O., Nakayama H., Doi K., Ohtomo Y., Satoh M., Nishida A., and Ogura A. 2002. Early death of mice cloned from somatic cells. *Nat. Genet.* **30:** 253.
Oswald J., Engemann S., Lane N., Mayer W., Olek A., Fundele R., Dean W., Reik W., and Walter J. 2000. Active demethylation of the paternal genome in the mouse zygote. *Curr. Biol.* **10:** 475.
Rehen S.K., McConnell M.J., Kaushal D., Kingsbury M.A., Yang A.H., and Chun J. 2001. Chromosomal variation in neurons of the developing and adult mammalian nervous system. *Proc. Natl. Acad. Sci.* **98:** 13361.
Reik W., Dean W., and Walter J. 2001. Epigenetic reprogramming in mammalian development. *Science* **293:** 1089.
Rideout W.M., Eggan K., and Jaenisch R. 2001. Nuclear cloning and epigenetic reprogramming of the genome. *Science* **293:** 1093.
Rideout W.M., III, Hochedlinger K., Kyba M., Daley G.Q., and Jaenisch R. 2002. Correction of a genetic defect by nuclear transplantation and combined cell and gene therapy. *Cell* **109:** 17.
Rossant J. 2002. A monoclonal mouse? *Nature* **415:** 967.
Santos F., Hendrich B., Reik W., and Dean W. 2002. Dynamic reprogramming of DNA methylation in the early mouse embryo. *Dev. Biol.* **241:** 172.
Tamashiro K.L., Wakayama T., Akutsu H., Yamazaki Y., Lachey J.L., Wortman M.D., Seeley R.J., D'Alessio D.A., Woods S.C., Yanagimachi R., and Sakai R.R. 2002. Cloned mice have an obese phenotype not transmitted to their offspring. *Nat. Med.* **8:** 262.
Terada N., Hamazaki T., Oka M., Hoki M., Mastalerz D.M., Nakano Y., Meyer E.M., Morel L., Petersen B.E., and Scott E.W. 2002. Bone marrow cells adopt the phenotype of other cells by spontaneous cell fusion. *Nature* **416:** 542.
Tian X.C., Xu J., and Yang X. 2000. Normal telomere lengths found in cloned cattle. *Nat. Genet.* **26:** 272.
Vassar R., Ngai J., and Axel R. 1993. Spatial segregation of odorant receptor expression in the mammalian olfactory epithelium. *Cell* **74:** 309.
Wakayama T. and Yanagimachi R. 1999. Cloning of male mice from adult tail-tip cells. *Nat. Genet.* **22:** 127.
———. 2001. Mouse cloning with nucleus donor cells of different age and type. *Mol. Reprod. Dev.* **58:** 376.
Wakayama T., Shinkai Y., Tamashiro K.L., Niida H., Blanchard D.C., Blanchard R.J., Ogura A., Tanemura K., Tachibana M., Perry A.C., Colgan D.F., Mombaerts P., and Yanagimachi R. 2000. Cloning of mice to six generations. *Nature* **407:** 318.
Wang X., Willenbring H., Akkari Y., Torimaru Y., Foster M., Al-Dhalimy M., Lagasse E., Finegold M., Olson S., and Grompe M. 2003. Cell fusion is the principal source of bone-marrow-derived hepatocytes. *Nature* **422:** 897.
Wilmut I., Schnieke A.E., McWhir J., Kind A.J., and Campbell K.H. 1997. Viable offspring derived from fetal and adult mammalian cells. *Nature* **385:** 810.
Yamazaki Y., Makino H., Hamaguchi-Hamada K., Hamada S., Sugino H., Kawase E., Miyata T., Ogawa M., Yanagimachi R., and Yagi T. 2001. Assessment of the developmental totipotency of neural cells in the cerebral cortex of mouse embryo by nuclear transfer. *Proc. Natl. Acad. Sci.* **98:** 14022.
Ying Q.L., Nichols J., Evans E.P., and Smith A.G. 2002. Changing potency by spontaneous fusion. *Nature* **416:** 545.
Young L.E., Sinclair K.D., and Wilmut I. 1998. Large offspring syndrome in cattle and sheep. *Rev. Reprod.* **3:** 155.
Zhang X. and Firestein S. 2002. The olfactory receptor gene superfamily of the mouse. *Nat. Neurosci.* **5:** 124.

Chromosome Loops, Insulators, and Histone Methylation: New Insights into Regulation of Imprinting in Clusters

W. REIK,*† A. MURRELL,†‡ A. LEWIS,† K. MITSUYA,† D. UMLAUF,§ W. DEAN,†
M. HIGGINS,∥ AND R. FEIL§

†*Laboratory of Developmental Genetics and Imprinting, The Babraham Institute, Cambridge CB2 4AT, United Kingdom;* ‡*The Wellcome Trust Sanger Institute, Wellcome Trust Genome Campus, Hinxton, Cambridge CB10 1SA, United Kingdom;* §*Institute of Molecular Genetics, CNRS, UMR-5535, and University of Montpellier-II, 34293 Montpellier cedex, France;* ∥*Department of Cancer Genetics, Roswell Park Cancer Institute, Buffalo, New York 14263*

Imprinted genes in mammals are organized into clusters in which genes share regulatory elements. *Igf2* and *H19* are separated by 100 kb (kilobases) of DNA, and both genes use enhancers that are located distal to *H19*. Alternate access to the enhancers by the two genes is in part regulated by a CTCF-dependent insulator located upstream of *H19*. We find that differentially methylated regions in both genes interact physically over the 100-kb distance. These interactions are epigenetically regulated and partition maternal and paternal chromatin into distinct loops. This creates a simple epigenetic switch for *Igf2* whereby it moves between an active and a silent chromatin domain. In the adjacent *Kcnq1ot1* cluster, by contrast, a noncoding RNA gene is flanked by several silent genes, which are marked by repressive histone modifications. Histone methylation is targeted directly or indirectly to the region by the noncoding RNA and is maintained in the absence of DNA methylation. We propose that imprinting regulation in clusters falls into different categories. The "insulator-loop" model may also be applicable to *Dlk1/Gtl2* and *Rasgrf1*. The "noncoding RNA" model is likely to be applicable to *Igf2r/Air*, to imprinted X-inactivation, and to other maternally imprinted loci.

Genomic imprinting was discovered 20 years ago (McGrath and Solter 1984; Surani et al. 1984). The realization of developmental nonequivalence of the two parental genomes (the maternal and the paternal one) was quickly followed by the insight that epigenetic mechanisms such as DNA methylation are key to this parental chromosomal identity (Sasaki et al. 1992; Brandeis et al. 1993; Stoger et al. 1993; Ferguson-Smith and Surani 2001; Reik and Walter 2001; Sleutels and Barlow 2002; Verona et al. 2003; Delaval and Feil 2004). At the same time, the "genetic conflict" hypothesis of evolution of imprinting was conceived (Haig and Westoby 1989), followed by the identification of the first imprinted genes in mice and humans (Barlow et al. 1991; Bartolomei et al. 1991; DeChiara et al. 1991). By the mid-1990s, therefore, some of the cornerstones of imprinting research were in place. Recent significant additions to this puzzle are experiments showing how imprints are introduced in the parental germ cells by DNA methyltransferases and their cofactors (Bourc'his et al. 2001; Hata et al. 2002; Kaneda et al. 2004). How imprints are erased in early germ cells is still a mystery that needs to be explored.

Meanwhile, a large amount of information has been accumulated on individual imprinted genes (currently about 80 in mouse) and their functions in fetal growth control, neonatal physiology, and adult behavior (http://www.mgu.har.mrc.ac.uk/research/imprinting/). Once a number of imprinted genes had been isolated, it quickly became clear that most imprinted genes are clustered in the genome. Some of the larger clusters extend over 1–2 Mb (megabases) and contain several imprinted genes. The reason why imprinted genes are clustered in this way and the evolution of clusters are still not understood. However, it can be imagined that genetic conflict occurs at the level of gene products as well as at different mechanistic levels. For example, maternal expression of the *H19* gene prevents maternal expression of the linked growth-enhancing *Igf2* gene (see below). Paternal expression of the noncoding *Air* transcript prevents paternal expression of the overlapping growth suppressing *Igf2r* gene. These mechanistic interactions ("battles") between imprinted genes are facilitated by physical linkage of the genes and shared regulatory elements.

Clustering and sharing of regulatory elements is indeed a common theme in imprinting. Such elements include enhancers, silencers, insulators, and activators, all of which have the potential to be epigenetically regulated. Of particular interest are the so called "imprinting centers" (ICs), which have the property that they regulate epigenetic modifications and imprinted expression of several genes throughout clusters. Their influence can extend over several hundred kilobases to megabases.

The purpose of this paper is to explore the nature of mechanistic interactions between imprinted genes in clusters. We describe recent insights into the *Igf2–H19* region suggesting that reciprocal access to a single set of enhancers is regulated by chromatin "looping." In the ad-

---

*To whom correspondence should be addressed. E-mail: wolf.reik@bbsrc.ac.uk.

jacent imprinted domain, we argue that an imprinted noncoding transcript, *Kcnq1ot1*, is pivotal for the recruitment of repressive histone modifications leading to a silent chromatin structure in the adjacent genes. We further explore the general applicability of these two models of imprinting regulation to other imprinting clusters.

THE *IGF2–H19* LOCUS (IC1 DOMAIN)

Structure of the Locus and Regulatory Elements

The fetal growth factor Insulin-like growth factor 2 gene (*Igf2*) is paternally expressed and lies upstream of the maternally expressed noncoding RNA gene *H19* (Fig. 1). The distance between *Igf2* and *H19* is ~100 kb. While the two genes are expressed from the opposite parental chromosomes, their spatial and temporal patterns of expression are strikingly similar. Both genes are expressed predominantly in mesodermal, endodermal, and extraembryonic tissues in the developing fetus, with a decline of expression during the first 3 weeks of postnatal life.

Lineage- and tissue-specific expression of both genes are governed by a number of different enhancers, most of which are located distal to *H19* (Leighton et al. 1995; Kaffer et al. 2001; Davies et al. 2002). A differentially methylated region (DMR) with paternally derived germ line methylation is located 2 kb upstream of *H19*. This region acts as an insulator or boundary element when unmethylated; it has multiple binding sites for the insulator protein CTCF, which is bound to the maternal allele (Bell and Felsenfeld 2000; Hark et al. 2000; Kanduri et al. 2000; Szabo et al. 2000; Arney 2003). CTCF binding is sensitive to DNA methylation, so the methylated paternal allele of the DMR does not bind CTCF. As a result, the insulator is inactive. The active insulator on the maternal allele is thought to restrict access of the enhancers exclusively to the *H19* promoter. On the paternal allele, by contrast, the *H19* promoter is methylated and so this gene is silent, but the insulator is now inactive, allowing the distal enhancers access to the *Igf2* gene. Notably, the *Igf2* promoters are not generally methylated on the inactive allele, and even show DNase hypersensitive sites, indicating that in principle the chromatin is transcriptionally permissive (Feil et al. 1995).

In addition to the distal enhancers and the insulator region, regulatory elements in *Igf2* are also important (Arney 2003). The *Igf2* gene has three DMRs: DMR0 is maternally methylated and overlaps with the placental specific *Igf2* promoter P0: DMR1 is paternally methylated and contains a methylation-sensitive silencer (Constancia et al. 2000; Eden et al. 2001); and the intragenic DMR2 is also paternally methylated and contains a methylation-sensitive activator (Murrell et al. 2001). These three DMRs are non-germ-line DMRs in that their differential methylation arises during early postimplantation development (Lopes et al. 2003). Importantly, deletion of the unmethylated DMR1 does lead to reactivation of the otherwise silent maternal *Igf2* allele, in the presence of an apparently intact insulator (Constancia et al. 2000). Additional silencer and enhancer sequences are present in the locus; a more complete summary can be found in Arney (2003).

Models for *Igf2-H19* Imprinting

An initial model for imprinting regulation in the locus was based on enhancer competition, whereby *H19* had privileged access to the enhancers, but when *H19* was methylated, *Igf2* could gain access to them (Bartolomei

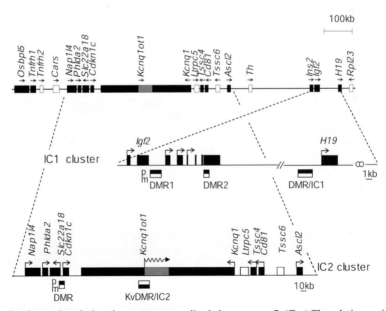

Figure 1. A map showing the two imprinting clusters on mouse distal chromosome 7. (*Top*) The relative positions of the IC1 and IC2 imprinting clusters and their respective ICs with imprinted genes indicated by *filled boxes*. (*Middle*) The IC1 cluster. The position of the DMRs is shown with a *filled black box* indicating the methylated allele and a *white box* for the unmethylated allele. The enhancers are represented by *unfilled circles*. (*Bottom*) The IC2 cluster with DMRs indicated as above. The *Kcnq1ot1* antisense transcript is represented by a *curly arrow*.

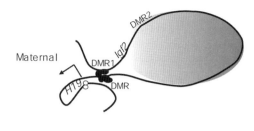

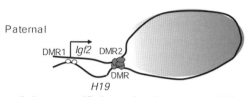

Figure 2. Parent-specific interactions between the DMRs provide an epigenetic switch for *Igf2*. On the maternal allele the unmethylated *H19* DMR, which is bound by CTCF and possibly other proteins (*stippled ovals*), and *Igf2* DMR1 interact, resulting in two chromatin domains, with the *H19* gene in an active domain with its enhancers (*small circles*) close to its promoter, and the *Igf2* gene in an inactive domain away from the enhancers (*shaded area*). On the paternal allele the methylated *H19* DMR associates with the methylated *Igf2* DMR2 through putative protein factors (*filled ovals*), moving *Igf2* into the active chromatin domain. The location of DMR2 at the end of the *Igf2* gene positions its promoters in close vicinity to the enhancers downstream of *H19*. *H19* remains in the active domain but is silenced by DNA methylation.

and Tilghman 1992). This model was superseded by the current one in which the *H19* DMR is shown to have methylation-sensitive insulator function, which depends on CTCF binding. How do insulators work? They are defined as elements that block promoter enhancer communication (Labrador and Corces 2002); recently CTCF has been found to bind nucleophosmin, which itself appears to be tethered to the nucleolar surface (Yusufzai et al. 2004). Such an anchoring of insulators to nuclear substructures could lead to the formation of chromatin loops that separate promoters and enhancers so that they can no longer interact (Fig. 2). Interestingly, a chromatin loop model was proposed some time ago for *Igf2* and *H19*, in which they come into close physical proximity, with different contact points on the maternal and paternal chromosomes (Fig. 2) (Banerjee and Smallwood 1995).

Another issue that needs to be considered in this context is the observation that epigenetic marks in clusters can be under the control of imprinting centers. Thus in Prader-Willi/Angelman syndrome, for example, mutations in the PWS IC on the paternal allele lead to methylation and silencing of multiple linked imprinted genes (Nicholls et al. 1998). In the mouse *Igf2–H19* locus, maternal deletion of the *H19* DMR (=IC1) leads to methylation of the maternal DMR1 and DMR2 in *Igf2* (Forne et al. 1997; Lopes et al. 2003). How this coordination of epigenetic marks in a cluster works and whether it is linked to the sharing of regulatory regions between the genes are unknown. However, some interesting observations have recently been made. First, in the case of the *H19* DMR, CTCF protein itself is apparently involved in protecting the maternal DMR (to which it binds) from de novo methylation during early postimplantation development (Schoenherr et al. 2003; Szabo et al. 2004). Second, while DMR1 and DMR2 in *Igf2* are actually differentially methylated in the germ line (paternal methylation) this methylation is lost soon after fertilization, but is then regained postimplantation. Importantly, if the maternal *H19* DMR is deleted, the maternal DMR1 and DMR2 become methylated at the same stage, suggesting that they are normally protected from de novo methylation by an intact *H19* DMR (Lopes et al. 2003) (whether this also involves CTCF is not known). From these observations, we developed a model in which on the maternal allele the *H19* DMR loops together with *Igf2* DMR1 (and perhaps also DMR2). When the loop is disrupted by deletion of the *H19* DMR, *Igf2* DMRs 1 and 2 are no longer protected from de novo methylation (Fig. 2) (Lopes et al. 2003). This model is complementary to the Banerjee and Smallwood model, but they make different predictions about where physical contacts are to be expected.

Testing the Loop Model

Chromosome looping was proposed many years ago to underlie the action of distant enhancer elements on proximal promoters, especially in more complex vertebrate genomes (Bulger and Groudine 1999). Only recently, however, have technological developments allowed these models to be directly tested. The first method, chromosome conformation capture (3C), is based on cutting cross-linked chromatin with defined restriction enzymes, and religating the ends under very dilute DNA concentrations (Dekker et al. 2002). Under these conditions, restriction sites from remote genomic regions can ligate to each other if they were in close physical proximity when the chromatin was cross-linked. Thus, successful PCR amplification with primers from the two remote regions would indicate physical proximity of the remote regions. In the mammalian system, this technique has been applied to study the developmental regulation of beta globin gene expression (Tolhuis et al. 2002). A remote enhancer (the LCR) has indeed been shown to come into close physical contact with the gene with the intervening DNA looping out. These interactions are dynamic and are regulated in a tissue and developmental fashion (Palstra et al. 2003).

The second method has also been developed to look at beta globin expression and is based on cross-linking locally an RNA FISH probe to the nuclear RNA as it is transcribed (of the globin gene in fetal blood cells in this case), followed by pulldown of the crosslinked protein–nucleic acid complex using a tag on the FISH probe. The complex is then analyzed to see which remote sequences are trapped. As with the 3C technique, areas of the enhancer (LCR) were found to contact the gene in tissues in which transcription occurs (Carter et al. 2002). The confirmation of "looping" with different techniques is important, and inspires confidence in these new technologies.

We chose to test the looping model for *Igf2-H19* by two different techniques. First, we reasoned that chromatin immunoprecipitation (ChIP) could be used with an antibody specific to a protein located on the *H19* DMR in order to see whether areas in *Igf2* would be coprecipitated. However, a protein such as CTCF (which is bound to the maternal *H19* DMR) is so ubiquitous that it cannot be used for this purpose. We thus decided to engineer a *H19* DMR with a unique protein tag. Three binding sequences for the yeast Gal4 transcriptional activator protein (termed UAS) were inserted into the *H19* DMR by homologous recombination in ES cells and knockin mice were generated (Murrell et al. 2004). These mice were then bred with transgenic mice ubiquitously expressing the DNA-binding domain of the Gal4 protein fused to a unique peptide tag (human MYC). Following ChIP with a MYC tag antibody, the DMRs and other regions were analyzed by Q-PCR. The *H19* DMR showed strong enrichment with the antibody, confirming that the knockin strategy was working and that the Gal4-MYC protein was indeed located at the *H19* DMR with the inserted UAS sequences. Interestingly, with maternal transmission of the modified DMR, DMR1 of *Igf2* was also enriched, but neither DMR2 nor intervening sequences between *Igf2* and *H19* were. With paternal transmission, by contrast, neither DMR1 nor intervening sequences were enriched, but DMR2 was. The preliminary conclusions from this experimental system were that on the maternal chromosome, the *H19* DMR was closely associated with *Igf2* DMR1, whereas on the paternal chromosome it was associated with DMR2.

We applied the 3C method in order to confirm these results. Ligation products were indeed detected between the *H19* DMR and DMR2, and the *H19* DMR and DMR1. For DMR2, it was possible to determine by sequence polymorphisms that it was the paternal allele that was associated with the *H19* DMR. Thus the combined evidence from the different techniques suggests that the *H19* DMR comes in close proximity to DMR1 on the maternal allele and to DMR2 on the paternal one. It should be noted that these analyses are not quantitative, and so the proportion of cells in any one tissue that contain the loop structures remains to be determined. The loops we see may be only transient and may or may not be present in nonexpressing cells. It will also be of interest to see whether other sequences between *Igf2* and *H19*, or those further away from the genes, particularly *Ins2*, which is also paternally expressed in some tissues, come into physical proximity with the DMRs.

An Epigenetic Switch

The model arising from these observations is simple (Fig. 2). On the maternal allele the unmethylated *H19* DMR and *Igf2* DMR1 come together; proteins involved in these interactions may well include CTCF and others. This places *Igf2* inside a loop and insulates it from the enhancers. On the paternal allele, *Igf2* moves out of the loop since the *H19* DMR and DMR2 are now interacting; this allows interaction between the enhancers and *Igf2*. While *H19* remains outside of the "silent" loop, it is inactivated by DNA methylation. CTCF cannot be involved in the interactions between the *H19* DMR and DMR2 because it does not bind to the methylated alleles. Whether proteins that bind preferentially to methylated DNA such as MBDs are involved in these interactions needs to be established. In human fibroblast cells the 3C method revealed further allele-specific interactions between restriction fragments containing the promoters of *IGF2* and *H19* and a fragment carrying their (presumptive) shared enhancers (Fig. 2) (Y.Yang and M. Higgins, unpubl.). Presumably these enhancer/promoter interactions are mediated by additional higher-order chromatin structures, perhaps involving a putative CTCF-dependent insulator (conserved between mouse and humans) that lies downstream of the enhancers (Ishihara and Sasaki 2002). Potential involvement of CTCF in looping can now be tested genetically using mutant DMRs that lack CTCF-binding sites (Pant et al. 2003; Schoenherr et al. 2003; Szabo et al. 2004), and whether methylation of the paternal DMR is needed for looping can be tested using a mutant DMR that lacks CpGs (M. Bartolomei, pers. comm.). These experiments may begin to reveal the *cis*- and *trans*-acting requirements for looping (Patrinos et al. 2004). It should be noted that our model is consistent with most knockout studies of the DMRs (Arney 2003). Knockouts of *H19* DMR or the *Igf2* DMRs may disrupt the looping structures. Upon maternal transmission of a *H19* DMR deletion, reduced *H19* expression accompanied by activated *Igf2* expression was observed, while upon paternal transmission of the deletion, reduced *Igf2* expression accompanied by activated *H19* expression was observed (Thorvaldsen et al. 1998). (It is noteworthy that the expression levels of the activated genes were still significantly lower than the wild type.) Thus, without the *H19* DMR, both the *Igf2* and the *H19* genes are not restricted to a particular silent or active domain. Deletion of the DMR1 of *Igf2* leads to biallelic *Igf2* expression upon maternal transmission (Constancia et al. 2000), while deletion of the DMR2 of *Igf2* leads to transcriptional down-regulation of *Igf2* upon paternal transmission (Murrell et al. 2001), which is consistent with DMR1 and DMR2 contributing to the looping structure on the maternal and paternal alleles, respectively.

A recent study shows that the paternal DMR2 is associated with the nuclear matrix in *Igf2*-expressing cells, so this may provide an anchor point for the base of the loop (Weber et al. 2003). The same has recently been observed for an insulator element upstream of the chicken beta globin genes (Yusufzai and Felsenfeld 2004), and CTCF has been found to bind to nucleophosmin, which can tether it to the nucleolar membrane (Yusufzai et al. 2004). Such interactions with subnuclear structures may well underlie the formation or maintenance of functional chromatin loops. In this way our observations would add to the growing body of evidence suggesting that regulatory elements and gene promoters in vertebrates are kept under tight control by higher-order chromatin structures, especially loops (Chubb and Bickmore 2003). The fascinating aspect of the observations on *Igf2-H19* is that these higher-order structures can be epigenetically regulated, thus providing simple epigenetic switches. It will be interesting to see whether other imprinted loci, or other epigenetically regulated loci, such as random monoallelically expressed genes (Chess 1998), possess similar epigenetic switches.

THE *Kcnq1ot1* LOCUS (IC2 DOMAIN)

Structure of the Locus and Regulatory Elements

The IC2 domain is located further telomeric from the IC1 domain but the two imprinted domains are closely linked (Fig. 1). Nevertheless, most of the genetic evidence available indicates that imprinting in the two domains is independently regulated. The IC2 domain contains the antisense RNA gene *Kcnq1ot1*, whose promoter region is currently defined as IC2. The *Kcnq1ot1* gene is located within an intron of the *Kcnq1* gene and its promoter overlaps with a DMR with germ line methylation arising from the oocyte (Engemann et al. 2000). The unmethylated paternal allele of *Kcnq1ot1* is therefore transcribed while the maternal allele remains transcriptionally silent. All other genes in the ~800-kb domain surrounding IC2 are maternally expressed (or not imprinted; Fig. 1). The *Kcnq1ot1* transcript partially overlaps with the large *Kcnq1* gene, but this overlap probably does not extend to the *Kcnq1* promoter region. A number of the genes in the IC2 domain are imprinted only in the placenta, attesting perhaps to the role in placental growth and function of a number of genes in the IC1 and IC2 domains. What is surprising is that most imprinted genes in the IC2 region do not have DMRs; this seems to be quite different from other larger imprinting clusters such as the PWS/AS region (Nicholls et al. 1998; A. Lewis et al., unpubl.).

The functional definition of the IC2 domain comes from knockout studies of the *Kcnq1ot1* promoter region, both in cell lines and in mice. These show that if this region is removed from the paternal chromosome, the otherwise silent genes flanking it are now expressed (Horike et al. 2000; Fitzpatrick et al. 2002). This has been tested as far as *Ascl2* on the centromeric side and *Phlda2* on the telomeric side. By this criterion, the imprinted domain regulated by IC2 is at least 800 kb in size. *Ascl2* is very likely the most centromeric imprinted gene in the domain, but the telomeric end has not been defined. It is unclear from these studies if it is the DNA sequence at the *Kcnq1ot1* promoter (that was deleted in the IC2 knockout), the *Kcnq1ot1* RNA, or the fact that the region is transcribed on the paternal chromosome that is responsible for *cis*-acting silencing. By transfection assays, the DNA segment itself appears to have methylation-sensitive silencer or insulator activity, depending on which cell line is used for the tests (Kanduri et al. 2002; Thakur et al. 2003). By analogy with the *Igf2r* antisense gene *Air*, which is also paternally expressed, it is possible that the *Kcnq1ot1* RNA (or the act of transcription) plays a role in silencing (Sleutels et al. 2002). This possibility is further strengthened by parallels with imprinted X-chromosome inactivation (below). What *cis*-acting sequences are responsible for imprinting and differential methylation of IC2 in the first place is unclear. A series of BAC and YAC transgenes containing IC2 and flanking sequences show that imprinting and the full tissue-specific pattern of expression (of *Cdkn1c*) depend upon the presence of remote sequence elements (John et al. 2001; F. Cerrato et al., unpubl.).

DNA Methylation Is Not Required for Maintenance of Imprinting: A Histone Methylation Imprint

Because of the relative scarcity of DMRs we were curious to see if DNA methylation was involved in regulating imprinting in the IC2 region, especially in the placenta. We used the *Dnmt1c* mutation, which removes the catalytic domain of Dnmt1 and results in a nonfunctional enzyme (Lei et al. 1996). Genomic methylation is dramatically reduced. Indeed, the maternal methylation of the *Kcnq1ot1* promoter was completely lost in both *Dnmt1*–/– embryos and *Dnmt1*–/– trophoblast (the precursor tissue of the placenta). This loss of DNA methylation causes a complete loss of imprinting at the IC2 cluster in the embryo. However, imprinted genes in the placenta that do not possess DMRs (such as *Ascl2*, *Obph1*, *Kcnq1*, etc.) do not lose imprinting (i.e., the paternal allele remains repressed) (A. Lewis et al., unpubl.). By contrast, the two genes with DMRs (*Kcnq1ot1* and *Cdkn1c*) do lose imprinting and are now biallelically expressed. Thus the maintenance of placental imprinting in this region does not apparently require DNA methylation. We therefore asked if the imprinted genes were marked by other types of epigenetic modifications. Chromatin immunoprecipitation was carried out in an allele-specific way using antibodies against acetyl and methyl modifications of histone tails. The Montpellier group (D. Umlauf and R. Feil, pers. comm.) thus found that in the placenta an 800-kb region (at least) surrounding *Kcnq1ot1*, and extending at least to *Ascl2* on the centromeric and *Obph1* on the telomeric side, was dimethylated at H3K9, and trimethylated at H3K27 (Fig. 3) (D. Umlauf et al., unpubl.). This is true not only of gene promoters, but also of all intra- and intergenic regions analyzed. By contrast, the maternal region in the placenta is marked with activating histone modifications (namely, H3K4 dimethylation and H3K9 and K14 acetylation). The only exception is the *Kcnq1ot1* DMR itself, in which histone methylation at K9 and K27 correlates with DNA methylation, and therefore marks the maternal chromosome, both in the placenta and in the embryo. In the embryo such allele-specific histone modifications are found only at confined regions comprising the DMRs. Other studies by D. Umlauf and coworkers (unpubl.) show that this regional histone methylation imprint is already present in ES cells, and thus may arise very early in development. It is then maintained in extraembryonic tissues, but apparently lost in embryonic ones.

The Repressive Histone Methylation Imprint Depends on IC2

We next asked if the histone methylation imprint depended on the IC2. Indeed, with paternal transmission of the IC2 deletion, histone modification differences as judged by allele-specific ChIP were eliminated at the several genes we analyzed. This indicates that the repressive histone methylation on the paternal chromosome is not established or is not maintained in mice carrying an IC2 deletion (A. Lewis et al., unpubl.). As stated before, it is

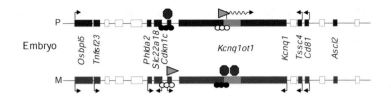

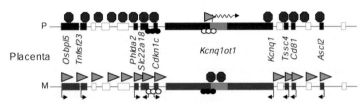

Figure 3. A summary of the differences in imprinted expression and histone modifications at the IC2 cluster and between the embryo and placenta. Imprinted genes are colored *blue* on the paternal chromosome and *red* on the maternal chromosome. Nonimprinted genes are *white*. *Arrows* show expression status in the embryo or placenta. DMRs are represented by *black circles* on the methylated allele and *white circles* on the unmethylated allele. *Green triangles* represent "active" histone modifications such as H3 acetylation and H3 K4 dimethylation. *Dark red hexagons* represent "repressive" histone modifications such as H3 K9 dimethylation and H3 K27 trimethylation. Differential histone modification in the embryo is limited to the DMRs possibly because DNA methylation is somehow required for its maintenance in the embryo (Fournier et al. 2002). In the placenta, however, the entire imprinted domain is marked by repressive modifications on the paternal allele and active modifications on the maternal allele.

not yet clear whether the DNA element at IC2, *Kcnq1ot1* transcription, or the *Kcnq1ot1* transcript is responsible for attracting repressive histone modifications to the region *in cis*. Our favored model is that the transcript itself plays some role. Thus we envisage (Fig. 4) that the *Kcnq1ot1* RNA is paternally expressed early in the preimplantation embryo. We then envisage that the noncoding *Kcnq1ot1* RNA coats the IC2 region, and that this recruits, directly or indirectly, histone methyltransferases to the paternal chromosome. Indeed Umlauf et al. (unpubl.) show that the Eed/Ezh2 polycomb complex is recruited to the paternal chromosome in ES cells. This complex has K27 and some K9 methyltransferase activity. Other methyltransferases, such as perhaps G9a, might be recruited to methylate the K9 residue. Indeed, it is interesting that in the *Eed* knockout mouse some of the paternally silenced genes analyzed here are derepressed, but not all of them are (Mager et al. 2003). Therefore differential, and perhaps also additive, effects of the different repressive histone modifications might be imagined. In addition, it is likely that histone methylation and DNA methylation cooperate in silencing. For example, the silent copies of *Kcnq1ot1* and of *Cdkn1c* (which possess DMRs) are considerably "more silent" than those of *Kcnq1* or *Cd81* (no DMRs), which show a bias only in parental expression.

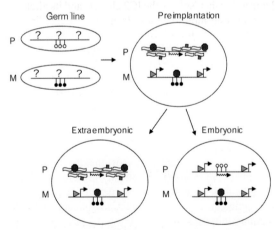

Figure 4. Model of the establishment of imprinting at the IC2 locus. *Arrows* show active gene transcription. Transcription of the antisense transcript is shown by a *curly arrow*. DMRs are represented by *black circles* on the methylated allele and *white circles* on the unmethylated allele. *Green triangles* represent histone acetylation or H3 K4 dimethylation and *red hexagons* represent histone K9 or K27 methylation. RNA coating the chromosome is shown by a *wavy rectangle*. Histone methyltransferases are shown as *red boxes*. The germ line histone modification status is as yet unknown. In preimplantation embryos we predict that *Kcnq1ot1* is expressed from the paternal allele and coats the chromosome. This coating may recruit histone modifications possibly using a mechanism similar to Xist RNA in X-inactivation. These modifications are then maintained in the extraembryonic lineages but lost in embryonic lineages.

We have shown that the histone methylation imprint, once established, is independent of DNA methylation. But if our model is correct the IC2 domain may require the continued presence of the *Kcnq1ot1* RNA to maintain imprinting. So in the *Dnmt1*–/– placenta, why does loss of methylation from the maternal *Kcnq1ot1* promoter (causing derepression of maternal *Kcnq1ot1* transcription) not result in silencing of the maternal chromosome? In heterozygous crosses between *Dnmt1*–/wt animals, the maternal oocyte still contains large amounts of Dnmt1 protein, which may become depleted only toward the blastocyst stage (at which stage the histone methylation imprint is already established; D. Umlauf et al., unpubl.). Thus activation of the maternal copy of *Kcnq1ot1* may occur only around the blastocyst stage at which perhaps a "window of opportunity" for *cis* establishment of the histone methylation imprint is already closed.

Why is histone methylation apparently important in the placenta, but less so in the embryo, for imprinting? It is possible that imprinting coevolved with the placenta in mammalian radiation, and that initially imprinted genes acted primarily in the placenta to regulate supply of maternal nutrients to the fetus (Reik et al. 2003). We speculate that during this phase of imprinting evolution, histone methylation may have been the primary epigenetic mark involved in imprinting. In marsupials, for example, the Insulin-like growth factor 2 receptor gene is imprinted as it is in mice, but the DMR2 region, which in mice carries the DNA methylation imprint, is not methylated (Killian et al. 2000). Similarly, inactivated genes on the X chromosome are not methylated in female marsupials, but they are in female mice (Wakefield et al. 1997). Thus an ancestral imprinting mechanism may have been based on histone modifications and was perhaps initially limited to the placenta. Since we believe a mechanism based on only histone modifications to be relatively "unstable" or "leaky," later evolution of imprinted gene expression in fetal tissues and after birth made necessary the recruitment of a more stable epigenetic mark, namely DNA methylation.

The Insulator-Loop Model and the Noncoding RNA Model: Application to Other Imprinted Regions

The noncoding RNA model is clearly applicable to other imprinted regions as well. First, there are overt similarities with imprinted X-chromosome inactivation. The current model of imprinted Xi (in the mouse placenta) envisages that *Xist* is expressed from the paternal X as soon as the two-cell stage, and begins to coat the inactivating X soon after. Histone methylation recruitment also begins during preimplantation development, and so does gene inactivation along the paternal X chromosome, possibly in a gradient from the *Xist* locus (Huynh and Lee 2003; Mak et al. 2004; Okamoto et al. 2004). What is still debated is whether, in addition to early postzygotic action of *Xist*, the paternal X is already marked or preinactivated in the germ line, perhaps because of the X-inactivation event that occurs during spermatogenesis (Huynh and Lee 2003). Notwithstanding this uncertainty, it is generally thought that in the morula and early blastocyst, the majority of cells, including those in the inner cell mass, carry an inactive paternal X chromosome, and that this silencing is erased (starting with down-regulation of Xist) in the ICM, followed by random X-inactivation at a slightly later stage in the epiblast (Mak et al. 2004; Okamoto et al. 2004).

We think that the noncoding RNA model is potentially applicable to a number of other imprinted loci, particularly those with paternal expression of noncoding antisense transcripts. The *Igf2r* locus is especially interesting since the *Igf2r* gene is overlapped by the noncoding antisense transcript *Air*, and the *Air* transcript or transcription is needed for *cis* inactivation of *Igf2r* as well as the two linked genes, *Slc22a2* and *Slc22a3*, which are only imprinted in the placenta (and do not have DMRs; Sleutels et al. 2002). Thus

our specific prediction for this locus would be that in the placenta there should be a histone methylation imprint in *Slc22a2* and *Slc22a3*, and that genetic removal of DNA methylation should not lead to loss of imprinting of these *Slc22a* genes. The same model may well be applicable to other clusters with maternal germ line methylation and a role in placental growth and function. For example, the *Snurf-Snrpn* promoter transcribes a very long paternal transcript which may have a role in *cis*-inactivation.

Conversely, we speculate that the looping-insulator model may be more widely applicable to clusters with paternal germ line methylation, such as *Dlk1-Gtl2*, and *Rasgrf1*. Maternal deletion of the (unmethylated) DMR upstream of *Gtl2* leads to reactivation of the normally silent *Dlk1* on the maternal chromosome, suggesting disruption of an insulator similar to that seen upstream of *H19* (Lin et al. 2003). *Rasgrf1* has a paternally methylated DMR with insulator activity (Yoon et al. 2002).

We must also consider the possibility that these two models are not necessarily mutually exclusive. Much of the regulatory machinery important for the epigenetic switch in the *Igf2-H19* region is also present in the IC2 cluster. For instance the *Kcnq1ot1* DMR has been shown to have methylation-sensitive insulator function in several somatic cell lines (Kanduri et al. 2002). Interestingly, in trophoblast-derived cells the DMR functions as a bidirectional silencer rather than a unidirectional insulator (Thakur et al. 2003). Perhaps these tissue-specific effects contribute to the differences in imprinted expression in embryonic and extraembryonic lineages. Perhaps the DMR region is also involved in a tissue-specific higher-order chromatin structure that isolates promoters and enhancers, or that isolates the IC2 cluster (and its allele-specific histone modifications) from the surrounding loci. The *Igf2-H19* cluster contains several noncoding transcripts including *H19* itself. None appear to be as extensive as the *Kcnq1ot1* transcript, which is at least 60 kb long, so a coating mechanism as described above seems unlikely. However, the process of transcription of these noncoding RNAs or the RNAs themselves may contribute to transcriptional, posttranscriptional, or RNAi-like mechanisms of gene regulation.

ACKNOWLEDGMENTS

We are grateful to many colleagues who have contributed to this work through discussions of ideas and comments on the manuscript, particularly P. Smith, J. Walter, G. Kelsey, P. Fraser, M. Constancia, F. Santos, A. Ferguson-Smith, and A. Riccio. The authors' work is funded by BBSRC, MRC, and CRUK. R.F. acknowledges grant funding from ARC. D.U. is a recipient of a Ministry of Research fellowship.

REFERENCES

Arney K.L. 2003. H19 and Igf2—Enhancing the confusion? *Trends Genet.* **19:** 17.
Banerjee S. and Smallwood A. 1995. A chromatin model of IGF2/H19 imprinting. *Nat. Genet.* **11:** 237.
Barlow D.P., Stoger R., Herrmann B.G., Saito K., and Schweifer

N. 1991. The mouse insulin-like growth factor type-2 receptor is imprinted and closely linked to the Tme locus. *Nature* **349:** 84.

Bartolomei M.S. and Tilghman S.M. 1992. Parental imprinting of mouse chromosome 7. *Semin. Dev. Biol.* **3:** 107.

Bartolomei M.S., Zemel S., and Tilghman S.M. 1991. Parental imprinting of the mouse H19 gene. *Nature* **351:** 153.

Bell A.C. and Felsenfeld G. 2000. Methylation of a CTCF-dependent boundary controls imprinted expression of the Igf2 gene. *Nature* **405:** 482.

Bourc'his D., Xu G.L., Lin C.S., Bollman B., and Bestor T.H. 2001. Dnmt3L and the establishment of maternal genomic imprints. *Science* **294:** 2536.

Brandeis M., Kafri T., Ariel M., Chaillet J.R., McCarrey J., Razin A., and Cedar H. 1993. The ontogeny of allele-specific methylation associated with imprinted genes in the mouse. *EMBO J.* **12:** 3669.

Bulger M. and Groudine M. 1999. Looping versus linking: Toward a model for long-distance gene activation. *Genes Dev.* **13:** 2465.

Carter D., Chakalova L., Osborne C.S., Dai Y.F., and Fraser P. 2002. Long-range chromatin regulatory interactions in vivo. *Nat. Genet.* **32:** 623.

Chess A. 1998. Expansion of the allelic exclusion principle? *Science* **279:** 2067.

Chubb J.R. and Bickmore W.A. 2003. Considering nuclear compartmentalization in the light of nuclear dynamics. *Cell* **112:** 403.

Constancia M., Dean W., Lopes S., Moore T., Kelsey G., and Reik W. 2000. Deletion of a silencer element in Igf2 results in loss of imprinting independent of H19. *Nat. Genet.* **26:** 203.

Davies K., Bowden L., Smith P., Dean W., Hill D., Furuumi H., Sasaki H., Cattanach B., and Reik W. 2002. Disruption of mesodermal enhancers for Igf2 in the minute mutant. *Development* **129:** 1657.

DeChiara T.M., Robertson E.J., and Efstratiadis A. 1991. Parental imprinting of the mouse insulin-like growth factor II gene. *Cell* **64:** 849.

Dekker J., Rippe K., Dekker M., and Kleckner N. 2002. Capturing chromosome conformation. *Science* **295:** 1306.

Delaval K. and Feil R. 2004. Epigenetic regulation of mammalian genomic imprinting. *Curr. Opin. Genet. Dev.* **14:** 188.

Eden S., Constancia M., Hashimshony T., Dean W., Goldstein B., Johnson A.C., Keshet I., Reik W., and Cedar H. 2001. An upstream repressor element plays a role in Igf2 imprinting. *EMBO J.* **20:** 3518.

Engemann S., Strodicke M., Paulsen M., Franck O., Reinhardt R., Lane N., Reik W., and Walter J. 2000. Sequence and functional comparison in the Beckwith-Wiedemann region: Implications for a novel imprinting centre and extended imprinting. *Hum. Mol. Genet.* **9:** 2691.

Feil R., Handel M.A., Allen N.D., and Reik W. 1995. Chromatin structure and imprinting: Developmental control of DNase-I sensitivity in the mouse insulin-like growth factor 2 gene. *Dev. Genet.* **17:** 240.

Ferguson-Smith A.C. and Surani M.A. 2001. Imprinting and the epigenetic asymmetry between parental genomes. *Science* **293:** 1086.

Fitzpatrick G.V., Soloway P.D., and Higgins M.J. 2002. Regional loss of imprinting and growth deficiency in mice with a targeted deletion of KvDMR1. *Nat. Genet.* **32:** 426.

Forne T., Oswald J., Dean W., Saam J.R., Bailleul B., Dandolo L., Tilghman S.M., Walter J., and Reik W. 1997. Loss of the maternal H19 gene induces changes in Igf2 methylation in both cis and trans. *Proc. Natl. Acad. Sci.* **94:** 10243.

Fournier C., Goto Y., Ballestar E., Delaval K., Hever A.M., Esteller M., and Feil R. 2002. Allele-specific histone lysine methylation marks regulatory regions at imprinted mouse genes. *EMBO J.* **21:** 6560.

Haig D. and Westoby M. 1989. Parent-specific gene expression and the triploid endosperm. *Am. Nat.* **134:** 147.

Hark A.T., Schoenherr C.J., Katz D.J., Ingram R.S., Levorse J.M., and Tilghman S.M. 2000. CTCF mediates methylation-sensitive enhancer-blocking activity at the H19/Igf2 locus. *Nature* **405:** 486.

Hata K., Okano M., Lei H., and Li E. 2002. Dnmt3L cooperates with the Dnmt3 family of de novo DNA methyltransferases to establish maternal imprints in mice. *Development* **129:** 1983.

Horike S., Mitsuya K., Meguro M., Kotobuki N., Kashiwagi A., Notsu T., Schulz T.C., Shirayoshi Y., and Oshimura M. 2000. Targeted disruption of the human LIT1 locus defines a putative imprinting control element playing an essential role in Beckwith-Wiedemann syndrome. *Hum. Mol. Genet.* **9:** 2075.

Huynh K.D. and Lee J.T. 2003. Inheritance of a pre-inactivated paternal X chromosome in early mouse embryos. *Nature* **426:** 857.

Ishihara K. and Sasaki H. 2002. An evolutionarily conserved putative insulator element near the 3´ boundary of the imprinted Igf2/H19 domain. *Hum. Mol. Genet.* **11:** 1627.

John R.M., Ainscough J.F., Barton S.C., and Surani M.A. 2001. Distant cis-elements regulate imprinted expression of the mouse p57(Kip2)(Cdkn1c) gene: Implications for the human disorder, Beckwith-Wiedemann syndrome. *Hum. Mol. Genet.* **10:** 1601.

Kaffer C.R., Grinberg A., and Pfeifer K. 2001. Regulatory mechanisms at the mouse Igf2/H19 locus. *Mol. Cell. Biol.* **21:** 8189.

Kanduri C., Fitzpatrick G., Mukhopadhyay R., Kanduri M., Lobanenkov V., Higgins M., and Ohlsson R. 2002. A differentially methylated imprinting control region within the Kcnq1 locus harbors a methylation-sensitive chromatin insulator. *J. Biol. Chem.* **277:** 18106.

Kanduri C., Pant V., Loukinov D., Pugacheva E., Qi C.F., Wolffe A., Ohlsson R., and Lobanenkov V.V. 2000. Functional association of CTCF with the insulator upstream of the H19 gene is parent of origin-specific and methylation-sensitive. *Curr. Biol.* **10:** 853.

Kaneda M., Okano M., Hata K., Sado T., Tsujimoto N., Li E., and Sasaki H. 2004. Essential role for de novo DNA methyltransferase Dnmt3a in paternal and maternal imprinting. *Nature* **429:** 900.

Killian J.K., Byrd J.C., Jirtle J.V., Munday B.L., Stoskopf M.K., MacDonald R.G., and Jirtle R.L. 2000. M6P/IGF2R imprinting evolution in mammals. *Mol. Cell* **5:** 707.

Labrador M. and Corces V.G. 2002. Setting the boundaries of chromatin domains and nuclear organization. *Cell* **111:** 151.

Lei H., Oh S.P., Okano M., Juttermann R., Goss K.A., Jaenisch R., and Li E. 1996. De novo DNA cytosine methyltransferase activities in mouse embryonic stem cells. *Development* **122:** 3195.

Leighton P.A., Saam J.R., Ingram R.S., Stewart C.L., and Tilghman S.M. 1995. An enhancer deletion affects both H19 and Igf2 expression. *Genes Dev.* **9:** 2079.

Lin S.P., Youngson N., Takada S., Seitz H., Reik W., Paulsen M., Cavaille J., and Ferguson-Smith A.C. 2003. Asymmetric regulation of imprinting on the maternal and paternal chromosomes at the Dlk1-Gtl2 imprinted cluster on mouse chromosome 12. *Nat. Genet.* **35:** 97.

Lopes S., Lewis A., Hajkova P., Dean W., Oswald J., Forne T., Murrell A., Constancia M., Bartolomei M., Walter J., and Reik W. 2003. Epigenetic modification in an imprinting cluster are controlled by a hierarchy of DMRs suggesting long-range chromatin interactions. *Hum. Mol. Genet.* **12:** 295.

Mager J., Montgomery N.D., de Villena F.P., and Magnuson T. 2003. Genome imprinting regulated by the mouse Polycomb group protein Eed. *Nat. Genet.* **33:** 502.

Mak W., Nesterova T.B., de Napoles M., Appanah R., Yamanaka S., Otte A.P., and Brockdorff N. 2004. Reactivation of the paternal X chromosome in early mouse embryos. *Science* **303:** 666.

McGrath J. and Solter D. 1984. Completion of mouse embryogenesis requires both the maternal and paternal genomes. *Cell* **37:** 179.

Murrell A., Heeson S., and Reik W. 2004. Physical contact between differentially methylated regions partitions the imprinted *Igf2* and *H19* genes into parent specific chromatin loops. *Nat. Genet.* **36:** 889.

Murrell A., Heeson S., Bowden L., Constancia M., Dean W.,

Kelsey G., and Reik W. 2001. An intragenic methylated region in the imprinted Igf2 gene augments transcription. *EMBO Rep.* **2:** 1101.

Nicholls R.D., Saitoh S., and Horsthemke B. 1998. Imprinting in Prader-Willi and Angelman syndromes. *Trends Genet.* **14:** 194.

Okamoto I., Otte A.P., Allis C.D., Reinberg D., and Heard E. 2004. Epigenetic dynamics of imprinted X inactivation during early mouse development. *Science* **303:** 644.

Palstra R.J., Tolhuis B., Splinter E., Nijmeijer R., Grosveld F., and de Laat W. 2003. The beta-globin nuclear compartment in development and erythroid differentiation. *Nat. Genet.* **35:** 190.

Pant V., Mariano P., Kanduri C., Mattsson A., Lobanenkov V., Heuchel R., and Ohlsson R. 2003. The nucleotides responsible for the direct physical contact between the chromatin insulator protein CTCF and the H19 imprinting control region manifest parent of origin-specific long-distance insulation and methylation-free domains. *Genes Dev.* **17:** 586.

Patrinos G.P., de Krom M., de Boer E., Langeveld A., Imam A.M., Strouboulis J., de Laat W., and Grosveld F.G. 2004. Multiple interactions between regulatory regions are required to stabilize an active chromatin hub. *Genes Dev.* **18:** 1495.

Reik W. and Walter J. 2001. Genomic imprinting: Parental influence on the genome. *Nat. Rev. Genet.* **2:** 21.

Reik W., Constancia M., Fowden A., Anderson N., Dean W., Ferguson-Smith A., Tycko B., and Sibley C. 2003. Regulation of supply and demand for maternal nutrients in mammals by imprinted genes. *J. Physiol.* **547:** 35.

Sasaki H., Jones P.A., Chaillet J.R., Ferguson-Smith A.C., Barton S.C., Reik W., and Surani M.A. 1992. Parental imprinting: Potentially active chromatin of the repressed maternal allele of the mouse insulin-like growth factor II (Igf2) gene. *Genes Dev.* **6:** 1843.

Schoenherr C.J., Levorse J.M., and Tilghman S.M. 2003. CTCF maintains differential methylation at the Igf2/H19 locus. *Nat. Genet.* **33:** 66.

Sleutels F. and Barlow D.P. 2002. The origins of genomic imprinting in mammals. *Adv. Genet.* **46:** 119.

Sleutels F., Zwart R., and Barlow D.P. 2002. The non-coding Air RNA is required for silencing autosomal imprinted genes. *Nature* **415:** 810.

Stoger R., Kubicka P., Liu C.G., Kafri T., Razin A., Cedar H., and Barlow D.P. 1993. Maternal-specific methylation of the imprinted mouse Igf2r locus identifies the expressed locus as carrying the imprinting signal. *Cell* **73:** 61.

Surani M.A., Barton S.C., and Norris M.L. 1984. Development of reconstituted mouse eggs suggests imprinting of the genome during gametogenesis. *Nature* **308:** 548.

Szabo P., Tang S.H., Rentsendorj A., Pfeifer G.P., and Mann J.R. 2000. Maternal-specific footprints at putative CTCF sites in the H19 imprinting control region give evidence for insulator function. *Curr. Biol.* **10:** 607.

Szabo P.E., Tang S.H., Silva F.J., Tsark W.M., and Mann J.R. 2004. Role of CTCF binding sites in the Igf2/H19 imprinting control region. *Mol. Cell. Biol.* **24:** 4791.

Thakur N., Kanduri M., Holmgren C., Mukhopadhyay R., and Kanduri C. 2003. Bidirectional silencing and DNA methylation-sensitive methylation-spreading properties of the Kcnq1 imprinting control region map to the same regions. *J. Biol. Chem.* **278:** 9514.

Thorvaldsen J.L., Duran K.L., and Bartolomei M.S. 1998. Deletion of the H19 differentially methylated domain results in loss of imprinted expression of H19 and Igf2. *Genes Dev.* **12:** 3693.

Tolhuis B., Palstra R.J., Splinter E., Grosveld F., and de Laat W. 2002. Looping and interaction between hypersensitive sites in the active beta-globin locus. *Mol. Cell* **10:** 1453.

Verona R.I., Mann M.R., and Bartolomei M.S. 2003. Genomic imprinting: Intricacies of epigenetic regulation in clusters. *Annu. Rev. Cell Dev. Biol.* **19:** 237.

Wakefield M.J., Keohane A.M., Turner B.M., and Graves J.A. 1997. Histone underacetylation is an ancient component of mammalian X chromosome inactivation. *Proc. Natl. Acad. Sci.* **94:** 9665.

Weber M., Hagege H., Murrell A., Brunel C., Reik W., Cathala G., and Forne T. 2003. Genomic imprinting controls matrix attachment regions in the Igf2 gene. *Mol. Cell. Biol.* **23:** 8953.

Yoon B.J., Herman H., Sikora A., Smith L.T., Plass C., and Soloway P.D. 2002. Regulation of DNA methylation of Rasgrf1. *Nat. Genet.* **30:** 92.

Yusufzai T.M. and Felsenfeld G. 2004. The 5′-HS4 chicken beta-globin insulator is a CTCF-dependent nuclear matrix-associated element. *Proc. Natl. Acad. Sci.* **101:** 8620.

Yusufzai T.M., Tagami H., Nakatani Y., and Felsenfeld G. 2004. CTCF tethers an insulator to subnuclear sites, suggesting shared insulator mechanisms across species. *Mol. Cell* **13:** 291.

Genomic Imprinting: Antagonistic Mechanisms in the Germ Line and Early Embryo

A.M. Fedoriw, N.I. Engel, and M.S. Bartolomei
*Howard Hughes Medical Institute and Department of Cell and Developmental Biology,
University of Pennsylvania School of Medicine, Philadelphia, Pennsylvania 19104*

Proper mammalian development requires the contributions of both a maternal and paternal genome (Barton et al. 1984; McGrath and Solter 1984). This unique constraint is due to a subset of genes whose monoallelic expression is dependent upon the parent of origin. Therefore, despite identical genetic information, alleles of such imprinted genes can be discriminated from one another by transcriptional activators or repressors in the nuclei of all somatic cells. Also implied by this phenomenon are mechanisms in each gamete that must establish a transcriptional state for an allele, one that is capable of being interpreted and stably maintained throughout the lifetime of the organism. Correct establishment and maintenance of these states is paramount, as misregulation of imprinted genes is associated with many developmental defects as well as with numerous cancers (Lalande 1996; Tycko 1999). Recent data have uncovered *cis*-elements and *trans*-factors that control the regulation of various imprinted genes in the soma. However, the mechanisms that govern establishment of imprints during gametogenesis remain incompletely understood.

The organization of imprinted genes within the genome provides several clues about their regulation. Approximately 70 imprinted genes have been identified in mammals (http://www.mgu.har.mrc.ac.uk/research/imprinted/). Not only is their imprinted expression well conserved between mouse and man, but their genomic organization is as well, as imprinted genes are often associated in clusters throughout the genome (Verona et al. 2003). The high level of conservation between imprinted expression patterns and clustered organization suggests a functional link between the two. Chromosomal rearrangements or deletions within these clusters often cause misregulation of linked genes, providing further evidence of a functional connection between their organization and regulation (Reik and Maher 1997). Central to their allele-specific expression, differential DNA methylation has been associated with almost all imprinted loci (Brannan and Bartolomei 1999). Targeted mutagenesis has shown that these differentially methylated domains (DMDs; also differentially methylated regions, DMRs) are critical to the regulation of neighboring imprinted genes (Thorvaldsen et al. 1998; Fitzpatrick et al. 2002; Lin et al. 2003). Because it is stable and heritable through cell division, DNA methylation is the leading candidate for the imprinting "mark," the epigenetic coding that distinguishes one allele from another.

The critical importance of DNA methylation to imprinted gene regulation has been demonstrated in mice that carry mutations in several members of the DNA methyltransferase (Dnmt) family, the enzymes responsible for methylating DNA in mammals. Hypomorphic mutations of the first such enzyme identified, Dnmt1, have demonstrated that this protein is primarily involved in maintaining methylation at both imprinted and nonimprinted loci (Li et al. 1993). An oocyte-specific, maternally loaded alternative splice form of Dnmt1, Dnmt1o, functions to maintain methylation during a brief window of preimplantation development (Howell et al. 2001).

The identification and characterization of three *Dnmt3* genes has linked these related proteins to establishment of imprints, as these genes are highly expressed during gametogenesis in both parents (La Salle et al. 2004). Compound mutations of *Dnmt3a* and *Dnmt3b* genes have uncovered a vital, yet possibly redundant, role in establishing methylation imprints (Okano et al. 1999). More recently, conditional null alleles of *Dnmt3a* have demonstrated an essential role for this de novo methyltransferase in the acquisition of many, but not all, methylation imprints in both parental germ lines (Kaneda et al. 2004). Alternative forms of *Dnmt3a* have also been identified, potentially adding greater complexity to the establishment of methylation imprints (Chen et al. 2002). Interestingly, the only member of this family that does not have DNA methyltransferase activity in vitro, *Dnmt3l*, is required for establishment of methylation at imprinted loci at least in oocytes (Bourc'his et al. 2001). Biochemical data suggest that Dnmt3l may exert its activity by interacting with Dnmt3a and 3b (Hata et al. 2002). Therefore, Dnmt3l may provide sequence specificity to the other Dnmt3 proteins.

Differentially methylated domains can be divided into those that are maternally or paternally hypermethylated. The majority of known DMDs, such as those for the *Snurf-snrpn*, *Igf2r*, *Kcnq1ot1*, and *Peg3* loci, are maternally hypermethylated, while the differentially methylated regions of *H19* and *Rasgrf1* are paternally hypermethylated (Stöger et al. 1993; Plass et al. 1996; Shemer et al. 1997; Tremblay et al. 1997; Li et al. 2000; Fitzpatrick et al. 2002). Recent work has focused on the paternally hypermethylated *H19* DMD as a paradigm not only for the regulation of imprinting, but for its establishment as well.

H19 and *Insulin-like growth factor 2* (*Igf2*) are two closely linked, yet oppositely imprinted genes (Bar-

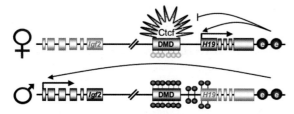

Figure 1. Imprinted expression of the *H19/Igf2* locus. Methylation of the DMD regulates the expression of *H19* and *Igf2*. The DMD is maternally hypomethylated (*open circles*) and paternally hypermethylated (*filled circles*). On the maternal allele, the DMD acts as an enhancer-blocking element through the binding of Ctcf. The shared enhancers (e) can only activate the *H19* promoter. On the paternal allele, methylation of the DMD prevents Ctcf binding, and *Igf2* is expressed. The methylation in the DMD spreads to the *H19* promoter during development, thereby preventing its expression.

tolomei et al. 1991; DeChiara et al. 1991). *H19* is located approximately 100 kb downstream of the *Igf2* promoter; the DMD lies 2 kb upstream of the *H19* transcriptional start site (Fig. 1). Analyses from several laboratories have demonstrated that the DMD is required on both parental chromosomes for proper imprinted expression. Various targeted deletions and mutations have shown that in somatic tissues, the DMD acts as a switch that directs transcriptional activity by regulating access of enhancers that are shared by both genes (Thorvaldsen et al. 1998; Srivastava et al. 2000). On the paternal chromosome, where the DMD is hypermethylated, *H19* is silent and *Igf2* is expressed. The repression of *H19* is thought to occur through the recruitment of repressive factors by the hypermethylated DMD. On the maternal allele, hypomethylation of the DMD results in the formation of an enhancer-blocking element that limits the action of the enhancers to the *H19* promoter, thereby preventing activation of *Igf2*. The function of the DMD on the maternal allele is mediated by the enhancer-blocking protein, Ctcf (Bell and Felsenfeld 2000; Hark et al. 2000; Kanduri et al. 2000; Szabo et al. 2000; Schoenherr et al. 2003). Recent work from our laboratory has shown that the requirement for Ctcf on the maternal allele extends to a critical period in oocyte growth when maternally hypermethylated DMDs acquire their characteristic levels of methylation (Fedoriw et al. 2004).

In this review, we summarize what is known about the establishment of DNA methylation on DMDs during gametogenesis. We also discuss mechanisms that are hypothesized to initiate and maintain differential methylation at imprinted loci.

METHYLATION OF DMRs IN THE GAMETES

One of the main implications of the existence of imprinted genes is that mechanisms must exist within each germ line to establish the future parental identity of a given locus. Presently, we know little about what molecules are involved in these processes and how the signature patterns are generated. Even the full nature of the marks remains elusive. As one of the first epigenetic modifications identified, DNA methylation has been most extensively described for several loci in the germ lines. Recently, the expression patterns of the Dnmt family of proteins has been shown to be quite different in the male and female germ line (La Salle et al. 2004). These data, along with the divergent processes that produce male and females gametes, provide ample opportunity for differential epigenetic modification. What remains to be shown, however, is what signals and factors cooperate to methylate a DMR exclusively in a single parental germ line.

Before a parental imprint can be conferred upon a given locus, any existing epigenetic modifications must be removed. The demethylation of various DMRs has been extensively studied in the developing mouse embryo. The allocation of the germ line occurs early in development; after their specification around 7 days postcoitum (dpc), primordial germ cells (PGCs) migrate to and populate the primitive gonad by 10.5 dpc (McLaren 2003). Erasure of DNA methylation at imprinted loci is complete before 12.5 dpc (Hajkova et al. 2002; Lee et al. 2002), coincident with the time when the PGC genome is globally undermethylated compared with the somatic genome (Monk et al. 1987). The paternal *H19* allele is hypomethylated by 12.5 dpc in both germ lines (Davis et al. 2000), an observation that may point to a common mechanism of erasure. Completion of erasure also coincides with cessation of PGC proliferation. Thus, methylation imprints are erased before further differentiation and proliferation of the male and female gametes.

At 12.5 dpc, gametogenesis becomes sex-specific and the processes that occur in the developing gonads differ greatly. In the developing male germ line, hypermethylation of select DMRs as well as intracisternal A particle (IAP) repetitive elements occurs by late gestation (Brandeis et al. 1993; Walsh et al. 1998; Davis et al. 2000). Concurrent with its proposed role in directing de novo methylation of imprinted regions (Hata et al. 2002), *Dnmt3l* expression peaks during the establishment of DNA methylation imprints before birth in the male germ line. Among DMRs that are paternally methylated, the acquisition of methylation has been best studied at the *H19* DMD (Davis et al. 1999, 2000; Ueda et al. 2000). The paternal allele is almost fully methylated by 13.5 dpc, shortly after demethylation is complete. Unexpectedly, significant levels of methylation are observed on the maternal DMD only after 16.5 dpc (Davis et al. 2000). This observation implies that although both alleles are hypomethylated, they continue to retain some parental identity. Additionally, this difference is sufficient to be recognized by Dnmts, or other factors, in the developing gamete. The nature of this difference is unknown, yet it may reflect differences in chromatin structure (such as bound Ctcf, histone modifications, etc.) that make the paternal allele more susceptible to methylation than the maternal allele (Davis et al. 2000; Ueda et al. 2000).

In contrast to the male germ line, oogonia proliferate and become meiotically arrested before birth. These incompetent, or nongrowing (ng), oocytes are globally undermethylated compared to the similar embryonic stage in the male. Additionally, DMRs are hypomethylated at this point (Lucifero et al. 2002). At birth, oocytes begin to grow and differentiate and gain competence to undergo the first meiotic division. During oocyte growth, and coincident with an increase in Dnmt3l mRNA levels, the DMRs of various maternally hypermethylated loci acquire their methylation (Lucifero et al. 2004).

The timing of DNA methylation at imprinted loci during oocyte growth has been characterized by several groups (Lucifero et al. 2002, 2004; Obata and Kono 2002). Their data show that not all DMRs acquire methylation simultaneously. The *Peg1* DMR, for example, does not become completely hypermethylated until the oocyte is nearly fully grown. Others, like the *Peg3* DMR and *Igf2r* DMR2, acquire methylation much earlier. Interestingly, de novo methylation of *Snurf-snrpn* DMR1 in the female germ line parallels that observed for the *H19* DMD in the male germ line with the maternal allele being hypermethylated earlier than the paternal allele (Lucifero et al. 2004). De novo methylation of maternally methylated DMRs is complete by the time the oocyte has fully matured.

MODELS FOR IMPRINT ESTABLISHMENT

Along with studies determining the timing of DNA methylation, *cis*-elements required to maintain imprints in the soma have also been well characterized. What remains to be elucidated are the signals required for establishment of a maternal or paternal identity in the germ line. Are the states actively determined in both germ lines or does one represent a default state for a particular locus? Two general models can be postulated to explain how these states are established. One possibility is the presence of sex-specific germ line factors that could target the de novo DNA methylation machinery to those DMRs that are to be hypermethylated (Fig. 2A). The other germ line would not express the same factor and, therefore, the DMR would remain unmethylated in that gamete. Alternate splice forms or differential expression of accessory factors would be sufficient to provide sequence specificity for Dnmt proteins in one germ line, but not the other. For example, the *Igf2r* DMR2 would remain hypomethylated during spermatogenesis because the protein(s) that direct its de novo methylation would be expressed only during oocyte growth. An alternative model posits that both germ lines could have similar potential for directing methylation to imprinted regions. However, sex-specific expression of factors that protect or exclude a locus from de novo methylation would be the primary determinant of a hypomethylated or hypermethylated DMR (Fig. 2B). Because this protective effect would be conferred in only one germ line, a differential state of methylation would be established.

ESTABLISHING THE MATERNAL IDENTITY OF THE *H19* DMD

Despite the large number of targeted mutations that have revealed essential *cis*-regulatory elements at the *H19* locus, none has identified an element that is required for establishment of the methylation pattern during gametogenesis. In contrast, deletion of an element adjacent to the *Rasgrf1* DMD results in a maternal-like hypomethylated state in sperm (Yoon et al. 2002). Therefore, it is unclear whether the *H19* locus harbors such an element. Alternatively, the current array of mutations may have abolished a signal that is required for *both* preventing and attracting methylation.

Several laboratories have proposed models for how the imprint is established at the *H19* DMD (Pfeifer 2000; Thorvaldsen and Bartolomei 2000). One possibility is that factors in the male germ line direct methylation to the *H19* DMD exclusively during spermatogenesis, overriding the default hypomethylated state. Here, the *H19* DMD would remain unmethylated during oocyte growth because of the absence of targeting factors. Furthermore, Ctcf would only be required on the maternal allele after fertilization to establish an insulator. Another model hypothesizes that Ctcf would be required to protect this locus from de novo methylation during oogenesis. In this case, Ctcf would be bound to the DMD during oocyte growth, establishing the enhancer-blocking chromatin structure prior to fertilization.

The latter model is particularly attractive because a protective role for a transcription factor has been described previously. The ubiquitous transcriptional activator, Sp1, has been shown to prevent de novo methylation of its binding site in a transgene model (Brandeis et al. 1994; Macleod et al. 1994). Could Ctcf play a similar role? Since Ctcf is expressed when methylation imprints are being established in the female germ line, we postulated that Ctcf is required for the *H19* DMD to remain hypomethylated during this period. If so, then oocytes that have been depleted of Ctcf protein should bear a hypermethylated *H19* DMD.

We used a transgenic-RNA-interference (RNAi)-based approach to target Ctcf mRNA for degradation specifically in growing oocytes (Fedoriw et al. 2004). We developed a series of transgenic lines that depleted Ctcf mRNA and protein to varying levels. This system provided us with a means to assess how vulnerable the *H19* DMD is during a particularly active time of methylation acquisition at other imprinted loci. In fact, oocytes from transgenic females with the greatest reduction of Ctcf did acquire significant methylation at the DMD, rivaling the

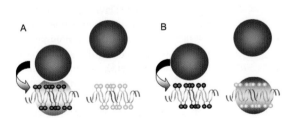

Figure 2. Models for establishment of methylation imprints. Two models for the differential methylation of imprinting control regions in the germ lines. (*A*) Sequence-specific factors (*green circle*) are expressed in only one germ line. These factors direct activity of de novo Dnmts (*orange circles*). Because the targeting factors are not expressed in the other germ line, the target sequence is not methylated. (*B*) Conversely, Dnmts recognize the CpG-rich sequences in both germ lines but protective factors (*red circles*) are only expressed in one developing gamete.

level seen on the wild-type paternal allele. This observation suggested that Ctcf is the critical factor in determining the maternal hypomethylated identity.

Since depletion of a broadly required transcription factor or essential chromatin component could lead to nonspecific responses, we also analyzed these oocytes for other potential defects in methylation. We have not observed abnormal methylation of other loci, nor has depletion of Ctcf led to loss of methylation at maternally hypermethylated DMRs (Fedoriw et al. 2004). We have also found that the *H19* DMD does not become methylated simply in the presence of other double-stranded RNAs (dsRNAs); in oocytes from a transgenic female carrying a similar transgene specific for the Mos mRNA (Stein et al. 2003), the *H19* DMD remained hypomethylated (A. Fedoriw, unpubl.). These data show that the hypermethylation observed at the *H19* DMD is not due to nonspecific effects; rather, the hypermethylation is the consequence of the DMD losing the protection it requires to repel the activity of Dnmts.

Recently, Loukinov and coworkers identified a Ctcf-like protein designated Boris (Brother of the Regulator of Imprinted Sites) (Loukinov et al. 2002). *Boris* is expressed exclusively in type A spermatocytes, which are reported to lack Ctcf. Additionally, these cells appear deficient for methylated DNA, as assayed by anti-5-methylcytosine antibody staining, although these data conflict with those reported by several other groups (Davis et al. 2000; Ueda et al. 2000). Loukinov and coworkers have hypothesized that Boris is responsible for the hypermethylated state of the DMD in the male germ line by either sequestering Ctcf or displacing it from its binding sites and allowing Dnmts to methylate those sites. In the models outlined in Figure 2, Boris may be an example of a targeting factor. We observed no ectopic expression of either Boris protein or mRNA in oocytes from transgenic females, suggesting that the methylation we observed in our system was unlikely to be mediated through this molecule (A. Fedoriw, unpubl.).

Our results showing ectopic *H19* methylation were rather surprising considering that targeted mutations of the DMD that ablate binding of Ctcf, as assayed by electrophoretic mobility shift experiments, resulted in methylation of the DMD only later during embryonic development (Pant et al. 2003; Schoenherr et al. 2003; Szabo et al. 2004). The DMD remained hypomethylated in fully grown oocytes and hypermethylated in sperm. Several possible explanations could account for this discrepancy. First, the four Ctcf binding sites that mediate the protection of the DMD in the soma may not exhibit the same function in the germ line. Other unidentified Ctcf binding sites located within or perhaps distant from the DMD may be responsible for conferring this function. Ctcf can bind a wide range of sequences, presumably through the use of distinct zinc fingers (Filippova et al. 1996). Therefore, the exact nucleotide sequences that Ctcf contacts in oocytes may be distinct from those required for the enhancer-blocking function in the soma. Second, it is possible that depletion of Ctcf results in misregulation of genes such as *Dnmts*, whose abnormal expression may lead to the observed phenotype. Although we are currently exploring this possibility, we do not think that it is a likely cause of ectopic methylation. The levels of transcripts for the *Dnmt* family are higher in the growing oocyte than at any other point during gametogenesis. Therefore, if increased RNA and protein caused nonspecific DNA methylation, it would likely have been detected at other loci, such as the *skeletal α-actin* promoter, which we did not observe (Fedoriw et al. 2004).

Finally, although the *H19* DMD acquires a high level of methylation in the Ctcf-depleted oocytes, many DNA strands remain hypomethylated. This may be a consequence of residual Ctcf that is present as the oocytes mature, leaving some protein that can protect the locus. However, as in the early male germ line, the hypomethylated paternal and maternal *H19* DMD alleles may be differentially regulated in the growing oocyte. We are currently investigating whether there is an allelic bias to the acquisition of methylation at the *H19* DMD in Ctcf-depleted oocytes. Perhaps only one of the alleles is susceptible to methylation at a given point during oogenesis, while the other has a chromatin conformation that is resistant. In this case, the maternally inherited DMD is protected by Ctcf, while the demethylated paternal allele is protected by other means. Therefore, mutant DMD alleles that cannot bind Ctcf would resemble the paternal alleles in the oocyte, and may have another mechanism whereby they remain protected from, or not targeted for, de novo methylation. The observed requirement for Ctcf would not occur in this case until the insulator is established during development. Together with previous observations from the paternal germ line, each allele of *H19* or any other imprinted gene may have a unique method for establishing its new parental identity.

THE PATERNAL *H19* DMD: ANTAGONISTIC FUNCTIONS WITHIN OVERLAPPING ELEMENTS

Although the maternal, hypomethylated state of the DMD is actively determined during oocyte growth, hypermethylation of the DMD in the paternal germ line is not necessarily a default state. Rather, the paternal identity of this locus may require specific factors or chromatin modifications to ensure that methylation is properly established, strictly maintained during each cell division, and correctly interpreted in the nucleus of every cell.

Data from our lab have demonstrated that the enhancer-blocking function of the DMD and the hypermethylated, silencing function of the DMD cannot coexist on the same allele (Engel et al. 2004). We generated a targeted allele of the DMD, $H19^{DMD-9CG}$, in which we mutated CpG dinucleotides in the Ctcf-binding sites. This allele was capable of binding Ctcf even in the presence of methylation (Fig. 3). When inherited paternally, these mutations did not interfere with the acquisition of methylation at this locus, as the mutant DMD is hypermethylated in sperm. However, and in contrast to the wild-type paternal allele where the hypermethylated state is strictly maintained, many paternal $H19^{DMD-9CG}$ chromosomes become gradually demethylated during development. The apparent instability of the methylation is suggestive

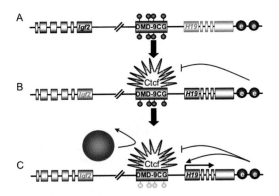

Figure 3. The paternal $H19^{DMD-9CG}$ allele activates *H19*. (*A*) In sperm, methylation of the mutant DMD is not affected. (*B*) However, since the mutant DMD can bind Ctcf even when methylated, the mutant DMD will still function as an enhancer blocker. (*C*) As a consequence, loss of methylation can be observed both at the DMD and at the promoter region. The instability of the hypermethylated state can occur through exclusion of maintenance methylation by Dnmt1 (*orange circle*).

of an antagonistic relationship between two mutually exclusive functions encoded within the same sequence. The targeted mutations result in an allele that cannot faithfully propagate the paternal epigenotype because it has acquired a property normally restricted to the maternal DMD, i.e., binding of Ctcf, a conclusion supported by studies in transfected embryonic stem cells (Rand et al. 2004). This is the only example of a DMD mutation that specifically affects the paternal allele.

It is quite surprising that mutation of only nine CpG dinucleotides results in a dramatic reduction in the ability of the paternal allele to retain its methylation state. One possible explanation is that loss of methylation occurs because of a decrease in CpG density. An alternative hypothesis is that the loss of methylation is dependent upon ectopic Ctcf binding. Ctcf may protect or exclude maintenance DNA methylation by Dnmt1 during embryonic divisions (Fig. 3).

While these two explanations may hold, the loss of methylation on the mutant allele may also reflect an incomplete establishment of the paternal state of the DMD. Perhaps the same sites that function in the maternal germ line as Ctcf-binding sites serve in the paternal germ line as docking sites for other factors. These factors would translate the hypermethylated state of the DMD into associated histone modifications and a resulting chromatin structure would lock in the methylation at this locus. Such an "epigenetic superstructure" could explain the resistance of the *H19* DMD to the genome-wide, active demethylation that occurs in the male pronucleus shortly after fertilization and the passive demethylation that occurs thereafter (Rougier et al. 1998; Mayer et al. 2000; Oswald et al. 2000; Santos et al. 2002). Therefore, loss of binding sites (or decrease in CpG density) would not affect acquisition of methylation at the DMD during spermatogenesis. It would, however, fail to propagate this state during embryogenesis and result in the gradual loss of methylation that we observe. As *Boris* seems the most likely candidate for such a role in spermatocytes, it will be of great interest to ascertain whether male mice that lack *Boris* faithfully establish a hypermethylated DMD that can be maintained in the offspring.

CONCLUSIONS

Imprinted genes are expressed in a parent-of-origin-specific manner. DNA methylation appears to be an epigenetic mark that allows two genetically identical alleles to be discriminated. Recently, research has demonstrated how DMDs control their associated imprinted genes. However, the mechanisms that establish a maternal or paternal identity of these DMDs in the germ lines remain poorly understood.

Our results argue that, at least for *H19*, the hypomethylated state of the maternal DMD is established by Ctcf binding during oogenesis. Thus, Ctcf appears to be a critical factor that distinguishes the *H19* DMD from its hypermethylated counterparts in the growing oocyte. It is not yet clear whether a similar mechanism can be extended to other loci. While it has been proposed that Ctcf may be a global regulator of imprint establishment, a role for Ctcf at other imprinted loci remains undefined. Other DMRs that contain Ctcf-binding sites, such as the KvDMR1, are maternally hypermethylated (Kanduri et al. 2002), in contrast to the maternally hypomethylated *H19* DMD. This observation suggests that either Ctcf is not itself sufficient to confer a maternally hypomethylated state or the mechanisms required for establishment of a methylation imprint may be unique to each DMR.

Just as a protective mechanism may not function at every imprinted locus, our results do not preclude the possibility that the hypermethylated state is also actively determined. In the *H19* paradigm, factors such as *Boris* may be required to generate a complete paternal imprint. This hypothesis is supported by the inability of the paternally inherited $H19^{DMD-9CG}$ allele to maintain a hypermethylated state. Certainly, the complex bipartite function of the *Snurf-snrpn* DMR2 argues that this locus may also be controlled through active modification in both germ lines (Nicholls and Knepper 2001). Chromosomal abnormalities, most frequently deletions, in the promoter region of the human *SNURF-SNRPN* promoter are associated with unique disease phenotypes, depending on the parent from which these lesions are inherited (Nicholls et al. 1999; Cassidy et al. 2000). Collectively, genetic data from these patients have refined the regions responsible for each phenotype, identifying critical domains for imprinting of this locus. The shortest regions of overlap of the different deletions for each syndrome do not themselves overlap, indicative of independent mechanisms functioning to establish the maternal and paternal identities.

Conversely, the imprint at other loci seems to be determined in only one germ line, while the other represents a default state. At such loci, deletion of the DMR would result in an allele reflecting the default state for the locus. This appears to be the case for the *Rasgrf1* and *Kcnq1ot1* DMRs. The *Rasgrf1* DMD is hypermethylated in sperm; however, deletion of adjacent repetitive sequences results in paternal hypomethylation of this domain (Yoon et al.

2002). Maternal inheritance of this deletion has no effect on *Rasgrf1* repression. At the *Kcnq1ot1* locus, the transcription of an antisense RNA (which is repressed by hypermethylation of the maternal KvDMR1) results in long-range repression on the paternal chromosome (Fitzpatrick et al. 2002). Paternal inheritance of a KvDMR1 deletion results in a maternal-like expression pattern of the associated imprinted genes. Maternal inheritance of this mutation has no effect on gene expression. Similarly, only maternal transmission of the IG-DMR deletion at the *Dlk1/Gtl2* locus results in a loss of imprinted expression on that chromosome (Lin et al. 2003).

Together, the above observations suggest that all imprints may not be created equal. While active mechanisms may be involved in the establishment of one—or both—imprinted states, it seems that each DMR may be governed by its own rules. Perhaps this diversity reflects the variety of mechanisms DMRs employ to exert long-range transcriptional activation or silencing: Boundaries, enhancer blockers, antisense-RNA, and functional noncoding RNAs have all been connected to imprinted loci. It would be expected that loss of methylation of these control regions would have unique consequences depending on the mode of regulation used by the DMR. At the *H19/Igf2* locus, the DMD serves as either an enhancer blocker or silencer, depending on its methylation state. Therefore, each gamete may have evolved a specific mechanism to generate the intended transcriptional state. Imprinted loci controlled by antisense RNA, such as *Igf2r*, may require only one establishing mechanism since antisense transcription is believed to be the key factor in repressing associated genes. While it remains to be seen whether the models and examples presented can be applied to other imprinted genes, the data described above should provide a better framework for dissecting mechanisms involved in establishing imprinted states.

ACKNOWLEDGMENTS

We are grateful for the helpful discussions and critical reading of the manuscript by the members of the laboratory. We are also indebted to our many collaborators, particularly Drs. Gary Felsenfeld, Adam West, and Richard Schultz, who contributed to the ideas expressed in this manuscript. This work was supported by NIH grants HD-42026 and GM-51279 and the Howard Hughes Medical Institute. A.M.F. was supported by an NIH predoctoral training grant (HD-07516) and N.I.E. is the recipient of a National Research Service Award (HD-41345).

REFERENCES

Bartolomei M.S., Zemel S., and Tilghman S.M. 1991. Parental imprinting of the mouse *H19* gene. *Nature* 351: 153.

Barton S.C., Surani M.A.H., and Norris M.L. 1984. Role of paternal and maternal genomes in mouse development. *Nature* 311: 374.

Bell A.C. and Felsenfeld G. 2000. Methylation of a CTCF-dependent boundary controls imprinted expression of the *Igf2* gene. *Nature* 405: 482.

Bourc'his D., Xu G.L., Lin C.S., Bollman B., and Bestor T.H. 2001. Dnmt3L and the establishment of maternal genomic imprints. *Science* 294: 2536.

Brandeis M., Kafri T., Ariel M., Chaillet J.R., McCarrey J., Razin A., and Cedar H. 1993. The ontogeny of allele-specific methylation associated with imprinted genes in the mouse. *EMBO J.* 12: 3669.

Brandeis M., Frank D., Keshet I., Siegfried Z., Mendelsohn M., Nemes A., Temper V., Razin A., and Cedar H. 1994. Sp1 elements protect a CpG island from de novo methylation. *Nature* 371: 435.

Brannan C.I. and Bartolomei M.S. 1999. Mechanisms of genomic imprinting. *Curr. Opin. Genet. Dev.* 9: 164.

Cassidy S.B., Dykens E., and Williams C.A. 2000. Prader-Willi and Angelman syndromes: Sister imprinted disorders. *Am. J. Med. Genet.* 97: 136.

Chen T., Ueda Y., Xie S., and Li E. 2002. A novel Dnmt3a isoform produced from an alternative promoter localizes to euchromatin and its expression correlates with active de novo methylation. *J. Biol. Chem.* 277: 38746.

Davis T.L., Yang G.J., McCarrey J.R., and Bartolomei M.S. 2000. The *H19* methylation imprint is erased and reestablished differentially on the parental alleles during male germ cell development. *Hum. Mol. Genet.* 9: 2885.

Davis T.L., Trasler J.M., Moss S.B., Yang G.J., and Bartolomei M.S. 1999. Acquisition of the *H19* methylation imprint occurs differentially on the parental alleles during spermatogenesis. *Genomics* 58: 18.

DeChiara T.M., Robertson E.J., and Efstratiadis A. 1991. Parental imprinting of the mouse insulin-like growth factor II gene. *Cell* 64: 849.

Engel N.I., West A.E., Felsenfeld G., and Bartolomei M.S. 2004. Antagonism between DNA hypermethylation and enhancer-blocking activity at the *H19* DMD is uncovered by CpG mutations. *Nat. Genet.* 36: 883.

Fedoriw A.M., Stein P., Svoboda P., Schultz R.M., and Bartolomei M.S. 2004. Transgenic RNAi reveals essential function for CTCF in *H19* gene imprinting. *Science* 303: 238.

Filippova G.N., Fagerlie S., Klenova E.M., Myers C., Dehner Y., Goodwin G., Neiman P.E., Collins S.J., and Lobanenkov V.V. 1996. An exceptionally conserved transcriptional repressor, CTCF, employs different combinations of zinc fingers to bind diverged promoter sequences of avian and mammalian c-myc oncogenes. *Mol. Cell. Biol.* 16: 2802.

Fitzpatrick G.V., Soloway P.D., and Higgins M.J. 2002. Regional loss of imprinting and growth deficiency in mice with a targeted deletion of *KvDMR1*. *Nat. Genet.* 32: 426.

Hajkova P., Erhardt S., Lane N., Haaf T., El-Maarri O., Reik W., Walter J., and Surani M.A. 2002. Epigenetic reprogramming in mouse primordial germ cells. *Mech. Dev.* 117: 15.

Hark A.T., Schoenherr C.J., Katz D.J., Ingram R.S., Levorse J.M., and Tilghman S.M. 2000. CTCF mediates methylation-sensitive enhancer-blocking activity at the *H19/Igf2* locus. *Nature* 405: 486.

Hata K., Okano M., Lei H., and Li E. 2002. Dnmt3L cooperates with the Dnmt3 family of de novo DNA methyltransferases to establish maternal imprints in mice. *Development* 129: 1983.

Howell C.Y., Bestor T.H., Ding F., Latham K.E., Mertineit C., Trasler J.M., and Chaillet J.R. 2001. Genomic imprinting disrupted by a maternal effect mutation in the Dnmt1 gene. *Cell* 104: 829.

Kanduri C., Fitzpatrick G., Mukhopadhyay R., Kanduri M., Lobanenkov V., Higgins M., and Ohlsson R. 2002. A differentially methylated imprinting control region within the Kcnq1 locus harbors a methylation-sensitive chromatin insulator. *J. Biol. Chem.* 277: 18106.

Kanduri C., Pant V., Loukinov D., Pugacheva E., Qi C.F., Wolffe A., Ohlsson R., and Lobanenkov V.V. 2000. Functional association of CTCF with the insulator upstream of the H19 gene is parent of origin-specific and methylation-sensitive. *Curr. Biol.* 10: 853.

Kaneda M., Okano M., Hata K., Sado T., Tsujimoto N., Li E., and Sasaki H. 2004. Essential role for *de novo* DNA methyltransferase 3a in paternal and maternal imprinting. *Nature* 429: 900.

Lalande M. 1996. Parental imprinting and human disease. *Annu. Rev. Genet.* 30: 173.

La Salle S., Mertineit C., Taketo T., Moens P.B., Bestor T.H., and Trasler J.M. 2004. Windows for sex-specific methylation marked by DNA methyltransferase expression profiles in mouse germ cells. *Dev. Biol.* **268:** 403.

Lee J., Inoue K., Ono R., Ogonuki N., Kohda T., Kaneko-Ishino T., Ogura A., and Ishino F. 2002. Erasing genomic imprinting memory in mouse clone embryos produced from day 11.5 primordial germ cells. *Development* **129:** 1807.

Li E., Beard C., and Jaenisch R. 1993. Role for DNA methylation in genomic imprinting. *Nature* **366:** 362.

Li L.L., Szeto I.Y., Cattanach B.M., Ishino F., and Surani M.A. 2000. Organization and parent-of-origin-specific methylation of imprinted Peg3 gene on mouse proximal chromosome 7. *Genomics* **63:** 333.

Lin S.P., Youngson N., Takada S., Seitz H., Reik W., Paulsen M., Cavaille J., and Ferguson-Smith A.C. 2003. Asymmetric regulation of imprinting on the maternal and paternal chromosomes at the Dlk1-Gtl2 imprinted cluster on mouse chromosome 12. *Nat. Genet.* **35:** 97.

Loukinov D.I., Pugacheva E., Vatolin S., Pack S.D., Moon H., Chernukhin I., Mannan P., Larsson E., Kanduri C., Vostrov A.A., Cui H., Niemitz E.L., Rasko J.E., Docquier F.M., Kistler M., Breen J.J., Zhuang Z., Quitschke W.W., Renkawitz R., Klenova E.M., Feinberg A.P., Ohlsson R., Morse H.C., III, and Lobanenkov V.V. 2002. BORIS, a novel male germ-line-specific protein associated with epigenetic reprogramming events, shares the same 11-zinc-finger domain with CTCF, the insulator protein involved in reading imprinting marks in the soma. *Proc. Natl. Acad. Sci.* **99:** 6806.

Lucifero D., Mann M.R., Bartolomei M.S., and Trasler J.M. 2004. Gene-specific timing and epigenetic memory in oocyte imprinting. *Hum. Mol. Genet.* **13:** 839.

Lucifero D., Mertineit C., Clarke H.J., Bestor T.H., and Trasler J.M. 2002. Methylation dynamics of imprinted genes in mouse germ cells. *Genomics* **79:** 530.

Macleod D., Charlton J., Mullins J., and Bird A.P. 1994. Sp1 sites in the mouse aprt gene promoter are required to prevent methylation of the CpG island. *Genes Dev.* **8:** 2282.

Mayer W., Niveleau A., Walter J., Fundele R., and Haaf T. 2000. Demethylation of the zygotic paternal genome. *Nature* **403:** 501.

McGrath J. and Solter D. 1984. Completion of mouse embryogenesis requires both the maternal and paternal genomes. *Cell* **37:** 179.

McLaren A. 2003. Primordial germ cells in the mouse. *Dev. Biol.* **262:** 1.

Monk M., Boubelik M., and Lehnert S. 1987. Temporal and regional changes in DNA methylation in the embryonic, extraembryonic and germ cell lineages during mouse embryo development. *Development* **99:** 371.

Nicholls R.D. and Knepper J.L. 2001. Genome organization, function, and imprinting in Prader-Willi and Angelman syndromes. *Annu. Rev. Genomics Hum. Genet.* **2:** 153.

Nicholls R.D., Ohta T., and Gray T.A. 1999. Genetic abnormalities in Prader-Willi syndrome and lessons from mouse models. *Acta Paediatr. Suppl.* **88:** 99.

Obata Y. and Kono T. 2002. Maternal primary imprinting is established at a specific time for each gene throughout oocyte growth. *J. Biol. Chem.* **277:** 5285.

Okano M., Bell D.W., Haber D.A., and Li E. 1999. DNA methyltransferases Dnmt3a and Dnmt3b are essential for de novo methylation and mammalian development. *Cell* **99:** 247.

Oswald J., Engemann S., Lane N., Mayer W., Olek A., Fundele R., Dean W., Reik W., and Walter J. 2000. Active demethylation of the paternal genome in the mouse zygote. *Curr. Biol.* **10:** 475.

Pant V., Mariano P., Kanduri C., Mattsson A., Lobanenkov V., Heuchel R., and Ohlsson R. 2003. The nucleotides responsible for the direct physical contact between the chromatin insulator protein CTCF and the H19 imprinting control region manifest parent of origin-specific long-distance insulation and methylation-free domains. *Genes Dev.* **17:** 586.

Pfeifer K. 2000. Mechanisms of genomic imprinting. *Am. J. Hum. Genet.* **67:** 777.

Plass C., Shibata H., Kalcheva I., Mullins L., Kotelevtseva N., Mullins J., Kato R., Sasaki H., Hirotsune S., Okazaki Y., Held W.A., Hayashizaki Y., and Chapman V. 1996. Identification of Grf1 on mouse chromosome 9 as an imprinted gene by RLGS-M. *Nat. Genet.* **14:** 106.

Rand E., Ben-Porath I., Keshet I., and Cedar H. 2004. CTCF elements direct allele-specific undermethylation at the imprinted H19 locus. *Curr. Biol.* **14:** 1007.

Reik W. and Maher E.R. 1997. Imprinting in clusters: Lessons from Beckwith-Wiedemann syndrome. *Trends Genet.* **13:** 330.

Rougier N., Bourc'his D., Gomes D.M., Niveleau A., Plachot M., Páldi A., and Viegas-Péquignot E. 1998. Chromosome methylation patterns during mammalian preimplantation development. *Genes Dev.* **12:** 2108.

Santos F., Hendrich B., Reik W., and Dean W. 2002. Dynamic reprogramming of DNA methylation in the early mouse embryo. *Dev. Biol.* **241:** 172.

Schoenherr C.J., Levorse J.M., and Tilghman S.M. 2003. CTCF maintains differential methylation at the Igf2/H19 locus. *Nat. Genet.* **33:** 66.

Shemer R., Birger Y., Riggs A.D., and Razin A. 1997. Structure of the imprinted mouse Snrpn gene and establishment of its parental-specific methylation pattern. *Proc. Natl. Acad. Sci.* **94:** 10267.

Srivastava M., Hsieh S., Grinberg A., Williams-Simons L., Huang S.-P., and Pfeifer K. 2000. H19 and Igf2 monoallelic expression is regulated in two distinct ways by a shared cis acting regulatory region upstream of H19. *Genes Dev.* **14:** 1186.

Stein P., Svoboda P., and Schultz R.M. 2003. Transgenic RNAi in mouse oocytes: A simple and fast approach to study gene function. *Dev. Biol.* **256:** 187.

Stöger R., Kubicka P., Liu C.-G., Kafri T., Razin A., Cedar H., and Barlow D.P. 1993. Maternal-specific methylation of the imprinted mouse Igf2r locus identifies the expressed locus as carrying the imprinting signal. *Cell* **73:** 61.

Szabo P.E., Tang S.-H., Rentsendorj A., Pfeifer G.P., and Mann J.R. 2000. Maternal-specific footprints at putative CTCF sites in the H19 imprinting control region give evidence for insulator function. *Curr. Biol.* **10:** 607.

Szabo P.E., Tang S.H., Silva F.J., Tsark W.M., and Mann J.R. 2004. Role of CTCF binding sites in the Igf2/H19 imprinting control region. *Mol. Cell. Biol.* **24:** 4791.

Thorvaldsen J.L. and Bartolomei M.S. 2000. Mothers setting boundaries. *Science* **288:** 2145.

Thorvaldsen J.L., Duran K.L., and Bartolomei M.S. 1998. Deletion of the H19 differentially methylated domain results in loss of imprinted expression of H19 and Igf2. *Genes Dev.* **12:** 3693.

Tremblay K.D., Duran K.L., and Bartolomei M.S. 1997. A 5′ 2-kilobase-pair region of the imprinted mouse H19 gene exhibits exclusive paternal methylation throughout development. *Mol. Cell. Biol.* **17:** 4322.

Tycko B. 1999. Genomic imprinting and cancer. *Results Probl. Cell Differ.* **25:** 133.

Ueda T., Abe K., Miura A., Yuzuriha M., Zubair M., Noguchi M., Niwa K., Kawase Y., Kona T., Matsuda Y., Fujimoto H., Shibata H., Hayashizaki Y., and Sasaki H. 2000. The paternal methylation imprint of the mouse H19 locus is acquired in the gonocyte stage during foetal testis development. *Genes Cells* **5:** 649.

Verona R.I., Mann M.R., and Bartolomei M.S. 2003. Genomic imprinting: Intricacies of epigenetic regulation in clusters. *Annu. Rev. Cell Dev. Biol.* **19:** 237.

Walsh C.P., Chaillet J.R., and Bestor T.H. 1998. Transcription of IAP endogenous retroviruses is constrained by cytosine methylation. *Nat. Genet.* **20:** 116.

Yoon B.J., Herman H., Sikora A., Smith L.T., Plass C., and Soloway P.D. 2002. Regulation of DNA methylation of Rasgrf1. *Nat. Genet.* **30:** 92.

Drosophila Su(Hw) Regulates an Evolutionarily Conserved Silencer from the Mouse *H19* Imprinting Control Region

S. SCHOENFELDER AND R. PARO
Center for Molecular Biology Heidelberg (ZMBH), University of Heidelberg, 69120 Heidelberg, Germany

Mammalian development requires the genomes from the maternal and the paternal side (McGrath and Solter 1984; Surani et al. 1984). Their genetic nonequivalence is mediated by genomic imprinting, an epigenetic mechanism resulting in a parent-of-origin-dependent expression of a small number of genes (Tilghman 1999). The best studied pair of imprinted genes are *H19* and *Igf2*, two closely linked, oppositely imprinted genes located on mouse chromosome 7. *H19* is transcribed from the maternally inherited allele (Bartolomei et al. 1991), whereas *Igf2* is paternally expressed (DeChiara et al. 1991). The coordinated expression of *H19* and *Igf2* is controlled by two regulatory elements: a shared enhancer downstream of the *H19* gene (Leighton et al. 1995), and an element referred to as imprinting control region (ICR) located upstream of the *H19* gene (Fig. 1). Much attention has focused on the *H19* ICR as a key element in the regulation of monoallelic expression of *H19* and *Igf2*, as its deletion results in activation of the normally silent *H19* allele and a concomitant reduction in *Igf2* expression upon paternal transmission, with a reciprocal effect upon maternal inheritance of the deletion (Thorvaldsen et al. 1998).

A METHYLATION-REGULATED INSULATOR IN THE *H19* IMPRINTING CONTROL REGION

Chromatin insulators prevent the interaction between a promoter and an enhancer when located between these two elements (Kellum and Schedl 1991). Enhancer-blocking properties characteristic of an insulator have been described for the *H19* ICR (Bell and Felsenfeld 2000; Hark et al. 2000; Kaffer et al. 2000). Parent-of-origin-specific DNA methylation patterns are a hallmark of most regulatory elements controlling the expression of imprinted genes (Reik and Walter 2001). Such a differentially methylated domain (DMD) has also been identified in the *H19* ICR, where exclusively the paternal allele is methylated (Bartolomei et al. 1993). A role for DNA methylation in genomic imprinting has been demonstrated by the finding that mouse embryos deficient for the DNA methyltransferase Dnmt1 show biallelic *H19* expression and no *Igf2* expression (Li et al. 1993). Together, these observations have led to the proposition that an insulator element upstream of *H19* regulated by DNA methylation is responsible for the reciprocal imprinted expression of *H19* and *Igf2* (Fig. 1a,b). Consistent with this idea, the unmethylated ICR on the maternal allele was found to be bound by the zinc finger protein CTCF (Szabo et al. 2000), a previously described insulator protein (Bell et al. 1999). According to the insulator model, this establishes a functional insulator that prevents the

a
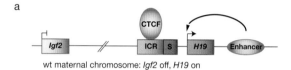
wt maternal chromosome: *Igf2* off, *H19* on

b
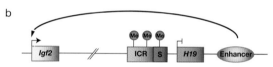
wt paternal chromosome: *Igf2* on, *H19* off

c
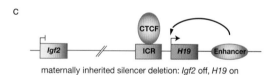
maternally inherited silencer deletion: *Igf2* off, *H19* on

d
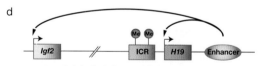
paternally inherited silencer deletion: *Igf2* on, *H19* on

Figure 1. The chromatin boundary model of *H19/Igf2* imprinted gene regulation. The imprinting control region (ICR) overlaps with the *H19* silencer (S) identified in *Drosophila* (Lyko et al. 1997). (*a*) CTCF binds to the unmethylated imprinting control region and establishes a chromatin insulator on the wild-type (wt) maternal chromosome. This prevents the downstream enhancer from interacting with *Igf2*, resulting in *H19* expression. (*b*) DNA methylation (Me) blocks the formation of a functional chromatin boundary on the wild-type paternal chromosome. The enhancer interacts with *Igf2*, whereas *H19* is silenced. (*c*) Maternal inheritance of the silencer deletion does not disrupt *H19* or *Igf2* expression (Drewell et al. 2000). (*d*) Paternal inheritance of the silencer deletion results in reactivation of *H19*, but does not change the methylation pattern at the locus.

downstream enhancer from activating the *Igf2* promoter. Thus, the enhancer is targeted to the *H19* promoter, leading to the expression of *H19* from the maternal chromosome. In contrast, on the paternal allele, methylation blocks the interaction of CTCF with the ICR, abrogating the formation of a functional insulator. As a consequence, the enhancer can activate the *Igf2* promoter, resulting in *Igf2* expression from the paternal allele (Bell and Felsenfeld 2000; Hark et al. 2000). Notably, the model predicts a dual function for DNA methylation on the paternal chromosome: prevention of functional insulator formation and silencing of the *H19* gene. However, a closer examination of the data suggests a more complex mode of regulation at the *H19/Igf2* locus than that implied by the insulator model (Arney 2003). Deletion of the *H19* ICR at different stages of development revealed that *Igf2* and *H19* are silenced by two distinct mechanisms (Srivastava et al. 2000). Remarkably, when the ICR was deleted in terminally differentiated cells, *H19* expression remains monoallelic, showing that the ICR is not continually required to silence *H19* expression. In contrast, continuous presence of the element throughout development is essential to repress maternal *Igf2* expression. Based on the mechanistic differences in silencing of *H19* and *Igf2*, the existence of two regulatory elements in the ICR has been postulated (Srivastava et al. 2000). Indeed, additional data support the idea of the ICR harboring two distinct regulatory sequences: an insulator and a silencer.

SILENCER ELEMENTS IN GENETIC IMPRINTING

At the *H19/Igf2* locus, three independent silencer elements act in concert with the *H19* ICR to assure monoallelic expression of *H19* and *Igf2*. Recently, a silencer has been identified within the imprinting control region of the *Kcnq1* gene (Mancini-DiNardo et al. 2003). The widespread occurrence of silencer elements in regulatory regions of imprinted genes is suggestive of their important role in the control of imprinting. At the *H19/Igf2* locus, three silencers have been proposed to fine-tune the activity of the *H19* insulator (Ferguson-Smith 2000). Two mesoderm-specific silencer elements lie within the differentially methylated region 1 upstream of *Igf2*, and 40 kb downstream of the gene, respectively (Ainscough et al. 2000; Constancia et al. 2000). In both cases, deletion of the repressive elements results in loss of imprinting of *Igf2* in mesodermally derived tissues, but has no effect on *H19* expression (Ainscough et al. 2000; Constancia et al. 2000; Erhardt et al. 2003). The third silencer element includes the 3′ end of the insulator locus within the *H19* imprinting control region (Fig. 1). A repressive activity of this element was first described in *Drosophila*, where the *H19* ICR functions as a bidirectional silencer in a transgene assay (Lyko et al. 1997). This finding came as a surprise, as *Drosophila* lacks genomic imprinting or detectable CpG methylation, yet in mouse, the unmethylated *H19* allele escapes silencing. Remarkably, the 1.2-kb silencer element identified in a genetic screen in *Drosophila* has subsequently been shown to be required to silence paternally transmitted *H19* mini-transgenes in mice (Brenton et al. 1999). More importantly, a targeted deletion of the silencer from the endogenous mouse locus resulted in loss of silencing on the paternal *H19* allele (Drewell et al. 2000). However, the deletion had no effect on *Igf2*, suggesting that the insulator sequences remained intact (Fig. 1c). Strikingly, the activation of *H19* caused by the deletion of the silencer occurred without a detectable change in the methylation status of the locus (Drewell et al. 2000), indicating that methylation at the ICR alone is not sufficient to silence *H19* expression (Fig. 1d). These findings suggest an evolutionarily conserved epigenetic silencing mechanism in flies and mice. However, it is unknown how the silencer fulfills its function, and the factors involved in this silencing process have not been characterized so far. We decided to take advantage of the existence of the huge number of characterized mutations and the powerful genetic tools available in *Drosophila* to identify *trans*-acting factors required for *H19* ICR-mediated gene silencing.

EXPERIMENTAL PROCEDURES

Generation of Transgenic Lines

Deletions of the Su(Hw)-responsive region were introduced into P(hzh) (Lyko et al. 1997) using an ET recombination cloning strategy to generate P(hzh)Δ72 and P(hzh)Δ256 (Zhang et al. 1998). Transgenic flies were generated using standard procedures (Spradling and Rubin 1982). Fly stocks were maintained at 25°C on standard medium.

Eye Pigment Extraction

For each line, ten male and ten female flies were collected within 4 hours of eclosion and aged for 2 days. Eye pigments were extracted by homogenizing heads in 30% EtOH-HCl pH 2.0, and measured at 480 nm. Mean values of triplicate determinations are reported.

EMSA Experiments

Nuclear extract was prepared as described by Zhao et al. (1995). DNA fragments were labeled with T4 polynucleotide kinase (NEB) and incubated with 2 μg nuclear extract at room temperature for 15 minutes in BSK-100 (25 mM HEPES pH 7.6, 100 mM KCl, 0.1 mM EDTA, 1 mM DTT, 10% glycerol, 0.1 mM PMSF) as described (Zhao et al. 1995). Reaction mixtures were electrophoresed on 5% polyacrylamide, 2.5% glycerol gels in 0.25× TBE.

Immunostaining on Polytene Chromosomes

The protocol has been described by Paro (2000). α-Su(Hw) antibody (a gift from Dirk-Henner Lankenau) was used in a 1:100 dilution. As a secondary antibody, we used Cy3 coupled anti-rat antibody (Dianova) in a 1:200 dilution.

SU(HW) IS A SUPPRESSOR OF *H19* ICR-MEDIATED SILENCING

In *Drosophila* transgenes containing the *H19* silencer, a down-regulation of the *mini-white* reporter gene is observed, resulting in yellow-eyed lines (Fig. 2a). In order to identify *trans*-acting factors involved in *H19* ICR-mediated gene silencing, we crossed a range of mutations in genes affecting chromatin regulation to these lines (Table 1). Only a few mutations resulted in a change in the expression of *mini-white*. Among these, a mutation in the *suppressor of hairy wing* (*su(Hw)*) gene shows the strongest derepression of the *mini-white* reporter gene (Table 1 and Fig. 2a).

The ability to function as a silencer in *Drosophila* is not a unique property of the *H19* ICR, but has also been demonstrated for an element from the imprinted human *SNRPN* locus (Lyko et al. 1998). However, we found that the loss of silencing caused by the (*su(Hw)*) mutation is specific to the *H19* ICR, as combining the mutation with control lines carrying a silencer element from the *SNRPN* locus did not result in significant changes of *mini-white* expression. In contrast, two independent lines carrying the 3.8-kb *H19* ICR transgene P(hzh) (Lyko et al. 1997) showed a significant loss of *mini-white* silencing when crossed to the $su(Hw)^{v/f}$ mutation (Fig. 2b). This result indicates that *su(Hw)* functions as a suppressor of *H19* ICR-mediated gene silencing. Interestingly, a mutation in the *mod(mdg4)* gene, whose gene product interacts with the Su(Hw) protein (Gerasimova et al. 1995), has a similar, though less pronounced, effect (Table 1 and Fig. 2b).

Table 1. Mutations Affecting Chromatin Regulation and Their Influence on the Expression of the *mini-white* Reporter Gene Used to Assay Silencing by the *H19* ICR in this Study

No effect on silencing	Silencing weakly relieved	Silencing strongly relieved
Wildtype +/+	$mod(mdg4)^{u1}$	$su(Hw)^{v/f}$
$brm^2\ trx^{E2}/+$	$mus\ 209^{B1}/+$	
$E(z)^{731}/+$	$Suvar\ 2\text{-}4^{01}/+$	
$E(z)^{2434}/+$	$Suvar\ 2\text{-}8^{01}/+$	
$e2F^{164}$	$Suvar\ 2\text{-}11^{01}/+$	
$mor^1/+$	$Suvar\ 2\text{-}13^{01}/+$	
$orc2^{\gamma 4e}/+$	$Suvar\ 2\text{-}14^{01}/+$	
$Pc^1/+$	$Suvar\ 2\text{-}15^{01}/+$	
$Pc^3/+$	$Suvar\ 3\text{-}9^{17}$	
$rfc4^{a18}/+$	$Suvar\ 3\text{-}9^{19}$	
$rpd3^1/+$	$su(z)\ 2^5/+$	
$Suvar\ 2\text{-}1^{01}/+$		
$Suvar\ 2\text{-}2^{01}/+$		
$Suvar\ 2\text{-}6^{01}/+$		
$Suvar\ 2\text{-}7^{01}/+$		
$Suvar\ 2\text{-}10^{01}/+$		
$Suvar\ 3\text{-}2^{01}/+$		
$Suvar\ 3\text{-}3^{02}/+$		
$Suvar\ 3\text{-}6^{01}/+$		
$Suvar\ 3\text{-}7^{14}/+$		
$Suvar\ 3\text{-}10^{01}/+$		
$su(z)12^3/+$		
$tara^{R/08}/+$		
$trl^{62}/+$		
trx^1		
$ttk^{D250}/+$		

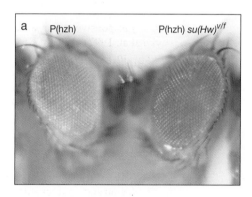

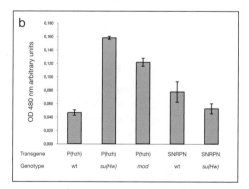

Figure 2. *su(Hw)* is a suppressor of *H19* ICR-mediated gene silencing. (*a*) Fly line carrying the P(hzh) transgene with a 3.8-kb element from the *H19* upstream region controlling the expression of the *mini-white* eye color reporter gene. Pale yellow eye pigmentation in wild-type (wt) genetic background (*left*) is indicative of silencing of *mini-white*, which is relieved by the $su(Hw)^{v/f}$ mutation (*right*). (*b*) Levels of eye pigment extracted from flies carrying the *H19* transgene P(hzh) or an element from the human SNRPN locus (SNRPN) upstream of *mini-white* in different genetic backgrounds: wild type (wt), $su(Hw)^{v/f}$ (*su(Hw)*), or $mod(mdg4)^{u1}$ (*mod*).

Su(Hw) BINDS THE *H19* TRANSGENE IN VIVO

We examined the possibility that Su(Hw) might exert its repressory role by directly binding the *H19* ICR using immunolocalization to polytene chromosomes. Indeed, we detected a signal at the site of transgene integration in flies that were homozygous (Fig. 3b) or heterozygous

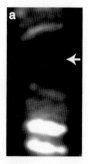 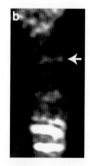 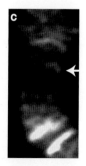

Figure 3. Localization of Su(Hw) protein on polytene chromosomes containing the *H19* transgene. Polytene chromosomes were prepared from wild-type larvae *(a)*, or from larvae homozygous *(b)* or heterozygous *(c)* for the *H19* transgene P(hzh). White arrows indicate the P(hzh) transgene integration site.

(Fig. 3c) for the 3.8-kb *H19* ICR transgene. Wild-type control polytene chromosomes did not show a comparable signal at this site (Fig. 3a). We conclude that Su(Hw) binds the *H19* transgene in vivo.

A *DROSOPHILA* NUCLEAR PROTEIN BINDS THE *H19* SILENCER IN VITRO

In order to determine which region of the *H19* transgene is bound by Su(Hw), we carried out electrophoretic mobility shift assays (EMSA). DNA fragments covering the 1.2-kb *H19* silencer region were incubated with nuclear extract obtained from *Drosophila* embryos (Fig. 4a). Using this assay, we detected an activity that bound a subdomain in the *H19* silencer region (Fig. 4b, lane 2). We observed binding to a fragment covering the base pairs –2257 to –2001 relative to the *H19* transcription start (EMSA 4 in Fig. 4a), and to two smaller fragments contained within this fragment (EMSA 4.1 and EMSA 4.1.1 in Fig. 4a, base pairs –2257 to –2115 and –2257 to –2185 relative to the *H19* transcription start, respectively). The binding activity was specific, as it could be competed with an excess of unlabeled probe, but not with the same molar concentration of an unspecific competitor probe obtained from the pBluescript vector (Fig. 4b, cf. lanes 3 and 4). Incubation of the reaction mixture with an antibody directed against the Su(Hw) protein abolished the shift, whereas a control antibody (α-Tubulin) had no effect (Fig. 4b, cf. lanes 5 and 6). A very similar result was obtained using a known Su(Hw) binding site from the *forked* locus as a probe (Lankenau et al. 2000): Addition of nuclear extract resulted in a bandshift (Fig. 4b, lane 8), which was abolished by incubating the reaction with an anti-Su(Hw) antibody (Fig. 4b, cf. lanes 12 and 13). Moreover, the bandshift obtained with the *forked* probe could be competed efficiently with an excess of both the *forked* and *H19* probes, but only to a lesser extent with the unspecific pBluescript competitor (Fig. 4b,

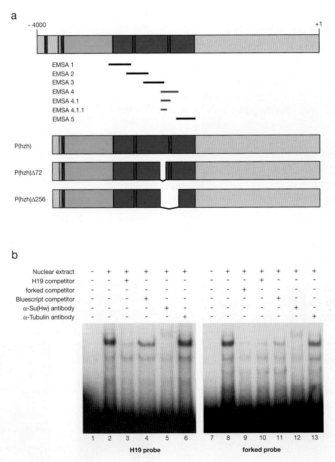

Figure 4. *H19* silencer electrophoretic mobility shift assays. (*a*) Schematic drawing of the *H19* 4-kb-upstream region (numbers relative to the *H19* transcription start site). The 1.2-kb silencer is shown as a *red rectangle*, the CTCF sites are shown as *dark gray boxes*. The *H19* ICR (*orange rectangle*) extends from –3.7 to –2.1 kb relative to the *H19* transcription start (Thorvaldsen et al. 1998) and overlaps with the silencer. *Bars* below show the fragments used as probes in electrophoretic mobility shift assays (EMSA1–EMSA5). *Red bars* represent probes to which a *Drosophila* nuclear protein bound. *Below:* Sequences from the *H19* upstream region contained in the P(hzh), P(hzh)Δ72, and P(hzh)Δ256 transgenic constructs. (*b*) EMSA experiment using EMSA 4.1 probe (*left*) and a probe from the *forked* locus (*right*).

lanes 9–11), indicating that both probes are bound by the same *Drosophila* nuclear protein. The finding that an antibody directed against the Su(Hw) protein blocks the specific shift strongly suggests that Su(Hw) is responsible for the bandshift we observed. To determine whether the Su(Hw) protein is also sufficient to bind to the *H19* silencer in the absence of other *Drosophila* proteins, we expressed Su(Hw) in *Saccharomyces cerevisiae* and repeated the EMSA experiments using yeast whole cell extract instead of *Drosophila* nuclear extract. As a control, we used a fragment derived from the gypsy insulator, which has been shown to possess high-affinity binding sites for Su(Hw) (Dorsett 1990; Kim et al. 1993). In contrast to the gypsy control probe, which was strongly bound by Su(Hw), we failed to detect an interaction of Su(Hw) protein expressed in *S. cerevisiae* with the *H19* silencer region (data not shown). We therefore conclude that Su(Hw) protein expressed in yeast is unable to bind the *H19* silencer. A possible explanation for the lack of binding by Su(Hw) expressed in yeast is that Su(Hw) needs additional interacting proteins or requires specific modifications not reproduced in the yeast expression system to achieve binding to the *H19* silencer.

DELETION OF THE PUTATIVE Su(Hw)-BINDING REGION RESULTS IN LOSS OF *H19* ICR-MEDIATED SILENCING

To address the in vivo role of the putative Su(Hw)-binding region identified by EMSA, we deleted the corresponding small regions from the original *H19* transgenic construct P(hzh) (Fig. 4a). Two different transgenic lines were generated: the first line (P(hzh)Δ256) lacks 256 bp of the putative Su(Hw)-binding region (EMSA 4 in Fig. 4a, base pairs –2257 to –2001 relative to the *H19* transcription start), and a second line (P(hzh)Δ72) lacking 72 bp (EMSA 4.1.1 in Fig. 4a, base pairs –2257 to –2185 relative to the *H19* transcription start). We established nine P(hzh)Δ256 and six P(hzh)Δ72 transgenic lines. We noted a significant loss of *mini-white* silencing in both P(hzh)Δ256 and P(hzh)Δ72 lines, indicating that the small fragment deleted was essential for *H19* ICR-mediated silencing (Fig. 5a). The degree of *mini-white* expression in both P(hzh)Δ256 and P(hzh)Δ72 lines showed some variability, in contrast to the uniform silencing of *mini-white* described for seven independent P(hzh) lines (Lyko et al. 1997). To determine whether the relief of silencing observed in the P(hzh)Δ256 and P(hzh)Δ72 lines was due to a removal of the Su(Hw)-binding region, we crossed them to the $su(Hw)^{v/f}$ mutation (three lines for P(hzh)Δ256 and two lines for P(hzh)Δ72). In both cases, the *su(Hw)* mutation led to a weak increase in *mini-white* expression (Fig. 5b). However, this increase is much less pronounced than in the case of the original line P(hzh). We conclude that the DNA elements deleted in P(hzh)Δ256 and P(hzh)Δ72 contain a Su(Hw)-responsive silencer. We do not rule out, however, the possibility that additional Su(Hw)-binding sites exist within the *H19* ICR, which could account for the weak increase in *mini-white* expression observed in P(hzh)Δ256 $su(Hw)^{v/f}$ and

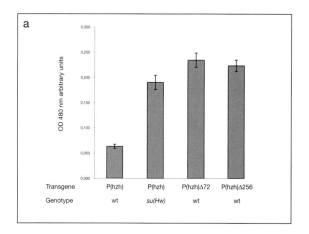

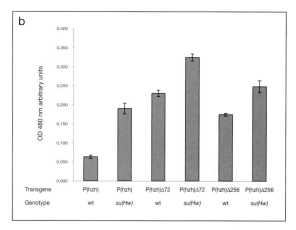

Figure 5. Deletion of the *H19* Su(Hw)-responsive element results in loss of silencing. (*a*) Levels of eye pigment from P(hzh), P(hzh)Δ72 (six lines), and P(hzh)Δ256 (nine lines) in a wild-type (wt) genetic background. The eye pigment level of P(hzh) is also shown in a $su(Hw)^{v/f}$ (*su(Hw)*) mutant genetic background. (*b*) Eye pigment levels from P(hzh), P(hzh)Δ72, and P(hzh)Δ256 in a wild-type (wt) or $su(Hw)^{v/f}$ (*su(Hw)*) mutant genetic background.

P(hzh)Δ72 $su(Hw)^{v/f}$ compared to P(hzh)Δ256 and P(hzh)Δ72, respectively. Alternatively, as well as acting through the interaction with the *H19* silencer directly, it is conceivable that Su(Hw) has an additional role in regulating the *H19* silencer by controlling the expression of other factors binding to the silencer. Such an indirect effect would not be eliminated by the removal of the Su(Hw)-responsive region within the *H19* silencer.

DISCUSSION

cis-acting DNA sequences involved in chromatin organization have been shown to operate by conserved mechanisms. A silencer element from the mouse *H19* locus mediates transcriptional repression of reporter genes in *Drosophila* (Lyko et al. 1997) and is required to silence the endogenous paternal *H19* allele in genomic imprinting (Drewell et al. 2000). The chicken β-globin insulator works in both human and *Drosophila* cells (Chung et al. 1993). The fact that these elements can function across a wide evolutionary spectrum suggests that the mechanisms revealed in *Drosophila* are relevant to other metazoans. It also implies that the *cis*-acting sequences may act through a conserved set of *trans*-acting factors. For these reasons, we used *Drosophila* as a model system to identify factors involved in imprinting associated silencing. In this study, we report the identification of *su(Hw)* as a suppressor of ICR-mediated gene silencing in *Drosophila*. We show that mutations in *su(Hw)* relieve *H19* ICR-associated silencing of the *mini-white* reporter gene and identify a Su(Hw)-responsive region within the *H19* silencer element.

Su(Hw): A *Drosophila* Insulator Protein

The *su(Hw)* gene encodes a zinc finger protein that is present at all stages of the development of *Drosophila*. Su(Hw) is localized in the nucleus, and immunofluorescence studies indicate that it binds approximately 200 sites on polytene chromosomes (Gerasimova and Corces 1998). Although endogenous Su(Hw)-binding sites have recently been described (Golovnin et al. 2003; Parnell et al. 2003), most of these Su(Hw) target sites are unknown. The gypsy retrotransposon contains Su(Hw)-binding sites in its transcribed, untranslated region and represents by far the most thoroughly investigated example of a Su(Hw)-binding sequence (Geyer and Corces 1992). Su(Hw) interacts with the gypsy element through its zinc finger domain, while a leucine zipper mediates interaction with its cofactor Mod(mdg4) (Ghosh et al. 2001). The gypsy Su(Hw)-binding site meets the two defining criteria of a chromatin insulator. First, it functions as a barrier to protect against position effects imposed by the chromosomal environment (Roseman et al. 1993). Second, if located between an enhancer and a promoter, the gypsy element can act as an "enhancer blocker" by interfering with enhancer-mediated gene activation (Holdridge and Dorsett 1991; Geyer and Corces 1992). To date, the gypsy insulator, composed of a 350-bp sequence of the retrotransposon and the protein components Su(Hw) and its cofactor Mod(mdg4), represents one of the best studied examples of a *Drosophila* insulator.

Su(Hw) and CTCF: Similar Mechanisms?

Intriguingly, the "insulator proteins" CTCF and Su(Hw) bind the *H19* ICR in mammals and *Drosophila*, respectively. Though their precise mode of action is unclear, insulators are proposed to divide the chromosome into functionally isolated domains, possibly by the formation of looped chromatin structures (West et al. 2002). Notably, both CTCF and Su(Hw) have been implicated in the formation of such loop structures (Gerasimova et al. 2000; Yusufzai et al. 2004). Although immunofluorescence studies have shown that Su(Hw) is present at hundreds of sites on *Drosophila* polytene chromosomes (Gerasimova and Corces 1998), recent work has demonstrated that in the interphase nuclei of diploid cells, Su(Hw) and its cofactor Mod(mdg4) are arranged in clusters near the nuclear periphery, referred to as insulator bodies (Gerasimova et al. 2000). The insertion of gypsy insulator sequences at two different sites normally found at separate nuclear locations forces these sites to colocalize, and the colocalization is dependent on the presence of Su(Hw). The interaction of Su(Hw) target sites may serve as a mechanism to create functionally independent chromosomal loops, which are held together at their base by interacting Su(Hw) complexes (Gerasimova et al. 2000). A similar role in chromosomal loop formation has been proposed for the scs and scs´ elements, two insulators flanking the *Drosophila hsp 70* locus (Blanton et al. 2003). Recently, CTCF, the protein that binds all mammalian insulators identified so far, has been demonstrated to interact with the nucleolar protein nucleophosmin, and it has been suggested that CTCF/nucleophosmin complexes create loop domain structures (Yusufzai et al. 2004). Interestingly, a looping mechanism has also been postulated for the *H19/Igf2* locus, though the precise role of CTCF in this process remains to be determined (Lopes et al. 2003). However, despite the involvement of both CTCF and Su(Hw) in creating higher-order chromatin structures through the interaction with insulators, important differences regarding their function in the regulation of gene expression at the *H19* ICR in mammals and *Drosophila* must be pointed out. Mutating the CTCF-binding sites within the *H19* ICR leads to loss of imprinting of *Igf2*, as the insulator model predicts, but also to a reduction of maternal *H19* expression (Schoenherr et al. 2003). This finding argues for a positive role for CTCF in *H19* transcription, while our studies are indicative of a role for Su(Hw) in ICR-mediated gene silencing in *Drosophila*. Interestingly, a CTCF paralog named BORIS (Brother of the Regulator of Imprinted Sites) has recently been identified (Loukinov et al. 2002). The DNA-binding specificities of CTCF and BORIS are indistinguishable, but BORIS is expressed exclusively in testis. The nonoverlapping expression patterns of BORIS and CTCF suggest a model in which BORIS, rather than CTCF, is involved in establishing a silencing imprint at the *H19* locus in the paternal germ line, whereas CTCF is required to protect the maternal *H19* ICR from de novo methylation (Fedoriw et al. 2004).

The Role of Su(Hw) in Gene Silencing

Numerous studies have addressed the role of the gypsy Su(Hw)-binding region as an insulator; however, a surprising versatility of the element has been demonstrated, including its ability to function as a silencer. In the absence of *mod(mdg4)*, the gypsy insulator can convert into a promoter-specific silencer (Cai and Levine 1997). Moreover, it has been postulated that Su(Hw) induces a heterochromatin-like chromatin structure that spreads bidirectionally from the Su(Hw)-binding region, and that the role of Mod(mdg4) is to impose a directionality on this effect by creating a chromatin insulator (Gerasimova et al. 1995). These studies indicate that the insulator function is not an intrinsic property of the Su(Hw)-binding region of the gypsy element, but rather that the activities of this element can be substantially modulated by the factors bound to it. Remarkably, we found that some mutations in *Su(var)* genes, whose gene products are components of heterochromatin, partially relieve *H19* ICR-mediated gene silencing (Table 1), consistent with the idea of a heterochromatin-like chromatin structure involved in gene silencing. Based on the observed differences in the interaction of Su(Hw) with the gypsy element and the *H19* silencer (data not shown), we propose the requirement of cofactors for Su(Hw) to interact with the *H19* silencer and exert its silencing function. Further studies are clearly needed to elucidate the role of the potential cofactors of Su(Hw) in the regulation of *H19* ICR-mediated gene silencing.

ACKNOWLEDGMENTS

We thank Dirk-Henner Lankenau for the gift of the α-Su(Hw) antibody and the P(w^+f^{asH}) plasmid containing the *forked* Su(Hw) binding site, and Dale Dorsett for the plasmids pSJ-su(Hw), pSJ-su(Hw)^{Zn}, and YEp24-BaBx. We are grateful to Gunter Reuter for *Su(var)* mutant lines. We thank Leonie Ringrose and Nara Lee for critical reading of the manuscript. This work was supported by grants from the Deutsche Forschungsgemeinschaft (DFG) and Fonds der Chemischen Industrie (FCI) to R.P.

REFERENCES

Ainscough J.F., John R.M., Barton S.C., and Surani M.A. 2000. A skeletal muscle-specific mouse Igf2 repressor lies 40 kb downstream of the gene. *Development* **127:** 3923.

Arney K.L. 2003. *H19* and *Igf2*—Enhancing the confusion? *Trends Genet.* **19:** 17.

Bartolomei M.S., Zemel S., and Tilghman S.M. 1991. Parental imprinting of the mouse *H19* gene. *Nature* **351:** 153.

Bartolomei M.S., Webber A.L., Brunkow M.E., and Tilghman S.M. 1993. Epigenetic mechanisms underlying the imprinting of the mouse *H19* gene. *Genes Dev.* **7:** 1663.

Bell A.C. and Felsenfeld G. 2000. Methylation of a CTCF-dependent boundary controls imprinted expression of the *Igf2* gene. *Nature* **405:** 482.

Bell A.C, West A.G., and Felsenfeld G. 1999. The protein CTCF is required for the enhancer blocking activity of vertebrate insulators. *Cell* **98:** 387.

Blanton J., Gaszner M., and Schedl P. 2003. Protein:protein interactions and the pairing of boundary elements *in vivo*. *Genes Dev.* **17:** 664.

Brenton J.D., Drewell R.A., Viville S., Hilton K.J., Barton S.C., Ainscough J.F., and Surani M.A. 1999. A silencer element identified in *Drosophila* is required for imprinting of *H19* reporter transgenes in mice. *Proc. Natl. Acad. Sci.* **96:** 9242.

Cai H.N. and Levine M. 1997. The gypsy insulator can function as a promoter-specific silencer in the *Drosophila* embryo. *EMBO J.* **16:** 1732.

Chung J.H., Whiteley M., and Felsenfeld G. 1993. A 5′ element of the chicken beta-globin domain serves as an insulator in human erythroid cells and protects against position effect in *Drosophila*. *Cell* **74:** 505.

Constancia M., Dean W., Lopes S., Moore T., Kelsey G., and Reik W. 2000. Deletion of a silencer element in *Igf2* results in loss of imprinting independent of *H19*. *Nat. Genet.* **26:** 203.

DeChiara T.M., Robertson E.J., and Efstratiadis A. 1991. Parental imprinting of the mouse insulin-like growth factor II gene. *Cell* **64:** 849.

Dorsett D. 1990. Potentiation of a polyadenylylation site by a downstream protein-DNA interaction. *Proc. Natl. Acad. Sci.* **87:** 4373.

Drewell R.A., Brenton J.D., Ainscough J.F., Barton S.C., Hilton K.J., Arney K.L., Dandolo L., and Surani M.A. 2000. Deletion of a silencer element disrupts *H19* imprinting independently of a DNA methylation epigenetic switch. *Development* **127:** 3419.

Erhardt S., Lyko F., Ainscough J.F., Surani M.A., and Paro R. 2003. Polycomb-group proteins are involved in silencing processes caused by a transgenic element from the murine imprinted *H19/Igf2* region in *Drosophila*. *Dev. Genes Evol.* **213:** 336.

Fedoriw A.M., Stein P., Svoboda P., Schultz R.M., and Bartolomei M.S. 2004. Transgenic RNAi reveals essential function for CTCF in *H19* gene imprinting. *Science* **303:** 238.

Ferguson-Smith A.C. 2000. Genetic imprinting: Silencing elements have their say. *Curr. Biol.* **10:** R872.

Gerasimova T.I. and Corces V.G. 1998. Polycomb and trithorax group proteins mediate the function of a chromatin insulator. *Cell* **92:** 511.

Gerasimova T.I., Byrd K., and Corces V.G. 2000. A chromatin insulator determines the nuclear localization of DNA. *Mol. Cell* **6:** 1025.

Gerasimova T.I., Gdula D.A., Gerasimov D.V., Simonova O., and Corces V.G. 1995. A *Drosophila* protein that imparts directionality on a chromatin insulator is an enhancer of position-effect variegation. *Cell* **82:** 587.

Geyer P.K. and Corces V.G. 1992. DNA position-specific repression of transcription by a *Drosophila* zinc finger protein. *Genes Dev.* **6:** 1865.

Ghosh D., Gerasimova T.I., and Corces V.G. 2001. Interactions between the Su(Hw) and Mod(mdg4) proteins required for gypsy insulator function. *EMBO J.* **20:** 2518.

Golovnin A., Birukova I., Romanova O., Silicheva M., Parshikov A., Savitskaya E., Pirrotta V., and Georgiev P. 2003. An endogenous Su(Hw) insulator separates the yellow gene from the Achaete-scute gene complex in *Drosophila*. *Development* **130:** 3249.

Hark A.T., Schoenherr C.J., Katz D.J., Ingram R.S., Levorse J.M., and Tilghman S.M. 2000. CTCF mediates methylation-sensitive enhancer-blocking activity at the *H19/Igf2* locus. *Nature* **405:** 486.

Holdridge C. and Dorsett D. 1991. Repression of hsp70 heat shock gene transcription by the suppressor of hairy-wing protein of *Drosophila melanogaster*. *Mol. Cell. Biol.* **11:** 1894.

Kaffer C.R., Srivastava M., Park K.Y., Ives E., Hsieh S., Batlle J., Grinberg A., Huang S.P., and Pfeifer K. 2000. A transcriptional insulator at the imprinted *H19/Igf2* locus. *Genes Dev.* **14:** 1908.

Kellum R. and Schedl P. 1991. A position-effect assay for boundaries of higher order chromosomal domains. *Cell* **64:** 941.

Kim J., Shen B., and Dorsett D. 1993. The *Drosophila melanogaster* suppressor of Hairy-wing zinc finger protein has minimal effects on gene expression in *Saccharomyces cerevisiae*. *Genetics* **135:** 343.

Lankenau D.H., Peluso M.V., and Lankenau S. 2000. The Su(Hw) chromatin insulator protein alters double-strand break repair frequencies in the *Drosophila* germ line. *Chromosoma* **109:** 148.

Leighton P.A., Saam J.R., Ingram R.S., Stewart C.L., and Tilghman S.M. 1995. An enhancer deletion affects both *H19* and *Igf2* expression. *Genes Dev.* **9:** 2079.

Li E., Beard C., and Jaenisch R. 1993. Role for DNA methylation in genomic imprinting. *Nature* **366:** 362.

Lopes S., Lewis A., Hajkova P., Dean W., Oswald J., Forne T., Murrell A., Constancia M., Bartolomei M., Walter J., and Reik W. 2003. Epigenetic modifications in an imprinting cluster are controlled by a hierarchy of DMRs suggesting long-range chromatin interactions. *Hum. Mol. Genet.* **12:** 295.

Loukinov D.I., Pugacheva E., Vatolin S., Pack S.D., Moon H., Chernukhin I., Mannan P., Larsson E., Kanduri C., Vostrov A.A., Cui H., Niemitz E.L., Rasko J.E., Docquier F.M., Kistler M., Breen J.J., Zhuang Z., Quitschke W.W., Renkawitz R., Klenova E.M., Feinberg A.P., Ohlsson R., Morse H.C., III, and Lobanenkov V.V. 2002. BORIS, a novel male germ-line-specific protein associated with epigenetic reprogramming events, shares the same 11-zinc-finger domain with CTCF, the insulator protein involved in reading imprinting marks in the soma. *Proc. Natl. Acad. Sci.* **99:** 6806.

Lyko F., Brenton J.D., Surani M.A., and Paro R. 1997. An imprinting element from the mouse *H19* locus functions as a silencer in *Drosophila*. *Nat. Genet.* **16:** 171.

Lyko F., Buiting K., Horsthemke B., and Paro R. 1998. Identification of a silencing element in the human 15q11-q13 imprinting center by using transgenic *Drosophila*. *Proc. Natl. Acad. Sci.* **95:** 1698.

Mancini-DiNardo D., Steele S.J., Ingram R.S., and Tilghman S.M. 2003. A differentially methylated region within the gene *Kcnq1* functions as an imprinted promoter and silencer. *Hum. Mol. Genet.* **12:** 283.

McGrath J. and Solter D. 1984. Completion of mouse embryogenesis requires both the maternal and paternal genomes. *Cell* **37:** 179.

Parnell T.J., Viering M.M., Skjesol A., Helou C., Kuhn E.J., and Geyer P.K. 2003. An endogenous suppressor of hairy-wing insulator separates regulatory domains in *Drosophila*. *Proc. Natl. Acad. Sci.* **100:** 13436.

Paro R. 2000. Mapping protein distributions on polytene chromosomes by immunostaining. In Drosophila *protocols* (ed. W. Sullivan et al.), p. 131. Cold Spring Harbor Laboratory Press, Cold Spring Harbor, New York.

Reik W. and Walter J. 2001. Genomic imprinting: Parental influence on the genome. *Nat. Rev. Genet.* **2:** 21.

Roseman R.R., Pirrotta V., and Geyer P.K. 1993. The su(Hw) protein insulates expression of the *Drosophila melanogaster white* gene from chromosomal position-effects. *EMBO J.* **12:** 435.

Schoenherr C.J., Levorse J.M., and Tilghman S.M. 2003. CTCF maintains differential methylation at the *Igf2/H19* locus. *Nat. Genet.* **33:** 66.

Spradling A.C. and Rubin G.M. 1982. Transposition of cloned P elements into *Drosophila* germ line chromosomes. *Science* **218:** 341.

Srivastava M., Hsieh S., Grinberg A., Williams-Simons L., Huang S.P., and Pfeifer K. 2000. *H19* and *Igf2* monoallelic expression is regulated in two distinct ways by a shared cis acting regulatory region upstream of *H19*. *Genes Dev.* **14:** 1186.

Surani M.A., Barton S.C., and Norris M.L. 1984. Development of reconstituted mouse eggs suggests imprinting of the genome during gametogenesis. *Nature* **308:** 548.

Szabo P., Tang S.H., Rentsendorj A., Pfeifer G.P., and Mann J.R. 2000. Maternal-specific footprints at putative CTCF sites in the *H19* imprinting control region give evidence for insulator function. *Curr. Biol.* **10:** 607.

Thorvaldsen J.L., Duran K.L., and Bartolomei M.S. 1998. Deletion of the *H19* differentially methylated domain results in loss of imprinted expression of *H19* and *Igf2*. *Genes Dev.* **12:** 3693.

Tilghman S.M. 1999. The sins of the fathers and mothers: Genomic imprinting in mammalian development. *Cell* **96:** 185.

West A.G., Gaszner M., and Felsenfeld G. 2002. Insulators: Many functions, many mechanisms. *Genes Dev.* **16:** 271.

Yusufzai T.M., Tagami H., Nakatani Y., and Felsenfeld G. 2004. CTCF tethers an insulator to subnuclear sites, suggesting shared insulator mechanisms across species. *Mol. Cell* **13:** 291.

Zhang Y., Buchholz F., Muyers J.P., and Stewart A.F. 1998. A new logic for DNA engineering using recombination in *Escherichia coli*. *Nat. Genet.* **20:** 123.

Zhao K., Hart C.M., and Laemmli U.K. 1995. Visualization of chromosomal domains with boundary element-associated factor BEAF-32. *Cell* **81:** 879.

The *Air* Noncoding RNA: An Imprinted *cis*-silencing Transcript

G. Braidotti,[*] T. Baubec,[†] F. Pauler,[†] C. Seidl,[†] O. Smrzka,[‡] S. Stricker,[†]
I. Yotova,[†] and D.P. Barlow[†]

[*]*AFI, Swinburne University of Technology, Hawthorn, Victoria 3122, Australia; [†]CeMM (Center of Molecular Medicine GmbH of the Austrian Academy of Science), c/o Institute of Microbiology and Genetics, A1030 Vienna, Austria; [‡]Axon Neuroscience GmbH, A-1030 Vienna, Austria*

Genomic imprinting is briefly defined as parental-specific gene expression. It is a feature of organisms with diploid cells that have inherited a complete haploid chromosome set from two parents and, thereby, have a maternal and paternal allele for each autosomal gene. Diploid cells would normally be expected to express both parental alleles equally but imprinted genes are different, because they acquire an epigenetic mark from one parental gamete that results in repression on one parental allele and expression from the other. Genomic imprinting uses an epigenetic mechanism (i.e., a reversible modification to DNA or chromatin) to regulate gene expression in *cis* that is assumed to be at the level of transcriptional, but this is not often tested.

Genomic imprinting has been so far identified in very diverse organisms: in angiosperm plants, in two types of invertebrates (sciarid flies, mealybugs), and lastly in mammals that have live-born young (marsupial and placental mammals) (Goday and Esteban 2001; Killian et al. 2001; Baroux et al. 2002). These diverse organisms lack a contemporary common ancestor that also shows imprinted gene expression, and this allows the possibility that imprinting arose independently at least three times in distantly related organisms. If this is so, the imprinting mechanism in these diverse organisms is less likely to share common features. The function of genomic imprinting is beginning to be understood following the functional analysis of a large number of imprinted genes in mice (see listing at http://www.mgu.har.mrc.ac.uk/research/imprinting/function.html). There is now sufficient information to suggest that genomic imprinting arose in mammals as a consequence of one parent carrying the total burden of provision for an embryo that carries the genomes of two parents. Thus, imprinted genes in mammals are able to influence embryonic growth in such a manner that imprinted growth suppressor genes are always maternally expressed, while imprinted growth promoter genes are always paternally expressed (Wilkins and Haig 2003). The reason imprinted genes exist in angiosperm plants is not yet clear, although it has been suggested that angiosperm plants resemble mammals because biparental seeds are also grown by one parent (Moore and Haig 1991). The two invertebrates with genomic imprinting appear to use this as a sex-determining mechanism, such that males are produced by inactivating the paternal genome in diploid offspring (Goday and Esteban 2001; Bongiorni and Prantera 2003).

The Harwell Mammalian Genetics Unit has for many years maintained a complete list of the known imprinted mouse genes and also lists mutant phenotypes for those imprinted genes so far analyzed (Beechey et al. 2003). Table 1 was drawn from this information and lists the 70 known imprinted genes by chromosomal subregions (note many imprinted genes have alternative names and these are listed on http://www.mgu.har.mrc.ac.uk/research/imprinting/). The imprinted genes are separated in this list according to parental expression and whether they are coding or noncoding. Several features of mammalian imprinted genes become apparent from this list.

a. Only 12 of 19 mouse autosomal chromosomes carry imprinted genes; however, four chromosomes (2, 6, 7, and 10) have multiple subregions carrying imprinted genes.

b. Imprinted genes are often clustered. There are 18 regions with imprinted genes and 12 have clusters containing between 2 and 9 imprinted genes.

c. Some genes inside the cluster escape imprinting.

d. Many imprinted genes are noncoding RNAs (ncRNAs) that show three types of pattern: antisense to one imprinted gene and expressed from the opposite parental allele (e.g., proximal 17: *Igf2r/Air*; distal 7a: *Kcnq1/Lit1*); antisense to one imprinted gene and expressed from the same chromosome (distal 7b: *Igf2/Igf2as*; central 7: *Zfp127/Zfp127as*); and sense to one imprinted gene and expressed from the opposite parental allele (distal 7b: *Igf2/H19*; distal 12: *Dlk1/Gtl2*).

e. There are approximately equal numbers of maternally and paternally expressed coding (mRNA) genes, but there are less maternally expressed ncRNAs than paternally expressed ncRNAs (note that the multiple ncRNAs from central 7 and distal 12 may arise from a single long processed ncRNA).

f. Reciprocal parental-specific expression between protein-coding genes and an ncRNA is often observed. Two patterns are seen: A simple type where all the protein-coding genes are expressed by one parental chromosome and the ncRNA from the other (e.g., distal 7a, distal 7b, distal 12, proximal 17), and a complex type where multiple protein coding genes are maternally or paternally expressed while the noncoding RNA is expressed from one parental chromosome (e.g., distal 2, subproximal 6, central 7).

Table 1. Imprinted Genes Are Mainly Organized into Clusters

Chr.	Region	Coding (mRNAs)		Noncoding (ncRNAs) as:antisense	
		Maternal expression	Paternal expression	Maternal expression	Paternal expression
2	central2	Gatm			
	distal2	Nesp, Gnas	Nnat, Gnasxl		Nespas
6	proximal6	Dlx5, Calcr, Neurabin, Pon3, Pon2, Asb4	Sgce, Peg10		
	subproximal6	Asb4, Cop2,	Peg1/Mest, Nap1l5		Cop2as, Mit1
7	proximal7	Zim1, Zim3,	Peg3/Pw1, Usp29, Zfp264,		
	central7	Atp10c, Ube3a, Nap1l4	Snrpn, Snurf, Magel2, Ndn, Zfp127/Mkrn3, Frat3		Zfp127/Mkrn3as, IC-Snurf-Snrpn processed, ncRNA, Ipw, Pwcr1, multiple MB11-snoRNAs, Ube3aas
	distal7a	Obph1, Tssc3, Slc22a1L, Msult, Cdkn1c, Kcnq1, Tssc4, Tapa1, Mash2			Lit1/Kcnq1ot1(as)
	distal7b		Ins2, Igf2	H19	Igf2as
9	distal9		A19, Rasgrf1		
10	proximal10		Zac1		
	distal10	Dcn1			
11	proximal11	Grb10/Meg1, Murr1	U2af1rs1		
12	distal12		Dlk1, Rtl1, Dio3	Gtl2, Mir126,136 Rain, C/D SnoRNA cluster	
14	distal14	Htr3a			
15	distal15		Peg13, Slc38a4		
17	proximal17	Slc22a3, Slc22a2, Igf2r			Air (Igf2ras)
18	proximal18		Impact		
19			Ins1		

Data from: Mouse imprinting map: http://www.mgu.har.mrc.ac.uk/research/imprinting/.

The observation that imprinted genes exist in clusters suggests that mechanisms to control imprinted expression act on the chromosomal domain rather than on individual genes (Reik and Walter 2001; Verona et al. 2003). If imprinting control mechanisms are specific to a domain this could explain why moving imprinted genes outside their endogenous locus on transgenes causes them to lose imprinted expression (Lee et al. 1993; Sleutels and Barlow 2001). The observation of reciprocal imprinted expression between the clustered protein-coding genes and the noncoding RNA is fascinating and naturally suggests a link between expression of a noncoding RNA and *cis*-induced silencing of the protein coding RNA. The data that tests the significance of this association between ncRNAs and *cis*-induced silencing is only beginning to appear. There are, however, two significant results so far (for review, see Verona et al. 2003). First, there was a clear demonstration that the *H19* ncRNA plays no role in silencing the reciprocally expressed *Ins2/Igf2* genes in *cis* (Jones et al. 1998; Schmidt et al. 1999). Instead a DNA sequence functions as a methyl-sensitive insulator that, when methylated, allows expression of *Igf2* and, when unmethylated, allows expression of *H19* (Bell and Felsenfeld 2000; Hark et al. 2000). The second result, and the topic of the remainder of this review, was the demonstration that the *Air* ncRNA plays a direct role in silencing a cluster of three genes spanning 300 kilobases (kb) in *cis* (Rougeulle and Heard 2002; Sleutels et al. 2002).

THE MOUSE CHROMOSOME 17 PROXIMAL IMPRINTED CLUSTER

The *Air* noncoding RNA (ncRNA) is part of an imprinted cluster on proximal mouse chromosome 17/band A2 (http://www.mgu.har.mrc.ac.uk/research/imprinting/). Figure 1 shows a map spanning 500 kb based on the public mouse and human sequence. Genes shown on each side of the sequence line have the same orientation, and it can be seen in the mouse map that the *Air* ncRNA has an antisense orientation with respect to *Igf2r* and *Mas*. The *Air* ncRNA promoter (marked by asterisks in Fig. 1) lies in the second intron of *Igf2r* and the transcript overlaps the 5′ part of the *Igf2r* gene for 29 kb (which includes *Igf2r* exons 1 and 2) and the 3′ part of the *Mas* gene (which

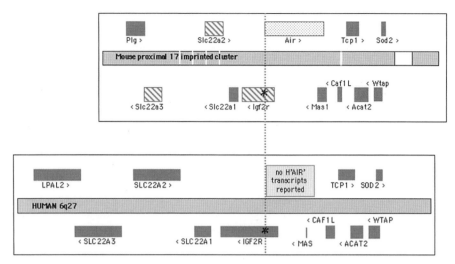

Figure 1. The mouse proximal 17 imprinted cluster (*top box*) and the syntenic human region on 6q27 (*bottom box*). Genes are positioned above or below the map according to the direction of transcription and their size indicated by the length of the box. Both regions are gene rich and show surprising conservation (n.b., *LPAL2* is a human-specific duplication of *Plg*); the *white boxes* in the mouse sequence indicate gaps. On the mouse map, the three maternally expressed imprinted genes are shown as *striped boxes*; the paternally expressed *Air* ncRNA is shown as a *dotted box*. The two regions are aligned (*vertical dotted line*) at the CpG island (*) in *Igf2r* intron 2 that is common to both mouse and human sequences, which in the mouse is the promoter for the *Air* ncRNA. The mouse *Air* ncRNA is 108-kb long and spans from *Igf2r* intron 2 to the *Mas* last intron. No human "H"AIR transcripts have been reported in the region between human *IGF2R* and *MAS*. Both maps were based on those shown in http://www.ensemble.org/.

includes the entire 1-kb coding exon of *Mas*). Gene overlaps are not rare in the mammalian genome and within this 500-kb region the *Acat2* gene overlaps the 3′ coding part of the *Tcp1* gene (Shintani et al. 1999). There are three maternally expressed genes indicated by striped boxes: *Igf2r* (a multifunctional transport receptor with embryonic growth suppressor function [Wang et al. 1994; Ludwig et al. 1996]); *Slc22a2* (solute cation transporter family 22a, an organic cation transporter [Jonker et al. 2003]); and *Slc22a3* (an organic cation transporter that may function as an extraneuronal monoamine transporter [Zwart et al. 2001b]). Only one paternally expressed gene, the *Air* ncRNA, indicated in Figure 1 by a dotted box, has been identified (Lyle et al. 2000). The remaining genes show biallelic expression in all tested tissues in embryos and adults.

The imprinted behavior of genes in this region is fully maintained in transgenes that span 300 kb (from *Mas* to the intergenic region between *Slc22a2* and *Slc22a3*) when they are integrated into autosomes (Wutz et al. 1997). "Short" transgenes that span 44 kb from 4 kb upstream of the *Igf2r* promoter to 8 kb downstream of the *Air* promoter are not reliably imprinted (Sleutels and Barlow 2001). This indicates that all elements and epigenetic marks for the imprinting mechanism lie within a 300-kb region. This rationale has formed the basis for generating epigenetic maps with the aim of identifying all parental-specific epigenetic modifications on the DNA and chromatin, which could contribute to the control of imprinting in this region. An enzyme-based DNA methylation scan that examined a 100-kb region spanning the *Igf2r* locus, only identified two parental-specific *d*ifferentially *m*ethylated *r*egions (DMR1 and DMR2), the remainder of the analyzed region showed various levels of high and low methylation that were the same on both parental alleles (Stoeger et al. 1993). Both DMRs were large CpG island promoters. DMR1 contained the *Igf2r* promoter and was specifically methylated on the silent paternal allele, but only after embryonic implantation. DMR2 contained the *Air* promoter and was specifically methylated on the silent maternal allele from late oocyte stages onward into adulthood (Stoeger et al. 1993). Methylation of DMR2 was also specifically maintained in preimplantation embryos during the process of genome-wide demethylation (Brandeis et al. 1993). While the identification of a DNA methylation imprint on the active maternal *Igf2r* allele was initially a surprise, it is now clear that the DMR2 methylation mark is the activator of maternal *Igf2r* expression and also serves as the primary imprint regulator of this cluster. This can be seen from analysis of mouse embryos deficient in genome-wide DNA methylation (through targeted inactivation of the DNA methylation enzyme system) that repress *Igf2r* on both parental chromosomes (Li 2002). Thus, DNA methylation acts to silence a *cis*-acting repressor of *Igf2r* but does not act as a direct silencer of *Igf2r*. Analysis of these methylation-deficient mice showed that the DMR1 methylation mark plays no role in initiating silencing of the paternal *Igf2r* promoter and, since this methylation mark occurs after embryonic implantation, it may be passively acquired after the promoter is silenced and may fulfill a maintenance role.

In addition to the surprise finding that the maternal *Igf2r* allele was imprinted by DNA methylation to be active, the lack of widespread DNA methylation on the silent paternal *Igf2r* allele was unexpected. Textbook models of the mammalian genome often view chromatin as organized into blocks of "euchromatin" containing active genes and blocks of "heterochromatin" containing silent genes. The *Igf2r* locus spans 86 kb and, although the *Air* ncRNA overlaps the silent paternal copy at the 5′ end, we had initially anticipated that parts of the *Igf2r* gene locus downstream of the *Air* promoter would show euchromatic features on the maternal allele and heterochromatic features on the paternal allele. However, in terms of DNA methylation, no other parental specific methylation marks, not even at the silent *Slc22a2*/*Slc22a3* promoters, were found at the resolution so far used (Stoeger et al. 1993; Zwart et al. 2001a). Histone modifications have been analyzed at specific regions within DMR1 and DMR2 and in the last *Igf2r* exon (Fournier et al. 2002; Yang et al. 2003). Chromatin immunoprecipitation experiments have shown that histone H3/H4 acetylation and histone H3 K4 methylation specifically distinguish the unmethylated DMRs that contain an active promoter (i.e., the maternal DMR1 and the paternal DMR2) from the methylated DMRs with a silent promoter (i.e., the paternal DMR1 and the maternal DMR2). The methylated DMRs were also marked by histone H3 K9 methylation. However, more detailed analysis of the whole cluster shown in Figure 1 is needed to test if the paternal allele that carries three silenced genes shows more widespread heterochromatin compared to the maternal allele that carries only one silenced gene.

A final chromatin feature associated with this locus is a widespread replication asynchrony that has been reported to cover the 700-kb mouse region shown in Figure 1, and extend approximately 1 Mbp (Mega base pair) upstream and downstream (Kitsberg et al. 1993). Once again this observed replication asynchrony does not follow textbook predictions; instead the early replicating allele is the paternal allele that is silent for *Igf2r*/*Slc22a2*/*Slc22a3* but active for the *Air* ncRNA. The early replicating paternal allele also lacks the critical DNA methylation imprint on DMR2; however, replication asynchrony has been shown to be independent of DNA methylation (Gribnau et al. 2003). Our work producing four different targeted alleles at DMR1 and DMR2 has highlighted another phenomenon that could also be related to chromatin structure, which is the preferential targeting in ES cells of the paternal allele (Sleutels et al. 2002, 2003; Wang et al. 1994; P. Latos, unpubl.).

THE *cis*-ACTING SILENCER IS THE *Air* NONCODING RNA

The analysis of epigenetic marks described above is limited but it already indicates that typical heterochromatic modifications, such as DNA methylation, histone deacetylation, or histone methylation, are not the primary cause of silencing *Igf2r*, *Slc22a2*, and *Slc22a3* on the paternal allele. Experiments using transgenes and targeted deletion of the endogenous locus, have, however, identified DMR2 as the imprint control element (ICE) for this cluster (Wutz et al. 1997, 2001; Spahn and Barlow 2003). Since DMR2 contained the *Air* promoter, the next logical step was to test for *Air* ncRNA involvement in silencing. An experiment was performed that inserted a PolyA signal 3 kb downstream of the *Air* transcription start (Sleutels et al. 2002). This truncated *Air* from 108 to 3 kb (in the truncated allele 80% of the transcripts were spliced from a donor at +53 bp to an acceptor in the PolyA signal cassette and only 20% were unspliced and shortened to 3 kb). Expression of the *Air* promoter from the truncated allele was equal and imprinted exactly as in the wild-type allele. Thus, the truncated allele was predicted to differ from the wild type only in the length of the *Air* ncRNA. Analysis of *Igf2r*, *Slc22a2*, and *Slc22a3* expression from the paternal chromosome carrying the truncated allele revealed a complete loss of repression of all three genes. This experiment confirms a direct role for the *Air* ncRNA transcript or its transcription, in the *cis*-silencing of three genes spread over a 300-kb region. A direct role for *Air* in silencing *Igf2r* was not unexpected in view of the sense/antisense overlap between the two genes. However, it was unexpected that *Air* would also silence the *Slc22a2* and *Slc22a3* genes. In view of this result, it was important to investigate if silencing of *Slc22a2* and *Slc22a3* arose by a two-step mechanism that initially silenced *Igf2r* via the sense/antisense overlap with *Air* and then established a silent chromatin state that spread to the flanking *Slc22a2* and *Slc22a3* genes. To test the role of the sense/antisense overlap between *Igf2r* and *Air*, a mouse was constructed that lacked the *Igf2r* promoter. In this mouse the *Air* ncRNA was not overlapped by any known transcript; despite this, the *Slc22a2* and *Slc22a3* genes, as well as the *Air* ncRNA itself, maintained their imprinted status (Sleutels et al. 2003). This experiment makes it less likely that a double-stranded RNA-based mechanism, such as RNAi, is involved in imprinting this cluster.

Thus gene-targeting experiments have shown that the *Air* ncRNA acts as a *cis*-acting silencer of a 300-kb domain of genes. There are still many interesting questions that need to be answered. It is, for example, fascinating why the *Air* ncRNA does not silence its own promoter yet can silence promoters that lie upstream and downstream. And it is also intriguing to consider why the effects of the *Air* ncRNA are limited to genes within a 300-kb region and why this region contains some genes that escape imprinting. Some clues to these questions can be obtained by noting that the imprinted expression of the four transcripts in this region shows a developmental and tissue-specific pattern. The *Igf2r* gene lacks imprinted expression in preimplantation embryos, undifferentiated ES cells, and also in testes, and adult and embryonic brain (Wang et al. 1994; Szabo and Mann 1995a,b; Lerchner and Barlow 1997; Hu et al. 1999; C. Seidl, unpubl.). ES cells do not express *Air*, and testes and brain express low levels compared to other organs (S. Stricker, unpubl.). However, all tissues that express *Air* ncRNA also maintain the correct imprinted paternal-specific expression. The *Slc22a2* and *Slc22a3* genes are not expressed in the embryo proper but both show imprinted expression in the placenta at 11.5 dpc (days postcoitum). However, at 15.5 dpc, while *Slc22a2* retains imprinted expression, *Slc22a3* shows biallelic expression. Adult tissues such as heart and kidney that retain imprinted expression of *Igf2r* and *Air* show biallelic expression of *Slc22a2* and *Slc22a3*. Of the nonimprinted genes displayed in Figure 1, the neighboring genes (*Slc22a1* and *Mas*) are not expressed in the embryo or placenta and show limited tissue-specific expression in adults. Thus, it may not be justified to call these genes "nonimprinted" because they are not expressed when imprinted expression is seen for the *Slc22a2* and *Slc22a3* genes. In view of the transcriptional overlap between *Mas* and the *Air* ncRNA it is of interest that *Mas* is only expressed in testes and brain, two tissues that lack imprinted *Igf2r* expression (Schweifer et al. 1997; Hu et al. 1999). *Plg*, *Tcp1*, and *Sod2* are expressed in the embryo and *Tcp1* and *Sod2* are also expressed in placenta, but all show biallelic expression (Zwart et al. 2001a). Of the imprinted genes in this cluster, *Igf2r* shows the strongest imprinted expression pattern and lies 29 kb from the *Air* promoter, while *Slc22a3* shows the weakest imprinted pattern and lies 240 kb distant (Fig. 1). This may indicate that the silencing effects of *Air* are reduced with distance.

IS THERE HUMAN "H"AIR?

The human *IGF2R* locus on 6q27 is embedded in a region with a similar organization to the mouse (Fig. 1). All tested human adult tissues appear to show a lack of imprinted *IGF2R* expression; however, some reports have described imprinted expression of human *IGF2R* restricted to 50% of the tested samples, in early embryos, in amniotic cell cultures, and in Wilms tumors (Giannoukakis et al. 1993; Kalscheuer et al. 1993; Xu et al. 1997; Oudejans et al. 2001). However, one report showed a fault with the RT reaction that results in a failure to reverse-transcribe one of the polymorphic alleles; thus, reports of imprinted expression from the human *IGF2R* gene must be confirmed by nonamplification techniques. A more recent report that used different polymorphisms to test the imprinting status of *IGF2R* in humans and lemurs clearly demonstrated that *IGF2R* was not imprinted and proposed that imprinted expression of *IGF2R* was lost in the primate lineage (Killian et al. 2001).

In view of the demonstrated lack of imprinted expression, it is surprising that the human *IGF2R* gene contains many sequence features and epigenetic marks associated with imprinting of the mouse *Igf2r* gene. For example, the human *IGF2R* gene contains a large CpG island in intron 2 in the same relative position to exon 3 as found for the mouse *Air* promoter (Smrzka et al. 1995; Riesewijk et al. 1996). The human intron 2 CpG island is also a differentially methylated region (DMR) that carries a maternal-specific methylation imprint and it is also contained in a region showing replication asynchrony. However, promoter prediction programs (http://www.gsf.de/biodv/matinspector.html) applied to the mouse and human intron 2 CpG islands predict promoter activity only for the mouse. The 108-kb mouse region spanned by *Air* contains more than 100 ESTs internally primed from A-rich stretches within the mature *Air* transcript, in contrast to the region predicted to correspond to a human "H"AIR ncRNA (http://www.ensembl.org/; S. Stricker and I. Yotova, unpubl.). This indicates the human CpG island in intron 2 does not normally act as promoter for an AIR-like transcript. However, the existence of a strong CpG island in intron 2 of the human *IGF2R* gene that is maintained as a DMR allows the possibility that the human gene could in some circumstances acquire imprinted maternal-specific expression by activating the putative "H"AIR promoter. It remains, therefore, to be tested if expression of a human AIR ncRNA can be found in some situations.

THE MOUSE *Air* ncRNA IS AN UNUSUAL RNA

The *Air* ncRNA was shown (Lyle et al. 2000) to span 108 kb from a promoter in *Igf2r* intron 2, to a PolyA tail in the last intron of *Mas*. The size of *Air* cannot be measured by standard Northern blots even in combination with Pulsed Field Gel Electrophoresis, as RNAs above 9 kb were not size-fractionated in formaldehyde gels (Lyle et al. 2000). Instead, the size of *Air* is assumed from mapping the 5' and 3' ends and the demonstration that truncating *Air* at 3 kb from the promoter removes all transcripts downstream from the truncation site (Sleutels et al. 2002). The major species of *Air* appears to be unspliced. However, the absence of splicing in the wild-type locus does not indicate that splicing is not possible in some conditions since insertion of a PolyA signal downstream from the *Air* promoter induced splicing at +53 bp in 80% of the truncated transcripts (Sleutels et al. 2002). In addition, mouse transgenes derived from the *Air* promoter consistently splice using the same +53-bp donor (Sleutels and Barlow 2001).

Mammalian genomes differ from those of other common experimental organisms (such as yeast, nematodes, fruit flies, and plants) because genes and repeats are comingled; thus, mammalian genes have extremely large introns relative to exons. As a result, unspliced precursor mRNA is rich in interspersed repeats; however, they would not represent a stable population since mRNAs are spliced as they are transcribed (Proudfoot et al. 2002). A recent bioinformatic-based survey indicated that a large number of cDNAs contain interspersed repeats in their 3'UTR (Nagashima et al. 2004); however, this would represent a small percentage of the transcript length. In contrast, the 108-kb *Air* analyzed by RepeatMasker (http://www.repeatmasker.org/) shows that interspersed repeats occupy 47% of the sequence. Thus the *Air* ncRNA is unusual because it is repeat rich as a mature transcript.

Seven of the imprinted noncoding RNAs listed in Table 1 show reciprocal parental-specific expression with respect to one or more imprinted protein-coding genes similar to that described for *Air* and *Igf2r*, and as a result have the potential to be involved in the imprinting mechanism. Table 2 lists a comparison of the features of these seven imprinted noncoding RNAs with respect to features noted for the *Air* ncRNA. Two out of seven are maternally expressed ncRNAs (*Gtl2* and *H19*), while five are paternally expressed (*Air*, *Kcnqot1*, *Nespas*, "IC-Snurf-Snurp long processed transcript," and *Cop2as*). Of the maternally expressed ncRNAs, the *H19* ncRNA has been excluded from a direct role in silencing (Jones et al. 1998; Schmidt et al. 1999), while *Gtl2* has not yet been tested. Of the paternally expressed ncRNAs, the *Air* ncRNA or its transcription has been shown to play a role in the imprinting mechanism (Rougeulle and Heard 2002; Sleutels et al. 2002, 2003). The other four paternally expressed ncRNAs have not yet been tested although dele-

Table 2. Characteristics of Imprinted Noncoding RNAs

Imprinted ncRNA	Parental expression	Size (kb)	Spliced or processed	References in addition to Ensembl build 30
H19	maternal	1.8	5 exons in 2.2 kb	
Gtl2	maternal	3.98	5 exons over 15.5 kb in ENSEMBL build 32, but 10 exons producing multiply spliced stable RNAs have been reported	Ensembl build 32; Paulsen et al. (2000); Lin et al. (2003)
Air	paternal	108	0 intron in 108 kb	Wutz et al. (1997); Lyle et al. (2000)
Kcnq1ot1	paternal	5′ mapped to Kcnq1 intron10, 3′ end not mapped	not known	Lee et al. (1999)
Nespas	paternal	>4.4 possibly longer	not known	Wroe et al. (2000)
IC-Snurf-Snrpn ncRNA	paternal	460 kb, *cis*tronic; mature RNA size unknown	>148 exons; long primary transcript possibly processed into multiple RNAs: IPW, PAR, PWCR1, multiple MB11 -snoRNAs, Ube3aas	Runte et al. (2001)
Cop2as	paternal	unknown	unknown, overlaps 3′ end of Cop2, possibly contains Mit1	Lee et al. (2000)

tions of the *Kcnq1ot1* and the "*IC-Snurf-Snurp long processed transcript*" promoters allow the possibility that these two ncRNAs may also have a silencing function.

A comparison of the ncRNAs listed in Table 2 with the features described above for *Air* does not yet allow clear conclusions to be drawn about similarities and differences. However, it does indicate trends that could be common to imprinted ncRNAs or trends that could be common either to the paternally or maternally expressed imprinted ncRNAs. For example, there is a tendency for both maternal and paternal imprinted ncRNAs to have few or short introns compared to normal mammalian mRNAs that have many and long introns. This tendency for imprinted transcripts per se to have few or short introns was noted before (Pfeifer and Tilghman 1994; Hurst et al. 1996). Interspersed repeats are likely to be present in the five paternally expressed imprinted ncRNAs and it is possible that there is tendency for these ncRNAs to be unusually long, but this will not be clear until more of these transcripts are finely mapped. One feature that may differentiate the maternally and paternally expressed ncRNAs is the presence of tandem direct repeats that map upstream of the transcription start for *H19* and *Gtl2* (Thorvaldsen et al. 1998; Takada et al. 2000) and downstream of the transcription start for *Air* and *Kcnq1* and thus potentially are present in the ncRNA (Lyle et al. 2000; Mancini-DiNardo et al. 2003).

FISHING FOR *Air*

RNA FISH (fluorescence in situ hybridization) is a procedure that applies fluorescent DNA hybridization probes to nondenatured cells, which detects single-stranded RNA but not double-stranded DNA (Dirks et al. 2003). This technique identifies a site close to a gene where nascent transcripts are accumulating and RNA processing is occurring. We have here used RNA FISH (Figs. 2–7) to test if the reciprocal parental-specific expression of *Air* and *Igf2r* observed in whole organs is also found in single cells.

Igf2r mRNA IS RESTRICTED TO THE MATERNAL CHROMOSOME IN MOST CELLS

A number of controls were performed to test the specificity of the *Igf2r* RNA FISH signals (Fig. 2). Note that the cells used in these experiments were not synchronized for cell cycle and the population contained cells with a mixed ploidy number but was dominated by diploid cells. To demonstrate that the observed fluorescent signal arose from RNA, control cells that were not denatured but were treated with Ribonuclease A produced little to no signal (Fig. 2A). Cells that were denatured and treated with RNase A produced two or four signals as expected for specific detection of DNA in diploid and tetraploid cells (Fig. 2B). Cells that were denatured without RNase A treatment showed an enhanced signal from one allele that is assumed to be *Igf2r* transcription from the maternal allele in addition to the DNA signal (Fig. 2C). These controls demonstrate that the hybridization target in nondenatured cells not treated with RNase A results from RNA (Fig. 2D). Figure 2D also shows that expression of *Igf2r* mRNA is generally restricted to one chromosome. In RNA FISH experiments with an *Igf2r*-specific probe, only 40–70% of nuclei in the examined population had any RNA FISH signal. The reason for the absence of RNA signals was not determined and may be due to the technical limits of the RNA FISH technique or may arise from stochastic gene expression. However, it should be noted that transcription would not be predicted to occur close to and during M phase. In cells positive for an RNA FISH signal, the majority of nuclei had a single signal (in one experiment with diploid NIH 3T3 cells, 80% of the positive cells had one RNA signal and 16% had two RNA signals). The reason for the presence of two RNA signals in a minority of the cell population was not determined and may be due to tetraploid cells containing two maternal chromosomes, to spontaneous denaturation allowing the probe access to DNA, or to expression of *Igf2r* from the paternal chromosome. Cells with more than one maternal chromosome, or with more than

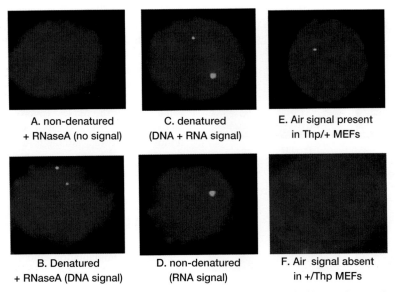

Figure 2. Control experiments for RNA FISH. (*A–D*) Diploid NIH 3T3 cells hybridized with an *Igf2r* strand-specific probe (*green*). (*E*) Thp/+ MEFs hybridized with an *Air* strand-specific probe (*green*). (*F*) +/Thp MEFs hybridized with an *Air* strand-specific probe. The RNA specific signal from *Igf2r* is seen in *D*. The paternal specific expression of *Air* is seen in *E*, since Thp/+ cells have a deletion on the maternal chromosome that removes the proximal Chr.17 imprinted cluster. In this and subsequent RNA FISH images, the probes were single stranded and derived from linear PCR. Two *Air*-strand-specific probes (FAS1 and 2) were amplified from cloned genomic DNA from *Igf2r* intron 1 (TGTCAAGGAGAACTGAGCGTCAT) or intron 2 (GGTAGGATGGTGTCTTACTC). The *Igf2r*-strand-specific probe (FcDNA) was amplified from the full-length mouse cDNA using an exon 28 primer (GTTTTC-GAAAGTCAGCTTCTGGC). Probes were labeled with either Digoxigenin or Biotin and detected by anti-Digoxigenin or Avidin antibody conjugated with FITC (*green*) or Rhodamine (*red*; see examples in Figs. 3–6). Nuclei were stained with DAPI. The full RNA FISH protocol and image capture has been described (Braidotti 2001).

one maternally inherited transgenic copy of *Igf2r*, will maintain correct maternal-specific expression, indicating that imprinted expression of *Igf2r* is not subject to a counting mechanism (Wutz et al. 1997; Vacik and Forejt 2003). It has also been reported that *Igf2r* can be expressed from the paternal allele in mouse embryos carrying a null *Igf2r* allele on the maternal chromosome, indicating a leaky repression of the paternal *Igf2r* promoter in embryos (Wang et al. 1994). In Thp/+ MEFs containing only the paternal *Igf2r* allele (the maternally inherited Thp chromosome carries a deletion of 6 Mbp that includes this imprinted cluster), a small subset of cells (<5%, data not shown) also produced an RNA signal in agreement with the suggestion that *Igf2r* mRNA can sometimes be expressed from a paternal allele in embryonic cells.

Air ncRNA IS RESTRICTED TO THE PATERNAL CHROMOSOME IN MOST CELLS

A similar RNA FISH hybridization series using probes for the *Air* ncRNA showed that *Air* expression is generally restricted to the paternal chromosome (Fig. 2E,F). In RNA FISH experiments using an *Air*-specific probe, ~50% of nuclei from diploid 3T3 fibroblasts were positive for any RNA signal. The reason for this increased number of nuclei negative for an *Air* RNA FISH signal was not determined, but may reflect reduced promoter activity or reduced RNA stability for *Air* compared with *Igf2r*. Of the 50% of nuclei with a positive signal for *Air*, 74% showed a single signal while 10% showed two signals (in 265 counted nuclei from NIH 3T3 cells). As with the results obtained with the *Igf2r* probe, a small proportion of NIH 3T3 cells also showed two *Air* transcription signals. Although double signals can arise for several reasons, possible *Air* expression from the maternal allele was seen in a small minority (4% of 71 counted nuclei) of +/Thp MEFs possessing just the maternal *Air* allele (in +/Thp cells the Thp deletion chromosome is paternally inherited). One difference from the results obtained with *Igf2r* was that *Air* showed some cells with a clustered signal pattern containing two to five distinct spots associated together, in ~16% of positive cells (Fig. 3B–D).

Air AND *Igf2r* ARE NOT COEXPRESSED IN *cis*

Figure 4A shows cohybridization with two probes, one recognizing the paternally derived *Air* transcript (red) and the second recognizing the maternally derived *Igf2r* mRNA (green). All positive cells showed distinct red or distinct green signals, indicating that expression of *Igf2r* and *Air* arose on different parental chromosomes. Since the *Igf2r* and *Air* probes are separated by ~45 kb, coexpression from the same chromosome would produce a yellow signal and none were observed. This data supports the claim that expression of *Igf2r* and *Air* is mutually exclusive in *cis*. Figure 4B shows a cell assumed to be tetraploid and this confirms other data that multiple copies of imprinted genes are not regulated by a counting mechanism (Wutz et al. 1997; Vacik and Forejt 2003).

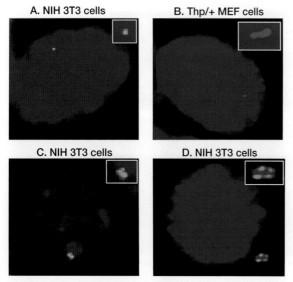

Figure 3. Four examples of *Air* RNA FISH showing (*A*) a single signal and (*B–D*) the clustered spots that appeared in 16% of diploid NIH 3T3 cells or Thp/+ MEFs. An enlarged image for the signals and clusters is shown in the top right of each picture.

THE *Air* ncRNA FORMS CLUSTERS THAT MIGRATE AWAY FROM CONDENSED CHROMOSOMES

Clustering of the *Air* ncRNA signal was seen in some cells assumed to be in interphase (Fig. 5A) but most often seen to be separate from the main body of condensing chromosomes in cells assumed from DAPI staining to be in early M phase. In Figure 5B,C, a cell was identified with a single transcription/accumulation signal on one focal plane within the spherical nucleus, plus a cluster of signals on a different focal plane that had migrated away from the condensed chromosomes. We cautiously interpret the behavior of these *Air* clusters as indicating that the *Air* transcript does not stay associated with the paternal *Igf2r* allele as the chromosomes condense in early M phase. In all cases where the *Air* ncRNA cluster was observed to dissociate from the chromosome, the signals remained grouped together in clusters. The possibility exists that these provisional results indicate that the *Air* ncRNA may form aggregates either with itself or with other proteins or RNAs. The *Air* clusters that were separated from metaphase chromosomes were often observed in an asymmetrical distribution relative to the metaphase plate (Fig. 5D), which suggested a link with the centrosome body or the MTOC (microtubule organizing center). However, cohybridization with a gamma tubulin antibody showed that although the *Air* cluster was often found associated with centrosomes, the cluster was also seen to be separate from the centrosomes (Fig. 6).

THE *Air* CLUSTERS ARE ABSENT IN ES CELLS THAT LACK IMPRINTED EXPRESSION OF *Igf2r* mRNA

Mouse preimplantation embryos express *Igf2r* from both parental chromosomes, with imprinted maternal-specific expression occurring between 4.5 and 6.5 dpc (Szabo and Mann 1995a; Lerchner and Barlow 1997). Embryonic stem (ES) cells show the same biallelic expression of *Igf2r* (Wang et al. 1994). *Air* ncRNA is weakly detected in the undifferentiated ES cell but is abundant in ES cells induced to differentiate (Fig. 7A). RNA FISH analysis of *Air* in undifferentiated ES cells identified only a few single very weak signals in ~5% of cells but no clusters were observed (Fig. 7B). Differentiated ES cells, however, produced an array of signal types similar to that observed in NIH 3T3 with a maximum of 34% of cells in the differentiated culture positive for any RNA FISH signal. Figure 7C shows a nucleus of a differentiated ES cell with a pronounced single signal and a cluster located at the edge of the nucleus. Three examples of *Air* ncRNA clusters in differentiated ES cells are shown magnified below.

The RNA FISH analysis of *Igf2r* and *Air* expression in MEFs shows that expression of both transcripts is parental specific. However, there is a possibility that requires further testing—that a minority of cells (<5%) can express *Igf2r* and *Air* from the opposite parental chromosome. The cohybridization experiments that assayed ex-

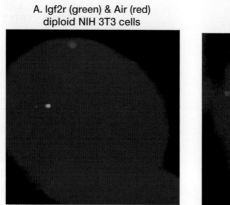
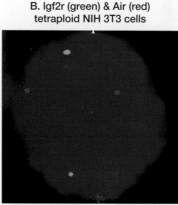

Figure 4. RNA FISH showing cohybridization of diploid and tetraploid NIH 3T3 cells with *Igf2r*-specific (*green*) and *Air*-specific (*red*) probes.

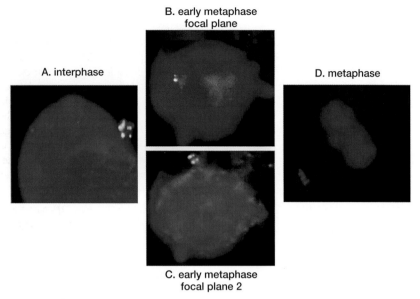

Figure 5. RNA FISH showing *Air* clustered signals (*red*) dissociated from chromosomes. *Air* clusters are seen in interphase nuclei (*A*), early metaphase nuclei (*B* and *C* show two focal planes in the same cell), and in metaphase (*D*).

pression of *Igf2r* and *Air* in the same cell, however, do clearly show that these two genes are not active in *cis*. This finding supports other data (Sleutels et al. 2002) that shows that *Air* expression is necessary to silence *Igf2r* (and the downstream flanking *Slc22a2* and *Slc22a3* genes) in *cis*. The RNA FISH analysis has also shown that the *Air* ncRNA, in contrast to *Igf2r*, can form clusters of two to five spots. *Air* clusters were seen in a small number of cells assumed from DAPI staining of chromosomes to be in interphase—the stage when *Igf2r* silencing must occur. However, in the majority of cases the *Air* ncRNA clusters were observed dissociated from chromosomes in cells assumed from DAPI staining to be in early

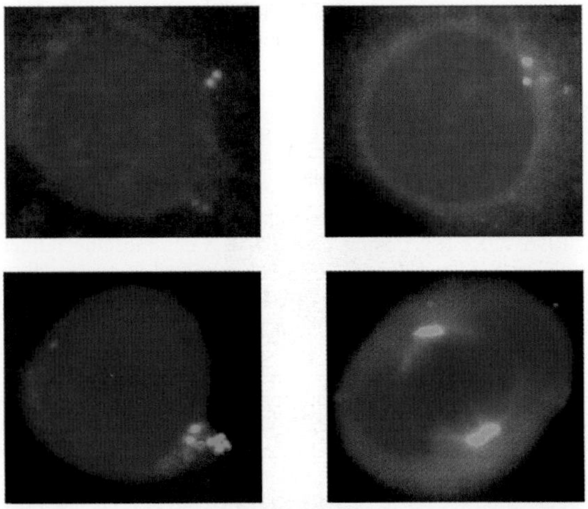

Figure 6. *Air* RNA FISH (*red*) combined with gamma tubulin immunofluorescence (*green*) in NIH 3T3 cells. Gamma tubulin antibody from Dr. Dagmar Ivanyi, NKI, Amsterdam. The *top left* image shows that *Air* is localized away from the centrosome, while the three other images show that *Air* is adjacent to the centrosome.

metaphase. We cautiously interpret this to indicate that *Air* clusters form in late interphase and metaphase. The meaning of these *Air* clusters and why they are also observed dissociated from chromosomes is not yet known. Since expression of *Air* is known to silence genes spread over 300 kb, it is possible that these clusters are relevant for the silencing function of *Air*. One speculation is that the clusters that show as multiple distinct bright spots at metaphase are derived from very tight aggregates present at interphase that cannot be resolved into separate spots. However, further work is needed to confirm that these clusters are consistently associated with *Air* expression or have any functional relevance. It should be noted that an earlier RNA FISH study on the imprinted maternally expressed *H19* ncRNA did not find evidence of RNA clusters (Jouvenot et al. 1999). Interestingly, the appearance of the human XIST ncRNA in diploid female cells has been described as an "accumulation of many, very distinct bright clusters" (Clemson et al. 1996). However, in contrast to the results for mouse *Air* reported here, the human XIST ncRNA clusters were specifically observed in interphase cells associated with the inactive chromosome.

SIMILARITIES BETWEEN GENOMIC IMPRINTING AND X-INACTIVATION

The *cis*-repression of a large subchromosomal domain by the *Air* ncRNA has clear parallels with the action of the *Xist* ncRNA in X-chromosome inactivation (Avner and Heard 2001; Wutz 2003). It has also been argued that X-chromosome inactivation arose from a localized form of genomic imprinting that initially affected only part of the chromosome (Graves 1996; Lyon 1999). Similarities between genomic imprinting and X-inactivation have been previously noted (Pfeifer and Tilghman 1994; Lyon 1999; Lee 2003) and the behavior of the *Air* ncRNA

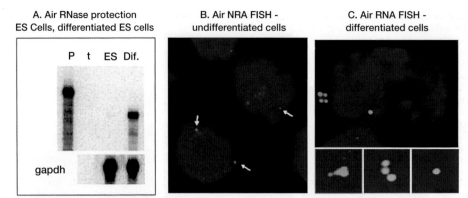

Figure 7. An RNase protection assay for *Air* (*A*) shows that Air is not expressed in ES cells but is abundant in ES cells differentiated by exposure for 5 days to retinoic acid. *gapdh* expression is used as the loading control (P, probe; t, tRNA). (*B*) The weak *Air* RNA FISH signal (*white arrows*) seen in ES cells that were not reproducible enough to count accurately. (*C*) The strong *Air* RNA FISH signal seen in 34% of differentiated cells. The three smaller images in *C* show examples of the *Air* signal clusters seen in these cells.

strengthens these similarities. The features in common between the two processes include (for review, see Lyon 1999) the following.

i. X-chromosome inactivation is imprinted such that the paternal X is preferentially inactivated in the extraembryonic tissues of placental mammals and in all organs of marsupials.

ii. The regulatory gene is a *cis*-silencing noncoding RNA that acts over a long range.

iii. Chromosome-specific methylation regulates expression of the noncoding RNA.

iv. Some genes escape repression.

The *Xist* noncoding RNA has some additional features that have not yet been characterized or may be absent for the *Air* ncRNA. First, experiments indicate that the *Xist* ncRNA has a sequence-specific function that may lie in the secondary structure (Wutz et al. 2002). Second, the *Xist* ncRNA has not been observed dissociated from chromosomes; instead it colocalizes with the repressed X chromosome in interphase in human cells (Clemson et al. 1996) and in interphase and metaphase in mouse cells (Sheardown et al. 1997). Third, expression of the *Xist* transcript is regulated in ES cells by a second antisense noncoding RNA, named *Tsix* (Lee 2000; Sado et al. 2001). Fourth, *Xist* recruits repressive chromatin proteins and modifications to the X chromosome (Avner and Heard 2001; Wutz 2003).

The similarities between genomic imprinting and X-inactivation do not imply that these are identical mechanisms. If, as suggested, the latter evolved from the former, it is likely that the *Xist* ncRNA may have also evolved extra functions when compared to the *Air* ncRNA, particularly with regard to its ability to spread along the entire X chromosome. While further experiments will be needed to discriminate between the possible modes of actions of the *Air* ncRNA and to refine the details of the repression mechanism, the high degree of similarity between *cis*-repression on autosomes by imprinted noncoding RNAs and *cis*-repression of the X chromosome by the noncoding *Xist* ncRNA is striking.

ACKNOWLEDGMENTS

We thank past and present members of the lab for contributing to the work and ideas described here, Dagmar Ivanyi for the gamma tubulin antibody, Roeland Dirks for advice on RNA FISH, and Laura Spahn for reading the manuscript. Research support from GEN-AU—Genomforschung in Österreich (F.P.), FWF—Fonds zur Förderung der wissenschaftlichen Forschung (I.Y. and C.S.), and KWF—the Dutch Cancer Society and the Netherlands Cancer Institute (G.B. and O.S.).

REFERENCES

Avner P. and Heard E. 2001. X-chromosome inactivation: Counting, choice and initiation. *Nat. Rev. Genet.* **2:** 59.
Baroux C., Spillane C., and Grossniklaus U. 2002. Genomic imprinting during seed development. *Adv. Genet.* **46:** 165.
Beechey C.V., Cattanach B.M., and Blake A. 2003. MRC Mammalian Genetics Unit, Harwell, Oxfordshire. World Wide Web site—Mouse Imprinting Data and References (http://www.mgu.har.mrc.ac.uk/research/imprinting/).
Bell A.C. and Felsenfeld G. 2000. Methylation of a CTCF-dependent boundary controls imprinted expression of the Igf2 gene. *Nature* **405:** 482.
Bongiorni S. and Prantera G. 2003. Imprinted facultative heterochromatization in mealybugs. *Genetica* **117:** 271.
Braidotti G. 2001. RNA-FISH to analyze allele-specific expression. *Methods Mol. Biol.* **181:** 169.
Brandeis M., Ariel M., and Cedar H. 1993. Dynamics of DNA methylation during development. *Bioessays* **15:** 709.
Clemson C.M., McNeil J.A., Willard H.F., and Lawrence J.B. 1996. XIST RNA paints the inactive X chromosome at interphase: Evidence for a novel RNA involved in nuclear/chromosome structure. *J. Cell Biol.* **132:** 259.
Dirks R.W., Molenaar C., and Tanke H.J. 2003. Visualizing RNA molecules inside the nucleus of living cells. *Methods* **29:** 51.
Fournier C., Goto Y., Ballestar E., Delaval K., Hever A.M., Esteller M., and Feil R. 2002. Allele-specific histone lysine methylation marks regulatory regions at imprinted mouse genes. *EMBO J.* **21:** 6560.
Giannoukakis N., Deal C., Paquette J., Goodyer C.G., and Polychronakos C. 1993. Parental genomic imprinting of the human IGF2 gene. *Nat. Genet.* **4:** 98.
Goday C. and Esteban M.R. 2001. Chromosome elimination in sciarid flies. *Bioessays* **23:** 242.
Graves J.A. 1996. Mammals that break the rules: Genetics of

marsupials and monotremes. *Annu. Rev. Genet.* **30:** 233.
Gribnau J., Hochedlinger K., Hata K., Li E., and Jaenisch R. 2003. Asynchronous replication timing of imprinted loci is independent of DNA methylation, but consistent with differential subnuclear localization. *Genes Dev.* **17:** 759.
Hark A.T., Schoenherr C.J., Katz D.J., Ingram R.S., Levorse J.M., and Tilghman S.M. 2000. CTCF mediates methylation-sensitive enhancer-blocking activity at the H19/Igf2 locus. *Nature* **405:** 486.
Hu J.F., Balaguru K.A., Ivaturi R.D., Oruganti H., Li T., Nguyen B.T., Vu T.H., and Hoffman A.R. 1999. Lack of reciprocal genomic imprinting of sense and antisense RNA of mouse insulin-like growth factor II receptor in the central nervous system. *Biochem. Biophys. Res. Commun.* **257:** 604.
Hurst L.D., McVean G., and Moore T. 1996. Imprinted genes have few and small introns. *Nat. Genet.* **12:** 234.
Jones B.K., Levorse J.M., and Tilghman S.M. 1998. Igf2 imprinting does not require its own DNA methylation or H19 RNA. *Genes Dev.* **12:** 2200.
Jonker J.W., Wagenaar E., Van Eijl S., and Schinkel A.H. 2003. Deficiency in the organic cation transporters 1 and 2 (Oct1/Oct2 [Slc22a1/Slc22a2]) in mice abolishes renal secretion of organic cations. *Mol. Cell. Biol.* **23:** 7902.
Jouvenot Y., Poirier F., Jami J., and Paldi A. 1999. Biallelic transcription of Igf2 and H19 in individual cells suggests a post-transcriptional contribution to genomic imprinting. *Curr. Biol.* **9:** 1199.
Kalscheuer V.M., Mariman E.C., Schepens M.T., Rehder H., and Ropers H.H. 1993. The insulin-like growth factor type-2 receptor gene is imprinted in the mouse but not in humans. *Nat. Genet.* **5:** 74.
Killian J.K., Nolan C.M., Wylie A.A., Li T., Vu T.H., Hoffman A.R., and Jirtle R.J. 2001. Divergent evolution in M6P/IGF2R imprinting from the Jurassic to the Quaternary. *Hum. Mol. Genet.* **10:** 1721.
Kitsberg D., Selig S., Brandeis M., Simon I., Keshet I., Driscoll D.J., Nicholls R.D., and Cedar H. 1993. Allele-specific replication timing of imprinted gene regions. *Nature* **364:** 459.
Lee J.E., Tantravahi U., Boyle A.L., and Efstratiadis A. 1993. Parental imprinting of an Igf-2 transgene. *Mol. Reprod. Dev.* **35:** 382.
Lee J.T. 2000. Disruption of imprinted X inactivation by parent-of-origin effects at Tsix. *Cell* **103:** 17.
———. 2003. Molecular links between X-inactivation and autosomal imprinting: X-inactivation as a driving force for the evolution of imprinting? *Curr. Biol.* **13:** R242.
Lee M.P., DeBaun M.R., Mitsuya K., Galonek H.L., Brandenburg S., Oshimura M., and Feinberg A.P. 1999. Loss of imprinting of a paternally expressed transcript, with antisense orientation to KVLQT1, occurs frequently in Beckwith-Wiedemann syndrome and is independent of insulin-like growth factor II imprinting. *Proc. Natl. Acad. Sci.* **96:** 5203.
Lee Y.J., Park C.W., Hahn Y., Park J., Lee J., Yun J.H., Hyun B., and Chung J.H. 2000. Mit1/Lb9 and Copg2, new members of mouse imprinted genes closely linked to Peg1/Mest(1). *FEBS Lett.* **472:** 230.
Lerchner W. and Barlow D.P. 1997. Paternal repression of the imprinted mouse Igf2r locus occurs during implantation and is stable in all tissues of the post-implantation mouse embryo. *Mech. Dev.* **61:** 141.
Li E. 2002. Chromatin modification and epigenetic reprogramming in mammalian development. *Nat. Rev. Genet.* **3:** 662.
Lin S.P., Youngson N., Takada S., Seitz H., Reik W., Paulsen M., Cavaille J., and Ferguson-Smith A.C. 2003. Asymmetric regulation of imprinting on the maternal and paternal chromosomes at the Dlk1-Gtl2 imprinted cluster on mouse chromosome 12. *Nat. Genet.* **35:** 97.
Ludwig T., Eggenschwiler J., Fisher P., D'Ercole A.J., Davenport M.L., and Efstratiadis A. 1996. Mouse mutants lacking the type 2 IGF receptor (IGF2R) are rescued from perinatal lethality in Igf2 and Igf1r null backgrounds. *Dev. Biol.* **177:** 517.
Lyle R., Watanabe D., te Vruchte D., Lerchner W., Smrzka O.W., Wutz A., Schageman J., Hahner L., Davie C., and Barlow D.P. 2000. The imprinted antisense RNA at the Igf2r locus overlaps but does not imprint Mas1. *Nat. Genet.* **25:** 19.
Lyon M.F. 1999. Imprinting and X-chromosome inactivation. *Results Probl. Cell Differ.* **25:** 73.
Mancini-DiNardo D., Steele S.J., Ingram R.S., and Tilghman S.M. 2003. A differentially methylated region within the gene Kcnq1 functions as an imprinted promoter and silencer. *Hum. Mol. Genet.* **12:** 283.
Moore T. and Haig D. 1991. Genomic imprinting in mammalian development: A parental tug-of-war. *Trends Genet.* **7:** 45.
Nagashima T., Matsuda H., Silva D.G., Petrovsky N., Konagaya A., Schonbac, C., Kasukawa T., Arakawa T., Carninci P., Kawai J., and Hayashizaki Y. 2004. FREP: A database of functional repeats in mouse cDNAs. *Nucleic Acids Res.* (database issue) **32:** D471.
Oudejans C.B., Westerman B., Wouters D., Gooyer S., Leegwater P.A., van Wijk I.J., and Sleutels F. 2001. Allelic IGF2R repression does not correlate with expression of antisense RNA in human extraembryonic tissues. *Genomics* **73:** 331.
Paulsen M., El-Maarri O., Engemann S., Strodicke M., Franck O., Davies K., Reinhardt R., Reik W., and Walter J. 2000. Sequence conservation and variability of imprinting in the Beckwith-Wiedemann syndrome gene cluster in human and mouse. *Hum. Mol. Genet.* **9:** 1829.
Pfeifer K. and Tilghman S.M. 1994. Allele-specific gene expression in mammals: The curious case of the imprinted RNAs. *Genes Dev.* **8:** 1867.
Proudfoot N.J., Furger A., and Dye M.J. 2002. Integrating mRNA processing with transcription. *Cell* **108:** 501.
Reik W. and Walter J. 2001. Genomic imprinting: Parental influence on the genome. *Nat. Rev. Genet.* **2:** 21.
Riesewijk A.M., Schepens M.T., Welch T.R., van den Berg-Loonen E.M., Mariman E.M., Ropers H.H., and Kalscheuer V.M. 1996. Maternal-specific methylation of the human IGF2R gene is not accompanied by allele-specific transcription. *Genomics* **31:** 158.
Rougeulle C. and Heard E. 2002. Antisense RNA in imprinting: Spreading silence through Air. *Trends Genet.* **18:** 434.
Runte M., Huttenhofer A., Gross S., Kiefmann M., Horsthemke B., and Buiting K. 2001. The IC-SNURF-SNRPN transcript serves as a host for multiple small nucleolar RNA species and as an antisense RNA for UBE3A. *Hum. Mol. Genet.* **10:** 2687.
Sado T., Wang Z., Sasaki H., and Li E. 2001. Regulation of imprinted X-chromosome inactivation in mice by Tsix. *Development* **128:** 1275.
Schmidt J.V., Levorse J.M., and Tilghman S.M. 1999. Enhancer competition between H19 and Igf2 does not mediate their imprinting. *Proc. Natl. Acad. Sci.* **96:** 9733.
Schweifer N., Valk P.J., Delwel R., Cox R., Francis F., Meier-Ewert S., Lehrach H., and Barlow D.P. 1997. Characterization of the C3 YAC contig from proximal mouse chromosome 17 and analysis of allelic expression of genes flanking the imprinted Igf2r gene. *Genomics* **43:** 285.
Sheardown S.A., Duthie S.M., Johnston C.M., Newall A.E., Formstone E.J., Arkell R.M., Nesterova T.B., Alghisi G.C., Rastan S., and Brockdorff N. 1997. Stabilization of Xist RNA mediates initiation of X chromosome inactivation. *Cell* **91:** 99.
Shintani S., O'hUigin C., Toyosawa S., Michalova V., and Klein J. 1999. Origin of gene overlap: The case of TCP1 and ACAT2. *Genetics* **152:** 743.
Sleutels F. and Barlow D.P. 2001. Investigation of elements sufficient to imprint the mouse Air promoter. *Mol. Cell. Biol.* **21:** 5008.
Sleutels F., Zwart R., and Barlow D.P. 2002. The non-coding Air RNA is required for silencing autosomal imprinted genes. *Nature* **415:** 810.
Sleutels F., Tjon G., Ludwig T., and Barlow D.P. 2003. Imprinted silencing of Slc22a2 and Slc22a3 does not need transcriptional overlap between Igf2r and Air. *EMBO J.* **22:** 3696.
Smrzka O.W., Fae I., Stoger R., Kurzbaue, R., Fischer G.F., Henn T., Weith A., and Barlow D.P. 1995. Conservation of a maternal-specific methylation signal at the human IGF2R locus. *Hum. Mol. Genet.* **4:** 1945.

Spahn L. and Barlow D.P. 2003. An ICE pattern crystallizes. *Nat. Genet.* **35:** 11.

Stoeger R., Kubicka P., Liu C G., Kafri T., Razin A., Cedar H., and Barlow D.P. 1993. Maternal-specific methylation of the imprinted mouse Igf2r locus identifies the expressed locus as carrying the imprinting signal. *Cell* **73:** 61.

Szabo P.E. and Mann J.R. 1995a. Allele-specific expression and total expression levels of imprinted genes during early mouse development: Implications for imprinting mechanisms. *Genes Dev.* **9:** 3097.

———. 1995b. Biallelic expression of imprinted genes in the mouse germ line: Implications for erasure, establishment, and mechanisms of genomic imprinting. *Genes Dev.* **9:** 1857.

Takada S., Tevendale M., Baker J., Georgiades P., Campbell E., Freeman T., Johnson M.H., Paulsen M., and Ferguson-Smith A.C. 2000. Delta-like and gtl2 are reciprocally expressed, differentially methylated linked imprinted genes on mouse chromosome 12. *Curr. Biol.* **10:** 1135.

Thorvaldsen J.L., Duran K.L., and Bartolomei M.S. 1998. Deletion of the H19 differentially methylated domain results in loss of imprinted expression of H19 and Igf2. *Genes Dev.* **12:** 3693.

Vacik T. and Forejt J. 2003. Quantification of expression and methylation of the Igf2r imprinted gene in segmental trisomic mouse model. *Genomics* **82:** 261.

Verona R.I., Mann M.R., and Bartolomei M.S. 2003. Genomic imprinting: Intricacies of epigenetic regulation in clusters. *Annu. Rev. Cell Dev. Biol.* **19:** 237.

Wang Z.Q., Fung M.R., Barlow D.P., and Wagner E.F. 1994. Regulation of embryonic growth and lysosomal targeting by the imprinted Igf2/Mpr gene. *Nature* **372:** 464.

Wilkins J.F. and Haig D. 2003. What good is genomic imprinting: The function of parent-specific gene expression. *Nat. Rev. Genet.* **4:** 359.

Wroe S.F., Kelsey G., Skinner J.A., Bodle D., Ball S.T., Beechey C.V., Peters J., and Williamson C.M. 2000. An imprinted transcript, antisense to Nesp, adds complexity to the cluster of imprinted genes at the mouse Gnas locus. *Proc. Natl. Acad. Sci.* **97:** 3342.

Wutz A. 2003. RNAs templating chromatin structure for dosage compensation in animals. *Bioessays* **25:** 434.

Wutz A., Rasmussen T.P., and Jaenisch R. 2002. Chromosomal silencing and localization are mediated by different domains of Xist RNA. *Nat. Genet.* **30:** 167.

Wutz A., Smrzka O.W., Schweifer N., Schellander K., Wagner E.F., and Barlow D.P. 1997. Imprinted expression of the Igf2r gene depends on an intronic CpG island. *Nature* **389:** 745.

Wutz A., Theussl H.C., Dausman J., Jaenisch R., Barlow D.P., and Wagner E.F. 2001. Non-imprinted Igf2r expression decreases growth and rescues the Tme mutation in mice. *Development* **128:** 1881.

Xu Y.Q., Grundy P., and Polychronakos C. 1997. Aberrant imprinting of the insulin-like growth factor II receptor gene in Wilms' tumor. *Oncogene* **14:** 1041.

Yang Y., Li T., Vu T.H., Ulaner G.A., Hu J.F., and Hoffman A.R. 2003. The histone code regulating expression of the imprinted mouse Igf2r gene. *Endocrinology* **144:** 5658.

Zwart R., Sleutels F., Wutz A., Schinkel A.H., and Barlow D.P. 2001a. Bidirectional action of the Igf2r imprint control element on upstream and downstream imprinted genes. *Genes Dev.* **15:** 2361.

Zwart R., Verhaagh S., Buitelaar M., Popp-Snijders C., and Barlow D.P. 2001b. Impaired activity of the extraneuronal monoamine transporter system known as uptake-2 in Orct3/Slc22a3-deficient mice. *Mol. Cell. Biol.* **21:** 4188.

The (Dual) Origin of Epigenetics

D. HAIG

*Department of Organismic and Evolutionary Biology, Harvard University,
Cambridge, Massachusetts 02138*

"Epigenetics" has different meanings for different scientists. Molecular biologists are probably most familiar with a definition of epigenetics as *"the study of mitotically and/or meiotically heritable changes in gene function that cannot be explained by changes in DNA sequence"* (Riggs et al. 1996). For them, epigenetic mechanisms would include DNA methylation and histone modification. Functional morphologists, however, would be more familiar with a definition such as that of Herring (1993), for whom epigenetics refers to *"the entire series of interactions among cells and cell products which leads to morphogenesis and differentiation."* She continues that "among the numerous epigenetic factors influencing the vertebrate face is mechanical loading" and that "epigenetic influences range from hormones and growth factors to ambient temperature and orientation in a gravitational field." In this note, I will argue that these disparate definitions have come about because "epigenetics" had at least two semi-independent origins during the 20th century.

The adjective "epigenetic" has a much longer history than the noun "epigenetics" because the adjective originally referred to a different noun, "epigenesis." Thus, the Oxford English Dictionary gives the primary sense of *epigenetic* as "[o]f pertaining to, or of the nature of epigenesis" and defines *epigenesis* as "the formation of an organic germ as a new product" with *the theory of epigenesis* defined as "the theory that the germ is brought into existence (by successive accretions), and not merely developed, in the process of reproduction." It is not my purpose here to delve into the history of theories of epigenesis, except to note that "epigenetic" is sometimes still used in this earlier sense as pertaining to epigenesis.

Epigenetics was coined by Waddington (1942) to refer to the study of the "causal mechanisms" by which "the genes of the genotype bring about phenotypic effects." Waddington (1939 [p. 156]) had earlier used epigenotype to refer to "the set of organizers and organizing relations to which a certain piece of tissue will be subject during development." He believed that genotype and phenotype referred to "differences between whole organisms . . . [and were] not adequate or appropriate for the consideration of differences within a single organism." Thus, epigenetics for Waddington referred to a subject similar to what we would now call developmental biology.

Waddington (1956) later provided some insight into his reasons for choosing epigenetics. He wrote, "The fact that the word 'epigenetics' is reminiscent of 'epigenesis' is to my mind one of the points in its favour. . . . We all realize that, by the time development begins, the zygote contains certain 'preformed' characters, but that these must interact with one another, in processes of 'epigenesis', before the adult condition is attained. The study of the 'preformed' characters nowadays belongs to the discipline known as 'genetics'; the name 'epigenetics' is suggested as the study of those processes which constitute the epigenesis which is also involved in development" (see also Waddington 1939 [pp. 154–155]).

Waddington targeted what he saw as the naive view of many geneticists that there was a simple correspondence between genes and characters. For Waddington, the course of development was determined by the interaction of many genes with each other and with the environment. Neo-Darwinism, he believed, involved "a breach between organism and nature as complete as the Cartesian dualism of mind and matter; an epigenetic consideration of evolution would go some way toward healing it" (Waddington 1953 [p. ix]). Waddington decried "the reigning modern view . . . [that] the direction of mutational change is entirely at random, and that adaptation results solely from the natural selection of mutations which happen to give rise to individuals with suitable characteristics." This he considered to be an "extremist" theory (Waddington 1953 [p. 151]). Because Waddington claimed to provide a richer paradigm for studying the interaction between organism and environment than the impoverished view of genetics, an epigenetic approach has appealed to critics of evolutionary "orthodoxy," whether these be biologists who feel that there is something lacking in the neo-Darwinian synthesis (see, e.g., Løvtrup 1972; Ho and Saunders 1979; Jablonka and Lamb 1989) or philosophers who favor a less gene-centric, more holistic, view of biology (see, e.g., articles in van de Vijver et al. 2002).

Waddington's term gained few converts prior to the 1960s. A notable exception was Huxley (1956) who encouraged others to use epigenetics to mean "the science of developmental process in general." Later, in a review of cancer biology, Huxley (1957) used epigenetics "to denote the analytic study of individual development (ontogeny) with its central problem of differentiation." For Huxley, "The method by which tissues and organs differentiate in the course of normal development is at the moment the main blank space in biology's map. . . . [W]e know little of the precise steps taken by epigenetic processes, of the biochemical factors involved, and above all, of what determines the replicable specificity of differentiated tissues." Thus, epigenetics was concerned with the processes by which a constant genotype gave rise to differentiated cell types and tissues, and perturbations of which could give rise to cancer.

Berry and Searle (1963) initiated an association of "epigenetics" with the study of variation in skeletal development. They used epigenetic "in Waddington's sense, to emphasize the developmental origin of the discontinuities being studied, with genetic factors determining the main features of the 'epigenetic landscape' but environmental forces influencing the final outcome." Their study of the rodent skeleton was later extended to a study of epigenetic variation in the human cranium (Berry and Berry 1967). Herring's (1993) definition of epigenetics (quoted in the first paragraph of this paper) is a direct descendant of this usage.

The second 20th-century derivation of "epigenetics" can be traced to Nanney (1958). At a conference in Gif-sur-Yvette on Extrachromosomal Heredity (March 1958), Nanney contrasted what he called genetic and *paragenetic* systems. However, Pontecorvo told Nanney of the criticism he had received from a professor of Greek for Pontecorvo's use of the term *parasexuality* (Ephrussi 1958). By the time the published version of Nanney's remarks appeared in July 1958, Nanney had substituted epigenetic for paragenetic and added references to Waddington (Nanney 1958). This history indicates that Nanney's concept of what he came to call epigenetic control systems was independent of Waddington's prior use of epigenetics, but that Nanney considered their two usages to be compatible.

Nanney (1958) started his comments by noting the great recent advances of chemical genetics. These advances had allowed a consistent hypothesis in which genetic control systems were based on a "template replicating mechanism" that determined the "library of specificities." However, he believed that "auxiliary mechanisms with different principles of operation are involved in determining which specificities are to be expressed in any particular cell." These auxiliary mechanisms he called epigenetic control systems. He saw them as accounting for the observation that cells with the same genotype could have different phenotypes. Moreover, epigenetic regulation could show properties of long-term persistence. Therefore, "the observation of indefinite persistence of differences does not distinguish persistent homeostasis due to DNA maintenance (genetic homeostasis) from persistent homeostasis due to epigenetic regulation (epigenetic homeostasis)." He contrasted "[t]he current concept of a primary genetic material (DNA), replicating by a template mechanism" with "a homeostatic system operating by, perhaps, self-regulating metabolic patterns."

It is no coincidence that Nanney made his remarks at a conference on Extrachromosomal Heredity. Nanney was a significant protagonist in what Sapp (1987) has styled the struggle for authority in genetics. On one side of this conflict were geneticists who ascribed a predominant role in heredity, evolution, and development to the genetic material of the nucleus. On the other side were biologists, such as Nanney, who wished to maintain an important role for the cytoplasm, or extranuclear factors, in hereditary. Nanney (1957 [p. 136]) had earlier expressed his views on the relative roles of the nucleus and cytoplasm in political terms, and these comments give some insight into his favored model of epigenetic control systems:

> Two concepts of genetic mechanisms have persisted side by side . . . The first of these we will designate the "Master Molecule" concept. . . . In its simplest form the concept places the "master molecules" in the chromosomes and attributes the characteristics of an organism to their specific construction; all other cellular constituents are considered relatively inconsequential except as obedient servants of the masters. This is in essence the Theory of the Gene, interpreted to suggest totalitarian government. . . .
>
> The second concept of a genetic mechanism is one which is more difficult to describe . . . This concept we will designate as the "Steady State" concept. By the term "Steady State" we envision a dynamic self-perpetuating organization of a variety of molecular species which owes its specific properties not to the characteristics of any one kind of molecule, but to the functional interrelationships of these molecular species. . . . In contrast to the totalitarian government by "master molecules," the "steady state" government is a more democratic organization, composed of interacting cellular fractions operating in self-perpetuating patterns.

By the time of the Cold Spring Harbor Symposium on Quantitative Biology, Nanney (1959) had refined his conception of epigenetic systems. These he characterized as "signal interpreting devices, yielding predictable results in response to specific stimuli from inside and outside the cell. They are conceived as the integrative systems regulating the expression of genetic potentialities; mutual exclusion, simultaneity of expression, and adaptive cellular transformation could scarcely be achieved without efficient triggering devices." He now conceded that epigenetic systems were "presumably limited by the information contained in the genetic library. . . . An epigenetic change should not result in a permanent loss of information and a return to a previous condition of expression is always theoretically possible."

Boris Ephrussi, one of the organizers of the conference in Gif-sur-Yvette, had long been interested in questions of cell heredity and was another proponent of an important role for extrachromosomal inheritance. Ephrussi championed Nanney's concept of epigenetic control at a conference on Genetic Approaches to Somatic Cell Variation held in Gatlinburg, Tennessee (April 1958). Ephrussi (1958) distinguished between "truly genetic mechanisms" based on "the transmission of particles carrying their own structural information" from "epigenetic mechanisms involving functional states of the nucleus." He adopted the terms genetic and epigenetic because they more clearly delimited "the notion of true genetic change. This is a concession on my part, and many of my geneticist friends will, I am sure, enjoy the shift of my stand. Unfortunately, I must remind them that, as a corollary, we must admit that not everything that is inherited is genetic."

In his summary comments to the Gatlinburg conference, Lederberg (1958) was not prepared to accept this olive branch without qualification. In Lederberg's view, Ephrussi wished to distinguish between genetic information that was structural from epigenetic information that was based on a dynamic flux equilibrium. But many properties that Ephrussi would ascribe to epigenetic mechanisms could also be explained by structural changes in the nucleus. Lederberg was prepared to "define a category of genetic information as being 'nucleic'; that is, depending on the *sequence of nucleotides* in a nucleic acid. By contrast, 'epinucleic' information is expressed in another form, e.g., as an aspect of nucleic acid configuration other than nucleotide sequence or in polypeptide or polyamine adjuncts to the polynucleotide." Offering his own olive branch, Lederberg was prepared to concede an adjunct role for "extranucleic information in molecules or reaction cycles not directly connected with nucleic acid." (Lederberg considered Ephrussi's choice of terms "confusing if only because 'epigenetic' is already widely current in a different sense, e.g., in Waddington's book 'The Epigenetics of Birds.'")

Lederberg's concept of epinucleic information is close to the concept of epigenetic modification used in modern molecular biology whereas Nanney's appeal to cytoplasmic steady states led nowhere. Ironically, Lederberg's terminology of nucleic, epinucleic, and extranucleic information obtained few converts whereas Nanney's dichotomy of genetic and epigenetic control was adopted by several researchers, especially those interested in cell heredity of somatic cells and cancer. For example, Harris (1964 [p. 1]) cited Nanney for a distinction between genetic changes that "depend on a recasting of hereditary determinants" and epigenetic changes that "take place against a constant cellular genome." In his view, "Truly genetic mechanisms are concerned with the preservation and replication of information in structural form" whereas epigenetic mechanisms "regulate the expression of genetic information. They serve to translate structural symbols into phenotypic reality."

During the 1960s, one can find a number of researchers working on cell culture or cancer who implicitly or explicitly derive their distinction between epigenetic and genetic processes from either Nanney or Harris (see, e.g., Waymouth 1967; Braun 1969 [p. 153]). The fact that Cahn and Cahn (1966) could refer to the epigenetics of development in clonal cell culture without providing an explicit definition or citation for "epigenetics" suggests that the authors expected the term would be understood by most of their intended audience. Markert (1968a) similarly provided no citation for "epigenetic" when he concluded that "there is probably no gene coding for cancer per se. Instead, normal gene activity is misprogrammed by epigenetic mechanisms to produce a neoplastic pattern of metabolism in which all of the individual components are normal."

Another early use of an epigenetic/genetic distinction occurred in the burgeoning field of protein electrophoresis. Soon after Nanney distinguished epigenetic from genetic control, Markert and Møller (1959) introduced "the term *isozyme* to describe the different molecular forms in which proteins may exist with the same enzymatic specificity. . . . The existence of isozymes raises the question of whether each of these also is controlled by a separate gene or whether they are all modifications of a single gene product." Nanney's terminology was ready-made to distinguish these alternatives. Thus, Markert (1968b) distinguished "the genetic control of final conformation through specifying amino acid sequences [from] epigenetic direction of conformational folding or subunit association" (also see Allen 1960; Ruddle and Roderick 1965; Munkres 1968).

Widespread use of epigenetics to refer to heritable changes that do not involve changes in DNA sequence did not occur until the 1990s, but I will leave my historical survey in the 1960s and limit myself to some brief comments on subsequent changes in usage based on a less systematic search of the literature. Since the 1960s, the evolving meaning of epigenetics in molecular and cellular biology has coexisted with continued use of epigenetics in a Waddingtonian sense, with the different definitions sometimes conflated and thought to represent a single coherent discipline (Jablonka and Lamb 2002; Müller and Olsson 2003). Holliday (1979) contains the earliest description of DNA methylation as an "epigenetic" process that I have found, but I suspect that "The inheritance of epigenetic defects" (Holliday 1987) was the critical paper that lit the fuse for the explosion in use of "epigenetic" in the 1990s.

REFERENCES

Allen S.L. 1960. Inherited variations in the esterases of *Tetrahymena*. *Genetics* **45**: 1051.
Berry A.C. and Berry R.J. 1967. Epigenetic variation in the human cranium. *J. Anat.* **101**: 361.
Berry R.J. and Searle A.G. 1963. Epigenetic polymorphism of the rodent skeleton. *Proc. Zool. Soc. Lond.* **1963**: 577.
Braun A.C. 1969. *The cancer problem. A critical analysis and modern synthesis*. Columbia University Press, New York.
Cahn R.D. and Cahn M.B. 1966. Heritability of cellular differentiation: Clonal growth and expression of differentiation in retinal pigment cells in vitro. *Proc. Natl. Acad. Sci.* **55**: 106.
Ephrussi B. 1958. The cytoplasm and somatic cell variation. *J. Cell. Comp. Physiol.* (suppl. 1) **52**: 35.
Harris M. 1964. *Cell culture and somatic variation*. Holt, Rinehart & Winston, New York.
Herring S.W. 1993. Formation of the vertebrate face: Epigenetic and functional influences. *Am. Zool.* **33**: 472.
Ho M.W. and Saunders P.T. 1979. Beyond neo-Darwinism—An epigenetic approach to evolution. *J. Theor. Biol.* **78**: 573.
Holliday R. 1979. A new theory of carcinogenesis. *Brit. J. Cancer* **40**: 513.
———. 1987. The inheritance of epigenetic defects. *Science* **238**: 163.
Huxley J. 1956. Epigenetics. *Nature* **177**: 807.
———. 1957. Cancer biology: Viral and epigenetic. *Biol. Rev.* **32**: 1.
Jablonka E. and Lamb M.J. 1989. The inheritance of acquired epigenetic variations. *J. Theor. Biol.* **139**: 69.
———. 2002. The changing concept of epigenetics. *Ann. N.Y. Acad. Sci.* **981**: 82.
Lederberg J. 1958. Genetic approaches to somatic cell variation: Summary comment. *J. Cell. Comp. Physiol.* (suppl. 1) **52**: 383.
Løvtrup S. 1972. *Epigenetics. A treatise on theoretical biology*. John Wiley & Sons, London.
Markert C.L. 1968a. Neoplasia: A disease of cell differentiation. *Cancer Res.* **28**: 1904.

———. 1968b. The molecular basis for isozymes. *Ann. N.Y. Acad. Sci.* **151:** 14.
Markert C.L. and Møller F. 1959. Multiple forms of enzymes: Tissue, ontogenetic, and species specific patterns. *Proc. Natl. Acad. Sci.* **45:** 753.
Munkres K.D. 1968. Genetic and epigenetic forms of malate dehydrogenase in *Neurospora. Ann. N.Y. Acad. Sci.* **151:** 294.
Müller G.B. and Olsson L. 2003. Epigenesis and epigenetics. In *Keywords and concepts in evolutionary developmental biology* (ed. B.K. Hall and W.M. Olson), p. 114. Harvard University Press, Cambridge, Massachusetts.
Nanney D.L. 1957. The role of the cytoplasm in heredity. In *The chemical basis of heredity* (ed. W.D. McElroy and B. Glass), p. 134. Johns Hopkins University Press, Baltimore, Maryland.
———. 1958. Epigenetic control systems. *Proc. Natl. Acad. Sci.* **44:** 712.
———. 1959. Epigenetic factors affecting mating type expression in certain ciliates. *Cold Spring Harbor Symp. Quant. Biol.* **23:** 327.
Riggs A.D., Martienssen R.A. and Russo V.E.A. 1996. Introduction. In *Epigenetic mechanisms of gene regulation* (ed. V.E.A. Russo et al.), p. 1. Cold Spring Harbor Laboratory Press, Cold Spring Harbor, New York.
Ruddle F.H. and Roderick T.H. 1965. The genetic control of three kidney esterases in C57BL/6J and RF/J mice. *Genetics* **51:** 445.
Sapp J. 1987. *Beyond the gene. Cytoplasmic inheritance and the struggle for authority in genetics.* Oxford University Press, New York.
van de Vijver G., van Speybroeck L. and de Waele D., Eds. 2002. *From epigenesis to epigenetics: The genome in context. Ann. N.Y. Acad. Sci.* **981**.
Waddington C.H. 1939. *An introduction to modern genetics.* Macmillan, New York.
———. 1942. The epigenotype. *Endeavour* **1:** 18.
———. 1953. *The strategy of the genes.* George Allen & Unwin, London.
———. 1956. Embryology, epigenetics and biogenetics. *Nature* **177:** 1241.
Waymouth C. 1967. Somatic cells *in vitro:* Their relationship to progenitive cells and to artificial milieux. *Natl. Cancer Inst. Monogr.* **26:** 1.

Sex and X-Chromosome-wide Repression in *Caenorhabditis elegans*

B.J. Meyer, P. McDonel, G. Csankovszki, and E. Ralston

Howard Hughes Medical Institute and Department of Molecular and Cell Biology, University of California, Berkeley, California 94720-3204

In many organisms, sex is determined by a precise chromosome counting mechanism that distinguishes one sex chromosome from two. In flies and worms, for example, XX embryos become females (or hermaphrodites), while XO or XY embryos become males (Fig. 1a) (Bridges 1916; Nigon 1951; Madl and Herman 1979). Sex can also be specified by the mere presence of a particular sex chromosome, such as the Y chromosome of mammals: XY or XXY embryos are males, and XX embryos females (Gubbay et al. 1990; Sinclair et al. 1990; Koopman et al. 1991). These sex-determining mechanisms cause the two sexes to differ in their dose of X chromosomes, yet both sexes require equivalent levels of X-chromosome gene products. A chromosome-wide regulatory process called dosage compensation neutralizes the difference in X-linked gene dose between males and females by equalizing X-chromosome transcript levels.

The strategies for dosage compensation are diverse (Fig. 2), but in all known cases, specialized dosage compensation complexes are targeted exclusively to the X chromosome(s) of one sex to modulate gene expression in only that sex (Fig. 1b,c). This selective recruitment of the dosage compensation machinery establishes the epigenetic regulation of X chromosomes that is maintained throughout the lifetime of the animal. Female mammals randomly inactivate one X chromosome (Huynh and Lee, this volume). Male flies double the transcription rate of

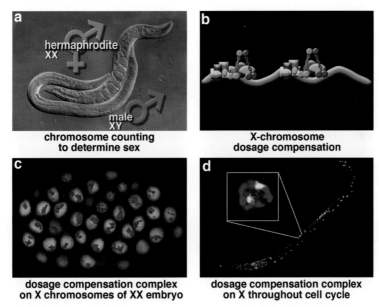

Figure 1. The dosage compensation process. (*a*) In *C. elegans*, sex is determined by an X-chromosome counting mechanism that distinguishes one X chromosome from two. XX animals are hermaphrodites, and XO animals are males. Shown is a male and hermaphrodite mating pair that harbors a mutation in the dosage compensation gene *dpy-28*. XX animals that escape lethality are dumpy in phenotype and retain their embryos; XO animals appear wild type. (*b*) The difference in X chromosome dose between sexes is neutralized by the process of dosage compensation. A dosage compensation complex (*multicolored shapes*) binds to both X chromosomes (*blue*) of hermaphrodites to repress transcript levels by half, thus equalizing the overall level of X chromosome gene expression between males and hermaphrodites. (*c*) The dosage compensation protein SDC-2 binds to both X chromosomes of XX embryos starting around the 40-cell stage. Shown is a multicell XX embryo stained with DAPI (*blue*) to identify nuclear DNA and SDC-2 antibodies (*red*) to monitor the dosage compensation complex. (*d*) Once bound, the dosage compensation complex remains localized to X chromosomes throughout the cell cycle. Shown is an XX L1 larva stained with an antibody to the dosage compensation protein DPY-26 (*green*) and one to histone H3 phosphorylated on serine 10 (*red*), which associates only with mitotic chromosomes. DPY-26 remains bound to both X chromosomes of the two mitotic cells in the larva.

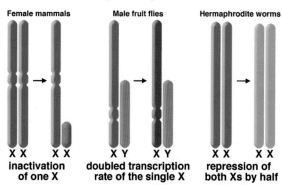

Figure 2. Diverse strategies for dosage compensation. Organisms use different strategies to equalize X-linked gene expression between males (XY or XO) and females or hermaphrodites (XX). Female mammals randomly inactivate one X chromosome. Male fruit flies double the transcription rate of their single X chromosome. Hermaphrodite worms halve the expression of both X chromosomes.

their single X chromosome (Oh et al., this volume). Hermaphrodite worms keep both X chromosomes active, but repress transcript levels from each X chromosome by half (Meyer and Casson 1986). In flies and mammals, this gene regulation is accompanied by the sex-specific modification of histones on the regulated X chromosomes (Bone et al. 1994; Hilfiker et al. 1997; Heard et al. 2001; Mermoud et al. 2002). In all three species, dosage compensation is essential, and failure to accomplish this global regulation causes either male- or female-specific lethality.

Fundamental questions are relevant to all forms of dosage compensation. First, what is the composition of the machinery that implements dosage compensation? Second, what are the sex-specific factors that activate the dosage compensation machinery in only one sex? Third, what are the *cis*-acting recruitment sites that target X chromosomes for regulation by the dosage compensation complex? Fourth, what is the molecular mechanism that underlies the coordinate regulation of gene expression along an entire X chromosome? Fifth, what is the explicit mechanism for fine-tuning X-linked gene expression by only twofold? We have addressed these basic questions in the nematode *Caenorhabditis elegans*, using integrated genetic, biochemical, and cell-biological approaches to dissect this regulatory process.

THE *C. ELEGANS* DOSAGE COMPENSATION MACHINERY

In *C. elegans*, the discovery of sex-specific lethal mutations that preferentially killed XX hermaphrodites was pivotal for the identification of dosage compensation proteins (Hodgkin 1983; Plenefisch et al. 1989). Mutations in eight genes—*sdc-1*, *sdc-2*, *sdc-3*, *dpy-21*, *dpy-26*, *dpy-27*, *dpy-28*, and *dpy-30*—reduced the viability of XX but not XO animals. All mutations disrupted dosage compensation, causing a twofold elevation in X-linked transcript levels in XX but not XO animals (Meyer and Casson 1986; Villeneuve and Meyer 1987; Nusbaum and Meyer 1989; DeLong et al. 1993; Hsu and Meyer 1994).

The *sdc* mutations not only interfere with X-chromosome gene regulation, they also disrupt sex determination, causing XX embryos to develop as males. The *sdc* genes coordinately control the hermaphrodite mode of both sex determination and dosage compensation, acting at a common step in the genetic regulatory hierarchy prior to the divergence of sex determination and dosage compensation genes into two separate pathways (Villeneuve and Meyer 1987; Nusbaum and Meyer 1989; DeLong et al. 1993).

The hermaphrodite-specific elevation in X-linked transcript levels caused by all dosage compensation mutations was consistent with either of two mechanisms for dosage compensation: random inactivation of a single hermaphrodite X chromosome or repression of both hermaphrodite X chromosomes by half. X-inactivation was unlikely since neither of the two genetic phenomena it would cause had been observed: First, hermaphrodites are not mosaic in phenotype for cell-autonomous X-linked markers; second, most X-linked loss-of-function mutations fail to behave as dominant alleles with variable penetrance and expressivity. The molecular architecture of the dosage compensation proteins and their localization to the two hermaphrodite X chromosomes (see below) confirmed that dosage compensation involves the repression of both X chromosomes by half.

Molecular analysis of the dosage compensation genes showed the similarity of DPY-26, DPY-27, and DPY-28 proteins to components of 13S condensin, a complex essential for the resolution and compaction of mitotic and meiotic chromosomes from yeast to man (Fig. 3a) (Chuang et al. 1994; Lieb et al. 1996; Swedlow and Hirano 2003; C. Tsai et al., in prep.). The prototypical condensin complex contains at least five subunits, including a pair of SMC (Structural Maintenance of Chromosomes) proteins (SMC2 and SMC4) and three non-SMC ancillary factors, which belong to the Chromosome-Associated Polypeptide CAP-D2, CAP-G, and CAP-H/Barren families (Hirano 2002; Swedlow and Hirano 2003). DPY-26 resembles CAP-H, DPY-28 resembles CAP-D2, and DPY-27 resembles SMC4. These three dosage compensation proteins form a complex that associates with both X chromosomes of hermaphrodites (Fig. 3a,b).

Genetic analysis failed to reveal the *C. elegans* homolog of SMC2, but biochemical analysis of the dosage compensation complex did (Lieb et al. 1996). MIX-1, the SMC2 homolog, immunoprecipitated with the complex and was identified by immunoreactivity to antibodies specific for the single *C. elegans* SMC2 protein. Mutations in *mix-1* disrupt not only dosage compensation, but also mitotic chromosome resolution, condensation, and segregation, causing the death of both XX and XO animals (Lieb et al. 1996; Hagstrom et al. 2002; Chan et al. 2004). The role of MIX-1 in this vital cellular process precluded its identification in genetic screens that exploited the hermaphrodite-specific action of dosage compensation genes.

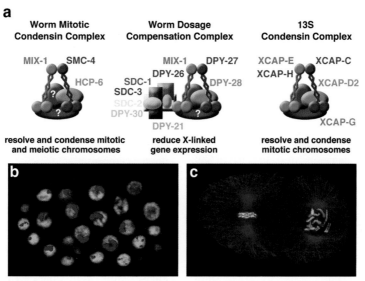

Figure 3. Two condensin complexes in *C. elegans*. (*a*) *C. elegans* has one condensin complex specialized for dosage compensation (*middle*) and one specialized for the resolution, compaction, and segregation of both mitotic and meiotic chromosomes (*left*). The canonical 13S condensin complex (*right*) is conserved from yeast to humans. Proteins in all three complexes are color-coded to denote homologous proteins. In addition to the condensin-like proteins, the dosage compensation complex contains proteins (SDC-2, SDC-3, and DPY-30) that are essential for the recruitment of the entire complex to X. The SMC2-like protein called MIX-1 is shared between the dosage compensation and mitotic condensin complexes. Preliminary biochemical analysis of the dosage compensation complex has identified an XCAP-G homolog (G. Csankovszki and B.J. Meyer, unpubl.). Preliminary biochemical analysis of the mitotic complex has identified an XCAP-H homolog and a different XCAP-G homolog (K.A. Hagstrom and B.J. Meyer, unpubl.). (*b*) The dosage compensation complex localizes to both X chromosomes of hermaphrodites. Shown is a multicell embryo stained with the DNA intercalating dye DAPI (*blue*) and an antibody to the dosage compensation protein SDC-2 (*red*). (*c*) The *C. elegans* mitotic condensin complex colocalizes with centromere proteins on the holocentric mitotic chromosomes. Shown is a two-celled embryo stained with the DNA dye DAPI (*red*) and antibodies to SMC-4 (*green*) and to tubulin (*blue*). One cell (*left*) is in metaphase and the other (*right*) is in prometaphase, where individual condensed chromosomes are readily apparent, and the centromeres appear as parallel tracts on the poleward sides of the sister chromatids. (*c*, Reprinted, with permission, from Hagstrom et al. 2002.)

MIX-1 binds to X chromosomes of hermaphrodites when it colocalizes with the dosage compensation protein DPY-27, and it colocalizes with the centromeres of mitosis chromosomes in both sexes when it associates with the *C. elegans* mitosis-specific SMC4 protein called SMC-4 (Fig. 3c) (Lieb et al. 1996; Hagstrom et al. 2002). That is, MIX-1 partitions its roles in two separate biological processes, gene expression and chromosome segregation, through its participation in two separate condensin-like complexes, a dosage compensation–specific complex and a mitotic complex (Fig. 3a) (Hagstrom et al. 2002; Chan et al. 2004). Together, these results suggest that the *C. elegans* dosage compensation process evolved by recruiting components used in other chromosome behaviors to the new task of regulating gene expression. The similarity of the dosage compensation complex to condensin and the participation of MIX-1 in both complexes suggest a common mechanism for repressing X-chromosome gene expression during dosage compensation and for establishing chromosome resolution and higher-order chromosome structure during mitotic chromosome segregation.

Other components of the dosage compensation complex have also retained their roles in conserved cellular processes. DPY-28 figures prominently in meiosis, where it regulates the number and distribution of crossovers between homologous chromosomes (C. Tsai et al., in prep.). DPY-30 not only binds X chromosomes to down-regulate gene expression, it associates with the COMPASS complex to methylate histones (C. Hassig and B. Meyer, unpubl.). DPY-30 is the *C. elegans* homolog of the *Saccharomyces cerevisiae* protein Sdc1 (genetic locus *YDR469W*) (Hsu and Meyer 1994; Nagy et al. 2002).

Recruitment of the DPY and MIX proteins to X chromosomes requires the activities of SDC-2 (a novel protein), SDC-3 (a protein with two zinc fingers), and DPY-30 (Klein and Meyer 1993; Chuang et al. 1996; Lieb et al. 1996; Davis and Meyer 1997; Lieb et al. 1998; Yonker and Meyer 2003). These three proteins recruit all other dosage compensation proteins to X, including DPY-21 (a novel protein) and SDC-1 (a protein with seven zinc fingers), through their own association with X (Davis and Meyer 1997; Dawes et al. 1999; Chu et al. 2002; Yonker and Meyer 2003). SDC-3 requires both SDC-2 and DPY-30 for its localization to X (Davis and Meyer 1997), and DPY-30 requires both SDC-2 and SDC-3 to bind X (C. Hassig and B. Meyer, unpubl.). SDC-2 is unique in that it can localize to X independently from all other dosage compensation proteins, suggesting that SDC-2 is pivotal for X-chromosome recognition and confers chromosome specificity to dosage compensation (Dawes et al. 1999). Once the dosage compensation machinery assembles on X chromosomes, it appears to remain localized to X throughout the entire cell cycle, including mitosis (Fig. 1d) (Lieb et al. 1998).

SEX-SPECIFIC TARGETING OF THE DOSAGE COMPENSATION COMPLEX TO HERMAPHRODITE X CHROMOSOMES

SDC-2 also confers hermaphrodite specificity to dosage compensation. All properties of SDC-2 are consistent with its role as the sex-specific switch that activates dosage compensation in hermaphrodites (Dawes et al. 1999). All other dosage compensation proteins are supplied maternally and are diffusely distributed throughout the nuclei of very young embryos (<30 cells) in both sexes and only later become specifically localized to X chromosomes of XX but not XO animals. SDC-2 differs in four important ways. First, SDC-2 is not maternally contributed and not expressed in very young embryos. Its initial expression occurs around the 40-cell stage, the stage in which the dosage compensation machinery assembles on X. Second, SDC-2 localizes to hermaphrodite X chromosomes from the onset of its expression. Third, SDC-2 is not expressed in wild-type XO embryos, indicating that SDC-2, unlike other dosage compensation proteins, is sex-specifically regulated. *sdc-2* is repressed in males by the male-specific gene *xol-1*, the master sex-determination switch gene and direct target of the primary sex-determination signal (Fig. 4) (Miller et al. 1988). When *xol-1* is active (in XO embryos), male development ensues; when *xol-1* is inactive (in XX embryos), hermaphrodite development ensues, including the activation of dosage compensation (Rhind et al. 1995). Finally, ectopic expression of *sdc-2* in XO animals using a ubiquitously expressed promoter triggers assembly of the dosage compensation complex on the single X, causing death. Thus, SDC-2 is the hermaphrodite-specific switch that is both necessary and sufficient to activate dosage compensation.

The mechanism by which *xol-1* controls *sdc-2* remains a mystery. Whether it acts directly or indirectly is also not known. Unexpectedly, the crystal structure of XOL-1 revealed it to be a GHMP kinase family member, despite having sequence identity of <10% (Luz et al. 2003). Small molecule kinases such as galactokinase and homoserine kinase are prototypic members of the GHMP kinase family. XOL-1 does not bind ATP under standard conditions, suggesting that XOL-1 may act by a mechanism distinct from that of GHMP small molecule kinases. However, possible clues to XOL-1 function come from the identity of GHMP kinase family members as regulators of gene expression. *S. cerevisiae* GAL3p, a protein

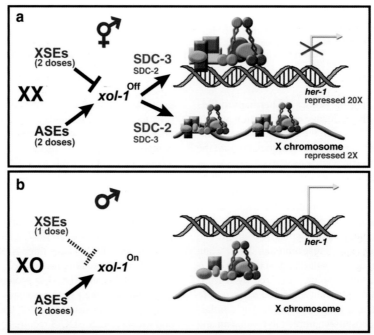

Figure 4. Genetic control of sex determination and dosage compensation in *C. elegans*. (*a*) *xol-1* is the master sex-determination switch gene that controls both sex determination and dosage compensation. It is the direct molecular target of the X-chromosome counting mechanism that determines sex. The twofold difference in X-chromosome dose between males and hermaphrodites is translated into the ON/OFF state of *xol-1*. In XX hermaphrodites, two doses of the X-signal elements (XSEs) repress *xol-1* by overcoming the *xol-1* activation achieved by autosomal signal elements (ASEs). Two mechanisms of *xol-1* repression are used: repression at the level of transcription and pre-mRNA splicing. When *xol-1* is repressed, the XX-specific gene SDC-2 is active and stabilizes SDC-3. SDC-2 acts with SDC-3 to target dosage compensation proteins to X, thereby repressing gene expression by half. SDC-2 plays the lead role in recognizing X-specific sequences. It is the only dosage compensation protein expressed solely in XX animals. SDC-2 also activates the hermaphrodite program of sexual development by repressing the male-specific sex-determination gene *her-1* by 20-fold. SDC-2 acts with SDC-3 to recruit dosage compensation proteins to *her-1*. In this case, SDC-3 plays the lead role in recognizing *her-1* DNA targets. The *her-1* and X complexes differ by one component: DPY-21 is present on X but not on *her-1*. (*b*) In XO males, the single dose of XSEs is insufficient to overcome the activating influence of ASEs. *xol-1* is active and promotes the male fate by repressing the activities of *sdc* genes. *her-1* is transcribed, and the single X is not repressed. In *xol-1* XO mutants, the dosage compensation complex localizes to the single X chromosome, killing XO animals by reducing X-linked gene expression.

that is structurally similar to galactokinase and can be converted to a galactokinase by the substitution of two amino acids, is a galactose- and ATP-dependent transcriptional inducer of *GAL* genes (Platt et al. 2000). Moreover, *GAL1* of *Kluyveromyces lactis* is a bifunctional protein that acts as both a galactokinase and a transcriptional inducer of *GAL* genes (Meyer et al. 1991). XOL-1 might also act with a small effector molecule to repress transcription of a downstream target gene such as *sdc-2*.

xol-1, in turn, is regulated by the primary sex-determination signal, the X:A signal, the ratio of X chromosomes to sets of autosomes (the ploidy) (Fig. 4) (Nigon 1951; Madl and Herman 1979; Miller et al. 1988; Rhind et al. 1995). *xol-1* is repressed by a group of dose-sensitive, X-linked genes called X-signal elements (XSEs) (Akerib and Meyer 1994; Nicoll et al. 1997; Carmi et al. 1998; Carmi and Meyer 1999) and activated by a set of dose-sensitive autosome-linked genes called Autosomal-signal elements (ASEs) (J. Powell and B. Meyer, in prep.). In XX animals, the double dose of XSEs opposes the double dose of ASEs, causing *xol-1* to be repressed. In XO animals, the single dose of XSEs cannot oppose the double dose of ASEs, allowing *xol-1* to be active (Fig. 4).

RECRUITMENT OF THE DOSAGE COMPENSATION COMPLEX FOR GENE-SPECIFIC VERSUS CHROMOSOME-WIDE REPRESSION

Not only does SDC-2 activate dosage compensation, it also induces hermaphrodite sexual differentiation in concert with SDC-1 and SDC-3 (Dawes et al. 1999; Chu et al. 2002). Proof that SDC-2 acts as the pivotal sex-specific factor to initiate the hermaphrodite program of sexual development came in part by the observation that overexpression of SDC-2 and SDC-1 in XO animals, made viable by a dosage compensation mutation, resulted in XO animals developing as hermaphrodites. SDC-2 activates hermaphrodite sexual development by binding directly to regulatory regions of the male sex-determination gene *her-1*, repressing its transcription 20-fold, thereby directing the sex-determination pathway to the hermaphrodite mode (Dawes et al. 1999). SDC-2 thus acts both as a strong gene-specific repressor and as a weaker chromosome-wide repressor. Unexpectedly, SDC-2, together with SDC-1 and SDC-3, recruits other dosage compensation components to *her-1*, directing this chromosome repression machinery to silence an individual, autosomal gene (Chu et al. 2002).

Functional dissection of *her-1* in vivo revealed three DNA recognition elements required for SDC binding, recruitment of the DCC, and transcriptional repression (Chu et al. 2002). These *her-1* elements differed in location (promoter and second intron), sequence, and strength of repression. Though no consensus sequence was found in the three *her-1* recognition elements, a 15-nucleotide (nt) repeat was shared between the two binding sites in the second intron. Scrambling of the 15-nt sequence prevented binding of the complex to the two recognition sites in vivo, causing derepression of *her-1* and thus masculinization of XX animals. The *her-1* binding sites bear no obvious resemblance to any sequences on X, leading to the speculation that the chromosome-wide repression complex can achieve different degrees of repression in part by associating with different DNA sequences. These results also imply that the dosage compensation complex regulates transcription along X chromosomes using diverse DNA recognition elements.

Further analysis revealed important molecular differences in the composition and targeting of the repression complex to *her-1* versus X (Yonker and Meyer 2003). The dosage compensation protein DPY-21, a novel protein, localizes to X but not to *her-1*. Furthermore, within the complex, different proteins play the lead role in recognizing DNA targets. SDC-2 recognizes X-chromosome targets, while SDC-3 recognizes *her-1* targets. These results begin to explain how closely related complexes can achieve uniformly weak repression of many genes in one context and strong repression of a specific gene in another.

RECRUITMENT AND SPREADING OF THE DOSAGE COMPENSATION COMPLEX ALONG HERMAPHRODITE X CHROMOSOMES

While most components of the dosage compensation complex have been identified, the *cis*-acting X-chromosome recognition elements that recruit the complex have been elusive. To define these recognition elements on X (*rex*), a chromosome-wide search was conducted for regions of X sufficient to recruit the complex when detached from X (Csankovszki et al. 2004). Regions were analyzed in 32-ploid intestinal cell nuclei of XX hermaphrodite strains carrying either free or autosome-attached X-chromosome duplications (Fig. 5). X-chromosome territories were marked by fluorescent in situ hybridization (FISH) probes; one identified the duplicated region (red), and a second identified the rest of X (blue). Localization of the dosage compensation complex was visualized by antibodies to a dosage compensation protein (green).

Provided that *C. elegans* X chromosomes contained discrete X-recognition elements that recruit the complex, four possibilities existed for how the complex might be targeted to X. First, a single site on X could recruit the complex and nucleate long-range spreading of the dosage compensation complex across the entire chromosome. Second, a limited number of recognition sites could recruit the complex, and some or all sites could nucleate short-range spreading. Third, a limited number of recognition sites could recruit the complex but the complex would not spread. It would occupy only sites that autonomously recruited it, influencing gene expression from a long distance, perhaps by altering chromosome structure. Fourth, a high density of X-recognition sites could recruit the complex, but no spreading would occur, implying direct, short-range regulation by the complex.

The first possibility, the presence of a single recruitment site, was eliminated by analysis of hermaphrodites carrying one copy of the X duplication *mnDp1* and two chromosomes deleted for the corresponding region (Fig.

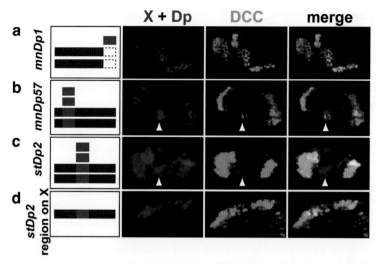

Figure 5. Multiple X regions recruit the dosage compensation complex and nucleate spreading. (*a-d*) Confocal images of individual 32-ploid intestinal cell nuclei with (*a–c*) and without (*d*) X chromosome duplications detached from X. Gut cells were stained with FISH probes to the duplicated region (*red*), FISH probes to the rest of X (*blue*), and antibodies to a dosage compensation protein (*green*). Cartoons on *left* represent the genotype of each nucleus, showing the copy number of each X duplication (*red*) and the location on X (*blue*) of the duplicated region (*red*). Shown are regions of X that strongly recruited the complex (*a*), weakly recruited the complex (*b*), or failed to recruit the complex (*c*). Regions that did not recruit the complex when detached from X exhibit robust recruitment of the dosage compensation when attached to the native X (*d*), indicating that the complexes spread to these regions from neighboring regions that contain X-recognition elements. *Arrows* mark duplications. (Reprinted, with permission, from Csankovszki et al. 2004 [©AAAS].)

5a and Fig. 6). The dosage compensation complex colocalized with both the duplication and the truncated X chromosomes, indicating that *mnDp1* harbors at least one X-recognition element, as does the rest of X. The interpretation of multiple, independent X-recognition elements was greatly reinforced by the identification of many other nonoverlapping, detached X regions that recruited the complex. Two classes of dosage compensation complex recruitment were found, strong recruitment typified by *mnDp1* and limited, but reproducible, recruitment typified by *mnDp57* (Fig. 5b and Fig. 6).

The fourth possibility, a high density of recruitment sites, was eliminated by the discovery of large segments of X incapable of recruiting the complex when detached from X (Fig. 5c and Fig. 6). Despite the inability to autonomously recruit the complex, these regions, typified by *stDp2*, harbor dosage compensated genes when present on the native X chromosome (Fig. 6). These results indicated that the dosage compensation complex regulates gene expression in X regions that lack recruitment sites either by spreading into the regions from neighboring X-recognition elements (second possibility) or by acting over a distance (third possibility). The second possibility predicts binding of the dosage compensation complex to the region on the native X corresponding to *stDp2*, and the third possibility predicts no binding. Ro-

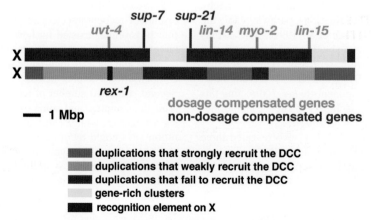

Figure 6. Map of X regions that recruit the dosage compensation complex when detached from the native X. Location of X regions that strongly (*dark green*), weakly (*light green*), or fail to (*red*) recruit the dosage compensation complex are shown relative to the gene-rich clusters (*yellow*) on X (*blue*) and the few known dosage compensated genes (*uvt-4*, *lin-14*, *myo-2*, *lin-15*) and non–dosage compensated genes (tRNA encoding genes *sup-7* and *sup-21*). Regions that apparently lack recruitment sites contain dosage compensated genes. This map summarizes DCC recruitment data obtained through experiments similar to those shown in Fig. 5.

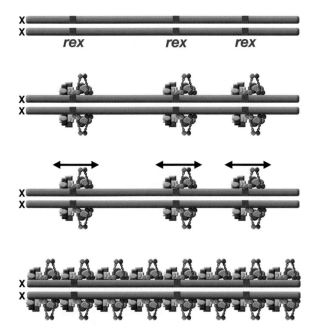

Figure 7. Spreading repression. The *C. elegans* X chromosome harbors multiple X-recognition elements (*rex* sites) that distinguish it from autosomes to recruit the dosage compensation complex. The complex binds to rex elements on both hermaphrodite X chromosomes and then spreads in *cis* to neighboring X regions that lack recruitment sites, thus establishing X-chromosome-wide repression that is maintained throughout the lifetime of the animal.

bust localization of the dosage compensation complex was found on the *stDp2* region (Fig. 5d), on other native X regions corresponding to duplications with no recruitment ability, and on regions corresponding to duplications with limited, but reproducible, binding. Thus, the *C. elegans* X chromosomes have discrete X-recognition elements that recruit the dosage compensation complex and nucleate spreading of the complex over short or long distances (Fig. 7) (Csankovszki et al. 2004).

MOLECULAR IDENTIFICATION OF DISCRETE X-RECOGNITION ELEMENTS

A general strategy was devised to identify small, discrete X-recognition elements responsible for recruitment activity (Csankovszki et al. 2004). Transgenic nematode lines carrying extrachromosomal DNA arrays with tandem copies of cosmids from a recruitment region were made, and arrays were assayed for dosage compensation complex binding. The arrays also carried *lac* operator repeats and *lac* repressor tagged with GFP. *lac* repressor binding to its operator sequences would allow the array to be identified by GFP autofluorescence or antibodies. Colocalization of dosage compensation protein antibodies with the array would indicate the presence of an X-recognition element (Fig. 8a). This approach identified the element *rex-1* from a region of X with limited recruitment ability (Fig. 6) (Csankovszki et al. 2004). Using a similar array approach, the recruitment activity was at-

tributed to a single cosmid and then ultimately a 790-base-pair subfragment (Fig. 8b) (Csankovszki et al. 2004).

Multiple copies of a bona fide X-recognition element would be expected to titrate dosage compensation complexes from X and thereby impair dosage compensation. Genetic assays indicated that dosage compensation was compromised in animals carrying arrays with X-recognition elements. Furthermore, X chromosomes could be detected easily by dosage compensation protein antibodies only in XX cells that lost arrays because of mitotic instability (Csankovszki et al. 2004). Adjacent cells harboring arrays showed robust localization of the dosage compensation complex to the array but not the X chromosomes (Fig. 8b). These results indicate that X regions with limited recruitment ability harbor high-affinity recruitment sites capable of competing with other recruitment sites on X for the dosage compensation complex (Csankovszki et al. 2004).

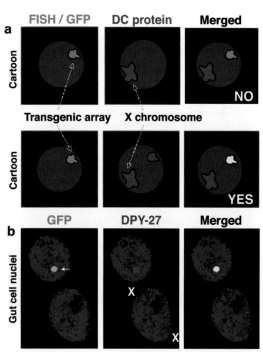

Figure 8. Molecular identification of an X-recognition element. (*a*) Strategy for isolating X-recognition elements that recruit the dosage compensation complex. Transgenic nematodes were created that carried extrachromosomal arrays made of X DNA, *lac* operator sequences, and a gene encoding the *lac* repressor-GFP bifunctional fusion protein. *lac* repressor binds the *lac* operator sequences, allowing arrays to be detected by GFP autofluorescence or antibodies (*green, arrow*). The dosage compensation complex is identified by antibodies to a dosage compensation protein (*red*), and the nuclear DNA by DAPI (*blue*). If X sequences contain an X-recognition element, dosage compensation protein antibodies will colocalize with the array. Within each nucleus, binding of the complex to both Xs serves as a staining control. (*b*) Identification of *rex-1*. This strategy identified an X-recognition element (*rex-1*) from a region of X with limited ability to recruit the complex (see map in Fig. 6). Shown are two gut cells, one with an array carrying a 4.5-kb piece of X DNA and one without. The dosage compensation complex binds to the array and titrates complexes from X. X chromosomes are readily detectable only in the cell that lacks the array.

A total of ten X-recognition elements have thus far been identified by this approach (J. Jans et al., unpubl.). Further dissection of these sites will reveal the features of X-recognition elements essential for recruitment of the dosage compensation complex. Those features could be primary DNA sequence, DNA structure, or a combination of both.

In summary, the strategy by which *C. elegans* X chromosomes attract the condensin-like dosage compensation complex is now known. Discrete X-recognition elements serve as entry sites to recruit the dosage compensation complex and to nucleate spreading of the complex to X regions that lack recruitment sites (Fig. 7). In this manner, a repressed chromatin state can be spread in *cis* over short or long distances, thus establishing the global regulation of X chromosomes that is maintained throughout the lifetime of hermaphrodites (Csankovszki et al. 2004).

CONCLUSIONS

Future endeavors will reveal whether the primary X-recognition protein SDC-2 binds X DNA directly or instead requires other X-bound cellular proteins to reach its target. Such endeavors will also define the mechanistic basis by which the dosage compensation complex spreads along the X chromosome and whether such spreading involves histone modifications or noncoding RNAs, as do other forms of epigenetic regulation across whole chromosomes or chromosomal domains (Allshire; Cam and Grewal; Oh et al.; Huynh and Lee; Martienssen et al.; all this volume). Ultimately, a detailed biochemical understanding of the twofold X-chromosome repression will emerge.

ACKNOWLEDGMENTS

This work was supported by NIH grant R37-GM30702 to B.J.M., NIH NRSA postdoctoral fellowship F32-GM065007 to G.C., and NIH pre-doctoral training grant GM07232 to P.M. B.J.M. is an investigator of the Howard Hughes Medical Institute.

REFERENCES

Akerib C.C. and Meyer B.J. 1994. Identification of X chromosome regions in *C. elegans* that contain sex-determination signal elements. *Genetics* **138:** 1105.
Bone J.R., Lavender J., Richman R., Palmer M.J., Turner B.M., and Kuroda M.I. 1994. Acetylated histone H4 on the male X chromosome is associated with dosage compensation in *Drosophila*. *Genes Dev.* **8:** 96.
Bridges C.B. 1916. Non-disjunction as proof of the chromosome theory of heredity. *Genetics* **1:** 1.
Carmi I. and Meyer B.J. 1999. The primary sex determination signal of *Caenorhabditis elegans*. *Genetics* **152:** 999.
Carmi I., Kopczynski J.B., and Meyer B.J. 1998. The nuclear hormone receptor SEX-1 is an X-chromosome signal that determines nematode sex. *Nature* **396:** 168.
Chan R.C., Severson A.F., and Meyer B.J. 2004. Condensin restructures chromosomes in preparation for meiotic divisions. *J. Cell Biol.* **167:** 613.
Chu D.S., Dawes H.E., Lieb J.D., Chan R.C., Kuo A.F., and Meyer B.J. 2002. A molecular link between gene-specific and chromosome-wide transcriptional repression. *Genes Dev.* **16:** 796.
Chuang P.-T., Albertson D.G., and Meyer B.J. 1994. DPY-27: A chromosome condensation protein homolog that regulates *C. elegans* dosage compensation through association with the X chromosome. *Cell* **79:** 459.
Chuang P.-T., Lieb J.D., and Meyer B.J. 1996. Sex-specific assembly of a dosage compensation complex on the nematode X chromosome. *Science* **274:** 1736.
Csankovszki G., McDonel P., and Meyer B.J. 2004. Recruitment and spreading of the *C. elegans* dosage compensation complex along X chromosomes. *Science* **303:** 1182.
Davis T.L. and Meyer B.J. 1997. SDC-3 coordinates the assembly of a dosage compensation complex on the nematode X chromosome. *Development* **124:** 1019.
Dawes H.E., Berlin D.S., Lapidus D.M., Nusbaum C., Davis T.L., and Meyer B.J. 1999. Dosage compensation proteins targeted to X chromosomes by a determinant of hermaphrodite fate. *Science* **284:** 1800.
DeLong L.D., Plenefisch J.D., Klein R.D., and Meyer B.J. 1993. Feedback control of sex determination by dosage compensation revealed through *Caenorhabditis elegans sdc-3* mutations. *Genetics* **133:** 875.
Gubbay J., Collignon J., Koopman P., Capel B., Economou A., Münsterberg A., Vivian N., Goodfellow P., and Lovell-Badge R. 1990. A gene mapping to the sex-determining region of the mouse *Y* chromosome is a member of a novel family of embryonically expressed genes. *Nature* **346:** 245.
Hagstrom K.A., Holmes V., Cozzarelli N., and Meyer B.J. 2002. *C. elegans* condensin promotes mitotic chromosome architecture, centromere organization, and sister chromatid segregation during mitosis and meiosis. *Genes Dev.* **16:** 729.
Heard E., Rougeulle C., Arnaud D., Avner P., Allis C.D., and Spector D.L. 2001. Methylation of histone H3 at Lys-9 is an early mark on the X chromosome during X inactivation. *Cell* **107:** 727.
Hilfiker A., Hilfiker-Kleiner D., Pannuti A., and Lucchesi J.C. 1997. *mof*, a putative acetyl transferase gene related to the Tip60 and MOZ human genes and to the SAS genes of yeast, is required for dosage compensation in *Drosophila*. *EMBO J.* **16:** 2054.
Hirano T. 2002. The ABCs of SMC proteins: Two-armed ATPases for chromosome condensation, cohesion, and repair. *Genes Dev.* **16:** 399.
Hodgkin J. 1983. *X* chromosome dosage and gene expression in *Caenorhabditis elegans:* Two unusual dumpy genes. *Mol. Gen. Genet.* **192:** 452.
Hsu D.R. and Meyer B.J. 1994. The *dpy-30* gene encodes an essential component of the *Caenorhabditis elegans* dosage compensation machinery. *Genetics* **137:** 999.
Klein R.D. and Meyer B.J. 1993. Independent domains of the *sdc-3* protein control sex determination and dosage compensation in *C. elegans*. *Cell* **72:** 349.
Koopman P., Gubbay J., Vivian N., Goodfellow P., and Lovell-Badge R. 1991. Male development of chromosomally female mice transgenic for *Sry*. *Nature* **351:** 117.
Lieb J.D., Albrecht M.R., Chuang P.-T., and Meyer B.J. 1998. MIX-1: An essential component of the *C. elegans* mitotic machinery executes X chromosome dosage compensation. *Cell* **92:** 265.
Lieb J.D., Capowski E.E., Meneely P., and Meyer B.J. 1996. DPY-26, a link between dosage compensation and meiotic chromosome segregation in the nematode. *Science* **274:** 1732.
Luz J.G., Hassig C.A., Pickle C., Godzik A., Meyer B.J., and Wilson I.A. 2003. XOL-1, primary determinant of sexual fate in *C. elegans* is a GHMP kinase family member and a structural prototype for a new class of developmental regulators. *Genes Dev.* **17:** 977.
Madl J.E. and Herman R.K. 1979. Polyploids and sex determination in *Caenorhabditis elegans*. *Genetics* **93:** 393.
Mermoud J.E., Popova B., Peters A.H., Jenuwein T., and Brockdorff N. 2002. Histone H3 lysine 9 methylation occurs rapidly at the onest of random X chromosome inactivation. *Curr. Biol.* **12:** 247.
Meyer B.J. and Casson L.P. 1986. *Caenorhabditis elegans* compensates for the difference in X chromosome dosage between the sexes by regulating transcript levels. *Cell* **47:** 871.

Meyer J., Walker-Jonah A., and Hollenberg C.P. 1991. Galactokinase encoded by GAL1 is a bifunctional protein required for induction of the GAL genes in *Kluyveromyces lactis* and is also able to suppress gal3 phenotype in *Saccharomyces cerevisiae*. *Mol. Cell. Biol.* **11:** 5454.

Miller L.M., Plenefisch J.D., Casson L.P., and Meyer B.J. 1988. *xol-1*: A gene that controls the male modes of both sex determination and X chromosome dosage compensation in *C. elegans*. *Cell* **55:** 167.

Nagy P.L., Griesenbeck J., Kornberg R.D., and Cleary M.L. 2002. A trithorax-group complex purified from *Saccharomyces cerevisiae* is required for methylation of histone H3. *Proc. Natl. Acad. Sci.* **99:** 90.

Nicoll M., Akerib C.C., and Meyer B.J. 1997. X-chromosome-counting mechanisms that determine nematode sex. *Nature* **388:** 200.

Nigon V. 1951. Polyploidie experimentale chez un Nematode libre, *Rhabditis elegans* Maupas. *Bull. Biol. Fr. Belg.* **85:** 187.

Nusbaum C. and Meyer B.J. 1989. The *Caenorhabditis elegans* gene *sdc-2* controls sex determination and dosage compensation in XX animals. *Genetics* **122:** 579.

Platt A., Ross H.C., Hankin S., and Reece R.J. 2000. The insertion of two amino acids into a transcriptional inducer converts it into a galactokinase. *Proc. Natl. Acad. Sci.* **97:** 3154.

Plenefisch J.D., DeLong L., and Meyer B.J. 1989. Genes that implement the hermaphrodite mode of dosage compensation in *Caenorhabditis elegans*. *Genetics* **121:** 57.

Rhind N.R., Miller L.M., Kopczynski J.B., and Meyer B.J. 1995. *xol-1* acts as an early switch in the *C. elegans* male/hermaphrodite decision. *Cell* **80:** 71.

Sinclair A.H., Berta P., Palmer M.S., Hawkins J.R., Griffiths B.L., Smith M.J., Foster J.W., Frischauf A.-M., Lovell-Badge R., and Goodfellow P.N. 1990. A gene from the human sex-determining region encodes a protein with homology to a conserved DNA-binding motif. *Nature* **346:** 240.

Swedlow J.R. and Hirano T. 2003. The making of the mitotic chromosome: Modern insights into classical questions. *Mol. Cell* **11:** 557.

Villeneuve A.M. and Meyer B.J. 1987. *sdc-1*: A link between sex determination and dosage compensation in *C. elegans*. *Cell* **48:** 25.

Yonker S.A. and Meyer B.J. 2003. Recruitment of *C. elegans* dosage compensation proteins for gene-specific versus chromosome-wide repression. *Development* **130:** 6519.

Targeting Dosage Compensation to the X Chromosome of *Drosophila* Males

H. OH,* X. BAI,*† Y. PARK,* J.R. BONE,‡ AND M.I. KURODA*†
*Howard Hughes Medical Institute, Harvard-Partners Center for Genetics & Genomics, Department of Genetics, Harvard Medical School, Boston, Massachusetts 02115; †Program in Developmental Biology, Baylor College of Medicine, Houston, Texas 77030; ‡Upstate USA, Charlottesville, Virginia 22903

RNA has attracted increasing attention from biologists because of its versatile talents in biological processes. RNA's capabilities are not confined just within traditional models, as messenger RNA or scaffolds of a ribosome, but have been extended to catalysis and regulation of diverse biological functions, supporting a primordial "RNA world hypothesis." One of RNA's fascinating roles is regulation of gene expression as an RNP (ribonucleoprotein) complex. Examples include dosage compensation in *Drosophila* by *roX* RNAs, Xist/Tsix RNAs in mammalian female X-inactivation, and siRNAs/miRNAs in posttranscriptional gene regulation in various organisms. *roX* RNAs are required for correct targeting of dosage compensation complexes to the *Drosophila* X chromosome (Meller and Rattner 2002). We previously proposed that targeting occurs by spreading *in cis* from multiple, discrete initiation sites (Kelley et al. 1999; Park et al. 2002; Oh et al. 2003). Here we emphasize that spreading cannot be the sole mechanism for targeting of dosage compensation and suggest that the MSL (Male-Specific Lethal) complex has evolved multiple ways to recognize genes on the X (Oh et al. 2004).

MSL complexes bind the male X chromosome and induce twofold hypertranscription so that the amount of gene product from one male X chromosome equals that from both female X chromosomes. This requires at least five proteins—MLE (*maleless*); MSL1, MSL2, and MSL3 (*male-specific lethal 1, 2*, and *3*, respectively); and MOF (*males absent on the first*)—and one of two *roX* (*RNA on X*) RNAs. MLE and MOF have enzymatic activities that are essential for dosage compensation: MLE is a DExH RNA helicase (Kuroda et al. 1991; Lee et al. 1997) and MOF is a MYST family histone acetyltransferase (Hilfiker et al. 1997; Akhtar and Becker 2000; Smith et al. 2000). JIL-1, a histone H3 kinase, also associates with the MSL proteins (Jin et al. 2000). The MSL proteins, JIL-1, and *roX* RNAs bind in a precise pattern along the length of the male X chromosome, resulting in enrichment of chromatin modifications associated with hypertranscription, such as histone H4 acetylated at lysine 16 and H3 phosphorylated at serine 10 (Turner et al. 1992; Wang et al. 2001).

The two noncoding RNAs, *roX1* and *roX2*, are functionally redundant (Meller and Rattner 2002) even though they have very little sequence homology and are distinct in size (3.7 kb for *roX*1 RNA vs. 0.5–1.2 kb for *roX*2 RNA) (Amrein and Axel 1997; Smith et al. 2000). Deletion of either *roX* gene has no effect on males. Missing both of them, however, results in male lethality (Meller et al. 1997; Meller and Rattner 2002). The MSL-binding pattern on the X chromosome is drastically disrupted in these *roX1 roX2* double mutant males, suggesting that *roX* RNAs are important for correctly targeting MSL complex to the X.

The targeting of hundreds of sites along the length of the X chromosome has been the subject of analysis and speculation for many years, but the mechanisms are still largely unknown. Initially, the prevailing view was that genes on the X evolved local *cis*-acting sequences for attracting the dosage compensation machinery. This was based on the observation that X-linked genes translocated to autosomes could retain dosage compensation. These studies demonstrated that the fly system is quite different from mammalian dosage compensation, in which a single region of the X (the X-inactivation center) controls the chromosome as a whole. Another key difference is that mammalian X-inactivation involves silencing one of two X chromosomes in females rather than up-regulation of the single X in males. However, several recent parallels suggest that the fly targeting mechanism might have mechanistic similarities to X-inactivation (Fig. 1). In both cases, noncoding RNAs (*roX1* and *roX2* in flies and *Xist* in mammals) are required for targeting the correct chromosome for regulation. Furthermore, in each case there is evidence for spreading of the dosage compensation process long distances along the chromosome from the sites of synthesis of those noncoding RNAs. Flies and mammals differ in that *Xist* is limited to action *in cis*, while *roX* RNAs and the MSL complex can clearly also act *in trans* (Meller et al. 1997).

MSL COMPLEX STRONGLY PREFERS TARGETING THE X CHROMOSOME

Large X to autosome transpositions retain both the ability to dosage compensate and the characteristic diffuse appearance of the male X chromosome (Offermann 1936; Aronson et al. 1954; Dobzhansky 1957). We analyzed several fly lines containing large X-ray-induced X to autosome transpositions obtained from the *Drosophila* stock center. The region of transposed X chromosome appears as wide as the paired autosomes that flank the in-

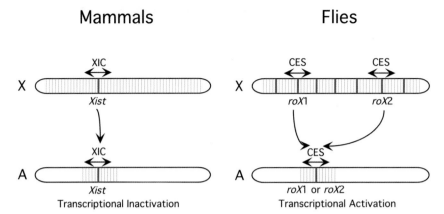

Figure 1. Spreading *in cis* of noncoding RNA from *Xist* and *roX* transgenes. (*Left*) In mammals, *Xist* RNA is transcribed from the XIC (X-inactivation center) and can spread *in cis* to "paint" the inactive X chromosome. An *Xist* transgene inserted on an autosome can cause *Xist* RNA to spread on the autosome. (*Right*) In flies, there are about 35 high-affinity sites (also called chromosome entry sites or CES) on the X chromosome including *roX1* and *roX2*. Like *Xist*, *roX1* and *roX2* transgenes also cause spreading *in cis* of noncoding RNAs on autosomes. (Adapted, with permission, from Park and Kuroda 2001 [©AAAS].)

sertion, suggesting that the transposed section of chromosome adopts a less compact chromatin structure similar to that of the intact male X chromosome. All X to A transposition stocks that we tested showed MSL binding within the transposed fragments (Fig. 2).

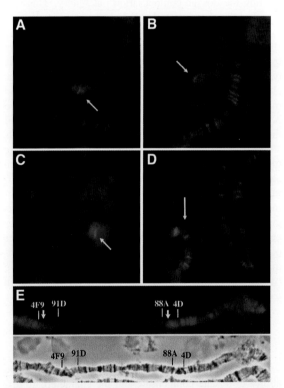

Figure 2. MSL complex strongly prefers binding to segments of the X chromosome, and avoids autosomes. Polytene chromosome squashes from male larvae containing large X to autosome (*A*–*D*) or autosome to X (*E*) transpositions, double stained with DAPI (*blue*) and antibodies specific for the MSL1 protein (*red*). (*A*) Tp(1;3) *sta* (1E1-2A:89B-C), (*B*) Tp(1;3) *JC153* (16E2-17A:99D), (*C*) Tp(1;2) $r^{+}75C$ (14B13-15A9:35D-E), (*D*) Tp(1;2) $rb^{+}71g$ (3F3-5E8:23A15), and (*E*) Tp(3;1) *O5* (88A-92C:4F2-4F6). *Arrows* indicate the site of transpositions. (Adapted, with permission, from Oh et al. 2004 [©Elsevier].)

Consistent with immunostaining of X to autosome transposition flies, an autosome to X transposition stock showed lack of MSL staining of a region of the third chromosome that was transposed to the X. MSL1 protein was detected at sites flanking the breakpoints of the transposition (Fig. 2, vertical arrows), but no staining was observed within the transposed section of the third chromosome. Thus, MSL immunostaining of translocation stocks supports a model featuring local control, suggesting that MSL complexes strongly prefer any piece of the X over the autosomes.

To characterize features of the X chromosome that attract the MSL complex, we isolated cosmids derived from X, inserted them on autosomes, and tested them for the ability to attract MSL complex to their ectopic site of insertion (Oh et al. 2004). In almost all cases, a 30-kb segment of X DNA appears sufficient to attract MSL complex, while smaller segments can be either positive or negative. Thus, it is possible that whatever marks the X might be dispersed on average every 20–30 kb, or that there is a cumulative effect of many very-low-affinity sites present on any 30-kb stretch of X-derived DNA. Ongoing transgenic studies should allow us to distinguish between these two possibilities.

Despite the global nature of regulation of the male X chromosome, not all genes on the X are dosage compensated (for review, see Baker et al. 1994), as is also the case for mammalian X-inactivation (Disteche 1995; Brown et al. 1997; Carrel et al. 1999). Genes that are sex-specific, present on both X and Y, or recently translocated to the X chromosome are not dosage compensated. This gene-by-gene adaptation for dosage compensation is likely to occur slowly, in response to the gradual degeneration of the Y chromosome. What could be the nature of *cis*-acting sequences? Molecular and evolutionary data have suggested that a unique sequence composition of the X chromosome correlates with dosage compensation (Huijser et al. 1987; Pardue et al. 1987; Lowenhaupt et al. 1989; Bachtrog et al. 1999; Gonzalez et al. 2002). Several repeated sequences (microsatellites), such as (CA)n/

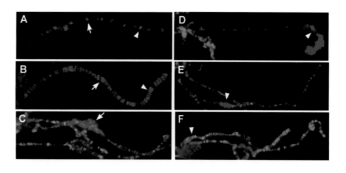

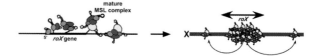

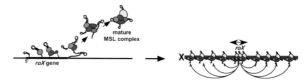

Figure 3. Local *cis* spreading of MSL complexes on the X chromosome. Polytene chromosomes were stained with anti-MSL1 antibodies, and visualized with Texas Red-conjugated secondary antibody (*red*). DNA was stained with DAPI (*blue*). *Arrowheads* and *arrows* indicate the locations of the *roX1* and *roX2* genes, respectively. The genotypes are (*A*) msl3[M1][M2] female, (*B*) wild-type male, (*C*) roX1^{ex6} ; [M1][M2] male, (*D*) Df(1)roX2^{52}; [M1][M2] male, (*E*) roX1^{ex6} Df(1)roX2^{52} [w$^+$ GMroX1-13A] ; [M1][M2] male, and (*F*) roX1^{ex6} Df(1)roX2^{52} [w$^+$ GMroX1-18C] male without extra MSL1 and MSL2. [M1][M2] stands for [w$^+$ H83M1][w$^+$ H83M2]/+ on the third chromosome. (*G,H*) Model for *cis*- vs. *trans*-interaction of MSL complexes with the X chromosome. (Adapted, with permission, from Oh et al. 2003.)

(GT)n, (CT)n/(GA)n, (TA)n, and (GC)n, are more abundant on the X chromosome than on the autosomes. However, no apparent consensus sequence has been found near X-linked genes that retain dosage compensation when moved to autosomes (for review, see Baker et al. 1994). Clearly, a genomic rather than piecemeal approach will be needed to identify features in DNA that attract the MSL complex.

LOCAL SPREADING OF MSL COMPLEX FROM *roX* GENES

In the absence of individual MSL proteins (MSL3, MLE, or MOF), the core subunits MSL1 and MSL2 bind approximately 35 high-affinity sites (Fig. 3A), which are dispersed along the X chromosome (Palmer et al. 1994; Lyman et al. 1997; Gu et al. 1998). Two of these are the *roX* genes themselves, which encode RNA components of the MSL complex (Amrein and Axel 1997; Meller et al. 1997; Kelley et al. 1999). The spreading model for MSL targeting to the X is based on the observation that MSL complexes spread *in cis* from small *roX* transgenes inserted on the autosomes, regardless of neighboring sequences (Kelley et al. 1999). Although the identities of the remaining high-affinity sites were not known, we proposed that they might also behave as spreading initiation sites based on the behavior of the two *roX* genes. In this model, MSL complexes would first recognize high-affinity sites in a sequence-specific manner, followed by spreading *in cis* along the neighboring chromatin to up-regulate flanking genes (Kelley et al. 1999). Regulation would still be relatively local, since there are approximately 35 dispersed high-affinity sites. The evolution of specific *cis*-acting sequences at each dosage compensated gene might not be necessary, as the complex might spread *in cis* from initiation sites to recognize something common to active genes.

Evidence for the *cis* spreading model came initially from studies of autosomal *roX* transgenes. Subsequently, we demonstrated that MSL complexes can spread locally from the endogenous *roX* genes on the X, the natural target of dosage compensation. We observed that wild-type males require a balance of MSL proteins and *roX* RNAs to evenly distribute MSL complexes both locally and at a distance along the X chromosome (Fig. 3B). When we artificially increased the amounts of MSL1 and MSL2, thought to be the limiting proteins (Kelley et al. 1997; Chang and Kuroda 1998; Park et al. 2002), and lowered the number of *roX* genes, MSL complexes concentrated predominantly over a local segment of the X surrounding a *roX* gene (Fig. 3C,D). More remote regions of the X were relatively depleted of MSL complex. This dramatically altered the morphology of polytene X chromosomes in many nuclei. When the *roX* genes were moved to new locations along the *roX*-deficient X chromosome, additional MSL1 and MSL2 produced bright MSL staining and diffuse chromosome morphology surrounding the relocated *roX*$^+$ gene (Fig. 3E). Again, more distant regions of the X had much less MSL staining. Because a strong bias for local MSL spreading was seen at several different locations of *roX* transgenes on the X chromosome, the sequences surrounding the endogenous *roX* genes are unlikely to contain special spreading elements. Moreover, some *roX* transgenes displayed a strong preference for regional spreading from their site of insertion with only wild-type levels of MSL1 and MSL2 (Fig. 3F). This indicates that changes in the MSL protein:*roX* RNA ratio or chromatin environment can produce large shifts in the pattern of local MSL spreading. A model based on these observations proposes that if MSL proteins are abundant and rapidly assemble onto growing *roX* transcripts, functional complexes will be completed before release of the nascent *roX* transcript from the DNA template. These complexes are postulated to immediately bind the flanking chromosome regardless of sequence and begin spreading *in cis* (Fig. 3G). When *roX* genes compete for a finite supply of MSL proteins, nascent *roX* RNA may be released from the template with an incomplete set of MSL subunits. After maturation is completed in solution, these complexes are postulated to diffuse through the nucleus until encountering the X (Fig. 3H).

These results suggest that the *roX* genes are the predominant spreading sites on the X chromosome. However, *roX*-deficient X chromosomes are painted by MSL complex as long as *roX* RNA is supplied from autosomes (Meller et al. 2000), suggesting that the nascent *roX* transcript can function as an assembly site for MSL proteins in any location (Park et al. 2003) and that other high-affinity sites might play a role in targeting MSL complex to the X chromosome.

TARGETING OF MSL COMPLEX THROUGH HIGH-AFFINITY SITES ON X: A COMPARISON OF *roX* GENES AND THE 18D REGION

We initially postulated that the approximately 35 high-affinity sites might contain a common sequence to attract the MSL complex. Therefore, we analyzed the MSL-binding sites within *roX* genes to look for targeting motifs. DNaseI hypersensitivity and transgenic deletion mapping identified an ~200-bp MSL-binding site in each *roX* gene, designated as DHS (DNaseI hypersensitive site) (Kageyama et al. 2001; Park et al. 2003). Sequence alignments revealed short stretches of evolutionarily conserved consensus elements in both sites and mutagenesis data have suggested that they are essential for MSL binding (Park et al. 2003). Unfortunately, this MSL-binding sequence has not led us to identify other high-affinity sites.

Surprisingly, we recently found that these strong MSL-binding sites within *roX* genes are not essential for spreading *in cis* from *roX* transgenes (Bai et al. 2004). Rather, the sites are utilized by MSL proteins to specifically up-regulate transcription of *roX* RNAs in males. How can spreading be initiated from a transgene in the absence of an observable MSL-binding site? Perhaps initial RNA binding can lead to a histone-binding/modification cycle in the absence of a specific DNA interaction, as proposed for HP1 spreading to form heterochromatin (Bannister et al. 2001; Lachner et al. 2001). This would be consistent with previous studies that *roX* RNA itself is the initial assembly target for MSL proteins and that copies of nontranscribed *roX* DNA cannot compete for assembly and spreading of MSL complexes (Park et al. 2002).

The idea that *roX* DHS sequences are dedicated to specific regulation of *roX* RNAs may explain why we were unable to find the MSL-binding consensus sequences at other sites on the X chromosome. Several other lines of evidence suggest that the majority of high-affinity sites are quite different from *roX* genes (Oh et al. 2003, 2004; and see below). First, the *roX* genes are highly dependent on MLE for recruiting MSL complex, unlike the other sites, which still display MSL binding in *mle*⁻ mutants (Meller et al. 2000; Kageyama et al. 2001). Second, the two *roX* RNAs seem to be the only RNA components of the MSL complex, suggested by male-specific SAGE analysis (Fujii and Amrein 2002) and subtractive RNA analysis from immunoprecipitated MSL complex (Oh et al. 2003). Third, the *roX* genes appear to be the two major MSL-spreading initiation sites, where the balance between local spreading *in cis* and diffusion of soluble complex *in trans* to the X may be determined by the rate of assembly of MSL proteins onto nascent *roX* transcripts (Oh et al. 2003).

To understand what role, if any, the approximately 35 high-affinity sites play in MSL targeting, we characterized an additional high-affinity MSL-binding site (18D10). We screened a cosmid library, constructed an overlapping cosmid contig around 18D10, created transgenic lines for each of the cosmids, and tested them for MSL binding at their new sites of insertion. In an *msl3*⁻ genetic background in which the high-affinity sites are most easily monitored, only one out of the three cosmids showed a strong MSL signal, comparable to the endogenous 18D10 region on the X chromosome. Further analysis, including male-specific DHS assays and chromatin immunoprecipitation (ChIP) assays, narrowed it down to a 510-bp fragment, which showed quite different characteristics from the *roX* DHS. First, the 18D entry site does not share any apparent sequence similarity with the *roX* DHS. Second, the 18D region does not seem to be transcribed nor encode a noncoding RNA. Third, the 510-bp fragment from the 18D region must be multimerized, while *roX* DHS (~200-bp) monomer was sufficient to recruit partial MSL complexes. Finally, MSL spreading from an 18D transgene was very rare and limited, consistent with a lack of RNA product for assembly of MSL complexes. These results suggest the possibility that interaction of MSL complexes with other high-affinity sites requires multiple, relatively weak, and dispersed *cis*-elements, instead of a strong interaction within a relatively short (~200-bp) *roX* DHS.

What could be the consequence of recruitment of the MSL complex to high-affinity sites? First, the other sites could function as male-specific enhancers to induce male-specific transcription of neighboring genes, like the *roX* DHS (Bai et al. 2004). Second, if we assume that several X-enriched *cis*-elements are required for MSL complex binding, those sequences may have evolved to cluster around genes which are haplo-insufficient, thus requiring dosage compensation. How many genes are dosage-compensated by MSL complexes is not known currently, so it is possible that the high-affinity sites are located around especially important target genes. Third, the other sites, evenly distributed along the X chromosome, may have evolved to act as physical facilitators of MSL spreading, to make it efficient and balanced. Finally, although spreading from 18D10 transgenes appears rare and modest, even MSL spreading from *roX* genes was rare in the presence of other *roX* genes in the genome, presumably because of the limited availability of protein components (Park et al. 2002; Oh et al. 2003). Therefore, it is still possible that the 18D10 site and other sites act as weak spreading nucleation sites to help paint the X chromosome with MSL complexes.

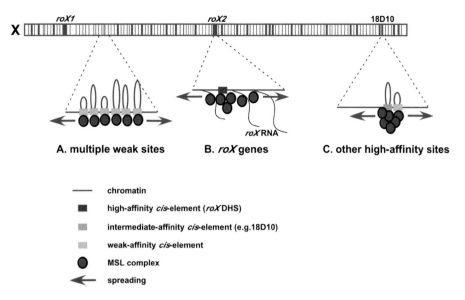

Figure 4. Genes on the X may have evolved multiple ways to attract the MSL complex. (*A*) Segments without a high-affinity site may contain weak-affinity sites that cooperate to recruit the MSL complex, followed by spreading only when complex reaches a threshold concentration. (*B*) Sites of *roX* RNA production recruit enough MSL complex for variable spreading, through a high local concentration of MSL complex. Current evidence suggests that the nascent roX RNA alone can be sufficient to reach this concentration even in the absence of the high-affinity DHS site. (*C*) An intermediate-affinity site requires another intermediate- or weak-affinity site for recruiting and spreading of MSL complex. (Adapted, with permission, from Oh et al. 2004 [©Elsevier].)

A MODEL FOR MSL BINDING AND SPREADING ON THE X CHROMOSOME

Our current data can be interpreted in the following framework (Fig. 4A–C). Perhaps there are diverse DNA recognition elements on the X chromosome that have different affinities for MSL complex: high, intermediate, or weak. High-affinity *cis*-elements, such as within the *roX* genes, would not require additional *cis*-elements for recruiting MSL complexes and might be involved in multifold gene activation instead of twofold hypertranscription. This interaction might be strengthened by *roX* RNA (Fig. 4B). An intermediate-affinity *cis*-element like the 18D10 site might require additional intermediate- and/or weak-affinity elements for robust binding and would have the ability to attract partial MSL complexes with a minimal MSL1/MSL2 composition (Fig. 4C). Weak-affinity *cis*-elements might require interaction with several additional weak-affinity *cis*-elements, which might explain occasional autosomal MSL signals and how X fragments on the autosomes attract wild-type MSL complexes even without a high-affinity site (Fig. 4A).

Regulation of MSL complex binding to target genes by multiple relatively weak *cis*-elements might be the most efficient way to achieve twofold activation, or could reflect the simplest way to evolve MSL activity on the X chromosome. MSL spreading may be mainly dependent on the local concentration of functional complex regardless of the presence of specific sequences.

SUMMARY AND CONCLUSION

A fundamental difference between *Xist* RNA and *roX* RNA is that transcription of *Xist* RNA can lead to inactivation of a whole chromosome, only *in cis*, whereas the presence of a *roX* gene is not sufficient to induce chromosome-wide gene regulation on a chromosome that it normally does not regulate. Instead, MSL complex prefers binding to the *roX*-deficient X chromosome, indicating that there is something else on the X chromosome that attracts MSL complexes. So far, we have identified a high-affinity site for MSL complex in the 18D10 region on the X chromosome showing male-specific DNaseI hypersensitivity, partial MSL complex recruitment in the absence of MSL3, and some MSL spreading from the high-affinity site. However, unlike our expectation that the other sites might share their own conserved sequence, no similar sequence to the 18D site was found in the rest of the genome. It is possible that conserved DNA sequences for MSL binding might be composed of short stretches of a few nucleotides, which makes homology search efforts very difficult. Another possibility is that MSL complex recognizes local chromatin structure, DNA topology, or a specific location within the nucleus, independent of nucleotide sequence. Given the fact that the Y chromosome has gradually degenerated over time and the X has evolved to overcome the deficiency of X-linked gene products in males, it is also possible that dosage-compensated genes evolved independently and ended up having a diversity of DNA sequences for MSL complex binding.

ISSUES FOR FUTURE STUDIES

Molecular and Functional Bases for Redundancy of *roX1* and *roX2*

Even though *roX1* (3.7 kb) and *roX2* (0.5 kb) are very different in size and sequence, they show a redundant function. However, the exact role of these *roX* RNAs in targeting and spreading of MSL complex to the male X chromosome is unknown. Therefore, characterization of the function of these *roX* RNAs will be a key to understanding how this RNP complex regulates chromatin organization.

It was reported that MSL3 and MOF have RNA-binding activity in vitro (Akhtar et al. 2000). In addition, MSL3, MOF, and MLE were shown to be dissociated from the MSL complex with RNase treatment in vivo, suggesting that the *roX* RNAs interact with each of those three MSL proteins to keep the MSL complex structure intact and functional. How MSL proteins and *roX* RNAs are associated with each other remains to be solved. Therefore, it is important to define the key functional domains of *roX* RNAs.

MSL proteins and *roX* RNAs are present in several other *Drosophila* species and are thought to localize to the male X chromosome. To find functional domains of *roX* RNAs, a phylogenetic comparison method could be employed to find conserved sequences and/or structure of *roX* RNAs in several different *Drosophila* species. Secondary or tertiary structures should be important for the function of noncoding *roX* RNAs, and therefore it is likely that the structure of essential domains in *roX* RNAs is conserved among different fly species, despite the substantial divergence in primary sequences of noncoding RNA through evolution. In this respect, it is much easier to map the functional domains within *roX2* RNA instead of *roX1*, since *roX2* (0.5 kb) is smaller than *roX1* (3.7 kb) RNA and nevertheless they show functional redundancy. Once the conserved secondary structures are defined, mutation, deletion, or domain-swapping experiments could be performed to draw a functional map between *roX* RNA and MSL proteins, which will help us understand the mechanism of gene regulation by the RNP complex.

A Delicate Twofold Up-Regulation of Gene Expression in Male X-linked Genes

Proper gene expression in the process of development is often achieved by dynamic reversible modification of histones to change chromatin structure at each gene locus. The *Drosophila* male X chromosome is acetylated at H4K16 by MOF protein. It is not known how the acetylated H4K16 is regulated to achieve just twofold activation of X-linked genes. Compared to X-inactivation, dealing with twofold up-regulation (in male flies) or down-regulation (in hemaphrodite *Caenorhabditis elegans*) requires much finer control of a dosage compensation complex. One possible way to keep the level of regulation twofold could be achieved by weak interactions between *cis*-acting DNA sequences and the *trans*-acting protein complex. Even though no conserved *cis*-element has been found around dosage compensated genes either in *Drosophila* or *C. elegans*, the *roX* genes in flies and the *her-1* gene in worms interact with dosage compensation complexes to activate (*roX*) or repress (*her-1*), not just twofold but in a sex-specific way. In each case, those high-affinity sequences were not found in other regions in either fly or *C. elegans*, indicating that the sequences for twofold regulation might be quite different. It is possible that some X-enriched repetitive nucleotide sequence might be responsible for weak interactions with the dosage compensation complex in both systems.

Another way of achieving just twofold regulation could be limiting the amount of dosage compensation complex in the nucleus. In metamales (1X3A), transcription of X-linked genes is increased and more diffuse X morphology is observed compared to the normal male (1X2A) (Lucchesi et al. 1977). This observation suggests that *trans*-acting factors encoded by autosomes (e.g., MSL2) may be normally limiting in diploids, and responsible for additional hyperactivation of the X chromosome when overexpressed. In *C. elegans*, the SDC complex recruitment to the X chromosome is reduced in the presence of multicopy arrays of putative SDC-binding fragments (Csankovszki et al. 2004). If dosage compensation complexes are limiting, how is the amount of complex controlled? It is possible that the stability or activity of protein components can be regulated by reversible protein modifications such as ubiquitination, phosphorylation, and acetylation. It was recently shown that MSL3 and MSL1 can be acetylated by MOF, which is a H4 acetyltransferase in the MSL complex (Buscaino et al. 2003; Morales et al. 2004). MSL1 also has several PEST domains (Chang and Kuroda 1998), which may be a target for phosphorylation, often followed by ubiquitination and degradation. MSL2 has a Ring-finger domain, which might act in the ubiquitination pathway as a E3 ligase to degrade interacting proteins or to modulate transcriptional activity of the interacting complex. Even though Jil-1 was initially found as an H3 kinase, it is possible that Jil-1 might phosphorylate other MSL proteins to modulate activity of MSL complex to achieve the proper level of gene expression.

Since MSL1 and MSL2 are less conserved among different species and function as two core components in the MSL complex, it is possible that those two proteins are the most ancient members of the MSL complex that have managed to achieve a primitive but inefficient mechanism of dosage compensation. As more genes degenerated on the Y chromosome, perhaps a more efficient mechanism of dosage compensation was adapted by adding general histone modifying enzymes to a core MSL1/MSL2 complex. Perhaps *roX* RNAs came even more recently, like a virus infection, and landed on the X chromosome by accident, giving the local region an advantage with regard to dosage compensation by a spreading mechanism.

cis-acting Elements on the X Chromosome

It seems that multiple classes of MSL-binding sites target dosage compensation to the X chromosome in fly. Many questions remain in the field of dosage compensation with regard to the identities of *cis*-acting elements on the X chromosome. Now that the fly genome sequence is available, more systematic approaches can be employed to find those *cis*-acting sequences on the X. Since it has been shown that there are X-enriched microsatellites such as (CA)n/(GT)n, (CT)n/(GA)n, (TA)n, and (GC)n (Huijser et al. 1987; Pardue et al. 1987; Lowenhaupt et al. 1989; Bachtrog et al. 1999; Gonzalez et al. 2002), the link between those repeated nucleotides and dosage compensation can be characterized by combining bioinformatics with molecular analyses such as in vivo MSL binding or chromatin immunoprecipitation assays. A microarray-based "ChIP on chip" analysis should be especially informative, although the availability of the proper microarray chips, which include intergenic regions, would be a key factor. Identification of a large number of small, specific *cis*-acting elements should greatly increase our understanding of how the MSL complex is recruited to achieve global twofold regulation of the male X chromosome.

ACKNOWLEDGMENTS

This work has been supported by the Welch Foundation (Q-1359), the National Institutes of Health (GM45744), and the Howard Hughes Medical Institute. M.I.K. is an HHMI Investigator.

REFERENCES

Akhtar A. and Becker P.B. 2000. Activation of transcription through histone H4 acetylation by MOF, an acetyltransferase essential for dosage compensation in *Drosophila. Mol. Cell* **5**: 367.

Akhtar A., Zink D., and Becker P.B. 2000. Chromodomains are protein-RNA interaction modules. *Nature* **407**: 405.

Amrein H. and Axel R. 1997. Genes expressed in neurons of adult male *Drosophila. Cell* **88**: 459.

Aronson J.F., Rudkin G.T., and Schultz J. 1954. A comparison of giant X-chromosomes in male and female *Drosophila melanogaster* by cytophotometry in the ultraviolet. *J. Histochem. Cytochem.* **2**: 458.

Bachtrog D., Weiss S., Zangerl B., Brem G., and Schlotterer C. 1999. Distribution of dinucleotide microsatellites in the *Drosophila melanogaster* genome. *Mol. Biol. Evol.* **16**: 602.

Bai X., Alekseyenko A.A., and Kuroda M.I. 2004. Sequence-specific targeting of MSL complex regulates transcription of the roX RNA genes. *EMBO J.* **23**: 2853.

Baker B.S., Gorman M., and Marin I. 1994. Dosage compensation in *Drosophila. Annu. Rev. Genet.* **28**: 491.

Bannister A.J., Zegerman P., Partridge J.F., Miska E.A., Thomas J.O., Allshire R.C., and Kouzarides T. 2001. Selective recognition of methylated lysine 9 on histone H3 by the HP1 chromo domain. *Nature* **410**: 120.

Brown C.J., Carrel L., and Willard H.F. 1997. Expression of genes from the human active and inactive X chromosome. *Am. J. Hum. Genet.* **60**: 1333.

Buscaino A., Kocher T., Kind J.H., Holz H., Taipale M., Wagner K., Wilm M., and Akhtar A. 2003. MOF-regulated acetylation of MSL-3 in the *Drosophila* dosage compensation complex. *Mol. Cell* **11**: 1265.

Carrel L., Cottle A.A., Goglin K.C., and Willard H.F. 1999. A first-generation X-inactivation profile of the human X chromosome. *Proc. Natl. Acad. Sci.* **96**: 14440.

Chang K.A. and Kuroda M.I. 1998. Modulation of MSL1 abundance in female *Drosophila* contributes to the sex specificity of dosage compensation. *Genetics* **150**: 699.

Csankovszki G., MacDonel P., and Meyer B.J. 2004. Recruitment and spreading of the *C. elegans* dosage compensation complex along X chromosomes. *Science* **303**: 1182.

Disteche C.M. 1995. Escape from X inactivation in human and mouse. *Trends Genet.* **11**: 17.

Dobzhansky T. 1957. The X-chromosome in the larval salivary glands of hybrids *Drosophila insularis* and *Drosophila tropicalis. Chromosoma* **8**: 691.

Fujii S. and Amrein H. 2002. Genes expressed in the *Drosophila* head reveal a role for fat cells in sex-specific physiology. *EMBO J.* **21**: 5353.

Gonzalez J., Ranz J.M., and Ruiz A. 2002. Chromosomal elements evolve at different rates in the *Drosophila* genome. *Genetics* **161**: 1137.

Gu W., Szauter P., and Lucchesi J.C. 1998. Targeting of MOF, a putative histone acetyl transferase, to the X chromosome of *Drosophila melanogaster. Dev. Genet.* **22**: 56.

Hilfiker A., Hilfiker-Kleiner D., Pannuti A., and Lucchesi J.C. 1997. mof, a putative acetyl transferase gene related to the Tip60 and MOZ human genes and to the SAS genes of yeast, is required for dosage compensation in *Drosophila. EMBO J.* **16**: 2054.

Huijser P., Hennig W., and Dijkhof R. 1987. Poly(dC-dA/dG-dT) repeats in the *Drosophila* genome: A key function for dosage compensation and position effects? *Chromosoma* **95**: 209.

Jin Y., Wang Y., Johansen J., and Johansen K.M. 2000. JIL-1, a chromosomal kinase implicated in regulation of chromatin structure, associates with the Male Specific Lethal (MSL) dosage compensation complex. *J. Cell Biol.* **149**: 1005.

Kageyama Y., Mengus G., Gilfillan G., Kennedy H.G., Stuckenholz C., Kelley R.L., Becker P.B., and Kuroda M.I. 2001. Association and spreading of the *Drosophila* dosage compensation complex from a discrete *roX1* chromatin entry site. *EMBO J.* **20**: 2236.

Kelley R.L., Wang J., Bell L., and Kuroda M.I. 1997. Sex lethal controls dosage compensation in *Drosophila* by a nonsplicing mechanism. *Nature* **387**: 195.

Kelley R.L., Meller V.H., Gordadze P.R., Roman G., Davis R.L., and Kuroda M.I. 1999. Epigenetic spreading of the *Drosophila* dosage compensation complex from roX RNA genes into flanking chromatin. *Cell* **98**: 513.

Kuroda M.I., Kernan M.J., Kreber R., Ganetzky B., and Baker B.S. 1991. The *maleless* protein associates with the X chromosome to regulate dosage compensation in *Drosophila. Cell* **66**: 935.

Lachner M., O'Carroll D., Rea S., Mechtler K., and Jenuwein T. 2001. Methylation of histone H3 lysine 9 creates a binding site for HP1 proteins. *Nature* **410**: 116.

Lee C.-G., Chang K.A., Kuroda M.I., and Hurwitz J. 1997. The NTPase/helicase activities of *Drosophila* Maleless, an essential factor in dosage compensation. *EMBO J.* **16**: 2671.

Lowenhaupt K., Rich A., and Pardue M.L. 1989. Nonrandom distribution on long mono- and dinucleotide repeats in *Drosophila* chromosomes: Correlation with dosage compensation, heterochromatin, and recombination. *Mol. Cell. Biol.* **9**: 1173.

Lucchesi J.C., Belote J.M., and Maroni G. 1977. X-linked gene activity in metamales (XY;3A) of *Drosophila. Chromosoma* **65**: 1.

Lyman L.M., Copps K., Rastelli L., Kelley R.L., and Kuroda M.I. 1997. *Drosophila* male-specific lethal-2 protein: Structure/function analysis and dependence on MSL-1 for chromosome association. *Genetics* **147**: 1743.

Meller V.H. and Rattner B.P. 2002. The roX RNAs encode redundant *male-specific lethal* transcripts required for targeting of the MSL complex. *EMBO J.* **21**: 1084.

Meller V.H., Wu K.H., Roman G., Kuroda M.I., and Davis R.L. 1997. *roX1* RNA paints the X chromosome of male *Drosophila* and is regulated by the dosage compensation system. *Cell* **88**: 445.

Meller V.H., Gordadze P.R., Park Y., Chu X., Stuckenholz C., Kelley R.L., and Kuroda M.I. 2000. Ordered assembly of *roX* RNAs into MSL complexes on the dosage compensated X chromosome in *Drosophila. Curr. Biol.* **10**: 136.

Morales V., Straub T., Neumann M.F., Mengus G., Akhtar A., and Becker P.B. 2004. Functional integration of the histone acetyltransferase MOF into the dosage compensation complex. *EMBO J.* **23**: 2258.

Offermann C.A. 1936. Branched chromosomes as symmetrical duplications. *J. Genet.* **32:** 103.

Oh H., Bone J.R., and Kuroda M.I. 2004. Multiple classes of MSL binding sites target dosage compensation to the X chromosome of *Drosophila*. *Curr. Biol.* **14:** 481.

Oh H., Park Y., and Kuroda M.I. 2003. Local spreading of MSL complexes from *roX* genes on the *Drosophila* X chromosome. *Genes Dev.* **17:** 1334.

Palmer M.J., Richman R., Richter L., and Kuroda M.I. 1994. Sex-specific regulation of the *male-specific lethal-1* dosage compensation gene in *Drosophila*. *Genes Dev.* **8:** 698.

Pardue M.L., Lowenhaupt K., Rich A., and Nordheim A. 1987. (dC-dA)$_n$-(dG-dT)$_n$ sequences have evolutionarily conserved chromosomal location in *Drosophila* with implications for roles in chromosome structure and function. *EMBO J.* **6:** 1781.

Park Y. and Kuroda M.I. 2001. Epigenetic aspects of X-chromosome dosage compensation. *Science* **293:** 1083.

Park Y., Kelley R.L., Oh H., Kuroda M.I., and Meller V.H. 2002. Extent of chromatin spreading determined by *roX* RNA recruitment of MSL proteins. *Science* **298:** 1620.

Park Y., Mengus G., Bai X., Kageyama Y., Meller V.H., Becker P.B., and Kuroda M.I. 2003. Sequence-specific targeting of *Drosophila roX* genes by the MSL dosage compensation complex. *Mol. Cell* **11:** 977.

Smith E.R., Pannuti A., Gu W., Steurnagel A., Cook R.G., Allis C.D., and Lucchesi J.C. 2000. The *Drosophila* MSL complex acetylates histone H4 at lysine 16, a chromatin modification linked to dosage compensation. *Mol. Cell. Biol.* **20:** 312.

Turner B.M., Birley A.J., and Lavender J. 1992. Histone H4 isoforms acetylated at specific lysine residues define individual chromosomes and chromatin domains in *Drosophila* polytene nuclei. *Cell* **69:** 375.

Wang Y., Zhang W., Jin Y., Johansen J., and Johansen K.M. 2001. The JIL-1 tandem kinase mediates histone H3 phosphorylation and is required for maintenance of chromatin structure in *Drosophila*. *Cell* **105:** 433.

Mammalian X-Chromosome Inactivation: An Epigenetics Paradigm

E. HEARD,* J. CHAUMEIL, O. MASUI, AND I. OKAMOTO

CNRS UMR218, Curie Institute, 75248 Paris Cedex 05, France

In mammals, dosage compensation for X-linked gene products between XX females and XY males is achieved through the transcriptional silencing of one of the two X chromosomes in the developing female embryo (Lyon 1961). This process of X-chromosome inactivation affects more than 1000 genes and is one of the most striking examples of long-range, monoallelic gene silencing in mammals. The choice of which X chromosome is inactivated is usually random in placental mammals, and the inactive state is then stably inherited, giving rise to adults that are mosaics for two cell types expressing one or the other X chromosome. However, in marsupials and during early embryogenesis of some placental mammals such as rodents, only the paternal X chromosome is inactivated because of a parent-specific imprint. In mice, this imprinted form of X-inactivation is subsequently reversed in cells of the embryo proper and random inactivation of either the paternally or maternally derived X chromosome occurs. Random X-inactivation also occurs when female embryonic stem (ES) cells are differentiated in vitro, thus providing a useful model system for mechanistic studies. Mechanistically, X-inactivation is usually described as being a multistep process, consisting of initiation from a unique region on the X chromosome, the X-inactivation center (Xic), followed by spread of inactivation over the X chromosome, and then maintenance of the inactive state through subsequent cell divisions. Initiation normally can occur only during a restricted developmental time window and is thought to involve the sensing of how many X chromosomes are present, the choice of X chromosome to inactivate, and the triggering of the silencing process itself. Spreading involves a *cis*-acting RNA molecule (Xist), but the factors that this RNA interacts with to mediate silencing and the regions of the X chromosome that it targets remain poorly understood. Maintenance appears to be independent of Xic function and to rely on a combination of epigenetic marks. Although the distinction of these three steps in X-inactivation facilitates our description of this process, they are in fact inextricably linked, as will become apparent below.

In this paper, we discuss a number of questions raised by X-inactivation that may also be relevant to other epigenetic processes. How does the inactivation of an entire X chromosome take place while its homolog, present in the same nucleus, remains genetically active? What are the epigenetic mark(s) that impose nonrandom X-inactivation in some species? What constitutes the cellular memory, whereby the inactive state is stably maintained, not only throughout the cell cycle, including disruptive S phase and during mitosis, but for hundreds of cell divisions during the lifetime of an organism? What are the epigenetic marks that underlie this cellular memory and how are they erased at the appropriate developmental times? Finally, could some aspects of the differential treatment of two genetically identical chromosomes be explained through specific nuclear localization?

Initiation: Defining the X-Inactivation Center

The key locus underlying the initial differential treatment of two X chromosomes is the Xic, which is involved in determining how many X chromosomes are present in a cell (counting) and which X chromosome will be inactivated (choice). Cells must register the presence of at least two Xics for X-inactivation to occur (Rastan 1983; Rastan and Robertson 1985) and only a single X chromosome will remain active in a diploid cell, all supernumerary X chromosomes being inactivated. The Xic is also the site from which inactivation spreads across the X chromosome *in cis*, or even across an autosome, in X-autosome translocations. Cytogenetic studies on such chromosomal rearrangements and deletions enabled the Xic region to be narrowed down over the years to ~1 Mbp of sequence on the human X (Fig. 1) (see Heard and Avner 1994). The corresponding syntenic region on the mouse X is approximately half this size (Chureau et al. 2002). Within this candidate region, the discovery of a remarkable gene, *Xist* (X-inactive-specific-transcript), in 1990 provided the first molecular insight into how inactivation is initiated (Brown et al. 1991). *Xist* is expressed exclusively from the inactive X chromosome, producing a 17,000-nucleotide-long, untranslated RNA that coats the X chromosome from which it is produced *in cis* (Brown et al. 1992; Clemson et al. 1998). Xist RNA is essential for the initiation and spread of X-inactivation *in cis* (Penny et al. 1996; Marahrens et al. 1997; Wutz and

---

*Corresponding author: Mammalian Developmental Epigenetics Group, CNRS UMR218, Curie Institute, 26 rue d'Ulm, 75248 Paris Cedex 05, France. E-mail: Edith.Heard@curie.fr.

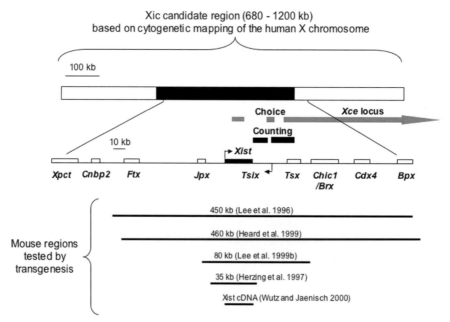

Figure 1. The candidate region for the X-inactivation center. Based on the candidate region defined by cytogenetic mapping of the human XIC region (*above*), the syntenic mouse region is shown (*below*). Elements defined to be involved in counting and choice by targeted deletion analysis or genetic mapping (Xce locus) are shown (for review, see Clerc and Avner 2003). Below the map of the murine Xic region, the extents of the mouse transgenes tested for Xic function are shown.

Jaenisch 2000), yet it is not involved in the counting function of the Xic (Penny et al. 1996). A complex combination of antisense transcription to *Xist* (in the form of Tsix) and *cis*-regulatory sequences located in the region 3´ to *Xist* appear to be involved in the choice and counting functions of the Xic (Fig. 1) (for reviews, see Boumil and Lee 2001; Clerc and Avner 2003). Several studies involving transgenesis, a stringent test for assessing the sequences that are not only necessary, but sufficient, for the function of a locus, have been performed for the Xic with the aim of defining the minimal region sufficient for Xic function at an ectopic (autosomal) site (Fig. 1) (Heard et al. 1996, 1999; Lee et al. 1996, 1999b; Herzing et al. 1997). In transgenic ES cells and mice, such transgenes are able to act as ectopic Xics (when they included the *Xist* gene); however, they can only do so when present as multicopy arrays. Indeed, single-copy *Xist* transgenes, up to 460 kb long (covering 130 kb of sequence 5´ and 300 kb of sequence 3´ to *Xist*) are unable to induce Xist RNA coating *in cis* or even to trigger inactivation of the endogenous X chromosome (counting) (Heard et al. 1999). Critical elements for autonomous Xic function must therefore still be lacking from these large single-copy transgenes, and multicopy arrays must compensate for this lack in some way. Indeed, comparison with the cytogenetic definition of the Xic region in humans (Fig. 1) suggests that missing sequences are likely to lie in the 5´ region of *Xist*.

These transgenes may be lacking sufficient numbers of repeat elements, or "way stations," as first proposed by Riggs (Riggs et al. 1985), necessary for the nucleation and propagation of the Xist RNA-mediated signal along the chromosome. Mary Lyon has proposed L1 repeat elements (LINEs) as possible way stations (Lyon 1998), given their enrichment on the X chromosome relative to autosomes. Strikingly, LINE density in the region covered by the *Xist* transgenes (Fig. 1) is not particularly high (Chureau et al. 2002), which would correlate well with a need for multiple copies of this region to compensate for this relative lack in LINEs. The overexpression of *Xist* can, however, compensate for whatever sequences might be lacking at ectopic sites, to produce *cis*-inactivation (Wutz and Jaenisch 2000). Thus, an alternative, but not mutually exclusive, possibility is that long-range elements that enhance *Xist* expression, or buffer it from position effects, are lacking in the single-copy genomic *Xist* transgenes. Consistent with this, recent studies have shown that unique characteristics define a region spanning hundreds of kilobases upstream of *Xist*. These include a 100-kb stretch of chromatin that is histone H4 hyperacetylated in XX but not XY cells (O'Neill et al. 1999), a >200-kb domain that is constitutively enriched for H3 Lys-9 dimethylation (Heard et al. 2001; Rougeulle et al. 2004), and a series of noncoding RNAs that are produced from this region (Chureau et al. 2002). The absence of at least part of the *Xist* upstream region in the genomic transgenes tested so far, especially the large H3 Lys-9 methylation domain, could explain the lack of single-copy Xic function (Heard et al. 2001; Rougeulle et al. 2004). Given that the region involved in counting, 3´ to Xist (Clerc and Avner 1998), lies intact within the large genomic transgenes tested so far (Heard et al. 1999), the complete absence of X-inactivation in these single-copy transgene lines could be related to an incorrect positioning of these transgenes in the nucleus, which would not permit their detection as a supplementary Xic and the initiation of inactivation of the endogenous X chromosome (see final section).

Spreading the Inactive State via Xist RNA

The earliest known event in the X-inactivation process, both in embryos and in differentiating ES cells (Fig. 2 and Fig. 4, below), consists of coating of the X chromosome by Xist RNA, which is rapidly followed by gene silencing along the length of the chromosome (Kay et al. 1993; Panning et al. 1997; Sheardown et al. 1997; Wutz and Jaenisch 2000). Xist RNA–induced transcriptional shutdown can occur only during an early developmental time window (Wutz and Jaenisch 2000). However, how this is achieved and the identity of the key protein or nucleic acid partners of Xist that are involved remain to be elucidated. In order to gain insight into Xist's functions, we set out to characterize early chromatin changes, such as covalent modifications of histones, on the X chromosome. Histone modifications, including acetylation, methylation, phosphorylation, and ubiquitination, represent an important means of regulating gene expression (Jenuwein and Allis 2001; Zhang and Reinberg 2001) and may even form a "histone code," through the binding of chromatin-associated proteins (Strahl and Allis 2000; Turner 2000). The inactive X chromosome is characterized by a number of histone modifications (see Fig. 3) (Jeppeson and Turner 1993; Boggs et al. 1996, 2002; Keohane et al. 1996), some of which are common to other regions of heterochromatin. We have shown that during female ES cell differentiation (Fig. 2), specific patterns of histone modifications, including H3 Lys-9 dimethylation and H3 Lys-4 hypomethylation, appear on the X chromosome soon after Xist RNA coating (Heard et al. 2001; Chaumeil et al. 2002). More recently, H3 Lys-27 trimethylation (Plath

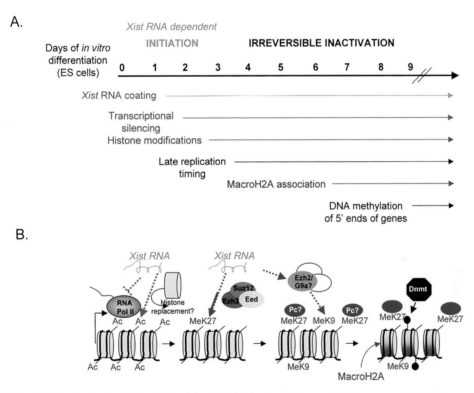

Figure 2. Kinetics of X-inactivation in differentiating ES cells. (*A*) The time of onset of different events during the onset of X-inactivation is shown, deduced thanks to numerous studies (Keohane et al. 1996; Penny et al. 1996; Panning et al. 1997; Sheardown et al. 1997; Wutz and Jaenisch 2000; Heard et al. 2001; Chaumeil et al. 2002; Plath et al. 2003; Silva et al. 2003; Kohlmaier et al. 2004). (*B*) Proposed mechanisms underlying X-inactivation. Xist RNA may first act by directly interfering with transcription initiation and/or elongation on the X chromosome (J. Chaumeil and E. Heard, unpubl.; for details, see text). In order to achieve the loss of euchromatic histone modifications such as H3 Lys-4 methylation (and given that no histone demethylases have so far been identified), Xist RNA may promote histone replacement or histone proteolysis. Finally, Xist RNA can recruit (either directly or via intermediate proteins) PRC2 complex polycomb group proteins (Ezh2/EnxI and Eed), which enable methylation of H3 Lys-27 (Plath et al. 2003; Silva et al. 2003; Kohlmaier et al. 2004); methylation of H3 Lys-9 may be mediated by Ezh2 or another histone methyltransferase such as G9a (Heard et al. 2001; Rougeulle et al. 2004). Although no PRC1 complex proteins have so far been found associated with the Xi, methylated H3 Lys-27 is likely to serve as a signal for the recruitment of the PRC1 complex, perhaps via a homolog of the Polycomb (Pc) protein itself (Fischle et al. 2003; Min et al. 2003), thus facilitating the maintenance of transcriptional silence. The histone variant, macroH2A, is Xist RNA–dependent, although it is not clear whether this is a direct recruitment or dependent on other changes induced by Xist (hence no arrow linking the two is shown). The mechanism by which DNA methyltransferases are recruited to the promoters of X-linked gene on the Xi remains unclear, but may be via the histone modifications or polycomb group proteins already bound. *Dashed arrows:* hypothetical links between Xist and various potential factors such as polycomb group proteins.

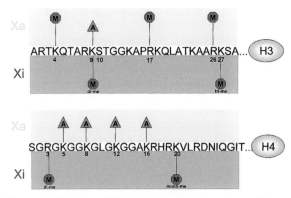

Figure 3. Histone modification patterns on the active and inactive X chromosomes. Histone modifications characterizing the inactive X chromosome and their equivalent state on the active X chromosome are shown (based on Jeppeson and Turner 1993; Boggs et al. 1996, 2002; Keohane et al. 1996; Heard et al. 2001; Chaumeil et al. 2002; Chadwick and Willard 2003; Plath et al. 2003; Silva et al. 2003; Kohlmeier et al. 2004).

et al. 2003; Silva et al. 2003) and H4 Lys-20 monomethylation (Kohlmaier et al. 2004) have been shown to appear within a similar time window. The early time of appearance of these histone changes and their presence on the Xi in somatic cells suggested that they could be involved in the transcriptional silencing process, or in its maintenance, or both. Furthermore, their appearance immediately following Xist RNA coating suggested that the Xist transcript might actually target histone-modifying enzymes to the X chromosome. Indeed, members of the Polycomb group PRC2 complex, Eed and Ezh2/Enx1, the latter being a histone methyltransferase capable of trimethylating H3 Lys-27 (Kuzmichev et al. 2002), were found to be recruited to the inactive X chromosome in what appears to be a Xist RNA–dependent fashion (Plath et al. 2003; Kohlmaier et al. 2004). The dimethylation of H3 Lys-9 on the Xi may not be Xist RNA–dependent and may be mediated by a different HMTase and/or complex, as its kinetics of appearance (Okamoto et al. 2004) and chromosomal distribution (Rougeulle et al. 2004) are different to H3 trimethyl Lys-27 (Fig. 2B).

Although Xist's role was considered to be restricted to transcriptional silencing up until recently, the recruitment of the PRC2 complex and H3 Lys-27 methylation by Xist does not seem to be involved in silencing (Kohlmaier et al. 2004), but rather to the early maintenance of the inactive state (Plath et al. 2003; Silva et al. 2003). A maintenance role for PRC2 complex is also supported by genetic evidence in Eed mutant embryos (Wang et al. 2001; Silva et al. 2003). Xist RNA may therefore have parallel roles, one in transcriptional silencing and the other in recruiting marks involved in early maintenance. Consistent with this, recent data from our laboratory, using immunofluorescence and green fluorescent protein (GFP)-tagged protein analyses, has shown that the exclusion of RNA polymerase II and associated transcription factors from the X chromosome occurs concurrently with Xist RNA coating and before histone modifications such as H3 Lys-27 and Lys-9 methylation (Fig. 2) (J. Chaumeil and E. Heard, unpubl.). Clearly, identification of the protein complexes involved in the various roles of Xist RNA, as well as the chromosomal regions that are targeted by Xist and its partners, represent critical questions for the future.

MAINTENANCE: EPIGENETIC MARKS AND CELLULAR MEMORY

Once silencing has been established on the X chromosome, it has to be maintained during the cell cycle and throughout the multiple cell divisions that occur during the lifetime of an organism. The inactive state of genes on the Xi is highly stable in somatic cells, with sporadic reactivation frequencies similar to mutation rates in wild-type cells ($<10^{-8}$). Furthermore, experimental reactivation of the entire chromosome in somatic cells has never been observed. What are the epigenetic marks responsible for this remarkable stability? Current thinking is that there are several layers of epigenetic modifications that could be involved in locking in the inactive state in a synergistic fashion (Csankovszki et al. 2001). Histone modifications, some of which may be recruited via Xist RNA as described above, represent one set of candidates. Such modifications are thought to participate in transcriptional repression on the one hand (Strahl and Allis 2000) and could be mitotically heritable chromatin marks on the other (Turner 2000). In order to represent true epigenetic marks, these modifications must also be heritable during DNA replication when nucleosomes are distributed randomly between the daughter strands, as well as being resistant to histone replacement processes that can occur at other times during the cell cycle (Henikoff et al. 2004; see Nakatani et al., this volume). Although many of these histone modifications appear to be detectable on the X chromosome at all times by immunofluorescence and by chromatin immunoprecipitation (ChIP), clearly this type of analysis cannot provide the resolution or dynamic vision necessary to address the issue of heritability. The purification of histones specifically associated with the Xi during different phases of the cell cycle, as well as the analysis of histone exchange rates associated with the Xi in living cells, should provide insights into this.

The polycomb group proteins are excellent candidates for participating in the cellular memory of inactivity, based on genetic studies in *Drosophila* showing their role in heritability of silent states (for recent review, see Cao and Zhang 2004). In the case of mammalian X-inactivation, members of the PRC2 complex and H3 Lys-27 methylation are recruited to the X, potentially via Xist, during the early stages of X-inactivation, as described above (Fig. 2B). However, recruitment to the the inactive X of PRC1 complex members and, in particular, of the Pc protein, which in *Drosophila* can bind H3 trimethyl K27, has not been reported so far, although it seems highly likely that at least some mouse Pc homologs (Tajul-Arifin et al. 2003) will indeed associate with the Xi. Our understanding of the role of polycomb group proteins in maintaining the inactive state of the X chromosome is still very rudimentary. Furthermore, significant differences have been found in the degree to which the maintenance of the Xi is dependent on PRC2 between extraembryonic and embryonic lineages (Wang et al. 2001). Biochemical and functional studies will provide important insights into this, although the situation is likely to be complicated, given the dynamic composition of these complexes and their functional redundancy in mammalian cells.

Aside from the histone modifications and polycomb group proteins, a number of other features that distinguish the inactive X chromosome from its active counterpart could be involved in maintenance of the inactive state. One such characteristic is asynchronous replication timing of the inactive X chromosome. Replication of the Xi at the very beginning or the very end of S phase may provide a temporal segregation of the two X chromosomes and minimize the exposure of the Xi to transcription factors, thus facilitating the maintenance of transcriptional silence. Indeed, asynchronous replication timing is one of the most conserved features of the inactive X chromosome across mammalian taxa (for review, see Heard et al. 1997). Its appearance correlates well with the transition period from the Xist RNA–dependent to Xist RNA–independent, irreversible silencing (Fig. 2A). Late replication of the Xi appears to occur after the changes in histone modifications described so far (Chaumeil et al. 2002), suggesting that the altered chromatin structure of the Xi may be necessary for replication to shift, although this has yet to be formally tested. The inactive X is also enriched in a histone H2A variant, macro-H2A.1 (Costanzi and Pehrson 1998). In vitro studies have shown that macro-H2A.1 can interfere with transcription factor binding and SWI/SNF nucleosome remodeling (Angelov et al. 2003), although its exact role in X-inactivation is unclear. Its relatively late appearance on the Xi during inactivation (Fig. 2A) suggests that it is a maintenance, as opposed to an initiation, mark, although intriguingly it seems to be dependent on Xist RNA coating (Csankovszki et al. 1999). Another apparently late modification of the inactive X chromosome is DNA (cytosine) methylation of CpG islands at the 5´ ends of Xi-linked genes. The recruitment of methyl binding proteins involved in silencing and the heritability of the methyl mark during DNA replication (thanks to DNA methyltransferases that can recognize hemimethylated DNA) makes DNA methylation probably the most "dependable" of epigenetic marks known to date. Analysis of $Dnmt1^{-/-}$ mutant embryos has shown that methylation is indeed required for stable maintenance of gene silencing on the Xi in the embryonic lineage (Sado et al. 2000). In extraembryonic lineages on the other hand, the 5´ ends of X-linked genes do not appear to be hypermethylated. Although this correlates with higher rates of sporadic reactivation of X-linked genes (Kratzer et al. 1983), it seems that the maintenance of the inactive state, at least in some extraembryonic tissues such as the visceral endoderm, can tolerate extensive demethylation in vivo (Sado et al. 2000). This could be consistent with an important role for other epigenetic marks, such as the polycomb group proteins, in extraembryonic lineages (Wang et al. 2001). In marsupials, on the other hand, CpG islands of genes are not methylated on the inactive X chromosome in any of the lineages (Kaslow and Migeon 1987). Perhaps as a consequence, higher frequencies of sporadic reactivation of X-linked genes are found in marsupials compared to placental mammals, although the rate of reactivation varies considerably between genes (for review, see Heard et al. 1997). Thus the relative roles of different epigenetic marks in maintaining the inactive state of the X chromosome seem to vary considerably across species, and even between cell lineages and different genes within the same species.

In order to assess the relative importance of different epigenetic marks in maintaining the inactive state of the X chromosome, studies aimed at preventing or reversing them have been performed. Removal of any one epigenetic mark (e.g., histone H3 Lys-27 methylation [using Eed mutant mice; Silva et al. 2003], hypoacetylation [following treatment with histone deacetylase inhibitors such as TSA], or DNA methylation [following 5-aza-cytosine treatment or in DNA methyltransferase $Dnmt1$ mutant cells]) can lead to a slightly higher sporadic reactivation rate of X-linked genes (Csankovszki et al. 2001). On the other hand, the combined absence of Xist RNA coating (which also leads to depletion of macroH2A.1 on the Xi), DNA methylation, and histone hypoacetylation in somatic cells does lead to a significantly increased rate of sporadic gene reactivation on the inactive X, implying that it is the multiplicity of epigenetic marks that ensures stability (Csankovszki et al. 2001). Nevertheless, even under these conditions, global reactivation of the X chromosome is never seen, although if it did occur it would be likely to induce rapid cell death owing to overexpression of X-linked genes. In fact, X-chromosome-wide reactivation normally takes place only during development (see below), in the inner cell mass of the blastocyst (4.5–5.5 dpc), and later in the primordial germ cells of the germ line (~12.5 dpc) (Fig. 4).

DEVELOPMENTAL DYNAMICS AND DIVERGENCE OF IMPRINTED AND RANDOM X-INACTIVATION

Evolutionary Considerations

In marsupials (Cooper et al. 1971; Sharman 1971) and in the extraembryonic tissues of some placental mammals, such as rodents (Takagi and Sasaki 1975; West et al. 1977) and cattle (Xue et al. 2002), X-inactivation is nonrandom, with exclusive inactivation of the paternal X chromosome (Xp). Based on phylogenetic analysis it has been suggested that imprinted X-inactivation may represent the ancestral state of this process, arising >130 million years ago in metatheria (Richardson et al. 1971). In this scenario, random X-inactivation would have evolved subsequently, only in placental mammals. Indeed, there are clear selective advantages for random X-inactivation, as the cellular mosaicism engendered allows survival of females even in the presence of deleterious X-linked gene mutations. In rodents, it has been proposed that the imprinted form of X-inactivation that occurs in the early embryo could represent a vestige of the ancestral form. In other placental mammals such as humans, however, this form appears to have been lost completely. Although the

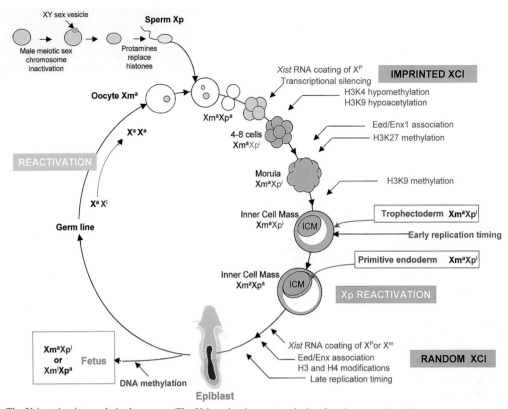

Figure 4. The X-inactivation cycle in the mouse. The X-inactivation events during female mouse development are summarized. Most evidence suggests that the Xp is initially active (X^a) in the early female embryo, like its maternal counterpart, although it has been proposed to arrive in a preinactivated state because of its passage through the male germ line (see text). Inactivation of the Xp chromosome initiates following the accumulation of Xist RNA from the four-cell stage onward (Huynh and Lee 2003; Okamoto et al. 2004). Xp-inactivation is maintained in the trophectoderm (3.5 dpc) and primitive endoderm (4.5 dpc), both of which go on to form extraembryonic tissues. Asynchronous replication timing of the Xi accompanies this transition. The Xp is reactivated in cells of the inner cell mass (ICM) of the blastocyst (Mak et al. 2004; Okamoto et al. 2004). Random X-inactivation of the paternal or maternal X chromosome then takes place at around 5.5 dpc, in cells derived from the ICM that go on to form the embryo proper, presumably following the kinetics shown based on studies in differentiating ES cells, which are derived from the ICM (see Fig. 2). The Xi is reactivated in the female germ line just prior to the onset of meiosis.

sequence of evolutionary steps described above is very appealing in its simplicity, an alternative hypothesis has to be considered, which is that imprinting of the X-inactivation process may have been imposed more than once during evolution, under selective pressures specific to different mammals. Indeed, despite the apparent similarities in imprinted X-inactivation between marsupials and the extraembryonic tissues of rodents, these mammals show fundamental differences in their early development and extraembryonic tissue formation (Selwood 2001), in their sexual differentiation and determination strategies (Watson et al. 2000), and possibly even in their regulation of X-inactivation: in fact, no homolog of the *Xist* gene has been identified in marsupials to date (Graves and Westerman 2002). The selection forces acting on X-linked gene expression during the development of different mammals are thus likely to be diverse and imprinted X-inactivation may have evolved independently in marsupials and eutherians (Olhsson et al. 2001). The strategies used to achieve imprinted X-inactivation may therefore be very different between the two taxa.

Whatever the reasons for the evolution of imprinted X-inactivation, an obvious question is why it always appears to be the paternal X that is inactivated. Although an imprinted mark on the maternal X chromosome preventing it from inactivation clearly exists, whether there is an imprint on the paternal X chromosome, predisposing it to be inactivated, is less clear (Jamieson et al. 1997). One hypothesis posits that preferential paternal X-inactivation could have evolved by exploiting a form of inactivation observed in male germ cells of many species (male meiotic sex chromosome inactivation or MMSI), which involves the transient inactivation of the Xp and Y chromosomes (Lifschytz and Lindsley 1972). The MMSI process presumably prevents deleterious, illegitimate recombination events between the unpaired regions of the sex chromosomes and other chromosomes during meiosis (Jablonka and Lamb 1988). If this meiotically induced inactivity of the Xp was somehow transmitted to the female zygote, it may have been selected for because of the dosage compensation it could provide in early XX embryos (for reviews, see Jamieson et al. 1997; Huynh and

Lee 2003). However, as discussed below, there is as yet little evidence to support this in rodents, and the question remains open in marsupials. Ultimately, our comprehension of the possible causes of imprinted X-inactivation, and of the evolutionary similarities or differences between marsupial and rodent imprinted X-inactivation, can come only from a knowledge of the epigenetic marks and the early events underlying this process in different species.

The Nature of the Imprint(s) Underlying Imprinted X-Inactivation in Mice

Conceptually, imprinted Xp-inactivation could be achieved in two ways: Either the paternal X chromosome is predisposed to inactivate or the maternal X chromosome (Xm) is marked to remain active (Lyon and Rastan 1984). The failure to develop extraembryonic tissues and the early death of mouse embryos carrying two Xm chromosomes suggests that there is initially a powerful maternal mark that prevents the Xm from being inactivated during early embryogenesis (Goto and Takagi 1998, 1999). On the other hand, mice with an XpO genotype are fully viable and normal (showing only a slight growth retardation in early development) (Burgoyne and Biggers 1976), which demonstrates that the paternal X is not irrevocably destined to inactivate during early development. Direct evidence that there is a maternal mark, which prevents the Xm from being inactivated, has come from an elegant study on embryos that were derived by combining maternal genomes from a fully grown (fg) oocyte and from an early nongrowing (ng) oocyte (Tada et al. 2000). In such embryos, the X chromosome derived from the ng oocyte was found to be inactivated in extraembryonic lineages, consistent with acquisition of an Xm mark on the fg X chromosome during oocyte maturation. However, what this mark is and where it lies have remained elusive. Analysis of mice carrying germ-line mutations of genes that encode factors potentially responsible for the epigenetic mark (such as DNA and histone methyltransferases) will be crucial for our understanding of its nature. In terms of its location, the imprinted mark is likely to act at the level of *Xist*, as the maternally inherited *Xist* gene is silent in the zygote and during early cleavage stages (Kay et al. 1994) and cannot be expressed, even in the face of strong selective pressure, such as when a paternally inherited *Xist* deletion prevents the Xp from undergoing inactivation (Marahrens et al. 1997). DNA methylation of *Xist*'s 5´ end has been found in oocytes (Ariel et al. 1995; Zuccotti and Monk 1995), but this methylation is mosaic (Goto and Monk 1998; McDonald et al. 1998). The antisense Tsix transcript has also been proposed as a candidate for the maternal imprint, given its repressive effect on Xist RNA accumulation in differentiating ES cells and blastocysts (Lee et al. 1999a; Lee 2000; Stavropoulos et al. 2001). However, *Tsix* is not normally expressed in cleavage-stage embryos at the time when the maternally inherited *Xist* allele is repressed (Debrand et al. 1999; I. Okamoto and E. Heard, unpubl.). Furthermore, no differential DNA methylation of the *Tsix* 5´ region has so far been identified during early development (Prissette et al. 2001). Deletions of the promoter region of *Tsix* do, nevertheless, lead to aberrant maternal *Xist* expression and embryonic lethality, presumably because of Xm-inactivation (Lee 2000; Sado et al. 2001). This could mean that the *Tsix* promoter region contains DNA elements responsible for the maternal silencing of *Xist*. However, *Tsix* mutants have so far been examined only at embryonic stages (blastocyst and beyond) where the maternal repressive mark would normally no longer be present (Lee 2000; Sado et al. 2001).

Although an imprinted mark on the maternal X chromosome preventing it from inactivation clearly exists, whether there is an imprint on the paternal X chromosome, predisposing it to be inactivated, is less clear. Recent work from our laboratory, as well as that of J. Lee (Huynh and Lee 2003; Okamoto et al. 2004), has shown that the Xp chromosome is, in fact, inactive in cleavage-stage embryos. Huynh and Lee have interpreted this finding as meaning that the Xp actually arrives in the zygote in a "preinactivated" state thanks to its passage through the male germ line, where MMSI takes place (Huynh and Lee 2003). Although, in an evolutionary sense, this interpretation may be very satisfying, several lines of evidence challenge it, at least in the mouse. First, using three different single-cell assays for transcriptional activity (the presence of RNA Polymerase II on the Xp, RNA fluorescence in situ hybridization [FISH] to detect nascent transcripts of X-linked genes, and Cot-1 repeat-specific RNA FISH), we have found that the Xp is, in fact, active in the two-cell embryo and becomes inactivated only from the four-cell stage onward (Fig. 4) following Xist RNA coating (Okamoto et al. 2004; I. Okamoto and E. Heard, in prep.). A second line of evidence is that the MMSI-induced inactivity of the Xp appears to be reversed prior to fertilization, as several X-linked genes are found to be reexpressed in postmeiotic spermatids (Hendriksen et al. 1995). Third, unlike imprinted X-inactivation, which is dependent on *Xist*, MMSI is *Xist*-independent (McCarrey et al. 2002; Turner et al. 2002) and occurs through a very different process (Fernandez-Capetillo et al. 2003). Finally, we have recently found that paternally transmitted autosomal *Xist* transgenes, which do not show any signs of MMSI in the male germ line, are still able to induce imprinted inactivation *in cis* in the early mouse embryo (I. Okamoto et al., in prep.). Based on these observations, we conclude that imprinted X-inactivation in rodents does not appear to depend on Xp-preinactivation as a result of MMSI in the male germ line. Rather, it is allele-specific *Xist* expression that appears to be the critical determinant for imprinted Xp-inactivation. The region 5´ to *Xist* becomes hypomethylated during spermatogenesis (Norris et al. 1994) and, following fertilization, the paternally inherited *Xist* allele is expressed, like other genes, at the time of zygotic gene activation (Fig. 5). In the face of the imprinted silence of the maternal *Xist* allele, paternal Xist RNA accumulation leads to the rapid silencing of the Xp. Indeed, consistent with *Xist* being the trigger for imprinted Xp-inactivation, silencing of X-linked genes appears to occur

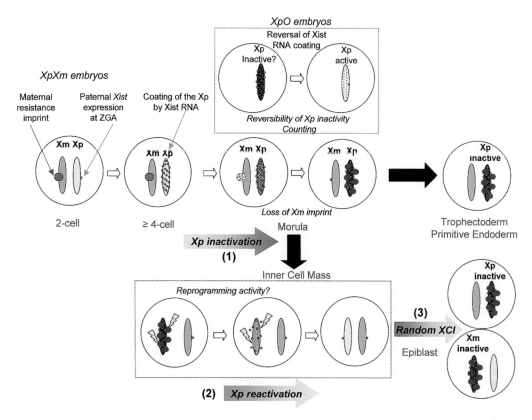

Figure 5. Proposed sequence of events underlying imprinted X-inactivation and X-reactivation in mice. (*1*) Initiation of Xp-inactivation: The maternal X chromosome (*pink*) initially carries a silent *Xist* gene in the zygote because of an unknown repressive imprint (red "stop" sign). The maternal imprint repressing *Xist* expression is lost from about the late morula to early blastocyst stage, thus the Xm chromosome is permissive to X-inactivation from this stage onward (see text). The *Xist* gene on the paternal X chromosome (*blue*) is switched on at zygotic gene activation (ZGA), presumably because its promoter becomes hypomethylated during spermatogenesis (Norris et al. 1994). As soon as paternal Xist RNA accumulates (*green spots*) and coats *in cis* (four-cell stage onward), it triggers transcriptional silencing (*darkened* chromosome). The polycomb group complex proteins Eed and Enx1 are only recruited to the Xp from the 16-cell stage or later (*purple dots*). By the late morula/early blastocyst stage, the Xp is inactive in all cells. Prior to this, there is presumably a window of time when, thanks to some form of counting, the initiation of X-inactivation can be halted or reversed if only a single Xp chromosome (XpO embryos, shown *boxed*) is present or on one of the two Xp chromosomes in androgenotes (not shown). During blastocyst growth, the Xp remains inactive and may be further locked-in by asynchronous replication timing in the trophectoderm and primitive endoderm lineages. (*2*) In the inner cell mass of the blastocyst the Xp is initially inactive but over the next 24 hr or so becomes reactivated by an unknown mechanism (*yellow flashes*) involving loss of Xist RNA coating and expression, loss of polycomb group protein association, and gradual reversal of the histone modifications (potentially via Tsix expression). (*3*) Initiation of random inactivation of the Xp or Xm: Both maternal and paternal alleles of *Xist* are expressed at low levels, presumably very transiently, prior to the onset of random X-inactivation in the epiblast and embryo proper (see Figs. 2 and 4 for kinetics).

as a gradient from the *Xist* locus during early development (Latham and Rambhatla 1995; Huynh and Lee 2003). In marsupials, on the other hand, where no *Xist* gene has been identified, the evolution of imprinted X-inactivation may well be due to a carryover effect from the male germ line.

It will be important to determine what happens to the Xp chromosome in a female embryo where the maternal *Xist* allele is not imprinted to be silent. Would inactivation during early development be random in this case? This would, of course, imply that counting can occur in preimplantation embryos and there is some evidence for this from XpXp androgenotes (Okamoto et al. 2000). Alternatively, in the absence of the imprint on the Xm, would the Xp still inactivate preferentially? Just after fertilization, the protamines with which the paternal genome is packaged are replaced by histones. This dramatic remodeling may provide an opening in which the paternal genome as a whole (Aoki et al. 1997), and perhaps the paternal *Xist* gene in particular, are transiently more highly transcribed than their maternal counterparts, thus providing a bias toward Xist RNA transcription and accumulation on the Xp compared to the Xm. In this case, the paternal X chromosome's predisposition to inactivate would be due to its hyperactivity rather than inactivity.

Kinetics of Imprinted Inactivation and Reactivation of the Paternal X Chromosome

Up until recently, based on classic cytogenetic and biochemical studies, imprinted paternal X-inactivation in mice was believed to initiate in the first lineages to differentiate the trophectoderm (3.5 dpc) and primitive endoderm (4.5 dpc) of the blastocyst. Prior to this, the two X chromosomes were thought to be active, although there was some evidence that the paternal X chromosome displayed a lower genetic activity compared to its maternal homolog (Singer-Sam et al. 1992; Latham and Rambhatla 1995). However, because of the techniques employed, which involved protein detection (allozyme analysis, detection of LacZ and GFP reporter genes) (West et al. 1977; Tam et al. 1994a; Hadjantonakis et al. 2001) or RNA detection by reverse transcriptase polymerase chain reaction (RT-PCR) (Singer-Sam et al. 1992; Latham and Rambhatla 1995; Lebon et al. 1995), interpretation of the results could be confounded by issues such as the unknown half-lives of the X-linked transcripts and proteins examined and the fact that only pools of cells or embryos, rather than single cells, could be examined. This prompted us to develop techniques that would allow Xp silencing to be examined in single cells at the transcriptional and chromatin levels during female preimplantation development. Xp chromosome activity and the timing of imprinted X-inactivation were investigated through the simultaneous detection of Xist RNA localization and nascent transcript detection of X-linked genes, as well as the association/depletion of RNA polymerase II (Okamoto et al. 2004; I. Okamoto and E. Heard, unpubl.). Further indicators of the inactive state came from the analysis of the PRC2 complex proteins Eed and Enx1, as well as characteristic histone modifications such as the trimethylation of Lys-27 (Plath et al. 2003) and the dimethylation of Lys-9 (Heard et al. 2001). Previous studies had shown that *Xist* is expressed from the two- to four-cell stage onward with exclusive expression from the paternally inherited allele only (Kay et al. 1993). Our RNA FISH and immuno-FISH analysis showed that the first signs of transcriptional inactivity of the Xp (exclusion of RNA PolII and down-regulation of the X-linked *Chic1* gene) can be detected at the four-cell stage, just after Xist RNA accumulation (Okamoto et al. 2004). The PRC2 complex and associated H3 Lys-27 methylation appear on the Xp from around the 16-cell stage (Okamoto et al. 2004), as does macroH2A (Costanzi et al. 2000), although the exact time of onset of these marks is quite variable from embyo to embryo. This variability in the time of onset of early maintenance marks could explain the ease with which the onset of Xp-inactivation can be reversed under special circumstances, such as in XpO embryos or XpXp androgenotes (see legend to Fig. 5). By the early blastocyst stage, however, the Xp is inactive in virtually all cells of normal female (XmXp) embryos. In the trophectoderm, this inactivity of the Xp is maintained, and presumably further locked in by the shift to asynchronous (early) replication timing, but what happens in the inner cell mass (ICM), which is the precursor to the embryo proper where random X-inactivation takes place? Our laboratory and that of Neil Brockdorff demonstrated that the Xp is, in fact, inactive in the ICM of early blastocysts (3.5 dpc), but during ICM growth the inactive state of the Xp is actually reversed, with cells rapidly losing their Xist RNA coating, Eed/Enx1 enrichment, and the histone modifications characteristic of X-inactivation (Fig. 4) (Mak et al. 2004; Okamoto et al. 2004). This reactivation permits the subsequent random inactivation of either the maternal or paternal X chromosome in epiblast cells following implantation. Here, the cascade of events and epigenetic marks that follow are presumably similar to those found during ES cell differentiation. The activity of the Xp chromosome is thus remarkably dynamic during preimplantation development. Furthermore, these results highlight the importance of the ICM in reprogramming epigenetic marks during early embryonic development. In a previous study (Eggan et al. 2000), it was shown that in cloned mouse embryos, the inactive X chromosome derived from a somatic cell carried over its inactive state to the extraembryonic tissues, but was subject to random X-inactivation in the embryo proper, presumably through the more recently uncovered reprogramming events in the ICM (Mak et al. 2004; Okamoto et al. 2004). Thus, ICM reprogramming of the X chromosome could have important general implications for cloning, where correct genome-wide erasure of epigenetic marks is critical for the establishment of a normal pattern of development. It will be interesting to investigate whether increasing the time span of the ICM (e.g., in diapause embryos) correlates positively with the efficiency of reprogramming and normal development in cloned embryos.

Reprogramming the Inactive X Chromosome

During female mouse embryogenesis, there are thus at least two rounds of X-inactivation followed by reactivation (see Fig. 5). First paternal X-inactivation occurs, followed by its reactivation in the inner cell mass; next random X-inactivation occurs in the epiblast, but a few days later, in the female germ line, the inactive X chromosome is reactivated. Epigenetic marks, such as histone modifications and polycomb group proteins, that ensure the cellular memory of the inactive state early on in development can be rapidly removed in the ICM. Loss of Xist RNA coating and Xist down-regulation appear to be accompanied by dissociation of the polycomb proteins from the Xp in this process and precede the loss of H3 Lys-27 and Lys-9 methylation (Mak et al. 2004; Okamoto et al. 2004) and the reactivation of X-linked genes (I. Okamoto and E. Heard, unpubl.). How Xist RNA becomes down-regulated on the Xp chromosome thus appears to be the critical question for understanding Xp-reactivation in the ICM. Given that *Tsix* is expressed at around this time and the repressive effect it has on *Xist* expression, this may be when *Tsix* actually plays a critical role in vivo.

DNA methylation of CpG islands of X-linked genes only occurs after the second wave of (random) X-inactivation. This mark is therefore present on the Xi, both in the soma and in the female germ line. How epigenetic marks, including DNA methylation, are reversed on the Xi in the germ line remains unknown. Both specific processes and common mechanisms that erase imprints and reprogram other loci are likely to be involved. Kinetic studies suggest that Xist RNA delocalization and macroH2A dissociation occur at 11.5 dpc during Xi-reactivation (Nesterova et al. 2002), when migrating germ cells enter the genital ridge (Tam et al. 1994). The exact relative timing of events requires further investigation.

Insights into the mechanisms and factors involved in the reactivation of the X chromosome should come from studies involving tissue-culture derivatives of the ICM and germ cells, namely ES and EG (embryonic germ) cells, respectively. Global reactivation of the Xi has been reported following fusion of female somatic cells with ES, embryonal carcinoma (EC), or EG cells (Yoshida et al. 1997; Tada et al. 2001). In one study, a minimum of five mitoses was required to reactivate the Xi at the level of replication timing, following fusion between a female thymocyte and an EC cell (Takagi 1988). This is consistent with the requirement of several rounds of DNA replication for the reversal of stable epigenetic marks such as DNA methylation (in the absence of any known DNA demethylases) and contrasts with the more rapid reversal of Xp-inactivity seen in the ICM, where no DNA methylation or asynchronous replication is present. Whether common reprogramming mechanisms are involved in reactivating the inactive X chromosome in the ICM and the germ line and the nature of the factors that are involved are exciting questions for future investigation.

X-INACTIVATION IN THE CONTEXT OF NUCLEAR ORGANIZATION

It is increasingly recognized that the localization of genes within specific nuclear compartments can be an important means of regulating gene expression (Cockell and Gasser 1999; Spector 2003). Subnuclear domains enriched in factors specialized in particular functions may provide a nuclear context, beyond that provided by flanking sequences, that influences chromatin architecture and gene expression. One of the most intriguing features of the X-inactivation process is the differential treatment of two identical chromosomes within the same nucleoplasm. How does one X chromosome become impervious to the same transcriptional machinery driving genetic activity of the second X chromosome? Spatial segregation of the two X chromosomes could be imagined to play a role both in the establishment and in the maintenance of X-inactivation. At the level of initiation, nuclear compartmentalization mediated by binding of one X chromosome to a single entity per nucleus was one of the earliest models proposed to explain the differential treatment of the two X chromosomes during X-inactivation (Comings 1968). In this context, one hypothesis for "counting" could be that the Xic on the X chromosome that will remain active is sequestered in a particular nuclear region, preventing it from exposure to factors involved in triggering X-inactivation. Transient colocalization or cross talk between Xics, as reported for imprinted loci (LaSalle and Lalande 1996), might also help the cell to count or "sense" the number of X chromosomes present (Fig. 6A). This hypothesis is particularly attractive in the light of the incapacity of single-copy transgenes to function as ectopic Xics and, more specifically, their inability to induce counting. An absence of long-range elements involved in directing the correct nuclear position of the Xic could result in these transgenes not being correctly "sensed" as supernumerary Xics because of their aberrant localization in the nucleus or with respect to the other Xic (Fig. 6A). The actual region of the Xic that has to be sensed could either be the counting region 3′ to *Xist* or else the *Xist* gene itself, as some kind of cross talk between *Xist* alleles has previously been proposed to be involved in the choice of X chromosome to inactivate during initiation (Marahrens et al. 1998).

The epigenetic modifications underlying X-chromosome-wide silencing and its maintenance might also extend beyond the level of chromatin structure and include changes in chromosome positioning within the nucleus and/or local changes in the nuclear compartment surrounding the X chromosome (Fig. 6B). Indeed, our recent data, showing that the exclusion of transcriptional machinery from the X chromosome is the earliest event induced by Xist RNA coating, prior to changes in histone modifications (J. Chaumeil and E. Heard, in prep.), could be consistent with this. Xist RNA coating may create a nuclear subcompartment depleted in members of the gene expression machinery. Transcriptional silencing of the X chromosome may thus initially be induced by exclusion of the transcription apparatus rather than by recruitment of repressors, although the two mechanisms are clearly not mutually exclusive.

To date, few studies have examined the role of nuclear compartmentalization in X-inactivation largely because of technical limitations. The recent advent of new technology such as the 3C technique (Dekker et al. 2002; Tolhuis et al. 2002) and of fluorescently tagged markers should open up the way to in vivo visualization and analysis of the spatial organization of the Xic and the Xist transcript and should enable the dynamic subnuclear localization of these key players in the X-inactivation process to be unraveled.

CONCLUSIONS

We have attempted to cover various aspects of X-chromosome inactivation, putting emphasis on those features of the process that may be relevant to other epigenetic phenomena and raising some of the numerous issues that we feel have not yet been fully addressed to date. The nature and extent of the Xic region that ensures the correct X-inactivation remains an outstanding question, the answer to which, we feel, may lie in the as yet unexplored dimension of nuclear organization. The nature of Xist RNA's silencing role is also a tantalizing issue, which will hopefully be resolved in the near future thanks to elegant studies dissecting function of different regions of this transcript (Wutz et al. 2002) and may also implicate nuclear segregation in addition to chromatin changes. The chromatin-modifying complexes involved in setting up the stability and mitotic heritability of the inactive X chromosome will no doubt be rapidly deciphered thanks

A. INITIATION: Regulation of counting and choice by nuclear localization

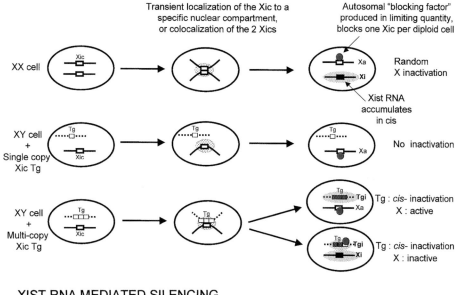

B. XIST RNA MEDIATED SILENCING

Xist RNA excludes transcription complexes, or sequesters the inactive X to a "silencing" compartment

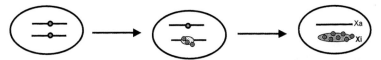

Xist RNA recruits silencing complexes and creates a "silent" compartment

Figure 6. Possible roles for nuclear compartmentalization during X-inactivation. (*A*) Nuclear localization during the initiation phase of X-inactivation: Models for counting usually evoke the action of an autosomally encoded blocking factor, produced in limited quantity, that binds a single Xic (per diploid autosome set) and prevents it from initiating X-inactivation. Transient localization of all of the Xics in a cell (XX cell shown, *top* row) within a compartment enriched in this limiting blocking factor (*red spots*), or even transient physical interaction between the Xics, may be required to count or sense the number of Xics present and ensure that only one will remain active per diploid cell. The active X has no Xist RNA accumulation and its Xic is blocked (*red circle*) within the counting region 3′ to *Xist* (see Fig. 1). All other Xics are free to accumulate Xist RNA and trigger inactivation *in cis*. In XY cells with a single-copy Xic transgene, however, the elements required to position the transgene in the correct nuclear location, or to interact with the Xic on the X chromosome, may be lacking and the transgene is not sensed as a second Xic and cannot trigger the counting process. Multiple copies of transgenes in a tandem array (but not at independent locations; see Heard et al. 1999) can override this lack of single-copy function and compensate for the missing elements. The latter are likely to lie in the 5′ region of Xist (see text for details). (*B*) Nuclear segregation during the propagation and maintenance phases of X-inactivation: Our recent data (J. Chaumeil and E. Heard, unpubl.) suggest that Xist RNA may trigger transcriptional silencing by excluding or impeding transcription factors (*top*). Alternatively, Xist RNA could act by recruiting silencing factors (*bottom*). These are not mutually exclusive possibilities (see text for details). Spatial and temporal segregation of the Xi relative to the active X chromosome and other active regions of the genome could then also participate in the highly efficient maintenance of the inactive state, once it has been established, throughout the cell cycle.

to the work of biochemists. The evolutionary origins of imprinted X-inactivation may be less easy to solve, given the complexity of the selective pressures that are likely to act on X-inactivation during early development, although clearly studies investigating the early events and players in X-inactivation in species other than mouse will be important to address this issue. Finally, the reprogramming events that the inactive X chromosome is subject to during development should provide us with important insights into the factors underlying the nature of stem cells and the mechanisms involved in cloning during the reprogramming of the genome.

ACKNOWLEDGMENTS

The authors wish to apologize for any omissions in references that may have been made and would like to thank Vincent Colot for critical input on the manuscript, as well as Géneviève Almouzni for her helpful comments. The work by our group described in this review was funded by the CNRS (ATIPE), the Human Frontier Science Program, the French Ministry of Research (ACI), and the Curie Institute (Program Incitatif et Collaboratif).

REFERENCES

Angelov D., Molla A., Perche P.Y., Hans F., Cote J., Khochbin S., Bouvet P., and Dimitrov S. 2003. The histone variant macroH2A interferes with transcription factor binding and SWI/SNF nucleosome remodeling. *Mol. Cell* **11:** 1033.

Aoki F., Worrard D.M., and Schultz R. 1997. Regulation of transcriptional activity during the first and second cell cycles in the preimplantation mouse embryos. *Dev. Biol.* **181:** 296.

Ariel M., Robinson E., McCarrey J.R., and Cedar H. 1995. Gamete-specific methylation correlates with imprinting of the murine Xist gene. *Nat. Genet.* **9:** 312.

Boggs B.A., Connors B., Sobel R.E., Chinault A.C., and Allis C.D. 1996. Reduced levels of histone H3 acetylation on the inactive X chromosome in human females. *Chromosoma* **105:** 303.

Boggs B.A., Cheung P., Heard E., Spector D.L., Chinault C.A., and Allis C.D. 2002. Differentially methylated forms of histone H3 show unique association patterns with inactive human X chromosomes. *Nat. Genet.* **30:** 73.

Boumil R.M. and Lee J. 2001. Forty years of decoding the silence in X-chromosome inactivation. *Hum. Mol. Genet.* **10:** 2225.

Brown C.J., Ballabio A., Rupert J.L., Lafreniere R.G., Grompe M., Tonlorenzi R., and Willard H.F. 1991. A gene from the region of the human X inactivation centre is expressed exclusively from the inactive X chromosome. *Nature* **349:** 38.

Brown C.J., Hendrich B.D., Rupert J.L., Lafreniere R.G., Xing Y., Lawrence J., and Willard H.F. 1992. The human XIST gene: Analysis of a 17 kb inactive X-specific RNA that contains conserved repeats and is highly localized within the nucleus. *Cell* **71:** 527.

Burgoyne P.S. and Biggers J.D. 1976. The consequences of X-dosage deficiency in the germ line: Impaired development *in vitro* of preimplantation embryos from X0 mice. *Dev. Biol.* **51:** 109.

Cao R. and Zhang Y. 2004. The functions of E(Z)/EZH2-mediated methylation of lysine 27 in histone H3. *Curr. Opin. Genet. Dev.* **14:** 155.

Chadwick B. and Willard H.F. 2003. Chromatin of the Barr body: Histone and non-histone proteins associated with or excluded from the inactive X chromosome. *Human Mol. Genet.* **12:** 2167.

Chaumeil J., Okamoto I., Guggiari M., and Heard E. 2002. Integrated kinetics of X chromosome inactivation in differentiating embryonic stem cells. *Cytogenet. Cell Genet.* **99:** 75.

Chureau C., Prissette M., Bourdet A., Barbe V., Cattolico L., Jones L., Eggen A., Avner P., and Duret L. 2002. Comparative sequence analysis of the X-inactivation center region in mouse, human, and bovine. *Genome Res.* **12:** 894.

Clemson C.M., Chow J.C., Brown C.J., and Lawrence J.B. 1998. Stabilization and localization of Xist RNA are controlled by separate mechanisms and are not sufficient for X inactivation. *J. Cell Biol.* **142:** 13.

Clerc P. and Avner P. 1998. Role of the region 3′ to *Xist* exon 6 in the counting process of X-chromosome inactivation. *Nat. Genet.* **19:** 249.

———. 2003. Multiple elements within the Xic regulate random X inactivation in mice. *Semin. Cell Dev. Biol.* **14:** 85.

Cockell M. and Gasser S. 1999. Nuclear compartments and gene regulation. *Curr. Opin. Genet. Dev.* **9:** 199.

Comings D.E. 1968. The rationale for an ordered arrangement of chromatin in the interphase nucleus. *Am. J. Hum. Genet.* **20:** 440.

Cooper D.W., VandeBerg J.L., Sharman G.B., and Poole W.E. 1971. Phosphoglycerate linase polymorphism in kangaroos provides further evidence for paternal X inactivation. *Nat. New Biol.* **230:** 155.

Costanzi C. and Pehrson J.R. 1998. Histone macroH2A.1 is concentrated on the inactive X chromosome of female mammals. *Nature* **393:** 599.

Costanzi C., Stein P., Worrad D.M., Schultz R.M., and Pehrson J.R. 2000. Histone macroH2A1 is concentrated in the inactive X chromosome of female preimplantation mouse embryos. *Development* **127:** 2283.

Csankovszki G., Nagy A., and Jaenisch R. 2001. Synergism of Xist RNA, DNA methylation, and histone hypoacetylation in maintaining X chromosome inactivation. *J. Cell Biol.* **153:** 773.

Csankovszki G., Panning B., Bates B., Pehrson J.R., and Jaenisch R. 1999. Conditional deletion of Xist disrupts histone macroH2A localization but not maintenance of X inactivation. *Nat. Genet.* **22:** 323.

Debrand E., Chureau C., Arnaud D., Avner P., and Heard E. 1999. Functional analysis of the *DXPas34* locus: A 3′ regulator of *Xist* expression. *Mol. Cell. Biol.* **19:** 8513.

Dekker J., Rippe K., Dekker M., and Kleckner N. 2002. Capturing chromosome conformation. *Science* **295:** 1306.

Eggan K., Akutsu H., Hochedlinger K., Rideout W., III, Yanagimachi R., and Jaenisch R. 2000. X-chromosome inactivation in cloned mouse embryos. *Science* **290:** 1578.

Fernandez-Capetillo O., Mahadevaiah S.K., Celeste A., Romanienko P.J., Camerini-Otero R.D., Bonner W.M., Manova K., Burgoyne P., and Nussenzweig A. 2003. H2AX is required for chromatin remodeling and inactivation of sex chromosomes in male mouse meiosis. *Dev. Cell.* **4:** 497.

Fischle W., Wang Y., Jacobs S.A., Kim Y., Allis C.D., and Khorasanizadeh S. 2003. Molecular basis for the discrimination of repressive methyl-lysine marks in histone H3 by Polycomb and HP1 chromodomains. *Genes Dev.* **17:** 1870.

Goto T. and Monk M. 1998. The regulation of X-chromosome inactivation in development in mouse and human. *Microbiol. Mol. Biol. Rev.* **62:** 362.

Goto Y. and Takagi N. 1998. Tetraploid embryos rescue embryonic lethality caused by an additional maternally inherited X chromosome in the mouse. *Development* **125:** 3353.

———. 1999. Maternally inherited X chromosome is not inactivated in mouse blastocysts due to parental imprinting. *Chromosome Res.* **7:** 101.

Graves J.A. and Westerman M. 2002. Marsupial genetics and genomics. *Trends Genet.* **18:** 517.

Hadjantonakis A.K., Cox L.L., Tam P.P., and Nagy A. 2001. An X-linked GFP transgene reveals unexpected paternal X-chromosome activity in trophoblastic giant cells of the mouse placenta. *Genesis* **29:** 133.

Heard E. and Avner P. 1994. Role play in X-inactivation. *Hum. Mol. Genet.* **3:** 1481.

Heard E., Clerc P., and Avner P. 1997. X chromosome inactivation in mammals. *Annu. Rev. Genet.* **31:** 571.

Heard E., Mongelard F., Arnaud D., and Avner P. 1999. *Xist* YAC transgenes function as Xics only in multicopy arrays and not as single copies. *Mol. Cell. Biol.* **19:** 3156.

Heard E., Rougeulle C., Arnaud D., Avner P., Allis C.D., and Spector D.L. 2001. Methylation of histone H3 at Lys-9 is an early mark on the X chromosome during X-inactivation. *Cell* **107:** 727.

Heard E., Kress C., Mongelard F., Courtier B., Rougeulle C., Ashworth A., Vourc'h C., Babinet C., and Avner P. 1996. Transgenic mice carrying an *Xist*-containing YAC. *Hum. Mol. Genet.* **5:** 441.

Hendriksen P.J., Hoogerbrugge J.W., Themmen A.P., Koken M.H., Hoeijmakers J.H., Oostra B.A., van der Lende T., and Grootegoed J.A. 1995. Postmeiotic transcription of X and Y chromosomal genes during spermatogenesis in the mouse. *Dev. Biol.* **170:** 730.

Henikoff S., Furuyama T., and Ahmad K. 2004. Histone variants, nucleosome assembly and epigenetic inheritance. *Trends Genet.* **20:** 320.

Herzing L.B., Romer J.T., Horn J.M., and Ashworth A. 1997. Xist has properties of the X-chromosome inactivation centre. *Nature* **386:** 272.

Huynh K.D. and Lee J.T. 2000. Imprinted X inactivation in eutherians: A model of gametic execution and zygotic relation.

Curr. Opin. Cell Biol. **13**: 690.
Huynh K.D. and Lee J.T. 2003. Inheritance of a pre-inactivated paternal X chromosome in early mouse embryos. *Nature* **426**: 857.
Jablonka E. and Lamb M.J. 1988. Meiotic pairing constraints and the activity of sex chromosomes. *J. Theor. Biol.* **133**: 23.
Jamieson R.V., Tam P.P.L., and Gardiner-Garden M. 1997. X-chromosome activity: Impact of imprinting and chromatin structure. *Int. J. Dev. Biol.* **40**: 1065.
Jenuwein T. and Allis C.D. 2001. Translating the histone code. *Science* **293**: 1074.
Jeppesen P. and Turner B.M. 1993. The inactive X chromosome in female mammals is distinguished by a lack of histone H4 acetylation, a cytogenetic marker for gene expression. *Cell* **74**: 281.
Kaslow D.C. and Migeon B.R. 1987. DNA methylation stabilizes X chromosome inactivation in eutherians but not in marsupials: Evidence for multistep maintenance of mammalian X dosage compensation. *Proc. Natl. Acad. Sci.* **84**: 6210.
Kay G.F., Barton S.C., Surani M.A., and Rastan S. 1994. Imprinting and X chromosome counting mechanisms determine Xist expression in early mouse development. *Cell* **77**: 639.
Kay G.F., Penny G.D., Patel D., Ashworth A., Brockdorff N., and Rastan S. 1993. Expression of Xist during mouse development suggests a role in the initiation of X chromosome inactivation. *Cell* **72**: 171.
Keohane A.M., O'Neill L.P., Belyaev N.D., Lavender J.S., and Turner B.M. 1996. X inactivation and histone H4 acetylation in ES cells. *Dev. Biol.* **180**: 618.
Kohlmaier A., Savarese F., Lachner M., Martens J., Jenuwein T., and Wutz A. 2004. A chromosomal memory triggered by xist regulates histone methylation in x inactivation. *PLoS Biol.* **2**: E171.
Kratzer P.G., Chapman V.M., Lambert H., Evans R.E., and Liskay R.M. 1983. Differences in the DNA of the inactive X chromosome of fetal and extraembryonic tissues of mice. *Cell* **33**: 37.
Kuzmichev A., Nishioka K., Erdjumant-Bromage H., Tempst P., and Reinberg D. 2002. Histone methyltransferase activity associated with a human multiprotein complex containing the Enhancer of Zeste protein. *Genes Dev.* **16**: 2893.
LaSalle J.M. and Lalande M. 1996. Homologous association of oppositely imprinted chromosomal domains. *Science* **272**: 725.
Latham K.E. and Rambhatla L. 1995. Expression of X-linked genes in androgenetic, cytogenetic and normal mouse preimplantation embryos. *Dev. Genet.* **17**: 212.
Lebon J.M., Tam P.P., Singer-Sam J., Riggs A.D., and Tan S.S. 1995. Mouse endogenous X-linked genes do not show lineage-specific delayed inactivation during development. *Genet. Res.* **65**: 223.
Lee J.T. 2000. Disruption of imprinted X inactivation by parent-of-origin effects at Tsix. *Cell* **103**: 17.
Lee J.T., Davidow L.S., and Warshawsky D. 1999a. Tsix, a gene antisense to Xist at the X-inactivation centre. *Nat. Genet.* **21**: 400.
Lee J.T., Lu N., and Han Y. 1999b. Genetic analysis of the mouse X inactivation center defines an 80-kb multifunction domain. *Proc. Natl. Acad. Sci.* **96**: 3836.
Lee J.T., Strauss W.M., Dausman J.A., and Jaenisch R. 1996. A 450 kb transgene displays properties of the mammalian X-inactivation center. *Cell* **86**: 83.
Lifschytz E. and Lindsley D.L. 1972. The role of X-chromosome inactivation during spermatogenesis (*Drosophila*-allocycly-chromosome evolution-male sterility-dosage compensation). *Proc. Natl. Acad. Sci.* **69**: 182.
Lyon M.F. 1961. Gene action in the X chromosome of the mouse (*Mus musculus* L.). *Nature* **190**: 372.
———. 1998. X-chromosome inactivation: A repeat hypothesis. *Cytogenet. Cell Genet.* **80**: 133.
Lyon M.F. and Rastan S. 1984. Parental source of chromosome imprinting and its relevance for X chromosome inactivation. *Differentiation* **26**: 63.
Mak W., Nesterova T.B., de Napoles M., Appanah R., Yamanaka S., Otte A.P., and Brockdorff N. 2004. Reactivation of the paternal X chromosome in early mouse embryos. *Science* **303**: 666.
Marahrens Y., Loring J., and Jaenisch R. 1998. Role of the Xist gene in X chromosome choosing. *Cell* **92**: 657.
Marahrens Y., Panning B., Dausman J., Strauss W., and Jaenisch R. 1997. Xist-deficient mice are defective in dosage compensation but not spermatogenesis. *Genes Dev.* **11**: 156.
McCarrey J.R., Watson C., Atencio J., Ostermeier G.C., Marahrens Y., Jaenisch R., and Krawetz S.A. 2002. X-chromosome inactivation during spermatogenesis is regulated by an Xist/Tsix-independent mechanism in the mouse. *Genesis* **34**: 257.
McDonald L.E., Paterson C.A., and Kay G.F. 1998. Bisulfite genomic sequencing-derived methylation profile of the *Xist* gene throughout early mouse development. *Genomics.* **54**: 379.
Min J., Zhang Y., and Xu R.M. 2003. Structural basis for specific binding of Polycomb chromodomain to histone H3 methylated at Lys 27. *Genes Dev.* **17**: 1823.
Nesterova T.B., Mermoud J.E., Brockdorff N., Hilton K., McLaren A., Surani M.A., and Pehrson J. 2002. Xist expression and macroH2A1.2 localisation in mouse primordial and pluripotent embryonic germ cells. *Differentiation* **69**: 216.
Norris D.P., Patel D., Kay G.F., Penny G.D., Brockdorff N., Sheardown S.A., and Rastan S. 1994. Evidence that random and imprinted *Xist* expression is controlled by preemptive methylation. *Cell* **77**: 41.
Ohlsson R., Paldi A., and Graves J.A. 2001. Did genomic imprinting and X chromosome inactivation arise from stochastic expression? *Trends Genet.* **17**: 136.
Okamoto I., Tan S., and Takagi N. 2000. X-chromosome inactivation in XX androgenetic mouse embryos surviving transplantation. *Development* **127**: 4137.
Okamoto I., Otte A.P., Allis C.D., Reinberg D., and Heard E. 2004. Epigenetic dynamics of imprinted X inactivation during early mouse development. *Science* **303**: 644.
O'Neill L.P., Keohane A.M., Levender J.S., McCabe V., Heard E., Avner P., Brockdorff N., and Turner B.M. 1999. A developmental switch in H4 acetylation upstream of Xist plays a role in X chromosome inactivation. *EMBO J.* **18**: 2897.
Panning B., Dausman J., and Jaenisch R. 1997. X-chromosome inactivation is mediated by *Xist* RNA stabilisation. *Cell* **90**: 907.
Penny G.D., Kay G.F., Sheardown S.S., Rastan S., and Brockdorff N. 1996. Requirement for *Xist* in X chromosome inactivation. *Nature* **379**: 131.
Plath K., Fang J., Mlynarczyk-Evans S.K., Cao R., Worringer K.A., Wang H., de la Cruz C.C., Otte A.P., Panning B., and Zhang Y. 2003. Role of histone H3 lysine 27 methylation in X inactivation. *Science* **300**: 131.
Prissette M., El-Marrii O., Arnaud D., Walter J., and Avner P. 2001. Methylation profiles of *DXPas34* during the onset of X inactivation. *Hum. Mol. Genet.* **10**: 31.
Rastan S. 1983. Non-random X-chromosome inactivation in mouse X-autosome translocation embryos: Location of the inactivation centre. *J. Embryol. Exp. Morphol.* **78**: 1.
Rastan S. and Robertson E.J. 1985. X chromosome deletions in embryo-derived EK cell lines associated with lack of X chromosome inactivation. *J. Embryol. Exp. Morphol.* **90**: 379.
Richardson B.J., Czuppon A.B., and Sharman G.B. 1971. Inheritance of glucose-6-phosphate dehydrogenase variation in kangaroos. *Nat. New Biol.* **230**: 154.
Riggs A.D., Singer S.J., and Keith D.H. 1985. Methylation of the PGK promoter region and an enhancer waystation model for X chromosome inactivation. In *Biochemistry and biology of DNA methylation* (ed. G.L. Cantoni and A. Razin), p. 211. Alan R. Liss, New York.
Rougeulle C., Chaumeil J., Sarma K., Allis C.D., Reinberg D., Avner P., and Heard E. 2004. Differential histone H3 Lys-9 and Lys-27 methylation profiles on the X chromosome. *Mol. Cell. Biol.* **24**: 5475.
Sado T., Wang Z., Sasaki H., and Li E. 2001. Regulation of imprinted X-chromosome inactivation in mice by Tsix. *Devel-*

opment **128:** 1275.

Sado T., Fenner M.H., Tan S.S., Tam P., Shioda T., and Li E. 2000. X inactivation in the mouse embryo deficient for Dnmt1: Distinct effect of hypomethylation on imprinted and random X inactivation. *Dev. Biol.* **225:** 294.

Selwood L. 2001. Mechanisms for pattern formation leading to axis formation and lineage allocation in mammals: A marsupial perspective. *Reproduction* **121:** 677.

Sharman G.B. 1971. Late DNA replication in the paternally derived X chromosome of female kangaroos. *Nature* **230:** 231.

Sheardown S.A., Duthie S.M., Johnston C.M., Newall A.E., Formstone E.J., Arkell R.M., Nesterova T.B., Alghisi G.C., Rastan S., and Brockdorff N. 1997. Stabilization of Xist RNA mediates initiation of X chromosome inactivation. *Cell* **91:** 99.

Silva J., Mak W., Zvetkova I., Appanah R., Nesterova T.B., Webster Z., Peters A.H., Jenuwein T., Otte A.P., and Brockdorff N. 2003. Establishment of histone h3 methylation on the inactive X chromosome requires transient recruitment of Eed-Enx1 polycomb group complexes. *Dev. Cell* **4:** 481.

Singer-Sam J., Chapman V., Lebon J.M., and Riggs A.D. 1992. Parental imprinting studied by allele-specific primer extension after PCR: Paternal X chromosome-linked genes are transcribed prior to preferential paternal X chromosome inactivation. *Proc. Natl. Acad. Sci.* **89:** 10469.

Spector D. 2003. The dynamics of chromosome organization and gene regulation. *Annu. Rev. Biochem.* **72:** 573.

Stavropoulos N., Lu N., and Lee J.T. 2001. A functional role for Tsix transcription in blocking Xist RNA accumulation but not in X-chromosome choice. *Proc. Natl. Acad. Sci.* **98:** 10232.

Strahl B.D. and Allis C.D. 2000. The language of covalent histone modifications. *Nature* **403:** 41.

Tada M., Takahama Y., Abe K., Nakatsuji N., and Tada T. 2001. Nuclear reprogramming of somatic cells by in vitro hybridization with ES cells. *Curr. Biol.* **11:** 1553.

Tada T., Obata Y., Tada M., Goto Y., Nakatsuji N., Tan S., Kono T., and Takagi N. 2000. Imprint switching for non-random X-chromosome inactivation during mouse oocyte growth. *Development* **127:** 3101.

Tajul-Arifin K., Teasdale R., Ravasi T., Hume D.A., Mattick J.S., RIKEN GER Group, and GSL Members. 2003. Identification and analysis of chromodomain-containing proteins encoded in the mouse transcriptome. *Genome Res.* **13:** 1416.

Takagi N. 1988. Requirement of mitoses for the reversal of X-inactivation in cell hybrids between murine embryonal carcinoma cells and normal female thymocytes. *Exp. Cell Res.* **175:** 363.

Takagi N. and Sasaki M. 1975. Preferential inactivation of the paternally derived X chromosome in the extraembryonic membranes of the mouse. *Nature* **256:** 640.

Tam P.P., Williams E.A., and Tan S.S. 1994a. Expression of an X-linked HMG-*lacZ* transgene in mouse embryos: Implication of chromosomal imprinting and lineage specific X-chromosome activity. *Dev. Genet.* **15:** 491.

Tam P.P., Zhou S.X., and Tan S.S. 1994b. X-chromosome activity of the mouse primordial germ cells revealed by the expression of an X-linked lacZ transgene. *Development* **120:** 2925.

Tolhuis B., Palstra R.J., Splinter E., Grosveld F., and de Laat W. 2002. Looping and interaction between hypersensitive sites in the active β-globin locus. *Mol. Cell* **10:** 1453.

Turner B.M. 2000. Histone acetylation and an epigenetic code. *Bioessays* **22:** 836.

Turner J.M., Mahadevaiah S.K., Elliott D.J., Garchon H.J., Pehrson J.R., Jaenisch R., and Burgoyne P.S. 2002. Meiotic sex chromosome inactivation in male mice with targeted disruptions of Xist. *J. Cell Sci.* **115:** 4097.

Wang J., Mager J., Chen Y., Schneider E., Cross J.C., Nagy A., and Magnuson T. 2001. Imprinted X inactivation maintained by a mouse Polycomb group gene. *Nat. Genet.* **28:** 371.

Watson C.M., Hughes R.L., Cooper D.W., Gemmell R.T., Loebel D.A., and Johnston P.G. 2000. Sexual development in marsupials: Genetic characterization of bandicoot siblings with scrotal and testicular maldevelopment. *Mol. Reprod. Dev.* **57:** 127.

West J.D., Frels W.I., Chapman V.M., and Papaioannou V.E. 1977. Preferential expression of the maternally derived X chromosome in the mouse yolk sac. *Cell* **12:** 873.

Wutz A. and Jaenisch R. 2000. A shift from reversible to irreversible X inactivation is triggered during ES cell differentiation. *Mol. Cell* **5:** 695.

Wutz A., Rasmussen T.P., and Jaenisch R. 2002. Chromosomal silencing and localization are mediated by different domains of Xist RNA. *Nat. Genet.* **30:** 167.

Xue F., Tian X.C., Du F., Kubota C., Taneja M., Dinnyes A., Dai Y., Levine H., Pereira L.V., and Yang X. 2002. Aberrant patterns of X chromosome inactivation in bovine clones. *Nat. Genet.* **31:** 216.

Yoshida I., Nishita Y., Mohandas T.K., and Takagi N. 1997. Reactivation of an inactive human X chromosome introduced into mouse embryonal carcinoma cells by microcell fusion with persistent expression of XIST. *Exp. Cell Res.* **230:** 208.

Zhang Y. and Reinberg D. 2001. Transcription regulation by histone methylation: Interplay between different covalent modifications of the core histone tails. *Genes Dev.* **15:** 2343.

Zuccotti M. and Monk M. 1995. Methylation of the mouse *Xist* gene in sperm and eggs correlates with imprinted *Xist* expression and paternal X-inactivation. *Nat. Genet.* **9:** 316.

A Continuity of X-Chromosome Silence from Gamete to Zygote

K.D. HUYNH AND J.T. LEE

Howard Hughes Medical Institute, Department of Molecular Biology, Massachusetts General Hospital, and Department of Genetics, Harvard Medical School, Boston, Massachusetts 02114

A PROBLEM OF DOSAGE

In all biological systems, gene expression is exquisitely controlled. The cell devotes vast resources to ensure that a given gene is expressed not only at the proper time but also at the appropriate level. Inappropriate or ectopic dosages often have grave consequences that cause developmental delay and organism death, as illustrated by disorders in human embryos with chromosome aneuploidies (e.g., Down's Syndrome in trisomy 21). In this regard, a problem faced by many sexually dimorphic organisms is coping with differences in sex-linked gene content arising from how sex is determined. In mammals, for example, sons (XY) inherit one X and one Y chromosome from their parents, while daughters (XX) inherit two X chromosomes (for review, see Zarkower 2001). Thus, female embryos have twice as much X-chromosome dosage as do male embryos. To deal with this dosage inequality, different organisms have evolved unique strategies (for review, see Cline and Meyer 1996; Park and Kuroda 2001). In the fruit fly (*Drosophila melanogaster*), the male X is hypertranscribed. In the round worm (*Caenorhabditis elegans*), expression from the two X's is reduced equally in the female. In mammals, one of the two X chromosomes is transcriptionally silenced through a process known as "X-chromosome inactivation" (XCI) (Lyon 1961). In all three organisms, dosage compensation is essential for proper development and the failure to achieve it results in early embryonic loss (Marahrens et al. 1997; Lee 2000; Sado et al. 2001; for review, see Cline and Meyer 1996).

Despite its essential nature, dosage compensation is apparently not necessary during the earliest stages of development. In the fruit fly and worm, dosage compensation does not take place until the blastoderm and 40-cell stages, respectively (Franke et al. 1996; Meyer 2000). In these systems, sex determination and dosage compensation are jointly regulated by the X-to-autosome ratio, so that equalizing any differences in X-chromosome dosage may not be desirable from the outset of development. In mammals, it is also thought that XCI does not take place until the implantation period. Because zygotic gene activation (ZGA) takes place almost immediately after fertilization in the mouse, this belief suggests that the first several cell divisions of the zygote proceed in the presence of a twofold imbalance of several thousand X-linked genes between males and females. Yet, unlike in the fruit fly and worm, mammalian sex is not determined by X-to-autosome dosage (Zarkower 2001). How and why, then, does the early mammalian embryo tolerate major deviations from normal gene dosage, especially given that key developmental decisions, such as the determination of the anatomic axes and formation of specific cell lineages, are made during early development?

Recent inquiries into this problem have led to some surprising and somewhat conflicting findings (Huynh and Lee 2003; Okamoto et al. 2004). Here, we will focus on recent developments in the understanding of when and how dosage compensation occurs in the mammal and also highlight major differences in current thinking.

TWO FORMS OF X-CHROMOSOME INACTIVATION IN THE MOUSE

XCI can take on two forms. In 1961, Mary Lyon described a form of silencing in which either X chromosome in a female cell can be selected for silencing (Lyon 1961), thus rendering every female a mosaic of cells expressing either the paternally or maternally inherited X chromosome. The random form of XCI is observed in somatic tissues of eutherian mammals such as humans and mice. In 1971, Sharman and colleagues documented one of the earliest examples of genomic imprinting in mammals when they described a form of XCI in which the paternal X chromosome (X^P) is preferentially silenced (Sharman 1971). Imprinted XCI takes place in the extraembryonic tissues of some eutherians such as the rodent (Takagi and Sasaki 1975) and cow (Xue et al. 2002). (In humans, the status of XCI in the placenta remains controversial [Ropers et al. 1978; Migeon and Do 1979; Zeng and Yankowitz 2003].) Interestingly, imprinted XCI can be traced back to some of the earliest mammals (metatherians), including the kangaroo and the opossum. This fact has led to the popular view that XCI evolved as an imprinted phenomenon, which was later modified in eutherians to a random mechanism (Graves 1996).

In support of this idea, imprinted and random XCI involve similar control elements in the mouse. Both are controlled by a *cis*-acting X-inactivation center (*Xic*) (Brown et al. 1991), a region that contains several noncoding RNA genes including *Xist* (Brockdorff et al. 1992; Brown et al. 1992) and the complementary *Tsix* gene (Lee et al. 1999). During imprinted XCI, silencing is associated with the spread of Xist RNA along the paternal X chromosome (Mak et al. 2002; Huynh and Lee 2003), while activity on the maternal X chromosome is preserved by expression of *Tsix* (Lee 2000; Sado et al. 2001). Random XCI is also initiated by Xist expression and blocked by *Tsix*. One key difference is that, because random XCI is not predetermined by parental choice, this form has the added complexity of incorporating an X-chromosome "counting" step, which determines whether the embryo is XX or XY, and then a "choice" step, which enables each cell to select one active X (Xa) and one inactive X (Xi) in a mutually exclusive way (for review, see Avner and Heard 2001). Once established, the Xi's of the extraembryonic and embryonic tissues both take on typical characteristics of heterochromatin and exist independently in the two anatomic compartments of the developing embryo.

THE CLASSICAL MODEL: TWO ACTIVE X CHROMOSOMES IN THE XX ZYGOTE

What is the ontogenic relationship between imprinted and random XCI in the early mouse embryo? For much of the past 30 years, the prevailing view has been one in which the XX zygote initiates development with two active X chromosomes (for review, see Grant and Chapman 1988; Jamieson et al. 1996; Heard et al. 1997; Goto and Monk 1998). It was thought that the imprints for preferential paternal X (X^P) silencing are laid down during gametogenesis, but that despite this silencing mark, the maternal and paternal X's are transmitted as fully active X chromosomes. Therefore, according to the conventional view, the early embryo cannot interpret the parental imprints, so that dosage compensation does not actually take place until implantation, at which point the differentiating trophectoderm lineage goes through imprinted XCI for the first time (embryonic day 3.5 [E3.5]) and the embryo proper goes through random inactivation a day or so later (E4.5–5.5) (Fig. 1; reviewed in Goto and Monk 1998).

This well-accepted view was formulated out of elegant studies from the 1970s and early 1980s. Biochemical analyses of the X-linked genes, *Hprt* (Epstein et al. 1978; Kratzer and Gartler 1978) and alpha galactosidase (Adler et al. 1977), indicate two peaks of expression, with a presumptive female peak at twice the activity level as compared to that of the presumptive male peak. Analyses of DNA replication timing reveal that the two X chromosomes replicated synchronously in XX embryos and that asynchronous replication does not occur until the blastocyst stage (Mukherjee 1976; Sugawara et al. 1985). This result has been taken to mean that X-inactivation does not take place prior to the blastocyst stage. Furthermore, cell differentiation studies suggest a tight linkage between XCI and the formation of the various germ lineages, events which take place around the morula–blastocyst transition (Monk and Harper 1979; Sugawara et al. 1985). Together, these studies led to the notion that female embryos initiate development with two fully active X chromosomes.

RIPPLES IN THE CLASSICAL MODEL

Despite its conformity to the bulk of published evidence, the classical model has not offered consistent explanations for some data in the literature. First, the model implies that mammalian embryos can tolerate a twofold imbalance of X-linked gene expression in the preimplantation embryo, despite conventional wisdom regarding the importance of dosage compensation for development (Takagi and Abe 1990; Goto and Takagi 1998, 2000). Given that preimplantation development represents a critical period of embryogenesis when the greatest demands might be placed on proper gene regulation, one naturally asks how early embryos can tolerate global dosage aberration. Relevant to this, there has been much anecdotal evidence of karyotypic instability and generally poor quality of female embryonic stem (ES) cells specific to their carriage of two active X's (in contrast, XO female ES cells rival XY ES cells in their genetic stability), further arguing for the essential nature of dosage compensation in undifferentiated cells. Moreover, if dosage compensation were truly dispensable during early development, why should XCI be required at all for the rest of development?

The two active X model also implies that the gametic imprint, which is laid down in the parental germ line, is essentially ignored by the early embryo and cannot be executed until the implantation stage. What is the nature of

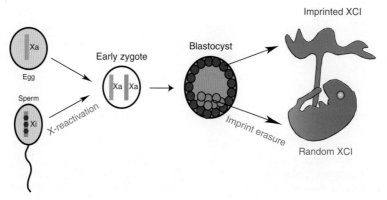

Figure 1. The classical model. In the classical model, the spermatic X is inactivated during spermatogenesis (indicated by *red circles*), but is believed to be reactivated prior to or soon after fertilization. The zygote thus maintains two active X's during preimplantation development and XCI commences for the first time with cellular differentiation at the implantation stage. It is thought that a molecular imprint(s) (e.g., DNA methylation), placed during gametogenesis, instructs cells that will form the extraembryonic tissues to undergo preferential paternal silencing (i.e., imprinted XCI), whereas cells giving rise to the embryo proper undergo imprint erasure and proceed with a random form of XCI.

the gametic signals? In principle, the imprint can either be a repressive mark placed on X^M to resist XCI or an activating mark placed on X^P to induce XCI. In fact, a cogent case can be made for the existence of both a maternal and a paternal imprint based on the results of a number of genetic manipulations. For example, $X^M X^M$ parthegenones and biparental embryos with supernumerary X^M's ($X^M X^M Y$, $X^M X^M X^P$) are surprisingly resistant to inactivation of the maternally inherited X (Shao and Takagi 1990; Goto and Takagi 1998, 2000; Takagi 2003), suggesting that the maternal germ line supplies a protective factor to prevent the inactivation of her X chromosome. In contrast, $X^P X^P$ androgenones have a propensity to inactivate all X's of paternal contribution (Kay et al. 1994; Latham 1996; Okamoto et al. 2000), suggesting that the paternal germ line actively imprints the X in the opposite manner to predispose to XCI. Given that clear marks are placed on one or both X's during gametogenesis, it is difficult to explain why and how there should be a lag in executing the imprinted program until the time of implantation.

Furthermore, the classical model has been difficult to reconcile with a number of recent experimental observations made in mice. Expression studies of the housekeeping gene, *Pgk1*, and an X-linked *LacZ* transgene indicated that, while the maternal alleles are active during preimplantation development, the paternal alleles are repressed until the blastocyst stage (Krietsch et al. 1982; Pravtcheva et al. 1991; Tam et al. 1994; Latham and Rambhatla 1995). These findings indicated that the X^P may not be uniformly active during preimplantation development. It has also been known for some time that *Xist* RNA can be detected at high levels at the two-cell stage (Latham and Rambhatla 1995; Sheardown et al. 1997; Matsui et al. 2001; Nesterova et al. 2001). Yet how there could be uncoupling of high *Xist* expression from gene silencing has never been explained satisfactorily.

In light of these discrepancies, we had previously postulated that the female mouse zygote may in fact carry an inactive X^P throughout preimplantation development and that this X may be inherited from the father as an already silent chromosome (Huynh and Lee 2001). This hypothesis extends thinking on the evolution of dosage compensation in which it is proposed that the ancestral mechanism seen in metatherians (marsupials) is one in which imprinted X^P silencing results directly from paternal germ line XY silencing (Lyon 1999; McCarrey 2001). It also extends the popular view that the nonrandom XCI pattern in the extraembryonic lineages is in part due to memory of the X^P having been inactive in the male germ line (Monk and McLaren 1981) and to obvious differences in chromatin structure between the X's of the maternal germ line, where both X's are euchromatic in character, and the paternal germ line, where the XY body is heterochromatic (Grant and Chapman 1988; Jamieson et al. 1996). These latter models, however, do not explicitly invoke an inheritance model for Xi^P.

Thus, while the classical view has been widely accepted for much of the past 30 years, the status of the early X^P was less than certain, caused in large part by the difficulty of experimentation on preimplantation mouse embryos and the lack of suitable molecular techniques for allelic expression analysis. Clearly, however, its status has been of paramount interest, as insight into the ontogeny of the X^P—whether it is inherited as a preinactivated X chromosome or instead silenced de novo in the embryo—is necessary to piece together the mechanistic details and perhaps also to drive thinking on the evolutionary sequence of X-chromosome inactivation. In the following sections, we will compare and contrast two most recent studies made possible by improvements in cytologic and genomic technologies over the past 10 years.

RECENT ADVANCES: EVIDENCE FOR A SILENT X^P IN PREIMPLANTATION EMBRYOS

The Preinactivation Hypothesis

In the study by our group (Huynh and Lee 2003), two techniques were used to examine the transcriptional status of the X chromosome in preimplantation embryos. To address whether nascent transcription occurs on the X^P, a two-color RNA fluorescence in situ hybridization (FISH) technique was used that simultaneously detects Xist RNA and the Cot-1 RNA fraction. Because the Cot-1 fraction contains highly repetitive sequences that occur within the introns of unprocessed RNA, hybridization to Cot-1 probes indicate nascent transcriptional activity. The results suggest that the Xist RNA domain, which delineates the X^P, is coincident with a Cot-1 "hole" and this Cot-1 exclusion is evident in XX embryos as early as the two-cell stage when the first major wave of ZGA takes place (Fig. 2). The Cot-1 hole can be seen throughout preimplantation development in a vast majority of blastomeres and persists into the extraembryonic lineage, as represented by trophoblasts. This result demonstrates that the X^P is transcription-poor and that this is the case essentially from the time of conception.

To examine the extent to which silencing occurs, direct analysis of gene expression was undertaken by allele-specific reverse transcriptase polymerase chain reaction (RT-PCR) analysis in the morula (8–16 cells) and revealed the imperfect nature of imprinted XCI in several ways. First, imprinted XCI is imperfect at the chromosomal level. RT-PCR analysis of a dozen X-linked genes revealed that, while most genes examined are subject to inactivation, genes residing at either end of the X^P tend to escape silencing (Huynh and Lee 2003). Of genes subject to inactivation, the degree of silencing varies depending on the location of the gene on the X^P—genes residing proximal to the *Xic* were subject to a higher degree of silencing than genes located more distal. The idea that X^P-inactivation "initiates" near the *Xic* and later "spreads" to the rest of the chromosome had been entertained previously (Latham 1996) and is consistent with other published studies (Singer-Sam et al. 1992; Latham and Rambhatla 1995).

Furthermore, expression is variable among blastomeres of a single embryo and among different embryos, the result of which is that X^M almost never contributes 100% of transcripts for any given gene (Huynh and Lee 2003). In general, there appears to be one paternal transcript for every three to four maternal transcripts. Inter-

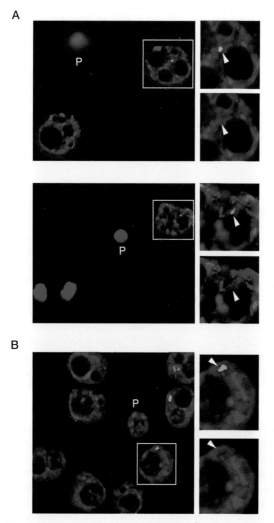

Figure 2. Transcriptional silencing of the X^P is detected as early as the two-cell stage. Cot-1 RNA FISH was performed on preimplantation mouse female embryos. In this experimental technique, Cot-1 sequences are used to probe for nascent transcription in the Xist RNA domain in undenatured nuclei. The results show that, whereas Cot-1 staining (*red*) is detected diffusely in the nucleoplasm, a clear absence of signal is observed within the Xist RNA domain (*green*), indicating a lack of transcription from one of the two X chromosomes. (*A*) Two-cell-stage embryos. (*B*) Eight-cell-stage embryo. Boxed nuclei are shown blown-up to the right; an *arrowhead* indicates Xist RNA domain and Cot-1 "hole." Dapi stains DNA (*blue*); P, polar body. (Reprinted, with permission, from Huynh and Lee 2003 [©Nature Publishing Group].)

estingly, ~20% of the blastomeres on average show markedly diminished or absent Xist coating as assessed by RNA FISH analysis, leading to the speculation that loss of imprinting in some blastomeres may account for a fraction of the paternal transcripts in any given embryo (Huynh and Lee 2003). Another idea worth entertaining is that blastomeres with down-regulated *Xist* expression might be precursors of epiblast cells. In other words, although they are initially imprinted in the zygote, some blastomeres may stochastically lose the imprint, resulting in a transient reactivation of the X^P and subsequent onset of random XCI in those cells.

The inabsolute nature of preimplantation XCI is consistent with conclusions reached by other genetic experiments in the mouse. For example, it explains the occasional surviving XO embryo carrying the imprinted X^P (X^PO) (Hunt 1991; Thornhill and Burgoyne 1993; Jamieson et al. 1998), the viability of some *Tsix* mutant mice that carry a maternally transmitted mutant allele (Lee 2000; Sado et al. 2001), and the occurrence of a small fraction of uniparental X^PX^P, X^PY, and X^MX^M embryos with apparently normal X-inactivation patterns (Okamoto et al. 2000; Matsui et al. 2001; Nesterova et al. 2001). We believe that the rare survivors arise from embryos in which a significant fraction of blastomeres have, by chance, a markedly weakened parental imprint that enables the embryo to use a predominantly random mechanism of XCI (for a more detailed discussion, see Huynh and Lee 2001). In light of the ability of some mutant embryos to overcome the parental imprints, imprinted XCI seems certain to be an inabsolute and reversible phenomenon.

We conclude that the preimplantation embryo is partially dosage compensated from the time of conception and propose that silencing is initiated in the paternal germ line. Importantly, this silencing mechanism is imperfect with much stochastic variation from gene to gene, blastomere to blastomere, and embryo to embryo. The lability of the silent X^P during cleavage-stage development may be a required property of a chromosome that undergoes X-reactivation, the dynamic process during which *Xist* must be down-regulated and heterochromatin along the entire X be converted to euchromatin. Thus, imprinted XCI must be reversible in early development and perhaps this very need to be reversible may result in a leaky mechanism. The fact that some hallmarks of classic heterochromatin (e.g., late DNA replication timing) do not appear until the peri-implantation stage may reflect the need to prevent a permanent "lockdown" of the chromatin during cleavage stage (Huynh and Lee 2003).

Dramatic changes take place in parallel in the two compartments of the embryo at the time of implantation (see Fig. 4 below). In the extraembryonic tissues, as represented by trophoblast stem (TS) cells, the gradient of silencing along the X^P disappears and all genes tested along the X^P exhibit complete silencing (Huynh and Lee 2003). A more global pattern of inactivation in postimplantation cells may naturally follow from increased recruitment of heterochromatic factors such as the polycomb group proteins (Mak et al. 2002, 2004; Okamoto et al. 2004), macroH2A1 (Costanzi et al. 2000), and DNA methylation, all of which serve to "lock in" the inactive state. Thus, imprinted XCI is characterized by two distinct phases—one in the preimplantation embryo, where dosage compensation favors maternal transcription without precluding leaky paternal expression, and one in the postimplantation embryo, where the imperfection of the earlier form gives way to a globalized paternal silencing mechanism.

While those events take place in the extraembryonic lineage, the X^P becomes transiently reactivated in the epiblast of the embryo proper in preparation for random silencing in the soma (Huynh and Lee 2003; Mak et al. 2004). Like imprinted XCI in the placenta, random XCI is also quite complete on a per gene and chromosomal basis relative to what is seen in the preimplantation embryo. There is evidence to suggest that the partial extent of inactivation in preimplantation embryos may reflect the spreading of Xist along the X^P. Xist RNA only partially coats metaphase X chromosomes of 8- to 16-cell-stage blastomeres, whereas Xist RNA coats the entire length of the chromosome in TS and somatic cells (Matsui et al. 2001; Mak et al. 2002; Huynh and Lee 2003). While these findings imply a role for *Xist*, whether *Xist* actually plays a role in preimplantation XCI has yet to be investigated.

The De Novo Inactivation Hypothesis (Okamoto et al. 2004)

Because the study of Okamoto et al. (2004) has been laid out in detail in Heard (this volume), this section will serve only to paraphrase their main conclusions as a prelude to discussing key differences. Using a slightly different approach, Okamoto and colleagues have also reported that signs of X^P-inactivation are evident during cleavage stage development (Okamoto et al. 2004). By combining Xist RNA FISH and immunostaining for characteristic signatures of heterochromatin, they showed that the X^P becomes increasingly enriched for histone H3 methylation at lysine 9 and 27, commencing at the four- to eight-cell stage and proceeding through implantation. In parallel, the polycomb group proteins, Eed and Ezh2, also show increasing recruitment to the X^P. Consistent with that, staining for RNA polII showed an exclusion in the Xist RNA domain, reportedly beginning at the four-cell stage and methylation at histone H3 lysine 4 (a mark of euchromatin) is largely absent throughout preimplantation development. In a significant departure from our conclusions, Okamoto et al. believe that the X^P is active at least during the first two cleavage stages. This conclusion is based on RNA FISH analysis of *Chic1* expression, in which putative *Chic1* signals lie near the Xist RNA domain.

Based on expression analysis of a single gene and chromatin dynamics during cleavage stages, the authors proposed that imprinted XCI is initiated no earlier than the four- to eight-cell stage. Relevant to this conclusion, one might issue the cautionary note that the expression status of the X chromosome cannot be concluded based on analysis of a single gene, as our study demonstrates obvious stochastic fluctuations in X^P repression from gene to gene, cell to cell, and embryo to embryo. Our work clearly indicates Cot-1 RNA exclusion in the X^P domain at the two-cell stage. This aside, the arguments in the next section lead us to favor a silent X^P from the time of conception.

PREINACTIVATION VERSUS DE NOVO X-INACTIVATION

Thus, using different approaches, the two studies both arrived at the conclusion that the mouse embryo exhibits hallmarks of dosage compensation long before implantation—a considerable shift from the prevailing view. An important difference between the recent studies, however, is that while we (Huynh and Lee 2003) observed evidence of XCI from the time of conception, Okamoto et al. believe that the X^P is initially active in the zygote. This discrepancy of one- to two-cell cycles has resulted in a significant interpretive difference. Because the signature modifications to chromatin are not present before, Okamoto et al. favor the idea that XCI takes place "de novo" at the four- to eight-cell stage (for supporting arguments, see Heard, this volume). On the other hand, we observe that the Cot-1-deficient Xist RNA domain can be seen in a majority of embryos beginning at the two-cell stage, leading to the conclusion that the X^P is silent from the time of conception and spurring the hypothesis that the X^P may be inherited from father as a preinactivated chromosome. Several excellent, well-balanced reviews have dealt with some of these differences (Cheng and Disteche 2004; Ferguson-Smith 2004; Hajkova and Surani 2004). Our purpose, in the remaining space, is to further build the argument that motivates us to support the preinactivation model.

First and foremost, there is an apparent continuity of silence from gamete to zygote. The "preinactivation model" integrates the well-established existence of paternal germ line X-inactivation and, according to our findings, the already silent state of the X^P in the early mouse embryo. During meiotic sex chromosome inactivation (MSCI) in the paternal germ line, the synapsed X and Y chromosomes become transcriptionally silenced at the pachytene stage of prophase I (Lifschytz and Lindsley 1972). Analyses of a limited number of X-linked genes have shown that *Pgk1*, *Pdha1*, *Hprt*, *Zfx*, *Phka1*, and *G6pd* all remain silent in the postmeiotic sperm (Singer-Sam et al. 1990; McCarrey et al. 1992a, 2000; Hendriksen et al. 1997). Nonetheless, some genes are clearly reexpressed in postmeiotic sperm (Hendriksen et al. 1995), a fact that has been used as a counterargument to the preinactivation hypothesis (Heard 2004). However, a more detailed examination indicates that the reactivated genes, *Ube1x* and *Ube2a*, reside at a considerable distance from the *Xic*, notably in a region where we observed escape from silencing in preimplantation embryos. (N.B., *Fmr1* and *Mage1/2* have also been examined, but the interpretation is complicated by the fact that *Fmr1* is not inactivated during pachytene and *Mage1/2* is part of a whole family of homologous genes that may have resulted in nonspecific PCR detection [Hendriksen et al. 1995; McCarrey et al. 2002].) Thus, the postmeiotic expression pattern of the X actually mirrors that in preimplantation embryos (Fig. 3), further supporting the hypothesis that partial X^P-inactivation results from inheritance of this expression pattern from the paternal germ line.

It is also worth noting that many X-linked genes have autosomal homologs that are expressed specifically during MSCI and in postmeiotic sperm (Dahl et al. 1990; McCarrey et al. 1992b; Hendriksen et al. 1997; Bradley et al. 2004; Emerson et al. 2004). As has been noted by others (McCarrey and Thomas 1987; Bradley et al. 2004), this phenomenon may be driven by the need to compen-

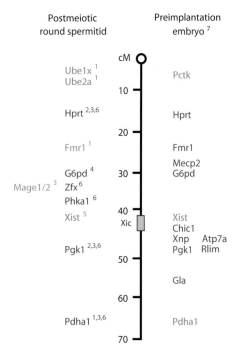

Figure 3. A continuity of X^P silencing from sperm to zygote. Shown is a comparision of X^P expression data in postmeiotic sperm and preimplantation embryos. *Green* genes, expressed; *red* genes, either not expressed or expression is severely reduced; *yellow* genes, expression not conclusive (see text). Sources: 1, Hendriksen et al. 1995; 2, Singer-Sam et al. 1990; 3, McCarrey et al. 2002; 4, Hendriksen et al. 1997; 5, McCarrey and Dilworth 1992; Richler et al. 1992; Salido et al. 1992; 6, McCarrey et al. 1992a; 7, Krietsch et al. 1982; Pravtcheva et al. 1991; Singer-Sam et al. 1992; Latham and Rambhatla 1995; Huynh and Lee 2003; Mak et al. 2004.

sate for the silencing of those X-linked genes whose continued expression is necessary throughout male meiosis. Throughout mammalian evolution, this need to compensate for MSCI may have spurred the "retrotransposition" of those genes to regions of the genome that are spared of meiotic silencing. The very existence of such X-to-autosome retrogenes indeed argues against wholesale reactivation of the X after male meiosis and indirectly supports the idea that the X^P is propagated as a partially silent X after male meiosis.

It is also important to emphasize that, while we believe that imprinted XCI may trace its origins to the paternal germ line, zygotic XCI and MSCI are clearly distinct in many ways. MSCI can occur independently of an intact *Xist* gene (McCarrey et al. 2002; Turner et al. 2002), but our data suggest that preimplantation XCI may depend on *Xist*, as judged by the correlation between how far Xist RNA spreads and the extent of silencing along the X^P (Huynh and Lee 2003). Some have argued that this apparent, though unproven, *Xist* dependence of zygotic XCI precludes MSCI as its origin (Heard 2004). Yet, this difference may simply reflect differences between the *initiation* phase of silencing through MSCI and the subsequent *maintenance* phase in the zygote. There is much precedent for initiation and maintenance requiring different factors. For example, while Polycomb proteins do not initiate heterochromatin formation, they are required to maintain heterochromatin once established (for review, see Francis and Kingston 2001). The preinactivation model, in fact, hypothesizes that the paternal germ line initiates X silencing in an *Xist*-independent fashion and that this silence is then actively maintained by the zygote, possibly through an *Xist*-dependent mechanism (Huynh and Lee 2003). Thus, if the recruitment of zygote-specific factors to maintain the silent X^P can be called a "de novo" mechanism, then the distinction between the preinactivation and de novo models might be partly a semantic one.

The preinactivation model is also a more parsimonious solution to the ontogeny of XCI. Indeed, it simplifies the sequence of silencing by eliminating multiple rounds of inactivation and reactivation during early development as required by the de novo model. According to Okamoto et al. (2004), the first round of inactivation takes place in the male germ line (MSCI), followed by a round of reactivation in the postmeiotic spermatid, then reinactivation of the X^P at the four- to eight-cell stage, followed by a second round of reactivation in the epiblast, and finally by yet another round of inactivation in the differentiating epiblast (Fig. 4). This view deviates from the conventional view only in that the second inactivation event takes place several cell cycles earlier than the blastocyst stage. In contrast, the preinactivation model posits a single initiation event in the paternal germ line through MSCI, followed by a single round of reactivation in the epiblast in preparation for random X-chromosome silencing in the soma (Fig. 4).

PREINACTIVATION FROM AN EVOLUTIONARY PERSPECTIVE

Our observations and the preinactivation model also fit neatly with various hypotheses for X-chromosome evolution in mammals. Many schemes share the proposal that the imprinted form of XCI is ancestral to the random form, based on the fact that metatherians—which predate eutherians by several million years—display an imprinted mechansim of XCI (for review, see Graves 1996). Since only female offspring inherit the X^P, a simple solution to the problem of dosage compensation in the early mammal would indeed have been to silence all X chromosomes of paternal origin.

This proposed evolutionary chronology is strikingly reflected in the ontogeny of XCI in the mouse, where the imprinted mechanism observed first in the cleavage stage embryo eventually gives rise to random XCI in the embryo proper (Huynh and Lee 2003). First, like marsupial XCI (Graves 1996), preimplantation XCI in the mouse shows a high degree of gene-to-gene and cell-to-cell variability and a propensity toward reactivation. Second, imprinted XCI in the mouse is biphasic, with a more localized and leaky form before preimplantation and a more globalized and complete form after implantation. Apropos to this, it has been proposed that, in its ancestral form, imprinted XCI may have been very limited in extent, perhaps involving only genes that have lost Y-chromosome homologs (Graves 1996; Lyon 1999). Later on, XCI could have spread to include more genes along the X

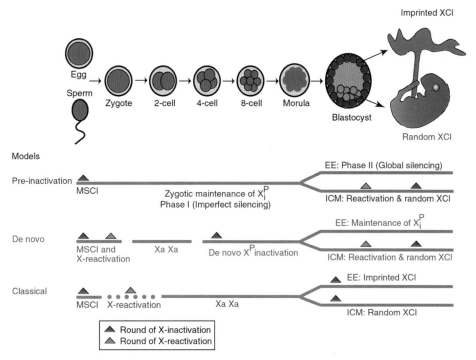

Figure 4. A comparison of the preinactivation, de novo, and classical models: a summary of the XCI models discussed in this review. The "preinactivation" model postulates just one round of inactivation and one round of reactivation: The female zygote inherits a preinactivated paternal X chromosome and maintains a partially silenced X^P throughout preimplantation development (phase I). This partial inactivation gives way to a more global silencing (phase II) in extraembryonic tissues (*blue circles*). Epiblast cells (*green circles*) of the inner cell mass (ICM), on the other hand, undergo a single round of reactivation followed by a random form of XCI. In contrast, the de novo model involves many rounds of inactivation and reactivation: In this model, the paternal germ line initiates MSCI, but the X is completely reactivated after meiosis. The zygote inherits two fully active X's and begins reinactivation at the four- to eight-cell stage. In the epiblast, yet another round of reactivation takes place in preparation for a final round of inactivation in the form of random XCI. This model is very similar to the classical model, the major difference being a shift to earlier zygotic inactivation by a couple of cell cycles.

chromosome in response to a rapidly degenerating Y chromosome.

If ontogeny is recapitulated by phylogeny in the case of XCI, one might ask whether the ancestral imprinted mechanism in marsupials might also involve "inheritance" of a preinactivated X chromosome from father. Indeed, others have suggested that paternal germ line X chromosome silencing might have been the original dosage compensation mechanism whose remnants are carried through into eutherians in the various forms we now observe (Cooper 1971; Graves 1996; Lyon 1999; McCarrey 2001). Meiotic sex chromosome inactivation with maintenance of that silent state in the zygote could have been a simple, though inefficient, system of ensuring dosage compensation in the earliest mammals. Millions of years of evolution would have provided enough time to evolve the more global and complete mechanism that we see today in eutherians.

Since MSCI is thought to be *Xist* independent, can we make predictions about whether there is an *Xist* ortholog in marsupials? To date, neither *Xist* nor *Tsix* sequences have been identified in metatherians, leading to some suggestion that marsupials may do XCI independently of *Xist*. On the other hand, preimplantation XCI in the mouse may require *Xist* (currently under testing). Given the similarities to marsupial XCI, it seems reasonable to think that the marsupial mechanism might also require *Xist* to inititate silencing on X^P and *Tsix* to block it on X^M. Sequencing of various marsupial genomes will provide the opportunity to answer these intriguing questions.

CONCLUDING STATEMENTS

In summary, recent advances have led to a considerable departure from the traditional view by reaching the consensus that the X^P is partially silent in the preimplantation mouse embryo. Thus, not only are the mechanisms of dosage compensation uniquely different among the fruit fly, worm, and mouse, but they are also different with respect to when dosage compensation takes place. Because sex determination and dosage compensation are linked by the X-to-autosome ratio in the fruit fly and the worm, it seems logical that dosage compensation would be delayed until a time when sex determination is initiated. With expressed X-linked genes playing a major role in setting up the X-to-autosome ratio, this delay in dosage compensation would enable X-dosage differences between XX and XY/XO individuals to be "read." Since mammalian sex is primarily determined by the Y chromosome, the constraint on timing of dosage compensation would not exist in the mouse. Whether other mammals show the same early XCI remains to be seen. The conclusions do raise several interesting questions: Is early XCI universally observed in mammals? If so, do they share an imprinted mechanism? In humans, several studies have strongly argued against the existence of imprinted XCI (Migeon and Do 1979; Migeon et al. 1985; Mohandas et al. 1989; Zeng and Yankowitz 2003). And, finally, how does the fruit fly or worm embryo tolerate dosage inequality between the sexes in light of the similarly essential nature of dosage compensation in those organisms?

Although further investigation into the exact timing of imprinted XCI is clearly warranted, we strongly favor the preinactivation model. Indeed, the molecular details may differ between paternal and zygotic XCI; the persistence of a partially silent X after MSCI and the later appearance of a similar X^P in the zygote suggest to us a direct inheritance of an Xi from father to daughter. To reiterate, it is the *continuity of silence* that motivates us to this conclusion irrespective of the molecular means.

A final consensus must await further improvements in technology and closer examination of gene expression during the critical gamete-to-zygote transition. Two questions are of foremost relevance. First, what is the status of the X chromosome after paternal meiosis—do X-linked genes remain silent or are they reactivated? Second, how does this pattern of expression differ from the X^P profile in the two-cell embryo? The answers to these questions will have distinct implications for the mechanism and evolution of XCI and also for any molecular links to the sister phenomenon of autosomal imprinting. With the rapid pace of technological innovation, the answers to these questions will not be long in coming.

ACKNOWLEDGMENTS

We thank all members of the Lee lab for discussion. This work was funded by the National Institutes of Health (GM58839), the Pew Scholar Program, and the Howard Hughes Medical Institute.

REFERENCES

Adler D.A., West J.D., and Chapman V.M. 1977. Expression of alpha-galactosidase in preimplantation mouse embryos. *Nature* **267:** 838.
Avner P. and Heard E. 2001. X chromosome inactivation: Counting, choice and initiation. *Nat. Rev. Genet.* **2:** 59.
Bradley J., Baltus A., Skaletsky H., Royce-Tolland M., Dewar K., and Page D.C. 2004. An X-to-autosome retrogene is required for spermatogenesis in mice. *Nat. Genet.* **36:** 872.
Brockdorff N., Ashworth A., Kay G.F., McCabe V.M., Norris D.P., Cooper P.J., Swift S., and Rastan S. 1992. The product of the mouse Xist gene is a 15 kb inactive X-specific transcript containing no conserved ORF and located in the nucleus. *Cell* **71:** 515.
Brown C.J., Hendrich B.D., Rupert J.L., Lafreniere R.G., Xing Y., Lawrence J., and Willard H.F. 1992. The human XIST gene: Analysis of a 17 kb inactive X-specific RNA that contains conserved repeats and is highly localized within the nucleus. *Cell* **71:** 527.
Brown C.J., Lafreniere R.G., Powers V.E., Sebastio G., Ballabio A., Pettigrew A.L., Ledbetter D.H., Levy E., Craig I.W., and Willard H.F. 1991. Localization of the X inactivation centre on the human X chromosome in Xq13. *Nature* **349:** 82.
Cheng M.K. and Disteche C.M. 2004. Silence of the fathers: Early X inactivation. *Bioessays* **26:** 821.
Cline T.W. and Meyer B.J. 1996. Vive la difference: Males vs females in flies vs worms. *Annu. Rev. Genet.* **30:** 637.
Cooper D.W. 1971. Directed genetic change model for X chromosome inactivation in eutherian mammals. *Nature* **230:** 292.
Costanzi C., Stein P., Worrad D.M., Schultz R.M., and Pehrson J.R. 2000. Histone macroH2A1 is concentrated in the inactive X chromosome of female preimplantation mouse embryos. *Development* **127:** 2283.
Dahl H.H., Brown R.M., Hutchison W.M., Maragos C., and Brown G.K. 1990. A testis-specific form of the human pyruvate dehydrogenase E1 alpha subunit is coded for by an intronless gene on chromosome 4. *Genomics* **8:** 225.
Emerson J.J., Kaessmann H., Betran E., and Long M. 2004. Extensive gene traffic on the mammalian X chromosome. *Science* **303:** 537.
Epstein C.J., Smith S., Travis B., and Tucker G. 1978. Both X chromosomes function before visible X chromosome inactivation in female mouse embryos. *Nature* **274:** 500.
Ferguson-Smith A.C. 2004. X inactivation: Pre- or post-fertilisation turn-off? *Curr. Biol.* **14:** R323.
Francis N.J. and Kingston R.E. 2001. Mechanisms of transcriptional memory. *Nat. Rev. Mol. Cell Biol.* **2:** 409.
Franke A., Dernburg A., Bashaw G.J., and Baker B.S. 1996. Evidence that MSL-mediated dosage compensation in *Drosophila* begins at blastoderm. *Development* **122:** 2751.
Goto T. and Monk M. 1998. Regulation of X chromosome inactivation in development in mice and humans. *Microbiol. Mol. Biol. Rev.* **62:** 362.
Goto Y. and Takagi N. 1998. Tetraploid embryos rescue embryonic lethality caused by an additional maternally inherited X chromosome in the mouse. *Development* **125:** 3353.
———. 2000. Maternally inherited X chromosome is not inactivated in mouse blastocysts due to parental imprinting. *Chromosome Res.* **8:** 101.
Grant S.G. and Chapman V.M. 1988. Mechanisms of X chromosome regulation. *Annu. Rev. Genet.* **22:** 199.
Graves J.A. 1996. Mammals that break the rules: Genetics of marsupials and monotremes. *Annu. Rev. Genet.* **30:** 233.
Hajkova P. and Surani M.A. 2004. Development. Programming the X chromosome. *Science* **303:** 633.
Heard E. 2004. Recent advances in X chromosome inactivation. *Curr. Opin. Cell Biol.* **16:** 247.
Heard E., Clerc P., and Avner P. 1997. X chromosome inactivation in mammals. *Annu. Rev. Genet.* **31:** 571.
Hendriksen P.J., Hoogerbrugge J.W., Baarends W.M., de Boer P., Vreeburg J.T., Vos E.A., van der Lende T., and Grootegoed J.A. 1997. Testis-specific expression of a functional retroposon encoding glucose-6-phosphate dehydrogenase in the mouse. *Genomics* **41:** 350.
Hendriksen P.J., Hoogerbrugge J.W., Themmen A.P., Koken M.H., Hoeijmakers J.H., Oostra B.A., van der Lende T., and Grootegoed J.A. 1995. Postmeiotic transcription of X and Y chromosomal genes during spermatogenesis in the mouse. *Dev. Biol.* **170:** 730.
Hunt P.A. 1991. Survival of XO mouse fetuses: Effect of parental origin of the X chromosome or uterine environment? *Development* **111:** 1137.
Huynh K.D. and Lee J.T. 2001. Imprinted X inactivation in eutherians: A model of gametic execution and zygotic relaxation. *Curr. Opin. Cell Biol.* **13:** 690.
———. 2003. Inheritance of a pre-inactivated paternal X chromosome in early mouse embryos. *Nature* **426:** 857.
Jamieson R.V., Tam P.P., and Gardiner-Garden M. 1996. X chromosome activity: Impact of imprinting and chromatin structure. *Int. J. Dev. Biol.* **40:** 1065.
Jamieson R.V., Tan S.S., and Tam P.P. 1998. Retarded postimplantation development of XO mouse embryos: Impact of the parental origin of the monosomic X chromosome. *Dev. Biol.* **201:** 13.
Kay G.F., Barton S.C., Surani M.A., and Rastan S. 1994. Im-

printing and X chromosome counting mechanisms determine Xist expression in early mouse development. *Cell* **77:** 639.

Kratzer P.G. and Gartler S.M. 1978. HGPRT activity changes in preimplantation mouse embryos. *Nature* **274:** 503.

Krietsch W.K., Fundele R., Kuntz G.W., Fehlau M., Burki K., and Illmensee K. 1982. The expression of X-linked phosphoglycerate kinase in the early mouse embryo. *Differentiation* **23:** 141.

Latham K.E. 1996. X chromosome imprinting and inactivation in the early mammalian embryo. *Trends Genet.* **12:** 134.

Latham K.E. and Rambhatla L. 1995. Expression of X-linked genes in androgenetic, gynogenetic, and normal mouse preimplantation embryos. *Dev. Genet.* **17:** 212.

Lee J.T. 2000. Disruption of imprinted X inactivation by parent-of-origin effects at Tsix. *Cell* **103:** 17.

Lee J.T., Davidow L.S., and Warshawsky D. 1999. Tsix, a gene antisense to Xist at the X-inactivation centre. *Nat. Genet.* **21:** 400.

Lifschytz E. and Lindsley D.L. 1972. The role of X chromosome inactivation during spermatogenesis (*Drosophila*-allocycly-chromosome evolution-male sterility-dosage compensation). *Proc. Natl. Acad. Sci.* **69:** 182.

Lyon M.F. 1961. Gene action in the X chromosome of the mouse (*Mus musculus* L.). *Naturwissenschaften* **190:** 372.

———. 1999. Imprinting and X chromosome inactivation. *Results Probl. Cell Differ.* **25:** 73.

Mak W., Baxter J., Silva J., Newall A.E., Otte A.P., and Brockdorff N. 2002. Mitotically stable association of polycomb group proteins eed and enx1 with the inactive x chromosome in trophoblast stem cells. *Curr. Biol.* **12:** 1016.

Mak W., Nesterova T.B., de Napoles M., Appanah R., Yamanaka S., Otte A.P., and Brockdorff N. 2004. Reactivation of the paternal X chromosome in early mouse embryos. *Science* **303:** 666.

Marahrens Y., Panning B., Dausman J., Strauss W., and Jaenisch R. 1997. Xist-deficient mice are defective in dosage compensation but not spermatogenesis. *Genes Dev.* **11:** 156.

Matsui J., Goto Y., and Takagi N. 2001. Control of Xist expression for imprinted and random X chromosome inactivation in mice. *Hum. Mol. Genet.* **10:** 1393.

McCarrey J.R. 2001. X chromosome inactivation during spermatogenesis: The original dosage compensation mechanism in mammals? In *Gene families: Studies of DNA, RNA, enzymes and proteins* (ed. G. Xue et al.), p. 59. World Scientific, Singapore.

McCarrey J.R. and Dilworth D.D. 1992. Expression of Xist in mouse germ cells correlates with X chromosome inactivation. *Nat. Genet.* **2:** 200.

McCarrey J.R. and Thomas K. 1987. Human testis-specific PGK gene lacks introns and possesses characteristics of a processed gene. *Nature* **326:** 501.

McCarrey J.R., Dilworth D.D., and Sharp R.M. 1992a. Semi-quantitative analysis of X-linked gene expression during spermatogenesis in the mouse: Ethidium-bromide staining of RT-PCR products. *Genet. Anal. Tech. Appl.* **9:** 117.

McCarrey J.R., Berg W.M., Paragioudakis S.J., Zhang P.L., Dilworth D.D., Arnold B.L., and Rossi J.J. 1992b. Differential transcription of Pgk genes during spermatogenesis in the mouse. *Dev. Biol.* **154:** 160.

McCarrey J.R., Watson C., Atencio J., Ostermeier G.C., Marahrens Y., Jaenisch R., and Krawetz S.A. 2002. X chromosome inactivation during spermatogenesis is regulated by an Xist/Tsix-independent mechanism in the mouse. *Genesis* **34:** 257.

Meyer B.J. 2000. Sex in the wormcounting and compensating X chromosome dose. *Trends Genet.* **16:** 247.

Migeon B.R. and Do T.T. 1979. In search of non-random X inactivation: Studies of fetal membranes heterozygous for glucose-6-phosphate dehydrogenase. *Am. J. Hum. Genet.* **31:** 581.

Migeon B.R., Wolf S.F., Axelman J., Kaslow D.C., and Schmidt M. 1985. Incomplete X chromosome dosage compensation in chorionic villi of human placenta. *Proc. Natl. Acad. Sci.* **82:** 3390.

Mohandas T.K., Passage M.B., Williams J.W., III, Sparkes R.S., Yen P.H., and Shapiro L.J. 1989. X chromosome inactivation in cultured cells from human chorionic villi. *Somat. Cell Mol. Genet.* **15:** 131.

Monk M. and Harper M.I. 1979. Sequential X chromosome inactivation coupled with cellular differentiation in early mouse embryos. *Nature* **281:** 311.

Monk M. and McLaren A. 1981. X chromosome activity in foetal germ cells of the mouse. *J. Embryol. Exp. Morphol.* **63:** 75.

Mukherjee A.B. 1976. Cell cycle analysis and X chromosome inactivation in the developing mouse. *Proc. Natl. Acad. Sci.* **73:** 1608.

Nesterova T.B., Barton S.C., Surani M.A., and Brockdorff N. 2001. Loss of xist imprinting in diploid parthenogenetic preimplantation embryos. *Dev. Biol.* **235:** 343.

Okamoto I., Tan S., and Takagi N. 2000. X chromosome inactivation in XX androgenetic mouse embryos surviving implantation. *Development* **127:** 4137.

Okamoto I., Otte A.P., Allis C.D., Reinberg D., and Heard E. 2004. Epigenetic dynamics of imprinted X inactivation during early mouse development. *Science* **303:** 644.

Park Y. and Kuroda M.I. 2001. Epigenetic aspects of X chromosome dosage compensation. *Science* **293:** 1083.

Pravtcheva D.D., Adra C.N., and Ruddle F.H. 1991. Timing of paternal Pgk-1 expression in embryos of transgenic mice. *Development* **111:** 1109.

Richler C., Soreq H., and Wahrman J. 1992. X inactivation in mammalian testis is correlated with inactive X-specific transcription. *Nat. Genet.* **2:** 192.

Ropers H.H., Wolff G., and Hitzeroth H.W. 1978. Preferential X inactivation in human placenta membranes: Is the paternal X inactive in early embryonic development of female mammals? *Hum. Genet.* **43:** 265.

Sado T., Wang Z., Sasaki H., and Li E. 2001. Regulation of imprinted X chromosome inactivation in mice by Tsix. *Development* **128:** 1275.

Salido E.C., Yen P.H., Mohandas T.K., and Shapiro L.J. 1992. Expression of the X-inactivation-associated gene XIST during spermatogenesis. *Nat. Genet.* **2:** 196.

Shao C. and Takagi N. 1990. An extra maternally derived X chromosome is deleterious to early mouse development. *Development* **110:** 969.

Sharman G.B. 1971. Late DNA replication in the paternally derived X chromosome of female kangaroos. *Nature* **230:** 231.

Sheardown S.A., Duthie S.M., Johnston C.M., Newall A.E., Formstone E.J., Arkell R.M., Nesterova T.B., Alghisi G.C., Rastan S., and Brockdorff N. 1997. Stabilization of Xist RNA mediates initiation of X chromosome inactivation. *Cell* **91:** 99.

Singer-Sam J., Chapman V., LeBon J.M., and Riggs A.D. 1992. Parental imprinting studied by allele-specific primer extension after PCR: Paternal X chromosome-linked genes are transcribed prior to preferential paternal X chromosome inactivation. *Proc. Natl. Acad. Sci.* **89:** 10469.

Singer-Sam J., Robinson M.O., Bellve A.R., Simon M.I., and Riggs A.D. 1990. Measurement by quantitative PCR of changes in HPRT, PGK-1, PGK-2, APRT, MTase, and Zfy gene transcripts during mouse spermatogenesis. *Nucleic Acids Res.* **18:** 1255.

Sugawara O., Takagi N., and Sasaki M. 1985. Correlation between X chromosome inactivation and cell differentiation in female preimplantation mouse embryos. *Cytogenet. Cell Genet.* **39:** 210.

Takagi N. 2003. Imprinted X chromosome inactivation: Enlightenment from embryos in vivo. *Semin. Cell Dev. Biol.* **14:** 319.

Takagi N. and Abe K. 1990. Detrimental effects of two active X chromosomes on early mouse development. *Development* **109:** 189.

Takagi N. and Sasaki M. 1975. Preferential inactivation of the paternally derived X chromosome in the extraembryonic membranes of the mouse. *Nature* **256:** 640.

Tam P.P., Williams E.A., and Tan S.S. 1994. Expression of an

X-linked HMG-lacZ transgene in mouse embryos: Implication of chromosomal imprinting and lineage-specific X chromosome activity. *Dev. Genet.* **15:** 491.

Thornhill A.R. and Burgoyne P.S. 1993. A paternally imprinted X chromosome retards the development of the early mouse embryo. *Development* **118:** 171.

Turner J.M., Mahadevaiah S.K., Elliott D.J., Garchon H.J., Pehrson J.R., Jaenisch R., and Burgoyne P.S. 2002. Meiotic sex chromosome inactivation in male mice with targeted disruptions of Xist. *J. Cell Sci.* **115:** 4097.

Xue F., Tian X.C., Du F., Kubota C., Taneja M., Dinnyes A., Dai Y., Levine H., Pereira L.V., and Yang X. 2002. Aberrant patterns of X chromosome inactivation in bovine clones. *Nat. Genet.* **31:** 216.

Zarkower D. 2001. Establishing sexual dimorphism: Conservation amidst diversity? *Nat. Rev. Genet.* **2:** 175.

Zeng S.M. and Yankowitz J. 2003. X-inactivation patterns in human embryonic and extra-embryonic tissues. *Placenta* **24:** 270.

Reading the DNA Methylation Signal

A. BIRD AND D. MACLEOD

*Wellcome Trust Centre for Cell Biology, University of Edinburgh,
Edinburgh EH9 3JR, Scotland, United Kingdom*

A cell's properties depend largely on its pattern of gene expression. During the development of complex multicellular organisms, multiple cell lineages emerge among the descendants of the single zygotic cell. Within each lineage, the range of gene expression programs is effectively restricted, culminating in the definitive program of the final differentiated cell. As part of this process, some genes are selected for future activity, whereas others are disqualified by long-term silencing. For a specific gene in one cell type, silence versus activity is not determined solely by the available combinations of transcripion factors in that cell, but also by local differentiation of chromatin structure near to or including the gene. A key feature of such local differentiation of the genome is its stability within and between cell generations. The study of these processes has come to be known as epigenetics. Contemporary epigenetics might therefore be defined as, "the study of the structural adaptation of chromosome regions so as to perpetuate local activity states."

Three sorts of epigenetic system have traditionally been studied: DNA methylation, the polycomb/trithorax system, and histone modification or variant substitution. Other epigenetic categories have emerged, however, and are the subject of intense study as reflected elsewhere in this volume. They include silencing based on noncoding RNAs, trancriptional silencing triggered by double-stranded RNA interference, and silencing or activity that responds to localization of genes within the nucleus. All of these processes appear to be closely interwoven with histone modification, which is itself a diverse and complex system of chromosome marking. For each system there is evidence that the trigger for altered chromatin is a preexisting transcriptional state. In other words, it appears to be the transcripion factors that primarily determine whether a gene is active or inactive, after which chromatin structure is adapted accordingly and the state is memorized. This "secondary" status of many epigenetic mechanisms does not diminish their importance. Stable, leakproof gene silencing requires sophisticated mechanisms for sensing the transcriptional status quo without perturbing it and then passing the memory between cell generations without disruption by DNA replication.

Obviously the most direct way of marking genes either for silence or activity would be to apply chemical tags to the DNA itself. In practice, however, the only widespread modification of DNA is methylation, either of the C5 position of the cytosine ring or of the N6 position of the adenine ring. Other modifications are found, but have limited phylogenetic distribution. In multicellular eukaryotes, only m^5C is convincingly reported. A plausible explanation for the narrow spectrum of DNA modifications is that chemical moieties, such as acetyl or phosphate groups, or methylation at other positions within the DNA structure would interfere with aspects of DNA function. Therefore the scope for chemically tagging genomic DNA, without compromising its essential information coding and storage properties, may be extremely limited. Even m^5C, which does not alter the base-pairing specificity of cytosine and has survived several billion years of scrutiny by natural selection, has drawbacks, as it is a mutagenic base in vertebrates and bacteria (Coulondre et al. 1978; Bird 1980). This trade-off between benefit and disadvantage may explain the evolutionary volatility of cytosine methylation across evolutionary time. Some organisms have dispensed with DNA methylation whereas their relatives retain it. For example, the fungus *Neurospora crassa* displays cytosine methylation, but the fission and budding yeasts—also fungi—do not.

The emergence of complex gene expression programs in multicellular organisms with functionally adapted cell types may have depended on the ability to epigenetically mark genes for either activity or silence during development. Without an effective memory system to enforce the activity state established at critical developmental stages, it would be necessary to constantly reiterate the initial conditions in order to reestablish the program. DNA is the obvious substrate for memorable marks of this kind, but, as pointed out above, has a limited capacity for chemical modification. This shortcoming of the genetic material may have provided a driving force behind the evolution of histones. Rather than marking the DNA itself, why not coat it uniformly with proteins that are susceptible to a plethora of chemical modifications (Jenuwein and Allis 2001)? A precondition of such a system is that histones do not move around. If they did, marks would not be locus specific and would lose their value. Histone loss by degradation or stochastic processes would also need to be compensated by insuring that replacement histones acquired modifications that are appropriate to the region. What we know about histones supports these requirements. The half-life of histones on DNA appears to be long and there is evidence for processes that transfer modification from one nucleosome to its (perhaps new) neighbor. An example is the interaction of HP1 with both methylated lysine 9 of histone H3 (H3K9) and the histone methyltransferase Suvar 3.9 (Bannister et al. 2001). Recruitment of H3K9 methyltransferase to an already methylated histone by HP1 may facilitate the spreading of this modification to neighboring H3 histones.

READING EPIGENETIC MARKS

For any epigenetic system, there are two key questions to be answered. (1) What creates patterns of chromatin marking? (2) How are the marks read by the cell to generate biological outcomes? Although this article focuses on DNA methylation and addresses the second question (how is the DNA methylation signal read?), similar questions apply to the decoding of histone marks. Two kinds of model have been advanced for the functional interpretation of histone modifications (methylation, phosphorylation, acetylation etc.): Either (1) the marks affect the properties of nucleosomal chromatin directly, for example, by changing histone DNA contacts; or (2) histone modifications attract proteins that recognize and bind to the modified (or unmodified) site and mediate downstream biological effects. Proteins that bind modified histones are comparable to methyl-CpG binding proteins, as both interpret chemical marks on chromatin. They include bromodomain proteins, which recognize acetylated lysine in histone tails (Winston and Allis 1999). Whereas bromodomain proteins are associated with gene activity, chromodomain proteins promote silencing and are found to interact with methylated lysine residues on K9 of histone H3 (Bannister et al. 2001; Lachner et al. 2001). At present, it is not known how many histone modifications have dedicated binding proteins to interpret their biological message.

As may be the case for repressive histone marks, methylation of cytosine in the dinucleotide CpG is known to repress transcription in two different ways: by repelling transcription factors or by attracting proteins that bring about transcriptional repression. There are examples of factors that cannot bind productively to DNA when there is a methyl-CpG within their binding site. An example is the protein CTCF that binds to chromatin boundaries. Its importance in the regulation of the Igf2-H19 imprinted domain is now well documented (Bell and Felsenfeld 2000; Hark et al. 2000). Likewise, we know of several proteins that can bind to methyl-CpG and bring about transcriptional repression in model reporter gene systems (Bird and Wolffe 1999), but have only begun to learn about endogenous genes that appear to be regulated by these proteins in normal cells.

DNA methylation has two well-documented biological consequences. On the negative side, it causes mutations due to inefficient repair of G:T mismatches that arise when m^5C becomes spontaneously deaminated. More positively, it causes local, heritable transcriptional silencing. Of the five known methyl-CpG-binding proteins—Mbd1, Mbd2, Mbd4, MeCP2, and Kaiso (Fig. 1)—Mbd4 has been clearly shown to mediate repair of m^5C mutations (Hendrich et al. 1999; Millar et al. 2002), whereas the other four proteins are implicated in transcriptional silencing (Bird and Wolffe 1999). This review will discuss current knowledge about these transcriptional repressors, based on work of this laboratory and others.

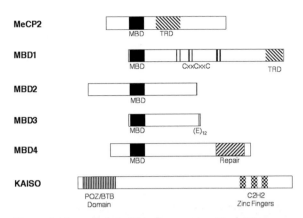

Figure 1. The methyl-CpG-binding proteins. Five of these proteins are related by possession of a methyl-CpG-binding domain (MBD), although in the case of MBD3 this domain does not show a strong preference for binding to methyl-CpG sites. Kaiso is an unrelated protein that binds to m^5CGm^5CG sites via its zinc finger domains. TRD indicates the transcriptional repression domains of MBD1 and MeCP2.

REPRESSION MEDIATED BY MBD PROTEINS

Evidence for "indirect" transcriptional repression by proteins that recognize methylated DNA was first obtained using in vitro transcription in nuclear extracts (Boyes and Bird 1991). In contrast to transfection of DNA into cultured cells, which invariably leads to silencing, methylated DNA could be transcribed as efficiently as nonmethylated DNA in extracts from these same cells. Specific repression did become apparent, however, when the amount of methylated DNA template was reduced below a threshold level. This repression was relieved by adding excess methylated DNA competitor, reinforcing the view that something in the extracts was able to bind to and silence methylated DNA. Prior to this study, an activity that bound multiple methylated sites was detected in nuclear extracts by bandshift assays (see Fig. 2) (Meehan et al. 1989). The properties of the putative repressor matched those of the "MeCP" (for Methyl-CpG-binding Protein) activity. In particular, F9 teratocarcinoma cells, which lacked the methyl-CpG binding activity, also showed unprecedentedly weak repression of methylated reporter genes (Boyes and Bird 1991; Levine et al. 1992). It was concluded that DNA methylation-mediated transcriptional repression depends heavily on methyl-CpG-binding activity.

Attempts to purify MeCP led to the isolation of a protein that was clearly different from the activity originally detected by bandshift assays (Lewis et al. 1992). The new protein was therefore christened MeCP2, to distinguish it from the initial activity, which by default became known as MeCP1. Deletion analysis showed that MeCP2 bound methylated sites via a methyl-CpG binding domain (MBD) of about 80 amino acids (Nan et al. 1993). The three-dimensional structure of the free MBD was determined by NMR (Ohki et al. 1999; Wakefield et al. 1999), soon to be followed by a structure of the MBD-DNA complex (Ohki et al. 2001). Database searching picked out proteins that contained domains related to the

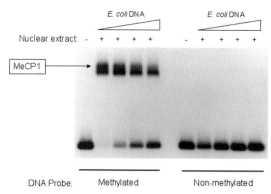

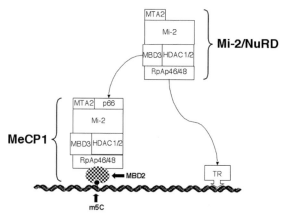

Figure 2. Detection of a methyl-CpG-binding activity in nuclear extracts from mouse liver. The upper DNA protein complex bands include the MeCP1 complex, whose DNA-binding component is Mbd2. The CG11 probe DNA is methylated at multiple CpG sites (*left* lanes) or is unmethylated (*right* lanes).

Figure 3. A diagramatic representation of the MeCP1 complex. The complex includes Mbd2 plus the NuRD complex. NuRD includes histone deacetylases and the chromatin remodeling protein Mi-2. NuRD is recruited to DNA by a number of alternative DNA binding proteins (TR), of which Mbd2 is just one.

MBD of MeCP2 and several of these (Mbd1, Mbd2, and Mbd4) were shown to bind methylated DNA (Cross et al. 1997; Hendrich and Bird 1998). Methylated DNA binding in vivo was demonstrated by assaying localization of exogenous MBD proteins to heterochromatic foci of mouse nuclei. These foci contain the mouse major satellite, which happens to possess ~40% of all methyl-CpG in the mouse nucleus. Mutations in the DNA methyltransferase gene *Dnmt1* that greatly reduce genomic CpG methylation (Li et al. 1992) prevent efficient localization of MeCP2 (Nan et al. 1996), Mbd1, Mbd2, and Mbd4 (Hendrich and Bird 1998), confirming the need for methyl-CpG as their in vivo chromosomal target.

What about the function of these proteins in the cell? Attempts to answer this question are still ongoing, but much is already known. As the first MBD protein to be purified, MeCP2 led the way in functional studies. DNA methylation-dependent transcriptional repression was established using in vitro extracts and transient transfections (Nan et al. 1997). In this way, a transcriptional repression domain was identified. A significant advance in understanding occurred when our group and the group of Alan Wolffe showed that the TRD of MeCP2 recruited the Sin3A-histone deacetylase corepressor complex (Jones et al. 1998; Nan et al. 1998). Repression by MeCP2 could be relieved by the histone deacetylase inhibitor trichostatin A. These studies provided a molecular explanation for the long suspected link between DNA methylation and a repressive chromatin structure. Since then there has been progress in characterizing all MBD protein complexes. Interestingly, each MBD protein appears to specialize by associating with a different corepressor complex. At the time of writing, the MBD1 complex is not fully described, but the involvement of MCAF suggests that a histone methyltransferase may be involved (Fujita et al. 2003). Unexpectedly, an isoform of Mbd1 turns out to have a second DNA-binding domain specific for nonmethylated CpG, suggesting that this protein can potentially interpret CpG as a repressive signal whether it is methylated or not (Jorgensen et al. 2004). The biological rationale for this dual DNA-binding specificity is currently unknown. MBD2 associates with the NuRD or Mi-2 complex (Fig. 3) which contains histone deacetylases and the ATP-dependent chromatin remodeling protein, Mi-2 (Ng et al. 1999; Wade et al. 1999; Feng and Zhang 2001). The latter complex corresponds to MeCP1 (Fig. 2), the methyl-CpG-binding activity originally detected in nuclear extracts (Meehan et al. 1989).

A recent addition to the list of methyl-CpG-binding proteins is Kaiso, which is not a member of the MBD family (Prokhortchouk et al. 2001). This protein was identified biochemically as the basis of a methyl-CpG-binding activity in nuclear extracts and found to be identical to the Kaiso protein that was initially identified as a binding partner of p120—itself a partner of the cell surface protein p120 (Daniel and Reynolds 1999). Kaiso is a POS-BTB domain protein in which the zinc fingers bind the sequence $5'm^5CGm^5CG$ with high affinity. It has also been reported that the isolated zinc finger domain of Kaiso is capable of binding to nonmethylated DNA sequences (Daniel et al. 2002). The involvement of Kaiso in repression of methylated genes in vivo was established by studies that identified the N-CoR corepressor complex as its binding partner (Yoon et al. 2003). The tantalizing link between Kaiso and cytoplasmic signaling via p120, as suggested by yeast two-hybrid and coimmunoprecipitation data, remains to be elucidated.

TARGET GENES OF METHYL-CpG-BINDING PROTEINS

Early studies of the function of methyl-CpG-binding proteins relied on model reporter gene systems. Next, it became important to assess their significance in cells of the organism by identifying target genes whose regulation is disrupted in the absence of a particular protein. Mouse gene knockouts are the system of choice for these studies. The first MBD protein gene to be disrupted was the X-linked *Mecp2* gene (Chen et al. 2001; Guy et al. 2001). Two laboratories found that *Mecp2*-null mice are born and develop normally for several weeks, but they acquire a variety of neurological symptoms at about 6 weeks of age leading to death at ~10 weeks. This delayed-onset phenotype, which is fully penetrant, recalls human Rett Syndrome, which is caused by *MECP2* mutations (Amir et al. 1999; see below). Interestingly, deletion of the *Mecp2* gene only in mouse brain cells using cre expression driven by the brain-specific nestin promoter caused the same symptoms as deletion in the whole mouse (Chen et al. 2001; Guy et al. 2001). Therefore, although MeCP2 is ubiquitously expressed in cells of the mouse, the *Mecp2*-null phenotype appears to be entirely due to its absence in the brain. Biochemical and immunocytochemical studies established that MeCP2 expression levels are highest in the brain; specifically in neurons (Shahbazian et al. 2002).

Knockout of the *Mbd2* gene in mice is compatible with viability and fertility, but some phenotypic effects were noticed (Hendrich et al. 2001). *Mbd2*–/– mothers failed to fully nurture their young, which consequently were underweight regardless of genotype. This weak maternal response to pups was also evident from the delayed retrieval of pups after they were removed from the nests of *Mbd2*–/– mothers. Other phenotypes were initially less obvious, but led to identification of the first target genes for a methyl-CpG binding protein. First, cells derived from *Mbd2*–/– mice were found to be unable to repress methylated reporter genes effectively. This finding matched early experiments using in vitro extracts, which showed that the MeCP1 complex—whose DNA-binding component is Mbd2—is largely responsible for the repression of methylated reporter genes.

This evidence that Mbd2-deficiency might compromise repression of methylated genes raised the possibility that a careful search might uncover misregulation of endogenous genes in these mice. Success came with an analysis of T cell differentiation in the mouse (Hutchins et al. 2002). Sorting of *Mbd2*–/– cells by fluorescence allowed detection of a small fraction of naive and differentiated T cells in which expression of both interleukin 4 (*Il4*) and interferon gamma (*Ifγ*) genes was derepressed. Expression of *Il4* and *Ifγ* is normally mutually exclusive and is initiated by triggering differentiation into either the Th2 or the Th1 pathway, respectively. In the absence of Mbd2, differentiated *Mbd2*–/– thymocytes often expressed both genes and this misexpression effect could be enhanced by addition of the histone deacetylase inhibitor trichostatin A. Chromatin immunoprecipitation showed that Mbd2 was normally present at the *Il4* promoter and was lost when the gene was transcriptionally activated. It is noteworthy that absence of Mbd2 does not lead to 100% reactivation of the inappropriate gene; the effect is instead stochastic in that only a fraction of all Th1 cells derepress *Ifγ*, but derepression in those cells appears to be complete. Thus the removal of Mbd2 appears to increase the normally very low probability that these silent genes will become fully reactivated. Similarly, reactivation of the aberrantly silenced pi-class glutathione S-transferase gene can be induced in cancer cell lines by Mbd2 deficiency (Lin and Nelson 2003). It seems likely that careful investigation of specific cell types will uncover further examples of the loosening of gene repression caused by absence of Mbd2.

The status of the *Mecp2*-null mouse as a model for Rett Syndrome has stimulated the search for MeCP2 target genes. Initial attempts to detect misregulated genes in the brains of the mutant animals offered partial success (Tudor et al. 2002). Deregulation of gene expression was apparent, but the degree of misexpression was subtle and only acquired statistical significance when groups of affected genes were considered together. The "candidate gene" approach proved more successful. Two groups (Chen et al. 2003; Martinowich et al. 2003) hypothesized that MeCP2 might play a role in activity-dependent expression of the gene for brain-derived neurotropic factor (*Bdnf*). Only one of several *Bdnf* promoters responds to calcium-dependent activation and this promoter was shown to associate with MeCP2 in cultured neurons. Upon stimulation of the neurons, MeCP2 became phosphorylated and was displaced from this promoter (Chen et al. 2003). Although the effect of MeCP2 deficiency on expression proved to be small in this in vitro system, it is possible that deregulation has important consequences in the animal itself. That MeCP2 deficiency can cause major gene expression changes was shown in a nonmammalian system, the embryo of the amphibian, *Xenopus laevis*. Here, MeCP2 was found associated with the promoter of the *Hairy2a* gene, whose product is important for limiting the number of embryonic cells that become neurons (Stancheva et al. 2003). Activation of the gene during development was accompanied by displacement of MeCP2, suggesting that, as in the case of the mouse *Bdnf* gene, MeCP2 is involved in the dynamic control of gene expression. These findings suggest that the view of methyl-CpG-binding proteins as long-term silencers of gene expression may need to be modified. It remains possible, however, that they may be involved in both stable and dynamic repression, depending on the locus concerned.

A target for Kaiso in cancer cell lines has clearly been shown to depend on recruitment of this protein plus the N-CoR corepressor (Yoon et al. 2003), but MBD1 target genes are, at the time of writing, yet to be reported. Immunoprecipitation experiments have identified multiple DNA sequences associated with various MBD proteins (Ballestar et al. 2003), but the regulatory consequences of MBD protein withdrawal at these loci have not yet been assessed. It is possible that all genomic regions that associate with one of these proteins are to some extent regulated by them. Alternatively, MBD proteins may spend time unproductively at methyl-CpG sites where their regulatory input is superfluous. In depth study of specific genes is required to distinguish these possibilities. A related issue concerns the redundancy or otherwise of these proteins. Do they compete for access to all methylated sites and therefore functionally back each other up? Or does each protein act at a subset of genes that are adapted to respond specifically to its particular regulatory influence? The answer to these questions is not yet clear, although we know that combining *Mecp2* and *Mbd2* mutations does not lead to an obvious "synthetic" enhancement of phenotype (Guy et al. 2001). Chromatin immunoprecipitation experiments identified some loci that bound to a single member of the family and other loci that appeared to bind multiple members (Ballestar et al. 2003).

METHYL-CpG-BINDING PROTEINS AND DISEASE

Dramatic demonstration of the medical relevance of methyl-CpG-binding proteins came with the discovery that at least 80% of patients with Rett Syndrome have new mutations in the *MECP2* gene (Amir et al. 1999; Shahbazian and Zoghbi 2002; Kriaucionis and Bird 2003). Rett Syndrome affects females that are heterozygous for the mutations. Because of X chromosome inactivation, the patients are mosaic for expression of either the mutant or the wild-type gene. They develop apparently normally for 6–18 months, at which time they show regression of motor skills, repetitive hand movements, abnormal breathing, microcephaly, and other symptoms. Given the role of MeCP2 as a transcriptional repressor, an obvious hypothesis to explain the disease is that genes in the brain that should be silenced by MeCP2 escape repression in its absence, leading to aberrant neuronal function. As mentioned above, the discovery that MeCP2 is involved in silencing the *Bdnf* gene, which encodes a neuronal growth factor, is compatible with this theory, although much remains to be done to connect the medical condition with misexpression of this particular gene. Meanwhile, attempts to define the molecular pathology of Rett Syndrome continue with the search for additional MeCP2 target genes.

The relationship between DNA methylation and cancer is the subject of much research, as well as animated debate (Baylin and Bestor 2002). At the experimental level, the connection is well illustrated by the demonstration that the Min mouse model of intestinal tumorigenesis depends on the DNA methyltransferase Dnmt1 (Laird et al. 1995; Eads et al. 2002). Reduced levels of this enzyme are accompanied by a significant drop in tumorigenesis, virtually abolishing the tumor-susceptible phenotype at the lowest enzyme levels compatible with viability. Parallel findings have been made in *Mbd2* mutant animals when crossed onto a Min background (Sansom et al. 2003). *Mbd2–/–* Min mice develop very few tumors and live considerably longer that *Mbd2+/+* controls. *Mbd2+/–* heterozygotes show an intermediate phenotype, stressing the sensitivity of the assay to full dosage of Mbd2. The requirements of tumorigenesis for a protein that methylates DNA at CpG and also for a protein that binds to the resulting methyl-CpGs begs an important question: Which pathways to cancer involve these proteins? At present we do not know, but the disturbance of the Wnt signalling pathway in cells that have lost both copies of the *Apc* gene (Fearnhead et al. 2001) raises the possibility that misregulation of downstream genes is somehow involved. A significant difference between *Dnmt1* and *Mbd2* mutant phenotypes is that the former is an embryonic lethal whereas the latter results in viable and fertile mice. Thus both the tumor and its host depend on Dnmt1 function, but only tumors depend on Mbd2. This distinction has therapeutic implications that deserve to be pursued.

CONCLUSIONS

DNA methylation can be considered together with histone modifications as a mechanism for adapting chromatin structure to local functional needs. Like histone modifications, DNA methylation can be read by binding proteins. Methyl-CpG-binding proteins were initially characterized in model reporter gene systems, but gene knockouts in mice have begun to functionally relate each protein to specific target genes. In doing so, the mouse studies have illuminated the relationships between methyl-CpG-binding proteins and human disease.

ACKNOWLEDGMENTS

Our research is supported by the Wellcome Trust, United Kingdom.

REFERENCES

Amir R.E., Van den Veyver I.B., Wan M., Tran C.Q., Francke U., and Zoghbi H.Y. 1999. Rett syndrome is caused by mutations in X-linked MECP2, encoding methyl-CpG-binding protein 2. *Nat. Genet.* **23:** 185.

Ballestar E., Paz M.F., Valle L., Wei S., Fraga M.F., Espada J., Cigudosa J.C., Huang T.H., and Esteller M. 2003. Methyl-CpG binding proteins identify novel sites of epigenetic inactivation in human cancer. *EMBO J.* **22:** 6335.

Bannister A.J., Zegerman P., Partridge J.F., Miska E.A., Thomas J.O., Allshire R.C., and Kouzarides T. 2001. Selective recognition of methylated lysine 9 on histone H3 by the HP1 chromo domain. *Nature* **410:** 120.

Baylin S. and Bestor T.H. 2002. Altered methylation patterns in cancer cell genomes: Cause or consequence? *Cancer Cell* **1:** 299.

Bell A.C. and Felsenfeld G. 2000. Methylation of a CTCF-dependent boundary controls imprinted expression of the Igf2 gene. *Nature* **405:** 482.

Bird A.P. 1980. DNA methylation and the frequency of CpG in animal DNA. *Nucleic Acids Res.* **8:** 1499.

Bird A. and Wolffe A.P. 1999. Methylation-induced repression—Belts, braces and chromatin. *Cell* **99:** 451.

Boyes J. and Bird A. 1991. DNA methylation inhibits transcription indirectly via a methyl-CpG binding protein. *Cell* **64:** 1123.

Chen R.Z., Akbarian S., Tudor M., and Jaenisch R. 2001. Deficiency of methyl-CpG binding protein-2 in CNS neurons results in a Rett-like phenotype in mice. *Nat. Genet.* **27:** 327.

Chen W.G., Chang Q., Lin Y., Meissner A., West A.E., Griffith E.C., Jaenisch R., and Greenberg M.E. 2003. Derepression of BDNF transcription involves calcium-dependent phosphorylation of MeCP2. *Science* **302:** 885.

Coulondre C., Miller J.H., Farabough P.J., and Gilbert W. 1978. Molecular basis of base substitution hotspots in *Escherichia coli*. *Nature* **274:** 775.

Cross S.H., Meehan R.R., Nan X., and Bird A. 1997. A component of the transcriptional repressor MeCP1 is related to mammalian DNA methyltransferase and trithorax-like protein. *Nat. Genet.* **16:** 256.

Daniel J.M. and Reynolds A.B. 1999. The catenin p120(ctn) interacts with Kaiso, a novel BTB/POZ domain zinc finger transcription factor. *Mol. Cell. Biol.* **19:** 3614.

Daniel J.M., Spring C.M., Crawford H.C., Reynolds A.B., and Baig A. 2002. The p120(ctn)-binding partner Kaiso is a bimodal DNA-binding protein that recognizes both a sequence-specific consensus and methylated CpG dinucleotides. *Nucleic Acids Res.* **30:** 2911.

Eads C.A., Nickel A.E., and Laird P.W. 2002. Complete genetic suppression of polyp formation and reduction of CpG-island hypermethylation in Apc(Min/+) Dnmt1-hypomorphic mice. *Cancer Res.* **62:** 1296.

Fearnhead N.S., Britton M.P., and Bodmer W.F. 2001. The ABC of APC. *Hum. Mol. Genet.* **10:** 721.

Feng Q. and Zhang Y. 2001. The MeCP1 complex represses transcription through preferential binding, remodeling, and deacetylating methylated nucleosomes. *Genes Dev.* **15:** 827.

Fujita N., Watanabe S., Ichimura T., Ohkuma Y., Chiba T., Saya H., and Nakao M. 2003. MCAF mediates MBD1-dependent transcriptional repression. *Mol. Cell. Biol.* **23:** 2834.

Guy J., Hendrich B., Holmes M., Martin J.E., and Bird A. 2001. A mouse *Mecp2*-null mutation causes neurological symptoms that mimic Rett syndrome. *Nat. Genet.* **27:** 322.

Hark A.T., Schoenherr C.J., Katz D.J., Ingram R.S., Levorse J.M., and Tilghman S.M. 2000. CTCF mediates methylation-sensitive enhancer-blocking activity at the H19/Igf2 locus. *Nature* **405:** 486.

Hendrich B. and Bird A. 1998. Identification and characterization of a family of mammalian methyl-CpG binding proteins. *Mol. Cell. Biol.* **18:** 6538.

Hendrich B., Guy J., Ramsahoye B., Wilson V.A., and Bird A. 2001. Closely related proteins Mbd2 and Mbd3 play distinctive but interacting roles in mouse development. *Genes Dev.* **15:** 710.

Hendrich B., Hardeland U., Ng H.-H., Jiricny J., and Bird A. 1999. The thymine glycosylase MBD4 can bind to the product of deamination at methylated CpG sites. *Nature* **401:** 301.

Hutchins A., Mullen A., Lee H., Barner K., High F., Hendrich B., Bird A., and Reiner S. 2002. Gene silencing quantitatively controls the function of a developmental trans-activator. *Mol. Cell* **10:** 81.

Jenuwein T. and Allis C.D. 2001. Translating the histone code. *Science* **293:** 1074.

Jones P.L., Veenstra G.J., Wade P.A., Vermaak D., Kass S.U., Landsberger N., Strouboulis J., and Wolffe A.P. 1998. Methylated DNA and MeCP2 recruit histone deacetylase to repress transcription. *Nat. Genet.* **19:** 187.

Jorgensen H.F., Ben-Porath I., and Bird A.P. 2004. Mbd1 is recruited to both methylated and nonmethylated CpGs via distinct DNA binding domains. *Mol. Cell. Biol.* **24:** 3387.

Kriaucionis S. and Bird A. 2003. DNA methylation and Rett syndrome. *Hum. Mol. Genet.* (Spec. No. 2) **12:** R221.

Lachner M., O'Carroll D., Rea S., Mechtler K., and Jenuwein T. 2001. Methylation of histone H3 lysine 9 creates a binding site for HP1 proteins. *Nature* **410:** 116.

Laird P.W., Jackson-Grusby L., Fazeli A., Dickinson S.L., Jung W.E., Li E., Weinberg R.A., and Jaenisch R. 1995. Suppression of intestinal neoplasia by DNA hypomethylation. *Cell* **81:** 197.

Levine A., Cantoni G.L., and Razin A. 1992. Methylation in the preinitiation domain suppresses gene transcription by an indirect mechanism. *Proc. Natl. Acad. Sci.* **89:** 10119.

Lewis J.D., Meehan R.R., Henzel W.J., Maurer-Fogy I., Jeppesen P., Klein F., and Bird A. 1992. Purification, sequence and cellular localisation of a novel chromosomal protein that binds to methylated DNA. *Cell* **69:** 905.

Li E., Bestor T.H., and Jaenisch R. 1992. Targeted mutation of the DNA methyltransferase gene results in embryonic lethality. *Cell* **69:** 915.

Lin X. and Nelson W.G. 2003. Methyl-CpG-binding domain protein-2 mediates transcriptional repression associated with hypermethylated GSTP1 CpG islands in MCF-7 breast cancer cells. *Cancer Res.* **63:** 498.

Martinowich K., Hattori D., Wu H., Fouse S., He F., Hu Y., Fan G., and Sun Y.E. 2003. DNA methylation-related chromatin remodeling in activity-dependent BDNF gene regulation. *Science* **302:** 890.

Meehan R.R., Lewis J.D., McKay S., Kleiner E.L., and Bird A.P. 1989. Identification of a mammalian protein that binds specifically to DNA containing methylated CpGs. *Cell* **58:** 499.

Millar C.B., Guy J., Sansom O.J., Selfridge J., MacDougall E., Hendrich B., Keightley P.D., Bishop S.M., Clarke A.R., and Bird A. 2002. Enhanced CpG mutability and tumorigenesis in MBD4-deficient mice. *Science* **297:** 403.

Nan X., Campoy J., and Bird A. 1997. MeCP2 is a transcriptional repressor with abundant binding sites in genomic chromatin. *Cell* **88:** 471.

Nan X., Meehan R.R., and Bird A. 1993. Dissection of the methyl-CpG binding domain from the chromosomal protein MeCP2. *Nucleic Acids Res.* **21:** 4886.

Nan X., Tate P., Li E., and Bird A.P. 1996. DNA methylation specifies chromosomal localization of MeCP2. *Mol. Cell. Biol.* **16:** 414.

Nan X., Ng H.-H., Johnson C.A., Laherty C.D., Turner B.M., Eisenman R.N., and Bird A. 1998. Transcriptional repression by the methyl-CpG-binding protein MeCP2 involves a histone deacetylase complex. *Nature* **393:** 386.

Ng H.-H., Zhang Y., Hendrich B., Johnson C.A., Burner B.M., Erdjument-Bromage H., Tempst P., Reinberg D., and Bird A. 1999. MBD2 is a transcriptional repressor belonging to the MeCP1 histone deacetylase complex. *Nat. Genet.* **23:** 58.

Ohki I., Shimotake N., Fujita N., Nakao M., and Shirakawa M. 1999. Solution structure of the methyl-CpG-binding domain of the methylation-dependent transcriptional repressor MBD1. *EMBO J.* **18:** 6653.

Ohki I., Shimotake N., Fujita N., Jee J., Ikegami T., Nakao M., and Shirakawa M. 2001. Solution structure of the methyl-CpG binding domain of human MBD1 in complex with methylated DNA. *Cell* **105:** 487.

Prokhortchouk A., Hendrich B., Jorgensen H., Ruzov A., Wilm M., Georgiev G., Bird A., and Prokhortchouk E. 2001. The p120 catenin partner Kaiso is a DNA methylation-dependent transcriptional repressor. *Genes Dev.* **15:** 1613.

Sansom O.J., Berger J., Bishop S.M., Hendrich B., Bird A., and Clarke A.R. 2003. Deficiency of Mbd2 suppresses intestinal tumorigenesis. *Nat. Genet.* **34:** 145.

Shahbazian M.D. and Zoghbi H.Y. 2002. Rett syndrome and MeCP2: Linking epigenetics and neuronal function. *Am. J. Hum. Genet.* **71:** 1259.

Shahbazian M.D., Antalffy B., Armstrong D.L., and Zoghbi H.Y. 2002. Insight into Rett syndrome: MeCP2 levels display tissue- and cell-specific differences and correlate with neuronal maturation. *Hum. Mol. Genet.* **11:** 115.

Stancheva I., Collins A.L., Van den Veyver I.B., Zoghbi H., and Meehan R.R. 2003. A mutant form of MeCP2 protein associated with human Rett syndrome cannot be displaced from methylated DNA by notch in *Xenopus* embryos. *Mol. Cell* **12:** 425.

Tudor M., Akbarian S., Chen R.Z., and Jaenisch R. 2002. Transcriptional profiling of a mouse model for Rett syndrome reveals subtle transcriptional changes in the brain. *Proc. Natl. Acad. Sci.* **99:** 15536.

Wade P.A., Gegonne A., Jones P.L., Ballestar E., Aubry F., and Wolffe A.P. 1999. Mi-2 complex couples DNA methylation to chromatin remodelling and histone deacetylation. *Nat. Genet.* **23:** 62.

Wakefield R.I.D., Smith B.P., Nan X., Free A., Soterious A., Uhrin D., Bird A., and Barlow P.N. 1999. The solution structure of the domain from MeCP2 that binds to methylated DNA. *J. Mol. Biol.* **291:** 1055.

Winston F. and Allis C.D. 1999. The bromodomain: A chromatin-targeting module? *Nat. Struct. Biol.* **6:** 601.

Yoon H.G., Chan D.W., Reynolds A.B., Qin J., and Wong J. 2003. N-CoR mediates DNA methylation-dependent repression through a methyl CpG binding protein Kaiso. *Mol. Cell* **12:** 723.

Genome Defense and DNA Methylation in *Neurospora*

E.U. SELKER

Department of Biology and Institute of Molecular Biology, University of Oregon, Eugene, Oregon 97403

The continuity of life is attributed to the faithful copying of DNA. It is therefore intriguing that some organisms chemically modify their DNA in ways that can lead to heritable changes. Soon after the discovery that a fraction of cytosines are methylated in DNA of some eukaryotes—including vertebrates, plants, and some fungi—evidence began to accumulate that DNA methylation is inherently mutagenic. For example, in mammals, where methylation is almost exclusively found at cytosines immediately preceding guanines ("CpG" dinucleotides), it was noticed that methylated chromosomal regions are typically unexpectedly poor in CpGs and unexpectedly rich in TpGs and CpAs, apparently as a result of C to T mutations at sites subject to methylation (Bird 1980). It was assumed that this reflected the fact that 5-methyl-cytosine, like cytosine, is somewhat unstable, spontaneously decomposing to 5-methyl-uridine (thymine) and that G:T mismatches are repaired less efficiently than G:U mismatches. Discovery of repeat-induced point mutation (RIP) in the filamentous fungus *Neurospora crassa* raised the possibility that DNA methyltransferases and similar enzymes may be directly responsible for such mutations (Selker 1990b; Freitag et al. 2002). What positive functions does DNA methylation serve? Although the control and function of DNA methylation are both still hotly debated, it is becoming increasingly clear that DNA methylation is one of several mechanisms that defends the eukaryotic genomes from mischievous DNA. *Neurospora* provided an early, and exceptionally clear, illustration of this and it continues to provide a convenient model to elucidate DNA methylation and other epigenetic processes.

THE METHYLATION LANDSCAPE

Although *Neurospora* was initially reported to have no methylated DNA (Antequera et al. 1984), observations in the 1980s revealed cytosine methylation in this model eukaryote. After methods became established to transform *Neurospora*, it was noticed that some transforming sequences are subject to de novo methylation, especially when they are repeated (Bull and Wootton 1984; Selker et al. 1987a; Orbach et al. 1988). At the same time, natural patches of methylation were found associated with several 5S rRNA pseudogenes (Selker and Stevens 1985; Margolin et al. 1998). No methylation has been reported in any of the hundreds of protein-coding genes that have been studied. Indeed, as discussed below, recent analyses suggest that most DNA methylation is associated with relics of transposons. We now know that ~1.5% of the cytosines in the DNA of *N. crassa* are methylated (Russell et al. 1987; Foss et al. 1993).

The first methylated region characterized in detail is the 1.6-kb zeta–eta (ζ–η) region (Selker and Stevens 1985; Selker et al. 1993b). This consists of a diverged tandem duplication of a 0.8-kb segment of DNA, including a 5S rRNA gene. Comparison of this region with the corresponding chromosomal region of strains lacking the duplication led to the idea that repeated sequences can somehow induce DNA methylation (Selker et al. 1985, 1987a; Grayburn and Selker 1989) and ultimately led to the discovery of the genome defense system that we named RIP (Selker et al. 1987b; Selker and Garrett 1988; Cambareri et al. 1989). Both the ζ–η region and the psi-63 (ψ63) region, the second methylated region discovered in *Neurospora* (Metzenberg et al. 1985; Foss et al. 1993; Miao et al. 1994; Margolin et al. 1998), are products of RIP.

THE PROTOTYPICAL GENOME DEFENSE SYSTEM: RIP

RIP detects duplicated sequences in the haploid genomes of special dikaryotic cells resulting from fertilization and then riddles both copies of the duplicated sequence with polarized transition mutations; C:G pairs are replaced with T:A pairs (Cambareri et al. 1989). RIP has a clear sequence preference (Cs preceding As are mutated most frequently), making it straightforward to recognize sequences that have been mutated by RIP (Margolin et al. 1998). In a single passage through the sexual cycle, up to ~30% of the G:C pairs in duplicated sequences can be changed to A:T pairs (Cambareri et al. 1991). After RIP, remaining cytosines are generally methylated, including those not in symmetrical sequences (e.g., CpGs) (Selker et al. 1993a). The methylation does not depend on the continued presence of sequence duplications and is normally stable through numerous rounds of DNA replication (Selker and Garrett 1988; Singer et al. 1995). DNA molecules show different patterns and levels of modification (Selker and Stevens 1985; Selker et al. 1993a). This heterogeneous methylation can extend beyond the mutated region and even beyond the edge of the segment that was originally duplicated (Cambareri et al. 1989; Selker et al. 1993a; Irelan and Selker 1997; Miao et al. 2000).

Results of genome-wide analyses of DNA methylation suggest that the vast majority of methylated residues are in relics of RIP (Galagan et al. 2003; Selker et al. 2003). Indeed, the only methylation not known to have resulted from RIP is that in the tandemly arranged rDNA (Perkins et al. 1986). A survey of methylated *Neurospora* sequences isolated by affinity chromatography using the methyl-binding domain (MBD) of MeCP2 revealed clear evidence of RIP in 47 of 51 sequenced fragments (Selker et al. 2003). Analysis of these sequences revealed significant similarities with a variety of transposable elements identified in other organisms, including both retrotransposons (e.g., lolligag, Tad, Tcen, and DAB1) and DNA-type transposons (e.g., Dodo1, Dodo2, Dodo3, listless, dPunt, Punt3, and Nogo) (Selker et al. 2003). Similar information came from large-scale genome sequencing projects. About 10% of the genome is composed of repetitive DNA and most of this is identifiable as transposon-like sequences that have been riddled with mutations by RIP (Galagan et al. 2003). Not a single intact transposable element was identifiable, consistent with failures to detect transpositions in standard *Neurospora* strains (Kinsey and Helber 1989). We conclude that RIP is an efficient genome defense mechanism. Does this mean that *Neurospora* cannot generate gene families by gene duplications? Nearly 20% of *N. crassa* genes are found in multigene families, but, interestingly, nearly all of the paralogs are sufficiently divergent and/or short that they should be invisible to RIP (Galagan et al. 2003). Thus, although *Neurospora* has gene families, RIP is presumably an obstacle to evolution of new genes through gene duplication (Galagan and Selker 2004).

DE NOVO AND MAINTENANCE METHYLATION IN *NEUROSPORA*

Laboratory experiments revealed that most products of RIP, but not their unmutated counterparts, are methylated (Selker et al. 1987a, 1993a; Cambareri et al. 1991; Singer et al. 1995; Miao et al. 2000). Similarly, predicted products of RIP identified in the genome based on their sequence composition are generally methylated (Galagan et al. 2003; Selker et al. 2003). In principle, such methylation could reflect signals for de novo methylation that work in vegetative cells. Alternatively, the mutated sequences may be methylated because of propagation of methylation established earlier (e.g., during RIP) and they could lack the capacity to trigger methylation de novo in vegetative cells. These possibilities were tested by determining whether a given sequence can induce methylation after it is stripped of its methylation by molecular cloning or by drug treatment (Singer et al. 1995). Such experiments revealed that, in general, sequences altered by RIP can trigger de novo methylation in vegetative cells. Most sequences mutated by RIP become remethylated both at their original genomic location and at arbitrary chromosomal positions (Selker et al. 1987a). Some products of RIP with relatively few mutations, however, do not normally become remethylated at their original site, suggesting that the observed methylation represents propagation of methylation established earlier (Singer et al. 1995). Although surprising, considering that this methylation is heterogeneous and is not limited to symmetrical sites, the capacity of *Neurospora* to perform such "maintenance methylation" on some sequences was experimentally demonstrated (Selker et al. 2002).

cis-ACTING SIGNALS THAT CONTROL DNA METHYLATION

As mentioned above, some transforming sequences are subject to de novo methylation without going through the sexual cycle (Bull and Wootton 1984; Selker et al. 1987a, 1993a; Orbach et al. 1988; Selker 1990a; Pandit and Russo 1992; Romano and Macino 1992). We have found that sequences differ in their susceptibility to methylation in vegetative cells. Native, unmutated sequences are not susceptible, while many foreign sequences with resemblance to sequences mutated by RIP are (M. Freitag and E. Selker, unpubl.). This methylation appears loosely correlated with the copy number of transforming DNA. Although single-copy sequences are less frequently methylated than multicopy sequences (Pandit and Russo 1992; Romano and Macino 1992; Selker et al. 1993b), multicopy sequences are not always methylated (Selker and Garrett 1988).

Observations on the specificity of methylation in *Neurospora* led to the "collapsed chromatin model," in which DNA methylation is the fate of sequences that are inert (i.e., completely inactive [Selker 1990a]). One prediction of the model was that short sequences should not have the ability to trigger methylation, for example, when inserted into an active gene. To explore this and other possibilities, we developed efficient gene-targeting systems that allowed us to test the methylation potential of single copies of sequences integrated precisely and without extraneous sequences at a common chromosomal position (at the *am* locus on LG VR [Miao et al. 1994] or the *his-3* locus on LG IR; Margolin et al. 1997). Using these systems, we demonstrated that mutations per se (e.g., numerous A:T to G:C mutations) do not trigger methylation and that products of RIP, such as the ζ–η region, effectively contain multiple, additive methylation signals, whose effects can spread hundreds of base pairs into flanking sequences. Fragments of the ζ–η region as short as 171 bp can trigger methylation (Selker et al. 1993b; Miao et al. 2000). We found that mutation density per se does not determine whether sequences become methylated and that neither A:T-richness nor high densities of TpA dinucleotides, typical attributes of methylated sequences in *Neurospora*, are essential features of methylation signals. Nevertheless, both A:T-richness and high densities of TpA dinucleotides appear to promote methylation in *Neurospora*. These and other findings from "transplanting" small fragments of genes, pieces of DNA mutated by RIP, and synthetic sequences led us to conclude that methylated sequences do not simply reflect the absence of signals that prevent methylation; they apparently contain positive signals that trigger methylation (Miao et al. 2000).

To better define the nature of these signals, we developed a more sensitive assay to test the capacity of short (25–100-bp) synthetic oligonucleotides to trigger methylation at a specific locus (Tamaru and Selker 2003). Our system used a *his-3* targeting vector carrying a 100-bp ζ–η segment surrounded by a lightly RIP-mutated allele of the *am* gene. We demonstrated that this mosaic construct does not trigger methylation itself at *his-3*, but does provide a sensitive context to assess the potential of various sequences to induce methylation. A variety of random sequences consisting of only A and T residues triggered methylation of nearby cytosines. Introduction of G:C pairs into the A:T-rich sequences was strongly inhibitory and both As and Ts were found to be required on the same strand to trigger significant methylation. Nevertheless, neither TpA nor ApT dinucleotides were essential. Tests of 20-, 25-, 40-, and 80-mer fragments of the most potent sequence, (TAAA)n, showed that longer tracts of this sequence act as stronger signals. It seems possible that an unidentified "A:T-hook"-type protein mediates methylation in *Neurospora*. Consistent with this possibility, we found that Distamycin A, an analog of the A:T-hook motif, interferes with de novo methylation in *Neurospora* (Tamaru and Selker 2003). Nevertheless, some of our findings do not support this hypothesis. In particular, we found that sequences with only two bp A:T tracts (e.g., (CTA)n) can induce methylation. In addition, using A:T-hook protein HMG-I (kindly provided by R. Reeves, Washington State University, Pullman), we showed that the sequence preference for HMG-I binding does not simply correlate with the sequence preferences for de novo methylation. Additional work will be required to determine how relics of RIP, and similar degenerate sequences, are recognized to trigger methylation in vegetative cells of *Neurospora*.

FORWARD GENETICS APPROACH TO ELUCIDATE CONTROL AND FUNCTION OF DNA METHYLATION

Neurospora is well suited to identify components of the methylation machinery by genetic approaches that do not rest on prior knowledge or preconceptions. We have successfully used several approaches to identify mutants defective in methylation (*dim*), but have not yet saturated the *Neurospora* genome for such mutations. DNA methylation can silence genes in *Neurospora* (Rountree and Selker 1997), allowing for the direct selection of mutations that affect DNA methylation. Nevertheless, most known *dim* mutants have been identified in other ways, such as in "brute force" screens using Southern hybridization to identify mutants that affect methylation (Foss et al. 1993, 1995). Two mutants, *dim-2* and *dim-5*, abolish all detectable DNA methylation. Mapping and complementation studies revealed that the *dim-2* gene encodes a DNA methyltransferase (DMTase) (Kouzminova and Selker 2001), while *dim-5* encodes a histone methyltransferase (HMTase) (Tamaru and Selker 2001).

Biochemical characterization of DIM-2 should be interesting. This predicted 1454-amino-acid protein is responsible for all detected DNA methylation in *Neurospora*, including methylation in a variety of sequence contexts. Mutations preventing all methylation have not been described in other eukaryotes, which may reflect the fact that DNA methylation results from multiple DMTases in most organisms that have been characterized and that DNA methylation is essential in some organisms. Although the DIM-2 carboxy-terminal domain includes all the well conserved motifs characteristic of DMTases, its amino-terminal domain shows no marked similarity to previously described proteins (Kouzminova and Selker 2001).

CONTROL OF DNA METHYLATION BY HISTONE METHYLATION

The discovery that *dim-5* encodes a HMTase provided the first indication that DNA methylation is regulated by histone methylation, at least in some organisms. Biochemical work showed that DIM-5 is specific for lysine 9 of histone H3 and that it efficiently trimethylates this residue in vitro and in vivo, unlike all previously characterized HMTases (Tamaru and Selker 2001; Tamaru et al. 2003). In order to determine whether histone H3 is the critical substrate for DIM-5 in vivo, we transformed a Dim+ *Neurospora* strain with engineered histone H3 genes carrying a mutation at lysine 9 that would preclude methylation. Amino acid substitutions at lysine 9 dramatically reduced DNA methylation, implicating histone H3 as the critical substrate for DNA methylation (Tamaru and Selker 2001). Evidence that some DNA methylation in *Arabidopsis* is also controlled by histone methylation came soon thereafter (Jackson et al. 2002; Malagnac et al. 2002). Interestingly, the critical mark in *Arabidopsis* appears to be di-, rather than tri-, methyl-lysine 9 (Jackson et al. 2004). Most recently, Jenuwein and colleagues have provided evidence in mice that trimethylation of lysine 9 of histone H3 directs DNA methylation to major satellite repeats at pericentric heterochromatin (Lehnertz et al. 2003). The extent to which DNA methylation is controlled by histones in eukaryotes is not yet known.

SEARCH FOR OTHER COMPONENTS OF DNA METHYLATION PATHWAY

In principle, forward genetics should reveal all components of the DNA methylation pathway except for those that are essential or redundant. However, use of reverse genetics and biochemical approaches can facilitate the identification of components in the methylation process. For example, evidence from other systems that a heterochromatin protein, HP1 (Eissenberg and Elgin 2000), binds methylated lysine 9 of histone H3 (Bannister et al. 2001; Jacobs et al. 2001; Lachner et al. 2001), led us to search for an HP1 homolog in *Neurospora* and to test its possible involvement in DNA methylation by disrupting the HP1 gene (*hpo*) by RIP. We found that HP1 is indeed essential for DNA methylation in *Neurospora*, implying that this protein directs DIM-2 to DNA associated with chromatin in which histone H3 is methylated at lysine 9 (Freitag et al. 2004a).

Clues that the RNAi machinery may be involved in DNA methylation (Wassenegger et al. 1994; Aufsatz et al. 2002; Chan et al. 2004) prompted us to test the components of the *Neurospora* RNAi machinery, including apparent RNA-dependent RNA polymerases, dicers, argonautes, and RecQ helicase homologs that had been identified by genetic or bioinformatics approaches (Galagan et al. 2003). We found no evidence that any of the RNAi machinery is involved in initiation or maintenance of DNA methylation (Freitag et al. 2004b). In contrast, mutational analyses of histone deacetylase genes of *Neurospora* revealed that at least one of these genes is involved in DNA methylation in *Neurospora* (K. Smith et al., unpubl.), consistent with the observation that the histone deacetylase inhibitor Trichostatin A reduces DNA methylation in some chromosomal regions of this organism (Selker 1998).

CONCLUSIONS

Why control DNA methylation through histone methylation? In the last few years it has become increasingly clear that histones are more than structural proteins; they are informational molecules (Jenuwein and Allis 2001). Particular combinations of posttranslational modifications of histones, including phosphorylation, methylation, acetylation, ubiquitination, and ADP ribosylation, can influence the function of the associated DNA and the modifications can depend on the presence or absence of other modifications. For example, methylation of lysine 9 on histone H3 is inhibited by phosphorylation of serine 10 and methylation of lysine 4, which are associated with active sequences (Rea et al. 2000; E. Berge et al., unpubl.) and there are indications that histone acetylation also influences methylation of lysine 9 (Selker 1998; Nakayama et al. 2001; K. Smith et al., unpubl.). Thus histones are well suited to integrate information bearing on the DNA that they are associated with.

Neurospora provided the first example of a genome defense system, RIP, and has more recently revealed two additional genetic mechanisms related to RNA interference (RNAi) mechanisms discovered in plants and animals that should also resist change in the genome. The first, "quelling," is specific to the vegetative phase of the *Neurospora* life cycle and is responsible for destroying RNA homologous to aberrant RNAs, for example, resulting from expression of tandem arrays of transgenes (Romano and Macino 1992; Catalanotto et al. 2002). The second, meiotic silencing by unpaired DNA (MSUD) (Aramayo and Metzenberg 1996; Shiu et al. 2001), scans paired homologs in meiosis for unpaired sequences and then destroys RNA matching unpaired DNA (Shiu and Metzenberg 2002; Lee et al. 2004). DNA methylation is commonly associated with both sequences mutated by RIP and with transgenes that trigger quelling, but it is not required for either process (Cogoni et al. 1996; Kouzminova and Selker 2001). Similarly, the RNAi processes are not required for DNA methylation. Thus, DNA methylation may be considered a fourth genome defense system, although there are undoubtedly additional connections between *Neurospora*'s genome defense systems that are yet to be elucidated.

It should be interesting to discover similarities and differences between the *Neurospora* genome defense systems and those in other organisms. For example, we need to better characterize how DNA methylation controls transposable elements in *Neurospora* and mammals. In mammals, DNA methylation appears to block initiation of transcription, but not transcription elongation (Jones 1999); while in *Neurospora*, DNA methylation does not appear to inhibit transcription initiation, but nonetheless prevents accumulation of transcripts (Rountree and Selker 1997). Results of in vivo labeling experiments in *Neurospora* and nuclear run-on assays suggest that methylation interferes with elongation, but it remains possible that methylation also triggers RNA degradation (e.g., by an RNAi pathway). Another difference between *Neurospora* and mammals pertains to how methylation is perceived. Whereas mammals possess a variety of protein complexes that specifically recognize methylated DNA and recruit chromatin modification factors (Bird and Wolffe 1999), *Neurospora* lacks MBD proteins, but preliminary investigations suggest that *Neurospora* uses different proteins to recognize methylated sequences (G. Kothe et al., unpubl.). Finally, until we have a better understanding about how different organisms "decide" which sequences to methylate, the possible generality of mechanisms involved in the function and control of DNA methylation will not be clear.

ACKNOWLEDGMENTS

I thank Michael Freitag, Keyur Adhvaryu, and Kristina Smith for comments on the manuscript and thank present and past members of my laboratory for their contributions to our research. I gratefully acknowledge funding from NIH (grant GM35690) and NSF (grant MCB0131383), which contributed to this work.

REFERENCES

Antequera F., Tamame M., Villanueva J.R., and Santos T. 1984. DNA methylation in the fungi. *J. Biol. Chem.* **259:** 8033.

Aramayo R. and Metzenberg R.L. 1996. Meiotic transvection in fungi. *Cell* **86:** 103.

Aufsatz W., Mette M.F., van der Winden J., Matzke A.J., and Matzke M. 2002. RNA-directed DNA methylation in *Arabidopsis*. *Proc. Natl. Acad. Sci.* (suppl. 4) **99:** 16499.

Bannister A.J., Zegerman P., Partridge J.F., Miska E.A., Thomas J.O., Allshire R.C., and Kouzarides T. 2001. Selective recognition of methylated lysine 9 on histone H3 by the HP1 chromo domain. *Nature* **410:** 120.

Bird A.P. 1980. DNA methylation and the frequency of CpG in animal DNA. *Nucleic Acids Res.* **8:** 1499.

Bird A.P. and Wolffe A.P. 1999. Methylation-induced repression—Belts, braces, and chromatin. *Cell* **99:** 451.

Bull J.H. and Wootton J.C. 1984. Heavily methylated amplified DNA in transformants of *Neurospora crassa*. *Nature* **310:** 701.

Cambareri E.B., Singer M.J., and Selker E.U. 1991. Recurrence of repeat-induced point mutation (RIP) in *Neurospora crassa*. *Genetics* **127:** 699.

Cambareri E.B., Jensen B.C., Schabtach E., and Selker E.U. 1989. Repeat-induced G-C to A-T mutations in *Neurospora*. *Science* **244:** 1571.

Catalanotto C., Azzalin G., Macino G., and Cogoni C. 2002. Involvement of small RNAs and role of the qde genes in the gene silencing pathway in *Neurospora*. *Genes Dev.* **16:** 790.

Chan S.W., Zilberman D., Xie Z., Johansen L.K., Carrington J.C., and Jacobsen S.E. 2004. RNA silencing genes control de novo DNA methylation. *Science* **303:** 1336.

Cogoni C., Irelan J.T., Schumacher M., Schmidhauser T.J., Selker E.U., and Macino G. 1996. Transgene silencing of the *al-1* gene in vegetative cells of *Neurospora* is mediated by a cytoplasmic effector and does not depend on DNA-DNA interactions or DNA methylation. *EMBO J.* **15:** 3153.

Eissenberg J.C. and Elgin S.C. 2000. The HP1 protein family: Getting a grip on chromatin. *Curr. Opin. Genet. Dev.* **10:** 204.

Foss H.M., Roberts C.J., Claeys K.M., and Selker E.U. 1993. Abnormal chromosome behavior in *Neurospora* mutants defective in DNA methylation. *Science* **262:** 1737.

———. 1995. Abnormal chromosome behavior in *Neurospora* mutants defective in DNA methylation (correction). *Science* **267:** 316.

Freitag M., Williams R.L., Kothe G.O., and Selker E.U. 2002. A cytosine methyltransferase homologue is essential for repeat-induced point mutation in *Neurospora crassa*. *Proc. Natl. Acad. Sci.* **99:** 8802.

Freitag M., Hickey P.C., Khlafallah T.K., Read N.D., and Selker E.U. 2004a. HP1 is essential for DNA methylation in *Neurospora*. *Mol. Cell* **13:** 427.

Freitag M., Lee D.W., Kothe G.O., Pratt R.J., Aramayo R., and Selker E.U. 2004b. DNA methylation is independent of RNA interference in *Neurospora*. *Science* **304:** 1939.

Galagan J.E. and Selker E.U. 2004. RIP: The evolutionary cost of genome defense. *Trends Genet.* **20:** 417.

Galagan J.E., Calvo S.E., Borkovich K.A., Selker E.U., Read N.D., Jaffe D., FitzHugh W., Ma L.J., Smirnov S., Purcell S., Rehman B., Elkins T., Engels R., Wang S., Nielsen C.B., Butler J., Endrizzi M., Qui D., Ianakiev P., Bell-Pedersen D., Nelson M.A., Werner-Washburne M., Selitrennikoff C.P., Kinsey J.A., and Braun E.L., et al. 2003. The genome sequence of the filamentous fungus *Neurospora crassa*. *Nature* **422:** 859.

Grayburn W.S. and Selker E.U. 1989. A natural case of RIP: Degeneration of DNA sequence in an ancestral tandem duplication. *Mol. Cell. Biol.* **9:** 4416.

Irelan J.T. and Selker E.U. 1997. Cytosine methylation associated with repeat-induced point mutation causes epigenetic gene silencing in *Neurospora crassa*. *Genetics* **146:** 509.

Jackson J.P., Lindroth A.M., Cao X., and Jacobsen S.E. 2002. Control of CpNpG DNA methylation by the KRYPTONITE histone H3 methyltransferase. *Nature* **416:** 556.

Jackson J.P., Johnson L., Jasencakova Z., Zhang X., Perez-Burgos L., Singh P.B., Cheng X., Schubert I., Jenuwein T., and Jacobsen S.E. 2004. Dimethylation of histone H3 lysine 9 is a critical mark for DNA methylation and gene silencing in *Arabidopsis thaliana*. *Chromosoma* **112:** 308.

Jacobs S.A., Taverna S.D., Zhang Y., Briggs S.D., Li J., Eissenberg J.C., Allis C.D., and Khorasanizadeh S. 2001. Specificity of the HP1 chromo domain for the methylated N-terminus of histone H3. *EMBO J.* **20:** 5232.

Jenuwein T. and Allis C.D. 2001. Translating the histone code. *Science* **293:** 1074.

Jones P.A. 1999. The DNA methylation paradox. *Trends Genet.* **15:** 34.

Kinsey J.A. and Helber J. 1989. Isolation of a transposable element from *Neurospora crassa*. *Proc. Natl. Acad. Sci.* **86:** 1929.

Kouzminova E.A. and Selker E.U. 2001. *Dim-2* encodes a DNA-methyltransferase responsible for all known cytosine methylation in *Neurospora*. *EMBO J.* **20:** 4309.

Lachner M., O'Carroll D., Rea S., Mechtler K., and Jenuwein T. 2001. Methylation of histone H3 lysine 9 creates a binding site for HP1 proteins. *Nature* **410:** 116.

Lee D.W., Seong K.Y., Pratt R.J., Baker K., and Aramayo R. 2004. Properties of unpaired DNA required for efficient silencing in *Neurospora crassa*. *Genetics* **167:** 131.

Lehnertz B., Ueda Y., Derijck A.A., Braunschweig U., Perez-Burgos L., Kubicek S., Chen T., Li E., Jenuwein T., and Peters A.H. 2003. Suv39h-mediated histone h3 lysine 9 methylation directs DNA methylation to major satellite repeats at pericentric heterochromatin. *Curr. Biol.* **13:** 1192.

Malagnac F., Bartee L., and Bender J. 2002. An *Arabidopsis* SET domain protein required for maintenance but not establishment of DNA methylation. *EMBO J.* **21:** 6842.

Margolin B.S., Freitag M., and Selker E.U. 1997. Improved plasmids for gene targeting at the *his-3* locus of *Neurospora crassa* by electroporation. *Fungal Genet. Newsl.* **44:** 34.

Margolin B.S., Garrett-Engele P.W., Stevens J.N., Yen-Fritz D., Garrett-Engele C., Metzenberg R.L., and Selker E.U. 1998. A methylated *Neurospora* 5S rRNA pseudogene contains a transposable element inactivated by RIP. *Genetics* **149:** 1787.

Metzenberg R.L., Stevens J.N., Selker E.U., and Morzycka-Wroblewska E. 1985. Identification and chromosomal distribution of 5S rRNA genes in *Neurospora crassa*. *Proc. Natl. Acad. Sci.* **82:** 2067.

Miao V.P., Freitag M., and Selker E.U. 2000. Short TpA-rich segments of the zeta-eta region induce DNA methylation in *Neurospora crassa*. *J. Mol. Biol.* **300:** 249.

Miao V.P.W., Singer M.J., Rountree M.R., and Selker E.U. 1994. A targeted replacement system for identification of signals for de novo methylation in *Neurospora crassa*. *Mol. Cell. Biol.* **14:** 7059.

Nakayama J., Rice J.C., Strahl B.D., Allis C.D., and Grewal S.I. 2001. Role of histone H3 lysine 9 methylation in epigenetic control of heterochromatin assembly. *Science* **292:** 110.

Orbach M.J., Schneider W.P., and Yanofsky C. 1988. Cloning of methylated transforming DNA from *Neurospora crassa* in *Escherichia coli*. *Mol. Cell. Biol.* **8:** 2211.

Pandit N.N. and Russo V.E.A. 1992. Reversible inactivation of a foreign gene, *hph*, during the asexual cycle in *Neurospora crassa* transformants. *Mol. Gen. Genet.* **234:** 412.

Perkins D.D., Metzenberg R.L., Raju N.B., Selker E.U., and Barry E.G. 1986. Reversal of a *Neurospora* translocation by crossing over involving displaced rDNA, and methylation of the rDNA segments that result from recombination. *Genetics* **114:** 791.

Rea S., Eisenhaber F., O'Carroll D., Strahl B.D., Sun Z.W., Schmid M., Opravil S., Mechtler K., Ponting C.P., Allis C.D., and Jenuwein T. 2000. Regulation of chromatin structure by site-specific histone H3 methyltransferases. *Nature* **406:** 593.

Romano N. and Macino G. 1992. Quelling: Transient inactivation of gene expression in *Neurospora crassa* by transformation with homologous sequences. *Mol. Microbiol.* **6:** 3343.

Rountree M.R. and Selker E.U. 1997. DNA methylation inhibits elongation but not initiation of transcription in *Neurospora crassa*. *Genes Dev.* **11:** 2383.

Russell P.J., Rodland K.D., Rachlin E.M., and McCloskey J.A. 1987. Differential DNA methylation during the vegetative life cycle of *Neurospora crassa*. *J. Bacteriol.* **169:** 2902.

Selker E.U. 1990a. DNA methylation and chromatin structure: A view from below. *Trends Biochem. Sci.* **15:** 103.

———. 1990b. Premeiotic instability of repeated sequences in *Neurospora crassa*. *Annu. Rev. Genet.* **24:** 579.

———1998. Trichostatin A causes selective loss of DNA methylation in *Neurospora*. *Proc. Natl. Acad. Sci.* **95:** 9430.

Selker E.U. and Garrett P.W. 1988. DNA sequence duplications trigger gene inactivation in *Neurospora crassa*. *Proc. Natl. Acad. Sci.* **85:** 6870.

Selker E.U. and Stevens J.N. 1985. DNA methylation at asymmetric sites is associated with numerous transition mutations. *Proc. Natl. Acad. Sci.* **82:** 8114.

Selker E.U., Fritz D.Y., and Singer M.J. 1993a. Dense non-symmetrical DNA methylation resulting from repeat-induced point mutation (RIP) in *Neurospora*. *Science* **262:** 1724.

Selker E.U., Jensen B.C., and Richardson G.A. 1987a. A portable signal causing faithful DNA methylation *de novo* in *Neurospora crassa*. *Science* **238:** 48.

Selker E.U., Stevens J.N., and Metzenberg R.L. 1985. Rapid evolutionary decay of a novel pair of 5S RNA genes. In *Molecular genetics of filamentous fungi* (ed. W. Timberlake), p. 309. Alan R. Liss, New York.

Selker E.U., Cambareri E.B., Jensen B.C., and Haack K.R. 1987b. Rearrangement of duplicated DNA in specialized cells of *Neurospora*. *Cell* **51**: 741.

Selker E.U., Richardson G.A., Garrett-Engele P.W., Singer M.J., and Miao V. 1993b. Dissection of the signal for DNA methylation in the ζ–η region of *Neurospora*. *Cold Spring Harbor Symp. Quant. Biol.* **58**: 323.

Selker E.U., Freitag M., Kothe G.O., Margolin B.S., Rountree M.R., Allis C.D., and Tamaru H. 2002. Induction and maintenance of nonsymmetrical DNA methylation in *Neurospora*. *Proc. Natl. Acad. Sci.* (suppl. 4) **99**: 16485.

Selker E.U., Tountas N.A., Cross S.H., Margolin B.S., Murphy J.G., Bird A.P., and Freitag M. 2003. The methylated component of the *Neurospora crassa* genome. *Nature* **422**: 893.

Shiu P.K. and Metzenberg R.L. 2002. Meiotic silencing by unpaired DNA: Properties, regulation and suppression. *Genetics* **161**: 1483.

Shiu P.K., Raju N.B., Zickler D., and Metzenberg R.L. 2001. Meiotic silencing by unpaired DNA. *Cell* **107**: 905.

Singer M.J., Marcotte B.A., and Selker E.U. 1995. DNA methylation associated with repeat-induced point mutation in *Neurospora crassa*. *Mol. Cell. Biol.* **15**: 5586.

Tamaru H. and Selker E.U. 2001. A histone H3 methyltransferase controls DNA methylation in *Neurospora crassa*. *Nature* **414**: 277.

———. 2003. Synthesis of signals for de novo DNA methylation in *Neurospora crassa*. *Mol. Cell. Biol.* **23**: 2379.

Tamaru H., Zhang X., McMillen D., Singh P.B., Nakayama J., Grewal S.I., Allis C.D., Cheng X., and Selker E.U. 2003. Trimethylated lysine 9 of histone H3 is a mark for DNA methylation in *Neurospora crassa*. *Nat. Genet.* **34**: 75.

Wassenegger M., Heimes S., Riedel L., and Sanger H.L. 1994. RNA-directed de novo methylation of genomic sequences in plants. *Cell* **76**: 567.

Role of De Novo DNA Methyltransferases in Initiation of Genomic Imprinting and X-Chromosome Inactivation

M. KANEDA,[*†] T. SADO,[*†‡] K. HATA,[*†] M. OKANO,[§‖] N. TSUJIMOTO,[‖#]
E. LI,[‖#] AND H. SASAKI[*†]

[*]Division of Human Genetics, Department of Integrated Genetics, National Institute of Genetics, Research Organization of Information and Systems (ROIS), Mishima 411-8540, Japan; [†]Department of Genetics, School of Life Science, Graduate University for Advanced Studies (SOKENDAI), Mishima 411-8540, Japan; [‡]PRESTO, Japan Science and Technology Agency (JST), Kawaguchi 332-0012, Japan; [§]Laboratory for Mammalian Epigenetic Studies, Center for Developmental Biology, RIKEN, Kobe 650-0047, Japan; [‖]Cardiovascular Research Center, Massachusetts General Hospital, Harvard Medical School, Charlestown, Massachusetts 02129

DNA methylation is an epigenetic mechanism that plays a key role in regulation of developmental gene expression, maintenance of genomic integrity, genomic imprinting, and X-chromosome inactivation (X-inactivation) in mammals. Methylation of mammalian genomic DNA occurs almost exclusively at the cytosine of CpG dinucleotides. The CpG methylation pattern of the mammalian genome is created and maintained by a combination of de novo DNA methyltransferases, *Dnmt3a* and *Dnmt3b*, and a maintenance DNA methyltransferase *Dnmt1*. Targeted disruption of these DNA methyltransferase genes in mice results in embryonic or early postnatal lethality, indicating that they are essential for normal mammalian development (Li et al. 1992; Okano et al. 1999).

Genomic imprinting and X-inactivation are the well-characterized, major epigenetic phenomena of mammals that regulate allelic expression of autosomal genes and X-linked genes, respectively (Lyon 1961; Reik and Walter 2001). Both phenomena are known to be crucial for normal mammalian development. Imprinting is initiated during male and female gametogenesis, marking a subset of autosomal genes (up to a few hundred) in a sex-specific way (paternal and maternal imprinting). The imprinted genes show either paternal-specific or maternal-specific monoallelic expression in the offspring (Reik and Walter 2001). Thus imprinting is dependent on the sex of the parent from which the gene is derived, but not on the sex of the individual that carries the gene. By contrast, X-inactivation is a dosage compensation mechanism found only in females, which equalizes the X-linked gene dosage between males (with one X and one Y chromosome) and females (with two X chromosomes) (Lyon 1961). In the embryo proper (the epiblast lineages), X-inactivation is initiated during early development, leading to random inactivation of either the paternal or the maternal X chromosome. However, in the extraembryonic lineages (trophoblast and primitive endoderm derivatives) of mice, preferential inactivation of the paternal X chromosome occurs (Takagi and Sasaki 1975). Thus X-inactivation can be subject to genomic imprinting (imprinted X-inactivation). Like the imprinting of autosomes, the imprinting of X chromosome is thought to occur in the parental germ line.

Previous studies with the mouse embryos and ES cells deficient for *Dnmt1* showed that DNA methylation plays an essential role in the maintenance of genomic imprinting and X-inactivation in the embryo proper (Table 1) (Li et al. 1993; Beard et al. 1995; Panning and Jaenisch, 1996; Sado et al. 2000). By contrast, in the trophoblast, the role of DNA methylation seems more relaxed (Table 1) (Caspery et al. 1998; Tanaka et al. 1999; Sado et al. 2000). However, whether DNA methylation is involved in their initiation has not been addressed. If DNA methylation were to play a role in the initiation step, *Dnmt3a* or *Dnmt3b* (or both) should be the key players because these are the enzymes that establish new genomic methylation patterns (Okano et al. 1999). We therefore asked whether *Dnmt3a* and/or *Dnmt3b* is involved in the initiation of autosomal imprinting and X-inactivation using the cells and embryos deficient for these genes. The Cre-loxP conditional gene knockout system was particularly useful because of the early lethality of conventional *Dnmt3a* or *Dnmt3b* knockout mice (Okano et al. 1999). In this article, we summarize the results obtained from these experiments and discuss the role of de novo DNA methylation in the initiation of the two epigenetic phenomena.

ROLE FOR DNA METHYLATION IN IMPRINTING OF AUTOSOMAL GENES IN THE PARENTAL GERM LINE

Several lines of evidence suggest a role for DNA methylation in the initiation of autosomal imprinting. Mice deficient for *Dnmt3L*, a protein sharing homology with *Dnmt3a* and *Dnmt3b* but lacking methyltransferase activity, showed a failure in establishment of oocyte-specific methylation imprints (Bourc'his et al. 2001; Hata et al. 2002). Offspring from such females showed loss of monoallelic expression of the maternally imprinted genes and died around embryonic day 10.5 (E10.5). Since *Dnmt3L* protein can interact with *Dnmt3a* and *Dnmt3b* enzymes in transfected cells (Hata et al. 2002), it has been

---

[#]Present address: Novartis Institute for Biomedical Research, Cambridge, Massachusetts 02139.

Table 1. DNA Methylation in Autosomal Imprinting and X-Chromosome Inactivation

Phenomenon	Step	Tissue	DNA methylation required?
Autosomal imprinting	Initiation	Germ line	Yes
	Maintenance	Embryo proper	Yes
	Maintenance	Extraembryonic	Yes/No[a]
Random X-inactivation	Initiation	Epiblast	No
	Maintenance	Embryo proper	Yes
Imprinted X-inactivation	Initiation	Germ line	No?/?[b]
	Maintenance	Extraembryonic	No

The findings described in this paper are underlined.
[a]DNA methylation is required for the maintenance of imprinting of some genes such as $p57^{kip2}$ but not for the other genes such as *Mash2*.
[b]DNA methylation is unlikely to be involved in the imprinting of Xm but its role in the imprinting of Xp has not been tested.

proposed that *Dnmt3L* may regulate maternal imprinting via the de novo methyltransferases. Consistent with this model, a [$Dnmt3a^{-/-}$, $Dnmt3b^{+/-}$] ovary transplanted into a wild-type female failed to establish the oocyte-specific (maternal) methylation imprints (Hata et al. 2002).

We wanted to investigate the role of *Dnmt3a* and/or *Dnmt3b* in the initiation step of imprinting in more detail by disrupting the genes in male and female germ cells. A big problem was that the conventional *Dnmt3b* knockout mice are embryonic lethal and the conventional *Dnmt3a* knockout mice die around 3–4 weeks of age, before reaching the reproductive stage (Okano et al. 1999). We therefore could not examine the gametes of these mice or the offspring derived from them. To overcome this problem, we took advantage of the Cre-loxP conditional knockout technology and disrupted the *Dnmt3* genes in a germ-cell-specific way, leaving the genes intact in most somatic cells (Fig. 1). We used the tissue nonspecific alkaline phosphatase (TNAP)-Cre knockin mice, which express the Cre recombinase in germ cells from E9.5 to late gestation (Lomeli et al. 2000). Although expression of TNAP-Cre was not strictly germ-cell-specific, we could derive conditional knockout mice ([$Dnmt3a^{2lox/1lox}$, TNAP-Cre] and [$Dnmt3b^{2lox/1lox}$, TNAP-Cre], where 2lox represents the functional allele and 1lox represents the nonfunctional allele) that can survive to adulthood (Fig. 1) (Kaneda et al. 2004).

When the [$Dnmt3a^{2lox/1lox}$, TNAP-Cre] females were crossed with wild-type males, no live pups were obtained. Subsequent studies revealed that all embryos died around E10.5 with various developmental defects, such as open neural tube, lack of branchial arches, and impediment of blood circulation (Kaneda et al. 2004). We examined the methylation status of the differentially methylated regions (DMRs) of the imprinted genes in the E10.5 embryos. The DMRs normally methylated on the maternal allele, such as those of *Snrpn*, *Igf2r*, and *Peg1*, were found to be unmethylated (Kaneda et al. 2004). By contrast, the methylation status of the paternally methylated *H19* and *Rasgrf1* DMRs was unaffected. We then examined the expression of the maternally imprinted genes in the same embryos and found that $p57^{kip2}$ (*Cdkn1c*) and *Igf2r* are silenced, consistent with a loss of expression from the normally active maternal alleles (Kaneda et al. 2004). We also found that expression of *Peg1*, *Snrpn*, and *Peg3* is increased, with a derepression of the normally silent maternal alleles (Kaneda et al. 2004). Thus, the *Dnmt3a* conditional mutant females fail to establish the oocyte-specific imprints at the maternally imprinted loci.

The [$Dnmt3a^{2lox/1lox}$, TNAP-Cre] males showed impaired spermatogenesis (Kaneda et al. 2004). Histological examinations showed that the testes from the mutant males contain a slightly reduced number of spermatogonia in the seminiferous tubules at postnatal day 11 (P11) (Kaneda et al. 2004). However, at 11 weeks of age, they contained only few spermatogonia and no spermatocytes, spermatids, or spermatozoa (Kaneda et al. 2004). Because of the azoospermia, we could not obtain offspring to examine. We therefore used the laser-microdissection technology to collect spermatogonia from histological sections of the P11 testes and analyzed three paternally methylated DMRs by bisulfite sequencing. We found that the spermatogonia from the *Dnmt3a* conditional mutant males lacked methylation at the normally methylated *H19* DMR and *Dlk1-Gtl2* intergenic DMR (Kaneda et al. 2004). The normally methylated DMR of *Rasgrf1* was slightly less methylated than in wild-type spermatogonia (Kaneda et al. 2004). These results indicate that *Dnmt3a*

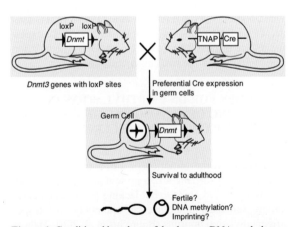

Figure 1. Conditional knockout of the de novo DNA methyltransferase genes *Dnmt3a* and *Dnmt3b* in mouse germ cells. The conditional mice with two loxP sites were produced by gene targeting and crossed with TNAP-Cre mice, which express the Cre recombinase in germ cells. The resulting conditional knockout mice survived to adulthood, and thus we could examine their germ cells (in both male and female mutants) and offspring (derived from the female mutants). For details, see Kaneda et al. (2004).

is required for the initiation of imprinting in both paternal and maternal germ lines.

We also generated [$Dnmt3b^{2lox/1lox}$, TNAP-Cre] mice, which were found to be phenotypically normal. When the [$Dnmt3b^{2lox/1lox}$, TNAP-Cre] males and females were crossed with wild-type partners, healthy pups were obtained (Kaneda et al. 2004). We analyzed the DMRs for the allelic methylation difference in these pups but all was found to be normal (Kaneda et al. 2004). Thus, so far there is no evidence that $Dnmt3b$ is involved in the initiation of imprinting in the parental germ line.

ROLE FOR DNA METHYLATION IN IMPRINTING OF $Mash2$ ($Ascl2$) IN THE MATERNAL GERM LINE

The mouse $Mash2$ ($Ascl2$) gene encodes a transcription factor of the basic helix-loop-helix class that is essential for extraembryonic development (Guillemot et al. 1994). $Mash2$ is located within a large imprinted cluster on mouse distal chromosome 7 and is exclusively expressed from the maternal allele (Guillemot et al. 1995). However, the imprinting of this gene is unique in that its maintenance is highly resistant to hypomethylation, as shown by the analysis of the $Dnmt1$-deficient embryos (Caspery et al. 1998; Tanaka et al. 1999). It is therefore interesting to ask whether the imprinting of this gene is initiated normally in the germ line of [$Dnmt3a^{2lox/1lox}$, TNAP-Cre] females. (It is known that $Mash2$ is a maternally imprinted gene.) Expression of $Mash2$ and $p57^{kip2}$ ($Cdkn1c$), as well as $Gapd$ (a nonimprinted control), was examined in the trophoblast of E9.5 embryos obtained from the [$Dnmt3a^{2lox/1lox}$, TNAP-Cre] females crossed with wild-type males. The results showed that both $Mash2$ and $p57^{kip2}$ are silenced in the trophoblast whereas $Gapd$ expression is unaffected (Fig. 2), indicating that the initiation of $Mash2$ imprinting in the maternal germ line does require $Dnmt3a$. This finding is consistent with the fact that the imprinting of $Mash2$ is controlled by an imprinting center located in an intron of $Kvlqt1$ ($Kcnq1$) (Fitzpatrick et al. 2002), which is methylated in oocytes but not in sperm.

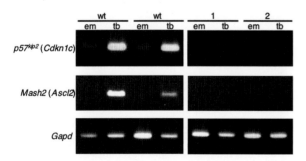

Figure 2. A maternally imprinted gene $Mash2$ is silenced in the trophoblast of embryos derived from [$Dnmt3a^{2lox/1lox}$, TNAP-Cre] mothers. Expression of $Mash2$, as well as $p57^{kip2}$, was examined by RT-PCR in the embryo proper (em) and trophoblast (tb) of two conceptuses (1, 2). The data suggest that the initiation of $Mash2$ imprinting in the maternal germ line requires DNA methylation. $Gapd$ is a nonimprinted housekeeping control. wt, wild-type.

ROLE FOR DNA METHYLATION IN INITIATION OF RANDOM X-CHROMOSOME INACTIVATION

We next asked whether the initiation of random X-inactivation in the embryo proper (epiblast lineages) and ES cells requires the de novo DNA methyltransferases. In the initiation step of X-inactivation, a noncoding RNA, X-inactive specific transcript ($Xist$), is upregulated on the future inactive X chromosome. The $Xist$ RNAs then coat the entire chromosome and presumably recruit factors required for heterochromatin formation (Brockdorff 2002). By contrast, $Xist$ is stably silenced on the active X chromosome. Although the maintenance of the silenced state of $Xist$ requires $Dnmt1$, monoallelic $Xist$ expression and subsequent X-inactivation can occur normally in ES cells deficient for $Dnmt1$ (Beard et al. 1995; Panning and Jaenisch 1996). We therefore examined whether X-inactivation can occur normally in the epiblast lineages of [$Dnmt3a^{-/-}$, $Dnmt3b^{-/-}$] embryos and in differentiating [$Dnmt3a^{-/-}$, $Dnmt3b^{-/-}$] ES cells (Sado et al. 2004).

We found by RNA fluorescence in situ hybridization (FISH) that monoallelic $Xist$ expression is appropriately initiated in most cells of the [$Dnmt3a^{-/-}$, $Dnmt3b^{-/-}$] female embryos at E9.5 (Sado et al. 2004). Furthermore, a cytological analysis showed that one of the two X chromosomes unanimously replicates late in S phase. In addition, one X chromosome was hypoacetylated at histone H4 in the mutant female embryos as in wild-type female embryos, as revealed by immunostaining. These results indicate that random X-inactivation occurred appropriately in the absence of the de novo methyltransferases (Sado et al. 2004). The $Xist$ promoter was extensively hypomethylated in these embryos.

In the above experiments, however, we observed ectopic accumulation of $Xist$ RNA (on the single X chromosome in males and on both X chromosomes in females) in a subset of cells (typically 4–5%) (Sado et al. 2004). Do these ectopic signals arise from inappropriate activation of $Xist$ at the onset of X-inactivation or from derepression of a once silenced $Xist$ locus? To address this question, we examined $Xist$ RNA accumulation in the course of differentiation of [$Dnmt3a^{-/-}$, $Dnmt3b^{-/-}$] male ES cells (Sado et al. 2004). Although the single X chromosome was never coated with $Xist$ RNA in undifferentiated state, ectopic $Xist$ accumulation was detected in 3.2% and 16.8% of cells at day 2 and day 5 of differentiation, respectively. At day 12 of differentiation, a surprisingly high percentage (68%) of cells from [$Dnmt3a^{-/-}$, $Dnmt3b^{-/-}$] embryoid bodies showed ectopic accumulation, suggesting a progressive derepression of the unmethylated $Xist$ locus. These observations establish that de novo DNA methylation is not required for the initial silencing of $Xist$ but is necessary for stabilizing the silenced state of $Xist$ (Sado et al. 2004).

ROLE FOR DNA METHYLATION IN INITIATION OF IMPRINTED X-CHROMOSOME INACTIVATION

In contrast to the random X-inactivation in the embryo proper, an imprinted X-inactivation occurs in the extraembryonic lineages (Takagi and Sasaki 1975). This seems to be due to an imprint on the maternal X chromosome (Xm) to remain active, as well as an imprint on the paternal X chromosome (Xp) to inactivate. The paternal imprint on the Xp can be reversed: XpO mice are developmentally retarded but viable and fertile (which indicates that the Xp is active) (Thornhill and Burgoyne 1993) and androgenetic embryos with two Xp chromosomes can undergo random X-inactivation in the extraembryonic tissues (Okamoto et al. 2000). The maternal imprint on the Xm may be more rigid: Genetic experiments using Robertsonian translocations showed that, in embryos carrying two Xm chromosomes, both remain active in the extraembryonic tissues and that such embryos die early because of poor development of the extraembryonic tissues (Goto and Takagi 1998). Furthermore, nuclear transplantation experiments showed that the maternal imprint is set on the Xm during oocyte growth (Tada et al. 2000), just as the maternal imprints on autosomes. Recently, it was found that imprinted X-inactivation is observed from the two-cell or four-cell stage (Huynh and Lee 2003; Okamoto et al. 2004) and it is proposed that the preinactivated state of the Xp is carried over from the paternal germ line (Huynh and Lee 2003).

We wanted to ask whether the de novo DNA methyltransferases plays a role in imprinting X chromosome in the male and female germ lines. Since disruption of Dnmt3a in the male germ line results in azoospermia (Kaneda et al. 2004), we could ask only whether its disruption in the female germ line has an effect on imprinted X-inactivation. As described above, embryos conceived by the [$Dnmt3a^{2lox/1lox}$, TNAP-Cre] females survived until E10.5 (Kaneda et al. 2004), indicating that the X-linked gene dosage was appropriately controlled, or nearly so, during early development. We recently obtained several embryos from [$Dnmt3a^{2lox/1lox}$, $Dnmt3b^{2lox/1lox}$, TNAP-Cre] females and found that their phenotype is almost identical with that of the above embryos. Two X-inactivation patterns are envisaged in these embryos. One is the normal imprinted X-inactivation, which suggests that the loss of the de novo DNA methyltransferases in the female germ line had no effect on the Xm. The other is a random X-inactivation, which suggests that the Xm had lost the imprint and then the counting and choice mechanisms were switched on. The previous findings that the Xm derived from nongrowing oocytes resembled the normal Xp (Tada et al. 2000) and the presence of two Xp chromosomes resulted in random X-inactivation (Okamoto et al. 2000) appear to support the latter idea.

To distinguish between the two possibilities, we made use of mice carrying an X-linked green fluorescent protein (GFP) transgene (X^{GFP}) (Takagi et al. 2002). [$Dnmt3a^{2lox/1lox}$, TNAP-Cre] and [$Dnmt3a^{2lox/1lox}$, $Dnmt3b^{2lox/1lox}$, TNAP-Cre] females were crossed with [X^{GFP}, Y] males, and embryos were recovered from the pregnant females at E8.5. In the embryo proper of female embryos derived from the mutant mothers, expression of the GFP transgene on the Xp was observed, just as in the embryo proper of control females, due to random X-inactivation (Fig. 3). By contrast, no GFP expression was ob-

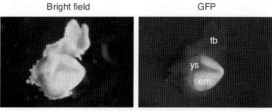

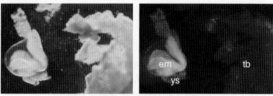

[$Dnmt3a^{1lox/+}$, $Dnmt3b^{1lox/+}$, XX^{GFP}]

Figure 3. Lack of expression of a GFP transgene in the trophoblast of E8.5 embryos derived from [$Dnmt3a^{2lox/1lox}$, $Dnmt3b^{2lox/1lox}$, TNAP-Cre] females crossed with X^{GFP}Y males. Expression of GFP from the Xp is observed in the embryo proper (as a result of random X-inactivation) but not in the trophoblast (because of preferential Xp-inactivation) of both wild-type (*top*) and mutant (*bottom*) embryos ([$Dnmt3a^{1lox/+}$, $Dnmt3b^{1lox/+}$, XX^{GFP}]). This suggests, although indirectly, that the Xm derived from the conditional mutant females has the imprint to be active in the trophoblast. em, embryo proper; ys, yolk sac; tb, trophoblast.

served from the Xp in the trophoblast of female embryos derived from the mutant mothers as well as those from the control mothers (Fig. 3). These results exclude the possibility of random X-inactivation in the trophoblast and suggest, although indirectly, that a loss of the de novo DNA methyltransferases in the female germ line does not affect the imprinted X-inactivation.

CONCLUSIONS

Based on the studies described here, we now begin to obtain a comprehensive view of the role of de novo DNA methyltransferases in the initiation of autosomal imprinting and X-inactivation. First, we found that *Dnmt3a* is required for the initiation of autosomal imprinting in both the paternal and maternal germ lines. Notably, we showed that a trophoblast-specific gene *Mash2*, which does not require *Dnmt1* for the maintenance of its imprinted state, does require *Dnmt3a* for the initiation of imprinting in the maternal germ line. Second, we found that the initiation of random X-inactivation can occur normally in the absence of *Dnmt3a* and *Dnmt3b*. Third, we showed that a disruption of both *Dnmt3a* and *Dnmt3b* in the maternal germ line does not affect the imprinted X-inactivation in the trophoblast. This seems to indicate that the Xm derived from the mutant females has the imprint to remain active. The role of *Dnmt3a* and *Dnmt3b* in the imprinting of Xp, however, remains an open question. These findings are summarized in Table 1, in terms of necessity for DNA methylation, together with the previous findings on the maintenance of these phenomena.

Autosomal imprinting and X-inactivation share a number of molecular features. These include *cis*-acting control centers, long-distance regulation, association with noncoding and antisense RNAs, involvement of histone modifications and chromatin-associated factors, and differential DNA methylation (Lee 2003). These features lead to the proposal that X-inactivation and autosomal imprinting may have a common origin (Lee 2003). Indeed, marsupials such as kangaroos show paternal-specific imprinted X-inactivation (Sharman 1971), just as the trophoblast of mice, and X-inactivation of this type is thought to be the early form of dosage compensation. Since de novo DNA methylation is required for the initiation of autosomal imprinting, it will be important to establish whether the same mechanism is involved in the initiation of imprinted X-inactivation. By contrast, we clearly showed that the initiation of random X-inactivation, which may be the more recent form of dosage compensation, does not require de novo DNA methylation. This indicates that, even if X-inactivation and autosomal imprinting have a common origin, a very different molecular mechanism evolved afterward to achieve random X-inactivation in the epiblast lineages.

ACKNOWLEDGMENTS

We thank A. Nagy and H. Lomeli (Samuel Lunenfeld Research Institute) for TNAP-Cre mice; M. Okabe (Osaka University) for X^{GFP} mice; K. Shiota and S. Tanaka (The University of Tokyo) for help in laser microdissection; T. Ohhata for photographing the X^{GFP} embryos; and all members of our laboratories for help, discussion, and encouragement throughout this work. The work was supported in part by Grants-in-Aids from the Ministry of Education, Culture, Sports, Science, and Technology of Japan to H.S. and Grants from the National Institutes of Health to E.L.

REFERENCES

Beard C., Li E., and Jaenisch R. 1995. Loss of methylation activates *Xist* in somatic but not in embryonic cells. *Genes Dev.* **9**: 2325.
Bour'chis D., Xu G.L., Lin C.S., Bollman B., and Bestor T.H. 2001. Dnmt3L and the establishment of maternal genomic imprints. *Science* **294**: 2536.
Brockdorff N. 2002. X-chromosome inactivation: Closing in on proteins that bind *Xist* RNA. *Trends Genet.* **18**: 352.
Caspery T., Cleary M.A., Baker C.C., Guan X.-J., and Tilghman S.M. 1998. Multiple mechanisms regulate imprinting of the mouse distal chromosome 7 gene cluster. *Mol. Cell. Biol.* **18**: 3466.
Fitzpatrick G.V., Soloway P.D., and Higgins M.J. 2002. Regional loss of imprinting and growth deficiency in mice with a targeted deletion of *KvDMR1*. *Nat. Genet.* **32**: 426.
Goto Y. and Takagi N. 1998. Tetraploid embryos rescue embryonic lethality caused by an additional maternally derived X chromosome in the mouse. *Development* **125**: 3353.
Guillemot F., Nagy A., Auerbach A., Rossant J., and Joyner A.L. 1994. Essential role of *Mash-2* in extraembryonic development. *Nature* **371**: 333.
Guillemot F., Caspary T., Tilghman S.M., Copeland N.G., Gilbert D.J., Jenkins N.A., Anderson D.J., Joyner, A.L., Rossant J., and Nagy A. 1995. Genomic imprinting of *Mash2*, a mouse gene required for trophoblast development. *Nat. Genet.* **9**: 235.
Hata K., Okano M., Lei H., and Li E. 2002. Dnmt3L cooperates with the Dnmt3 family of de novo DNA methyltransferases to establish maternal imprints in mice. *Development* **129**: 1983.
Huynh K.D. and Lee J.T. 2003. Inheritance of a pre-inactivated paternal X chromosome in early mouse embryos. *Nature* **426**: 857.
Kaneda M., Okano M., Hata K., Sado T., Tsujimoto N., Li E., and Sasaki H. 2004. Essential role for de novo DNA methyltransferase Dnmt3a in paternal and maternal imprinting. *Nature* **429**: 900.
Lee J.T. 2003. Molecular links between X-inactivation and autosomal imprinting: X-inactivation as a driving force for the evolution of imprinting? *Curr. Biol.* **13**: R242.
Li E., Beard C., and Jaenisch R. 1993. Role for DNA methylation in genomic imprinting. *Nature* **366**: 362.
Li E., Bestor T., and Jaenisch R. 1992. Targeted mutation of the DNA methyltransferase gene results in embryonic lethality. *Cell* **69**: 915.
Lomeli H., Ramos-Mejia V., Gertsenstein M., Lobe C.G., and Nagy A. 2000. Targeted insertion of Cre recombinase into the TNAP gene: Excision in primordial germ cells. *Genesis* **26**: 116.
Lyon M.F. 1961. Gene action in the X-chromosome of the mouse (*Mus musculus L*). *Nature* **190**: 372.
Okamoto I., Tan S.-S., and Takagi N. 2000. X-chromosome inactivation in XX androgenetic mouse embryos surviving implantation. *Development* **127**: 4137.
Okamoto I., Otte A.P., Allis C.D., Reinberg D., and Heard E. 2004. Epigenetic dynamics of imprinted X-inactivation during early mouse development. *Science* **303**: 644.
Okano M., Bell D.W., Haber D.A., and Li E. 1999. DNA methyltransferases *Dnmt3a* and *Dnmt3b* are essential for de novo methylation and mammalian development. *Cell* **99**: 247.
Panning B. and Jaenisch R. 1996. DNA hypomethylation can activate *Xist* expression and silence X-linked genes. *Genes Dev.* **10**: 1991.
Reik W. and Walter J. 2001. Genomic imprinting: Parental influence on the genome. *Nature Rev. Genet.* **2**: 21.
Sado T., Okano M., Li E., and Sasaki H. 2004. De novo methylation is dispensable for the initiation and propagation of X chromosome inactivation. *Development* **131**: 957.
Sado T., Fenner M.H., Tan S.-S., Tam P., Shioda T., and Li E. 2000. X-inactivation in the mouse embryo deficient for Dnmt1: Distinct effect of hypomethylation on imprinted and random X inactivation. *Dev. Biol.* **225**: 294.
Sharman G.B. 1971. Late DNA replication in the paternally derived X chromosome of female kangaroos. *Nature* **230**: 231.
Tada T., Obata Y., Tada M., Goto Y., Nakatsuji N., Tan S.-S., Kono T., and Takagi N. 2000. Imprint switching for non-random X-chromosome inactivation during mouse oocyte growth. *Development* **127**: 3101.
Takagi N. and Sasaki M. 1975. Preferential inactivation of the paternally derived X chromosome in the extraembryonic membranes of the mouse. *Nature* **256**: 640.
Takagi N., Sugimoto M., Yamaguchi S., Ito M., Tan S.-S., and Okabe M. 2002. Nonrandom X chromosome inactivation in mouse embryos carrying Searle's T(X;16)16H translocation visualized using X-linked *lacZ* and *GFP* transgenes. *Cytogenet. Genome Res.* **99**: 52.
Tanaka M., Puchyr M., Gertsenstein M., Harpal K., Jaenisch R., Rossant J., and Nagy A. 1999. Parental-origin-specific expression of Mash2 is established at the time of implantation with its imprinting mechanism highly resistant to genome-wide demethylation. *Mech. Dev.* **87**: 129.
Thornhill A.R. and Burgoyne P.S. 1993. A paternally imprinted X chromosome retards the development of the early mouse embryo. *Development* **118**: 171.

Gene Repression Paradigms in Animal Cells

L. Lande-Diner, J. Zhang, T. Hashimshony, A. Goren, I. Keshet, and H. Cedar

Department of Cellular Biochemistry, Hebrew University, Jerusalem, 91120 Israel

In simple organisms with relatively small genomes almost all of the genes are expressed in every cell. Silenced genes are usually controlled by specific repressor proteins that recognize and bind *cis*-acting elements in the vicinity of the gene, thereby preventing transcription. This organization contrasts dramatically with the program of repression in higher multicellular organisms. In these cells, one observes a bimodal pattern of expression. Housekeeping genes are transcribed constitutively while other genes are tissue specific and thus turned off in most cell types. These repressed genes probably make up >60% of the genome. Although repression mechanisms for individual genes have been studied in animal cells, little is known about the overall organization and programming of gene silencing.

One of the key mechanisms involved in gene repression in animal cells is DNA methylation. The genome itself appears to be modified in a bimodal pattern with most of the DNA being methylated, while CpG islands at the 5′ end of housekeeping genes are constitutively unmethylated. Genes that do not contain CpG islands are generally repressed by the presence of DNA methylation in the vicinity of their promoter sequences. It should be noted that, unlike the genetic information contained in the DNA sequence, the methylation profile is not inherited from the germ line (Kafri et al. 1992).

Almost all DNA methylation coming from the germ cells is erased from the genome in the morula, and the bimodal pattern is then reestablished anew in each individual at the time of implantation (Kafri et al. 1992). At this stage the embryo expresses powerful de novo methylases capable of modifying the entire genome (Okano et al. 1999). In parallel, *cis*-acting sequences within CpG islands are recognized by specific *trans*-acting factors that serve to protect these regions from modification (Brandeis et al. 1994; Macleod et al. 1994). Once established, this bimodal pattern is preserved intact following each round of cell division through the action of a maintenance methylase that copies the CpG methylation profile present on the native DNA strand once the replication fork has passed (Cedar 1988). It should be noted that this repression mechanism operates without the need to recognize or identify specific gene sequences and, in this sense, is truly global. In addition, this represents a genuine epigenetic regulatory mechanism because it is preprogrammed to be preserved as part of the genome sequence itself.

METHYLATION MODELS HISTONE MODIFICATION

In order to better understand how methylation contributes to gene repression, we designed an experiment in which the methylation state of a transgene could be controlled in vivo, and then used these animals to examine the effect of DNA methylation on chromatin structure (Fig. 1a). The test gene (Hashimshony et al. 2003) was composed of a normal human β-globin sequence driven by its own natural minimal promoter placed near a 120-base pairs (bp) segment (island element [IE]) of the mouse Aprt CpG island that has been shown to protect against de novo methylation at the time of implantation (Siegfried et al. 1999). This IE is actually part of a cassette flanked by two LoxP sites. When this mouse is crossed with mice that express Cre from early stages of embryogenesis (Lallemand et al. 1998), the presence of this *trans*-acting factor causes the IE to be deleted prior to implantation, and as a result the nearby β-globin promoter region undergoes de novo methylation, which is then maintained in all cells of the offspring.

Alternatively, when the adult mouse is mated with animals that express an interferon inducible Cre gene (Mx) (Kuhn et al. 1995), the IE element remains intact at the time of implantation, thus protecting against de novo methylation in the promoter region. When adult mice are then treated with the inducer, newly synthesized Cre protein brings about deletion of the IE, while still leaving the promoter region in its unmethylated state. Using these crosses, it is thus possible to generate identical transgenic mice that carry either a methylated or nonmethylated promoter. Previous studies (Siegfried et al. 1999) have demonstrated that while the level of transcription is low in both animals, the presence of DNA methylation at the promoter brings about a 50–100-fold inhibition of β-globin basal expression in many different cell types.

We next asked whether DNA methylation influences chromatin structure at the level of local histone modification. To this end, we isolated lymphoid cells from the spleen and prepared mononucleosomes that were then subjected to chromatin immunoprecipitation (ChIP) analysis using antibodies to acetylated histones (Fig. 1b). The bound and input DNA were then used to assay nucleosomes in the β-globin promoter region by means of semiquantitative polymerase chain reaction (PCR). As a con-

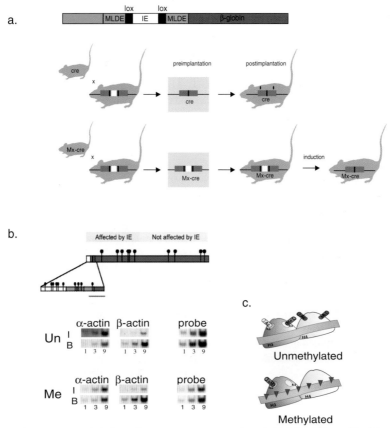

Figure 1. (*a*) Programmed transgene methylation. The transgene construct (not shown to scale) used in this study is made up of human β-globin (*HBB*) sequences (*pink*), including a 48-bp minimal promoter, two 30-bp MLDE sequences (*blue*), two 34-bp *lox*P elements (*black*), the 120-bp island element (IE; *white*), and 100 bp of flanking buffer DNA sequences (*green*). Mouse bML120β contains a single copy of this transgene. To generate mice carrying transgenes identical in sequence but either methylated or unmethylated, the *lox*P-flanked island element was deleted by mating carriers with two different Cre-expressing lines. In the first line, Cre is expressed before implantation. In mice carrying this construct (Cre mice), the island element is deleted before the wave of de novo methylation, and surrounding CpG sites thus become methylated (*black circles*). The second Cre-expressing line carries interferon-inducible *cre*. In these mice (Mx-Cre), the island element remains present during implantation and protects adjacent regions from methylation. By treating adult mice with polyI-polyC, Cre activity is induced and the island element is removed, generating an unmethylated version of the transgene flanked by *lox*P sites (Hashimshony et al. 2003). (*b*) Mononucleosomes were prepared from spleens taken from mice with either the methylated (Me) or the unmethylated (Un) ML$^\beta$ transgene (Hashimshony et al. 2003) and subjected to ChIP analysis using antibody to acetylated H4. The map shows the locations of CpG residues (*black circles*) in the construct. All of the CpG sites were constitutively modified in mice carrying the methylated transgene. In mice with the unmethylated transgene, the region whose methylation is affected by the IE is >90% unmodified. Input (I) and bound (B) DNA fractions were used for quantitative PCR on samples of 1, 3, or 9 μl using primer sets specific for the region shown on the map (*red line*). The input DNA was diluted at least 1:100 to allow comparison to bound DNA. For each ChIP preparation, α-actin was assayed as a negative control and β-actin as a positive control (Hashimshony et al. 2003). (*c*) The histone modification pattern and DNA structure (*ribbon*) of the unmethylated and methylated (*red triangles*) transgene. Unmethylated DNA is packaged in an open configuration, where histone H3 is acetylated in K9 and methylated in K4, and histone H4 is acetylated. Methylated DNA is packaged in a closed chromatin structure, where histone H3 is methylated in the K9 residue and demethylated in lysine 4, and histone H4 is deacetylated.

trol, we first showed that an active gene like β-actin is indeed enriched for Ac-H3, while an inactive gene such as α actin is depleted for this modification. Interestingly, the unmethylated β-globin promoter on the transgene was also found to be in an active acetylated conformation. In contrast, when methylated, this same region is depleted for histone acetylation. By analyzing additional transgene founders, we found that this effect is highly reproducible and clearly differentiates between methylated and nonmethylated DNA over a defined region surrounding the IE, but not in areas unaffected by this element. Additional experiments employing antibodies to other histone modifications showed that the presence of DNA methylation has a big influence on histone structure, causing chromatin to be packaged with nucleosomes that have deacetylated histones H3 and H4, which are also unmethylated at histone H3(Lys4). It thus appears that DNA methylation represents an epigenetic mark that causes DNA to be packaged in "closed" nucleosomes after every round of replication.

GENE REPRESSION PATTERNS

While these studies clearly demonstrate that DNA methylation serves as a global repressor of gene structure and activity, it must be remembered that the genome utilizes additional gene repression mechanisms, as well. In order to evaluate the relative contribution of DNA modification to the overall process of gene silencing, it is thus necessary to remove methyl moieties from the DNA, and then determine how this effects gene expression. To this end, we prepared primary mouse embryo fibroblasts that are Dnmt1$^-$ (Li et al. 1992) on a p53$^-$ background. Since the amount of Dnmt1 in these cells is limiting, DNA methylation is diluted out, and after about ten generations the cells have lost 95% of their methyl groups as shown by nearest neighbor analysis (Gruenbaum et al. 1981).

In order to evaluate the effect of this undermethylation on gene expression, we prepared RNA from p53$^{-/-}$ control cultures and from p53$^{-/-}$ Dnmt1$^{-/-}$ cells and hybridized to microarray chips (Affymetrix) containing 22,000 mouse gene cDNA segments. Since DNA methylation is thought to work through its effect on histone acetylation (Razin 1998), we also examined the expression pattern induced by treatment of these cells with Trichostatin A (TSA), a potent inhibitor of histone deacetylation. This analysis revealed the presence of four distinct gene repression paradigms: genes that are unaffected by TSA and Dnmt1$^{-/-}$, genes induced by TSA treatment alone, genes induced by demethylation alone, and genes that could be expressed only following a combination of demethylation and TSA treatment.

GENE STRUCTURE PATTERNS

We next addressed ourselves to understanding the relationship between these expression paradigms and gene structure. To this end, we chose about 40 tissue-specific genes representative of the four categories and analyzed the methylation state and histone modification pattern in their promoter regions. The DNA methylation status of each gene was determined by methylated DNA immunoprecipitation (mDIP), a new technique that utilizes an antibody specific to 5mC (Reynaud et al. 1992) to identify methylated sequences by PCR analysis. Histone modifications, including Ac-H4, Ac-H3, 2me-H3(K9), and me-H3(K4), were assessed by ChIP. In all cases the degree of enrichment was determined by semiquantitative PCR assays of the bound and input DNA fractions following immunoprecipitation. The overall structural features of each gene category are summarized in diagrammatic form in Figure 2. These studies reveal a striking correlation between expression patterns and gene structure.

GROUP I GENES

Group I tissue-specific genes are essentially very similar to housekeeping genes. They are expressed in fibroblasts at relatively high levels, have an unmethylated CpG island promoter, and are packaged with an "open" nucleosome structure containing acetylated histones enriched for me-H3(K4) but depleted for me-H3(K9). These genes are considered tissue specific because they are expressed preferentially in one or a number of different cell types. Our studies clearly demonstrate that in terms of structure and expression, these genes are not repressed in the true sense of the word. Rather, they are packaged in an open nucleosome structure and their main

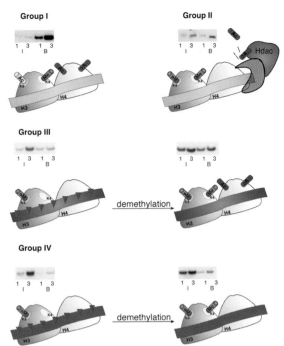

Figure 2. Epigenetic structures of tissue-specific genes. Histone modification patterns and DNA structure are shown for each category of tissue-specific genes. Group I and group II genes have CpG island (*yellow ribbon*) promoters that are constitutively unmethylated. Group I genes have acetylated histones and are methylated at H3(K4). ChIP analysis shows that the bound (B) fraction is enriched over the input (I) fraction. Group II genes are inhibited by repressors (*gray*) that recruit histone deacetylase (Hdac). H3(K4) remains methylated, and once deacetylated H3(K9) can become methylated. Group III and group IV genes have non-CpG island (*blue ribbon*) promoters that are generally methylated (*red triangles*). In the presence of DNA methylation, histones are hypoacetylated (B is low as compared to I) and unmethylated at H3(K4). When the methyl groups are removed (Dnmt1$^-$), however, the histones undergo reacetylation and become methylated at H3(K4). For group IV genes, removal of methyl groups has no effect on histone acetylation (B is depleted as compared to I), but H3(K4) becomes methylated. Gene activity is not affected because these genes have an additional layer of repression that also works through histone acetylation.

mode of regulation is probably through interactions with tissue-specific transcriptional enhancers that boost their expression even more in the cell type of choice.

GROUP II GENES

Group II genes are normally expressed at low levels in fibroblasts, but can be readily induced simply by treating with TSA. Like group I genes, they contain CpG island promoter regions that are constitutively unmethylated, thus explaining why their level of transcription does not respond to undermethylation. In keeping with their repressed state, the promoter regions of these genes are uniformly packaged in nucleosomes containing deacetylated histones and me-H3(K9). In light of these structural details, it is very likely that these genes are inhibited by *trans*-acting repression factors, which work by bringing about local histone deacetylation. It is for this reason that transcription can be induced simply by reversing this state using TSA. In keeping with this idea, we found that many of the genes in this category are nervous tissue–specific and carry upstream *cis*-acting elements that bind the well characterized repressor, REST (Chong et al. 1995; Lunyak et al. 2002). This protein, which is expressed in all cell types except neurons, is known to work by recruiting histone deacetylases to the gene promoter (Huang et al. 1999).

GROUP III GENES

Almost all of the genes automatically induced in cells carrying the Dnmt1 knockout were indeed found to be methylated over their promoter sequences in normal fibroblasts. While these genes are underacetylated in wild-type cells, the removal of methyl groups converts nucleosomes at these promoters to an acetylated form. This is consistent with the idea that methylated DNA binds methyl-binding proteins, which then recruit histone deacetylases (Jones et al. 1998; Nan et al. 1998). Thus, once the methyl groups are removed, as in the Dnmt1$^-$ cells, these deacetylases will no longer be present at the promoter, allowing the DNA to get repackaged in an acetylated form. Since methyl-mediated repression is partially due to deacetylation (Eden et al. 1998), we find that TSA also has a small inductive effect on these genes. In addition to being packaged with underacetylated histones, these methylated genes also carry histones depleted for me-H3(K4). When methyl moieties are removed, however, the promoters get reassembled with histone H3 that is highly methylated at the lysine 4 position. Overall, it appears that genes in group III are repressed predominantly because of DNA methylation, and once this block is removed, the gene can now adopt an open chromatin conformation conducive to transcription.

It should be noted that some of the genes that are automatically induced in Dnmt1$^{-/-}$ cells actually have nonmethylated CpG island promoters (Jackson-Grusby et al. 2001) and appear to be constitutively packaged with acetylated histones H3 and H4 in wild-type cells. Although the mechanism for this repression has not been worked out, it should be noted that Dnmt1 itself has been shown to be capable of repressing transcription through a mechanism that is independent of DNA methylation (Fuks et al. 2000; Robertson et al. 2000). One possibility is that this protein recruits a histone methylase that modifies H3(K9) (Fuks et al. 2003), a mark that is clearly present on this category of genes.

GROUP IV GENES

The largest number of tissue-specific genes fall into group IV. These genes are expressed at extremely low levels and can be induced only by a combination of undermethylation and TSA treatment. Like group II genes, these sequences were found to be methylated in their promoter region and packaged in a closed nucleosome structure. In contrast to group III, however, group IV genes retain their closed nucleosome structure even after DNA methylation has been removed, suggesting that these genes are silenced by multiple layers of repression. Thus, while taking away DNA methylation would normally lead to an alteration in chromatin structure, these genes must be subject to additional repression mechanisms that also work by bringing about histone deacetylation.

It should be noted that induction of group IV genes by TSA in Dnmt1$^-$ cells is only transient, and once the drug is removed, full gene repression is restored. This suggests that each layer of repression must be inherently programmed in the genome and is constantly capable of regenerating the closed deacetylated chromatin conformation even after nucleosomes have been artificially acetylated by TSA treatment. In other words, the histone acetylation pattern itself does not appear to be autonomously maintained.

REPLICATION TIMING

One potential candidate for another basic global repression mechanism is late replication timing. The genome as a whole is divided into replication time zones that are programmed to replicate at specific times in S phase (Goren and Cedar 2003). This pattern is an integral part of the overall chromosome structure characterized by a striking division into distinct structural and functional bands. At the level of individual genes, there appears to be a correlation between early replication and gene expression. Housekeeping genes all replicate early in S phase, while many tissue-specific genes are developmentally regulated to be late replicating in most tissues, but early replicating in their cell type of expression. Although this correlation has been known for a long time, it was only recently that experiments were done to test whether replication timing itself plays a role in the regulation of expression (Zhang et al. 2002).

Previous models had suggested that the transcriptional competence of each gene may actually be set up at the time of replication, while chromatin structure is somewhat relaxed. It was hypothesized that the factors necessary to impose gene activation could be cell cycle regulated and expressed only in early S (Gottesfeld and Bloomer 1982; Brown 1984). On the basis of this idea, we reasoned that naked DNA introduced into the nucleus early in S phase would become transcriptionally competent, while the same DNA would be inactive if introduced in late S cells. In order to test this simple concept, we inserted reporter genes into nuclei from early or late S phase using microinjection. To this end, Rat-1 cells were placed on slides carrying a well-marked grid that allowed each cell to be uniquely identified, and were then injected with reporters containing various different promoters linked to GFP or β-Gal coding sequences.

By following the division progress of each cell microscopically, it was possible to determine when in S phase each cell was injected (Fig. 3a). From a large series of these experiments using a wide variety of different promoters, we were able to show that plasmids injected in early S nuclei are ten times more active than when the same plasmid was injected into late S phase cells (Zhang et al. 2002). Once set, this pattern remains fixed for a long time even though the cells themselves may continue cycling. Since this phenomenon was also observed with

REPRESSION PARADIGMS

a.

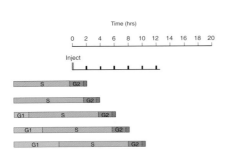

b.

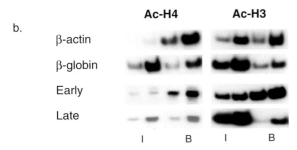

Figure 3. (*a*) Cells were plated on a Cellocate grid that contains markings that enable one to identify each cell in order to follow its cell cycle progression (Zhang et al. 2002). Cells were injected (*red vertical bar*) at zero time with a promoter-driven β-Gal reporter gene and followed every 2 hr (short bars) to determine whether mitosis (*red*) had occurred. Mitosis at 10 hr indicates that injection took place in early S. Mitosis at 2 hr indicates that injection took place in late S. At 20 h after the injection, the cells were stained to visualize β-Gal, and the percentage of positive divided cells recorded. (*b*) Histone acetylation of early-replicating and late-replicating chromosomal Rat-1 cells injected in early S or late S with plasmid pBluescript and harvested 3 hr later. ChIP (anti-Ac-H4 or anti-Ac-H3) was performed on mononucleosomes and enrichment was determined for total plasmid sequences (which should be the same in both cell populations), as well as for β-actin and β-globin (Zhang et al. 2002). Note that early injected plasmids are enriched for acetylation while late injected DNA is depleted.

H4, while late-injected plasmids are automatically packaged with histones depleted for acetylation.

These studies provide an elementary picture of how replication timing may play a role in regulating gene expression. According to this model, replication timing is a basic property of each chromosomal band that is set up during early development in the embryo. Once established, each region replicates at a fixed time in S phase in every division cycle. As part of the process of DNA replication, nucleosomal structure becomes disrupted. Once the replication fork has passed by, this structure must then be restored. If replication takes place in early S, the newly synthesized DNA will be assembled with nucleosomes containing acetylated histones. On the other hand, genes replicating in late S will get packaged with deacetylated histones. Although the precise mechanism for this is not known, it has been shown that late replication foci are specifically associated with histone deacetylase 2 (Hdac2) (Rountree et al. 2000), and this could clearly serve as a basis for late S repression. In this scheme, replication timing serves as a global mechanism for maintaining gene expression patterns through cell division.

REPRESSION BY LATE REPLICATION

With this picture in mind, we then asked whether late replication timing may represent one of the mechanisms involved in the repression of group IV genes. To this end, we used fluorescent in situ hybridization (FISH) to assay the replication timing pattern (Selig et al. 1992) of genes in each tissue-specific expression category. In this procedure, bacterial artificial chromosome (BAC) clones containing each individual gene sequence are labeled with biotinylated nucleotides and used as hybridization probes on interphase wild-type fibroblast nuclei labeled with BrdU. Each S-phase nucleus (BrdU positive) reveals either two single hybridization dots if the gene has not yet undergone replication, or two double dots if it has. By counting the number of nuclei containing single or double signals one can calculate the replication time. Genes that yield a relatively high number of double dots must replicate early in S, while genes showing a relatively high number of singles replicate in late S (Fig. 4a).

Using this method, we obtained a full snapshot of replication timing in the genome (Figs. 4b and 5). All of the genes in groups I, II, and III appear to replicate in early S. The significance of this is that these genes would have a normal tendency to be packaged with acetylated histones following replication. In the case of group II, the presence of repressor proteins would counteract this by actively causing local deacetylation. In a similar manner, group III genes would become deacetylated by virtue of the presence of DNA methylation. Since group I genes do not have any inherent repression mechanisms based on histone deacetylation, they would retain the acetylation pattern that is predetermined by early replication.

MULTIPLE LAYERS OF REPRESSION

As predicted from our model, >60% of all group IV genes replicate late in S phase, and thus could represent

replicating plasmids as well, these results strongly suggest that late replication may indeed represent an endogenous mechanism for repressing gene expression.

In order to understand the molecular mechanism of this effect, we next asked whether transcriptional competence may be set up through effects on nucleosome structure. To this end, we injected plasmids into either early or late S phase cells and then used ChIP to determine their histone modification pattern. As shown in Figure 3b, plasmids injected into early S phase cells are highly enriched in nucleosomes containing acetylated histones H3 and

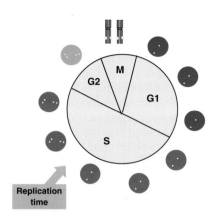

a.

b.

Group	Gene Name	% singles
HK	β-actin	34 (E)
	Dhfr	38 (E)
I	Mea	37 (E)
	CD24	38 (E)
II	Int-1	46 (E)
III	KC	39 (E)
	Pem	36 (E)
IV	Oct-4	37 (E)
	Y1-globin	73 (L)
	β-globin	64 (L)
	Thy 19.4	66 (L)

Figure 4. (*a*) How gene sequences are detected by FISH throughout the cell cycle. Each allele is represented by a single dot. In G1 prior to replication, each homologous chromosome generates a single dot. In S phase, until the gene replicates there are two single dots. After the specific gene has replicated, however, one will observe two sets of double dots. The time point where the transition from two single dots to two doublets occurs depends on the replication time of each gene. In G2 all genes have two sets of double dots. In metaphase there are two chromosomes, each composed of two sister chromatids, so that they yield four dots altogether. (*b*) Representable examples from the four gene groups, assayed for their replication timing by FISH (Selig et al. 1992). Replication time is shown as percentage of single dots. Genes with ≤50% singlets are early replicating, whereas genes whose percentage of single dots is more than 60 are late replicating.

Figure 5. Summary of the mechanisms responsible for the repression status of tissue-specific genes. Group I genes are not repressed and have an open chromatin structure. Group II genes are probably repressed by *trans*-acting repressors and are therefore packed in a closed chromatin conformation. Group III genes are repressed by DNA methylation. All of these groups replicate early in the cell cycle. Group IV genes are repressed by DNA methylation and by additional mechanisms that serve to maintain the repressed state even after the methyl moieties have been removed. Late replication timing may represent one of these effectors. Many of these genes are late replicating, and this could explain the layered effect. Other genes in group IV may be influenced both by methylation and by specific repressors, as in group II.

one of the multiple mechanisms employed for gene repression. Since these genes are also methylated, they carry not one, but two fundamental repression layers. Thus, when DNA methylation is removed (Dnmt1⁻ cells), methyl groups are no longer present to recruit deacetylase activity, but the gene promoters remain unacetylated by virtue of late replication timing, which automatically packages this DNA in a closed conformation following replication. It should be noted that some group IV genes do not replicate in late S and thus must have additional layers of repression that have not yet been characterized. One possibility is that these genes, like those in group II, are subject to factor-mediated repression. Thus, these genes would also have two independent mechanisms for packaging genes in a closed nucleosome structure, DNA methylation (like group III) and transcriptional repressors capable of recruiting histone deacetylases (like group II). One example of this is Oct4, which has been shown to be inactivated by both DNA methylation (Gidekel and Bergman 2002) and, more recently, Polycomb-mediated repression (N. Feldman, unpubl.).

HISTONE CODE HYPOTHESIS

It has been suggested that modifications on histone tail lysines and arginines are used as a molecular code for affecting nucleosome structure and in this way regulate different expression paradigms (Jenuwein and Allis 2001). Currently, most of the data describing histone modification states for different gene sequences reveal a more conservative picture where active genes are always packaged with nucleosomes containing histones that are generally acetylated and methylated at H3(K4). In the same manner, inactive genes are also characterized by a standard structure in which the "activating modifications" are absent, but H3 is methylated on lysine 9 (Vermaak et al. 2003).

In our study we demonstrate for the first time that there is indeed a code that relates epigenetic mechanisms at the level of DNA to nucleosome structure. The most striking example of this is the effect of DNA methylation on me-H3(K4). Genes that contain unmethylated CpG island promoters are all packaged with histone H3 methylated at lysine 4 regardless of the activity state of the gene or the level of histone acetylation. In contrast, this lysine residue is always unmodified in nucleosomes covering methylated DNA. Removal of DNA methylation appears to bring about an automatic reversal that allows this lysine 4 to become methylated. Although the mechanism for this phenomenon has not yet been elucidated, it is reasonable to assume that the presence of methyl moieties on the DNA inhibits histone H3(K4) methylation. It should be noted that in our studies, H3(K9) methylation also seems to behave like an independent variable with many genes remaining modified at this position even within the context of general histone acetylation. Although our experiments do not reveal the role of each histone modification in the regulation of gene expression, they do indicate that these modifications help to define different gene repression paradigms.

STEPWISE GENE ACTIVATION

The fact that many tissue-specific genes are repressed by multiple layers of gene repression has a significant influence on the way in which these genes ultimately become activated in their cell type of expression. While group I genes can evidently be induced automatically once they are exposed to tissue-specific *trans* activators, all of the other groups are probably packaged in an inaccessible form that prevents these interactions, and this structure must be opened before high level expression can be attained. A good example of how this may occur is the human β-globin gene locus on chromosome 11 in man. This region contains three globin genes (ε, γ, β), which are specifically expressed in different cell types during development (Fu et al. 2002).

In nonexpressing tissues, the locus is late replicating and is in a DNaseI-insensitive chromatin conformation with all of the gene promoters being methylated. In all erythroid cells, the first change in structure occurs early in the differentiation process when the entire globin region becomes early replicating (Epner et al. 1988; Simon et al. 2001) and DNaseI sensitive (Forrester et al. 1986). Despite this overall partial opening, only one of the genes actually undergoes demethylation and becomes active (Busslinger et al. 1983). This could be the ε gene in embryonic blood, the γ gene in fetal liver, or the β gene in adult erythroid precursors. It appears that only the demethylated gene is fully accessible to tissue-specific *trans*-acting factors. While the other genes in the cluster have already shed some of their epigenetic repression mechanisms, they still remain inactive because of the closed structure imposed on them by the presence of DNA methylation. In the immune system, as well, repression is removed in a stagewise manner with demethylation being required as the last step that makes the locus accessible to the rearrangement machinery of the cell (Bergman and Cedar 2004).

ACKNOWLEDGMENTS

This work was supported by grants from the National Institutes of Health, the Israel Science Foundation, the Israel Cancer Research Fund, and the Belfer Foundation.

REFERENCES

Bergman Y. and Cedar H. 2004. A stepwise epigenetic process controls immunoglobulin allelic exclusion. *Nat. Rev. Immunol.* **4:** 753.

Brandeis M., Frank D., Keshet I., Siegfried Z., Mendelsohn M., Nemes A., Temper V., Razin A., and Cedar H. 1994. Sp1 elements protect a CpG island from de novo methylation. *Nature* **371:** 435.

Brown D.D. 1984. The role of stable complexes that repress and activate eucaryotic genes. *Cell* **37:** 359.

Busslinger M., Hurst J., and Flavell R.A. 1983. DNA methylation and the regulation of the globin gene expression. *Cell* **34:** 197.

Cedar H. 1988. DNA methylation and gene activity. *Cell* **53:** 3.

Chong J.A., Tapia-Ramirez J., Kim S., Toledo-Aral J.J., Zheng Y., Boutros M.C., Altshuller Y.M., Frohman M.A., Kraner S.D., and Mandel G. 1995. REST: A mammalian silencer protein that restricts sodium channel gene expression to neurons. *Cell* **80:** 949.

Eden S., Hashimshony T., Keshet I., Thorne A.W., and Cedar H. 1998. DNA methylation models histone acetylation. *Nature* **394:** 842.

Epner E., Forrester W.C., and Groudine M. 1988. Asynchronous DNA replication within the human β-globin gene locus. *Proc. Natl. Acad. Sci.* **85:** 8081.

Forrester W.C., Thompson C., Elder J.T., and Groudine M. 1986. A developmentally stable chromatin structure in the human beta-globin gene cluster. *Proc. Natl. Acad. Sci.* **83:** 1359.

Fu X.H., Liu D.P., and Liang C.C. 2002. Chromatin structure and transcriptional regulation of the beta-globin locus. *Exp. Cell Res.* **278:** 1.

Fuks F., Hurd P.J., Deplus R., and Kouzarides T. 2003. The DNA methyltransferases associate with HP1 and the SUV39H1 histone methyltransferase. *Nucleic Acids Res.* **31:** 2305.

Fuks F., Burgers W.A., Brehm A., Hughes-Davies L., and Kouzarides T. 2000. DNA methyltransferase Dnmt1 associates with histone deacetylase activity. *Nat. Genet.* **24:** 88.

Gidekel S. and Bergman Y. 2002. A unique developmental pattern of Oct-3/4 DNA methylation is controlled by a cis-demodification element. *J. Biol. Chem.* **277:** 34521.

Goren A. and Cedar H. 2003. Replicating by the clock. *Nat. Rev. Mol. Cell Biol.* **4:** 25.

Gottesfeld J. and Bloomer L.S. 1982. Assembly of transcriptionally active 5S RNA gene chromatin in vitro. *Cell* **28:** 781.

Gruenbaum Y., Stein R., Cedar H., and Razin A. 1981. Methylation of CpG sequences in eukaryotic DNA. *FEBS Lett.* **123:** 67.

Hashimshony T., Zhang J., Keshet I., Bustin M., and Cedar H. 2003. The role of DNA methylation in setting up chromatin structure during development. *Nat. Genet.* **34:** 187.

Huang Y., Myers S.J., and Dingledine R. 1999. Transcriptional repression by REST: Recruitment of Sin3A and histone deacetylase to neuronal genes. *Nat. Neurosci.* **2:** 867.

Jackson-Grusby L., Beard C., Possemato R., Tudor M., Fambrough D., Csankovszki G., Dausman J., Lee P., Wilson C., Lander E., and Jaenisch R. 2001. Loss of genomic methylation causes p53-dependent apoptosis and epigenetic deregulation. *Nat. Genet.* **27:** 31.

Jenuwein T. and Allis C.D. 2001. Translating the histone code. *Science* **293:** 1074.

Jones P.L., Veenstra G.J.C., Wade P.A., Vermaak D., Kass S.U., Landsberg N., Strouboulis J., and Wolffe A.P. 1998. Methylated DNA and MeCP2 recruit histone deacetylase to repress transcription. *Nat. Genet.* **19:** 187.

Kafri T., Ariel M., Brandeis M., Shemer R., Urven L., McCarrey J., Cedar H., and Razin A. 1992. Developmental pattern of gene-specific DNA methylation in the mouse embryo and germline. *Genes Dev.* **6:** 705.

Kuhn R., Schwenk F., Aguet M., and Rajewsky K. 1995. Inducible gene targeting in mice. *Science* **269:** 1427.

Lallemand Y., Luria V., Haffner-Krausz R., and Lonai P. 1998. Maternally expressed PGK-Cre transgene as a tool for early and uniform activation of the Cre site specific recombinase. *Transgenic Res.* **7:** 105.

Li E., Bestor T.H., and Jaenisch R. 1992. Targeted mutation of the DNA methyltransferase gene results in embryonic lethality. *Cell* **69:** 915.

Lunyak V.V., Burgess R., Prefontaine G.G., Nelson C., Sze

S.-H., Chenoweth J., Schwartz P., Pevzner P.A., Glass C., Mandel G., and Rosenfeld M.G. 2002. Corepressor-dependent silencing of chromosomal regions encoding neuronal genes. *Science* **298:** 1747.

Macleod D., Charlton J., Mullins J., and Bird A.P. 1994. Sp1 sites in the mouse Aprt gene promoter are required to prevent methylation of the CpG island. *Genes Dev.* **8:** 2282.

Nan X., Ng H.-H., Johnson C.A., Laherty C.D., Turner B.M., Eisenman R.N., and Bird A. 1998. Transcriptional repression by the methyl-CpG-binding protein MeCP2 involves a histone deacetylase complex. *Nature* **393:** 386.

Okano M., Bell D.W., Haber D.A., and Li E. 1999. DNA methyltransferases Dnmt3a and Dnmt3b are essential for de novo methylation and mammalian development. *Cell* **99:** 247.

Razin A. 1998. CpG methylation, chromatin structure and gene silencing—A three-way connection. *EMBO J.* **17:** 4905.

Reynaud C., Bruno C., Boullanger P., Grange J., Barbesti S., and Niveleau A. 1992. Monitoring of urinary excretion of modified nucleosides in cancer patients using a set of six monoclonal antibodies. *Cancer Lett.* **61:** 255.

Robertson K.D., Ait-Si-Ali S., Yokochi T., Wade P.A., Jones P.L., and Wolffe A.P. 2000. DNMT1 forms a complex with Rb, E2F1 and HDAC1 and represses transcription from E2F-responsive promoters. *Nat. Genet.* **25:** 338.

Rountree M.R., Bachman K.E., and Baylin S.B. 2000. DNMT1 binds HDAC2 and a new co-repressor, DMAP1, to form a complex at replication foci. *Nat. Genet.* **25:** 269.

Selig S., Okumura K., Ward D.C., and Cedar H. 1992. Delineation of DNA replication time zones by fluorescence in situ hybridization. *EMBO J.* **11:** 1217.

Siegfried Z., Eden S., Mendelsohn M., Feng X., Tzubari B., and Cedar H. 1999. DNA methylation represses transcription in vivo. *Nat. Genet.* **22:** 203.

Simon I., Tenzen T., Mostoslavsky R., Fibach E., Lande L., Milot E., Gribnau J., Grosveld F., Fraser P., and Cedar H. 2001. Developmental regulation of DNA replication timing at the human β globin locus. *EMBO J.* **20:** 6150.

Vermaak D., Ahmad K., and Henikoff S. 2003. Maintenance of chromatin states: An open-and-shut case. *Curr. Opin. Cell Biol.* **15:** 266.

Zhang J., Feng X., Hashimshony T., Keshet I., and Cedar H. 2002. The establishment of transcriptional competence in early and late S-phase. *Nature* **420:** 198.

Control of Development and Transposon Movement by DNA Methylation in *Arabidopsis thaliana*

T. KAKUTANI, M. KATO, T. KINOSHITA, AND A. MIURA
Department of Integrated Genetics, National Institute of Genetics, Mishima, Shizuoka 411-8540, Japan

In both vertebrates and plants, mutations abolishing DNA methylation induce developmental abnormalities, which often accompany perturbation of transcription (for review, see Richards 1997; Habu et al. 2001; Li 2002). A unique feature in plants is that the induced epigenetic changes are often inherited over multiple generations even after introduction into a wild-type background (for review, see Martienssen 1998; Kakutani 2002). This feature has facilitated identification of plant genes controlled by DNA methylation through conventional genetic linkage analysis. Tissue-specific as well as parent-of-origin-specific transcriptional control by DNA methylation can be studied further using *Arabidopsis* mutants affecting epigenetic states.

Control of transcription of genes involved in development is not the only function of DNA methylation. In vertebrates, plants, and some fungi, methylated cytosine is often found in repeated sequences including transposons. One model to explain this distribution proposes that the primary function of eukaryotic DNA methylation is to protect the genome from the deleterious effects of transposons (Yoder et al. 1997; Matzke et al. 1999; Selker et al. 2003). Consistent with such a "genome defense" hypothesis, endogenous *Arabidopsis* transposons are mobilized by mutations affecting genomic DNA methylation (Miura et al. 2001; Singer et al. 2001). In this paper, we summarize our genetic approaches to understand the roles of DNA methylation in the control of development and behavior of transposons.

DEVELOPMENTAL ABNORMALITIES INDUCED BY *ARABIDOPSIS* DNA METHYLATION MUTANTS

Arabidopsis provides a genetically tractable system to examine the role of DNA methylation, because viable mutants defective in DNA methylation are available. MET1 (METHYLTRANSFERASE1, an ortholog of the mammalian DNA methyltransferase Dnmt1) is necessary for maintaining genomic cytosine methylation at 5′-CG-3′ sites (Finnegan et al. 1996; Ronemus et al 1996). *Arabidopsis* additionally methylates non-CG sites using CHROMOMETHYLASE3 (CMT3) (Bartee et al. 2001; Lindroth et al. 2001). A third *Arabidopsis* gene necessary for DNA methylation is *DDM1* (*DECREASE IN DNA METHYLATION1*), which encodes a chromatin remodeling factor (Vongs et al. 1993; Jeddeloh et al. 1999; Brzeski and Jerzmanowski 2002).

A striking feature of the *ddm1* DNA hypomethylation mutation is that it results in a variety of developmental abnormalities by inducing heritable changes in other loci (Fig. 1) (Kakutani et al. 1996; Kakutani 1997). Target loci have been identified for three of *ddm1*-induced heritable developmental abnormalities. Two of them, *FWA* and *BAL*, are caused by epigenetic changes in transcription, and one of them, *CLAM*, is due to a genetic change, a transposon insertion. Each of them will be discussed in the following sections.

FWA: INHERITANCE OF IMPRINTING OVER MULTIPLE GENERATIONS

The *ddm1*-induced phenotype observed most frequently is delay in flowering initiation (Kakutani et al. 1996; Kakutani 1997). Linkage analysis and molecular analysis revealed that this phenotype was due to ectopic transcription of a homeobox gene *FWA* (Koornneef et al. 1991; Kakutani 1997; Soppe et al. 2000). The *FWA* gene is transcriptionally silent in wild-type adult tissues, but the transcript accumulates in the *ddm1*-induced late-flowering

Figure 1. Developmental abnormalities found in *ddm1* DNA hypomethylation mutant lines. Each of the *ddm1* lines has been self-pollinated seven times independently. (*a*) Flowers of a wild-type Col plant. (*b*) Flowers with thin sepals in a *ddm1* line. (*c*) Homeotic transformation of ovule into other flower organs observed in a *ddm1* line. (*d*) *Right*, upward leaf curling in a *ddm1* line. *Left*, control wild-type Columbia plant. (*e*) *bonsai* phenotype in a *ddm1* line.

lines (Soppe et al. 2000). Interestingly, the *ddm1*-induced transcriptional *FWA* activation and late-flowering phenotypes are heritable over multiple generations, which allowed the identification of *FWA* locus by the linkage analysis (Kakutani 1997; A. Miura and T. Kakutani, unpubl.). A similar phenotype is also induced by a loss of function in the CG methyltransferase gene *MET1*, indicating that DNA methylation is critical for the silencing of the *FWA* gene (Ronemus et al. 1996; Kankel et al. 2003).

Although the *FWA* transcript is undetectable in wild-type adult tissues, further characterization of the *FWA* transcription during earlier developmental stages revealed that even in wild-type plants the *FWA* transcript accumulates in endosperm, a tissue supporting embryo growth in the seed (Fig. 2a) (Kinoshita et al. 2004). In addition, *FWA* transcription in the endosperm is imprinted; only the maternal-origin transcript was detected (Fig. 2b) (Kinoshita et al. 2004). The silencing of *FWA* in embryonic tissues and silencing of paternal-origin copy depend on the *MET1* gene function (Kinoshita et al 2004). Thus CG methylation is necessary for maintenance of *FWA* imprinting, as is the case in many of imprinted mammalian genes (Li et al 1993). Unlike mammals, however, the maternal-specific *FWA* expression was not established by paternal-specific de novo methylation; it is established by maternal-specific activation in the central cell of female gametophyte, which depends on the DNA glycosylase gene, *DEMETER* (Choi et al. 2002; Kinoshita et al. 2004). Thus the silent methylated state is the default for this class of imprinted genes. Resilencing of the activated *FWA* gene is not necessary, because endosperm does not contribute to the next generation. When methylation is lost in the embryonic lineage (e.g., by *ddm1* or *met1* mutation), the *fwa* epigenetic mutation and the associated late-flowering phenotype can be stably inherited over multiple generations.

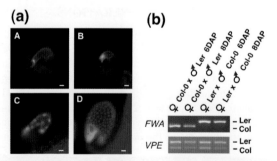

Figure 2. Endosperm-specific and imprinted transcription of *FWA* gene during early development in wild-type Col plants. (*a*) Expression pattern of the *FWA* gene examined by fluorescence images of the *pFWA::NLS-GFP* transgenic plants. (*A*) Green fluorescent protein (GFP) expression visible in mature central cell nucleus before fertilization. (*B*) Primary endosperm nucleus after fertilization. (*C*) 16–32 nuclei endosperm stage. (*D*) Heart embryo stage corresponding to 5 days after pollination. When the entire *FWA* coding region was fused to the GFP, the signal disappeared at an earlier stage (Kinoshita et al. 2004). (*b*) Imprinted transcription of *FWA* in endosperm. Wild-type Col and Ler strains were used as parents for bidirectional crosses. RT-PCR shows maternal-allele-specific expression in the endosperm at 6 and 8 days after pollination (DAP). A *VPE* gene was used as a control. (Reprinted, with permission, from Kinoshita et al. 2004 [©AAAS].)

OTHER EPIGENETIC ALLELES: *bal*, *sup*-LIKE, AND *ag*-LIKE

Another *ddm1*-induced developmental abnormality, *bal*, maps to a locus with clustered disease resistance genes; and it is also associated with overexpression of sequences within this region (Stokes et al. 2002). Like *fwa*, this phenotype is stably inherited over multiple generations even after introduction into wild-type *DDM1* background (Stokes et al. 2002). Details of the inheritance of *bal* phenotype are described in another chapter (Yi et al., this volume).

Both *bal* and *fwa* phenotypes are due to overexpression induced by the loss of DNA methylation. On the other hand, some of the *ddm1*-induced phenotypes are due to transcriptional suppression. Floral pattern phenotypes similar to those of *superman* (*sup*) or *agamous* (*ag*) mutants have been induced in *ddm1* or *met1* mutant background. Each of the phenotypes is associated with transcriptional suppression and hypermethylation of the *SUP* or *AG* gene (Jacobsen et al. 2000). Interestingly, a global decrease in DNA methylation level in these mutants is associated with local hypermethylation in *SUP* and *AG* locus. Although the mechanism is unknown, a similar phenomenon is seen in cancer cells; they can simultaneously show genome-wide hypomethylation and hypermethylation of specific genes (Jacobsen et al. 2000).

OTHER *ddm1*-INDUCED DEVELOPMENTAL ABNORMALITIES

Other types of abnormalities in flower pattern formation have been found in the self-pollinated *ddm1* lines, such as decrease in sepal number, thin sepals, and homeotic transformation of ovules into other flower organs (Fig. 1b,c) (Kakutani et al 1996). Upward curling of leaves was also induced in several independent *ddm1* lines (Fig. 1d). This phenotype was inherited as a dominant trait over generations in the presence of wild-type *DDM1* gene (T. Kakutani, unpubl.). Altered phyllotaxy, clustered siliques, and dwarfism were also found in multiple independent *ddm1* lines (Fig. 1e). This combination of phenotypes, which was named *bonsai* (Kakutani 1997), was inherited as an unstable recessive trait, and it was genetically mapped to the bottom arm of chromosome 1 (T. Kakutani, unpubl.). Target loci for these abnormalities have not been identified yet at the molecular level.

On the other hand, the characterization of another example of heritable *ddm1*-induced abnormality, *clam*, lead us to identify a transposon-induced genome rearrangement in DNA hypomethylation backgrounds, as will be described in the next section.

MOBILIZATION OF *CACTA* TRANSPOSONS BY DNA HYPOMETHYLATION MUTATIONS

The *ddm1*-induced *clam* phenotype was unstable; phenotypically normal sectors were occasionally observed (Fig. 3). Through genetic characterization of the *clam* phenotype, we identified an endogenous *Arabidopsis* transposon, named *CACTA1*. The *CACTA1* has features

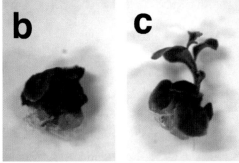

Figure 3. *clam* phenotype and its reversion. (*a*) A wild-type Columbia plant showing expanded leaves (*right*), and a *clam* homozygous plant with defects in leaf and stem elongation (*left*). Both are 35-day old. (*b–f*) The *clam* homozygous plants with (*c,e,f*) or without (*b,d*) reversion sectors. Plants were grown on soil (*a*) or nutrient agar plates (*b–f*). White bar: 10 mm. (Reprinted, with permission, from Miura et al. 2001 [©Nature Publishing Group (http://www.nature.com/)].)

Table 1. Transposition of *CACTA* Elements in *ddm1*, *met1*, *cmt3*, and *met1-cmt3* Double Mutants Detected by Changes in the Banding Pattern by Southern Analysis

Parental genotype	F$_2$ Genotype	F$_3$: self-pollinated progeny from F$_2$	
		Number examined[a]	Number of transposants[b]
DDM1/ddm1	*ddm1/ddm1*	29	17 (59%)
	DDM1/DDM1	35	0
MET1/met1	*met1/met1*	42	0
	MET1/MET1	12	0
CMT3/cmt3	*cmt3/cmt3*	64	0
	CMT3/CMT3	48	0
MET1/met1-	*cmt3-met1*[c]	39	38 (97%)
CMT3/cmt3	*CMT3-met1*[c]	45	0
	cmt3-MET1[c]	42	4 (10%)
	CMT3-MET1[c]	43	0

[a]Number of F$_3$ plants examined by Southern analysis.
[b]Number of plants with bands in new positions detectable by the Southern analysis.
[c]Homozygotes for the *CMT3* and *MET1* loci.
This table summarizes, with permission, the results shown in Table 1 and Supplementary Figure S3 of Kato et al. 2003 (©Elsevier).

typical of the CACTA family DNA-type transposons found in many plant species; the 5′ and 3′ ends of which have the conserved terminal inverted sequence CACTA-CAA. The *clam* phenotype was due to insertion of a CACTA transposon into *DWF4*, a gene involved in regulation of plant growth. The phenotypic reversion of *clam* was associated with excision of the transposon. *CACTA1* is silent in wild-type Columbia (Col) strain, but transposes and increases in the copy number in a *ddm1* DNA hypomethylation background (Miura et al. 2001).

The CACTA family transposons are found in many plant species. The CACTA member identified first and studied most extensively is the maize *Suppressor-mutator* (*Spm*)/*Enhancer* (*Spm*) transposons (for review, see Fedoroff 1996). McClintock found in the 1950s that the activity of *Spm* transposon changes heritably and reversibly (McClintock 1958). Subsequent molecular characterization of the *Spm* revealed that changes in transposon activity correlate with epigenetic changes in DNA methylation of element sequences (Fedoroff 1996).

Indeed, loss of DNA methylation seems to be sufficient for mobilization of the *Arabidopsis CACTA1*, because this element was mobilized in mutants of DNA methyltransferase genes, *MET1* (maintenance CG methylase) and *CMT3* (non-CG methylase). High-frequency transposition of *CACTA* elements was detected in *cmt3-met1* double mutants. However, single mutants in either *met1* or *cmt3* were much less effective in the mobilization (Table 1) (Kato et al. 2003). Transposition was not detected by Southern analysis in either single mutant segregating from the respective heterozygotes. This observation was confirmed by a polymerase chain reaction (PCR)–based method, in which the original *CACTA1* locus with somatic excision less than 1/100 was detectable (Fig. 4).

Interestingly, low frequency of transposition was detected in *cmt3* single mutant homozygotes segregating from self-pollinated progeny of *MET1/met1 CMT3/cmt3* double heterozygotes (Table 1) (Kato et al. 2003). This correlates with the accumulation of *CACTA1* transcript in

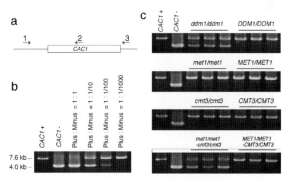

Figure 4. Excision of *CACTA1* detected by PCR. (*a*) Schematic drawing of the strategy. Products from primers 1 and 3 and primers 1 and 2 will be amplified from template genomic DNA with or without the excision of *CACTA1* from this locus, respectively. (*b*) Sensitivity of this system. The 4.0-kb product was amplified more efficiently than the 7.6-kb product. (*c*) Excision of *CACTA1* detected in *ddm1*, *met1*, *cmt3*, or *met1-cmt3* mutants segregating in self-pollinated progeny of heterozygotes. Each of the genotypes is compared to wild-type homozygotes in the corresponding segregating families. The primer sequences and PCR conditions will be provided on request.

these plants, which was undetectable in *cmt3* single mutants segregating in the progeny of *CMT3/cmt3* single heterozygotes (not shown). The *met1* mutation, even in heterozygous state, seems to affect the *CACTA1* transcription of the progeny. That may be because *met1* mutations heritably activate transcription of silent sequences during the haploid gametophyte generation (Kankel et al. 2003; Saze et al. 2003)

The mobilization of *CACTA* elements may be mediated by changes in chromatin structure, because histone modifications are also affected by the DNA methyltransferase mutants (Johnson et al. 2002; Soppe et al. 2002; Lippman et al. 2003). Although downstream mechanisms remain to be clarified, our results suggest that CG and non-CG methylation systems redundantly function as heritable epigenetic tags for immobilization of *CACTA* transposons.

FACTORS DETERMINING TRANSPOSON DISTRIBUTION WITHIN GENOME: INTEGRATION SPECIFICITY VERSUS NATURAL SELECTION

Transposons are not distributed randomly within genomes. Transposon-related sequences are often enriched in centromeric and pericentromeric regions in many organisms including *Arabidopsis* (Arabidopsis Genome Initiative 2000). The clustering of the transposons in these heterochromatic regions can be due to preferential integration of transposons into these regions (Boeke and Devine 1998) or can be caused by purifying selection against transposon insertion into gene-rich chromosomal arm regions in natural populations. To evaluate the possible contribution of natural selection, we examined the distribution of the *CACTA* transposons in genomes of 19 natural *Arabidopsis* variants (ecotypes) in comparison to new integration sites induced in the laboratory.

Sequences similar to mobile *CACTA1* are distributed among the ecotypes and show a strikingly high polymorphism in genomic localization (Miura et al. 2004). This suggests that *CACTA1* has transposed in the recent past in natural populations. Despite the high polymorphism, the copy number of *CACTA1* was low in all the examined ecotypes (Miura et al. 2004). In addition, they are localized preferentially in pericentromeric and transposon-rich regions. This contrasts to transposition induced in laboratory in the *ddm1* mutant background, in which the integration sites are less biased and the copy number frequently increases (Miura et al. 2001). The differences in the distribution of integration sites could be due to loss of the heterochromatin marks in the *ddm1* mutants, because this mutation affects DNA and histone modification in the heterochromatic regions (Gendrel et al. 2002; Johnson et al. 2002; Soppe et al. 2002). However, when the *CACTA1* element mobilized by *ddm1* mutation was introduced into a *DDM1* wild-type background, new integration events still were not targeted to the pericentromeric heterochromatin regions (Kato et al. 2004).

Because each of the transposition events does not seem to be biased toward pericentromeric heterochromatin, other factor(s) should be responsible for the accumulation of *CACTA1* elements in these regions. One possible mechanism is that if transposon excision frequency from these heterochromatic regions is low, net accumulation of the "cut-and-paste"-type of transposons in these regions would result. This possibility can be tested experimentally.

Another possible factor is natural selection against deleterious insertion into chromosomal arm regions. This mechanism is consistent with our observation that the copy number of *CACTA1* is low in all the examined ecotypes (Miura et al. 2004). If natural selection against deleterious insertions is a significant factor determining transposon distribution, the next question is how the selection operates. An obvious mechanism is that transposon insertions into gene-rich regions sometimes reduce host fitness by direct gene disruption. Other unknown selection mechanisms might also be working, as transposon insertion sometimes have phenotypic effects much stronger than loss of the entire chromosomal region around the insertion site (McClintock 1956).

PERSPECTIVES

Here we reported that loss of DNA methylation in *Arabidopsis* induces developmental abnormalities through both transposon mobilization and perturbation of transcription. The change in transcription and associated developmental variation was often heritable over multiple generations. Such an inheritance of differential epigenetic state is an enigmatic phenomenon found often in plants (Jacobsen and Meyerowitz 1997; Cubas et al. 1999; Kakutani et al. 1999; Soppe et al. 2000, 2002; Stokes et al. 2002). Similar inheritance of epigenetic variations has also been reported for some alleles of mammalian genes (Whitelaw and Martin 2001; Rakyan et al. 2003). Interestingly, the affected mammalian alleles often have transposon insertions compared to the wild-type allele. Inheritance of epigenetic silencing may be used, at least for some sequences, for transgenerational genome defense against deleterious genome rearrangement induced by transposon movement. For the genome defense by DNA methylation (Yoder et al. 1997; Matzke et al. 1999), the heritable property might be advantageous, because the silencing is maintained at every stage of development, including early development before the de novo silencing is established (Kato et al 2004).

Components controlling epigenetic states are being clarified in *Arabidopsis*. In addition, a variety of inbred accessions from natural populations are available, as well as genetic and genomic tools to exploit these resources. *Arabidopsis* will provide a good system to understand control of transposons, not only of their transcription, but also of their effects on genome integrity and chromosomal evolution.

ACKNOWLEDGMENTS

Tetsuji Kakutani thanks Eric Richards for advice in the initial stages of the project. Thanks also to Eric Richards and Jun-ichi Tomizawa for critically reading the manuscript. Supported by Grant-in-Aid for Creative Scientific Research 14GS0321 from Japan Society for the Promotion of Science.

REFERENCES

Arabidopsis Genome Initiative. 2000. Analysis of the genome sequence of the flowering plant *Arabidopsis thaliana*. *Nature* **408:** 796.

Bartee L., Malagnac F., and Bender J. 2001. *Arabidopsis* cmt3 chromomethylase mutations block non-CG methylation and silencing of an endogenous gene. *Genes Dev.* **15:** 1753.

Boeke J.D. and Devine S.E. 1998. Yeast retrotransposons: Finding a nice quiet neighborhood. *Cell* **93:** 1087.

Brzeski J. and Jerzmanowski A. 2002. Deficient in DNA methylation 1 (DDM1) defines a novel family of chromatin remodeling factors. *J. Biol. Chem.* **278:** 823.

Choi Y., Gehring M., Johnson L., Hannon M., Harada J.J., Goldberg R.B., Jacobsen S.E., and Fischer R.L. 2002. DEMETER, a DNA glycosylase domain protein, is required for endosperm gene imprinting and seed viability in *Arabidopsis*. *Cell* **110:** 33.

Cubas P., Vincent C., and Coen E. 1999. An epigenetic mutation responsible for natural variation in floral symmetry. *Nature* **401:** 157.

Fedoroff N. 1996. Epigenetic regulation of the maize *Spm* transposable element. In *Epigenetic mechanisms of gene regulation* (ed. V.E.A. Russo et al.), p 575. Cold Spring Harbor Laboratory Press, Cold Spring Harbor, New York.

Finnegan E., Peacock J., and Dennis E. 1996. Reduced DNA methylation in *Arabidopsis thaliana* results in abnormal plant development. *Proc. Natl. Acad. Sci.* **93:** 8449.

Gendrel A.V., Lippman Z., Yordan C., Colot V., and Martienssen R.A. 2002. Dependence of heterochromatic histone H3 methylation patterns on the *Arabidopsis* gene DDM1. *Science* **297:** 1871.

Habu Y., Kakutani T., and Paszkowski J. 2001. Epigenetic developmental mechanisms in plants: Molecules and targets of plant epigenetic regulation. *Curr. Opin. Genet. Dev.* **11:** 215.

Jacobsen S.E. and Meyerowitz E.M. 1997. Hypermethylated SUPERMAN epigenetic alleles in *Arabidopsis*. *Science* **277:** 1100.

Jacobsen S.E., Sakai H., Finnegan E.J., Cao X., and Meyerowitz E.M. 2000. Ectopic hypermethylation of flower-specific genes in *Arabidopsis*. *Curr. Biol.* **10:** 179.

Jeddeloh J.A., Stokes T.L., and Richards E.J. 1999. Maintenance of genomic methylation requires a SWI2/SNF2-like protein. *Nat. Genet.* **22:** 94.

Johnson L., Cao X., and Jacobsen S. 2002. Interplay between two epigenetic marks. DNA methylation and histone H3 lysine 9 methylation. *Curr. Biol.* **12:** 1360.

Kakutani T. 1997. Genetic characterization of late-flowering traits induced by DNA hypomethylation mutation in *Arabidopsis thaliana*. *Plant J.* **12:** 1447.

———. 2002. Epi-alleles in plants: Inheritance of epigenetic information over generations. *Plant Cell Physiol.* **43:** 1106.

Kakutani T., Munakata K., Richards E.J., and Hirochika H. 1999. Meiotically and mitotically stable inheritance of DNA hypomethylation induced by ddm1 mutation of *Arabidopsis thaliana*. *Genetics* **151:** 831.

Kakutani T., Jeddeloh J., Flowers S., Munakata K., and Richards E. 1996. Developmental abnormalities and epimutations associated with DNA hypomethylation mutations. *Proc. Natl. Acad. Sci.* **93:** 12406.

Kankel M.W., Ramsey D.E., Stokes T.L., Flowers S.K., Haag J.R., Jeddeloh J.A., Riddle N.C., Verbsky M.L., and Richards E.J. 2003. *Arabidopsis* MET1 cytosine methyltransferase mutants. *Genetics* **163:** 1109.

Kato M., Takashima K., and Kakutani T. 2004. Epigenetic control of CACTA transposon mobility in *Arabidopsis thaliana*. *Genetics* **168:** 961.

Kato M., Miura A., Bender J., Jacobsen S.E., and Kakutani T. 2003. Role of CG and non-CG methylation in immobilization of transposons in *Arabidopsis*. *Curr. Biol.* **13:** 421.

Kinoshita T., Miura A., Choi Y., Kinoshita Y., Cao X., Jacobsen S.E., Fischer R., and Kakutani T. 2004. One-way control of FWA imprinting in *Arabidopsis* endosperm by DNA methylation. *Science* **303:** 521.

Koornneef M., Hanhart C.J., and van der Veen J.H. 1991. A genetic and physiological analysis of late flowering mutants in *Arabidopsis thaliana*. *Mol. Gen. Genet.* **229:** 57.

Li E. 2002. Chromatin modification and epigenetic reprogramming in mammalian development. *Nat. Rev. Genet.* **3:** 662.

Li E., Beard C., and Jaenisch R. 1993. Role for DNA methylation in genomic imprinting. *Nature* **366:** 362.

Lindroth A.M., Cao X., Jackson J.P., Zilberman D., McCallum C.M., Henikoff S., and Jacobsen S.E. 2001. Requirement of CHROMOMETHYLASE3 for maintenance of CpXpG methylation. *Science* **292:** 2077.

Lippman Z., May B., Yordan C., Singer T., and Martienssen R. 2003. Distinct mechanisms determine transposon inheritance and methylation via small interfering RNA and histone modification. *PLoS Biol.* **1:** E67.

Martienssen R. 1998. Chromosomal imprinting in plants. *Curr. Opin. Genet. Dev.* **8:** 240.

Matzke M.A., Mette M.F., Aufsatz W., Jakowitsch J., and Matzke A.J. 1999. Host defenses to parasitic sequences and the evolution of epigenetic control mechanisms. *Genetica* **107:** 271.

McClintock B. 1956. Controlling elements and the gene. *Cold Spring Harbor Symp. Quant. Biol.* **21:** 197.

———. 1958. The Suppressor-mutator system of control of gene action in maize. *Carnegie Inst. Wash. Year Book* **57:** 415.

Miura A., Kato M., Watanabe K., Kawabe A., Kotani H., and Kakutani T. 2004. Genomic localization of endogenous mobile CACTA family transposons in natural variants of *Arabidopsis thaliana*. *Mol. Gen. Genomics* **270:** 524.

Miura A., Yonebayashi S., Watanabe K., Toyama T., Shimada H., and Kakutani T. 2001. Mobilization of transposons by a mutation abolishing full DNA methylation in *Arabidopsis*. *Nature* **411:** 212.

Rakyan V.K., Chong S., Champ M.E., Cuthbert P.C., Morgan H.D., Luu K.V., and Whitelaw E. 2003. Transgenerational inheritance of epigenetic states at the murine Axin(Fu) allele occurs after maternal and paternal transmission. *Proc. Natl. Acad. Sci.* **100:** 2538.

Richards E.J. 1997. DNA methylation and plant development. *Trends Genet.* **13:** 319.

Ronemus M., Galbiati M., Ticknor C., Chen J., and Dellaporta S. 1996. Demethylation-induced developmental pleiotropy in *Arabidopsis*. *Science* **273:** 654.

Saze H., Scheid O.M., and Paszkowski J. 2003. Maintenance of CpG methylation is essential for epigenetic inheritance during plant gametogenesis. *Nat. Genet.* **34:** 65.

Selker E., Tountas N., Cross S., Margolin B., Murphy J., Bird A., and Freitag M. 2003. The methylated components of the *Neurospora* genome. *Nature* **422:** 893.

Singer T., Yordan C., and Martienssen R. A. 2001. Robertson's Mutator transposons in *A. thaliana* are regulated by the chromatin-remodeling gene Decrease in DNA Methylation (DDM1). *Genes Dev.* **15:** 591.

Soppe W.J., Jacobsen S.E., Alonso-Blanco C., Jackson J.P., Kakutani T., Koornneef M., and Peeters A.J. 2000. The late flowering phenotype of fwa mutants is caused by gain-of-function epigenetic alleles of a homeodomain gene. *Mol. Cell* **6:** 791.

Soppe W., Jasencakova Z., Hauben A., Kakutani T., Meister A., Huang M., Jacobsen S., Schubert I., and Fransz P. 2002. DNA methylation controls histone H3 lysine 9 methylation and heterochromatin assembly in *Arabidopsis*. *EMBO J.* **21:** 6549.

Stokes T.L., Kunkel B.N., and Richards E.J. 2002. Epigenetic variation in *Arabidopsis* disease resistance. *Genes Dev.* **16:** 171.

Vongs A., Kakutani T., Martienssen R.A., and Richards E.J. 1993. *Arabidopsis thaliana* DNA methylation mutants. *Science* **260:** 1926.

Whitelaw E. and Martin D. I. 2001. Retrotransposons as epigenetic mediators of phenotypic variation in mammals. *Nat. Genet.* **27:** 361.

Yoder J.A., Walsh C.P., and Bestor T.H. 1997. Cytosine methylation and the ecology of intragenomic parasites. *Trends Genet.* **13:** 335.

DNA Methylation of the Endogenous *PAI* Genes in *Arabidopsis*

J. BENDER

Department of Biochemistry and Molecular Biology, Johns Hopkins University
Bloomberg School of Public Health, Baltimore, Maryland 21205

In eukaryotes, the association of genomic DNA with histone proteins to form chromatin allows packaging of DNA into the nucleus. However, the degree of chromatin compaction varies along the length of each chromosome, with some regions present in relatively open transcriptionally active euchromatin, and some regions present in more compact transcriptionally silent heterochromatin. In particular, repetitive arrays associated with centromeres and ribosomal RNA-encoding genes (rDNA), as well as transposable element sequences (both DNA transposons and retrotransposons), are assembled into heterochromatin. In the case of repeat arrays, heterochromatin is likely to serve as a means of stabilizing these structures against rearrangement, whereas in the case of transposable elements heterochromatin serves as a means of genome defense against deleterious effects of their transcription and movement. Thus, the integrity of chromatin organization is critical for correct patterns of gene expression and genome stability. Key questions are how chromatin organization patterns are established, and how these patterns are maintained through cell divisions.

In mammals and plants, as well as some species of fungi such as *Neurospora crassa*, heterochromatic regions of the genome are marked by cytosine methylation. Because this covalent DNA modification can be inherited as hemimethylated DNA after each round of replication, it provides an epigenetic memory of heterochromatin patterning in the previous generation. In mammals, cytosine methylation occurs almost exclusively in CG contexts, whereas in plants and *Neurospora* both CG and non-CG cytosines can be methylated. These differences reflect the different substrate specificities of DNA methyltransferase (DMTase) enzymes present in each organism.

Heterochromatin is also associated with particular patterns of posttranslational modifications on the amino-terminal "tails" of histone proteins, which extend outward from the globular core of the histone octamer. For example, in many eukaryotes heterochromatin-associated modification patterns include methylation of histone H3 at lysine 9 (H3 mK9) and lack of acetylation (deacetylation) of lysines on H3 and H4. Conversely, euchromatin is associated with methylation of H3 at lysine 4 (H3 mK4) and hyperacetylation of lysines on H3 and H4. These and other posttranslational histone modifications are thought to constitute a "histone code" that guides the formation of heterochromatin, euchromatin, or specialized chromatin structures, most likely through recruitment of appropriate chromatin remodeling factors to histones carrying particular combinations of modifications (Jenuwein and Allis 2001).

A striking recent discovery in *Neurospora*, *Arabidopsis*, and mammals is that cytosine methylation patterns can be guided by one of the heterochromatin-associated histone modifications, methylation of H3 at lysine 9. *Neurospora* presents the simplest example of this relationship: Loss of H3 mK9, either by mutation of the H3 K9 histone methyltransferase (HMTase) DIM-5 or by mutation of the K9 residue of the single H3-encoding gene, abolishes cytosine methylation, similar to mutation of the DMTase DIM-2 (Kouzminova and Selker 2001; Tamaru and Selker 2001). In *Arabidopsis*, mutation of the H3 K9 HMTase SUVH4/KYP (hereafter called SUVH4) reduces non-CG but not CG patterns of cytosine methylation, similar to mutation of the DMTase CMT3 (Jackson et al. 2002; Malagnac et al. 2002). In the mouse, double mutation of the hSUVH391 and hSUV392 H3 K9 HMTases causes demethylation of major satellite repeat sequences (Lehnertz et al. 2003). Similarly, mutation of the mouse G9a H3 K9 HMTase causes demethylation of an imprinted locus (Xin et al. 2003). Taken together, these findings suggest that DMTases can be recruited to appropriate target sites by recognizing the H3 mK9 mark.

Another emerging area in our understanding of how cytosine methylation and heterochromatin are targeted is the finding that RNA molecules can guide these epigenetic modifications. For example, in plant systems that express double-stranded RNA (dsRNA) and "diced" small interfering RNAs (siRNAs) associated with RNA interference (RNAi), DNA sequences with identity to the aberrant RNAs can acquire cytosine methylation (Bender 2004). This RNA-directed DNA methylation (RdDM) is typically very specifically targeted, and results in dense CG and non-CG methylation.

Interestingly, an examination of transgene and transposon targets of RdDM in *Arabidopsis* suggests that different loci are differentially sensitive to methylation mutations. Some silenced loci are demethylated and reactivated by mutation of the MET1 CG-specific DMTase and the HDA6 histone deacetylase, whereas other silenced loci are demethylated and reactivated by mutations in CMT3 and SUVH4, which control CNG and other patterns of non-CG methylation (Lindroth et al. 2001; Murfett et al. 2001; Aufsatz et al. 2002; Cao and Jacobsen 2002a; Jackson et al. 2002; Cao et al. 2003;

Kankel et al. 2003; Kato et al. 2003; Lippman et al. 2003; Probst et al. 2004). In general, these differences correspond to the CG versus non-CG content of the relevant expression sequences. In addition, some loci are sensitive to mutations in factors thought to be involved in converting aberrant single-stranded RNA templates into dsRNA such as RNA-dependent RNA polymerases (RdRPs) or argonaute (AGO) proteins (Dalmay et al. 2000, 2001; Fagard et al. 2000; Mourrain et al. 2000; Morel et al. 2002; Zilberman et al. 2003; Xie et al. 2004). The emerging view is that different RNA processing systems and perhaps different RNA species can trigger RdDM. Thus, plants might contain more than one RdDM pathway.

THE *ARABIDOPSIS PAI* GENES

Although the primary targets of DNA methylation in plant genomes are transposon-derived sequences, rDNA repeats, and centromere repeats, some low- or single-copy sequences can also be methylated. These simple targets present tractable systems for understanding the establishment and maintenance of methylation, which can then be extrapolated to methylation affecting complex sequences. Because they have evolved along with the rest of the host genome, simple endogenous targets can also provide unique insights that might not be obvious from studies with transgene-based reporters.

The phosphoribosylanthranilate isomerase (*PAI*) tryptophan biosynthetic genes in *Arabidopsis* provide a model system to study DNA methylation of relatively low-expression endogenous genes. In different backgrounds of *Arabidopsis* isolated from the wild, these genes can be found in either of two general arrangements—one associated with an absence of *PAI* DNA methylation and one associated with dense *PAI* methylation.

Most *Arabidopsis* strains, including the commonly used laboratory strains Columbia (Col) and Landsberg *erecta* (Ler), carry three unlinked unmethylated *PAI* genes—*PAI1* on the upper arm of chromosome 1, *PAI2* on the upper arm of chromosome 5, and *PAI3* near the center of chromosome 1 (Fig. 1) (Bender and Fink 1995; Melquist et al. 1999). In Col, *PAI1* and *PAI2* are 98% identical to each other over a region of 2.3 kilobases (kb) extending from 350 base pairs (bp) upstream of the translational start codon through the five exons and four introns of the coding region until 470 bp downstream of the translational stop codon. A divergently transcribed fatty acid desaturase gene *FAD8* lies in full length just downstream of *PAI2*, and the 3′ end of the *FAD8* sequence is included in the duplicated *PAI1* downstream sequences. This arrangement implies that *PAI2* was the progenitor locus for the *PAI1* duplication. The *PAI3* gene is only 90% identical to *PAI1* and *PAI2* because of polymorphisms throughout the gene coding region. Based on cDNA abundance studies, *PAI1* and *PAI2* transcripts are present at similar levels, whereas *PAI3* is less well expressed. Moreover, whereas *PAI1* and *PAI2* encode functional PAI enzyme, a *PAI3* polymorphism renders its gene product inactive.

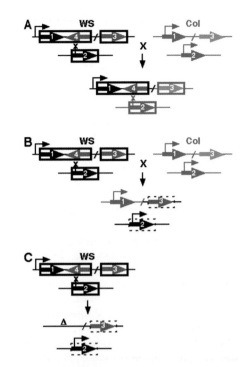

Figure 1. *PAI* methylation in the WS strain of *Arabidopsis* is triggered by a transcribed inverted repeat of *PAI* genes. *Arrows* indicate duplicated *PAI* gene sequences including the proximal promoter regions: *solid black or red arrows* indicate that the gene encodes functional PAI enzyme, and *gray or pink arrows* indicate that the gene encodes nonfunctional PAI enzyme. The transcriptional activity of the functional *PAI* genes is shown by either a *small arrow* (expressed) or an x (silenced). DNA methylation is indicated by boxes around the affected *PAI* genes: A *solid box* indicates dense CG plus non-CG methylation patterning, and a *dashed box* indicates reduced mostly CG methylation patterning. The *diagonal line* between the *PAI1* locus and the *PAI3* locus indicates that these loci are on the same chromosome but at unlinked positions. (*A*) The WS *PAI1–PAI4* locus can trigger dense de novo methylation of Col *PAI2* and *PAI3* target genes in hybrid progeny plants. (*B*) WS × Col hybrid plants that lack the WS *PAI1–PAI4* locus cannot maintain dense methylation on WS *PAI2* and *PAI3*. (*C*) A deletion mutation (Δ) that removes the WS *PAI1–PAI4* locus results in reduced methylation on *PAI2* and *PAI3*.

In contrast to the "normal" *PAI* gene arrangement found in Col and most other *Arabidopsis* strains, a few strains including Wassilewskija (WS) carry an unusual duplication at the *PAI1* locus (Fig. 1) (Melquist et al. 1999). In these strains, rather than a singlet *PAI1* gene, the locus carries the *PAI1* gene plus another full-length gene *PAI4* arranged as a tail-to-tail inverted repeat. The *PAI4* gene is highly identical to *PAI1* and *PAI2*, except that in WS and most other strains with the *PAI1–PAI4* inverted repeat arrangement, *PAI4* does not encode a functional enzyme because of inactivating polymorphisms. Another feature of the *PAI1–PAI4* locus in WS is that the duplicated *PAI* sequences are flanked by 2.9 kb of almost-perfect direct repeat sequences. In fact, the same 2.9-kb sequence lies upstream of the Col *PAI1* gene, but only the distal 0.8 kb of this sequence is present downstream of *PAI1*. This arrangement suggests that the WS-like structure might have given rise to the Col-like structure by an internal deletion of the *PAI4* gene and part of the downstream direct repeat.

Besides the different structure at the *PAI1* locus, a striking feature of WS and other strains with a *PAI1–PAI4* inverted repeat is that they carry cytosine methylation on all four *PAI* genes (Bender and Fink 1995; Luff et al. 1999; Melquist et al. 1999). This methylation is precisely targeted to the regions of shared *PAI* sequence identity including both exon and intron sequences and is densely patterned over both CG and non-CG cytosines (Luff et al. 1999). Given this *PAI*-specific methylation patterning, plus the perfect correlation between a *PAI1–PAI4* inverted repeat and *PAI* methylation in a survey of *Arabidopsis* strains (Melquist et al. 1999), we formed the hypothesis that the inverted repeat produces a signal for *PAI* methylation. To test this hypothesis, we crossed WS and Col and isolated hybrid progeny plants where the WS *PAI1–PAI4* putative initiator locus was combined with unmethylated Col *PAI* genes, and analyzed the *PAI* methylation patterns during an inbreeding regime (Fig. 1) (Luff et al. 1999). This experiment revealed that the Col *PAI* genes become methylated de novo once combined in the same genome with the WS *PAI1–PAI4* locus. The de novo methylation requires several generations of inbreeding to accumulate on *PAI2* and *PAI3* to a density similar to that observed in parental WS, with *PAI2* methylation occurring more rapidly than *PAI3* methylation. Interestingly, the Col *PAI1* locus can acquire substantial de novo methylation during the F1 generation when made heterozygous with the WS *PAI1–PAI4* locus, but only in plants where the WS *PAI1–PAI4* locus was inherited from the female parent in the cross. Taken together, these experiments strongly support the model that *PAI1–PAI4* produces a *PAI*-specific methylation signal.

Another line of evidence supporting this model comes from analysis of WS × Col hybrid progeny plants carrying the WS *PAI2* and *PAI3* genes, but the Col singlet *PAI1* gene in place of *PAI1–PAI4* (Fig. 1) (Luff et al. 1999). In hybrid plants with this genotype, residual methylation persists on *PAI2* and *PAI3*, but this methylation is reduced in density and occurs mainly at CG cytosines. Similarly, in a derivative of WS where the *PAI1–PAI4* inverted repeat is deleted by recombination between the flanking direct repeats, reduced mainly CG methylation persists on *PAI2* and *PAI3* (Fig. 1) (Bender and Fink 1995; Jeddeloh et al. 1998). Furthermore, this residual *PAI* methylation is not completely stable, such that progeny plants with demethylated *PAI2* and *PAI3* genes segregate from the Δ*pai1–pai4* parent at 1–5% per generation (Bender and Fink 1995). These observations suggest that the *PAI1–PAI4* inverted repeat not only acts to establish the *PAI* methylation imprint, but must be continuously present to maintain dense CG plus non-CG methylation on *PAI* sequences. In the absence of the inverted repeat, the residual CG methylation is likely to be maintained as a relic of the previous dense imprint by the MET1 CG DMTase. The loss of residual CG methylation presumably reflects inefficiency of this maintenance process.

TRANSCRIPTION OF THE METHYLATED *PAI* GENES IN WS

In strains like Col with unmethylated *PAI* genes, *PAI1* and *PAI2* express transcripts that initiate approximately 130 bp upstream of the translational start codon (Melquist et al. 1999; Melquist and Bender 2003). In strains like WS with methylated *PAI* genes, these proximal promoter regions are densely methylated and therefore predicted to be transcriptionally silenced. However, WS does not display any phenotypes diagnostic of tryptophan pathway deficiencies, and in fact contains steady-state levels of *PAI* transcripts that are slightly higher than those observed in Col (Bender and Fink 1995). A detailed analysis of WS *PAI* transcripts revealed the explanation for this apparent paradox: all the detectable transcripts correspond to the *PAI1* gene and initiate from a unique fortuitous promoter in upstream unmethylated sequences with a transcription start site that lies approximately 500 bp beyond the *PAI* methylation boundary (Fig. 1) (Melquist et al. 1999; Melquist and Bender 2003).

This fortuitous upstream promoter is derived from the 2.9-kb direct repeat sequences that flank the *PAI1–PAI4* locus. At the end of the direct repeat proximal to *PAI4*, the promoter drives expression of a putative ribosomal protein-encoding gene *S15a*. In fact, the first exon and first intron as well as the promoter of *S15a* are included in the direct repeat segment. At the end of the direct repeat proximal to *PAI1*, these sequences are fused to *PAI1* upstream sequences instead of to *S15a*. This *S15a–PAI1* fusion yields three splice variants at the 5′ ends of WS *PAI1* transcripts upstream of the coding region (Melquist and Bender 2003). The two more abundant species remove ~700 nucleotides (nt) of upstream sequences such that the resulting transcripts are of similar length to the Col *PAI* transcripts that originate from the *PAI* proximal promoter. A third less-abundant species fails to splice the upstream sequences. Interestingly, even though the Col *PAI1* gene carries the same upstream *S15a*-derived sequences as the WS *PAI1* gene, the unmethylated proximal *PAI1* promoter is used preferentially to drive Col *PAI1* expression (Melquist and Bender 2004). The greater accumulation of the *S15a*-driven *PAI1* transcripts in WS than of the combined *PAI1*, *PAI2*, and *PAI3* transcripts in Col implies that the *S15a* promoter is stronger than the *PAI* promoter.

Given the precedent for dsRNA-derived species to promote DNA methylation in plants, an attractive view is that transcription through the *PAI1–PAI4* inverted repeat from the upstream unmethylated *S15a* promoter produces a *PAI* dsRNA methylation signal. To test this hypothesis, we constructed a transgenic derivative of WS where the *S15a* promoter sequences are methylated and silenced (Fig. 2) (Melquist and Bender 2003). This effect was achieved by transforming WS with a transgene where the strong Cauliflower Mosaic Virus 35S promoter drives expression through an inverted repeat of *S15a* promoter sequences (S15aIR) to trigger RdDM of the endogenous promoter sequence targets in the direct repeats flanking the *PAI1–PAI4* locus. The resulting WS(S15aIR) strain

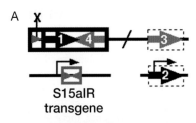

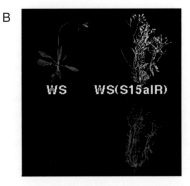

Figure 2. Transgene-induced methylation and silencing of the upstream promoter that drives transcription through the WS *PAI1–PAI4* inverted repeat results in reduced methylation on the *PAI2* and *PAI3* target genes. (*A*) *PAI* gene arrangements, transcriptional status, and methylation patterning in the WS(S15aIR) transgenic strain are indicated as described in Fig. 1. (*B*) Reduced *PAI* expression in the WS(S15aIR) transgenic strain causes PAI-deficient phenotypes including blue fluorescence under UV light (*lower* panel).

has dense methylation of *S15a* promoter sequences, strongly reduced steady-state levels of both *S15a* and *PAI1* transcripts, and conspicuous PAI-deficient phenotypes including blue fluorescence under UV light caused by accumulation of a tryptophan precursor compound.

Analysis of *PAI* methylation patterning in the WS(S15aIR) strain versus parental WS showed that *PAI* methylation is indeed impaired by suppression of transcription through the *PAI1–PAI4* locus (Melquist and Bender 2003). In WS(S15aIR), the *PAI2* and *PAI3* singlet genes display reduced mostly CG methylation similar to strains where the *PAI1–PAI4* locus is removed from the genome (Fig. 2) (Bender and Fink 1995; Jeddeloh et al. 1998; Luff et al. 1999). This finding strongly supports the view that *PAI* methylation is mediated by an RNA signal. However, in WS(S15aIR) the *PAI1–PAI4* locus maintains dense CG and non-CG patterning similar to the patterning in parental WS. A possible explanation for this difference is that the *PAI1–PAI4* locus might be more susceptible to an RNA methylation signal than the outlying singlet genes such that, in the WS(S15aIR) strain where the signal is suppressed but not eliminated, enough RNA persists to maintain methylation at this sensitive site. Alternatively, the *PAI1–PAI4* inverted repeat could maintain methylation by an RNA-independent mechanism, such as presenting a novel structure that directly recruits the methylation machinery.

In addition to this genetic evidence for an RNA-based *PAI* methylation signal, molecular analysis of WS *PAI* transcripts indicates that a low level (<10%) of transcripts extends beyond a major polyadenylation site at the end of the *PAI1* gene into palindromic *PAI4* sequences to make *PAI* dsRNA (Melquist and Bender 2003). However, diced *PAI* siRNAs derived from this dsRNA are not detectable by standard gel blot methods. Thus, the *PAI* methylation signal, whether it is precursor dsRNA or processed siRNAs, acts efficiently even at very low levels.

An indirect line of evidence suggests that there might be a low level of RNAi directed against full-length *PAI* transcripts, implying the presence of *PAI* siRNAs. This evidence comes from a mutant derivative of WS, *invpai1-Δpai4*, where the central sequences in *PAI1–PAI4* are rearranged such that there are now novel strong polyadenylation signals that suppress readthrough transcription from *PAI1* into *PAI4* (Melquist and Bender 2004). In this mutant, *PAI1* transcripts accumulate to 2.5-fold higher levels than in wild-type WS, suggesting that the WS transcripts might be partially destabilized by inefficient RNAi. Alternatively, the increased stability of *PAI1* transcripts in the *invpai1-Δpai4* mutant could be accounted for by the novel 3′ end structures in these transcripts.

Consistent with suppression of *PAI* dsRNA in the *invpai1-Δpai4* rearrangement mutant, the *PAI* genes carry only residual mostly CG methylation (Melquist and Bender 2004). But in contrast to the WS(S15aIR), where *PAI* dsRNA production is blocked by silencing of the upstream promoter (Melquist and Bender 2003), the rearrangement mutant displays reduced methylation at the *invpai1-Δpai4* locus as well as at the unlinked singlet genes *PAI2* and *PAI3* (Melquist and Bender 2004). This difference could be accommodated by either of the explanations provided above for the retention of dense *PAI1–PAI4* methylation in the WS(S15aIR) strain. If *PAI1–PAI4* methylation depends on an RNA-based signal, then there could be a stronger suppression of this signal in *invpai1-Δpai4* than in WS(S15aIR). If *PAI1–PAI4* methylation depends on a DNA structure–based signal, then the rearrangement mutant could disrupt critical structural elements.

If *PAI* siRNAs constitute the *PAI* RdDM signal, then strain backgrounds with impaired processing of these species should display a reduction in *PAI* methylation as well as an increase in *PAI* steady-state message levels. The *Arabidopsis* genome encodes four Dicer-like (DCL) proteins with sequences related to animal dicer ribonucleases required for siRNA cleavage (Schauer et al. 2002; Finnegan et al. 2003). Mutations in *DCL1* have emerged from many different mutant screens because of their pleiotropic developmental defects: strong loss-of-function alleles are embryo-lethal and weaker alleles such as "*carpel factory*" (*dcl1-9*) have aberrant morphology including defects in flower architecture that confer complete sterility. These developmental defects have been traced to defects in the processing of microRNAs (miRNAs), a specialized class of small RNAs produced from endogenous short dsRNA precursors and used in the regulation of a wide variety of developmental genes (Park et al. 2002; Reinhart et al. 2002; Finnegan et al. 2003;

Kasschau et al. 2003; Papp et al. 2003). However, production of transgene or transposon-derived siRNAs is not obviously impaired in *dcl1* mutant backgrounds, suggesting that the other DCL proteins might be involved in processing other types of precursor aberrant RNAs. Consistent with this view, a mutation in *dcl2* has been shown to impair accumulation of siRNAs derived from infecting viruses, and a mutation in *dcl3* has been shown to impair accumulation of siRNAs derived from some endogenous sequences including a short interspersed nucleotide element (SINE) retrotransposon (Xie et al. 2004). The *dcl3* mutant also displays a partial reduction of DNA methylation on its target sequences. In contrast to *dcl1* mutants, insertional disruptions in *DCL2, DCL3*, and *DCL4* do not confer morphological abnormalities (Xie et al. 2004; J. Bender, unpubl.).

We found that each of the single *dcl* mutants has no obvious effect on *PAI* transcript accumulation or *PAI* DNA methylation (Melquist and Bender 2003; J. Bender, unpubl.). Thus, either (1) the DCL proteins act redundantly and multiple *dcl* mutants will be required to reveal *PAI* phenotypes; (2) DCL1 is the key factor for processing *PAI* dsRNA, but the weak *dcl1-9* allele we used for our analysis was not sufficient to reveal *PAI* phenotypes; or (3) *PAI* dsRNA is not significantly processed by dicer ribonucleases, and the unprocessed longer species constitute the *PAI* DNA methylation signal. Future studies of multiple *dcl* mutants might allow us to discriminate among these possibilities. Another potentially informative approach will be to express viral proteins that suppress RNAi (Chapman et al. 2004; Dunoyer et al. 2004) in the WS background and monitor effects on *PAI* RNA accumulation and *PAI* DNA methylation.

THE WS *PAI* GENES AS REPORTERS FOR EPIGENETIC CHANGES

Because *PAI1* is the only expressed *PAI* gene in WS, mutations that disrupt this gene have the potential to confer a blue fluorescent PAI-deficient phenotype. For example, as described above, the WS(S15aIR) strain is blue fluorescent as a result of transcriptional silencing of the upstream promoter that drives *PAI1* expression (Fig. 2) (Melquist and Bender 2003). Loss of *PAI1* expression by itself would be predicted to be lethal because of the defect in tryptophan synthesis, but because the suppression of *PAI* dsRNA in WS(S15aIR) causes partial demethylation and reexpression of the other functional *PAI* gene, *PAI2*, enough enzyme is produced for the plant to be viable and relatively healthy. Similarly, in the *invpai1-Δpai4* rearrangement mutant strain or the Δ*pai1–pai4* complete deletion mutant strain the *PAI1* gene is inactivated, causing a blue fluorescent phenotype (Bender and Fink 1995; Melquist and Bender 2004). But in these strains the *PAI1* defect is compensated for by the suppression of *PAI* dsRNA and the partial demethylation and reactivation of *PAI2* to yield a relatively healthy plant.

Although WS(S15aIR), *invpai1-Δpai4*, and Δ*pai1–pai4* all display similar *PAI1*-deficient phenotypes and methylation patterning on *PAI2* and *PAI3*, a striking difference among these strains is that whereas WS(S15aIR) and *invpai1-Δpai4* stably retain fluorescence and residual *PAI2* methylation (Melquist and Bender 2004), the Δ*pai1–pai4* strain gives rise to nonfluorescent *PAI2*-demethylated progeny at a frequency of 1–5% per generation (Bender and Fink 1995). As mentioned above, the instability of residual mostly CG *PAI2* methylation in Δ*pai1–pai4* is presumably caused by inefficiency in the maintenance of this methylation together with the complete absence of the initiating RNA signal. Accordingly, the stability of the residual mostly CG methylation in WS(S15aIR) and *invpai1-Δpai4* suggests that these strains still produce a low level of the RNA signal that is sufficient, when combined with maintenance methylation systems, to perpetuate a partial methylation imprint. These results should be borne in mind when studying the relationship between aberrant RNA species and methylation patterning for other methylated loci in plant genomes: the RNA species expressed during the maintenance of a preexisting methylation imprint might have altered substantially because of epigenetic changes, mutations, or segregation of unlinked repetitive sequences from the RNA species that originally established the imprint.

Because the blue fluorescent phenotype is a facile and sensitive reporter for *PAI2* promoter methylation and transcriptional silencing, we isolated a *pai1* missense mutant derivative of WS where this phenotype is revealed to use as a reporter strain (Fig. 3) (Bartee and Bender 2001).

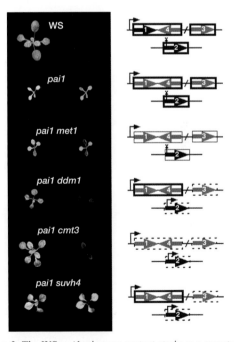

Figure 3. The WS *pai1* missense mutant strain as a reporter for mutations that reduce *PAI2* DNA methylation and transcriptional silencing. (*Left*) Photographs of the indicated strains are shown under visible (*left*) or UV (*right*) light. (*Right*) The *PAI* gene arrangements, transcriptional status, and methylation patterning as described in Fig. 1. The methylation boxes drawn with a *fine solid line* in the *pai1 met1* diagram indicate a reduction mainly in CG methylation.

In contrast to the Δpai1–pai4, invpai1-Δpai4, and WS(S15aIR) lesions where the defect in PAI1 is accompanied by a defect in the PAI RdDM signal, the WS pai1 missense mutation only affects the function of the PAI1 enzyme. Thus, the WS pai1 mutant maintains dense CG and non-CG methylation on all three PAI loci including the PAI2 promoter region, and the strain is severely deficient in functional PAI enzyme. In consequence, the WS pai1 strain is brightly blue fluorescent in all parts of the plant at all developmental stages, as well as having reduced size and fertility. In this background, trans-acting mutations that release PAI2 promoter methylation and transcriptional silencing can be easily identified by suppression of blue fluorescence.

As a proof of concept for the WS pai1 reporter strain, we crossed two mutations known to cause demethylation of centromere repeats into this background: met1-1, with a missense mutation in the MET1 CG DMTase (Kankel et al. 2003), and ddm1-2, with a splice junction mutation in the DDM1 chromatin remodeling helicase (Jeddeloh et al. 1999). In both cases, the newly segregated homozygous methylation mutation in the pai1 background did not obviously suppress fluorescence (Bartee and Bender 2001). However, each methylation mutation conferred a distinct pattern of PAI demethylation upon inbreeding (Fig. 3). The met1 mutation caused a reduction mainly in CG methylation on all three PAI loci; because the PAI2 promoter sequence carries relatively few cytosines in the CG context, the met1-induced demethylation resulted in only a subtle suppression of the PAI-deficient fluorescent phenotype. In contrast, the ddm1 mutation caused a strong loss of both CG and non-CG methylation on the PAI2 and PAI3 singlet genes, but only a slight reduction in methylation on the PAI1–PAI4 inverted repeat; the strong demethylation of the PAI2 promoter resulted in an almost-complete suppression of the PAI-deficient fluorescent phenotype. The observation that PAI1–PAI4 methylation is not dependent on DDM1 chromatin remodeling helicase function could reflect an unusual chromatin composition for this locus. Alternatively, this locus could more efficiently recruit DMTases than the other PAI loci such that it is able to maintain methylation even without the facilitating function of DDM1.

MAINTENANCE OF NON-CG METHYLATION ON THE PAI GENES

We also conducted a forward genetic screen for ethyl methane sulfonate (EMS)-induced mutations in the WS pai1 background that suppress the PAI-deficient fluorescent phenotype. This screen yielded 11 loss-of-function mutations in the CMT3 DMTase (Bartee et al. 2001) and seven loss-of-function mutations in the SUVH4 H3 K9 HMTase (Fig. 3) (Malagnac et al. 2002). CMT3 is a member of a plant-specific class of DMTases that contains a chromodomain interaction motif in addition to canonical DMTase catalytic motifs. The cmt3 mutations confer a strong loss of methylation in non-CG contexts on all three PAI loci (Bartee et al. 2001). Thus, CMT3 is the major DMTase involved in maintenance of non-CG methylation associated with PAI RdDM. The suvh4 mutations also confer a loss of non-CG methylation on the PAI2 reporter locus (Malagnac et al. 2002). Furthermore, a cmt3 suvh4 double mutant has similar methylation patterning and suppression phenotypes to the cmt3 single mutant, arguing that cmt3 and suvh4 are part of a common pathway for maintenance of non-CG cytosine methylation (Malagnac et al. 2002). Given that most of the cytosines in the PAI2 promoter occur in non-CG contexts, it is not surprising that this locus is most sensitive to mutations that affect non-CG methylation.

The effects of cmt3 and suvh4 mutations are not limited to the PAI genes. Both mutations also reduce CNG methylation at centromere repeats and transposon-derived sequences (Bartee et al. 2001; Lindroth et al. 2001; Jackson et al. 2002; Malagnac et al. 2002; Lippman et al. 2003). However, in contrast to the PAI genes, some of these target loci are able to maintain methylation of cytosines in asymmetric contexts in the cmt3 mutant background (Lindroth et al. 2001; Cao and Jacobsen 2002a). The domains-rearranged methyltransferase (DRM) class of DMTase, represented in Arabidopsis by DRM1 and DRM2 gene products, has been implicated through both genetic and in vitro studies in catalyzing this residual asymmetric methylation (Cao and Jacobsen 2002a; Wada et al. 2003). The loss of asymmetric methylation on the PAI genes in a cmt3 mutant background could therefore be explained as an indirect effect on DRM activity caused by the loss of CNG methylation. A WS pai1 drm1 drm2 mutant shows no obvious alteration in PAI methylation patterns or PAI-deficient phenotypes (J. Bender, unpubl.), but since PAI asymmetric methylation is less dense than CG or CNG methylation (Luff et al. 1999), we cannot rule out the possibility that the drm mutations confer subtle effects on this patterning.

The observation that suvh4 and cmt3 mutations both affect CNG methylation can be accounted for by proposing that the H3 mK9 mark catalyzed by SUVH4 recruits CMT3 to appropriate genomic target regions. Moreover, CMT3 carries a chromodomain motif related to the chromodomain of the animal heterochromatin protein 1 (HP1), which is known to bind to H3 mK9 (Jacobs and Khorasanizadeh 2002; Nielsen et al. 2002), suggesting that CMT3 might directly bind to the SUVH4-catalyzed histone modification. It has also been proposed that the Arabidopsis HP1-related protein LHP1 acts as a bridge between H3 mK9 and CMT3 (Jackson et al. 2002), but this model is unlikely because an lhp1 mutation has no effect on PAI or centromere repeat methylation (Malagnac et al. 2002). However, it should be noted that in the fungus Neurospora, genomic DNA methylation requires the DMTase DIM-2 (Kouzminova and Selker 2001), the H3 K9 HMTase DIM-5 (Tamaru and Selker 2001), and an HP1-related protein (Freitag et al. 2004). Thus, both Arabidopsis and Neurospora maintain cytosine methylation using H3 mK9, a chromodomain, and a DMTase, but differ in whether the chromodomain is contained in the DMTase polypeptide.

Interestingly, mutations in cmt3 and suvh4 do not confer obvious morphological phenotypes despite genome-wide loss of CNG methylation (Bartee et al. 2001; Lindroth et al. 2001; Jackson et al. 2002; Malagnac et al.

2002). In contrast, mutations in the CG DMTase MET1 confer a number of abnormal phenotypes including late flowering due to demethylation and ectopic expression of the *FWA* gene (Lindroth et al. 2001; Kankel et al. 2003) and fertility defects caused by dysregulation of flower development genes (Jacobsen et al. 2000). These findings suggest that CMT3-mediated non-CG methylation serves primarily as a reinforcement to the basal level of CG methylation maintained by MET1. Consistent with this view, the densely methylated *CACTA* transposons in *Arabidopsis* cannot transpose when they are only partially demethylated by single *cmt3* or *met1* mutations; only in the *cmt3 met1* double mutant background are the *CACTA* elements fully demethylated and mobilized (Kato et al. 2003).

The sensitivity of the PAI-deficient fluorescence phenotype and our ability to monitor methylation changes on each of the three *PAI* loci allowed us to determine that *cmt3* mutations confer stronger effects on *PAI* methylation and silencing than *suvh4* mutations (Malagnac et al. 2002). First, *pai1 cmt3* mutant seedlings are almost completely nonfluorescent whereas *pai1 suvh4* mutant seedlings retain partial fluorescence (Fig. 3). Correspondingly, the *cmt3* mutant has a stronger loss of non-CG methylation on the *PAI2* promoter than the *suvh4* mutant. Second, the *cmt3* mutation also confers a strong loss of non-CG methylation on the *PAI1–PAI4* inverted repeat, but this locus retains dense CG plus non-CG methylation in the *suvh4* mutant background (Fig. 3). One possible explanation for this discrepancy is to propose that SUVH4 is partially redundant with other H3 K9 HMTases. Consistent with this possibility, the *Arabidopsis* genome encodes eight other related SUVH proteins (Baumbusch et al. 2001). Furthermore, at least one of these gene products, SUVH6, has H3 mK9 activity in vitro (Jackson et al. 2004). An alternative possibility is that CMT3 can maintain a subset of non-CG methylation independently of the H3 mK9 mark. We plan to explore these models by testing the histone modifications on the *PAI* genes in wild-type versus *suvh4* mutant backgrounds using chromatin immunoprecipitation, and by testing the effects of mutations in other *SUVH* genes on *PAI* epigenetic modifications.

Strikingly, the *pai1* fluorescence suppressor screen did not yield mutations in factors involved in RNA processing, even though the results with the WS(S15aIR) transgenic strain implicate a *PAI* dsRNA-based signal in maintenance of *PAI* non-CG methylation (Fig. 2). But as discussed for the *DCL* genes, such mutations might be lethal or affect redundant genes. In addition, because the *PAI1–PAI4* inverted repeat locus provides a direct source of dsRNA, it is unlikely to require factors that convert aberrant RNAs into dsRNA, such as RdRP enzymes. In fact, we have explicitly determined that maintenance of *PAI* methylation does not require any of the SDE RNA processing factors including the SDE1/SGS2 RdRP (Melquist and Bender 2003).

The *pai1* fluorescence suppressor screen also did not yield new alleles of *met1* and *ddm1*. From our explicit tests with *met1* and *ddm1* mutations we know that their *PAI* demethylation effects are progressive (Bartee and Bender 2001) and are thus unlikely to be apparent in newly segregated mutant seedlings such as we scored in our suppressor screen. Furthermore, in the case of *met1*, the partial loss of CG methylation conferred by this mutation on the *PAI2* promoter only slightly suppresses the fluorescent silenced phenotype (Fig. 3).

THE ESTABLISHMENT OF *PAI* METHYLATION

An area that remains to be more fully explored is the factors that contribute to the establishment of *PAI* methylation. As discussed above, we can trigger de novo methylation of a previously unmethylated *PAI2* or *PAI3* gene from the Col strain by crossing WS and Col and isolating hybrid progeny where the *PAI1–PAI4* inverted repeat trigger locus from WS is combined with the unmethylated Col *PAI* targets in the same genome (Fig. 1) (Luff et al. 1999). In addition, if we perform this experiment with the WS *pai1* strain rather than wild-type WS, we can follow the progressive methylation and silencing of the target functional *PAI2* gene by the acquisition of the blue fluorescent PAI-deficient phenotype (Malagnac et al. 2002; Melquist and Bender 2004).

When this assay is performed with parent plants that both carry *suvh4* mutations, de novo methylation can still occur on the *PAI2* gene inherited from the unmethylated parent, although in this case the methylation accumulates with reduced density mostly at CG cytosines because of the *suvh4* methylation maintenance defect (Malagnac et al. 2002). However, given the possibility that *Arabidopsis* encodes other H3 K9 HMTases (Baumbusch et al. 2001; Jackson et al. 2004), this experiment does not rule out H3 mK9 as an establishing epigenetic mark.

In contrast to the *suvh4* experiment, when the assay is performed with parent plants that both carry *cmt3* mutations the *PAI2* target gene remains unmethylated even after several generations of inbreeding (Malagnac et al. 2002). Three possible interpretations of this result are (1) CMT3 makes the initial epigenetic mark on the *PAI* genes as well as maintaining this mark in non-CG contexts; (2) the DRM enzymes, which have been implicated in de novo methylation (Cao and Jacobsen 2002b; Cao et al. 2003; Wada et al. 2003), make the initial epigenetic mark, but CMT3 is needed to propagate this mark into an appropriate patterning or density to allow recognition of the target region by the MET1 CG maintenance DMTase; or (3) the partial demethylation of *PAI1–PAI4* in the *cmt3* mutant background attenuates the RNA signal for DNA methylation produced from this locus. An analysis of the *drm1 drm2* double mutant in our *PAI* de novo methylation assay is currently under way, and will allow us to determine whether these enzymes are involved in the initial *PAI* cytosine methylation mark, indicating distinct pathways for establishment versus maintenance of *PAI* methylation patterning.

CONCLUSIONS

The endogenous *PAI* genes in *Arabidopsis* provide an especially facile system for understanding the establishment and maintenance of DNA methylation in plant genomes. Although the four WS *PAI* genes are similar to transposable element sequences in that they have duplicated to multiple sites in the genome, they present a uniquely simple arrangement where only a single promoter drives the production of aberrant RNAs that trigger *PAI* DNA methylation. In addition, silencing of *PAI* expression can be sensitively monitored by the PAI-deficient blue fluorescent phenotype. These features have allowed us to characterize the CMT3/SUVH4 pathway for maintenance of non-CG methylation. These features have also allowed us to determine that the *PAI1–PAI4* inverted repeat trigger locus can maintain methylation even in strain backgrounds where the other *PAI* genes are demethylated. In future work we plan to dissect further the nature of the RNA signal for *PAI* DNA methylation and the relationship between this signal, CMT3, and SUVH4 at the *PAI* loci.

ACKNOWLEDGMENTS

Thanks to members of the laboratory including Lisa Bartee, Michelle Ebbs, Bradley Luff, Fabienne Malagnac, Stacey Melquist, and Laura Pawlowski for their contributions to this work. Thanks also to the Arabidopsis Biological Resouce Center, David Baulcombe, Jim Carrington, Jean Finnegan, Steve Jacobsen, Ortrun Mittelsten Scheid, Jane Murfett, Eric Richards, and Hervé Vaucheret for contributions of mutant strains. This work has been supported by grants from the Searle Scholars Program, the March of Dimes Birth Defects Foundation, and the National Institutes of Health.

REFERENCES

Aufsatz W., Mette M.F., van der Winden J., Matzke M., and Matzke A.J. 2002. HDA6, a putative histone deacetylase needed to enhance DNA methylation induced by double-stranded RNA. *EMBO J.* **21:** 6832.

Bartee L. and Bender J. 2001. Two *Arabidopsis* methylation-deficiency mutations confer only partial effects on a methylated endogenous gene family. *Nucleic Acids Res.* **29:** 2127.

Bartee L., Malagnac F., and Bender J. 2001. *Arabidopsis cmt3* chromomethylase mutations block non-CG methylation and silencing of an endogenous gene. *Genes Dev.* **15:** 1753.

Baumbusch L.O., Thorstensen T., Krauss V., Fischer A., Naumann K., Assalkhou R., Schulz I., Reuter G., and Aalen R.B. 2001. The *Arabidopsis thaliana* genome contains at least 29 active genes encoding SET domain proteins that can be assigned to four evolutionarily conserved classes. *Nucleic Acids Res.* **29:** 4319.

Bender J. 2004. DNA methylation and epigenetics. *Annu. Rev. Plant Biol.* **55:** 41.

Bender J. and Fink G.R. 1995. Epigenetic control of an endogenous gene family is revealed by a novel blue fluorescent mutant of *Arabidopsis*. *Cell* **83:** 725.

Cao X. and Jacobsen S.E. 2002a. Locus-specific control of asymmetric and CpNpG methylation by the *DRM* and *CMT3* methyltransferase genes. *Proc. Natl. Acad. Sci.* (suppl. 4) **99:** 16491.

———. 2002b. Role of the *Arabidopsis DRM* methyltransferases in de novo DNA methylation and gene silencing. *Curr. Biol.* **12:** 1138.

Cao X., Aufsatz W., Zilberman D., Mette M.F., Huang M.S., Matzke M., and Jacobsen S.E. 2003. Role of the *DRM* and *CMT3* methyltransferases in RNA-directed DNA methylation. *Curr. Biol.* **13:** 2212.

Chapman E.J., Prokhnevsky A.I., Gopinath K., Dolja V.V., and Carrington J.C. 2004. Viral RNA silencing suppressors inhibit the microRNA pathway at an intermediate step. *Genes Dev.* **18:** 1179.

Dalmay T., Horsefield R., Braunstein T.H., and Baulcombe D.C. 2001. *SDE3* encodes an RNA helicase required for post-transcriptional gene silencing in *Arabidopsis*. *EMBO J.* **20:** 2069.

Dalmay T., Hamilton A., Rudd S., Angell S., and Baulcombe D.C. 2000. An RNA-dependent RNA polymerase gene in *Arabidopsis* is required for posttranscriptional gene silencing mediated by a transgene but not by a virus. *Cell* **101:** 543.

Dunoyer P., Lecellier C.H., Parizotto E.A., Himber C., and Voinnet O. 2004. Probing the microRNA and small interfering RNA pathways with virus-encoded suppressors of RNA silencing. *Plant Cell* **16:** 1235.

Fagard M., Boutet S., Morel J.B., Bellini C., and Vaucheret H. 2000. AGO1, QDE-2, and RDE-1 are related proteins required for post-transcriptional gene silencing in plants, quelling in fungi, and RNA interference in animals. *Proc. Natl. Acad. Sci.* **97:** 11650.

Finnegan E.J., Margis R., and Waterhouse P.M. 2003. Posttranscriptional gene silencing is not compromised in the *Arabidopsis CARPEL FACTORY* (*DICER-LIKE1*) mutant, a homolog of Dicer-1 from *Drosophila*. *Curr. Biol.* **13:** 236.

Freitag M., Hickey P.C., Khlafallah T.K., Read N.D., and Selker E.U. 2004. HP1 is essential for DNA methylation in *Neurospora*. *Mol. Cell* **13:** 427.

Jackson J.P., Lindroth A.M., Cao X., and Jacobsen S.E. 2002. Control of CpNpG DNA methylation by the KRYPTONITE histone H3 methyltransferase. *Nature* **416:** 556.

Jackson J.P., Johnson L., Jasencakova Z., Zhang X., PerezBurgos L., Singh P.B., Cheng X., Schubert I., Jenuwein T., and Jacobsen S.E. 2004. Dimethylation of histone H3 lysine 9 is a critical mark for DNA methylation and gene silencing in *Arabidopsis thaliana*. *Chromosoma* **112:** 308.

Jacobs S.A. and Khorasanizadeh S. 2002. Structure of HP1 chromodomain bound to a lysine 9-methylated histone H3 tail. *Science* **295:** 2080.

Jacobsen S.E., Sakai H., Finnegan E.J., Cao X., and Meyerowitz E.M. 2000. Ectopic hypermethylation of flower-specific genes in *Arabidopsis*. *Curr. Biol.* **10:** 179.

Jeddeloh J.A., Bender J., and Richards E.J. 1998. The DNA methylation locus *DDM1* is required for maintenance of gene silencing in *Arabidopsis*. *Genes Dev.* **12:** 1714.

Jeddeloh J.A., Stokes T.L., and Richards E.J. 1999. Maintenance of genomic methylation requires a SWI2/SNF2-like protein. *Nat. Genet.* **22:** 94.

Jenuwein T. and Allis C.D. 2001. Translating the histone code. *Science* **293:** 1074.

Kankel M.W., Ramsey D.E., Stokes T.L., Flowers S.K., Haag J.R., Jeddeloh J.A., Riddle N.C., Verbsky M.L., and Richards E.J. 2003. *Arabidopsis MET1* cytosine methyltransferase mutants. *Genetics* **163:** 1109.

Kasschau K.D., Xie Z., Allen E., Llave C., Chapman E.J., Krizan K.A., and Carrington J.C. 2003. P1/HC-Pro, a viral suppressor of RNA silencing, interferes with *Arabidopsis* development and miRNA unction. *Dev. Cell* **4:** 205.

Kato M., Miura A., Bender J., Jacobsen S.E., and Kakutani T. 2003. Role of CG and non-CG methylation in immobilization of transposons in *Arabidopsis*. *Curr. Biol.* **13:** 421.

Kouzminova E. and Selker E.U. 2001. *dim-2* encodes a DNA methyltransferase responsible for all known cytosine methylation in *Neurospora*. *EMBO J.* **20:** 4309.

Lehnertz B., Ueda Y., Derijck A.A., Braunschweig U., Perez-Burgos L., Kubicek S., Chen T., Li E., Jenuwein T., and Peters A.H. 2003. Suv39h-mediated histone H3 lysine 9 methylation directs DNA methylation to major satellite repeats at pericentric heterochromatin. *Curr. Biol.* **13:** 1192.

Lindroth A.M., Cao X., Jackson J.P., Zilberman D., McCallum C.M., Henikoff S., and Jacobsen S.E. 2001. Requirement of *CHROMOMETHYLASE3* for maintenance of CpXpG methylation. *Science* **292:** 2077.

Lippman Z., May B., Yordan C., Singer T., and Martienssen R. 2003. Distinct mechanisms determine transposon inheritance and methylation via small interfering RNA and histone modification. *PLoS Biol.* **1:** E67.

Luff B., Pawlowski L., and Bender J. 1999. An inverted repeat triggers cytosine methylation of identical sequences in *Arabidopsis*. *Mol. Cell* **3:** 505.

Malagnac F., Bartee L., and Bender J. 2002. An *Arabidopsis* SET domain protein required for maintenance but not establishment of DNA methylation. *EMBO J.* **21:** 6842.

Melquist S. and Bender J. 2003. Transcription from an upstream promoter controls methylation signaling from an inverted repeat of endogenous genes in *Arabidopsis*. *Genes Dev.* **17:** 2036.

——— 2004. An internal rearrangement in an *Arabidopsis* inverted repeat locus impairs DNA methylation triggered by the locus. *Genetics* **166:** 437.

Melquist S., Luff B., and Bender J. 1999. *Arabidopsis PAI* gene arrangements, cytosine methylation and expression. *Genetics* **153:** 401.

Morel J.B., Godon C., Mourrain P., Béclin C., Boutet S., Feuerbach F., Proux F., and Vaucheret H. 2002. Fertile hypomorphic *ARGONAUTE* (*ago1*) mutants impaired in post-transcriptional gene silencing and virus resistance. *Plant Cell* **14:** 629.

Mourrain P., Béclin C., Elmayan T., Feuerbach F., Godon C., Morel J.B., Jouette D., Lacombe A.M., Nikic S., Picault N., Remoue K., Sanial M., Vo T.A., and Vaucheret H. 2000. *Arabidopsis SGS2* and *SGS3* genes are required for posttranscriptional gene silencing and natural virus resistance. *Cell* **101:** 533.

Murfett J., Wang X.J., Hagen G., and Guilfoyle T.J. 2001. Identification of *Arabidopsis* histone deacetylase HDA6 mutants that affect transgene expression. *Plant Cell* **13:** 1047.

Nielsen P.R., Nietlispach D., Mott H.R., Callaghan J., Bannister A., Kouzarides T., Murzin A.G., Murzina N.V., and Laue E.D. 2002. Structure of the HP1 chromodomain bound to histone H3 methylated at lysine 9. *Nature* **416:** 103.

Papp I., Mette M.F., Aufsatz W., Daxinger L., Schauer S.E., Ray A., van der Winden J., Matzke M., and Matzke A.J. 2003. Evidence for nuclear processing of plant micro RNA and short interfering RNA precursors. *Plant Physiol.* **132:** 1382.

Park W., Li J., Song R., Messing J., and Chen X. 2002. CARPEL FACTORY, a Dicer homolog, and HEN1, a novel protein, act in microRNA metabolism in *Arabidopsis thaliana*. *Curr. Biol.* **12:** 1484.

Probst A.V., Fagard M., Proux F., Mourrain P., Boutet S., Earley K., Lawrence R.J., Pikaard C.S., Murfett J., Furner I., Vaucheret H., and Scheid O.M. 2004. *Arabidopsis* histone deacetylase HDA6 is required for maintenance of transcriptional gene silencing and determines nuclear organization of rDNA repeats. *Plant Cell* **16:** 1021.

Reinhart B.J., Weinstein E.G., Rhoades M.W., Bartel B., and Bartel D.P. 2002. MicroRNAs in plants. *Genes Dev.* **16:** 1616.

Schauer S.E., Jacobsen S.E., Meinke D.W., and Ray A. 2002. *DICER-LIKE1*: Blind men and elephants in *Arabidopsis* development. *Trends Plant Sci.* **7:** 487.

Tamaru H. and Selker E.U. 2001. A histone H3 methyltransferase controls DNA methylation in *Neurospora crassa*. *Nature* **414:** 277.

Wada Y., Ohya H., Yamaguchi Y., Koizumi N., and Sano H. 2003. Preferential de novo methylation of cytosine residues in non-CpG sequences by a domains rearranged DNA methyltransferase from tobacco plants. *J. Biol. Chem.* **278:** 42386.

Xie Z., Johansen L.K., Gustafson A.M., Kasschau K.D., Lellis A.D., Zilberman D., Jacobsen S.E., and Carrington J.C. 2004. Genetic and functional diversification of small RNA pathways in plants. *PLoS Biol.* **2:** E104.

Xin Z., Tachibana M., Guggiari M., Heard E., Shinkai Y., and Wagstaff J. 2003. Role of histone methyltransferase G9a in CpG methylation of the Prader-Willi syndrome imprinting center. *J. Biol. Chem.* **278:** 14996.

Zilberman D., Cao X., and Jacobsen S.E. 2003. *ARGONAUTE4* control of locus-specific siRNA accumulation and DNA and histone methylation. *Science* **299:** 716.

Induced and Natural Epigenetic Variation

H. YI, N.C. RIDDLE*, T.L. STOKES†, H.-R. WOO, AND E.J. RICHARDS
Department of Biology, Washington University, St. Louis, Missouri 63130

The molecular nature of the inherited variation that underlies morphological and biochemical diversity generally is assumed to be genetic. In other words, the range of form among individuals and species is caused by changes in primary nucleotide sequence. This view has been cemented by countless studies defining the molecular genetic basis of important changes driving alternative developmental programs or characterizing evolutionary lineages. However, it is becoming increasingly evident that nongenetic alterations can have a striking impact on the morphology and physiology of cells and organisms. Such epigenetic alterations, which do not change nucleotide sequence but do affect the expression of this genetic information, can involve diverse mechanisms, including chromatin structural modifications and steady-state regulatory networks. A growing list of epigenetic alterations have been induced and studied in laboratory settings, and a few examples of spontaneous epigenetic alleles in a natural context have been reported.

Is epigenetic variation an important component of the raw material acted on by artificial and natural selection? The view that epigenetic variation plays a role in evolutionary change has been challenged, in part, because of the perceived similarity between epigenetic variation and Lamarckian evolution. The parallels between this widely discredited view of evolution by the inheritance of acquired characters and epigenetic inheritance stems from the possibility that environmentally induced adaptive variation in an individual can be passed on to future generations. While there is nothing controversial about the notion that environmental conditions (e.g., mutagens) can influence the generation of heritable variation, conventional neo-Darwinian models assume that variation is random and not adapted to the particular inducing environment. However, epigenetic variation is frequently nonrandom and might lead to inheritance of gene expression programs that are induced by the environment.

Significant objections based on mechanism also are leveled against the notion that epigenetic variation could fuel evolution (Maynard Smith 1990). The first objection is that epigenetic variation cannot be important in transmission genetics or evolution because such variation is confined to a single generation as a result of an erasure/resetting event during the organism's life cycle. This view incorporates a central aspect of the original definition of epigenetics put forward by Waddington (1939), a definition we would currently equate with developmental mechanisms of differential gene expression, signal transduction, and morphogenesis. The more specialized usage of the term epigenetics focuses on the acquisition, elaboration, and inheritance of biochemical marks that are superimposed on genetic information (Russo et al. 1996; Rice and Allis 2001; Richards and Elgin 2002). These marks need not be confined to a single generation. For example, whereas a widespread erasure and resetting of cytosine methylation patterns occurs in early mammalian embryogenesis (Reik et al. 2001; Santos et al. 2002), DNA methylation patterns can be inherited across generations in many organisms, including mammals (Silva and White 1988; Morgan et al. 1999; Rakyan et al. 2003), because some cytosine methylation sites can escape erasure (Kakutani et al. 1999; Macleod et al. 1999).

Even if it is conceded that epigenetic information can be inherited between organismal generations, including between sexual generations traversing meiosis, another serious objection arises—metastability. Indeed, instability and reversibility are among the features that continue to draw researchers' attention to classic epigenetic phenomena, such as nucleolar dominance (Pikaard 2000), paramutation (Chandler and Stam 2004), transposon changes in phase (McClintock 1952), and X-chromosome inactivation (Plath et al. 2002). It is argued that epigenetic changes cannot be relevant in an evolutionary context, or even in a breeding context, because inherited epigenetic variation is too unstable. Although metastability is a frequent feature of epigenetic variation, stable epigenetic alterations also exist that can be inherited with a high degree of fidelity mimicking that of traditional genetic Mendelian factors (Holliday and Ho 1990).

To learn more about the characteristics of epigenetic variation and the mechanisms that underlie this variation, our lab has been pursuing two different approaches using the flowering plant *Arabidopsis thaliana*. These two approaches will be summarized below in brief, and we will draw connections between our findings and some of the related work from other groups.

INDUCED EPIGENETIC VARIATION

The first approach grew out of our work on *Arabidopsis* mutants with dramatically reduced cytosine methylation. These mutants include plants deficient in either the

---

*Present address: Division of Biological Sciences, University of Missouri—Columbia, Columbia, Missouri 65211.
†Present address: New York University, Department of Biology, New York, New York 10003.

DDM1 SWI2/SNF2 nucleosome remodeling factor (Vongs et al. 1993; Jeddeloh et al. 1999) or the Dnmt1-class cytosine-DNA-methyltransferase MET1 (Kankel et al. 2003). Given the prevailing view at the time (early 1990s), it was expected that such mutants should be severely developmentally compromised. This view was solidified by the landmark paper published in 1992 by Li et al. (1992) demonstrating that Dnmt1 cytosine-DNA-methyltransferase knockout mice were inviable. In contrast, *Arabidopsis* DNA hypomethylation mutants did not exhibit significant developmental abnormalities (Vongs et al. 1993). Still more puzzling was the stochastic occurrence of morphologically aberrant individuals in *ddm1* mutant lines that had been propagated through several generations by self-pollination (Kakutani et al. 1996). Indeed, progressively more severe phenotypes were seen in inbred *ddm1* lines as the number of self-pollinations increased. Subsequent genetic analysis demonstrated that the morphological defects were caused by the accumulation of stable alterations at a limited number of genomic loci. Our lab and others have characterized a handful of these alterations during the past several years. One of the best-understood alterations induced by *ddm1* or *met1* is hypomethylated epigenetic alleles (Kakutani 1997; Kankel et al. 2003) of the *FWA* locus that lead to a delay in flowering time associated with ectopic expression of the putative transcription factor encoded by the locus (Soppe et al. 2000).

Our lab has focused on another *ddm1*-induced alteration, called *bal*, which causes dwarfing and twisted leaf morphology (Stokes et al. 2002). This developmental syndrome stems from an alteration mapping to a complex ~90-kb cluster of NBS-LRR class (nucleotide binding site-leucine rich repeat) *Resistance* or *R* genes and transposable elements on the lower arm of chromosome 4 (Fig. 1). These *R* genes encode signal transduction components that act early in pathogen recognition and defense pathways (Dangl and Jones 2001). The *bal* defect causes an overexpression of *At4g16890*, the most highly expressed gene in the cluster in wild-type plants (Stokes et al. 2002). Overexpression of *At4g16890* in transgenic plants recapitulates the phenotype. This overexpression leads to constitutive signaling through the salicylic acid (SA) pathway, which induces pathogenesis-related (PR) proteins, eventually leading to dwarfing and twisted leaf morphology. The sequence of *At4g16890* in the *bal* variant is identical to that seen in wild-type plants (strain Columbia), suggesting that the genomic change causing the *bal* syndrome is epigenetic or resides within the *R*-gene cluster at some distance from *At4g16890*, or possibly both.

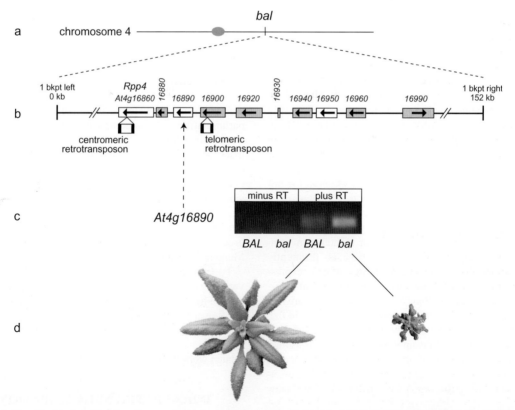

Figure 1. The *ddm1*-induced *bal* dwarfing syndrome is caused by misregulation of a pathogen resistance gene cluster. The *bal* allele maps to *Arabidopsis* chromosome 4 (*a*) in a 152-kb interval (*b*) containing a cluster of ten *R* genes or gene fragments and transposable elements (in strain Columbia). The direction of gene transcription is indicated by the *arrow*. *Open boxes*, functional or potentially functional genes; *gray boxes*, genes predicted to be nonfunctional. (*c*) A reverse transcriptase polymerase chain reaction (RT-PCR) experiment (loading controls not shown) demonstrating overexpression of *At4g16890* in *bal* plants. (*d*) The phenotypes of a wild-type (*BAL*) individual and *bal* dwarf (strain Columbia).

Support for the hypothesis that the *bal* syndrome is caused by an epigenetic allele comes from the unusual metastability of the *bal* alteration. We have not observed reversion or instability of the *bal* defect under standard greenhouse conditions, suggesting that the *bal* allele is very stable. Yet, attempts to isolate suppressor mutations by mutagenizing a *bal* homozygote yielded wild-type *BAL* revertants at a frequency approaching 10% in the first generation after mutagenesis with ethylmethanesulfonate (EMS) (Stokes et al. 2002; H. Yi and E.J. Richards, unpubl.). This frequency is orders of magnitude higher than that expected from reversion of point mutations. No rearrangements or transposition events were observed in the locus, arguing against high-frequency genetic reversion mechanisms. As expected, reversion to the *BAL* state is accompanied by the loss of *At4g16890* overexpression.

The chromosome 4 *R*-gene cluster acts as an island of heterochromatin and appears to be a hot spot for epigenetic variation. At least two other metastable alterations mapping to or near the same chromosome 4 *R* gene cluster have been reported. Both alterations, *ssi1* (Shah et al. 1999) and *cpr1* (Bowling et al. 1994), display (1) constitutive pathogen signaling and dwarfing, (2) metastability, and (3) overexpression of at least one *R* gene related to *At4g16890* (Stokes and Richards 2002; Stokes et al. 2002; Yang and Hua 2004). One objective of our current efforts is to understand the molecular mechanisms that underlie the formation and metastability of *bal* and related alterations as a model for the types of epigenetic mechanisms that exercise control in complex genomic regions typical of eukaryotic genomes.

Interestingly, both *ssi1* and *cpr1* were generated from EMS mutageneses (Bowling et al. 1994; Shah et al. 1999). Two of the best studied *Arabidopsis* epigenetic alleles—hypomethylated *fwa* flowering time alleles (Soppe et al. 2000) and hypermethylated *sup* flower development alleles (Jacobsen and Meyerowitz 1997)—were also isolated from screens of chemically mutagenized plants, demonstrating that traditional mutagenesis protocols can generate both genetic and epigenetic alleles.

NATURAL EPIGENETIC VARIATION

If it is well established that epigenetic variation can be generated at a high frequency in plants in response to traditional mutageneses or as a consequence of pharmacological or genetic manipulation of epigenetic modifications such as DNA methylation, the relevance of analogous epigenetic variants in natural plant populations is less clear. One striking example of an epigenetic variant in a wild population was described by Enrico Coen and colleagues in 1999—a variant form of toadflax (in the genus *Linaria*) with radially rather than bilaterally symmetric flowers (Cubas et al. 1999). This morphologically distinct *Linaria* variant was recognized by Linnaeus over 250 years earlier. The mechanistic basis of the alternate forms in flower morphology is epigenetic silencing, associated with DNA hypermethylation, of the *Linaria CYCLOIDEA* homolog, *Lcyc*.

Our lab began an approach to study the prevalence and stability of epigenetic variation among natural accessions of *Arabidopsis*. These accessions are laboratory lines propagated from seeds originally collected from throughout the natural range of *Arabidopsis* and are often referred to as "ecotypes," suggesting that they represent different genotypes fine-tuned to specific ecological habitats. These lines capture a wide variety of variation that has been exploited as a source of new genetic alleles (Alonso-Blanco and Koornneef 2000). We asked the simple question, "How much epigenetic variation in the form of differential cytosine methylation is found among these accessions?" (also see Cervera et al. 2002). In our initial study (Riddle and Richards 2002), we found that a number of genomic sequences were invariably highly methylated, such as the 180-base pair (bp) centromeric repeats and the pericentromeric *Athila* transposable element. However, we found certain genomic loci were differentially methylated among the accessions surveyed. Particularly striking was the variation in DNA methylation on the highly repeated ribosomal RNA genes found in large arrays at the top of chromosome 2 and chromosome 4, the so-called nucleolus organizer regions (NORs) (Fig. 2).

The different NOR methylation phenotypes characteristic of the different accessions result from a combination of sources. Two lines of evidence indicate that differential NOR methylation is partially due to genetic variation (Riddle and Richards 2002). First, the number of rRNA genes strongly correlates with total NOR methylation content, in keeping with previous observations suggest-

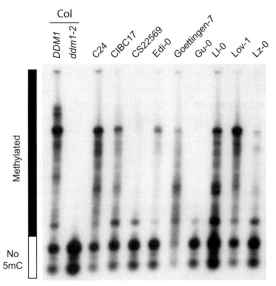

Figure 2. Natural variation in rRNA gene methylation among different *Arabidopsis* accessions. Southern blot assay to monitor genomic cytosine methylation among different *Arabidopsis* strains (http://walnut.usc.edu/2010.html) and controls (Columbia wild-type = Col *DDM1*; plus a *ddm1-2* homozygote in a Columbia background). The hybridization signal indicates the extent of methylation in the rRNA gene clusters; unmethylated fragments are cleaved by the methylation-sensitive endonuclease *Hpa*II to small fragments (*open box*), whereas methylated genomic fragments remain uncleaved or partially cleaved and migrate as larger fragments (*solid box*).

ing that excess rRNA gene repeats are archived in a silenced, methylated state. A second line of evidence comes from quantitative trait loci (QTL) analysis using recombinant inbred (RI) lines generated from crosses between the high NOR methylation accession Ler and the low NOR methylation accession Cvi. At least two significant *trans*-acting QTL influence the variation of NOR methylation among the different RI lines. But total NOR methylation is not under strict genetic control; QTL analysis and supporting studies demonstrate that the inheritance and maintenance of differential NOR methylation states are the most important factors explaining the variation in NOR methylation among the RI lines. These findings suggest that the unique NOR methylation content characteristic of different *Arabidopsis* accessions results from a combination of genetic factors (i.e., rRNA gene number and *trans*-acting modifiers) pushing against the inertia of the parental NOR methylation state.

These results provide some insight into the forces and dynamics that shape cytosine methylation patterns in natural populations. In the short timescale represented by the transmission genetic experiments described above (i.e., a handful of generations), epigenetic inheritance is paramount in defining genomic cytosine methylation patterns. Over longer time frames, genetic control of epigenetic modification is expected to play the dominant role. Nonetheless, the importance of stochastic changes that shift cytosine methylation patterns cannot be discounted. For example, there is precedent for abiotic stresses (e.g., temperature; Sherman and Talbert 2002) or biotic stresses (e.g., pathogen infection; Guseinov and Vanyushin 1975) altering cytosine methylation levels. Such changes would establish a new methylation default state that would be inherited over time because of the self-propagating nature of epigenetic codes. Thus, epigenetic variation need not march in lockstep with genetic variation at loci that are targets of epigenetic modification or loci that encode the machinery that executes or guides this modification.

CONCLUSIONS

The modes of induction, meiotic transmission, and genetic behavior of alternative cytosine methylation alleles in *Arabidopsis* suggest that epigenetic alleles will be frequently encountered in plant genetics and often mistaken for genetic mutations. The larger questions concerning the relevance of epigenetic variation in natural populations and in an evolutionary context persist. The semi-independent nature of epigenetic variation—the possibility of formation and transmission of epigenetic alleles independent of genetic variation controlling these events—suggests a plausible role for epigenetic variation in evolution. A single genotype can express a range of phenotypes based on alternative epigenotypes. This mechanism incorporates an element beyond that implied by phenotypic plasticity because epigenetic information can be inherited. Selection for one alternative epigenotype may serve as an intermediate bridge to a greater stabilization of an adaptive phenotypic outcome by genetic variation.

This chain of events is a type of "genetic assimilation" mechanism, a process through which apparently environmentally induced phenotypic variation can be inherited. As originally formulated and demonstrated by Waddington (1942, 1953), genetic assimilation involves the selection of cryptic genetic variation uncovered by environmental stress. Selection for stress-induced phenotypes leads to the accumulation of genetic alleles, the effects of which are now exposed in the new environment, that favor the expression of the selected phenotype. After several generations of selection the variants express the abnormal phenotype in the absence of the inducing environment. Inherited epigenetic variation might play two roles in the genetic assimilation of adaptive phenotypes. In one scenario, selection would initially act on alternative heritable epigenetic states rather than on newly exposed genetic variation. Subsequent selection for genetic alleles that buttress or supplant the epigenetic changes would complete the assimilation. Alternatively, selection may be operating on both genetic and epigenetic variation (Ruden et al. 2003; Sollars et al. 2003), with inherited epigenetic variation stabilizing the selected phenotype in transition to complete genetic assimilation.

ACKNOWLEDGMENTS

We thank Justin Rincker for comments on the manuscript. Research in our group has been supported by grants from the National Science Foundation (DBI-9975930, MCB-9985348, and MCB-0321990) and the Ministry of Agriculture, Forestry and Fisheries, Japan.

REFERENCES

Alonso-Blanco C. and Koornneef M. 2000. Naturally occurring variation in *Arabidopsis:* An underexploited resource for plant genetics. *Trends Plant Sci.* **5:** 22.

Bowling S.A., Guo A., Cao H., Gordon A.S., Klessig D.F., and Dong X. 1994. A mutation in *Arabidopsis* that leads to constitutive expression of systemic acquired resistance. *Plant Cell* **6:** 1845.

Cervera M.T., Ruiz-Garcia L., and Martinez-Zapater J.M. 2002. Analysis of DNA methylation in *Arabidopsis thaliana* based on methylation-sensitive AFLP markers. *Mol. Genet. Genomics* **268:** 543.

Chandler V.L. and Stam M. 2004. Chromatin conversations: Mechanisms and implications of paramutation. *Nat. Rev. Genet.* **5:** 532.

Cubas P., Vincent C., and Coen E. 1999. An epigenetic mutation responsible for natural variation in floral symmetry. *Nature* **401:** 157.

Dangl J.L. and Jones J.D. 2001. Plant pathogens and integrated defence responses to infection. *Nature* **411:** 826.

Guseinov V.A. and Vanyushin B.F. 1975. Content and localisation of 5-methylcytosine in DNA of healthy and wilt-infected cotton plants. *Biochim. Biophys. Acta* **395:** 229.

Holliday R. and Ho T. 1990. Evidence for allelic exclusion in Chinese hamster ovary cells. *New Biol.* **2:** 719.

Jacobsen S.E. and Meyerowitz E.M. 1997. Hypermethylated SUPERMAN epigenetic alleles in *Arabidopsis. Science* **277:** 1100.

Jeddeloh J.A., Stokes T.L., and Richards E.J. 1999. Maintenance of genomic methylation requires a SWI2/SNF2-like protein. *Nat. Genet.* **22:** 94.

Kakutani T. 1997. Genetic characterization of late-flowering traits induced by DNA hypomethylation mutation in *Arabidopsis thaliana. Plant J.* **12:** 1447.

Kakutani T., Munakata K., Richards E.J., and Hirochika H. 1999. Meiotically and mitotically stable inheritance of DNA hypomethylation induced by ddm1 mutation of *Arabidopsis thaliana*. *Genetics* **151:** 831.

Kakutani T., Jeddeloh J.A., Flowers S.K., Munakata K., and Richards E.J. 1996. Developmental abnormalities and epimutations associated with DNA hypomethylation mutations. *Proc. Natl. Acad. Sci.* **93:** 12406.

Kankel M.W., Ramsey D.E., Stokes T.L., Flowers S.K., Haag J.R., Jeddeloh J.A., Riddle N.C., Verbsky M.L., and Richards E.J. 2003. *Arabidopsis* MET1 cytosine methyltransferase mutants. *Genetics* **163:** 1109.

Li E., Bestor T.H., and Jaenisch R. 1992. Targeted mutation of the DNA methyltransferase gene results in embryonic lethality. *Cell* **69:** 915.

Macleod D., Clark V.H., and Bird A. 1999. Absence of genome-wide changes in DNA methylation during development of the zebrafish. *Nat. Genet.* **23:** 139.

Maynard Smith J. 1990. Models of a dual inheritance system. *J. Theor. Biol.* **143:** 41.

McClintock B. 1952. Chromosome organization and genic expression. *Cold Spring Harbor Symp. Quant. Biol.* **16:** 13.

Morgan H.D., Sutherland H.G., Martin D.I., and Whitelaw E. 1999. Epigenetic inheritance at the agouti locus in the mouse. *Nat. Genet.* **23:** 314.

Pikaard C.S. 2000. The epigenetics of nucleolar dominance. *Trends Genet.* **16:** 495.

Plath K., Mlynarczyk-Evans S., Nusinow D.A., and Panning B. 2002. Xist RNA and the mechanism of X chromosome inactivation. *Annu. Rev. Genet.* **36:** 233.

Rakyan V.K., Chong S., Champ M.E., Cuthbert P.C., Morgan H.D., Luu K.V., and Whitelaw E. 2003. Transgenerational inheritance of epigenetic states at the murine Axin(Fu) allele occurs after maternal and paternal transmission. *Proc. Natl. Acad. Sci.* **100:** 2538.

Reik W., Dean W., and Walter J. 2001. Epigenetic reprogramming in mammalian development. *Science* **293:** 1089.

Rice J.C. and Allis C.D. 2001. Code of silence. *Nature* **414:** 258.

Richards E.J. and Elgin S.C. 2002. Epigenetic codes for heterochromatin formation and silencing: Rounding up the usual suspects. *Cell* **108:** 489.

Riddle N.C. and Richards E.J. 2002. The control of natural variation in cytosine methylation in *Arabidopsis*. *Genetics* **162:** 355.

Ruden D.M., Garfinkel M.D., Sollars V.E., and Lu X. 2003. Waddington's widget: Hsp90 and the inheritance of acquired characters. *Semin. Cell Dev. Biol.* **14:** 301.

Russo V.E.A., Martienssen R.A., and Riggs A.D., eds. 1996. *Epigenetic mechanisms of gene regulation*. Cold Spring Harbor Laboratory Press, Cold Spring Harbor, New York.

Santos F., Hendrich B., Reik W., and Dean W. 2002. Dynamic reprogramming of DNA methylation in the early mouse embryo. *Dev. Biol.* **241:** 172.

Shah J., Kachroo P., and Klessig D.F. 1999. The *Arabidopsis* ssi1 mutation restores pathogenesis-related gene expression in npr1 plants and renders defensin gene expression salicylic acid dependent. *Plant Cell* **11:** 191.

Sherman J.D. and Talbert L.E. 2002. Vernalization-induced changes of the DNA methylation pattern in winter wheat. *Genome* **45:** 253.

Silva A.J. and White R. 1988. Inheritance of allelic blueprints for methylation patterns. *Cell* **54:** 145.

Sollars V., Lu X., Xiao L., Wang X., Garfinkel M.D., and Ruden D.M. 2003. Evidence for an epigenetic mechanism by which Hsp90 acts as a capacitor for morphological evolution. *Nat. Genet.* **33:** 70.

Soppe W.J., Jacobsen S.E., Alonso-Blanco C., Jackson J.P., Kakutani T., Koornneef M., and Peeters A.J. 2000. The late flowering phenotype of fwa mutants is caused by gain-of-function epigenetic alleles of a homeodomain gene. *Mol. Cell* **6:** 791.

Stokes T.L. and Richards E.J. 2002. Induced instability of two Arabidopsis constitutive pathogen-response alleles. *Proc. Natl. Acad. Sci.* **99:** 7792.

Stokes T.L., Kunkel B.N., and Richards E.J. 2002. Epigenetic variation in *Arabidopsis* disease resistance. *Genes Dev.* **16:** 171.

Vongs A., Kakutani T., Martienssen R.A., and Richards E.J. 1993. *Arabidopsis thaliana* DNA methylation mutants. *Science* **260:** 1926.

Waddington C.H. 1939. *An introduction to modern genetics*. Allen and Unwin, London.

———. 1942. Canalization of development and the inheritance of acquired characters. *Nature* **150:** 563.

———. 1953. Genetic assimilation of an acquired character. *Evolution* **7:** 118.

Yang S. and Hua J. 2004. A haplotype-specific Resistance gene regulated by BONZAI1 mediates temperature-dependent growth control in *Arabidopsis*. *Plant Cell* **16:** 1060.

Linking Covalent Histone Modifications to Epigenetics: The Rigidity and Plasticity of the Marks

Y. Wang,[*†] J. Wysocka,[*†] J.R. Perlin,[*] L. Leonelli,[*]
C.D. Allis,[†] and S.A. Coonrod[*]

[*]Department of Genetic Medicine, Weill Medical College of Cornell University, New York, New York 10021;
[†]Laboratory of Chromatin Biology, The Rockefeller University, New York, New York 10021

In higher eukaryotes, genetic information, encoded by the DNA double helix, is organized into a complex chromatin structure with the assistance of histone and nonhistone proteins. As such, chromatin is the physiologically relevant form of eukaryotic genomes. While it is largely uncontested that different genetic compositions of individual species underlie their unique physiology and final morphology, chromatin provides an additional layer for dictating the translation of genetic information into meaningful biological readouts (for review, see Strahl and Allis 2000; Jenuwein and Allis 2001; Felsenfeld and Groudine 2003). Increasing evidence underscores an emerging theme in biology wherein "epigenetic" mechanisms, defined as mechanisms operating outside of changes in DNA sequence, govern gene function by creating potentially heritable alternative states of chromatin. DNA methylation and covalent modifications of histones, such as methylation, acetylation, and phosphorylation, are thought to contribute to these chromatin states by serving as a platform that integrates cell differentiation signals and environmental cues as well as regulating the timely release of the appropriate genetic information (Cheung et al. 2000; Schreiber and Bernstein 2002). Several of the above histone modifications are readily reversible through enzymatic means; others are potentially static (Bannister et al. 2002; see below). One central problem that remains to be resolved is how these mechanisms are regulated either during normal developmental conditions or under various pathological conditions.

EPIGENETIC MECHANISMS UNDERLYING DIFFERENTIATION DURING MAMMALIAN EMBRYONIC DEVELOPMENT

Gametogenesis gives rise to two different types of terminally differentiated cells, the sperm and the egg, which fuse during fertilization to initiate a new life cycle. During early developmental stages, the zygote and the blastomeres in two-, four-, and eight-cell embryos are totipotent, meaning that they possess an unrestricted developmental potential and are capable of differentiating into all cell types (Matzuk and Lamb 2002; Sassone-Corsi 2002). By the 16-cell stage, a gradual restriction in the developmental potential of cells takes place resulting in the production of two distinct lineages: (1) the trophectoderm, which eventually develops into the extraembryonic tissues; and (2) the inner cell mass, which develops into the embryo proper. The inner cell mass then gives rise to ectoderm, endoderm, and mesoderm. These different germ layers further develop and are organized to form various somatic tissues and organs (Fig. 1). This developmental process is associated with dramatic changes in gene expression profiles and is often referred to as cellular differentiation.

While mechanisms underlying the differentiation process remain unclear, several lines of evidence suggest the involvement of epigenetic regulation (as defined above; Kelly and Trasler 2004). For example, nuclear cloning experiments show that the developmental potential of a terminally differentiated somatic cell nucleus can be

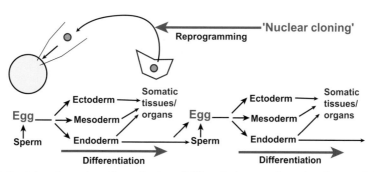

Figure 1. Cell differentiation is based on epigenetic mechanisms. Differentiation, germ layer (ectoderm, endoderm, or mesoderm) formation, and organogenesis in a developing embryo are associated with heritable changes in gene expression profiles. However, the change of gene expression is caused not by the changes in the genetic information per se, but by epigenetic mechanisms. This notion is strongly supported by animal cloning experiments, which also suggest that eggs are rich with machineries capable of erasing the epigenetic information of a somatic nucleus.

fully restored to totipotency following transfer into an enucleated egg (Campbell et al. 1996; Wakayama et al. 1998). However, the low success rate of cloning (1–3% for adult somatic cell nuclei and 10–30% for embryonic nuclei) and the high rate of developmental defects associated with cloning strongly argue that epigenetic information carried by the transferred nucleus can be reprogrammed only on a limited basis (Hiendleder et al. 2004). On the other hand, these experiments suggest that poorly defined machinery stored in the egg is capable of reversing the biological clock back to its primary state. Identification of these suspected regulatory factors and the elucidation of their role during normal and abnormal developmental processes would offer important insights toward our understanding of epigenetic regulation of gene activity.

FUNDAMENTAL ROLES OF HISTONES IN ORCHESTRATING THE EPIGENETIC INFORMATION

The nucleosome is the basic structural unit of chromatin, consisting of about 146 bp of DNA wrapped around an octamer of core histones (including two each of histone H3, H4, H2B, and H2A) (van Holde 1989; Luger et al. 1997). Nucleosomal arrays are further stabilized by members of the linker histone H1/H5 family whose precise roles in chromatin function remains unclear (Wolffe 1998). While initially thought to be an inert structure involved in packaging DNA into the confines of the nucleus, significant progress has been made in documenting a more dynamic view of chromatin, particularly during transcriptional activation and repression (for review, see Felsenfeld and Groudine 2003; Khorasanizadeh 2004). This includes experiments with test "designer" genes where remarkable transitions between decondensed and condensed chromatin have been documented in living cells (Strukov et al. 2003; Janicki et al. 2004).

While there are many ways to introduce physiologically relevant variation into the chromatin polymer (Felsenfeld and Groudine 2003), numerous reports have documented that covalent histone modifications, working in a sequential and/or combinatorial manner, regulate DNA and chromatin templated cellular events, such as gene expression, DNA repair, and chromosome condensation (see Strahl and Allis 2000; Jenuwein and Allis 2001; Fischle et al. 2003). Well-documented histone modifications include acetylation, methylation, phosphorylation, ubiquitination, and ADP-ribosylation (for reviews, see Grunstein 1997; Cheung et al. 2000; Roth et al. 2001; Zhang and Reinberg 2001; Kouzarides 2002; Lachner et al. 2003). Other modifications, such as sumoylation (Shiio and Eisenman 2003), or the newly discovered citrullination (Hagiwara et al. 2002; this paper), are less well understood, but are likely to play key regulatory roles that will require further study. Although it remains unclear as to whether a true "histone code" exists, and, if so, the precise form and extent to which it is utilized (Agalioti et al. 2002; Schreiber and Bernstein 2002; Kurdistani et al. 2004; Schubeler et al. 2004), this hypothesis offers a useful framework for the study of histone modifications, modifying enzymes, and effector proteins, as well as the role of these machineries and pathways in regulating important cellular functions.

HISTONE METHYLATION: STORIES OF Lys AND Arg

For the purpose of the discussion below, we highlighted several relevant modification sites in histone H3 and H4 in Figure 2A. Methylation of histone Lys residues, which can be mono-, di-, or trimethyl, is associated with actively expressed genes, as well as with silenced genes or large heterochromatin regions. For example, methylation at Lys4 and Lys79 of H3 is often found to be associated with actively expressed genes (Strahl et al. 1999; Santos-Rosa et al. 2003; Schubeler et al. 2004). On the other hand, methylation at Lys9 and Lys27 of H3 is closely correlated with the formation of repressive chromatin structure at both centromeric regions and silenced genes such as homoetic genes (for references, Schubeler et al. 2004; see below). Multiple histone H3 Lys9 methyltransferases have been reported, including G9a, Suv39h1, h2, and ESET (for review, see Lachner et al. 2003). Mice carrying mutations of these enzymes die at early embryonic stages (Peters et al. 2001; Dodge et al. 2004). EZH2, the well-studied methyltransferase of H3 Lys27, is also important for early mouse embryonic development (Erhardt et al. 2003). These results indicate that "setting" the epigenetic marks of histone Lys methylation ensures the normal embryonic development.

In addition to Lys methylation, protein arginine methyltransferases (PRMTs) modify multiple Arg residues in the amino-terminal tails of H3 and H4 (Fig. 2A). PRMT1 methylates histone H4 Arg3 residues, while histone H3 Arg17 is a preferred target site for CARM1 (Strahl et al. 2001; Wang et al. 2001; Bauer et al. 2002). Furthermore, methylation of Arg residues in H3 and H4 functions synergistically with histone Lys acetylation during activation of nuclear hormone regulated genes (Wang et al. 2001; Daujat et al. 2002). Recently, it was found that PRMT1, CARM1, and histone acetyltransferase p300 cooperate with p53 to activate gene expression (An et al. 2004). Taken together, these studies suggest that Arg methylation of histones is intimately involved in gene activation from chromatin templates.

CROSS-TALK OF DNA METHYLATION AND Lys METHYLATION IN HISTONES

For many years, DNA methylation, namely the 5-methylcytosine (5mC) modification at CpG islands of the genome, has been the main focus of the epigenetic gene regulation field (for review, see Jaenisch and Bird 2003; Feinberg and Tycko 2004). The findings that histone modifications can regulate DNA methylation patterns suggest that histone modifications, particularly Lys methylation, are important regulatory mechanisms of epigenetic phenomena, such as X-chromosome inactivation, imprinting, and cancer etiology. In *Neurospora crassa*, DIM5, a methyltransferase of histone H3 Lys9, mediates

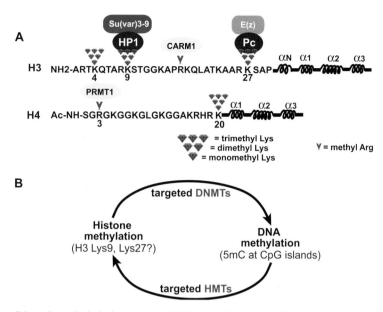

Figure 2. Components of the epigenetic indexing system. (*A*) Covalent histone modifications, such as methylation of Lys and Arg residues, are marks associated with the genome to indicate gene expression potential of a chromatin region. These marks are "written" by enzymes that covalently modify histones. They exert their function by interacting with binding "effector" proteins, such as HP1, which "reads" H3 Lys9 methyl mark, and polycomb (Pc), which "reads" H3 Lys27 methyl marks. However, without clear binding effector proteins, methyl marks on some histone residues, including H3 Lys4, H4 Lys20, H3 Arg17, and H4 Arg3, remain as "orphan methyl marks." (*B*) DNA methyl marks and histone methyl marks cross-talk with each other (see text for details). HMT, histone methyltransferase; DNMT, DNA methyltransferase.

DNA methylation (Tamaru et al. 2003). In *Arabidopsis thaliana*, KRYPTONITE, another histone H3 Lys9 methyltransferase, is also required for DNA methylation (Jackson et al. 2002). These studies suggest a regulatory mechanism whereby DNA methylation is targeted by histone methylation (Fig. 2B).

While the above-mentioned data supports the model that histone methylation guides DNA methylation, other reports suggest that DNA methylation may, in fact, regulate histone methylation as well. For example, DNA hypomethylation due to compromised activity of a maintenance DNA methyltransferase, MET1, causes defects in H3 Lys9 methylation in *Arabidopsis thaliana* (Soppe et al. 2002). Biochemical studies found that the methyl-DNA binding protein (MeCP2) interacts with H3 Lys9 methyltransferase (Fuks et al. 2003). These results suggest that there may be a continous interplay between histone methylation and DNA methylation in certain biological systems. The possibility that DNA methylation may guide histone methylation is especially intriguing in the context of maintaining histone modification patterns following DNA replication.

BASIC CHARACTERISTICS OF EPIGENETIC MARKS: RIGIDITY AND PLASTICITY

Epigenetic gene expression patterns can be inherited by daughter cells following mitotic and sometimes even meiotic cell division (Cavalli and Paro 1999; Wolffe and Matzke 1999). Therefore, mechanisms must exist to establish and maintain these epigenetic marks in order to faithfully perpetuate gene expression patterns throughout cellular generations.

Establishing the Mark

What are the signals that establish an epigenetic state on a region of chromatin or on a particular gene? While some of the protein components involved in these pathways have been identified, it is less clear as to (1) how these marks are initially written onto the chromatin template and (2) how a multitude of epigenetic events are temporally coordinated and ordered. Moreover, the "trigger" of a given epigenetic event may vary from case to case.

In genomic imprinting, for example, the difference between two homologous alleles is often their developmental history (i.e., whether the allele is of maternal or paternal origin). During mouse embryonic development, the silencing of maternal or paternal X chromosome is random in the inner cell mass cells and their descendants. However, in the trophectoderm, the paternal X chromosome is selectively inactivated. Multiple epigenetic events, such as Xist RNA coating and changes in histone-modification patterns, are involved in these two X-inactivation processes (Okamoto et al. 2004).

On the other hand, in *Drosophila*, the formation of the dosage compensation complex is directly linked with the sex determination process. Translation of male specific lethal 2 (*msl2*) mRNA is repressed by the sex lethal

(SXL) protein in females (Kelley et al. 1997). Therefore, the MSL dosage compensation complex is formed to upregulate the gene expression of the single male X chromosome by twofold. The mechanism of targeting the MSL complex remains poorly understood, but also involves production of noncoding RNAs and histone hyperacetylation (for review, see Park and Kuroda 2001).

In the case of establishing epigenetic marks on an individual gene, the transcriptional status of a given gene seems to determine the downstream epigenetic events. For example, in the process of homeotic gene silencing in Drosophila, the expression pattern is initially established by upstream transcription factors, which is then faithfully maintained in the subsequent development stages by the polycomb group of genes (for review, see Simon and Tamkun 2002). This finding underscores the interesting possibility that the absence of active gene transcription from a specific promoter may trigger downstream epigenetic events as witnessed by a recent study on $p16^{INK4a}$ (Bachman et al. 2003).

Maintaining the Mark

To impart their functions as epigenetic marks, histone modifications must be maintained during the cell cycle. The paradigm of how DNA methylation marks are propagated during DNA replication and cell division serves as a framework for a discussion on the maintenance of histone modifications, about which much less is known (Jaenisch and Bird 2003). Initial DNA methylation marks are generated by de novo methyltransferases, DNMT3a and DNMT3b. Since DNA methylation is often on the CpG islands of the DNA double helix, it can be inherited by semiconservative DNA replication. However, after DNA replication, the DNA methylation mark is absent from the newly synthesized daughter strands and therefore could be lost during subsequent rounds of DNA replication. To offset possible dilution effects, the maintenance methyltransferase, DNMT1, methylates the newly synthesized daughter strands of DNA according to the methylation patterns of the parental strands.

However, there are several significant differences in the way that histones are transmitted or inherited as compared to DNA methylation. First, although DNA is wrapped around histones to form the nucleosome, histones can be readily dissociated from DNA during the cell cycle. In extreme cases, such as mammalian spermatogenesis, the majority of histones are removed from the chromatin template and replaced by protamine (Sassone-Corsi 2002). Second, DNA methylation is retained on the parental strands after DNA replication, but histone modifications are distributed in a different manner: (1) In S phase, parental histones are distributed randomly onto the two newly duplicated DNA double helices (Wolffe and Matzke 1999 and references therein). Currently, it is not known whether the core histone octamer is incorporated as a whole entity or whether there is some dynamic change of the core histone components. Interestingly, based on biochemical data in Hela cells, a semiconservative nucleosome assembly model was recently proposed (Tagami et al. 2004). (2) Although the majority of histones are deposited into chromatin during S phase when DNA is replicated, interphase or transcription-associated histone replacement can also occur. For example, the histone H3 variant, H3.3, is actively incorporated into chromatin during interphase (Ahmad and Henikoff 2002).

In sum, mechanisms that govern the inheritance of histones and histone modifications are poorly understood. We suspect that incorporation of histone-modifying enzymes and binding effector proteins may serve to retain the pattern of chromatin marks in much the same way that spreading of heterochromatin, brought about by Lys methylation and HP1 binding, has been proposed to occur (Bannister et al. 2001). Determining how localized enzymatic activities recognize and propagate the patterns of covalent marks to nearby nucleosomes to compensate for the dilution effect of the cell cycle will be an important area of future investigation.

RESETTING EPIGENETIC MARKS IN GERM CELLS AND EARLY EMBRYOS

Dramatic resetting of the epigenetic marks occurs in germ line cells as well as in preimplantation embryos. DNA methylation marks are first erased and then regenerated during both male and female germ cell development (Hajkova et al. 2002; Lee et al. 2002). Shortly after fertilization, paternal DNA is actively demethylated by an unidentified DNA demethylase, while the maternal DNA methylation pattern is reset passively during subsequent cell divisions (Mayer et al. 2000; Li 2002). However, there are particular loci, especially the imprinted genes, that can escape active DNA demethylation.

Since most histones are replaced by protamine during spermatogenesis, it seems unlikely that the vast majority of "somatic" histone modifications can be transmitted to the next generation through sperm. The maternal genome in the egg, however, does appear to be decorated with a variety of histone modifications whose fate during early development is only beginning to be examined. To begin to test the hypothesis that the reprogramming of histone modifications may play a role in activation of the embryonic genome, we investigated global changes in levels of a series of histone tail modifications during oocyte maturation and preimplantation mouse development (Sarmento et al. 2004). We found that certain "static" histone modifications, such as methylation of H3 Lys 9 and H3 Lys 4, appear to be relatively stable epigenetic marks during early development without exhibiting global change in our assays. However, other modifications, such as acetylation of H3 and H4 and methylation of H3 Arg 17 and H4 Arg 3, appear to be dynamically regulated. While the significance for the global resetting of histone Arg methylation marks remains unclear, we favor the view that this process may be involved in the dramatic changes in gene expression patterns observed during the egg-to-embryo transition.

HISTONE (PROTEIN) Arg METHYLATION: IS THIS A "PLASTIC" SIGNAL?

The finding that histone Arg methylation marks are dynamically regulated in early mouse eggs and embryos prompted us to search for enzymatic activities that can remove methyl groups from histone methyl-Arg residues. Although the fate of methyl group on protein Arg is not known, the modified amino acid, methyl-Arg, can be metabolized by dimethylarginine dimethylaminohydrolase (DDAH) to produce citrulline (Cit) and methylamine as reaction products (depicted in Fig. 3A) (Ogawa et al. 1989; Boger 2003). To investigate the possibility that an enzyme(s) exists that can catalyze a similar reaction on protein substrates like that of DDAH, we scanned the database for DDAH homologous proteins using the Fugue sequence-structure homology-analysis program (www-cryst.bioc.cam.ac.uk/fugue). Significant sequence and secondary structure similarity was found between DDAH and peptidylarginine deiminase (hereafter PAD) (Fig. 3B). PADs are a family of enzymes known to covert Arg residues in proteins to Cit via a deimination reaction, releasing ammonium (Vossenaar et al. 2003). There are five isoforms of PADs that have been identified in human and mouse genomes, including PAD1, -2, -3, -4, and -6. Interestingly, several of these enzymes are correlated with genetic diseases, including multiple sclerosis (PAD2) and rheumatoid arthritis (PAD4) (De Keyser et al. 1999; Suzuki et al. 2003).

Relevant to this discussion are recent findings that one of the most abundant polypeptides in the ovulated mouse egg appears to be a novel PAD family member that is specifically expressed in eggs and early embryos (Wright et al. 2003). In addition, PAD4 is also expressed in the nucleus of eggs and early embryos (Y. Wang and S.A. Coonrod, unpubl.). Taken together, these findings have prompted us to begin testing the hypothesis that PADs can regulate histone Arg methylation levels during early mammalian developmental processes.

BACK FROM THE FUTURE: TO WHAT EXTENT DOES THE "CHROMATIN STATE" INFLUENCE CLONING AND CELLULAR DIFFERENTIATION AND DEDIFFERENTIATION?

As mentioned above, the success of nuclear cloning experiments in mammals provided a "proof of principle" example that cellular differentiation is regulated by epigenetic mechanisms. The steps involved in cellular differentiation are numerous and are associated with a continual refinement of the pools of cellular mRNAs and proteins, thus ultimately allowing the cell to reach a state in which it can successfully carry out its unique physiological functions. During this process, there are many cues that a cell must respond to and interpret in order to decide which differentiation pathway to choose.

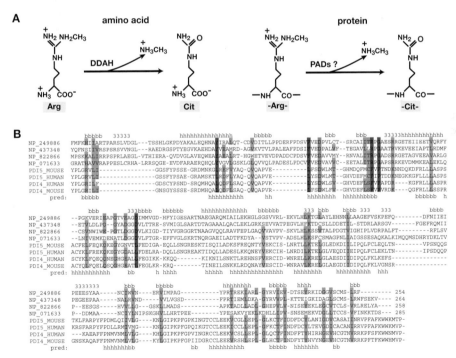

Figure 3. Comparison of DDAH and PAD proteins. (*A*) The biochemical reaction of dimethylarginine dimethylaminohydrolase (DDAH) on methyl-Arg amino acid and a likely biochemical reaction of peptidylarginine deiminase (PAD) on methyl-Arg residues in proteins. (*B*) Sequence and structure homology between DDAH and PAD. Identical residues are highlighted by *green*. Similar residues, such as basic: K, R, H; acidic: D, E; aromatic: F, Y, W, H; beta-branched: T, V, I; aliphatic: L, I, V; hydrogen bonding: C, S, T; or D/N or EQ, are highlighted in *blue*. Completely identical residues in both DDAH and PAD families but not identical to each other are highlighted in *red*. DDAH: NP_249886 (*Pseudomonas aeruginosa*), NP_437348 (*Sinorhizobium meliloti*), NP_822866 (*Stroptomyces avermitilis*), NP_071633 (*Rattus norvegicus*).

From an epigenetic viewpoint, we favor the idea that the developmental life history of a cell is engraved on chromatin by posttranslational histone modifications and DNA methylation. If we expand upon and broaden this concept, then in fact all of the structural proteins and enzymes responsible for the generation and maintenance of these epigenetic modifications can also be considered as a part of the "epigenetic information." For example, a certain structure formed on a chromatin region, such as condensed structures on pericentric heterochromatin or silenced homeotic genes, is inherited through cell cycles or memorized by the next generation. Thus, we would like to introduce the concept of "chromatin state," in which DNA and histone modifications interact with their binding or "effector" proteins and affect the local chromatin conformation in order to accommodate different chromatin/DNA templated events, such as transcription (Fig. 4).

Differentiation would be a process of continually refining the chromatin state leading to different sets of gene expression. We can consider the differentiation signals or cues as input, which then leads to the fine adjustment of the chromatin state in a process of "editing" and "writing." The adjustment of the chromatin state can bring about changes from state A to state B, leading to cell differentiation during embryonic development. Multiple machineries, such as the transcriptional complexes, would respond to this change of chromatin state, read the epigenetic information on chromatin, and translate it into meaningful biological readouts. Although differentiation has been considered to be a unidirectional commitment, recent data demonstrated that the forced expression of C/EBPα and C/EBPβ (transcription factors involved in hematopoietic cell differentiation) can reprogram B cells into macrophages, suggesting that the manipulation of upstream master control genes can dramatically change the chromatin state of one cell type into another, thus altering the differentiation pathway (Xie et al. 2004).

In the context of the cloning experiments, the chromatin state in a given cell type, such as a neuron, might be reprogrammed to the primary chromatin state (i.e., the zygote) from which all of the other chromatin states arise (Eggan et al. 2004). The mammalian egg's ability to reprogram terminally differentiated nuclei suggests that eggs have all the machineries required to erase the historic "records" on histones and DNA. New proteomic techniques, combined with the completed human and mouse genome sequences, will likely allow us to identify many of the factors involved in these early steps of development. Ultimately, biochemical and genetic approaches should help us obtain an idea of how these protein components work together to orchestrate this dramatic reprogramming event.

HISTONE METHYLTRANSFERASES AND EPIGENETIC TUMORIGENESIS

The imbalance of stem cell proliferation and differentiation can lead to cancer. For example, in patients with acute promyelocytic leukemia (APL), promyelocytes keep dividing and fail to differentiate into granulocytes. The ability of all-trans retinoic acid (ATRA) to induce the differentiation of APL cells has offered a successful chemotherapeutic agent to treat APL patients for more than a decade (Fenaux et al. 1992). Therefore, changes of gene expression profiles and epigenetic histone modifications in response to ATRA switch a "cancer" chromatin state to another chromatin state of a short-lived granulocyte. In the following discussion, we will focus on the epigenetic aspect of tumorigenesis.

Enzymes that modify different Lys sites in H3 have numerous links with human cancers. For example, human MLL1 (also termed ALL1, HRX, or HTRX) has been shown to methylate H3 Lys4 (Milne et al. 2002; Nakamura et al. 2002). The translocations and fusions of MLL1 with dozens of other genes result in acute leukemia. It is intriguing to speculate that MLL normally functions as a tumor "epigenetic" suppressor (see Fig. 5). Chromosome translocations between the *mll* locus and many other loci result in the loss of the SET domain of MLL. Therefore, the set of genes epigenetically regulated by MLL through marking/indexing with H3 Lys4 methylation is perturbed, which might led to carcinogenesis. The lacking of the H3 Lys9 methyltransferases, Suv39h1 and -2, in mice leads to telomere abnormalities, chromosome instability, and increased rate of B cell lymphomas (Peters et al. 2001). Although the effect on chromosome instability and telomere length should be well considered, we speculate an epigenetic gene-regulation mechanism similar to that of MLL underlying the high incidence of B cell lymphoma. Interestingly, RIZ1, a Rb interacting H3 Lys9 methyltransferase, has been well characterized as a tumor suppressor

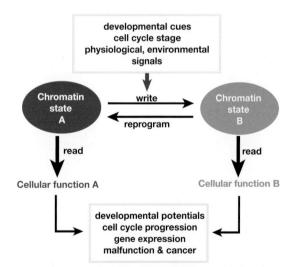

Figure 4. The concept of a chromatin state. The epigenetic information and the binding effector protein of epigenetic marks form an index system of gene expression potentials and regional chromatin conformation that determines a chromatin state. In response to extracellular cues (input), new epigenetic information is "written," and chromatin state "A" can be modified to become chromatin state "B." The change of chromatin state is translated (read) to cellular functions, which underlie(s) various processes such as cell differentiation, gene activity, and etiology of some diseases. Change(s) of chromatin state in the opposite direction (reprogramming) can happen during cloning or dedifferentiation.

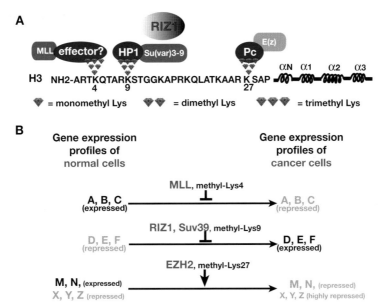

Figure 5. Histone Lys methyltransferases and epigenetic cause of tumorigenesis. (*A*) In this model, MLL (targeting Lys4), RIZ1, and Su(var)3-9 (targeting Lys9) are considered as tumor "epigenetic" suppressors. Members of the E(z) pathway (EZH2, Bmi1) are considered as "epigenetic" oncogenes. (*B*) MLL marks the expression potential of genes (A, B, C) by methylation of H3 Lys4 in normal cells. Loss of the normal function of MLL by chromosome translocations causes the repression of these genes thereby inducing cancers. On the other hand, RIZ1 and Su(var)3-9 index the silencing of their target genes (D, E, F); the loss of the repressive marks (H3 Lys9 methylation) leads to overexpression of these genes, thus leading to carcinogenesis. In contrast, the E(z) pathway maintains the cell proliferation potential by silencing a set of genes (X, Y, Z). The ultra-active E(z) function can cause cancers either by super-silencing of its target genes (X, Y, Z) or by silencing a new set of genes (M, N), which is normally expressed to prevent cancer development.

(Kim et al. 2003). The low expression and point mutations of RIZ1 are correlated with various human cancers, including breast cancer, liver cancer, colon cancer, neuroblastoma, melanoma, lung cancer, and osteosarcoma, suggesting that RIZ1 may suppress tumorigenesis via an epigenetic mechanism to ensure gene expression profiles of normal cells (Steele-Perkins et al. 2001).

On the other hand, human H3 Lys27 methyltransferase EZH2 was often overexpressed in metastatic prostate cancer and Hodgkin and non-Hodgkin B cell lymphoma (van Kemenade et al. 2001; Varambally et al. 2002). Similarly, Bmi-1, the human homolog of *Drosophila* Psc, which was shown to regulate the proliferative capacity of normal and leukemic stem cells, is upregulated in cancer cells (Lessard and Sauvageau 2003; Molofsky et al. 2003). Therefore, the EZH2 complex may be overexpressed in cancer cells to hyperrepress cell cycle inhibitor genes, thereby promoting tumorigenesis in a manner like an "epigenetic" oncogene.

In summary, proteins controlling epigenetic gene expression are important regulators of normal cell division, and differentiation, as well as tumorigenesis. We and others favor the idea that histone- and chromatin-modifying enzymes represent excellent targets for exploring pharmaceutical compounds to counteract cancers and other epigenetic disorders (Huang 2002).

ACKNOWLEDGMENTS

The authors would like to thank members of the Allis lab (The Rockefeller University), and Drs. X. Zhang and X. Chen (Emory University) for insightful discussions and comments. We appreciate Dr. F. Campagne's help on bioinformatical analyses of PAD4 and DDAH. We thank Upstate Biotech. Inc. for antibody development. Works discussed here are supported by a NIH MERIT Award GM R01 50659 (C.D.A.) and funding provided by The Rockefeller University (C.D.A.), NIH grant R01 HD38353 (S.A.C.), and Damon Runyon Cancer Research Foundation Fellowship (J.W.).

REFERENCES

Agalioti T., Chen G., and Thanos D. 2002. Deciphering the transcriptional histone acetylation code for a human gene. *Cell* **111:** 381.

Ahmad K. and Henikoff S. 2002. The histone variant H3.3 marks active chromatin by replication-independent nucleosome assembly. *Mol. Cell* **9:** 1191.

An W., Kim J., and Roeder R.G. 2004. Ordered cooperative functions of PRMT1, p300, and CARM1 in transcriptional activation of p53. *Cell* **117:** 735.

Bachman K.E., Park B.H., Rhee I., Rajagopalan H., Herman J.G., Baylin S.B., Kinzler K.W., and Vogelstein B. 2003. Histone modifications and silencing prior to DNA methylation of a tumor suppressor gene. *Cancer Cell* **3:** 89.

Bannister A.J., Schneider R., and Kouzarides T. 2002. Histone methylation: Dynamic or static? *Cell* **109:** 801.

Bannister A.J., Zegerman P., Partridge J.F., Miska E.A., Thomas J.O., Allshire R.C., and Kouzarides T. 2001. Selective recognition of methylated lysine 9 on histone H3 by the HP1 chromo domain. *Nature* **410:** 120.

Bauer U.M., Daujat S., Nielsen S.J., Nightingale K., and Kouzarides T. 2002. Methylation at arginine 17 of histone H3 is linked to gene activation. *EMBO Rep.* **3:** 39.

Boger R.H. 2003. Asymmetric dimethylarginine ADMA modulates endothelial function-therapeutic implications. *Vasc. Med.* **8**: 149.

Campbell K.H., McWhir J., Ritchie W.A., and Wilmut I. 1996. Sheep cloned by nuclear transfer from a cultured cell line. *Nature* **380**: 64.

Cavalli G. and Paro R. 1999. Epigenetic inheritance of active chromatin after removal of the main transactivator. *Science* **286**: 955.

Cheung P., Allis C.D., and Sassone-Corsi P. 2000. Signaling to chromatin through histone modifications. *Cell* **103**: 263.

Daujat S., Bauer U.M., Shah V., Turner B., Berger S., and Kouzarides T. 2002. Crosstalk between CARM1 methylation and CBP acetylation on histone H3. *Curr. Biol.* **12**: 2090.

De Keyser J., Schaaf M., and Teelken A. 1999. Peptidylarginine deiminase activity in postmortem white matter of patients with multiple sclerosis. *Neurosci. Lett.* **260**: 74.

Dodge J.E., Kang Y.K., Beppu H., Lei H., and Li E. 2004. Histone H3-K9 methyltransferase ESET is essential for early development. *Mol. Cell. Biol.* **24**: 2478.

Eggan K., Baldwin K., Tackett M., Osborne J., Gogos J., Chess A., Axel R., and Jaenisch R. 2004. Mice cloned from olfactory sensory neurons. *Nature* **428**: 44.

Erhardt S., Su I.H., Schneider R., Barton S., Bannister A.J., Perez-Burgos L., Jenuwein T., Kouzarides T., Tarakhovsky A., and Surani M.A. 2003. Consequences of the depletion of zygotic and embryonic enhancer of zeste 2 during preimplantation mouse development. *Development* **130**: 4235.

Feinberg A.P. and Tycko B. 2004. The history of cancer epigenetics. *Nat. Rev. Cancer* **4**: 143.

Felsenfeld G. and Groudine M. 2003. Controlling the double helix. *Nature* **421**: 448.

Fenaux P., Castaigne S., Dombret H., Archimbaud E., Duarte M., Morel P., Lamy T., Tilly H., Guerci A., and Maloisel F., et al. 1992. All-transretinoic acid followed by intensive chemotherapy gives a high complete remission rate and may prolong remissions in newly diagnosed acute promyelocytic leukemia: A pilot study on 26 cases. *Blood* **80**: 2176.

Fischle W., Wang Y., and Allis C.D. 2003. Binary switches and modification cassettes in histone biology and beyond. *Nature* **425**: 475.

Fuks F., Hurd P.J., Wolf D., Nan X., Bird A.P., and Kouzarides T. 2003. The methyl-CpG-binding protein MeCP2 links DNA methylation to histone methylation. *J. Biol. Chem.* **278**: 4035.

Grunstein M. 1997. Histone acetylation in chromatin structure and transcription. *Nature* **389**: 349.

Hagiwara T., Nakashima K., Hirano H., Senshu T., and Yamada M. 2002. Deimination of arginine residues in nucleophosmin/B23 and histones in HL-60 granulocytes. *Biochem. Biophys. Res. Commun.* **290**: 979.

Hajkova P., Erhardt S., Lane N., Haaf T., El-Maarri O., Reik W., Walter J., and Surani M.A. 2002. Epigenetic reprogramming in mouse primordial germ cells. *Mech. Dev.* **117**: 15.

Hiendleder S., Mund C., Reichenbach H.D., Wenigerkind H., Brem G., Zakhartchenko V., Lyko F., and Wolf E. 2004. Tissue-specific elevated genomic cytosine methylation levels are associated with an overgrowth phenotype of bovine fetuses derived by in vitro techniques. *Biol. Reprod.* **71**: 217.

Huang S. 2002. Histone methyltransferases, diet nutrients and tumor suppressors. *Nat. Rev. Cancer* **2**: 469.

Jackson J.P., Lindroth A.M., Cao X., and Jacobsen S.E. 2002. Control of CpNpG DNA methylation by the KRYPTONITE histone H3 methyltransferase. *Nature* **416**: 556.

Jaenisch R. and Bird A. 2003. Epigenetic regulation of gene expression: How the genome integrates intrinsic and environmental signals. *Nat. Genet.* (suppl.) **33**: 245.

Janicki S.M., Tsukamoto T., Salghetti S.E., Tansey W.P., Sachidanandam R., Prasanth K.V., Ried T., Shav-Tal Y., Bertrand E., Singer R.H., and Spector D.L. 2004. From silencing to gene expression: Real-time analysis in single cells. *Cell* **116**: 683.

Jenuwein T. and Allis C.D. 2001. Translating the histone code. *Science* **293**: 1074.

Kelley R.L., Wang J., Bell L., and Kuroda M.I. 1997. Sex lethal controls dosage compensation in *Drosophila* by a non-splicing mechanism. *Nature* **387**: 195.

Kelly T.L. and Trasler J.M. 2004. Reproductive epigenetics. *Clin. Genet.* **65**: 247.

Khorasanizadeh S. 2004. The nucleosome: From genomic organization to genomic regulation. *Cell* **116**: 259.

Kim J.H., Yoon S.Y., Kim C.N., Joo J.H., Moon S.K., Choe I.S., Choe Y.K., and Kim K.C., Geng L., and Huang S. 2003. Inactivation of a histone methyltransferase by mutations in human cancers. *Cancer Res.* **63**: 7619.

Kouzarides T. 2002. Histone methylation in transcriptional control. *Curr. Opin. Genet. Dev.* **12**: 198.

Kurdistani S.K., Tavazoie S., and Grunstein M. 2004. Mapping global histone acetylation patterns to gene expression. *Cell* **117**: 721.

Lachner M., O'Sullivan R.J., and Jenuwein T. 2003. An epigenetic road map for histone lysine methylation. *J. Cell Sci.* **116**: 2117.

Lee J., Inoue K., Ono R., Ogonuki N., Kohda T., Kaneko-Ishino T., Ogura A., and Ishino F. 2002. Erasing genomic imprinting memory in mouse clone embryos produced from day 11.5 primordial germ cells. *Development* **129**: 1807.

Lessard J. and Sauvageau G. 2003. Bmi-1 determines the proliferative capacity of normal and leukaemic stem cells. *Nature* **423**: 255.

Li E. 2002. Chromatin modification and epigenetic reprogramming in mammalian development. *Nat. Rev. Genet.* **3**: 662.

Luger K., Mader A.W., Richmond R.K., Sargent D.F., and Richmond T.J. 1997. Crystal structure of the nucleosome core particle at 2.8 Å resolution. *Nature* **389**: 251.

Matzuk M.M. and Lamb D.J. 2002. Genetic dissection of mammalian fertility pathways. *Nat. Cell Biol.* (suppl.) **4**: s41.

Mayer W., Niveleau A., Walter J., Fundele R., and Haaf T. 2000. Demethylation of the zygotic paternal genome. *Nature* **403**: 501.

Milne T.A., Briggs S.D., Brock H.W., Martin M.E., Gibbs D., Allis C.D., and Hess J.L. 2002. MLL targets SET domain methyltransferase activity to Hox gene promoters. *Mol. Cell* **10**: 1107.

Molofsky A.V., Pardal R., Iwashita T., Park I.K., Clarke M.F., and Morrison S.J. 2003. Bmi-1 dependence distinguishes neural stem cell self-renewal from progenitor proliferation. *Nature* **425**: 962.

Nakamura T., Mori T., Tada S., Krajewski W., Rozovskaia T., Wassell R., Dubois G., Mazo A., Croce C.M., and Canaani E. 2002. ALL-1 is a histone methyltransferase that assembles a supercomplex of proteins involved in transcriptional regulation. *Mol. Cell* **10**: 1119.

Ogawa T., Kimoto M., and Sasaoka K. 1989. Purification and properties of a new enzyme, NG,NG-dimethylarginine dimethylaminohydrolase, from rat kidney. *J. Biol. Chem.* **264**: 10205.

Okamoto I., Otte A.P., Allis C.D., Reinberg D., and Heard E. 2004. Epigenetic dynamics of imprinted X inactivation during early mouse development. *Science* **303**: 644.

Park Y. and Kuroda M.I. 2001. Epigenetic aspects of X-chromosome dosage compensation. *Science* **293**: 1083.

Peters A.H., O'Carroll D., Scherthan H., Mechtler K., Sauer S., Schofer C., Weipoltshammer K., Pagani M., Lachner M., Kohlmaier A., Opravil S., Doyle M., Sibilia M., and Jenuwein T. 2001. Loss of the Suv39h histone methyltransferases impairs mammalian heterochromatin and genome stability. *Cell* **107**: 323.

Roth S.Y., Denu J.M., and Allis C.D. 2001. Histone acetyltransferases. *Annu. Rev. Biochem.* **70**: 81.

Santos-Rosa H., Schneider R., Bernstein B.E., Karabetsou N., Morillon A., Weise C., Schreiber S.L., Mellor J., and Kouzarides T. 2003. Methylation of histone H3 K4 mediates association of the Isw1p ATPase with chromatin. *Mol. Cell* **12**: 1325.

Sarmento O.F., Digilio L.C., Wang Y., Perlin J., Herr J.C., Allis C.D., and Coonrod S.A. 2004. Dynamic alterations of specific histone modifications during early murine development. *J. Cell Sci.* **117**: 4449.

Sassone-Corsi P. 2002. Unique chromatin remodeling and transcriptional regulation in spermatogenesis. *Science* **296:** 2176.

Schreiber S.L. and Bernstein B.E. 2002. Signaling network model of chromatin. *Cell* **111:** 771.

Schubeler D., MacAlpine D.M., Scalzo D., Wirbelauer C., Kooperberg C., van Leeuwen F., Gottschling D.E., O'Neill L.P., Turner B.M., Delrow J., Bell S.P., and Groudine M. 2004. The histone modification pattern of active genes revealed through genome-wide chromatin analysis of a higher eukaryote. *Genes Dev.* **18:** 1263.

Shiio Y. and Eisenman R.N. 2003. Histone sumoylation is associated with transcriptional repression. *Proc. Natl. Acad. Sci.* **100:** 13225.

Simon J.A. and Tamkun J.W. 2002. Programming off and on states in chromatin: Mechanisms of Polycomb and trithorax group complexes. *Curr. Opin. Genet. Dev.* **12:** 210.

Soppe W.J., Jasencakova Z., Houben A., Kakutani T., Meister A., Huang M.S., Jacobsen S.E., Schubert I., and Fransz P.F. 2002. DNA methylation controls histone H3 lysine 9 methylation and heterochromatin assembly in *Arabidopsis*. *EMBO J.* **21:** 6549.

Steele-Perkins G., Fang W., Yang X.H., Van Gele M., Carling, T., Gu J., Buyse I.M., Fletcher J.A., Liu J., Bronson R., Chadwick R.B., de la Chapelle A., Zhang X., Speleman F., and Huang S. 2001. Tumor formation and inactivation of RIZ1, an Rb-binding member of a nuclear protein-methyltransferase superfamily. *Genes Dev.* **15:** 2250.

Strahl B.D. and Allis C.D. 2000. The language of covalent histone modifications. *Nature* **403:** 41.

Strahl B.D., Ohba R., Cook R.G., and Allis C.D. 1999. Methylation of histone H3 at lysine 4 is highly conserved and correlates with transcriptionally active nuclei in Tetrahymena. *Proc. Natl. Acad. Sci.* **96:** 14967.

Strahl B.D., Briggs S.D., Brame C.J., Caldwell J.A., Koh S.S., Ma H., Cook R.G., Shabanowitz J., Hunt D.F., Stallcup M.R., and Allis C.D. 2001. Methylation of histone H4 at arginine 3 occurs in vivo and is mediated by the nuclear receptor coactivator PRMT1. *Curr. Biol.* **11:** 996.

Strukov Y.G., Wang Y., and Belmont A.S. 2003. Engineered chromosome regions with altered sequence composition demonstrate hierarchical large-scale folding within metaphase chromosomes. *J. Cell Biol.* **162:** 23.

Suzuki A., Yamada R., Chang X., Tokuhiro S., Sawada T., Suzuki M., Nagasaki M., Nakayama-Hamada M., Kawaida R., Ono M., Ohtsuki M., Furukawa H., Yoshino S., Yukioka M., Tohma S., Matsubara T., Wakitani S., Teshima R., Nishioka Y., Sekine A., Iida A., Takahashi A., Tsunoda T., Nakamura Y., and Yamamoto K. 2003. Functional haplotypes of PADI4, encoding citrullinating enzyme peptidylarginine deiminase 4, are associated with rheumatoid arthritis. *Nat. Genet.* **34:** 395.

Tagami H., Ray-Gallet D., Almouzni G., and Nakatani Y. 2004. Histone H3.1 and H3.3 complexes mediate nucleosome assembly pathways dependent or independent of DNA synthesis. *Cell* **116:** 51.

Tamaru H., Zhang X., McMillen D., Singh P.B., Nakayama J., Grewal S.I., Allis C.D., Cheng X., and Selker E.U. 2003. Trimethylated lysine 9 of histone H3 is a mark for DNA methylation in *Neurospora crassa*. *Nat. Genet.* **34:** 75.

van Holde K.E. 1989. *Chromatin*. Springer-Verlag, New York.

van Kemenade F.J., Raaphorst F.M., Blokzijl T., Fieret E., Hamer K.M., Satijn D.P., Otte A.P., and Meijer C.J. 2001. Coexpression of BMI-1 and EZH2 polycomb-group proteins is associated with cycling cells and degree of malignancy in B-cell non-Hodgkin lymphoma. *Blood* **97:** 3896.

Varambally S., Dhanasekaran S.M., Zhou M., Barrette T.R., Kumar-Sinha C., Sanda M.G., Ghosh D., Pienta K.J., Sewalt R.G., Otte A.P., Rubin M.A., and Chinnaiyan A.M. 2002. The polycomb group protein EZH2 is involved in progression of prostate cancer. *Nature* **419:** 624.

Vossenaar E.R., Zendman A.J., van Venrooij W.J., and Pruijn G.J. 2003. PAD, a growing family of citrullinating enzymes: Genes, features and involvement in disease. *Bioessays* **25:** 1106.

Wakayama T., Perry A.C., Zuccotti M., Johnson K.R., and Yanagimachi R. 1998. Full-term development of mice from enucleated oocytes injected with cumulus cell nuclei. *Nature* **394:** 369.

Wang H., Huang Z.Q., Xia L., Feng Q., Erdjument-Bromage H., Strahl B.D., Briggs S.D., Allis C.D., Wong J., Tempst P., and Zhang Y. 2001. Methylation of histone H4 at arginine 3 facilitating transcriptional activation by nuclear hormone receptor. *Science* **293:** 853.

Wolffe A. 1998. *Chromatin*. Academic Press, London.

Wolffe A.P. and Matzke M.A. 1999. Epigenetics: Regulation through repression. *Science* **286:** 481.

Wright P.W., Bolling L.C., Calvert M.E., Sarmento O.F., Berkeley E.V., Shea M.C., Hao Z., Jayes F.C., Bush L.A., Shetty J., Shore A.N., Reddi P.P., Tung K.S., Samy E., Allietta M.M., Sherman N.E., Herr J.C., and Coonrod S.A. 2003. ePAD, an oocyte and early embryo-abundant peptidylarginine deiminase-like protein that localizes to egg cytoplasmic sheets. *Dev. Biol.* **256:** 73.

Xie H., Ye M., Feng R., and Graf T. 2004. Stepwise reprogramming of B cells into macrophages. *Cell* **117:** 663.

Zhang Y. and Reinberg D. 2001. Transcription regulation by histone methylation: Interplay between different covalent modifications of the core histone tails. *Genes Dev.* **15:** 2343.

Steps Toward Understanding the Inheritance of Repressive Methyl-Lysine Marks in Histones

D. Reinberg,* S. Chuikov,* P. Farnham,† D. Karachentsev,‡ A. Kirmizis,†
A. Kuzmichev,* R. Margueron,* K. Nishioka,* T.S. Preissner,*,§ K. Sarma,*
C. Abate-Shen,‖ R. Steward,‡ AND A. Vaquero*

*Howard Hughes Medical Institute, Division of Nucleic Acids Enzymology, Department of Biochemistry, Robert Wood Johnson Medical School, Piscataway, New Jersey 08854; †UC Davis Genome Center and Biomedical Sciences Facility, Davis, California 95616-8816; ‡Waksman Institute, Rutgers University, Piscataway, New Jersey 08854; §X Inactivation Group, MRC Clinical Sciences Centre, ICSM, Hammersmith Hospital, London W12 0NN, United Kingdom; ‖Center for Advanced Biotechnology and Medicine, Department of Medicine and Cancer Institute of New Jersey, Robert Wood Johnson Medical School, Piscataway, New Jersey 08854

The establishment of cellular identity in multicellular organisms proceeds according to a pattern of gene expression that is adjusted temporally and spatially. It has become evident that changes within the chromatin structure brought about by covalent modifications of histones are of crucial importance in this as well as in many other biological processes.

A focus of our studies involves the identity and functional roles of the cellular proteins that enable the transcriptional machinery to access and move through the nucleosomal DNA. This entails alterations within the nucleosomal structure. To get to this stage, the overall chromatin structure must unfold. Our recent studies have focused on the properties of the nucleosomal proteins that are determinant to the overall organization of compacted chromatin. It is these proteins that proscribe or enable DNA accessibility. The complex chromosome structure around silent genes is tightly wound (heterochromatin) prohibiting access to the transcriptional machinery, while it is more loosely wrapped around the genes active in RNA synthesis (euchromatin). Over the past few years, researchers have uncovered that the key to these two different chromatin structures resides in the histone tails that embody unstructured, hooklike projections extending outward from each nucleosome. Posttranslational modifications in these histone tails determine whether they lock together tightly or loosely and thus whether the genes spooled around them are silent or active.

Several discoveries in 1996 heralded a momentous year for members of the transcription and chromatin communities. Two reports provided the long awaited connection between chromatin and the regulation of gene expression. In one report, Allis and coworkers (Brownell et al. 1996) described a histone acetyltransferase and discovered that it was GCN5, whose gene was previously identified in yeast using a genetic screen scoring for transcriptional regulators. In a second report, Schreiber and coworkers (Taunton et al. 1996) uncovered that a histone deacetylase resides in a mammalian homolog of yeast RPD3, originally isolated in genetic screens scoring for transcriptional repressors. These were groundbreaking discoveries. Subsequent developments enhanced their impact by showing that the enzymatic activities of Gcn5 and Rpd3 are regulated through their association with other proteins (Sterner and Berger 2000; Kuzmichev and Reinberg 2001). These findings galvanized the field of chromatin research, within the context of transcriptional regulation and gene expression. We now know that many regulators of transcription contain activities that covalently modify the histone tails (Struhl 1998; Sterner and Berger 2000).

Another important finding that intersected histone methylation with heterochromatin formation came about 4 years later. Jenuwein and colleagues (Rea et al. 2000) uncovered that the mammalian homolog of a gene initially isolated in *Drosophila* as a suppressor of variegation, Suv39H1, and its *Schizosaccharomyces pombe* homolog, CLR4, encode proteins with intrinsic histone lysine methyltransferase (HKMT) activity. These respective HKMT activities exhibited a strict specificity for lysine-9 of histone H3, a residue previously identified as being methylated in vivo (DeLange et al. 1973). Lysine residues can potentially be fully methylated generating trimethylated moieties; however, HKMTs can incorporate mono-, di-, or trimethyl groups (Zhang and Reinberg 2001), providing another level of diversity in regulation. The bond formed between the side chain of lysine and the methyl group is not easily reversed (Bannister et al. 2002; Kubicek and Jenuwein 2004). The stability of this lysine-methyl bond is a unique property of this modification. Similar to clusters of CpG methylation in DNA, histone lysine methylations are now appreciated as stably modified chromatin constituents whose transfer from mother to daughter cells qualifies the epigenetic process. Thus, histone lysine methylations are categorized as "epigenetic marks" (Jenuwein and Allis 2001; Sims et al. 2003). Importantly, numerous studies have provided evidences that the enzymes responsible for the modifications of histones function in a coordinated pattern to control gene expression (Berger 2002; Ng et al. 2002; Sun and Allis

2002; Krogan et al. 2003; Wood et al. 2003). This ranges from short- to long-term effects, with the latter involving the heritable transferral of stable modifications from mother to daughter cells.

The molecular means by which the transcriptional machinery deals with the repressive nature of unfolded chromatin is one level of complexity being investigated by many laboratories including ours. How higher-order chromatin unfolds to allow access of transcription factors to DNA represents the next level. Our current studies are geared toward the identity and biochemical properties of the proteins required for the formation of higher-order chromatin (heterochromatin) and its unfolding to euchromatin, the template of the transcriptional machinery. In this chapter, we describe our studies on two HKMTs that target sites within the histone tails that function in the establishment of repressive facultative heterochromatin. The sites independently targeted for methylation are lysine-20 of histone H4 and lysine-27 of histone H3. We describe our studies toward understanding how these "repressive marks" might be transmitted from mother to daughter cells.

PR-SET7

A big step toward analyzing how dividing cells can retain tissue identity came about with the isolation of the human protein, PR-Set7, which specifically methylates lysine-20 of histone H4 (Nishioka et al. 2002). PR-Set7 was found to be nucleosomal specific. Our initial biochemical and immunofluorescence experiments using tissue culture cells confirmed that PR-Set7 catalyzed methylation of H4 lysine-20 within nucleosomes (Nishioka et al. 2002). Recent structural studies demonstrated that PR-Set7 is a monomethylase (B. Xiao et al., unpubl.).

We uncovered that methylation of histone H4-lysine-20 correlates with the presence of repressive chromatin in vivo (Nishioka et al. 2002). Moreover, other studies uncovered that monomethylation of H4-K20 is present in the inactive female X chromosome (Kohlmaier et al. 2004). This finding together with our observations led us to conclude that monomethylation of H4-K20 is a repressive mark associated with facultative heterochromatin. Consistently, the "mark" set by PR-Set7 was found to be present in the euchromatic arms of *Drosophila* polytene chromosomes, but excluded from loci associated with the transcriptionally active form of RNA polymerase II (Fig. 1). Most important to our goals was the discovery that the expression of PR-Set7 was regulated during cell division (Rice et al. 2002). PR-Set7 was found to be expressed early during mitosis in human cells (Fig. 2) and to be bound to mitotic human chromosomes during prometaphase, prior to separation of the chromosomes (Fig. 3). These findings placed PR-Set7 at a time and position appropriate for propagating the methyl mark to newly generated chromatin templates.

From the studies described above, at least two important questions remained unanswered: (1) Although PR-Set7 is a monomethylase, does it somehow participate in di- and trimethylation of H4-K20 that occurs in vivo? And (2) what are the in vivo evidences demonstrating that methylation of H4-K20 is a stable mark that can be propagated from mother to daughter cells?

Immunofluorescence studies using highly specific antibodies that distinguish between the three different states of H4-K20 methylation, together with the expression of various SET domain–containing proteins in tissue culture cells, uncovered two novel genes, distinct from PR-SET7, that mediate di- and trimethylation of H4-K20 in mammals (Schotta et al. 2004). Interestingly, di- and trimethylation of H4-K20 was dependent on the expression of the Suv3-9H1/2 enzymes, which catalyze trimethylation of H3-K9 (Rea et al. 2000). This finding underscores that the enzymes responsible for the modifications of histone tails function coordinately to control gene expression and chromatin structure, as observed by others (Berger 2002; Ng et al. 2002; Sun and Allis 2002; Krogan et al. 2003; Wood et al. 2003). Di- and trimethylation of H4-K20 was found to be catalyzed by SUV4-20h1/2 in mammals and the "mark" established by these enzymes was found to be primarily pericentromeric (Schotta et al. 2004); i.e., corresponding to constitutive heterochro-

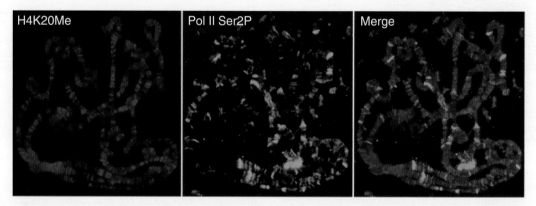

Figure 1. H4 lysine-20 methylation is a mark for repression. *Drosophila* polytene chromosomes were costained with polyclonal antibodies to H4 methyl-K20 (*red*) and monoclonal antibodies recognizing the elongating, Ser2-phosphorylated form of RNA polymerase II (*green*). H4 methyl-K20 and actively transcribing RNA polymerase II are mutually exclusive suggesting that the methyl mark at H4 lysine-20 is repressive. (Adapted from Nishioka et al. 2002.)

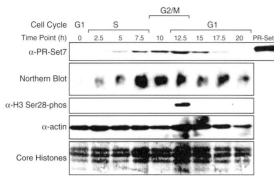

Figure 2. PR-Set7 is up-regulated in mitosis. HeLa cells were synchronized by cell cycle arrest in the G1 phase and released. Cells were collected every 2.5 hr. RNA levels were analyzed by northern blot with a probe corresponding to the open reading frame of PR-Set7. Protein levels and H3-Ser28 phosphorylation, which is a mitotic marker, were analyzed by western blot. Actin as well as the Ponceau staining of histones served as loading controls. PR-Set7 protein levels are maximal in the G2/M phase of the cell cycle and overlap with H3-Ser28 phosphorylation, confirming that PR-Set7 is a mitotic HKMT. (Adapted from Rice et al. 2002.)

matin. This is distinct from monomethylation at H4-K20, a mark present within facultative heterochromatin (Nishioka et al. 2002; Kohlmaier et al. 2004). Thus, disparate enzymes catalyze the monomethylation versus the di-and trimethylation of H4-K20 in mammalian cells.

The expression of PR-Set7 during early mitosis and its physical association with mitotic chromosomes in mammalian tissue culture cells suggested that monomethyl H4-K20 might be a transferable chromatin mark. Strong evidence for this epigenetic feature comes from studies in *Drosophila*. By employing P element mobilization, we isolated a null allele in *Drosophila* PR-SET7 such that the entire protein coding region is missing ($PR\text{-}Set7^{20}$). Interestingly, while PR-Set7 was maternally deposited in the homozygous $PRSet7^{20}$ embryo, it did not persist into the first-instar larval stage (Fig. 4). Yet the monomethyl mark was detectable until late larval stages (data not shown, see below) indicating that once set, this mark is stable over several subsequent cell generations. The absence of PR-Set7 resulted in lethality at the larval to pupal transition (Nishioka et al. 2002; Karachentsev et al. 2005). $PR\text{-}Set7^{20}$ now allowed us to define the functional role of PR-Set7 and of H4-K20 monomethylation in the organism.

Position effect variegation (PEV) monitors the spread of silent heterochromatin through an inserted euchromatic, active region. The extent of PEV is reflected by a phenotypic change, such as eye color, associated with the levels of expression of the inserted gene. PEV is suppressed in $Pr\text{-}Set7^{20}$ when insertions into the centromeric and telomeric heterochromatin of the fourth chromosome are analyzed. Thus, PR-Set7 functions, like the enzymes that trimethylate H4-K20 and H3-K9, as a suppressor of variegation (Su(var)), confirming its role in gene silencing. Surprisingly, and in contrast with results obtained in mammalian tissue culture cells, the levels of all three methyl H4-K20 marks were reduced in $PR\text{-}Set7^{20}$ mutant larvae (Fig. 5). This may be related to the temporal sequence of the marks attained during early wild-type embryogenesis with monomethyl preceding the di- and trimethyl marks in *Drosophila*. Of note, there are two enzymes that catalyze di- and trimethylation of H4-K20 in mammals, while only Suv4-20 performs this function in *Drosophila* (Schotta et al. 2004).

The levels of PR-Set7 cycle and are highest during the late G2 phase through the early M phase. Notably, we found that PR-Set7 is important to the mitotic process. $Pr\text{-}Set7^{20}$ larval imaginal leg and eye discs contained fewer cells than did their wild-type counterparts and these cells had elevated DNA levels (Fig. 6).

These results collectively led us to conclude that (1) monomethylation of histone H4-K20 localizes with transcriptionally inactive, condensed chromatin and (2) PR-Set7 expression occurs at the time during mitosis when fresh DNA templates are available to be marked and must be so modified for sustained cell division. PR-Set7 may well be determinant to restoring the repression of cell cycle regulatory genes, a possibility that awaits further investigation. We postulate that binding of PR-Set7 to mitotic chromosomes establishes the basis for propagation of this stable mark through cell divisions. PR-Set7 would recognize monomethylation of histone H4-lysine-20 in the mother chromosomes and transmit this mark to the daughter chromosomes before their separation. This is indeed a very exciting proposition because the modification can be passed down to daughter cells prescribing the silence or expression of genes (in other words, determining the cell's identity).

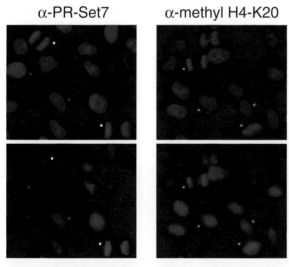

Figure 3. PR-Set7 localizes to mitotic chromosomes. Immunofluorescence staining of PR-Set7 (*left, red*) and methyl H4-K20 (*right, red*) in HeLa cells shows that PR-Set7 is up-regulated in the mitotic phase (*yellow dots*) and localizes to the chromosomes. In interphase cells (*pink dots*), the levels of PR-Set7 are low or not detectable. The staining of methyl H4-K20 (*right, red*) shows that this modification is also associated with mitotic chromosomes (*green dots*), but it is still present in interphase cells (*orange dots*). This might be attributable to the stability of the methyl-lysine mark over several cell generations. The mitotic phases were determined by DAPI staining (*blue*). (Adapted from Rice et al. 2002.)

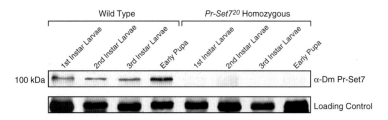

Figure 4. Maternally deposited PR-Set7 does not perdure into late larval stages in PR-Set7 homozygous null flies (*PR-Set7[20]*). Different developmental stages of wild-type and *PR-Set7[20]* flies were analyzed for levels of PR-Set7 protein. Although maternal PR-Set7 is deposited into eggs and is present in early embryogenesis (data not shown), it is absent from the first-instar larval stages. In contrast, the H4 monomethyl mark is detectable, albeit much reduced, even in the third-instar larval stage, as seen in the polytene chromosome staining in Fig. 5. (Adapted from Karachentsev et al. 2005.)

While PR-Set7 may establish heritable cellular identity during mitosis, the histone lysine methyltransferase discussed next, Ezh2, may do so during DNA replication.

EZH2

Ezh2 is an HKMT that exists in at least three different large protein complexes that share common subunits (Fig. 7A). Remarkably, many of these subunits exhibit altered expression levels during cellular transformation. Particularly noteworthy is that one component is enriched in undifferentiated and transformed cells. The resultant change in the complexes harboring Ezh2 directly alters the histone substrates that are targeted. As discussed below, we postulate that this alteration in histone substrate specificity has profound effects on the pattern of Ezh2-targeted gene expression.

As in many cases, the history of Ezh2 is an interesting one. Early studies in *Drosophila* capitalized on two distinct processes functionally relevant to gene repression. One such phenomenon was the long-term transcriptional silencing of homeotic genes whose regulation was determinant to proper embryonic development (Pirrotta and Rastelli 1994). The Polycomb group (PcG) of proteins was identified as being crucial to this process (Jacobs and van Lohuizen 1999). How these proteins repress transcription is a field of active investigation (Lund and van Lohuizen 2004), but it is known that they exist in different multiprotein complexes. The second phenomenon is PEV, whereby the repositioning of an active reporter gene within a heterochromatic region results in its repression. This was insightfully employed as an assay to identify genes that, when mutated or altered in expression, may suppress or enhance PEV. Such candidates were recognized as somehow maintaining the integrity of heterochromatin. The characterization of Ezh2 discussed below demonstrates that it is an HKMT whose activity bridges both phenomena.

Ezh2 was one of the first to be identified as a member of the SET domain family of proteins that includes PR-Set7. In most cases the SET domain is associated with lysine-specific methyltransferase activity. Overexpression of Ezh2 was known to enhance PEV (Laible et al. 1997). We set out to examine whether this Ezh2-associated gene repression correlates with its putative HKMT activity.

Our first evidence that the characterization of Ezh2 would prove to be intriguing was that recombinant Ezh2 isolated from bacteria was devoid of HKMT activity (data not shown). A clue that other proteins may be important in eliciting Ezh2 activity was contained in a previous report demonstrating that Ezh2 exists in a complex with other proteins. Only two components of this com-

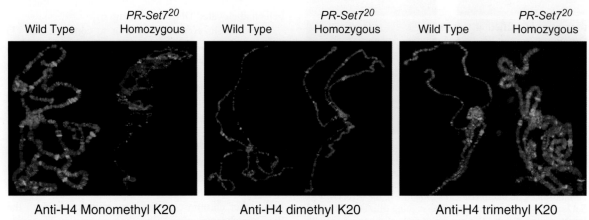

Figure 5. Mono-, di-, and trimethyl levels of H4-K20 are reduced in PR-Set7 homozygous null flies. Immunofluorescence staining of polytene chromosomes from wild-type and *PR-Set7[20]* flies show that the mutant flies have very low levels of monomethyl K20 (*left*). The di- and trimethyl levels (*middle* and *right*, respectively) are also reduced. The staining of RNA polymerase II (*green*) is not altered in the mutants, indicating that the loss of methyl marks from the polytene chromosomes is not a result of chromosome degradation. (Adapted from Karachentsev et al. 2005.)

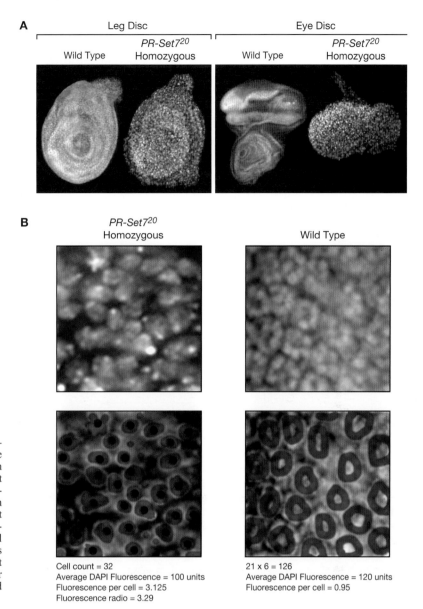

Figure 6. Mitosis is affected in *PR-Set7²⁰* flies. (*A*) Hoescht staining of eye (*left*) and leg (*right*) imaginal discs from wild-type and *PR-Set7²⁰* flies show that the cell number in the mutant flies is reduced. (*B*) Quantification of the DNA in these cells confirmed that the mutant cells had a larger DNA content as compared to wild type. The decrease in cell number and increase in DNA levels strongly suggests that the cells deficient for PR-Set7 are compromised in their ability to undergo cell division. (Adapted from Karachentsev et al. 2005.)

plex were identified at that time, Ezh2 and the PcG protein ESC (*Drosophila* homolog of mammalian Eed) (Ng et al. 2000). Nonetheless, given that Ezh2 did not function in isolation, this suggested to us that its association with other proteins might regulate its activity. As described below, this was indeed the case. Not only did associated proteins elicit Ezh2 activity but several such complexes were identified. While these complexes shared many subunits, such as Ezh2, they differed in others and this correlated with the differential substrate specificity exhibited by the enzyme (see Fig. 7A).

Using standard chromatographic techniques, we initially detected two complexes that contained Ezh2 in mammalian tissue culture cells. Purification of one of these complexes followed by mass spectrometry in conjunction with immunoblotting revealed that, in addition to Ezh2 and ESC (mammalian Eed), the complex contained another PcG protein, JJAZ1 (the human homolog of *Drosophila* Su(z)12), as well as the histone-binding proteins RbAp46 and RbAp48 (Fig. 7) (Kuzmichev et al. 2002). We designated this complex PRC2, as another complex containing a completely different set of PcG proteins had been previously designated Polycomb Repressive Complex 1 or PRC1 (Shao et al. 1999). The PRC2 complex was also isolated from cells that were stably transformed with ESC containing a FLAG tag (Kuzmichev et al. 2002).

Analysis of the second Ezh2-containing complex revealed it to be similar in polypeptide composition to PRC2; however, the Eed species contained in this complex differed to that present in PRC2 (Fig. 7). We referred to this complex as PRC3.

Further investigation using tissue culture cells revealed that Eed protein is actually present in four different isoforms (Eed1–4) with the largest species being Eed1 (Fig. 7B; and see below). These isoforms appear to arise from

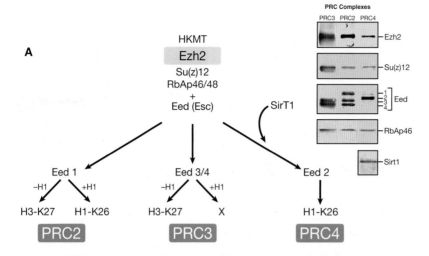

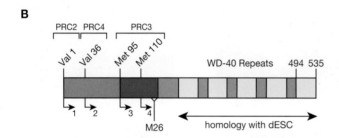

Figure 7. PRC complexes and Eed protein. (*A*) *Left*: Summary of the different PRC complexes showing their composition and substrate specificities. *Right*: Western blot for the PRC components with partially purified PRC2/3/4. (*B*) Schematic representation of the Eed protein. (Adapted from Kuzmichev et al. 2004.)

the use of different translational start sites present in Eed mRNA (Fig. 7B) (Denisenko et al. 1998; Kuzmichev et al. 2004). PRC2 was enriched in Eed1 and PRC3 contained exclusively Eed3 and Eed4 (Fig. 7A). Neither complex contained Eed2. Were these differences in the Eed isoform determinant to the function/activity of the PRC2/3 complexes?

In contrast to Ezh2 in isolation, PRC2 and PRC3 now exhibited HKMT activity and this was dependent upon the SET domain of Ezh2 (Kuzmichev et al. 2002, 2004). Given octamers in isolation, PRC2 and PRC3 both targeted histone H3-lysine-27 (H3-K27) for methylation (Kuzmichev et al. 2002, 2004). However, when histone H3 was present in the form of oligonucleosomes containing histone H1, presumably the natural substrate in vivo, differences in PRC2 and PRC3 activity were detected. PRC2 additionally targeted lysine-26 of histone H1 (H1-K26) for methylation, but the presence of H1 repressed PRC3 activity completely (Fig. 7A). As histone H1 had not been documented as a substrate for lysine methylation, the first issue to address was whether or not H1-K26 is methylated in vivo. Indeed, by generating antibodies specific for me-H1-K26, we observed immunoreactivity against native, but not recombinant, histone H1. Importantly, the stretches of amino acid sequence immediately surrounding H1-K26 and H3-K27 are remarkably similar (Fig. 8) and likely contribute to their targeting by Ezh2. We hypothesize that the disparity in Eed makeup of PRC2 and PRC3 is determinant to the differences in their substrate specificity. Indeed, as described below, we isolated a novel PRC that exhibited yet another difference in

substrate specificity. This complex contained the elusive Eed2 isoform and the conditions that spurred its formation were quite notable (Fig. 7A; and see below).

Given the biological relevance of methylated histone H1, instead of focusing on Ezh2-containing complexes, we next sought activities other than PRC2 that may methylate lysine residues of histone H1 (H1-HKMT) from crude HeLa cell extract. We found at least two such activities (Kuzmichev et al. 2005). One of them was a mixture of PRC2 and PRC3. Interestingly, another species displayed the highest level of activity and had a larger molecular mass, and yet its composition was similar to PRC2 and PCR3 as it included Ezh2, Su(z)12, Eed, and RbAp46/48 polypeptides. However, we noticed a distinct difference between this complex and PRC2/3 at this point in our analyses. This larger complex specifically contained the elusive Eed2 isoform (Figs. 7A and 9). Thus, our search for additional H1-HKMT activity

Figure 8. Alignment of sequence surrounding H3-K9, H3-K27, and H1-K26. Alignment of the human sequence surrounding H3-K9, H3-K27, and H1-K26 from different H1 isoforms and the corresponding sequence from H1° is shown. The two existing nomenclatures for H1 isoforms are indicated.

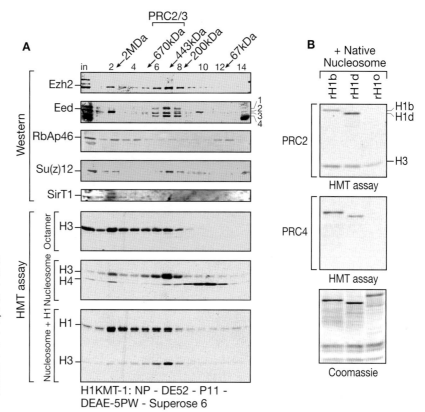

Figure 9. PRC4 identification and substrate specificity. (*A*) Gel filtration analysis of the PRC4 complex. Fraction numbers and corresponding molecular weight standards are indicated at the top. *Top:* Western blot of column fractions using indicated antibodies. *Bottom:* HKMT assays of column fractions performed with substrates indicated on the *left*. (*B*) Comparison of PRC2 and PRC4 methylation of H1 isoforms. (Adapted from Kuzmichev et al. 2005.)

other than that associated with PRC2 revealed a novel PRC as judged by its composition. We designated this complex PRC4.

The substrate specificity of PRC4 was most similar to that of PRC2. In the absence of H1, PRC4 methylated histone H3-K27 when present in oligonucleosomes. However, in contrast to PRC2, PRC4-mediated methylation of H3 was greatly reduced when H1-containing oligonucleosomes were the substrates (Fig. 9A). Similar to PRC2, PRC4 now preferentially methylated H1-K26. Further analyses did reveal a distinction in PRC2 and PRC4 substrate preferences. While PRC2 preferentially methylated the histone H1d isoform (human H1.2), PRC4 preferentially targeted H1b (human H1.4) (Fig. 9B). None of the complexes methylated H1°, which lacks lysine-26 (Fig. 8). Thus PRC2, PRC3, and PRC4 differed in the specific Eed isoform(s) they contained and this correlated with their differential substrate preferences.

Further examination of PRC4 revealed another significant difference in its composition and associated activity relative to PRC2 and PRC3. We thought it possible that SirT1 might be associated with one or more of the PRCs, based on the following findings. Our studies had identified SirT1 as a NAD$^+$-dependent histone deacetylase. It exhibited specificity for H1-K26, the same residue that is methylated by PRC4 and PRC2 (Vaquero et al. 2004). We noted that methylation of H3-K9 is known to be facilitated by its prior deacetylation through HDAC1 in *Drosophila* (Czermin et al. 2001) and by deacetylation through the NAD$^+$-dependent deacetylase spSir2 in *S. pombe* (Shankaranarayana et al. 2003). Thus, SirT1 may

function analogously to HDAC1 or spSir2, but with respect to H1-K26. Moreover, a connection between SirT1 and Ezh2 was revealed by their independent participation in PEV (Laible et al. 1997; Rosenberg and Parkhurst 2002). Following this hunch, we found that SirT1 was indeed associated, specifically with PRC4 (Figs. 9A and 10A), in a manner dependent on SirT1 enzymatic activity (Fig. 10A). Moreover this association was specific because the related SirT2 failed to interact with the PRC4 complex (Fig. 10). In fact, SirT1 interacts directly with Su(z)12 (Fig. 11). As Su(z)12 is a common subunit of PRC2–4, the fact that SirT1 associates solely with PRC4 is intriguing. Perhaps the specific Eed isoforms present in the PRCs determine their HKMT substrate specificity and also the availability of the Su(z)12 epitope necessary for SirT1 interaction. This remains to be examined. Determining if SirT1 was relevant to PRC-regulated gene expression in vivo was of high priority.

We initially identified PRC-regulated genes through a combination of approaches. We first sought genes whose expression may change upon down-regulation of one of the PRC common subunits, Su(z)12. RNA isolated from colon cancer cells that were either untreated or treated with siRNA against Su(z)12 was scored using DNA arrays containing CpG islands (Kirmizis et al. 2004). The latter are enriched for promoter-containing DNA. A group of responsive genes was identified. We then tested for the presence of PRC components at the promoters of these genes in vivo, in wild-type cells. Chromatin immunoprecipitation (ChIP) experiments identified that several of the genes whose expression had responded to

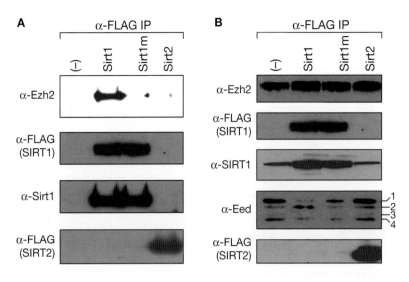

Figure 10. SirT1 interacts with the PRC4 complex in mammalian cells. (*A*) Anti-FLAG immunoprecipitation was performed with extracts from 293 cells transfected with expression vectors encoding untagged Ezh2 and FLAG-tagged SirT proteins (wild-type SirT1 ["SirT1" lane], active site mutant SirT1 ["SirT1m" lane], or wild-type SirT2 protein ["SirT2" lane]). The immunoprecipitates were analyzed by western blot using the indicated antibodies. (*B*) Inputs for immunoprecipitation were analyzed by western blot using antibodies as indicated on the *left*. (Adapted from Kuzmichev et al. 2005.)

Su(z)12 deprivation contained Su(z)12, Ezh2, and Eed at their promoters in the wild-type case, in vivo. We later found that SirT1 was present at these same PRC-regulated genes (Fig. 11B). Thus, SirT1 is an in vivo constituent of a PRC and our biochemical studies revealed this to be PRC4.

Given the differential substrate specificity exhibited by the distinct PRCs, conditions in the cell that may promote formation of a particular complex would be expected to have marked consequences to the histones that are modified and, thus, to chromatin structure and gene expression. At this stage in our investigations a new report provided an important clue to such conditions, conditions that we found to favor PRC4 formation.

Ezh2 was found to be expressed at increasingly elevated levels during the progression of prostate cancer (Varambally et al. 2002) and the same was true in breast cancer (Kleer et al. 2003). In the context of our biochem-

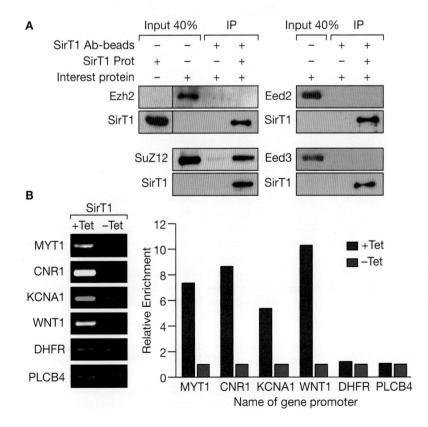

Figure 11. SirT1 interacts with Su(z)12 and is present on PRC target promoters. (*A*) Recombinant Ezh2, Su(z)12, Eed (30-535), or Eed (95-535) derived from Sf9 cells infected with recombinant baculoviruses or from *Escherichia coli* (Eed proteins) were incubated separately with or without recombinant SirT1 protein and immunoprecipitated using SirT1-specific antibody (2G1; see Vaquero et al. 2004) coupled to beads. Inputs and elutions were analyzed by western blot using antibodies as indicated on the *left*. (*B*) Chromatin immunoprecipitation experiments using antibodies to Gal4 were performed in 293f cells expressing Gal4-SirT1 under the control of tetracycline in its presence (+Tet) or its absence (–Tet). The precipitated chromatin was analyzed with PCR using primers specific to the promoters targeted by PRC2/3 (MYT1, CNR1, KCNA1, and WNT1) and to the negative control promoters (DHFR and PLCB4) as indicated. The relative enrichment of the signal in the absence or in the presence of tetracycline (which causes the induction of Gal4-SirT1) is shown. (Adapted from Kuzmichev et al. 2005.)

ical analyses of Ezh2, such overexpression was intriguing. As Ezh2-associated HKMT activity requires its association with the other PRC components, Ezh2 would not be expected to be functional were it to be solely overexpressed. To approach this quandary, we first examined the levels of Ezh2, SirT1, and Eed isoforms during differentiation of mouse embryonic stem (ES) cells in vitro as cancer cells exhibit at least partial dedifferentiation. Previous studies had shown that Ezh2 and Eed levels overall decline during ES cell differentiation (Silva et al. 2003). We further noted that the levels of SirT1 also declined (Fig. 12A). We could now distinguish the four Eed isoforms, all of which were present in undifferentiated cells. Remarkably, Eed1–4 did not decline in unison. The PRC4-specific Eed2 isoform was down-regulated immediately upon induction to differentiate (Fig. 12A). Over time, the only isoform predominately remaining was Eed3. This change in pattern of the Eed isoform levels upon ES cell differentiation was consistent with the profile of Eed expression in normal mouse tissue; the levels of Eed2 are undetectable and Eed3 is predominant.

The state of Ezh2 activity upon its elevation in cancer cells became less perplexing when we observed elevated levels not only of Ezh2, but also of other PRC components: SirT1, Su(z)12, and all four Eed isoforms were elevated in human cancer specimens relative to normal tissue (Fig. 12B). These conditions would favor formation of any of the PRC2–4 candidates, including the SirT1/Eed2-specific PRC4. Similar to differentiated mouse ES cells and normal mouse cells, Eed2 was the only Eed isoform that was undetectable in normal human cells derived from the colon (Fig. 12B). Normal cells appear to favor PRC2 and PRC3 and to be deficient in PRC4. This survey of ES cells as a function of differentiation and of human cells as a function of cellular transformation strongly suggested that there were indeed cellular conditions that discriminate among the PRCs. This was evident in the differential expression profiles of the Eed isoforms that distinguished PRC2–4.

We next investigated the affects on PRC integrity when Ezh2 or SirT1 is overexpressed, this time in tissue culture cells. Biochemical analyses revealed that overexpression of Ezh2 drove PRC4 formation as evidenced by the accumulation of a large complex containing the PRC4-specific components, SirT1 and Eed2 (Fig. 13). Remarkably, overexpression of SirT1 resulted in an altered pattern of Eed isoform expression in the cells with Eed2 now being prominent, thus suggesting that SirT1 overexpression may also lead to PRC4 formation (Fig. 10B). These findings underscore the fact that overexpression of the PRC components can have a direct impact on the composition of the PRCs and can promote formation of the distinctive PRC4. With this basis, we next investigated the levels of PRC components and gene expression profiles during

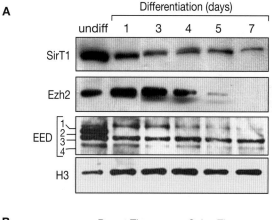

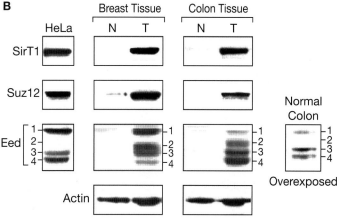

Figure 12. Expression of PRC components is modulated as a function of differentiation and cellular transformation. (*A*) Western blot analysis of nuclear extracts from PGK12.1 ES cells. Days of differentiation are indicated above each lane. The western blot was probed with antibodies against SirT1, Ezh2, and Eed as well as control antibody directed against histone H3. (*B*) Western blot analysis of PRC components using HeLa cell nuclear extracts or whole tissue extracts prepared from normal (N) and tumor (T) tissues obtained from a breast cancer patient and a colon cancer patient. Tissue samples were obtained from the Cooperative Human Tissue Network (CHTN). The blots were probed with antibodies to Su(z)12, SirT1, Eed, and actin, which served as a loading control. A long exposure of the same panel containing the normal colon extract probed with Eed antibody is also shown. (Adapted from Kuzmichev et al. 2005.)

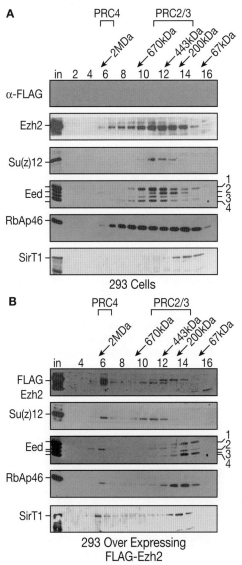

Figure 13. PRC complexes in 293f cells overexpressing Ezh2. Gel filtration of nuclear extracts derived from 293f (*A*) and 293-FLAG-Ezh2 (*B*) cell lines on Sephacryl-400 columns. Fraction numbers and molecular weight standards are indicated on *top*; "in" corresponds to column input. Western blots were probed with antibodies as indicated on the *left*. The amount of protein analyzed in *A* was approximately twice that analyzed in *B*. (Adapted from Kuzmichev et al. 2005.)

discrete stages of carcinogenesis using a well-established mouse model for prostate cancer.

A unique mouse model for prostate cancer was generated based on haplo-insufficiency in two genes that are candidate targets for prostate cancer in humans. Pten is a ubiquitously expressed tumor suppressor gene and Nkx3.1 is a prostate-specific homeobox gene. Mice heterozygotic for these genes ($Pten^{+/-}$; $Nkx3.1^{+/-}$) constitute an ideal system for studies covering the range of prostate cancer progression (Abate-Shen and Shen 2002)—an investigation not currently feasible in the case of humans. These mice develop prostatic intraepithelial neoplasia (PIN) lesions by 6 months of age and adenocarcinoma by 12 months. Following androgen deprivation by castration, these mice develop metastases to the lung and lymph nodes (Fig. 14A) (Abate-Shen et al. 2003). We examined the levels of PRC2, PRC3, and PRC4 components as a function of prostate cancer progression using immunohistochemical analyses. Low levels of Ezh2 and SirT1 were evident in normal prostate. PIN lesions exhibited a moderate elevation and further elevation in Ezh2 and SirT1 levels were evident in cancerous tissue (Fig. 14B). This recapitulated and expanded the previous findings with human specimens. The Ezh2, Su(z)12, and Eed expression levels were further investigated using microarray gene expression profiling. In this case, RNA was isolated after laser-capture microdissection of prostatic lesions through different stages of cancer. These PRC components were all elevated relative to normal tissue (Fig. 14, C and D).

The validity of this mouse model was further buttressed. As discussed above, several PRC-regulated genes had been identified. Some of these genes were down-regulated and others were up-regulated as a consequence of depriving human colon cancer cells of the common PRC component, Su(z)12. We now found that these same PRC-regulated genes exhibited a steady, inverse change in expression profiles with cancer progression in the mouse model, relative to normal cells (Fig. 14D).

The elevated levels of PRC components in human cancer cells and during prostate cancer progression in the mouse model has not been shown to be causal to carcinogenesis. This can, however, be tested in the mouse model. For example, the mouse model can be engineered to express the Eed isoforms excluding the PRC4-specific Eed2 and then surveyed for repercussions on prostate cancer progression.

The Ezh2-associated PRCs have proven to be fascinating in that their constitution is a dynamic one reflecting the expression levels of the disparate isoforms of the common Eed subunit. By exchanging one isoform for another, the histone substrate specificity of Ezh2 is alternately regulated in the cell. Given the disparate histone substrate specificities exhibited by PRC2–4, conditions that promote the formation of a particular PRC were suspected to impact on gene expression. Such an outcome has now been substantiated as cellular transformation favors the constitution of PRC4. Consistent with this, PRC-targeted gene expression is demonstrably altered between normal and cancer cells.

ACKNOWLEDGMENTS

We thank Dr. A. Otte for providing anti-Eed antibodies (M26) and Dr. Lynne Vales for valuable comments on the manuscript. We also thank Xiaohui Sun for immunohistochemistry and Jayshree Rao and Yong Lin for assistance with microarray analyses. We thank Dr. Michael Shen for advice and Dr. John Lis for providing Figure 1. This work was supported by NIH grants to D.R., C.A.-S., and P.F. D.R. is an Investigator of the Howard Hughes Medical Institute.

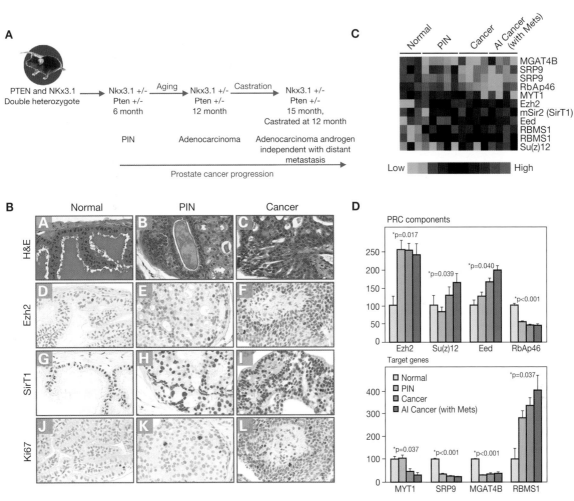

Figure 14. PRC4 components in a mouse model of prostate cancer. (*A*) Mouse model of prostate cancer. Double heterozygote mice ($Nkx3.1^{+/-}$; $Pten^{+/-}$) develop prostate intraepithelial neoplasia (PIN) at 6 months and cancer at 12 months. Tumors from mice castrated at 12 months become androgen independent and develop metastases to the lung and lymph nodes. (*B*) Immunohistochemical detection of SirT1 and Ezh2 in PIN and cancer lesions. Sections from anterior prostates of wild-type and $Nkx3.1^{+/-}$; $Pten^{+/-}$ mice were processed for hematoxylin-eosin (H&E) staining (*A–C*), or immunostained using Ezh2 (*D–F*), SirT1 (*G–I*), or Ki67 (*J–L*) antisera. (*A,D,G,J*) Sections from a wild-type $Nkx3.1^{+/+}$; $Pten^{+/+}$ mouse at 9 months of age. (*B,E,H,K*) Sections from a $Nkx3.1^{+/-}$; $Pten^{+/-}$ mouse at 12 months. (*C,F,I,L*) Sections from a $Nkx3.1^{+/-}$; $Pten^{+/-}$ mouse at 15 months of age. Note that Ezh2 immunostaining is infrequent in normal tissue, but is more common in PIN and carcinoma, while SirT1 is broadly expressed in normal epithelium and stroma, but is up-regulated in epithelial cells in PIN and carcinoma. (*C,D*) Expression profiling of PRC components and target genes from normal to cancer lesions of prostate from compound mutant mice described in *A* and performed using RNA obtained by laser-capture microdissection. (*C*) Tree representation. (*D*) Quantitative representation of microarray results. (Adapted from Kuzmichev et al. 2005.)

REFERENCES

Abate-Shen C. and Shen M.M. 2002. Mouse models of prostate carcinogenesis. *Trends Genet.* **18:** S1.

Abate-Shen C., Banach-Petrosky W.A., Sun X., Economides K.D., Desai N., Gregg J.P., Borowsky A.D., Cardiff R.D., and Shen M.M. 2003. Nkx3.1; Pten mutant mice develop invasive prostate adenocarcinoma and lymph node metastases. *Cancer Res.* **63:** 3886.

Bannister A.J., Schneider R., and Kouzarides T. 2002. Histone methylation: Dynamic or static? *Cell* **109:** 801.

Berger S.L. 2002. Histone modifications in transcriptional regulation. *Curr. Opin. Genet. Dev.* **12:** 142.

Brownell J.E., Zhou J., Ranalli T., Kobayashi R., Edmondson D.G., Roth S.Y., and Allis C.D. 1996. Tetrahymena histone acetyltransferase A: A homolog to yeast Gcn5p linking histone acetylation to gene activation. *Cell* **84:** 843.

Czermin B., Schotta G., Hulsmann B.B., Brehm A., Becker P.B., Reuter G., and Imhof A. 2001. Physical and functional association of SU(VAR)3-9 and HDAC1 in *Drosophila*. *EMBO Rep.* **2:** 915.

DeLange R.J., Hooper J.A., and Smith E.L. 1973. Histone 3. 3. Sequence studies on the cyanogen bromide peptides; complete amino acid sequence of calf thymus histone 3. *J. Biol. Chem.* **248:** 3261.

Denisenko O., Shnyreva M., Suzuki H., and Bomsztyk K. 1998. Point mutations in the WD40 domain of Eed block its interaction with Ezh2. *Mol. Cell. Biol.* **18:** 5634.

Jacobs J.J. and van Lohuizen M. 1999. Cellular memory of transcriptional states by Polycomb-group proteins. *Semin. Cell Dev. Biol.* **10:** 227.

Jenuwein T. and Allis C.D. 2001. Translating the histone code. *Science* **293:** 1074.

Karachentsev D., Sarma K., Reinberg D., and Steward R. 2005. PR-Set7-dependent methylation of histone H4 Lys 20 functions in repression of gene expression and is essential for mitosis. *Genes Dev.* **19:** 431.

Kirmizis A., Bartley S.M., Kuzmichev A., Margueron R., Reinberg D., Green R., and Farnham P.J. 2004. Silencing of human polycomb target genes is associated with methylation of histone H3 Lys 27. *Genes Dev.* **18:** 1592.

Kleer C.G., Cao Q., Varambally S., Shen R., Ota I., Tomlins S.A., Ghosh D., Sewalt R.G., Otte A.P., Hayes D.F., Sabel M.S., Livant D., Weiss S.J., Rubin M.A., and Chinnaiyan A.M. 2003. EZH2 is a marker of aggressive breast cancer and promotes neoplastic transformation of breast epithelial cells. *Proc. Natl. Acad. Sci.* **100:** 11606.

Kohlmaier A., Savarese F., Lachner M., Martens J., Jenuwein T., and Wutz A. 2004. A chromosomal memory triggered by xist regulates histone methylation in x inactivation. *PLoS Biol.* **2:** E171.

Krogan N.J., Kim M., Tong A., Golshani A., Cagney G., Canadien V., Richards D.P., Beattie B.K., Emili A., Boone C., Shilatifard A., Buratowski S., and Greenblatt J. 2003. Methylation of histone H3 by Set2 in *Saccharomyces cerevisiae* is linked to transcriptional elongation by RNA polymerase II. *Mol. Cell. Biol.* **23:** 4207.

Kubicek S. and Jenuwein T. 2004. A crack in histone lysine methylation. *Cell* **119:** 903.

Kuzmichev A. and Reinberg D. 2001. Role of histone deacetylase complexes in the regulation of chromatin metabolism. *Curr. Top. Microbiol. Immunol.* **254:** 35.

Kuzmichev A., Jenuwein T., Tempst P., and Reinberg D. 2004. Different EZH2-containing complexes target methylation of histone H1 or nucleosomal histone H3. *Mol. Cell* **14:** 183.

Kuzmichev A., Nishioka K., Erdjment-Bromage H., Tempst P., and Reinberg D. 2002. Histone methyltransferase activity associated with a human multiprotein complex containing the Enhancer of Zeste protein. *Genes Dev.* **16:** 2893.

Kuzmichev A., Margueron R., Vaquero A., Preissner T., Scher M., Kirmizis A., Ouyang X., Brockdorff N., Abate-Shen C., Farnham P.J., and Reinberg D. 2005. Composition and histone substrates of polycomb regressive group complexes change during cellular differentiation. *Proc. Natl. Acad. Sci.* **102:** 1859.

Laible G., Wolf A., Dorn R., Reuter G., Nislow C., Lebersorger A., Popkin D., Pillus L., and Jenuwein T. 1997. Mammalian homologues of the Polycomb-group gene Enhancer of zeste mediate gene silencing in *Drosophila* heterochromatin and at *S. cerevisiae* telomeres. *EMBO J.* **16:** 3219.

Lund A.H. and van Lohuizen M. 2004. Polycomb complexes and silencing mechanisms. *Curr. Opin. Cell Biol.* **16:** 239.

Ng H.H., Xu R.M., Zhang Y., and Struhl K. 2002. Ubiquitination of histone H2B by Rad6 is required for efficient Dot1-mediated methylation of histone H3 lysine 79. *J. Biol. Chem.* **277:** 34655.

Ng J., Hart C.M., Morgan K., and Simon J.A. 2000. A *Drosophila* ESC-E(Z) protein complex is distinct from other polycomb group complexes and contains covalently modified ESC. *Mol. Cell. Biol.* **20:** 3069.

Nishioka K., Rice J.C., Sarma K., Erdjment-Bromage H., Werner J., Wang Y., Chuikov S., Valenzuela P., Tempst P., Steward R., Lis J.T., Allis C.D., and Reinberg D. 2002. PR-Set7 is a nucleosome-specific methyltransferase that modifies lysine 20 of histone H4 and is associated with silent chromatin. *Mol. Cell* **9:** 1201.

Pirrotta V. and Rastelli L. 1994. White gene expression, repressive chromatin domains, and homeotic gene regulation in *Drosophila*. *Bioessays* **16:** 549.

Rea S., Eisenhaber F., O'Carroll D., Strahl B.D., Sun Z.W., Schmid M., Opravil S., Mechtler K., Ponting C.P., Allis C.D., and Jenuwein T. 2000. Regulation of chromatin structure by site-specific histone H3 methyltransferases. *Nature* **406:** 593.

Rice J.C., Nishioka K., Sarma K., Steward R., Reinberg D., and Allis C.D. 2002. Mitotic-specific methylation of histone H4 Lys 20 follows increased PR-Set7 expression and its localization to mitotic chromosomes. *Genes Dev.* **16:** 2225.

Rosenberg M.I. and Parkhurst S.M. 2002. *Drosophila* Sir2 is required for heterochromatic silencing and by euchromatic Hairy/E(Spl) bHLH repressors in segmentation and sex determination. *Cell* **109:** 447.

Schotta G., Lachner M., Sarma K., Ebert A., Sengupta R., Reuter G., Reinberg D., and Jenuwein T. 2004. A silencing pathway to induce H3-K9 and H4-K20 trimethylation at constitutive heterochromatin. *Genes Dev.* **18:** 1251.

Shankaranarayana G.D., Motamedi M.R., Moazed D., and Grewal S.I. 2003. Sir2 regulates histone H3 lysine 9 methylation and heterochromatin assembly in fission yeast. *Curr. Biol.* **13:** 1240.

Shao Z., Raible F., Mollaaghababa R., Guyon J.R., Wu C.T., Bender W., and Kingston R.E. 1999. Stabilization of chromatin structure by PRC1, a Polycomb complex. *Cell* **98:** 37.

Silva J., Mak W., Zvetkova I., Appanah R., Nesterova T.B., Webster Z., Peters A.H., Jenuwein T., Otte A.P., and Brockdorff N. 2003. Establishment of histone h3 methylation on the inactive X chromosome requires transient recruitment of Eed-Enx1 polycomb group complexes. *Dev. Cell* **4:** 481.

Sims R.J., III, Nishioka K., and Reinberg D. 2003. Histone lysine methylation: A signature for chromatin function. *Trends Genet.* **19:** 629.

Sterner D.E. and Berger S.L. 2000. Acetylation of histones and transcription-related factors. *Microbiol. Mol. Biol. Rev.* **64:** 435.

Struhl K. 1998. Histone acetylation and transcriptional regulatory mechanisms. *Genes Dev.* **12:** 599.

Sun Z.W. and Allis C.D. 2002. Ubiquitination of histone H2B regulates H3 methylation and gene silencing in yeast. *Nature* **418:** 104.

Taunton J., Hassig C.A., and Schreiber S.L. 1996. A mammalian histone deacetylase related to the yeast transcriptional regulator Rpd3p. *Science* **272:** 408.

Vaquero A., Scher M., Lee D., Erdjment-Bromage H., Tempst P., and Reinberg D. 2004. Human SirT1 interacts with histone H1 and promotes formation of facultative heterochromatin. *Mol. Cell* **16:** 93.

Varambally S., Dhanasekaran S.M., Zhou M., Barrette T.R., Kumar-Sinha C., Sanda M.G., Ghosh D., Pienta K.J., Sewalt R.G., Otte A.P., Rubin M.A., and Chinnaiyan A.M. 2002. The polycomb group protein EZH2 is involved in progression of prostate cancer. *Nature* **419:** 624.

Wood A., Krogan N.J., Dover J., Schneider J., Heidt J., Boateng M.A., Dean K., Golshani A., Zhang Y., Greenblatt J.F., Johnston M., and Shilatifard A. 2003. Bre1, an E3 ubiquitin ligase required for recruitment and substrate selection of Rad6 at a promoter. *Mol. Cell* **11:** 267.

Zhang Y. and Reinberg D. 2001. Transcription regulation by histone methylation: Interplay between different covalent modifications of the core histone tails. *Genes Dev.* **15:** 2343.

Noncovalent Modification of Chromatin: Different Remodeled Products with Different ATPase Domains

H.-Y. Fan, G.J. Narlikar,* and R.E. Kingston

Department of Molecular Biology, Massachusetts General Hospital, Boston, Massachusetts 02114;
Department of Genetics, Harvard Medical School, Boston, Massachusetts 02115

The ability to maintain the same master regulatory gene in an "on" state in one cell lineage and in an "off" state in another cell lineage is fundamental to proper development. There is general agreement that modification of chromatin structure can contribute to this form of epigenetic regulation. Thus, a locus that must be repressed in a heritable fashion might have a specialized chromatin structure that is repressive to transcription, and the same locus in another cell lineage where it is heritably active might have a permissive chromatin structure.

THE DYNAMIC RANGE OF NUCLEOSOME STRUCTURE

To understand the roles that chromatin structure can play in epigenetic regulation, it is essential to understand the dynamic range of chromatin structure. The basic building block of chromatin, the nucleosome, is known to be dynamic: Covalent modification and noncovalent events that either move or lock the nucleosome in place can alter the stability of the nucleosome. An explosion of data generated using specific antisera and the technique of chromatin immunoprecipitation (ChIP) has shown that large regions of chromatin can be covalently modified in a manner that correlates with regulatory state and that therefore is likely to contribute to epigenetic regulatory events (Turner 2002; Fischle et al. 2003; Weissmann and Lyko 2003).

The role that noncovalent modification of the genome plays is not as well characterized. Noncovalent modification of chromatin has been strongly implicated in epigenetic events by genetic studies. Examples of this include the Polycomb and trithorax systems (Simon and Tamkun 2002). Polycomb-Group (PcG) and trithorax-Group (trxG) genes are required for maintaining expression patterns of master regulatory genes (Ringrose and Paro 2004). The expression pattern of these master regulators is established by one set of mechanisms early in embryogenesis, but must be maintained in spatially restricted patterns throughout the lifetime of any organism. Most PcG and trxG gene products that have been characterized are able to modify chromatin structure. Several of these products perform covalent modification events such as methylation or acetylation (Milne et al. 2002; Fischle et al. 2003; Schotta et al. 2004). Others modify chromatin structure without covalent modification, by creating chromatin structures refractory to transcription (e.g., PRC1) or by using the energy of ATP hydrolysis to create access to chromatin (e.g., SWI/SNF family complexes) (McCall and Bender 1996; Fitzgerald and Bender 2001; King et al. 2002; Simon and Tamkun 2002; Francis et al. 2004).

A key question concerns the dynamic range of these noncovalent modifications to the nucleosome: Do these modifications involve changes in nucleosome position, changes in nucleosome constitution, and/or changes in nucleosome conformation? It has long been appreciated that chromatin structure can be altered noncovalently. Theoretical considerations led to the hypothesis 30 years ago that transcription through the nucleosome might require or create alterations in position. Several subsequent studies have shown that transcribed genes have changes in histone content and nuclease sensitivity in vivo (Macleod and Bird 1982; Sweet et al. 1982; Weisbrod 1982). In vitro systems have shown that transcription of nucleosomal templates by RNA polymerase II can result in removal of a histone H2A/H2B dimer, providing one defined alteration in nucleosome constitution that can impact regulation (Kireeva et al. 2002).

The discovery of ATP-dependent nucleosome remodeling complexes 10 years ago demonstrated that there are complexes whose primary function is to alter nucleosome structure noncovalently (Cote et al. 1994; Imbalzano et al. 1994; Kwon et al. 1994). This class of complexes contains numerous families of complexes with distinct biological roles, prominently the SWI/SNF family and the ISWI family (Table 1). These complexes are abundant; for example, it is estimated that mammalian nuclei contain more than 100,000 copies of ISWI-family complexes and 25,000 copies of SWI/SNF-family complexes. Both of these families of complexes can create access to DNA-binding factors on nucleosomal templates (Cote et al. 1994; Imbalzano et al. 1994; Kwon et al. 1994; Tsukiyama and Wu 1995). One way that they can do this is by changing the position of nucleosomes, thereby moving a specific sequence in DNA from a position where it is inaccessible because of histone contacts to a position where it is accessible because it is in a region of free linker DNA between nucleosomes. Complexes in the ISWI family are known to be able to "slide" nucleosomes along a template, which allows them to alter spacing and also allows them

---

*Present address: Department of Biochemistry and Biophysics, University of California at San Francisco, San Francisco, California 94143.

Table 1. Functions of ATP-dependent Chromatin Remodeling Families

Family	Motor proteins	Functions
SWI/SNF	hBRG1, hBRM, dBRM, ySth1p, ySwi2p/Snf2p	Transcription DNA replication Recombination Higher-order chromatin structure?
ISWI	hSNF2h, hSNF21, dISWI, xISWI, yIsw1p, yIsw2p	Chromatin assembly Transcription Higher-order chromatin structure DNA replication Nucleotide excision repair
Mi-2	Mi2α/CHD3, Mi2β/CHD4, Chd1p, Hrp1, Hrp3	Transcription histone deacetylation
INO80	yIno80p, hINO80	Transcription DNA repair
SWR1	ySwr1p, hSCRAP, human p400, dDomino	Transcription Histone exchange DNA repair
CSB[a]	CSB/ERCC6, yRad26p	Transcription-coupled DNA repair
Rad54[a]	hRad54, hRad54B, dORK, yRad54p, hATRX, ARIP4, DRD1	Recombination Transcription DNA methylation
DDM1	DDM1, LSH1	DNA methylation

[a]Some members of these families demonstrate ATP-dependent chromatin remodeling activities in vitro; nonetheless, it is unclear if these activities represent their authentic biological functions.

to function in regulation by closing or opening regions of chromatin (Narlikar et al. 2002). In addition to regulating access and altering spacing, other ATP-dependent remodeling complexes (not the focus of this paper) are able to promote exchange of histone H2A/H2B dimers (Bruno et al. 2003; Mizuguchi et al. 2004).

While it is clear that noncovalent modification can involve changes in nucleosome position and changes in nucleosome constitution, it is not clear what structural alterations occur to the nucleosome upon ATP-dependent remodeling and whether these alterations lead to stable conformational changes in the nucleosome. In theory, there are several ways that the energy of ATP hydrolysis might alter histone–DNA contacts to create access to sites (see Fig. 7). These complexes might twist the DNA and that twisting might propagate through the nucleosome. DNA could be peeled off the edge of the nucleosome by movement of the complex into the edge of the nucleosome, or a writhe could be created that would form a bulge of DNA on the nucleosome that would propagate (see Fig. 7A). These events could transiently alter structure and lead to changes in position of a canonical nucleosome, they could transiently alter structure to create quasistable structures that are identical to canonical nucleosomes except for the inclusion of looped regions of DNA, or they could more fundamentally alter nucleosome conformation.

Based on the above considerations, two classes of hypotheses have been proposed for how noncovalent modification of nucleosome structure might occur. At one extreme, it is possible that noncovalent modification always creates remodeled products that are identical or highly related to the canonical nucleosome structure as defined crystallographically. At the other extreme, it is possible that noncovalent modification can create altered conformations of the nucleosome in which either histone–histone or histone–DNA interactions are demonstrably altered in nucleosomal structures that differ significantly from the canonical structure. Defining the dynamic range of noncovalently modified nucleosome structures is key to understanding the potential roles for the nucleosome in epigenetic regulation.

COMPARISON OF COMPLEXES IN THE SWI/SNF AND ISWI FAMILIES

There has been ongoing discussion concerning the ISWI and SWI/SNF families of ATP-dependent remodeling complexes around whether these complexes function by similar or distinct mechanisms. While it is too early to resolve this issue, the purpose here is to summarize recent data, with an emphasis on experiments done in our laboratory, concerning the characteristic behavior of these complexes in remodeling nucleosomes. We show that the ATPase domain itself plays a primary role in determining the outcome of the remodeling reaction (see Fig. 5). We use this observation and previous observations to argue in favor of the existence of nucleosomal structures that differ significantly from the canonical nucleosomal structure (Fan et al. 2003). Thus, the dynamic range of nucleosome architecture might be greater than is widely appreciated, which, if true, would significantly expand the regulatory capabilities of chromatin structural changes.

The premise of this article is that the nucleosome should not be assumed to have an unwavering stable conformation. There are a significant number of experiments whose results are difficult to reconcile with the hypothesis that dynamic changes to nucleosome structure consist solely of changes in position and/or changes in histone constitution. We begin with a summary of published experimental results that demonstrate that there are stable remodeled nucleosomal structures, and then we present data from our own laboratory that bear on the nature and genesis of these structures. The best studied example of a family of complexes that might create altered nucleosomal structures is the SWI/SNF family.

SWI/SNF-family remodeling complexes were originally identified in yeast, first via genetic studies using yeast strains that are *s*ucrose *n*on-*f*ermenters or that are defective in mating type *swi*tching (Carlson et al. 1981; Stern et al. 1984). Subsequent biochemical studies demonstrated that certain of the proteins encoded by the SWI and SNF genes form a large complex (Cote et al. 1994; Wang et al. 1996). There are several forms of this complex, most organisms having at leaast two versions; these complexes have a mass greater than 1 MD and are composed of at least 11–15 subunits (Fig. 1). The central subunit of these complexes, homologous to SWI2/SNF2,

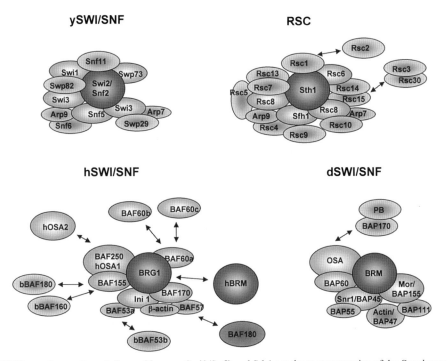

Figure 1. SWI/SNF complexes of yeast, fly, and human. Swi2/Snf2 and Sth1 are the motor proteins of the *Saccharomyces cerevisiae* SWI/SNF complexes. Human SWI/SNF complexes have either BRG1 or hBRM as the central ATPase subunit. *Drosophila* has only one SWI/SNF complex, and brama is the motor.

is the ATPase subunit, which has remodeling activity as an isolated protein (see Fig. 1). These complexes are involved in a wide variety of regulatory events on numerous genes and can be responsible for either activation or repression.

EVIDENCE FOR STABLY REMODELED STRUCTURES FORMED BY SWI/SNF

Two papers published in 1998 demonstrated that SWI/SNF-family complexes could create a stably remodeled product when a mononucleosome was used as substrate. This product had altered mobility on a native gel and could revert to a standard nucleosome with either incubation under specific conditions or through the ATP-dependent action of SWI/SNF (Lorch et al. 1998; Schnitzler et al. 1998). Both human and yeast members of the family were shown to create this structure; in subsequent work, it has been proposed that the structure consists of DNA strands that bridge two nucleosomes (Lorch et al. 1999). Similar structures have not been observed with other ATP-dependent remodeling complexes, suggesting that the ability to form this structure might represent a special aspect of SWI/SNF function.

A second unusual property of the SWI/SNF family is the ability to create significant topological changes in closed circular nucleosomal templates. These changes are also stable, reverting to standard topology with a half-life measured in hours (Guyon et al. 1999, 2001). The characteristics of these topologically altered products imply that there is a significant energy barrier between the stably remodeled state and a standard state. The effects of SWI/SNF-family complexes on topology and the ability to form stable remodeled products from mononucleosomes might be related to each other mechanistically. Both properties could be explained by an altered nucleosomal conformation induced by SWI/SNF that has distinct topology and that is prone to aggregation. Alternatively, both properties could be explained by modeling the standard nucleosome in a manner that has adjacent nucleosomes sharing DNA strands (Lorch et al. 1998; Schnitzler et al. 1998). These experiments indicated that SWI/SNF complexes can create distinct products from other remodeling complexes that might involve altered nucleosomal structures.

DIFFERENT REMODELED PRODUCTS PRODUCED BY BRG1 AND SNF2h

To follow up on these studies, we have performed a series of experiments that compare the remodeled products that are formed by a SWI/SNF-family complex and an ISWI-family complex. The goal of these ongoing studies is to understand the capabilities of these remodeling complexes that might relate to their biological activity and to determine characteristics of each complex that will facilitate a detailed mechanistic and structural understanding of the different remodeling reactions. We focused on the major remodeling ATPases in humans in each class: BRG1 (SWI/SNF family) and SNF2h (ISWI family). Each individual remodeling protein can perform a reaction that has similar characteristics to the reaction performed by the intact complexes that form around these proteins (Phelan et al. 1999; Aalfs et al. 2001).

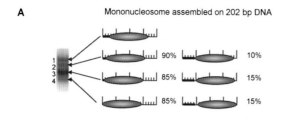

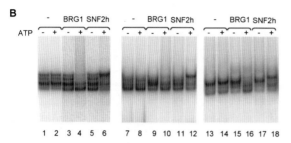

Figure 2. BRG1 and SNF2h generate different remodeled products. (*A*) 202-bp mononucleosomes were resolved as four bands on a 5% native polyacrylamide gel. The histone octamer positions in each band were mapped according to standard procedure and are indicated as *ovals* (Hamiche et al. 1999; Langst and Becker 2001). The *solid box* represents a 40-bp GT nucleosome phasing sequence. Distance between ticks is 10 bp. (*B*) Three different glycerol gradient nucleosome fractions were used in independent remodeling reactions.

Native gel electrophoresis can be used to analyze changes in nucleosome position. Mononucleosomes that have been formed on DNA that is longer than 146 base pairs (bp) (we have used a 202-bp DNA fragment in these studies) will run with different mobility during native gel electrophoresis depending upon whether the histone octamer is near the center of the DNA or near the end of the DNA fragment (Fig. 2A). Incubation of mononucleosomes with remodeling proteins results in ATP-dependent changes in mobility (Fig. 2B, lanes 4, 6, 10, 12, 16, and 18). SNF2h generates primarily slowly migrating species, while BRG1 generates more rapidly migrating species. Thus, these remodelers create qualitatively different products.

To understand how these products differ, the BRG1- and SNF2h-remodeled products were digested with MNase. The size of the resultant DNA fragments was determined by denaturing gel electrophoresis (Fig. 3A, lanes 3, 4, 6, and 7). Once again, there was a distinct difference between the SNF2h and BRG1 products: The SNF2h product ran with a discrete size appropriate to a 146-bp mononucleosome while the BRG1 product contained fragments across a spectrum of size. The location of the fragments that were protected from MNase cleavage was determined by excising the DNA fragment produced by MNase digestion and cleaving it with restriction enzymes. The primary SNF2h product mapped to a central position (Fig. 3B), while the BRG1 product had a large number of MNase fragments that mapped to a variety of positions (Fig. 3C).

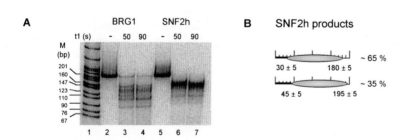

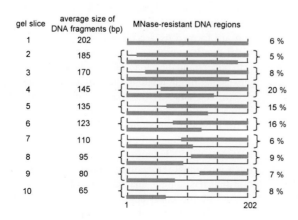

Figure 3. Mapping of the BRG1- and SNF2h-remodeled products. (*A*) BRG1- and SNF2h-remodeled products were treated with MNase, deproteinized, and resolved on an 8% polyacrylamide gel. (*B*) Mapping the major SNF2h products. (*C*) Mapping BRG1-remodeled mononucleosomes. *Bars* represent DNA regions protected by the histone octamer from MNase digestion. DNA fragments ranging from the average sizes (as shown) ± 10 bp were mapped. DNA fragments with an average size of 65, 95, 170, or 185 bp are more spread out and thus are less visible in *A*.

The surprising finding from these studies was that the major BRG1 product migrated as a discrete band when analyzed by native gel electrophoresis, but displayed a wide spectrum of differently sized and positioned MNase fragments. Why this happened is not clear. One possible explanation is that each BRG1 product had a similar amount of DNA that was associated with the histone octamer, but that there were MNase-sensitive regions where the DNA was looped away from the mononucleosome. By this hypothesis, the spectrum of MNase fragments formed would result from loops in different positions on the mononucleosome.

BRG1 CREATES MORE ACCESSIBLE DNA SITES THAN SNF2h

These studies demonstrate that there are measurable differences between the products that are created by BRG1 and SNF2h. We were interested in determining whether these differences also reflected differences in what is believed to be a key biological function of these remodeling proteins—the ability to create access of DNA-binding proteins to nucleosomal DNA. To study this problem, we constructed a series of mononucleosomal templates that were engineered to have restriction sites at a variety of different positions on the nucleosome. We then made use of protocols that have been shown to measure the rate of opening of a specific site by the rate of restriction enzyme digestion (Polach and Widom 1995; Logie and Peterson 1997; Narlikar et al. 2001). It has been shown previously that, under the conditions used here, the restriction enzyme is at high enough concentration to "report" whether a site has opened; the restriction enzyme will cleave every nucleosome that has been remodeled to have an open site and does not participate in the remodeling reaction.

Consistent with previous data on *Drosophila* ISWI-family complexes, we found that SNF2h opened sites near the edge of a nucleosome significantly faster than it opened sites near the center of the nucleosome (Fig. 4). This is consistent with a requirement for SNF2h to "slide" nucleosomes. If sliding is required, then it will prove dif-

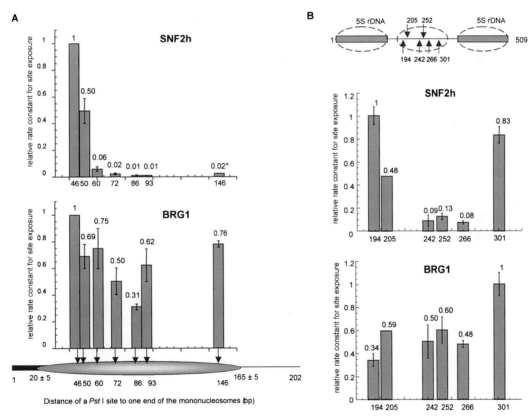

Figure 4. SNF2h and BRG1 create different profiles of accessible DNA sites on mononucleosomes and trinucleosomes. (*A*) Mononucleosome remodeling profiles of SNF2h and BRG1 monitored by continuous restriction enzyme accessibility assays. The rate constants for cutting each *Pst* I position were normalized to that of position 46 for BRG1 (0.2–0.3 min^{-1}) and SNF2h (0.2–0.4 min^{-1}): The normalized values are shown above the bars. At positions 46 and 50, SNF2h increased the rate of *Pst* I exposure by at least 30-fold relative to reactions without ATP. BRG1 increased the rate of *Pst* I exposure at all positions by at least 30-fold relative to reactions without ATP. (*, No increase in DNA exposure relative to the reaction with no ATP.) (*B*) A schematic illustration of the 509-bp DNA templates used to assemble trinucleosomes. The *Pst* I sites in the different templates are indicated by *arrows*. The rate constants for opening each *Pst* I position were measured relative to that for opening up position 205 for SNF2h (0.1–0.3 min^{-1}) and BRG1 (0.1–0.2 min^{-1}); the relative values were then normalized with respect to the highest remodeling rate constants (positions 194, 301, and 252, for SNF2h and BRG1, respectively). Normalized values are shown above the bars. SNF2h opened up positions 194, 205, and 301 at least 30-fold faster than reactions without ATP. BRG1 opened up the different positions at least tenfold faster than reactions without ATP.

ficult to open sites at the center of a fragment that has insufficient room for a nucleosome to slide. In contrast, BRG1 was able to open sites throughout the fragment at roughly similar rates. This could be explained by an ability of BRG1 to promote a sliding reaction where the histone octamer slides off of the end of the DNA fragment. Alternatively, this might result from the trapping of loops of DNA on the surface of the histone octamer.

These marked differences in the ability to open sites throughout a nucleosome provided a straightforward way to compare mutant remodeling proteins to see whether they had characteristics of SWI/SNF-family proteins or ISWI-family proteins. In the experiments described below, we focused on access to sites at positions 50 and 93, as ISWI-family proteins show a significantly lower ability to open site 93 than site 50, while SWI/SNF-family proteins open both sites efficiently.

THE IDENTITY OF THE ATPase DOMAIN DEFINES REMODELING CAPABILITY

The experiments described above demonstrate that the SNF2h and BRG1 proteins create products with different characteristics and have different abilities to create access to nucleosomal DNA sites. To understand these differences, we designed chimeric proteins that contained mixtures of BRG1 and SNF2h in order to map which region determined the characteristic activity of each remodeling protein (Fig. 5). The first set of constructs that were examined swapped ATPase domains. (The boundaries of the ATPase domain in these experiments were defined by homology between BRG1, SNF2h, and their orthologs; deletion analysis; and published genetic studies [Khavari et al. 1993; Elfring et al. 1994].) Surprisingly, we found that the ATPase domain itself determined both the nature of the remodeled products that were formed and the ability of the remodeling protein to create access to a series of sites.

When restriction enzyme access was used to characterize these chimeric proteins, the construct containing the BRG1 ATPase domains with SNF2h flanking regions (called "S-B-S"; Fig. 5A) showed significant ability to create access to site 93 (Fig. 5B). In contrast, the construct containing the SNF2h ATPase domain surrounded by BRG1 flanking regions ("B-S-B"; Fig. 5A) was not able to create efficient access to site 93 (Fig. 5B). Both of these chimeric remodeling proteins showed activity at

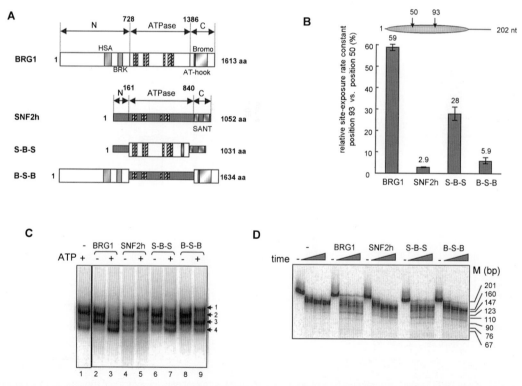

Figure 5. Remodeling activities of BRG1/SNF2h chimeric proteins. (*A*) Schematic representation of the proteins used in this study. BRG1 and SNF2h are divided into three regions: a homologous central ATPase domain (ATPase) and nonhomologous amino-terminal (N) and carboxy-terminal (C) regions. BRG1 and SNF2h contain seven conserved helicase motifs (*striped boxes*). BRG1 also contains HSA, BRK, AT-hook, and bromodomains, while SNF2h contains two SANT domains. Exchanging the central ATPase domains (BRG1 residues 728–1386 and SNF2h residues 161–840) forms the chimeric proteins B-S-B and S-B-S. The detailed breakpoints for construction of chimeras are B(1-727L)-S(161A-840Q)-B(1387G-end) and S(1-160K)-B(728Q-1386A)-S(841G-end). (*B*) Restriction enzyme accessibility assays using 202-bp mononucleosomes containing a *Pst* I site at either position 50 or 93. Site-exposure rate constants were determined. Results are expressed as the ratio of the site-exposure rate constants of positions 93 vs. 50. Relative rate constants are averages of at least three independent experiments. (*C*) Nucleosome mobility assays. Remodeled products as well as unremodeled nucleosomes were resolved on a 5% native polyacrylamide gel. (*D*) MNase sensitivity assays. Mononucleosomes remodeled by BRG1, SNF2h, S-B-S, and B-S-B were treated with 0.03 units of MNase for 0, 1, 2, and 3 min, deproteinized, and resolved on an 8% PAGE.

opening site 50 that was similar to intact BRG1 and intact SNF2h, demonstrating that swapping the ATPase domains does not generally impair remodeling activity.

These rate experiments suggest that the characteristic remodeling behavior of BRG1 and SNF2h is largely determined by the ATPase domain. To buttress these rate studies, we also performed qualitative measurements of remodeling function. We used native gel electrophoresis to show that B-S-B protein created remodeled products with similar migration to the products of the SNF2h reaction and S-B-S protein created products with similar migration to those of the BRG1 reaction (Fig. 5C). We characterized the products of these remodeling reactions using MNase digestion, and found that the B-S-B protein created a primarily a single protected band centered at 150, while the S-B-S protein created a spectrum of bands of varying lengths. We conclude from these studies that the characteristic differences in outcome of the remodeling reactions that are catalyzed by BRG1 and SNF2h are determined by the BRG1 and SNF2h ATPase domains.

To determine whether swapping the ATPase domain could alter function of a remodeling complex, we assembled a minimal SWI/SNF remodeling complex using both BRG1 and B-S-B. In addition to the ATPase subunit, these minimal complexes contain the human homologs of yeast SWI3 (BAF170 and BAF155) and SNF5 (INI1) proteins. Both BRG1 and the B-S-B chimeric protein were able to form complexes with these subunits with similar stoichiometries (Fig. 6A). Both complexes were active for remodeling site 50 of a mononucleosome. As was seen with the isolated ATPase subunits, the complex that contained BRG1 was able to efficiently open site 93, while the complex that contained B-S-B was not (Fig. 6B). Thus, swapping the SNF2h ATPase domain into a minimal SWI/SNF complex created remodeling activity that mirrored that of SNF2h. As above, we used native gel electrophoresis and MNase cleavage analysis to analyze the products of the remodeling reactions. The minimal complex containing BRG1 created products that ran on a native gel with different characteristics than the products created by the minimal complex containing B-S-B (Fig. 6C). Similarly, swapping the ATPase domain into the minimal complex caused a change in the MNase pattern of the remodeled products (Fig. 6D; compare "BRG1 minimal complex" to "B-S-B minimal complex").

Similar results were obtained when we isolated intact SWI/SNF complexes from SW13 cells that had been transfected with expression constructs for BRG1 and B-

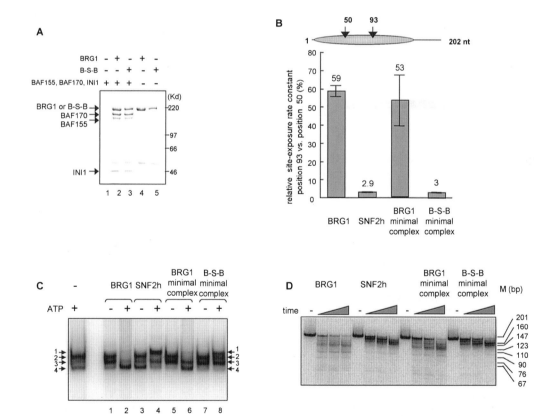

Figure 6. In vitro characterization of the reconstituted BRG1 and B-S-B minimal complex. (*A*) SF9 cell nuclear extracts containing Flag-tagged BRG1 or B-S-B were mixed with an SF9 cell nuclear extract containing BAF170, BAF155, and INI1, and anti-Flag M2 beads were used to purify BRG1 or B-S-B complexes. The purified complexes were eluted with the Flag peptide, resolved by 8% SDS-PAGE. (*B*) Comparison of the relative site-exposure rate constants at position 93 vs. position 50 of mononucleosomes remodeled by the BRG1 complex, the B-S-B complex, BRG1, or SNF2h. Relative rate constants are averages of at least three independent experiments. (*C*) Gel mobility assays to compare mononucleosomes remodeled by the BRG1 complex, the B-S-B complex, BRG1, and SNF2h. (*D*) The remodeled products from *C* were treated with 0.03 units of MNase for 0, 0.75, 1.5, and 3 min, deproteinized, and resolved on an 8% PAGE.

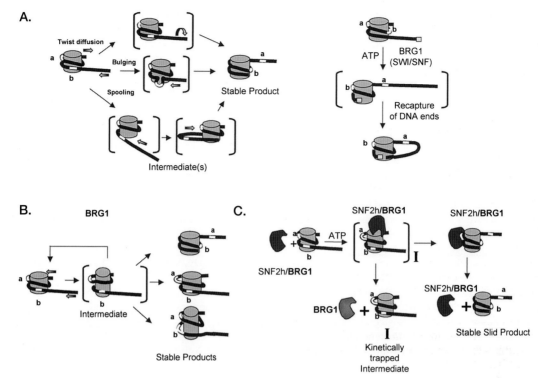

Figure 7. Models accounting for different products generated by SNF2h and BRG1. (*A*) Both SNF2h and BRG1 slide histone octamers. SNF2h may slide histone octamers via twist diffusion, bulging, or spooling to expose a DNA site such as "**a**." BRG1 and SWI/SNF may expose site "**a**" by sliding the histone octamer off the DNA ends. The exposed DNA end subsequently rebinds the histone octamer to form a stable loop (Kassabov et al. 2003). (*B*) SNF2h and BRG1 use different mechanisms to remodel nucleosomes. While SNF2h slides histone octamer to remodel nucleosome (*A*), BRG1 may create an altered nucleosome conformation; in this model, repositioning of a histone octamer is not a necessary outcome for exposure of site "**a**" or "**b**." (*C*) BRG1 (and hSWI/SNF) and SNF2h remodel nucleosomes by the same mechanism and the two reactions proceed though similar intermediates such as "**I**." In this model, BRG1 and hSWI/SNF release "**I**" more often than SNF2h does and thus create a kinetically trapped intermediate (**I′**) with site "**a**" exposed within the histone bounds. The structures depicted for the intermediate and final products in *A* and *B* are hypothetical and could involve changes in the conformation of DNA, histones, or both.

S-B (data not shown). Therefore, we conclude that the ATPase domain itself plays a critical role in determining the outcome of the remodeling reaction; in fact, simply swapping this domain changes the measured characteristics of the remodeling reaction even in the context of a full remodeling complex.

SPECULATIONS ON THE ROLE OF THE ATPase DOMAIN IN DEFINING THE REMODELING REACTION

These studies demonstrate that there is a direct link between the nature of the ATPase domain and the outcome of the remodeling reaction. The surprising finding that the ATPase domain is the primary determinant for outcome, even in the context of intact complexes, raises the possibility that the differences in function between complexes are determined by differences in the manner in which ATP hydrolysis is coupled to remodeling. This finding is emphasized by the observation that the isolated BRG1 ATPase domain is functional for remodeling (data not shown). Thus, the ATPase domain serves as the central component of the engine that drives remodeling.

Do the engines of BRG1 and SNF2h differ solely in efficiency, or do they differ in the mechanism by which they harness the energy of ATP hydrolysis? One possible explanation for the data above is that both BRG1 and SNF2h perform precisely the same function (e.g., sliding the nucleosome) (Fig. 7A), and that BRG1 opens up a greater spectrum of sites because it is more potent at promoting the sliding reaction. A greater potency might allow BRG1 to slide the nucleosome off of the ends of the DNA or into adjacent nucleosomes. This ability could lead to the increased restriction enzyme access that is observed, and folding back of the DNA onto the nucleosome could result in the formation of loops of DNA (Fig. 7A). A second possibility is that BRG1 uses a distinct mechanism that creates access in the middle of the nucleosome; for example, BRG1 might use energy to push DNA toward the nucleosome dyad from both the entry and exit points, thereby inducing a strain that creates an altered conformation (Fig. 7B).

There is no data that allows one to rule out either of the above hypotheses for the differences in BRG1 and SNF2h function. We argue, however, that the characteristics of the SWI/SNF remodeling reaction might be most simply explained by the hypothesis that BRG1 differs fundamentally from SNF2h in mechanism.

To frame these arguments, it is necessary to define the types of mechanisms that might be involved (Fig. 7). One prominent hypothesis for sliding a nucleosome by ISWI-family complexes involves the creation of a segment of DNA that dissociates from the nucleosome, creating a "bulge." This bulge might then propagate through the entirety of the nucleosome to cause histone displacement in the size of the bulge (thus sliding the nucleosome by an amount determined by the amount of DNA in the bulge) (Fig. 7A). It has been proposed that all ATP-dependent remodeling enzymes share this common mechanism, and that the differences in outcome of the remodeling reaction such as those highlighted above are caused either by sliding the octamer off of the edge of the histone, followed by strand recapture to create loops of DNA with altered topology, or by having the bulge arrest in the center of the nucleosome to create loops of DNA that are accessible to restriction enzyme access. The alternative hypothesis considered here is that the energy of ATP hydrolysis is used differently for different classes of complexes, in that some such as ISWI might induce a bulge that propagates, and others such as SWI/SNF-family complexes might either push DNA into the structure from both the entry and exit points or otherwise induce a strain on the canonical structure that favors the formation of quasistable nucleosome structures that have exposed sites near the center of the nucleosome (Fig. 7B).

Two considerations argue against a model in which the mechanism that induces sliding is also used to create access to sites that are centrally located. The first concerns energetics. If SWI/SNF were doing something more energetically unfavorable like moving the octamer off the end of the DNA, we might expect it to use more ATP per remodeling event than ISWI complexes. Instead it uses similar or less ATP suggesting SWI/SNF action is not a simple extension of ISWI action (Narlikar et al. 2001; Fyodorov and Kadonaga 2002). In addition, this simplest of notions of sliding as a mechanism of creating access would necessitate an energy gradient, where, for example, sites near the entry/exit point required less energy to open than sites near the dyad. Instead, with SWI/SNF-family enzymes, similar amounts of ATP hydrolysis are needed to open sites near or at the nucleosome dyad as are needed at sites away from the dyad.

A variant on this hypothesis is that SWI/SNF-family enzymes create access by arresting a propagating bulge as it traverses the nucleosome (Fig. 7C). By this model, both ISWI complexes and SWI/SNF complexes function by creating propagating bulges, but these bulges arrest during SWI/SNF remodeling. This would require that the components of the complex that maintain contact with the nucleosome to either cause dissociation from the nucleosomes partway through the remodeling reaction, or actively arrest the bulge. The finding that ATPase domains determine the outcome of the remodeling reaction, combined with this and previous studies that show that many of the domains that interact with the nucleosome are outside the ATPase domain, argues against this model.

Thus, we argue that current data do not support the hypothesis that ISWI and SWI/SNF remodeling complexes function by fundamentally similar mechanisms. While these data by no means disprove these hypotheses, they favor the consideration of the alternative hypothesis that fundamentally distinct mechanisms are used by the SWI/SNF family of remodeling complexes to create access to nucleosomal sites. Previous data has shown that SWI/SNF complexes create stable remodeled structures with long half-lives (Guyon et al. 1999, 2001). In addition, SWI/SNF complexes create dramatic changes in topology of nucleosomal arrays (Guyon et al. 1999; Gavin et al. 2001). The nature of these stable remodeled structures and these topological shifts is not known. The finding that the ATPase domain is central to function of SWI/SNF complexes implies that the remodeling parameters are tightly connected to ATP hydrolysis. These considerations are all consistent with the possibility that a specialized reaction is performed by SWI/SNF in which ATP hydrolysis creates stably altered nucleosomal structures. To test this hypothesis, it will be essential to identify the nature of these putative structures and the relationship of these structures to that of the canonical nucleosome.

SUMMARY AND PERSPECTIVES ON THE ROLE OF REMODELING IN EPIGENETIC REGULATION

It is generally agreed that different ATP-dependent remodeling complexes create different products. We have argued here that these differences might reflect a diverse range of dynamic changes in nucleosome structure. Identifying the nature of the products of these different remodeling reactions is a matter that is central to determining the range of possible chromatin structures that might contribute to epigenetic regulation. If the nucleosome is more dynamic than currently demonstrated, and if complexes have evolved to regulate that dynamic state, then the pallet of changes that can occur to chromatin structure to create stable epigenetic states is increased. Any changes in the nucleosomal structure would also be anticipated to impact the range of potential higher-order structures. An important and interesting frontier is the elucidation of the full spectrum of structures that can be formed in chromatin.

REFERENCES

Aalfs J.D., Narlikar G.J., and Kingston R.E. 2001. Functional differences between the human ATP-dependent nucleosome remodeling proteins BRG1 and SNF2H. *J. Biol. Chem.* **276:** 34270.

Bruno M., Flaus A., Stockdale C., Rencurel C., Ferreira H., and Owen-Hughes T. 2003. Histone H2A/H2B dimer exchange by ATP-dependent chromatin remodeling activities. *Mol. Cell* **12:** 1599.

Carlson M., Osmond B.C., and Botstein D. 1981. Mutants of yeast defective in sucrose utilization. *Genetics* **98:** 25.

Cote J., Quinn J., Workman J.L., and Peterson C.L. 1994. Stimulation of GAL4 derivative binding to nucleosomal DNA by the yeast SWI/SNF complex. *Science* **265:** 53.

Elfring L.K., Deuring R., McCallum C.M., Peterson C.L., and Tamkun J.W. 1994. Identification and characterization of *Drosophila* relatives of the yeast transcriptional activator SNF2/SWI2. *Mol. Cell. Biol.* **14:** 2225.

Fan H.Y., He X., Kingston R.E., and Narlikar G.J. 2003. Distinct strategies to make nucleosomal DNA accessible. *Mol. Cell* **11:** 1311.

Fischle W., Wang Y., and Allis C.D. 2003. Histone and chromatin cross-talk. *Curr. Opin. Cell Biol.* **15:** 172.

Fitzgerald D.P. and Bender W. 2001. Polycomb group repression reduces DNA accessibility. *Mol. Cell. Biol.* **21:** 6585.

Francis N.J., Kingston R.E., and Woodcock C.L. 2004. Chromatin compaction by a polycomb group protein complex. *Science* **306:** 1574.

Fyodorov D.V. and Kadonaga J.T. 2002. Dynamics of ATP-dependent chromatin assembly by ACF. *Nature* **418:** 897.

Gavin I., Horn P.J., and Peterson C.L. 2001. SWI/SNF chromatin remodeling requires changes in DNA topology. *Mol. Cell* **7:** 97.

Guyon J.R., Narlikar G.J., Sif S., and Kingston R.E. 1999. Stable remodeling of tailless nucleosomes by the human SWI-SNF complex. *Mol. Cell. Biol.* **19:** 2088.

Guyon J.R., Narlikar G.J., Sullivan E.K., and Kingston R.E. 2001. Stability of a human SWI-SNF remodeled nucleosomal array. *Mol. Cell. Biol.* **21:** 1132.

Hamiche A., Sandaltzopoulos R., Gdula D.A., and Wu C. 1999. ATP-dependent histone octamer sliding mediated by the chromatin remodeling complex NURF. *Cell* **97:** 833.

Imbalzano A.N., Kwon H., Green M.R., and Kingston R.E. 1994. Facilitated binding of TATA-binding protein to nucleosomal DNA. *Nature* **370:** 481.

Kassabov S.R., Zhang B., Persinger J., and Bartholomew B. 2003. SWI/SNF unwraps, slides, and rewraps the nucleosome. *Mol. Cell* **11:** 391.

Khavari P.A., Peterson C.L., Tamkun J.W., Mendel D.B., and Crabtree G.R. 1993. BRG1 contains a conserved domain of the SWI2/SNF2 family necessary for normal mitotic growth and transcription. *Nature* **366:** 170.

King I.F., Francis N.J., and Kingston R.E. 2002. Native and recombinant polycomb group complexes establish a selective block to template accessibility to repress transcription in vitro. *Mol. Cell. Biol.* **22:** 7919.

Kireeva M.L., Walter W., Tchernajenko V., Bondarenko V., Kashlev M., and Studitsky V.M. 2002. Nucleosome remodeling induced by RNA polymerase II: Loss of the H2A/H2B dimer during transcription. *Mol. Cell* **9:** 541.

Kwon H., Imbalzano A.N., Khavari P.A., Kingston R.E., and Green M.R. 1994. Nucleosome disruption and enhancement of activator binding by a human SW1/SNF complex. *Nature* **370:** 477.

Langst G. and Becker P.B. 2001. ISWI induces nucleosome sliding on nicked DNA. *Mol. Cell* **8:** 1085.

Logie C. and Peterson C.L. 1997. Catalytic activity of the yeast SWI/SNF complex on reconstituted nucleosome arrays. *EMBO J.* **16:** 6772.

Lorch Y., Zhang M., and Kornberg R.D. 1999. Histone octamer transfer by a chromatin-remodeling complex. *Cell* **96:** 389.

Lorch Y., Cairns B.R., Zhang M., and Kornberg R.D. 1998. Activated RSC-nucleosome complex and persistently altered form of the nucleosome. *Cell* **94:** 29.

Macleod D. and Bird A. 1982. DNAase I sensitivity and methylation of active versus inactive rRNA genes in *Xenopus* species hybrids. *Cell* **29:** 211.

McCall K. and Bender W. 1996. Probes of chromatin accessibility in the *Drosophila* bithorax complex respond differently to Polycomb-mediated repression. *EMBO J.* **15:** 569.

Milne T.A., Briggs S.D., Brock H.W., Martin M.E., Gibbs D., Allis C.D., and Hess J.L. 2002. MLL targets SET domain methyltransferase activity to Hox gene promoters. *Mol. Cell* **10:** 1107.

Mizuguchi G., Shen X., Landry J., Wu W.H., Sen S., and Wu C. 2004. ATP-driven exchange of histone H2AZ variant catalyzed by SWR1 chromatin remodeling complex. *Science* **303:** 343.

Narlikar G.J., Fan H.Y., and Kingston R.E. 2002. Cooperation between complexes that regulate chromatin structure and transcription. *Cell* **108:** 475.

Narlikar G.J., Phelan M.L., and Kingston R.E. 2001. Generation and interconversion of multiple distinct nucleosomal states as a mechanism for catalyzing chromatin fluidity. *Mol. Cell* **8:** 1219.

Phelan M.L., Sif S., Narlikar G.J., and Kingston R.E. 1999. Reconstitution of a core chromatin remodeling complex from SWI/SNF subunits. *Mol. Cell* **3:** 247.

Polach K.J. and Widom J. 1995. Mechanism of protein access to specific DNA sequences in chromatin: A dynamic equilibrium model for gene regulation. *J. Mol. Biol.* **254:** 130.

Ringrose L. and Paro R. 2004. Epigenetic regulation of cellular memory by the polycomb and trithorax group proteins. *Annu. Rev. Genet.* **38:** 413.

Schnitzler G., Sif S., and Kingston R.E. 1998. Human SWI/SNF interconverts a nucleosome between its base state and a stable remodeled state. *Cell* **94:** 17.

Schotta G., Lachner M., Peters A.H., and Jenuwein T. 2004. The indexing potential of histone lysine methylation. *Novartis Found. Symp.* **259:** 22.

Simon J.A. and Tamkun J.W. 2002. Programming off and on states in chromatin: Mechanisms of Polycomb and trithorax group complexes. *Curr. Opin. Genet. Dev.* **12:** 210.

Stern M., Jensen R., and Herskowitz I. 1984. Five SWI genes are required for expression of the HO gene in yeast. *J. Mol. Biol.* **178:** 853.

Sweet R.W., Chao M.V., and Axel R. 1982. The structure of the thymidine kinase gene promoter: Nuclease hypersensitivity correlates with expression. *Cell* **31:** 347.

Tsukiyama T. and Wu C. 1995. Purification and properties of an ATP-dependent nucleosome remodeling factor. *Cell* **83:** 1011.

Turner B.M. 2002. Cellular memory and the histone code. *Cell* **111:** 285.

Wang W., Cote J., Xue Y., Zhou S., Khavari P.A., Biggar S.R., Muchardt C., Kalpana G.V., Goff S.P., Yaniv M., Workman J.L., and Crabtree G.R. 1996. Purification and biochemical heterogeneity of the mammalian SWI-SNF complex. *EMBO J.* **15:** 5370.

Weisbrod S. 1982. Active chromatin. *Nature* **297:** 289.

Weissmann F. and Lyko F. 2003. Cooperative interactions between epigenetic modifications and their function in the regulation of chromosome architecture. *Bioessays* **25:** 792.

Acetylation of Yeast Histone H4 Lysine 16: A Switch for Protein Interactions in Heterochromatin and Euchromatin

C.B. MILLAR, S.K. KURDISTANI, AND M. GRUNSTEIN

Department of Biological Chemistry, UCLA School of Medicine and the Molecular Biology Institute, Los Angeles, California 90095

Histone acetylation is an evolutionarily conserved phenomenon that can alter chromatin structure and activity. The basic unit of chromatin, the nucleosome, contains two molecules of each of the core histones, H3, H4, H2A, and H2B, wrapped by DNA. Each core histone has a central hydrophobic domain that contributes to the histone–histone and histone–DNA contacts that form the basis of the structural organization of the nucleosome (Luger et al. 1997). Extending from the nucleosomal core, the charged amino- and carboxy-terminal tail domains of the histones are free to interact with linker DNA, adjacent nucleosomes, and other chromosomal proteins (Zheng and Hayes 2003). These tail domains can be modified by acetyltransferases, methyltransferases, and kinases, thus altering their chemical characteristics and, consequently, those of the chromatin fiber.

Acetylation of lysine residues by histone acetyltransferases (HATs) is the most prevalent type of histone modification (Zhang et al. 2003) and can be rapidly reversed through the action of histone deacetylases (HDACs). Addition of an acetyl group neutralizes the positive charge of a lysine, and this change in charge can affect the in vitro interactions of histone tails with DNA, adjacent nucleosomes, and other chromatin components, ultimately altering the higher-order folding of chromatin (Tse et al. 1998; Zheng and Hayes 2003). However, the extent to which such charge-induced structural changes occur in vivo is unclear. A complementary view of the function of acetylation is based on its role in regulating the binding of proteins to the histone tails. It has been shown in genetic and biochemical studies that interactions of certain chromosomal regulatory proteins with histone tails are dependent on the histones' acetylatable lysines (Johnson et al. 1990; Hecht et al. 1995; Jacobson et al. 2000; Corona et al. 2002; Kasten et al. 2004). For example, Sir3, a chromosomal regulator that binds to histones and organizes heterochromatin structure in yeast, requires the deacetylation of H4-K16 for its interaction with chromatin (Johnson et al. 1990; Hecht et al. 1995) and preferentially binds to a completely deacetylated H4 amino terminus in vitro (Carmen et al. 2002). In contrast, proteins containing bromodomains interact preferentially with acetylated histone tails via a conserved hydrophobic pocket within the bromodomain that selectively binds acetyl-lysine residues (Dhalluin et al. 1999). SANT domains also interact with histones and may prefer to bind unacetylated histone tails, although the structural basis for this discrimination is unknown (Boyer et al. 2002; Yu et al. 2003). Numerous chromatin remodeling complexes contain at least one bromo- and/or SANT domain, indicating that the regulation of binding by acetylation and deacetylation may be a common mechanism for controlling protein–protein interactions within chromatin. In this paper we will discuss the role of H4-K16 acetylation in yeast heterochromatin and euchromatin as a switch to affect the binding of regulatory proteins to chromatin.

HISTONE H4 ACETYLATION SITES ARE NONREDUNDANT

The lysine residues at positions 5, 8, 12, and 16 (K5, K8, K12, and K16) within the amino-terminal tail of H4 are acetylated in *Saccharomyces cerevisiae* and all other species examined to date. In yeast, this region of H4 is important for a number of cellular processes including chromatin assembly (Ma et al. 1998), heterochromatic silencing (Kayne et al. 1988), activation of inducible genes (Durrin et al. 1991), repression of basal transcription (Lenfant et al. 1996), and double-strand break repair (Bird et al. 2002). In all of these processes, there is a distinction between the roles of sites K5, K8, and K12 (and their acetylation and/or deacetylation) and that of K16. For example, sites K5, K8, and K12 are redundant with each other, but not with K16, for chromatin assembly (Ma et al. 1998). Correspondingly, mutations in K16 have a uniquely strong effect on repression of heterochromatin at the silent mating loci (Johnson et al. 1990; Megee et al. 1990; Park and Szostak 1990) and telomeres (Aparicio et al. 1991). The roles for acetylated sites in DNA double-strand break repair are not well-characterized, but mutants in K5, K8, and K12 behave similarly, while K16 is less important in this process (Bird et al. 2002). Taken together, these results suggest that K16 has functions that are distinct from the other acetylated sites on H4.

ROLE OF H4-K16 ACETYLATION AND DEACETYLATION IN THE FORMATION AND SPREADING OF TELOMERIC HETEROCHROMATIN

The H4 amino-terminal tail plays an important role in heterochromatic silencing at yeast telomeres and mating-type loci (Kayne et al. 1988; Aparicio et al. 1991). Mutations of individual lysine residues in the H4 tail show that,

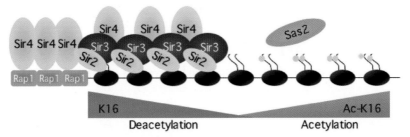

Figure 1. Initiation and spreading of telomeric heterochromatin and its regulation by H4-K16 (de)acetylation. Rap1 determines where heterochromatin initiates by binding to *cis*-acting DNA sequences and recruiting Sir4. This step is independent of H4-K16, but all subsequent events require the unacetylated K16 residue (Hoppe et al. 2002; Luo et al. 2002). These events include recruitment of Sir2 and Sir3 to the initiation site and subsequent spreading of Sir2, Sir3, and Sir4 along the chromosome. The nucleosomes are shown as *black circles*, with the H4 amino termini extending from them. The interaction between Sir3 and the histones requires H4-K16 deacetylation, which is achieved through the action of Sir2. The acetylation of H4-K16 (*orange circles*) by Sas2 prevents Sir protein binding and therefore establishes a boundary to the spreading of heterochromatin.

of the four acetylated sites, K16 is critical for silencing whereas K5, K8, and K12 are much less important (Kayne et al. 1988; Johnson et al. 1990; Megee et al. 1990; Park and Szostak 1990). The H4 amino terminus, and K16 in particular, is required for the binding of Sir3, one of the components of the SIR (silent information regulator) silencing complex that spreads from initiation sites to form heterochromatin (Johnson et al. 1990; Hecht et al. 1996). At the right arm of chromosome VI, the spreading of the SIR complex extends only as far as ~3 kilobases from the telomere, despite the absence of a discrete DNA boundary element. In fact, specific changes in chromatin components, both within the telomeric heterochromatin and in the subtelomeric euchromatin, regulate the extent of spreading of the SIR proteins from the telomeres into the subtelomeric chromatin (van Leeuwen et al. 2002; Kristjuhan et al. 2003; Ladurner et al. 2003; Meneghini et al. 2003; Ng et al. 2003). One such change is the deacetylation (in telomeric heterochromatin) and acetylation (in subtelomeric euchromatin) of K16 by the Sir2 HDAC and Sas2 HAT, respectively (Kimura et al. 2002; Suka et al. 2002). The role of H4-K16 in initiation and spreading of telomeric heterochromatin is illustrated in Figure 1.

K16 is part of a larger stretch of basic amino acids (residues 16–19) that are required for Sir3 binding in vivo and in vitro (Johnson et al. 1990; Hecht et al. 1995, 1996). The charge of K16 is crucial, since substitution with an uncharged residue or increased acetylation of K16 reduces Sir3 binding to the H4 amino terminus (Johnson et al. 1990; Carmen et al. 2002; Kimura et al. 2002). In this context, K16 acetylation and deacetylation may function as an electrostatic switch that prevents or enables binding of Sir3 to a larger domain and thus controls the formation and boundaries of heterochromatin.

GLOBAL H4-K16 ACETYLATION

Heterochromatin accounts for a relatively small proportion of the yeast genome (if the amount of heterochromatin at chromosome VI-R is indicative of other chromosomes). However, H4-K16 acetylation is a highly abundant modification, with ~80% of histone H4 molecules having an acetyl group on K16 (Clarke et al. 1993; Smith et al. 2003). The function of K16 acetylation is therefore unlikely to be restricted to the formation of a boundary between telomeric heterochromatin and euchromatin. Indeed, in *Drosophila*, H4-K16 acetylation plays a key role in transcriptional activation of the male X chromosome for dosage compensation (Akhtar and Becker 2000). In yeast, K16 may similarly modulate transcription regulation in euchromatic regions. It is known that the acetylation sites at the amino terminus of H4 are required for the activation of a number of inducible promoters (Durrin et al. 1991) and for the steady-state expression of many genes (Sabet et al. 2003), but the specific role of K16 in transcription is otherwise unclear. As part of a systematic analysis of the acetylation status of lysines in the four core histones, we have obtained information about the genome-wide acetylation profile of 11 of these sites, including K16. This information has allowed us to examine the relationships between the acetylated sites and between each site and transcriptional activity (Kurdistani et al. 2004).

PATTERNS OF HISTONE ACETYLATION THROUGHOUT THE GENOME

The development of antibodies that are highly specific to each acetylated lysine has allowed us to examine the relative acetylation levels of specific sites at a number of heterochromatic and euchromatic loci by chromatin immunoprecipitation (ChIP) (Suka et al. 2001; Wu et al. 2001). Moreover, the fusion of ChIP analysis and microarrays (Ren et al. 2000; Iyer et al. 2001) has enabled genome-wide studies of the targets of histone deacetylases in yeast and, most recently, of the patterns of acetylation in wild-type cells (Kurdistani et al. 2002, 2004; Robyr et al. 2002).

Using antibodies specific for 11 different acetylated lysines on the four core histones, we isolated fragments of DNA cross-linked to histones acetylated on particular sites. These DNA fragments were then amplified and used to probe microarrays spotted with either 6200 open reading fames (ORFs) or 6700 intergenic regions (IGRs) from the yeast genome. We analyzed genomic loci for which we had data for all 11 acetylated sites. In addition

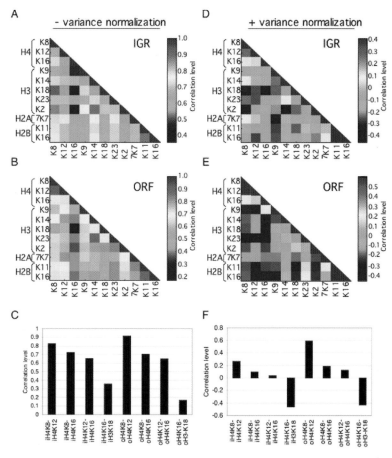

Figure 2. Correlation of acetylation levels at individual lysine residues with each other. The matrices (*A,B,D,E*) show the degree of correlation (indicated by the color scale) between acetylation levels of different lysines. The diagonal is the correlation of each site to itself. (*A, B*) Data from IGR and ORF arrays without variance normalization. (*C*) Correlation values for the acetylation levels of H4 lysine pairs on IGR ("i") and ORF ("o") arrays, depicted as bar graphs. The correlation of H4-K16 and H3-K18 is also shown. (*D, E*) Correlation between acetylation levels of different sites after variance normalization of the acetylation profile across the 11 sites. These values reflect the relative acetylation level of each site compared to all other sites. (*F*) Correlation values for the variance-normalized acetylation levels of H4 lysine pairs and H4-K16/H3-K18, depicted as bar graphs. (*A,B,D,E*, Reprinted, with permission, from Kurdistani et al. 2004 [©Cell Press].)

to comparing the acetylation levels of individual lysines, we variance-normalized the data across the acetylation sites to compare relative acetylation levels of each lysine to all other lysines (for a more detailed description of methods and normalization procedures, see Kurdistani et al. 2004). Variance normalization emphasizes the local relationship among the various lysines rather than the more global (background) acetylation levels.

The correlation values obtained, before and after variance normalization, when the 11 acetylated sites were compared in a pair-wise manner are shown in Figure 2. Overall, acetylated lysines are positively correlated with each other. This means that, for a particular genomic region, if a single site is acetylated the probability of acetylation at other sites is high. In contrast to this general trend, the acetylation of H4-K16 has a distinctly lower correlation with other sites (Fig. 2A–C), and it is evident after variance normalization that H4-K16 acetylation is, in fact, negatively correlated with most other sites (Fig. 2D–F). In particular, H4-K16 is strongly anticorrelated with H3-K18 (Fig. 2F). H4-K16 acetylation correlates best with the other two H4 sites that were examined, K8 and K12 (Fig. 2C), although after variance normalization it is clear that K16 acetylation correlates poorly even with these two sites (Fig. 2F), mirroring the discrete functions of K16 and the other H4 acetylation sites. We believe that the high correlations we see before variance normalization are due to the pervasive global acetylation and deacetylation of lysines (Kuo et al. 2000; Vogelauer et al. 2000; Waterborg 2001), which would mask smaller local changes given the low resolution of the current generation of DNA microarrays.

ACETYLATION AND TRANSCRIPTIONAL ACTIVITY

Historically, hyperacetylation has been associated with transcription, while inactive regions have been characterized as hypoacetylated. This view is too simplistic, however, given the finding that the HDAC Hos2 acts prefer-

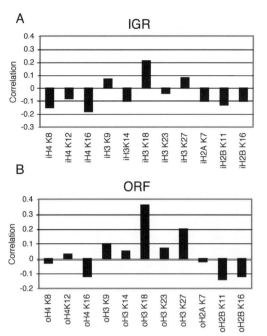

Figure 3. Correlation values between transcriptional activity and variance-normalized lysine acetylation levels on (*A*) IGRs and (*B*) ORFs.

entially on the coding regions of active genes genome-wide, and is, in fact, required for gene activation (Wang et al. 2002). In order to assess how acetylation of particular lysines is correlated with gene expression, we compared our acetylation data with data on the transcriptional activity of genes throughout the genome under the experimental conditions we used (Causton et al. 2001). For both IGRs and ORFs, we conclude that the sites for which acetylation is best correlated with gene activity are on H3 (K18, followed by K9 and K27; Fig. 3A,B). Acetylation of lysines on H2A, H2B, and H4, on the other hand, show either a low or negative correlation with transcription. Notably, the acetylation of H2B sites K11 and K16, as well as H4-K16, shows a marked anticorrelation with active genes, lending further weight to the idea that deacetylation, as well as acetylation, can be important for normal gene activity.

GENE CLUSTERS WITH SIGNATURE ACETYLATION PATTERNS

The above correlations compare acetylation states for individual lysines with gene expression throughout the genome. The positive and negative correlations between acetylation sites and gene activity suggest that combinations of acetylated and deacetylated lysines may generate distinct modification patterns, each of which may be common to genes with related functions. To address this question systematically and without bias, we applied the *k*-means partitional clustering method (Hartigan 1975) to the variance-normalized acetylation data. Each IGR or ORF is included in only one cluster, based solely on the similarity of its acetylation patterns to other cluster members (Fig. 4A,C). In any one cluster, the correlation between patterns of acetylation for genes is 0.7 or higher. In this manner, we identified 53 and 68 groups of genes from IGRs and ORFs, respectively. Examination of the average acetylation patterns of the IGR clusters reveals a diverse array of acetylation patterns on the promoter regions. Remarkably, independent clustering of the ORF acetylation data also uncovered groups of genes with patterns as diverse as those of IGRs. We have shown by four different criteria that the genes in the acetylation clusters are biologically related to each other. Genes in the same cluster show significant coexpression, are significantly grouped within the same functional classes or physiological processes, have common upstream DNA sequence motifs, and show significant enrichment for binding of particular transcription factors (Kurdistani et al. 2004).

We find that the whole-genome relationships between individual lysines and their correlations to gene activity are further amplified in the acetylation clusters. This is perhaps due to the fact that these clusters are comprised of biologically cohesive groups of genes, and thus have a better signal to noise ratio. For instance, we find that H4-K16 is hypoacetylated and H3-K18 is hyperacetylated in IGR cluster 1 and, conversely, in IGR cluster 30, H3-K18 is hypoacetylated and H4-K186 is hyperacetylated (Fig. 4B). Consistent with the genome-wide correlations, cluster 1 is highly expressed whereas cluster 30 is highly repressed under the conditions of growth used. The same is true for ORF clusters 1 and 6 (Fig. 4D). In fact, the inverse acetylation states of H4-K16 and H3-K18 are correlated strongly with expression state for a number of different clusters, indicating that, both globally and in the identified clusters, H3-K18 and H4-K16 are acetylated in an inverse manner with respect to each other. Therefore, for many clusters, the acetylation state of these sites can be a mark for transcriptional status. It will be interesting to determine whether such a mark exists in other eukaryotes as well.

HOW IS K16 ACETYLATION REGULATED DURING GENE ACTIVITY?

The acetylation and deacetylation of H4-K16 must be achieved by the enzymes that show specificity for this site within H4, making Esa1 and Sas2 the candidate HATs and Rpd3 and Hos2 the potential HDACs that target K16 in euchromatin (Allard et al. 1999; Kimura et al. 2002; Robyr et al. 2002; Suka et al. 2002). Rpd3 can be recruited to promoters by transcription factors such as Ume6, but also binds and deacetylates chromatin globally (Vogelauer et al. 2000; Kurdistani et al. 2002). Hos2, on the other hand, acts primarily on the coding region of active genes (Wang et al. 2002). While Rpd3 is strongly required for the deacetylation of H4-K5, K8, and K12, Rpd3 deletion has the least effect on H4-K16 (Wu et al. 2001). Interestingly, the effect of Hos2 deletion on K16 acetylation is greater than that of Rpd3 (Wang et al. 2002). Since Hos2 acts on active coding regions, we must consider the possibility that this enzyme is responsible for the observed underacetylation of K16 on active genes

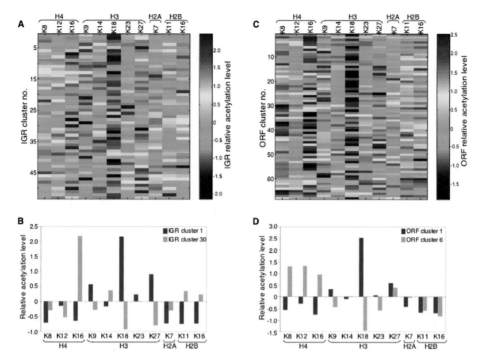

Figure 4. Groups of genes clustered based on their acetylation profile. (*A*) Average acetylation patterns of genes in the IGR clusters. Each row represents one cluster and, for each cluster, the average acetylation of each lysine is color-coded according to the scale shown. (*B*) Average acetylation patterns of genes in IGR clusters 1 and 30 as bar graphs. (*C*) Average acetylation patterns of genes in the ORF clusters presented as in *A*. (*D*) Average acetylation patterns of genes in ORF clusters 1 and 6 as bar graphs. (Reprinted, with permission, from Kurdistani et al. 2004 [©Cell Press].)

genome-wide. Conversely, by acetylating K16, Esa1 and/or Sas2 may negatively regulate gene activity.

ACETYLATION OF THE HISTONE H4 TAIL AND BDF1 BINDING

The acetylation microarrays give us a picture of the genome-wide patterns of acetylation in logarithmically growing yeast cells. We wondered whether this information could be used to predict the binding preferences of chromatin proteins whose interaction with histones is affected by acetylation. Because the bromodomain has been identified as an acetyl-lysine binding domain, we looked at the genome-wide binding of Bdf1, a chromosomal protein that has two bromodomains and that binds preferentially to increasingly acetylated isoforms of H4 in vitro (Chua and Roeder 1995; Ladurner et al. 2003; Matangkasombut and Buratowski 2003).

Comparison of the binding profile of HA-tagged Bdf1 to the individual acetylation sites indicates that Bdf1 binding correlates positively with acetylation at all sites examined (Fig. 5A). The poorest correlation was with H4-K16 acetylation on ORFs. When the acetylation levels of the 11 lysines are normalized to each other (variance-normalized) and then compared to the genome-wide Bdf1 binding, the correlation with most sites remained positive, apart from H3-K9 and K27, H2B-K11, and H4-K16. Relative acetylation of H4-K16 showed the strongest anticorrelation with Bdf1 binding (Fig. 5B).

Since Bdf1 is known to interact with the acetylated H4 tail, we looked in vivo at the consequences of point mutations in the H4 tail on Bdf1 binding (Fig. 5C,D). We found that mutation of K12 to arginine reduced binding at three out of the four promoters tested, arguing that, as predicted from the in vitro data, acetylation of H4-K12 is important for Bdf1 binding. In contrast, mutation of K16 to arginine had little effect on Bdf1 binding, suggesting that K16 acetylation is not required for Bdf1 binding at these promoters. In fact, substitution of K16 by glutamine (approximating the acetylated state) strongly reduced binding of Bdf1 to the four promoters in vivo. We conclude that optimal binding of Bdf1 to these promoters requires a positive charge at H4 site K16, indicating that the specific deacetylation of K16 may enhance Bdf1 binding to the otherwise acetylated H4 tail.

H4-K16 ACETYLATION/DEACETYLATION IS A SWITCH FOR PROTEIN BINDING

The amino terminus of H4 can be bound by a number of chromosomal proteins, and the acetylation status of K16 is a key regulator of these interactions. In most cases documented so far, deacetylation of K16 is a prerequisite for protein binding, but this deacetylation can affect binding to different subdomains in the H4 tail (Fig. 6). Sir3 binds the region carboxy-terminal to unacetylated K16 in the histone H4 tail (Fig. 6A). This region consists of a highly basic region (residues 16–19) and a relatively un-

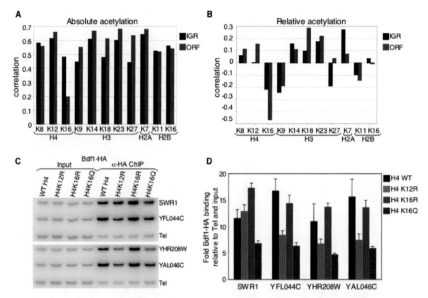

Figure 5. Genome-wide Bdf1 binding correlates with hypoacetylated H4-K16. (*A*) Bdf1 binding shows significant correlations with individual sites of acetylation on IGR (*black bars*) and ORF (*red bars*) regions, indicating that it binds to generally hyperacetylated regions of the genome. (*B*) Correlation of Bdf1 binding levels with variance-normalized acetylation levels. (*C, D*) Effect of mutations in histone H4 on Bdf1 binding in vivo at the promoter regions of four randomly selected Bdf1-target genes. The intensity of Bdf1 enrichment is normalized to a region 500 bp from the end of chromosome VI-R (Tel), which acts as an internal loading control. The fold binding is the ratio of Tel-normalized values of immunoprecipitated DNA to its input, averaged across three independent ChIP experiments. A representative gel is shown in *C*, and a bar graph plotting the enrichment values is shown in *D*. (K-R) Lysine to arginine substitution; (K-Q) lysine to glutamine mutation. (Reprinted, with permission, from Kurdistani et al. 2004 [©Cell Press].)

charged domain (residues 21–29), both of which are required for Sir3 binding in vivo and in vitro (Johnson et al. 1990, 1992; Hecht et al. 1995, 1996). The *Drosophila* remodeling factor ISWI has similar binding characteristics to Sir3. ISWI is also dependent on residues 16–19 of the H4 tail for its interaction with chromatin (Hamiche et al. 2001; Clapier et al. 2002) and acetylation of K16 reduces the ATPase activity of ISWI (Fig. 6B) (Clapier et al. 2002; Corona et al. 2002). The SLIDE (SANT-like-ISWI-domain) is responsible for the interaction between ISWI and this region of H4 (Grune et al. 2003). Unexpectedly, K16 deacetylation is also favorable for the binding of the euchromatic bromodomain protein Bdf1. In this case, however, the acetylation of the other H4 amino-terminal lysines is also critical, indicating that K16 acetylation status may be important for the binding of Bdf1 to the region that is amino-terminal to K16 (Fig. 6C). Because the deacetylation of K16 is important for Sir3, ISWI, and Bdf1 binding, it is possible that K16 deacetylation enhances the accessibility of the entire amino terminus of H4 in vivo, while acetylation of K16 restricts access. However, at least one protein is known to depend on K16 acetylation for its interaction with the H4 amino terminus. The crystal structure of the Gcn5 bromodomain complexed with residues 15–29 of H4 shows that Gcn5 makes contacts with the same region of the H4 tail involved in Sir3 binding (Owen et al. 2000), but K16 acetylation is favorable for binding in this case (Fig. 6D). It will be interesting to see if Gcn5, or other factors, rely on K16 acetylation in order to interact with the H4 amino-terminal tail in vivo.

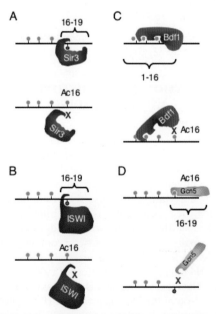

Figure 6. The acetylation state of H4-K16 controls the binding of multiple protein regulators. The H4 amino terminus is shown as a *black line*, with lysines 5, 8, 12, and 16 indicated as *lollipops* (*blue* for acetylated; *orange* for unacetylated). (*A*) Sir3 interacts with a region of H4 (residues 16–19) including K16. Acetylation of K16 decreases this interaction. (*B*) ISWI also interacts with residues 16–19 of H4 when K16 is deacetylated. (*C*) Bdf1 binding is favored by K16 deacetylation, but acetylation of other sites is important, indicating that Bdf1 interacts with the region amino-terminal to K16. (*D*) Gcn5 makes contacts with residues 16–19 of H4 in vitro, but this interaction, unlike those in *A–C*, is favored by acetylation of K16.

In conclusion, the acetylation state of H4-K16 may act a switch that determines the binding of numerous regulatory factors to different H4 regions, amino- or carboxy-terminal to K16. The ability of K16 acetylation and deacetylation to affect multiple protein interactions with the H4 amino terminus may explain the unique status of K16 in the regulation of both heterochromatin and euchromatin.

ACKNOWLEDGMENTS

We thank Mona Shahbazian for helpful comments on the manuscript. C.B.M. is a recipient of a Wellcome Trust International Research Fellowship (ref: 069856). S.K.K. is a Howard Hughes Medical Institute Physician Postdoctoral Fellow. This work was supported by Public Service grants of NIH to M.G.

REFERENCES

Akhtar A. and Becker P.B. 2000. Activation of transcription through histone H4 acetylation by MOF, an acetyltransferase essential for dosage compensation in *Drosophila*. *Mol. Cell* **5**: 367.

Allard S., Utley R.T., Savard J., Clarke A., Grant P., Brandl C.J., Pillus L., Workman J.L., and Cote J. 1999. NuA4, an essential transcription adaptor/histone H4 acetyltransferase complex containing Esa1p and the ATM-related cofactor Tra1p. *EMBO J.* **18**: 5108.

Aparicio O.M., Billington B.L., and Gottschling D.E. 1991. Modifiers of position effect are shared between telomeric and silent mating-type loci in *S. cerevisiae*. *Cell* **66**: 1279.

Bird A.W., Yu D.Y., Pray-Grant M.G., Qiu Q., Harmon K.E., Megee P.C., Grant P.A., Smith M.M., and Christman M. 2002. Acetylation of histone H4 by Esa1 is required for DNA double-strand break repair. *Nature* **419**: 411.

Boyer L.A., Langer M.R., Crowley K.A., Tan S., Denu J.M., and Peterson C.L. 2002. Essential role for the SANT domain in the functioning of multiple chromatin remodeling enzymes. *Mol. Cell* **10**: 935.

Carmen A.A., Milne L., and Grunstein M. 2002. Acetylation of the yeast histone H4 N terminus regulates its binding to heterochromatin protein SIR3. *J. Biol. Chem.* **277**: 4778.

Causton H.C., Ren B., Koh S.S., Harbison C.T., Kanin E., Jennings E.G., Lee T.I., True H.L., Lander E.S., and Young R.A. 2001. Remodeling of yeast genome expression in response to environmental changes. *Mol. Biol. Cell* **12**: 323.

Chua P. and Roeder G.S. 1995. Bdf1, a yeast chromosomal protein required for sporulation. *Mol. Cell. Biol.* **15**: 3685.

Clapier C.R., Nightingale K.P., and Becker P.B. 2002. A critical epitope for substrate recognition by the nucleosome remodeling ATPase ISWI. *Nucleic Acids Res.* **30**: 649.

Clarke D.J., O'Neill L.P., and Turner B.M. 1993. Selective use of H4 acetylation sites in the yeast *Saccharomyces cerevisiae*. *Biochem. J.* **294**: 557.

Corona D.F., Clapier C.R., Becker P.B., and Tamkun J.W. 2002. Modulation of ISWI function by site-specific histone acetylation. *EMBO Rep.* **3**: 242.

Dhalluin C., Carlson J.E., Zeng L., He C., Aggarwal A.K., and Zhou M.M. 1999. Structure and ligand of a histone acetyltransferase bromodomain. *Nature* **399**: 491.

Durrin L.K., Mann R.K., Kayne P.S., and Grunstein M. 1991. Yeast histone H4 N-terminal sequence is required for promoter activation *in vivo*. *Cell* **65**: 1023.

Grune T., Brzeski J., Eberharter A., Clapier C.R., Corona D.F., Becker P.B., and Muller C.W. 2003. Crystal structure and functional analysis of a nucleosome recognition module of the remodeling factor ISWI. *Mol. Cell* **12**: 449.

Hamiche A., Kang J.G., Dennis C., Xiao H., and Wu C. 2001. Histone tails modulate nucleosome mobility and regulate ATP-dependent nucleosome sliding by NURF. *Proc. Natl. Acad. Sci.* **98**: 14316.

Hartigan J. 1975. *Clustering algorithms*. Wiley, New York.

Hecht A., Strahl-Bolsinger S., and Grunstein M. 1996. Spreading of transcriptional repressor SIR3 from telomeric heterochromatin. *Nature* **383**: 92.

Hecht A., Laroche T., Strahl-Bolsinger S., Gasser S.M., and Grunstein M. 1995. Histone H3 and H4 N-termini interact with SIR3 and SIR4 proteins: A molecular model for the formation of heterochromatin in yeast. *Cell* **80**: 583.

Hoppe G.J., Tanny J.C., Rudner A.D., Gerber S.A., Danaie S., Gygi S.P., and Moazed D. 2002. Steps in assembly of silent chromatin in yeast: Sir3-independent binding of a Sir2/Sir4 complex to silencers and role for Sir2-dependent deacetylation. *Mol. Cell. Biol.* **22**: 4167.

Iyer V.R., Horak C.E., Scafe C.S., Botstein D., Snyder M., and Brown P.O. 2001. Genomic binding sites of the yeast cell-cycle transcription factors SBF and MBF. *Nature* **409**: 533.

Jacobson R.H., Ladurner A.G., King D.S., and Tjian R. 2000. Structure and function of a human TAFII250 double bromodomain module. *Science* **288**: 1422.

Johnson L.M., Kayne P.S., Kahn E.S., and Grunstein M. 1990. Genetic evidence for an interaction between SIR3 and histone H4 in the repression of the silent mating loci in *Saccharomyces cerevisiae*. *Proc. Natl. Acad. Sci.* **87**: 6286.

Johnson L.M., Fisher-Adams G., and Grunstein M. 1992. Identification of a non-basic domain in the histone H4 N-terminus required for repression of the yeast silent mating loci. *EMBO J.* **11**: 2201.

Kasten M., Szerlong H., Erdjument-Bromage H., Tempst P., Werner M., and Cairns B.R. 2004. Tandem bromodomains in the chromatin remodeler RSC recognize acetylated histone H3 Lys14. *EMBO J.* **23**: 1348.

Kayne P.S., Kim U.J., Han M., Mullen J.R., Yoshizaki F., and Grunstein M. 1988. Extremely conserved histone H4 N terminus is dispensable for growth but essential for repressing the silent mating loci in yeast. *Cell* **55**: 27.

Kimura A., Umehara T., and Horikoshi M. 2002. Chromosomal gradient of histone acetylation established by Sas2p and Sir2p functions as a shield against gene silencing. *Nat. Genet.* **32**: 370.

Kristjuhan A., Wittschieben B.O., Walker J., Roberts D., Cairns B.R., and Svejstrup J.Q. 2003. Spreading of Sir3 protein in cells with severe histone H3 hypoacetylation. *Proc. Natl. Acad. Sci.* **100**: 7551.

Kuo M.H., vom Baur E., Struhl K., and Allis C.D. 2000. Gcn4 activator targets Gcn5 histone acetyltransferase to specific promoters independently of transcription. *Mol. Cell* **6**: 1309.

Kurdistani S.K., Tavazoie S., and Grunstein M. 2004. Mapping global histone acetylation patterns to gene expression. *Cell* **117**: 721.

Kurdistani S.K., Robyr D., Tavazoie S., and Grunstein M. 2002. Genome-wide binding map of the histone deacetylase Rpd3 in yeast. *Nat. Genet.* **31**: 248.

Ladurner A.G., Inouye C., Jain R., and Tjian R. 2003. Bromodomains mediate an acetyl-histone encoded antisilencing function at heterochromatin boundaries. *Mol. Cell* **11**: 365.

Lenfant F., Mann R.K., Thomsen B., Ling X., and Grunstein M. 1996. All four core histone N-termini contain sequences required for the repression of basal transcription in yeast. *EMBO J.* **15**: 3974.

Luger K., Mader A.W., Richmond R.K., Sargent D.F., and Richmond T.J. 1997. Crystal structure of the nucleosome core particle at 2.8 Å resolution. *Nature* **389**: 251.

Luo K., Vega-Palas M.A., and Grunstein M. 2002. Rap1-Sir4 binding independent of other Sir, yKu, or histone interactions initiates the assembly of telomeric heterochromatin in yeast. *Genes Dev.* **16**: 1528.

Ma X.J., Wu J., Altheim B.A., Schultz M.C., and Grunstein M. 1998. Deposition-related sites K5/K12 in histone H4 are not required for nucleosome deposition in yeast. *Proc. Natl. Acad. Sci.* **95**: 6693.

Matangkasombut O. and Buratowski S. 2003. Different sensitivities of bromodomain factors 1 and 2 to histone H4 acetyla-

tion. *Mol. Cell* **11:** 353.
Megee P.C., Morgan B.A., Mittman B.A., and Smith M.M. 1990. Genetic analysis of histone H4: Essential role of lysines subject to reversible acetylation. *Science* **247:** 841.
Meneghini M.D., Wu M., and Madhani H.D. 2003. Conserved histone variant H2A.Z protects euchromatin from the ectopic spread of silent heterochromatin. *Cell* **112:** 725.
Ng H.H., Dole S., and Struhl K. 2003. The Rtf1 component of the Paf1 transcriptional elongation complex is required for ubiquitination of histone H2B. *J. Biol. Chem.* **278:** 33625.
Owen D.J., Ornaghi P., Yang J.C., Lowe N., Evans P.R., Ballario P., Neuhaus D., Filetici P., and Travers A.A. 2000. The structural basis for the recognition of acetylated histone H4 by the bromodomain of histone acetyltransferase Gcn5p. *EMBO J.* **19:** 6141.
Park E.C. and Szostak J.W. 1990. Point mutations in the yeast histone H4 gene prevent silencing of the silent mating type locus HML. *Mol. Cell. Biol.* **10:** 4932.
Ren B., Robert F., Wyrick J.J., Aparicio O., Jennings E.G., Simon I., Zeitlinger J., Schreiber J., Hannett N., Kanin E., Volkert T.L., Wilson C.J., Bell S.P., and Young R.A. 2000. Genome-wide location and function of DNA binding proteins. *Science* **290:** 2306.
Robyr D., Suka Y., Xenarios I., Kurdistani S.K., Wang A., Suka N., and Grunstein M. 2002. Microarray deacetylation maps determine genome-wide functions for yeast histone deacetylases. *Cell* **109:** 437.
Sabet N., Tong F., Madigan J.P., Volo S., Smith M.M., and Morse R.H. 2003. Global and specific transcriptional repression by the histone H3 amino terminus in yeast. *Proc. Natl. Acad. Sci.* **100:** 4084.
Smith C.M., Gafken P.R., Zhang Z., Gottschling D.E., Smith J.B., and Smith D.L. 2003. Mass spectrometric quantification of acetylation at specific lysines within the amino-terminal tail of histone H4. *Anal. Biochem.* **316:** 23.

Suka N., Luo K., and Grunstein M. 2002. Sir2p and Sas2p opposingly regulate acetylation of yeast histone H4 lysine16 and spreading of heterochromatin. *Nat. Genet.* **32:** 378.
Suka N., Suka Y., Carmen A.A., Wu J., and Grunstein M. 2001. Highly specific antibodies determine histone acetylation site usage in yeast heterochromatin and euchromatin. *Mol. Cell* **8:** 473.
Tse C., Sera T., Wolffe A.P., and Hansen J.C. 1998. Disruption of higher-order folding by core histone acetylation dramatically enhances transcription of nucleosomal arrays by RNA polymerase III. *Mol. Cell. Biol.* **18:** 4629.
van Leeuwen F., Gafken P.R., and Gottschling D.E. 2002. Dot1p modulates silencing in yeast by methylation of the nucleosome core. *Cell* **109:** 745.
Vogelauer M., Wu J., Suka N., and Grunstein M. 2000. Global histone acetylation and deacetylation in yeast. *Nature* **408:** 495.
Wang A., Kurdistani S.K., and Grunstein M. 2002. Requirement of Hos2 histone deacetylase for gene activity in yeast. *Science* **298:** 1412.
Waterborg J.H. 2001. Dynamics of histone acetylation in *Saccharomyces cerevisiae*. *Biochemistry* **40:** 2599.
Wu J., Suka N., Carlson M., and Grunstein M. 2001. TUP1 utilizes histone H3/H2B-specific HDA1 deacetylase to repress gene activity in yeast. *Mol. Cell* **7:** 117.
Yu J., Li Y., Ishizuka T., Guenther M.G., and Lazar M.A. 2003. A SANT motif in the SMRT corepressor interprets the histone code and promotes histone deacetylation. *EMBO J.* **22:** 3403.
Zhang L., Eugeni E.E., Parthun M.R., and Freitas M.A. 2003. Identification of novel histone post-translational modifications by peptide mass fingerprinting. *Chromosoma* **112:** 77.
Zheng C. and Hayes J.J. 2003. Intra- and inter-nucleosomal protein-DNA interactions of the core histone tail domains in a model system. *J. Biol. Chem.* **278:** 24217.

Histone Deposition Proteins: Links between the DNA Replication Machinery and Epigenetic Gene Silencing

A.A. Franco and P.D. Kaufman

Lawrence Berkeley National Laboratory and Department of Molecular and Cell Biology, University of California, Berkeley, California 94720

The information encoded in all eukaryotic genomes is organized into a nucleoprotein complex called chromatin. Chromatin is essential for compacting genomic DNA and plays a primary role in governing the expression status of genes. Chromatin is a diverse and dynamic polymer whose composition is modulated to build specialized chromosome structures, such as those found at centromeres and telomeres (for review, see Grewal and Moazed 2003). The fundamental repeating unit of chromatin is the nucleosome, containing an octamer of histone proteins, two each of H2A, H2B, H3, and H4, around which 146 bp of DNA wraps 1.7 times. This review will focus on how nucleosomes are assembled by histone chaperones that deposit histones during DNA replication or RNA transcription. We will begin by describing the biophysical properties governing nucleosome formation and the mechanisms of de novo nucleosome formation at sites of DNA synthesis. Then, we will discuss the functional overlap between histone chaperones involved in DNA replication-coupled and replication-independent nucleosome assembly. Finally, we will review recent data demonstrating the role of histone chaperones in the tight temporal coupling of histone deposition and DNA replication.

MOLECULAR MECHANISMS OF NUCLEOSOME ASSEMBLY

Nucleosome Architecture

Nucleosome assembly occurs in two steps. First, histones H3 and H4 are deposited onto DNA, followed by the addition of histones H2A and H2B (for review, see Verreault 2000). This order of addition reflects the structure of the nucleosome (Luger et al. 1997) and the biophysical properties of histone subcomplexes (Eickbush and Moudrianakis 1978). In the nucleosome, an $(H3/H4)_2$ tetramer organizes the central 120 bp of the DNA and is flanked by two histone H2A/H2B dimers. In the absence of DNA, the histone octamer is stable only when salt concentrations exceed 2 M and the temperature is maintained near 4°C. At more physiological conditions, histone octamers dissociate into two H2A/H2B dimers and an $(H3/H4)_2$ tetramer (Kornberg and Thomas 1974; Eickbush and Moudrianakis 1978). These stable subcomplexes had previously been thought to be the only intermediates in nucleosome formation (for review, see Verreault 2000); as discussed below, this view has been challenged by recent data (Tagami et al. 2004).

The primary forces governing nucleosome formation are electrostatic. Combining pure histones and DNA at physiological salt concentrations results in the formation of insoluble aggregates. However, high concentrations of anionic molecules facilitate in vitro nucleosome formation by shielding the positively charged histones. Indeed, gradual reduction of salt concentration also promotes nucleosome formation because $(H3/H4)_2$ tetramers associate with DNA more avidly then H2A/H2B dimers (for review, see Kaufman and Almouzni 2000). Because charge shielding is sufficient to promote nucleosome assembly in vitro, verification of the physiological relevance of biochemically identified histone chaperones has been important. This had previously been most easily achieved in genetically tractable organisms. However, the recent introduction of siRNA techniques has facilitated the study of deposition factors in human cell culture experiments (Hoek and Stillman 2003; Nabatiyan and Krude 2004).

DNA Replication Creates Platforms for Epigenetic Marks

The deposition of a genome's worth of nucleosomes occurs concomitant with DNA replication during S phase. To understand how histone chaperones coordinate with the DNA replication machinery we will first review some of the fundamentals of eukaryotic DNA replication. A prominent theme of this discussion will be the cascade of protein–protein interactions that guide the assembly and disassembly of replisome components and thereby order events during DNA replication. This will lead to consideration of how replication proteins regulate epigenetic states.

DNA replication. To dissect eukaryotic DNA synthesis in vitro, biochemists have taken advantage of viral replication systems to circumvent the complexity of cellular origins. Plasmids harboring the Simian Virus 40 (SV40) DNA replication origin are replicated in human cell extracts upon addition of a single virally encoded protein, the SV40 large tumor antigen (T antigen) (for review, see Fanning and Knippers 1992). T antigen is both a DNA sequence-specific replication initiation protein and a DNA helicase. All other factors required for DNA synthesis were isolated from human cell extracts using this assay system (Waga and Stillman 1998). These include the single-stranded binding protein, RPA, which stabilizes unwound DNA during initiation and plays a central role in regulating replisome formation (Brill and Stillman 1989). For example, RPA interacts with two other essential components of the replisome, polymerase α (Polα) primase and the clamp-loading complex Replication Factor C (RFC) (see below).

Most polymerases cannot initiate synthesis de novo. Thus, after DNA unwinding, DNA synthesis begins when the four-subunit DNA Polα uses both its RNA primase and DNA polymerase activities to synthesize short (35–50-nt) RNA–DNA primers (Murakami et al. 1992). Polα is recruited to single-stranded DNA via an interaction with RPA (for review, see Waga and Stillman 1998). After primer synthesis, DNA replication elongation proceeds continuously on the leading strand and discontinuously on the lagging strand at each Okazaki fragment. Elongation begins with polymerase switching, which occurs when Polα is replaced with the more processive DNA polymerase δ (Polδ) (Tsurimoto and Stillman 1991). Polδ processivity is greatly enhanced by association with the accessory factor Proliferating-Cell Nuclear Antigen (PCNA). PCNA is a toroidal homotrimer that encircles DNA, topologically tethering polymerases to the DNA template and thus greatly increasing their processivity. PCNA is therefore referred to as a "clamp" (for review, see Jeruzalmi et al. 2002). The three-dimensional structures, though not the primary sequences, of these clamps are conserved among eubacteria, archaebacteria, bacteriophage T4, and eukaryotes (Jeruzalmi et al. 2002). Besides polymerases, PCNA also binds to a large number of other proteins involved in DNA replication, repair, and modification (for review, see Warbrick 2000). Additionally, PCNA directs a nucleosome assembly protein to replicated DNA templates (see below) (Shibahara and Stillman 1999; Zhang et al. 2000; Krawitz et al. 2002).

The task of topologically loading processivity clamps onto DNA is performed by evolutionarily conserved protein complexes termed "clamp loaders" (for review, see Jeruzalmi et al. 2002). The eukaryotic PCNA-loading complex is RFC, a DNA- and PCNA-stimulated ATPase that loads PCNA onto 3′OH-primer-template junctions in an ATP-dependent manner. RFC promotes the switch from replication initiation to elongation by competing with Polα for interaction with RPA at primer–template junctions prior to PCNA loading (Yuzhakov et al. 1999). This polymerase switch need happen only once on the leading strand, but on the lagging strand occurs at every Okazaki fragment (Waga et al. 1994). Once RFC has loaded PCNA onto DNA, Polδ binds to PCNA, displacing RFC. Interestingly, RFC may remain associated with the elongating replisome though interactions with Polδ and RPA (Yuzhakov et al. 1999). This association would leave RFC available for interaction with factors that regulate DNA replication.

Finally, after elongation, Okazaki fragment maturation occurs. RNA primers are removed by RNase HI and FEN1, the resulting gaps are filled in by the polδ replisome, and the DNA fragments are covalently joined by DNA ligase I (for review, see Waga and Stillman 1998). FEN1 and DNA ligase I also bind PCNA, highlighting PCNA's central role in coordinating DNA replication events. Similar coordination of PCNA-mediated events operates during strand resealing associated with DNA repair (for review, see Maga and Hubscher 2003).

Histone deposition during DNA synthesis: Relation to epigenetic marks. Covalent histone modifications, such as methylation, acetylation, and ubiquitination, regulate local chromosome structure and gene expression status (for review, see Jenuwein and Allis 2001). These modifications can function as epigenetic marks defining chromosome regions independent of DNA sequence. The new histones deposited during DNA synthesis are not marked in the same way as the parental histones (Ruiz-Carrillo et al. 1975; Jackson et al. 1976) and thus might erase or dilute existing epigenetic marks. Therefore, a major question regarding replication-linked nucleosome assembly had been the fate of parental nucleosomes after passage of the replication fork. To address this, pulse-chase experiments were performed using isotope and density labels to mark parental histones. Upon cross-linking of histones in nucleosomes and separation of labeled species by density gradient sedimentation, two distinct species of nucleosomes were detected. The first contained tetramers of new H3/H4 molecules in the same nucleosome as two dimers of old H2A/H2B. The second contained tetramers of old H3/H4 associated with one dimer of new and one dimer of old H2A/H2B (Jackson 1987). These data are consistent with the structure and biophysical properties of the nucleosome, and suggest that the (H3/H4)$_2$ tetramers and H2A/H2B dimers of parental nucleosomes are separately inherited behind the replication fork. Furthermore, (H3/H4)$_2$ tetramers are segregated to the daughter strands as intact units, while H2A/H2B dimers randomly reassociated with new and old (H3/H4)$_2$ tetramers on both daughter strands (Fig. 1) (Jackson 1987, 1990; Gruss et al. 1993). These data are consistent with other experiments demonstrating that incorporation of H3 and H4 precedes H2A and H2B in vivo (Worcel et al. 1978) and in vitro (Smith and Stillman 1991). Intriguingly, recent data suggest that histone chaperones may bind dimers of H3/H4 instead of tetramers (Tagami et al. 2004) suggesting that two chaperone molecules may act in concert to form new (H3/H4)$_2$ tetramers.

Notably, models for the lateral spreading of chromatin states easily accommodate interspersed new and parental histones (for review, see Richards and Elgin 2002). These models rely on recent data describing the self-reinforcing nature of epigenetic marks, in which a particular mark recruits the enzymes responsible for generating that mark. For instance, methylation of H3 lysine 9 (H3-K9Me) is an epigenetic mark recognized by the HP1 protein. HP1, in turn, interacts with the H3 K9 methyltransferase, Su(var)3-9, thus recruiting the marking enzyme to areas already containing the mark. Additionally, like HP1, Su(var)3-9 contains a conserved chromodomain that contributes to targeting the methyltransferase to heterochromatin via the H3-K9Me mark (Melcher et al. 2000).

Epigenetic marks on histones, then, generally are renewed after DNA replication. However, replication may

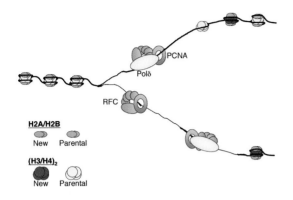

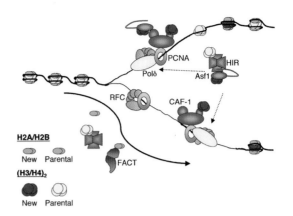

Figure 1. Segregation of histones at replication forks. Nucleosomes are detected within ~250 bp of the site of DNA synthesis. During fork passage, parental nucleosomes separate into H2A/H2B dimers (*blue*) and (H3/H4)$_2$ tetramers (*yellow*) that randomly reassociate with newly synthesized histones (*red* and *green*) on each daughter strand. DNA synthesis on the leading strand (*top*) is continuous and accomplished by Polδ tethered to the DNA substrate via interaction with the sliding clamp PCNA. On the lagging strand (*bottom*), RNA/DNA primers synthesized by Polα are extended by Polδ discontinuously. At each Okazaki fragment, a polymerase-switching event catalyzed by the PCNA-loading complex RFC displaces Polα and recruits Polδ. RFC may remain associated with the replisome via Polδ and the single-strand binding protein RPA (not shown).

Figure 2. Histone deposition proteins at sites of DNA replication. CAF-1 is recruited to replication forks though an interaction with PCNA. Asf1 and CAF-1 cooperate to deposit newly synthesized histones. However, Asf1 and the Hir proteins also contribute to nucleosome formation in a manner independent of CAF-1. Chaperones are also likely to be involved in redepositing parental nucleosomes after replication fork passage. Likely candidates include the FACT complex and the Hir protein complex.

provide a window of opportunity where these marks can be erased or altered. Consistent with this idea, mutations in PCNA alter the extent and stability of epigenetic gene silencing in budding yeast (Zhang et al. 2000; Sharp et al. 2001). As discussed below, this role for PCNA has been associated with its interaction with the DNA replication-linked chromatin assembly factor, CAF-1.

Histone chaperones that function during DNA replication. A variety of data from multiple organisms indicate close coordination between DNA synthesis and histone deposition. Nucleosome formation occurs rapidly following DNA replication in vivo (Worcel et al. 1978; Jackson 1990). Additionally, in both human and yeast cells, histone deposition during S phase is required for viability (Han et al. 1987; Kim et al. 1988; Nelson et al. 2002; Nabatiyan and Krude 2004), and in human cells concomitant histone deposition is required for completing DNA replication (Hoek and Stillman 2003; Ye et al. 2003).

Several histone chaperones have been implicated in nucleosome assembly at sites of DNA synthesis (Fig. 2). Biochemical experiments designed to isolate replication-coupled chromatin assembly factors from human cell extracts led to the discovery of Chromatin Assembly Factor-1 (CAF-1) (Smith and Stillman 1989). CAF-1 is an evolutionarily conserved heterotrimeric protein complex that binds histones H3/H4 and delivers them to replicating DNA through an interaction with PCNA (Shibahara and Stillman 1999; Zhang et al. 2000; Krawitz et al. 2002). The biochemical activity and subunit structure of CAF-1 is conserved among yeast, plants, frogs, and flies (Kamakaka et al. 1996; Kaufman et al. 1997; Kaya et al. 2001; Quivy et al. 2001). In mammalian cells, CAF-1 colocalizes with DNA replication foci during S phase, and is recruited to sites of DNA repair at other times (Krude 1995; Green and Almouzni 2003), consistent with a function at sites of DNA synthesis. Recent work shows that inactivation of CAF-1 in human cells blocks cell cycle progression via S phase checkpoint activation (Hoek and Stillman 2003; Ye et al. 2003). Additionally, shutting off histone synthesis during S phase blocks DNA synthesis in human cells (Nelson et al. 2002; Ye et al. 2003). Together, these data demonstrate a requirement for coordinating nucleosome assembly and DNA replication in human cells. In contrast, in *Saccharomyces cerevisiae*, DNA replication can be completed in the absence of new histone synthesis, although the resulting chromosomes are unable to segregate and viability is lost (Han et al. 1987; Kim et al. 1988). These data suggest that human cells are more sensitive to perturbation of histone deposition during S phase.

Data from multiple organisms demonstrate that replication-linked histone deposition contributes to epigenetic control of gene expression. In budding yeast, the genes encoding the CAF-1 subunits (*CAC1*, *CAC2*, and *CAC3*) are not essential and are not required for cell cycle progression (Kaufman et al. 1997). However, cells lacking CAF-1 are defective in position-dependent gene silencing (Enomoto et al. 1997, 1998; Kaufman et al. 1997; Monson et al. 1997). Likewise, *Arabidopsis* mutants lacking CAF-1 display loss of transcriptional gene silencing and developmental defects in meristematic tissue (Kaya et al. 2001; Takeda et al. 2004). Additionally, human cells overexpressing a dominant-negative fragment of the large subunit of CAF-1 display reduced transcriptional silencing of a reporter gene (Tchenio et al. 2001). Together, these data demonstrate that CAF-1 plays a role in regulating chromatin-mediated transcriptional states, and that this function is conserved throughout eukaryotic organisms.

However, CAF-1 is not the only route for delivery of histones H3/H4 to chromatin. In budding yeast, simultaneous disruption of CAF-1 and a second histone chaperone protein complex, the HIR proteins, causes much more severe defects in heterochromatic silencing than observed in cells lacking either complex alone (Kaufman et al. 1998). Additionally, yeast cells lacking both CAF-1 and HIR proteins also have altered centromeric chromatin and impaired kinetochore function (Sharp et al. 2002). Together, these data demonstrate significant functional overlap between CAF-1 and the HIR proteins in budding yeast. There are four *HIR* genes in yeast (*HIR1*, *HIR2*, *HIR3*, and *HPC2*) that were identified in genetic screens for factors required for repression of histone gene transcription upon replication arrest (Osley and Lycan 1987; Xu et al. 1992). In humans, a single polypeptide, HIRA, shares homology with both the yeast Hir1 and Hir2 proteins. In budding and fission yeasts and humans, the HIR proteins repress histone transcription outside of S phase (Osley and Lycan 1987; Nelson et al. 2002; Blackwell et al. 2004). Additionally, in vitro studies of human and *Xenopus* HIRA show that HIRA facilitates replication-independent histone deposition (Ray-Gallet et al. 2002; Tagami et al. 2004). Thus, the HIR proteins are conserved eukaryotic histone deposition factors that supplement CAF-1 function.

In addition to the HIR proteins, Asf1 is another conserved histone H3/H4-binding and deposition factor. Asf1 was isolated from *Drosophila* extracts as a factor that stimulates CAF-1 activity in vitro. Specifically, addition of Asf1 stimulates substoichiometric amounts of CAF-1 to produce more nucleosomes (Tyler et al. 1999). Asf1 proteins from yeast and humans also display this activity (Sharp et al. 2001; Mello et al. 2002). Stimulation of CAF-1 by Asf1 likely is related to the direct interaction between Asf1 and the Cac2/p60 subunit of CAF-1 (Tyler et al. 2001; Krawitz et al. 2002; Mello et al. 2002). Like CAF-1, Asf1 is not essential for viability in *S. cerevisiae*, and deletion of *ASF1* does not significantly alter heterochromatic gene silencing. However, Asf1 is essential for efficient silencing and cell cycle progression in the absence of CAF-1 (Tyler et al. 1999; Sharp et al. 2001; Sutton et al. 2001). Genetic epistasis experiments demonstrated that Asf1 is required for the HIR protein-mediated silencing pathway, and Asf1 directly binds to HIR proteins (Sharp et al. 2001; Tagami et al. 2004). Additionally, Asf1 becomes essential for CAF-1 activity both in vivo and in vitro when the interaction between CAF1 and PCNA is disrupted (Krawitz et al. 2002). These data suggest that Asf1 provides an alternative mechanism for coupling histone deposition to DNA replication in addition to the PCNA–CAF-1 interaction.

Native protein complexes containing either CAF-1 or Asf1 also contain newly synthesized histones H3 and H4 (Kaufman et al. 1995; Verreault et al. 1996; Tyler et al. 1999) identified by the conserved pattern of acetylation on K5 and K12 of histone H4. These data are consistent with the view that these protein complexes target newly synthesized histones to nascent DNA. However, the functional role for histone acetylation in nucleosome assembly is not yet clear. Histone acetylation is not essential for histone deposition by either CAF-1 or Asf1 in vitro (Shibahara et al. 2000; Sharp et al. 2001). Indeed, the histone H3/H4 amino-termini can be entirely deleted without affecting deposition by CAF-1 (Shibahara et al. 2000). Therefore, the biological role for these modifications may instead be important for transport of newly translated histones into the nucleus. Consistent with this idea, an enzyme complex that performs synthesis-related histone modification has recently been identified in both the cytoplasmic and nuclear compartments (Ai and Parthun 2004; Poveda et al. 2004).

The ability of CAF-1 and Asf1 to assemble histones regardless of acetylation status suggests an additional possibility regarding histone inheritance during DNA replication. Histones in front of replication forks have different modification patterns than do newly synthesized molecules. Therefore, the lack of modification specificity of CAF-1 and Asf1 would allow them to participate in the transient dissociation and redeposition of parental histones during replication fork movement. Development of efficient in vitro systems for the replication of chromatin templates will be required to test for this activity.

Functional Overlap between Replication-coupled and Replication-independent Histone Deposition

In addition to nucleosome assembly factors that operate during DNA replication, other histone deposition proteins are likely to function during both DNA replication and transcription. Elegant studies recently established the FACT (Facilitates Chromatin Transcription/Transactions) complex as a bona fide histone deposition protein (Belotserkovskaya et al. 2003). FACT binds histone H2A/H2B dimers, moves them out of the way of elongating RNA polymerase, and subsequently promotes reformation of the nucleosome (Belotserkovskaya et al. 2003; Saunders et al. 2003). Human FACT was isolated based on its ability to promote transcription elongation on a chromatin template in vitro (Orphanides et al. 1998). Additionally, protein components of yeast FACT were isolated as Polα binding factors (Wittmeyer et al. 1999), and genetic studies in yeast and biochemical studies in *Xenopus* indicate that FACT facilitates DNA replication as well as transcription (for review, see Formosa 2003).

In budding yeast, the HIR proteins and Asf1 become important for growth in the absence of FACT (Formosa et al. 2002). A simple explanation for these genetic interactions is that the HIR/Asf1 proteins, in their capacity as histone chaperones, compensate for defects in FACT-mediated histone deposition (Formosa et al. 2002). Because FACT appears to act during both replication and transcription, these genetic data raise the possibility that the HIR/Asf1 proteins function during both replication-coupled and replication-independent nucleosome assembly (Fig. 3).

The extensive functional overlap between replication-coupled and replication-independent histone deposition in yeast may also be facilitated by the existence of a sin-

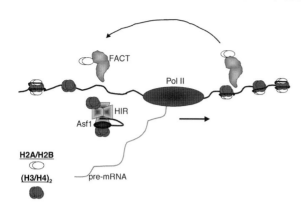

Figure 3. Transcription-coupled histone deposition. RNA transcript elongation requires disruption of nucleosomes ahead of RNA polymerase and reformation of chromatin structure in the wake of the polymerase. FACT performs both of these activities. Additionally, the HIR proteins and Asf1 might also assist reformation of nucleosomes in the wake of RNA polymerase.

gle isoform of H3 in this organism. Histone H3 in yeast is most closely related to the constitutively expressed H3.3 isoform of higher eukaryotes found in actively transcribed chromatin (Baxevanis and Landsman 1998; Ahmad and Henikoff 2002). In human cells, H3.3 copurifies with HIRA but not CAF-1. Human H3.1, in contrast, is expressed exclusively during S phase (Franklin and Zweidler 1977), and copurifies with CAF-1 but not HIRA (Tagami et al. 2004). Thus, in humans there appears to be a clear division of labor between CAF-1-mediated replication-coupled and HIRA-mediated replication-independent histone deposition. However, Asf1 is present in both the H3.1- and H3.3-containing complexes. Thus, in both single-celled yeasts and multicellular eukaryotes, replication-linked and replication-independent histone deposition pathways share Asf1 as a common factor.

Active Coupling of Histone Deposition DNA Replication

In vivo, histone deposition and DNA replication are temporally coupled (Worcel et al. 1978; Sogo et al. 1986; Jackson 1990), and inhibition of either CAF-1 or new histone synthesis in human cells inhibits DNA replication (Nelson et al. 2002; Hoek and Stillman 2003; Ye et al. 2003). Why does DNA synthesis halt when histone deposition is disrupted? One possibility is that replication in the absence of chromatin assembly results in DNA damage and activation of checkpoints halts fork movement. Indeed, disruption of CAF-1 activates the DNA Damage/S phase checkpoint (Hoek and Stillman 2003; Ye et al. 2003). Subsequent inactivation of the checkpoint either by the addition of caffeine or inhibition of both the ATM and ATR checkpoint kinases allowed DNA synthesis to proceed, but reduced cell viability (Ye et al. 2003). In yeast, the lethality caused by uncoupling histone and DNA synthesis cannot be reversed by restoring histone protein synthesis once cells have completed S phase (Han et al. 1987; Kim et al. 1988). Together, these data suggest that irreparable damage occurs to chromosomes when DNA is synthesized in the absence of concomitant histone deposition.

What kind of signal leads to checkpoint activation in human cells when CAF-1 function is disrupted? Because DNA synthesis continues in the absence of CAF-1 only when the S-phase checkpoint is removed, it is unlikely that lack of histone deposition causes a topological or physical barrier to DNA replication. Rather, we hypothesize that the observed S-phase arrest occurs because replication forks pause or stall in response to CAF-1 depletion, thereby activating the checkpoint. Consistent with this idea, replication forks themselves act as sensors of DNA damage and are required for the activation of the S-phase checkpoint in response to the DNA alkyating agent MMS (Tercero et al. 2003).

What proteins are likely to be involved in this chromatin assembly surveillance mechanism? The most well-studied mechanism for coupling histone deposition and DNA replication is the physical interaction between CAF-1 and PCNA (Shibahara and Stillman 1999; Sharp et al. 2001; Krawitz et al. 2002). Thus, one possibility is that CAF-1 via PCNA signals to the replisome that histone deposition is occurring. Additionally, the checkpoint kinase, Rad53, has recently been implicated in regulating nonnucleosomal histone protein degradation (Gunjan and Verreault 2003) and in this capacity may function as a sensor of histone levels.

A variety of data suggest that Asf1 is also important for coupling histone deposition to DNA synthesis. In budding yeast, $asf1\Delta$ cells, unlike cells lacking CAF-1, are hypersensitive to the replication inhibitor hydroxyurea (HU) (Le et al. 1997; Tyler et al. 1999; Emili et al. 2001; Hu et al. 2001). Reentry into the cell cycle after HU arrest is poor in cells lacking Asf1 (Tyler et al. 1999), suggesting the presence of irreversible damage on replication fork arrest when Asf1 is absent. Asf1 physically interacts with the DNA damage and S phase checkpoint kinase, Rad53 (Emili et al. 2001; Hu et al. 2001; Daganzo et al. 2003; Schwartz et al. 2003). Additionally, deletion of *ASF1* causes synthetic growth phenotypes in combination with mutations in a number of genes implicated in DNA replication. These include *SGS1*, encoding a RecQ-family helicase required for maintaining stalled replication fork integrity (Tong et al. 2001; Cobb et al. 2003), *TOP1*, encoding DNA topoisomerase 1, and *CDC45*, encoding a replication initiation and elongation factor (Tercero et al. 2000; Tong et al. 2004). Finally, deletion of *ASF1* increases the frequency of gross chromosomal rearrangements (Myung et al. 2003), a common phenotype for cells with DNA replication defects, causes a modest increase in plasmid loss rates that could be indicative of DNA replication or chromosome segregation defects (Le et al. 1997) and also results in increased rates of sister-chromatid exchanges (Prado et al. 2004). Together, these data suggest that Asf1 helps to actively couple histone deposition and DNA replication.

CONCLUSIONS

Histones are deposited onto DNA during DNA synthesis and at other times; some histone chaperones likely function in both pathways. Mounting evidence suggests that DNA replication and RNA transcription occur at discrete sites within the nucleus and are accomplished by large macromolecular machines that coordinate the multiple tasks necessary to complete these complex processes (for review, see Cook 1999). We are just beginning to uncover mechanisms that target histone chaperones to sites of DNA or RNA synthesis. One example of such a targeting mechanism is the interaction between the replication-coupled histone deposition complex CAF-1 and the DNA polymerase processivity protein PCNA (Shibahara and Stillman 1999; Krawitz et al. 2002). However, genetic data in yeast suggest that other pathways exist to target histone chaperones, like Asf1, to sites of DNA synthesis (Sharp et al. 2001; Krawitz et al. 2002). Much remains to be discovered about the molecular mechanisms that couple histone deposition to DNA metabolic processes.

Additionally, recent work demonstrates that concomitant histone deposition and DNA replication are essential for S phase progression in human cells (Hoek and Stillman 2003; Ye et al. 2003), raising the possibility that cells possess mechanisms for monitoring histone deposition during DNA replication. Understanding this chromatin surveillance and the replication fork defects that occur in the absence of histone deposition will be important contributions to our comprehension of genome stability.

ACKNOWLEDGMENTS

We thank Erin Green for comments on the manuscript. This work was supported by NIH grant GM55712, NSF grant MCB-0234014, and Department of Energy Field Work Proposal KP110301, and by Department of Energy funds administered through the Lawrence Berkeley National Laboratory. A.A.F. was supported by an NSF predoctoral training grant and an NIH predoctoral training grant.

REFERENCES

Ahmad K. and Henikoff S. 2002. The histone variant H3.3 marks active chromatin by replication-independent nucleosome assembly. *Mol. Cell* **9:** 1191.

Ai X. and Parthun M.R. 2004. The nuclear Hat1p/Hat2p complex: A molecular link between type B histone acetyltransferases and chromatin assembly. *Mol. Cell* **14:** 195.

Baxevanis A.D. and Landsman D. 1998. Histone sequence database: New histone fold family members. *Nucleic Acids Res.* **26:** 372.

Belotserkovskaya R., Oh S., Bondarenko V.A., Orphanides G., Studitsky V.M., and Reinberg D. 2003. FACT facilitates transcription-dependent nucleosome alteration. *Science* **301:** 1090.

Blackwell C., Martin K.A., Greenall A., Pidoux A., Allshire R.C., and Whitehall S.K. 2004. The *Schizosaccharomyces pombe* HIRA-like protein Hip1 is required for the periodic expression of histone genes and contributes to the function of complex centromeres. *Mol. Cell. Biol.* **24:** 4309.

Brill S.J. and Stillman B. 1989. Yeast replication factor-A functions in the unwinding of the SV40 origin of DNA replication. *Nature* **342:** 92.

Cobb J.A., Bjergbaek L., Shimada K., Frei C., and Gasser S.M. 2003. DNA polymerase stabilization at stalled replication forks requires Mec1 and the RecQ helicase Sgs1. *EMBO J.* **22:** 4325.

Cook P.R. 1999. The organization of replication and transcription. *Science* **284:** 1790.

Daganzo S.M., Erzberger J.P., Lam W.M., Skordalakes E., Zhang R., Franco A.A., Brill S.J., Adams P.D., Berger J.M., and Kaufman P.D. 2003. Structure and function of the conserved core of histone deposition protein Asf1. *Curr. Biol.* **13:** 2148.

Eickbush T.H. and Moudrianakis E.N. 1978. The histone core complex: An octamer assembled by two sets of protein-protein interactions. *Biochemistry* **17:** 4955.

Emili A., Schieltz D.M., Yates J.R., III, and Hartwell L.H. 2001. Dynamic interaction of DNA damage checkpoint protein Rad53 with chromatin assembly factor Asf1. *Mol. Cell* **7:** 13.

Enomoto S., Berman J., McCune-Zierath P.D., Gerami-Nejad M., and Sanders M.A. 1998. Chromatin assembly factor I contributes to the maintenance, but not the re-establishment, of silencing at the yeast silent mating loci. *Genes Dev.* **12:** 219.

Enomoto S., McCune-Zierath P.D., Gerami-Nejad M., Sanders M.A., and Berman J. 1997. RLF2, a subunit of yeast chromatin assembly factor-I, is required for telomeric chromatin function in vivo. *Genes Dev.* **11:** 358.

Fanning E. and Knippers R. 1992. Structure and function of simian virus 40 large tumor antigen. *Annu. Rev. Biochem.* **61:** 55.

Formosa T. 2003. Changing the DNA landscape: Putting a SPN on chromatin. *Curr. Top. Microbiol. Immunol.* **274:** 171.

Formosa T., Ruone S., Adams M.D., Olsen A.E., Eriksson P., Yu Y., Rhoades A.R., Kaufman P.D., and Stillman D.J. 2002. Defects in SPT16 or POB3 (yFACT) in *Saccharomyces cerevisiae* cause dependence on the Hir/Hpc pathway: Polymerase passage may degrade chromatin structure. *Genetics* **162:** 1557.

Franklin S.G. and Zweidler A. 1977. Non-allelic variants of histones 2a, 2b and 3 in mammals. *Nature* **266:** 273.

Green C.M. and Almouzni G. 2003. Local action of the chromatin assembly factor CAF-1 at sites of nucleotide excision repair in vivo. *EMBO J.* **22:** 5163.

Grewal S.I. and Moazed D. 2003. Heterochromatin and epigenetic control of gene expression. *Science* **301:** 798.

Gruss C., Wu J., Koller T., and Sogo J.M. 1993. Disruption of the nucleosomes at the replication fork. *EMBO J.* **12:** 4533.

Gunjan A. and Verreault A. 2003. A Rad53 kinase-dependent surveillance mechanism that regulates histone protein levels in *S. cerevisiae*. *Cell* **115:** 537.

Han M., Chang M., Kim U.J., and Grunstein M. 1987. Histone H2B repression causes cell-cycle-specific arrest in yeast: Effects on chromosomal segregation, replication, and transcription. *Cell* **48:** 589.

Hoek M. and Stillman B. 2003. Chromatin assembly factor 1 is essential and couples chromatin assembly to DNA replication in vivo. *Proc. Natl. Acad. Sci.* **100:** 12183.

Hu F., Alcasabas A.A., and Elledge S.J. 2001. Asf1 links Rad53 to control of chromatin assembly. *Genes Dev.* **15:** 1061.

Jackson V. 1987. Deposition of newly synthesized histones: New histones H2A and H2B do not deposit in the same nucleosome with new histones H3 and H4. *Biochemistry* **26:** 2315.

———. 1990. In vivo studies on the dynamics of histone-DNA interaction: Evidence for nucleosome dissolution during replication and transcription and a low level of dissolution independent of both. *Biochemistry* **29:** 719.

Jackson V., Shires A., Tanphaichitr N., and Chalkley R. 1976. Modifications to histones immediately after synthesis. *J. Mol. Biol.* **104:** 471.

Jenuwein T. and Allis C.D. 2001. Translating the histone code. *Science* **293:** 1074.

Jeruzalmi D., O'Donnell M., and Kuriyan J. 2002. Clamp loaders and sliding clamps. *Curr. Opin. Struct. Biol.* **12:** 217.

Kamakaka R.T., Bulger M., Kaufman P.D., Stillman B., and

Kadonaga J.T. 1996. Postreplicative chromatin assembly by *Drosophila* and human chromatin assembly factor 1. *Mol. Cell. Biol.* **16:** 810.

Kaufman P.D. and Almouzni G. 2000. DNA replication, nucleotide excision repair, and nucleosome assembly. In *Chromatin structure and gene expression* (ed. S.C.R. Elgin and J.L. Workman), p. 24. Oxford University Press, Oxford, United Kingdom.

Kaufman P.D., Cohen J.L., and Osley M.A. 1998. Hir proteins are required for position-dependent gene silencing in *Saccharomyces cerevisiae* in the absence of chromatin assembly factor I. *Mol. Cell. Biol.* **18:** 4793.

Kaufman P.D., Kobayashi R., and Stillman B. 1997. Ultraviolet radiation sensitivity and reduction of telomeric silencing in *Saccharomyces cerevisiae* cells lacking chromatin assembly factor-I. *Genes Dev.* **11:** 345.

Kaufman P.D., Kobayashi R., Kessler N., and Stillman B. 1995. The p150 and p60 subunits of chromatin assembly factor I: A molecular link between newly synthesized histones and DNA replication. *Cell* **81:** 1105.

Kaya H., Shibahara K.I., Taoka K.I., Iwabuchi M., Stillman B., and Araki T. 2001. FASCIATA genes for chromatin assembly factor-1 in *Arabidopsis* maintain the cellular organization of apical meristems. *Cell* **104:** 131.

Kim U.J., Han M., Kayne P., and Grunstein M. 1988. Effects of histone H4 depletion on the cell cycle and transcription of *Saccharomyces cerevisiae*. *EMBO J.* **7:** 2211.

Kornberg R.D. and Thomas J.O. 1974. Chromatin structure; oligomers of the histones. *Science* **184:** 865.

Krawitz D.C., Kama T., and Kaufman P.D. 2002. Chromatin assembly factor I mutants defective for PCNA binding require Asf1/Hir proteins for silencing. *Mol. Cell. Biol.* **22:** 614.

Krude T. 1995. Chromatin assembly factor 1 (CAF-1) colocalizes with replication foci in HeLa cell nuclei. *Exp. Cell Res.* **220:** 304.

Le S., Davis C., Konopka J.B., and Sternglanz R. 1997. Two new S-phase-specific genes from *Saccharomyces cerevisiae*. *Yeast* **13:** 1029.

Luger K., Mader A.W., Richmond R.K., Sargent D.F., and Richmond T.J. 1997. Crystal structure of the nucleosome core particle at 2.8 Å resolution. *Nature* **389:** 251.

Maga G. and Hubscher U. 2003. Proliferating cell nuclear antigen (PCNA): A dancer with many partners. *J. Cell Sci.* **116:** 3051.

Melcher M., Schmid M., Aagaard L., Selenko P., Laible G., and Jenuwein T. 2000. Structure-function analysis of SUV39H1 reveals a dominant role in heterochromatin organization, chromosome segregation, and mitotic progression. *Mol. Cell. Biol.* **20:** 3728.

Mello J.A., Sillje H.H., Roche D.M., Kirschner D.B., Nigg E.A., and Almouzni G. 2002. Human Asf1 and CAF-1 interact and synergize in a repair-coupled nucleosome assembly pathway. *EMBO Rep.* **3:** 329.

Monson E.K., de Bruin D., and Zakian V.A. 1997. The yeast Cac1 protein is required for the stable inheritance of transcriptionally repressed chromatin at telomeres. *Proc. Natl. Acad. Sci.* **94:** 13081.

Murakami Y., Eki T., and Hurwitz J. 1992. Studies on the initiation of simian virus 40 replication in vitro: RNA primer synthesis and its elongation. *Proc. Natl. Acad. Sci.* **89:** 952.

Myung K., Pennaneach V., Kats E.S., and Kolodner R.D. 2003. *Saccharomyces cerevisiae* chromatin-assembly factors that act during DNA replication function in the maintenance of genome stability. *Proc. Natl. Acad. Sci.* **100:** 6640.

Nabatiyan A. and Krude T. 2004. Silencing of chromatin assembly factor 1 in human cells leads to cell death and loss of chromatin assembly during DNA synthesis. *Mol. Cell. Biol.* **24:** 2853.

Nelson D.M., Ye X., Hall C., Santos H., Ma T., Kao G.D., Yen T.J., Harper J.W., and Adams P.D. 2002. Coupling of DNA synthesis and histone synthesis in S phase independent of cyclin/cdk2 activity. *Mol. Cell. Biol.* **22:** 7459.

Orphanides G., LeRoy G., Chang C.H., Luse D.S., and Reinberg D. 1998. FACT, a factor that facilitates transcript elongation through nucleosomes. *Cell* **92:** 105.

Osley M.A. and Lycan D. 1987. Trans-acting regulatory mutations that alter transcription of *Saccharomyces cerevisiae* histone genes. *Mol. Cell. Biol.* **7:** 4204.

Poveda A., Pamblanco M., Tafrov S., Tordera V., Sternglanz R., and Sendra R. 2004. Hif1 is a component of yeast histone acetyltransferase B, a complex mainly localized in the nucleus. *J. Biol. Chem.* **279:** 16033.

Prado F., Cortes-Ledesma F., and Aguilera A. 2004. The absence of the yeast chromatin assembly factor Asf1 increases genomic instability and sister chromatid exchange. *EMBO Rep.* **5:** 497.

Quivy J.P., Grandi P., and Almouzni G. 2001. Dimerization of the largest subunit of chromatin assembly factor 1: Importance in vitro and during *Xenopus* early development. *EMBO J.* **20:** 2015.

Ray-Gallet D., Quivy J.P., Scamps C., Martini E.M., Lipinski M., and Almouzni G. 2002. HIRA is critical for a nucleosome assembly pathway independent of DNA synthesis. *Mol. Cell.* **9:** 1091.

Richards E.J. and Elgin S.C. 2002. Epigenetic codes for heterochromatin formation and silencing: Rounding up the usual suspects. *Cell* **108:** 489.

Ruiz-Carrillo A., Wangh L.J., and Allfrey V.G. 1975. Processing of newly synthesized histone molecules. *Science* **190:** 117.

Saunders A., Werner J., Andrulis E.D., Nakayama T., Hirose S., Reinberg D., and Lis J. T. 2003. Tracking FACT and the RNA polymerase II elongation complex through chromatin in vivo. *Science* **301:** 1094.

Schwartz M.F., Lee S., Duong J.K., Eminaga S., and Stern D.F. 2003. FHA domain-mediated DNA checkpoint regulation of Rad53. *Cell Cycle* **2:** 384.

Sharp J.A., Fouts E.T., Krawitz D.C., and Kaufman P.D. 2001. Yeast histone deposition protein Asf1p requires Hir proteins and PCNA for heterochromatic silencing. *Curr. Biol.* **11:** 463.

Sharp J.A., Franco A.A., Osley M.A., and Kaufman P.D. 2002. Chromatin assembly factor I and Hir proteins contribute to building functional kinetochores in *S. cerevisiae*. *Genes Dev.* **16:** 85.

Shibahara K. and Stillman B. 1999. Replication-dependent marking of DNA by PCNA facilitates CAF-1-coupled inheritance of chromatin. *Cell* **96:** 575.

Shibahara K., Verreault A., and Stillman B. 2000. The N-terminal domains of histones H3 and H4 are not necessary for chromatin assembly factor-1-mediated nucleosome assembly onto replicated DNA in vitro. *Proc. Natl. Acad. Sci.* **97:** 7766.

Smith S. and Stillman B. 1989. Purification and characterization of CAF-I, a human cell factor required for chromatin assembly during DNA replication in vitro. *Cell* **58:** 15.

———. 1991. Stepwise assembly of chromatin during DNA replication in vitro. *EMBO J.* **10:** 971.

Sogo J.M., Stahl H., Koller T., and Knippers R. 1986. Structure of replicating simian virus 40 minichromosomes. The replication fork, core histone segregation and terminal structures. *J. Mol. Biol.* **189:** 189.

Sutton A., Bucaria J., Osley M.A., and Sternglanz R. 2001. Yeast Asf1 protein is required for cell cycle regulation of histone gene transcription. *Genetics* **158:** 587.

Tagami H., Ray-Gallet D., Almouzni G., and Nakatani Y. 2004. Histone H3.1 and H3.3 complexes mediate nucleosome assembly pathways dependent or independent of DNA synthesis. *Cell* **116:** 51.

Takeda S., Tadele Z., Hofmann I., Probst A.V., Angelis K.J., Kaya H., Araki T., Mengiste T., Scheid O.M., Shibahara K.-I., Scheel D., and Paszkowski J. 2004. BRU1, a novel link between responses to DNA damage and epigenetic gene silencing in *Arabidopsis*. *Genes Dev.* **18:** 782.

Tchenio T., Casella J.F., and Heidmann T. 2001. A truncated form of the human CAF-1 p150 subunit impairs the maintenance of transcriptional gene silencing in mammalian cells. *Mol. Cell. Biol.* **21:** 1953.

Tercero J.A., Labib K., and Diffley J.F. 2000. DNA synthesis at individual replication forks requires the essential initiation factor Cdc45p. *EMBO J.* **19:** 2082.

Tercero J.A., Longhese M.P., and Diffley J.F. 2003. A central role for DNA replication forks in checkpoint activation and response. *Mol. Cell* **11:** 1323.

Tong A.H., Evangelista M., Parsons A.B., Xu H., Bader G.D., Page N., Robinson M., Raghibizadeh S., Hogue C.W., Bussey H., Andrews B., Tyers M., and Boone C. 2001. Systematic genetic analysis with ordered arrays of yeast deletion mutants. *Science* **294:** 2364.

Tong A.H., Lesage G., Bader G.D., Ding H., Xu H., Xin X., Young J., Berriz G.F., Brost R.L., Chang M., Chen Y., Cheng X., Chua G., Friesen H., Goldberg D.S., Haynes J., Humphries C., He G., Hussein S., Ke L., Krogan N., Li Z., Levinson J.N., Lu H., Menard P., Munyana C., Parsons A.B., Ryan O., Tonikian R., Roberts T., Sdicu A.M., Shapiro J., Sheikh B., Suter B., Wong S.L., Zhang L.V., Zhu H., Burd C.G., Munro S., Sander C., Rine J., Greenblatt J., Peter M., Bretscher A., Bell G., Roth F.P., Brown G.W., Andrews B., Bussey H., and Boone C. 2004. Global mapping of the yeast genetic interaction network. *Science* **303:** 808.

Tsurimoto T. and Stillman B. 1991. Replication factors required for SV40 DNA replication in vitro. II. Switching of DNA polymerase alpha and delta during initiation of leading and lagging strand synthesis. *J. Biol. Chem.* **266:** 1961.

Tyler J.K., Adams C.R., Chen S.R., Kobayashi R., Kamakaka R.T., and Kadonaga J.T. 1999. The RCAF complex mediates chromatin assembly during DNA replication and repair. *Nature* **402:** 555.

Tyler J.K., Collins K.A., Prasad-Sinha J., Amiott E., Bulger M., Harte P.J., Kobayashi R., and Kadonaga J.T. 2001. Interaction between the *Drosophila* CAF-1 and Asf1 chromatin assembly factors. *Mol. Cell. Biol.* **21:** 6574.

Verreault A. 2000. De novo nucleosome assembly: New pieces in an old puzzle. *Genes Dev.* **14:** 1430.

Verreault A., Kaufman P.D., Kobayashi R., and Stillman B. 1996. Nucleosome assembly by a complex of CAF-1 and acetylated histones H3/H4. *Cell* **87:** 95.

Waga S. and Stillman B. 1998. The DNA replication fork in eukaryotic cells. *Annu. Rev. Biochem.* **67:** 721.

Waga S., Bauer G., and Stillman B. 1994. Reconstitution of complete SV40 DNA replication with purified replication factors. *J. Biol. Chem.* **269:** 10923.

Warbrick E. 2000. The puzzle of PCNA's many partners. *Bioessays* **22:** 997.

Wittmeyer J., Joss L., and Formosa T. 1999. Spt16 and Pob3 of *Saccharomyces cerevisiae* form an essential, abundant heterodimer that is nuclear, chromatin-associated, and copurifies with DNA polymerase alpha. *Biochemistry* **38:** 8961.

Worcel A., Han S., and Wong M.L. 1978. Assembly of newly replicated chromatin. *Cell* **15:** 969.

Xu H., Kim U.J., Schuster T., and Grunstein M. 1992. Identification of a new set of cell cycle-regulatory genes that regulate S-phase transcription of histone genes in *Saccharomyces cerevisiae*. *Mol. Cell. Biol.* **12:** 5249.

Ye X., Franco A.A., Santos H., Nelson D.M., Kaufman P.D., and Adams P.D. 2003. Defective S phase chromatin assembly causes DNA damage, activation of the S phase checkpoint, and S phase arrest. *Mol. Cell* **11:** 341.

Yuzhakov A., Kelman Z., Hurwitz J., and O'Donnell M. 1999. Multiple competition reactions for RPA order the assembly of the DNA polymerase delta holoenzyme. *EMBO J.* **18:** 6189.

Zhang Z., Shibahara K., and Stillman B. 2000. PCNA connects DNA replication to epigenetic inheritance in yeast. *Nature* **408:** 221.

Trilogies of Histone Lysine Methylation as Epigenetic Landmarks of the Eukaryotic Genome

M. LACHNER, R. SENGUPTA, G. SCHOTTA, AND T. JENUWEIN
Research Institute of Molecular Pathology (IMP), The Vienna Biocenter, A-1030 Vienna, Austria

The past several years have been highlighted by the landmark descriptions of the genomes of several model organisms. The significance of these findings becomes even greater following the nearly full sequence assembly of the human genome. Together, these "genome projects" have shown that more complex eukaryotic model organisms have a much bigger genome than unicellular eukaryotes, although the increased "biocomplexity" is not reflected by an equivalent expansion in the number of protein-coding genes (e.g., ~40,000 in humans vs. ~6,000 in *Saccharomyces cerevisiae*). These results strongly suggest that biocomplexity is only in part regulated by overall gene number, but largely depends on combinatorial control triggering a vast number of gene expression patterns. In addition, mechanisms other than DNA sequence information have been used during evolution to better index and regulate the complex developmental programs and key regulatory processes, such as chromosome segregation and cell division of eukaryotic genomes.

In the nuclei of almost all eukaryotic cells, genomic DNA is highly folded and compacted with histone and nonhistone proteins in a dynamic polymer called chromatin. The discoveries that DNA methylation, nucleosome remodeling, histone modification, and noncoding RNAs can organize chromatin into accessible ("euchromatic") and inaccessible ("heterochromatic") subdomains reveal epigenetic mechanisms that considerably extend the information potential of the genetic code. Thus, one genome can generate many "epigenomes" (Fig. 1), as the fertilized egg progresses through development and translates its information into a multitude of cell fates. These epigenetic mechanisms are crucial for the function of most, if not all, chromatin-templated processes and link alterations in chromatin structure to allele-specific expression differences, developmental programming of cell lineages, chromosome segregation, DNA repair, and genome stability. The implications of epigenetic research for human biology and disease, including cancer and aging, are far reaching.

The basic repeating unit of chromatin is the nucleosome, consisting of 147 bp of DNA wrapped around an octamer of the core histones H2A, H2B, H3, and H4 (Luger et al. 1997). Posttranslational modifications of the protruding histone amino-termini (histone "tails") were proposed 40 years ago to affect gene expression (Allfrey et al. 1964) and have since been demonstrated as important modulators of chromatin structure, culminating in the "histone code" hypothesis (Strahl and Allis 2000; Turner 2000; Jenuwein and Allis 2001).

Histone modifications include acetylation, phosphorylation, methylation (arginine and lysine), ubiquitination, and ADP ribosylation (van Holde 1988). In the last 4 years, histone lysine methylation has emerged as a central epigenetic modification (Jenuwein 2001; Zhang and

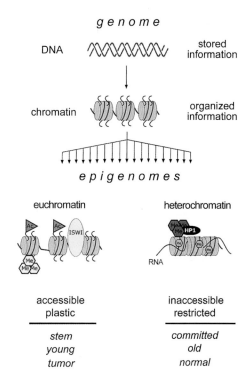

Figure 1. Genome versus epigenomes. The figure illustrates the distinction between genome (DNA sequence) and epigenome (the collective modification pattern of chromatin). Various mechanisms, including noncoding RNAs, histone modification, DNA methylation, and chromatin-remodeling activities, contribute to the generation of >1000 "epigenomes" from one genome template, as the fertilized egg progresses during development. Differences in the ratio between accessible ("euchromatic") and inaccessible ("heterochromatic") chromatin domains allow or restrict transcription from the underlying DNA sequence and may represent an organizing principle to regulate the use of the genetic information. The various epigenetic modifications are indicated as histone acetylation (*blue flag*), histone lysine methylation (*green hexagons*, active marks; *red hexagons*, repressive marks), DNA methylation (*small orange hexagons*), and remodeling factors (e.g., ISWI). See text for further explanation.

Reinberg 2001; Kouzarides 2002; Fischle et al. 2003a; Lachner et al. 2003). To date there are five well described and prominently methylated lysine positions in the histone H3 and H4 tails. According to their associated functions, histone H3 lysine 4 (H3-K4) and lysine 36 (H3-K36) methylation are classified as transcriptional "ON" marks, while histone H3 lysine 9 (H3-K9), lysine 27 (H3-K27), and histone H4 lysine 20 (H4-K20) methylation represent "OFF" marks, which are mainly involved in the organization of repressive chromatin structures. Each of these individual lysine positions can exist in three distinct methylation states: mono-, di, and trimethylation (the "trilogies").

Here, we describe a comprehensive analysis of the presence and abundance of all possible methylation states for H3-K9, H3-K27, and H4-K20 in mammalian chromatin and in several eukaryotic model organisms. We show that selective combinations between distinct methylation states can discriminate constitutive from facultative heterochromatin. Yet other combinations of repressive methyl-lysine marks appear to be involved in Polycomb-mediated gene silencing. We also discuss implications of why there are differences in mono-, di-, and trimethylation and how distinct histone lysine methylation systems are functionally connected to impart a combinatorial histone lysine methylation pattern.

In mammalian chromatin, monomethylation and, to a larger extent, trimethylation are focally enriched at chromosomal subdomains, whereas dimethylation is not very informative for regional chromatin organization. H3-K9 methylation is a hallmark of constitutive heterochromatin and is evolutionarily conserved, although there are significant differences between its abundance in *Schizosaccharomyces pombe* and mammalian chromatin. H3-K27 methylation appears to indicate the emergence of multicellularity and of the Polycomb system; it is not detectable in *S. pombe* chromatin. H4-K20 methylation is highly abundant and the most conserved of the three repressive methyl-lysine marks. Based on its broad chromosomal presence, it may serve several functions, including gene silencing and chromatin condensation. Together, these data allow the assignment of distinct roles for repressive histone lysine methylation states and underscore the indexing potential of histone lysine methylation as epigenetic landmarks in eukaryotic chromatin.

EPIGENETIC CONTROL OF EUKARYOTIC GENOMES

Identical twins are used as the cover illustration for this 69th Cold Spring Harbor Symposium—*Epigenetics*—and as such underscore the power of genetics over the subtle differences that are imparted by epigenetic control. However, "*You can inherit something beyond the DNA sequence. That's where the real excitement in genetics is now*" (Watson 2003). Indeed, alterations in DNA methylation patterns (Bird 2002), misregulated chromatin remodeling (Klochendler-Yeivin et al. 2002; Narlikar et al. 2002), and changes in histone modifications (Schneider et al. 2002) have been linked with perturbed development and tumorigenesis (Jones and Baylin 2002) and with inefficient reprogramming of cloned mammalian embryos after nuclear transfer (Jaenisch and Bird 2003).

Genome size, and how the genetic information is partitioned in more complex organisms, is another aspect to be considered when gauging the importance of epigenetic control. A comparison between the genomes of *S. cerevisiae* (Yeast Genome Directory 1997), *S. pombe* (Wood et al. 2002), *Drosophila melanogaster* (Adams et al. 2000), and *Mus musculus* (Waterston et al. 2002) indicates that genome size significantly expands with the complexity of the respective organism (≥1000-fold between *S. cerevisae* and mouse), despite an only modest increase in overall gene number (~6–7-fold between *S. cerevisiae* and mouse) (see Fig. 2). In contrast to the largely "open" genomes of the unicellular fungi, multicellular organisms have accumulated repetitive elements and noncoding regions, which, for example, in the mouse account for the majority of its DNA sequence (52% noncoding and 44% repetitive DNA). Only ~4% of the mouse genome encodes for protein function. This massive expansion of repetitive and noncoding sequences in multicellular organisms is most likely due to the incorpo-

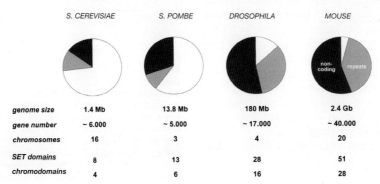

Figure 2. Genome partitioning of eukaryotic model organisms. Overview on the relative partitioning of coding (*white*), repetitive (*gray*), and noncoding (*black*) sequences within the genomes of *S. cerevisiae, S. pombe, D. melanogaster,* and *M. musculus*. Additionally, the respective genome sizes (in base pairs), approximate gene numbers, and chromosome numbers are indicated. All data were retrieved from published sequences (Yeast Genome Directory 1997; Adams et al. 2000; Waterston et al. 2002; Wood et al. 2002). Also listed are the approximate number of SET-domain and chromodomain proteins that are encoded by the various genomes and that have been derived by blast searches from public databases.

ration of invasive elements, such as DNA and retrotransposons (Kazazian 2004). Although these represent a burden for coordinated gene expression programs, they also allow for genome plasticity and a certain degree of stochastic gene regulation (Han et al. 2004).

S. cerevisiae with its "open" genome lacks almost all of the known components for epigenetic gene silencing, such as the RNAi machinery, repressive histone lysine methylation systems (see below), and DNA methylation. Intriguingly, during evolution from *S. cerevisiae* to mouse, there is a similar increase (~6–7- fold) in SET-domain and chromodomain genes, as there is for overall gene number. Putative HMTases (i.e., proteins that carry the signature SET domain) expand from 8 in *S. cervisiae* to 51 in mouse (see Fig. 2). For chromodomain (i.e., a module that has been shown to confer methyl-binding activity) proteins, there is a similar increase from 4 in *S. cerevisiae* to ~28 in mice.

Histone lysine methylation systems generally consist of a SET-domain enzyme and a chromodomain adaptor. Since there are only five prominent methyl-lysine residues in the histone H3 and H4 tails, each methylatable lysine can potentially be targeted by multiple HMTases. For example, the H3-K9 position is a substrate for the activity of at least three distinct HMTase systems in mammalian chromatin (Suv39h1 and Suv39h2: O'Carroll et al. 2000; Rea et al. 2000; G9a and Glp1: Ogawa et al. 2002; Tachibana et al. 2002; ESET: Yang et al. 2002). Thus, for the "closed" genomes of complex multicellular organisms, various repressive HMTase systems may interact with the other silencing mechanisms, such as the RNAi machinery and DNA methylation, to restrict aberrant transcriptional activity in a very effective manner. This silencing synergy is also important to prevent remobilization of transposons and contributes to the specialized chromatin organization at pericentric and telomeric regions in these highly partitioned genomes.

FUNCTIONAL ROLES OF HISTONE LYSINE MONO-, DI-, AND TRIMETHYLATION

Early biochemical analyses have indicated that histone lysine positions can be mono-, di-, or trimethylated (Paik and Kim 1971; DeLange and Smith 1973). These findings have recently been substantiated by functional studies in *S. cerevisiae* (Santos-Rosa et al. 2002) and by mass-spectrometry analyses of bulk histone preparations (Peters et al. 2003; Rice et al. 2003; L. Zhang et al. 2003). Additionally, the structural resolutions of the SET domains of various HMTases have revealed the molecular mechanism of substrate and product specificity for these enzymes (B. Xiao et al. 2003; X. Zhang et al. 2003). Based on these three-dimensional structures, putative HMTases can be predicted to be mono-, di-, or trimethylating enzymes by comparing the primary amino acid sequence of their associated SET domain. Further, directed mutational analyses allow the conversion of a monomethylating enzyme to become a trimethylating HMTase by engineering a slightly larger catalytic pocket via the exchange of a Tyr with a Phe (X. Zhang et al. 2003). Despite these significant advances, the question as to the physiological relevance of why there are three distinct histone lysine methylation states remains largely unsolved.

One possibility is that the distinct methylation states would generate different affinities for chromatin-associated proteins. The structural resolution of several chromodomain proteins, such as HP1 (Jacobs and Khorasanizadeh 2002; Nielsen et al. 2002) and Polycomb (Fischle et al. 2003c; Min et al. 2003) have allowed predictions that discriminate position-specific binding toward H3-K9 or H3-K27 methylation. HP1 displays a similar in vitro affinity for H3-K9 dimethylation and trimethylation, while its K_d is significantly reduced for H3-K9 monomethylation (Fischle et al. 2003c). An H3-K9 monomethyl mark is not sufficient for in vivo HP1 binding to pericentric heterochromatin (Peters et al. 2003). Chromatin recruitment of HP1 may be facilitated by nucleosomal surfaces (Nielsen et al. 2001; Meehan et al. 2003) and an RNA affinity (Maison et al. 2002; Muchardt et al. 2002). Similarly, the K_d of Polycomb for H3-K27 trimethylation is significantly stronger than for H3-K27 dimethyl or monomethyl peptides (Fischle et al. 2003c). These in vitro data reflect only potential in vivo affinities and, in particular for Polycomb binding, additional marks, such as H3-K9 (Czermin et al. 2002; L. Ringrose and R. Paro, pers. comm.) and possibly histone H1b-K26 methylation (Kuzmichev et al. 2004), may be required. Dimerization of HP1 (Brasher et al. 2000) or of Polycomb (Min et al. 2003) proteins could further stabilize histone tail interactions. Thus, there are known chromatin binders for dimethyled and trimethylated histone lysine residues, but none for a monomethylated substrate. Also, no mutational analyses have been described that would convert chromodomain affinity from tri- to di- to monomethyl binding.

A second possibility is that the distinct methylation states could reflect the stability of histone lysine methylation. There is currently no biochemical data on differential turnover rates of mono-, di-, and trimethylation of the various histone lysine positions. Immunofluorescence analyses with position- and state-specific methyl-lysine histone antibodies (see below) have suggested the trimethylated state as a robust mark, because it appears mitotically stable at pericentric heterochromatin (H3-K9 and H4-K20 trimethylation) (Peters et al. 2003; Schotta et al. 2004) or at the Xi (H3-K27 trimethylation) (Plath et al. 2003; Silva et al. 2003; Okamoto et al. 2004). Recent kinetic studies using an inducible *Xist* transgene have indicated prolonged association of H3-K27 trimethylation at the Xi after removal of *Xist* (Kohlmaier et al. 2004). In addition, nuclear transfer experiments revealed the persistent presence of somatic-specific histone lysine trimethylation patterns in cloned embryos (Santos et al. 2003).

There is yet another intriguing possibility for a more stable propagation of the trimethylated state. H3-K9 dimethylation and H3-S10 phosphorylation have been shown to be mutually exclusive (Rea et al. 2000). Since histone modifications do not appear to occur in isolation (Fischle et al. 2003b), H3-K9 trimethylation could be

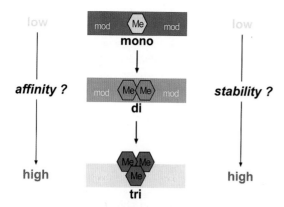

Figure 3. Potential roles of histone lysine mono-, di-, and trimethylation. Selective lysine positions in histone tails can be mono-, di-, or trimethylated, as illustrated by increasing color contrast of methyl-lysine marks (*red hexagons*). The distinct methylation states could modulate the affinity toward methyl-lysine binding proteins, differ in their turnover rates and stability, or display various potentials to interfere with antagonistic modifications (*yellow mod*) at a histone modification cassette (Fischle et al. 2003b) or in a given histone tail segment (*gray bars*). Decreasing shades of *gray bars* would reflect that a trimethylated state may impair antagonistic modifications at a higher efficiency as compared to monomethylation or dimethylation.

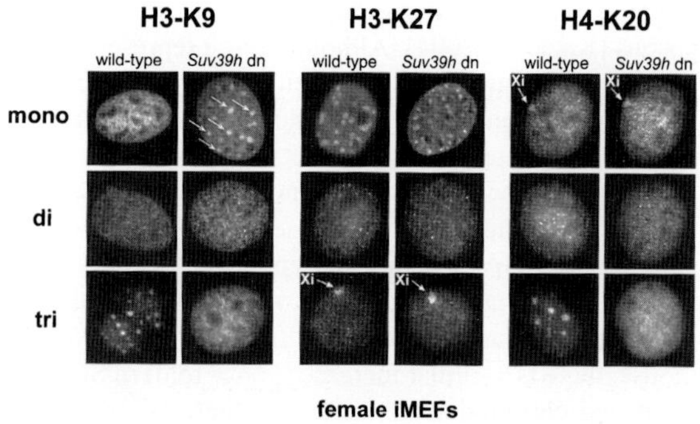

Figure 4. Repressive histone lysine methylation states are epigenetic landmarks in mouse interphase chromatin. Female wild-type and *Suv39h* dn iMEFs were stained with methyl-lysine histone antibodies that discriminate mono-, di-, and trimethylation of H3-K9 (Peters et al. 2003; Perez-Burgos et al. 2004), H3-K27 (Peters et al. 2003; Perez-Burgos et al. 2004), and H4-K20 (Schotta et al. 2004). Foci of pericentric heterochromatin that were visualized by costaining with DAPI (not shown) are decorated by H3-K9 tri-, H3-K27 mono-, and H4-K20 trimethylation in wild-type nuclei. In addition, the inactive X chromosome (Xi) is enriched for H3-K27 trimethylation and H4-K20 monomethylation. The occurrence of H3-K9 monomethylation at pericentric heterochromatin in *Suv39h* dn nuclei is indicated by multiple *arrows*.

more efficient in preventing antagonistic marks such as acetylation or phosphorylation within the same modification cassette on a given histone segment (Fig. 3). A methyl/phospho switch for HP1 binding has been proposed, predicting impaired HP1 association if the H3-K9 methylated histone tail is additionally phosphorylated at H3-S10 (Fischle et al. 2003b). It is conceivable that H3-K9 trimethylation could inhibit substrate recognition of H3-S10-specific kinases more efficiently than H3-K9 monomethylation or dimethylation (see Fig. 3) (e.g., by presenting a stronger sterical hindrance). According to this model, H3-K9 trimethylated chromosomal regions would maintain the recruitment signal for HP1 and hence propagate a distinct chromatin structure in a more stable manner. Recently, phosphoacetylation of histone H3 peptides has been shown to weaken the affinity of HP1 toward an H3-K9 dimethylated position (Mateescu et al. 2004).

HISTONE LYSINE METHYLATION PATTERNS ARE EPIGENETIC LANDMARKS OF EUKARYOTIC CHROMATIN

To examine the in vivo distribution of distinct histone lysine methylation states in native chromatin, we developed a series of position-specific methyl-lysine antibodies against H3-K9, H3-K27, and H4-K20 that additionally discriminate between mono-, di-, and trimethylation (Peters et al. 2003; Perez-Burgos et al. 2004; Schotta et al. 2004). Immunofluorescence analysis of wild-type and *Suv39h* double null (*Suv39h* dn) female mouse embryonic fibroblasts (iMEFs) (Fig. 4) indicated a rather speckled and noninformative distribution for the dimethylated state throughout the interphase nucleus. By contrast, the monomethyl and trimethyl specific antibodies revealed selective enrichment with distinct subnuclear chromatin domains.

Using DAPI staining (data not shown) to visualize the foci of pericentric heterochromatin, wild-type iMEFs display enrichment for H3-K9 tri-, H3-K27 mono-, and H4-K20 trimethylation at constitutive heterochromatin (Fig. 4). In addition, H3-K9 monomethylation is excluded from these pericentric regions. In *Suv39h* dn cells, H3-K9 trimethylation is lost from pericentric heterochromatin, which instead accumulates prominent H3-K9 monomethyl signals (see multiple arrows in Fig. 4). These data suggest that H3-K9 monomethylation (mediated by a currently unidentified H3-K9 monomethylase) could serve as an intermediate in vivo substrate for the trimethylating Suv39h enzymes. Strikingly, the loss of Suv39h activity also abolishes the staining for H4-K20 trimethylation at pericentric heterochromatin, although the H4-K20 position is not an intrinsic substrate for these enzymes. Recent data have identified a silencing pathway in which nucleosome-specific H4-K20 trimethylating enzymes (Suv4-20h HMTases) act downstream of Suv39h-dependent H3-K9 trimethylation (Schotta et al. 2004).

In these female iMEFs, the Barr body that represents the inactive X chromosome (Xi) can also be visualized as a single focus that is largely localized at the nuclear periphery (see arrows indicating Xi in Fig. 4). Intriguingly, the Xi is decorated by a different combination of methyl-lysine marks, which include H3-K27 trimethylation and H4-K20 monomethylation. This staining pattern remains unaltered in wild-type and *Suv39h* dn cells, consistent with distinct HMTase systems to control epigenetic imprints of the Xi (Plath et al. 2003; Silva et al. 2003; Kohlmaier et al. 2004; Okamoto et al. 2004; see below).

Thus, our comprehensive analysis on all nine repressive histone lysine methylation states (the three "trilogies") has revealed selective patterns to index distinct chromosomal subdomains. Constitutive heterochromatin at pericentric regions is characterized by a combination of H3-K9 tri-, H3-K27 mono-, and H4-K20 trimethylation, whereas the signature for facultative heterochromatin at the Xi consists of the two prominent marks H3-K27 trimethylation and H4-K20 monomethylation.

HMTase NETWORKS INDEX HETEROCHROMATIN, THE Xi AND Polycomb SILENCING

Based on the above data, induction of the observed histone lysine methylation pattern at constitutive heterochromatin would involve interplay of (at least) four different HMTases. Whereas the function of the predicted H3-K27 monomethylase is unresolved, another monomethylase appears required to prepare the histone H3 tail via H3-K9 monomethylation for subsequent substrate recognition by the trimethylating Suv39h enzymes (Peters et al. 2003). Suv39h-dependent H3-K9 trimethylation then provides a binding platform for the HP1 proteins and recruitment of the nucleosome-specific, trimethylating Suv4-20h HMTases (Schotta et al. 2004). None of the proposed H3-K27 and H3-K9 monomethylating enzymes are currently identified, nor do we know the nature of any of the putative H3-K27 monomethyl and H4-K20 trimethyl-lysine binders (Fig. 5).

At the Xi, the presence of H4-K20 monomethylation requires the activity of an unidentified H4-K20 monomethylase (Kohlmaier et al. 2004). In addition, H3-K9 dimethylation may accumulate at the Xi during the cell cycle (K. Plath and B. Panning, pers. comm.), where it could provide a transient signal for HP1 (Chadwick and Willard 2003). The most robust modification at facultative heterochromatin, however, is H3-K27 trimethylation, which is mediated by the Polycomb group and SET-domain enzyme Ezh (Plath et al. 2003; Silva et al. 2003; Okamoto et al. 2004). There is currently no reported association of other Polycomb group proteins with the Xi, although not all components of the Polycomb system have been examined. It is also unclear how the predicted H4-K20 monomethylase would interact with an Xi-specific Ezh2 complex to induce a combinatorial histone lysine methylation pattern.

For Polycomb-mediated gene silencing, the initial data suggested Ezh2-dependent H3-K9 and H3-K27 methylation as important signals, with a preference for H3-K27 trimethylation (Cao et al. 2002; Czermin et al. 2002; Kuzmichev et al. 2002; Müller et al. 2002). Ezh2 is an HMTase that displays activity only upon complex formation with the other Polycomb-group proteins Su(z)12 and Eed. Biochemical fractionation has recently identified several distinct Ezh2 complexes that contain different amino-terminal isoforms of Eed: PRC2 (Eed1) and PRC3 (Eed 3/4) (Kuzmichev et al. 2004). Strikingly, these complexes target mutually exclusive histone substrates when tested in a nucleosomal context containing linker histones. In this setting, PRC2 is inactive toward histone H3, and instead trimethylates the linker histone H1b at the K26 position. In contrast, PRC3 has a preference for H3-K27 trimethylation, but only if the nucleosomal templates lack the linker histones. From these data, Ezh2 appears as a very promiscuous HMTase whose substrate specificity would be context dependent. It would thus not be inconceivable that distinct Ezh complexes at different developmental options could target H3-K9, H3-K27, and/or H1b-K26 positions, which together may generate a robust combinatorial signal to recruit Polycomb complexes for stable gene silencing.

	histone methyl-mark		methyl binder	HMTase mouse	S. pombe
heterochromatin shRNAs	H3-K27	mono	?	?	n.a.
	(H3-K9)	(mono)	none	?	?
	H3-K9	tri	HP1	Suv39h	clr4
	H4-K20	tri	?	Suv4-20h	SET9
Xi Xist	H4-K20	mono	?	?	?
	H3-K9	di ?			
	H3-K27	tri	?	Ezh-?	n.a.
Polycomb "aberrant" RNAs	H3-K9	tri	Polycomb ?	Ezh-?	
	H3-K27	tri		Ezh-eed3/4	n.a.
	H1b-K26	tri ?		Ezh-eed1	n.a.

Figure 5. HMTase networks for heterochromatin, the Xi, and Polycomb. This figure summarizes combinations of mammalian histone lysine methylation marks that are enriched at distinct chromatin domains, such as pericentric heterochromatin or the inactive X chromosome (see Fig. 4), or that could provide a combinatorial signal to stabilize Polycomb complexes at their target regions. Confirmed histone lysine-methyl marks are shown in bold. Also indicated are known methyl-lysine binding partners and the relevant HMTases (with their *S. pombe* homologs, where appropriate). The Polycomb group HMTase Ezh only works in a complex and appears as a promiscuous enzyme whose substrate specificity is directed via interaction with different eed isoforms (Kuzmichev et al. 2004). Whether additional eed isoforms, eed modifications, or other Ezh complex components may alter substrate specificity toward H3-K9 or facilitate recruitment at the Xi is currently unresolved (Ezh-?). n.a.: not applicable.

HISTONE LYSINE METHYLATION STATES IN MOUSE VERSUS S. POMBE

At mouse and *Drosophila* (Schotta et al. 2004) heterochromatin, we have uncovered a sequential silencing pathway for H3-K9 and H4-K20 trimethylation. To address whether this link may also be operational in unicellular eukaryotes, we analyzed histone lysine methylation states in *S. pombe*. In this organism, the Suv39h/HP1 methylation system is represented through the Clr4/Swi6 homologs, and the Suv4-20h HMTases are conserved via the SET9 enzyme (see Fig. 5, and above). Further, the enrichment of Clr4-mediated H3-K9 methylation at the outer centromeric repeats and at the mating type loci has been demonstrated by ChIP data (Nakayama et al. 2001; Noma et al. 2001; Volpe et al. 2003).

To examine the relative abundance of repressive histone lysine methylation marks, we probed bulk histone preparations from wild-type and *clr4Δ S. pombe* strains with our panel of position- and state-specific methyl-lysine histone antibodies (Fig. 6). Nuclear extracts from wild-type and *Suv39h* dn iMEFs were used as a reference. To demonstrate the quality of the histone preparations from *S. pombe*, we first hybridized the protein blots with H3-K4-specific antibodies. Similar and abundant levels for H3-K4 mono-, di-, and trimethylation could be visualized, consistent with this active mark being a prominent modification in fission yeast (Noma and Grewal 2002). Surprisingly, there were no detectable signals with any of the H3-K9 state-specific antibodies, despite the documented presence of H3-K9 dimethylation by ChIP analyses (Nakayama et al. 2001; Noma et al. 2001; Volpe et al. 2003). These data indicate a very low abundance for H3-K9 methylation and suggest that this mark may be locally enriched only in *S. pombe* chromatin. For the three H3-K27 methylation states, we also failed to detect any signals. This would be consistent with the absence of the Polycomb system in unicellular eukaryotes, and there is no fission yeast SET-domain gene with apparent sequence homology to the Ezh HMTases.

In striking contrast, however, H4-K20 mono-, di-, and trimethylation states are very prominent in the *S. pombe* histone preparations. Interestingly, there is a moderate reduction in H4-K20 tri-, but not mono- or dimethylation levels, in *clr4Δ* mutants, similar to that observed in wild-type versus *Suv39h* dn nuclear extracts (Fig. 6, right panel). These data suggest that a fraction of H4-K20 trimethylation could be dependent on Clr4 function, indicating a possible evolutionary conservation between H3-K9 and H4-K20 methylation systems from *S. pombe* to mammals.

EVOLUTIONARY CONSERVATION OF HISTONE LYSINE METHYLATION STATES

The surprising differences in the selective abundance of repressive histone lysine methylation states in *S. pombe* versus mouse prompted us to compile a summary of histone lysine methylation "trilogies" in several other model organisms, including *S. cerevisiae*, *Arabidopsis thaliana*, and *Drosophila* (Fig. 7). For simplification, we focused only on prominent methyl-lysine positions in the histone tails and divided this summary into "ON" marks (H3-K4 and H3-K36) and "OFF" marks (H3-K9, H3-K27, and H4-K20).

S. cerevisiae chromatin does not carry any of the currently described "OFF" marks. This is consistent with the lack of the other components that have been implicated in the epigenetic hierarchy of gene silencing, such as the RNAi machinery and DNA methylation. For its few repressed chromatin domains (telomeres and the mating-type loci), *S. cerevisiae* has developed a unique silencing system that is mainly based on the SIR proteins (Grunstein 1997; Gasser and Cockell 2001). Consistent with the "activated" genome of *S. cerevisiae* (see Fig. 2), the "ON" marks are present (Briggs et al. 2001), including H3-K4 trimethylation (Santos-Rosa et al. 2002), H3-K36 methylation (Strahl et al. 2002; T. Xiao et al. 2003), and the antisilencing function imparted by H3-K79 methylation (Feng et al. 2002; van Leeuwen et al. 2002).

Although the *S. pombe* epigenome also has only a limited amount of repressed chromosomal regions (centromeres, telomeres, and mating-type loci), it contains a

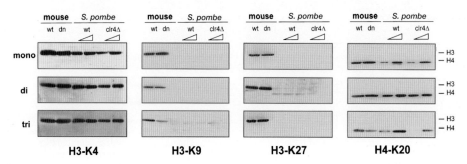

Figure 6. Protein blot analysis for histone lysine methylation states in mouse versus *S. pombe* extracts. Protein blots of nuclear extracts (12 μg) from wild-type (wt) and *Suv39h* dn iMEFs and of acid-extracted histone preparations from wt and *clr4Δ S. pombe* strains (25 μg and 50 μg, respectively) were probed with methyl-lysine histone antibodies directed against mono-, di-, and trimethylation of H3-K4 (Upstate Biotech), H3-K9 (Peters et al. 2003; Perez-Burgos et al. 2004), H3-K27 (Peters et al. 2003; Perez-Burgos et al. 2004), and H4-K20 (Schotta et al. 2004). The relative positions of histone H3 and H4 are indicated on the *far right*.

			S. cerevisiae	S. pombe	Arabidopsis	Drosophila	mouse
ON	H3-K4	mono	+	+	+	+	+
		di	+	+	+	+	+
		tri	+	+	+	+	+
	H3-K36	mono	+	nd	nd	+	+
		di	+	+	+	+	+
		tri	nd	+	+	+	+
OFF	H3-K9	mono	-	?	+	+	+
		di	-	(+)	+	+	+
		tri	-	(+)	?	(+)	+
	H3-K27	mono	-	-	+	+	+
		di	-	-	+	+	+
		tri	-	-	nd	+	+
	H4-K20	mono	-	+	+	+	+
		di	-	+	+	+	+
		tri	-	+	+	+	+

Figure 7. Evolutionary conservation of histone lysine methylation trilogies. Summary of available data on the presence and abundance of mono-, di-, and trimethylation (the "trilogies") at the H3-K4 and H3-K36 ("ON" marks) or at the H3-K9, H3-K27, and H4-K20 ("OFF" marks) positions in S. cerevisiae, S. pombe, A. thaliana, D. melanogaster, and M. musculus chromatin. +: abundant presence; –: undetectable levels; (+): this given methyl-lysine mark has been detected only by ChIP analyses or is underrepresented with respect to the other methylation states at this specific lysine position; ?: the data on this methyl-lysine mark are not confirmed. Information on some distinct methylation states was not available. nd: not done.

This summary integrates data from immunofluorescence, protein blot, ChIP, and mass spectrometry analyses, based on the following references: S. cerevisiae; H3-K4 (Briggs et al. 2001; Santos-Rosa et al. 2002); H3-K36 (Strahl et al. 2002; B. Strahl, pers. comm.); S. pombe, H3-K4 (see Fig. 6); H3-K36 (R. Sengupta and T. Jenuwein, unpubl.); H3-K9 (Nakayama et al. 2001; Noma et al. 2001; Volpe et al. 2003; M. Portoso and R. Allshire, pers. comm.; S. Grewal, pers. comm.); H3-K27 and H4-K20 (see Fig. 6); A. thaliana, H3-K4; and H3-K36 (A. Fischer and G. Reuter, pers. comm.); H3-K9 (Jasencakova et al. 2003; Jackson et al. 2004); H3-K27 (Bastow et al. 2004; Sung and Amasino 2004; A. Fischer and G. Reuter, pers. comm.); H4-K20 (A. Fischer and G. Reuter, pers. comm.); D. melanogaster; H3-K4 (G. Reuter, pers. comm.); H3-K36 (S. Kubicek and T. Jenuwein, unpubl.); H3-K9 and H3-K27 (A. Ebert et al., in prep.); H4-K20 (Schotta et al. 2004); M. musculus and H3-K4 (see Fig. 6); H3-K36 (S. Kubicek and T. Jenuwein, unpubl.); H3-K9 and H3-K27 (Peters et al. 2003); and H4-K20 (Schotta et al. 2004). Additional data used to compile this summary are available from R. Sengupta and T. Jenuwein (unpubl.).

subset of the "OFF" marks. In addition to a local enrichment of H3-K9 dimethylation at these silenced domains (Nakayama et al. 2001; Noma et al. 2001; Volpe et al. 2003), some signals for H3-K9 trimethylation may also be present at the outer centromeric regions (S. Grewal, pers. comm.). In contrast, there is no proven evidence for H3-K9 monomethylation (M. Portoso and R. Allshire, pers. comm.). Also, none of the three H3-K27 methylation states can be detected, consistent with the absence of Polycomb function in fission yeast. In *Neurospora crassa*, H3-K9 trimethylation has been shown to be important to direct DNA methylation (Tamaru et al. 2003). Intriguingly, H4-K20 methylation represents the most prominent "OFF" mark in *S. pombe*, including all three methylation states at similar abundance. However, H4-K20 methylation may not be restricted to a role in gene repression, since loss of H4-K20 methylation in *set9Δ* mutants does not appear to cause silencing defects, but rather results in increased sensitivity toward DNA damage (S. Saunders, R. Allshire, and T. Kouzarides, pers. comm.).

In *A. thaliana* chromatin, all "OFF" marks, including H3-K27 are subject to methylation, which is consistent with described Polycomb function and H3-K27-mediated gene repression at developmentally regulated loci (Bastow et al. 2004; Sung and Amasino 2004). However, there may be significant differences as compared to mammalian chromatin with respect to the full definition of the trimethylated state, as *A. thaliana* apparently lacks H3-K9 trimethylation at heterochromatic subdomains (Jasencakova et al. 2003; Jackson et al. 2004), and there are no reported data for the presence of H3-K27 trimethylation. This does not exclude local accumulation of H3-K9 and/or H3-K27 trimethylation at distinct targets or selected chromatin regions (A. Fischer and G. Reuter, pers. comm.), particularly if synergisms between some of the corresponding HMTase systems may be impaired.

Drosophila chromatin displays the full spectrum of all repressive histone lysine methylation states. Although H3-K9 dimethylation is most prominent at pericentric heterochromatin and dominates the trimethylated state, H3-K9 trimethylation is focally enriched at the core of the chromocenter (A. Ebert et al., in prep.). All three H3-K27 methylation states are present at the chromosomal arms and their accumulation is collectively governed by the Polycomb group HMTase E(z). H4-K20 trimethylation is broadly dispersed over many chromosomal regions but selectively enriched at the chromocenter, where it is required for modulation of position effect variegation (PEV) (Schotta et al. 2004).

Finally, mammalian chromatin exhibits abundant mono-, di-, and trimethylation of all examined repressive histone lysine positions (see Fig. 4). This probably reflects the need to strenghten functional interactions between the major silencing systems of the RNAi machinery, histone lysine methylation, and DNA methylation in the efficient epigenetic control of large parts of their "closed" genomes (see Fig. 2). Mass-spectrometry analyses have indicated that the sum of mono-, di-, and trimethylation accounts for >50% of bulk histones in mouse embryonic stem cell extracts to be methylated at the H3-K9 or H3-K27 positions (Peters et al. 2003).

SUMMARY AND OUTLOOK

We have examined the three repressive "trilogies" for H3-K9, H3-K27, and H4-K20 mono-, di-, and trimethylation. We have shown that all of these histone lysine methylation states are present in mammalian chromatin and that they can index selective chromatin regions in a combinatorial manner. In addition, we provided evidence for a functional interplay of the associated HMTase systems. Together with the data on the evolutionary conservation, it is possible to assign the following major roles for the distinct histone lysine methylation marks.

H3-K9 methylation is largely, although not abundantly, conserved from *S. pombe* to mammals and serves as a hallmark for gene silencing and the formation of heterochromatic regions. H3-K9 methylation can induce a repressed chromatin domain primarily at repetitive sequence elements, where its HMTase systems are likely to be recruited via short heterochromatic RNAs (shRNAs, see Fig. 5). At pericentric heterochromatin, the interplay of shRNAs and H3-K9 methylation (and of DNA methylation in the appropriate organisms) is important to safeguard centromere function and chromosome segregation.

H3-K27 methylation could indicate the emergence of multicellularity and cell-type differentiation, as it is absent in both budding and fission yeast. The presence of H3-K27 methylation signals appears tightly coupled to the existence of the Polycomb system, which is involved in lineage commitment and in coregulating the stability of gene expression programs. H3-K27 trimethylation is also a prominent epigenetic imprint for facultative heterochromatin at the Xi, where it is targeted via the noncoding *Xist* RNA. Based on the recurrent parallels of RNA-dependent recruitment of HMTase systems, it is conceivable that Polycomb-mediated "transcriptional memory" could also be triggered by an RNA moiety that may comprise "aberrant" or stalled transcripts (see Fig. 5) from developmentally regulated promoters.

H4-K20 methylation is highly conserved and most abundant across epigenetic model organisms ranging from *S. pombe* to mammals. Although H4-K20 trimethylation is enriched at pericentric heterochromatin, where it appears to be dependent on preexisting H3-K9 trimethylation, it is also present at many other chromosomal locations. Based on its high conservation and broad distribution, H4-K20 methylation is probably involved in several important functions, ranging from mitotic chromatin condensation (Fang et al. 2002; Nishioka et al. 2002; Rice et al. 2002) to transcriptional regulation and to sensing chromatin damage and probably also DNA repair. A crucial function for H4-K20 methylation in the structural organization of chromatin is further suggested by the positioning of H4-K20 at the outer boundary of the nucleosome (Luger et al. 1997; Luger 2003), where targeted modifications or recruitment of chromatin-associated factors and linker histones could trigger transitions from the 11-nm to a 30-nm chromatin fiber.

Although the above examples highlight significant advances in our understanding of the functional implications of histone lysine methylation, many questions remain. These include demonstrations on whether distinct histone marks can in general be assigned with specific biological roles, what their modification-dependent binding partners are, and how combinations of these marks may affect downstream events—all of these notions represent predictions from the "histone code" hypothesis. To examine combinatorial signals of histone lysine methylation and of other histone modifications, high-resolution mass spectrometry analyses on purified nucleosomes or on enriched chromosomal subdomains will be required, but will also pose a major challenge. Further, there are crucial questions in defining the mechanisms that confer inheritance and propagation of epigenetic information, important lines of research that may possibly lead to a molecular explanation for the distinction between germ cells versus somatic cells and to a better understanding of the nature of pluripotency. Will there be an enzyme(s) that actively removes histone lysine methyl marks, or what are the alternative mechanisms to erase or stabilize chromatin imprints during continuous rounds of cell division? The functional connections between the RNAi machinery, histone lysine methylation, and DNA methylation will continue to provide exciting insights into normal and perturbed development. It is also conceivable that differences in the relative abundance between distinct histone lysine methylation states, such as the apparent underrepresentation of H3-K9 and H3-K27 trimethylation in *S. pombe* and *A. thaliana* (see Fig. 7), may reflect the greater proliferative and regenerative potential in these organisms, as compared to the more restricted developmental programs of metazoan systems.

Thus, the detailed analysis on the three repressive "trilogies" of H3-K9, H3-K27, and H4-K20 methylation has not only broadened our views on the range of chromatin modifications, but also promises to yield future insights into many basic questions of epigenetic control, ranging from cell differentiation, stem cell plasticity, and regeneration to tumorigenesis and even aging.

ACKNOWLEDGMENTS

We are particularly indebted to Manuela Portoso and Robin Allshire for *S. pombe* strains and advice on protein blot analysis with fission yeast extracts; to Andreas Fischer, Anja Ebert, and Gunter Reuter for their expert help in defining histone lysine methylation states in *Drosophila* and *A. thaliana* chromatin; and to Upstate Biotechnology for development and exchange of some of the described methyl-lysine histone antibodies. We would like to thank Danny Reinberg, Robin Allshire, Tony Kouzarides, Shiv Grewal, Brian Strahl, Barbara Panning, and Gunter Reuter for allowing us to cite work prior to its publication. We are further grateful to Joost Martens, Stefan Kubicek, and Roddy O'Sullivan for contributing some of their unpublished data, and to all other members of the Jenuwein laboratory for their continuous interest and enthusiasm on histone lysine methylation systems. Research in the laboratory of T.J. is supported by the IMP through Boehringer Ingelheim and by grants from the Vienna Economy Promotion Fund, the European Union (EU-network HPRN-CT 2000-00078), and the Austrian GEN-AU initiative, which is financed by the Austrian Ministry of Education, Science, and Culture.

REFERENCES

Adams M.D., Celniker S.E., Holt R.A., Evans C.A., Gocayne J.D., Amanatides P.G., Scherer S.E., Li P.W., Hoskins R.A., Galle R.F., George R.A., Lewis S.E., Richards S., Ashburner M., Henderson S.N., Sutton G.G., Wortman J.R., Yandell M.D., Zhang Q., Chen L.X., Brandon R.C., Rogers Y.H., Blazej R.G., Champe M., Pfeiffer B.D., et al. 2000. The genome sequence of *Drosophila melanogaster*. *Science* **287**: 2185.

Allfrey V.G., Faulkner R., and Mirsky A.E. 1964. Acetylation and methylation of histones and their possible role in the regulation of RNA synthesis. *Proc. Natl. Acad. Sci.* **51**: 786.

Bastow R., Mylne J.S., Lister C., Lippman Z., Martienssen R.A., and Dean C. 2004. Vernalization requires epigenetic silencing of FLC by histone methylation. *Nature* **427**: 164.

Bird A. 2002. DNA methylation patterns and epigenetic memory. *Genes Dev.* **16**: 6.

Brasher S.V., Smith B.O., Fogh R.H., Nietlispach D., Thiru A., Nielsen P.R., Broadhurst R.W., Ball L.J., Murzina N.V., and Laue E.D. 2000. The structure of mouse HP1 suggests a unique mode of single peptide recognition by the shadow chromo domain dimer. *EMBO J.* **19**: 1587.

Briggs S.D., Bryk M., Strahl B.D., Cheung W.L., Davie J.K., Dent S.Y., Winston F., and Allis C.D. 2001. Histone H3 lysine 4 methylation is mediated by Set1 and required for cell growth and rDNA silencing in *Saccharomyces cerevisiae*. *Genes Dev.* **15**: 3286.

Cao R., Wang L., Wang H., Xia L., Erdjument-Bromage H., Tempst P., Jones R.S., and Zhang Y. 2002. Role of histone H3 lysine 27 methylation in Polycomb-group silencing. *Science* **298**: 1039.

Chadwick B.P. and Willard H.F. 2003. Chromatin of the Barr body: Histone and non-histone proteins associated with or excluded from the inactive X chromosome. *Hum. Mol. Genet.* **12**: 2167.

Czermin B., Melfi R., McCabe D., Seitz V., Imhof A., and Pirrotta V. 2002. *Drosophila* Enhancer of Zeste/ESC complexes have a histone H3 methyltransferase activity that marks chromosomal Polycomb sites. *Cell* **111**: 185.

DeLange R.J. and Smith E.L. 1973. Histone 3. I. Isolation and sequences of the tryptic peptides from the maleylated calf thymus protein. *J. Biol. Chem.* **248**: 3248.

Fang J., Feng Q., Ketel C.S., Wang H., Cao R., Xia L., Erdjument-Bromage H., Tempst P., Simon J.A., and Zhang Y. 2002. Purification and functional characterization of SET8, a nucleosomal histone H4-lysine 20-specific methyltransferase. *Curr. Biol.* **12**: 1086.

Feng Q., Wang H., Ng H.H., Erdjument-Bromage H., Tempst P., Struhl K., and Zhang Y. 2002. Methylation of H3-lysine 79 is mediated by a new family of HMTases without a SET domain. *Curr. Biol.* **12**: 1052.

Fischle W., Wang Y., and Allis C.D. 2003a. Histone and chromatin cross-talk. *Curr. Opin. Cell Biol.* **15**: 172.

———. 2003b. Binary switches and modification cassettes in histone biology and beyond. *Nature* **425**: 475.

Fischle W., Wang Y., Jacobs S.A., Kim Y., Allis C.D., and Khorasanizadeh S. 2003c. Molecular basis for the discrimination of repressive methyl-lysine marks in histone H3 by Polycomb and HP1 chromodomains. *Genes Dev.* **17**: 1870.

Gasser S.M. and Cockell M.M. 2001. The molecular biology of the SIR proteins. *Gene* **279**: 1.

Grunstein M. 1997. Molecular model for telomeric heterochromatin in yeast. *Curr. Opin. Cell Biol.* **9**: 383.

Han J.S., Szak S.T., and Boeke J.D. 2004. Transcriptional disruption by the L1 retrotransposon and implications for mammalian transcriptomes. *Nature* **429**: 268.

Jackson J.P., Johnson L., Jasencakova Z., Zhang X., PerezBurgos L., Singh P.B., Cheng X., Schubert I., Jenuwein T., and Jacobsen S.E. 2004. Dimethylation of histone H3 lysine 9 is a critical mark for DNA methylation and gene silencing in *Arabidopsis thaliana*. *Chromosoma* **112**: 308.

Jacobs S.A. and Khorasanizadeh S. 2002. Structure of HP1 chromodomain bound to a lysine 9-methylated histone H3 tail. *Science* **295**: 2080.

Jaenisch R. and Bird A. 2003. Epigenetic regulation of gene expression: How the genome integrates intrinsic and environmental signals. *Nat. Genet.* **33**: 245.

Jasencakova Z., Soppe W.J., Meister A., Gernand D., Turner B.M., and Schubert I. 2003. Histone modifications in *Arabidopsis*- high methylation of H3 lysine 9 is dispensable for constitutive heterochromatin. *Plant J.* **33**: 471.

Jenuwein T. 2001. Re-SET-ting heterochromatin by histone methyltransferases. *Trends Cell Biol.* **11**: 266.

Jenuwein T. and Allis C.D. 2001. Translating the histone code. *Science* **293**: 1074.

Jones P.A. and Baylin S.B. 2002. The fundamental role of epigenetic events in cancer. *Nat. Rev. Genet.* **3**: 415.

Kazazian H.H., Jr. 2004. Mobile elements: Drivers of genome evolution. *Science* **303**: 1626.

Klochendler-Yeivin A., Muchardt C., and Yaniv M. 2002. SWI/SNF chromatin remodeling and cancer. *Curr. Opin. Genet. Dev.* **12**: 73.

Kohlmaier A., Savarese F., Lachner M., Martens J.H., Jenuwein T., and Wutz A. 2004. A chromosomal memory triggered by Xist regulates histone methylation in X inactivation. *PloS Biol.* **2**: 991.

Kouzarides T. 2002. Histone methylation in transcriptional control. *Curr. Opin. Genet. Dev.* **12**: 198.

Kuzmichev A., Jenuwein T., Tempst P., and Reinberg D. 2004. Different EZH2-containing complexes target methylation of histone H1 or nucleosomal histone H3. *Mol. Cell* **14**: 183.

Kuzmichev A., Nishioka K., Erdjument-Bromage H., Tempst P., and Reinberg D. 2002. Histone methyltransferase activity associated with a human multiprotein complex containing the Enhancer of Zeste protein. *Genes Dev.* **16**: 2893.

Lachner M., O'Sullivan R.J., and Jenuwein T. 2003. An epigenetic road map for histone lysine methylation. *J. Cell Sci.* **116**: 2117.

Luger K. 2003. Structure and dynamic behavior of nucleosomes. *Curr. Opin. Genet. Dev.* **13**: 127.

Luger K., Mader A.W., Richmond R.K., Sargent D.F., and Richmond T.J. 1997. Crystal structure of the nucleosome core particle at 2.8 Å resolution. *Nature* **389**: 251.

Maison C., Bailly D., Peters A.H., Quivy J.P., Roche D., Taddei A., Lachner M., Jenuwein T., and Almouzni G. 2002. Higher-order structure in pericentric heterochromatin involves a distinct pattern of histone modification and an RNA component. *Nat. Genet.* **30**: 329.

Mateescu B., England P., Halgand F., Yaniv M., and Muchardt C. 2004. Tethering of HP1 proteins to chromatin is relieved by phosphoacetylation of histone H3. *EMBO Rep.* **5**: 490.

Meehan R.R., Kao C.F., and Pennings S. 2003. HP1 binding to native chromatin in vitro is determined by the hinge region and not by the chromodomain. *EMBO J.* **22**: 3164.

Min J., Zhang Y., and Xu R.M. 2003. Structural basis for specific binding of Polycomb chromodomain to histone H3 methylated at Lys 27. *Genes Dev.* **17**: 1823.

Muchardt C., Guilleme M., Seeler J.S., Trouche D., Dejean A., and Yaniv M. 2002. Coordinated methyl and RNA binding is required for heterochromatin localization of mammalian HP1alpha. *EMBO Rep.* **3**: 975.

Müller J., Hart C.M., Francis N.J., Vargas M.L., Sengupta A., Wild B., Miller E.L., O'Connor M.B., Kingston R.E., and Simon J.A. 2002. Histone methyltransferase activity of a *Drosophila* Polycomb group repressor complex. *Cell* **111**: 197.

Nakayama J., Rice J.C., Strahl B.D., Allis C.D., and Grewal S.I. 2001. Role of histone H3 lysine 9 methylation in epigenetic control of heterochromatin assembly. *Science* **292**: 110.

Narlikar G.J., Fan H.Y., and Kingston R.E. 2002. Cooperation between complexes that regulate chromatin structure and transcription. *Cell* **108**: 475.

Nielsen A.L., Oulad-Abdelghani M., Ortiz J.A., Remboutsika E., Chambon P., and Losson R. 2001. Heterochromatin formation in mammalian cells: Interaction between histones and HP1 proteins. *Mol. Cell* **7**: 729.

Nielsen P.R., Nietlispach D., Mott H.R., Callaghan J., Bannister A., Kouzarides T., Murzin A.G., Murzina N.V., and Laue E.D. 2002. Structure of the HP1 chromodomain bound to histone H3 methylated at lysine 9. *Nature* **416**: 103.

Nishioka K., Rice J.C., Sarma K., Erdjument-Bromage H., Werner J., Wang Y., Chuikov S., Valenzuela P., Tempst P., Steward R., Lis T.J., Allis C.D., and Reinberg D. 2002. PR-Set7 is a nucleosome-specific methyltransferase that modifies lysine 20 of histone H4 and is associated with silent chromatin. *Mol. Cell* **9:** 1201.

Noma K. and Grewal S.I. 2002. Histone H3 lysine 4 methylation is mediated by Set1 and promotes maintenance of active chromatin states in fission yeast. *Proc. Natl. Acad. Sci.* **99:** 16438.

Noma K., Allis C.D., and Grewal S.I. 2001. Transitions in distinct histone H3 methylation patterns at the heterochromatin domain boundaries. *Science* **293:** 1150.

O'Carroll D., Scherthan H., Peters A.H., Opravil S., Haynes A.R., Laible G., Rea S., Schmid M., Lebersorger A., Jerratsch M., Sattler L., Mattei M.G., Denny P., Brown S.D., Schweizer D., and Jenuwein T. 2000. Isolation and characterization of Suv39h2, a second histone H3 methyltransferase gene that displays testis-specific expression. *Mol. Cell. Biol.* **20:** 9423.

Ogawa H., Ishiguro K., Gaubatz S., Livingston D.M., and Nakatani Y. 2002. A complex with chromatin modifiers that occupies E2F- and Myc-responsive genes in G0 cells. *Science* **296:** 1132.

Okamoto I., Otte A.P., Allis C.D., Reinberg D., and Heard E. 2004. Epigenetic dynamics of imprinted X inactivation during early mouse development. *Science* **303:** 644.

Paik W.K. and Kim S. 1971. Protein methylation. *Science* **174:** 114.

Perez-Burgos L., Peters A.H., Opravil S., Kauer M., Mechtler K., and Jenuwein T. 2004. Generation and characterization of methyl-lysine histone antibodies. *Methods Enzymol.* **376:** 234.

Peters A.H., Kubicek S., Mechtler K., O'Sullivan R.J., Derijck A.A., Perez-Burgos L., Kohlmaier A., Opravil S., Tachibana M., Shinkai Y., Martens J.H., and Jenuwein T. 2003. Partitioning and plasticity of repressive histone methylation states in mammalian chromatin. *Mol. Cell* **12:** 1577.

Plath K., Fang J., Mlynarczyk-Evans S.K., Cao R., Worringer K.A., Wang H., de la Cruz C.C., Otte A.P., Panning B., and Zhang Y. 2003. Role of histone H3 lysine 27 methylation in X inactivation. *Science* **300:** 131.

Rea S., Eisenhaber F., O'Carroll D., Strahl B.D., Sun Z.W., Schmid M., Opravil S., Mechtler K., Ponting C.P., Allis C.D., and Jenuwein T. 2000. Regulation of chromatin structure by site-specific histone H3 methyltransferases. *Nature* **406:** 593.

Rice J.C., Nishioka K., Sarma K., Steward R., Reinberg D., and Allis C.D. 2002. Mitotic-specific methylation of histone H4 Lys 20 follows increased PR-Set7 expression and its localization to mitotic chromosomes. *Genes Dev.* **16:** 2225.

Rice J.C., Briggs S.D., Ueberheide B., Barber C.M., Shabanowitz J., Hunt D.F., Shinkai Y., and Allis C.D. 2003. Histone methyltransferases direct different degrees of methylation to define distinct chromatin domains. *Mol. Cell* **12:** 1591.

Santos F., Zakhartchenko V., Stojkovic M., Peters A., Jenuwein T., Wolf E., Reik W., and Dean W. 2003. Epigenetic marking correlates with developmental potential in cloned bovine preimplantation embryos. *Curr. Biol.* **13:** 1116.

Santos-Rosa H., Schneider R., Bannister A.J., Sherriff J., Bernstein B.E., Emre N.C., Schreiber S.L., Mellor J., and Kouzarides T. 2002. Active genes are tri-methylated at K4 of histone H3. *Nature* **419:** 407.

Schneider R., Bannister A.J., and Kouzarides T. 2002. Unsafe SETs: Histone lysine methyltransferases and cancer. *Trends Biochem. Sci.* **27:** 396.

Schotta G., Lachner M., Sarma K., Ebert A., Sengupta R., Reuter G., Reinberg D., and Jenuwein T. 2004. A silencing pathway to induce H3-K9 and H4-K20 trimethylation at constitutive heterochromatin. *Genes Dev.* **18:** 1251.

Silva J., Mak W., Zvetkova I., Appanah R., Nesterova T.B., Webster Z., Peters A.H., Jenuwein T., Otte A.P., and Brockdorff N. 2003. Establishment of histone H3 methylation on the inactive X chromosome requires transient recruitment of Eed-Enx1 polycomb group complexes. *Dev. Cell* **4:** 481.

Strahl B.D. and Allis C.D. 2000. The language of covalent histone modifications. *Nature* **403:** 41.

Strahl B.D., Grant P.A., Briggs S.D., Sun Z.W., Bone J.R., Caldwell J.A., Mollah S., Cook R.G., Shabanowitz J., Hunt D.F., and Allis C.D. 2002. Set2 is a nucleosomal histone H3-specific methyltransferase that mediates transcriptional repression. *Mol. Cell. Biol.* **22:** 1298.

Sung S. and Amasino R.M. 2004. Vernalization in *Arabidopsis thaliana* is mediated by the PHD finger protein VIN3. *Nature* **427:** 159.

Tachibana M., Sugimoto K., Nozaki M., Ueda J., Ohta T., Ohki M., Fukuda M., Takeda N., Niida H., Kato H., and Shinkai Y. 2002. G9a histone methyltransferase plays a dominant role in euchromatic histone H3 lysine 9 methylation and is essential for early embryogenesis. *Genes Dev.* **16:** 1779.

Tamaru H., Zhang X., McMillen D., Singh P.B., Nakayama J., Grewal S.I., Allis C.D., Cheng X., and Selker E.U. 2003. Trimethylated lysine 9 of histone H3 is a mark for DNA methylation in *Neurospora crassa*. *Nat. Genet.* **34:** 75.

Turner B.M. 2000. Histone acetylation and an epigenetic code. *Bioessays* **22:** 836.

van Holde K.E. 1988. *Chromatin*. Springer Verlag, New York.

van Leeuwen F., Gafken P.R., and Gottschling D.E. 2002. Dot1p modulates silencing in yeast by methylation of the nucleosome core. *Cell* **109:** 745.

Volpe T., Schramke V., Hamilton G.L., White S.A., Teng G., Martienssen R.A., and Allshire R.C. 2003. RNA interference is required for normal centromere function in fission yeast. *Chromosome Res.* **11:** 137.

Waterston R.H., Lindblad-Toh K., Birney E., Rogers J., Abril J.F., Agarwal P., Agarwala R., Ainscough R., Alexandersson M., An P., Antonarakis S.E., Attwood J., Baertsch R., Bailey J., Barlow K., Beck S., Berry E., Birren B., Bloom T., Bork P., Botcherby M., Bray N., Brent M.R., Brown D.G., and Brown S.D., et al. (Mouse Genome Sequencing Consortium). 2002. Initial sequencing and comparative analysis of the mouse genome. *Nature* **420:** 520.

Watson J.D. (Interview). 2003. Celebrating the genetic jubilee: A conversation with James D. Watson. *Sci. Am.* (April issue) **2003:** 67.

Wood V., Gwilliam R., Rajandream M.A., Lyne M., Lyne R., Stewart A., Sgouros J., Peat N., Hayles J., Baker S., Basham D., Bowman S., Brooks K., Brown D., Brown S., Chillingworth T., Churcher C., Collins M., Connor R., Cronin A., Davis P., Feltwell T., Fraser A., Gentles S., and Goble A., et al. 2002. The genome sequence of *Schizosaccharomyces pombe*. *Nature* **415:** 871.

Xiao B., Jing C., Wilson J.R., Walker P.A., Vasisht N., Kelly G., Howell S., Taylor I.A., Blackburn G.M., and Gamblin S.J. 2003. Structure and catalytic mechanism of the human histone methyltransferase SET7/9. *Nature* **421:** 652.

Xiao T., Hall H., Kizer K.O., Shibata Y., Hall M.C., Borchers C.H., and Strahl B.D. 2003. Phosphorylation of RNA polymerase II CTD regulates H3 methylation in yeast. *Genes Dev.* **17:** 654.

Yang L., Xia L., Wu D.Y., Wang H., Chansky H.A., Schubach W.H., Hickstein D.D., and Zhang Y. 2002. Molecular cloning of ESET, a novel histone H3-specific methyltransferase that interacts with ERG transcription factor. *Oncogene* **21:** 148.

Yeast Genome Directory. 1997. The yeast genome directory. *Nature* (suppl. 6632) **387:** 5.

Zhang L., Eugeni E.E., Parthun M.R., and Freitas M.A. 2003. Identification of novel histone post-translational modifications by peptide mass fingerprinting. *Chromosoma* **112:** 77.

Zhang X., Yang Z., Khan S.I., Horton J.R., Tamaru H., Selker E.U., and Cheng X. 2003. Structural basis for the product specificity of histone lysine methyltransferases. *Mol. Cell* **12:** 177.

Zhang Y. and Reinberg D. 2001. Transcription regulation by histone methylation: Interplay between different covalent modifications of the core histone tails. *Genes Dev.* **15:** 2343.

Histone H3 Amino-Terminal Tail Phosphorylation and Acetylation: Synergistic or Independent Transcriptional Regulatory Marks?

C.J. Fry,[*] M.A. Shogren-Knaak,[*] and C.L. Peterson
Program in Molecular Medicine, University of Massachusetts Medical School, Worcester, Massachusetts 01605

Eukaryotic cells package their large genomes into higher-order nucleoprotein complexes termed chromatin, creating a barrier to many nuclear processes that require access to the DNA. The fundamental unit of chromatin is the nucleosome, which consists of an octamer of core histone proteins (two each of histone H2A, H2B, H3, and H4). Each histone octamer contains an $(H3-H4)_2$ tetramer flanked by two H2A-H2B dimers that form a nucleosome core that binds to and wraps 1.65 turns (147 bp) of superhelical DNA (Luger et al. 1997). This wrapping of DNA around the nucleosome core partially restricts access to DNA-binding proteins. In addition, histone proteins also contain nonstructured amino- and carboxy-terminal tail domains that protrude from the nucleosome core and may interact with neighboring nucleosomes and nucleosomal DNA to further restrict access to DNA-binding proteins. Individual nucleosomes are connected by 10–60 bp of "linker" DNA to form contiguous nucleosomal arrays that are condensed into highly compacted 100–400-nm-thick fibers in interphase chromosomes. Chromosomes are further compacted during mitosis, condensing the DNA up to 10,000-fold (Hayes and Hansen 2001; Woodcock and Dimitrov 2001). How then do cells regulate the accessibility of their genomic DNA?

In the last several decades, we have begun to understand how cells regulate chromatin structure to allow accessibility of the DNA. One mechanism is provided by the ATP-dependent chromatin remodeling enzymes, which have been shown to increase the mobility of nucleosomes and to increase the accessibility of nucleosomal DNA to DNA-binding factors (Becker and Horz 2002). A second mechanism involves the posttranslational modification of the histone proteins. The amino-terminal tails of histones H2A, H2B, H3, and H4, in addition to the carboxy-terminal tail of histone H2A, are subject to numerous posttranslational modifications including phosphorylation, acetylation, ubiquitination, methylation, sumoylation, and ribosylation (Fischle et al. 2003b and references within). Recent studies have even identified modifications of residues within the nucleosome core that are involved in regulating chromatin structure (Ng et al. 2002; van Leeuwen et al. 2002; Zhang et al. 2003). Posttranslational histone modifications may directly regulate DNA/chromatin compaction by altering histone–DNA contacts or disrupting nucleosome–nucleosome interactions that mediate chromatin folding (Hayes and Hansen 2001). Alternatively, these modifications may regulate the recognition/binding motifs for other chromatin enzymes or binding proteins, providing a regulatory scaffold for transcription and other cellular processes that access DNA.

Here we focus on one of the earliest and least understood modifications linked to transcriptional regulation, phosphorylation of the histone H3 amino-terminal tail, and discuss several mechanisms by which phosphorylation is thought to regulate transcription. Specifically, we will present experiments addressing a recent model proposing that histone H3 phosphorylation and acetylation are coupled and discuss several alternative models by which H3 phosphorylation may contribute to the regulation of chromatin structure and transcription.

EARLY EVIDENCE FOR HISTONE PHOSPHORYLATION: ACTIVATION OF IMMEDIATE-EARLY GENES

Studies on immediate-early gene expression in mammalian cells provided some of the earliest evidence that chromatin plays a major regulatory role in transcription. When stimulated with growth factors, cytokines, or specific pharmacological agents, quiescent cells elicit a rapid response, involving the activation of intracellular signaling cascades leading to altered gene expression and cell growth. The first phase of gene activation following stimulation with mitogens involves the expression of immediate-early genes c-fos and c-jun. Upon stimulation, the MAP kinase cascade targets the phosphorylation of downstream transcriptional regulatory proteins that activate transcription of these genes. Early studies by Allfrey and colleagues found that the chromatin structure of the c-fos and c-myc genes assumes a more open confirmation upon gene activation (Allegra et al. 1987; Chen and Allfrey 1987), suggesting that the modification of chromatin structure may be a key step in activation; however, it was not clear whether the observed chromatin changes were a cause or consequence of transcription.

Mahadevan and coworkers sought to identify downstream components of the MAP kinase signaling cascade by phosphate labeling mouse fibroblasts and purifying proteins that become phosphorylated after induction with epidermal growth factor (EGF) and 12-0-tetrade-

---

[*]These authors contributed equally to this work.

canoylphorbol 13-acetate (TPA) (Mahadevan et al. 1991). Surprisingly, the authors identified histone H3 as a key downstream target of the MAP kinase cascade, suggesting that histone H3 phosphorylation contributes to the activation of immediate-early genes. Phosphorylation of histone H3 was resistant to the transcriptional inhibitor α-amanitin, suggesting that phosphorylation of H3 was not a consequence of transcription, but rather a regulatory signal delivered to chromatin upon stimulation with mitogens. This was later confirmed by studies showing that pretreatment of cells with kinase inhibitors or deletion of the histone H3 kinase reduced activation of c-fos and c-jun upon mitogen stimulation (Soloaga et al. 2003). These studies provided the first evidence linking histone H3 phosphorylation to transcriptional activation.

Histone H3, in addition to histone H1, had long been known to be globally phosphorylated during mitosis when chromosomes are condensed (for a recent review, see Prigent and Dimitrov 2003). Unlike mitosis, histone H3 phosphorylation during immediate-early gene activation occurs on only a small fraction of nucleosomes and is targeted to active genes. Studies using fibroblasts established from Coffin–Lowry Syndrome (CLS) patients suggested that Rsk-2, a member of the pp90^{rsk} family of kinases implicated in cell growth control, was the downstream effector kinase responsible for phosphorylating histone H3 during activation of immediate-early genes (Sassone-Corsi et al. 1999). However, more recent studies strongly implicate the mitogen-stimulated kinases Msk1 and Msk2 (Wiggin et al. 2002; Soloaga et al. 2003). How does histone H3 phosphorylation, a mark associated with condensed chromosomes, facilitate transcriptional activation? The answers came much later with the advent of histone modification-specific antibodies and chromatin immunoprecipitation techniques.

HISTONE H3 PHOSPHORYLATION AND ACETYLATION: EVIDENCE FOR COUPLED VERSUS INDEPENDENT MARKS

In addition to phosphorylation, early studies showed that histone H3 was also acetylated at immediate-early genes during transcriptional activation (Allegra et al. 1987; Chen and Allfrey 1987). This acetylation is likely to be mediated by histone acetyltransferase (HAT) proteins such as CBP/p300 or P/CAF, which have been shown to interact with numerous transcriptional activator proteins. Recent chromatin immunoprecipitation (ChIP) studies identified serine 10 as the phosphorylation site and lysines 9 and 14 as acetylation sites on the histone H3 amino-terminal tail that accompany activation of the c-fos and c-jun genes (Cheung et al. 2000; Thomson et al. 2001). More importantly, these studies showed that phosphorylation and acetylation occur on the same histone H3 tails, suggesting that the two modifications might be linked.

There has been much debate over the last several years whether H3-S10 phosphorylation and K9/K14 acetylation are coupled or whether they have independent functions during transcriptional activation. One model suggests a synergistic mechanism for the two modifications, where prior S10 phosphorylation enhances subsequent K9/K14 acetylation (Cheung et al. 2000; Lo et al. 2000). This model is based in part on studies of the immediate-early genes. Cheung and coworkers found that after stimulation of C3H cells with EGF, bulk histone H3 S10 phosphorylation precedes K9/K14 acetylation, suggesting a temporal pattern of these two modifications (Cheung et al. 2000). Furthermore, the authors suggest that pretreatment of cells with an inhibitor of the MAP kinase signaling pathway (PD98059) reduces both phosphorylation and acetylation of histone H3 at the c-fos promoter, suggesting that the two modifications may be linked. Strong support for the synergistic model comes from in vitro studies showing that H3-S10 phosphorylation stimulates the acetylation of a small peptide corresponding to the H3 amino-terminal tail by the catalytic HAT domain of yeast Gcn5p (Cheung et al. 2000; Lo et al. 2000). A very modest stimulation was also reported for the catalytic domains of PCAF and p300, and purified yeast NuA3 and SAGA HAT complexes (Lo et al. 2000).

Although H3-S10 phosphorylation was found to stimulate the activity of HAT proteins in vitro, it is still not clear whether histone H3 phosphorylation and acetylation are coupled during the activation of immediate-early genes. In an extensive study of c-jun activation, Thomson and colleagues show that prior treatment of cells with MAP kinase inhibitors SB 203580 (specific for MAP kinase 38) or H89 (specific for Msk1), or deletion of MSK1 and MSK2, severely reduces H3-S10 phosphorylation levels but has no effect on acetylation of histone H3 at the c-jun promoter following anisomycin stimulation (Thomson et al. 2001; Soloaga et al. 2003). Furthermore, ChIP experiments using modification-specific antibodies did not find phosphorylation preceding acetylation of histone H3 at the c-jun promoter during activation, rather H3-S10 phosphorylation and K9/K14 acetylation occurred simultaneously. Taken together, these later studies strongly suggest that H3-S10 phosphorylation and K9/K14 acetylation are marks deposited on histone H3 independently during activation of immediate-early genes.

Although the function of H3-S10 phosphorylation may be to enhance K9/K14 acetylation in some instances, other studies clearly show that phosphorylation has additional acetylation-independent functions in transcriptional activation. Nowak and Corces (2000) show that during the heat shock response in *Drosophila*, H3-S10 phosphorylation, but not K9/K14 acetylation, increases at the heat shock genes during activation. Furthermore, non–heat shock regions of polytene chromosomes show a decrease in H3-S10 phosphorylation upon transcriptional repression during heat shock. Taken together, these results suggest that in *Drosophila*, histone H3 phosphorylation plays a global role in transcriptional activation that is independent of acetylation.

Further support for a synergistic phosphorylation/acetylation model comes primarily from studies in yeast. Mutation of histone H3 serine 10 to alanine reduces transcription levels of a few genes in yeast, including HO, HIS3, and INO1 (Lo et al. 2000; Lo et al. 2001); however, we have found no changes in cell-cycle-dependent HO expression in an S10A mutant strain (C.J. Fry et al., unpubl.). During their studies of the INO1 gene, Lo and coworkers (2001) found that Snf1p, a protein kinase previously implicated in the regulation of INO1 gene expression, was able to phosphorylate histone H3 on S10 in vitro. Deletion of the Snf1 kinase, or the Gcn5p HAT, or mutation of histone H3 S10 or K14 reduces the activation of INO1 when yeast cells are shifted from high- to low-inositol medium. Furthermore, deletion of Snf1p or mutation of histone H3 S10A greatly reduces the acetylation of K9/K14 by Gcn5p during activation. Taken together, these results invoke the synergistic phosphorylation/acetylation model for the yeast INO1 gene. However, it is unclear whether histone H3 S10 phosphorylation plays a role in the activation of any other Gcn5p-regulated genes.

MECHANISM OF COUPLING ACETYLATION TO PHOSPHORYLATION: BINDING OF PHOSPHORYLATED TAILS

Does histone phosphorylation directly enhance histone acetylation? If so, what is the mechanism for the coupling of these two events? At least three different models exist to explain transcriptional activation by H3-S10 phosphorylation, some of which are independent of acetylation (Fig. 1). In the first model (Fig. 1A), histone tail phosphorylation creates a binding site for a HAT complex, thereby facilitating histone acetylation. In the second model (Fig. 1B), H3 serine 10 phosphorylation disrupts higher-order chromatin structure to facilitate the binding of transcriptional regulators. In the third model (Fig. 1C), histone tail phosphorylation disrupts binding of inhibitor/repressor proteins, allowing the binding of positive regulators of transcription.

Initial studies focused on the ability of H3-S10 phosphorylation to directly recruit acetyltransferase enzymes. Many chromatin-associated proteins bind histone tails at acetyl-lysine and methyl-lysine residues using structurally conserved bromo- and chromodomains, respectively (Dhalluin et al. 1999; Nielsen et al. 2002). Furthermore, a number of phosphoserine- and phosphothreonine-binding domains exist, such as 14-3-3 proteins, WW domains, and forkhead-associated domains (Yaffe and Elia 2001). While these domains have not been shown to bind histone tails, it was expected that a similar phosphoserine, histone tail-binding domain would be found. Based on the cocrystal structure of Tetrahymena Gcn5 bound to an unphosphorylated histone H3 peptide (Rojas et al. 1999), it was proposed that arginine 164 of Gcn5 might interact electrostatically with the anionic oxygens of a phosphoserine 10 moiety, thereby increasing the affinity of Gcn5 for phosphorylated histone H3 tails (Cheung et al. 2000; Lo et al. 2000). In support of this hypothesis, kinetic analysis using the catalytic domain of yeast Gcn5 and phosphorylated and unphosphorylated histone H3 tail peptides show that phosphorylation does enhance peptide acetylation by a factor of 6–10 (Cheung et al. 2000; Lo et al. 2000). This enhancement is due to an increase in the binding affinity, not turnover efficiency, of Gcn5 toward phosphorylated peptides. Furthermore, mutagenesis of the putative phosphate-binding residue, arginine 164, eliminated this phosphate-mediated stimulation of acetylation, suggesting that arginine 164 does contact the phosphoserine 10 residue. Genetic studies in yeast show that mutation of arginine 164 results in slow growth under some conditions, and a reduction in levels of transcription of some genes, including HIS3 and HO, lending in vivo support for a coupled phosphorylation and acetylation mechanism (Lo et al. 2000).

In cells, histones exist within chromatin, and this context has the potential to modulate the interactions with histone tails. For example, histone tails mediate interaction between nucleosomes in folded nucleosomal arrays, and these interactions change the ability of histone tails to be acetylated (Hayes and Hansen 2001). Furthermore, in a cellular context, histone acetyltransferases do not function in isolation, but are instead part of larger protein complexes that influence histone acetyltransferase activity. It is known that Gcn5-containing complexes bind more tightly to nucleosomal substrates and acetylate a larger range of lysine residues on nuclesomal histone tails

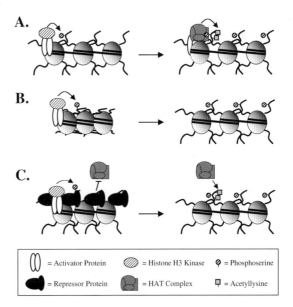

Figure 1. Possible mechanisms by which histone H3 phosphorylation may regulate chromatin structure and transcription. (*A*) Phosphorylation of H3 serine 10 by an activator-recruited histone H3 kinase such as MSK1 (*left*) may directly increase the binding affinity of the histone H3 amino-terminal tail for histone acetyltransferase proteins such as Gcn5, and enhance the acetylation of lysines 9 and 14 (*right*), leading to transcriptional activation. (*B*) Phosphorylation of histone H3 serine 10 (*left*) may disrupt histone–DNA or histone–histone interactions required for chromatin folding, leading to chromatin decondensation (*right*) and increased accessibility of the DNA to the transcriptional machinery. (*C*) H3 Serine 10 phosphorylation may remove a transcriptional repressor protein such as HP1 or IN-HAT from chromatin (*left*) by modifying a recognition motif on the histone H3 amino-terminal tail. Removal of the repressor protein may increase the accessibility of histone tails for further modification by histone acetyltransferase proteins (*right*), leading to transcriptional activation.

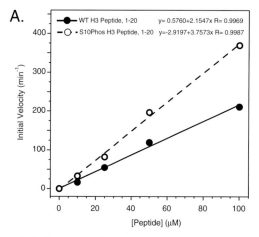 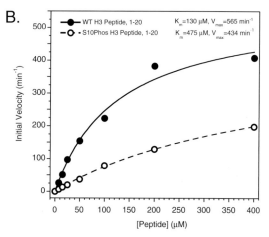

Figure 2. Serine 10 phosphorylation inhibits acetylation of a histone H3 peptide by the Gcn5p-containing SAGA complex. A comparison of the HAT activities of yeast Gcn5p (*A*) and SAGA complex (*B*) on a serine 10 phosphorylated (*open circles*) or unphosphorylated (*closed circles*) histone H3 amino-terminal peptide (amino acids 1–20). Histone acetyltransferase assays were performed with histone H3 peptide (0–400 μM) and recombinant Gcn5p (20–100 nM) or purified yeast SAGA complex (10–20 nM), as described previously (Shogren-Knaak et al. 2003). For each concentration of histone H3 peptide, the initial velocities were calculated by plotting the average incorporation of tritiated acetate as a function of time. Each assay was repeated at least twice and the average initial velocities are plotted as a function of peptide concentration. Recombinant yeast Gcn5p (amino acids 99–262) and purified yeast SAGA complex were prepared as described previously (Shogren-Knaak et al. 2003).

than Gcn5 alone (Tse et al. 1998b; Grant et al. 1999; Sendra et al. 2000). To assess the effects of chromatin context and Gcn5-associated proteins on the mechanism of acetylation of phosphorylated H3 amino-terminal tails, we performed a thorough in vitro kinetic analysis.

To test physiologically relevant HAT complexes, we compared the ability of recombinant yeast Gcn5 and the native SAGA complex to acetylate phosphorylated H3 amino-terminal tail peptides. In agreement with published results (Cheung et al. 2000; Lo et al. 2000), we found that serine 10 phosphorylation of H3 histone-tail peptides stimulated the rate of histone acetylation by Gcn5, although to a lesser degree than reported (Fig. 2A). However, similar analyses with peptides and the Gcn5-containing complex, SAGA, proved different than Gcn5 alone. Phosphorylation of the histone H3 tail peptide on serine 10 actually decreased the rate of SAGA-mediated acetylation (Fig. 2B). This decrease is observed both at subsaturating and saturating substrate concentrations, with a 3.6-fold increase in K_m for peptide, and a 1.3-fold reduction in k_{cat}, contradicting the mild stimulation reported previously for SAGA at a single concentration and time point (Lo et al. 2000).

To test more physiologically relevant chromatin substrates, we generated histone H3 homogeneously phosphorylated at serine 10 using native chemical ligation (Shogren-Knaak et al. 2003; Shogren-Knaak and Peterson 2004) and incorporated it with other histones into nucleosomal arrays by salt dialysis (Carruthers et al. 1999). Like the peptide substrates, the rate of acetylation of the arrays by Gcn5 is stimulated by serine 10 phosphorylation (Fig. 3A). However, with the SAGA complex, serine 10 phosphorylation does not stimulate acetylation, and this lack of stimulation is observed over both saturating and subsaturating concentrations (Fig. 3B). Thus, our results suggest that serine 10 phosphorylation of histone H3 does not stimulate acetylation by a native histone acetyltransferase complex on a chromatin substrate.

Recent structural studies of the catalytic domain of Tetrahymena Gcn5 in complex with a phosphorylated peptide and coenzyme A suggest that phosphopeptide binding occurs in a different manner than originally proposed and may be less stimulatory than originally reported (Clements et al. 2003). In the crystal structure of this ternary complex, the phosphate group adopts two separate orientations and does not interact with the arginine 164 residue, as originally proposed. Instead the phosphate group appears to facilitate a number of changes in the mode of peptide binding both distal and proximal to the phosphoserine residue, with threonine 11 adopting a number of new interactions. Altogether, there is a net increase in buried surface area for the phosphopeptide, and this would be expected to translate into increased binding affinity. Indeed, in kinetic analysis of this protein, there is roughly a threefold decrease in K_m for the phosphopeptide, which denotes augmented phosphopeptide binding. Although in agreement with our own results with Gcn5 and peptides, this increase is less than the six- to tenfold originally reported. Interestingly, in their kinetic analysis of Tetrahymena Gcn5, the authors find that the maximal catalytic turnover rate, k_{cat}, also decreases fourfold for the phosphopeptide. Since the enzymatic specificity constant is determined by k_{cat}/K_m, the net result of these changes would not be expected to augment selectivity for phosphorylated peptides, which is contrary to results seen with yeast Gcn5.

Altogether, the available in vitro data shows that the acetyltransferase activity of recombinant Gcn5 may be stimulated by histone H3 serine 10 phosphorylation, while the native SAGA HAT complex is not stimulated

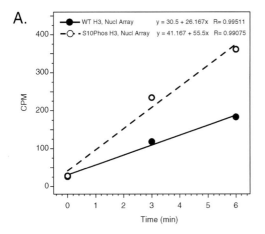
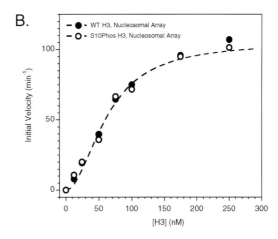

Figure 3. Histone H3 serine 10 phosphorylation stimulates the acetylation of nucleosomal arrays by Gcn5p, but not by the yeast Gcn5p-containing SAGA complex. (*A*) Comparison of Gcn5p HAT activity on nucleosomal arrays containing unphosphorylated histone H3 (*closed circles*) or histone H3 phosphorylated on serine 10 (*open circles*). Histone acetyltransferase assays were performed with recombinant Gcn5p (2.45 μM) and nucleosomal arrays (100-nM histone H3), as described previously (Shogren-Knaak et al. 2003). The extent of histone acetylation was determined at multiple time points by counting the incorporation of tritiated acetate (counts per minute) using a scintillation counter. Each assay was repeated at least three times and the average counts per minute (CPM) is plotted as a function of time. (*B*) Comparison of SAGA HAT activity on phosphorylated (*open circles*) and unphosphorylated (*closed circles*) nucleosomal arrays. Histone acetyltransferase assays were performed with nucleosomal arrays (12.5–250-nM histone H3) and purified SAGA complex (4.9–9.8 nM). For each concentration of histone H3, the initial velocities were calculated by plotting the average incorporation of tritiated acetate as a function of time. Each assay was repeated at least three times and the average initial velocities are plotted as a function of histone H3 concentration. Nucleosomal arrays containing unphosphorylated histone H3 or histone H3 phosphorylated on S10 were prepared as described previously, using a native peptide ligation technique (Shogren-Knaak et al. 2003).

by histone phosphorylation of a peptide or a chromatin substrate. Why then does phosphorylation of serine 10 appear to enhance acetylation of H3 at some promoters in vivo? This coupling could be facilitated through a mechanism that does not rely on direct recruitment of Gcn5 or activator proteins.

MODULATING CHROMATIN STRUCTURE

Histone phosphorylation at serine 10 could facilitate transcriptional activation by disrupting higher-order chromatin structure, thereby increasing the ability of transcription-related proteins to bind. Beyond the basic structure of the nucleosome, neighboring nucleosomes can interact to form a 30-nm chromatin fiber. These fibers can associate to generate 100–400-nm fibers, which then may adopt even higher orders of structure (Hayes and Hansen 2001). The charged and bulky nature of the phosphate group could lead to disruption of chromatin structure at any of these levels, either directly or indirectly. Because of the relatively high positive charge of the H3 histone tail (+14), any effect of phosphorylation (–2 charges) would probably result from the disruption or generation of specific binding contacts, instead of a bulk change in charge.

In vitro studies of nucleosomal arrays support the idea that changes to histone tails can influence chromatin structure. Nucleosomal arrays can be assembled from histones lacking amino-terminal tails, and these arrays prove defective in their ability both to fold into 30-nm fibers and to self-associate into 100–400-nm fibers (Fletcher and Hansen 1995; Schwarz et al. 1996). Furthermore, posttranslational modifications can disrupt higher-order structure. Nucleosomal arrays generated from hyperacetylated histones are also defective in folding into 30-nm fibers and are stimulatory toward transcription in vitro (Tse et al. 1998a).

Evidence that histone phosphorylation might directly influence chromatin structure comes from studies of mitotic chromatin. During mitosis, when H3 is highly phosphorylated, the amount of DNA that can be cross-linked to histone H3 by UV irradiation is decreased relative to interphase levels. This loss is partially rescued by treating the mitotic cells with a kinase inhibitor, suggesting that kinase activity somehow leads to structural changes at the chromatin level. Moreover, the effect of inhibiting kinase activity on chromatin structure is apparent at a gross level, as disruption of mitotic condensation is visible microscopically (Sauve et al. 1999).

To more directly test the effects of histone H3 serine 10 phosphorylation on chromatin structure, we have assembled nucleosomal arrays containing homogeneously phosphorylated H3 histones and characterized these arrays biophysically using sedimentation analysis. In TE buffer, conditions where nucleosomal arrays are expected to adopt an extended "beads-on-a-string" conformation, arrays with or without H3 serine 10 phosphorylation display a similar, relatively tight distribution of sedimentation coefficients (Fig. 4A). These data reflect most directly on the similarity of the nucleosome structure and

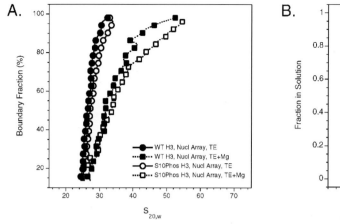
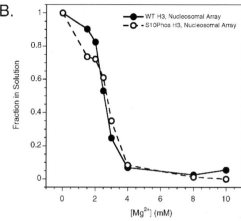

Figure 4. Histone H3-S10 phosphorylation does not effect higher-order chromatin structure. Experiments were performed with nucleosomal arrays assembled from 208-12 5S rDNA template, recombinant *Xenopus* H2A, H2B, and H4 histone protein, and either recombinant *Xenopus* H3 histone or ligated H3 histone-containing serine phosphorylated at position 10 (Shogren-Knaak et al. 2003). (*A*) Intramolecular folding of a nucleosomal array is not changed by histone H3 serine 10 phosphorylation. The sedimentation velocities of phosphorylated (*open circles*) and nonphosphorylated (*closed circles*) nucleosomal arrays were determined at 20°C in TE in the presence or absence of 1.75 mM $MgCl_2$ (Horn et al. 2002). The $G(s)$ distributions were calculated by the method of van Holde and Weischet (van Holde and Weischet 1978) and corrected to water at 20°C. (*B*) Intermolecular association of a nucleosomal array is not changed by histone H3 serine 10 phosphorylation. Nucleosomal arrays were incubated in TE containing increasing amounts of $MgCl_2$. After 15 min at room temperature, samples were centrifuged for 10 min at 14,000g, and the concentration of array in the supernatant was determined by measurement of absorption at 260 nm. Shown is the fraction of array in solution as a function of $MgCl_2$ concentration.

argue that serine 10 phosphorylation of histone H3 does not dramatically change the manner in which DNA is wrapped around the histone octamer nor how histone octamers become distributed onto the DNA. With the addition of 1.75-mM divalent magnesium cation, which promotes intramolecular nucleosome–nucleosome interactions, phosphorylated and unphosphorylated arrays show similar sedimentation properties (Fig. 4A). Both arrays adopt a distribution of S values from 29–55S, with 55S reflecting fully folded, 30-nm fibers. Thus, H3 serine 10 phosphorylation does not appear to disrupt tail-mediated chromatin folding. Finally, under concentrations of divalent magnesium above 2 mM, conditions that promote reversible intermolecular nucleosomal array interactions, phosphorylated and nonphosphorylated arrays demonstrate similar sedimentation properties (Fig. 4B), suggesting that H3 serine 10 phosphorylation also does not disrupt tail-mediated internucleosomal array interactions.

Altogether, our sedimentation studies suggest that phosphorylation of serine 10 on histone H3 does not have a discernible effect on chromatin structure. Consistent with this conclusion is our previous work showing that SWI-SNF-mediated remodeling of nucleosomal arrays is not sensitive to histone phosphorylation (Shogren-Knaak et al. 2003), and recent results from the Richmond lab demonstrating that loss of histone H3 amino-terminal tails is not sufficient to disrupt chromatin folding (Dorigo et al. 2003). Nonetheless, histone H3 serine 10 phosphorylation may still exert a disruptive effect on chromatin structure in the context of additional histone modifications, including the high level of basal tail acetylation, or in combination with other tail phosphorylation marks (H3-S28, H3-T11, H3-T3, H4-S1, H2A-S1, H2B-S33) that occur during mitosis, meiosis, or transcriptional activation.

DECREASING THE BINDING AFFINITY OF TRANSCRIPTIONAL REPRESSORS

Several recent studies suggest an additional mechanism by which histone H3 phosphorylation may mediate transcriptional activation, by directly disrupting the binding of transcriptional repressor proteins. During mitosis, removal of the heterochromatin protein HP1 is preceded by histone H3 serine 10 phosphorylation, and phosphorylation is believed to directly promote loss of HP1 binding (Fischle et al. 2003a). However, the factors required to remove HP1 appear to be more complicated than just phosphorylation, as it seems that both serine 10 phosphorylation and lysine 14 acetylation are required for HP1 loss (Mateescu et al. 2004). The importance of acetylation hints at a potential connection to transcriptional activation. Indeed it has been found that HP1 plays a role in Rb-mediated transcriptional repression of inducible genes (Nielsen et al. 2001), and phosphorylation in addition to acetylation may play a direct role in relieving this repression.

This strategy of derepressing genes by phosphorylation of the histone H3 tail may also apply to other transcriptional repressor proteins. Kouzarides and coworkers have shown that acetylation of histone H3 peptides at multiple sites, or phosphorylation of threonine 3, serine 10, or threonine 11, is sufficient to disrupt binding of the IN-HAT repressor complex (Schneider et al. 2004). The INHAT complex is a chromatin-associated complex that may inhibit acetylation of histones by physically masking acetylating sites (Seo et al. 2002). While loss of INHAT binding due to histone phosphorylation has not been confirmed in vivo, it does provide a mechanism in which phosphorylation and acetylation are coupled, and thus provides an appealing model for how histone phosphorylation might facilitate transcription.

CONCLUSION

Histone H3 phosphorylation at serine 10 is an important means of facilitating transcription of a number of inducible genes. However, the mechanism for this coupling remains controversial. Initial in vitro studies with Gcn5 and peptides suggested that transcriptional stimulation might be accomplished by augmented binding of Gcn5 to H3-S10 phosphorylated histone tails. However, studies with more physiologically relevant Gcn5-containing complexes and nucleosomal arrays do not show an increase in binding affinity toward H3-S10-phosphorylated histone, and alternative mechanisms are now being uncovered. Structural studies with H3-S10-phosphorylated nucleosomal arrays demonstrate that histone H3 phosphorylation on serine 10 is not sufficient to disrupt higher-order chromatin structure. Alternatively, preliminary studies suggest that H3-S10 phosphorylation may disrupt binding of transcriptional repressors and provides an attractive mechanism by which H3-S10 phosphorylation might facilitate histone acetylation and transcription. The complex interplay of phosphorylation, acetylation, and methylation of histone tails remains a rich area of inquiry and continued studies will provide a deeper understanding of transcription and other crucial biological processes.

ACKNOWLEDGMENTS

Our studies on histone phosphorylation were supported by a grant from the NCI to C.L.P. (CA82834), and postdoctoral fellowships from the NIH to M.S.-K. (F32 AI10611) and from the Leukemia and Lymphoma Society of America to C.J.F.

REFERENCES

Allegra P., Sterner R., Clayton D.F., and Allfrey V.G. 1987. Affinity chromatographic purification of nucleosomes containing transcriptionally active DNA sequences. *J. Mol. Biol.* **196:** 379.
Becker P.B. and Horz W. 2002. ATP-dependent nucleosome remodeling. *Annu. Rev. Biochem.* **71:** 247.
Carruthers L.M., Tse C., Walker K.P., III, and Hansen J.C. 1999. Assembly of defined nucleosomal and chromatin arrays from pure components. *Methods Enzymol.* **304:** 19.
Chen T.A. and Allfrey V.G. 1987. Rapid and reversible changes in nucleosome structure accompany the activation, repression, and superinduction of murine fibroblast protooncogenes c-fos and c-myc. *Proc. Natl. Acad. Sci.* **84:** 5252.
Cheung P., Tanner K.G., Cheung W.L., Sassone-Corsi P., Denu J.M., and Allis C.D. 2000. Synergistic coupling of histone H3 phosphorylation and acetylation in response to epidermal growth factor stimulation. *Mol. Cell* **5:** 905.
Clements A., Poux A.N., Lo W.S., Pillus L., Berger S.L., and Marmorstein R. 2003. Structural basis for histone and phosphohistone binding by the GCN5 histone acetyltransferase. *Mol. Cell* **12:** 461.
Dhalluin C., Carlson J.E., Zeng L., He C., Aggarwal A.K., and Zhou M.M. 1999. Structure and ligand of a histone acetyltransferase bromodomain. *Nature* **399:** 491.
Dorigo B., Schalch T., Bystricky K., and Richmond T.J. 2003. Chromatin fiber folding: Requirement for the histone H4 N-terminal tail. *J. Mol. Biol.* **327:** 85.
Fischle W., Wang Y., and Allis C.D. 2003a. Binary switches and modification cassettes in histone biology and beyond. *Nature* **425:** 475.
———. 2003b. Histone and chromatin cross-talk. *Curr. Opin. Cell Biol.* **15:** 172.
Fletcher T.M. and Hansen J.C. 1995. Core histone tail domains mediate oligonucleosome folding and nucleosomal DNA organization through distinct molecular mechanisms. *J. Biol. Chem.* **270:** 25359.
Grant P.A., Eberharter A., John S., Cook R.G., Turner B.M., and Workman J.L. 1999. Expanded lysine acetylation specificity of Gcn5 in native complexes. *J. Biol. Chem.* **274:** 5895.
Hayes J.J. and Hansen J.C. 2001. Nucleosomes and the chromatin fiber. *Curr. Opin. Genet. Dev.* **11:** 124.
Horn P.J., Crowley K., Carruthers L.M., Hansen J.C., and Peterson C.L. 2002. The SIN domain of the histone octamer is essential for intramolecular folding of nucleosomal arrays. *Nat. Struct. Biol.* **9:** 167.
Lo W.S., Duggan L., Emre N.C., Belotserkovskya R., Lane W.S., Shiekhattar R., and Berger S.L. 2001. Snf1—A histone kinase that works in concert with the histone acetyltransferase Gcn5 to regulate transcription. *Science* **293:** 1142.
Lo W.S., Trievel R.C., Rojas J.R., Duggan L., Hsu J.Y., Allis C.D., Marmorstein R., and Berger S.L. 2000. Phosphorylation of serine 10 in histone H3 is functionally linked in vitro and in vivo to Gcn5-mediated acetylation at lysine 14. *Mol. Cell* **5:** 917.
Luger K., Mader A.W., Richmond R.K., Sargent D.F., and Richmond T.J. 1997. Crystal Structure of the nucleosome core particle at 2.8 Å resolution. *Nature* **389:** 251.
Mahadevan L.C., Willis A.C., and Barratt M.J. 1991. Rapid histone H3 phosphorylation in response to growth factors, phorbol esters, okadaic acid, and protein synthesis inhibitors. *Cell* **65:** 775.
Mateescu B., England P., Halgand F., Yaniv M., and Muchardt C. 2004. Tethering of HP1 proteins to chromatin is relieved by phosphoacetylation of histone H3. *EMBO Rep.* **5:** 490.
Ng H.H., Feng Q., Wang H., Erdjument-Bromage H., Tempst P., Zhang Y., and Struhl K. 2002. Lysine methylation within the globular domain of histone H3 by Dot1 is important for telomeric silencing and Sir protein association. *Genes Dev.* **16:** 1518.
Nielsen P.R., Nietlispach D., Mott H.R., Callaghan J., Bannister A., Kouzarides T., Murzin A.G., Murzina N.V., and Laue E.D. 2002. Structure of the HP1 chromodomain bound to histone H3 methylated at lysine 9. *Nature* **416:** 103.
Nielsen S.J., Schneider R., Bauer U.M., Bannister A.J., Morrison A., O'Carroll D., Firestein R., Cleary M., Jenuwein T., Herrera R.E., and Kouzarides T. 2001. Rb targets histone H3 methylation and HP1 to promoters. *Nature* **412:** 561.
Nowak S.J. and Corces V.G. 2000. Phosphorylation of histone H3 correlates with transcriptionally active loci. *Genes Dev.* **14:** 3003.
Prigent C. and Dimitrov S. 2003. Phosphorylation of serine 10 in histone H3, what for? *J. Cell Sci.* **116:** 3677.
Rojas J.R., Trievel R.C., Zhou J., Mo Y., Li X., Berger S.L., Allis C.D., and Marmorstein R. 1999. Structure of *Tetrahymena* GCN5 bound to coenzyme A and a histone H3 peptide. *Nature* **401:** 93.
Sassone-Corsi P., Mizzen C.A., Cheung P., Crosio C., Monaco L., Jacquot S., Hanauer A., and Allis C.D. 1999. Requirement of Rsk-2 for epidermal growth factor-activated phosphorylation of histone H3. *Science* **285:** 886.
Sauve D.M., Anderson H.J., Ray J.M., James W.M., and

Roberge M. 1999. Phosphorylation-induced rearrangement of the histone H3 NH2-terminal domain during mitotic chromosome condensation. *J. Cell Biol.* **145:** 225.

Schneider R., Bannister A.J., Weise C., and Kouzarides T. 2004. Direct binding of INHAT to H3 tails disrupted by modifications. *J. Biol. Chem.* **279:** 23859.

Schwarz P.M., Felthauser A., Fletcher T.M., and Hansen J.C. 1996. Reversible oligonucleosome self-association: Dependence on divalent cations and core histone tail domains. *Biochemistry* **35:** 4009.

Sendra R., Tse C., and Hansen J.C. 2000. The yeast histone acetyltransferase A2 complex, but not free Gcn5p, binds stably to nucleosomal arrays. *J. Biol. Chem.* **275:** 24928.

Seo S.B., Macfarlan T., McNamara P., Hong R., Mukai Y., Heo S., and Chakravarti D. 2002. Regulation of histone acetylation and transcription by nuclear protein pp32, a subunit of the INHAT complex. *J. Biol. Chem.* **277:** 14005.

Shogren-Knaak M.A. and Peterson C.L. 2004. Creating designer histones by native chemical ligation. *Methods Enzymol.* **375:** 62.

Shogren-Knaak M.A., Fry C.J., and Peterson C.L. 2003. A native peptide ligation strategy for deciphering nucleosomal histone modifications. *J. Biol. Chem.* **278:** 15744.

Soloaga A., Thomson S., Wiggin G.R., Rampersaud N., Dyson M.H., Hazzalin C.A., Mahadevan L.C., and Arthur J.S. 2003. MSK2 and MSK1 mediate the mitogen- and stress-induced phosphorylation of histone H3 and HMG-14. *EMBO J.* **22:** 2788.

Thomson S., Clayton A.L., and Mahadevan L.C. 2001. Independent dynamic regulation of histone phosphorylation and acetylation during immediate-early gene induction. *Mol. Cell* **8:** 1231.

Tse C., Sera T., Wolffe A.P., and Hansen J.C. 1998a. Disruption of higher-order folding by core histone acetylation dramatically enhances transcription of nucleosomal arrays by RNA polymerase III. *Mol. Cell. Biol.* **18:** 4629.

Tse C., Georgieva E.I., Ruiz-Garcia A.B., Sendra R., and Hansen J.C. 1998b. Gcn5p, a transcription-related histone acetyltransferase, acetylates nucleosomes and folded nucleosomal arrays in the absence of other protein subunits. *J. Biol. Chem.* **273:** 32388.

van Holde K.E. and Weischet W.O. 1978. Boundary analysis of sedimentation-velocity experiments with monodisperse and paucidisperse solutes. *Biopolymers* **17:** 1387.

van Leeuwen F., Gafken P.R., and Gottschling D.E. 2002. Dot1p modulates silencing in yeast by methylation of the nucleosome core. *Cell* **109:** 745.

Wiggin G.R., Soloaga A., Foster J.M., Murray-Tait V., Cohen P., and Arthur J.S. 2002. MSK1 and MSK2 are required for the mitogen- and stress-induced phosphorylation of CREB and ATF1 in fibroblasts. *Mol. Cell. Biol.* **22:** 2871.

Woodcock C.L. and Dimitrov S. 2001. Higher-order structure of chromatin and chromosomes. *Curr. Opin. Genet. Dev.* **11:** 130.

Yaffe M.B. and Elia A.E. 2001. Phosphoserine/threonine-binding domains. *Curr. Opin. Cell Biol.* **13:** 131.

Zhang L., Eugeni E.E., Parthun M.R., and Freitas M.A. 2003. Identification of novel histone post-translational modifications by peptide mass fingerprinting. *Chromosoma* **112:** 77.

Structural Characterization of Histone H2A Variants

S. Chakravarthy,* Y. Bao,* V.A. Roberts,† D. Tremethick,‡ and K. Luger*

Department of Biochemistry and Molecular Biology, Colorado State University, Fort Collins, Colorado 80523-1870; †Department of Molecular Biology, The Scripps Research Institute, La Jolla, California 92037; ‡The John Curtin School of Medical Research, The Australian National University, Australian Capital University, Canberra ACT, 2601 Australia

Eukaryotic DNA associates with an equal amount of protein to form chromatin, the fundamental unit of which is the nucleosome core particle (NCP). An NCP consists of two copies each of the four core histones H2A, H2B, H3, and H4. This histone octamer binds 147 base pairs of DNA around its outer surface in 1.65 tight superhelical turns (Fig. 1A) (Luger et al. 1997; Richmond and Davey 2003). Linker histones and other nonhistone proteins promote or stabilize the folding of nucleosomal arrays into superstructures of increasing complexity and largely unknown architecture (Hansen 2002). Covalent modification of the core histones and variations in the fundamental biochemical composition of nucleosomes distinguish transcriptionally active from inactive chromatin regions, by either changing the structure of the nucleosomes, altering their ability to interact with other protein factors, or modifying their propensity to fold into varying degrees of higher-order structures (or by any combination of the above). Studying the mechanism for establishing distinct chromatin domains is essential to understanding differential regulation of gene expression and all other DNA-dependent processes. Much progress has been made in this direction in the past few years.

Substitution of one or more of the core histones with the corresponding histone variants has the potential to exert considerable influence on the structure and function of chromatin. Histone variants are distinct nonallelic forms of conventional, major-type histones that form the bulk of nucleosomes during replication and whose synthesis is tightly coupled to S phase. Histone variants are characterized by a completely different expression pattern that is not restricted to S phase. They are found in most eukaryotic organisms and are expressed in all tissue types (unlike some H2B isoforms that are found only in specialized tissues such as testes). Compared to their major-type counterparts, histone variants exhibit moderate to significant degrees of sequence homology (Fig. 1B). H2A.X (82%) and H3.3 (~96%) are the least divergent of all histone variants. H2A.Z (~60%), macroH2A (~65%), H2A.Bbd (40%), and CenpA, which has a 93 amino acid domain that is 62% identical to H3 (Palmer et al. 1991; Sullivan et al. 1994), are increasingly divergent in their

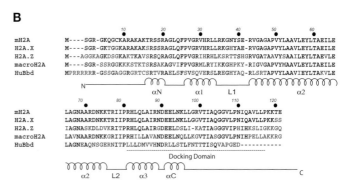

Figure 1. Histone H2A variants. (*A*) Overview of nucleosome structure. Only 74 base pairs of DNA and associated proteins are shown. *Yellow*: H2A; *red*: H2B; *blue*: H3; *green*: H4. The axis of noncrystallographic symmetry is indicated by a *dashed line*. The four-helix bundle structure formed by the two H3 molecules and the H2A docking domain are *boxed*; other structural features are indicated. (*B*) Sequence alignment of the histone domain of human H2A.X, mouse H2A.Z, mouse macroH2A, and human H2A.Bbd with major-type mouse H2A. *Bullets*: every tenth residue in major H2A; *black*: identical residues; *blue*: similar residues; *red*: different residues. Also indicated are the secondary structure elements of the histone fold (α1, α2, and α3) and the loops and extensions (L1, L2, αN, and αC).

histone moiety from H2A and H3, respectively. As is the case with histones in general, the structured regions of the histones (encompassing histone folds and extensions) are more conserved than the histone tails. The structured region of H2A.X is 97% conserved to its major-type H2A counterpart, that of macroH2A, 70%, of H2A.Z, 66%, and of H2A.Bbd, 48%. MacroH2A is unique in that it contains an additional nonhistone-like domain that is connected to the histone-homology domain by a flexible linker (Pehrson and Fried 1992). All histone variants are highly conserved between different species. In many cases, they are even more conserved than their major-type paralogs (Sullivan et al. 2002), indicating that they all have evolved to fulfill important functions that cannot be accomplished by major-type H2A and H3, as has been demonstrated for H2A.Z (van Daal and Elgin 1992; Clarkson et al. 1999; Faast et al. 2001).

While the modus operandi of most of histone variants remains unknown, they are all characterized by unique in vivo localization patterns, which in turn shed light on their putative function. H2A.X is distributed throughout the genome. It is implicated in double-stranded DNA repair (Celeste et al. 2003; Rothkamm and Lobrich 2003) and is necessary for programming DNA breakage that occurs in developing lymphocytes (Bassing et al. 2003). The presence of H2A.Z within euchromatin plays a role in preventing silencing from spreading into regions of the chromosome that are normally transcriptionally active (Meneghini et al. 2003). Interestingly, H2A.Z can coexist with Sir proteins at the telomere (Krogan et al. 2003). Most recently, H2A.Z has also been shown to play a role in chromosome segregation (Rangasamy et al. 2004). MacroH2A is found at the inactive X chromosome of adult female mammals, which consists predominantly of heterochromatin and is transcriptionally inactive (Costanzi and Pehrson 1998), while H2A.Bbd colocalizes with hyperacetylated histone H4, indicating that it might be associated with actively transcribed chromatin (Chadwick and Willard 2001). The histone H3 variant H3.3 is thought to be associated with active chromatin, and CENP-A is a major component of centromeric heterochromatin (Vermaak and Wolffe 1998; Ahmad and Henikoff 2002a,b).

Here we will summarize and review available structural information on nucleosomes and chromatin containing histone H2A variants, and will attempt to explain how structure relates to their varied function. We note that an exhaustive review of all biological and functional data would clearly exceed the scope of this manuscript. We will present a hypothesis why "true" histone variants have been identified for only histone H2A and H3, and we will show data in support of our hypothesis that particular regions in the H2A amino acid sequence appeared to have been targets during the evolution of H2A histone variants.

WHY ARE THERE NO H4 AND H2B HISTONE VARIANTS?

All known true histone variants are replacements for either histone H3 or H2A (for review, see Malik and Henikoff 2003; Henikoff et al. 2004). From a structural vantage point, we hypothesize that this is the case because only H3 and H2A are engaged in homotypic interactions (Fig. 2). In contrast, neither H4 nor H2B interact with the other H4 or H2B molecule within the histone octamer. A four-helix bundle formed by residues from the two H3 chains holds together the (H3-H4)$_2$ tetramer, which is stable in the absence of DNA under physiological conditions (Fig. 2A,B). This interface, which is characterized by a combination of salt bridges, ionic interactions, and some hydrophobic contacts (Luger et al. 1997), had been proposed many years ago to be conformationally flexible, especially in the absence of the (H2A-H2B) dimers (Chen et al. 1991; Hamiche et al. 1996; but see also Protacio and Widom 1996). In contrast, the interface formed between two H2A molecules is quite small and only exists in the context of a folded nucleosome (Fig. 2C,D). It is, however, the only point of contact between the two (H2A-H2B) dimers in a nucleosome, and thus may be responsible in part for highly cooperative incorporation of the two H2A-H2B dimers, as well as for tethering the two gyres of the DNA superhelix together. Intriguingly, major sequence differences between H3 and the centromeric H3 variant are found in this four-helix bundle region (Shelby et al. 1997; Black et al. 2004). Similarly, variability among the many H2A variants themselves and differences between variants and major-type H2A are found in the L1 loop (Fig. 1B). This sug-

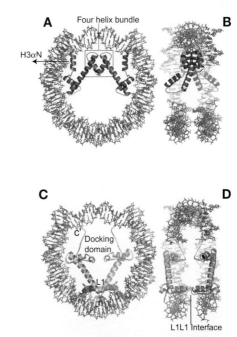

Figure 2. H3 and H2A fulfill special roles within the NCP. (*A*) Structure of the NCP viewed down the superhelical axis. Only the two H3 chains are shown. (*B*) The same structure is shown after rotation by 90° around the *y*-axis, with parts of the DNA omitted for clarity. (*C*) Structure of the NCP viewed down the superhelical axis. Only the two H2A chains are shown. The structures of NCPs containing major-type H2A from *Xenopus laevis* (*yellow*; pdb entry code 1AOI), H2A.Z (*wheat*; pdb entry code 1F66), and macro-H2A (*gray*; S. Chakravarthy, unpubl.) have been superimposed. (*D*) The structures displayed in *C*, shown in the same orientation as *B*.

gests that sequence variability in these regions of self-interactions may serve to ensure that only nucleosomes with two identical H2A or H3 "flavors" are formed. Equally likely, these interfaces may preclude the formation of such nucleosomes and may instead favor the incorporation of a major-type histone H3 with variant H3, or major-type H2A with variant H2A, respectively, as will be shown below.

A second common feature of H3 and H2A not shared by H4 and H2B is the fact that they both are involved in the organization of more than the requisite ~70 base pairs of DNA that are bound by a canonical histone fold dimer (Luger and Richmond 1998a), because of quite extended regions in both H3 and H2A outside the histone fold. The amino-terminal helix of H3 (αN) is positioned to interact with the penultimate turn of the DNA double helix before it exits the confines of the nucleosome (Figs. 1A and 2A). It is held in position by the H2A docking domain (Bao et al. 2004). The carboxy-terminal tail of H2A is poised to interact with linker DNA that extends beyond the 147 nucleosomal base pairs (Figs. 1A and 2C).

Thus, unlike H2B and H4, replacement of H3 and H2A with histone variants has the potential to affect DNA organization (or exit angle) at the penultimate 15 base pairs of nucleosomal DNA, with potential implications for higher-order structure. As pointed out above, sequence variations that are specific to either H3 or H2A variants have the potential to increase the pool of theoretically possible nucleosomes, in being able to choose their interaction partners. Finally, in light of the current models for linker histone binding to the nucleosome (Crane-Robinson 1997; Hayes and Hansen 2001), histone H1 and its variants and isoforms are more likely to interact with H3 and H2A than with H2B and H4. One intriguing (but as yet untested) model for how histone variants exert their function in chromatin may be by modulating the interaction between the nucleosome and linker histones or non-histone architectural chromatin proteins, such as HP1. For example, it is conceivable that nucleosomes harboring certain H2A or H3 variants may be unable to interact with linker histones or other architectural proteins, or that they may prefer certain H1 isoforms over others, with profound effects on gene regulation and/or chromatin higher-order structure.

STRUCTURAL CHARACTERISTICS OF NUCLEOSOMES AND CHROMATIN CONTAINING HISTONE H2A VARIANTS

H2A.X

Amino acids (1–120) of H2A.X are very similar in sequence to major H2A; indeed, with an only two amino acid difference in the structured domain it is safe to assume that the structural properties of a mononucleosome are likely to remain unaffected. The two sequence changes in the structured domain of H2A.X are in the L1 loop and in the docking domain, respectively, and are also found to distinguish macroH2A from major-type H2A (H2A N38 to H, and R99 to G, Fig. 1B).

The carboxy-terminal domain of H2A.X is phosphorylated, and it is this phosphorylated form of this variant that is implicated in DNA repair (Celeste et al. 2003). One possible mode of H2A.X action could be via recruitment of repair proteins to the site of DNA damage. It is indeed found that formation of Nbs1, 53bp1, and Brca1 foci at the damage sites is severely impaired in H2A.X–/– cells, as well as suppressing genomic instability (Celeste et al. 2003).

H2A.Z

H2A.Z, which is essential in *Drosophila*, mouse, and *Tetrahymena* (van Daal and Elgin 1992; Liu et al. 1996; Faast et al. 2001), is the histone variant whose structure and function is perhaps best studied among all histone variants. While not being essential in budding yeast, H2A.Z (Htz1) functions to prevent the spread of heterochromatin into euchromatin (Meneghini et al. 2003) and plays a role in transcription with a function that is partially redundant with specific chromatin remodeling complexes.

The crystal structure of an NCP in which major H2A is replaced by H2A.Z reveals no major differences in the path of the DNA superhelix or in the nature of protein–DNA interactions (Suto et al. 2000). While tyrosine quenching revealed no major change in nucleosome stability in vitro, a more sensitive fluorescence resonance energy transfer (FRET) approach (Park et al. 2004) revealed that the sequence changes of H2A.Z result in subtle stabilization of the (H2A-H2B) dimer interaction with the (H3-H4)$_2$ tetramer-DNA complex upon increasing ionic strength (Fig. 3).

Amino acid changes in the H2A.Z docking domain also contribute to an extended acidic patch, which is a prominent feature of the nucleosome surface (Fig. 4) (Suto et al. 2003). The docking domain is the region that

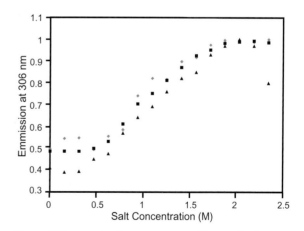

Figure 3. The tyrosine fluorescence quenching profile shows no pronounced differences in the overall stability of macro-NCP, H2A.Z-NCP, and major-NCP. Tyrosine quenching profiles in response to increased ionic strength for nucleosomes in which the histone domain of macroH2A, or H2A.Z, has replaced major H2A. Values have been normalized using the formula $A_n = A/A_h$, where A_n is the normalized value, A is the observed signal, and A_h is the highest signal observed. *Diamonds:* major-NCP; *squares:* macro-NCP; *triangles:* H2A.Z-NCP.

is essential for *Drosophila* development (Clarkson et al. 1999). This region of H2A.Z (and the amino-terminal tail) has been shown to interact with the pericentric heterochromatin-binding protein INCENP (Rangasamy et al. 2003), a protein critical for proper chromosome segregation. Thus, it appears that subtle sequence differences between major-type H2A and H2A.Z bear a direct relevance to its ability to interact with nonhistone proteins.

Restriction digest analysis of an in vitro chromatin model system (H2A or H2A.Z nucleosome arrays assembled on a DNA template that contains 12 repeats of a 208-base pair nucleosome positioning sequence) in combination with sedimentation velocity studies established that H2A and H2A.Z nucleosomes assemble with comparable affinity (Fan et al. 2002). Consistent with the increase in nucleosome core stability, H2A.Z arrays are more regularly spaced than H2A arrays. Sedimentation velocity analysis at elevated salt concentrations was used to determine that H2A.Z facilitates intramolecular folding of nucleosome arrays resulting in the formation of a folded state (55S). This folded state is thought to reflect the canonical 30-nm fiber. On the other hand, H2A.Z is found to hinder the oligomerization mediated by interactions between nucleosome arrays (Fan et al. 2002). Therefore, H2A.Z arrays form higher-order chromatin structures distinctly different from H2A arrays. Consistent with this, H2A.Z is located at constitutive heterochromatin in mammalian cells and is required for faithful chromosome segregation (Rangasamy et al. 2003, 2004). Taken together, H2A.Z may have a dual role in modulating the recruitment of nonhistone proteins and in chromatin fiber folding. These features would enable H2A.Z to play a role in a number of different processes that include gene expression and mitosis.

macroH2A

macroH2A is perhaps the most unusual of all histone variants because of its large size (~370 amino acids and a molecular weight of 42 kD) and tripartite structural organization. The amino-terminal third (amino acid 1–122) is ~64% identical to major H2A (Pehrson and Fried 1992). This region is followed by a highly basic stretch (amino acid 132–160), which is homologous to the carboxyl terminus of linker histone H1 (57% identical to sea urchin H1γ over 30 residues; Pehrson and Fried 1992). Amino acids 161–370 form a tightly folded domain whose high-resolution structure we have determined recently (G.S.Y.K. Swamy and S. Chakravarthy, unpubl.). The structure of an NCP in which H2A has been replaced by the histone domain of macroH2A shows that the significant differences in sequence are accommodated with surprisingly minor structural changes (Fig. 2C,D) (S. Chakravarthy et al., unpubl.). The most pronounced structural divergence is found in the L1 region of macroH2A. Changes in the electrophoretic behavior of mononucleosomes containing the H2A-like domain of macroH2A, in the stability of the histone octamer at decreased ionic strength, and in the absence of the stabilizing influence of the DNA were observed. A mutant of major-type H2A in which four amino acids in the L1 loop of major-type H2A (38–NYAE–41) were replaced with the corresponding amino acids from macroH2A (38–HPKY–41) was shown to confer all in vitro characteristics of macroH2A-containing octamer and NCP onto major-type H2A (S. Chakravarthy et al., unpubl.).

Despite the pronounced effect on histone octamer stability, preliminary data show that the overall stability of the nucleosome in response to increased ionic strength is not affected (Fig. 3). However, given our results with H2A.Z-containing nucleosomes (Park et al. 2004), it is possible and even likely that more subtle changes exist. It should be pointed out that since the first and foremost task of histone variants is to form nucleosomes, subtle changes are expected at most, but clearly these differences are fundamental to the function of histone variants.

macroH2A-containing chromatin has been shown to be inhibitory to transcription in vivo (Perche et al. 2000). Furthermore, macroH2A-containing mononucleosomes are not remodeled by SWI/SNF and are unable to bind transcription factors (Angelov et al. 2003). More precisely, the nonhistone domain or the flexible linker (or both) were responsible for the inability of these variant nucleosomes to bind the transcription factor NF-κB. In contrast, the histone-like domain of macroH2A was shown to be responsible for the inability of macroH2A-containing nucleosomes to be remodeled (Angelov et al. 2003). It will be of interest to see whether the L1 loop alone can confer this inhibition of chromatin remodeling and transcription to nucleosomes. The key location of the L1 loop within the NCP structure (where it seemingly holds together the two gyres of the DNA double helix) certainly makes this region a prime candidate to affect the dynamic motions of the NCP associated with chromatin remodeling.

macroH2A displays a very unique nuclear localization pattern. While it is expressed in equal quantities in both males and females (Rasmussen et al. 1999), in almost all the higher eukaryotes it is enriched in the inactive X chromosome (Xi) of adult female mammals manifested as a Macro Chromatin Body (MCB) (Costanzi and Pehrson 1998; Hoyer-Fender et al. 2000). Various GFP-tagged constructs have been used to determine the importance of the histone and the nonhistone domain in the localization of macroH2A to the inactive X chromosome (Chadwick et al. 2001). The histone domain alone (without amino- and carboxy-terminal tails) fused to GFP was found to form the Xi-associated MCB with efficiencies comparable to full-length macroH2A. Of the 19 amino acids that are different in macroH2A compared to major H2A, no single amino acid could be attributed this function by point mutation studies. On the basis of preliminary analysis of the crystal structure of macroH2A mononucleosomes and biochemical studies of L1 mutants of major H2A, it seemed likely that the L1 loop is crucial for the localization on the Xi. This hypothesis has been tested experimentally (S. Chakravarthy et al., unpubl.).

macroH2A is gradually emerging as a "family" of variants. At least two different genes are known to encode this protein with significant variations in sequence but the same basic structural organization (H2A1 and macroH2A2, respectively). macroH2A1 has two splice variants (macroH2A1.1 and macroH2A1.2), which are nonidentical in a very small region of the nonhistone region starting at amino acid 195 (Pehrson et al. 1997). macroH2A2 is overall 68% identical to macroH2A1.2. The histone region is 84% identical to that of macroH2A1 and only 66% identical to major H2A. The sequence of the macroH2A2 L1-loop (TFKY) seems to be a convolution between that of major H2A (NYAE) and macroH2A (HPKY). The basic region is the most varied and is only 25% identical to that of macroH2A1, whereas the nonhistone region is 64% identical to that of macroH2A1.2. macroH2A1.2 and macroH2A2 display very similar (and sometimes overlapping) nuclear localization patterns, at least at a global level, and the functional relevance of the sequence differences remains unclear (Chadwick et al. 2001; Costanzi and Pehrson 2001).

H2A.Bbd

The structured region of H2A.Bbd is only 48% identical to that of major H2A, making H2A.Bbd the most divergent histone H2A variant known to date. So far, this histone variant has been identified in only humans and mice. Major hallmarks of the amino acid sequence of H2A.Bbd as compared to that of major H2A are (1) the presence of a continuous stretch of five arginines and the conspicuous absence of lysines in its amino-terminal tail; (2) the absence of a carboxy-terminal tail and the very last segment of the docking domain; (3) major sequence differences in the docking domain of H2A; (4) the presence of only one lysine in H2A.Bbd compared to 14 in major H2A, resulting in a slightly less basic protein (pI 10.7, compared to a pI of 11.2 for major H2A); and (5) the absence of the "acidic patch" (Luger and Richmond 1998b) on the docking domain (Fig. 1B).

We found that mononucleosomes containing H2A.Bbd had a more relaxed structure with less tightly bound DNA ends (Bao et al. 2004). Only 118 ± 2 bp of DNA are protected against digestion with micrococcal nuclease, in contrast to 146 bp in canonical nucleosomes. These results are consistent with the observed more rapid exchange of GFP-H2A.Bbd in vivo (Gautier et al. 2004). Intriguingly, we also found a lower repeat length in micrococcal nuclease digestion of nucleosomes reconstituted onto plasmids using a recombinant in vitro assembly system (Georges et al. 2002) (125 bp as opposed to ~160 bp for major nucleosome arrays), which suggests that the H2A.Bbd nucleosomes are deposited at a higher density. At this high density, H2A.Bbd represses transcription comparable to H2A. Intriguingly, domain swap experiments (in which the H2A docking domain was exchanged with that of Bbd) show that the H2A.Bbd docking domain is largely responsible for its behavior. The conservation of the histone fold together with the nuclear localization pattern suggests that H2A.Bbd alters chromatin structure at the nucleosomal level, giving rise to transcriptionally active domains (Chadwick and Willard 2001).

EVOLUTIONARY TARGETS IN THE HISTONE FOLD OF H2A VARIANTS

Sequence comparisons (Fig. 1B) together with analysis of the two available crystal structures of nucleosomes containing histone variants (Suto et al. 2000; S. Chakravarthy, unpubl.) and analysis of biochemical and biophysical data from our and several other laboratories show that histone variants are true replacement histones in that they can form functional nucleosomes and chromatin. The majority of structural and functional changes in histone H2A variants reside in the docking domain and in the L1 loops, with the latter being more structurally divergent than the former.

The Multifunctional H2A Docking Domain

The docking domain is involved in interactions between the H2A-H2B dimer and the (H3-H4)$_2$ tetramer and harbors three of the seven residues that form the "acidic patch" on the surface of the nucleosome (Luger and Richmond 1998b). Thus, sequence divergence in this region may affect the stability of the H2A variant-H2B dimer/(H3-H4)$_2$ tetramer interface, as has been observed for H2A.Z (Park et al. 2004), which will have effects on chromatin remodeling and transcription. The inefficient organization of the penultimate ~15–20 base pairs of nucleosomal DNA can also be a consequence of relatively minor sequence changes in this domain, as described for H2A.Bbd (Bao et al. 2004). Changes in amino acid sequence may also alter the surface of the nucleosome, with important implications for the ability of nucleosomes to interact with other factors or to form more compact higher-order structures. It is interesting to note that while the acidic patch is decreased in H2A.Bbd (not shown), its size is actually increased in H2A.Z (Suto et al. 2000), and expanded towards the carboxy-terminal end of the docking domain in macroH2A (Fig. 4).

The H2A L1 Loop May Select for the Second H2A-H2B Dimer

The L1 loops of the two H2A moieties within the nucleosome are involved in the formation of the L1L1 interface, which is the only site of interaction between the two H2A-H2B dimers in the nucleosome. The L1L1 interface is responsible for the cooperative incorporation of the two H2A-H2B dimers. It may also stabilize the two gyres of the nucleosome core particle (Fig. 2D). Altering the biochemical nature of this interface should therefore result in an altered response to the transcriptional machinery and to chromatin remodeling factors, as well as determine the histone composition of the nucleosome.

There is no experimental evidence for the tacit assumption that one nucleosome contains, for example, two H2A.Z-H2B dimers. It is theoretically possible to have a nucleosome in which only one of the H2A moieties has been replaced by its corresponding variant, resulting in a nucleosome with one (H3-H4)$_2$ tetramer, one major H2A-H2B dimer, and one variant H2A-H2B dimer. From

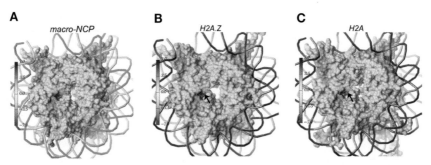

Figure 4. Subtle changes in the molecular surface of macro-NCP compared to H2A.Z-NCP and major-NCP. (*A*) macro-NCP; (*B*) H2A.Z-NCP; (*C*) major-NCP. The electrostatic potential ranges from +7.0 (*blue*) to –7.0 (*red*) kcal/mole/e (*color bar*); shown is the value at the solvent-accessible surface (out 1.4 Å, the radius of a water molecule, from the molecular surface) mapped back on to the molecular surface. Molecular surfaces were calculated with the program MSMS (Sanner et al. 1996) using a 1.4-Å probe sphere. The acidic patch that provides an essential crystal contact with the amino-terminal tail of histone H4 of a neighboring nucleosome core particle is indicated by a *black arrow*. (*B,C*, Reprinted, with permission, from Suto et al. 2000 [©Nature Publishing Group; http://www.nature.com/].)

a nucleosomal viewpoint, the primary determinant of the composition of a given nucleosome must be the compatibility of the L1 loop of major H2A with those of different H2A variants. To investigate this possibility in vitro, we performed salt-gradient reconstitutions with mixtures of (H3-H4)$_2$ tetramer, (H2A-H2B) dimers, and H2A.Z-H2B dimers (Fig. 5); or with H2A.Bbd-H2B dimers (Fig. 6). We used either nickel-affinity chromatography to isolate nucleosomes containing his-tagged H2A (Fig. 5), or gel elution of "mixed nucleosome bands" (Fig. 6), followed by analysis of the histone content by SDS PAGE. Using these two approaches, we could show that while all histone H2A variants are capable of forming hybrid nucleosomes (Figs. 5 and 6), the propensity to do so clearly differs between macroH2A, H2A.Z, and H2A.Bbd.

Intriguingly, hybrid nucleosomes reconstituted from a mixture of H2A-H2B dimers and H2A.Bbd-H2B dimers together with (H3-H4)$_2$ tetramer protect only ~130 bp of DNA against digestion with micrococcal nuclease, that is, ~12 base pairs more than nucleosomes containing two H2A.Bbd chains and ~15 base pairs less than canonical nucleosome (not shown). The ability of H2A.Bbd to form hybrid nucleosomes is independent of the assembly pathway, since the same end result is obtained by yNAP-1-dependent assembly under physiological ionic strength (Fig. 6C).

Our results clearly indicate the possibility that variants may be combined with major-type histone H2A in a single nucleosome, thus generating yet another level of structural and functional heterogeneity. The specific nature of the different L1L1 interfaces and their potential influence on the accessibility of nucleosomal DNA will become more obvious in light of future structural studies. How would such nucleosomes be assembled in vivo? Histone H2A-H2B dimers are in rapid exchange even in the absence of transcription and replication (Louters and Chalkley 1985; Kimura and Cook 2001). Recently discovered histone-variant specific assembly factors for H2A.Z (Krogan et al. 2003; Kobor et al. 2004; Mizuguchi et al. 2004) promote the replacement of one or both H2A-H2B dimer with a H2A.Z-H2B dimer. ATP-dependent chromatin remodeling factors are also capable of actively exchanging histone variant dimers into folded nucleosomes (Bruno et al. 2003). Since it is unlikely that both dimers are exchanged at the same time, the final "equilibrium" makeup of a particular nucleosome will be determined by the L1 loop, and by the relative availability of major-type and variant histones.

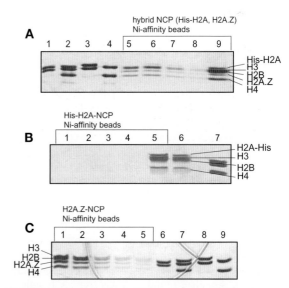

Figure 5. H2A.Z can form hybrid nucleosomes with major-type H2A. (*A*) Lane *1*: H2A.Z-H2B dimer. Lane *2*: Histone octamer with untagged H2A. Lane *3*: His-tagged H2A-H2B dimer. Lane *4*: H3-H4 tetramer. Lanes *5–8*: Flowthrough and washes with 5-mM imidazole of NCP reconstituted from a mixture of H2A.Z, His-tagged H2A, H2B, H3, and H4 after a 2-hr incubation with Ni-NTA beads. Lane *9*: Elution with 1 M imidazole. (*B*) Control with his-tagged H2A NCP. Lanes *1–4*: Flowthrough and washes with 5-mM imidazole after a 2-hr incubation of His-tagged H2A NCP to Ni-NTA beads. Lane *5*: Elution with 1 M imidazole. Lane *6*: Onput. Lane *7*: Histone octamer with untagged H2A. (*C*) Control with untagged H2A.Z nucleosomes. Lanes *1–4*: Flowthrough and washes with 5-mM imidazole after 2-hr binding of H2A.Z NCPs to Ni-NTA beads. Lane *5*: Elution with 1 M imidazole. Lane *6*: H2A.Z-H2B dimer. Lane *7*: Histone octamer with non His-tagged H2A. Lane *8*: His-tagged H2A-H2B dimer (His-tagged H2A, *upper band*; H2B, *lower band*). Lane *9*: H3-H4 tetramer. All gels shown here are 18% SDS-PAGE gels stained with Coomassie brilliant blue.

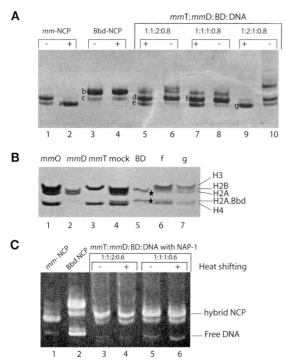

Figure 6. H2A.Bbd can form hybrid nucleosomes with major-type H2A. (*A*) Salt gradient reconstituted mouse NCP (*mm*-NCP; lanes *1* and *2*), Bbd-NCP (lanes *3* and *4*), and hybrid NCPs (lanes *5–10*), before (–) and after (+) a 1-hr incubation at 37°C, were analyzed by 5% native gel and stained by Coomassie brilliant blue. Ratios between mouse (H3-H4)$_2$ tetramer (*mm*T), mouse (H2A-H2B) dimer (*mm*D), H2A.Bbd-H2B dimer (BD), and DNA for "hybrid NCP" reconstitution are indicated. Bands that were subsequently excised from the gel for further analysis are labeled *a–g*. (*B*) Bands *f*, *g* (from *A*) were analyzed by 18% SDS-PAGE, stained with Coomassie brilliant blue. Mouse H2A and H2A.Bbd are indicated by *stars*. Mouse octamer (MO, lane *1*), MD (lane *2*), *mm*T (lane *3*), *mm*T:*mm*D:BD mixtures of 1:1:1 (mock, lane *4*), and BD (lane *5*) are shown as controls. (*C*) Hybrid NCPs are also obtained with yNAP-1-dependent nucleosome assemblies. yNAP-1-reconstituted NCP (reconstituted at indicated ratios) was compared with NCP reconstituted over a salt gradient on a 5% native gel. The gel was stained with ethidium bromide. *Asterisks:* hybrid NCP.

CONCLUSIONS AND OUTLOOK

Histone variants are receiving an ever increasing amount of attention, but much needs to be done before we reach a complete understanding of their role in the complex biology of chromatin. In vitro, structural and biophysical analysis of nucleosomes containing H2A.Bbd and CenpA (in addition to the already available structures for H2A.Z- and macroH2A-containing nucleosomes) will provide further insight into the ways in which nucleosome structure and dynamics is affected by histone variants. These studies will also have to be expanded to take into the account the possible existence of hybrid nucleosomes. Second, we will need to extend our investigations to include model nucleosomal arrays, as demonstrated for H2A.Z (Fan et al. 2002). Third, the interaction of variant-containing chromatin with various linker histones and nonhistone proteins needs to be investigated. Fourth, more systematic analyses are needed to study the dynamic behavior of variant-containing nucleosomes and chromatin in the presence of various chromatin remodeling factors (see, e.g., Flaus et al. 2004). Finally, the transcriptional properties of variant-containing nucleosomes from several model promoters will have to be scrutinized carefully in vitro.

In vivo, the challenges are perhaps even greater. For example, knockout models are available for only a few histone variants. Second, we need to investigate the distribution of histone variants at the single-nucleosome (and subnucleosome) level at a variety of chromatin loci. This represents a technical challenge that may not be easily overcome. Third, the pathways by which variant nucleosomes (or hybrid nucleosomes) are assembled need to be further investigated. Finally, the ability of histone variants to be posttranslationally modified must be scrutinized. Combined, these studies are likely to yield a picture of mind-boggling complexity that probably will approach or even exceed that of the histone modification network. Doubtlessly, histone variants add an entirely new dimension to the "histone code."

REFERENCES

Ahmad K. and Henikoff S. 2002a. Histone H3 variants specify modes of chromatin assembly. *Proc. Natl. Acad. Sci.* (suppl. 4) **99:** 16477.

———. 2002b. The histone variant H3.3 marks active chromatin by replication-independent nucleosome assembly. *Mol. Cell* **9:** 1191.

Angelov D., Molla A., Perche P.Y., Hans F., Cote J., Khochbin S., Bouvet P., and Dimitrov S. 2003. The histone variant macroH2A interferes with transcription factor binding and SWI/SNF nucleosome remodeling. *Mol. Cell* **11:** 1033.

Bao Y., Konesky K., Park Y.J., Rosu S., Dyer P.N., Rangasamy D., Tremethick D.J., Laybourn P.J., and Luger K. 2004. Nucleosomes containing the histone variant H2A.Bbd organize only 118 base pairs of DNA. *EMBO J.* **23:** 3314.

Bassing C.H., Suh H., Ferguson D.O., Chua K.F., Manis J., Eckersdorff M., Gleason M., Bronson R., Lee C., and Alt F.W. 2003. Histone H2AX: A dosage-dependent suppressor of oncogenic translocations and tumors. *Cell* **114:** 359.

Black E.B., Foltz D.R., Chakravarthy S., Luger K., Woods V.L.J., and Cleveland D.W. 2004. Structural determinants for generating centromeric chromatin. *Nature* **430:** 578.

Bruno M., Flaus A., Stockdale C., Rencurel C., Ferreira H., and Owen-Hughes T. 2003. Histone H2A/H2B dimer exchange by ATP-dependent chromatin remodeling activities. *Mol. Cell* **12:** 1599.

Celeste A., Fernandez-Capetillo O., Kruhlak M.J., Pilch D.R., Staudt D.W., Lee A., Bonner R.F., Bonner W.M., and Nussenzweig A. 2003. Histone H2AX phosphorylation is dispensable for the initial recognition of DNA breaks. *Nat. Cell Biol.* **5:** 675.

Chadwick B.P. and Willard H.F. 2001. A novel chromatin protein, distantly related to histone H2A, is largely excluded from the inactive X chromosome. *J. Cell Biol.* **152:** 375.

Chadwick B.P., Valley C.M., and Willard H.F. 2001. Histone variant macroH2A contains two distinct macrochromatin domains capable of directing macroH2A to the inactive X chromosome. *Nucleic Acids Res.* **29:** 2699.

Chen T.A., Smith M.M., Le S.Y., Sternglanz R., and Allfrey V.G. 1991. Nucleosome fractionation by mercury affinity chromatography. Contrasting distribution of transcriptionally active DNA sequences and acetylated histones in nucleosome fractions of wild-type yeast cells and cells expressing a histone H3 gene altered to encode a cysteine 110 residue. *J. Biol. Chem.* **266:** 6489.

Clarkson M.J., Wells J.R., Gibson F., Saint R., and Tremethick D.J. 1999. Regions of variant histone His2AvD required for *Drosophila* development. *Nature* **399**: 694.

Costanzi C. and Pehrson J.R. 1998. Histone macroH2A1 is concentrated in the inactive X chromosome of female mammals. *Nature* **393**: 599.

———. 2001. MacroH2A2, a new member of the MacroH2A core histone family. *J. Biol. Chem.* **276**: 21776.

Crane-Robinson C. 1997. Where is the globular domain of linker histone located on the nucleosome? *Trends Biochem. Sci.* **22**: 75.

Faast R., Thonglairoam V., Schulz T.C., Beall J., Wells J.R., Taylor H., Matthaei K., Rathjen P.D., Tremethick D.J., and Lyons I. 2001. Histone variant H2A.Z is required for early mammalian development. *Curr. Biol.* **11**: 1183.

Fan J.Y., Gordon F., Luger K., Hansen J.C., and Tremethick D.J. 2002. The essential histone variant H2A.Z regulates the equilibrium between different chromatin conformational states. *Nat. Struct. Biol.* **19**: 172.

Flaus A., Rencurel C., Ferreira H., Wiechens N., and Owen-Hughes T. 2004. Sin mutations alter inherent nucleosome mobility. *EMBO J.* **23**: 343.

Gautier T., Abbott D.W., Molla A., Verdel A., Ausio J., and Dimitrov S. 2004. Histone variant H2ABbd confers lower stability to the nucleosome. *EMBO Rep.* **5**: 715.

Georges S.A., Kraus W.L., Luger K., Nyborg J.K., and Laybourn P.J. 2002. p300-mediated tax transactivation from recombinant chromatin: Histone tail deletion mimics coactivator function. *Mol. Cell. Biol.* **22**: 127.

Hamiche A., Carot V., Alilat M., De Lucia F., O'Donohue M.F., Revet B., and Prunell A. 1996. Interaction of the histone (H3-H4)2 tetramer of the nucleosome with positively supercoiled DNA minicircles: Potential flipping of the protein from a left- to a right-handed superhelical form. *Proc. Natl. Acad. Sci.* **93**: 7588.

Hansen J.C. 2002. Conformational dynamics of the chromatin fiber in solution: Determinants, mechanisms, and functions. *Annu. Rev. Biophys. Biomol. Struct.* **31**: 361.

Hayes J.J. and Hansen J.C. 2001. Nucleosomes and the chromatin fiber. *Curr. Opin. Genet. Dev.* **11**: 124.

Henikoff S., Furuyama T., and Ahmad K. 2004. Histone variants, nucleosome assembly and epigenetic inheritance. *Trends Genet.* **20**: 320.

Hoyer-Fender S., Costanzi C., and Pehrson J.R. 2000. Histone macroH2A1.2 is concentrated in the XY-body by the early pachytene stage of spermatogenesis. *Exp. Cell Res.* **258**: 254.

Kimura H. and Cook P.R. 2001. Kinetics of core histones in living human cells: Little exchange of H3 and H4 and some rapid exchange of H2B. *J. Cell Biol.* **153**: 1341.

Kobor M.S., Venkatasubrahmanyam S., Meneghini M.D., Gin J.W., Jennings J.L., Link A.J., Madhani H.D., and Rine J. 2004. A protein complex containing the conserved Swi2/Snf2-related ATPase Swr1p deposits histone variant H2A.Z into euchromatin. *PLoS Biol.* **2**: E131.

Krogan N.J., Keogh M.C., Datta N., Sawa C., Ryan O.W., Ding H., Haw R.A., Pootoolal J., Tong A., Canadien V., Richards D.P., Wu X., Emili A., Hughes T.R., Buratowski S., and Greenblatt J.F. 2003. A Snf2 family ATPase complex required for recruitment of the histone H2A variant Htz1. *Mol. Cell* **12**: 1565.

Liu X., Li B., and Gorovsky M.A. 1996. Essential and nonessential histone H2A variants in *Tetrahymena thermophila*. *Mol. Cell. Biol.* **16**: 4305.

Louters L. and Chalkley R. 1985. Exchange of histones H1, H2A, and H2B in vivo. *Biochemistry* **24**: 3080.

Luger K. and Richmond T.J. 1998a. DNA binding within the nucleosome core. *Curr. Opin. Struct. Biol.* **8**: 33.

———. 1998b. The histone tails of the nucleosome. *Curr. Opin. Genet. Dev.* **8**: 140.

Luger K., Maeder A.W., Richmond R.K., Sargent D.F., and Richmond T.J. 1997. X-ray structure of the nucleosome core particle at 2.8 Å resolution. *Nature* **389**: 251.

Malik H.S. and Henikoff S. 2003. Phylogenomics of the nucleosome. *Nat. Struct. Biol.* **10**: 882.

Meneghini M.D., Wu M., and Madhani H.D. 2003. Conserved histone variant H2A.Z protects euchromatin from the ectopic spread of silent heterochromatin. *Cell* **112**: 725.

Mizuguchi G., Shen X., Landry J., Wu W.H., Sen S., and Wu C. 2004. ATP-driven exchange of histone H2AZ variant catalyzed by SWR1 chromatin remodeling complex. *Science* **303**: 343.

Palmer D.K., O'Day K., Trong H.L., Charbonneau H., and Margolis R.L. 1991. Purification of the centromere-specific protein CENP-A and demonstration that it is a distinctive histone. *Proc. Natl. Acad. Sci.* **88**: 3734.

Park Y.J., Dyer P.N., Tremethick D.J., and Luger K. 2004. A new fluorescence resonance energy transfer approach demonstrates that the histone variant H2AZ stabilizes the histone octamer within the nucleosome. *J. Biol. Chem.* **279**: 24274.

Pehrson J.R. and Fried V.A. 1992. MacroH2A, a core histone containing a large nonhistone region. *Science* **257**: 1398.

Pehrson J.R., Costanzi C., and Dharia C. 1997. Developmental and tissue expression patterns of histone macroH2A1 subtypes. *J. Cell. Biochem.* **65**: 107.

Perche P.Y., Vourc'h C., Konecny L., Souchier C., Robert-Nicoud M., Dimitrov S., and Khochbin S. 2000. Higher concentrations of histone macroH2A in the Barr body are correlated with higher nucleosome density. *Curr. Biol.* **10**: 1531.

Protacio R.U. and Widom J. 1996. Nucleosome transcription studied in a real-time synchronous system: Test of the lexosome model and direct measurement of effects due to histone octamer. *J. Mol. Biol.* **256**: 458.

Rangasamy D., Greaves I., and Tremethick D.J. 2004. RNA interference demonstrates a novel role for H2A.Z in chromosome segregation. *Nat. Struct. Mol. Biol.* **11**: 650.

Rangasamy D., Berven L., Ridgway P., and Tremethick D.J. 2003. Pericentric heterochromatin becomes enriched with H2A.Z during early mammalian development. *EMBO J.* **22**: 1599.

Rasmussen T.P., Huang T., Mastrangelo M.A., Loring J., Panning B., and Jaenisch R. 1999. Messenger RNAs encoding mouse histone macroH2A1 isoforms are expressed at similar levels in male and female cells and result from alternative splicing. *Nucleic Acids Res.* **27**: 3685.

Richmond T.J. and Davey C.A. 2003. The structure of DNA in the nucleosome core. *Nature* **423**: 145.

Rothkamm K. and Lobrich M. 2003. Evidence for a lack of DNA double-strand break repair in human cells exposed to very low x-ray doses. *Proc. Natl. Acad. Sci.* **100**: 5057.

Sanner M.F., Olson A.J., and Spehner J.C. 1996. Reduced surface: An efficient way to compute molecular surfaces. *Biopolymers* **38**: 305.

Shelby R.D., Vafa O., and Sullivan K.F. 1997. Assembly of CENP-A into centromeric chromatin requires a cooperative array of nucleosomal DNA contact sites. *J. Cell Biol.* **136**: 501.

Sullivan K.F., Hechenberger M., and Masri K. 1994. Human CENP-A contains a histone H3 related histone fold domain that is required for targeting to the centromere. *J. Cell Biol.* **127**: 581.

Sullivan S., Sink D.W., Trout K.L., Makalowska I., Taylor P.M., Baxevanis A.D., and Landsman D. 2002. The histone database. *Nucleic Acids Res.* **30**: 341.

Suto R.K., Clarkson M.J., Tremethick D.J., and Luger K. 2000. Crystal structure of a nucleosome core particle containing the variant histone H2A.Z. *Nat. Struct. Biol.* **7**: 1121.

Suto R.K., Edayathumangalam R.S., White C.L., Melander C., Gottesfeld J.M., Dervan P.B., and Luger K. 2003. Crystal structures of nucleosome core particles in complex with minor groove DNA-binding ligands. *J. Mol. Biol.* **326**: 371.

van Daal A. and Elgin S.C. 1992. A histone variant, H2AvD, is essential in *Drosophila melanogaster*. *Mol. Biol. Cell* **3**: 593.

Vermaak D. and Wolffe A.P. 1998. Chromatin and chromosomal controls in development. *Dev. Genet.* **22**: 1.

Epigenetics, Histone H3 Variants, and the Inheritance of Chromatin States

S. Henikoff,* E. McKittrick,* and K. Ahmad[†]

*Howard Hughes Medical Institute, Fred Hutchinson Cancer Research Center, Seattle, Washington 98109;
[†]Department of Biological Chemistry and Molecular Pharmacology, Harvard Medical School, Boston,
Massachusetts 02115

The 68th Cold Spring Harbor Symposium celebrated an extraordinary accomplishment in molecular biology: determination of the human genomic sequence. This landmark achievement highlights the rapid and steady progress that has occurred over the past 50 years beginning with the Watson–Crick model of DNA and culminating in the nearly complete specification of our genetic inheritance. In contrast, the 69th Symposium deals with a topic, epigenetics, that was already 25 years old at the time of the Watson–Crick model, and yet the salient features of epigenetic phenomena have remained almost as mysterious now as when they were first described.

The experimental study of epigenetics began when H.J. Muller discovered a remarkable class of mutations that appeared following X-ray irradiation of flies (Muller 1930). This phenomenon, called position-effect variegation (PEV), is the unstable expression of a gene that results when chromosome breaks juxtapose it to heterochromatin, the cytologically condensed regions of chromosomes (Spofford 1976). Although Muller first observed the phenomenon in rearrangements that affect the white eye color gene, similar effects have been found for many other types of genes and in organisms ranging from animals to fungi to plants (Richards and Elgin 2002).

Three striking features that are apparent from PEV phenotypes continue to intrigue us today. First, PEV appears to be an all-or-none phenomenon. This is readily apparent in a modern example of classical PEV seen in Figure 1, which shows a fly that harbors a variegating allele of a green fluorescent protein (GFP) gene that had been juxtaposed to heterochromatin by a chromosomal rearrangement. The gene shows its normal level of expression in some tissues of the eye, but appears to be completely inactive in other tissues. Second, inactivation is patchy, as though a decision was made as to whether or not to be active, and that decision was inherited through multiple rounds of cell division. Finally, the patterns of expression are completely different between the two eyes of this fly, as if such decisions could occur essentially at random throughout development. These three features, all-or-none expression, heritability, and random decisions between states, are characteristic of epigenetic phenomena in general.

We have attempted to understand these prominent features of PEV by applying modern visualization tools and reporters to the problem. We first describe the use of a *Drosophila* genetic system for probing the accessibility of a promoter that is subject to PEV (Ahmad and Henikoff 2001b). This experimental system has provided insights into the all-or-none and stochastic nature of PEV. Next, we describe a *Drosophila* cytological system for observing dynamic chromatin changes during the cell cycle that might be involved in epigenetic phenomena (Ahmad and Henikoff 2001a). We have applied this system to the study of variant H3 histones, where we have uncovered a distinct pathway for nucleosome assembly that takes place at active genes (Ahmad and Henikoff 2002c). This discovery motivated a biochemical approach to ascertain the relationship between H3 variants and their covalent modifications (McKittrick et al. 2004). We propose that nucleosome assembly pathways represent fundamental means of differentiating chromatin states that can account for precise inheritance of chromatin memory and PEV through multiple rounds of cell division.

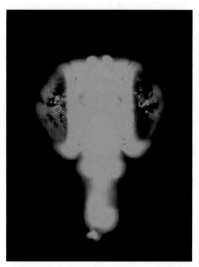

Figure 1. PEV illustrates features of epigenetic inheritance. Image of the head of a fruit fly, showing the stochastic epigenetic silencing of a transgene encoding GFP in individual ommatidia of the eyes. See text for details.

EPIGENETIC PHENOTYPES AND NUCLEOSOME DYNAMICS

Probing Heterochromatin during *Drosophila* Development

To study PEV during development, we introduced a novel in vivo reporter system (Fig. 2A). Our starting line harbored a *P* transposon that contained a set of five tandem yeast GAL4 binding sites upstream of a gene encoding GFP (*UASGFP*), inserted into a euchromatic site (Ahmad and Henikoff 2001b). The transposon also contained a *mini-white* eye color gene, which provided a reporter for selection of PEV mutants. After X-ray irradiation to induce chromosomal rearrangements, we screened for progeny that showed classical white variegation. We then subjected lines to *P*-element mobilization, which allowed us to select for more severe variegators (Dorer and Henikoff 1994). The PEV nature of these mutant lines was confirmed by showing that variegation was suppressed at elevated temperature and in the presence of a *Su(var)205* mutation, a suppressor of PEV that encodes Heterochromatin-associated Protein 1 (HP1) (Weiler and Wakimoto 1995).

When two genes are present on a transposon and one is subject to PEV, it is generally observed that the other gene variegates as well. To test for variegation of the *UASGFP* gene in lines showing *mini-white* variegation, we crossed each line to flies that produce GAL4, which activates the *UASGFP* gene when bound to its upstream activation sequence (Ahmad and Henikoff 2001b). Expression was assayed in eye discs of third instar larvae. Using a "driver" (*GawB*) that expresses GAL4 beginning in the early larval period and continuing throughout development, we observed patchy expression relative to the uniform expression that is characteristic of wild-type, indicating PEV of *UASGFP* (Fig. 2B). Patches appeared clonal, often crossing the morphogenetic furrow that separates proliferating from differentiated cells as it moves from posterior to anterior of the disc. Interestingly, silencing in the same *UAS-GFP* lines was less pronounced in the presence of GAL4 produced by a driver (*A5C*) that began to express during early embryonic development, and silencing was more severe using a driver (*GMR*) that expressed behind the morphogenetic furrow during pupal development (Fig. 2B). Such differences could not be accounted for by differences in the amount of GAL4 produced, because these levels were estimated to be highest in tissues expressing GAL4 only in differentiated cells (*GMR*), which showed the strongest degree of silencing (Fig. 2B).

Activation Depends on Transcription Factor Abundance

A simple explanation for different degrees of silencing produced by drivers that turn on at different times during development is that heterochromatin usually, but not always, prevents GAL4 from gaining access to its binding site. However, once GAL4 binds, then the gene remains active throughout the remainder of development. In this way, the longer that GAL4 is available in developing tissue, the higher the probability that it will bind and permanently activate expression. This hypothesis predicts that the probability of expression will also be increased by higher concentrations of GAL4. Indeed, this is the case; we found that two drivers that begin expression at the same early time in development but at approximately tenfold different levels (*A5C* and *arm*) show corresponding differences in the probability of activation (Fig. 2B). Further support for this hypothesis came from the detection of spontaneous activation in cultured discs on both sides of the morphogenetic furrow (Ahmad and Henikoff 2001b). These observations imply that heterochromatin is in a dynamic state, because yeast GAL4, which can act in *Drosophila* only once it is bound, appears to gain access to its binding site in heterochromatin in a stochastic manner at any time during development.

Our conclusion that heterochromatin is dynamic (Ahmad and Henikoff 2001b, 2002a) later received strong support from fluorescence recovery after photobleaching

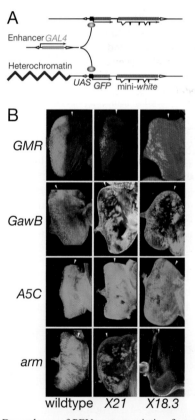

Figure 2. Dependence of PEV on transcription factor concentration (Polach and Widom 1995). (*A*) Schematic diagram of a system for probing heterochromatin during *Drosophila* development. GAL4 controls expression of the *UAS-GFP* reporter gene located either in its normal position (*top*) or juxtaposed to constitutive heterochromatin (*bottom*). (*B*) Eye-antennal imaginal discs expressing *UAS-GFP* under control of GAL4 driven by the *GMR* eye- and differentiation-specific enhancer (*GMR*), by the *GawBT80* larval enhancer (*GawB*), by the strong Actin5C (*A5C*) embryonic enhancer, and by the weak armadillo (*arm*) embryonic enhancer. The reporter gene is on a wild-type chromosome (*left* column) and is subject to PEV in two different lines (*center* and *right* column). *Arrowheads*, the location of the morphogenetic furrow.

studies of the key component of heterochromatin, HP1 (Cheutin et al. 2003; Festenstein et al. 2003). HP1 associates with chromatin with a residence time of only several seconds. The dynamic nature of heterochromatin is surprising given the perception that it reflects a condensed state, indicating the need to consider dynamic models of heterochromatin to explain PEV. An attractive possibility comes from a proposal by Widom and colleagues based on their in vitro studies of the accessibility of a DNA-binding site when it is wrapped up in a nucleosome (Polach and Widom 1995). They hypothesized that an equilibrium exists between a conformation in which a binding site is inaccessible because it is wrapped and a conformation in which the site is accessible because it is unwrapped. We expect that in the wild-type euchromatic situation, this equilibrium would not be directly observable, because the unwrapped conformation would also be in equilibrium with an unwrapped factor-bound conformation, and this bound state would be strongly favored (Fig. 3). However, if juxtaposition to heterochromatin increases the frequency of a nucleosome being wrapped, then there will be two relatively stable conformations (wrapped and factor-bound) separated by a transient unwrapped and unbound intermediate (Ahmad and Henikoff 2001b). The existence of a distinct intermediate state is crucial, because in a two-conformation competition model (Dillon and Festenstein 2002), each cell should show a time-averaged intermediate level of expression, not stochastic variegation.

Our dynamic equilibrium model fits with several known features of heterochromatin and PEV. HP1 is a dimer with two chromodomains that bind to methylated histone H3 lysine-9 (H3K9me), a modification that is especially abundant in heterochromatin (Jenuwein and Allis 2001). H3 tails exit between the DNA gyres on either side of the nucleosomal dyad axis and associate with linker histones and linker DNA (Luger et al. 1997; Zheng and Hayes 2003). Therefore, it is expected that the effect of HP1 binding, even briefly, to pairs of H3 tails would be to reduce the frequency of unwrapping, thus disfavoring intermediate formation and reducing the availability of factor-binding sites. The model also can explain the dependence of activation frequency on transcription factor concentration observed in our in vivo assay (Fig. 2B); examination of the equilibrium equations reveals that the ratio of active to silent tissue should be directly proportional to the concentration of transcription factor (see the Fig. 3 legend). Finally, the model provides an intuitive explanation for both the on–off and the stochastic nature of PEV. So, although PEV is notorious as a long-range effect, the ultimate nature of silencing (but not "spreading") is a process that may be understood in molecular detail at the level of a single nucleosome.

NUCLEOSOME ASSEMBLY PATHWAYS DISTINGUISH CHROMATIN STATES

How Are Chromatin States Inherited?

The dynamic nature of heterochromatin makes the problem of propagating the heterochromatic state through rounds of cell division seem even more mysterious. If the silent state is unstable, with HP1 and other chromatin-associated proteins rapidly exchanging, why is PEV so often clonal? This question of epigenetic memory is a general one that appears to underlie most cases of developmental decisions that are remembered through multiple rounds of cell division.

One possible way that epigenetic memory might be inherited is by replication of histone modifications. This possibility has generated much excitement in the chromatin field in recent years with the realization that active and silent chromatin are differentially modified by enzymes that have been implicated in the specification of chromatin states (Jenuwein and Allis 2001; Grewal and Elgin 2002). Furthermore, the binding of HP1 to H3K9me provides a direct link between this particular histone modification and the heterochromatic state. However, the process whereby modifications are replicated has remained unclear. This is a serious problem, because old nucleosomes are randomly distributed intact behind the replication fork (Leffak et al. 1977; Yamasu and Senshu 1990), leading to dilution of epigenetic signals by a factor of 2 and loss of precision with each cell division. Chromatin inheritance appears to be unlike DNA methylation, which is maintained by the Dnmt1 DNA methyltransferase with base-pair precision behind the replication fork (Leonhardt et al. 1992). Thus far, no mechanism is known that can replicate modifications on histones following DNA replication.

Our work on histone variants provides a potential solution to this enigma. We have discovered that a histone 3 variant, H3.3, replaces canonical H3 at active loci via a distinct pathway of replication-independent nucleosome assembly (Ahmad and Henikoff 2002c). The fact that H3 is not used by this pathway provides a distinction between active and inactive chromatin states. Replication-independent deposition may serve to replenish a histone mark by replacing histones at active genes throughout the cell cycle.

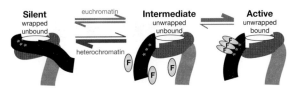

Figure 3. A site-exposure model for gene silencing (Ahmad and Henikoff 2001b). When DNA containing factor-binding sites (*stars*) is wrapped around a nucleosome, the sites are inaccessible. Transient unwrapping of the DNA exposes the site, which can then be bound by factor (F). Let [S] = the concentration of silent nucleosomes, [A] = the concentration of active nucleosomes, [I] = the concentration of the transient intermediate, and [F] = the concentration of factor. Define rate constants as k_1 (S→I), k_2 (I→S), k_3 (A→I+F), and k_4 (I+F→A). Then $d[S]/dt = -k_1[S] + k_2[I]$, and $d[A]/dt = -k_3[A] + k_4[I][F]$. At equilibrium, $d[S]/dt = d[A]/dt = 0$, so $[I] = (k_1/k_2)[S] = (k_3/k_4)[A]/[F]$, or $[A]/[S] = (k_1k_4/k_2k_3)[F]$. That is, the ratio of active to silent tissue should be directly proportional to transcription factor concentration. (Reprinted, with permission, from Ahmad and Henikoff 2001b [©Elsevier].)

A Cytological Assay Reveals Alternative Nucleosome Assembly Pathways

To study changes that occur in nuclei during the cell cycle, we took advantage of the *Drosophila Kc* cell line in which the difference between early replicating euchromatin and late replicating heterochromatin is especially pronounced (Ahmad and Henikoff 2001a). Brief pulses of nucleoside triphosphates or probing with an antibody against a component of the DNA replication apparatus provides a snapshot of replication forks at the time that cells are fixed for cytological observations. The formation of a chromocenter by the coalescence of all heterochromatin in a cell reveals a clear spatial distinction between early and late replication, thus identifying cell cycle stages in an unsynchronized population. To follow the behavior of particular nuclear proteins during the cell cycle, we constructed plasmids encoding fusions to GFP under control of a heat shock promoter, introduced the plasmids into *Kc* cells by transient transfection, and produced chimeric protein with a heat shock pulse.

When this cytological assay system was applied to the centromere-specific H3 variant, CID, we were surprised to find that CID-GFP localized to centromeres throughout the cell cycle, indicating replication-independent (RI) assembly (Ahmad and Henikoff 2001a). Similar conclusions were reached by Sullivan and coworkers (Shelby et al. 2000) for CENP-A, the human centromere-specific H3 variant. In contrast, canonical H3 only deposited at replication forks in our assay (Ahmad and Henikoff 2001a).

A control experiment led us to realize that there is a distinct RI pathway for nucleosome assembly that acts at active loci. We had expected that histone H4-GFP would deposit at all sites of H3 and CID localization in gap phase cells when produced in the same assay. It did, but it also localized elsewhere (Ahmad and Henikoff 2002c). By far the most prominent site of RI deposition of H4-GFP was an rDNA array (confirmed by fluorescence in situ hybridization using an rDNA probe), with weaker labeling throughout the chromosome arms, but not in heterochromatin. Thus, H4, but not H3, is deposited during gap phase at particular loci.

H3 and H4 are folded together by histone chaperones upon synthesis and are always deposited as a unit (Mello and Almouzni 2001), so we reasoned that the partner for H4 deposition at particular loci must be the replacement variant H3.3. Indeed, like H4, H3.3 was deposited in a replication-coupled (RC) manner during S phase and localized to an rDNA array and euchromatin during gap phases (Fig. 4) (Ahmad and Henikoff 2002c). This differential behavior of H3 and H3.3 was unexpected, insofar as the RI form of histone 3 in Tetrahymena was shown to be interchangeable with the RC form (Yu and Gorovsky 1997). It was assumed that H3 and H3.3 would be likewise interchangeable, with only four amino acid differences between them. However, these differences are responsible for completely different deposition behavior of the histones when produced in the same way during gap phase (cf. Fig. 5A and Fig. 5B). The fact that H3.3 is 100% identical throughout its length in organisms

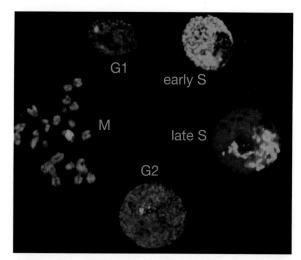

Figure 4. RC and RI deposition of H3.3 (Ahmad and Henikoff 2002c). Nuclei labeled with DAPI (*gray*) show the location of H3.3-GFP (*green*) when it is produced in a pulse at different times in the cell cycle. A mitotic figure is shown on the *left* (M), with H3.3 prominently deposited at the rDNA locus within X chromosome heterochromatin and throughout the euchromatic arms but not the heterochromatin. Proceeding *clockwise* are nuclei in G1 phase showing a bright rDNA focus, in early S phase showing RC euchromatic deposition, in late S phase showing RC heterochromatic deposition at the chromocenter, and in G2 phase showing bright rDNA foci and faint RI euchromatic deposition.

as diverse as flies, humans, and clams suggested that this difference in nucleosome assembly behavior is a general feature of animal chromatin, and a comparable situation appears to hold in plants (Waterborg 1993).

Distinct Substrate Requirements for Alternative Nucleosome Assembly Pathways

Three of the four amino acids that distinguish H3 from H3.3 are found in a cluster on the core (residues 87–90) and one lies on the amino-terminal tail (residue 31) (Malik and Henikoff 2003). Our gap phase incorporation assay revealed that H3 does not deposit into chromosomes at all when produced during G2 phase, but rather can be seen to accumulate throughout the cell at mitosis (Fig. 5B). Swapping these residues from H3.3 into H3 led to partial deposition of the GFP fusion at the rDNA array for each of the three core residues (Fig. 5C), but swapping the tail residue or swapping any single H3 residue into H3.3 had no effect. This suggests that something interacts with the three core residues of H3 and actively prevents its RI deposition.

Another sequence requirement that distinguishes RC from RI nucleosome assembly is a requirement for the amino-terminal tail. RC but not RI deposition of either H3 or H3.3 is gradually lost as more and more of the amino-terminal tail is removed (Ahmad and Henikoff 2002c). Although we do not understand the basis for this restriction, it is interesting that yeast lack this requirement, and, like other ascomycetes, yeast lack the H3 (RC)

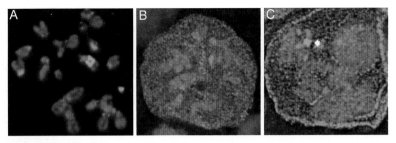

Figure 5. Mapping determinants of RC and RI assembly (Ahmad and Henikoff 2002c). (*A*) G2 phase incorporation of H3.3-GFP at the X-long rDNA array (*green*). Centromeres are marked with anti-CID antibody (*red*). (*B*) H3-GFP fails to incorporate into the chromosomes when produced during G2, and accumulates in the cell volume at mitosis. (*C*) Changing a single H3 amino acid residue on the core to the corresponding H3.3 residue (Met to Gly at position 90) causes partial incorporation of GFP at the rDNA (*arrow*). Similar results were obtained for the other two core residues (A87S and I89V), but no effect on incorporation was seen for the tail residue (A31S) or for any single change from H3.3 to H3. (Reprinted, with permission, from Ahmad and Henikoff 2002c [©Elsevier].)

form, having lost the canonical variant after splitting from basidiomycete fungi (Malik and Henikoff 2003). Thus, ascomycetes have only the H3.3 form, which is ironic considering that most of our knowledge of chromatin comes from studies in budding yeast, whereas almost all studies of higher eukaryotic chromatin have used the more abundant H3 form.

Our discovery of alternative chromatin assembly pathways based on substrate requirements has recently received direct biochemical support. Using affinity purification of tagged H3 and H3.3, Tagami et al. (2004) have found that soluble H3 is present in a complex with the RC-specific chromatin assembly factor 1 (CAF-1), and soluble H3.3 with the RI-specific histone regulator A (HIRA).

H3.3 Replaces H3K9me at Epigenetically Activated Loci

In tetraploid *Drosophila Kc* cells, RI assembly of H3.3 typically occurs at only one rDNA array (Ahmad and Henikoff 2002c), despite the presence of multiple X chromosomes, each with a single rDNA array. The X chromosomes in *Kc* cells are of two different types based on the size of their rDNA arrays (Echalier 1997). One X chromosome has undergone a large expansion of the rDNA locus relative to wild type, evidently an adaptation to culture conditions after the lines were established. We noticed that in quiescent cells, this "X-long" chromosome was always labeled with H3.3-GFP when produced during G2 (Fig. 6A). The absence of RI deposition of H3.3-GFP at "X-short" does not appear to result from any permanent defect in X-short, because when cells are induced to grow, about half of them show RI deposition of H3.3-GFP at the rDNA locus on at least one of the X-short chromosomes (Fig. 6B). This epigenetic change in an rDNA array is reminiscent of the well-known phenomenon of nucleolar dominance, where the rDNA array from one species in an interspecific hybrid is active while that from the other species is inactive (Pikaard 2000). The similarity of RI deposition of H3.3 to epigenetic activation suggests that H3.3 is depositing at active, but not inactive rDNA arrays.

We could ascertain whether H3.3 deposits at sites of active or inactive rDNA arrays using an antibody to H3K9me, which decorates heterochromatin on all of the chromosome arms. At mitosis, the prominent labeling of X-long by H3.3-GFP falls into a large gap in the H3K9me pattern (Fig. 6). Close examination of the X-short rDNA arrays that display H3.3-GFP staining after induction reveals that these spots always fall within a gap in the H3K9me pattern (Fig. 6B, inset). No such gaps were seen in uninduced cells, leading to the conclusion that gaps in H3K9me appeared as a consequence of induction, with concomitant deposition of H3.3 (Ahmad and Henikoff 2002c). This replacement process must have occurred within the 16-hour time window between induction by feeding and fixation of the cells. Recently Spector and coworkers (Janicki et al. 2004) confirmed and extended this observation by inducing a large array for transcription by RNA Polymerase II in a human cell line and showing that H3K9me and HP1 disappear while H3.3 accumulates to high levels during an approximately one-hour period. Therefore, the replacement of heterochromatin by H3.3 appears to be a general process that can occur rapidly upon transcriptional induction.

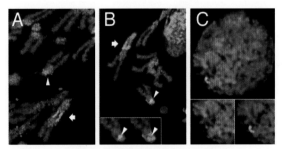

Figure 6. Replacement of H3K9me by H3.3 at rDNA arrays (Ahmad and Henikoff 2002c). (*A*) Mitotic figures from stationary phase cells show intense H3.3-GFP labeling (*green*) only on the large rDNA array on the elongated X-long chromosome (*arrow*), but not on the X-short chromosomes (*arrowhead*), which also carry rDNA genes. (*B*) Mitotic figures from induced cells show H3.3-GFP labeling on both X-long and X-short chromosomes. (*Inset*) An enlargement of the proximal part of X-short. Wherever H3.3-GFP is present, there is a gap in heterochromatin (antibody to H3K9me, *red*). (*C*) RI H3.3-GFP labeling shows little overlap with heterochromatin in interphase nuclei. DNA staining (DAPI) is in *gray*. (*Insets*) Separated channels. (Reprinted, with permission, from Ahmad and Henikoff 2002c [©Elsevier].)

H3.3 IS ENRICHED IN "ACTIVE" LYSINE MODIFICATIONS

If H3.3 marks active chromatin, then it should be enriched in covalent modifications that have been correlated with the active state. To test this, we fractionated histone 3 forms from *Kc* cells using reverse-phase high performance liquid chromatography (HPLC) (McKittrick et al. 2004), which had previously been used by Waterborg to resolve alfalfa H3.1 (the plant RC form) and H3.2 (the RI form) (Waterborg 1990). When acid-extracted nuclear proteins were chromatographed, histone 3 resolved into two peaks (Fig. 7). After ArgC protease digestion of HPLC fractions, we performed mass spectrometry. Based on the encoded difference at position 31, which is alanine in H3 and serine in H3.3, mass analysis of peptide 27–40 identified the first peak as H3 and the second peak as H3.3 (McKittrick et al. 2004).

Repeated measurements of relative peak heights revealed that H3.3 accounts for 24±3% of bulk histone 3 in *Kc* cells. This percentage is not significantly different from the fraction (26%) of the *Drosophila* genome that is transcribed in *Kc* cells ([176 Mb total genomic DNA – 59 Mb heterochromatic – 49 Mb intergenic]/176 Mb total genomic [Hoskins et al. 2002; Misra et al. 2002] = 0.386, multiplied by the fraction [68%] of genes that are transcribed in *Kc* cells [0.39 × 0.68 = 0.26] [Schübeler et al. 2002]). Therefore, H3.3 is sufficiently abundant to package all actively transcribed regions of the *Drosophila* genome.

The separation of H3 and H3.3 allowed us to ask whether they show differences in covalent modifications. To do this, we performed mass analysis of ArgC-digested H3 and H3.3 peptides using liquid chromatography with electrospray ionization mass spectrometry (LC/ESI MS) and tandem mass spectrometry (McKittrick et al. 2004). We also measured modifications on H3 and H3.3 using modification-specific antibodies, assaying on western and slot blots and by enzyme-linked immunoabsorbent assays (ELISA). In all, we detected methyl and acetyl group on nine lysines (positions 4, 9, 14, 18, 23, 28, 36, 37, and 79). All of these modifications are of residues that are identical between H3 and H3.3, and in no case did we detect a modification on one variant by LC/ESI MS but not on the other. However, there were numerous quantitative differences between H3 and H3.3 in the relative abundance of modifications. Relative to H3, H3.3 is enriched twofold to fivefold in di- and tri-methyl lysine 4, mono- and di-methyl lysine 79, and acetyl lysines 9, 14, and 23+28, whereas H3 is enriched in dimethyl lysine 9 (Table 1). In general, our results on *Drosophila Kc* cells correspond well to those reported for alfalfa H3.1 and H3.2 by Waterborg (1990), who assayed modifications using Edman degradation analysis (Table 1). In both cases, hyperacetylation of tail lysines and enrichment of methylated K4 are found for the RI form and enrichment of dimethylated K9 is found for the RC forms of histone 3. Therefore, differences between RC and RI forms of histone 3 in the relative abundances of histone modifications are consistent between plants and animals. Lysine modifications are also highly abundant in plants and animals, with usually only a minority of each lysine residue on one or the other form of histone 3 remaining unmodified when measured in bulk (Waterborg 1990; McKittrick et al. 2004).

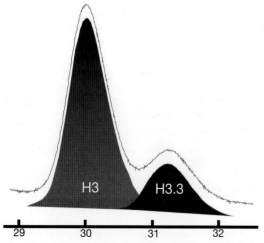

Figure 7. Abundance of H3.3 in a *Drosophila* cell line (McKittrick et al. 2004). Acid-extracted histones from *Kc* cells were injected onto a reverse-phase HPLC column in 30% acetonitrile 0.1% trifluoroacetic acid. All histone 3 eluted in the two peaks shown here, and the peaks were determined to be composed of pure histone 3 by SDS PAGE and Coomassie staining. The relative abundance of the two peaks was determined by integrating the area under each curve (*gray* and *black* areas) and averaging over multiple measurements.

Table 1. Differential Enrichment of Modifications on Alternative Forms of Histone 3

Residue	Modification	*Drosophila*[a] H3 (RC)	*Drosophila*[a] H3.3 (RI)	Alfalfa[b] H3.1 (RC)	Alfalfa[b] H3.2 (RI)
K4	monomethyl	+	+	+	++
	dimethyl	+	++	+	++
	trimethyl	+	++	+	+
K9	monomethyl	+	+	++	+
	dimethyl	++	+	++	+
	trimethyl				++
	acetyl	+	++		+
K14	monomethyl	+	+		
	dimethyl	+	+		
	trimethyl			+	+
	acetyl	+	++	+	++
K18 + K23	acetyl	+	++	+	++
K27	monomethyl	+	+	+	+
	dimethyl	+	+	++	
	trimethyl	+	+	+	+
	acetyl			+	+
K36	monomethyl	+	+		
	dimethyl	+	+		
K37	monomethyl	+	+		
	dimethyl	+	+		
K79	monomethyl	+	++		
	dimethyl	+	++		

+, observed; ++, relatively enriched.
[a]Determined by LC/ESI or tandem MS, ELISA, slot blot, and/or western blot analysis (McKittrick et al. 2004).
[b]Determined by Edman analysis (Waterborg 1990).

The modifications that are enriched on H3.3 relative to H3 are the same ones that many studies have shown to be enriched in active chromatin (Turner 2002), whereas dimethyl K9, which is enriched on H3 relative to H3.3, marks heterochromatin (e.g., Fig. 6). This biochemical correspondence strengthens our conclusions based on cytological observations that H3.3 marks active chromatin and H3 marks silent chromatin. Further support for this two-state model comes from a genome-wide chromatin immunoprecipitation analysis showing that actively transcribed genes are consistently enriched in acetylated H3 and H4 and in methylated H3K4 and H3K79 in *Kc* cells in an all-or-none fashion (Schübeler et al. 2004). Taken together, these observations suggest that histones are primarily modified as part of the nucleosome assembly process (Vermaak et al. 2003; McKittrick et al. 2004; Workman and Abmayr 2004). Modification enzymes would travel along with the RC and RI assembly complexes, acting just before, during, or just after assembly of the H3-H4 units into core nucleosomes. Replication and transcription are spatially and temporally separate processes that use chemically distinct substrates, and the same appears to be the case for their accompanying nucleosome assembly and modification complexes.

RI NUCLEOSOME ASSEMBLY AND CHROMATIN INHERITANCE

The relative abundance of covalent modifications that are associated with active chromatin on H3.3 may have profound implications for the relationship between chromatin and transcription in general. The assembly of H3.3-H4 at active, but not inactive, loci throughout the cell cycle (Fig. 4) and the sufficient abundance of H3.3 to package essentially all active chromatin (Fig. 7) suggest that RI assembly is a transcription-coupled process that results in the continual turnover of nucleosomes. RNA polymerase transit, together with associated chromatin remodeling machines (Kingston and Narlikar 1999), might provide the force to expel a nucleosome, either removing it entirely or exchanging it with new H3.3-H4 in a stepwise fashion (Henikoff et al. 2004). As a result, transcription units would become fully packaged in H3.3-containing nucleosomes over the course of a cell cycle, resulting in enrichment of histone modifications known to be associated with active chromatin. Genes would be inactivated by recruitment of histone deacetylases and H3K9 methyltransferases (Richards and Elgin 2002), which would act on H3.3-containing nucleosomes to promote a silent conformation.

Each round of cell division deposits H3-containing nucleosomes, which are relatively deficient in modifications associated with active chromatin. This is likely to be important for the majority of the genome, which must be kept in a transcriptionally silent and condensed state in typical higher eukaryotic cells. At transcriptionally active regions, RC assembly would cause a factor-of-2 dilution of old H3.3-containing nucleosomes with those containing H3 (Fig. 8). But an abundance of active modifications on the remaining H3.3-containing nucleo-

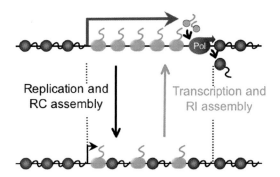

Figure 8. Model for chromatin inheritance (Henikoff et al. 2004). During active transcription, H3.3 is continually inserted, replacing H3 from the previous replication event. In this way, the DNA transited by RNA polymerases (*magenta arrow*) defines the boundaries for regions of active chromatin. Nucleosomes with histone tails that lack active modifications (*gray circles*) bind avidly to DNA and neighboring nucleosomes. Nucleosomes with active modifications are more mobile (depicted as *blurry green circles*) and their tails do not package chromatin tightly as a result of acetylations of lysines and other modifications enriched on H3.3. During DNA replication, H3-containing nucleosomes are deposited, diluting H3.3-containing nucleosomes by half, only partially reducing mobility in the transcribed region compared to the rest of the genome. The presence of these mobile nucleosomes would allow transcription to proceed.

somes could be sufficient to maintain the transcriptional process.

There are several ways in which the presence of H3.3-containing nucleosomes might open chromatin for transcription. For example, the more acetylated tails of H3.3 might reduce contact with linker DNA, thus allowing the DNA to be unwrapped for transcription during RNA polymerase transit (van Holde et al. 1992). In any case, the presence of some H3.3 in chromatin may be sufficient to maintain a positive feedback loop, where ongoing transcription-coupled replacement of H3- with H3.3-containing nucleosomes returns the active gene to a highly acetylated H3.3-enriched state during the remainder of the cell cycle (Fig. 8). In this way, the process of RI nucleosome assembly can resolve the enigma of chromatin inheritance by restoring activating epigenetic signals following replication over a precisely defined interval, the span of the transcription unit (Ahmad and Henikoff 2002b).

CONCLUSIONS

Despite tremendous progress in recent years in understanding the details of eukaryotic transcription and chromatin enzymology, we are just beginning to understand the molecular mechanisms that are responsible for PEV and developmental memory. Here we have described evidence that leads to surprisingly simple explanations for these phenomena, based on what we have learned about the dynamic behavior and assembly of nucleosomes. We find that the classical two-state model of active and silent chromatin based on nuclease accessibility (Weintraub and Groudine 1976) holds up remarkably well. The active chromatin state would be perpetuated despite the dilution of epigenetic information after DNA replication by the coupling of RI assembly to transcription.

Other processes that act on chromatin, including the mobilization of nucleosomes by remodeling complexes, the covalent modifications of histone tails, and the movement of polymerases, might also regulate the dynamic behavior of nucleosomes. Nucleosome assembly complexes that are specific for histone variants provide other means of affecting chromatin dynamics by altering the composition of nucleosomes (Krogan et al. 2003; Mizuguchi et al. 2003; Kobor et al. 2004; Tagami et al. 2004). Whereas RC deposition of nucleosomes behind the replication fork would assemble chromatin into a relatively immobile default state, the small fraction of a higher eukaryotic genome that is exposed for transcription is subject to RI assembly, which is potentially a gene-regulatory process. So despite differentiation of nucleosomes resulting from histone modifications, binding of chromatin-associated proteins, and incorporation of histone variants, the coupling of these events to assembly of nucleosomes at replication forks and transcription units provides a fundamentally simple framework for understanding epigenetic processes.

ACKNOWLEDGMENTS

We thank Danielle Vermaak, Jim Smothers, Harmit Malik, Také Furuyama, Yamini Dalal, and Yoshiko Mito for helpful discussions. This work was supported by the Howard Hughes Medical Institute.

REFERENCES

Ahmad K. and Henikoff S. 2001a. Centromeres are specialized replication domains in heterochromatin. *J. Cell Biol.* **153:** 101.
——. 2001b. Modulation of a transcription factor counteracts heterochromatic gene silencing in *Drosophila*. *Cell* **104:** 839.
——. 2002a. Epigenetic consequences of nucleosome dynamics. *Cell* **111:** 281.
——. 2002b. Histone H3 variants specify modes of chromatin assembly. *Proc. Natl. Acad. Sci.* (suppl. 4) **99:** 16477.
——. 2002c. The histone variant H3.3 marks active chromatin by replication-independent nucleosome assembly. *Mol. Cell* **9:** 1191.
Cheutin T., McNairn A.J., Jenuwein T., Gilbert D.M., Singh P.B., and Misteli T. 2003. Maintenance of stable heterochromatin domains by dynamic HP1 binding. *Science* **299:** 721.
Dillon N. and Festenstein R. 2002. Unravelling heterochromatin: Competition between positive and negative factors regulates accessibility. *Trends Genet.* **18:** 252.
Dorer D.R. and Henikoff S. 1994. Expansions of transgene repeats cause heterochromatin formation and gene silencing in *Drosophila*. *Cell* **77:** 993.
Echalier G. 1997. Drosophila *cells in culture*. Academic Press, New York.
Festenstein R., Pagakis S.N., Hiragami K., Lyon D., Verreault A., Sekkali B., and Kioussis D. 2003. Modulation of Heterochromatin Protein 1 dynamics in primary mammalian cells. *Science* **299:** 719.
Grewal S.I. and Elgin S.C. 2002. Heterochromatin: New possibilities for the inheritance of structure. *Curr. Opin. Genet. Dev.* **12:** 178.
Henikoff S., Furuyama T., and Ahmad A. 2004. Histone variants, nucleosome assembly and epigenetic inheritance. *Trends Genet.* **20:** 320.
Hoskins R.A., Smith C.D., Carlson J.W., Carvalho A.B., Halpern A., Kaminker J.S., Kennedy C., Mungall C.J., Sullivan B.A., Sutton G.G., Yasuhara J.C., Wakimoto B.T., Myers E.W., Celniker S.E., Rubin G.M., and Karpen G.H. 2002. Heterochromatic sequences in a *Drosophila* whole-genome shotgun assembly. *Genome Biol.* **3:** RESEARCH0085.
Janicki S.M., Tsukamoto T., Salghetti S.E., Tansey W.P., Sachidanandam R., Prasanth K.V., Ried T., Shav-Tal Y., Bertrand E., Singer R.H., and Spector D.L. 2004. From silencing to gene expression: Real-time analysis in single cells. *Cell* **116:** 683.
Jenuwein T. and Allis C.D. 2001. Translating the histone code. *Science* **293:** 1074.
Kingston R.E. and Narlikar G.J. 1999. ATP-dependent remodeling and acetylation as regulators of chromatin fluidity. *Genes Dev.* **13:** 2339.
Kobor M.S., Venkatasubrahmanyam S., Meneghini M.D., Gin J.W., Jennings J.L., Link A.J., Madhani H.D., and Rine J. 2004. A protein complex containing the conserved Swi2/Snf2-related ATPase Swr1p deposits histone variant H2A.Z into euchromatin. *PLoS Biol.* **2:** E131.
Krogan N.J., Keogh M.C., Datta N., Sawa C., Ryan O.W., Ding H., Haw R.A., Pootoolal J., Tong A., Canadien V., Richards D.P., Wu X., Emili A., Hughes T.R., Buratowski S., and Greenblatt J.F. 2003. A Snf2 family ATPase complex required for recruitment of the histone H2A variant Htz1. *Mol. Cell* **12:** 1565.
Leffak I.M., Grainger R., and Weintraub H. 1977. Conservative assembly and segregation of nucleosomal histones. *Cell* **12:** 837.
Leonhardt H., Page A.W., Weier H.U., and Bestor T.H. 1992. A targeting sequence directs DNA methyltransferase to sites of DNA replication in mammalian nuclei. *Cell* **71:** 865.
Luger K., Mader A.W., Richmond R.K., Sargent D.F., and Richmond T.J. 1997. Crystal structure of the nucleosome core particle at 2.8 Å resolution. *Nature* **389:** 251.
Malik H.S. and Henikoff S. 2003. Phylogenomics of the nucleosome. *Nat. Struct. Biol.* **10:** 882.
McKittrick E., Gafken P.R., Ahmad K., and Henikoff S. 2004. Histone H3.3 is enriched in covalent modifications associated with active chromatin. *Proc. Natl. Acad. Sci.* **101:** 1525.
Mello J.A. and Almouzni G. 2001. The ins and outs of nucleosome assembly. *Curr. Opin. Genet. Dev.* **11:** 136.
Misra S., Crosby M.A., Mungall C.J., Matthews B.B., Campbell K.S., Hradecky P., Huang Y., Kaminker J.S., Millburn G.H., et al. 2002. Annotation of the *Drosophila melanogaster* euchromatic genome: A systematic review. *Genome Biol.* **3:** RESEARCH0083.
Mizuguchi G., Shen X., Landry J., Wu W.H., Sen S., and Wu C. 2003. ATP-driven exchange of histone H2AZ variant catalyzed by SWR1 chromatin remodeling complex. *Science* **303:** 343.
Muller H.J. 1930. Types of visible variations induced by X-rays in *Drosophila*. *J. Genet.* **22:** 299.
Pikaard C.S. 2000. The epigenetics of nucleolar dominance. *Trends Genet.* **16:** 495.
Polach K.J. and Widom J. 1995. Mechanism of protein access to specific DNA sequences in chromatin: A dynamic equilibrium model for gene regulation. *J. Mol. Biol.* **254:** 130.
Richards E.J. and Elgin S.C. 2002. Epigenetic codes for heterochromatin formation and silencing: Rounding up the usual suspects. *Cell* **108:** 489.
Schübeler D., Scalzo D., Kooperberg C., van Steensel B., Delrow J., and Groudine M. 2002. Genome-wide DNA replication profile for *Drosophila melanogaster:* A link between transcription and replication timing. *Nat. Genet.* **32:** 438.
Schübeler D., MacAlpine D.M., Scalzo D., Wirbelauer C., Kooperberg C., van Leeuwen F., Gottschling D.E., O'Neill L.P., Turner B.M., Delrow J., Bell S.P., and Groudine M. 2004. The histone modification pattern of active genes revealed through genome-wide chromatin analysis of a higher eukaryote. *Genes Dev.* **18:** 1263.
Shelby R.D., Monier K., and Sullivan K.F. 2000. Chromatin assembly at kinetochores is uncoupled from DNA replication. *J. Cell Biol.* **151:** 1113.
Spofford J.B. 1976. Position-effect variegation in *Drosophila*. In *Genetics and biology of* Drosophila (ed. M. Ashburner and E. Novitski), p. 955. Academic Press, London.

Tagami H., Ray-Gallet D., Almouzni G., and Nakatani Y. 2004. Histone H3.1 and H3.3 complexes mediate nucleosome assembly pathways dependent or independent of DNA synthesis. *Cell* **116:** 51.

Turner B.M. 2002. Cellular memory and the histone code. *Cell* **111:** 285.

van Holde K.E., Lohr D.E., and Robert C. 1992. What happens to nucleosomes during transcription? *J. Biol. Chem.* **267:** 2837.

Vermaak D., Ahmad K., and Henikoff S. 2003. Maintenance of chromatin states: An open-and-shut case. *Curr. Opin. Cell Biol.* **15:** 266.

Waterborg J.H. 1990. Sequence analysis of acetylation and methylation in two histone H3 variants of alfalfa. *J. Biol. Chem.* **265:** 17157.

———. 1993. Histone synthesis and turnover in alfalfa: Fast loss of highly acetylated replacement histone variant H3.2. *J. Biol. Chem.* **268:** 4912.

Weiler K.S. and Wakimoto B.T. 1995. Heterochromatin and gene expression in *Drosophila*. *Annu. Rev. Genet.* **29:** 577.

Weintraub H. and Groudine M. 1976. Chromosomal subunits in active genes have an altered conformation. *Science* **193:** 848.

Workman J.L. and Abmayr S.M. 2004. Histone H3 variants and modifications on transcribed genes. *Proc. Natl. Acad. Sci.* **101:** 1429.

Yamasu K. and Senshu T. 1990. Conservative segregation of tetrameric units of H3 and H4 histones during nucleosome replication. *J. Biochem.* **107:** 15.

Yu L. and Gorovsky M.A. 1997. Constitutive expression, not a particular primary sequence, is the important feature of the H3 replacement variant hv2 in *Tetrahymena thermophila*. *Mol. Cell. Biol.* **17:** 6303.

Zheng C. and Hayes J.J. 2003. Structures and interactions of the core histone tail domains. *Biopolymers* **68:** 539.

Chromatin Boundaries and Chromatin Domains

G. Felsenfeld,* B. Burgess-Beusse,* C. Farrell,* M. Gaszner,* R. Ghirlando,*
S. Huang,* C. Jin,* M. Litt,* F. Magdinier,* V. Mutskov,* Y. Nakatani,[†]
H. Tagami,[†] A. West,*[‡] and T. Yusufzai*

*Laboratory of Molecular Biology, NIDDK, National Institutes of Health, Bethesda, Maryland 20892-0540;
[†]Dana Farber Cancer Institute, Boston, Massachusetts 02115

Insulator elements were first described in *Drosophila*, but subsequent studies have shown that they are present in vertebrates as well (for review, see West et al. 2002). Over the past several years we have focused our attention on the properties of an insulator at the 5′ end of the chicken β-globin locus that has begun to provide an understanding of how such elements function. This work, as well as studies in other laboratories, has revealed that there are two distinct kinds of insulator activities, which are different in their function. The first of these is the *enhancer-blocking activity*, which can prevent interaction between a distal enhancer and a promoter when placed between them (Fig. 1A). This has the effect of preventing an incorrect interaction between regulatory elements in adjacent, but separately regulated, gene systems. The second insulator function is connected with *barrier activity*, which prevents condensed heterochromatin from extending into adjacent chromatin domains carrying transcriptionally active genes (Fig. 1B).

The chicken β-globin locus extends over 30 kb. It contains four members of the globin gene family, with different programs of expression during development (Fig. 2), which have been studied extensively (Felsenfeld 1993). Strong positive regulatory elements, components of the locus control region (LCR), are distributed both upstream of the gene cluster and within it. Further upstream is a DNase I "hypersensitive site," 5′HS4, which, unlike others in the locus, is not erythroid-specific, but is nuclease sensitive in all cells that have been tested (Reitman and Felsenfeld 1990). It seemed an attractive possibility that this marked the 5′ end of the open chromatin domain; in fact, it was shown not long afterward that immediately upstream of 5′HS4 there is an abrupt decrease of nuclease sensitivity and histone acetylation in globin-expressing cells, consistent with a transition from the open chromatin of the globin locus to a more inactive, condensed chromatin structure (Hebbes et al. 1994). We explored the possibility that this element might have the properties of an insulator (Chung et al. 1997).

ENHANCER-BLOCKING INSULATION AND CTCF

The analysis of the gypsy element in *Drosophila* had provided the first example of enhancer-blocking action, and we began our studies by testing whether a 1.2-kb fragment containing 5′HS4 could similarly prevent enhancer–promoter interaction. The assay we devised placed the element between a strong erythroid-specific enhancer and promoter that were driving expression of a drug resistance gene. The number of colonies able to grow under selective conditions was used as a measure of enhancer-blocking activity. We showed that the 1.2-kb element reduced colony number by an order of magnitude, but only when placed between the enhancer and promoter (Chung et al. 1997). A 250-bp "core" element derived from this fragment retained the enhancer-blocking function and allowed us to dissect the activity further. We report here on subsequent results that followed from these observations.

DNase footprinting experiments with nuclear extracts showed that there were five discrete protected regions within the 250-bp core, which could be tested individually for enhancer-blocking activity (Fig. 3) (Chung et al. 1997). Footprint II was necessary and sufficient for activity in our assay and provided a DNA sequence that could be used in gel retardation assays to follow purification of the protein that bound to it. This protein, CTCF, had been described earlier, but had not been associated with insulator activity (Bell et al. 1999).

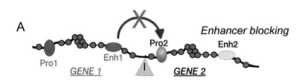

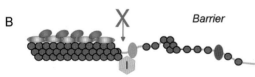

Figure 1. Two kinds of insulator function. (*A*) Enhancer blocking: The insulator (I) prevents Enhancer 1, belonging to gene system 1, from acting inappropriately on Promoter 2 in gene system 2. (*B*) Barrier function: The insulator (I) prevents the advance of heterochromatic regions on the *left* into transcriptionally active chromatin on the *right*.

[‡]Present address: Division of Cancer Sciences and Molecular Pathology, Gene Regulation and Mechanisms of Disease Section, Pathology Department, Glasgow University, Glasgow G11 6NT, Scotland, United Kingdom.

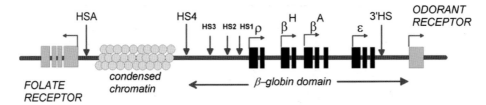

Figure 2. A map (not to scale) of the chicken β-globin domain and its neighbors. The distance between HS4, the globin 5′ insulator element, and 3′HS, the CTCF site at the 3′ end of the domain, is about 30 kb.

CTCF is a member of the zinc finger family, with the unusually large number of 11 fingers. Since its identification as the active agent in the enhancer-blocking activity of the 5′HS4 chicken β-globin element, it has been implicated in similar functions within a wide variety of vertebrate loci, including the 3′ end of the chicken β-globin locus (Saitoh et al. 2000). CTCF binding sites with enhancer-blocking action have been found within the mouse and human β-globin loci (Farrell et al. 2002; Tanimoto et al. 2003), and associated with the Tsix antisense gene in the mouse choice/imprinting center involved in control of X chromosome inactivation (Chao et al. 2002). Recent studies have identified a single CTCF site (Fig. 4) located between the mouse T cell receptor α (*TCR* α) gene and the downstream *Dad 1* gene, where it may play a role in preventing inappropriate interaction between enhancers committed to one gene and promoter controlling the other (Magdinier et al. 2004).

Perhaps the most convincing evidence for the biological role of such insulators is provided by studies of the regulation of the mouse and human *Igf2/H19* imprinted locus (Bell and Felsenfeld 2000; Hark et al. 2000; Kanduri et al. 2000). The *H19* gene is expressed only from the maternally transmitted allele, and *Igf2* only from the paternal (Fig. 5). *Igf2* expression is shut down on the maternal allele because of the presence of an insulator element with multiple binding sites for CTCF, which blocks the action of downstream endodermal enhancers that normally activate *Igf2*. These DNA sites are methylated on the imprinted paternal allele, which prevents CTCF binding, inactivates the insulator, and activates *Igf2* expression.

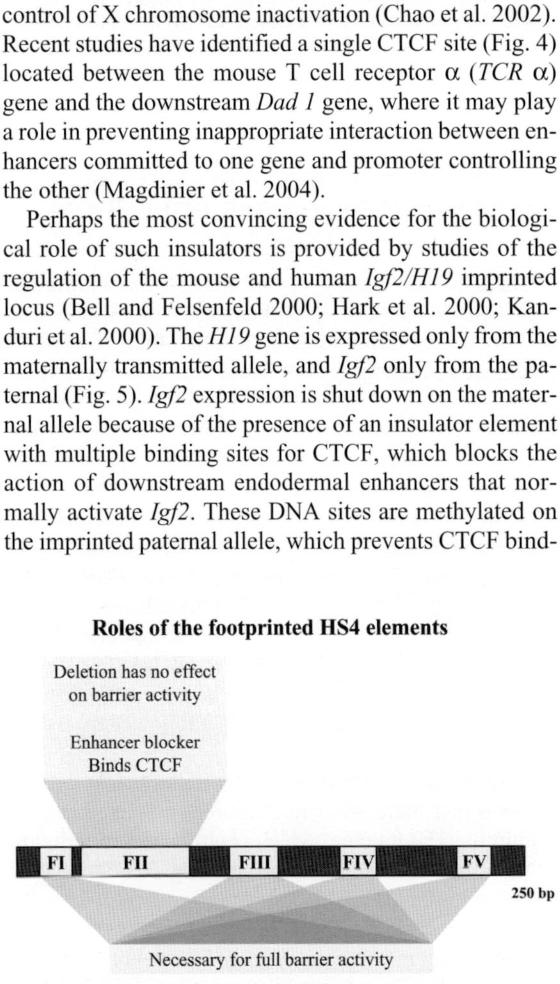

Since it seemed unlikely that CTCF could carry out this function without the intervention of other proteins, the next step was to search for factors with which it interacted. We expressed a double epitope-tagged version of the protein and immunopurified complexes containing candidate cofactors (Yusufzai et al. 2004). Among the proteins purified in this way were poly ADP-ribose polymerase, Set1, H2A, H2A.Z, and, most notably, the nucleolar protein, nucleophosmin, which was present in the greatest abundance. When a singly purified complex was centrifuged on a glycerol gradient, some of the CTCF cosedimented with nucleophosmin in two fractions, indicating that a well-defined complex had formed between them. Chromatin immunoprecipitation (ChIP) using antibodies against CTCF and nucleophosmin showed that these proteins also colocalized in vivo over the two known CTCF sites at either end of the chicken β-globin locus (Fig. 6A). Fluorescence in situ hybridization analysis of cell lines carrying multiple copies of the insulator showed localization of the insulator elements at the surface of the nucleolus, where nucleophosmin is also concentrated. Mutation of the CTCF site abolished localization (Yusufzai et al. 2004). In other experiments, it was shown also that CTCF can form homodimers and probably higher-order oligomers as well.

The class of models suggested by these results is quite similar to one proposed earlier by Corces and his collaborators to explain the action of the gypsy element in *Drosophila* (Gerasimova et al. 2000). These models invoke the generation of separate loop domains, mediated by the protein bound to the insulators, to generate structures in which enhancer and promoter occupy separate topologically independent loops. The β-globin insulator is tethered to the nucleolar surface by its interaction with nucleophosmin, and this would be sufficient to create such domains (Fig. 6B). However, the ability of CTCF molecules to interact with one another would also result in the formation of loops. In the case of the gypsy element, both clustering and tethering to the nuclear envelope have been observed.

There are two kinds of enhancer mechanisms arising from these models that could serve to explain insulation. Either the enhancer can no longer make a necessary physical contact with the promoter, or some processive signal that normally passes from enhancer to promoter is blocked at the point where the insulator helps to form the

Figure 3. The five footprinted regions of the β-globin HS4 "core" insulator element, showing the separate contributions of each element to enhancer-blocking or barrier activity.

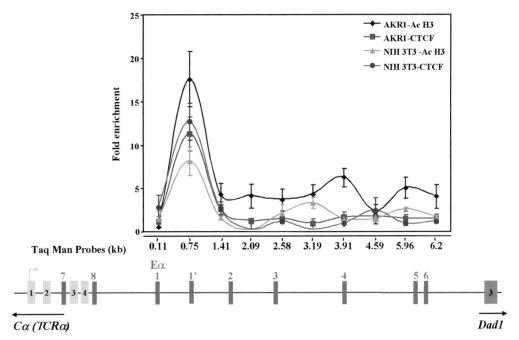

Figure 4. The distribution of histone H3 acetylation and CTCF binding in vivo, determined by chromatin immunoprecipitation, for the region between the mouse *T cell receptor* α gene and the downstream *Dad1* gene. Experiments were carried out both in the fibroblast line NIH-3T3 and in a lymphocyte line, AKR1. Hypersensitive sites 1–6 previously described (Zhong and Krangel 1999; Ortiz et al. 2001) are shown below as dark vertical bars. The peak at 0.75 kb corresponds to hypersensitive site 1′. (Reprinted, with permission, from Magdinier et al. 2004.)

base of the loop. Evidence for such processive mechanisms has been reported (Hatzis and Talianidis 2002). It should be recalled, however, that there are other proteins that interact with CTCF (Yusufzai et al. 2004), and that their possible role in enhancer blocking remains to be explored.

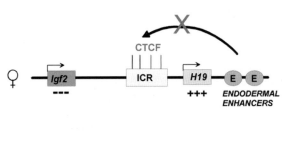
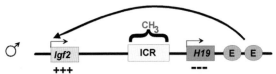

Figure 5. The imprinted *Igf2/H19* locus in mouse. The imprinted control region (ICR) of the maternally transmitted allele binds CTCF and acts as an enhancer-blocking insulator to prevent the downstream endodermal enhancers (E) from activating *Igf2* expression. In the paternal allele, the DNA of the ICR is methylated, CTCF does not bind, and the enhancers can activate *Igf2* expression (Bell and Felsenfeld 2000; see also Hark et al. 2000; Kanduri et al. 2000). (Reprinted, with permission, from Bell and Felsenfeld 2000.)

BOUNDARY FUNCTIONS AND THE SEPARATION OF INSULATOR ACTIVITIES

The second kind of insulator serves to provide a barrier against invasion of an open chromatin region by heterochromatin. We devised an assay in which we measured the expression in erythroid cell lines of an integrated transgene surrounded by the element to be tested for barrier activity. In most lines carrying unprotected constructs, expression is extinguished by 20–80 days in culture. This probably reflects integration into endogenous sites that are silent; some lines do continue to express the reporter gene, in many cases presumably by trapping a strong enhancer (Pikaart et al. 1998). We tested the ability of both the full 1.2-kb globin HS4 insulator element and the 250-bp "core" to protect against extinction of expression. When the reporter is flanked on each side by two copies of either of these elements, silencing does not occur. Thus the globin insulator also possesses the second, "barrier" activity that protects against position effect.

The obvious next step was to dissect the core element to isolate the components responsible for this activity. We undertook successive deletion of each of the five footprinted regions of the core described above. To our surprise, deletion of the CTCF binding site had no effect on the barrier activity. In contrast, deletion of any one of the other four footprint regions severely inhibited barrier function. The HS4 insulator enhancer-blocking function is therefore separable from the barrier function; HS4 is a compound element with interspersed regulatory sequences (Recillas-Targa et al. 2002).

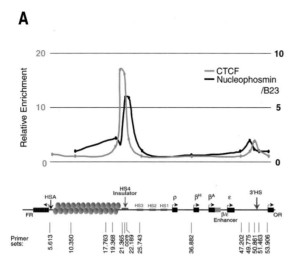

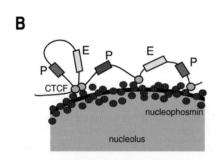

Figure 6. (*A*) Chromatin immunoprecipitation over the chicken β-globin domain with antibodies to CTCF and nucleophosmin show that nucleophosmin colocalizes with CTCF over the insulators at each end of the domain. (*B*) A possible model for enhancer-blocking action, in which CTCF is tethered to the nucleolar surface to form topologically isolated loop domains. Equivalent structures can be generated simply by pair-wise or clustered interactions between CTCF molecules (see text). (Reprinted, with permission, from Yusufzai et al. 2004.)

HISTONE MODIFICATIONS AND BARRIER FUNCTION

A clue as to how barriers might be established and maintained comes from studies of the distribution of histone modifications over the β-globin locus. Chromatin immunoprecipitation studies show that there are peaks of histone acetylation over 5′HS4 that are present in every cell type that has been studied (Fig. 7) (Litt et al. 2001b). Immediately upstream of 5′HS4 and the β-globin locus is a region of condensed chromatin (see below), marked by dimethylation of histone H3 lysine 9 (Litt et al. 2001a). This methylation mark is characteristic of heterochromatin. Schemes have been proposed in which the presence of this modification on one nucleosome can recruit the methylating enzyme necessary to modify an adjacent nucleosome, leading to propagation of the inactivation signal. We have suggested that the presence of a high level of H3 lysine 9 acetylation at HS4 could prevent the adjacent histones from being methylated at lysine 9, preventing the advance of the propagating silencing signal. When the HS4 insulator flanks a transgene, the targeted histone acetylation by the insulators may prevent the influence of surrounding endogenous condensed chromatin in the same way.

Recent evidence from our laboratories suggests that other modifications induced by HS4 may similarly block the advance of other silencing signals. A similar mechanism has been proposed for the barrier function at the end of the mating type locus in yeast (Donze and Kamakaka 2001).

In this connection it is interesting to note that silencing of the integrated but uninsulated reporter used in our barrier assays can be prevented by growing the cells in the constant presence of Trichostatin A, which inhibits histone deacetylation (Mutskov et al. 2002). In the absence of such inhibition, the kinetics of loss of histone H3 and H4 acetylation closely parallel loss of transcript; promoter DNA methylation, usually associated with silencing, occurs only afterward. The kinetics of silencing exclude DNA methylation as the primary causative event in this transgene system (Mutskov and Felsenfeld 2004).

We are presently dissecting the HS4 region further to identify the proteins responsible for the barrier behavior. It is already clear that multiple proteins are involved, and that these include histone-modifying enzymes that correspond to the histone modifications observed in the neighborhood of the insulator.

CONDENSED CHROMATIN

The condensed chromatin region that begins immediately upstream of 5′HS4 extends for about 16 kb and is followed at its 5′ end by a gene for an erythroid-specific folate receptor (Prioleau et al. 1999). It is important to understand the structure of such regions, which in vivo may well affect the expression of nearby genes. Our early studies of this region showed that it was composed of regularly arrayed nucleosomes, and that the DNA within it was highly methylated at CpG sites. We therefore took advantage of accessible and unmethylated HpaII restriction sites at the borders of the region to excise it from the nucleus. Sedimentation in a sucrose gradient revealed an essentially monodisperse fragment as revealed either by Southern blotting or polymerase chain reaction (PCR) analysis (Fig. 8). The individual gradient fractions could be studied in the analytical ultracentrifuge, yielding a precise value for the sedimentation coefficient of the heterochromatin particle. We were also able to measure the buoyant density of the particle. The combined density and sedimentation information gave its frictional coefficient, a measure of particle shape. The measured value corresponds to an extended rod-like particle, and is consistent with results obtained from electron micrograph and diffraction studies of chromatin fibers that detect a rod-like structure about 30 nm in diameter (Ghirlando et al. 2004).

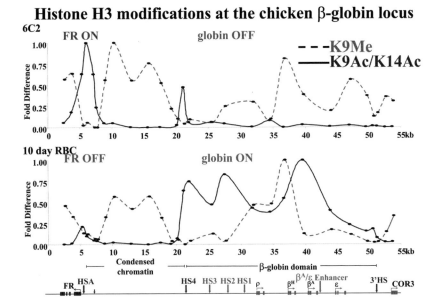

Figure 7. Distribution of histone H3 lysine 9 methylation and lysine 4 methylation over the β-globin region in the chicken erythroleukemia cell line 6C2 (*top*) and 10-day chick embryonic red blood cells (*bottom*). (*Top*, Reprinted, with permission, from Litt et al. 2001a; *bottom*, reprinted, with permission, from Litt et al. 2001b.)

CONCLUSIONS

The original purpose of this study was to investigate the structure of chromatin at the boundaries of the chicken β-globin locus and to understand how longer-range chromatin structure might mediate globin gene expression. During the ensuing studies we identified a complex insulator element at the 5′ end of the locus that separated it from an extended adjacent heterochromatic region and, beyond that, a folate receptor gene with a different program of expression from that of the globin genes. Dissection of the insulator showed that it had separate elements capable both of serving as a barrier to heterochromatinization and of blocking distal enhancers. The latter activity was attributable to binding of the protein CTCF, and we were able to show that CTCF insulator sites are present elsewhere in the genome, notably at the Igf2/H19 imprinted locus. Recent results show that CTCF can interact both with itself and with a nucleolar protein, nucleophosmin; this suggests possible mechanisms of enhancer-blocking action. Barrier activity is also mediated by the globin insulator, but recent results in our laboratory show that it involves a different set of binding sites and proteins, appropriate to what we believe is its task of preventing the immediately adjacent, compact, and inactive 16-kb heterochromatin domain from advancing.

This series of investigations also has revealed a lot about the role of chromatin structure in regulating gene expression. The analysis of the developmentally regulated distribution of histone modifications over the β-globin locus provides important correlations between these modifications and the state of expression over the globin and folate receptor genes, as well as the inactive 16-kb heterochromatic region. Recent data from our laboratory (G. Felsenfeld et al., unpubl.) extends this work to a variety of other modifications, with the ultimate goal of determining the time course of activation and inactivation events. The studies of enhancer-blocking activity take us in yet another direction, calling attention to the likely importance of higher-order structures and interaction with components of the nuclear architecture in regulating expression. The study of insulators has thus provided a powerful way of studying problems of chromatin structure and gene expression that might otherwise be difficult to address.

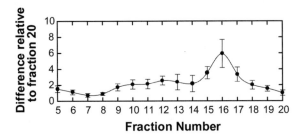

Figure 8. Sucrose gradient sedimentation of a *Hpa*II digest of 6C2 cell nuclei. The digestion releases an ~16-kb condensed chromatin fragment that lies between the 5′ end of the chicken β-globin domain and the folate receptor gene further upstream (see Fig. 2). The fragment was detected by PCR, and the chromatin in the peak fraction was subjected to analytical ultracentrifugation. (Reprinted, with permission, from Ghirlando et al. 2004.)

REFERENCES

Bell A.C. and Felsenfeld G. 2000. Methylation of a CTCF-dependent boundary controls imprinted expression of the Igf2 gene. *Nature* **405:** 482.

Bell A.C., West A.G., and Felsenfeld G. 1999. The protein CTCF is required for the enhancer blocking activity of vertebrate insulators. *Cell* **98:** 387.

Chao W., Huynh K.D., Spencer R.J., Davidow L.S., and Lee J.T. 2002. CTCF, a candidate trans-acting factor for X-inactivation choice. *Science* **295:** 345.

Chung J.H., Bell A.C., and Felsenfeld G. 1997. Characterization of the chicken beta-globin insulator. *Proc. Natl. Acad. Sci.* **94:** 575.

Donze D. and Kamakaka R.T. 2001. RNA polymerase III and RNA polymerase II promoter complexes are heterochromatin barriers in *Saccharomyces cerevisiae*. *EMBO J.* **20:** 520.

Farrell C.M., West A.G., and Felsenfeld G. 2002. Conserved CTCF insulator elements flank the mouse and human beta-globin loci. *Mol. Cell. Biol.* **22:** 3820.

Felsenfeld G. 1993. Chromatin structure and the expression of globin-encoding genes. *Gene* **135:** 119.

Gerasimova T.I., Byrd K., and Corces V.G. 2000. A chromatin insulator determines the nuclear localization of DNA. *Mol. Cell* **6:** 1025.

Ghirlando R., Litt M.D., Prioleau M.N., Recillas-Targa F., and Felsenfeld G. 2004. Physical properties of a genomic condensed chromatin fragment. *J. Mol. Biol.* **336:** 597.

Hark A.T., Schoenherr C.J., Katz D.J., Ingram R.S., Levorse J.M., and Tilghman S.M. 2000. CTCF mediates methylation-sensitive enhancer-blocking activity at the H19/Igf2 locus. *Nature* **405:** 486.

Hatzis P. and Talianidis I. 2002. Dynamics of enhancer-promoter communication during differentiation-induced gene activation. *Mol. Cell* **10:** 1467.

Hebbes T.R., Clayton A.L., Thorne A.W., and Crane-Robinson C. 1994. Core histone hyperacetylation co-maps with generalized DNase I sensitivity in the chicken beta-globin chromosomal domain. *EMBO J.* **13:** 1823.

Kanduri C., Pant V., Loukinov D., Pugacheva E., Qi C.F., Wolffe A., Ohlsson R., and Lobanenkov V.V. 2000. Functional association of CTCF with the insulator upstream of the H19 gene is parent of origin-specific and methylation-sensitive. *Curr. Biol* **10:** 853.

Litt M.D., Simpson M., Gaszner M., Allis C.D., and Felsenfeld G. 2001a. Correlation between histone lysine methylation and developmental changes at the chicken beta-globin locus. *Science* **293:** 2453.

Litt M.D., Simpson M., Recillas-Targa F., Prioleau M.N., and Felsenfeld G. 2001b. Transitions in histone acetylation reveal boundaries of three separately regulated neighboring loci. *EMBO J.* **20:** 2224.

Magdinier F., Yusufzai T.M., and Felsenfeld G. 2004. Both CTCF-dependent and -independent insulators are found between the mouse T cell receptor alpha and Dad1 genes. *J. Biol. Chem.* **279:** 25381.

Mutskov V. and Felsenfeld G. 2004. Silencing of transgene transcription precedes methylation of promoter DNA and histone H3 lysine 9. *EMBO J.* **23:** 138.

Mutskov V.J., Farrell C.M., Wade P.A., Wolffe A.P., and Felsenfeld G. 2002. The barrier function of an insulator couples high histone acetylation levels with specific protection of promoter DNA from methylation. *Genes Dev.* **16:** 1540.

Ortiz B.D., Harrow F., Cado D., Santoso B., and Winoto A. 2001. Function and factor interactions of a locus control region element in the mouse T Cell receptor-α/Dad1 gene locus. *J. Immunol.* **167:** 3836.

Pikaart M.J., Recillas-Targa F., and Felsenfeld G. 1998. Loss of transcriptional activity of a transgene is accompanied by DNA methylation and histone deacetylation and is prevented by insulators. *Genes Dev.* **12:** 2852.

Prioleau M.N., Nony P., Simpson M., and Felsenfeld G. 1999. An insulator element and condensed chromatin region separate the chicken beta-globin locus from an independently regulated erythroid-specific folate receptor gene. *EMBO J.* **18:** 4035.

Recillas-Targa F., Pikaart M.J., Burgess-Beusse B., Bell A.C., Litt M.D., West A.G., Gaszner M., and Felsenfeld G. 2002. Position-effect protection and enhancer blocking by the chicken beta-globin insulator are separable activities. *Proc. Natl. Acad. Sci.* **99:** 6883.

Reitman M. and Felsenfeld G. 1990. Developmental regulation of topoisomerase II sites and DNase I-hypersensitive sites in the chicken beta-globin locus. *Mol. Cell. Biol.* **10:** 2774.

Saitoh N., Bell A.C., Recillas-Targa F., West A.G., Simpson M., Pikaart M., and Felsenfeld G. 2000. Structural and functional conservation at the boundaries of the chicken beta-globin domain. *EMBO J.* **19:** 2315.

Tanimoto K., Sugiura A., Omori A., Felsenfeld G., Engel J.D., and Fukamizu A. 2003. Human beta-globin locus control region HS5 contains CTCF- and developmental stage-dependent enhancer-blocking activity in erythroid cells. *Mol. Cell. Biol.* **23:** 8946.

West A.G., Gaszner M., and Felsenfeld G. 2002. Insulators: Many functions, many mechanisms. *Genes Dev.* **16:** 271.

Yusufzai T.M., Tagami H., Nakatani Y., and Felsenfeld G. 2004. CTCF tethers an insulator to subnuclear sites, suggesting shared insulator mechanisms across species. *Mol. Cell* **13:** 291.

Zhong X. and Krangel M.S. 1999. Enhancer-blocking activity within the DNase I hypersensitive site 2 to 6 region between the TCR α and Dad1 genes. *J. Immunol.* **163:** 295.

Do Higher-Order Chromatin Structure and Nuclear Reorganization Play a Role in Regulating *Hox* Gene Expression during Development?

W.A. BICKMORE, N.L. MAHY, AND S. CHAMBEYRON

MRC Human Genetics Unit, Edinburgh EH4 2XU, Scotland, United Kingdom

It is quite clear that nucleosome structure affects gene expression and contributes to epigenetic inheritance of gene expression states. The best understood examples are the correlations of histone acetylation and methylation to either active transcription or gene repression (Fischle et al. 2003). There is a growing realization that variant histones, such as histones H2A.Z and H3.3, are also critical in demarcating active and silent chromatin (Meneghini et al. 2003; McKittrick et al. 2004). But can all epigenetic information about gene expression be accounted for by chromatin structure at the level of the nucleosome and its modifications, or could poorly understood higher-order chromatin structures and nuclear organization also have a role? To answer this question we have focused here on chromatin and nuclear structure at the murine *HoxB* locus.

Despite decades of analysis of *cis*-regulatory elements that control *Hox* gene expression (see, e.g., Marshall et al. 1994; Popperl et al. 1995; Maconochie et al. 1997; Oosterveen et al. 2003), very little is known about the chromatin structure of *Hox* loci. The correct spatial and temporal regulation of *Hox* expression is important for anteroposterior patterning of the embryo (Krumlauf 1994). As with the homologous Antennapaedia (ANT) and Bithorax (BX) complexes in *Drosophila*, the time and place of mammalian *Hox* gene activation is colinear with the gene order along the chromosome (Kmita and Duboule 2003). Genes at the 3′ end of the clusters are expressed earlier, and in more anterior parts of the embryo, than more 5′ genes (Fig. 1). Manipulations of the mouse genome that reposition *Hox* genes within a cluster, for example moving a 3′ gene to a more 5′ position, have suggested that a transition from an inactive to an active chromatin state is propagated through *Hox* clusters from 3′ to 5′ (van der Hoeven et al. 1996; Kondo and Duboule 1999; Roelen et al. 2002). Is the spatiotemporal regulation of *Hox* loci therefore controlled by, for example, a progressive sweep of histone modification through the complex?

Recapitulating Early *Hox* Gene Activation ex Vivo

Analyzing chromatin structures in specific cells/tissues of the developing embryo is difficult. Therefore we sought initially to study the temporal activation of *Hox* genes ex vivo. Retinoids mediate *Hox* expression in vivo, and retinoic acid (RA) also induces the temporal program of *Hox* expression in murine embryonic stem (ES) cells (Papalopulu et al. 1991). There is no *HoxB* expression in undifferentiated ES cells. After 2–4 days of differentiation (day 2–4), *Hoxb1* expression is detected, but is then extinguished. The more 5′ *Hoxb9* is expressed later, at day 10, and is also then silenced. No activation of *Hoxb13* expression is seen even by day 12 of differentiation (Fig. 2) (Chambeyron and Bickmore 2004).

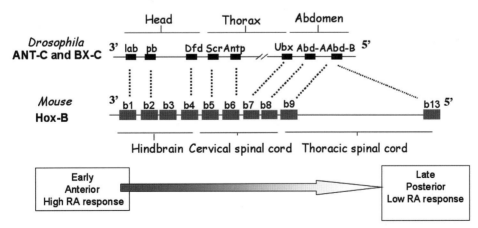

Figure 1. Colinear organization of homeotic genes. In both flies and mammals, the genomic organization of homeotic gene clusters—Antennapaedia (ANT-C) and Bithorax (BX-C) in *Drosophila*, and *HoxA-D* in mammals—is colinear with the site of gene expression in the animal. In mammals, genes at the 3′ end of the clusters also switch on their expression earlier during development than more 5′ genes.

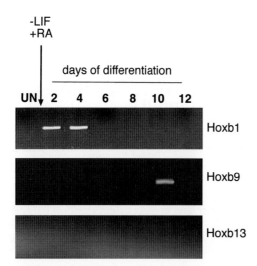

Figure 2. Hox gene expression ex vivo. The temporal pattern of *Hox* gene expression, induced by the addition of retinoic acid (RA) to ES cells, mimics that seen in vivo. The 3′ *Hoxb1* gene is expressed first (day 2–4), a more 5′ gene, *Hoxb9*, is expressed later at day 10, and expression of the most 5′ gene, *Hoxb13*, is not detected at all by RT-PCR. (Adapted from Chambeyron and Bickmore 2004.)

Histone Modifications and the *HoxB* Temporal Program of Gene Expression

What changes in histone modification occur at *HoxB* during this program of transcriptional activation? We investigated this question by using chromatin immunoprecipitation (ChIP) to examine the promoters and exons of *Hoxb1* and *Hoxb9*, as well as two *cis*-regulatory elements. The retinoic acid response element (RARE) 3′DR2 mediates the early response to RA and is 1.2 kb 3′ of *Hoxb1* (Marshall et al. 1994). The r4 enhancer is required for the rhombomere 4 restricted expression of *Hoxb1* in the embryo and is just upstream of the *Hoxb1* promoter (Fig. 3) (Popperl et al. 1995). We first chose to examine histone H3 acetylation, and methylation of H3 on lysine 4 (metH3-K4), because these modifications have been associated with transcriptionally active genes (Lachner et al. 2003).

The acquisition of these histone marks at *Hoxb1* and its regulatory elements nicely parallels the induction of gene expression (Chambeyron and Bickmore 2004). The levels of both AcH3-K9 and met$_2$H3-K4 at *Hoxb1* are low in undifferentiated ES cells and increase by day 4 of differentiation (Fig. 3A). However, histone acetylation is also acquired simultaneously at *Hoxb9*, even though this gene is not expressed until day 10—6 days later (Figure 3B). AcH3 K9 persists at *Hoxb9* until day 10, whereas both acH3 K9 and met$_2$H3 K4 are lost from *Hoxb1* as the gene switches off. The ES cells continue to divide rapidly during the induction process and therefore acetylation and K4 methylation of H3 precede the transcriptional activation of *Hoxb9* by many cell cycles. This suggests that these histone modifications cannot encode the epigenetic information for transcriptional activation of *Hoxb9*. Rather they may be necessary, but not sufficient, for *Hox* gene activation by, for example, establishing a transcrip-

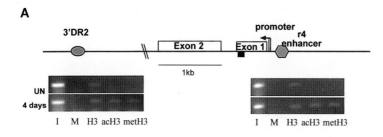

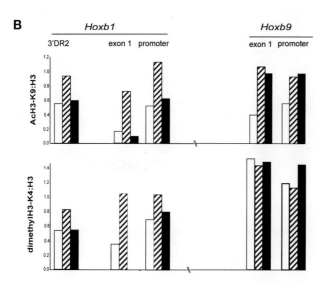

Figure 3. Changes in histone H3 acetylation and methylation upon induction. (*A*) PCR amplification of the *Hoxb1* 3′DR2 and the promoter/r4 enhancer in input (I), and mock IP (M) chromatin samples and in chromatin immunoprecipitated with antibodies that recognize H3, H3 acetylated on K9 (acH3), and H3 dimethylated on K4 (metH3). Chromatin was prepared from ES cells before (UN) and 4 days after induction with RA. The position of these elements at *Hoxb1* is shown. (*B*) AcH3 K9 and Met$_2$H3 K4 levels, relative to those of H3, at *Hoxb1* and *Hoxb9*, measured by quantitative real-time PCR of ChIP samples, from chromatin prepared at 0 (*open bars*), 4 days (*hatched bars*), and 10 days (*filled bars*) of differentiation. (Adapted from Chambeyron and Bickmore 2004.)

tionally poised state. Surprisingly, high levels of met$_2$H3 K4 seem to be present at *Hoxb9* in undifferentiated ES cells. Interestingly, a microRNA (mir-196a-1) has recently been described just upstream of the *Hoxb9* promoter, but its transcriptional status in undifferentiated ES cells is unknown (Yekta et al. 2004).

The acquisition of histone acetylation at *HoxB* during differentiation is not just a reflection of generally low levels of histone acetylation in undifferentiated ES cells, because chromatin of X-linked genes, and the *Oct4* locus, is known to be hyperacetylated in these cells (O'Neill et al. 2003; Hattori et al. 2004).

A Visible "Opening" of *HoxB*

Since histone modifications do not seem to provide a simple explanation for the temporal program of *Hox* gene expression, we may need to look at higher-order chromatin structures. This is especially pertinent given the models that raise the possibility of an (implicitly structural) opening of chromatin structure propagated through *Hox* loci (van der Hoeven et al. 1996; Kondo and Duboule 1999; Roelen et al. 2002). Visible chromatin decondensation accompanies transcriptional activation in some reporter systems (Tumbar et al. 1999; Tsukamoto et al. 2000; Muller et al. 2001; Ye et al. 2001; Nye et al. 2002), but evidence for decondensation at endogenous mammalian loci has been lacking.

We took a cytological approach to investigate chromatin condensation at *HoxB*. At probe separations of <2 Mb, there is a linear relationship between the mean-square interphase distances between them (d^2) and their genomic separation (kb) (van den Engh et al. 1992). This type of analysis can reveal different levels of chromatin compaction at different genomic regions (Yokota et al. 1997). However, we reasoned that it should also be able to detect different levels of chromatin compaction at the same genomic region, in cells at different states of differentiation.

Since *Hoxb1* and *Hoxb9* are only 90 kb apart, we were not surprised that in the nuclei of undifferentiated ES cells their fluorescence in situ hybridization (FISH) signals were superimposed on one another (i.e., the two genes are not spatially separable). However, by day 4 of differentiation we saw a 10-fold decondensation of *HoxB* ($d^2 = 0.88$ μm^2 for the 90-kb region between *Hoxb1* and *Hoxb9*) (Chambeyron and Bickmore 2004). Since the length of 90 kb of DNA double helix is ~31 μm, and folding into a 30-nm fiber is thought to compact this almost 50-fold (Wolffe 1998), then this level of compaction is approximately consistent with unwinding of the *Hoxb1* to *b9* region down to a 30-nm chromatin fiber, and perhaps also beyond this. This is several-fold more decondensation than has been reported at exogenous reporters in mammalian cells (see, e.g., Muller et al. 2001; Nye et al. 2002).

CHROMATIN DECONDENSATION CORRELATES WITH ALTERED NUCLEAR ORGANIZATION

Does chromatin decondensation just represent a local unfolding of chromatin structure without further consequences for larger-scale organization, or could it correspond to gross changes in the way that the chromatin is organized in the nucleus? An analysis of the organization of *HoxB* relative to its chromosome territory (CT) suggests the latter.

Using FISH we found that upon differentiation *Hoxb1* moves from the surface of the MMU11 CT to a position ~1 μm outside of the territory by day 4 (Fig. 4). *Hoxb9* remains inside the MMU11 territory at this time. *Hoxb1* partially relocalizes back toward the CT by day 10, when its expression is switched off. *Hoxb9* can only be detected outside of the CT at day 10, coincident with its expression, and its looping out never reaches the same extent as that seen for *Hoxb1* at day 4 (Chambeyron and Bickmore 2004). There is therefore a choreographed extrusion of *Hoxb1* and *b9* outside of CTs that occurs in synchrony with the execution of the gene expression program.

Domains of high transcriptional activity have previously been correlated with a looping out from CTs. Extrusion of some large domains of coordinately regulated genes from their CTs correlates with the upregulation of gene expression (Volpi et al. 2000; Williams et al. 2002). In other parts of the human genome, regions that contain a high density of generally widely expressed, but not necessarily coordinately regulated, genes are also looped out of their CTs (Mahy et al. 2002). The fact that looping at *HoxB* parallels the temporal expression of the genes in the complex (at least for *Hoxb1* and *b9*), and that the pattern of looping is modulated during cell differentiation, strongly suggests that these large-scale chromatin structure changes are important in the regulation of gene expression.

Where do the genes extruded from CTs go? In this case the extruded Hox genes appear to be moving toward the center of the nucleus during induction, relative to the MMU11 CT (Fig. 5A) (Chambeyron and Bickmore 2004). We also saw recruitment toward the nuclear cen-

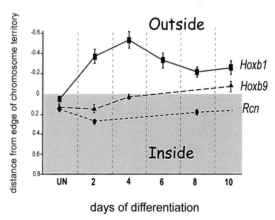

Figure 4. Extrusion from chromosome territory. Localization of *Hoxb1* and *Hoxb9*, relative to the MMU11 chromosome territory during 10 days of RA-induced differentiation. A control gene (*Rcn*), that is on MMU2 and not induced by RA, is included as a control. Mean position (± S.E.M.) of genes is shown, normalized to the size of the chromosome territory, to account for changes in territory size during differentiation. 1, center of chromosome territory; 0, territory edge; negative values, localization outside of the chromosome territory. $n = 100$ territories.

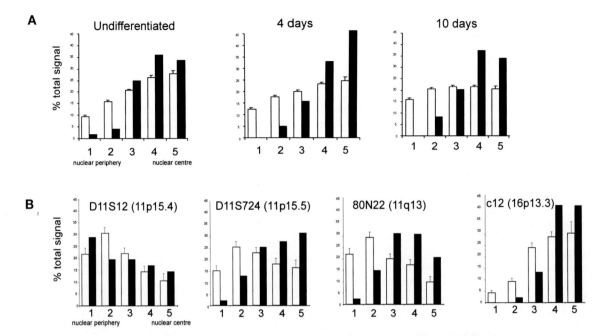

Figure 5. Movement of transcriptionally active regions toward the center of the nucleus. (*A*) Mean localization of *Hoxb1* signals (*filled bars*) and the distribution of MMU11 CT signal (*open bars*) within shells eroded from the edge (shell 1) to the center (shell 5) of the nucleus from undifferentiated murine ES cells, and ES cells differentiated with RA for 4 or 10 days. $n = 100$ territories. (*B*) Mean localization of specific loci (*filled bars*) and their respective human CT signals (*open bars*) within shells eroded from the edge (shell 1) to the center (shell 5) of human nuclei. $n = 100$ territories. The 11p15.4 has been shown to be located inside of the CT, whereas the other three loci are known to be located outside of their respective CTs in a high proportion of cells (Mahy et al. 2002).

ter, of genes that are looped out of CTs in human cells (Fig. 5B) (Mahy et al. 2002). A similar differential subnuclear localization of imprinted regions in murine ES cells has been reported (Gribnau et al. 2003).

The functional significance of localizing active loci toward the center of the nucleus remains unclear. It might be that the central volume of the nucleus provides an environment that favors high levels of transcription, for example, by enhancing transcription rates. Alternatively, very active regions may be being recruited to distinct nuclear compartments that enhance transcription and that themselves have a biased distribution toward the nuclear center. For example, recently it was shown that very transcriptionally active regions of the human genome are preferentially associated with promyelocytic leukemia (PML) bodies (Wang et al. 2004).

CAN CHROMATIN DECONDENSATION BE SEPARATED FROM HISTONE ACETYLATION?

We are suggesting that higher-order and large-scale changes in chromatin structure underpin the regulated expression of, at least, *Hox* loci during development. Is it possible that these cytological levels of chromatin structure are just a visible readout of what is happening at the level of the nucleosome and its modifications? Histone modifications may directly affect higher-order chromatin structure and condensation (Tse et al. 1998; Wolffe and Hayes 1999; Carruthers and Hansen 2000; Wang et al. 2001).

To confirm that simple histone acetylation is not sufficient to induce the temporal program of *HoxB* gene expression, we used the histone deacetylase inhibitor Trichostatin A (TSA) to increase histone acetylation in undifferentiated ES cells, in the absence of RA. ChIP showed that this treatment increased AcH3-K9 levels at *Hoxb1* and its regulatory elements to the same extent as that seen when untreated cells are differentiated with RA (Fig. 6A). But there was no visual decondensation of the locus and no movement outside of CTs—indeed, *Hoxb1* seemed to be further inside its CT in TSA-treated ES cells as compared with untreated cells (Fig. 6B). Despite the acetylation of H3 at and around *Hoxb1*, there was no induction of *Hoxb1* expression in TSA-treated cells (Fig. 6C) (Chambeyron and Bickmore 2004). It has also been shown that TSA cannot rescue the expression of a *Hoxb1lacZ* transgene in a polycomb mutant embryo (de Graaff et al. 2003).

CONCLUSIONS

During embryogenesis, expression of *Hoxb1* is first seen at the onset of gastrulation in the most posterior part of the primitive streak (E7.0). The domain of expression then spreads anterior to this during the next day of development (Forlani et al. 2003). However, *Hoxb1* is poised to respond to exogenous RA earlier than this (by E6.0), before the primitive streak is formed (Roelen et al. 2002). The rapid response of *Hoxb1* to RA in ES cells (Fig. 1B) (Papalopulu et al. 1991) suggests that *Hoxb1* located at

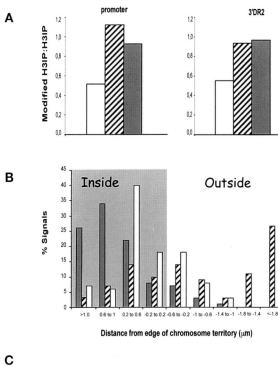

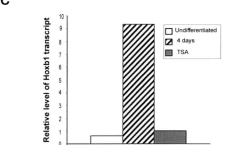

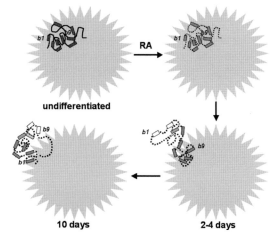

Figure 7. Model for the progress activation of genes within *HoxB*. Within undifferentiated ES cells the genes of the *HoxB* complex are inactive (*filled gray boxes*) and are packaged in condensed unacetylated chromatin that resides within the MMU11 chromosome territory (*light gray shape*). Upon induction by RA there is widespread acetylation of histones at the locus (*dotted line*), but the locus remains compact and within the CT. However at day 2 of differentiation the chromatin at *HoxB* decondenses, and the active (*white box*) *Hoxb1* gene moves out of the CT. The inactive *Hoxb9* remains within the CT. This loop is then remodeled so that by day 10, the now inactivated *Hoxb1* is recondensing and moving back into the CT, while *Hoxb9* moves out of the CT and becomes active.

Figure 6. Histone acetylation does not induce changes in higher-order chromatin structure. (*A*) ChIP to determine the levels of AcH3 K9 at the promoter and 3´DR2 of *Hoxb1*, in control (*open bars*) and TSA-treated (*filled bars*) undifferentiated ES cells and untreated ES cells induced to differentiate with RA for 4 days (*hatched bars*). (*B*) Histogram of the position of *Hoxb1*, relative to the surface of the CT in untreated (*open bars*), and TSA-treated (*filled bars*) ES cells and in ES cells differentiated with RA for 4 days (*hatched bars*). Negative values indicate localization outside of the CT. (*C*) Quantitative RT-PCR of *Hoxb1* expression in RNA prepared from untreated (*open bars*) and TSA-treated (*filled bars*) ES cells and ES cells differentiated with RA for 4 days (*hatched bars*). Levels of *Hoxb1* are normalized relative to those of actin.

the surface of its CT is similarly poised for transcription in these cells that are derived from the inner cell mass of the blastocyst. A prediction therefore is that *Hoxb1* will be similarly localized at the CT surface of nuclei of epiblast cells in E6.0 mouse embryos.

The early temporal expression pattern of *Hox* genes is essential for their better known functions later in development, though it is not clear why. By studying the expression of *Hox* transgenes, and transpositions that rearrange *Hox* genes within and between clusters, several models have been put forward to explain the colinearity between *Hox* gene order and the time and place of their expression (Kmita and Duboule 2003). Most, but not all, of these models have changes in chromatin structure at their heart. In the first, it is suggested that there is a polarized physical "opening" of chromatin from the 3´ to 5´ end of the cluster that might, for example, allow progressive access of the genes to the transcriptional machinery. In a similar vein there might be a progressive change in chromatin structure that relieves the genes from an active silencing mechanism. Alternatively, there might be sequential interaction of the genes within a cluster with distal control elements, akin to the regulation of the β-globin cluster by the LCR.

These models are not necessarily mutually exclusive, and they could each operate at different times during development, or in different ways for the different *Hox* clusters. For example, there is good evidence for a global enhancer located outside of the *HoxD* cluster that controls colinear expression in developing digits (Spitz et al. 2003). However, our analysis of *HoxB* is most consistent with the idea of a progressive opening of chromatin structure through the cluster and manifest in our experiments as visible chromatin decondensation (Fig. 7).

What is the trigger mechanism of chromatin decondensation and looping outside of the CT at *HoxB*? In answering this question we are faced with the fact that we still do not even know what the structure of chromatin fibers is (Thoma et al. 1979; McGhee et al. 1983; Woodcock et al. 1984; van Holde and Zlatanova 1996; Wolffe 1998), let alone the mechanisms that may modulate them. It has previously been predicted that Hox clusters gradually move from inside of CTs to a space outside of them (interchromatin domain) during development (Papageorgiou 2004). It was suggested that the force for this movement would be a differential concentration of transcription factors within and outside of the CT. However, we think this unlikely since the studies of transcription factor mobility in the nucleus using fluorescence recovery after photobleaching (FRAP) indicate that there is no diffusional barrier for proteins imposed by chromatin in the nucleus (Misteli 2001).

Whatever the mechanism of chromatin opening at *HoxB*, it seems likely that the cis-elements responsible for initiating it lie in and around *Hoxb1*, since placing of a *Hoxb1lacZ* transgene at the 5′ end of the HoxD complex resulted in early mesodermal expression not only of the transposed *Hoxb1*, but also of the adjacent *Hoxd13* (Kmita et al. 2000). Identifying this element and the mechanisms of chromatin decondensation will be the focus of future studies.

ACKNOWLEDGMENTS

S.C. was supported by fellowships from La Ligue Contre le Cancer and the Wellcome Trust (GR071481). W.A.B. is a Centennial fellow of the James S. McDonnell foundation. This work was supported by the Medical Research Council of the United Kingdom.

REFERENCES

Carruthers L.M. and Hansen J.C. 2000. The core histone N termini function independently of linker histones during chromatin condensation. *J. Biol. Chem.* **275**: 37285.

Chambeyron S. and Bickmore W.A. 2004. Chromatin decondensation and nuclear reorganization of the *HoxB* locus upon induction of transcription. *Genes Dev.* **18**: 1119.

de Graaff W., Tomotsune D., Oosterveen T., Takihara Y., Koseki H., and Deschamps J. 2003. Randomly inserted and targeted *Hox*/reporter fusions transcriptionally silenced in *Polycomb* mutants. *Proc. Natl. Acad. Sci.* **100**: 13362.

Fischle W., Wang Y., and Allis C.D. 2003. Histone and chromatin cross-talk. *Curr. Opin. Cell Biol.* **15**: 172.

Forlani S., Lawson K.A., and Deschamps J. 2003. Acquisition of Hox codes during gastrulation and axial elongation in the mouse embryo. *Development* **130**: 3807.

Gribnau J., Hochedlinger K., Hata K., Li E., and Jaenisch R. 2003. Asynchronous replication timing of imprinted loci is independent of DNA methylation, but consistent with differential subnuclear localization. *Genes Dev.* **17**: 759.

Hattori N., Nishino K., Ko Y.G., Hattori N., Ohgane J., Tanaka S., and Shiota K. 2004. Epigenetic control of mouse Oct-4 gene expression in embryonic stem cells and trophoblast stem cells. *J. Biol. Chem.* **279**: 17063.

Kmita M. and Duboule D. 2003. Organizing axes in time and space; 25 years of colinear tinkering. *Science* **301**: 331.

Kmita M., van der Hoeven F., Zakany J., Krumlauf R., and Duboule D. 2000. Mechanisms of Hox gene colinearity: Transposition of the anterior Hoxb1 gene into the posterior HoxD complex. *Genes Dev.* **14**: 198.

Kondo T. and Duboule D. 1999. Breaking colinearity in the mouse HoxD complex. *Cell* **97**: 407.

Krumlauf R. 1994. Hox genes in vertebrate development. *Cell* **78**: 191.

Lachner M., O'Sullivan R.J., and Jenuwein T. 2003. An epigenetic road map for histone lysine methylation. *J. Cell Sci.* **116**: 2117.

Maconochie M.K., Nonchev S., Studer M., Chan S.K., Popperl H., Sham M.H., Mann R.S., and Krumlauf R. 1997. Cross-regulation in the mouse HoxB complex: The expression of Hoxb2 in rhombomere 4 is regulated by Hoxb1. *Genes Dev.* **11**: 1885.

Mahy N.L., Perry P.E., and Bickmore W.A. 2002. Gene density and transcription influence the localization of chromatin outside of chromosome territories detectable by FISH. *J. Cell Biol.* **159**: 753.

Marshall H., Studer M., Popperl H., Aparicio S., Kuroiwa A., Brenner S., and Krumlauf R. 1994. A conserved retinoic acid response element required for early expression of the homeobox gene Hoxb-1. *Nature* **370**: 567.

McGhee J.D., Nickol J.M., Felsenfeld G., and Rau D.C. 1983. Higher order structure of chromatin: Orientation of nucleosomes within the 30nm chromatin solenoid is independent of species and spacer length. *Cell* **33**: 831.

McKittrick E., Gafken P.R., Ahmad K., and Henikoff S. 2004. Histone H3.3 is enriched in covalent modifications associated with active chromatin. *Proc. Natl. Acad. Sci.* **101**: 1525.

Meneghini M.D., Wu M., and Madhani H.D. 2003. Conserved histone variant H2A.Z protects euchromatin from the ectopic spread of silent heterochromatin. *Cell* **112**: 725.

Misteli T. 2001. Protein dynamics: Implications for nuclear architecture and gene expression. *Science* **291**: 843.

Muller W.G., Walker D., Hager G.L., and McNally J.G. 2001. Large-scale chromatin decondensation and recondensation regulated by transcription from a natural promoter. *J. Cell Biol.* **154**: 33.

Nye A.C., Rajendran R.R., Stenoien D.L., Mancini M.A., Katzenellenbogen B.S., and Belmont A.S. 2002. Alteration of large-scale chromatin structure by estrogen receptor. *Mol. Cell. Biol.* **22**: 3437.

O'Neill L.P., Randall T.E., Lavender J., Spotswood H.T., Lee J.T., and Turner B.M. 2003. X-linked genes in female embryonic stem cells carry an epigenetic mark prior to the onset of X inactivation. *Hum. Mol. Genet.* **12**: 1783.

Oosterveen T., Niederreither K., Dolle P., Chambon P., Meijlink F., and Deschamps J. 2003. Retinoids regulate the anterior expression boundaries of 5′ Hoxb genes in posterior hindbrain. *EMBO J.* **22**: 262.

Papageorgiou S. 2004. A cluster translocation model may explain the collinearity of Hox gene expressions. *Bioessays* **26**: 189.

Papalopulu N., Lovell-Badge R., and Krumlauf R. 1991. The expression of murine Hox-2 genes is dependent on the differentiation pathway and displays a collinear sensitivity to retinoic acid in F9 cells and *Xenopus* embryos. *Nucleic Acids Res.* **19**: 5497.

Popperl H., Bienz M., Studer M., Chan S.K., Aparicio S., Brenner S., Mann R.S., and Krumlauf R. 1995. Segmental expression of Hoxb-1 is controlled by a highly conserved autoregulatory loop dependent upon exd/pbx. *Cell* **81**: 1031.

Roelen B.A.J., de Graaff W., Forlani S., and Deschamps J. 2002. Hox cluster polarity in early transcriptional availability: A high order regulatory level of clustered Hox genes in the mouse. *Mech. Dev.* **119**: 81.

Spitz F., Gonzalez F., and Duboule D. 2003. A global control region defines a chromosomal regulatory landscape containing the *HoxD* cluster. *Cell* **113**: 405.

Thoma F., Koller T., and Klug A. 1979. Involvement of histone H1 in the organization of the nucleosome and of the salt-dependent superstructures of chromatin. *J. Cell Biol.* **83**: 403.

Tse C., Sera T., Wolffe A.P., and Hansen J.C. 1998. Disruption of higher-order folding by core histone acetylation dramatically enhances transcription of nucleosomal arrays by RNA polymerase III. *Mol. Cell. Biol.* **18**: 4629.

Tsukamoto T., Hashiguchi N., Janicki S.M., Tumbar T., Belmont A.S., and Spector D.L. 2000. Visualization of gene activity in living cells. *Nat. Cell Biol.* **2**: 871.

Tumbar T., Sudlow G., and Belmont A.S. 1999. Large-scale chromatin unfolding and remodeling induced by VP16 acidic activation domain. *J. Cell Biol.* **145**: 1341.

van den Engh G., Sachs R., and Trask B.J. 1992. Estimating genomic distance from DNA sequence location in cell nuclei by a random walk. *Science* **257**: 1410.

van der Hoeven F., Zakany J., and Duboule D. 1996. Gene transpositions in the HoxD complex reveal a hierarchy of regulatory controls. *Cell* **85:** 1025.

van Holde K. and Zlatanova J. 1996. What determines the folding of the chromatin fiber? *Proc. Natl. Acad. Sci.* **93:** 10548.

Volpi E.V., Chevret E., Jones T., Vatcheva R., Williamson J., Beck S., Campbell R.D., Goldsworthy M., Powis S.H., Ragoussis J., Trowsdale J., and Sheer D. 2000. Large-scale chromatin organisation of the major histocompatibility complex and other regions of human chromosome 6 and its response to interferon in interphase nuclei. *J. Cell Sci.* **113:** 1565.

Wang J., Shiels C., Sasieni P., Wu P.J., Islam S.A., Freemont P.S., and Sheer D. 2004. Promyelocytic leukemia nuclear bodies associate with transcriptionally active genomic regions. *J. Cell Biol.* **164:** 515.

Wang X., He C., Moore S.C., and Ausio J. 2001. Effects of histone acetylation on the solubility and folding of the chromatin fiber. *J. Biol. Chem.* **276:** 12764.

Williams R.R.E., Broad S., Sheer D., and Ragoussis J. 2002. Subchromosomal positioning of the epidermal differentiation complex (EDC) in keratinocyte and lymphoblast interphase nuclei. *Exp. Cell Res.* **272:** 163.

Wolffe A.P. 1998. *Chromatin: Structure and function.* Academic Press, San Diego, California.

Wolffe A.P. and Hayes J.J. 1999. Chromatin disruption and modification. *Nucleic Acids Res.* **27:** 711.

Woodcock C.L., Frado L.L., and Rattner J.B. 1984. The higher-order structure of chromatin: Evidence for a helical ribbon arrangement. *J. Cell Biol.* **99:** 42.

Ye Q., Hu Y.F., Zhong H., Nye A.C., Belmont A.S., and Li R. 2001. BRCA1-induced large-scale chromatin unfolding and allele-specific effects of cancer-predisposing mutations. *J. Cell Biol.* **155:** 911.

Yekta S., Shih I.H., and Bartel D.P. 2004. MicroRNA-directed cleavage of HOXB8 mRNA. *Science* **304:** 594.

Yokota H., Singer M.J., van den Engh G.J., and Trask B.J. 1997. Regional differences in the compaction of chromatin in human G_0/G_1 interphase nuclei. *Chromosome Res.* **5:** 157.

SIR1 and the Origin of Epigenetic States in *Saccharomyces cerevisiae*

L. Pillus[*] AND J. Rine[†]

[*]Division of Biological Sciences, Section of Molecular Biology and UCSD Cancer Center,
University of California, San Diego, La Jolla, California 92093-0347; [†]Molecular and Cell Biology,
University of California, Berkeley, Berkeley, California 94720-3202

Silencing transcription of the majority of the genome in every cell is critical for normal growth, development, and regulation in multicellular eukaryotes. Current models of silencing build on the emerging details of chromatin modifications, as described in many of the papers in this volume. Modification states of histones, transcription factors, and other cellular proteins can strongly influence the recruitment and processivity of RNA polymerases.

Our goal in this short and decidedly noncomprehensive piece is to recap some insights from what has become a classic example of epigenetic control of transcription, that of silencing in the budding yeast *Saccharomyces cerevisiae*. Thorough and timely reviews may be found elsewhere (see, e.g., Rusche et al. 2003). Here, our historical and affectionate touchstone is the Silent Information Regulator gene, *SIR1*, which is pivotal in causing heterochromatin to assemble at and spread from silencers.

Most key players of silencing in yeast were initially identified mutationally. The subsequent molecular identification of some of their contributions to chromatin modification is mechanistically satisfying. Yet questions remain about other regulators, underscoring the probability of surprises in store as more is learned about how functionally different chromosomal states are established, maintained, and interconverted.

MUTANTS OF *SIR1* REVEAL EPIGENETIC SILENCING STATES

Historically, a popular approach to transcriptional studies was the biochemical identification of molecules and machines required for activation. These analyses provided much detail as in vitro reconstitution of transcription on purified DNA templates became increasingly refined. However, contemporaneously, genetic analyses provided examples in which the absence of transcription, rather than its activation, is essential and most relevant for normal cell function. One such key example in yeast is the existence of silent copies of the mating-type information. The purpose of these copies is not to be expressed. Rather, these genes serve as sources of genetic information that, when transposed elsewhere, determine cell identity.

The elegant "cassette hypothesis" developed by Ira Herskowitz and his colleagues explained many aspects of the formal patterns through which cells could switch their identity by regulated access to the extra, although silent, copies of mating-type information (Herskowitz et al. 1977). Early versions of the hypothesis entertained the notion that something special about the mating-type locus, perhaps its promoter, caused expression at that locus. However, genetic studies soon turned the problem inside out with the discovery of recessive mutations that could activate the otherwise silent mating-type genes (Herskowitz et al. 1977; Haber and George 1979; Klar et al. 1979; Rine et al. 1979). What remained to be found, though, were the proteins necessary to prevent expression of the silent mating-type cassettes. Extended systematic analysis defined four major complementation groups of silent information regulators, the *SIR* genes, some alleles of which were also uncovered early on as *STE*, *MAR*, or *CMT* genes (Rine and Herskowitz 1987).

Even before molecular identification, the silent information regulator mutants fit into two categories: those with complete defects in silencing and one unusual case, *sir1*. In *sir1* mutants, silencing appeared to be only moderately affected in populations of mutant cells (Rine et al. 1979). Several different tests supported these distinctions between the mutants.

One way of assessing silencing is through mating tests in plate assays (Fig. 1). If haploid cells have normal silencing of the extra mating-type genes at *HML* and *HMR*, they mate with cells of the opposite mating type. If silencing is disrupted, haploids have many characteristics of diploid cells and fail to mate because they express mating-type information of both cell types. Under standard lab conditions, mating assays are performed so that the strains being tested have nonoverlapping auxotrophic markers. Therefore, mating is detected by selecting for prototrophic growth, which can occur only if a diploid has successfully formed by mating of the two haploids. In the example shown, *sir2Δ* mutants are completely mating defective, whereas, *sir1Δ* mutants appear to mate with an efficiency comparable to that of wild-type cells. However, this assay is blind to partial reductions in mating efficiency.

At first blush, a simple but incorrect explanation for the apparently incomplete effect on mating is that the *sir1* mutant alleles tested were only partial loss-of-function mutants, rather than null alleles that might completely destroy silencing. Therefore, when null alleles of *SIR1* were

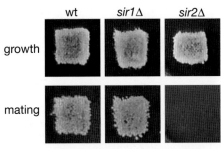

Figure 1. Silencing in populations of cells is easily assayed in patch mating tests. Wild-type and *sir*Δ mutant cells grow well, as shown in the control growth patches on rich medium in the *top* row. In mating tests for silencing, cells grow only if they successfully mate and form diploids with a wild-type cell of the mating type. In this example, as shown on the *bottom* row, *sir2*Δ mutants are completely mating defective in comparison to wild-type and *sir1*Δ cells. By other assays, *sir1* mutants were initially identified as silencing defective (Rine et al. 1979; Rine and Herskowitz 1987), although these defects are masked in population assays such as these. In quantitative mating assays or single-cell analyses (see text and Pillus and Rine 1989), 20–40% of *sir1*Δ cells are silencing competent, whereas the remainder of the population appears as defective as *sir2*Δ mutants.

found to have the same phenotype, all models based upon leaky alleles had to be discarded. Instead, two different views were equally compatible: (1) Sir1p played only a supporting role in silencing such that in all mutant cells the silent mating-type genes were activated to a slight extent or (2) in the absence of Sir1p, some cells silenced the *HML* and *HMR* loci and some did not.

In 1989, after many happy and, at times, perplexing hours and days in the company of single *sir1*Δ mutant cells at the micromanipulator, we concluded that *SIR1* provided insight into important aspects of transcriptional regulation. Notably, epigenetic transcription states could be a fundamental part of the biology of single eukaryotic cells.

In normal haploid cells, with functional silencing, in the presence of mating pheromone of the opposite cell type, the cells undergo cell-cycle arrest and characteristic morphological alterations in preparation for mating. This acutely sensitive assay for transcriptional status allows silencing to be measured in single mutant cells, their siblings, and their progeny simply by examining them microscopically.

In "Epigenetic inheritance of transcriptional states in *S. cerevisiae*," we reported that seemingly identical cells from an isogenic population of *sir1*Δ mutants had two decidedly distinct transcriptional states: those that appeared to be completely normal and silenced, and those that were as defective in silencing as their *sir2*, *sir3*, or *sir4* mutant brethren (Pillus and Rine 1989).

Importantly, cells of either *sir1*Δ phenotypic class were mitotically stable with respect to their transcriptional silencing of mating-type information. This stability could last for many generations. Yet cells of either state, silenced or expressed, could also "switch" to assume the opposite phenotype at a low frequency, on the order of once in every 250 cell divisions. Therefore, it appeared that rather than existing as some sort of continuum or threshold phenomenon, silencing was an either/or, on/off biological switch.

Furthermore, either transcriptional state could be propagated through meiosis. There was not any requisite, wholesale reprogramming of silenced states to move successfully through this major developmental process. Because either on or off transcriptional state could be sustained in *sir1* mutants, we concluded that Sir1p was not necessary for the maintenance of silencing, but rather was more likely to play a key role in its establishment. This point has caused some confusion over the years. After all, the two populations observed in *sir1*Δ cells could result from defects in either the establishment or the maintenance of the repressed state. In principle, one could imagine that establishment of silencing is ordinarily slow or inefficient in wild-type cells, but that maintenance is efficient, whereas maintenance is inefficient in *sir1*Δ cells.

The link of *SIR1* function to establishment is supported by two additional observations. First, in the small subset of cells in which the silenced state is established in *sir1*Δ cells, all cells in that generation are silenced rather than a small fraction. Second, under conditions in which *SIR1* function can be restored, such as with regulated promoter fusions, silencing is efficiently restored to all the *sir1*Δ cells in the population (Fox et al. 1997; Kirchmaier and Rine 2001).

ADDITIONAL EPIGENETICALLY REGULATED LOCI REVEAL COMMON FEATURES

As the molecular identification of *SIR1* was being finalized (Stone et al. 1991), independent observations were made that the epigenetic transcriptional states revealed by *sir1*Δ cells were not likely to be restricted to loss of *SIR1*. Indeed, regulatory site mutations at the silent mating-type loci, and expression of reporter genes at telomere-proximal sites revealed further fundamental possibilities for dynamic epigenetic changes in gene expression in yeast (Gottschling et al. 1990; Mahoney et al. 1991). The early studies of telomeric position effects paralleled those of *sir1* mutants and silent mating-type control. Indeed, both mitotic stability and distinct switches between states were observed in normal yeast cells. Telomeric silencing was also dependent on *SIR2*, *SIR3*, and *SIR4*, underscoring the potential coordinate function of these genes, although the epigenetic character of telomeric silencing is *SIR1* independent (Aparicio et al. 1991).

The possibility of an expanded role for *SIR* genes was also recognized in studies defining *SIR2* as a suppressor of recombination between the repetitive array of rDNA repeats within the nucleolus (Gottlieb and Esposito 1989). Although yeast cells are ordinarily characterized by high levels of homologous recombination in both meiotic and mitotic cells, suppression of recombination in the rDNA helps assure optimal numbers of the repeats and thereby coordinates regulation of ribosomal inventories, protein synthesis, and growth. A recombinational silencing role in the nucleolus is restricted to *SIR2*, since rDNA recombination is not affected in *sir1*, *sir3*, or *sir4* mutants. *SIR2* suppresses not just recombination, but also transposition, and transcription of pol II-driven reporter genes within the rDNA (Bryk et al. 1997; Fritze et al. 1997; Smith and Boeke 1997). Like recombination, these functions are intact in other *sir* mutant cells.

SIR2 is required for silencing at all three major genomic regions in yeast subject to epigenetic control, a role now known to be tied to its catalytic properties as an NAD+-dependent protein deacetylase of histones and other substrates. Significant progress has been made understanding the mechanisms of Sir2p's activities and the broader biological roles for *SIR2* and its relatives in yeast and in organisms ranging from bacteria to vertebrates (for recent reviews, see Blander and Guarente 2004; North and Verdin 2004).

THINKING ABOUT *SIR1* MECHANISTICALLY

Unlike *SIR2*, *SIR3*, or *SIR4*, which act in more than one region of the genome, from early studies, *SIR1* was distinguished by its apparent restriction to function at the silent mating-type loci. The *SIR1* gene can encode a 678 amino acid open reading frame (ORF), with no telltale structural domains or predicted catalytic activities (Stone et al. 1991). However, functional studies define an interaction between Sir1p and the DNA-binding Origin Recognition Complex (ORC; Triolo and Sternglanz 1996; Fox et al. 1997). The discovery of this relationship was particularly appealing because of the longstanding connection between the requirement for passage through S phase for the reestablishment of silencing after its disruption (Miller and Nasmyth 1984), the role of ORC in silencing (Foss et al. 1993), and the role of Sir1p in establishment. The requirement for both S phase and ORC for silencing suggested early notions of a requirement for DNA replication in silencing, but these proved incorrect (Kirchmaier and Rine 2001; Li et al. 2001). Models emerging from these observations and others (Chien et al. 1993) suggested that targeting Sir1p to silenced regions either through tethering to site-specific DNA-binding proteins or through the naturally targeted ORC complex is a critical feature of silent chromatin formation.

Indeed, both a genetic screen and molecular dissection of *SIR1* reveal a carboxy-terminal region of Sir1p that is critical for its interaction with Orc1p, the largest subunit of ORC, and with Sir4p (Gardner et al. 1999; Bose et al. 2004). These specific interactions, together with the low abundance of Sir1p (Gardner and Fox 2001), may help explain its restricted genomic functions. Furthermore, although Sir1p does not appear to be conserved at the level of primary amino acid sequence in multicellular eukaryotes, the regions identified as critical for interaction with ORC are conserved in other *Saccharomyces* species, as are two-hybrid interactions between the *Saccharomyces bayanus* and *Saccharomyces mikatae* Sir1 proteins and *S. cerevisiae* Orc1p. Tellingly, the most divergent ortholog is from *Saccharomyces castellii*, and in this case, where five amino acid substitutions are found in the critical interaction region, the two-hybrid interaction is lost (Bose et al. 2004).

Roles beyond transcriptional silencing for Sir1p have been revealed by its occupancy in centromeric chromatin (Sharp et al. 2003). Interestingly, Sir1p appears to act independently of the other Sir proteins, because they are not found in centric chromatin. Furthermore, Sir1p binds to the Cac1p chromatin assembly factor and stabilizes its centromeric association. This binding may be functionally significant. Although chromosome stability is normal in single mutants, the *sir1Δ cac1Δ* double mutant has elevated nondisjunction rates that are further exacerbated when Hir1p, another Cac1p-associated factor, is also mutant.

Unlike larger eukaryotes, *Saccharomyces* has active genes very close to centromeres. Hence, Sir1p function at the centromeres does not reflect silencing in the sense that *HML* and *HMR* are silenced. However, recombination is suppressed near centromeres, and transcription through centromeres can destroy their function (Panzeri et al. 1984; Snyder et al. 1988). Thus perhaps Sir1p works with other as-yet-undefined proteins to repress recombination at centromeres, possibly in partnership with one of the *HST* cousins of *SIR2*. Alternatively, perhaps it nucleates a nonspreading chromatin structure that helps insulate centromeres from an occasional errant RNA polymerase transit.

ENHANCERS OF *sir1Δ* MUTANTS MAKE MORE CHROMATIN CONNECTIONS

Even in the absence of a detailed understanding of *SIR1*'s molecular mechanism, it seemed likely that the phenotype of the null mutants could give more insight into the epigenetic control of silencing. For example, what protein or process fulfills Sir1p's role in establishment in the fraction of cells that silence *HML* and *HMR* in *sir1Δ* mutants?

One approach to answering this question was to identify additional mutants that themselves were mating competent, but in combination with the *sir1Δ* mutation would become completely mating defective (Reifsnyder et al. 1996). In this screen for enhancers of the *sir1Δ* phenotype, nonnull alleles of *SIR2*, *SIR3*, and *SIR4* were identified, some of which have been studied in detail and give insight into locus-specific requirements for these genes (Stone et al. 2000; Garcia and Pillus 2002). Significantly, an allele of the *SAS2* gene was recovered. Other alleles of *SAS2* had been identified independently as suppressors that restore mating to strains bearing a defective, sensitized *HMR* cis-silencer sequence (Ehrenhofer-Murray et al. 1997). Both studies found that *sas2* mutants completely eliminate silencing of telomeric reporter genes. The fact that Sas2p functions may be positive in the case of *sir1Δ* mutants and at telomeres, but antagonistic at *HMR*, underscores that it has distinct roles depending on the region of the genome where it acts.

Sas2p is part of the MYST family, so named for its founding members, the leukemia-associated Moz protein, the Yeast Sas proteins, and Tip-60, a human protein first found as an interactor with the HIV Tat transactivator.

The MYST family is broadly conserved, with two closely related proteins in yeast, Esa1p and Sas3p, and multiple orthologs in every eukaryote for which sequence information is available. Relatively weak sequence similarity to proteins with acetyltransferase activity led to the discovery of histone and nucleosome acetyltransferase activity for all three yeast paralogs and many multicellular orthologs. These orthologs contribute to such diverse processes as dosage compensation, apoptosis, and response to DNA damage (for review, see Carrozza et al. 2003; Utley and Côté 2003).

MYST family proteins, including Sas2p, are found in complexes whose subunits contribute to activity and specificity. In the case of Sas2p's partners, Sas4p and Sas5p, the mating defect of $sir1\Delta$ cells is enhanced by mutants of either subunit (Xu et al. 1999a,b).

Other enhancers of $sir1\Delta$ also have connections to chromatin modification. *NAT1* and *ARD1* are required for mating in *sir1* mutant backgrounds (Whiteway et al. 1987; Stone et al. 1991). Nat1p and Ard1p are components of an amino-terminal protein acetyltransferase (NatA) with many cellular substrates (for review, see Polevoda and Sherman 2003), including a recently described role acetylating Sir3p and Orc1p (Wang et al. 2004). Further, the $sir1\Delta$ $cac1\Delta$ $hir1\Delta$ mutant introduced above not only affects centromeres, but is also completely mating defective (Kaufman et al. 1998). The loss of mating in these additional mutant backgrounds underscores the idea that Sir1p activity becomes critical for silencing in many genetically compromised circumstances, including in cells with synthetic versions of silencers (McNally and Rine 1991).

We discovered a circumstance requiring *SIR1* that points to a previously unsuspected role of higher-order structure or topological constraints in silencing. Both of the silent mating-type loci, *HML* and *HMR*, and the actively transcribed *MAT* locus reside on chromosome III. Homologous recombination between *HML* and *HMR* results in circular derivatives of the chromosome (Strathern et al. 1979; Newlon et al. 1991) in which only genetic information between the silent loci and telomeres is lost. Fortunately, there are no essential genes distal to these loci. Thus, strains carrying a large circular chromosome III are fully viable and mating competent, yet have no free chromosome ends or telomeres on this chromosome. We observed that in such strains, mutation of *SIR1* results in loss of mating (Fig. 2). Thus, changing chromosomal structure renders silencing completely dependent on *SIR1*. The mechanism of this dramatic effect is not yet clear, but may result from topological changes, altered localization, altered replication timing, or other higher-order influences on chromosome structure.

ANALOGS BUT NOT HOMOLOGS OF *SIR1*

Given that heterochromatin is common to all eukaryotes, and that *SIR2* family members seem to participate in heterochromatin in many if not all of these, it is striking that *SIR1* homologs have not been found outside of the *Saccharomyces* genera. If it is so central to the establishment of heterochromatin in *Saccharomyces*, what plays this role in other species?

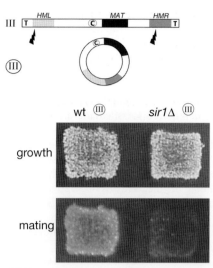

Figure 2. *SIR1* is required for mating-type silencing when chromosome III is circular. Unlike the normal genomic context of a linear chromosome III (as in Fig. 1), $sir1\Delta$ mutants become completely mating-defective when chromosome III is circularized by recombination between homologous sequences of the silent mating-type loci. Cartoons of chromosome III (*top*) highlight the telomeres (T), the centromere (C), and the silent mating-type loci *HML* and *HMR*, on the left and right arms of the chromosome. Transcription of the centromere-proximal *MAT* locus defines cell type. In the circular chromosome, recombination has occurred between the sites highlighted by lightning bolts. Although wild-type and $sir1\Delta$ mutants are healthy with this circular derivative (*top*), the $sir1\Delta$ mutants fail to mate (*bottom*).

A reasonable argument can be made for HP1 being a Sir1p analog. HP1, so named as the first discovered heterochromatin protein, found initially in *Drosophila*, is required for heterochromatin function in both flies and in *Schizosaccharomyces pombe* (for review, see Eissenberg and Elgin 2000; Kellum 2003; Maison and Almouzni 2004). In *Drosophila*, HP1 binds to the amino terminus of the Orc1 subunit of ORC analogously to the way that Sir1p binds to ORC, and *orc* mutants in *Drosophila* function disrupt heterochromatin (Shareef et al. 2001; Prasanth et al. 2004). Moreover, the heterochromatin function of ORC subunits is functionally interchangeable between yeast and *Drosophila* (Ehrenhofer-Murray et al. 1995; Pak et al. 1997). Hence, there seems to be a deep and conserved connection between proteins that can initiate heterochromatin and ORC. It is curious that a link between *S. pombe's* ORC and heterochromatin has not been made, especially given Sir2p's requirement for heterochromatin formation in *S. pombe* (Shankaranarayana et al. 2003; Freeman-Cook et al. 2005).

From a different perspective, perhaps it is more surprising that *Saccharomyces* does use Sir1 and its paralogs to enhance the efficiency of silencing. After all, a mechanism of repression that has no way for being turned on and off would not be useful for regulating genes that need to be turned on under some conditions but not others. This consideration leads us to suggest that Sir protein–based silencing may have evolved without Sir1 to allow for the variable yet heritable expression states of certain genes such as those turned on under some conditions but not others, including subtelomeric genes, such as the *FLO* genes of *Saccharomyces* and adhesion genes of *Candida* (B. Cormack, pers. comm.). Sir1 would then be a protein whose function is to push the equilibrium between the assembly of silenced chromatin versus active chromatin decidedly in the direction of silencing for genes whose purpose is not to be expressed, but to serve as a donor in the repair of double-stranded breaks at *MAT*. It will be interesting to see if HP1 serves a similar linchpin role in recruiting intercalary heterochromatin proteins to sites of ORC localization in *S. pombe* and *Drosophila*.

MORE *SIR1*s, MORE EPIGENETICS

The accumulating evidence from $sir1\Delta$ mutants and their enhancers demonstrates that loss of *SIR1* sensitizes transcriptional silencing to require other structural and catalytic regulators of chromatin for efficient silencing. Elements of Sir1p's functions at the *HML* and *HMR* silencer elements are clearly related to its recruitment to those sites by Orc1p and Sir4p. Precisely what does Sir1p do when bound? Most current models propose kinetic roles in the assembly of silent chromatin. For example, by providing an extra kilocalorie or so of free energy, Sir1p's interactions with ORC and Sir4p may overcome a kinetic barrier to the formation of silenced chromatin.

Recent studies cited above suggest that Sir1p has a broader spectrum of function than previously suspected. Sir1p's contribution to centromere function is independent of the other Sir proteins (Sharp et al. 2003). Enhancer screens in $sir1\Delta$ mutants focused on chromosome loss or segregation phenotypes may yield deeper insight into epigenetic aspects of centromere functions in yeast.

It has not yet been established whether the same regions of Sir1p defined for silencer binding are required for its centromeric functions or for functions with the circular derivative of chromosome III. Any differences might help define specificity of targeting and functionally distinct genomic interactions. Indeed, the *SIR1* gene may encode more than one species of protein. Specifically, *SIR1* produces two transcripts, the shorter of which is twice as abundant as the longer (Ivy et al. 1986; Stone et al. 1991). The more abundant transcript, if it is translated, would yield an alternative Sir1p isoform with a disrupted Orc1p-Sir4p binding domain (Sparks and Dieckmann 1998). Multiple in-frame methionine codons near the 5´ end of the *SIR1* ORF amplify the diversity of potential proteins from the locus. Determining whether other Sir1p isoforms have distinct or interfering functions in silencing may provide new insights into previously defined epigenetic states or those that control other less well understood biological processes.

Indeed, epigenetic control of pseudohyphal switching was recently reported to be influenced by the Hst1p and Hst2p members of the Sir2p family (Halme et al. 2004). Genes regulated by these deacetylases are relatively telomere proximal, regions known from several genome-wide studies to be particularly sensitive to histone acetylation state (see, e.g., the review by Millar et al., this volume). It seems reasonable to expect that as understanding of physiological states becomes more fully integrated with knowledge of transcriptional regulation, additional roles will be uncovered for molecules like Sir1p that may have the capacity to buffer epigenetic states. Moreover, the *Saccahromyces* genus offers unusually favorable opportunities to uncover new roles for Sir1p-like functions as up to four paralogs of *SIR1* can be found within its species.

Although the epigenetic aspects of silencing revealed by *sir1* mutations have been the focus of this discussion, the underlying on-or-off nature of silencing is also of interest as it bears on whether repression mechanisms have analog or digital qualities. This issue has received some attention. For example, the *GAL1* gene of *S. cerevisiae* is repressed in medium containing glucose by two different repressors, Gal80p and Mig1p. In the absence of Gal80p, glucose repression of *GAL1* is graded, whereas in the absence of Mig1p, it is on or off (Biggar and Crabtree 2001). Hence the quality of gene repression in response to a physiological stimulus can be either analog or digital, depending upon which repressor mediates the signal. The studies of *SIR1* suggest that mutants with "leaky" phenotypes can be particularly informative regarding mechanisms of repression, but capitalizing on such opportunities requires assays that can be performed robustly at the single-cell level. The versatility of mating-type regulation makes it supremely amenable to such studies.

ACKNOWLEDGMENTS

We thank J. Game for providing the parental strain with the circular derivative of chromosome III shown in Figure 2. We have greatly enjoyed continuing discussions with our colleagues and their contributions to the development of our ideas and we appreciate comments from S. Jacobson and J. Heilig on this manuscript. It is a particular delight to acknowledge the help of Ira Herskowitz in the early phases of our work on *SIR1* and the continuing impact his legacy makes. Work in our laboratories is occasionally funded by the National Institutes of Health. We appreciate that support.

REFERENCES

Aparicio O.M., Billington B.L., and Gottschling D.E. 1991. Modifiers of position effect are shared between telomeric and silent mating-type loci in *S. cerevisiae*. *Cell* 66: 1279.

Biggar S.R. and Crabtree G.R. 2001. Cell signaling can direct either binary or graded transcriptional responses. *EMBO J.* 20: 3167.

Blander G. and Guarente L. 2004. The Sir2 family of protein deacetylases. *Annu. Rev. Biochem.* 73: 417.

Bose M.E., McConnell K.H., Gardner-Aukema K.A., Muller U., Weinreich M., Keck J.L., and Fox C.A. 2004. The origin recognition complex and Sir4 protein recruit Sir1p to yeast silent chromatin through independent interactions requiring a common Sir1p domain. *Mol. Cell. Biol.* 24: 774.

Bryk M., Banerjee M., Murphy M., Knudsen K.E., Garfinkel D.J., and Curcio M.J. 1997. Transcriptional silencing of Ty1 elements in the *RDN1* locus of yeast. *Genes Dev.* 11: 255.

Carrozza M.J., Utley R.T., Workman J.L., and Côté J. 2003. The diverse functions of histone acetyltransferase complexes. *Trends Genet.* **19:** 321.

Chien C.-T., Buck S., Sternglanz R., and Shore D. 1993. Targeting of SIR1 protein establishes transcriptional silencing at *HM* loci and telomeres in yeast. *Cell* **75:** 531.

Ehrenhofer-Murray A., Rivier D., and Rine J. 1997. The role of Sas2, an acetyltransferase homolog, in silencing and ORC function in *Saccharomyces cerevisiae*. *Genetics* **145:** 923.

Ehrenhofer-Murray A.E., Gossen M., Pak D.T.S., Botchan M.R., and Rine J. 1995. Separation of origin recognition complex functions by cross-species complementation. *Science* **270:** 1671.

Eissenberg J.C. and Elgin S.C. 2000. The HP1 protein family: Getting a grip on chromatin. *Curr. Opin. Genet. Dev.* **10:** 204.

Foss M., McNally F.J., Laurenson P., and Rine J. 1993. Origin recognition complex (ORC) in transcriptional silencing and DNA replication in *S. cerevisiae*. *Science* **262:** 1838.

Fox C.A., Ehrenhofer-Murray A.E., Loo S., and Rine J. 1997. The origin recognition complex, *SIR1*, and the S phase requirement for silencing. *Science* **276:** 1547.

Freeman-Cook L.L., Gómez E.B., Spedale E.J., Marlett J., Forsburg S., Pillus L., and Laurenson P. 2005. Conserved locus-specific silencing functions of *Schizosaccharomyces pombe* $sir2^+$. *Genetics* **169:** (in press).

Fritze C.E., Verschueren K., Strich R., and Esposito R.E. 1997. Direct evidence for *SIR2* modulation of chromatin structure in yeast rDNA. *EMBO J.* **16:** 6495.

Garcia S.N. and Pillus L. 2002. A unique class of conditional *sir2* mutants displays distinct silencing defects in *Saccharomyces cerevisiae*. *Genetics* **162:** 721.

Gardner K.A. and Fox C.A. 2001. The Sir1 protein's association with a silenced chromosome domain. *Genes Dev.* **15:** 147.

Gardner K.A., Rine J., and Fox C.A. 1999. A region of the Sir1 protein dedicated to recognition of a silencer and required for interaction with the Orc1 protein in *Saccharomyces cerevisiae*. *Genetics* **151:** 31.

Gottlieb S. and Esposito R.E. 1989. A new role for a yeast transcriptional silencer gene, *SIR2*, in regulation of recombination in ribosomal DNA. *Cell* **56:** 771.

Gottschling D.E., Aparicio O.M., Billington B.L., and Zakian V.A. 1990. Position effect at *S. cerevisiae* telomeres: Reversible repression of Pol II transcription. *Cell* **63:** 751.

Haber J.E. and George J.P. 1979. A mutation that permits the expression of normally silent copies of mating type information in *Saccharomyces cerevisiae*. *Genetics* **93:** 13.

Halme A., Bumgarner S., Styles C., and Fink G.R. 2004. Genetic and epigenetic regulation of the *FLO* gene family generates cell-surface variation in yeast. *Cell* **116:** 405.

Herskowitz I., Strathern J.N., Hicks J.B., and Rine J. 1977. Mating type interconversion in yeast and its relationship to development in higher eukaryotes. *ICN-UCLA Symp. Mol. Cell. Biol.* (Eukaryotic Genetics Systems) **111:** 193.

Ivy J.M., Klar A.J.S., and Hicks J.B. 1986. Cloning and characterization of four SIR genes of *Saccharomyces cerevisiae*. *Mol. Cell. Biol.* **6:** 688.

Kaufman P.D., Cohen J.L., and Osley M.A. 1998. Hir proteins are required for position-dependent gene silencing in *Saccharomyces cerevisiae* in the absence of chromatin assembly factor I. *Mol. Cell. Biol.* **18:** 4793.

Kellum R. 2003. HP1 complexes and heterochromatin assembly. *Curr. Top. Microbiol. Immunol.* **274:** 53.

Kirchmaier A.L. and Rine J. 2001. DNA replication-independent silencing in *S. cerevisiae*. *Science* **291:** 646.

Klar A.J.S., Fogel S., and Macleod K. 1979. *MAR1*, a regulator of *HM*a and *HM*α loci in *Saccharomyces cerevisiae*. *Genetics* **93:** 37.

Li Y.C., Cheng T.H., and Gartenberg M.R. 2001. Establishment of transcriptional silencing in the absence of DNA replication. *Science* **291:** 650.

Mahoney D., Marquardt R., Shei G., Rose A., and Broach J. 1991. Mutations in the *HML* E silencer of *Saccharomyces cerevisiae* yield metastable inheritance of transcriptional repression. *Genes Dev.* **5:** 605.

Maison C. and Almouzni G. 2004. HP1 and the dynamics of heterochromatin maintenance. *Nat. Rev. Mol. Cell Biol.* **5:** 296.

McNally F.J. and Rine J. 1991. A synthetic silencer mediates *SIR*-dependent functions in *Saccharomyces cerevisiae*. *Mol. Cell. Biol.* **11:** 5648.

Miller A.M. and Nasmyth K.A. 1984. Role of DNA replication in the repression of silent mating type loci in yeast. *Nature* **312:** 247.

Newlon C.S., Lipschitz L.R., Collins I., Deshpande A., Devenish R.J., Green R.P., Klein H.L., Palzkill T.G., Ren R., Synn S., and Woody S.T. 1991. Analysis of a circular derivative of *Saccharomyces cerevisiae* chromosome III: A physical map and identification and location of *ARS* elements. *Genetics* **129:** 343.

North B.J. and Verdin E. 2004. Sirtuins: Sir2-related NAD-dependent protein deacetylases. *Genome Biol.* **5:** 224.

Pak D.T.S., Pflumm M., Chesnokov I., Huang D.W., Kellum R., Marr J., Romanowski P., and Botchan M.R. 1997. Association of the origin recognition complex with heterochromatin and HP1 in higher eukaryotes. *Cell* **91:** 311.

Panzeri L., Groth-Clausen I., Shepard J., Stotz A., and Phillippsen P. 1984. Centromeric DNA in yeast. *Chromosomes Today* **8:** 46.

Pillus L. and Rine J. 1989. Epigenetic inheritance of transcriptional states in *S. cerevisiae*. *Cell* **59:** 637.

Polevoda B. and Sherman F. 2003. Composition and function of the eukaryotic N-terminal acetyltransferase subunits. *Biochem. Biophys. Res. Commun.* **308:** 1.

Prasanth S.G., Prasanth K.V., Siddiqui K., Spector D.L., and Stillman B. 2004. Human Orc2 localizes to centrosomes, centromeres and heterochromatin during chromosome inheritance. *EMBO J.* **23:** 2651.

Reifsnyder C., Lowell J., Clarke A., and Pillus L. 1996. Yeast *SAS* silencing genes and human genes associated with AML and HIV-1 Tat interactions are homologous with acetyltransferases. *Nat. Genet.* **14:** 42.

Rine J. and Herskowitz I. 1987. Four genes responsible for a position effect on expression from *HML* and *HMR* in *Saccharomyces cerevisiae*. *Genetics* **116:** 9.

Rine J., Strathern J.N., Hicks J.B., and Herskowitz I. 1979. A suppressor of mating-type locus mutations in *Saccharomyces cerevisiae*: Evidence for and identification of cryptic mating-type loci. *Genetics* **93:** 877.

Rusche L.N., Kirchmaier A.L., and Rine J. 2003. The establishment, inheritance, and function of silenced chromatin in *Saccharomyces cerevisiae*. *Annu. Rev. Biochem.* **72:** 481.

Shankaranarayana G.D., Motamedi M.R., Moazed D., and Grewal S.I. 2003. Sir2 regulates histone H3 lysine 9 methylation and heterochromatin assembly in fission yeast. *Curr. Biol.* **13:** 1240.

Shareef M.M., King C., Damaj M., Badagu R., Huang D.W., and Kellum R. 2001. *Drosophila* heterochromatin protein 1 (HP1)/origin recognition complex (ORC) protein is associated with HP1 and ORC and functions in heterochromatin-induced silencing. *Mol. Biol. Cell* **12:** 1671.

Sharp J.A., Krawitz D.C., Gardner K.A., Fox C.A., and Kaufman P.D. 2003. The budding yeast silencing protein Sir1 is a functional component of centromeric chromatin. *Genes Dev.* **17:** 2356.

Smith J.S. and Boeke J.D. 1997. An unusual form of transcriptional silencing in yeast ribosomal DNA. *Genes Dev.* **11:** 241.

Snyder M., Sapolsky R.J., and Davis R.W. 1988. Transcription interferes with elements important for chromosome maintenance in *Saccharomyces cerevisiae*. *Mol. Cell. Biol.* **8:** 2184.

Sparks K.A. and Dieckmann C.L. 1998. Regulation of poly(A) site choice of several yeast mRNAs. *Nucleic Acids Res.* **26:** 4676.

Stone E.M., Reifsnyder C., McVey M., Gazo B., and Pillus L. 2000. Two classes of *sir3* mutants enhance the *sir1* mutant mating defect and abolish telomeric silencing in *Saccharomyces cerevisiae*. *Genetics* **155:** 509.

Stone E.M., Swanson M.J., Romeo A.M., Hicks J.B., and Sternglanz R. 1991. The *SIR1* gene of *Saccharomyces cerevisiae* and its role as an extragenic suppressor of several mating-defective mutants. *Mol. Cell. Biol.* **11:** 2253.

Strathern J.N., Newlon C.S., Herskowitz I., and Hicks J.B. 1979. Isolation of a circular derivative of yeast chromosome III: Implication for the mechanism of mating type interconversion. *Cell* **18:** 309.

Triolo T. and Sternglanz R. 1996. Role of interactions between the origin recognition complex and SIR1 in transcriptional silencing. *Nature* **381:** 251.

Utley R.T. and Côté J. 2003. The MYST family of histone acetyltransferases. *Curr. Top. Microbiol. Immunol.* **274:** 203.

Wang X., Connelly J.J., Wang C.L., and Sternglanz R. 2004. Importance of the Sir3 N terminus and its acetylation for yeast transcriptional silencing. *Genetics* **168:** 547.

Whiteway M., Freedman R., Arsdell S.V., Szostak J.W., and Thorner J. 1987. The yeast *ARD1* gene product is required for repression of cryptic mating-type information at the *HML* locus. *Mol. Cell. Biol.* **7:** 3713.

Xu E.Y., Kim S., and Rivier D.H. 1999a. *SAS4* and *SAS5* are locus-specific regulators of silencing in *Saccharomyces cerevisiae*. *Genetics* **153:** 25.

Xu E.Y., Kim S., Replogle K., Rine J., and Rivier D.H. 1999b. Identification of *SAS4* and *SAS5*, two genes that regulate silencing in *Saccharomyces cerevisiae*. *Genetics* **153:** 13.

Analyzing Heterochromatin Formation Using Chromosome 4 of *Drosophila melanogaster*

K.A. HAYNES, B.A. LEIBOVITCH, S.H. RANGWALA, C. CRAIG, AND S.C.R. ELGIN
Department of Biology, Washington University, St. Louis, Missouri 63130

While chromosomes provide a broad level of genomic organization within the nuclei of higher eukaryotes, the chromosome itself is further organized into multiple domains with distinct properties. This level of organization is visible in interphase nuclei as areas of condensed heterochromatin and regions of more dispersed euchromatin. Chromatin domains are biochemically distinguished by the types of histone modification and associated nonhistone chromosomal proteins. In most higher eukaryotes, domains of constitutive heterochromatin are normally restricted to pericentric and telomeric DNA. A remarkable property of heterochromatin in fungi, flies, and mammals is the ability to spread in *cis*, in response to loss of boundary constraints or to changes in dosage or activity of chromatin components; this results in silencing of euchromatic genes that are abnormally juxtaposed to heterochromatic domains by chromosome rearrangement or transposition, referred to as Position Effect Variegation (PEV) (for review, see Grewal and Elgin 2002). Recent studies in fungi and plants suggest that heterochromatin formation is targeted to repetitious elements through an RNAi mechanism, resulting in a domain of silenced chromatin (Volpe et al. 2002; Matzke et al. 2004; Schramke and Allshire 2004).

The small fourth chromosome of *Drosophila melanogaster* exhibits a rather unusual chromatin organization compared to the other chromosome arms. The distal 20–25% of chromosome 4 is amplified in polytene nuclei to a similar extent as the other euchromatic arms, and contains 82 genes in 1.2 Mb of DNA (Flybase Consortium 2003), a gene density comparable to that found in other euchromatic regions. At the same time, chromosome 4 exhibits characteristics of heterochromatin throughout its length. Chromosome 4 is rich in dispersed repetitious sequences (Kaminker et al. 2002), shows no detectable meiotic recombination (Bridges 1935), and is late-replicating (Barigozzi et al. 1966), all well-established characteristics of heterochromatic regions. Further, immunofluorescent staining of polytene chromosomes shows that Heterochromatin Protein 1 (HP1), known to play a key role in heterochromatin-induced silencing, is localized both in the pericentric heterochromatin and across the whole of the fourth chromosome (Fig. 1A) (James et al. 1989; Eissenberg and Elgin 2000).

CHROMOSOME 4: INTERSPERSED HETEROCHROMATIC AND EUCHROMATIC DOMAINS DEMONSTRATED BY IMMUNOCYTOLOGY

Unlike the fairly homogenous staining visible throughout the chromocenter, the fourth possesses contrasting

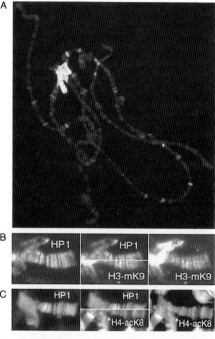

Figure 1. Immunofluorescent staining of polytene chromosomes of *Drosophila melanogaster* using antibodies specific for HP1 and antibodies specific for a modified histone. While the distribution pattern for histone H3 methylated at lysine 9 is essentially congruent with that for HP1, the distribution of histone H4 acetylated at lysine 8 is often opposing, suggesting distinct heterochromatic and euchromatic domains. (*A*) An entire chromosome set; (*B,C*) an enlargement of the fourth chromosome. (*A*) Immunofluorescent staining of polytene chromosomes from third instar larvae was carried out as described (James et al. 1989) using antibodies specific for HP1 (mouse monoclonal C1A9, James et al. 1989). (*B*) Chromosomes were stained simultaneously with anti-HP1 (mouse C1A9) and anti-H3-mK9 (rabbit, Upstate Biotechnology); secondary antibodies (Molecular Probes) were labeled with Alexa Fluor red or green. (*C*) Chromosomes were stained simultaneously with anti-HP1 (mouse C1A9) and rabbit anti-H4-acK8 (Suka et al. 2001).

HP1-rich and HP1-lacking bands, suggesting the presence of alternating domains of heterochromatin and euchromatin. This idea can be tested by examining the pattern of histone modification. Histone modifications have been shown to play a critical role in establishing alternative forms of chromatin packaging. Methylation of histone H3 at lysine 9 (producing H3-mK9) has been linked to HP1 interaction and heterochromatin formation, while acetylation of most lysines in the H3 and H4 tails is associated with gene activation (for review, see Berger 2002; Grewal and Elgin 2002). Immunocytology of polytene chromosomes indicates that the fourth chromosome includes domains with modifications indicative of both heterochromatin and euchromatin (Fig. 1). Distribution of the H3-mK9 isoform aligns with the distribution of HP1, as anticipated (Fig. 1B) (see also Schotta et al. 2002), while the H3-acK8 isoform has a distribution in opposition to that seen for HP1 (Fig. 1C). Thus, while the prominent association of HP1 and H3-mK9 indicates a high percentage of heterochromatin, there are interspersed domains within the fourth that exhibit modifications associated with euchromatin, suggesting a pattern in which the active genes are clustered, flanked by heterochromatic domains.

These observations illustrate the need to organize distinct domains within the context of a single chromosome. The interaction of HP1 both with H3-mK9 and with the H3 histone methyltransferases able to generate that modification suggests a model for maintenance and spreading of heterochromatin (for review, see Grewal and Elgin 2002). On binding to the modified histone H3-mK9, HP1 can recruit H3 methyltransferase, facilitating modification of adjacent nucleosomes. This model is supported by recent findings in *Schizosaccharomyces pombe*. Spreading of heterochromatin occurs upon removal of a putative boundary of the silent mating type domain; this requires the yeast homologs of HP1 and SU(VAR)3-9, an H3-K9 methyltransferase (Hall et al. 2002). In flies, both mutations in HP1 and in SU(VAR)3-9 result in suppression of PEV, i.e., in loss of silencing at a variegating gene (Eissenberg et al. 1990; Schotta et al. 2002). Thus, the maintenance and spread of heterochromatin are explained as a cycle of biochemical interactions between a histone signal, a structural protein, and the enzyme that generates the signal. However, the question of how this heterochromatin formation is targeted to specific domains remains unanswered.

INTERSPERSED HETEROCHROMATIC AND EUCHROMATIC DOMAINS REPORTED BY *hsp70-white*

We have investigated fourth chromosome domain organization using a transposable P element P[*hsp26-pt, hsp70-w*] (Fig. 2) as a phenotypic probe. Earlier studies have shown that while this P element in a euchromatic domain bestows a full red eye phenotype on a *white* mutant line, transposition into pericentric or telomeric heterochromatin results in variegated expression of *white*, reduced nuclease accessibility in the *hsp26* regulatory region, and a shift to a more regularly spaced nucleosome array across the transgene (Wallrath and Elgin 1995; Cry-

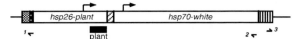

Figure 2. A map of the P[*hsp26-pt, hsp70-w*] construct drawn to scale (Wallrath and Elgin 1995). The *hsp26* sequences from –1917 to +490 were fused to a 740-bp fragment of barley cDNA (plant probe), followed by *hsp70* transcription termination sequences (*slashed box*). This fusion gene was cloned into the P element vector A412 possessing an *hsp70*-driven copy of the *white* gene. Transcription start sites for *hsp26-pt* and *hsp70-w* are marked by *bent arrows*. (*Left*) The 3´ P end (*dotted box*); (*right*) the 5´ P end (*striped box*). Primers used for inverse or direct PCR to identify flanking DNA (*small arrows below the diagram*) are *1* [5´-aactcgaggcctcgaggt-3´]; *2* [5´-gacgaaatgacc cactcgg-3´]; *3* [5´-gcttcggctatcgacgggaccacc-3´].

derman et al. 1999a; Sun et al. 2001). Insertion of this element at various sites in the fourth chromosome has identified a few euchromatic domains (resulting in a red eye) interspersed with heterochromatic domains (causing a variegating eye) (Sun et al. 2000). Insertion sites have now been precisely mapped onto the published *D. melanogaster* genome sequence (Flybase Consortium 2003) using genomic DNA sequences flanking the 5´ P element end, recovered by inverse PCR. The distribution of insertion sites in lines from this study resulting in a variegating or red eye phenotype relative to the positions of genes and transposable elements is shown in Figure 3. The pattern of reporter gene expression reveals two euchromatic domains in a largely heterochromatic chromosome. The present map is most likely not saturated for euchromatic domains; in practice, it is much easier to recover fourth chromosome P element insertion events from among variegating lines than from among red eye lines, as the bulk of the red eye lines represent insertion into the major euchromatic domains on the other chromosomes.

While one might anticipate that variegating inserts would be associated with gene-poor regions of the chromosome, potentially containing tandem arrays of repetitious elements, this proved not to be the case. In fact, most (eight of ten) of the variegating P elements shown lie within 2 kb of a gene, five lying within the transcribed portions of four different genes. In all cases tested, the variegating phenotype is suppressed (loss of silencing) by a mutation in HP1 (Sun et al. 2000). Recently we have recovered and analyzed an additional 11 lines carrying the P element on the fourth. The results support the conclusions discussed above, while identifying an additional euchromatic domain associated with the *sv* gene (Sun et al. 2004).

Taken together, the banded pattern of HP1 distribution and the position-dependent expression of the P element reporter inserted at different sites across the fourth chromosome strongly support a model of interspersed euchromatic and heterochromatic domains. Rather than finding the fourth chromosome genes restricted to euchromatic domains, the experimental results described above point to the conclusion that many fourth chromosome genes lie in heterochromatic domains, defined as regions inducing a variegating *white* phenotype. One can infer that a significant number of fourth chromosome genes are packaged with HP1. This is not without prece-

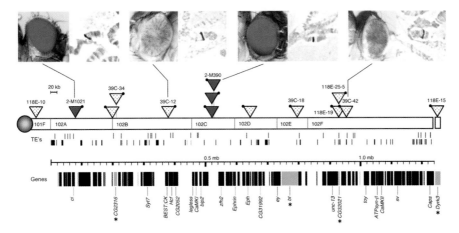

Figure 3. Chromosome 4 is made up of interspersed heterochromatic and euchromatic domains, as reported by *hsp70-white*. Characteristics of representative *P* insert lines are shown at the top: photographs of the eye phenotypes (*left*) are adjacent to photographs showing the cytological position of the *P* element mapped by in situ hybridization of third instar larvae polytene chromosomes (*right*) for a given line. The map of the fourth chromosome (regions 101F-102F) shows the positions of known and predicted genes, and of transposable elements (TEs) (Flybase Consortium 2003). *Solid triangles* and *dotted triangles* mark *P* element insertion sites that result in a solid red eye or induce a variegating phenotype, respectively. The black dot indicates the 5′ end of the *P* element. In cases where a line has been recovered with a *P* element inserted within a gene, the gene (in *green*) is indicated by an asterisk. Figure based on data from Sun et al. (2000, 2004).

dent. While much of the *Drosophila* heterochromatin at centromeres and telomeres is made up of tandem repeats, classical genetic analysis has identified several genes within the pericentric heterochromatin, and data from genome sequencing suggest that several hundred genes may reside in these regions in *D. melanogaster* (Hoskins et al. 2002). Many of the genes on the fourth have specific developmental functions. It will be of interest to determine how these genes function within a heterochromatic environment.

THE SPATIAL POSITION OF CHROMOSOME 4, CLOSE TO PERICENTRIC HETEROCHROMATIN, IS CRITICAL TO MAINTAIN SILENCING

Many lines of evidence from studies in yeast, flies, and mammals (for review, see Gasser 2001) demonstrate a link between gene silencing and nuclear location. We have used X-ray-induced translocation to show that the position of the *P* element reporter in the nucleus, in addition to its position in the genome, can impact chromatin packaging. Transgenes that map to the distal portion of the fourth chromosome, such as that present in line 118E-15 (see Fig. 3), show a variegating phenotype with similar responses to genetic modifiers as observed for the pericentric and other variegating fourth chromosome transgenes, including a loss of silencing on depletion of HP1 (Wallrath and Elgin 1995). By using X-ray mutagenesis we can generate translocations in line 118E-15 to address whether nuclear position influences silencing (Cryderman et al. 1999b). Translocations resulting in a displacement of the fourth further away from the centromeric heterochromatin (stocks X-2 and X-4) lead to a dramatic loss of silencing (Fig. 4). However, when a translocation places the fourth in a new locus proximal to the pericentric heterochromatin (stock X-10), silencing is maintained. Variegation in all these cases is suppressed by $Su(var)2-5^{02}$, a mutation in the gene encoding HP1. These results indicate that the local chromatin structure is impacted by changes in nuclear organization and suggest that the high concentration of HP1 in the pericentromeric heterochromatin may generate a local nuclear environment that facilitates heterochromatin formation and maintenance.

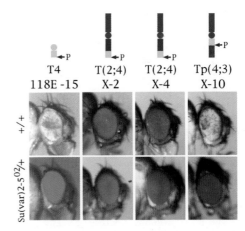

Figure 4. The spatial position of chromosome 4 is critical to maintain silencing. Diagrams of the relevant chromosomes (the fourth highlighted in *green*, second and third in *blue*) are shown at the top; a *red arrow* labeled "P" marks the *P* element insertion 118E-15 (see Fig. 3). Translocation of ~1 Mb of the distal portion of the fourth chromosome (carrying the *P* element close to its telomere) to the distal end of chromosome 2 [stocks labeled T(2;4)] causes a loss of silencing of the *white* reporter. Transposition of approximately the same fourth chromosome fragment to a proximal region of the third chromosome [stock Tp(4;3)] allows silencing. Variegation observed in these cases is suppressed by a mutation in the gene encoding HP1, $Su(var)2-5^{02}$. (Reprinted, with permission, from Cryderman et al. 1999b [©Nature Publishing Group] [http://www.nature.com/].)

HETEROCHROMATIC SILENCING, H3-K9 METHYLATION, AND HP1 LOCALIZATION IN *DROSOPHILA* ARE DEPENDENT UPON THE RNAi MACHINERY

The presence of euchromatic domains on chromosome 4 argues against the simple idea that proximity to pericentromeric heterochromatin is sufficient to induce formation of heterochromatin, and suggests that *cis*-acting DNA sequence elements are required to target heterochromatin formation in specific regions. Repetitious elements have been implicated as targeted sites for heterochromatin formation in fungi and plants via an RNAi mechanism (Volpe et al. 2002; Hall et al. 2002; for reviews, see Matzke et al. 2004; Schramke and Allshire 2004). Region 101F-102F of chromosome 4 is enriched in repetitious sequences (see Fig. 3) compared to similar intervals on the other euchromatic chromosome arms (Kaminker et al. 2002). The average transposable element density is 10–15/Mb in the major chromosome arms, but <82/Mb for chromosome 4, because of an order-of-magnitude increase in remnants of LINE-like and TIR elements (elements that transpose via a DNA intermediate, flanked by short inverted repeats). A recent effort to generate a developmental profile of small RNAs from *D. melanogaster* has yielded a cloned and characterized nonredundant collection of 62 miRNAs and 178 repeat-associated small interfering RNAs (rasiRNAs). rasiRNAs were recovered from 38 different transposable elements (corresponding to 40% of the known elements), as well as from satellite DNA and the subterminal minisatellite at the 2L telomere. Of these, 15 are represented in the 1.2-Mb banded portion of chromosome 4. The rasiRNAs, 16–28 nucleotides, were most abundant in testis and early embryo, the latter suggesting a role in establishing heterochromatin structure (Aravin et al. 2003).

To examine whether heterochromatin assembly in *Drosophila* might be targeted by siRNA, we tested the potential of available mutations in RNAi components *piwi*, *aubergine*, and *homeless* to impact heterochromatic silencing using two test systems, tandem mini-*white* arrays and *white* transgenes in either pericentric heterochromatin or a heterochromatin domain on the fourth chromosome. In both test systems, the *white* reporter gene shows a variegating phenotype that is suppressed (loss of silencing) in the presence of mutations in HP1 and exhibits other characteristics typical of heterochromatin-induced silencing. Both systems exhibited a loss of silencing (suppression of PEV) in the presence of mutations in the RNAi components (Fig. 5). The loss of silencing is accompanied by a decrease in the amount of H3-mK9 (in *homeless* mutant flies, to ca. 25% that observed without the mutation) and by a redistribution of HP1 to sites within the euchromatic arms (Pal-Bhadra et al. 2004).

DOES RNAi TARGET HISTONE MODIFICATION OR HP1 INTERACTIONS?

A major question that remains is the connecting link between RNAi and heterochromatin formation in

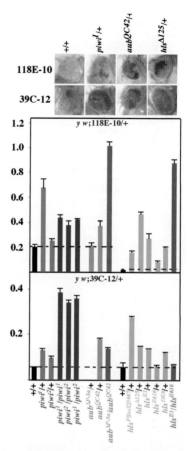

Figure 5. Loss of silencing is associated with mutations in components of the RNAi system. Homozygous or heteroallelic mutations in *piwi* result in an increase in *white* gene expression (loss of silencing) in line 118E-10 (*white* transgene in pericentric heterochromatin). The heteroallelic mutant combination of *aubergine* produces strong suppression. The *homeless* mutations have a dominant phenotype with a several-fold increase in expression, depending on the allele. Similar, but less dramatic results were obtained using stock 39C-12 (transgene in the fourth chromosome). Eye phenotypes are shown in photographs above. Pigment values (below) are given relative to the control $y\ w^{67c23}$ stock carrying the respective transgene. (Adapted, with permission, from Pal-Bhadra et al. 2004 [©AAAS].)

Drosophila. The recent characterization of the RITS complex (RNA-induced initiation of transcriptional gene silencing) in *S. pombe* provides some insight into this question. RITS is composed of siRNA (22–25 nucleotides); Chp1, a chromodomain protein that binds to centromeric repeats; Ago1, the fission yeast Argonaute homolog; and Tas3, a novel protein (Verdel et al. 2004). Gene silencing, H3-K9 methylation, and Swi6 (HP1) localization are dependent upon the function of all three RITS components.

Is there a similar targeting complex in *Drosophila*? A speculative model for targeting heterochromatin formation is presented in Figure 6. The model is based on the reported work in *S. pombe*, our studies in flies, work by others describing the components of the RNAi system (particularly the RISC complex) in *Drosophila* (for re-

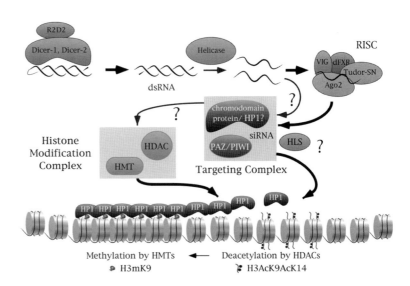

Figure 6. A speculative model of targeted heterochromatin formation. Long dsRNA is processed by the Dicer/R2D2 complex. The resulting siRNA (probably in a single-stranded form) may be complexed with RISC or another loading complex. We postulate a targeting complex that includes an siRNA from repetitious DNA, a chromo-domain protein (arguably HP1), and a PAZ/PIWI group member (possibly AUB or PIWI); HLS plays a critical role, but its placement is uncertain. At the heterochromatin initiation site, the targeting complex presumably interacts with a histone-modification complex, likely including HDAC1/ SU(VAR)3-9. We hypothesize that siRNA-dependent targeting of HP1 is the primary event, followed by histone modification to stabilize and maintain HP1 association (*thick arrows*). Alternative pathways are indicated by *thin arrows*.

view, see Hannon 2002), and work describing histone-modification enzymes and heterochromatin complexes in *Drosophila* (for review, see Grewal and Elgin 2002). The model points to a number of important questions that need to be resolved. Is the RISC complex, known to be critical for posttranscriptional gene silencing, a required component for transcriptional gene silencing? What role does the *homeless* product play? Can one identify a "targeting complex" similar to RITS? While *Drosophila* has several members of the PAZ/PIWI family of proteins (AGO1, AGO2, PIWI, AUB), there are no homologs for Chp1 (aside from HP1) or Tas3. *homeless* activity is critical for proper localization of HP1 (Pal-Bhadra et al. 2004) and encodes a protein containing structural motifs that suggest the capacity to bind nucleic acids, to unwind double helices, and to interact with methylated peptides (Flybase Consortium 2003; Maurer-Stroh et al. 2003). We propose that HLS plays a critical role in the processing of siRNAs involved in transcriptional gene silencing, perhaps facilitating their interaction with RISC or perhaps participating directly in a targeting complex.

What sort of target might be present on the fourth chromosome? As noted above, the fourth chromosome is particularly rich in repetitious sequences, mostly fragments of transposable elements. We have used a genetic approach to search for *cis*-acting DNA sequences that specify heterochromatic domains. Genetic screens for a switch in phenotype on mobilization of the *P* element reporter described above have demonstrated that local deletions or duplications of 5–80 kb of DNA flanking the transposon reporter can lead to loss or acquisition of variegation, pointing to short-range *cis*-acting determinants of silencing. We have mapped the distance from the *P* element to the surrounding repetitious DNA fragments for a series of lines derived from 39C-12 (see Fig. 3), inserted close to the *Hcf* gene in region 102B. The results suggest that if the *P* element with *hsp70-w* lies within 10 kb of a fragment of the *1360* repetitious element, it will have a variegating phenotype, indicating heterochromatic packaging, whereas if it lies farther away, it will have a uniform red eye, indicating euchromatic packaging (Sun et al. 2004). This suggests that *1360*, and perhaps other repetitious elements present in *D. melanogaster*, could serve as the source of the siRNA in the putative Targeting Complex. Once targeted, heterochromatin formation apparently spreads over a distance of about 10 kb, a result similar to that found in *S. pombe* at the mating type locus (Hall et al. 2002). The results to date do not identify any discrete boundaries to the spreading of HP1-dependent silencing in this case, but rather suggest a competition based on alternative histone modifications to establish the mode of chromatin packaging (Sun et al. 2004).

Once a locus is targeted for silencing via the RNAi pathway, how is heterochromatin packaging nucleated? Localization of HP1 and the histone-modification machinery appears interdependent in many systems, but "who's on first" may differ. Genetic analysis in *S. pombe* has indicated that H3-K9 methylation precedes localization of Swi6, the HP1 homolog (for review, see Grewal and Elgin 2002). However, work by Schotta et al. (2002) demonstrates that the normal restriction of SU(VAR)3-9 (the major H3-K9 methyl transferase) and H3-mK9 to heterochromatic domains is dependent upon HP1 in *Drosophila*. Loss of HP1 results in mislocalization (presence throughout the euchromatic arms) of SU(VAR)3-9; loss of SU(VAR)3-9 results in a loss of HP1 in the pericentric heterochromatin, but a general delocalization is not observed. One can suggest that siRNA targeting of HP1 may be the initial event in heterochromatin formation, although histone modification is likely to be essential to generate a stable chromatin state. Both alternatives (HP1 on first vs. H3-K9 methylation first) appear plausible, and are indicated as possible routes to heterochromatin formation in Figure 6. The exciting challenge for the next several years is to determine which pathways are the correct ones and which players have lead roles.

ACKNOWLEDGMENTS

We thank Michael Grunstein (UCLA) for making antibodies directed against modified histones available, and Gabriella Farkas (Washington University) for assistance in the preparation of figures. This work was supported by a grant from NIGMS, NIH, to S.C.R.E. (GM068388).

REFERENCES

Aravin A.A., Lagos-Quintana M., Yalcin A., Zavolan M., Marks D., Snyder B., Gaasterland T., Meyer J., and Tuschl T. 2003. The small RNA profile during *Drosophila melanogaster* development. *Dev. Cell* **5**: 337.

Barigozzi C., Dolfini S., Fracacaro M., Rezzonico-Raimondi G., and Tiepolo L. 1966. In vitro study of the DNA replication patterns of somatic chromosomes of *Drosophila melanogaster*. *Exp. Cell Res.* **43**: 231.

Berger S. 2002. Histone modifications in transcriptional regulation. *Curr. Opin. Genet. Dev.* **12**: 142.

Bridges C.B. 1935. The mutants and linkage data of chromosome four of *Drosophila melanogaster*. *Biol. Zh.* **4**: 401.

Crydernan D.E., Tang H., Gilmour D.S., and Wallrath L.L. 1999a. Heterochromatic silencing of *Drosophila* heat shock genes acts at the level of promoter potentiation. *Nucleic Acids Res.* **27**: 3364.

Cryderman D.E., Morris E.J., Biessmann H., Elgin S.C., and Wallrath L.L. 1999b. Silencing at *Drosophila* telomeres: Nuclear organization and chromatin structure play critical roles. *EMBO J.* **18**: 3724.

Eissenberg J.C. and Elgin S.C.R. 2000. The HP1 protein family: Getting a grip on chromatin. *Curr. Opin. Genet. Dev.* **10**: 204.

Eissenberg J.C., James T.C., Foster-Hartnett D.M., Hartnett T., Ngan V., and Elgin S.C.R. 1990. A mutation in a heterochromatin-specific chromosomal protein is associated with suppression of position effect variegation in *Drosophila melanogaster*. *Proc. Natl. Acad. Sci.* **87**: 9923.

FlyBase Consortium. 2003. The FlyBase database of the *Drosophila* genome projects and community literature (http://flybase.org/[release 3]). *Nucleic Acids Res.* **31**: 172.

Gasser S.M. 2001. Positions of potential: Nuclear organization and gene expression. *Cell* **104**: 639.

Grewal S. and Elgin S.C.R. 2002. Heterochromatin: New possibilities for the inheritance of structure. *Curr. Opin. Genet. Dev.* **12**: 178.

Hall I.H., Shankaranarayana G.D., Noma K., Ayoub N., Cohen A., and Grewal S.I. 2002. Establishment and maintenance of a heterochromatin domain. *Science* **297**: 2232.

Hannon G. 2002. RNA interference. *Nature* **418**: 244.

Hoskins R.A., Smith C.D., Carlson J.W., Carvalho A.B., Halpern A., Kaminker J.S., Kennedy C., Mungall C.J., Sullivan B.A., Sutton G.G., Yasuhara J.C., Wakimoto B.T., Myers E.W., Celniker S.E., Rubin G.M., and Karpen G.H. 2002. Heterochromatic sequences in a *Drosophila* whole-genome shotgun assembly. *Genome Biol.* **3**: RESEARCH0085.

James T.C., Eissenberg J.C., Craig C., Dietrich V., Hobson A., and Elgin S.C.R. 1989. Distribution patterns of HP1, a heterochromatin-associated nonhistone chromosomal protein of *Drosophila*. *Eur. J. Cell Biol.* **50**: 170.

Kaminker J., Bergman V., Kronmileer B., Carlson J., Sviskas R., Patel S., Frise E., Wheeler D.A., Lewis S., Rubin G.M., Ashburner M., and Celniker S. 2002. The transposable elements of the *Drosophila melanogaster* euchromatin: A genomic perspective (http://genomebiology.com/2002/3/12/research/0084). *Genome Biology* **3**: RESEARCH0084.

Matzke M., Aufsatz W., Kanno T., Daxinger L., Papp I., Mette M.F., and Matzke A.J.M. 2004. Genetic analysis of RNA-mediated transcriptional gene silencing. *Biochim. Biophys. Acta* **1677**: 129.

Maurer-Stroh S., Dickens N.J., Hughes-Davies L., Kouzarides T., Eisenhaber F., Ponting C.P. 2003. The Tudor domain 'Royal Family': Tudor, plant Agenet, Chromo, PWWP and MBT domains. *Trends Biochem. Sci.* **28**: 69.

Pal-Bhadra M., Leibovitch B.A., Gandhi S.G., Rao M., Bhadra U., Birchler J.A., and Elgin S.C.R. 2004. Heterochromatic silencing and HP1 localization in *Drosophila* are dependent on the RNAi machinery. *Science* **303**: 669.

Schotta G., Ebert A., Krauss V., Fischer A., Hoffmann J., Rea S., Jenuwein T., Dorn R., and Reuter G. 2002. Central role of *Drosophila* SU(VAR)3-9 in histone H3-K9 methylation and heterochromatic gene silencing. *EMBO J.* **21**: 1121.

Schramke V. and Allshire R. 2004. Those interfering little RNAs! Silencing and eliminating chromatin. *Curr. Opin. Genet. Dev.* **14**: 174.

Suka N., Suka Y., Carmen A.A., Wu J., and Grunstein M. 2001. Highly specific antibodies determine histone acetylation site usage in yeast heterochromatin and euchromatin. *Mol. Cell* **8**: 473.

Sun F.-L., Cuaycong M.H., and Elgin S.C.R. 2001. Long-range nucleosome ordering is associated with gene silencing in *Drosophila melanogaster* pericentric heterochromatin. *Mol. Cell. Biol.* **21**: 2867.

Sun F.-L., Cuaycong M.H., Craig C.A., Wallrath L.L., Locke J., and Elgin S.C.R. 2000. The fourth chromosome of *Drosophila melanogaster*: Interspersed euchromatic and heterochromatic domains. *Proc. Natl. Acad. Sci.* **97**: 5340.

Sun F.-L., Haynes K., Simpson C.L., Lee S.D., Collins L., Wuller J., Eissenberg J.C., and Elgin S.C.R. 2004. Cis-acting determinants of heterochromatin formation on *Drosophila melanogaster* chromosome four. *Mol. Cell. Biol.* **24**: 18.

Verdel A., Jia S., Gerber S., Sugiyama T., Gygi S., Grewal S.I., and Moazed D. 2004. RNAi-mediated targeting of heterochromatin by the RITS complex. *Science* **303**: 672.

Volpe T.A., Kidner C., Hall I.M., Teng G., Grewal S.I., and Martienssen R.A. 2002. Regulation of heterochromatic silencing and histone H3 lysine-9 methylation by RNAi. *Science* **297**: 1833.

Wallrath L.L. and Elgin S.C.R. 1995. Position effect variegation in *Drosophila* is associated with an altered chromatin structure. *Genes Dev.* **9**: 1263.

Two Distinct Nucleosome Assembly Pathways: Dependent or Independent of DNA Synthesis Promoted by Histone H3.1 and H3.3 Complexes

Y. Nakatani,* D. Ray-Gallet,[†] J.-P. Quivy,[†] H. Tagami,*[‡] and G. Almouzni[†]

*Dana-Farber Cancer Institute and Harvard Medical School, Boston, Massachusetts 02115;
[†]Section de Recherche, Institut Curie, UMR 218 du CNRS, 75248 Paris Cedex 05, France

Covalent modifications of histones, such as acetylation, phosphorylation, and methylation, have been shown to contribute to the formation and maintenance of transcriptionally active and inactive chromatin (Jenuwein and Allis 2001; Grewal and Elgin 2002; Turner 2002; Felsenfeld and Groudine 2003; Kurdistani and Grunstein 2003). In addition to histone modifications, histone variants that mark specific chromatin loci could play important roles in formation and maintenance of epigenetic memory (Henikoff et al. 2004). In mammals, three isotypes of histone H3 (H3.1, H3.2, and H3.3) have been identified in addition to the centromere-specific histone H3 variant, CENP-A. Histone H3.1 is the major histone H3, which is predominantly synthesized and assembled in nucleosomes during S phase. Histone H3.2 is closely related to H3.1 and has been shown to belong to the family of S-phase subtypes (Franklin and Zweidler 1977). In contrast to H3.1 and H3.2, H3.3 is expressed in proliferating cells at all stages of the cell cycle as well as in quiescent cells (Wu et al. 1982). This latter observation is consistent with the fact that H3.3 can be incorporated into nucleosomes in the absence of DNA synthesis. Given these properties, H3.3-H4 has been suggested to replace H3-H4 in nucleosomes, in particular during transcriptional activation (Ahmad and Henikoff 2002; Henikoff et al. 2004). Although, we are only at an early stage in our understanding of the functional role of these histone variants, it is clear that their mode of incorporation represents a crucial event with major implications for cell fate and stability of expression programs. In this context, it is remarkable to realize that while the principles for a mechanism to ensure the faithful propagation of genetic information through DNA replication were discovered a half-century ago (Watson and Crick 1953), the molecular events leading to the maintenance and transmission of epigenetic information remain a real puzzle. Much of our knowledge concerning chromatin assembly and nucleosome formation has dealt so far mainly with bulk histones (Verreault 2000; Mello and Almouzni 2001), and the issue of how to deal with histone variants, in particular H3, has remained an enigma. To gain insights into how histones H3.1 and H3.3 are assembled into nucleosomes, we decided to isolate and characterize the protein complexes containing predeposition forms of these histones. Significantly, we found that histone H3.1 associates with the DNA synthesis–dependent chaperone CAF-1, while histone H3.3 associates with the DNA synthesis–independent chaperone HIRA (Tagami et al. 2004). Consistent with these observations, the H3.1 and H3.3 complexes mediate nucleosome assembly in a DNA synthesis–dependent and –independent manner, respectively, in vitro. Most strikingly, our data point to the existence of histones H3 and H4 in both predeposition complexes as a dimer (Tagami et al. 2004). In the present paper, we will summarize these recent findings to highlight the potential of the experimental strategies we developed. We will then discuss how we can envisage the relevance of a H3-H4 dimer, rather than a tetramer, as a basic "building unit" for chromatin dynamics. Finally, possible implications of our findings for mechanisms involved in nucleosome assembly and epigenetic inheritance will be considered to help define some promising new avenues and to stimulate future developments in the field.

HISTONE DEPOSITION COMPLEXES

To explore the mechanisms of how histones H3.1 and H3.3 are assembled into chromatin, we purified preassembled histones H3.1 and H3.3 from the nuclear extract fraction of HeLa cells expressing epitope-tagged H3.1 or H3.3, respectively (Tagami et al. 2004). For immuno-affinity purification (Nakatani and Ogryzko 2003), we have fused the FLAG (DYKDDDDK) and HA (YPYDVPDYA) epitope sequences to the carboxy-terminal end of histones H3.1 (e-H3.1) or H3.3 (e-H3.3). To optimize accessibility to the epitope tag, and to avoid major interference with normal functions of histones H3.1 and H3.3, we inserted a "flexible" linker, AAAGG, between the carboxy-terminal end of histones H3.1 or H3.3 and the epitope tag. Immunofluorescent staining of the mitotic cells expressing e-H3.1 or e-H3.3 reveals that most of the tagged histones colocalize with chromosomes, indicating that the epitope-tagged H3.1 and H3.3 could be effectively assembled into nucleosomes in vivo.

[‡]Present address: Division of Biological Science, Graduate School of Science, Nagoya University, Chikusa, Nagoya, Aichi 464-8602, Japan.

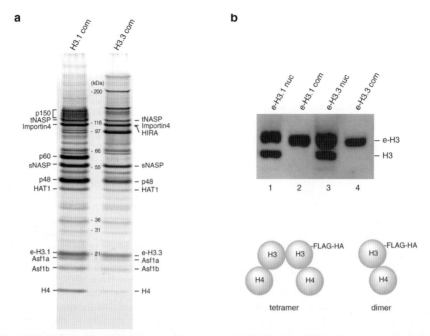

Figure 1. (*a*) Compositions of the H3.1 and H3.3 complexes. Silver staining of the H3.1 and H3.3 complexes (H3.1 com and H3.3 com). The polypeptides identified by mass spectrometric analyses are indicated. (*b*) Histones H3 and H4 in the H3.1 and H3.3 complexes exist as a dimer. (*Top*) Immunoblotting of mononucleosomes and nucleosome assembly complexes containing epitope-tagged histone H3.1 and H3.3. e-H3.1-containing (lane *1*) and e-H3.3-containing (lane *3*) mononucleosomes and the H3.1 (lane *2*) and H3.3 (lane *4*) complexes were analyzed by immunoblotting with anti-histone H3 antibody. The positions for epitope-tagged histone H3 (e-H3.1 or e-H3.3) and native H3 are indicated. (*Bottom*) Given that untagged H3 is far more abundant than epitope-tagged H3 in the cell lines employed, the epitope-tagged H3 could preferentially pair with untagged H3 to form the H3-H4 tetramer. Accordingly, mononucleosomes contain epitope-tagged and untagged H3 (lanes *1* and *3*). In contrast, the H3.1 and H3.3 complexes have no detectable untagged H3, indicating that histones H3 and H4 in these complexes exist as a dimer. (Adapted, with permission, from Tagami et al. 2004 [©Cell Press].)

The H3.1 complex resolution is shown in Figure 1a as revealed after analysis by PAGE and silver staining. Table 1 summarizes the subunit composition of the different complexes that we characterized, with a list of the polypeptides identified so far. The H3.1 complex contains two distinct histone chaperones for H3-H4; the entire set of CAF-1 subunits (p150, p60, and p48) (Smith and Stillman 1989) and the two isoforms of Asf1 that exist in human cells (Asf1a and b) (Munakata et al. 2000; Sillje and Nigg 2001; Tyler et al. 2001; Mello et al. 2002).

Table 1. Nucleosomal and Nonnucleosomal Complexes Containing Epitope-Tagged H3

Epitope-tagged histone e-H3	Histones associated with nucleosomal e-H3 (1)	Proteins associated with nonnucleosomal e-H3 (2)		
e-H3.1	H3.1 H2A H2B H4	p150 p60 p48	} CAF-1	Smith and Stillman (1989); Gaillard et al. (1996)
		tNASP, sNASP importin 4 HAT1 Asf1a, Asf1b H4 and others		Richardson et al. (2000); Alekseev et al. (2003) Jakel et al. (2002) Verreault (2000) Sillje and Nigg (2001)
e-H3.3	H3.3 H2A H2B H4	HIRA Cabin 1 Ubinuclein p48 tNASP, sNASP importin 4 HAT1 Asf1a, Asf1b H4 and others		Lamour et al. (1995); Ray-Gallet et al. (2002) Sun et al. (1998) Aho et al. (2000) Ridgway and Almouzni (2000) Richardson et al. (2000); Alekseev et al. (2003) Jakel et al. (2002) Verreault (2000) Sillje and Nigg (2001)

Affinity purification was carried out using (1) mononucleosomes isolated after glycerol fractionation of MNase digestion products, and (2) nuclear extracts prepared according to Dignam et al. (1983).

Moreover, the complex contains NASP, which is a candidate for histone H1 chaperone (Richardson et al. 2000; Alekseev et al. 2003). In addition to these histone-binding proteins, the complex contains the histone-acetylating enzyme HAT1, which can modify predeposited histones H3 and H4 (Verreault 2000), as well as importin 4, which presumably mediates nuclear translocation (Jakel et al. 2002). Glycerol gradient sedimentation of the purified complex suggests that most histone H3.1 belongs to a major complex.

We also purified preassembled histone H3.3 from the nuclear extract and the subunit composition of the H3.3 complex is shown in Figure 1a and Table 1. Glycerol gradient sedimentation profiles of the purified complex show that H3.3 belongs to at least three subcomplexes. Importantly, among these three complexes, one of them contains HIRA, a DNA synthesis–independent histone chaperone (Lamour et al. 1995; Ray-Gallet et al. 2002). This important feature was a first demonstration of a link between histone H3.3 and HIRA. Western blotting analysis confirmed that HIRA is specific to the H3.3 complex, while CAF-1 is specific to the H3.1 complex. As described later, this difference could explain why H3.1 and H3.3 are deposited into nucleosomes through distinct pathways. Moreover, Cabin 1 (Sun et al. 1998) and ubinuclein (Aho et al. 2000) are found in the HIRA-containing complex. Subunits in common with the H3.1 complex were Asf1, importin 4, HAT1, and p48 (Ridgway and Almouzni 2000), which suggests that they could play related roles in both complexes.

HISTONE H3-H4 EXISTS AS A DIMER IN THE NUCLEOSOME ASSEMBLY COMPLEXES

Although histones H3 and H4 are known to form tetramers under physiological conditions, several lines of evidence support the fact that the dimeric form of histones H3 and H4 could exist within predeposition complexes associated with chaperones. Western blotting analysis of the H3.1 and H3.3 complexes with anti-histone H3 antibody shows that these complexes contain epitope-tagged histone H3 (e-H3.1 or e-H3.3), but not untagged H3 (Fig. 1b, lanes 2 and 4). Yet, we should consider that the epitope-tagged histones are far less abundant than the endogenous histones in our cell lines. Thus, as illustrated in Figure 1b, if histone H3-H4 were to exist as a tetramer in a predeposition form, statistically one would expect the complex to contain both epitope-tagged and untagged H3, which is not the case. In contrast, mononucleosome core particles purified by anti-FLAG antibody immunoprecipitation contain both epitope-tagged and untagged H3 (Fig. 1b, lanes 1 and 3). These data lead us to conclude that histones H3 and H4 in the predepostion nucleosome assembly complexes are present as dimers, rather than tetramers. The existence of H3-H4 dimers as intermediates in chromatin dynamics will be discussed later. It is also important to note that, within the mixed population as tagged and untagged versions of histone H3 in our mononucleosomes, the variants involved were identical, as shown by the mass spectrometry analysis. It is still unclear how some kind of "recognition" can bring together the same kind of variants within the same particle. However, this can provide interesting information concerning how the differences in sequences between H3.1 and H3.3 may have implications for their association as tetramers.

HIRA- AND CAF-1-DEPENDENT ASSEMBLY PATHWAYS

The H3.1 and H3.3 complexes isolated from nuclear extracts as shown above contain distinct histone chaperones. Most remarkably, all three subunits of CAF-1—p150, p60, and p48—are associated with H3.1, consistent with deposition coupled to DNA replication as expected, while HIRA is associated with the H3.3 complex (Fig. 1 and Table 1). Indeed, these two chaperones have been shown to promote distinct nucleosome assembly pathways in higher eukaryotes. HIRA is critical for a DNA synthesis–independent nucleosome assembly process (Ray-Gallet et al. 2002), while CAF-1 mediates nucleosome formation coupled to DNA synthesis during DNA replication (Smith and Stillman 1989) or DNA repair (Gaillard et al. 1996). To test the capacity of our complexes in each of these assembly pathways, the potent nucleosome assembly capacity of *Xenopus* egg extract system (HSE) is a convenient system (Ray-Gallet and Almouzni 2004). Pioneering work using *Xenopus* egg extracts (Laskey et al. 1977) provided the basis for analyzing different assembly pathways by immunodepleting HIRA or p150 CAF-1 from HSE, as shown in Figure 2a.

HIRA-depleted HSE supports only nucleosome assembly coupled to DNA synthesis using UV-treated plasmid DNA (DNA UV), to follow nucleosome assembly in the presence of DNA synthesis during nucleotide excision repair, whereas CAF-1-depleted HSE supports only nucleosome formation in the absence of DNA synthesis using intact plasmid DNA (DNA 0) (Fig. 2b). With such a system in hand, we could test our different purified H3.1 and H3.3 complexes (and subcomplexes) for their ability to rescue the defect in nucleosome assembly of the depleted extracts. Results from these data showed that the histone H3.1 and H3.3 complexes mediate DNA synthesis–dependent and DNA synthesis–independent nucleosome assembly, respectively. Although the specificity of each pathway could be dictated by the two chaperones—CAF-1 (for H3.1) and HIRA (for H3.3), respectively (Loyola and Almouzni 2004)—the basis of the specific interaction of the distinct deposition complexes with H3.1 and H3.3 is still unclear. Given that specific interactions of HIRA with H3.3 and CAF-1 with H3.1 could not be revealed in assays using recombinant proteins (not shown), formation of specific complexes is likely to involve additional aspects, in vivo, which will have to be understood in future studies. Furthermore, CAF-1 and HIRA are not necessarily found only in histone-containing complexes. This has been recently illustrated by the fact that CAF-1 was found in a complex containing HP1α, which was devoid of histone H3 (Quivy et al. 2004). In addition, the function of HIRA has been associated with repression of histone transcription (Nelson et al. 2002), as found initially in *Saccharomyces cerevisiae* for Hir1 and Hir2 (Sherwood et al. 1993). Whether this is related to the function of the com-

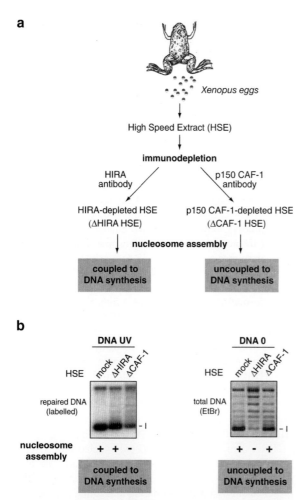

Figure 2. Two distinct chromatin assembly pathways: one coupled to DNA synthesis and CAF-1 mediated, and another uncoupled to DNA synthesis and HIRA dependent. (*a*) Immunodepletion strategy used to monitor these two distinct chromatin pathways individually. *Xenopus* egg extract (HSE), supporting both pathways, is depleted either of HIRA or of the p150 subunit of CAF-1 by using specific antibodies. HIRA-depleted HSE (ΔHIRA HSE) supports only chromatin assembly coupled to DNA synthesis, whereas p150 CAF-1–depleted HSE (ΔCAF-1 HSE) supports only chromatin assembly uncoupled to DNA synthesis. (*b*) Analysis of the nucleosome assembly activities coupled and uncoupled to DNA synthesis in ΔHIRA and ΔCAF-1 HSEs. Mock, ΔHIRA, or ΔCAF-1 HSEs were used in nucleosome assembly assays that were carried with either UV-C-treated plasmid (DNA UV) in the presence of α-^{32}P(dCTP) to follow the nucleosome assembly pathway coupled to DNA synthesis or nonirradiated plasmid (DNA 0) to follow the DNA synthesis–independent pathway. The repaired and nonrepaired assembled DNA were analyzed by supercoiling assay and visualized by autoradiography and by ethidium bromide staining, respectively. Migration position of DNA plasmid form I (supercoiled) corresponding to the assembled DNA is indicated. (Adapted, with permission, from Ray-Gallet and Almouzni 2004 [©Elsevier].)

plex that we have studied here or to another one remains to be elucidated. Thus, the functions of CAF-1 and HIRA in histone deposition are thus likely to be important in multiple contexts. In any case, distinct pathways can be promoted by the distinct histone-containing complexes in which we found, respectively, CAF-1 and HIRA.

HISTONES H3.1 AND H3.3 ARE DEPOSITED THROUGH DISTINCT NUCLEOSOMAL ASSEMBLY PATHWAYS

Having shown that nuclear complexes containing histones H3.1 and H3.3 could promote distinct nucleosomal assembly pathways, it was important to determine if histones H3.1 and H3.3 are indeed specifically deposited onto DNA through such distinct pathways. Cytosolic extracts from HeLa cells expressing e-H3.1 or e-H3.3, which contain no detectable CAF-1 or HIRA, provided us with a source of free-tagged histones for nucleosomal assembly reactions. We then used a DNA linked to bead assay, as shown in Figure 3a, to analyze histone deposition "de novo," i.e., to test nucleosome formation on a template free of histone. To examine DNA synthesis–dependent nucleosomal assembly, UV-damaged DNA immobilized onto magnetic beads was incubated with HIRA-depleted HSE in the presence of cytosolic extracts containing e-H3.1 or e-H3.3. In these HIRA-depleted extracts, which are still functional for the CAF-1-dependent assembly pathway, histones bound to immobilized DNA could be analyzed by immunoblotting. e-H3.1 was indeed recovered onto immobilized UV-damaged DNA in a CAF-1 (from the HSE)-dependent manner (Fig. 3b, left, lanes 2 and 4). In contrast, no e-H3.3 was detected on the immobilized DNA (Fig. 3b, left, lane 8). Thus, histone H3.1, but not H3.3, is utilized by the DNA synthesis–dependent nucleosome assembly pathway.

In parallel, we also examined the deposition of histones H3.1 and H3.3 in a DNA synthesis–independent nucleosomal assembly pathway. This time, we used immobilized intact DNA as a template and a CAF-1 p150–depleted HSE in which only the HIRA-dependent pathway is functional. In contrast to the DNA synthesis–dependent deposition, e-H3.3, but not e-H3.1, was recovered onto immobilized intact DNA (Fig. 3b, right, lanes 4 and 8), indicating that histone H3.3 is specifically deposited in DNA synthesis–independent nucleosome assembly.

We conclude that in this system it is possible to reveal a specific loading of histone H3.1 and H3.3 onto DNA that occurs using DNA synthesis–dependent and –independent "de novo" nucleosome assembly pathways, respectively (Fig. 3c). In such a scheme one would predict H3.1 loading during DNA replication in S phase or during repair synthesis. H3.3 loading would occur throughout the cell cycle. Given that we have not been able to detect a loading of H3.3 coupled to DNA synthesis in vitro, a postreplicative mode of deposition could perhaps be considered for H3.3 during S phase.

"DE NOVO" DEPOSITION THROUGH TWO DISTINCT PATHWAYS USING DIMERS

Importantly, the purification scheme of the distinct H3 (H3.1 or H3.3) complexes suggested that they were present in dimeric form together with histone H4 in the nuclear extract. A simple way to explain the mixed population of H3 (tagged and nontagged) produced in mononucleosomes isolated from our cells could be according to the model in Figure 4 with two nonexclusive options. In the first case, two dimers of one type are

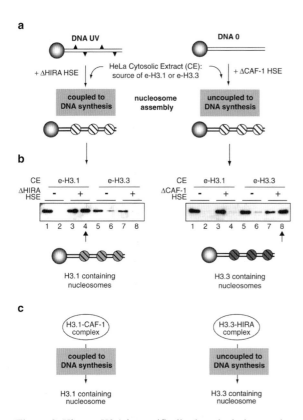

Figure 3. Histone H3.1 is specifically deposited via a nucleosome assembly pathway coupled to DNA synthesis, whereas histone H3.3 is specifically deposited via a nucleosome assembly pathway independent of DNA synthesis. (*a*) Scheme of the experiment: Magnetic bead–linked UV-irradiated DNA (DNA UV) or intact DNA (DNA 0) were incubated with cytosolic extracts from HeLa cells expressing either e-H3.1 or e-H3.3 complemented with ΔHIRA HSE for the UV-irradiated DNA or ΔCAF-1 HSE for the intact DNA. The nucleosome assembly pathway coupled to DNA synthesis supported by the ΔHIRA HSE extract occurs during the DNA repair synthesis of the UV-irradiated DNA. The nucleosome assembly pathway uncoupled to DNA synthesis supported by the ΔCAF-1 HSE occurs on the intact DNA. (*b*) Analysis by Western blot of the histones H3.1 and H3.3 deposition during the nucleosome assembly pathways either coupled or uncoupled to DNA synthesis. Bead-DNA was incubated with cytosolic extract containing e-H3.1 (lanes *1–4*) or e-H3.3 (lanes *5–8*) in the absence (lanes *1*, *2*, *5*, and *6*) or presence (lanes *3*, *4*, *7*, and *8*) of immunodepleted HSE. Histones H3.1 and H3.3 were detected by anti-HA antibodies as 10% input of the reaction (*odd* lanes) and bound material (*even* lanes). (*c*) Hypothetical nucleosome assembly line where the H3.1-H4 dimer specifically associated with CAF-1 helps to form H3.1-containing nucleosome by a DNA synthesis–dependent nucleosome assembly pathway. The H3.3-H4 dimer specifically associated with HIRA helps to form H3.3-containing nucleosome by a DNA synthesis–independent nucleosome assembly pathway. (Adapted, with permission, from Tagami et al. 2004 [©Cell Press].)

loaded in a coordinated manner by two identical chaperone complexes that play an equivalent role in the specific loading reaction (Fig. 4a). Alternatively, one dimer could be targeted onto DNA through the specific dedicated chaperone (complex A) and another complex (complex B) would provide the second dimer (the origin of which is discussed later) (Fig. 4b). In the latter scenario the specificity for one pathway or another would be dictated by a component in complex A, such as CAF-1 for a pathway coupled to DNA synthesis to promote H3.1 loading, or HIRA for the pathway independent of DNA synthesis to promote H3.3 deposition. Obviously some cross talk between complexes A and B would be needed, which could perhaps be achieved by a shuttling between the complexes of certain subunits. An attractive candidate for such a histone-shuttling activity could be Asf1. Present in both cytosolic and nuclear fractions in HeLa cells (Mello et al. 2002), Asf1 could potentially act both as a histone acceptor, as recently proposed in yeast (Adkins et al. 2004), and as histone donor in an assembly line (Koundrioukoff et al. 2004; Loyola and Almouzni 2004). This would be consistent with the fact that both the H3.1 and H3.3 complexes possess this additional histone H3-H4 chaperone, Asf1 (Fig. 1 and Table 1). Given that both CAF-1 and Asf1 have been independently shown to interact with H3-H4, an additional H3-H4 dimer could potentially be used to provide the two H3-H4 dimers necessary to form a tetramer. A similar mechanism could be at work with the H3.3 complex.

Thus, a "de novo" assembly pathway as followed in our in vitro system using naked DNA as a substrate may also be used in vivo in situations where DNA becomes available in a "histone-free" form, subsequent to some disruptive event, such as is found during DNA replication, transcription, repair, or recombination.

H3-H4 DIMERS AS INTERMEDIATES DURING CELLULAR LIFE: HISTONE FATE AND NUCLEOSOME DYNAMICS

Our data collectively point to the existence of H3-H4 dimers as intermediates that have to be considered during cellular life. In principle, dimeric forms of H3-H4 could originate from various possible sources, as summarized in

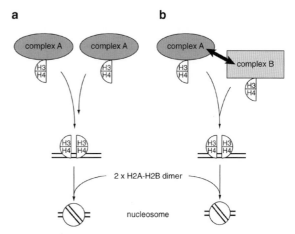

Figure 4. Hypothetical model for nucleosome assembly from H3-H4 dimer–containing complexes. (*a*) Two H3-H4 dimers are deposited onto DNA by two identical complexes (A) in one step or in two subsequent steps. Following association of two H2A-H2B dimers, a nucleosome is formed. (*b*) Two distinct H3-H4-containing complexes (A and B) are used.

our scheme in Figure 5. The first option could be that all dimers originate from newly synthesized histones that are to be incorporated into nucleosomes "de novo" (Fig. 5, left panel). In such a case, it will be particularly important to examine which modifications are associated with the histone-containing complexes and when they are imposed. The relative enrichment in chromatin of specific combination of histone modifications associated with each of the variants to distinguish active from silent chromatin found in *Alfafa* and *Drosophila* cells (Waterborg 1990; McKittrick et al. 2004) could be strictly dependent on the nature of the variants and the corresponding specific complexes. In this way specific patterns could be dictated at rather early stages soon after synthesis. Although this is a simple solution at first glance, it remains unclear how subtle distinct patterns for different types of silent regions can be established. In mammals, for example, how can the distinction be made between constitutive and facultative heterochromatin as found on the inactive X chromosome in females (Heard 2004; Maison and Almouzni 2004) and how can specific developmentally regulated expression be tightly regulated in a more subtle way than just on/off? Surely, in addition to a certain degree of specificity provided by histone variants and corresponding associated complexes, one will have to also consider the contribution of additional modifications targeted by other means through specific modifying enzymes.

A second potential source of H3-H4 dimers would be when parental histones (i.e., initially already present in a nucleosomal form) are recycled (Fig. 5, middle and right panels). At one extreme, if the recycling does not involve disruption of a tetramer, or only a transient dissociation followed by rapid reassociation, no mixing with newly synthesized histones would be expected (right panel). Such a scenario would be in agreement with experiments in the late 1980s involving the analysis of bulk histones without any particular focus on a specific genomic region (Prior et al. 1980; Jackson 1987, 1988; Annunziato 1990; Yamasu and Senshu 1990; Wolffe 1998; Henikoff et al. 2004). In such a scheme, parental H3-H4 tetramers that contain epigenetic information that do not necessarily evenly segregate onto the two daughter strands may dilute parental information that could be used in a templating mechanism. This may not necessarily be a major issue in genome regions that are organized in a repetitive manner and in which modification (such as amino-terminal histone modifications or associated proteins) of the neighbor can propagate the information. However, it is perhaps more difficult to envision how local marks spanning over a limited number of nucleosomes could be reestablished exactly in the same manner on the new DNA strands. In this context, it may be interesting to consider an alternative scenario. For example, if disruption of the tetramer can be promoted under specific circumstances, then a mixing event between parental and newly synthesized histones may occur (Fig. 5, middle panel). In this way, nucleosome core particles formed on newly synthesized DNA would inherit parental H3-H4 dimers with their encoded epigenetic information, an option that could solve the problem of inheritance of nucleosomal patterns restricted to a limited region.

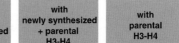
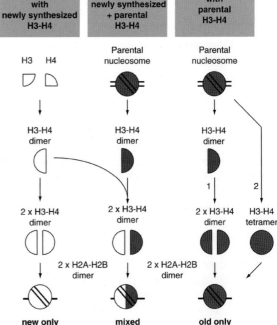

Figure 5. Dimer intermediates during nucleosome assembly. Three hypothetical and nonexclusive models of nucleosome assembly using H3-H4 dimer intermediates can be envisaged. *Left:* "De novo" assembly uses newly synthesized H3 and H4 histones to form two H3-H4 dimers, which, after association with two H2A-H2B dimers, leads to a nucleosome containing only new H3-H4. *Middle:* Assembly using a H3-H4 dimer of newly synthesized histones and a H3-H4 dimer recycled from a disrupted parental nucleosome. This could lead, after association with two H2A-H2B dimers, to a mixed nucleosome containing both newly synthesized and parental H3-H4. *Right:* Assembly using either (1) two H3-H4 dimers recycled from a transiently disrupted parental nucleosome that self-reassociate or (2) a H3-H4 tetramer stably inherited from parental nucleosome. After association with two H2A-H2B dimers, a nucleosome containing only old H3-H4 is formed.

This is purely hypothetical at this stage, and, of course, much work needs to be done to examine how preexisting nucleosomes can be disrupted and which machineries are capable of promoting such events associated with either replication, repair, or transcription. The stability of the tetramer entity, largely supported by many biophysical studies (D'Anna and Isenberg 1974; van Holde 1989; Karantza et al. 1996), although recently challenged (Banks and Gloss 2003), does not exclude the possibility of in vivo mechanisms to deal with the problem. The recent example of histone eviction as a mechanism underlying transcriptional activation at the PHO5 promoter in yeast (Boeger et al. 2003, 2004; Reinke and Horz 2003) is one illustration. Furthermore, the idea of histone exchange has been supported by recent data; in the case of variants of H2A, such as H2AZ, the combined action of histone chaperone and remodeling factors could facilitate the reaction (Mizuguchi et al. 2004). Evidence in live cells using an elegant system to follow transcriptional activation has also shown appearance of H3.3 while detection of methylation on H3K9 was lost (Janicki et al. 2004). However, whether the presence of H3.3 is a cause or a consequence of transcription remains an open issue. Given that, somehow, histone exchange may be at work in the examples mentioned above, to incorporate H3.3, remodeling factors that could use ATP to disrupt preexisting histone interaction may act in combination with specific chaperones like HIRA (or CAF-1). It will thus be interesting in this respect to examine whether remaining polypeptides among the complexes containing H3.3 or H3.1 correspond to known remodeling factors. The connection with transcription is particularly tempting, especially because the idea of a switch from "closed to open configuration" in chromatin is thought to occur with profound changes in nucleosomal conformation, where structural elements initially hidden become exposed. For instance, cysteine residue 110 on histone H3 can be chemically modified in transcribed regions (Prior et al. 1983). Our scheme summarizes possibilities that can be considered theoretically, and hopefully adequate systems will be developed to test them, taking advantage of all the recent advances in the chromatin field.

In conclusion, we found the existence of two distinct nucleosome assembly pathways that promote specific deposition of histone H3.1 or H3.3. How many more pathways remain to be identified is still an open question. The issue of dimeric histone intermediates, still provocative at this stage, is likely to stimulate future work and provides a working model that may enable our vision of nucleosome dynamics to evolve. It will be fascinating over the next few years to learn more about the many faces of nucleosome dynamics. Hopefully this will help in finding a solution to the puzzle of the inheritance of epigenetic information.

ACKNOWLEDGMENTS

We thank E. Heard and A. Loyola for critical reading. Y.N. is supported by grants from NIH (GM065939-02) and G.A. is supported by la Ligue Nationale contre le Cancer (Equipe labellisée la Ligue), Euratom (FIGH-CT-1999-00010, FIGH-CT-2002-00207), Commissariat à l'Energie Atomique (LRC #.26), Curie Program on epigenetic parameters, and NoE "Epigenome" (LSHG-CT-2004-503433).

REFERENCES

Adkins M.W., Howar S.R., and Tyler J.K. 2004. Chromatin disassembly mediated by the histone chaperone Asf1 is essential for transcriptional activation of the yeast PHO5 and PHO8 genes. *Mol. Cell* **14:** 657.

Ahmad K. and Henikoff S. 2002. The histone variant H3.3 marks active chromatin by replication-independent nucleosome assembly. *Mol. Cell* **9:** 1191.

Aho S., Buisson M., Pajunen T., Ryoo Y.W., Giot J.F., Gruffat H., Sergeant A., and Uitto J. 2000. Ubinuclein, a novel nuclear protein interacting with cellular and viral transcription factors. *J. Cell Biol.* **148:** 1165.

Alekseev O.M., Bencic D.C., Richardson R.T., Widgren E.E., and O'Rand M.G. 2003. Overexpression of the Linker histone-binding protein tNASP affects progression through the cell cycle. *J. Biol. Chem.* **278:** 8846.

Annunziato A.T. 1990. Chromatin replication and nucleosome assembly. In *Eukaryotic nucleus: Molecular biochemistry and macromolecular assemblies* (ed. P.R. Strauss and S. Wilson), vol. 2, p. 687. Telford Press, Caldwell, New Jersey.

Banks D.D. and Gloss L.M. 2003. Equilibrium folding of the core histones: The H3-H4 tetramer is less stable than the H2A-H2B dimer. *Biochemistry* **42:** 6827.

Boeger H., Griesenbeck J., Strattan J.S., and Kornberg R.D. 2003. Nucleosomes unfold completely at a transcriptionally active promoter. *Mol. Cell* **11:** 1587.

———. 2004. Removal of promoter nucleosomes by disassembly rather than sliding in vivo. *Mol. Cell* **14:** 667.

D'Anna J.A., Jr. and Isenberg I. 1974. A histone cross-complexing pattern. *Biochemistry* **13:** 4992.

Dignam J.D., Lebovitz R.M., and Roeder R.G. 1983. Accurate transcription initiation by RNA polymerase II in a soluble extract from isolated mammalian nuclei. *Nucleic Acids Res.* **11:** 1475.

Felsenfeld G. and Groudine M. 2003. Controlling the double helix. *Nature* **421:** 448.

Franklin S.G. and Zweidler A. 1977. Non-allelic variants of histones 2a, 2b and 3 in mammals. *Nature* **266:** 273.

Gaillard P.H., Martini E.M., Kaufman P.D., Stillman B., Moustacchi E., and Almouzni G. 1996. Chromatin assembly coupled to DNA repair: A new role for chromatin assembly factor I. *Cell* **86:** 887.

Grewal S.I. and Elgin S.C. 2002. Heterochromatin: New possibilities for the inheritance of structure. *Curr. Opin. Genet. Dev.* **12:** 178.

Heard E. 2004. Recent advances in X-chromosome inactivation. *Curr. Opin. Cell Biol.* **16:** 247.

Henikoff S., Furuyama T., and Ahmad K. 2004. Histone variants, nucleosome assembly and epigenetic inheritance. *Trends Genet.* **20:** 320.

Jackson V. 1987. Deposition of newly synthesized histones: New histones H2A and H2B do not deposit in the same nucleosome with new histones H3 and H4. *Biochemistry* **26:** 2315.

———. 1988. Deposition of newly synthesized histones: Hybrid nucleosomes are not tandemly arranged on daughter DNA strands. *Biochemistry* **27:** 2109.

Jakel S., Mingot J.M., Schwarzmaier P., Hartmann E., and Gorlich D. 2002. Importins fulfill a dual function as nuclear import receptors and cytoplasmic chaperones for exposed basic domains. *EMBO J.* **21:** 377.

Janicki S.M., Tsukamoto T., Salghetti S.E., Tansey W.P., Sachidanandam R., Prasanth K.V., Ried T., Shav-Tal Y., Bertrand E., Singer R.H., and Spector D.L. 2004. From silencing to gene expression: Real-time analysis in single cells. *Cell* **116:** 683.

Jenuwein T. and Allis C.D. 2001. Translating the histone code. *Science* **293:** 1074.

Karantza V., Freire E., and Moudrianakis E.N. 1996. Thermodynamic studies of the core histones: pH and ionic strength effects on the stability of the (H3-H4)/(H3-H4)2 system. *Biochemistry* **35:** 2037.

Koundrioukoff S., Polo S., and Almouzni G. 2004. Interplay between chromatin and cell cycle checkpoints in the context of ATR/ATM-dependent checkpoints. *DNA Repair* **3:** 969.

Kurdistani S.K. and Grunstein M. 2003. Histone acetylation and deacetylation in yeast. *Nat. Rev. Mol. Cell Biol.* **4:** 276.

Lamour V., Lecluse Y., Desmaze C., Spector M., Bodescot M., Aurias A., Osley M.A., and Lipinski M. 1995. A human homolog of the *S. cerevisiae* HIR1 and HIR2 transcriptional repressors cloned from the DiGeorge syndrome critical region. *Hum. Mol. Genet.* **4:** 791.

Laskey R.A., Mills A.D., and Morris N.R. 1977. Assembly of SV40 chromatin in a cell-free system from *Xenopus* eggs. *Cell* **10:** 237.

Loyola A. and Almouzni G. 2004. Histone chaperones, a supporting role in the limelight. *Biochim. Biophys. Acta* **15:** 1.

Maison C. and Almouzni G. 2004. HP1 and the dynamics of heterochromatin maintenance. *Nat. Rev. Mol. Cell Biol.* **5:** 296.

McKittrick E., Gafken P.R., Ahmad K., and Henikoff S. 2004.

Histone H3.3 is enriched in covalent modifications associated with active chromatin. *Proc. Natl. Acad. Sci.* **101:** 1525.

Mello J.A. and Almouzni G. 2001. The ins and outs of nucleosome assembly. *Curr. Opin. Genet. Dev.* **11:** 136.

Mello J.A., Sillje H.H., Roche D.M., Kirschner D.B., Nigg E.A., and Almouzni G. 2002. Human Asf1 and CAF-1 interact and synergize in a repair-coupled nucleosome assembly pathway. *EMBO Rep.* **3:** 329.

Mizuguchi G., Shen X., Landry J., Wu W.H., Sen S., and Wu C. 2004. ATP-driven exchange of histone H2AZ variant catalyzed by SWR1 chromatin remodeling complex. *Science* **303:** 343.

Munakata T., Adachi N., Yokoyama N., Kuzuhara T., and Horikoshi M. 2000. A human homologue of yeast anti-silencing factor has histone chaperone activity. *Genes Cells* **5:** 221.

Nakatani Y. and Ogryzko V. 2003. Immunoaffinity purification of mammalian protein complexes. *Methods Enzymol.* **370:** 430.

Nelson D.M., Ye X., Hall C., Santos H., Ma T., Kao G.D., Yen T.J., Harper J.W., and Adams P.D. 2002. Coupling of DNA synthesis and histone synthesis in S phase independent of cyclin/cdk2 activity. *Mol. Cell. Biol.* **22:** 7459.

Prior C.P., Cantor C.R., Johnson E.M., and Allfrey V.G. 1980. Incorporation of exogenous pyrene-labeled histone into *Physarum* chromatin: A system for studying changes in nucleosomes assembled in vivo. *Cell* **20:** 597.

Prior C.P., Cantor C.R., Johnson E.M., Littau V.C., and Allfrey V.G. 1983. Reversible changes in nucleosome structure and histone H3 accessibility in transcriptionally active and inactive states of rDNA chromatin. *Cell* **34:** 1033.

Quivy J.P., Roche D., Kirschner D., Tagami H., Nakatani Y., and Almouzni G. 2004. A CAF-1 dependent pool of HP1 during heterochromatin duplication. *EMBO J.* **12:** 12.

Ray-Gallet D. and Almouzni G. 2004. DNA synthesis-dependent and -independent chromatin assembly pathways in *Xenopus* egg extracts. *Methods Enzymol.* **375:** 117.

Ray-Gallet D., Quivy J.P., Scamps C., Martini E.M., Lipinski M., and Almouzni G. 2002. HIRA is critical for a nucleosome assembly pathway independent of DNA synthesis. *Mol. Cell* **9:** 1091.

Reinke H. and Horz W. 2003. Histones are first hyperacetylated and then lose contact with the activated PHO5 promoter. *Mol. Cell* **11:** 1599.

Richardson R.T., Batova I.N., Widgren E.E., Zheng L.X., Whitfield M., Marzluff W.F., and O'Rand M.G. 2000. Characterization of the histone H1-binding protein, NASP, as a cell cycle-regulated somatic protein. *J. Biol. Chem.* **275:** 30378.

Ridgway P. and Almouzni G. 2000. CAF-1 and the inheritance of chromatin states: At the crossroads of DNA replication and repair. *J. Cell Sci.* **113:** 2647.

Sherwood P.W., Tsang S.V., and Osley M.A. 1993. Characterization of HIR1 and HIR2, two genes required for regulation of histone gene transcription in *Saccharomyces cerevisiae*. *Mol. Cell. Biol.* **13:** 28.

Sillje H.H. and Nigg E.A. 2001. Identification of human Asf1 chromatin assembly factors as substrates of Tousled-like kinases. *Curr. Biol.* **11:** 1068.

Smith S. and Stillman B. 1989. Purification and characterization of CAF-I, a human cell factor required for chromatin assembly during DNA replication in vitro. *Cell* **58:** 15.

Sun L., Youn H.D., Loh C., Stolow M., He W., and Liu J.O. 1998. Cabin 1, a negative regulator for calcineurin signaling in T lymphocytes. *Immunity* **8:** 703.

Tagami H., Ray-Gallet D., Almouzni G., and Nakatani Y. 2004. Histone H3.1 and H3.3 complexes mediate nucleosome assembly pathways dependent or independent of DNA synthesis. *Cell* **116:** 51.

Turner B.M. 2002. Cellular memory and the histone code. *Cell* **111:** 285.

Tyler J.K., Collins K.A., Prasad-Sinha J., Amiott E., Bulger M., Harte P.J., Kobayashi R., and Kadonaga J.T. 2001. Interaction between the *Drosophila* CAF-1 and ASF1 chromatin assembly factors. *Mol. Cell. Biol.* **21:** 6574.

van Holde K. 1989. *Chromatin*. Springer-Verlag, New York.

Verreault A. 2000. De novo nucleosome assembly: New pieces in an old puzzle. *Genes Dev.* **14:** 1430.

Waterborg J.H. 1990. Sequence analysis of acetylation and methylation in two histone H3 variants of alfalfa. *J. Biol. Chem.* **265:** 17157.

Watson J.D. and Crick F.H.C. 1953. Molecular structure of nucleic acids. *Nature* **171:** 737.

Wolffe A. 1998. *Chromatin: Structure and function,* 3rd edition. Academic Press, San Diego, California.

Wu R.S., Tsai S., and Bonner W.M. 1982. Patterns of histone variant synthesis can distinguish G0 from G1 cells. *Cell* **31:** 367.

Yamasu K. and Senshu T. 1990. Conservative segregation of tetrameric units of H3 and H4 histones during nucleosome replication. *J. Biochem.* **107:** 15.

The Chromatin Accessibility Complex: Chromatin Dynamics through Nucleosome Sliding

P.B. BECKER

*Adolf-Butenandt-Institut, Molekularbiologie, Ludwig-Maximilians-Universität,
80336 München, Germany*

Understanding chromatin structure and function in all its facets has been a fascinating challenge for generations of researchers. The organization of eukaryotic genomes into chromatin serves three functions. First and foremost, the packaging of DNA as chromatin involves neutralization of the negative charges in the phosphodiester backbone of DNA and its bending around molecular spools, the histone octamers, to form nucleosomes. Not only is this organization necessary for fitting a genome into the nuclear interior, but it also serves to protect the valuable genetic information from damage and instability. At the same time, chromatin organization hinders the interaction of DNA-binding regulators, thereby creating a globally repressed ground state of gene activity. The majority of all genes have to remain switched off in any given cell and primary mechanisms of tight repression build on a basic chromatin infrastructure. Finally, this organization creates opportunities for regulation: Utilization of the DNA—be it to transcribe genes, to replicate the genome, to recombine chromosomal segments, or to repair DNA damage—involves changing the occlusive chromatin environment, thereby allowing regulators to gain access to the substrate. Although nucleosomes are dynamic entities due to the very nature of the histone–DNA interactions (Widom 1999), the opening and closing of chromatin is not left to chance: A large number of enzymes dedicated to this task have been identified during recent years. Besides those that covalently modify histones to change their interaction properties (Jenuwein and Allis 2001), one interesting class of enzymes is fueled by chemical energy: They couple ATP hydrolysis to changing chromatin organization (Becker and Hörz 2002; Narlikar et al. 2002; Peterson 2002; Lusser and Kadonaga 2003). In broadest terms, the process of changing histone–DNA interactions in order to modulate the access of DNA is called "nucleosome remodeling." About 10 years ago we discovered a nucleosome remodeling entity, the Chromatin Accessibility Complex (CHRAC), during our search for an energy-dependent chromatin opening (Varga-Weisz et al. 1997). Since then, we and several other groups have studied the molecular anatomy of CHRAC and its mechanism of nucleosome remodeling. Although its physiological function has not been clearly established, all available results point to crucial functions of CHRAC for the assembly and maintenance of dynamic chromatin structure.

IN SEARCH OF ATP-DEPENDENT NUCLEOSOME REMODELING ACTIVITIES

Because of its potential for genetic analysis, *Drosophila melanogaster* is an excellent model organism suited for deciphering molecular mechanisms of eukaryotic gene function during development. Although less appreciated, *Drosophila* also provides a very good opportunity for biochemical studies. *Drosophila* embryos, which can be harvested in kilogram amounts from mass populations of fruit flies, are among the most active tissues. Extracts from embryos are rich in enzymatic activities, notably factors governing all aspects of gene expression. Relevant for the current discussion are the large quantities of chromatin assembly and nucleosome remodeling activities present in very early embryos. The first ten nuclear divisions during embryonic development occur very rapidly in the absence of zygotic transcription and protein synthesis. The assembly of the more than 1000 genomes arising into chromatin relies entirely on large stockpiles of histones, histone chaperones, and other assembly activities. Extracts from early embryos contain all these activities in proper stoichiometry and are, therefore, very competent in the assembly of experimental DNA into long nucleosomal arrays with physiological spacing. The assembly of such arrays also involves an ATP-dependent step. In the absence of ATP, nucleosomes will form, but their succession on DNA will be unordered and interrupted by gaps of free DNA (Becker and Wu 1992).

Chromatin reconstituted in this way contains many nonhistone proteins and is surprisingly transparent for DNA-binding proteins, despite its physiological nucleosome density. The discovery that the interaction of DNA-binding proteins with their target sequences in crude chromatin was much facilitated by the presence of hydrolyzable ATP (Pazin et al. 1994; Tsukiyama et al. 1994; Varga-Weisz et al. 1995; Wall et al. 1995) initiated searches for ATP-dependent nucleosome remodeling enzymes in diverse laboratories. The harvest was impressive. The Wu lab found the Nucleosome Remodeling Factor (NURF), a complex that can be targeted to promoters in chromatin to derepress transcription (Tsukiyama and Wu 1995; Xiao et al. 2001; Badenhorst et al. 2002). Focusing on chromatin assembly factors, Kadonaga and colleagues found ACF, an ATP-dependent remodeling factor that can improve chromatin assembly in vitro and in vivo (Ito et al. 1997; Fyodorov and Kadonaga 2002b; Fyodorov et al. 2004). Since we had hints about an untargeted nucleosome remodeling activity (see below), we devised an assay that made use of prokaryotic restriction enzymes as probes for access to DNA in chromatin and purified CHRAC from the system (Varga-Weisz et al. 1995, 1997). Interestingly, all three protein complexes purified from fly embryo chromatin contained a common ATPase, ISWI, as the primary constituent of the nucleosome remodeling activity (Längst and Becker 2001b).

NUCLEOSOME SLIDING CATALYZED BY CHRAC AND OTHER ISWI-CONTAINING COMPLEXES

As the name suggests, the Chromatin Accessibility Complex generates access to DNA in chromatin allowing the interaction of DNA-binding proteins including the replication initiator T-antigen (Alexiadis et al. 1998) suggesting that chromatin might be disrupted or disassembled during the remodeling reaction. Surprisingly, however, it turned out that this is not the case. Rather, the ATP-dependent action of CHRAC improves the quality of chromatin, as judged from the regularity and extent of nucleosomal arrays (Varga-Weisz et al. 1997), consistent with a role for CHRAC as a chromatin assembly factor, rather than a disassembly factor. Chromatin assembly in the absence of ATP results in irregular successions of nucleosomes, which can be converted into arrays with regular spacing, a hallmark of physiological chromatin by CHRAC and ATP (Fig. 1). Interestingly, a long-range nucleosome spacing in vivo has been observed in conjunction with recruitment of CHRAC/ACF to the IL2R-α locus through interaction with the scaffolding protein SATB1 in thymocytes (Yasui et al. 2002).

The only way an increased accessibility of DNA in chromatin could be reconciled with this CHRAC activity was to assume that intact nucleosomes could be rendered mobile on DNA. This mobility would bring nucleosomal DNA into the more accessible linker between histone octamers for a fraction of time, exposing it to potential DNA-binding proteins.

We envision that nucleosome mobilization by CHRAC contributes to the assembly of chromatin fibers with regular nucleosome spacing, a requirement for folding of the nucleosomal chain into higher-order structures. Conceivably, its activity would not come to rest once this task has been achieved, but would continue to move nucleosomes, thereby endowing chromatin with fluidity and the flexibility required to respond to changing molecular environments (Becker 2002).

Evidence for the mobility of nucleosomes in *Drosophila* embryo chromatin had been obtained earlier. The type of chromatin reconstituted in this system depends on the experimental conditions. For example, increasing cation concentrations during assembly will lead to a gradual increase in internucleosomal distances, such that nucleosomal arrays with repeat lengths covering the entire physiological range from 160 to >200 base pairs

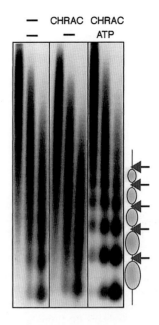

Figure 1. CHRAC catalyzes nucleosome spacing. Chromatin was assembled in an extract from *Drosophila* embryos in the absence of ATP. It was then treated with detergent in order to remove bound nucleosome remodeling activities. During further incubations, CHRAC and ATP were added as indicated. Treatment of the chromatin with Micrococcal Nuclease, which cleaves the internucleosomal linker DNA preferentially, reveals a regular "laddering" of fragments after gel electrophoresis only if the nucleosomes in the array are regularly spaced. A "smear" reveals the lack of ordering, irregular distances, and nucleosome-free gaps in chromatin. Details can be found in Varga-Weisz et al. (1997).

(bp) are obtained (Blank and Becker 1995). Evidently, chromatin structures are flexible to accommodate various linker lengths during assembly. But how flexible are they afterward? In order to address this question we assembled arrays at high-salt conditions resulting in long linkers, followed by buffer change to lower ionic strength. Remarkably, we observed that the nucleosomal array changed in response to the altered conditions to accommodate a shorter repeat length (Fig. 2) (Varga-Weisz et al. 1995). The nucleosomal array was only flexible to respond to changes in the milieu if hydrolyzable ATP was present, pointing to an active, catalyzed process (Fig. 2). Since this assay monitored arrays of up to ten nucleosomes, all of which changed their positions relative to their neighbors, some nucleosomes must have moved over more than 100 base pairs. These and other experiments (Längst and Becker 2001b) with purified enzymes demonstrated that even extended arrays are still sufficiently flexible to allow nucleosomes to change their positions. Recently, Tsukiyama and colleagues provided the first evidence to suggest that ISWI-induced nucleosome sliding may also occur in vivo (Fazzio and Tsukiyama 2003). The discovery of ATP-dependent nucleosome mobility provided an elegant solution to the problem of how to ensure that regulatory proteins gain access to their binding sites while chromatin disruption is minimized (Becker 2002).

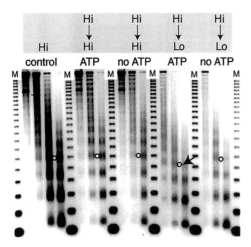

Figure 2. Nucleosomal arrays are dynamic. Chromatin was assembled in vitro at high-salt conditions, yielding long nucleosomal repeat lengths (NRLs) followed by buffer change to lower salt buffer. In the presence of ATP (and nucleosome remodeling factors) the positions of nucleosomes in the arrays relative to each other adjust to the altered ionic strength: All nucleosomes move closer together as seen by a shorter NRL. For details, see the text and Varga-Weisz et al. (1995). (Adapted, with permission, from Varga-Weisz et al. 1995 [©Nature Publishing Group].)

The analysis of nucleosome mobility in more quantitative and mechanistic terms was facilitated by the development of an assay that allows observing the relocation of single histone octamers on DNA (Hamiche et al. 1999; Längst et al. 1999; Eberharter et al. 2004a). Wrapping of short DNA fragments around a single histone octamer frequently leads to different positions of the nucleosome with respect to the fragment ends. Particles that contain a more central nucleosome can be distinguished from those where the nucleosome is close to one fragment end by a simple electrophoretic mobility shift assay (Fig. 3). Because of the inherent stability of nucleosomes at low-salt conditions these different particles can be isolated and used as substrates for remodeling enzymes. CHRAC, ACF, NURF, and other ISWI-containing enzymes can catalyze movement of histone octamers on these short DNA fragments (Fig. 3) and the detailed characterization of this reaction demonstrated that histone octamers were moved as intact entities (nucleosome sliding) rather than being rapidly dis- and reassembled (Hamiche et al. 1999; Längst et al. 1999; Eberharter et al. 2001). Interestingly, although NURF can also slide nucleosomes, it is unable to function as a nucleosome spacing factor (Tsukiyama and Wu 1995).

MOLECULAR ANATOMY OF CHRAC

The enzyme responsible for the nucleosome remodeling activity of CHRAC is the ATPase ISWI, which also drives several other nucleosome remodeling complexes with distinct subunit composition (Längst and Becker 2001b; Corona and Tamkun 2004). In CHRAC, ISWI is associated with ACF1, a BAZ/WAL family member characterized by prominent PHD fingers and a bromodomain, and two histone-fold proteins of 14 and 16 kD

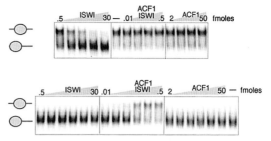

Figure 3. Interaction of ACF1 with ISWI modulates nucleosome mobilization. The nucleosome sliding assay monitors relocation of nucleosomes between central to peripheral positions on small DNA fragments. The cartoons to the left indicate which gel electrophoresis band corresponds to which type of particle. Association of ACF1 with ISWI changes the directionality of nucleosome sliding and at the same time increases the efficiency of the reaction: Far fewer femtomoles of remodeling complex are required to relocate nucleosomes in presence of ACF1 (for details see Eberharter et al. 2001). (Adapted, with permission, from Eberharter et al. 2001 [©Nature Publishing Group].)

(CHRAC14 and CHRAC16) (Corona et al. 2000; Eberharter et al. 2001). Although topoisomerase II was initially thought to be an integral part of CHRAC (Varga-Weisz et al. 1997), we later realized that this enzyme could be separated from CHRAC without affecting the activity or size of the complex (Eberharter et al. 2001). CHRAC is highly related to ACF, which solely consists of ISWI and ACF1 (Ito et al. 1999). CHRAC is a highly conserved entity: It has been isolated from human and frog cells (Poot et al. 2000; MacCallum et al. 2002) and recently it was found that the yeast ISW2 complex resembles CHRAC in its subunit composition (Iida and Araki 2004; McConnell et al. 2004). Metazoan CHRAC is organized by the large ACF1 subunit, which contacts ISWI at a central domain (Collins et al. 2002; Fyodorov and Kadonaga 2002a; Eberharter et al. 2004b) and binds the histone-fold proteins with an amino-terminal domain containing a sequence conservation termed the "WAC homology" (Kukimoto et al. 2004). The yeast protein Itc1p, which is equivalent to ACF1 in the ISW2 complex, is dissimilar to ACF1 except for the WAC sequence in the amino terminus where the two histone-fold proteins interact (McConnell et al. 2004).

MECHANISTIC CONSIDERATIONS

The ATPase ISWI is a nucleosome remodeling factor by itself, which responds to the presence of the nucleosomal substrate with increased rates of ATP hydrolysis and induces the sliding of histone octamers on DNA (Corona et al. 1999; Längst et al. 1999; Clapier et al. 2001; Grüne et al. 2003). Currently, our favorite model for how ISWI-containing nucleosome remodeling factors induce nucleosome sliding (Fig. 4) presumes that they function analogously to DNA translocases, such as helicases or restriction enzymes of type I, which move relative to DNA in a series of conformational changes triggered by nucleotide binding and/or hydrolysis (Längst and Becker 2004). If the enzyme was attached to a histone octamer, it could use analogous conformational choreography to

Figure 4. Hypothetical model of how nucleosome remodeling factors may mobilize histone octamers. The enzyme is depicted as two globular domains connected by a flexible hinge. The remodeler interacts with the histone body (*large sphere*) and the linker DNA. A conformational change in the enzyme pushes a segment of linker DNA (*lighter gray*) over the surface of the histone octamer.

"bulge up" the adjacent linker DNA and to "push" it into the realm of the nucleosome particle. Propagation of this loop over the histone surface, either by diffusion or through an active process (Haushalter and Kadonaga 2003), until it emerges at the opposite side of the particle, would effectively move the nucleosome relative to the DNA sequence (Fig. 4). While this model is consistent with many experimental observations (Havas et al. 2000; Längst and Becker 2001a, 2004; Becker and Hörz 2002), formal proof will require knowledge of the atomic structure of the remodeling machinery in contact with the nucleosome substrate as well as the characterization of reaction intermediates. Alternative potential outcomes of nucleosome remodeling reactions of the kinds observed for SWI/SNF-type remodelers, such as the disruption of histone–DNA interactions in the absence of nucleosome relocation (Fan et al. 2003), are less frequently observed with ISWI-containing enzymes.

The structure of a prominent substrate recognition domain within the carboxy-terminal half of ISWI, consisting of a DNA-binding SLIDE domain and a SANT domain, has recently been determined (Grüne et al. 2003). The SLIDE domain resembles classical helix-loop-helix DNA-binding domains in general and more specifically the DNA-binding modules of c-myb. ISWI preferentially interacts with the linker DNA adjacent to a nucleosome particle, a strategic position to initiate nucleosome remodeling since the DNA at its entry into the nucleosome is only loosely attached to the histone surface. SANT modules have been suggested to function as histone-tail-binding units in the context of other chromatin regulators (Boyer et al. 2004). However, even though ISWI requires the histone H4 tail for efficient nucleosome remodeling (Clapier et al. 2001, 2002; Hamiche et al., 2001), a strong, direct interaction of the enzyme with a histone tail has not yet been described. Interaction with the nucleosome substrate is likely to be complex since DNA and histone interactions of ACF1 have also been observed (Fyodorov and Kadonaga 2002a; Eberharter et al. 2004b) and the CHRAC14-CHRAC16 histone-fold dimer can also bind to DNA, albeit with modest affinity (Poot et al. 2000; Kukimoto et al. 2004; F. Hartlepp et al., in prep.). Furthermore, because of the nature of the remodeling process, these interactions are supposedly dynamic and presumably partially dependent on nucleotide-induced conformational changes of the enzyme.

REGULATION OF ISWI ACTIVITY BY OTHER CHRAC SUBUNITS

Although recombinant ISWI can carry out basic nucleosome sliding reactions, the enzyme is not very efficient. Association of other CHRAC subunits profoundly improves the effectiveness of the reaction (i.e., the extent of remodeling per constant amount of ATP hydrolyzed). Interaction of ISWI with ACF1 improves nucleosome sliding efficiency by an order of magnitude (Fig. 3) (Ito et al. 1999; Eberharter et al. 2001). We recently found that this improved efficiency correlated with interactions of the ACF1 PHD fingers with the histone moiety within nucleosomes (Eberharter et al. 2004b). Conceivably, ISWI is a relatively inefficient remodeling enzyme because it lacks a stable anchoring point on the histone body. ACF1-histone contacts may provide such an anchor for the enzyme, assuring that the presumed conformational changes triggered by nucleotide binding and hydrolysis are efficiently converted into positional shifts of DNA relative to histones. Interestingly, association of ACF1 with ISWI not only boosts the efficiency of the enzyme, but it also changes the outcome of the remodeling reaction qualitatively (see Fig. 3). While ISWI alone moves nucleosomes from the center of a fragment to more peripheral positions (and not back), ACF (i.e., ISWI+ACF1) and CHRAC direct the converse relocation of end-positioned nucleosomes to more central positions (Fig. 3). While the significance of these differences in highly synthetic in vitro reactions remains to be established, it is nevertheless clear that the approach of ISWI to the nucleosome differs whether or not ACF1 is present.

The two small subunits CHRAC14 and CHRAC16 that differentiate CHRAC from ACF are of considerable interest, since their predicted (Corona et al. 2000) and experimentally determined structures (F. Hartlepp et al., in prep.) resemble that of a histone H2A-H2B dimer pair. CHRAC14-CHRAC16 or their human counterparts, CHRAC17-CHRAC15, bind DNA with modest affinity (Poot et al. 2000; Kukimoto et al. 2004; F. Hartlepp et al., in prep.). Their dynamic interactions with structured DNA suggest that they may function as DNA chaperones in analogy to HMGB1 (see below). In vitro, the small subunits improve the activity of human and *Drosophila* ACF under conditions of limiting enzyme.

REGULATION OF CHRAC ACTIVITY— TIP OF THE ICEBERG

Very little is known about how CHRAC (or other ISWI-containing nucleosome remodeling factors) are regulated, but we assume that evolution has not missed the opportunity to control the level of chromatin dynamics. Among the many possibilities are posttranslational modifications of the remodeling machinery, which are being worked out at the moment. One other important parameter of chromatin dynamics is obviously the state of the substrate, chromatin, itself. The packaging of a nucleosomal fiber is mainly determined by interacting proteins: the linker histone H1, abundant nonhistone proteins, such as the HMG proteins, or more specialized chromatin regulators, such as the polycomb group proteins. The interaction of histone H1 with the DNA at the entry/exit of the nucleosome is known to impede the action of nucleosome remodeling machines of the SWI2/SNF2 type (Hill and Imbalzano 2000; Horn et al. 2002). However, early experiments with crude, reconstituted chromatin suggested that nucleosomal arrays retain their dynamic properties in the presence of physiological H1 levels (Varga-Weisz et al. 1995). The effect of linker histones on ISWI-dependent nucleosome remodeling has not been determined so far. Even if it turns out to be inhibitory, there are obvious ways of how it may be modulated. The interaction of H1 with nucleosomes may be weakened by phosphorylation of the linker histone (Horn et al. 2002), or H1 may be competed off the nucleosome by HMGB1 (Varga-Weisz et al. 1994). HMGB1 interacts with the nucleosomal linker much more dynamically than H1. Under conditions of limiting remodeling activity the association of HMGB1 with mononucleosomal substrates activates ACF- and CHRAC-dependent nucleosome mobilization (Bonaldi et al., 2002). This stimulatory effect of HMGB1 may be due to the "DNA chaperone" function of HMGB1, which could conceivably help to flexibilize the DNA at its entry into the nucleosome such that it can be more readily distorted and bound by the remodeling factor (Bonaldi et al. 2002; Travers 2003). We thus predict that substitution of HMGB1 for H1 should result in dramatic stimulation of nucleosome mobility.

Very little is known about the effect of histone modifications on nucleosome remodeling activity. ACF1 contains a bromodomain, which might serve to contact acetylated lysine residues and hence to discriminate between histones of different acetylation status within the nucleosome (Strahl and Allis 2000). But even the recombinant ATPase ISWI itself is able to sense certain modifications. For unknown reasons, efficient completion of the ISWI ATPase cycles requires a basic, DNA-bound patch on the amino terminus of histone H4. On peptide substrates, lysine acetylation close to this "ISWI response determinant" dampens the stimulation of its ATPase activity (Clapier et al., 2001, 2002; Hamiche et al. 2001).

NUCLEOSOME REMODELING MACHINE IN SEARCH OF A FUNCTION

Although mutation of the ATPase ISWI itself leads to a rich phenomenology of chromosome aberrations and changes in gene expression (Corona and Tamkun 2004), the fact that ISWI resides in at least one other complex (NURF) besides CHRAC/ACF precludes a straightforward correlation of effects and complexes (Längst and Becker 2001b; Tsukiyama 2002; Corona and Tamkun 2004). Monitoring the ACF1 subunit does not allow discrimination between functions of CHRAC or ACF, but nevertheless yields important information on a combined activity profile. Kadonaga and colleagues recently described their analysis of a genetic deficiency of ACF1 in flies (Fyodorov et al. 2004). A large fraction of the animals harboring a homozygous null mutation of the ACF1 gene die during the larval–pupal transition, but those that do survive show chromatin assembly defects and corresponding alleviation of chromatin-based repression mechanisms, such as polycomb-mediated silencing and position effect variegation (Fyodorov et al. 2004). The data are most consistent with a global role of CHRAC/ACF for the assembly of chromatin and chromosomes, which is expected to indirectly affect most nuclear functions. However, the flies lacking ACF1 also show phenotypes reminiscent of defects in trithorax group genes, which are involved in the regulation of homeotic genes during development, and thus more specific roles of the remodeling complex in gene regulation are still possible (Fyodorov et al. 2004).

The acceleration of S phase in embryonic and larval cells, which may be explained by a lesser hindrance of DNA synthesis in a defective and hence less repressive chromatin environment, could be an example for such an indirect effect (Fyodorov et al. 2004). However, direct roles for CHRAC/ACF in replication have been invoked from the analysis of other systems as well. In unresolved contrast to the acceleration of S phase in ACF1-deficient flies, Varga-Weisz and colleagues (Collins et al. 2002) found that ablation of ACF1 by RNA interference or expression of a dominant negative ACF in human cells resulted in a delay of cell-cycle progression, which was explained by a slowed replication of pericentric heterochromatin. This scenario would be more consistent with a role for CHRAC/ACF as chromatin accessibility factors, which facilitate the progress of the enzyme through a resistant chromatin environment. The effects were not seen if the cells were treated with 5-aza-deoxycytidine, a drug that, among other effects, inhibits the maintenance DNA methyltransferase (Collins et al. 2002). Since fly chromatin does not carry DNA methylation marks outside of a very narrow window in early embryonic development, the tasks of fly and human CHRAC/ACF may differ according to their chromatin environment.

Proper functional discrimination of CHRAC and ACF relies entirely on monitoring the role of the two small histone-fold subunits. The antibodies currently available against the metazoan subunits do not allow unambiguous localization of these factors in nuclei. However, the genetic analysis of the histone-like CHRAC subunits in yeast, Dls1p and Dpb4p, revealed that they are required for proper functioning of the ISW2 complex, which involves site-specific nucleosome positioning and repression of selected genes in synergisms with histone deacetylases (McConnell et al. 2004). In an interesting analogy, targeting of ISWI and ACF1 to the IL2R-α gene locus correlates with extensive ordering of chromatin and repression of the gene in mouse thymocytes (Yasui et al. 2002).

Since Isw2p is targeted to sites of action in the absence of the small subunits, these must play a role in the remodeling process itself, rather than in recruiting the remodeling complex. Interestingly, Dpb4p is also a subunit of a DNA polymerase ε complex, which stabilizes repressive chromatin environment at telomers, whereas the ISW2 complex (yCHRAC) maintains open chromatin at telomers (Iida and Araki 2004). The human DNA polymerase ε has been colocalized to late-replicating heterochromatin, like hACF1 (Collins et al. 2002; Fuss and Linn 2002). Whether there is a functional interaction between the two complexes in heterochromatin and, if so, whether this is mediated by their shared subunit Dpb4p remain to be seen.

Several rather fundamental questions concerning CHRAC function remain to be answered. Do the two histone-fold subunits that distinguish CHRAC from ACF provide the remodeling complex with a different quality? Does CHRAC have global, untargeted roles in chromatin assembly and the maintenance of dynamic chromatin structure or is its physiological action limited locally by recruitment of the complex to sites of action such as replication forks or selected promoters? We have evidence from different model systems to support either scenario: The defect of ACF1 in the mutant fly points to a rather global function (Fyodorov et al. 2004), whereas experiments in yeast and mammalian cells suggest targeting principles (Yasui et al. 2002; McConnell et al. 2004). How does nucleosome mobilization work in an environment, where most nucleosomes are part of highly folded nucleosomal fibers? Does CHRAC/ACF affect higher-order chromatin structures, or is its action limited to mobilization of individual nucleosomes in a decondensed nucleosomal fiber? And if so, what are the activities and principles that allow productive interaction of the nucleosome remodeling machinery itself with its substrate? The rich phenomenology that has emerged from the analysis of nucleosome remodeling by CHRAC and ACF promises exciting answers to all these questions.

ACKNOWLEDGMENTS

I thank my colleagues who during recent years spent long hours studying CHRAC and chromatin dynamics in my laboratory: Thiemo Blank, Patrick Varga-Weisz, Davide Corona, Simona Ferrari, Karl Nightingale, Cedric Clapier, Edgar Bonte, Gernot Längst, Jan Brzeski, Anton Eberharter, Felix Hartlepp, and Roger Ferreira. I am also grateful for trust- and fruitful collaborations with John Tamkun, Marco Bianchi, Tiziana Bonaldi, Christoph Müller, Tim Grüne, John Widom, and Rein Aasland. I thank P. Varga-Weisz, G. Längst, and A. Eberharter for comments on the manuscript; P. Varga-Weisz and G. Längst for Figures 1 and 4, respectively. Work on CHRAC was supported by Deutsche Forschungsgemeinschaft through grants Be 1140/2-4, SFB 190, and SFB 594; the European Union through Network Grant HPRN-CT-2000-00078; and Fonds of the Chemische Industrie.

REFERENCES

Alexiadis V., Varga-Weisz P.D., Bonte E., Becker P.B., and Gruss C. 1998. In vitro chromatin remodelling by chromatin accessibility complex (CHRAC) at the SV40 origin of DNA replication. *EMBO J.* **17:** 3428.
Badenhorst P., Voas M., Rebay I., and Wu C. 2002. Biological functions of the ISWI chromatin remodeling complex NURF. *Genes Dev.* **16:** 3186.
Becker P.B. 2002. Nucleosome sliding: Facts and fiction. *EMBO J.* **21:** 4749.
Becker P.B. and Hörz W. 2002. ATP-dependent nucleosome remodeling. *Annu. Rev. Biochem.* **71:** 247.
Becker P.B. and Wu C. 1992. Cell-free system for assembly of transcriptionally repressed chromatin from *Drosophila* embryos. *Mol. Cell. Biol.* **12:** 2241.
Blank T.A. and Becker P.B. 1995. Electrostatic mechanism of nucleosome spacing. *J. Mol. Biol.* **252:** 305-313.
Bonaldi T., Längst G., Strohner R., Becker P.B., and Bianchi M.E. 2002. The DNA chaperone HMGB1 facilitates ACF/CHRAC-dependent nucleosome sliding. *EMBO J.* **21:** 6865.
Boyer L.A., Latek R.R., and Peterson C.L. 2004. The SANT domain: A unique histone-tail-binding module? *Nat. Rev. Mol. Cell Biol.* **5:** 158.
Clapier C.R., Nightingale K.P., and Becker P.B. 2002. A critical epitope for substrate recognition by the nucleosome remodeling ATPase ISWI. *Nucleic Acids Res.* **30:** 649.
Clapier C.R., Längst G., Corona D.F., Becker P.B., and Nightingale K.P. 2001. Critical role for the histone H4 N terminus in nucleosome remodeling by ISWI. *Mol. Cell. Biol.* **21:** 875.
Collins N., Poot R.A., Kukimoto I., Garcia-Jimenez C., Dellaire G., and Varga-Weisz P.D. 2002. An ACF1-ISWI chromatin-remodeling complex is required for DNA replication through heterochromatin. *Nat. Genet.* **32:** 627.
Corona D.F. and Tamkun J.W. 2004. Multiple roles for ISWI in transcription, chromosome organization and DNA replication. *Biochim. Biophys. Acta* **1677:** 113.
Corona D.F., Längst G., Clapier C.R., Bonte E.J., Ferrari S., Tamkun J.W., and Becker P.B. 1999. ISWI is an ATP-dependent nucleosome remodeling factor. *Mol. Cell* **3:** 239.
Corona D.F., Eberharter A., Budde A., Deuring R., Ferrari S., Varga-Weisz P., Wilm M., Tamkun J., and Becker P.B. 2000. Two histone fold proteins, CHRAC-14 and CHRAC-16, are developmentally regulated subunits of chromatin accessibility complex (CHRAC). *EMBO J.* **19:** 3049.
Eberharter A., Längst G., and Becker P.B. 2004a. A nucleosome sliding assay for chromatin remodeling factors. *Methods Enzymol.* **377:** 344.
Eberharter A., Vetter I., Ferreira R., and Becker P.B. 2004b. ACF1 improves the effectiveness of nucleosome mobilization by ISWI through PHD-histone contacts. *EMBO J.* **23:** 4029.
Eberharter A., Ferrari S., Längst G., Straub T., Imhof A., Varga-Weisz P., Wilm M., and Becker P.B. 2001. Acf1, the largest subunit of CHRAC, regulates ISWI-induced nucleosome remodelling. *EMBO J.* **20:** 3781.
Fan H.Y., He X., Kingston R.E., and Narlikar G.J. 2003. Distinct strategies to make nucleosomal DNA accessible. *Mol. Cell* **11:** 1311.
Fazzio T.G. and Tsukiyama T. 2003. Chromatin remodeling in vivo: Evidence for a nucleosome sliding mechanism. *Mol. Cell* **12:** 1333.
Fuss J. and Linn S. 2002. Human DNA polymerase epsilon colocalizes with proliferating cell nuclear antigen and DNA replication late, but not early, in S phase. *J. Biol. Chem.* **277:** 8658.
Fyodorov D.V. and Kadonaga J.T. 2002a. Binding of Acf1 to DNA involves a WAC motif and is important for ACF-mediated chromatin assembly. *Mol. Cell. Biol.* **22:** 6344.
———. 2002b. Dynamics of ATP-dependent chromatin assembly by ACF. *Nature* **418:** 897.
Fyodorov D.V., Blower M.D., Karpen G.H., and Kadonaga J.T. 2004. Acf1 confers unique activities to ACF/CHRAC and promotes the formation rather than disruption of chromatin in vivo. *Genes Dev.* **18:** 170.
Grüne T., Brzeski J., Eberharter A., Clapier C.R., Corona D.F., Becker P.B., and Müller C.W. 2003. Crystal structure and functional analysis of a nucleosome recognition module of the remodeling factor ISWI. *Mol. Cell* **12:** 449.

Hamiche A., Sandaltzopoulos R., Gdula D.A., and Wu C. 1999. ATP-dependent histone octamer sliding mediated by the chromatin remodeling complex NURF. *Cell* **97:** 833.

Hamiche A., Kang J.G., Dennis C., Xiao H., and Wu C. 2001. Histone tails modulate nucleosome mobility and regulate ATP-dependent nucleosome sliding by NURF. *Proc. Natl. Acad. Sci.* **98:** 14316.

Haushalter K.A. and Kadonaga J.T. 2003. Chromatin assembly by DNA-translocating motors. *Nat. Rev. Mol. Cell Biol.* **4:** 613.

Havas K., Flaus A., Phelan M., Kingston R., Wade P.A., Lilley D.M., and Owen-Hughes T. 2000. Generation of superhelical torsion by ATP-dependent chromatin remodeling activities. *Cell* **103:** 1133.

Hill D.A. and Imbalzano A.N. 2000. Human SWI/SNF nucleosome remodeling activity is partially inhibited by linker histone H1. *Biochemistry* **39:** 11649.

Horn P.J., Carruthers L.M., Logie C., Hill D.A., Solomon M.J., Wade P.A., Imbalzano A.N., Hansen J.C., and Peterson C.L. 2002. Phosphorylation of linker histones regulates ATP-dependent chromatin remodeling enzymes. *Nat. Struct. Biol.* **9:** 263.

Iida T. and Araki H. 2004. Noncompetitive counteractions of DNA polymerase epsilon and ISW2/yCHRAC for epigenetic inheritance of telomere position effect in *Saccharomyces cerevisiae*. *Mol. Cell. Biol.* **24:** 217.

Ito T., Bulger M., Pazin M.J., Kobayashi R., and Kadonaga J.T. 1997. ACF, an ISWI-containing and ATP-utilizing chromatin assembly and remodeling factor. *Cell* **90:** 145.

Ito T., Levenstein M.E., Fyodorov D.V., Kutach A.K., Kobayashi R., and Kadonaga J.T. 1999. ACF consists of two subunits, Acf1 and ISWI, that function cooperatively in the ATP-dependent catalysis of chromatin assembly. *Genes Dev.* **13:** 1529.

Jenuwein T. and Allis C.D. 2001. Translating the histone code. *Science* **293:** 1074.

Kukimoto I., Elderkin S., Grimaldi M., Oelgeschlager T., and Varga-Weisz P.D. 2004. The histone-fold protein complex CHRAC-15/17 enhances nucleosome sliding and assembly mediated by ACF. *Mol. Cell* **13:** 265.

Längst G. and Becker P.B. 2001a. ISWI induces nucleosome sliding on nicked DNA. *Mol. Cell* **8:** 1085.

———. 2001b. Nucleosome mobilization and positioning by ISWI-containing chromatin remodeling factors. *J. Cell Sci.* **114:** 2561.

———. 2004. Nucleosome remodeling: one mechanism, many phenomena? *Biochim. Biophys. Acta* **1677:** 58.

Längst G., Bonte E.J., Corona D.F.V., and Becker P.B. 1999. Nucleosome movement by CHRAC and ISWI without disruption or trans-displacement of the histone octamer. *Cell* **97:** 843.

Lusser A. and Kadonaga J.T. 2003. Chromatin remodeling by ATP-dependent molecular machines. *Bioessays* **25:** 1192.

MacCallum D.E., Losada A., Kobayashi R., and Hirano T. 2002. ISWI remodeling complexes in *Xenopus* egg extracts: Identification as major chromosomal components that are regulated by INCENP-aurora B. *Mol. Biol. Cell* **13:** 25.

McConnell A.D., Gelbart M.E., and Tsukiyama T. 2004. Histone fold protein Dls1p is required for Isw2-dependent chromatin remodeling in vivo. *Mol. Cell. Biol.* **24:** 2605.

Narlikar G.J., Fan H.Y., and Kingston R.E. 2002. Cooperation between complexes that regulate chromatin structure and transcription. *Cell* **108:** 475.

Pazin M.J., Kamakaka R., and Kadonaga J.T. 1994. ATP-dependent nucleosome reconfiguration and transcriptional activation from preassembled chromatin templates. *Science* **266:** 2007.

Peterson C.L. 2002. Chromatin remodeling enzymes: Taming the machines. Third in review series on chromatin dynamics. *EMBO Rep.* **3:** 319.

Poot R.A., Dellaire G., Hulsmann B.B., Grimaldi M.A., Corona D.F., Becker P.B., Bickmore W.A., and Varga-Weisz P.D. 2000. HuCHRAC, a human ISWI chromatin remodelling complex contains hACF1 and two novel histone-fold proteins. *EMBO J.* **19:** 3377.

Strahl B.D. and Allis C.D. 2000. The language of covalent histone modifications. *Nature* **403:** 41.

Travers A.A. 2003. Priming the nucleosome: A role for HMGB proteins? *EMBO Rep.* **4:** 131.

Tsukiyama T. 2002. The in vivo functions of ATP-dependent chromatin-remodelling factors. *Nat. Rev. Mol. Cell Biol.* **3:** 422.

Tsukiyama T. and Wu C. 1995. Purification and properties of an ATP-dependent nucleosome remodeling factor. *Cell* **83:** 1011.

Tsukiyama T., Becker P.B., and Wu C. 1994. ATP-dependent nucleosome disruption at a heat-shock promoter mediated by binding of GAGA transcription factor. *Nature* **367:** 525.

Varga-Weisz P.D., Blank T.A., and Becker P.B. 1995. Energy-dependent chromatin accessibility and nucleosome mobility in a cell-free system. *EMBO J.* **14:** 2209.

Varga-Weisz P.D., van Holde K., and Zlatanova J. 1994. Competition between linker histones and HMG1 for binding to four-way junction DNA: Implications for transcription. *Biochem. Biophys. Res. Commun.* **203:** 1904.

Varga-Weisz P.D., Wilm M., Bonte E., Dumas K., Mann M., and Becker P.B. 1997. Chromatin-remodelling factor CHRAC contains the ATPases ISWI and topoisomerase II (erratum in *Nature* [1997] **389:** 1003). *Nature* **388:** 598.

Wall G., Varga-Weisz P.D., Sandaltzopoulos R., and Becker P.B. 1995. Chromatin remodeling by GAGA factor and heat shock factor at the hypersensitive *Drosophila hsp26* promoter in vitro. *EMBO J.* **14:** 1727.

Widom J. 1999. Equilibrium and dynamic nucleosome stability. *Methods Mol. Biol.* **119:** 61.

Xiao H., Sandaltzopoulos R., Wang H., Hamiche A., Ranallo R., Lee K., Fu D., and Wu C. 2001. Dual functions of the largest NURF subunit NURF301 in nucleosome sliding and transcription factor interactions. *Mol. Cell* **8:** 531.

Yasui D., Miyano M., Cai S., Varga-Weisz P., and Kohwi-Shigematsu T. 2002. SATB1 targets chromatin remodelling to regulate genes over long distances. *Nature* **419:** 641.

Histone H2B Ubiquitylation and Deubiquitylation in Genomic Regulation

N.C.T. EMRE AND S.L. BERGER

Gene Expression and Regulation Program, The Wistar Institute, Philadelphia, Pennsylvania 19104

DNA is organized and packaged in the eukaryotic nucleus via association with octamers of the core histone proteins, comprising two H2A/H2B dimers bound to a tetramer of H3/H4. The nucleosome core is formed when 146 base pairs of DNA are wrapped approximately twice around this histone cylinder (Luger 2003). Alterations of the repeating nucleosome structure occur during DNA transactions that include transcription, DNA repair, heterochromatic silencing, and large-scale changes in chromosome compaction during mitosis/apoptosis. The major nucleosome changes are posttranslational covalent modifications of constituent histones and remodeling of the nucleosome. Remodeling encompasses altered nucleosome position and composition, as well as disruption (Flaus and Owen-Hughes 2004). Histone modifications that regulate DNA-based processes include acetylation, phosphorylation, and methylation (Berger 2002). In addition to these relatively small covalent modifications, histones are also modified by much larger ubiquitin (ub) and the related sumo polypeptides (for review, see Jason et al. 2002; Nathan et al. 2003; Zhang 2003).

Several key principles have emerged over a decade of intensive investigation of specific enzymes and their histone substrates. One central paradigm is the notion that addition of and subtraction of a modification are functionally opposed. Thus, histone acetylase enzymes are typically activators of transcription, while histone deacetylase enzymes are generally transcriptional repressors. Another main idea is that acetylation is highly dynamic, being added soon after a transcriptional inducing signal occurs and then being removed following the cessation of the signal (Katan-Khaykovich and Struhl 2002).

A second major theme is the genomic organization of histone modifications, a major dichotomy being that they occur either locally or more globally (Fischle et al. 2003). Thus, during processes that are limited in distance, such as gene-specific transcriptional activation, the scope of the modification may be just one or two nucleosomes. This localization is achieved by sequence-specific DNA-binding proteins, such as transcriptional activators, that recruit the enzymes to the promoter (see, e.g., Brown et al. 2001). More global modifications are accomplished through spreading, which permits the modification and its functional consequence to encompass a broad region, even to the size of chromosomes in the cases of chromosome condensation or X chromosome dosage compensation.

The histone-modifying enzymes are typically constituents of large multimolecular complexes, which include multiple mechanisms through organization into separate protein modules that carry out and regulate specific functions. For example, the transcriptional cofactor and histone acetylation complex SAGA in the yeast *Saccharomyces cerevisiae* is composed of several modules (Grant et al. 1998). One key function is its ability to associate with DNA-bound activators to provide recruitment for the entire complex. A second function is its enzymatic acetylation activity, which is provided by a catalytic subunit, Gcn5, and substrate specificity is fine-tuned by additional proteins. A third module both positively and negatively regulates TATA-binding protein association with the proximal promoter (Grant et al. 1998; Sterner et al. 1999).

In this paper, we discuss the historical context of histone ubiquitylation and describe recent studies in *S. cerevisiae* from our laboratory focusing on the role of reversible histone H2B ubiquitylation in genomic regulation. Each of the themes touched upon above will be discussed in greater detail. The interesting points that emerge are areas where the role of H2B ubiquitylation and deubiquitylation either follow the paradigms that have arisen for other small covalent modifications or, more strikingly, where the role of ubiquitylation diverges from previous general themes.

UBIQUITYLATION AND HISTONES

Ubiquitin is a 76-residue polypeptide conserved across eukaryotes and, through covalent attachment to substrate proteins, is involved either in proteosome-mediated degradation (Hochstrasser 1996) or in intracellular signaling (Aguilar and Wendland 2003). Ubiquitin is conjugated via an isopeptide bond involving its carboxy-terminal glycine residue through sequential action of three classes of enzymes. These are the E1-activating enzymes and E2-conjugating enzymes working with E3 ligases to transfer ub to specific substrates (Hochstrasser 1996; Wilkinson 2000). The fate of the protein depends on the number of molecules in the ub chain and on which residue within ub is used to form chains. Poly-ub chains longer than four molecules formed through residue Lys48 of ub correlate with proteosomal degradation (Hochstrasser 1996; Wilkinson 2000). However, proteins associated with chains less than four ub molecules, and especially with a single (mono) ub, or chains formed through Lys63 linkages are destined for alternate functions. These span diverse processes such as vesicular trafficking, DNA repair, and intracellular localization (Hicke 2001; Aguilar and Wendland 2003). Thus, ub and other ub-like modifications such as sumo are versatile signaling modules that are functionally reminiscent of small molecule protein posttranscriptional modifications like phosphorylation or acetylation.

Most of the histone proteins occur in ub-conjugated forms (Table 1). More than 25 years ago, histone H2A was the first protein shown to be ubiquitylated. Histones constitute the most abundant ub proteins in higher eukaryotes, including 5–15% of H2A and a few percent of H2B (Jason et al. 2002; Zhang 2003; Osley 2004). Ubiquitylation of H3 may have a role in mammalian spermatogenesis (Chen et al. 1998). The largest subunit of the TFIID coactivator complex, TAF1, monoubiquitylates the linker histone H1, which may have a role in gene activation in *Drosophila* (Pham and Sauer 2000).

Overall, despite a long history, the detailed mechanisms and physiological significance of histone ubiquitylation has remained relatively poorly characterized until recently. Interestingly, histone ubiquitylation does not cause a dramatic effect on nucleosome compaction in vitro, although its effect on higher-order chromatin formation in vivo remains an open and important question (Jason et al. 2002; Moore et al. 2002). While highly prevalent, ubiquitylation apparently does not mark histones for proteosomal degradation. Regulatory roles of ubiquitylated histones include transcription (as described in detail below), spermatogenesis, and cell cycle (Jason et al. 2002; Zhang 2003; Osley 2004).

HISTONE UBIQUITYLATION SITES AND CROSS-TALK TO HISTONE METHYLATION

Both sites of histone ubiquitylation and relevant enzymes are currently under investigation (Table 1). Mammalian H2A is ubiquitylated at Lys119 and, while yeast *S. cerevisiae* H2A can be ubiquitylated in vitro, it has not been detected in vivo (Robzyk et al. 2000). Lys123 on H2B is probably the major yeast histone ubiquitylation site in vivo in rapidly growing cells and is monoubiquitylated by the Rad6/Ubc2 E2 conjugase. Loss of histone H2B ubiquitylation causes slow growth in mitotically growing cells and meiotic defects (Robzyk et al. 2000).

Table 1. Histone Ubiquitylation in *Saccharomyces cerevisiae* and Mammals

Histone	Site	Process	Enzyme
H2A	Lys119 (m)	gene repression	hPRC1L/Ring2 (E3)
H2B	Lys120 (m)	gene repression	Mdm2 (E3)
	Lys123 (y)	gene activation[a]	Rad6 (E2)/Bre1(E3)
H3	NK (m)	spermatogenesis	NK
H4	NR	NR	NR
H1	NK (m)	gene activation	Taf1(E1/E2)

(m), mammals; (y), yeast *S. cerevisiae*; NK, not known; NR, not reported.
[a]Indirect role in silencing.

Bre1, a RING finger protein, is the E3 ligase partner of Rad6 (Hwang et al. 2003; Wood et al. 2003b) involved in histone ubiquitylation (HUB) function. The homologous mammalian residue (Lys120) is also ubiquitylated in vivo in human cells (Jason et al. 2002; Zhang 2003; Minsky and Oren 2004). Ubiquitylation sites have not yet been mapped in cellular H3 and H1.

Aside from roles in DNA repair (Matunis 2002) and meiosis (Yamashita et al. 2004), Rad6-mediated ubiquitylation is implicated in transcription, including inducible gene activation (Kao et al. 2004) and repression (Turner et al. 2002; Carvin and Kladde 2004), as well as telomeric and *HMR* gene silencing (Huang et al. 1997). The histone H2B ubiquitylating activity of Rad6 is specifically required for these transcriptional activities (Huang et al. 1997; Sun and Allis 2002; Turner et al. 2002; Henry et al. 2003; Hwang et al. 2003; Wood et al. 2003b; Carvin and Kladde 2004; Kao et al. 2004). Rad6 is recruited to gene promoters of constitutive (Wood et al. 2003b) or inducible genes in a Bre1-dependent manner (Kao et al. 2004) coinciding with increased ubH2B and preceding mRNA accumulation (Henry et al. 2003; Kao et al. 2004).

The functional outcome of H2B Lys123 ubiquitylation is gene specific. For example, in yeast, transcription of certain genes is activated by ubiquitylation (Henry et al. 2003; Kao et al. 2004), while transcription of other genes is repressed (Turner et al. 2002; Carvin and Kladde 2004). How ubH2B exerts opposing effects on gene transcription is currently not clear. However, the outcome parallels gene-specific positive or negative roles of the SAGA coactivator/acetylase complex itself, and SAGA acts in the same pathways as does ubH2B (Turner et al. 2002; Henry et al. 2003; Kao et al. 2004). In addition, ubiquitylation of mammalian H2A and H2B has recently been correlated with transcriptional repression rather than activation. First, Mdm2 is known to ubiquitylate p53 to mediate its degradation under normal conditions, that is, in the absence of DNA damage, which activates p53 (Michael and Oren 2003). It now appears that, under normal conditions, Mdm2 associates with p53-inducible genes in a p53-dependent fashion to repress them via mono-ub of H2B at Lys120 and/or Lys125 (Minsky and Oren 2004). It is not clear how repression via p53 recruitment of Mdm2 leading to subsequent H2B ubiquitylation is coordinated with Mdm2-mediated p53 degradation, but there is a precedent in the occurrence of overlapping activation and degradation motifs within transcriptional activators (Muratani and Tansey 2003). A second set of recent studies identified an E3 ligase complex composed of several Polycomb-group proteins that link H2A ubLys119 to gene repression (Wang et al. 2004) and to the inactive X chromosome (de Napoles et al. 2004; Fang et al. 2004). Thus, the transcriptional outcome of histone ubiquitylation is not simple and may depend upon many factors, such as the nucleosome structure of the promoter or whether a specific ub site is tied to other modifications, as described below.

In yeast, ubH2B at Lys123 has an intriguing effect on other histone modifications, which may underlie its transcriptional activation role. Modification of histone H3 with methyl (me) groups on both Lys4 (Dover et al.

2002; Sun and Allis 2002) and Lys79 (Briggs et al. 2002; Ng et al. 2002a) depends on ubH2B; that is, either H2B Lys123Arg substitution or RAD6/BRE1 deletion (Hwang et al. 2003; Wood et al. 2003b) abolishes detectable methylation. However, although having a related function, meLys36 in H3 does not require ubH2B (Briggs et al. 2002; Ng et al. 2002a). Cross-talk between modifications in *cis*, on the same histone tail, has been characterized in both H3 and H4 (for example Lo et al. 2001; see Fischle et al. 2003), caused by increased binding of effector enzymes/proteins (Clements et al. 2003). However, cross-talk in *trans*, among modifications residing on different molecules (i.e., ubH2B and meH3) is unique. *Trans*-tail cross-talk may actually be direct since ubH2B and meLys4 may coexist on the same or neighboring nucleosomes (Sun and Allis 2002). Although the molecular mechanism creating the linkage is not understood (Henry and Berger 2002), methylation may be the functional readout of ubiquitylation in transcriptional activation, since there are strong parallels between ubH2B and meH3. For example, similar to ubH2B, Dot1-dependent meLys79 and Set1-dependent meLys4 are implicated in silencing (Nislow et al. 1997; Singer et al. 1998; Briggs et al. 2001; Bryk et al. 2002; Feng et al. 2002; Krogan et al. 2002; Ng et al. 2002b; van Leeuwen et al. 2002), activated gene transcription (Noma and Grewal 2002; Santos-Rosa et al. 2002), and repression (Carvin and Kladde 2004). These data strongly suggest that ubH2B acts in the same pathway with meLys4 and meLys79 in transcription.

Recent studies have identified two potential mechanisms that link ubiquitylation and methylation in this *trans*-histone pathway. First, components of the Paf1 transcriptional elongation complex are implicated in the cross-talk (Gerber and Shilatifard 2003). The Paf1 complex is associated primarily with elongating forms of RNA polymerase II (RNAPII), and thus Paf1 mutations exhibit defects in transcriptional elongation (Shilatifard 2004). Interestingly, mutations in the Paf1 complex also have very low levels of ubH2B, meLys4, and meLys79 and are defective in silencing. Indeed, Rad6/Bre1 and Set1 are all associated with early elongating RNAPII and loss of Paf1 leads to loss of the histone modification enzymes (Krogan et al. 2003a; Ng et al. 2003a,b; Wood et al. 2003a). In this setting, the Paf1 complex may provide a "platform" for interaction of the modification enzymes with RNAPII (Gerber and Shilatifard 2003). A second fascinating connection between histone ubiquitylation and methylation is related to previous observations that nondegradative portions of the proteosome are associated with gene promoters (Gonzalez et al. 2002). ATPase components of the proteosome were recently shown to be required for efficient meLys4 and meLys79, and ubH2B is required for recruitment of the ATPases (Ezhkova and Tansey 2004). ATPase mutations, similar to ubiquitylation and methylation deficiencies, cause both silencing and gene activation defects. It is possible that the subproteosomal module may alter the nucleosomal template to mediate methylation of histone H3. Thus, although the precise steps and physical associations are not yet known, there are several potential direct recruitment steps in this intricate pathway.

H2B DEUBIQUITYLATION BY Ubp8 IS INVOLVED IN TRANSCRIPTION OF SAGA/Gcn5-DEPENDENT GENES

Our recent data indicate that H2B deubiquitylation also regulates transcriptional activation. Many posttranslational modifications of proteins are reversed by specific enzymes. For example, HDACs (histone deacetylases) remove acetyl groups from histones. In most cases, removal results in the opposing function, e.g., histone acetyl transferases (HATs) activate genes and HDACs repress them (Kurdistani and Grunstein 2003). Ubiquitylation is similar in that ub conjugation is countered by the action of deubiquitylating enzymes (DUBs; for review, see Wilkinson 1997, 2000; Soboleva and Baker 2004). DUBs are thiol proteases that cleave the isopeptide bond at the carboxy-terminal-most Gly76 residue of ub, either removing ub from the substrate or processing the poly-ub chain. DUBs have been traditionally classified into two families, although there may be additional families (Soboleva and Baker 2004). One family, called ub carboxy-terminal hydrolases (UCHs), cleave ub from small adducts, such as amides or short peptides. The second family comprises the ub-specific proteases (abbreviated as UBPs for yeast proteins and as USPs for mammalian proteins). These are composed of 16 putative members in yeast and more than 80 in humans, making this family the largest in all ubiquitylation-related processes. The defining characteristic of UBPs is the presence of conserved domains, called the Cys-box and His-box, which are involved in formation of a "catalytic triad" for proteolytic cleavage and are reminiscent of the papain family of cysteine proteases (Wilkinson 1997; Hu et al. 2002, and references therein). UBPs mainly differ by insertions between the catalytic boxes or having divergent amino- or carboxy-terminal extensions, all of which may be involved in substrate selectivity (Wilkinson 1997; Soboleva and Baker 2004). DUBs are involved in several biological processes, such as cell growth, differentiation, immune function, and memory consolidation; DUB mutants are implicated in disease states, such as cancer (for review, see Wilkinson 1997; Chung and Baek 1999; Soboleva and Baker 2004). Nevertheless, other than in proteosomal degradation, little was known about deubiquitylation until recently, especially with regard to mono ub signaling.

One revealing finding was that Ubp8, a putative ub hydrolase, is a novel component of the SAGA coactivator complex (Sanders et al. 2002). Because we linked H2B ubiquitylation with SAGA complex–dependent gene activity (Henry et al. 2003; Kao et al. 2004), we investigated whether Ubp8-mediated H2B deubiquitylation occurs in the context of SAGA (reviewed in Wyce et al. 2004). We and others found that Ubp8 is a stable and stoichiometric subunit of SAGA and has a role in H2B deubiquitylation in vitro and in vivo (Henry et al. 2003; Daniel et al. 2004). Ubp8 is recruited to the SAGA-dependent *GAL1-10* gene promoter upon induction (Henry et al. 2003; Daniel et al. 2004), coincident with the Gcn5 HAT subunit of SAGA, leading to decreased ubH2B levels (Henry et al. 2003). As discussed above, antagonistic actions of histone-modifying enzymes generally result in opposite functional outcomes. However, we found that both ubiquitylation and deubiquitylation of H2B are required for the proper activation of SAGA-dependent genes (Henry et al. 2003). We examined the basis of this unusual requirement and found that the absence of H2B ubiquitylation compared to the absence of H2B deubiquitylation exhibits different defects in meH3. Targeted H3 Lys4 trimethylation occurs at the 5′ end of the ORF early during transcript elongation (Santos-Rosa et al. 2002; Ng et al. 2003b), while H3 Lys36 methylation occurs primarily later during transcript elongation (Krogan et al. 2003b; Xiao et al. 2003). As described above, ubH2B is required for H3 Lys4 and Lys79 methylation, but is not required for H3 Lys36 methylation. We found that ubiquitylation is required for meLys4 while deubiquitylation is required for meLys36, leading to a proposal that sequential ubiquitylation and deubiquitylation of H2B establishes the correct balance among H3 methylations required for proper gene activation (Fig. 1) (Henry et al. 2003; Wyce et al. 2004).

The catalytic domain of Ubp8 is similar to other putative UBPs. Indeed, we found that yeast Ubp10 also deubiquitylates H2B, but has a distinct genomic function and localization (see below). It appears that correct targeting is crucial for distinct roles of catalytically indistinguishable (at least with regard to H2B) Ubp8 or Ubp10. We investigated the basis for Ubp8 targeting and noted an amino-terminal cysteine/histidine-rich region. This region is predicted to bind zinc, forming a type of zinc finger called ZnF_UBP (Pfam domain ID# PF02148; Bateman et al. 2004). ZnF_UBP is also present in other UBPs and in mHDAC6 (a histone deacetylase), where, interestingly, it is involved in ub binding (Seigneurin-Berny et al. 2001; Hook et al. 2002). Since zinc fingers are known to be versatile modules for DNA and protein interaction (Laity et al. 2001), we tested a possible role in interaction with SAGA using a panel of deletion and substitution mutants. We found that the Ubp8 ZnF is important for association of Ubp8 with SAGA and for Ubp8's deubiquitylating activity in vivo (i.e., Ubp8 must be associated with SAGA for its enzymatic activity). In contrast, mutations in the catalytic domain reduce enzyme activity on ubH2B, as expected, but not SAGA association (Ingvarsdottir et al. 2005). Furthermore, when combined with deletion of *GCN5*, either catalytic or zinc finger mutants are defective in pathways requiring SAGA, suggesting that Ubp8 exerts its effects through histone H2B deubiquitylation and as a part of SAGA complex (Ingvarsdottir et al. 2005).

We and others recently identified a second novel component of SAGA, called Sgf11 (SAGA-associated factor 11 kD) (Lee et al. 2004; Powell et al. 2004; Ingvarsdottir

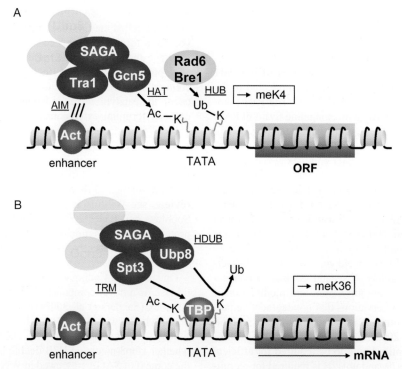

Figure 1. Model of SAGA modularity during transcriptional gene activation. SAGA is depicted in *red*, with one representative component from each module highlighted (i.e., Gcn5 of the HAT module; Tra1 of the activator interaction module [AIM]; Spt3 of the TBP [TATA-binding protein] regulation module [TRM]; Ubp8 of the histone deubiquitylase [HDUB] module). The roles of targeted action of SAGA in controlling transcriptional initiation and the transition from initiation to elongation are represented in two panels. (*A*) Recruitment of SAGA through AIM by a DNA-bound activator, and consequent acetylation of histones by the HAT module works in parallel with Rad6/Bre1-dependent histone H2B ubiquitylation and consequent Set1-dependent histone H3 methylation at Lys4. (*B*) This leads to preinitiation complex formation at the promoter through recruitment of TBP by the TRM. The transition to efficient transcriptional elongation depends on HDUB-mediated histone deubiquitylation and consequent Set2-dependent histone H3 Lys36 methylation. SAGA functions in these two stages need not be mutually exclusive, such as a possible requirement of the HAT module in the later elongation stages.

et al. 2005). Interestingly, Sgf11 is important for association of Ubp8 with SAGA (Powell et al. 2004; Ingvarsdottir et al. 2005; Lee et al. 2005) and for histone H2B deubiquitylation (Ingvarsdottir et al. 2005). Ubp8 interacts with Sgf11 in vivo through the Ubp8 ZnF (Ingvarsdottir et al. 2005). We have found that the functions of Ubp8 and Sgf11 are related and separable from other components of SAGA, such as Gcn5/Ada2/Ada3 (HAT module) or Spt3/Spt7/Spt8 (TBP regulatory module, TRM). In particular, the profiles of *UBP8* and *SGF11* deletions are remarkably similar in global gene expression analyses and large-scale synthetic genetic interactions (Ingvarsdottir et al. 2005). These results indicate that Ubp8 and Sgf11 represent a new functional module within SAGA ("histone deubiquitylase" [HDUB] module) involved in gene regulation through H2B deubiquitylation and consistent with the modular construction of SAGA (Grant et al. 1998). Whether Ubp8 and Sgf11 represent a separable, physical module remains to be established.

In summary, SAGA possesses a histone deubiquitylation module in addition to a histone acetylation module and these have quite distinct functions within the complex. Both enzymatic activities are required for SAGA's role in transcriptional activation of certain genes. One novel aspect to dynamic histone ubiquitylation is that it appears to increase early in gene activity and then, in contrast to histone acetylation, is rapidly cleaved. The failure to remove ub blocks the sequence of steps that occurs during gene activation and results in the failure to progress from H3 Lys4 methylation to H3 Lys36 methylation (Henry et al. 2003). Finally, Ubp8 is targeted to genes in a local fashion, through zinc finger domain and Sgf11-mediated association with SAGA (Ingvarsdottir et al. 2005), which is recruited to promoters through activator interaction with its Tra1 subunit (Brown et al. 2001).

SILENCING AND Ubp10 AS A HISTONE H2B DEUBIQUITYLATING ENZYME IMPLICATED IN SILENCING

In contrast to activator-mediated local recruitment of chromatin modification complexes to regulate transcription, other DNA-templated processes require broader recruitment, to either certain regions of the genome or even to whole chromosomes. We have identified a second H2B deubiquitylating enzyme, Ubp10, that indeed has a broader role than Ubp8, helping to maintain quiescence of certain regions within the yeast genome (Emre et al. 2005). These quiescent regions are similar in some respects to densely staining, late replicating heterochromatic regions of higher eukaryotic genomes. The chromatin within these regions tends to repress RNA polymerase II transcription, a phenomenon called "heterochromatic gene silencing." For example, transgenes inserted into silenced loci are repressed in a gene-nonspecific manner. In yeast these regions famously include telomeres, silent mating type loci (*HMR* and *HML*), and the rDNA locus (Huang 2002; Rusche et al. 2003).

Covalent modifications of histone tails are involved in silencing (Grunstein 1997). Telomeric silencing, also referred to as telomere position effect (TPE) in budding yeast (Gottschling et al. 1990), requires the SIR (Silent Information Regulatory) complex, composed of Sir3, Sir4, and the Sir2 HDAC. The Sir complex is recruited to telomere tips through interactions with sequence-specific DNA-binding proteins such as Rap1. Sir proteins bind to hypoacetylated tails and spread inward from telomeres as Sir2 deacetylates histone tails of neighboring nucleosomes (Huang 2002; Rusche et al. 2003; Moazed et al. 2004). Similar mechanisms act at the silent mating loci. Recent studies suggest spreading is counteracted at boundary regions by acetylation through the HAT Sas2 (Kimura et al. 2002; Suka et al. 2002) and by the presence of the histone variant H2A.Z (Meneghini et al. 2003) and other proteins, including bromodomain-containing Bdf1 (Ladurner et al. 2003) at boundary regions.

Methylation of lysine residues in histone tails has also been implicated in heterochromatin related silencing both in budding yeast (van Leeuwen and Gottschling 2002; Ng et al. 2003c; Santos-Rosa et al. 2004) and in other eukaryotes, however with notable differences: H3 Lys9 and Lys27 methylation silence through direct recruitment of heterochromatic chromodomain-containing proteins HP1 and Polycomb (Pc), respectively (Grewal and Moazed 2003). On the other hand, yeast apparently lack repressive histone methylation and homologs of the binding partners. Instead H3 Lys4 and Lys79 methylation are required, but indirectly, because these modifications are found primarily in active chromatin and may provide a relatively stable (compared to acetylation) memory mark for genes per se (Bernstein et al. 2002; Ng et al. 2003b). Furthermore, ubH2B has also been implicated in silencing, as described above, in part probably because of its requirement for H3 Lys4 and Lys79 methylation.

Our study of Ubp8 in histone deubiquitylation during gene activation prompted us to investigate whether other ub proteases in yeast may target H2B in transcriptional regulation. We considered two candidates in regulating gene silencing, Ubp3 and Ubp10 (Moazed and Johnson 1996; Singer et al. 1998). We found that the level of ubH2B is increased in a strain deleted for *UBP10*, similar to the effect of *UBP8* deletion, but not in the absence of *UBP3* (Henry et al. 2003; Emre et al. 2005). We confirmed that H2B is a substrate for Ubp10 in vitro using purified recombinant Ubp10, suggesting that ubH2B is a direct target of Ubp10. In addition, ChIP assays demonstrate that deletion of *UBP10* leads to increased ubH2B, specifically at a telomere proximal region, and Ubp10 is present at the affected regions. Further, in the absence of Ubp10, expression of a normally strongly silenced gene within this affected region is increased (Emre et al. 2005). These data provide a clear model that Ubp10 is directed to telomere proximal regions to maintain low ubiquitylation leading to low gene expression.

We further investigated molecular mechanisms of Ubp10 in silencing. We hypothesized that histone H3 methylation levels may be increased and Sir protein localization may be lowered at the telomere upon loss of Ubp10. The reasoning followed from evidence that ubH2B is required for H3 Lys4 (Dover et al. 2002; Sun and Allis 2002) and Lys79 (Briggs et al. 2002; Ng et al. 2002a) methylation in euchromatic regions of the genome, and especially in open reading frames (ORFs). Mutations that prevent these methylations cause silencing defects (see above), likely because of Sir protein relocalization from silenced to active regions (van Leeuwen and Gottschling 2002). These observations and earlier studies on Sir complex binding to undermodified histones (Carmen et al. 2002) prompted a model to explain the relationship between global histone modifications and silencing (van Leeuwen and Gottschling 2002): Lack of global modifications on histone tails (such as lack of H3 Lys4 or Lys79 methylation, or lysine acetylation) lead to loss of Sir proteins (whose levels are limited in the cell) away from silenced regions, causing silencing defects. These observations and models led us to test whether the increase in ubH2B in the absence of Ubp10 leads to an increase in meH3. We found that overexpression of Ubp10 leads to loss of global ubH2B and loss of H3 meLys4/meLys79 as well, although certain other unrelated modifications were not altered. Chromatin immunoprecipitation (ChIP) analysis showed that loss of Ubp10 also causes increased H3 meLys4 and meLys79 methylations close to the telomere but not at internal, active regions (Emre et al. 2005). Interestingly, loss of Ubp10 lowers the level of Sir2 association at the telomere and, reciprocally, loss of Sir2 lowers Ubp10. Further, using ChIP analysis we detected Ubp10 association with rDNA regions where Ubp10 deletion also leads to increased H3 methylation (Emre et al. 2005).

These observations suggest the following model (Fig. 2): Ubp10 is preferentially associated with and deubiquitylates H2B within silent chromatin, contributing to low methylation and helping to promote Sir association. Deletion of *UBP10* causes an increase in ubH2B, leading, in turn, to increased meH3; methylation of histone H3 is unfavorable for Sir protein association with the chromatin (van Leeuwen and Gottschling 2002; Ng et al. 2003c; Santos-Rosa et al. 2004), and presumably ubiquitylation is also incompatible with Sir binding. Decreased Sir2 binding in turn leads to decreased Ubp10 binding. Thus, there is a feedback loop between the HDAC and HDUB to maintain low modifications and silencing. On the other hand, overexpression of Ubp10 leads to global deubiquitylation of histone H2B (and global hypomethylation of H3 Lys4 and Lys79), which results in promiscuous binding of Sir proteins to normally active chromatin, causing the limited pools of Sirs to escape from silenced regions.

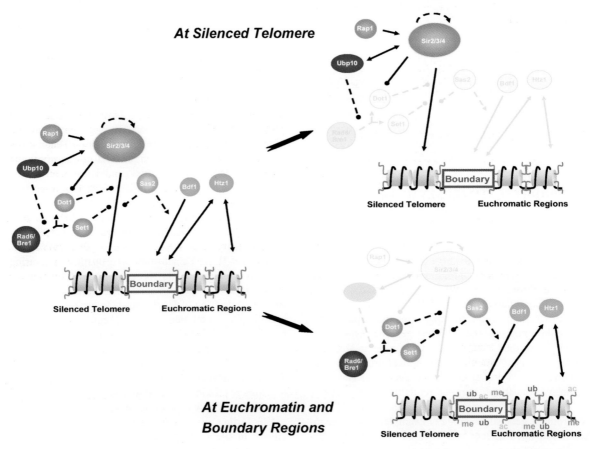

Figure 2. Model for yeast telomeric silencing incorporating the major factors known to act through histone modifications. Silencing Sir complex (Sir2/3/4), and factors that limit silencing in yeast (e.g., Htz1 and Bdf1) are shown competing for chromatin binding (*left*). *Solid lines* denote molecular interactions (with *arrowheads*), or inhibition of molecular interactions (with *spheres*). Dashed lines denote enzymatic activities on histones, either promoting a step in the pathway (with *arrowheads*) or inhibiting (with *spheres*). Typical outcomes are silencing at the telomere-proximal regions (*top right*), or protection from silencing at the euchromatin boundary (*bottom right*). Only relevant factors are drawn (not to scale).

Ubp8 VERSUS Ubp10: SAME HISTONE SUBSTRATE, DIFFERENT GENOMIC PROCESSES

Our characterization of Ubp8 and Ubp10 as histone H2B deubiquitylating enzymes raised the possibility that there might be an overlap or redundancy in their functions, in spite of the fact that they were discovered in very different functional contexts (Singer et al. 1998; Sanders et al. 2002; Henry et al. 2003). However, numerous observations (Table 2) suggest that Ubp10 and Ubp8 engage in nonoverlapping, distinct molecular processes, i.e., regional silencing and locus-specific activated gene transcription, respectively. First, loss of both Ubp10 and Ubp8 exhibit far higher levels of ubH2B compared to loss of either single enzyme, consistent with each enzyme targeting a separate subpopulation of ubH2B. Second, while Ubp8 and Gcn5 work cooperatively to activate genes that require SAGA function (Henry et al. 2003), Ubp10 does not display cooperativity with Gcn5 (Emre et al. 2005). Third, Ubp8 does not function similarly to Ubp10 in regulation of silencing: Ubp8 is not preferentially localized to telomere-proximal regions; deletion of Ubp8 does not increase telomere proximal levels of ubH2B, meH3, or gene expression (Emre et al. 2005). Finally, deletion of Ubp10, but not Ubp8, shows genetic interactions with the components of SWR-C, the nucleosome remodeling complex implicated in the disposition of silencing-related histone variant H2A.Z (reviewed in Korber and Horz 2004). Taken together, these data suggest that Ubp10 and Ubp8 are involved in distinct molecular processes, although they both target histone H2B for deubiquitylation.

SUMMARY AND PERSPECTIVES

Our observations considered along with recent studies from other laboratories indicate that histone H2B ubiquitylation and deubiquitylation, similar to acetylation and deacetylation, regulate two distinct aspects of transcriptional regulation: gene-specific expression (Fig. 1) and heterochromatic-like silencing in yeast (Fig. 2). Thus, in the first case, transcription is regulated by localized, dynamic ubiquitylation and deubiquitylation. H2B ubiquitylation is catalyzed by recruitment of the E2/E3 partners Rad6/Bre1, which, in a *trans*-tail relationship, triggers H3 methylation of Lys4 and Lys79. However, in apparent contrast to the typical situation, such as with acetylation/deacetylation causing activation/repression, H2B deubiquitylation is also required for productive transcription (Henry et al. 2003). This is catalyzed by activator-dependent recruitment of SAGA-associated ub protease Ubp8, which causes a transition of the predominant H3 methylation from Lys4/Lys79 to Lys36. Ubp8 is associated with Sgf11 as a functional module within SAGA that is mechanistically distinct from the acetylase Gcn5/Ada2/Ada3 or TBP-regulatory Spt3/Spt7/Spt8 modules (Ingvarsdottir et al. 2005).

Certain larger regions of the yeast genome, such as telomeres, are maintained in special "gene silencing" conformations, probably to prevent recombination and promote genome stability (Huang 2002; Rusche et al. 2003). It appears that histone acetylation/ubiquitylation, and thus H3 methylation, are constitutively low in these regions relative to transcriptionally poised ORF-rich euchromatin, which in contrast appears to bear constitutive H3 methylation to mark genes. Our data suggest that the deacetylase Sir2 and the deubiquitylase Ubp10 cooperatively maintain the hypomodified state through binding/spreading along the chromatin (Emre et al. 2005). Although the Sir complex is initially recruited by DNA sequence–specific binding proteins, the binding thereafter is less specific, with preference to undermodified histone tails. The complex is believed to extend along the chromatin fiber until a boundary element is reached, which is composed of H2A.Z, tRNA genes, and other marks. Sir binding, in turn, directly or indirectly recruits Ubp10 to the silenced regions. As previously suggested (Ng et al. 2003c), positive feedback loops may contribute to the epigenetic nature of telomere position effect, which is similar to position effect variegation (PEV) where distinct quasi-stable states, such as transcriptionally active or repressed telomeres, can be maintained for many generations among genetically identical cells.

Some general questions arise in considering the roles of histone ubiquitylation and deubiquitylation. One issue is whether there is a general role for ubiquitylation. In yeast, ubiquitylation is transcriptionally activating, whereas, in higher eukaryotes, it has both activating and

Table 2. Comparison of Ubp8 and Ubp10 Functions

Function	Ubp8	Ubp10
HDUB activity		
in vitro	yes	yes
in vivo	yes	yes
HDUB genomic targets and localization	SAGA-inducible gene promoters	silenced regions (*TEL*, *HML/R*, rDNA)
Recruitment	DNA-bound activators[a]	Sir complex
Effect on silenced genes	no	yes
Effect on gene induction	yes	no[b]
Effect on H3 methylation	regulatory: lowers Lys4[c] raises Lys36	lowers Lys4/Lys79

[a]As a component of SAGA, through its ZnF and its association with Sgf11.
[b]May have an indirect effect on the expression of euchromatic genes.
[c]Lys79 not tested.

repressing functions, as described above. How are these diverse outcomes achieved? One possibility is that, similar to methylation and acetylation, various effector proteins bind to ubiquitylated histones at different promoters, and these effectors lead to different functional readouts. One interesting example is that ubH2A may bind the linker histone H1 more strongly than H2A resulting in repression through chromatin compaction (Jason et al. 2004). A second model is suggested by the ubiquitylation/methylation *trans*-tail connection that exists in yeast. Similar ub/me patterns may exist in higher eukaryotes, but where methylation is both activating (e.g., H3 meLys4) and repressing (e.g., H3 meLys9), leading to dramatically divergent final functional readouts. However, as yet there is no reported evidence for a ubiquitylation/methylation *trans*-tail pattern outside of yeast (*S. cerevisiae* and *Schizosaccharomyces pombe* [Roguev et al. 2003]). A third hypothesisis is that ubiquitylation leads to alterations in the nucleosomal template itself and, depending on the nucleosome structure at the promoter, opens or compacts the chromatin. Finally, histone depletion has been detected around genes in yeast (Boeger et al. 2004; Korber et al. 2004; Schwabish and Struhl 2004). Ubiquitylation could lead to histone depletion through more conventional proteosomal degradation, which would activate or repress depending on subsequent binding of new regulatory proteins.

Considering both the significance of the reversibility concept in protein posttranslational modifications and the large number of putative deubiquitylating enzymes, it is not surprising that histone deubiquitylation is emerging as a means to control transcriptional processes, such as gene activation and silencing. However, it is not clear whether histone deubiquitylation has a general role during transcription, as appears to be true for Ubp8 in yeast, to serve as a gatekeeper or checkpoint. Our data suggest that deubiquitylation primes the elongation process to proceed to the next step in a series. Thus, early elongation (characterized by H3 Lys4 methylation) requires ubiquitylation; the switch to later elongation (characterized by H3 Lys36 methylation) requires deubiquitylation. Many questions arise here: What is the mechanism by which ubiquitylation blocks the elongation progress? Does ubiquitylation serve a similar checkpoint function during transcriptional repression? It is also completely unknown whether higher eukaryotes have a similar system at authentic heterochromatin as Ubp10's role in heterochromatin-like silencing in yeast in maintaining constitutively low ubiquitylation/methylation.

Histone ubiquitylation has a role in processes beyond transcription and silencing. The ubH2B-meH3 *trans*-tail modification pathway is involved in DNA double strand break (DSB) formation and repair. Rad6-dependent histone H2B ubiquitylation is important for meiosis (Robzyk et al. 2000), possibly because of its role in DSBs (Yamashita et al. 2004). Similarly, Set1 (Sollier et al. 2004) and Dot1 (San-Segundo and Roeder 2000; Game et al. 2005) function during meiosis and/or DSB formation and, in mammalian cells, meLys79 may provide a recognition mark for binding of DNA damage sensor proteins to DSBs (Huyen et al. 2004). It will be interesting to learn whether histone deubiquitylating enzymes also regulate these processes.

Thus, histone ubiquitylation and deubiquitylation have emerged as key regulatory signals in eukaryotes during transcription, silencing, and beyond. Similar to histone acetylation, ubiquitylation largely correlates with open, active chromatin in yeast and thus is involved in transcription and DNA repair and is absent from heterochromatin-like regions. In contrast to the role of histone acetylation in transcription, both adding and removing ub is required for productive RNA synthesis in yeast. Understanding the function of transcriptional repressive ubiquitylation in higher eukaryotes may provide key insight into possible general mechanisms.

ACKNOWLEDGMENTS

We thank K. Ingvarsdottir, A. Wyce, K.W. Henry, A. Wood, N.J. Krogan, K. Li, R. Marmorstein, J.F. Greenblatt, and A. Shilatifard for their contributions to the studies on ubiquitin proteases that were in press at the time of this review. These studies are supported by grants from the National Institutes of Health and the National Science Foundation to S.L.B.

REFERENCES

Aguilar R.C. and Wendland B. 2003. Ubiquitin: Not just for proteasomes anymore. *Curr. Opin. Cell Biol.* **15:** 184.

Bateman A., Coin L., Durbin R., Finn R.D., Hollich V., Griffiths-Jones S., Khanna A., Marshall M., Moxon S., Sonnhammer E.L., Studholme D.J., Yeats C., and Eddy S.R. 2004. The Pfam protein families database. *Nucleic Acids Res.* **32:** database issue D138.

Berger S.L. 2002. Histone modifications in transcriptional regulation. *Curr. Opin. Genet. Dev.* **12:** 142.

Bernstein B.E., Humphrey E.L., Erlich R.L., Schneider R., Bouman P., Liu J.S., Kouzarides T., and Schreiber S.L. 2002. Methylation of histone H3 Lys 4 in coding regions of active genes. *Proc. Natl. Acad. Sci.* **99:** 8695.

Boeger H., Griesenbeck J., Strattan J.S., and Kornberg R.D. 2004. Removal of promoter nucleosomes by disassembly rather than sliding in vivo. *Mol. Cell* **14:** 667.

Briggs S.D., Bryk M., Strahl B.D., Cheung W.L., Davie J.K., Dent S.Y., Winston F., and Allis C.D. 2001. Histone H3 lysine 4 methylation is mediated by Set1 and required for cell growth and rDNA silencing in *Saccharomyces cerevisiae*. *Genes Dev.* **15:** 3286.

Briggs S.D., Xiao T., Sun Z.W., Caldwell J.A., Shabanowitz J., Hunt D.F., Allis C.D., and Strahl B.D. 2002. Gene silencing: Trans-histone regulatory pathway in chromatin. *Nature* **418:** 498.

Brown C.E., Howe L., Sousa K., Alley S.C., Carrozza M.J., Tan S., and Workman J.L. 2001. Recruitment of HAT complexes by direct activator interactions with the ATM-related Tra1 subunit. *Science* **292:** 2333.

Bryk M., Briggs S.D., Strahl B.D., Curcio M.J., Allis C.D., and Winston F. 2002. Evidence that Set1, a factor required for methylation of histone H3, regulates rDNA silencing in *S. cerevisiae* by a Sir2-independent mechanism. *Curr. Biol.* **12:** 165.

Carmen A.A., Milne L., and Grunstein M. 2002. Acetylation of the yeast histone H4 N terminus regulates its binding to heterochromatin protein SIR3. *J. Biol. Chem.* **277:** 4778.

Carvin C.D. and Kladde M.P. 2004. Effectors of lysine 4 methylation of histone H3 in *Saccharomyces cerevisiae* are negative regulators of PHO5 and GAL1-10. *J. Biol. Chem.* **279:** 33057.

Chen H.Y., Sun J.M., Zhang Y., Davie J.R., and Meistrich M.L. 1998. Ubiquitination of histone H3 in elongating spermatids

of rat testes. *J. Biol. Chem.* **273:** 13165.
Chung C.H. and Baek S.H. 1999. Deubiquitinating enzymes: Their diversity and emerging roles. *Biochem. Biophys. Res. Commun.* **266:** 633.
Clements A., Poux A.N., Lo W.S., Pillus L., Berger S.L., and Marmorstein R. 2003. Structural basis for histone and phosphohistone binding by the GCN5 histone acetyltransferase. *Mol. Cell* **12:** 461.
Daniel J.A., Torok M.S., Sun Z.W., Schieltz D., Allis C.D., Yates J.R., III, and Grant P.A. 2004. Deubiquitination of histone H2B by a yeast acetyltransferase complex regulates transcription. *J. Biol. Chem.* **279:** 1867.
de Napoles M., Mermoud J.E., Wakao R., Tang Y.A., Endoh M., Appanah R., Nesterova T.B., Silva J., Otte A.P., Vidal M., Koseki H., and Brockdorff N. 2004. Polycomb group proteins Ring1A/B link ubiquitylation of histone H2A to heritable gene silencing and X inactivation. *Dev. Cell* **7:** 663.
Dover J., Schneider J., Tawiah-Boateng M.A., Wood A., Dean K., Johnston M., and Shilatifard A. 2002. Methylation of histone H3 by COMPASS requires ubiquitination of histone H2B by Rad6. *J. Biol. Chem.* **277:** 28368.
Emre N.C., Ingvarsdottir K., Wyce A., Wood A., Krogan N.J., Henry K.W., Li K., Marmorstein R., Greenblatt J.R., Shilatifard A., and Berger S.L. 2005. Maintenance of low histone ubiquitylation by Ubp10 correlates with telomere proximal Sir2 association and gene silencing. *Mol. Cell* **17:** 585.
Ezhkova E. and Tansey W.P. 2004. Proteasomal ATPases link ubiquitylation of histone H2B to methylation of histone H3. *Mol. Cell* **13:** 435.
Fang J., Chen T., Chadwick B., Li E., and Zhang Y. 2004. Ring1b-mediated H2A ubiquitination associates with inactive X chromosomes and is involved in initiation of X inactivation. *J. Biol. Chem.* **279:** 52812.
Feng Q., Wang H., Ng H.H., Erdjument-Bromage H., Tempst P., Struhl K., and Zhang Y. 2002. Methylation of H3-lysine 79 is mediated by a new family of HMTases without a SET domain. *Curr. Biol.* **12:** 1052.
Fischle W., Wang Y., and Allis C.D. 2003. Histone and chromatin cross-talk. *Curr. Opin. Cell Biol.* **15:** 172.
Flaus A. and Owen-Hughes T. 2004. Mechanisms for ATP-dependent chromatin remodelling: Farewell to the tuna-can octamer? *Curr. Opin. Genet. Dev.* **14:** 165.
Game J.C., Williamson M.S., and Baccari C. 2005. X-ray survival characteristics and genetic analysis for nine *Saccharomyces* deletion mutants that affect radiation sensitivity. *Genetics* **169:** 51.
Gerber M. and Shilatifard A. 2003. Transcriptional elongation by RNA polymerase II and histone methylation. *J. Biol. Chem.* **278:** 26303.
Gonzalez F., Delahodde A., Kodadek T., and Johnston S.A. 2002. Recruitment of a 19S proteasome subcomplex to an activated promoter. *Science* **296:** 548.
Gottschling D.E., Aparicio O.M., Billington B.L., and Zakian V.A. 1990. Position effect at *S. cerevisiae* telomeres: Reversible repression of Pol II transcription. *Cell* **63:** 751.
Grant P.A., Sterner D.E., Duggan L.J., Workman J.L., and Berger S.L. 1998. The SAGA unfolds: Convergence of transcription regulators in chromatin-modifying complexes. *Trends Cell Biol.* **8:** 193.
Grewal S.I. and Moazed D. 2003. Heterochromatin and epigenetic control of gene expression. *Science* **301:** 798.
Grunstein M. 1997. Molecular model for telomeric heterochromatin in yeast. *Curr. Opin. Cell Biol.* **9:** 383.
Henry K.W. and Berger S.L. 2002. Trans-tail histone modifications: Wedge or bridge? *Nat. Struct. Biol.* **9:** 565.
Henry K.W., Wyce A., Lo W.S., Duggan L.J., Emre N.C., Kao C.F., Pillus L., Shilatifard A., Osley M.A., and Berger S.L. 2003. Transcriptional activation via sequential histone H2B ubiquitylation and deubiquitylation, mediated by SAGA-associated Ubp8. *Genes Dev.* **17:** 2648.
Hicke L. 2001. Protein regulation by monoubiquitin. *Nat. Rev. Mol. Cell Biol.* **2:** 195.
Hochstrasser M. 1996. Ubiquitin-dependent protein degradation. *Annu. Rev. Genet.* **30:** 405.
Hook S.S., Orian A., Cowley S.M., and Eisenman R.N. 2002. Histone deacetylase 6 binds polyubiquitin through its zinc finger (PAZ domain) and copurifies with deubiquitinating enzymes. *Proc. Natl. Acad. Sci.* **99:** 13425.
Hu M., Li P., Li M., Li W., Yao T., Wu J.W., Gu W., Cohen R.E., and Shi Y. 2002. Crystal structure of a UBP-family deubiquitinating enzyme in isolation and in complex with ubiquitin aldehyde. *Cell* **111:** 1041.
Huang H., Kahana A., Gottschling D.E., Prakash L., and Liebman S.W. 1997. The ubiquitin-conjugating enzyme Rad6 (Ubc2) is required for silencing in *Saccharomyces cerevisiae*. *Mol. Cell. Biol.* **17:** 6693.
Huang Y. 2002. Transcriptional silencing in *Saccharomyces cerevisiae* and *Schizosaccharomyces pombe*. *Nucleic Acids Res.* **30:** 1465.
Huyen Y., Zgheib O., Ditullio R.A., Jr., Gorgoulis V.G., Zacharatos P., Petty T.J., Sheston E.A., Mellert H.S., Stavridi E.S., and Halazonetis T.D. 2004. Methylated lysine 79 of histone H3 targets 53BP1 to DNA double-strand breaks. *Nature* **432:** 406.
Hwang W.W., Venkatasubrahmanyam S., Ianculescu A.G., Tong A., Boone C., and Madhani H.D. 2003. A conserved RING finger protein required for histone H2B monoubiquitination and cell size control. *Mol. Cell* **11:** 261.
Ingvarsdottir K., Krogan N.J., Emre N.C.T., Wyce A., Thompson N.J., Emili A., Hughes T.R., Greenblatt J., and Berger S.L. 2005. H2B ubiquitin protease Ubp8 and Sgf11 constitute a discrete functional module within the *Saccharomyces cerevisiae* SAGA complex. *Mol. Cell. Biol.* **25:** 1162–1172.
Jason L.J., Finn R.M., Lindsey G., and Ausio J. 2005. Histone H2A ubiquitination does not preclude histone H1 binding, but it facilitates its association with the nucleosome. *J. Biol. Chem.* **280:** 4975.
Jason L.J., Moore S.C., Lewis J.D., Lindsey G., and Ausio J. 2002. Histone ubiquitination: A tagging tail unfolds? *Bioessays* **24:** 166.
Kao C.F., Hillyer C., Tsukuda T., Henry K., Berger S., and Osley M.A. 2004. Rad6 plays a role in transcriptional activation through ubiquitylation of histone H2B. *Genes Dev.* **18:** 184.
Katan-Khaykovich Y. and Struhl K. 2002. Dynamics of global histone acetylation and deacetylation in vivo: Rapid restoration of normal histone acetylation status upon removal of activators and repressors. *Genes Dev.* **16:** 743.
Kimura A., Umehara T., and Horikoshi M. 2002. Chromosomal gradient of histone acetylation established by Sas2p and Sir2p functions as a shield against gene silencing. *Nat. Genet.* **32:** 370.
Korber P. and Horz W. 2004. SWRred not shaken; mixing the histones. *Cell* **117:** 5.
Korber P., Luckenbach T., Blaschke D., and Horz W. 2004. Evidence for histone eviction in trans upon induction of the yeast PHO5 promoter. *Mol. Cell. Biol.* **24:** 10965.
Krogan N.J., Dover J., Khorrami S., Greenblatt J.F., Schneider J., Johnston M., and Shilatifard A. 2002. COMPASS, a histone H3 (Lysine 4) methyltransferase required for telomeric silencing of gene expression. *J. Biol. Chem.* **277:** 10753.
Krogan N.J., Dover J., Wood A., Schneider J., Heidt J., Boateng M.A., Dean K., Ryan O.W., Golshani A., Johnston M., Greenblatt J.F., and Shilatifard A. 2003a. The Paf1 complex is required for histone H3 methylation by COMPASS and Dot1p: Linking transcriptional elongation to histone methylation. *Mol. Cell* **11:** 721.
Krogan N.J., Kim M., Tong A., Golshani A., Cagney G., Canadien V., Richards D.P., Beattie B.K., Emili A., Boone C., Shilatifard A., Buratowski S., and Greenblatt J. 2003b. Methylation of histone H3 by Set2 in *Saccharomyces cerevisiae* is linked to transcriptional elongation by RNA polymerase II. *Mol. Cell. Biol.* **23:** 4207.
Kurdistani S.K. and Grunstein M. 2003. Histone acetylation and deacetylation in yeast. *Nat. Rev. Mol. Cell Biol.* **4:** 276.
Ladurner A.G., Inouye C., Jain R., and Tjian R. 2003. Bromodomains mediate an acetyl-histone encoded antisilencing function at heterochromatin boundaries. *Mol. Cell* **11:** 365.
Laity J.H., Lee B.M., and Wright P.E. 2001. Zinc finger proteins: New insights into structural and functional diversity.

Curr. Opin. Struct. Biol. **11:** 39.

Lee K.K., Florens L., Swansen S.K., Washburn M.P., and Workman J.L. 2005. The deubiquitylation activity of Ubp8 is dependent upon Sgf11 and its association with the SAGA complex. *Mol. Cell. Biol.* **25:** 1173–1182.

Lee K.K., Prochasson P., Florens L., Swanson S.K., Washburn M.P., and Workman J.L. 2004. Proteomic analysis of chromatin-modifying complexes in *Saccharomyces cerevisiae* identifies novel subunits. *Biochem. Soc. Trans.* **32:** 899.

Lo W.S., Duggan L., Emre N.C.T., Belotserkovskya R., Lane W.S., Shiekhattar R., and Berger S.L. 2001. Snf1—A histone kinase that works in concert with the histone acetyltransferase Gcn5 to regulate transcription. *Science* **293:** 1142.

Luger K. 2003. Structure and dynamic behavior of nucleosomes. *Curr. Opin. Genet. Dev.* **13:** 127.

Matunis M.J. 2002. On the road to repair: PCNA encounters SUMO and ubiquitin modifications. *Mol. Cell* **10:** 441.

Meneghini M.D., Wu M., and Madhani H.D. 2003. Conserved histone variant H2A.Z protects euchromatin from the ectopic spread of silent heterochromatin. *Cell* **112:** 725.

Michael D. and Oren M. 2003. The p53-Mdm2 module and the ubiquitin system. *Semin. Cancer Biol.* **13:** 49.

Minsky N. and Oren M. 2004. The RING domain of Mdm2 mediates histone ubiquitylation and transcriptional repression. *Mol. Cell* **16:** 631.

Moazed D. and Johnson D. 1996. A deubiquitinating enzyme interacts with SIR4 and regulates silencing in *S. cerevisiae*. *Cell* **86:** 667.

Moazed D., Rudner A.D., Huang J., Hoppe G.J., and Tanny J.C. 2004. A model for step-wise assembly of heterochromatin in yeast. *Novartis Found. Symp.* **259:** 48.

Moore S.C., Jason L., and Ausio J. 2002. The elusive structural role of ubiquitinated histones. *Biochem. Cell Biol.* **80:** 311.

Muratani M. and Tansey W.P. 2003. How the ubiquitin-proteasome system controls transcription. *Nat. Rev. Mol. Cell Biol.* **4:** 192.

Nathan D., Sterner D.E., and Berger S.L. 2003. Histone modifications: Now summoning sumoylation. *Proc. Natl. Acad. Sci.* **100:** 13118.

Ng H.H., Dole S., and Struhl K. 2003a. The Rtf1 component of the Paf1 transcriptional elongation complex is required for ubiquitination of histone H2B. *J. Biol. Chem.* **278:** 33625.

Ng H.H., Robert F., Young R.A., and Struhl K. 2003b. Targeted recruitment of Set1 histone methylase by elongating Pol II provides a localized mark and memory of recent transcriptional activity. *Mol. Cell* **11:** 709.

Ng H.H., Xu R.M., Zhang Y., and Struhl K. 2002a. Ubiquitination of histone H2B by Rad6 is required for efficient Dot1-mediated methylation of histone H3 lysine 79. *J. Biol. Chem.* **277:** 34655.

Ng H.H., Ciccone D.N., Morshead K.B., Oettinger M.A., and Struhl K. 2003c. Lysine-79 of histone H3 is hypomethylated at silenced loci in yeast and mammalian cells: A potential mechanism for position-effect variegation. *Proc. Natl. Acad. Sci.* **100:** 1820.

Ng H.H., Feng Q., Wang H., Erdjument-Bromage H., Tempst P., Zhang Y., and Struhl K. 2002b. Lysine methylation within the globular domain of histone H3 by Dot1 is important for telomeric silencing and Sir protein association. *Genes Dev.* **16:** 1518.

Nislow C., Ray E., and Pillus L. 1997. SET1, a yeast member of the trithorax family, functions in transcriptional silencing and diverse cellular processes. *Mol. Biol. Cell* **8:** 2421.

Noma K. and Grewal S.I. 2002. Histone H3 lysine 4 methylation is mediated by Set1 and promotes maintenance of active chromatin states in fission yeast. *Proc. Natl. Acad. Sci.* (suppl. 4) **99:** 16438.

Osley M.A. 2004. H2B ubiquitylation: The end is in sight. *Biochim. Biophys. Acta* **1677:** 74.

Pham A.D. and Sauer F. 2000. Ubiquitin-activating/conjugating activity of TAFII250, a mediator of activation of gene expression in *Drosophila*. *Science* **289:** 2357.

Powell D.W., Weaver C.M., Jennings J.L., McAfee K.J., He Y., Weil P.A., and Link A.J. 2004. Cluster analysis of mass spectrometry data reveals a novel component of SAGA. *Mol. Cell. Biol.* **24:** 7249.

Robzyk K., Recht J., and Osley M.A. 2000. Rad6-dependent ubiquitination of histone H2B in yeast. *Science* **287:** 501.

Roguev A., Schaft D., Shevchenko A., Aasland R., Shevchenko A., and Stewart A.F. 2003. High conservation of the Set1/Rad6 axis of histone 3 lysine 4 methylation in budding and fission yeasts. *J. Biol. Chem.* **278:** 8487.

Rusche L.N., Kirchmaier A.L., and Rine J. 2003. The establishment, inheritance, and function of silenced chromatin in *Saccharomyces cerevisiae*. *Annu. Rev. Biochem.* **72:** 481.

Sanders S.L., Jennings J., Canutescu A., Link A.J., and Weil P.A. 2002. Proteomics of the eukaryotic transcription machinery: Identification of proteins associated with components of yeast TFIID by multidimensional mass spectrometry. *Mol. Cell. Biol.* **22:** 4723.

San-Segundo P.A. and Roeder G.S. 2000. Role for the silencing protein Dot1 in meiotic checkpoint control. *Mol. Biol. Cell* **11:** 3601.

Santos-Rosa H., Bannister A.J., Dehe P.M., Geli V., and Kouzarides T. 2004. Methylation of H3 lysine 4 at euchromatin promotes Sir3p association with heterochromatin. *J. Biol. Chem.* **279:** 47506.

Santos-Rosa H., Schneider R., Bannister A.J., Sherriff J., Bernstein B.E., Emre N.C.T., Schreiber S.L., Mellor J., and Kouzarides T. 2002. Active genes are tri-methylated at K4 of histone H3. *Nature* **419:** 407.

Schwabish M.A. and Struhl K. 2004. Evidence for eviction and rapid deposition of histones upon transcriptional elongation by RNA polymerase II. *Mol. Cell. Biol.* **24:** 10111.

Seigneurin-Berny D., Verdel A., Curtet S., Lemercier C., Garin J., Rousseaux S., and Khochbin S. 2001. Identification of components of the murine histone deacetylase 6 complex: Link between acetylation and ubiquitination signaling pathways. *Mol. Cell. Biol.* **21:** 8035.

Shilatifard A. 2004. Transcriptional elongation control by RNA polymerase II: A new frontier. *Biochim. Biophys. Acta* **1677:** 79.

Singer M.S., Kahana A., Wolf A.J., Meisinger L.L., Peterson S.E., Goggin C., Mahowald M., and Gottschling D.E. 1998. Identification of high-copy disruptors of telomeric silencing in *Saccharomyces cerevisiae*. *Genetics* **150:** 613.

Soboleva T.A. and Baker R.T. 2004. Deubiquitinating enzymes: Their functions and substrate specificity. *Curr. Protein Pept. Sci.* **5:** 191.

Sollier J., Lin W., Soustelle C., Suhre K., Nicolas A., Geli V., and De La Roche Saint-Andre C. 2004. Set1 is required for meiotic S-phase onset, double-strand break formation and middle gene expression. *EMBO J.* **23:** 1957.

Sterner D.E., Grant P.A., Roberts S.M., Duggan L.J., Belotserkovskaya R., Pacella L.A., Winston F., Workman J.L., and Berger S.L. 1999. Functional organization of the yeast SAGA complex: Distinct components involved in structural integrity, nucleosome acetylation, and TATA-binding protein interaction. *Mol. Cell. Biol.* **19:** 86.

Suka N., Luo K., and Grunstein M. 2002. Sir2p and Sas2p opposingly regulate acetylation of yeast histone H4 lysine16 and spreading of heterochromatin. *Nat. Genet.* **32:** 378.

Sun Z.W. and Allis C.D. 2002. Ubiquitination of histone H2B regulates H3 methylation and gene silencing in yeast. *Nature* **418:** 104.

Turner S.D., Ricci A.R., Petropoulos H., Genereaux J., Skerjanc I.S., and Brandl C.J. 2002. The E2 ubiquitin conjugase Rad6 is required for the ArgR/Mcm1 repression of ARG1 transcription. *Mol. Cell. Biol.* **22:** 4011.

van Leeuwen F. and Gottschling D.E. 2002. Genome-wide histone modifications: Gaining specificity by preventing promiscuity. *Curr. Opin. Cell Biol.* **14:** 756.

van Leeuwen F., Gafken P.R., and Gottschling D.E. 2002. Dot1p modulates silencing in yeast by methylation of the nucleosome core. *Cell* **109:** 745.

Wang H., Wang L., Erdjument-Bromage H., Vidal M., Tempst P., Jones R.S., and Zhang Y. 2004. Role of histone H2A ubiquitination in Polycomb silencing. *Nature* **431:** 873.

Wilkinson K.D. 1997. Regulation of ubiquitin-dependent processes by deubiquitinating enzymes. *FASEB J.* **11:** 1245.
———. 2000. Ubiquitination and deubiquitination: Targeting of proteins for degradation by the proteasome. *Semin. Cell Dev. Biol.* **11:** 141.
Wood A., Schneider J., Dover J., Johnston M., and Shilatifard A. 2003a. The Paf1 complex is essential for histone monoubiquitination by the Rad6-Bre1 complex, which signals for histone methylation by COMPASS and Dot1p. *J. Biol. Chem.* **278:** 34739.
Wood A., Krogan N.J., Dover J., Schneider J., Heidt J., Boateng M.A., Dean K., Golshani A., Zhang Y., Greenblatt J.F., Johnston M., and Shilatifard A. 2003b. Bre1, an E3 ubiquitin ligase required for recruitment and substrate selection of Rad6 at a promoter. *Mol. Cell* **11:** 267.
Wyce A., Henry K.W., and Berger S.L. 2004. H2B ubiquitylation and de-ubiquitylation in gene activation. *Novartis Found. Symp.* **259:** 63.
Xiao T., Hall H., Kizer K.O., Shibata Y., Hall M.C., Borchers C.H., and Strahl B.D. 2003. Phosphorylation of RNA polymerase II CTD regulates H3 methylation in yeast. *Genes Dev.* **17:** 654.
Yamashita K., Shinohara M., and Shinohara A. 2004. Rad6-Bre1-mediated histone H2B ubiquitylation modulates the formation of double-strand breaks during meiosis. *Proc. Natl. Acad. Sci.* **101:** 11380.
Zhang Y. 2003. Transcriptional regulation by histone ubiquitination and deubiquitination. *Genes Dev.* **17:** 2733.

Polycomb Silencing Mechanisms in *Drosophila*

Y.B. Schwartz, T.G. Kahn, G.I. Dellino, and V. Pirrotta
Department of Zoology, University of Geneva, CH-1211 Geneva, Switzerland

Homeotic genes are a preeminent target for epigenetic mechanisms that program and maintain a chromatin state. The study of their function in *Drosophila* originally allowed the identification of Polycomb Group (PcG) genes, although we now know that PcG mechanisms regulate many other genes. Homeotic genes must be expressed in specific segmental domains of the *Drosophila* body plan throughout development. The expression domains are set in the earliest stages of embryonic development by transient regulators localized in specific regions of the embryo by maternal cues. Shortly after blastoderm, PcG complexes take over to maintain the repressed state in those cells in which the target genes had not been activated in the first 3 hours of development. The descendants of these cells maintain this silent state of the target genes for the rest of development.

PcG proteins are present in all nuclei at all stages in which they have been examined, including the very early embryo, where they are both maternally and zygotically supplied. Remarkably, however, they neither prevent the initial activation of homeotic genes nor their continued expression in the cells in which activation had initially occurred. Both the silenced and the nonsilenced state are therefore epigenetic states that are maintained for the rest of development. After each round of cell division, the gene preserves an epigenetic memory of its chromatin state in the previous cycle. Critical for the establishment and maintenance of this state is the Polycomb Response Element (PRE), the site of action of PcG proteins in the regulatory region of the target genes (Chan et al. 1994). In the case of the *Ultrabithorax* (*Ubx*) gene, the best known, the PRE is located 24 kilobases (kb) upstream of the *Ubx* promoter. When the PRE in transgenic construct is excised during development, the repression is lost, showing that it is required not only to establish but to maintain the silenced chromatin state (Busturia et al. 1997).

Despite the attractiveness of this two-state, on-or-off epigenetic model, it is important to bear in mind that the function of PcG complexes does not always adhere to this scheme, which is derived from the regulation of the homeotic genes. A large number of other PcG targets are unlikely to be regulated the same way. Although we know little about most of these other targets, genetic and molecular evidence shows that at least two of them, the *engrailed* and *hedgehog* genes, must be reprogrammed at some stages during development (Maurange and Paro 2002). In at least two other cases, the *polyhomeotic* (*ph*) and *Posterior sex combs* (*Psc*) genes, themselves encoding PcG proteins, the PcG regulation does not define a simple on-or-off alternative. In these genes, the PcG mechanisms down-regulate the expression rather than permanently switching it off (Fauvarque and Dura 1993; Rastelli et al. 1993). How PcG complexes up- and down-regulate transcription is even less understood than the all-or-none silencing mechanism and most likely depends on the structure of the different PREs and the dynamics of the proteins that interact with them. In this discussion, we will confine ourselves to the behavior of the PREs found in homeotic genes.

THE MAINTENANCE OF THE EPIGENETIC STATE

An illustration of how the PRE maintains the silenced or nonrepressed state is given in Figure 1, which summarizes the results of reporter constructs containing the *Ubx* promoter fused to lacZ under control of different combinations of early enhancers, imaginal disc enhancers, and PRE (Pirrotta et al. 1995; Poux et al. 1996). The early enhancers of the *Ubx* gene respond to segmentation genes and are activated before the cellular blastoderm stage in the 3-hour-old embryo. They are repressed by the Hunchback segmentation gene product in the anterior half of the embryo. This repression in regions (for simplicity, anterior to parasegment 6) must be maintained since *Ubx* expression in more anterior segments would cause segmental transformations. However, since the Hunchback product disappears shortly after blastoderm, a reporter gene containing only segmental enhancers becomes derepressed and expression occurs in all segments. No expression occurs in the larval imaginal discs since the segmentation genes do not function in larval stages. When an early enhancer is combined with a PRE, permanent repression sets in after blastoderm wherever the reporter gene was inactive at that stage but no repression occurs wherever the gene was active. As a consequence, the segmental domain of expression defined by the early regulatory proteins (posterior to parasegment 6) is maintained.

Imaginal disc enhancers do not function in the early embryo but become activated when the disc primordia are determined. They direct expression in imaginal discs regardless of segmental origin. These enhancers can function therefore in the haltere disc (posterior thorax) but also in the wing (mesothorax) and the eye/antenna (head) discs. If a PRE is added to the imaginal enhancer, there is no expression at any stage because at blastoderm, when

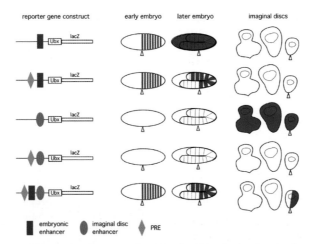

Figure 1. Epigenetic transmission of the repressed state mediated by the PRE. (*Left column*) Diagrams of reporter constructs containing combinations of early embryonic segmental enhancers, larval imaginal enhancers, and PRE, controlling the expression of the *Ubx-lacZ* reporter gene. (*Right columns*) The expression patterns (*blue*) in the early embryo, later embryo, and in the larval eye antenna, wing, and haltere disk, in anterior–posterior order. (*Arrowhead*) The parasegment 6 boundary.

silencing is established, the imaginal enhancers are not active. However, if all three elements are present together, the early enhancers activate the gene in the posterior segments, silencing is established only in the anterior segments, and, when the imaginal enhancers become active, they will drive expression only in the segmental domain defined by the early enhancers.

In this way, the PRE could be said to convey the positional information from the early enhancers (which are repressed in the anterior segments) to the larval enhancers (which do not distinguish head and anterior thorax from posterior thorax). There is a gap between the time when the early enhancers cease functioning and when the imaginal enhancers become active. What prevents PcG silencing from setting in during that time? Current evidence indicates that this is the function of the Trithorax system. Interaction with Trithorax is also mediated by the PRE, which contains a subdomain necessary for Trithorax recruitment (Tillib et al. 1999). While Trithorax also stimulates the expression of the *Ubx* promoter, it appears to maintain a memory of an earlier state of activity that antagonizes the establishment of silencing.

PcG PROTEINS ARE ASSEMBLED IN TWO DISTINCT COMPLEXES

The effectors of PcG silencing are the multiprotein complexes exemplified by the biochemically purified PRC1 complex, which includes a core quartet of PcG proteins PC, PH, PSC, and dRING (Saurin et al. 2001), and the PRC2 complex, which includes ESC and the histone methyltransferase E(Z) (Cao et al. 2002; Czermin et al. 2002; Kuzmichev et al. 2002 Müller et al. 2002). Both of these complexes are recruited to PREs and both appear to be essential for PcG silencing. Various kinds of genetic, cytogenetic, and biochemical evidence indicate that there is probably considerable variation in the composition of these complexes at different target sites, tissues, or developmental stages.

The targets of the E(Z)/ESC complex are lysine 9 and lysine 27 (K9, K27), which become trimethylated in vitro. Whether both are also methylated in vivo is not clear, but experiments with different antibodies in different laboratories agree that K27 is certainly methylated in vivo at genomic PcG target sites. By analogy with the roles of Suvar3-9 and HP1, it has been proposed that the histone methylation serves to recruit PC by increasing the affinity of the PC chromodomain for histone H3. In vitro experiments confirm that histone H3 K27 methylation, but not K9 methylation, increases the in vitro binding of the PC chromodomain to a peptide containing the K27 region about 100-fold (Fischle et al. 2003). The increase in binding to full-length recombinant H3 me3K27 is less dramatic (Czermin et al. 2002), probably because PC has additional affinity for intact H3 not mediated by the chromodomain (Breiling et al. 1999).

One current model envisions the early recruitment of the E(Z) complex in the early embryo as the first step in the establishment of PcG silencing. This recruitment would depend on DNA-binding proteins that recognize specific sequences in the PRE. The E(Z)-dependent histone H3 methylation at K27 would then be necessary and sufficient for the subsequent recruitment of the PC-containing complex by the affinity of PC for the methylated H3. Alternatively, DNA-binding recruiters have been suggested to act cooperatively to recruit various components of the E(Z) complex and of the PC complex.

EVIDENCE FOR RECRUITMENT

PREs are DNA sequences spread over 500–1000 base pairs (bp) within which a functional minimum core region of a few hundred base pairs is capable of recapitulating most of the silencing activities when included in a reporter construct (Chan et al. 1994; Horard et al. 2000). The additional flanking sequences appear to contribute to the stability of the silenced state. Transposon constructs containing PRE core sequences bind PcG proteins that can be visualized at the site of insertion in polytene chromosomes and can be detected by chromatin immunoprecipitation. These core sequences contain certain recurring motifs, some of which are known to be consensus sequences for DNA-binding proteins, but there is otherwise little sequence homology between different PREs. Two such motifs are the GAGAG consensus binding sites for GAGA factor and the GCCAT consensus for Pleiohomeotic (PHO), the fly homolog of the mammalian YY1 factor (Brown et al. 1998; Horard et al. 2000). A third motif frequently associated with PREs is the YGAGYG consensus that binds Zeste (Benson and Pirrotta 1987, 1988). However, Zeste function is not required for PcG silencing and Zeste binding sites can be separated from PREs without impairing their silencing activity. On the contrary, the addition of Zeste binding sites generally stimulates the expression of PRE-containing reporter constructs in a Zeste-dependent manner (Horard et al. 2000), consistent with the fact that Zeste has been shown to recruit the Brahma chromatin-remodeling complex (Kal et al. 2000). Because of their prominent presence in multiple copies in most PRE sequences, the PHO and GAGA binding sites have been proposed to bind proteins that contribute to the recruitment of PcG components that have themselves no specific DNA-binding activity.

GAGA Factor

GAGAG motifs contained in PREs have been found to bind GAGA factor but also Pipsqueak, both BTB domain proteins important for PcG silencing (Hagstrom et al. 1997; Busturia et al. 2001; Hodgson et al. 2001; Americo et al. 2002; Huang et al. 2002). Both of these proteins have been found to be localized at homeotic loci on polytene chromosomes, and mutations in either enhance homeotic phenotypes of PcG mutations, implying that they contribute to PcG silencing. However, GAGAG motifs are found at many promoters, including those of homeotic genes but also those of many other genes such as heat shock genes, which are not known to be repressed by PcG mechanisms.

A PcG complex biochemically purified from *Drosophila* embryos was not found to contain GAGA factor or Pipsqueak although it apparently included a number of other proteins in addition to the core PcG components. However, coimmunoprecipitation experiments with embryonic nuclear extracts have shown that GAGA factor isoforms can associate with protein complexes containing PC. Furthermore, PC-containing complexes present in the nuclear extracts can bind in vitro to DNA fragments from the *bxd* PRE and the binding is specifically competed by oligonucleotides containing GAGA factor binding sites (Horard et al. 2000). We conclude that a variety of PcG complexes exist in vivo, a subset of which does contain GAGA factor which could act as a recruiter of PcG components to the PRE. It is not sufficient, however. GAGA factor binding sites are present at a very large number of genomic sites in contexts that bear no relationship to PcG silencing. We suppose therefore that GAGA factor is one of several recruiters whose concerted action brings together and eventually stabilizes a PcG complex to the PRE.

PHO Factor

PHO and its related protein PHO-like are necessary for PcG silencing (Fritsch et al. 1999; Brown et al. 2003). Coimmunoprecipitation experiments have detected PHO in association with E(Z)/ESC complexes but not with PC complexes in most embryonic nuclear extracts (Poux et al. 2001a; but see also Poux et al. 2001b). However, PHO is not found in the best characterized form of the E(Z)/ESC complex, the 600-kD complex. Here too, however, other and higher molecular weight complexes have been identified and it is possible that they include PHO. Other studies have also detected physical interactions between PHO and PC (Mohd-Sarip et al. 2002). These discrepant reports raise again the possibility that different types of complexes and different interactions may exist in different cells at different developmental stages or may be detected under different in vitro conditions.

MULTIPLE DNA-BINDING RECRUITERS

The analysis of PRE sequences indicates that different DNA-binding proteins play a role in their function. The fact that two proteins bind to each of the two best characterized motifs could explain why mutations in the GAGA factor, Pipsqueak, PHO, and PHO-like have relatively mild effects on PcG silencing, if each can be at least partly replaced by its related DNA-binding protein. Other sequence motifs are likely to be important since synthetic combinations of GAGA and PHO binding sites are not sufficient to act as PREs. Different PREs share no extended sequence homology but Ringrose et al. (2003), using a whole genome approach, found that an algorithm that searches for clusters of the GAGA, PHO, and Zeste binding motifs identified most known PREs and predicted successfully many others. An analysis of these sequences showed that a few other motifs are frequently associated with these sites, suggesting that additional DNA-binding proteins might be implicated. It is possible therefore that the loss of one of several DNA-binding recruiters of PcG proteins might have only weak effects in the overall recruitment process.

CONSERVATION OF PRE SEQUENCES ACROSS SPECIES

Another approach often used to identify functionally important sequence elements is a phylogenetic comparison. The sequences of the *bxd* PREs from four *Drosophila* species—*D. melanogaster*, *D. eugracilis*, *D. virilis* (Dellino et al. 2002), and *D. pseudoobscura*—show that conservation is extensive and remarkable over an interval of almost one kilobase, in view of the fact that *D. virilis* and *D. pseudoobscura* are thought to have diverged from *D. melanogaster* between 40 and 60 million years ago. Figure 2 compares the core PRE region from

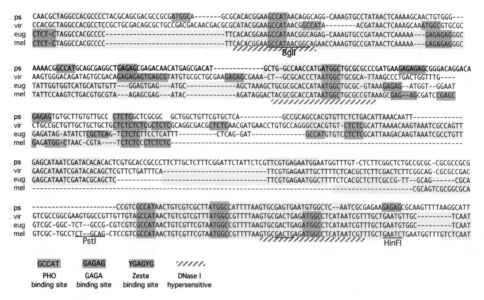

Figure 2. Conservation of *bxd* PRE sequences among *Drosophila* species. The sequences of the core PRE region from *D. melanogaster*, *D. eugracilis*, *D. virilis*, and *D. pseudoobscura* are aligned for maximum homology. (*Dotted lines*) Sequences that have been inserted or deleted in different species. Consensus sequences for known DNA-binding proteins are shown in *blue* (GAGA), *red* (PHO), and *green* (Zeste). (*Gray blocks*) Highly conserved blocks; (*brown barred lines*) the DNase I hypersensitive regions. For reference, three restriction sites present in the *D. melanogaster* sequence are indicated: *Bgl*I, *Pst*I, and *Hin*FI. The *D. pseudoobscura* sequence was obtained from the Baylor Drosophila Genome Project (http://www.hgsc.bcm.tmc.edu/projects/drosophila/). The three other sequences are from Dellino et al. (2002).

the four species, showing that the GAGA and PHO binding sites are either strictly conserved or where one is lost through mutation, it is recreated nearby by compensatory mutations. Even more remarkable is the complete conservation of some sequence intervals over more than 100 nt although they do not contain any identified DNA-binding motifs. The sequence conservation is reflected by the conservation of a set of four prominent DNase I hypersensitive sites within a conserved sequence context. These results support the idea that PREs are targets for a number of DNA-binding proteins, of which we know at present only a few, but which act together to recruit PcG components. However, the extensive sequence conservation makes it difficult to identify potential DNA-binding motifs other than those already known.

The four DNase I hypersensitive sites are clustered within <400 bp in the PRE core, making it unlikely that the PRE core region could be assembled in nucleosomes. The presence of ten GAGA factor binding sites in this interval is also incompatible with nucleosome formation, in view of the cooperative binding properties of GAGA factor (Katsani et al. 1999) and its known propensity to reposition nucleosomes (Tsukiyama et al. 1994). The PRE core region is also highly accessible to restriction enzymes (Dellino et al. 2004).

To obtain more detailed information on the chromatin structure of the PRE core region, we carried out genomic footprinting using ligation-mediated PCR. DNase I footprinting was done in parallel with chromatin from wild-type embryos or from *esc–/esc–* embryos produced by *esc–/esc–* mothers, hence lacking both maternal and zygotic ESC and unable to establish PcG silencing. Figure 3 shows that the hypersensitive sites are preserved in the mutant embryos but a pattern of relatively more protected, footprint-like regions becomes more DNase I sensitive in the mutant. A similar change, though less pronounced, was seen with chromatin from embryos carrying the homozygous $E(z)^{S2}$ temperature-sensitive mutation, raised at nonpermissive temperature. The fact that neither the *esc* nor the *E(z)* mutation affect the DNase I hypersensitive sites, suggests that they are due to the DNA-binding proteins such as GAGA factor or PHO. The changes induced by the mutations probably indicate sequences protected by the recruited PcG proteins.

TARGETING OF PcG COMPLEXES BY DNA-BINDING DOMAIN FUSIONS

Dissection of the recruitment process has been attempted using fusions of different PcG proteins with exogenous DNA-binding domains to target the proteins to a reporter gene containing the corresponding DNA-binding sites (Müller 1995; Poux et al. 2001a,b). The rationale of these experiments was to bypass part of the recruitment process and ask which protein, when targeted to a *Ubx-lacZ* reporter gene, can recruit the remaining components necessary to establish functional silencing. These experiments showed that PC, PH, PSC, or ESC can all silence the reporter gene to which they are targeted in a manner dependent on the other PcG components, implying that

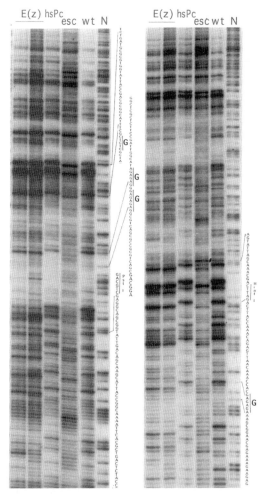

Figure 3. Genomic footprinting of the *bxd* PRE core region. Embryonic chromatin was partially digested with DnaseI and the DNA was then amplified by ligation-mediated PCR. The product was analyzed by short (*left*) or long (*right*) runs of gel electrophoresis. The lanes show the wild-type (wt), $E(z)^{S2}$ temperature-sensitive mutants treated at nonpermissive temperature 4 hr or overnight (E(z)), *esc* embryos from *esc* mothers (esc), or embryos overexpressing PC protein (hsPc). N indicates free DNA. The sequence is approximately aligned with the bands. PstI and HinF1 restriction sites and GAGA consensus sequences (G) are indicated. Note that the *E(z)* and *esc* sequences contain an insertion of two nucleotides relative to the others.

they can recruit the other necessary proteins. These experiments revealed two surprising facts. One is that, although the reporter gene was correctly repressed in the embryo, the silenced state was not maintained during larval development. The second was that targeted ESC and targeted PC establish silencing dependent on the endogenous *Pc* or *esc* gene, respectively, implying that each can recruit the other although they appear to belong to different, noncoimmunoprecipitating complexes.

These results imply that there must be an early stage at which PC-containing complexes interact with ESC-containing complexes. When we prepared nuclear extracts from 0–3-hour-old embryos, we found, in fact, that PC coimmunoprecipitates with ESC, E(Z), and PHO (Poux et al. 2001b). In contrast, at later embryonic stages, PC no longer appears to interact directly with the ESC/E(Z) complex. If both complexes must be present for silencing to occur, these results would explain why tethered PC or tethered ESC can silence a reporter gene at earlier stages but not during later development. They also imply that the presence of the ESC/E(Z) histone methyltransferase complex is not sufficient to recruit PC complexes at later stages.

The ESC/E(Z) complex is essential for the establishment silencing. Genetic experiments show that *esc* function is needed only in the early embryo, within the first 4 hours of development (Struhl and Brower 1982; Simon et al. 1995), while during later embryonic development *esc* transcription is said to disappear. A sufficient maternal endowment of ESC protein can support the development of homozygous *esc* embryos and lead to nearly normal adults. Since ESC is a constituent of the E(Z) histone H3 methyltransferase complex and is required for its enzymatic activity, how is methylation maintained in the absence of ESC in later development? Current work indicates that although *esc* expression ceases during embryonic development, a second *esc* gene, *esc-like*, takes over to ensure continued H3 K27 methylation (D. McCabe et al., unpubl.).

DISTRIBUTION OF PcG PROTEINS AND HISTONE H3 METHYLATION IN THE REPRESSED *Ubx* GENE

To test the role of histone methylation in the recruitment and localization of PcG proteins at a repressed gene, we determined the distribution of H3 me3K27 and of PcG proteins on various target genes or transgene constructs in embryos and in tissue culture cells (Y.B. Schwartz et al., in prep.). In brief, these chromatin immunoprecipitation experiments showed that PC, PSC, and E(Z) proteins are primarily localized over the PRE of the silenced *Ubx* gene, although a weak but significant association of PC can be detected over much of the gene. In contrast, H3 me3K27 is broadly distributed over the entire 75 kb of the *Ubx* transcription unit and at least 40 kb of upstream regulatory region. Only the PRE core itself appears to be unmethylated and, in fact, depleted of histone H3, in agreement with the idea that it is probably nucleosome-free. These observations support the idea that DNA-binding proteins are the primary recruiters of PcG complexes at the PRE and that H3 methylation plays an important but subsequent role. They raise the question of what this role might be and how methylation takes place over such an extended domain if the E(Z) methyltransferase is localized at the PRE.

H3 methylation is important for the stability and repressive function of the PcG complex. Loss of the E(Z) methyltransferase function during development eventually causes the derepression of homeotic genes and the dissociation of PcG proteins from most of their binding sites on polytene chromosomes (Rastelli et al. 1993; Beuchle et al. 2001). It is not simply needed for binding of the PC complex since a requirement for ESC and E(Z) is still observed even when PC is directly recruited to a reporter gene by fusion with a DNA-binding domain.

The fact that the PRE is essential for maintenance of silencing implies that the repressed state is not just perpetuated by modification of a silenced chromatin region or by the persistence of a protein complex that acts as a template. The broad distribution of the H3 K27 methylation over the silenced gene also implies that methylation is not sufficient for the stable recruitment of PcG complexes. Histone methylation might have several possible roles. It might be necessary to stabilize the association of the PcG complex with the PRE. It might allow the spreading of PcG complex assembly, extending from the PRE to cover a larger chromatin domain. It might induce a switch in the PcG complex that permits it to repress promoter function. It might simply act as the carrier of the epigenetic memory by facilitating the reassembly of the complex at sites that were methylated in the previous cell cycle. Chromatin immunoprecipitation results do not support a uniform, continuous spreading of the PcG complex fed in from an entry site (the PRE) and coating the chromatin over an extensive domain, such as is thought to occur in the case of the yeast SIR complex (Hecht et al. 1996). However, they would be consistent with a more transient interaction of PcG complex mediated by histone methylation according to two scenarios (Fig. 4). One is a looping mechanism by which the complex bound at the PRE loops over to interact transiently with nucleosomes. This interaction would be facilitated by the affinity of PC for the methylated nucleosomes without stable binding to any one sequence for any length of time. At the same time, it would allow a large domain to be constantly surveyed by the E(Z) complex bound at the PRE, thus maintaining the fully methylated state. In another view, the interaction of PC with the methylated histone would contribute to retain a local concentration of PC in the vicinity of the PRE. The binding of chromatin proteins such as HP1, like that of DNA-binding proteins, has been shown by photobleaching techniques to be short-lived with residence times of the order of seconds (Cheutin et al. 2003; Phair et al. 2004). The transient binding of PC to a methylated chromatin domain could maintain a local high concentration of PC protein to ensure rebinding to the PRE. Both mechanisms might, in fact, help to stabilize the presence of the PcG complex and to mediate the widespread methylation. The large methylated domain in turn contributes to the stability of the PcG complex mediated by the interaction of the PC chromodomain with the methylated histone and ensures the persistence of the epigenetic mark from one round of DNA replication to the next if the old nucleosomes are randomly distributed between the two daughter DNA molecules.

SILENCING MECHANISMS

The traditional view of chromatin silencing presupposed that the affected chromatin would be highly condensed. In this view, silencing would result from the inability of transactivators and transcriptional machinery to have access to the condensed chromatin. This type of mechanism implies that condensation is responsible for silencing rather than the reverse. To test this idea, we have recently used a reporter construct in which the bxd PRE was placed next to the hsp26 promoter (Dellino et al. 2004). As a result, the *hsp26* promoter becomes repressed and unable to be induced in a high proportion of the cells of animals carrying the transgene.

The heat shock promoter is normally preset by the cooperative binding of GAGA factor, TBP, and RNA Pol II, which together position a nucleosome and leave DNase I-sensitive sites over the heat shock elements. The polymerase is known to initiate transcription but to be unable to elongate beyond 25–50 nt in the absence of HSF. At the repressed *hsp26* promoter, the nucleosome is still correctly positioned and chromatin immunoprecipitation showed that both TBP and RNA polymerase were able to bind to the promoter and, upon heat shock, the HSF had normal access to the heat shock elements. The analysis of the promoter by permanganate sensitivity showed that, although the polymerase is bound to the promoter even without heat shock, it fails to open the strands and initiate transcription. Therefore, these results imply that, at least at this promoter, PcG silencing does not prevent access of the promoter factors but acts directly on the promoter complex to prevent initiation. This does not exclude the possibility that in other promoter contexts the PcG complex might also inhibit certain kinds of chromatin remodeling necessary to make the promoter accessible to the transcription machinery, as has been proposed by Shao et al. (1999). A similar direct effect on the promoter complex has been found in yeast, where the SIR-silencing complexes do not prevent the binding of

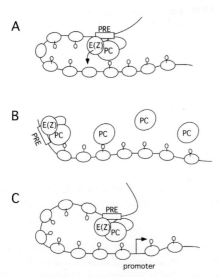

Figure 4. Three possible roles of histone methylation. (*A*) The chromodomain of PC bound at the PRE mediates transient interactions with methylated nucleosomes by a looping mechanism that allows the PRE complex to scan a large region, maintaining the fully methylated state. (*B*) The affinity of PC for methylated nucleosomes maintains a local high concentration of PC protein, stabilizing the PRE complex. A similar mechanism helps to rebind to between previously repressed genes every cell cycle. (*C*) The transient looping mediated by PC interaction with methylated nucleosomes also brings the PRE complex in the vicinity of the promoter, allowing it to block the promoter complex, if present, or prevent its recruitment by blocking nucleosome remodeling. The ball and stick indicates a methylated H3 lysine 27.

promoter factors and RNA polymerase but inhibit productive transcription (Sekinger and Gross 2001)

To account for our observations we envision two possible scenarios. Both depend on the looping mechanism described above. One scenario would view the methylated nucleosomes over the chromatin domain as providing a ready access for the PcG complex to interact with the promoter region. This interaction would be transient unless the PcG complex meets with promoter components for which it has specific affinities. Evidence for such affinities is the finding of several TAFs copurifying with the PRC1 complex as well as the observation that PcG components coimmunoprecipitate with TBP (Breiling et al. 2001; Saurin et al. 2001). This would be the situation at promoters that have independent mechanisms for recruiting general transcription factors, such as heat shock promoters, where the chromatin configuration is permanently preset by the cooperative binding of GAGA factor, TBP, and RNA polymerase. However, different promoters recruit promoter factors in different ways. In the case of the *Ubx* promoter, in particular, we know nothing about this mechanism except that, as a TATA-less promoter, it is presumed to require help from activators and other mechanisms. Inhibition of chromatin remodeling might also play a role at such sites.

CONCLUSIONS

The study of PcG silencing mechanisms has made enormous progress over the last 10 years and has revealed an extraordinary degree of complexity in the molecular mechanisms that recruit the complexes, establish silencing, and maintain the epigenetic silent state. More complexity is certain to be revealed in the future. The phylogenetically most ancient elements of the mechanism are the E(Z)/ESC components that appear to antedate all others and function in epigenetic silencing in plants (Goodrich et al. 1997; Köhler et al. 2003). In animals, the PC and associated components were added to work in concert with the histone methyltransferase complex and their complex interactions have been conserved at least from insects to mammals.

We might wonder why this large array of components and complex recruiting mechanisms is necessary. If it is sufficient to provide a PcG protein with a DNA-binding domain in order to assemble the functional silencing complex, why has nature not provided this simple solution? One explanation might be that the assembly of the repressive complexes must be carefully regulated. The PcG proteins are present at all times but they must not prevent the activation of the target genes in their appropriate domains. The assembly must be designed to be exquisitely sensitive to the chromatin state of the target gene so that an active gene does not become silenced. Other explanations might be that different components of the silencing mechanism may have special roles, not necessarily in association with the others, but participate in the PcG complexes to provide stability, flexibility, and additional regulatory features. The role of PcG proteins in the regulation of many other genes in addition to their classical targets, the homeotic genes, will be the object of research for the future.

ACKNOWLEDGMENTS

This work was supported by grants from the Human Frontiers Science Program, from the Swiss National Science Foundation, and by the "Frontiers in Genetics" Pôle de Recherche National of the Swiss National Science Foundation.

REFERENCES

Americo G., Whiteley M., Brown J.L., Fujioka M., Jaynes J.B., amd Kassis J.A. 2002. A complex array of DNA-binding complexes required for pairing-sensitive silencing by a polycomb group response element from the *Drosophila engrailed* gene. *Genetics* **160:** 1561.

Benson M. and Pirrotta V. 1987. The product of the *Drosophila zeste* gene binds to specific DNA sequences in *white* and *Ubx*. *EMBO J.* **6:** 1387.

———. 1988. The *Drosophila* zeste protein binds cooperatively to sites in many gene regulatory regions: Implications for transvection and gene regulation. *EMBO J.* **7:** 3907.

Beuchle D., Struhl G., and Müller J. 2001. Polycomb group proteins and heritable silencing of *Drosophila Hox* genes. *Development* **128:** 993.

Breiling A., Turner B.M., Bianchi M.E., and Orlando V. 2001. General transcription factors bind promoters repressed by Polycomb group proteins. *Nature* **412:** 651.

Breiling A.E., Bonte E., Ferrari S., Becker P.B., and Paro R. 1999. The *Drosophila* polycomb protein interacts with nucleosomal core particles in vitro via its repression domain. *Mol. Cell. Biol.* **19:** 8451.

Brown J.L., Fritsch C., Müller J., and Kassis J.A. 2003. The *Drosophila pho-like* gene encodes a YY1-related DNA binding protein that is redundant with *pleiohomeotic* in homeotic gene silencing. *Development* **130:** 285.

Brown J.L., Mucci D., Whiteley M., Dirksen M.L., and Kassis J.A. 1998. The *Drosophila* Polycomb group gene *pleiohomeotic* encodes a DNA binding protein with homology to the transcription factor YY1. *Mol. Cell* **4:** 1057.

Busturia A., Wightman C.D., and Sakonju S. 1997. A silencer is required for maintenance of transcriptional repression throughout *Drosophila* development. *Development* **124:** 4343.

Busturia A., Lloyd A., Bejarano F., Zavortink M. Xin H., and Sakonju S. 2001. The MCP silencer of the *Drosophila Abd-B* gene requires both *Pleiohomeotic* and GAGA factor for the maintenance of repression. *Development* **128:** 2163.

Cao R., Wang L., Wang H., Xin L., Erdjument-Bromage H., Tempst P., Jones R.S., and Zhang Y. 2002. Role of histone H3 lysine 27 methylation in Polycomb-group silencing. *Science* **298:** 1039.

Chan C.-S., Rastelli L., and Pirrotta V. 1994. A *Polycomb* response element in the *Ubx* gene that determines an epigenetically inherited state of repression. *EMBO J.* **13:** 2553.

Cheutin T., McNairn A., Jenuwein T., Gilbert D.M., Singh P.B., and Misteli T. 2003. Maintenance of stable heterochromatin domains by dynamic HP1 binding. *Science* **299:** 721.

Czermin B., Melfi R., McCabe D., Seitz V., Imhof A., and Pirrotta V. 2002. *Drosophila* enhancer of Zeste/ESC complexes have a histone H3 methyltransferase activity that marks chromosomal Polycomb sites. *Cell* **111:** 185.

Dellino G.I., Tatout C., and Pirrotta V. 2002. Conservation of sequences and chromatin structure of the bxd polycomb response element among *Drosophila* species. *Int. J. Dev. Biol.* **46:** 133.

Dellino G.I., Schwartz Y.B., Farkas G., McCabe D., Elgin S.C.R., and Pirrotta V. 2004. Polycomb silencing blocks transcription initiation. *Mol. Cell* **13:** 887.

Fauvarque M.-O. and Dura J.-M. 1993. polyhomeotic regulatory sequences induce developmental regulator-dependent variegation and targeted P-element insertions in *Drosophila*. *Genes Dev.* **7:** 1508.

Fischle W., Wang Y., Jacobs S.A., Kim Y., Allis C.D., and Khorasanizadeh S. 2003. Molecular basis for the discrimination of repressive methyl-lysine marks in histone H3 by Polycomb and HP1 chromodomains. *Genes Dev.* **17:** 1870.

Fritsch C., Brown J.L., Kassis J.A., and Müller J. 1999. The DNA-binding polycomb group protein pleiohomeotic mediates silencing of a *Drosophila* homeotic gene. *Development* **126:** 3905.

Goodrich J., Puangsomlee P., Martin M., Long D., Meyerowitz E., and Coupland G. 1997. A Polycomb-group gene regulates homeotic gene expression in *Arabidopsis*. *Nature* **386:** 44.

Hagstrom K., Müller M., and Schedl P. 1997. A Polycomb and GAGA dependent silencer adjoins the Fab7 boundary in the *Drosophila* bithorax complex. *Genetics* **146:** 1365.

Hecht A., Strahl-Bolsinger S., and Grunstein M. 1996. Spreading of transcriptional repressor SIR3 from telomeric heterochromatin. *Nature* **383:** 92.

Hodgson J.W., Argiropoulos B., and Brock H.W. 2001. Site-specific recognition of a 70-base-pair element containing d(GA)(n) repeats mediates bithoraxoid polycomb group response element-specific silencing. *Mol. Cell. Biol.* **21:** 4528.

Horard D., Tatout C., Poux S., and Pirrotta V. 2000. Structure of a polycomb response element and in vitro binding of polycomb group complexes containing GAGA factor. *Mol. Cell. Biol.* **20:** 3187.

Huang D.-H., Chang Y.-L., Yang C.-C., Pan I.-C., and King B. 2002. *pipsqueak* encodes a factor essential for sequence-specific targeting of a polycomb group protein complex. *Mol. Cell. Biol.* **22:** 6261.

Kal A.J., Mahmoudi T., Zak N.B., and Verrijzer C.P. 2000. The *Drosophila* brahma complex is an essential coactivator for the trithorax group protein zeste. *Genes Dev.* **14:** 1058.

Katsani K.R., Hajibagheri N., and Verrijzer C.P. 1999. Co-operative DNA binding by GAGA transcription factor requires the conserved BTB/POZ domain and reorganizes promoter topology. *EMBO J.* **18:** 698.

Kuzmichev A., Nishioka K., Erdjument-Bromage H., Tempst P., and Reinberg D. 2002. Histone methyltransferase activity associated with a human multiprotein complex containing the Enhancer of Zeste protein. *Genes Dev.* **22:** 2893.

Köhler C., Hennig L., Spillane C., Pien S., Gruissem W., and Grossniklaus U. 2003. The Polycomb-group protein MEDEA regulates seed development by controlling expression of the MADS-box gene PHERES1. *Genes Dev.* **17:** 1540.

Maurange C. and Paro R. 2002. A cellular memory module conveys epigenetic inheritance of hedgehog expression during *Drosophila* wing imaginal disc development. *Genes Dev.* **16:** 2672.

Mohd-Sarip A.F., Venturini F., Chalkley G.E., and Verrijzer C.P. 2002. Pleiohomeotic can link polycomb to DNA and mediate transcriptional repression. *Mol. Cell. Biol.* **22:** 7473.

Müller J. 1995. Transcriptional silencing by the Polycomb protein in *Drosophila* embryos. *EMBO J.* **14:** 1209.

Müller J., Hart C.M., Francis N.J., Vargas M.L., Sengupta A., Wild B., Miller E.L., O'Connor M.B., Kingston R.E., and Simon J.A. 2002. Histone methyltransferase activity of a *Drosophila* Polycomb group repressor complex. *Cell* **111:** 197.

Phair R.D., Scaffidi P., Elbi C., Vecerova J., Dey A., Ozato K., Brown D.T., Hager G., Bustin M., and Misteli T. 2004. Global nature of dynamic protein-chromatin interactions in vivo: Three-dimensional genome scanning and dynamic interaction networks of chromatin proteins. *Mol. Cell. Biol.* **24:** 6393.

Pirrotta V., Chan C.-S., McCabe D., and Qian S. 1995. Upstream parasegmental enhancers and the establishment of the expression pattern of the *Ubx* gene. *Genetics* **141:** 1439.

Poux S., Kostic C., and Pirrotta V. 1996. Hunchback-independent silencing of late *Ubx* enhancers by a Polycomb Group Response Element. *EMBO J.* **15:** 4713.

Poux S., McCabe D., and Pirrotta V. 2001a. Recruitment of components of Polycomb Group chromatin complexes in *Drosophila*. *Development* **128:** 75.

Poux S., Melfi R., and Pirrotta V. 2001b. Establishment of Polycomb silencing requires a transient interaction between PC and ESC. *Genes Dev.* **15:** 2509.

Rastelli L., Chan C.S., and Pirrotta V. 1993. Related chromosome binding sites for zeste, suppressors of zeste and Polycomb group proteins in *Drosophila* and their dependence on Enhancer of zeste function. *EMBO J.* **12:** 1513.

Ringrose L., Rehmsmeier M., Dura J.M., and Paro R. 2003. Genome-wide prediction of Polycomb/Trithorax response elements in *Drosophila melanogaster*. *Dev. Cell* **5:** 759.

Saurin A.J., Shao Z., Erdjument-Bromage H., Tempst P., and Kingston R.E. 2001. A *Drosophila* Polycomb group complex includes Zeste and dTAFII proteins. *Nature* **412:** 655.

Sekinger E.A. and Gross D.S. 2001. Silenced chromatin is permissive to activator binding and PIC recruitment. *Cell* **105:** 403.

Shao Z., Raible F., Mollaaghababa R., Guyon J.R., Wu C.T., Bender W., and Kingston R.E. 1999. Stabilization of chromatin structure by PRC1, a Polycomb complex. *Cell* **98:** 37.

Simon J., Bornemann D., Lunde K., and Schwartz C. 1995. The *extra sex combs* product contains WD40 repeats and its time of action implies a role distinct from other Polycomb group products. *Mech. Dev.* **53:** 197.

Struhl G. and Brower D. 1982. Early role of the *esc+* gene product in the determination of segments in *Drosophila*. *Cell* **31:** 285.

Tsukiyama T., Becker P.B., and Wu C. 1994. ATP-dependent nucleosome disruption at a heat-shock promoter mediated by binding of GAGA transcription factor. *Nature* **367:** 525.

Tillib S., Petruk S., Sedkov Y., Kuzin A., Fujioka M., Goto T., and Mazo A. 1999. Trithorax- and Polycomb-group response elements within an *Ultrabithorax* transcription maintenance unit consist of closely situated but separable sequences. *Mol. Cell. Biol.* **19:** 5189.

Mechanism of Polycomb Group Gene Silencing

Y. ZHANG,* R. CAO,* L. WANG,† AND R.S. JONES†

Department of Biochemistry and Biophysics, Lineberger Comprehensive Cancer Center, University of North Carolina at Chapel Hill, Chapel Hill, North Carolina 27599-7295; †Department of Biological Sciences, Southern Methodist University, Dallas, Texas 75275

The *Drosophila* trithorax-group (trxG) and Polycomb-group (PcG) proteins function in an antagonistic manner to maintain the transcriptionally active and silence states of target genes, respectively. Although they regulate numerous genes, mutant alleles of most trxG and PcG genes were first identified on the basis of homeotic phenotypes resulting from misexpression of Hox genes of the Antennapedia and bithorax gene complexes. *Drosophila* Hox genes, which encode transcription factors that regulate numerous downstream genes, must be continuously expressed in appropriate patterns throughout embryonic and larval development in order to assign segmental identities to cells along the anterior–posterior body axis. The expression patterns of the Hox genes are initially established in early embryos by activators and repressors encoded by gap and pair rule genes, but soon after Hox gene expression is initiated, these activators and repressors decay. It is during this window of time that trxG and PcG proteins somehow recognize the transcriptionally active or repressed states of Hox genes and become responsible for maintaining their expression states in cell lineages throughout embryonic and larval development. Thus, trxG and PcG proteins serve as molecular memory systems central to the process of cellular determination (Francis and Kingston 2001; Simon and Tamkun 2002).

Here we discuss our recent progress in understanding the mechanisms of PcG silencing, but, because PcG proteins function antagonistically to the trxG, we will first briefly describe the trxG and the mechanisms by which they help maintain transcriptional activity. The trxG comprises approximately 20 genes. Several encode components of the 2-MD Brahma (BRM) complex, which is a member of the SWI/SNF family of nucleosome remodeling complexes (Papoulas et al. 1998; Kal et al. 2000), and others encode proteins that are members of the SWI2/SNF2 family of ATPases, but are physically independent of the BRM complex (Daubresse et al. 1999; Ruhf et al. 2001). Two members of the trxG, Trithorax (Trx) and Abnormal small or homeotic discs-1 (ASH-1), contain SET domains [Su(var)3-9, Enhancer of zeste, Trx], conserved domains present in numerous chromatin proteins that possess histone lysine methyltransferase (HMTase) activity (Jenuwein et al. 1998; Rea et al. 2000). Both Trx and ASH-1 methylate histone H3 at lysine 4 (H3-K4) (Beisel et al. 2002; Byrd and Shearn 2003; Smith et al. 2004), a modification generally associated with gene activation (Bernstein et al. 2002). Trx coexists in the 1-MD TAC1 complex with dCBP, a histone acetyltransferase, and dSbf1 (Petruk et al. 2001).

Originally identified as regulators of *Drosophila* Hox genes, PcG homologs have since been identified across a wide phylogenetic spectrum, including *Caenorhabditis elegans*, *Arabidopsis thalania*, and mammals. A list of *Drosophila* PcG proteins and their mammalian homologs are provided in Table 1. Thus, PcG proteins appear to be an evolutionarily conserved gene-silencing system that has been adapted for the regulation of different genes and developmental purposes. The *Drosophila* PcG comprises approximately 15 genes, many of which encode components of multiprotein complexes. The Polycomb repressive complex 1 (PRC1) contains the PcG proteins Polycomb (Pc),

Table 1. A List of Known PcG Proteins in *Drosophila* and Mammalians

Drosophila proteins	Human proteins	Mouse proteins
Sequence-specific DNA-binding proteins		
Pho	YY1	Yy1
Phol	YY1	Yy1
Esc-E(z) complex		
Esc	EED	Eed
E(z)	EZH1	Ezh1/Enx2
	EZH2	Ezh2/Enx1
Su(z)12	SUZ12	Suz12
PRC1 complex		
Pc	HPC1/CBX2	M33/Cbx2
	HPC2/CBX4	Mpc2/Cbx4
	HPC3/CBX8	
Ph	HPH1/EDR1	Mph1/Rae28/Rae28
	HPH2/EDR2	Mph2/Edr2
	HPH3/EDR3	
dRing/Sce	RING1/RNF1/RING1A	Ring1/Ring1a
	RING1B/RNF2	Ring1b/Rnf2
Psc	BMI1	Bmi1
	ZFP144/RNF110	Mel18/Zfp144/Rnf110
	ZNF134	Znf134/Mblr
Undefined function		
Asx	ASXL1	
	ASXL2	
Crm		
Mxc		
Scm	SCML1	Scmh1
	SCML2	Scmh2
Pcl	hMTF2	MTF2
	PHF1	
Sxc		

310 ZHANG ET AL.

polyhomeotic (Ph), Posterior sex combs (Psc), dRing1 (also known as Sex combs extra, Sce; Fritsch et al. 2003), in addition to Zeste (which has been also classified as a trxG protein), dSbf1, HSC4, and five general transcription factors (dTAFIIs 250, 110, 85, 62, and 42) (Saurin et al. 2001). A second complex, referred to as Esc-E(z), contains the PcG proteins Extra sex combs (Esc), Enhancer of zeste [E(z)], and Suppressor 12 of zeste [Su(z)12], in addition to the histone-binding protein NURF-55. The histone deacetylase HDAC1 (Rpd3) has been identified in some forms of the complex (Tie et al. 2001; Czermin et al. 2002), but is absent from others (Müller et al. 2002). The human counterparts of both complexes have been purified and the core components are found to be conserved (Cao et al. 2002; Levine et al. 2002).

To fully understand the molecular mechanism of PcG-mediated gene silencing, several major questions must be addressed. (1) How is the repressed state of target genes initially recognized? (2) What are the mechanisms by which PcG proteins repress transcription? (3) How is the silenced state faithfully transmitted through many cycles of cell division? Here, we describe our recent studies aimed at addressing the latter two questions. In particular, we will discuss the role of sequence-specific DNA-binding PcG proteins Pleiohomeotic (Pho) and Pho-like (Phol) and H3-K27 methylation by ESC-E(Z)/EED-EZH2 complexes in maintenance of transcriptional silencing. We will also examine the roles of both catalytic and noncatalytic subunits of this HMTase complex and how they contribute to H3-K27 methylation. In addition, we will discuss the mechanisms by which PcG proteins may repress transcription, including the contribution of a novel enzymatic activity associated with the PRC1 complex.

MATERIALS AND METHODS

All materials used and methods described in the studies presented here have been previously described as indicated throughout the text.

RESULTS

Purification and Characterization of the EED-EZH2 HMTase Complex

Histone tails are rich in covalent modifications that include acetylation, methylation, ubiquitination, and phosphorylation (van Holde 1988). While acetylation on lysine residues generally correlates with gene activation, methylation on lysine residues results in either gene activation or repression depending on the particular lysine residues that are methylated (Zhang and Reinberg 2001; Lachner et al. 2003). In an attempt to understand the function of histone methylation, we have been using a systematic biochemical approach to purify and characterize histone methyltransferases from HeLa cells (Fang et al. 2003). Of the six HMTases that we have characterized so far, the EED-EZH2/ESC-E(Z) complex is of particular interest because of its roles in diverse biological processes including PcG silencing, X-inactivation, germ-line development, stem cell pluripotency, and cancer (Cao and Zhang 2004a).

By following a nucleosomal histone H3 methyltransferase activity, we had previously purified a protein complex of about 500 kD from HeLa cells (Cao et al. 2002). The complex is composed of five subunits including EZH2, SUZ12, AEBP2, EED, and RbAp48 (Fig. 1a). A similar protein complex was also purified independently by several other groups (Czermin et al. 2002; Kuzmichev et al. 2002; Müller et al. 2002). RbAp48 is a WD40-repeat protein initially identified as a Rb-binding protein (Qian et al. 1993). Subsequent studies revealed the presence of this protein in many protein complexes involved in histone modification and nucleosome remodeling, consistent with the notion that this protein is a histone-binding protein (Verreault et al. 1998). AEBP2 is a zinc finger transcriptional repressor that may contribute to targeting of the complex to specific genes (He et al. 1999). EZH2, EED, and SUZ12 are PcG proteins (Table 1). Since, with the exception of AEBP2, the composition

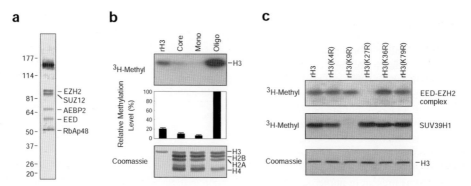

Figure 1. Purification and characterization of the EED-EZH2 histone methyltransferase complex. (*a*) Coomassie-stained polyacrylamide-SDS gel containing the purified EED-EZH2 complex. The identity of the proteins in the complex is indicated. The largest prominent protein is a contaminant. The protein size markers are indicated. (*b*) The EED-EZH2 HMTase complex prefers oligonucleosomal histone substrate. Equal amounts of the enzyme complex were used to methylate equal amounts of histone H3 alone, in octamer, mono-, and oligonucleosome forms (*bottom* panel). The *top* panel is an autoradiography of the *bottom* panel. Quantification of the autoradiography is presented in the *middle* panel. (*c*) EED-EZH2 complex methylates H3 at lysine 27. Equal amounts of wild-type and mutant histone H3 (*bottom* panel) were methylated by EED-EZH2 complex (*top* panel) and SUV39H1 (*middle* panel), respectively. The lysines that were mutated are indicated on top of the panel. (Adapted, with permission, from Cao et al. 2002 [©AAAS].)

of this complex is conserved in the *Drosophila* ESC-E(Z) complex (Ng et al. 2000; Czermin et al. 2002; Kuzmichev et al. 2002; Müller et al. 2002), we refer to it as the EED-EZH2 complex. The facts that most subunits of the complex belong to the PcG proteins and that EZH2 contains a SET domain suggest a potential link between the intrinsic HMTase activity of the complex and PcG silencing.

To understand the relationship between PcG silencing and the HMTase activity, we characterized the enzymatic activity further by determining its substrate specificity and the lysine residue on H3 that the complex methylates. Toward this end, equivalent amounts of isolated histone H3, histone H3 assembled with other core histones, and mono- or oligonucleosomes were subjected to methylation by equal amounts of the enzyme complex. Results shown in Figure 1b indicate that the enzyme complex has a strong preference for H3 in oligonucleosome form. To identify the lysine residue that the complex methylates, we generated H3 mutants in which each of the five potential methylation sites (K4, K9, K27, K36, and K79) was individually mutated. The effect of the mutations on the ability of H3 to serve as substrates for the enzyme complex was evaluated. As a control, the ability of these H3 mutants to serve as substrates for the H3-K9 methyltransferase SUV39H1 was also analyzed. Results shown in Figure 1c (top panel) indicate that mutation on K27 completely abolished the ability of H3 to serve as a substrate, whereas mutations on other sites had little effect. As expected, only mutation of K9 affected the SUV39H1-mediated H3 methylation (Fig. 1c, middle panel). These results strongly suggest that H3-K27 is the target site of methylation for the complex. To further verify the result, oligonucleosomes were subjected to methylation. After purification, the methylated H3 was subjected to microsequencing followed by liquid scintillation counting. This again revealed that K27 is the target site (Cao et al. 2002). Therefore, we conclude that the EED-EZH2 complex prefers oligonucleosomal substrates and methylates H3-K27.

H3-K27 Methylation Is Required for PRE Binding by PC and *Ubx* Gene Silencing

To study the function of H3-K27 methylation in vivo, we generated a polyclonal antibody that recognizes methylated, but not nonmethylated, H3-K27 (Cao et al. 2002). Using this antibody, we evaluated whether the *Drosophila* ESC-E(Z) complex is responsible for H3-K27 methylation in vivo. Previous studies have identified an *E(z)* temperature-sensitive allele, $E(z)^{61}$, which contains a Cys-to-Tyr substitution (C603Y) in the cysteine-rich region immediately preceding the SET domain (Carrington and Jones 1996). At 18°C (permissive temperature), the protein functions normally and $E(z)^{61}$ homozygotes exhibit no detectable mutant phenotype and maintain wild-type expression patterns of Hox genes, such as *Ubx* (Jones and Gelbart 1990; Carrington and Jones 1996). However, at 29°C (restrictive temperature), $E(Z)61$ protein fails to bind to chromatin leading to disruption of chromosome binding by Polycomb (PC) and other PRC1 components (Rastelli et al. 1993; Platero et al. 1996). As a result, $E(z)^{61}$ produces multiple homeotic phenotypes because of derepression of Hox genes (Jones and Gelbart 1990). Therefore, if E(Z) is responsible for H3-K27 methylation in vivo, we expect partial or complete loss of H3-K27 methylation when $E(Z)^{61}$ mutants are shifted from 18°C to 29°C. Results shown in Figure 2a confirm this prediction and demonstrate that H3-K27 methylation is dramatically decreased in the $E(z)^{61}$ embryos at 29°C (middle panel). However, these conditions do not affect H3-K9 methylation (top panel). Therefore, we conclude that functional E(Z) protein is required for H3-K27 methylation in vivo.

Previous studies have demonstrated that transcriptional silencing of the *Ubx* gene requires both the ESC-E(Z) and the PRC1 complexes, in addition to a *cis*-acting Polycomb response element (PRE), to which the two complexes bind. To understand the functional relationship between E(Z)-mediated H3-K27 methylation and Hox gene silencing, we analyzed E(Z) binding, H3-K27 methylation, and recruitment of PC, a core component of the PRC1 complex (Francis et al. 2001), to the major *Ubx* PRE (PRE_D) by chromatin immunoprecipitation (ChIP) (Fig. 2b). Analysis of S2 tissue culture cells revealed a precise colocalization of E(Z), H3-K27 methylation, and PC binding to the PRE_D region (Cao et al. 2002). Importantly, disruption of the ESC-E(Z) complex by RNAi resulted in greatly reduced E(Z) binding, H3-K27 methylation, and concomitant loss of PC binding to the PRE (Cao et al. 2002), suggesting that ESC-E(Z)-mediated H3-K27 methylation contributes to PRE binding by PC. We also performed similar experiments using dissected wing imaginal discs from homozygous $E(z)^{61}$ larvae, which had been either reared continuously at 18°C or shifted from 18°C to 29°C ~48 hours prior to dissection. Results shown in Figure 2c (left panels) demonstrate that at permissive temperatures, as in S2 cells, $E(Z)^{61}$ binding, H3-K27 methylation, and PC binding colocalize at the PRE_D region. At restrictive temperatures, however, loss of $E(Z)^{61}$ binding is concomitant with loss of H3-K27 methylation and PC binding (Fig. 2c, right panels). In contrast, similar changes in H3-K9 methylation were not observed under the same conditions (Fig. 2c). Similar inactivation of an *E(z)* temperature-sensitive allele during larval development has been shown to result in significant derepression of *Ubx* in wing discs (LaJeunesse and Shearn 1996). Collectively, these data suggest that H3-K27 methylation plays an important role in the maintenance of *Ubx* gene silencing.

PC Chromodomain Recognizes Methyl-K27 of H3

The "histone code" hypothesis predicts that single or combinational histone modifications may serve as molecular marks that can be recognized by specific protein modules or domains that in turn direct the functional consequence of the modification (Strahl and Allis 2000; Turner 2000). Consistent with this hypothesis, the chromodomain of the heterochromatin protein HP1 has been demonstrated to specifically bind to H3 tails that are methylated at K9 by the HMTase SUV39H1 (Bannister et al. 2001; Lachner et al. 2001). Several lines of evidence

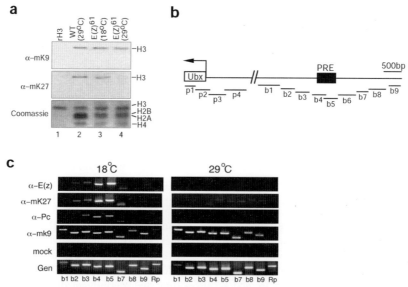

Figure 2. Loss of E(Z) function abolishes H3-K27 methylation, PC binding, and Ubx gene silencing. (*a*) H3-K27 methylation depends on functional E(Z) protein. Equivalent amounts (*bottom* panel) of histones purified from wild-type (lane 2) and mutant E(z)[61] (lanes 3 and 4) *Drosophila* embryos were probed with H3-2mK9- (*top* panel) or H3-2mK27- (*middle* panel) specific antibodies. (*b*) Schematic representation of the Ubx promoter and bxd PRE regions. The regions amplified by PCR in these and subsequent ChIP assays, p1–p4 and b1–b9, are depicted as horizontal lines below. (Adapted from Wang et al. 2004.) (*c*) ChIP assays demonstrate colocalization of E(Z) binding and H3-K27 methylation in *E(z)[61]* wing imaginal discs at 18°C (*left* panel) and loss of binding in wing discs at 29°C (*right* panel). Antibodies used in each assay are indicated on the *left*. Genomic DNA from pooled collection of wing imaginal discs was PCR amplified as controls for efficiencies of PCR primers. Numbers below the panels indicate the PCR primers used in each ChIP assay. Lanes *1–9* corresponding to the regions are as indicated in *b*; lane *10* is a PCR product of RpII140 promoter, which served as a negative control. (Adapted, with permission, from Cao et al. 2002 [©AAAS].)

suggest that the chromodomain of PC may recognize H3 tails methylated at K27, analogous to that of the HP1 binding to H3 tails methylated on K9. First, the chromodomain of PC is both necessary and sufficient for targeting PC, as well as other components of the PRC1 complex, to specific chromosomal locations in vivo (Messmer et al. 1992; Platero et al. 1995). Second, loss of E(Z) function abolishes H3-K27 methylation as well as PC binding to the *Ubx* PRE (Fig. 2c). Third, all the amino acids in HP1 chromodomain that are involved in methyl-lysine binding are conserved in the PC chromodomain. These lines of evidence prompted us to test the *Drosophila* PC protein, generated using the rabbit reticulocyte transcription/translation system, for its ability to bind to biotinylated H3 peptides with or without K27 methylation. Results shown in Figure 3a (top panel) indicated that methylation on K27 facilitates binding of PC to the H3 peptide. This binding is mediated through the chromodomain as mutations in two of the highly conserved amino acids within the chromodomain (W47A, W50A) abolished preferential binding of PC to the methylated peptide (Fig. 3a, middle panel). Binding of PC to the peptides is specific because the chromodomain-containing protein HP1 failed to bind to the same peptides under the same conditions (Fig. 3a, bottom panel).

The above in vitro binding results were recently confirmed by structural studies in which the PC chromodomain in complex with an H3 peptide trimethylated on K27 was crystallized and the structure solved (Fischle et al. 2003; Min et al. 2003). The study revealed a conserved mode of methyl-lysine binding and provided structural basis for specific recognition of PC chromodomain to histone H3 methylated on K27, but not K9. As shown in Figure 3b, the *Drosophila* PC chromodomain consists of three β strands (β1–β3) and a carboxy-terminal helix (αA). The histone H3 peptide is bound in a cleft formed between the PC amino terminal to β1 and the loop connecting β3 and αA. Although the overall structures of PC and HP1 chromodomains are very similar (Jacobs and Khorasanizadeh 2002; Nielsen et al. 2002), differences between the two chromodomains are noticeable. For example, while the methyl-lysine-binding pocket of HP1 interacts with methyl-K9 via hydrophobic interaction, the corresponding aromatic residues on PC interact with methyl-K27 through cation–π interactions. In addition, unique interactions between Leu 20, Thr 22 of histone H3, and Arg 67 of PC were noticed. However, these interactions cannot account for the binding specificity of PC chromodomain to methyl-K27, but not methyl-K9, because only the main-chain atoms of histone H3 are involved in the interaction (Min et al. 2003).

A careful examination of the cocrystal structure identified a potential chromodomain dimer that can account for the binding specificity of PC chromodomain to methyl-K27. As depicted in Figure 3c, the chromodomain dimer interacts via intermolecular hydrogen bonds between the main-chain atoms of Leu 64 and Arg 66, which appear to be specific to the PC family of proteins. An additional hydrogen bond can also form between Arg 66 and Val 61. The chromodomain dimer juxtaposes the two H3-binding clefts in an antiparallel fashion and results in histone–histone interactions involving Leu 20, Thr 22, and Ala 24

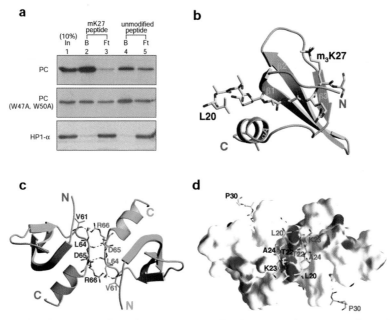

Figure 3. H3-K27 methylation facilitates binding of PC to H3 through its chromodomain. (*a*) Autoradiographs of peptide pulldown experiments. ^{35}S-labeled PC, PC mutant (W47A, W50A), and HP1-α were incubated with biotinylated H3 peptides (aa 19–35), which were either methylated or unmethylated at K27, in the presence of streptavidin-conjugated Sepharose beads. After extensive washing, the beads were boiled with SDS loading buffer and resolved in SDS-polyacrylamide gels. In: 10% of the total input used for the pulldown assays; B: bound; Ft: flowthrough. (*b*) Overall structure of the *Drosophila* PC chromodomain (aa 23–77) in complex with a histone H3 peptide (aa 19–33) trimethylated on K27. The choromodomain is shown in a ribbon diagram (*brown*), and the H3 peptide is shown as a ball-and-stick model (*red,* oxygen; *blue,* nitrogen; and *yellow,* carbon). (*c*) The PC chromodomain dimer. The PC chromodomain are shown in a ribbon representation (*brown* and *cyan*). Key residues involved in dimerization are shown in a bond model. Hydrogen bonds involving these residues are indicated with *broken lines*. (*d*) The PC chromodomain dimer juxtaposes the two binding sites of methyl-K27 of H3. The PC chromodomain dimer is shown as surface representation (*red,* negatively charged area; *blue,* positively charged area; *white,* neutral). Two bound H3-3mK27 peptides are shown in a ball-and-stick model. (Courtesy of Dr. Rui-Ming Xu.)

(Fig. 3d). This recognition mode can effectively exclude the binding of a histone H3 peptide encompassing methylated Lys 9, as the residues corresponding to Leu 20, Thr 22, and Ala 24 of H3 would be Arg 2, Lys 4, and Thr 5, respectively. Therefore, the key determinants that confer specific recognition of methyl-K27 by PC chromodomain are both the histone H3 sequence (Leu 20, Thr 22, and Ala 24) and the dimerization of the PC chromodomain.

H3-K27 Methylation Contributes to PC/PRC1 Recruitment

As described above, loss of E(Z) results in rapid loss of H3-K27 methylation and PC binding to the PRE$_D$ region. Previous studies also suggest that the E(Z) complex can transiently interact with components of the PRC1 complex (Poux et al. 2001). Therefore, results from the above study cannot distinguish between the contribution of H3-K27 methylation and the physical interaction between the ESC-E(Z) complex and the PRC1 components in PC recruitment. However, ChIP analysis indicate that E(Z), PC, and trimethyl H3-K27 are also present near the *Ubx* promoter in wing imaginal discs (Fig. 4a). Following inactivation of E(Z)[61] and loss of the HMTase complex, H3-K27 methylation is maintained near the *Ubx* promoter for ~24 hours. PC also remains near the *Ubx* pro-

moter in the absence of E(Z), but is finally lost when H3-K27 methylation is no longer detectable (Fig. 4b). Thus, PC binding correlates with H3-K27 methylation, but not with the physical presence of E(Z)-containing complex, consistent with H3-K27 methylation serving as a tag that is primarily responsible for recruiting PC-containing complexes.

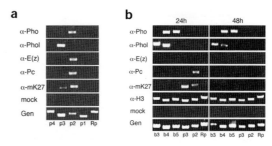

Figure 4. Pc binding at the *Ubx* promoter and PRE$_D$ regions is dependent on H3-K27 methylation. (*a*) ChIP assays showing distribution of PcG proteins and H3-3mK27 in the *Ubx* promoter region in wing imaginal discs. Wing imaginal discs were dissected from *E(z)*[61] larvae reared continuously at 18°C. (*b*) ChIP assays of wing imaginal discs from *E(z)*[61] larvae shifted from 18°C to 29°C (*left*) 24 hr or (*right*) 48 hr prior to dissection. α-H3, anti-histone H3 was used as a positive control in the ChIP assays shown in this figure and in Fig. 5. (Adapted, with permission, from Wang et al. 2004 [©Elsevier].)

Hierarchical Recruitment of PcG Complexes

PcG proteins maintain the transcriptionally silenced state of target genes through many cell cycles. Both initiation and maintenance of transcriptional silence require a *cis*-acting PRE. Reporter genes contained within P element constructs become derepressed after one to a few cell generations following deletion of a flanking PRE (Busturia et al. 1997; Sengupta et al. 2004). Several components of the PRC1 dissociate from chromosomes in tissue culture cells during mitosis (Buchenau et al. 1998). This suggests that proteins capable of binding directly to sites within PREs play important roles in repeatedly recruiting PcG proteins to PREs following mitosis. PREs contain binding sites for several sequence-specific DNA-binding proteins. These include GAGA factor, Pipsqueak (Psq), Pleiohomeotic (Pho), and Pho-like (Phol). GAGA factor and Psq both bind to GAGAG repeats and mutation of these sequences in PRE_D within the context of a P element transgene results in partial derepression of a reporter gene in embryos (Horard et al. 2000; Hodgson et al. 2001). However, GAGA factor and Psq may not be required for maintenance of PcG silencing in larvae, since mutation of PRE GAGAG sites does not affect repression of a reporter gene in imaginal discs (Fritsch et al. 1999).

Pho and Phol are homologs of human Yin Yang 1 (YY1), and are identical in their sequence-specific DNA-binding activities in vitro (Brown et al. 1998, 2003). Maternally expressed Pho is needed early in embryogenesis in order to establish PcG silencing, but individuals that are homozygous for null *pho* alleles (derived from heterozygous mothers) die as late pupae with relatively mild homeotic phenotypes and show only moderate derepression of Ubx in wing imaginal discs (Brown et al. 2003). The relatively mild zygotic phenotypes of *pho* mutants appear to be due to functional redundancy with *phol*. Although *phol* nulls are homozygous viable, *phol;pho* double mutants exhibit extensive *Ubx* derepression in wing imaginal discs and die as late larvae/early pupae (Brown et al. 2003). This suggests that Pho and Phol play important roles in maintaining PcG silencing during larval development and that their functions may be partially redundant. We have recently demonstrated that PHO and PHOL directly interact with E(Z) and/or ESC (Wang et al. 2004), which suggests that the ESC-E(Z) complex may be recruited to PREs through protein–protein interactions with PHO and/or PHOL. ChIP analysis of wing imaginal discs revealed the presence of PHO and PHOL in the PRE_D and *Ubx* promoter regions at sites that overlapped those of E(Z) and PC (Figs. 4a, 5a). Binding by neither E(Z) nor PC was affected in *phol* or *pho* mutant wing imaginal discs. However, binding by both E(Z) and PC is lost in *phol;pho* double mutants (Fig. 5b). Taken together with the role of H3-K27 methylation in PC recruitment described above, we propose the following hierarchical pathway of PcG recruitment. PHO and/or PHOL bind to sites within PREs and directly recruit ESC-E(Z) complex, which then methylates H3 at K27. The PC chromodomain then binds to the methylated H3-K27 tag, facilitating recruitment of PC-containing complexes such as PRC1 (Fig. 6).

In addition to their presence at the PRE_D region, Pc, E(z), Pho, and Phol are associated with discrete regions near the transcription start site (Fig. 4a). In the absence of

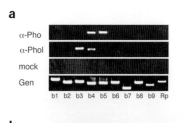

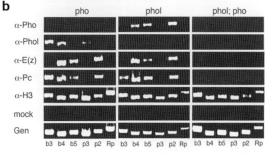

Figure 5. Pho and Phol are redundant for recruitment of E(z)- and Pc-containing complexes. (*a*) ChIP assays showing distribution of Pho and Phol in the PRE_D region in wing imaginal discs. (*b*) ChIP assays of wing imaginal discs dissected from pho^1 (*left*), $phol^{81A}$ (*middle*), or $pho^1;phol^{81A}$ (*right*) larvae. (Adapted, with permission, from Wang et al. 2004 [©Elsevier].)

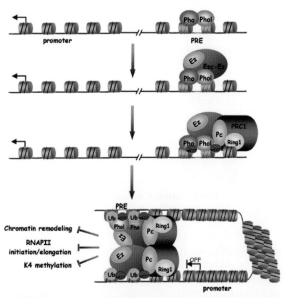

Figure 6. Model depicting the mechanism of PcG silencing. Binding of transcription factors, such as Pho and Phol, to PRE initiates the recruitment of the ESC-E(Z) complex, which methylates H3-K27. H3-3mK27, recognized by the chromodomain of PC, serves as a marker for the recruitment of the PRC1 complex. A yet-to-be-identified mechanism mediates loop formation bringing the PRE and associated PcG proteins into proximity of the transcriptional start site, which inhibits transcription by (1) interfering with chromatin remodeling, (2) directly inhibiting transcription initiation or elongation by RNAPII, and/or (3) ubiquitinating histone H2A, thereby inhibiting H3-K4 methylation by Trx and/or Ash1 complexes.

E(z), Pho and Phol remain at the PRE, but are no longer detected near the *Ubx* promoter (Fig. 4b). This is consistent with a model in which PcG proteins assemble at the PRE followed by the formation of a loop that brings them into contact with the promoter (Fig. 6). Assuming this model is correct, it is not clear what may mediate loop formation. PRC1 has been shown to be able to recruit chromatin templates *in trans* (Lavigne et al. 2004). Alternatively, the sequence-specific DNA-binding protein Zeste has been shown to be a component of PRC1 (Saurin et al. 2001), raising the possibility that Zeste may mediate loop formation.

Evolutionary Conservation of PcG Gene Silencing

As listed in Table 1, PcG proteins have been structurally and functionally conserved during evolution. In addition, the core components of the ESC-E(Z)/EED-EZH2 and the PRC1 complexes are conserved from *Drosophila* to human (Francis et al. 2001; Cao et al. 2002; Levine et al. 2002; Muller et al. 2002). One of the conserved functions of PcG proteins is their involvement in Hox gene silencing. For example, PcG mutations in *Drosophila* or mice result in homeotic transformation because of derepression of Hox genes (Kmita and Duboule 2003). Data presented above illustrate the importance of ESC-E(Z)-mediated H3-K27 methylation in *Ubx* gene silencing. To examine whether the function of H3-K27 methylation is conserved in mammalian cells, we reconstituted the human EED-EZH2 complex and demonstrated that the HMTase activity requires a minimum of three components, including EZH2, EED, and SUZ12. Addition of RbAp48 and AEBP2 stimulated the enzymatic activity (Cao and Zhang 2004b).

To evaluate the role of SUZ12 in H3-K27 methylation in vivo, we generated a stable SUZ12 knockdown cell line that expresses ~25% of the normal levels of SUZ12 protein and ~35% of the normal levels of SUZ12 mRNA (Fig. 7a). Compared with the control empty vector knockdown cells, SUZ12-targeted knockdown resulted in a significant decrease on the trimethyl-K27 level but had little effect on the trimethyl-K9 level (Fig. 7b, third and fourth panels). Interestingly, an increase in monomethyl-K27 and a moderate decrease in dimethyl-K27 were also observed (Fig. 7b, top two panels). The fact that SUZ12 knockdown does not affect EZH2 level (Fig. 7a) in combination with the requirement of SUZ12 for H3-K27 methyltransferase activity in vitro (Fig. 7b) allows us to conclude that SUZ12 directly contributes to H3-K27 methylation in vivo.

Previous studies in *Drosophila* have established a critical role for Su(z)12 in Hox gene silencing (Birve et al. 2001). The fact that SUZ12 is required for H3-K27 methylation in combination with the fact that H3-K27 methylation is critical in Hox gene silencing (Cao et al. 2002; Muller et al. 2002) predict that SUZ12 knockdown will result in derepression of at least some Hox genes. Analysis of HoxC6, HoxC8, and HoxA9 in the knockdown cells and the parallel control cells revealed derepression of HoxC8 and HoxA9 genes in the knockdown cells (Fig. 7c). These data support the notion that, like most other PcG proteins, the function of SUZ12/Su(z)12 in Hox gene silencing is conserved from human to *Drosophila*.

CONCLUSIONS AND FUTURE DIRECTIONS

As a result of these and other studies, we can now begin to assign molecular/biochemical activities to more than half of the known PcG proteins. We propose that PcG proteins may be placed in either of two categories: Recruiters or Effectors. Proteins such as Pho, Phol, or their mammalian homolog YY1 and components of the Esc-E(z)/EED-EZH2 complex primarily function as Recruiters. The sequence-specific DNA-binding Pho and Phol bind to sites within PREs and directly recruit ESC-E(Z) complexes, which in turn methylates H3 at K27 in the immediate vicinity of the PRE. The PC chromodomain binds to the methylated H3-K27 tag, facilitating recruitment of PRC1, or related complexes (Fig. 6). Thus,

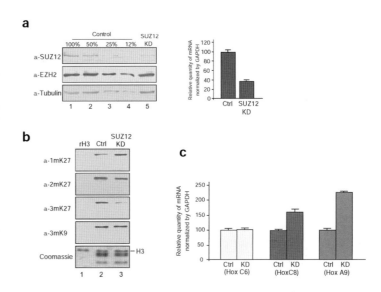

Figure 7. SUZ12 knockdown affects H3-K27 methylation and Hox gene expression. (*a*) Western blot (*left* panel) and quantitative RT-PCR (*right* panel) analysis of a SUZ12 stable knockdown cell line and a parallel mock knockdown cell line. Tubulin serves as a loading control for Western blotting. GAPDH serves as control for normalization in the quantitative RT-PCR. (*b*) Western blot analysis of histones extracted from control and knockdown HeLa cells with antibodies specific for mono-, di-, or trimethylated K27 and trimethylated K9. Equal loading of histone H3 was verified by Coomassie staining of a parallel gel (*bottom* panel). (*c*) Quantitative RT-PCR analysis of HoxC6, HoxC8, and HoxA9 expression in SUZ12 knockdown and mock knockdown cells. GAPDH was used as a control for normalization. Quantification is an average of two independent experiments with error bars. (Adapted from Cao and Zhang 2004b.)

the primary function of H3-K27 methylation in PcG silencing appears to be recruitment of PC-containing complexes. In vitro studies suggest that PRC1 may be classified as an Effector of transcriptional repression, which may inhibit transcription by any of several possible mechanisms. For example, PRC1 inhibits nucleosome remodeling by SWI/SNF complexes (Shao et al. 1999; Francis et al. 2001). Therefore, it may antagonize the nucleosome remodeling activity of the trxG BRM complex, thus interfering with activator binding or assembly of the preinitiation complex. In addition, PRC1 has been shown to be able to block transcription of chromatin or naked DNA templates by RNA polymerase II or T7 RNA polymerase (King et al. 2002). PRC1 does not appear to block activator binding in these assays, but instead seems to act upon the template to interfere with transcription initiation or elongation. These observations are consistent with in vivo studies in which RNA polymerase II and basal transcription factors were shown to be present at promoters under conditions of PcG repression (Dellino et al. 2004) and the presence of Pc- and E(z)-containing complexes at a discrete site just downstream of a silenced endogenous target gene, Ubx, in wing imaginal discs (Wang et al. 2004). In addition, our recent studies indicate that a PRC1-like complex possesses H2A ubiquitin ligase activity. Human Ring 2, a homolog of dRing/Sce, was identified as the catalytic subunit (data not shown). Although the mechanism by which this activity affects transcription has not been determined, it nevertheless suggests that PRC1 may interfere with transcription by multiple mechanisms.

Among the questions to be addressed in the near future is whether PRE-promoter loops actually form, what is the mechanistic basis for loop formation, and how are PcG complexes targeted to a site just downstream of the transcription start site. Once positioned downstream of the transcription start site, what is the mechanism by which transcription is prevented? Does PRC1 directly act upon the DNA template to prevent duplex melting, or might it interfere with some step in initiation such as RNA polymerase II CTD phosphorylation? What is the effect of H2A ubiquitination by dRing/hRing2? It is also important to point out that of the 15 genetically identified PcG genes, the products of only 9 have been identified either as sequence-specific DNA-binding proteins (Pho and Phol) or components of the PRC1 (Pc, Ph, Psc, dRing) or Esc-E(z) (Esc, E(z), Su(z)12) complexes. The remaining PcG proteins also play important roles in transcriptional silencing, but their activities are yet to be defined. In addition, other proteins, which may have pleiotropic functions and therefore are not easily classifiable as members of the PcG on the basis of genetic studies, also contribute to PcG silencing. Full understanding of this epigenetic gene regulation system will require an understanding of these other players in addition to those that have received the bulk of our attention to date.

ACKNOWLEDGMENTS

This work was supported by NIH grants GM068804 (Y.Z.) and GM46567 (R.S.J.).

REFERENCES

Bannister A.J., Zegerman P., Partridge J.F., Miska E.A., Thomas J.O., Allshire R.C., and Kouzarides T. 2001. Selective recognition of methylated lysine 9 on histone H3 by the HP1 chromo domain. *Nature* **410**: 120.

Beisel C., Imhof A., Greene J., Kremmer E., and Sauer F. 2002. Histone methylation by the *Drosophila* epigenetic transcriptional regulator Ash1. *Nature* **419**: 857.

Bernstein B.E., Humphrey E.L., Erlich R.L., Schneider R., Bouman P., Liu J.S., Kouzarides T., and Schreiber S.L. 2002. Methylation of histone H3 Lys 4 in coding regions of active genes. *Proc. Natl. Acad. Sci.* **99**: 8695.

Birve A., Sengupta A.K., Beuchle D., Larsson J., Kennison J.A., Rasmuson-Lestander A., and Muller J. 2001. Su(z)12, a novel *Drosophila* Polycomb group gene that is conserved in vertebrates and plants. *Development* **128**: 3371.

Brown J.L., Fritsch C., Mueller J., and Kassis J.A. 2003. The *Drosophila* pho-like gene encodes a YY1-related DNA binding protein that is redundant with pleiohomeotic in homeotic gene silencing. *Development* **130**: 285.

Brown J.L., Mucci D., Whiteley M., Dirksen M.L., and Kassis J.A. 1998. The *Drosophila* Polycomb group gene pleiohomeotic encodes a DNA binding protein with homology to the transcription factor YY1. *Mol. Cell* **1**: 1057.

Buchenau P., Hodgson J., Strutt H., and Arndt-Jovin D.J. 1998. The distribution of polycomb-group proteins during cell division and development in *Drosophila* embryos: Impact on models for silencing. *J. Cell Biol.* **141**: 469.

Busturia A., Wightman C.D., and Sakonju S. 1997. A silencer is required for maintenance of transcriptional repression throughout *Drosophila* development. *Development* **124**: 4343.

Byrd K.N. and Shearn A. 2003. ASH1, a *Drosophila* trithorax group protein, is required for methylation of lysine 4 residues on histone H3. *Proc. Natl. Acad. Sci.* **100**: 11535.

Cao R. and Zhang Y. 2004a. The functions of E(Z)/EZH2-mediated methylation of lysine 27 in histone H3. *Curr. Opin. Genet. Dev.* **14**: 155.

———. 2004b. SUZ12 is required for both the histone methyltransferase activity and the silencing function of the EED-EZH2 complex. *Mol. Cell* **15**: 57.

Cao R., Wang L., Wang H., Xia L., Erdjument-Bromage H., Tempst P., Jones R.S., and Zhang Y. 2002. Role of histone H3 lysine 27 methylation in Polycomb-group silencing. *Science* **298**: 1039.

Carrington E.A. and Jones R.S. 1996. The *Drosophila Enhancer of zeste* gene encodes a chromosomal protein: Examination of wild-type and mutant protein distribution. *Development* **122**: 4073.

Czermin B., Melfi R., McCabe D., Seitz V., Imhof A., and Pirrotta V. 2002. *Drosophila* Enhancer of Zeste/ESC complexes have a histone H3 methyltransferase activity that marks chromosomal Polycomb sites. *Cell* **111**: 185.

Daubresse G., Deuring R., Moore L., Papoulas O., Zakrajsek I., Waldrip W.R., Scott M.P., Kennison J.A., and Tamkun J.W. 1999. The *Drosophila* kismet gene is related to chromatin-remodeling factors and is required for both segmentation and segment identity. *Development* **126**: 1175.

Dellino G.I., Schwartz Y.B., Farkas G., McCabe D., Elgin S.C., and Pirrotta V. 2004. Polycomb silencing blocks transcription initiation. *Mol. Cell* **13**: 887.

Fang J., Wang H., and Zhang Y. 2003. Purification of histone methyltransferases from HeLa cells. *Methods Enzymol.* **377**: 213.

Fischle W., Wang Y., Jacobs S.A., Kim Y., Allis C.D., and Khorasanizadeh S. 2003. Molecular basis for the discrimination of repressive methyl-lysine marks in histone H3 by Polycomb and HP1 chromodomains. *Genes Dev.* **17**: 1870.

Francis N.J. and Kingston R.E. 2001. Mechanisms of transcriptional memory. *Nat. Rev. Mol. Cell Biol.* **2**: 409.

Francis N.J., Saurin A.J., Shao Z., and Kingston R.E. 2001. Reconstitution of a functional core polycomb repressive complex. *Mol. Cell* **8**: 545.

Fritsch C., Beuchle D., and Muller J. 2003. Molecular and genetic analysis of the Polycomb group gene Sex combs extra/Ring in *Drosophila*. *Mech. Dev.* **120:** 949.

Fritsch C., Brown J.L., Kassis J.A., and Muller J. 1999. The DNA-binding polycomb group protein pleiohomeotic mediates silencing of a *Drosophila* homeotic gene. *Development* **126:** 3905.

He G.P., Kim S., and Ro H.S. 1999. Cloning and characterization of a novel zinc finger transcriptional repressor. A direct role of the zinc finger motif in repression. *J. Biol. Chem.* **274:** 14678.

Hodgson J.W., Argiropoulos B., and Brock H.W. 2001. Site-specific recognition of a 70-base-pair element containing d(GA)(n) repeats mediates bithoraxoid polycomb group response element-dependent silencing. *Mol. Cell. Biol.* **21:** 4528.

Horard B., Tatout C., Poux S., and Pirrotta V. 2000. Structure of a polycomb response element and in vitro binding of polycomb group complexes containing GAGA factor. *Mol. Cell. Biol.* **20:** 3187.

Jacobs S.A. and Khorasanizadeh S. 2002. Structure of HP1 chromodomain bound to a lysine 9-methylated histone H3 tail. *Science* **295:** 2080.

Jenuwein T., Laible G., Dorn R., and Reuter G. 1998. SET domain proteins modulate chromatin domains in eu- and heterochromatin. *Cell. Mol. Life Sci.* **54:** 80.

Jones R.S. and Gelbart W.M. 1990. Genetic analysis of the enhancer of zeste locus and its role in gene regulation in *Drosophila melanogaster*. *Genetics* **126:** 185.

Kal A.J., Mahmoudi T., Zak N.B., and Verrijzer C.P. 2000. The *Drosophila* brahma complex is an essential coactivator for the trithorax group protein zeste. *Genes Dev.* **14:** 1058.

King I.F., Francis N.J., and Kingston R.E. 2002. Native and recombinant polycomb group complexes establish a selective block to template accessibility to repress transcription in vitro. *Mol. Cell. Biol.* **22:** 7919.

Kmita M. and Duboule D. 2003. Organizing axes in time and space; 25 years of colinear tinkering. *Science* **301:** 331.

Kuzmichev A., Nishioka K., Erdjument-Bromage H., Tempst P., and Reinberg D. 2002. Histone methyltransferase activity associated with a human multiprotein complex containing the Enhancer of Zeste protein. *Genes Dev.* **16:** 2893.

Lachner M., O'Sullivan R.J., and Jenuwein T. 2003. An epigenetic road map for histone lysine methylation. *J. Cell Sci.* **116:** 2117.

Lachner M., O'Carroll D., Rea S., Mechtler K., and Jenuwein T. 2001. Methylation of histone H3 lysine 9 creates a binding site for HP1 proteins. *Nature* **410:** 116.

LaJeunesse D. and Shearn A. 1996. E(z): A polycomb group gene or a trithorax group gene? *Development* **122:** 2189.

Lavigne M., Francis N.J., King I.F., and Kingston R.E. 2004. Propagation of silencing; recruitment and repression of naive chromatin in trans by polycomb repressed chromatin. *Mol. Cell* **13:** 415.

Levine S.S., Weiss A., Erdjument-Bromage H., Shao Z., Tempst P., and Kingston R.E. 2002. The core of the polycomb repressive complex is compositionally and functionally conserved in flies and humans. *Mol. Cell. Biol.* **22:** 6070.

Messmer S., Franke A., and Paro R. 1992. Analysis of the functional role of the Polycomb chromo domain in *Drosophila melanogaster*. *Genes Dev.* **6:** 1241.

Min J., Zhang Y., and Xu R.M. 2003. Structural basis for specific binding of Polycomb chromodomain to histone H3 methylated at Lys 27. *Genes Dev.* **17:** 1823.

Müller J., Hart C.M., Francis N.J., Vargas M.L., Sengupta A., Wild B., Miller E.L., O'Connor M.B., Kingston R.E., and Simon J.A. 2002. Histone methyltransferase activity of a *Drosophila* Polycomb group repressor complex. *Cell* **111:** 197.

Ng J., Hart C.M., Morgan K., and Simon J.A. 2000. A *Drosophila* ESC-E(Z) protein complex is distinct from other polycomb group complexes and contains covalently modified ESC. *Mol. Cell. Biol.* **20:** 3069.

Nielsen P.R., Nietlispach D., Mott H.R., Callaghan J., Bannister A., Kouzarides T., Murzin A.G., Murzina N.V., and Laue E.D. 2002. Structure of the HP1 chromodomain bound to histone H3 methylated at lysine 9. *Nature* **416:** 103.

Papoulas O., Beek S.J., Moseley S.L., McCallum C.M., Sarte M., Shearn A., and Tamkun J.W. 1998. The *Drosophila* trithorax group proteins BRM, ASH1 and ASH2 are subunits of distinct protein complexes. *Development* **125:** 3955.

Petruk S., Sedkov Y., Smith S., Tillib S., Kraevski V., Nakamura T., Canaani E., Croce C.M., and Mazo A. 2001. Trithorax and dCBP acting in a complex to maintain expression of a homeotic gene. *Science* **294:** 1331.

Platero J.S., Hartnett T., and Eissenberg J.C. 1995. Functional analysis of the chromo domain of HP1. *EMBO J.* **14:** 3977.

Platero J.S., Sharp E.J., Adler P.N., and Eissenberg J.C. 1996. In vivo assay for protein-protein interactions using *Drosophila* chromosomes. *Chromosoma* **104:** 393.

Poux S., Melfi R., and Pirrotta V. 2001. Establishment of Polycomb silencing requires a transient interaction between PC and ESC. *Genes Dev.* **15:** 2509.

Qian Y.-W., Wang Y.-C.J., Hollingsworth R.E.J., Jones D., Ling N., and Lee E.Y.-H.P. 1993. A retinoblastoma-binding protein related to a negative regulator of Ras in yeast. *Nature* **364:** 648.

Rastelli L., Chan C.S., and Pirrotta V. 1993. Related chromosome binding sites for zeste, suppressors of zeste and Polycomb group proteins in *Drosophila* and their dependence on Enhancer of zeste function. *EMBO J.* **12:** 1513.

Rea S., Eisenhaber F., O'Carroll D., Strahl B.D., Sun Z.W., Schmid M., Opravil S., Mechtler K., Ponting C.P., Allis C.D., and Jenuwein T. 2000. Regulation of chromatin structure by site-specific histone H3 methyltransferases. *Nature* **406:** 593.

Ruhf M.L., Braun A., Papoulas O., Tamkun J.W., Randsholt N., and Meister M. 2001. The domino gene of *Drosophila* encodes novel members of the SWI2/SNF2 family of DNA-dependent ATPases, which contribute to the silencing of homeotic genes. *Development* **128:** 1429.

Saurin A.J., Shao Z., Erdjument-Bromage H., Tempst P., and Kingston R.E. 2001. A *Drosophila* Polycomb group complex includes Zeste and dTAFII proteins. *Nature* **412:** 655.

Sengupta A.K., Kuhrs A., and Muller J. 2004. General transcriptional silencing by a Polycomb response element in *Drosophila*. *Development* **131:** 1959.

Shao Z., Raible F., Mollaaghababa R., Guyon J.R., Wu C.T., Bender W., and Kingston R.E. 1999. Stabilization of chromatin structure by PRC1, a Polycomb complex. *Cell* **98:** 37.

Simon J.A. and Tamkun J.W. 2002. Programming off and on states in chromatin: Mechanisms of Polycomb and trithorax group complexes. *Curr. Opin. Genet. Dev.* **12:** 210.

Smith S.T., Petruk S., Sedkov Y., Cho E., Tillib S., Canaani E., and Mazo A. 2004. Modulation of heat shock gene expression by the TAC1 chromatin-modifying complex. *Nat. Cell Biol.* **6:** 162.

Strahl B.D. and Allis C.D. 2000. The language of covalent histone modifications. *Nature* **403:** 41.

Tie F., Furuyama T., Prasad-Sinha J., Jane E., and Harte P.J. 2001. The *Drosophila* Polycomb Group proteins ESC and E(Z) are present in a complex containing the histone-binding protein p55 and the histone deacetylase RPD3. *Development* **128:** 275.

Turner B.M. 2000. Histone acetylation and an epigenetic code. *Bioessays* **22:** 836.

van Holde K.E. 1988. *Histone modifications*. Springer, New York.

Verreault A., Kaufman P.D., Kobayashi R., and Stillman B. 1998. Nucleosomal DNA regulates the core-histone-binding subunit of the human Hat1 acetyltransferase. *Curr. Biol.* **8:** 96.

Wang L., Brown J.L., Cao R., Zhang Y., Kassis J.A., and Jones R.S. 2004. Hierarchical recruitment of polycomb group silencing complexes. *Mol. Cell* **14:** 637.

Zhang Y. and Reinberg D. 2001. Transcription regulation by histone methylation: Interplay between different covalent modifications of the core histone tails. *Genes Dev.* **15:** 2343.

Emerging Roles of Polycomb Silencing in X-Inactivation and Stem Cell Maintenance

I. Muyrers-Chen, I. Hernández-Muñoz, A.H. Lund, M.E. Valk-Lingbeek,
P. van der Stoop, E. Boutsma, B. Tolhuis, S.W.M. Bruggeman, P. Taghavi,
E. Verhoeven, D. Hulsman, S. Noback, E. Tanger, H. Theunissen,
and M. van Lohuizen

*The Netherlands Cancer Institute, Division of Molecular Genetics,
1066 CX Amsterdam, The Netherlands*

Maintenance of cell identity and cell fate depends on the tight regulation of gene expression patterns in correct time and space. Two families of proteins, the trithorax group (trxG) and the Polycomb group (PcG), use epigenetic mechanisms to faithfully ensure that designated genes are maintained on or off throughout the life of the organism. This maintenance function is imperative to allow the proper development of an organism from a single cell to an organized combination of multifunctional cells. Here, we briefly review the advances achieved in recent years aimed at understanding how members of PcG and trxG function (for more in-depth reviews, see Otte and Kwaks 2003; Pirrotta et al. 2003; Lund and van Lohuizen 2004; Valk-Lingbeek et al. 2004). Particularly, we will discuss methods that can be employed to uncover additional target genes regulated by PcG and/or trxG families. Additionally, we will focus on recent results linking PcG regulation with X-inactivation and with stem cell biology.

INTRODUCTION

The Polycomb group (PcG) and trithorax group (trxG) of gene families were first identified in *Drosophila* as important regulators of homeotic genes, responsible for the development of the body plan. The functional conservation of these protein families from plants and worms to mammals illustrates how basic and essential PcG and trxG regulation is for development. PcG and trxG use epigenetic mechanisms to properly modulate gene expression and to maintain appropriate levels of gene activity over several mitotic divisions. Such epigenetic mechanisms require chromatin structure alterations, independent of a specific DNA sequence, as their mode of action. Both PcG and trxG families act within large protein complexes to execute their function. The initiation complex isolated in *Drosophila*, designated PRC2 (Polycomb Repressive Complex 2), is 400–600 kD in size and consists of Esc, E(z), Su(z)12, and RbAp48/Nurf-55 (Cao et al. 2002; Kuzmichev et al. 2002; Muller et al. 2002). Recently, variants of this complex, dubbed PRC3, have been characterized, harboring different isoforms of the Esc/EED subunit (Kuzmichev et al. 2004). The second complex, PRC1, contains Pc, Ph, Psc, and dRing as its core members (Saurin et al. 2001). In addition, a variety of proteins have been copurified with the PRC1 core complex giving rise to several subcomplexes, which are most probably required for regulation of diverse sets of target genes. Comparable PRC1 and PRC2 complexes have also been isolated in mammals, emphasizing conservation of this ancient family not only in protein structure homology but also in the functional mechanism employed by PcG/trxG members to regulate their target sites (Lund and van Lohuizen 2004). However, owing to duplications of part of the mammalian genome involving PcG genes, dissecting the function of mammalian PcG/trxG complexes has proven to be complex because of partial functional redundancy.

In the fly, recent studies showed that both trxG and PcG members can interact with the same CIS elements to insure the proper status of gene activity. The best noted example is the study reporting that both Polycomb (Pc) and GAGA factor (a trxG member) can bind to close but separable entities within the classically defined Fab-7 polycomb response element (PRE) (Francis and Kingston 2001). PREs and TREs (trithorax response elements) have therefore been renamed as Cellular Memory Modules (CMMs) (Cavalli and Paro 1998). Moreover, RNA polymerase and TBP-associated factors (TAFs) can be found simultaneously with Polycomb group members in addition to trxG proteins at a silent locus (Breiling et al. 2001). This supports the idea that the status of gene activity is not merely dependent on the presence or the ability of PcG or trxG members to change chromatin configuration to prevent activators or repressors from binding (Dellino et al. 2004). Additional levels of regulation certainly must exist to dictate the action of bound PcG or trxG members in keeping a gene either on or off depending on its developmental context. Based on recent studies in *Drosophila*, the prevailing model for how PcG silencing works appears to be a looping model, in which the distantly located PREs bound by PcG proteins can interact in a dynamic fashion through multiple low-affinity interactions, with a subset of proteins located at or close to the transcriptional start of homeotic genes (see Schoenfelder and Paro; Schwartz et al.; both this volume). The binding of PcG members to distinct elements within PREs/CMMs and the ability of PcG complexes to repress artificial chromatin templates by template bridging support looping models for PcG silencing (Lavigne et al. 2004). Alternative models have recently been proposed in recent reports linking PcG proteins with noncoding RNAs. For instance, it has been suggested that PcG members, rather than altering chromatin structure to block transcription factor accessibility (Francis and Kingston 2001), somehow directly inhibit the ability of the RNA polymerase to transcribe (Dellino et al. 2004). Moreover, it is imaginable that PcG/trxG members may use their ability to change the packaging of higher-order chromatin to allow RNA polymerase activity at intergenic regions (such as at regulatory regions), generating noncoding RNAs (Hogga and Karch 2002; Rank et al. 2002). The detection of noncoding RNAs in *Drosophila*, in yeast, and in mammals and the link between chromatin modifiers and the RNAi machinery further substantiate the idea that RNA polymerase activity along the DNA may prevent activator binding (Pal-Bhadra et al. 2002; Lund and van Lohuizen 2004). Whether the noncoding RNAs, generated artificially or physiologically present (microRNAs, siRNAs), have a specific function in targeting PcG/trxG protein complexes or in regulating PcG/trxG function is an intriguing possibility that deserves to be tested.

In mammals, thus far, PREs/TREs/CMMs have not been described. It is formally possible that because of the diversification of the genome throughout evolution, such as reflected in the vast increase of repetitive elements in mammalian genomes, a different, as of yet elusive targeting mechanism for PcG/trxG exists in mammals. However, given their strong structural and functional conservation, this possibility would be surprising. In this regard, the tethering of PcG/trxG complexes onto chromatin has been best understood in *Drosophila*. *Drosophila* pho and pho-like are two PcG proteins that recognize specific DNA sequences and that have been shown recently to recruit E(z)/PRC2 and PRC1complexes to their target sites (Mohd-Sarip et al. 2002; Wang et al. 2004). Importantly, the mammalian homolog of pho, YY1, is also a DNA-specific binding protein and has been described to associate with PRC1 and PRC2 members (Brown et al. 1998; Satijn et al. 2001). Knockdown experiments of YY1 in Xenopus resulted in anterior–posterior patterning defects reminiscent of a classical PcG/trxG mutant (Kwon and Chung 2003). Moreover, the ability of human YY1 to repress target genes in transgenic flies and to partially rescue pho fly mutants further substantiates YY1 as a bona fide PcG member (Atchinson et al. 2003). In summary, a main question in the field is how mammalian PcG protein complexes are recruited to specific genes and what the identity and genome-wide distribution of these PcG target genes is.

SEARCHING FOR MAMMALIAN PREs/CMMs

To be able to uncover target genes regulated by PcG/trxG, an in vivo formaldehyde cross-linking and chromatin immunoprecipitation technique (ChIP) was developed and successfully applied in *Drosophila* (Orlando and Paro 1993). However, because of the complexity of the mammalian genome and the increase in repetitive elements relative to the *Drosophila* genome, using ChIP to identify mammalian PREs has proven to be difficult, especially when applied to screen in a genome-wide fashion (ChIP on DNA microarrays or ChIP on Chip). An additional limitation is the requirement of high-quality antisera, suitable to work after harsh cross-linking conditions. Only very recently, the first successful PRC2 ChIP on Chip with mammalian CpG island arrays has been reported (Kirmizis et al. 2004). Two alternative techniques have recently been developed that aim at uncovering how various regulatory elements separated over hundreds of kilobases can physically interact to dictate the level of gene expression. The manner in which these regulatory elements come together through "looping" type of interactions reflects the chromatin structure in which a particular gene, whether active or inactive, is embedded. The first technique, called RNA TRAP (tagging and recovery of associated proteins), detects chromatin fragments close to a nascent transcript via RNA fluorescence in situ hybridization (FISH) (Carter et al. 2002). The second method, Chromatin Conformation Capture (3C), depends on computing the frequency of interaction of distant chromatin fragments after formaldehyde cross-linking (Dekker et al. 2002). The 3C method, unlike the RNA TRAP technique, has the advantage that it does not depend on a nascent transcript to uncover chromatin fragment association. The RNA TRAP and 3C techniques have been used to unravel the molecular nature of how the globin gene cluster is regulated (Carter et al. 2002; Tolhuis et al. 2002). In light of this success, it will be interesting to utilize both RNA TRAP and 3C technology to study in more detail how various PREs/TREs/CMMs may regulate the expression of gene activity and to find support for the proposed looping-type models. However, both RNA TRAP and 3C require the prior knowledge of target loci for PcG/trxG binding, posing a problem for mammalian applications.

In order to find mammalian PcG/trxG target genes, an attractive alternative technique to ChIP on Chip is the DamID (Dam identification) methodology (van Steensel et al. 2001). The technique is based on generating fusions of proteins of interest with the *Escherichia coli* DAM methylase. The exogenous deposition of methylation marks on adenosine, which does not occur naturally in mammals, can be detected using specific restriction enzymes followed by amplification of DNA fragments enriched for methyl-Adenosines. The resulting pool of amplified DNA would reflect the binding sites of the protein of interest. This pool of DNA is then hybridized to appropriate microarrays, allowing outliers specifically associated with the Dam-fusion protein to be identified on a genome-wide scale. This technique, when applied to PcG proteins in mammalian cells, would provide in theory a complete picture of PcG target genes at a genomic level, but would also allow the analysis for common motifs present in the isolated sequences. Could a common motif identified by this method reflect a true mammalian PRE? A precedent in this respect is the recent demonstration by the Paro lab that a sequence-based search tool could be generated, based on ChIP on Chip data, which identified over 100 PREs/TREs in *Drosophila*, some of which have been functionally verified (Ringrose et al. 2003). As a first step toward more comprehensive analyses, an array encompassing the genomic loci of HOX gene clusters (representing conserved PcG/trxG target genes) would provide a good starting point in mammals. Ultimately, further comparison of DamID profiles with data generated by alternative methods, such as ChIP on Chip experiments, will undoubtedly contribute to our understanding of the mechanism(s) that PcG (and trxG) proteins employ to regulate their target genes.

PcG AND trxG COMPLEXES POSSESS ENZYMATIC ACTIVITIES THAT MODIFY HISTONES/NUCLEOSOMES

A main finding over the last few years was the demonstration that complexes containing both PcG and trxG members exhibit enzymatic activities. The targets of these activities have been shown to be histones, although it is conceivable that nonhistone targets may also exist. The various modifications on histones, including phosphorylation, acetylation, methylation, and ubiquitination on different residues, act in a combinatorial manner to regulate gene activity. This so-called "histone code" is thought to provide an epigenetic tag, which serves to recruit appropriate regulators as well as to dictate chromatin compaction (Jenuwein and Allis 2001). The mammalian PRC2 protein Ezh2 sets the repressive methylation mark on H3 at lysine 27, which is dependent on the association of Ezh2 with PRC2 members Eed and Su(z)12 (Cao et al. 2002; Kuzmichev et al. 2002; Muller et al. 2002). In turn, the meK27 mark has been shown to recruit Polycomb via its chromodomain and, thus, initiate sustained silencing by PRC1 (Cao et al. 2002; Czermin et al. 2002; Kuzmichev et al. 2002). An additional level of regulation by the Ezh2 HMT activity is determined by the type of Eed isoforms with which Ezh2 associates. The four Eed isoforms generated by alternate translation sites result in various complexes that direct Ezh2 HMT activity via PRC2 to the H3 at lysine 27 or to H1 at lysine 26 (Kuzmichev et al. 2004). This recent finding is intriguing given the important role of H1-linker histones in mediating higher-order chromatin folding. Both H1 and H3 methylation marks are important for transcriptional repression by Ezh2. These reports are pivotal in our understanding of PcG mechanisms as they provide the first link as to how PRC1 is recruited to target genes via the activity of PRC2.

In contrast, methylation of H3 at lysine 4 by TRX/MLL and of lysine 4, 9, and 20 by Ash1 is generally correlated with active genes (Beisel et al. 2002; Milne et al. 2002; Nakamura et al. 2002). The recruitment of the remodeling Brahma trxG complex, via the methylation of lysine 4 and double methylation of lysine 9 epigenetic tags and the consequent opening of chromatin by nucleosome repositioning, may be one way in which TRX/MLL and Ash1 enzymatic activities help to induce gene activation (Beisel et al. 2002). An alternative mechanism resulting from the deposition of me-H3K4 marks by trxG members may also include prevention of setting the meK27 mark by PcG and, thus, prohibition of the establishment of silencing rather than recruiting activators. The finding that double mutants for Trx or Ash1 and PcG members can reconstitute misexpression of Hox genes suggests that at least Trx and Ash1 maintain active genes by indeed functioning as antirepressors rather than classical coactivators (Klymenko and Muller 2004).

IMPLICATIONS OF PcG IN MULTILAYERED GENE SILENCING: THE X-INACTIVATION EXAMPLE

Arguably the best example of multilayer silencing involving PcG is the recent demonstration that EzH2/PRC2 activity is an important step in the early phases of X-inactivation (Plath et al. 2003; Silva et al. 2003). In mammals, dosage compensation of gene activity of the X-chromosome is normalized between males and females by inactivating one of the two X chromosomes in female cells (Heard 2004). This process takes place during early development where one of the X chromosomes is randomly inactivated in the inner cell mass, which gives rise to the embryo proper. The inactivation process is highly dependent on the coating of the inactive X chromosome by a large noncoding RNA, named Xist (X-inactive-specific transcript), and is influenced by the state of histone acetylation and by DNA methylation. Upon differentiation of embryonic stem (ES) cells, initiation of X-inactivation can be studied, where Xist noncoding RNA is transcribed and recruited to the prospective Xi, and modification of chromatin on the amino-terminal tails of histones H3 and H4 is observed. Interestingly, this early stage in the X-inactivation process is still reversible, perhaps reflecting the inherent plasticity of precursor cells. However, later in the differentiation process, inactivation of the X becomes largely Xist-independent indicating that other silencing mechanisms contribute to assure stable repression of the Xi during subsequent development.

It was recently demonstrated that at least part of the histone tags on the inactive X is mediated by the HMT activity of Ezh2/Eed complex PRC2 polycomb (Plath et al. 2003; Silva et al. 2003). The transient binding of Ezh2/Eed complex at the inactive X during the early stages of X-inactivation fits well with its putative function as an initiating repressive complex. Importantly, setting the repressive meK27 mark on the Xi by EzH2/PRC2 was found to be dependent on Xist, although so far no direct binding to the noncoding RNA has been reported (Plath et al. 2003). The trimethylation of H3K27 is localized within distinct domains on the inactive X, which appears to not overlap with the dimethylation of H3K9, a mark correlated with facultative heterochromatin (Heard 2004). Worthy to note is that although recruitment of Ezh2/Eed to the inactive X may be essential for the onset of X-inactivation, it is not enough to maintain silencing. Additional layers of regulation, such as histone acetylation and DNA methylation, must act to synergize with the Xist RNA to silence the X and/or to contribute in creating a compact chromatin structure that is able to maintain its silencing state over several mitotic divisions. The fact that

Ezh2/Eed complex binds to the Xi would logically imply that the deposition of trimethylation of H3K27 at the inactive X should recruit members of the PRC1 complex to the inactive X. This turned out to be correct as we recently detected Bmi1, Ring1B/Rnf2, and hPC2 localization on the inactive X in mouse and human cells (I. Hernández-Muñoz et al., in prep.). The recruitment of these PRC1 members occurred during the initial stage of X-inactivation following closely the recruitment of Ezh2/Eed. Interestingly, PRC1 binding appeared highly dynamic and dependent on the cell cycle stage, as Bmi-1, Rnf2, and HPC2 were found on the Xi chromosome primarily in early and mid S phase (I. Hernández-Muñoz et al., in prep.). The dynamic kinetics of the recruitment of PRC2 and PRC1 was found in somatic, extraembryonic, and embryonic cells. Such a phenomenon suggests that upon each cell division the reestablishment of the compact silent Xi chromatin needs to be reformed de novo, requiring the recruitment of both PRC2 and PRC1.

The inactive X has been described to be enriched in histone-variant macroH2A, as is illustrated by specific staining of the Barr body by macroH2A antisera. macroH2A has recently been shown to inhibit accessibility of chromatin to transcription and remodeling factors such as Brahma, thus possibly helping to ensure stable X-inactivation (Angelov et al. 2003). However, a direct role of macroH2A in silencing of the Xi remained obscure. New insight came with the discovery that macroH2A binds to the SPOP protein (Takahashi et al. 2002), and our observation that SPOP also binds to the PRC1 protein Bmi1 (I. Hernández-Muñoz et al., in prep.). Significantly, SPOP is comprised of both a BTB and a POZ domain, suggesting that SPOP may interact with CULLIN3 to form a functional E3 ubiquitin ligase (see Fig. 1) (Nagai et al. 1997; Furukawa et al. 2003). Subsequent experiments confirmed that SPOP and Bmi1 as well as SPOP and macroH2A can interact and form active E3 ubiquitin ligase complexes with CULLIN3 under physiological conditions. Both macroH2A and Bmi1 themselves are subject to poly- and monoubiquitination by these ligases. Importantly, ubiquitination appeared not to influence the overall stability of Bmi1 and macroH2A, suggesting that the ubiquitination mark on these proteins serves a regulatory function and that the proteins are not targeted for proteasomal for degradation (I. Hernández-Muñoz et al., in prep.). What the precise molecular function of ubiquitination of Bmi1 and macroH2A is remains to be fully elucidated. However, the importance of these novel SPOP/CUL3 E3 ligases for X-inactivation was illustrated by a series of RNAi experiments and by using an in vivo assay where a silent GFP reporter is embedded in the inactive X. Knockdown of CULLIN3 and SPOP resulted in a profound delocalization of macroH2A from the inactive X. It thus appears that ubiquitination is a vital feature during the X-inactivation process, which directly or indirectly assists in correct deposition of macroH2A to the inactive X (Fig. 1). Moreover, knockdown of either macroH2A or CULLIN3 and SPOP in cells carrying a silent GFP reporter on the inactive X allowed partial reactivation of the GFP reporter. The RNAi-mediated reactivation of the GFP reporter, however, was only observed after mild TSA and 5-azadC treatment of the cells. Taken together, these results suggest that macroH2A, regulated through ubiquitination, represents another level of silencing to the inactive X, acting on top of DNA methylation and histone code changes (I. Hernández-Muñoz et al., in prep.). This multilayered silencing of the inactive X most probably assures a tight silent chromatin structure that can be stably propagated over several cell generations.

Although functional data demonstrating if and how ubiquitination of Bmi1 contributes to the silencing of the inactive X remain to be shown, it is tempting to speculate that modification of Bmi1 by ubiquitin can help to enhance the formation of silencing complexes required to

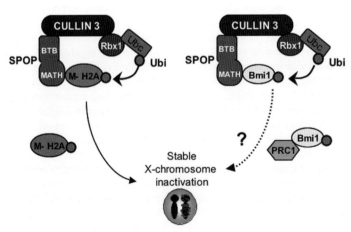

Figure 1. Ubiquitination of MacroH2A represents an additional level of regulation in X-chromosome inactivation. Bmi1 and macroH2A are ubiquitinated by the SPOP/Cul3 E3 ligase. The ubiquitination of macroH2A is important for its proper localization onto the inactive X. The localization of macroH2A is required for the silencing of the inactive X. The PRC1 complex is recruited to the inactive X by the HMT activity of the Ezh2/Eed complex. The exact role(s) of PRC1 and of ubiquitination of Bmi1 in the inactive X process remains unclear, although it can be speculated that PRC1/Bmi1 complexes may be required to stably maintain the compact chromatin of the inactive X over several mitotic divisions.

maintain the X inactive by either changes in Bmi1 conformation/localization or by recruiting proteins carrying ubiquitin-binding domains (Fig. 1). Moreover, it will be important to investigate if such posttranslation modifications may be a general theme employed by Bmi1 (and possibly other PcG members) to modulate the extent of its repressive function at other target sites. In this regard, the *Caenorhabditis elegans* SOP-2 (the corresponding mammalian homolog is Rae28/Mph1) has been recently reported to be sumoylated (an ubiquitin-like molecule) and that this sumoylation is required for SOP-2's ability to localize to nuclear bodies and, most importantly, to enable repression of HOX genes (Zhang et al. 2004). It will be important to further delineate the functional significance of ubiquitinated Bmi1 as well as to explore Bmi1's (and other PRC1 members') role in X-inactivation.

CONNECTIONS AMONG PcG REGULATION, STEM CELL BIOLOGY, AND CANCER FORMATION

It has often been suggested that stem cells share similarities with cancer cells, most importantly, in their ability to continuously proliferate and in the presence of subsets of cells with undifferentiated characteristics within a heterogeneous tumor cell population (Pardal et al. 2003). Best evidence for the presence of such "cancer stem cells" has been found in several classes of leukemia (Bonnet and Dick 2001; Singh et al. 2003). Given that sustained abnormal expression of Bmi1 together with c-Myc potently induces leukemia, is there evidence linking Bmi-1 regulation (and possibly other PcG members) with stem cell biology? Several recent observations indeed have highlighted this. Deficiency for several PRC1 members, including Bmi1, Mph1/Rae28, M33, and Mel-18, cause severe and progressive reductions in T and B cell lymphocytes (van der Lugt et al. 1994; Akasaka et al. 1997; Core et al. 1997; Takihara et al. 1997; Tokimasa et al. 2001). These defects can be understood in light of the recently discovered essential role for both Bmi1 and Mph1/Rae28 in sustaining the self-renewal capacity of adult hematopoietic stem cells (HSCs) (Ohta et al. 2002; Park et al. 2003). Importantly with respect to its role in cancer formation, similar observations were made with mouse leukemic cells, where Bmi1 was essential for the maintenance of leukemia following transplantation of primary leukemic cells into secondary recipients (Lessard and Sauvageau 2003). This indicates that a subpopulation of leukemic cancer stem cells rely on Bmi1/PRC1 function for their continuous proliferation.

Extending Bmi-1's role in stem cell biology, Bmi-1 was reported to also be essential for the self-renewal capacity of neural stem cells from the central and peripheral nervous systems and for the proliferation of cerebellar granule neuron precursor cells (CGNPs), which are essential for formation of the cerebellum (Molofsky et al. 2003; Leung et al. 2004). High expression of Bmi1 was observed in the external granular layer (EGL), a thin layer of cells residing at the surface of the developing cerebellum, and was strongest in proliferating granule cell precursors within the EGL at postnatal days 5–8. The previously observed dramatic and progressive reduction in cerebellum size of Bmi1 null mice, giving rise to ataxia, tremors, and paralysis due to motorneuron defects, correlates well with the observed reduced proliferation capacity of the CGNPs in vivo and in vitro (Valk-Lingbeek et al. 2004).

Given these observations, obvious following questions are how does Bmi-1 control the ability of stem cells to self-renew, which target genes are implicated, and how in turn is Bmi1 itself regulated in stem cells? A first clue came from our previous observations, indicating that the Ink4a/Arf tumor-suppressor locus was an important in vivo relevant target for Bmi1/PRC1, when it was observed that the early senescent phenotype exhibited by cells lacking Bmi1 can be rescued by repressing or deleting the Ink4A/Arf locus. Indeed, Bmi1 deficiency leads to derepression of the Ink4a/Arf locus, which in turn controls both the Rb and p53 tumor-suppressor pathways and acts as an important tumor-prevention mechanism (for review, see Jacobs and van Lohuizen 2002). Conversely, overexpression of Bmi1 was shown to hyperrepress the Ink4a/Arf locus, phenocopying loss of these tumor suppressors, explaining why overexpression of Bmi1 is oncogenic. Hence, the Ink4a/Arf locus is an ideal candidate through which Bmi1 could exert its effects on stem cell self-renewal. Indeed, through genetic epistasic experiments and functional studies aberrant derepression of the Ink4a/Arf locus was found to contribute to the reduction in self-renewal of stem cells in Bmi1 knockouts (Jacobs et al. 1999; Molofsky et al. 2003; Park et al. 2003). This suggests that Bmi1 acts at least in part by repressing the p16/Ink4a and p19/Arf loci postnatally, thus allowing continued proliferation of adult hematopoietic and neural stem cells (Fig. 2).

Apart from implicating the Ink4a/Arf locus as an important downstream target in stem cells, studying the cerebellar phenotypes in Bmi1-deficient mice also provided new insight as to how, in turn, Bmi1/PRC1 is regulated. As uncontrolled proliferation of cerebellar precursor cells is thought to be the culprit in causing medulloblastoma, a major childhood brain tumor with stem/precursor cell characteristics, and the same precursor cells require Bmi1 for sustained proliferation, BMI1 levels in primary human medulloblastoma samples were investigated. We found high BMI1 overexpression in 8 out of 12 medulloblastomas. Interestingly, these same tumors displayed a constitutive active Sonic Hedgehog (Shh) signaling pathway (Leung et al. 2004). The Shh morphogen and stem cell growth factor has been shown to be required for cerebellar precursor cell proliferation, and mutations in Shh pathway components leading to constitutive signaling have been found in ±25% of medulloblastomas. We subsequently showed that Shh itself and the downstream Gli1 transcription factor of the Shh pathway induce Bmi1 expression, providing a direct link between a cell-extrinsic stem cell growth factor and cell-intrinsic epigenetic gene silencing by Bmi1/PRC1. Since the proliferation of CGNPs is induced by the Shh, the reduction in cerebellar granule neurons in the cerebel-

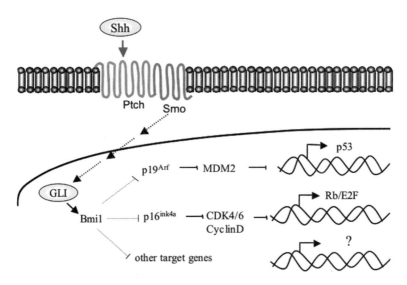

Figure 2. Developmental morphogen Sonic Hedgehog (Shh) acts via Bmi1 to control stem cell fate. Shh via its receptors, Patched (Ptch) and Smoothened (Smo), signals to downstream transcription factors, GLI, to control the protein level of Bmi1. Bmi1 is required to modulate $p16^{ink4a}$ and $p19^{Arf}$ protein levels to allow stem cells to self-renew via the Rb and p53 pathways, respectively. The connection among Shh pathway, Bmi1, and the ink4a/Arf locus provides the first example of a developmental signaling pathway impinging on a PcG protein to allow normal stem cell function.

lum of Bmi1 knockout appears to be a result of attenuated Shh signaling. The perfect correlation between Bmi1 overexpression and sustained activation of the Shh pathway in medulloblastoma suggests that partial activation of the hedgehog pathway may cause high levels of Bmi1, which in turn acts together with other Shh responses such as induction of Nmyc, in full proliferation (and perhaps concomitant blocking of differentiation) of cerebellar precursor cells (for review, see Valk-Lingbeek et al. 2004). The Ink4a/Arf locus was shown to be important for the ability of stem cells to self-renew in a Bmi1-dependent manner. Taken together, it is plausible that the subset of medulloblastomas that exhibit overexpression of Bmi1 results at least in part from repression of both the RB (via p16/Ink4a) and the p53 (via p19/Arf) pathways leading to disturbance in the differentiation and proliferation capacity of cerebellar precursors. This hypothesis will be tested in new mouse models.

CONCLUSIONS

The emerging dynamic role of both PRC1 and PRC2 polycomb protein complexes in X-chromosome inactivation under control of a novel ubiquitin ligase provides new insight in multilayered silencing processes needed for stable repression of larger parts of the genome (Fig. 1). In addition, the role of Bmi1 in stem cell biology provides the first example of a developmental signaling pathway (Shh) controlling a PcG gene (Bmi1), through which downstream PcG targets (such as Ink4a/Arf) mediate stem cell self-renewal potential (Fig. 2). It is tempting to speculate that additional external signaling pathways may indeed exert their effects in normal cell biology as well as in tumorigenesis via downstream PcG targets. The Shh and Indian hedgehog pathways have been implicated in the regulation of hematopoietic stem cells. Whether Bmi-1 is a downstream target of these pathways in other (tumor) cell contexts awaits further study.

REFERENCES

Akasaka T., Tsuji K., Kawahira H., Kanno M., Harigaya K., Hu L., Ebihara Y., Nakahata T., Tetsu O., Taniguchi M., and Koseki H. 1997. The role of mel-18, a mammalian Polycomb group gene, during IL-7-dependent proliferation of lymphocyte precursors. *Immunity* **7:** 135.

Angelov D., Molla A., Perche P.Y., Hans F., Cote J., Khochbin S., Bouvet P., and Dimitrov S. 2003. The histone variant macroH2A interferes with transcription factor binding and SWI/SNF nucleosome remodeling. *Mol. Cell* **11:** 1033.

Atchinson L., Ghias A., Wilkinson F., Bonini N., and Atchinson M. 2003. Transcription factor YY1 functions as a PcG protein in vivo. *EMBO J.* **22:** 1347.

Beisel C., Imhof A., Greene J., Kremmer E., and Sauer F. 2002. Histone methylation by the *Drosophila* epigenetic transcriptional regulator Ash1. *Nature* **419:** 857.

Bonnet D. and Dick J. 2001. Human acute myeloid leukaemia is organized as a hierarchy that originates from a primitive hematopoietic cell. *Nat. Med.* **3:** 730.

Breiling A., Turner B., Bianchi M., and Orlando V. 2001. General transcription factors bind promoters repressed by Polycomb group proteins. *Nature* **412:** 651.

Brown J., Mucci D., Whiteley M., Dirksen M., and Kassis J. 1998. The *Drosophila* polycomb group gene pleiohomeotic encodes a DNA binding protein with homology to the transcription factor YY1. *Mol. Cell* **1:** 1057.

Cao R., Wang L., Wang H., Xia L., Erdjument-Bromage H., Tempst P., Jones R.S., and Zhang Y. 2002. Role of histone H3 lysine 27 methylation in Polycomb-group silencing. *Science* **298:** 1039.

Carter D., Chakalova L., Osborne C., Dai Y., and Fraser P. 2002. Long-range chromatin regulatory interactions in vivo. *Nat. Genet.* **4:** 623.

Cavalli G. and Paro R. 1998. The *Drosophila* Fab-7 chromosomal element conveys epigenetic inheritance during mitosis and meiosis. *Cell* **93:** 505.

Core N., Bel S., Gaunt S.J., Aurrand-Lions M., Pearce J., Fisher A., and Djabali M. 1997. Altered cellular proliferation and mesoderm patterning in Polycomb-M33-deficient mice. *Development* **124:** 721.

Czermin B., Melfi R., McCabe D., Seitz V., Imhof A., and Pirrotta V. 2002. *Drosophila* enhancer of Zeste/ESC complexes have a histone H3 methyltransferase activity that marks chromosomal Polycomb sites. *Cell* **111:** 185.

Dekker J., Rippe K., Dekker M., and Kleckner N. 2002. Capturing chromosome conformation. *Science* **295:** 1306.

Dellino G., Schwartz Y., Farkas G., McCabe D., Elgin S., and Pirrotta V. 2004. Polycomb silencing blocks transcription initiation. *Mol. Cell* **13:** 887.

Francis N. and Kingston R. 2001. Mechanisms of transcriptional memory. *Nat. Rev. Mol. Cell Biol.* **2:** 409.

Furukawa M., He Y., Borchers C., and Xiong Y. 2003. Targeting of protein ubiquitination by BTB-Cullin 3-Roc1 ubiquitin ligases. *Nat. Cell Biol.* **5:** 1001.

Heard E. 2004. Recent advance in X-chromosome inactivation. *Curr. Opin. Cell Biol.* **16:** 247.

Hogga I. and Karch F. 2002. Transcription through the iab-7 cis-regulatory domain of the bithorax complex interferes with maintenance of polycomb-mediated silencing. *Development* **129:** 4915.

Jacobs J.J. and van Lohuizen M. 2002. Polycomb repression: From cellular memory to cellular proliferation and cancer. *Biochim. Biophys. Acta* **1602:** 151.

Jacobs J.J., Kieboom K., Marino S., DePinho R.A., and van Lohuizen M. 1999. The oncogene and Polycomb-group gene bmi-1 regulates cell proliferation and senescence through the ink4a locus. *Nature* **397:** 164.

Jenuwein T. and Allis C. 2001. Translating the histone code. *Science* **293:** 1074.

Kirmizis A., Bartley S., Kuzmichev A., Margueron R., Reinberg D., Green R., and Farnham P. 2004. Silencing of human polycomb target genes is associated with methylation of histone H3 Lys 27. *Genes Dev.* **18:** 1592.

Klymenko T. and Muller J. 2004. The histone methyltransferases Trithorax and Ash1 prevent transcriptional silencing by Polycomb group proteins. *EMBO Rep.* **5:** 373.

Kuzmichev A., Jenuwein T., Tempst P., and Reinberg D. 2004. Different EZH2-containing complexes target methylation of histone H1 or nucleosomal histone H3. *Mol. Cell* **14:** 183.

Kuzmichev A., Nishioka K., Erdjument-Bromage H., Tempst P., and Reinberg D. 2002. Histone methyltransferase activity associated with a human multiprotein complex containing the Enhancer of Zeste protein. *Genes Dev.* **16:** 2893.

Kwon H. and Chung H. 2003. Yin Yang 1, a vertebrate polycomb group gene, regulates antero-posterior neural patterning. *Biochem. Biophys. Res. Commun.* **306:** 1008.

Lavigne M., Francis N., King I., and Kingston R. 2004. Propagation of silencing: Recruitment and repression of native chromatin *in trans* by polycomb repressed chromatin. *Mol. Cell* **13:** 415.

Lessard J. and Sauvageau G. 2003. Bmi-1 determines the proliferative capacity of normal and leukaemic stem cells. *Nature* **423:** 255.

Leung C., Lingbeek M., Shakhova O., Liu J., Tanger E., Saremaslani P., van Lohuizen M., and Marino S. 2004. Bmi1 is essential for cerebellar development and is overexpressed in human medulloblastomas. *Nature* **428:** 337.

Lund A.H. and van Lohuizen M. 2004. Polycomb complexes and silencing mechanisms. *Curr. Opin. Cell Biol.* **16:** 239.

Milne T., Briggs S., Brock H., Martin M., Gibbs D., Allis C., and Hess J. 2002. MLL targets SET domain methyltransferase activity to Hox gene promoters. *Mol. Cell* **10:** 1107.

Mohd-Sarip A., Venturini F., Chalkley G.E., and Verrijzer C.P. 2002. Pleiohomeotic can link polycomb to DNA and mediate transcriptional repression. *Mol. Cell. Biol.* **22:** 7473.

Molofsky A.V., Pardal R., Iwashita T., Park I.K., Clarke M.F., and Morrison S.J. 2003. Bmi-1 dependence distinguishes neural stem cell self-renewal from progenitor proliferation. *Nature* **425:** 962.

Muller J., Hart C.M., Francis N.J., Vargas M.L., Sengupta A., Wild B., Miller E.L., O'Connor M.B., Kingston R.E, and Simon J.A. 2002. Histone methyltransferase activity of a *Drosophila* Polycomb group repressor complex. *Cell* **111:** 197.

Nagai Y., Kojima T., Muro Y., Hachiya T., Nishizawa Y., Wakabayashi T., and Hagiwara M. 1997. Identification of a novel nuclear speckle-type protein, SPOP. *FEBS Lett.* **418:** 23.

Nakamura T., Mori T., Tada S., Krajewski W., Rozovskaia T., Wassell R., Dubois G., Mazo A., Croce C., and Canaani E. 2002. ALL-1 is a histone methyltransferase that assembles a supercomplex of proteins involved in transcriptional regulation. *Mol. Cell* **10:** 1119.

Ohta H., Sawada A., Kim J.Y., Tokimasa S., Nishiguchi S., Humphries R.K., Hara J., and Takihara Y. 2002. Polycomb group gene rae28 is required for sustaining activity of hematopoietic stem cells. *J. Exp. Med.* **195:** 759.

Orlando V. and Paro R. 1993. Mapping Polycomb-repressed domains in the bithorax complex using in vivo formaldehyde cross-linked chromatin. *Cell* **75:** 1187.

Otte A.P. and Kwaks T.H. 2003. Gene repression by Polycomb group protein complexes: A distinct complex for every occasion? *Curr. Opin. Genet. Dev.* **13:** 448.

Pal-Bhadra M., Bhadra U., and Birchler J. 2002. RNAi related mechanisms affect both transcriptional and posttranscriptional transgene silencing in *Drosophila*. *Mol. Cell* **9:** 315.

Pardal R., Clarke M., and Morrison S. 2003. Applying the principles of stem-cell biology to cancer. *Nat. Rev. Cancer* **12:** 895.

Park I.K., Qian D., Kiel M., Becker M.W., Pihalja M., Weissman I.L., Morrison S.J., and Clarke M.F. 2003. Bmi-1 is required for maintenance of adult self-renewing haematopoietic stem cells. *Nature* **423:** 302.

Pirrotta V., Poux S., Melfi R., and Pilyugin M. 2003. Assembly of polycomb complexes and silencing mechanisms. *Genetica* **117:** 191.

Plath K., Fang J., Mlynarczyk-Evans S.K., Cao R., Worringer K.A., Wang H., de la Cruz C.C., Otte A.P., Panning B., and Zhang Y. 2003. Role of histone H3 lysine 27 methylation in X inactivation. *Science* **300:** 131.

Rank G., Prestel M., and Paro R. 2002. Transcription through intergenic chromosomal memory elements of the *Drosophila* bithorax complex correlates with an epigenetic switch. *Mol. Cell. Biol.* **22:** 8026.

Ringrose L., Rehmsmeier M., Dura J., and Paro R. 2003. Genome-wide prediction of polycomb/trithorax response elements in *Drosophila melanogaster*. *Dev. Cell* **5:** 759.

Satijn D., Hamer K., den Blaauwen J., and Otte A. 2001. The polycomb group protein EED interacts with YY1, and both proteins induce neural tissue in *Xenopus* embryos. *Mol. Cell. Biol.* **21:** 1360.

Saurin A., Shao Z., Erdjument-Bromage H., Tempst P., and Kingston R. 2001. A *Drosophila* Polycomb group complex includes Zeste and dTAFII proteins. *Nature* **412:** 655.

Silva J., Mak W., Zvetkova I., Appanah R., Nesterova T.B., Webster Z., Peters A.H., Jenuwein T., Otte A.P., and Brockdorff N. 2003. Establishment of histone h3 methylation on the inactive X chromosome requires transient recruitment of Eed-Enx1 polycomb group complexes. *Dev. Cell* **4:** 481.

Singh S., Clarke I., Terasaki M., Bonn V., Hawkins C., Squire J., and Dirks P. 2003. Identification of a cancer stem cell in human brain tumors. *Cancer Res.* **63:** 5821.

Takahashi I., Kameoka Y., and Hashimoto K. 2002. MacroH2A1.2 binds the nuclear protein Spop. *Biochim. Biophys. Acta* **1591:** 63.

Takihara Y., Tomotsune D., Shirai M., Katoh-Fukui Y., Nishii K., Motaleb M.A., Nomura M., Tsuchiya R., Fujita Y., Shibata Y., Higashinakagawa T., and Shimada K. 1997. Targeted disruption of the mouse homologue of the *Drosophila* polyhomeotic gene leads to altered anteroposterior patterning and neural crest defects. *Development* **124:** 3673.

Tokimasa S., Ohta H., Sawada A., Matsuda Y., Kim J.Y., Nishiguchi S., Hara J., and Takihara Y. 2001. Lack of the Polycomb-group gene rae28 causes maturation arrest at the early B-cell developmental stage. *Exp. Hematol.* **29:** 93.

Tolhuis B., Palstra R., Splinter E., Grosvel F., and de Laat W. 2002. Looping and interaction between hypersensitive sites in the active β-globin locus. *Mol. Cell* **10:** 1453.

Valk-Lingbeek M., Bruggeman S., and van Lohuizen M. 2004. Stem cells and cancer: The polycomb connection. *Cell* **118:** 409.

van der Lugt N.M., Domen J., Linders K., van Roon M., Robanus-Maandag E., te Riele H., van der Valk M., Deschamps J., Sofroniew M., van Lohuizen M., and Berns A. 1994. Posterior transformation, neurological abnormalities, and severe hematopoietic defects in mice with a targeted deletion of the bmi-1 proto-oncogene. *Genes Dev.* **8:** 757.

van Steensel B., Delrow J., and Henikoff S. 2001. Chromatin profiling using targeted DNA adenine methyltransferase. *Nat. Genet.* **27:** 304.

Wang L., Brown J.L., Cao R., Zhang Y., Kassis J.A., and Jones R.S. 2004. Heirarchical recruitment of polycomb group silencing complexes. *Mol. Cell* **14:** 637.

Zhang H., Smolen G., Palme R., Christoforou A., van den Heuvel S., and Haber D. 2004. SUMO modification is required for *in vivo* Hox gene regulation by the *Caenorhabditis elegans* Polycomb group protein SOP-2. *Nat. Genet.* **36:** 507.

The Function of Telomere Clustering in Yeast: The Circe Effect

S.M. Gasser,*† F. Hediger,* A. Taddei,* F.R. Neumann,*
and M.R. Gartenberg‡

Department of Molecular Biology and Frontiers in Genetics NCCR Program, University of Geneva, CH-1211 Geneva, Switzerland; ‡Department of Pharmacology University of Medicine and Dentistry of New Jersey–Robert Wood Johnson Medical School, Piscataway, New Jersey 08854

Recent work has shown convincingly that telomeres of unicellular organisms, such as budding and fission yeast, *Trypanosoma* and *Plasmodia*, are spatially clustered in the nucleus. Live imaging (Heun et al. 2001b; Tham et al. 2001; Hediger et al. 2002b) and controlled in situ hybridization experiments (Gotta et al. 1996; Hediger et al. 2002a) in budding yeast indicate that telomere clusters are often adjacent to the nuclear envelope (NE), although they are distributed independently of nuclear pores (Fig. 1A). Rather than through pores, telomeres are bound through two redundant mechanisms: one that requires the end-binding complex yKu, and a second that makes use of the "partitioning and anchoring domain" (PAD; Ansari and Gartenberg 1997) of the silent information regulator Sir4 (Hediger et al. 2002b; Taddei et al. 2004). This domain, which is adjacent to the lamin-like carboxyl terminus of Sir4, binds specifically to the protein Esc1 (Establishes Silent Chromatin; Andrulis et al. 2002). Esc1 is large acidic protein localized along the inner nuclear membrane at interpore spaces (Fig. 1B) (Taddei et al. 2004). The anchorage mediated by the yKu heterodimer is not fully characterized, yet, when targeted to DNA, the yKu80 subunit is able to relocalize DNA to the nuclear periphery in a manner independent of Sir4 or Esc1 (Taddei et al. 2004). The efficiency of yKu80-mediated anchorage varies between G1 and S phases of the cell cycle, being stronger in G1. This mechanism may reflect a link between telomere anchoring and telomere end protection. In contrast to earlier reports (Galy et al. 2000; Feuerbach et al. 2002), we find that yKu does not anchor telomeres through interaction with the two Myosin-like proteins Mlp1 or Mlp2. Deletion of both *MLP* genes has no effect on telomere position, telomere silencing, or the localization of silent information regulatory proteins in budding yeast (Fig. 2) (Hediger et al. 2002a,b). Nonetheless, Mlp proteins do have a minor effect on telomere length (Hediger et al. 2002a).

BIOLOGICAL FUNCTIONS OF TELOMERE CLUSTERING

In wild-type cells, the clustering of telomeres creates zones in which silent information regulatory proteins, Sir2,
Sir3, and Sir4, accumulate (Gotta et al. 1996). These proteins form a complex that is involved in the repression of transcription in a domain-specific, rather than promoter-specific, manner. The complex binds chromatin through interaction with histone amino-terminal tails. A point mutation in histone H4 K16 can displace the Sir complex and compromise telomeric repression (Hecht et al., 1995). Whereas Sir2 is a universally conserved NAD (nicotinamide adenine)-dependent histone deacetylase, Sir3 and Sir4 are neither conserved nor display any recognizable enzymatic activity (for review, see Gasser and Cockell 2001). The complex is thought to bind cooperatively along the nucleosomal fiber, a process that can be counteracted by barrier proteins (for review, see Bi and Broach 2001) or by Sas2, a histone acetylase that targets histone H4 K16 (Kimura et al. 2002; Suka et al. 2002). It is assumed, but has not yet been shown, that Sir complex binding folds the nucleosomal fiber into a higher-order structure.

Colocalization of the Sir protein pools with subtelomeric sequences and silent-mating-type loci was demonstrated by Chromatin IP and by double FISH-IF (fluorescence in situ hybridization immunofluorescence) localization (Gotta et al. 1996; Strahl-Bolsinger et al. 1997). The binding of Sir proteins to nucleosomes in these telomeric foci correlates strongly with subtelomeric transcriptional repression, called TPE (Hecht et al. 1995, 1996). Recent studies presented below have shown that if one disperses Sir pools throughout the nucleoplasm by eliminating the telomere anchors yKu and Esc1, silencer-flanked reporter genes can be silenced without being proximal to telomeres (Gartenberg et al. 2004; see below). Similar effects can be achieved by overexpressing Sir proteins, although this can also have toxic effects (Holmes et al. 1997).

GENETIC DIVERSITY THROUGH TELOMERE-ASSOCIATED VARIEGATION

Insight into the relationships between telomere clustering, anchoring, and silencing have emerged from studies of pathogens such as *Plasmodium falciparum*, *Trypanosoma brucei*, or *Candida glabrata*. In all three species, telomeres are grouped at the NE as they are in budding yeast (Chung et al. 1990; Freitas-Junior et al. 2000). Furthermore, orthologs of yeast Sir proteins have been identified in *Plasmodia* and *Candida* (Scherf et al.

---

†Corresponding author. Current address: Friedrich Miescher Institute, Maulbeerstrasse 66, CH-4058 Basel, Switzerland. E-mail: susan.gasser@fmi.ch.

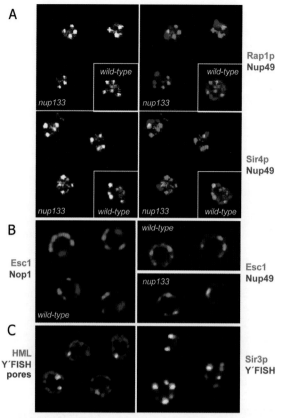

Figure 1. Subnuclear localization of telomeres, silencing factors, Esc1, and the silent *HML* locus. (*A*) Mutation of *nup133* does not affect telomere position, as detected by immunofluorescence (IF) for telomere-bound Rap1 and Sir4. IF was performed on fixed cells using affinity-purified polyclonal antibodies specific for Rap1 or Sir4, along with the monoclonal Mab414 (anti-pore). Cells were grown at 24°C and were prepared for IF as described in Heun et al. (2001a). Strains used were either wild type (shown in *insets*) or an isogenic strain deleted for *nup133*. In the merged image, Rap1 or Sir4 signals are in *green*, while nuclear pore is in *red*. Overlap is in *yellow*. Bar, 2 μm. (*B*) Esc1 is localized at the nuclear periphery independently of pores (Taddei et al. 2004). Live visualization of endogenous Esc1-GFP tagged and the nucleolar protein Nop1 (*left* panel), or the nuclear pore protein Nup49 (*right* panels) fused to CFP (shown in *red*) in a wild-type or *nup133* strain, as indicated. Shown are equatorial sections of nuclei after deconvolution of wide-field Z-stacks of GFP and CFP signals. (*C*) *Left:* Two-color FISH with probes specific for the telomeric Y′ repeat (*red*) and the silent *HML* locus in a diploid cell, which was arrested in G1 by alpha factor and counterstained with antinuclear pore Mab414 (*blue*). Shown are equatorial confocal sections of four cells. Labeling and microscopy as described in Heun et al. (2001a). *Right:* Colocalization of FISH for the telomeric Y′ probe (*red*) and Sir3, detected with an affinity purified polyclonal antibody (*green*). Only ~70% of foci colocalize (overlap in *white*; Gotta et al. 1996). (*A*, Reprinted, with permission, from Hediger et al. 2002a [©Elsevier]; *B*, reprinted from Taddei et al. 2004 [©Nature Publishing Group; http://www.nature.com/].)

2001; De Las Penas et al. 2003), increasing the likelihood for conserved anchoring and silencing pathways. In these pathogens, telomere clustering is linked to the organism's virulence (De Las Penas et al. 2003). High virulence in *Plasmodia* or *Trypanosoma* requires a sequential activation of variant-specific surface glycoprotein genes (vsg or var) at subtelomeric expression sites, allowing the parasite to escape the host immune system. The clustering of different *var* genes in groups, which correspond to telomeric foci, is thought to enhance recombination efficiency and thereby increases the parasites' virulence by increasing antigenic variability. Although telomere-anchoring proteins are not yet characterized, Figueiredo and colleagues showed that the clustering of *Plasmodia* telomeres depends on subtelomeric elements (Figueiredo et al. 2002). It is not known as yet whether virulence can be compromised by interfering with telomere clustering.

In the fungal pathogen, *Candida glabrata*, virulence depends on the adherence of the parasite to the intestine epithelia through the lectin EPA-1 (Cormack et al. 1999). EPA-1 is member of a family of related *Candida* genes found in subtelomeric clusters, most of which are not expressed at any given time. Their repression resembles the reversible position effect found at budding yeast telomeres and is mediated by homologs of the ScSir3 and ScRap1 proteins (De Las Penas et al. 2003). Highly analogous is a budding yeast gene family of *FLO* genes, which encode cell-wall glycoproteins involved in regulating cell adherence and a developmental transition. One family member, the *FLO11* gene, is nonsubtelomeric, yet shows a variegated pattern of repression that requires a promoter-bound factor and the histone deacetylase Hda1 (Halme et al. 2004). This silencing does not require the *SIR* gene products. In contrast, *FLO10*, like most other *FLO* genes, is located in a subtelomeric zone of reduced gene expression, called the HAST domain (Hda1-affected subtelomeric regions; Robyr et al. 2002). These genes show variegated, position-dependent repression that does require telomere-binding factors yKu and Sir3. Surprisingly, neither Sir4 nor Sir2 is implicated in this regulation, although the Sir2 homolog Hst1 is. The variegated expression of the yeast *FLO* genes is highly reminiscent of mechanisms used to ensure phenotypic variation of cell surface markers in pathogenic microorganisms, like *Plasmodia*, *Trypanosoma*, and *Candida*. Thus we propose that unicellular organisms may use the perinuclear clustering of telomeres and its associated recombination and variegation phenotypes to expand their repertoire of gene expression patterns improving survival in adverse situations.

SILENT CHROMATIN DIRECTS ITS OWN ANCHORAGE INDEPENDENTLY OF A CHROMOSOMAL CONTEXT

Our studies of yeast telomere anchoring raised the following question: Does silent chromatin that has no chromosomal end nearby also associate with the nuclear periphery? Does repressed chromatin per se have a subnuclear "address"? Because the targeting of a Sir4^{PAD} fusion relocalizes an internal chromosomal domain to the NE (Taddei et al. 2004), we hypothesized that silent chromatin may be anchored directly. We therefore tested whether full-length Sir4, as an integral component of silent chromatin, could also mediate anchorage of a chromosomal domain.

In situ hybridization studies showed that the two silent yeast mating-type loci, *HML* and *HMR*, which are posi-

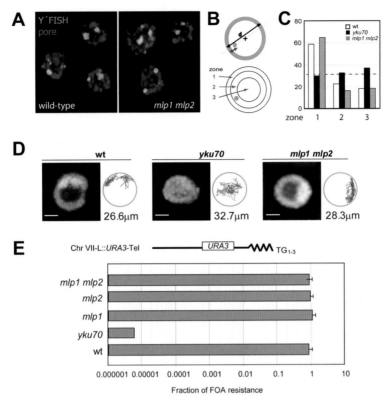

Figure 2. *MLP1* and *MLP2* deletion does not alter telomere distribution nor TPE. (*A*) In situ hybridization with a probe recognizing $TG_{(1-3)}$ and Y′ sequences and immunofluorescence with affinity-purified polyclonal antibodies specific for Sir4 were performed on wild-type LM11 (wt; Maillet et al. 2001) and an isogenic strain deleted for *mlp1* and *mlp2* (GA1471, *mlp1 mlp2*). Telomeres are in *green*, while nuclear pore staining (Mab414) is in *red*. Shown are confocal images as published in Hediger et al. (2002a) showing equatorial sections of yeast nuclei that are intact after the staining procedure. Spread or flattened nuclei were not analyzed. Bar, 2 mm. (*B*) A scheme showing an equatorial section of fluorescence for cells bearing Nup49-GFP and a tagged genomic locus. Distances between the lacI-GFP-tagged locus and Nup49-GFP at the nuclear envelope (x) are normalized to the nuclear diameter (y) and binned according to three zones of equal surface area. In a 2-μm-diameter nucleus, zone 1 corresponds to the outermost ring of ~190 nm width. Random distribution implies 33% in each zone. (*C*) Telomere 6R is not delocalized in the *mlp1mlp2* mutant. Position of telomere 6R in G1 cells is represented in bar graphs as percentage of spot (y axis) per zone (x axis) for wild type, *yku70*, or *mlp1mlp2* strains (from Hediger et al. 2002b). Values >33% show an enrichment of Tel6R at the nuclear periphery. Cells analyzed are wt, n=110; *yku70*, n = 168; for *mlp1mlp2*, n= 124. (*D*) 250 sequential confocal images (at 1.5-sec intervals) from a typical time-lapse series of G1 phase nuclei bearing the Tel 6R tag were first aligned based on their nuclear pore signals, and then the absolute position of focus was marked using the AIM tool of the Zeiss LSM 510 software (rel. 2.8). The locus' trajectory is projected in *red* on a single focal section of the nucleus. Strains used are wild-type (GA-1459), a *yku70* deletion (GA-1489), and the *mlp1 mlp2* double deletion (GA-1731). The mean length of the path in millimeters for a 5-min movie (200 frames) averaged over eight movies is indicated at the lower right of each image (for details, see Hediger et al. 2002b). (*E*) *mlp1 mlp2* mutants show normal levels of repression at telomeres. The wild-type LM11 background containing the *URA3* reporter at Tel VII-L and the same strain with either *mlp1*, *mlp2*, or *mlp1mlp2* (*mlp1/2*) deletions were grown overnight in YPAD, and tenfold serial dilutions were plated onto synthetic complete medium (SC) and SC containing 0.1% 5-FOA. Similar results were obtained for the *ADE2* gene on TelV-R and for *URA3* in another strain background (Hediger et al. 2002a). (*A,E*, Reprinted from Hediger et al. 2002a [© Elsevier]; *D*, reprinted from Hediger et al. 2002b [©Elsevier].)

tioned within 25 kb of telomeres, are indeed found at or near the NE, and *HML* was frequently found clustered with other telomeres (Fig. 1C) (Laroche et al. 2000). The contour length of DNA sequence (in base pairs [bp]) separating the telomeres and silent loci is below the resolution of the light microscope. Thus, it is impossible to determine whether or not *HM* loci are simply held at the nuclear periphery because of anchorage of the adjacent telomere. To eliminate the influence of *cis*-linked telomeres on silent chromatin localization, we used inducible site-specific recombination to uncouple the *HMR* locus from its normal chromosomal context (Fig. 3A) (Cheng et al. 1998). A genomic region of 17 kb, containing *HMR* with its silencers and an integrated lac operator array (lac^{op}), was flanked by a pair of target sites for the R site–specific recombinase. Galactose-induced expression of the recombinase yielded 95% excision within 2 hours in a silencing-competent strain. The locus could be monitored relative to GFP-tagged nuclear pores and, by comparing strains that had or had not undergone excision, the impact of chromosomal context on location and dynamics of *HMR* could be evaluated (Fig. 3).

In a strain that lacks recombinase (SIR+/Rec–), the unexcised *HMR* remains closely linked to the nuclear envelope (Fig. 3B). In a silencing-defective strain (*sir3* deletion; *sir3*/Rec-), the perinuclear fraction of *HMR* dropped only slightly (91% to 81%; Fig. 3B). This was expected, since a yKu-dependent mechanism is known to anchor telomeres in the absence of silencing (Hediger et al. 2002b). Intriguingly, when *HMR* was uncoupled from the chromosome as an excised ring, it remained at the nuclear envelope in 86% of silencing-competent cells (SIR+/Rec+; Fig. 3B). This shows that *HMR* association with the nuclear periphery is an intrinsic property of the locus and does not depend on linkage to a telomere. Furthermore, when the ring was excised in a *sir3* deletion strain, it lost anchorage (*sir3*/Rec+; Fig. 3B). The unconstrained mobility of a nonsilent ring can be demonstrated by time-lapse tracking (Fig. 3C; see 5-min trajectories of excised rings). This allows us to conclude that immobilization of the excised *HMR* locus at the NE depends on its silent state, unlike the situation at telomeres.

Two Types of Perinuclear Anchorage: Independent of and Dependent on Silencing

Because yeast has no intermediate filament proteins like the nuclear lamina, and because neither yKu nor Sir proteins have membrane-spanning domains, we screened for NE components that tether yKu and silent chromatin. By exploiting an in vivo chromatin relocalization assay (Fig. 4A) (Taddei et al. 2004), we have been able to assay protein domains for their ability to direct chromatin to the nuclear periphery independently of their interaction with silencing factors. As shown here (Fig. 4B), minimal protein domains that are sufficient to relocate a tagged locus to the NE include yKu80, Sir4^{PAD}, the carboxy-terminal domain of Esc1, and a transmembrane protein like Yif1 (Taddei et al. 2004). Esc1 interacts with Sir4 and is required for the stable mitotic partitioning of a Sir4-bound plasmid (Andrulis et al. 2002). Importantly, as Esc1 is positioned at the nuclear periphery in strains lacking *YKU* and *SIR* genes, it appears to be a structural element that does not require chromatin for its localization. Nonetheless, Esc1 contributes to efficient telomeric silencing (Taddei et al. 2004). Most importantly, the relocalization of a chromatin domain through the targeting of the Sir4^{PAD} domain requires either Esc1 or Ku (Fig. 4C,D), arguing that these ligands constitute necessary but redundant pathways for the anchoring of both telomeres and silent chromatin by Sir4.

Mlp1 and Mlp2 were previously proposed to form a bridge between yKu and nuclear pores (Galy et al. 2000). If this were true, telomeres would be released from their perinuclear anchorage in the double *mlp1 mlp2* deletion strain, similar to their delocalization in *yku*-deficient strains. Based both on an analysis of live cells and on FISH studies, this is not the case: As shown above, telo-

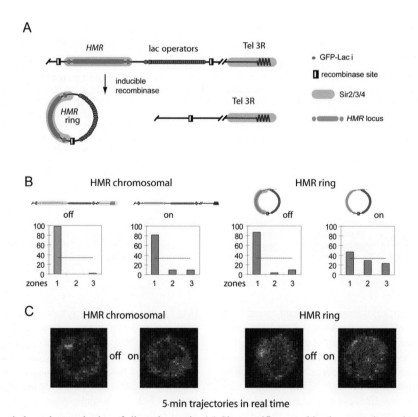

Figure 3. Telomere independent anchoring of silent chromatin. (*A*) Site-specific recombination at engineered target sites (*RS* sites) uncouples the lac^{op}-tagged *HMR* locus from the chromosome. The strains carry Lac-GFP, Nup49-GFP, and a galactose-inducible R recombinase, and are either competent for repression (Sir+) or deficient (*sir3* deletion). (*B*) Distribution of *HMR* in the three zones is plotted as a percentage of the total number of cells counted for each strain, as described in Fig. 2B. Zone 1 is the outermost zone and 33% would be a random distribution. Strains used and cells counted are MRG2253 SIR+/chromo, *n* = 140; MRG2251 *sir3*/chromo, *n* = 232; MRG2249 SIR+/ring, *n* = 173; MRG2250 *sir3*/ring, *n* =192. (*C*) The trajectory of the tagged *HMR* locus is traced in a representative cell from each relevant strain. A 3D stack of images was taken at 1.5-sec intervals for 5 min and projected onto one plane for tracking and quantitation (see Gartenberg et al. 2004).

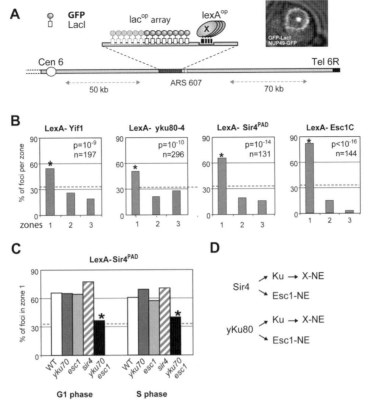

Figure 4. A chromatin relocalization assay identifies yKu- and Sir4-dependent anchoring pathways. (*A*) A yeast strain bearing *lac*^{op} repeats and LexA binding sites at an internal locus on the right arm of Chr 6 (Chr6^{int}) and a GFP-Nup49 fusion is transformed with plasmids expressing different LexA-protein fusions (Taddei et al. 2004). The position of the locus is scored as in Fig. 2B. (*B*) LexA.Yif1, LexA-Sir4^{PAD} LexA-yku80, and LexA-Esc1^C are able to relocate the locus to the NE. Bar graphs present the percentage of spots (*y* axis; *n* = number of cells analyzed) per zone (*x* axis). *P*-values are calculated by comparing actual proportion of cells bearing a spot within zone 1 to a hypothetical random distribution. (*C*) LexA-Sir4^{PAD} was expressed in GA-1461 bearing complete disruptions of *yku70*, *esc1*, *sir4*, and *yku70 esc1*, as indicated. Chr6^{int} position was scored as in Fig. 2B, and zone 1 values are compared among strains for G1 and S phase (for statistics, see Taddei et al. 2004). The black * indicates that values are indistinguishable from a random distribution ($p > 0.05$). LexA-Sir4^{PAD} chromatin relocation activity was unaffected by loss of either Esc1 or yKu70 individually, yet we see a complete loss of the Sir4^{PAD} tethering activity in the absence of both Esc1 and Ku, throughout the cell cycle. This result strongly suggests that Esc1 and Ku provide two parallel anchorage pathways for Sir4^{PAD}. (*D*) We show the dual pathways of anchoring mediated by Sir4-Esc1 and by Ku through an unknown protein (x) at the nuclear envelope, which allows relocalization of internal sequences in the absence of silencing. During G1 phase in wild-type yeast cells the yKu-mediated pathway appears to be favored while the Sir-Esc1 anchoring pathway functions efficiently in S phase (for details, see Taddei et al. 2004). (Reprinted, with permission, from Taddei et al. 2004 [©Nature Publishing Group; http://www.nature.com/].)

meres remain completely anchored in the double *mlp1 mlp2* mutant based on both live imaging and Y′ FISH (Fig. 2A,C). Moreover, while some but not all telomeres are released from the periphery in the *yku70* single mutant, we could show that all lose anchorage in *yku70 esc1* double mutants (Fig. 5A,B). Thus, the anchoring from yKu and from Sir4PAD-Esc1 provides parallel mechanisms that tether telomeres to the NE. Intriguingly, different telomeres show a differential dependence on the two pathways, possibly reflecting the efficiency of subtelomeric heterochromatin formation (Fig. 5A,B) (Tham et al. 2001; Hediger et al. 2002b; Taddei et al. 2004).

TESTING THE ROLE OF PERINUCLEAR ANCHORING IN REPRESSION

It has been proposed that the anchorage of telomeres at the nuclear periphery sequesters silencing factors and thereby promotes repression of adjacent genes (Maillet et al. 1996). This model is based on several observations: The first is that normal cellular Sir protein concentrations are limiting for repression. This notion was supported by the observation that balanced overexpression of Sir2, Sir3, and Sir4 (Maillet et al. 1996) or of Sir3 alone (Renauld et al. 1993) improves TPE, although *HM* repression appears to be resistant to this variability. Moreover, mutations were also discovered that enhance the ability of the *HM* loci and telomeres to compete with one another for Sir factors (Buck and Shore 1995). Second, we know from the work of several laboratories that silencer-flanked reporter genes are more efficiently repressed when inserted near the telomere. Although this shows that linear distance from a telomere can influence repression, it was unclear whether this reflected a positive effect of the nuclear envelope on repression or an effect of Sir factor concentration in telomeric pools. Indeed, the spatial or subnuclear position of these translocated silencing cassettes was never determined. Third, it was shown that

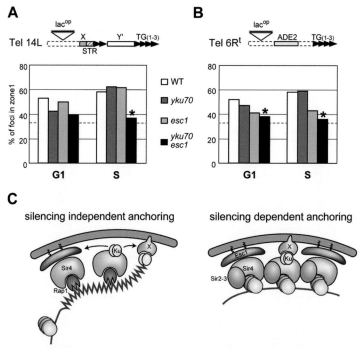

Figure 5. YKu and Esc1 anchor yeast telomeres in vivo. (*A*) Tel 14L position was determined in G1 and S phase cells of GA-1985, *yku70* (GA-1983), *esc1* (GA-2074), and *yku70 esc1* (GA-2082) strains, as in Fig. 4. For statistics see Taddei et al. (2004). (*B*) Truncated Tel 6R[t] localization was determined in WT (GA-1917), *yku70* (GA-1918), *esc1* (GA-2229), and *yku70 esc1* (GA-2230) strains as in A. (*C*) Scheme of dual telomere anchoring pathways: one through Esc1 and the other through yKu. This can occur for natural telomeres through yKu even in the absence of silencing (Hediger et al. 2002a; Taddei et al. 2004). In the presence of silent chromatin, Sir4 is available to bind Esc1 and yKu, thereby anchoring repressed chromatin in the absence of a nearby telomere. (Reprinted, with permission, from Taddei et al. 2004 [©Nature Publishing Group; http://www.nature.com/].)

the delocalization of Sir factors from telomeres, provoked either by mutation or overexpression, was able to restore repression at loci distant from telomeres (Lustig et al. 1996; Maillet et al. 1996, 2001; Marcand et al. 1996). This confirms that telomeric foci sequester repressors from other sites of action. Again, it was not shown whether the "internal" silenced loci relocate to the periphery once repressed. Such analysis is complicated by the fact that internal repression efficiency is often low.

In support of a functional role for telomere anchoring, the Sternglanz laboratory showed that the artificial "tethering" of a reporter gene to the yeast NE through a membrane-spanning protein or motif could favor transcriptional repression (Andrulis et al. 1998). This required the presence of at least one silencer element, a specific *cis*-acting sequence that nucleates Sir-dependent repression. It was proposed that this NE "tether" would promote repression because it places the weak silencer near a zone of high Sir protein concentration. However, other interpretations were not only possible, but likely. Notably, three results suggested that the nuclear periphery contributes to repression through something other than an increase in proximity to Sir proteins. First, Andrulis et al. (1998) showed that overexpression of Sir proteins did not obviate the positive effect conferred by the transmembrane tethering of the reporter construct. Thus, it would seem that the NE provides something necessary for repression in addition to the high Sir concentrations. Consistently, Galy et al. (2000) argued that loss of the pore-associated proteins Mlp1 and Mlp2 disrupted telomeric silencing and telomere anchoring. Although Mlp proteins are not needed for telomeric repression (Fig. 2E) (Andrulis et al. 2002), the notion that perturbation of nuclear pore architecture influences repression is often repeated (Feuerbach et al. 2002). From these arguments arose the notion that the nuclear envelope/pore environment is needed to promote repression.

Making use of the excised *HM* ring described in Figure 3, we were able to test this hypothesis. Knowing that Sir4[PAD] can anchor through both yKu and Esc1, we examined the locus dynamics and the repression status of the excised *HMR* ring in a strain lacking the two anchorage pathways. We ask whether the double *esc1 ku70* deletion would release the *HMR* ring from its perinuclear anchorage and, if so, whether this would compromise silencing. Although telomeric repression is eliminated by the deletion of yKu subunits, *yku70* deletion does not weaken mating repression at the endogenous *HML* (Laroche et al. 1998) or *HMR* loci (Boulton and Jackson 1998; Mishra and Shore 1999) to any significant degree. To monitor repression at the silent mating-type loci, the active *MAT* locus was deleted in these strains and the transcriptional state of the *a1* gene at *HMR*, the sole remaining copy of this gene, was determined by northern blot analysis. We note that although *a1* mRNA was repressed as expected in the single *ku70* and *esc1* mutants and derepressed in the *sir3* strain, it was surprisingly completely silent in the *esc1 yku70* double mutant, irrespective of whether *HMR* was excised or not (Fig. 6A) (see also Gartenberg et al. 2004).

We then monitored the dynamics of the silent *HMR* ring in the *esc1 yku70* mutant. Rapid time-lapse imaging of the ring in intact cells reveals that the excised *HMR* locus is completely mobile, showing no preferential association with the nuclear envelope. Its dynamics and radii of constraint (for description of imaging and Mean Squared Displacement analysis, see Gartenberg et al. 2004) are indistinguishable from a nonsilent ring (Fig. 6B). This allows us to functionally separate silencing from perinuclear anchorage, and to conclude that the NE per se is not necessary for maintenance of the repressed state.

At the same time we have probed for Sir protein distribution by immunostaining. Sir3 and Sir4 are almost completely dispersed from subtelomeric foci in the *esc1 yku70* double mutant, while they remain largely perinuclear when only *esc1* or *yku70* is deleted (Fig. 6C for Sir4) (Gartenberg et al. 2004). Thus, Sir-mediated repression does not have to occur at the nuclear envelope, at least when Sir proteins are no longer sequestered by telomeres. This does not argue, however, that Sir pools

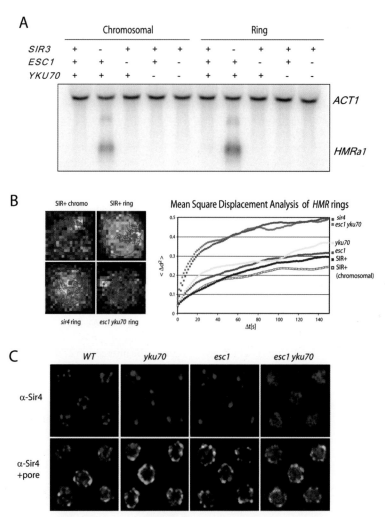

Figure 6. (*A*) The excised *HMR* ring remains transcriptionally silent. Cultures were grown to induce excision of the *HMR* locus as shown in Fig. 3. Samples were harvested immediately before galactose addition (labeled chromosomal) or 4 hr thereafter (labeled ring). Lanes *1* and *6*, MRG2262 SIR+; lanes *2* and *7*, MRG2263 *sir3*; lanes *3* and *8*, MRG2264 *esc1*; lanes *4* and *9*, MRG2265 *yku70*; lanes *5* and *10*, MRG2266 *esc1 yku70*. Blot was hybridized simultaneously with probes to the *a1* gene at *HMR* and *ACT1* as an internal control (data from Gartenberg et al. 2004). (*B*) The 5-min trajectories of the movement of the excised *HMR* ring in MRG2197 SIR+/chromo, MRG2201 SIR+/ring, MRG2267 *sir4*/ring, and MRG2255 *esc1 yku70*/ring. Images were taken as 3D stacks every 1.5-sec over a 5-min interval, and are projected onto one plane for tracking and quantitation. Mean Squared Displacement analysis of the excised *HMR* ring was performed by computing the square of the displacement, $\Delta d^2 = [d(t) - d(t + \Delta t)]^2$, over increasing time intervals 1.5–150-sec; see Gartenberg et al. 2004). The average of all Δd^2 *values for each* Δt value was plotted against Δt. Over 65 min of projected 3D stacks taken at 1.5-sec intervals are analyzed per strain. All data are after induced excision except SIR+ chromosomal, which shows the constraint of mobility imposed by the chromosomal fiber and the repressed state. (*C*) Immunolocalization of Sir4 (*red*) and nuclear pore (*green*) on strains used in *A* and *B*. Affinity-purified anti-Sir4 rabbit serum and Mab414 (mouse anti-pore) were applied after growth and fixation with formaldehyde in YPD (Gartenberg et al. 2004).

have no role in repression in a wild-type situation. Rather, one could argue that a local increase in Sir protein concentration may be the *only* essential contribution of the nuclear envelope to repression, since silent chromatin is maintained internally after ablation of the two perinuclear anchors for Sir4. We propose that if critical Sir concentrations are achieved by other means, the NE will have no further role in repression. This does not, however, rule out a possible role for the NE in nucleation events. Indeed, the weaker the silencer, the more important positioning at the NE and proximity to telomere clusters will be.

These results also provide a plausible explanation for why improved silencing in wild-type cells correlates with improved anchoring (Tham et al. 2001; Hediger et al. 2002b). Because silent chromatin can itself target the repressed locus to the periphery (as long as either yKu80 or Esc1 is present), then silent domains will accumulate at the periphery as a result of repression. Ironically, the general limitation on Sir protein availability, which is the reason one cannot repress internally, stems in part from the sequestering of Sir proteins by telomeres. This chicken-and-egg situation is resolved by noting that telomere anchoring at the NE can occur in the absence of silencing, providing a means to initiate the formation of these repressive subcompartments. Thus telomere anchoring probably precedes repression.

A corollary to these findings is the expectation that relocalization to the NE alone will not be sufficient to silence a locus. Support for this, data showing that perinuclear anchoring can occur without silencing, comes from many sources. First, Andrulis et al. (1998) showed that reporter gene anchoring through a membrane protein did not repress the reporter if no silencer was present. Second, truncated telomeres remain fully anchored in *yku* mutants, although TPE is compromised by 10^4-fold (Tham et al. 2001) and internal sequences can be dragged to the nuclear envelope by targeted protein domains, in the absence of silencing (Taddei et al. 2004). This conclusion was actually evident from the very first descriptions of TPE and telomere anchoring. Gottschling et al. (1990) showed that subtelomeric silencing is stochastic and occurs at a frequency ranging from 0.5% to maximally 30% of cells (depending on the reporter and the telomere), while telomeres are on average about 70% tightly anchored (Gotta et al. 1996). Further, in a *sir*-deficient strain, most telomeres remain at the periphery despite the fact that there is no silencing. It follows that transcribed telomeres are able to remain anchored, a phenomenon we now know to be mediated by yKu (Hediger et al. 2002b; Taddei et al. 2004).

SPONTANEOUS AND SELF-PROPAGATING Sir POOLS

These results lead to a model that explains how a Sir-rich compartment can form spontaneously in the nucleus (Fig. 7). Central to this model are two facts: Silencing-independent anchorage can be achieved through yKu (and possibly also Sir4; see Fig. 5C), and silencing-dependent anchorage occurs through Sir4. The yKu-mediated anchoring provides a means for telomeres to accumulate at the NE prior to the assembly of repressed chromatin. The anchoring of telomeric repeats creates a large number of potential Sir4 protein binding sites within a restricted volume, because of the presence of 20-25 Rap1 consensuses on each telomere end (Gilson et al. 1993), and the ability of Rap1 to bind both Sir3 and Sir4 (Moretti et al. 1994). The telomere-associated yKu dimer also binds Sir4 tightly (Roy et al. 2004). Subtelomeric repeats could further contribute to the efficiency of Sir4 accumulation through ORC- and Abf1-binding sites. The resulting high density of Sir4-binding chromatin would in turn recruit Sir2 and Sir3, again through protein–protein interactions (Moretti et al. 1994; for review, see Gasser and Cockell 2001). Once repressed chromatin is nucleated, spreading would be favored by the high local concentration of silencing factors. As silent chromatin spreads, it augments the amount of stably bound Sir complex and in turn reinforces the interaction of the telomere (or of another locus) with the NE. This creates a feedback loop, which can be initiated by binding a very limited number of yKu or Sir4

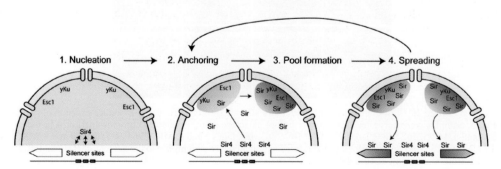

Figure 7. Model for the role of chromatin anchoring in the promotion of silencing. We propose that Sir4 is first recruited at the nucleation center by DNA binding proteins that can also bind Sir proteins (step 1; Nucleation). These include Rap1, ORC, Abf1, and yKu. The presence of Sir4 at the locus will then bring it to the nuclear periphery through one of the two Sir4 anchoring pathways (step 2; yKu or Esc1) where the high local concentrations of Sir proteins will be created, in part by the clustering of telomeres (step 3). This, in turn, helps silencing complexes assemble and spread (step 4). The anchoring of silent loci at the periphery will increase the concentration of Sir proteins and reinforce the silencing of other loci within this region. This can be self-propagating. In addition, the multiple telomere-associated Rap1 sites create a sink that maintains a pool of Sir factors for incorporation into repressed chromatin—the so-called "Circe effect."

proteins at given site (summarized in Fig. 7). It is not clear what the minimal number of anchoring molecules may be, but in a targeting assay, four double lexA sites were sufficient (Taddei et al. 2004). We envision such mechanisms as also being relevant for the clustering of centromeric satellite and heterochromatin sequences in higher eukaryotes, which may similarly promote the sequestration of heterochromatin factors like HP1.

THE CIRCE EFFECT: CREATING COMPARTMENTS FOR GENERAL REPRESSORS

This apparent sequestration of Sir proteins in telomeric pools can be described by a term from enzymology—the "Circe effect." (In Greek mythology, Circe was an enchantress who detained Odysseus on her island and turned his men into swine. Jencks [1975], a brilliant enzymologist, defined the Circe effect as "the utilization of strong attractive force to lure a substrate into a site in which it undergoes an extraordinary transformation.") This term was initially used to describe the attractive effect that weak nonproductive binding sites would have on the association of a substrate with the active site of an enzyme (Jencks 1975). The presence of numerous "nonproductive" binding sites are described as "luring" the substrate into the active site of an enzyme, where it would undergo transformation "in form and structure." It could be argued that the substrate would not be in equilibrium with its concentration as determined for a large pool (the nucleus), but would be in equilibrium with a local volume that is enriched for the ligand, because of the spatial proximity of multiple weak interaction sites. Here we apply this notion to describe how the pools of "nonengaged" Sir proteins held by clustered yeast telomeres and their multiple Rap1 sites might favor the assembly of a repressed chromatin state. Once Sir proteins form a stable complex with chromatin, we argue they are "transformed in form and structure."

To apply this concept to general repressors such as Sir proteins we would have to postulate first that Sir proteins are limiting for assembly into silent chromatin. The well-documented effects of Sir protein overexpression are consistent with this idea (Renauld et al. 1993; Maillet et al. 1996). Second, we need to postulate that there are two types of binding sites for Sir proteins. Again this seems to be the case. There are weak, unstable interactions of Sir3 and Sir4 with Rap1, which itself binds with high affinity to the TG repeats at telomeres. The instability of Rap1-Sir interactions is attested by the fact that these proteins cannot be immunoprecipitated without cross-linking reagents, whereas the SIR-nucleosome complexes can be. The average spacing of one Rap1 site per 18 bp within the TG repeat leads to 20–25 weak binding sites for Sir3 and Sir4 per chromosomal end. Assuming that telomeres cluster in groups of 6, this results in 180 weak binding sites. Indeed, ChIP (chromatin immunoprecipitation) data indicate that Sir4 can be recovered with telomeres in the absence of silencing (Luo et al. 2002). We would argue that within a given focus of telomeres, the Sir proteins stay sequestered even when they are not incorporated into repressive chromatin, thanks to their weak affinity for Rap1. This is further supported by the fact that Rif1 and Rif2 compete for Sir interaction with Rap1, and that the loss of Rif proteins improves silencing (Mishra and Shore 1999). The second type of "Sir interaction" that occurs in telomeric foci would be that resulting from the incorporation of the complex into a higher-order chromatin structure. FRAP (fluorescence recovery after photobleaching) experiments show this interaction to be much more stable than that of Rap1 with DNA (T. Cheutin et al., unpubl.). We note further that the assembly of Sirs into silent chromatin may be cooperative. Although this remains to be proven, the linear spreading of silent chromatin would be consistent with cooperativity in the assembly of this higher-order structure (Renauld et al. 1993). For a cooperative event, local reserves of Sir proteins should indeed favor repression. The fact that unlinked or imbalanced Sir2 or Sir4 expression derepresses TPE suggests that factor dosage is also important within the complex or else that secondary, nonproductive complexes can be formed by two of the three Sir partners.

Given the different types of binding sites, one being more stable than the other, and the silencing-independent clustering of yeast telomeres, it would seem that the "Circe effect" may indeed accurately describe the sequestering of Sir proteins in perinuclear pools by telomeric repeats. These then serve to favor the formation of repressed chromatin, which is proposed to occur in at least two steps (Hoppe et al. 2002), the sum of which may well be considered equivalent to "transformation by an enzyme's active site."

Why should a cell bother to create repressive subcompartments, when the *esc1 yku70* mutant shows that release of Sirs allows silencing without specific localization? We propose that telomere anchoring, and its encumbant sequestering of Sir factors, serves to prevent promiscuous repression throughout the rest of the active genome. The creation of subcompartments enriched for Sir proteins allows the cell to ensure that most other loci remain active, while nonetheless exploiting the power of a general, promoter-independent, transcriptional repression mechanism. This protective function has also been proposed for the methylation of lysine 79 in histone H3, a modification that is found throughout the active genome and that appears to be incompatible with silencing (van Leeuwen et al. 2002). Similarly the histone variant Htz1 is thought to help insulate the rest of the genome from inappropriate silencing events (Meneghini et al. 2003). By creating subcompartments that favor repression through the sequestering of Sir proteins, the cellular concentration of these general repressors can be maintained at relatively low levels to protect the genome from the deleterious action of runaway repression.

EXPERIMENTAL PROCEDURES

Strain Construction and Yeast Techniques

All strains are described elsewhere in the cited references, including Gartenberg et al. (2004), Hediger et al. (2002a,b), and Taddei et al. (2004). Standard techniques were used for immunofluorescence, FISH (Laroche et al. 2000; Hediger et al. 2004), Northern analysis, and silencing assays.

Single Z-Stack Microscopy and Time-Lapse Analysis

For the excision experiments, freshly streaked cells grown in SC-trp media containing 2% dextrose were diluted into SC-his + 2% raffinose. When cultures reached ~0.25 OD_{600} after well-aerated overnight growth, galactose was added to 2%. After 2 hours, 1 ml of cell culture was harvested by centrifugation and either placed on microscope slides bearing 1.4% agarose plugs containing either 4% galactose (pH 5.8) or 4% dextrose (pH 7.0), as needed, or were mounted in a Ludin chamber flushed with appropriate media (Hediger et al. 2004). Data collection was limited to 2 hours after mounting. For position data, Z-stacks were collected on an Olympus IX70 fluorescence microscope coupled to a TillVision imaging system using 475-nm excitation, an FITC (fluorescein isothiocyanate) filter, and a 200-ms acquisition time for each image. The stack range was 4.5 µm with 0.25 µm z-axis increments (total of 19 images/stack). The statistical significance of distributions was compared with random distributions by χ^2 analysis, and Student t-tests determined the similarity of zone 1 values.

Time-lapse imaging was performed on a Zeiss LSM510 confocal microscope with three-dimensional (3D) imaging over time, taking six optical slices of 450 nm with a Hyperfine HRZ 200 motor, at 1.5-second time intervals. For high-throughput analysis the 3D stacks were projected onto a single x,y plane by maximum intensity projection prior to analysis, so that data sets from either the Nup49-GFP or the tetR-GFP background essentially represent 2D projections of 3D information. 3D tracking of focus movement in x,y,z after deconvolution was performed with Imaris Time module (Rel 4.0; Bitplane, Zürich). The quantity of data analyzed and the efficiency of our computing algorithms limits our 3D positional analysis for large numbers of time-lapse series.

ACKNOWLEDGMENTS

We thank T. Laroche, K. Dubrana, G. Van Houwe, and K. Bystricky for thoughtful discussions, technical advice, and experimental support. This work was funded by the NIH (GM51402) and the Swiss National Science Foundation and its NCCR Frontiers in Genetics program. The Novartis Foundation supported M.R.G. while on sabbatical in the Gasser lab.

REFERENCES

Andrulis E., Neiman A.M., Zappulla D.C., and Sternglanz R. 1998. Perinuclear localization of chromatin facilitates transcriptional silencing. *Nature* **394:** 592.

Andrulis E.D., Zappulla D.C., Ansari A., Perrod S., Laiosia C.V., Gartenberg M.R., and Sternglanz R. 2002. Esc1, a nuclear periphery protein required for Sir4-based plasmid anchoring and partitioning. *Mol. Cell. Biol.* **22:** 8292.

Ansari A. and Gartenberg M.R. 1997. The yeast silent information regulator Sir4p anchors and partitions plasmids. *Mol. Cell. Biol.* **17:** 7061.

Bi X. and Broach J.R. 2001. Chromosomal boundaries in *S. cerevisiae. Curr. Opin. Genet. Dev.* **11:** 199.

Boulton S.J. and Jackson S.P. 1998. Components of the Ku-dependent non-homologous end-joining pathway are involved in telomeric length maintenance and silencing. *EMBO J.* **17:** 1819.

Buck S.W. and Shore D. 1995. Action of a RAP1 carboxy-terminal silencing domain reveals an underlying competition between HMR and telomeres in yeast. *Genes Dev.* **9:** 370.

Cheng T.-H., Li Y.-C., and Gartenberg, M.R. 1998. Persistence of an alternate chromatin structure at silenced loci in the absence of silencers. *Proc. Natl. Acad. Sci.* **95:** 5521.

Chung H.M., Shea C., Fields S., Taub R.N., Van der Ploeg L.H., and Tse D.B. 1990. Architectural organization in the interphase nucleus of the protozoan *Trypanosoma brucei:* Location of telomeres and mini-chromosomes. *EMBO J.* **9:** 2611.

Cockell M. and Gasser S.M. 1999. Nuclear compartments and gene regulation. *Curr. Opin. Genet. Dev.* **9:** 199.

Cormack B.P., Ghori N., and Falkow S. 1999. An adhesin of the yeast pathogen *Candida glabrata* mediating adherence to human epithelial cells. *Science* **285:** 578.

De Las Penas A., Pan S.J., Castano I., Alder J., Cregg R., and Cormack B.P. 2003. Virulence-related surface glycoproteins in the yeast pathogen *C. glabrata* are encoded in subtelomeric clusters and subject to RAP1- and SIR-dependent transcriptional silencing. *Genes Dev.* **17:** 2245.

Feuerbach F., Galy V., Trelles-Sticken E., Fromont-Racine M., Jacquier A., Gilson E., Olivo-Marin J.C., Scherthan H., and Nehrbass U. 2002. Nuclear architecture and spatial positioning help establish transcriptional states of telomeres in yeast. *Nat. Cell Biol.* **4:** 214.

Figueiredo L.M., Freitas-Junior L.H., Bottius E., Olivo-Marin J.C., and Scherf A. 2002. A central role for *Plasmodium falciparum* subtelomeric regions in spatial positioning and telomere length regulation. *EMBO J.* **21:** 815.

Freitas-Junior L.H., Bottius E., Pirrit L.A., Deitsch K.W., Scheidig C., and Scherf A. 2000. Frequent ectopic recombination of virulence factor genes in telomeric chromosome clusters of *P. falciparum. Nature* **407:** 1018.

Galy V., Olivo-Marin J.-C., Scherthan H., Doye V., Rascalou N., and Nehrbass U. 2000. Nuclear pore complexes in the organization of silent telomeric chromatin. *Nature* **403:** 108.

Gartenberg M.R., Neumann F.R., Laroche T., Blaszczyk M., and Gasser S.M. 2004. Sir-mediated repression can occur independently of chromosomal and subnuclear contexts. *Cell* **119:** 955.

Gasser S.M. and Cockell M.M. 2001. The molecular biology of SIR proteins. *Gene* **279:** 1.

Gilson E., Roberge M., Giraldo R., Rhodes D., and Gasser S.M. 1993. Distortion of the DNA double helix by RAP1 at silencers and multiple telomeric binding sites. *J. Mol. Biol.* **231:** 293.

Gotta M., Laroche T., Formenton A., Maillet L., Scherthan H., and Gasser S.M. 1996. The clustering of telomeres and colocalization with Rap1, Sir3, and Sir4 proteins in wild-type *S. cerevisiae. J. Cell Biol.* **134:** 1349.

Gottschling D.E., Aparicio O.M., Billington B.L., and Zakian V.A. 1990. Position effect at *S. cerevisiae* telomeres: Reversible repression of Pol II transcription. *Cell* **63:** 751.

Halme A., Bumgarner S., Styles C., and Fink G.R. 2004. Genetic and epigenetic regulation of the FLO gene family generates cell-surface variation in yeast. *Cell* **116:** 405.

Hecht A., Strahl-Bolsinger S., and Grunstein M. 1996. Spreading of transcriptional repressor SIR3 from telomeric heterochromatin. *Nature* **383:** 92.

Hecht A., Laroche T., Strahl-Bolsinger S., Gasser S.M., and Grunstein M. 1995. Histone H3 and H4N-termini interact with SIR3 and SIR4 proteins: A molecular model for the formation of heterochromatin in yeast. *Cell* **80:** 583.

Hediger F., Dubrana K., and Gasser S.M. 2002a. Myosin-like proteins 1 and 2 are not required for silencing or telomere anchoring but act in the Tel1 pathway of telomere length control. *J. Struct. Biol.* **140:** 79.

Hediger F., Taddei A., Neumann F.R., and Gasser S.M. 2004. Methods for visualizing chromatin dynamics in living yeast. *Methods Enzymol.* **375:** 345.

Hediger F., Neumann F.R., Van Houwe G., Dubrana K., and Gasser S.M. 2002b. Live imaging of telomeres: yKu and Sir proteins define redundant telomere-anchoring pathways in yeast. *Curr. Biol.* **12:** 2076.

Heun P., Laroche T., Raghuraman M.K., and Gasser S.M. 2001a. The positioning and dynamics of origins of replication in the budding yeast nucleus. *J. Cell Biol.* **152:** 385.

Heun P., Laroche T., Shimada K., Furrer P., and Gasser S.M. 2001b. Chromosome dynamics in the yeast interphase nucleus. *Science* **294:** 2181.

Holmes S.G., Rose A.B., Steuerle K., Saez E., Sayegh S., Lee Y.M., and Broach J.R. 1997. Hyperactivation of the silencing proteins, Sir2p and Sir3p, causes chromosome loss. *Genetics* **145:** 605.

Hoppe G.J., Tanny J.C., Rudner A.D., Gerber S.A., Danaie S., Gygi S.P., and Moazed D. 2002. Steps in assembly of silent chromatin in yeast: Sir3-independent binding of a Sir2/Sir4 complex to silencers and role for Sir2-dependent deacetylation. *Mol. Cell. Biol.* **22:** 4167.

Jencks W.P. 1975. Binding energy, specificity, and enzymic catalysis: The circe effect. *Adv. Enzymol. Relat. Areas Mol. Biol.* **43:** 219.

Kimura A., Umehara T., and Horikoshi M. 2002. Chromosomal gradient of histone acetylation established by Sas2p and Sir2p functions as a shield against gene silencing. *Nat. Genet.* **32:** 370.

Laroche T., Martin S.G., Tsai-Pflugfelder M., and Gasser S.M. 2000. The dynamics of yeast telomeres and silencing proteins through the cell cycle. *J. Struct. Biol.* **129:** 159.

Laroche T., Martin S.G., Gotta M., Gorham H.C., Pryde F.E., Louis E.J., and Gasser S.M. 1998. Mutation of yeast Ku genes disrupts the subnuclear organization of telomeres. *Curr. Biol.* **8:** 653.

Luo K., Vega-Palas M.A., and Grunstein M. 2002. Rap1-Sir4 binding independent of other Sir, yKu, or histone interactions initiates the assembly of telomeric heterochromatin in yeast. *Genes Dev.* **16:** 1528.

Lustig A.J., Liu C., Zhang C., and Hanish J.P. 1996. Tethered Sir3p nucleates silencing at telomeres and internal loci in *Saccharomyces cerevisiae*. *Mol. Cell. Biol.* **16:** 2483.

Maillet L., Boscheron C., Gotta M., Marcand S., Gilson E., and Gasser S.M. 1996. Evidence of silencing compartments within the yeast nucleus: A role for telomere proximity and Sir protein concentration in silencer-mediated repression. *Genes Dev.* **10:** 1796.

Maillet L., Gaden F., Brevet V., Fourel G., Martin S.G., Dubrana K., Gasser S.M., and Gilson E. 2001. Ku-deficient yeast strains exhibit alternative states of silencing competence. *EMBO Rep.* **2:** 203.

Marcand S., Buck S.W., Moretti P., Gilson E., and Shore D. 1996. Silencing of genes at nontelomeric sites in yeast is controlled by sequestration of silencing factors at telomeres by Rap1 protein. *Genes Dev.* **10:** 1297.

Meneghini M.D., Wu M., and Madhani H.D. 2003. Conserved histone variant H2A.Z protects euchromatin from the ectopic spread of silent heterochromatin. *Cell* **112:** 725.

Mishra K. and Shore D. 1999. Yeast Ku protein plays a direct role in telomeric silencing and counteracts inhibition by Rif proteins. *Curr. Biol.* **9:** 1123.

Moretti P., Freeman K., Coodly L., and Shore D. 1994. Evidence that a complex of SIR proteins interacts with the silencer and telomere-binding protein RAP1. *Genes Dev.* **8:** 2257.

Renauld H., Aparicio O.M., Zierath P.D., Billington B.L., Chhablani S.K., and Gottschling D.E. 1993. Silent domains are assembled continuously from the telomere and are defined by promoter distance and strength, and by SIR3 dosage. *Genes Dev.* **7:** 1133.

Robyr D., Suka Y., Xenarios I., Kurdistani S.K., Wang A., Suka N., and Grunstein M. 2002. Microarray deacetylation maps determine genome-wide functions for yeast histone deacetylases. *Cell* **109:** 437.

Roy R., Meier B., McAinsh A.D., Feldmann H.M., and Jackson S.P. 2004. Separation-of-function mutants of yeast Ku80 reveal a Yku80p-Sir4p interaction involved in telomeric silencing. *J. Biol. Chem.* **279:** 86.

Scherf A., Figueiredo L.M., and Freitas-Junior L.H. 2001. Plasmodium telomeres: A pathogen's perspective. *Curr. Opin. Microbiol.* **4:** 409.

Strahl-Bolsinger S., Hecht A., Luo K., and Grunstein M. 1997. SIR2 and SIR4 interactions differ in core and extended telomeric heterochromatin in yeast. *Genes Dev.* **11:** 83.

Suka N., Luo K., and Grunstein M. 2002. Sir2p and Sas2p opposingly regulate acetylation of yeast histone H4 lysine16 and spreading of heterochromatin. *Nat. Genet.* **32:** 378.

Taddei A., Hediger F., Neumann F.R., Bauer C., and Gasser S.M. 2004. Separation of silencing from perinuclear anchoring functions in yeast Ku80, Sir4 and Esc1 proteins. *EMBO J.* **23:** 1301.

Tham W.H., Wyithe J.S., Ferrigno P.K., Silver P.A., and Zakian V.A. 2001. Localization of yeast telomeres to the nuclear periphery is separable from transcriptional repression and telomere stability functions. *Mol. Cell* **8:** 189.

van Leeuwen F., Gafken P.R., and Gottschling D.E. 2002. Dot1p modulates silencing in yeast by methylation of the nucleosome core. *Cell* **109:** 745.

Genetic Instability in Aging Yeast: A Metastable Hyperrecombinational State

M.A. McMurray and D.E. Gottschling

Division of Basic Sciences, Fred Hutchinson Cancer Research Center, Seattle, Washington 98109

We are all aware of changes that occur as a person reaches middle age and beyond—wrinkled skin, gray hair, poor vision, etc. But one particularly intriguing phenomenon is the dramatic rise in incidence of cancer with increasing age: About 75% of all cancers are diagnosed after the age of 55 (ACS 2004). Because cancer is typically considered a genetic disease—genetic alterations are a hallmark of tumors and the inactivation of tumor suppressor genes and/or activation of oncogenes facilitate oncogenesis (for review, see Hanahan and Weinberg 2000)—it has been suggested that the link between aging and increased incidence of cancer may simply be a steady accumulation of genetic changes over the course of a person's life. The increased chance of attaining a sufficient number of changes to elicit oncogenesis may explain the exponential increase in cancer incidence with advancing age (Armitage and Doll 1954, 1957; Frank 2004).

However, this idea is likely too simple. For instance, the rates of spontaneous mutation observed in human tissue culture cells cannot account for the amount of genetic change observed in most tumors (Loeb et al. 2003). Furthermore, a number of nongenetic changes occur as a cell ages that could have an impact on oncogenesis. DNA methylation at gene promoters, which could inactivate tumor suppressor genes (Jones and Laird 1999), increases with age (DePinho 2000). Similarly, changes in histone modification levels, which also can impact gene expression, are observed with increasing age (DePinho 2000; Bandyopadhyay and Medrano 2003). Oxidative damage to DNA accumulates with age (Bohr 2002) and may contribute to the acquisition of mutations during carcinogenesis (Jackson and Loeb 2001).

Even if we consider only how genetic alterations contribute to the age-related increase in cancer, we can appreciate the enormous complexity of this issue. For instance, do mutations arise at a constant rate throughout an individual's life, or do mutation rates increase with age? Many biological and clinical issues confound this determination, including clonal expansion, by which an early mutation provides a growth advantage and increases opportunities for subsequent mutations; tissue-specific effects, in terms of both the frequency of tumor detection and the inherent susceptibility of certain cell types to tumorigenesis; environmental challenges, including extrinsic (e.g., carcinogens) and intrinsic (e.g., hormones) factors that arise over a lifetime and can affect tumor formation; and the continual improvement of diagnostic methods.

The budding yeast *Saccharomyces cerevisiae* has been developed as a model system for studying many aspects of cellular aging. Because of the asymmetric nature of its cell divisions, it has been particularly useful to study replicative life span. Replicative aging in yeast is defined by the number of mitotic cell divisions a yeast mother cell undergoes before she ceases dividing and ultimately lyses (for review, see Bitterman et al. 2003). In typical laboratory strains of haploid yeast, wild-type life spans average 20–30 divisions, during which characteristic changes affect the aging mother cell, such as increased cell size, wrinkled cell surface, and decreased division rates.

The daughters of very old mothers display all of the cellular phenotypes of old cells mentioned above, as well as a reduced life span (Egilmez and Jazwinski 1989; Kennedy et al. 1994), consistent with the daughter cells inheriting age-induced alterations. However, these traits are not stably inherited; the great-granddaughters of old mothers appear rejuvenated, with full life spans and a normal cell size and division rate (Egilmez and Jazwinski 1989; Kennedy et al. 1994). Thus, the appearance of old cell phenotypes in the daughters of old mothers is not the result of a permanent genetic change.

Nevertheless, the first proposed cause of yeast aging was centered on a genetic event: the formation of extrachromosomal rDNA circles (ERCs), which accumulate to high levels in old mother cells and are inherited by the daughters of old mothers (Sinclair and Guarente 1997). It was put forth that the ERC phenomenon represents a role for "genetic instability" in yeast aging (Jazwinski 2001; Bitterman et al. 2003). However, the actual chromosomal events creating ERCs are likely age independent. A "stochastic trigger" of recombination-mediated ERC formation early in a cell's life span, followed by exponential accumulation to a lethal threshold, can explain the sigmoidal shape of survival curves of populations of *S. cerevisiae* (Sinclair et al. 1998). Thus, the accumulation of ERCs is not a true example of age-associated genetic instability.

In an attempt to understand whether there are fundamental aspects of cellular aging that contribute to oncogenesis, we developed an assay to examine genomic instability as a function of cellular age in *S. cerevisiae*. Here we describe our surprising observations, which suggest that aged cells enter a metastable epigenetic state of hyperrecombination. Finally, we propose a molecular model to explain these results, with implications for the effects of aging on genomic stability in higher organisms.

GENETIC INSTABILITY AND YEAST AGING

Monitoring Genome Stability throughout a Life Span: Yeast Pedigree Analysis

In setting up a system for examining genomic instability, we chose to follow loss of heterozygosity (LOH) events, an important mechanism of tumor suppressor gene inactivation in the development of cancer (Lengauer et al. 1998). LOH may be the result of chromosome loss, recombination, or deletion (see Fig. 1). To this end we engineered a set of diploid yeast strains with heterozygosity at several different loci. Loss of the functional allele at each locus was easily assayed by colony growth under selective conditions or a change in colony color.

These marked strains were then subjected to pedigree analysis, in which a mother cell is isolated on agar media, and each of her newborn daughter cells is successively moved to a different location and allowed to form a colony. These "daughter colonies" provide a retrospective life history of the mother cell: If a genetic change such as LOH occurs during the mother's life span, it is detected by a color change, or a lack of growth when the daughter colony is transferred to selective media. If an LOH event occurred in the mother cell during her life span, it is evident as LOH in every daughter colony until the end of her life (e.g., Fig. 2B, fourth line from bottom). However, if the mother remained wild type and LOH occurred in a single daughter cell, it is evident as a single daughter colony with the LOH phenotype and subsequent wild-type daughter colonies (e.g., Fig. 2B, third line from bottom).

In our initial pedigree analyses, strains carried marker genes at various heterozygous loci, together representing six different marked chromosomes (II, III, IV, V, XII, and XV). Among these, LOH at the *MET15* locus on chromosome XII was observed most often—3.2×10^{-3} per division of a mother cell. This value contrasted with the rate of spontaneous LOH at *MET15* in young (growing in culture) cells, which was significantly lower, $\sim 7 \times 10^{-4}$ per cell division, in the same strain. Thus, we found evidence of increased genetic instability in the cell divisions of aging mother cells.

More significantly though, the *MET15* LOH events tended to occur in cell divisions nearer the end of life span rather than the beginning. Among the pedigrees of 16 independent mother cells with such LOH events, there were 31 ± 3.3 (median ± S.E.M.) divisions before the first LOH event was observed, compared to 11 ± 1.4 divisions after the LOH event and until the end of the life span ($P < 0.0001$, one-tailed unpaired *t*-test). Thus, replicative aging in yeast is associated with an increased rate of *MET15* LOH events in the cell divisions of old mothers.

Inheriting Instability: LOH in the Progeny of Old Mothers

In addition to LOH events affecting the mother or her buds, we observed an unexpected increase in the frequency of daughter colonies containing sectors of cells

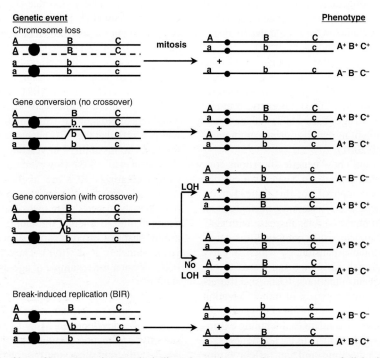

Figure 1. Mechanisms of loss of heterozygosity. For the indicated genetic event, the arrangement of alleles in each cell after mitotis (separated by "+") is given for a generic chromosome, with double-stranded DNA represented by *lines*, centromeres by *circles*, and dominant (uppercase) or recessive (lowercase) alleles at each of three loci. To the right, the phenotypes of each cell after mitosis.

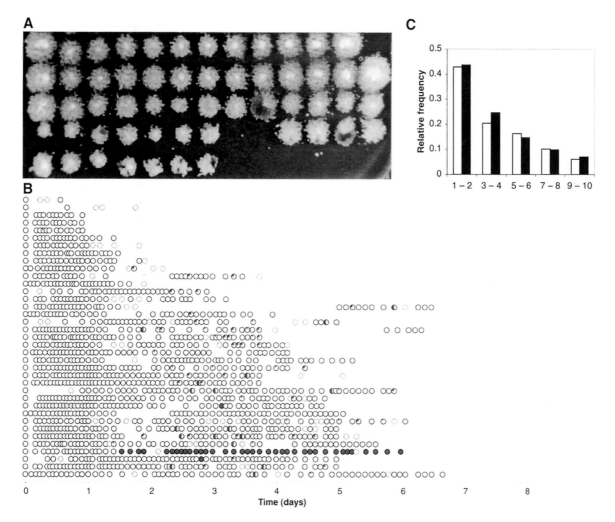

Figure 2. Pedigree analysis reveals an age-induced switch to a state of frequent LOH. (*A*) The pedigree of a single *MET15/met15Δ ADE2/ade2Δ* mother cell. Colonies were formed from the individual daughter cells of a single mother cell throughout her life span of 57 cell divisions and replica plated to lead nitrate-containing media, on which *met15Δ/met15Δ* cells—resulting from *MET15* LOH—turn brown (Cost and Boeke 1996). The age of the mother when each daughter was produced increases *left* to *right*, *top* to *bottom*. (*B*) Thirty seven mother cells of the strain shown in *A* were analyzed by pedigree analysis. Each daughter colony from a single mother is represented as a *circle* in each row. *Black*, *met15Δ/met15Δ*; *gray*, *ade2Δ/ade2Δ*. *Dashed outlines* indicate daughters that failed to form visible colonies. (*C*) The data represented in *B* were analyzed by counting the number of cell divisions between daughter colonies containing *MET15* LOH events, defined as an eighth sector or larger of *met15Δ/met15Δ* cells. *White bars* show a frequency distribution of these intervals, in bins of two cell divisions. *Black bars* represent the expected (binomial) distribution if the probability per cell division of generating a daughter colony with a *MET15* LOH event is 0.25.

with LOH: Part of the colony remained wild type, but a significant fraction showed the phenotype associated with marker loss (Fig. 2A) (McMurray and Gottschling 2003). These events were readily detectable by exploiting the brown colony color phenotype of *met15* cells on lead nitrate media (Cost and Boeke 1996), and the red colony color phenotype of *ade2* cells on media with low adenine content (Roman 1957). Colored sectors within colonies indicate LOH events that occurred after the daughter cell was removed from the mother.

In the pedigree analysis, sectored daughter colonies representing LOH at each of the marked chromosomes were observed. These sectored colonies were most frequent when the daughter came from older mother cells. In addition, they were detected far more commonly than LOH events in the aging mothers themselves (Fig. 2B) (McMurray and Gottschling 2003; data not shown). This surprising observation indicates that old mother cells produce progeny that are prone to LOH, raising the possibility that the chromosomal instability manifested as increased LOH in the mother's divisions might be inherited by her progeny.

Countdown to Instability: A Two-State Switch Running on Its Own Aging Clock

Mother cells of the strain shown in Figure 2 divided on average 43 times (median value, $n = 37$). (Note that these diploid cells are substantially longer-lived than other reported wild-type laboratory strains, most of which are haploid [Kaeberlein et al. 2004].) The first LOH event—scored as a wholly brown or brown/white daughter colony with an eighth or larger brown sector—was observed in a mother's life span after 25 divisions without an event (median value, $n = 22$). Thereafter, LOH events were much more frequent, occurring on average in every third daughter colony ($n = 49$; see Fig. 2). Interestingly, this frequency did not increase as mother cells aged further (Spearman correlation coefficient $r \geq 0$), meaning that the rate of LOH did not continue to increase as the end of the life span approached. Furthermore, after the first LOH event in each mother's life span, the distribution of cell divisions between subsequent LOH events for all the mothers appeared to fit a binomial distribution, in which each cell division by the mother is associated with a constant probability (~0.25) of observing a daughter colony with LOH (Fig. 2C). Thus, if we consider the first appearance of an LOH event in a mother's pedigree to represent the onset of age-induced chromosomal instability, then the subsequent rate of LOH events is constant throughout the remainder of the life span. Taken together, the data suggest that after approximately 25 cell divisions by an aging mother, there is a switch-like increase in the rate of LOH, manifested as a high frequency of LOH events in daughter colonies.

Among individual mother cells of the same strain, there is considerable variation in life span (Fig. 2B) (Mortimer and Johnston 1959). If the onset of increased LOH is related to the process that brings about the end of the life span, then short-lived mothers should display an earlier age of onset, whereas the onset of LOH should be delayed in long-lived mothers. Instead, we noticed that the onset of *MET15* LOH occurred after approximately 25 cell divisions regardless of the life span of the mother (see Fig. 2B). This is perhaps best illustrated by the fact that the mothers whose pedigrees were devoid of LOH were significantly shorter-lived (mean life span 22 ± 4.2) than mothers with pedigrees including at least one daughter colony with LOH (mean life span 44 ± 2.7; one-tailed unpaired *t*-test $P < 0.0001$). Thus, if a mother cell failed to achieve 25 cell divisions before dying, she was unlikely to have experienced the switch to high LOH. Furthermore, the onset of age-induced *MET15* LOH was equivalent (median 23 divisions) in other LOH detection strains with significantly shorter life spans (median ~30) (McMurray and Gottschling 2003; and data not shown).

We next wanted to examine more closely whether the switch to higher levels of LOH is causally related to the number of cell divisions from the beginning of the life span. If so, then manipulations that alter life span would not alter the age of onset of age-induced LOH. As expected, deletion of both copies of *FOB1*, whose gene product causes the intrachromosomal rDNA recombination events that create ERCs (Defossez et al. 1999), increased the life span of diploid mother cells by 80% compared to wild type (McMurray and Gottschling 2003). However, the kinetics of the *MET15* LOH events were unchanged. The median number of cell divisions in *FOB1*-deleted mother cells before a first LOH event was 25 cell divisions, just as it was in shorter-lived wild-type cells (McMurray and Gottschling 2003). Also unchanged was the increased frequency with which subsequent LOH events occurred (McMurray and Gottschling 2003). Thus, prolonging life span by reducing the level of ERCs did not delay the onset nor decrease the frequency of age-induced genomic instability. Furthermore, these data suggest that the onset of age-induced LOH operates on a different "clock" than does yeast replicative life span.

THE MECHANISM OF LOH IN OLD CELLS

Break-induced Replication Is Responsible for Most Age-induced LOH Events

We next determined by which pathway LOH occurred in aging yeast cells. In considering the possibilities (see Fig. 1), we found that chromosome loss was not responsible for age-induced LOH, because single-copy markers located on the opposite arm of chromosomes with an LOH event remained intact (McMurray and Gottschling 2003). Thus, age-induced LOH is likely initiated by chromosomal damage and is not the result of chromosome nondisjunction.

It appears that age-induced LOH proceeds via repair of random genome-wide damage: We found that markers located further from a centromere underwent LOH more frequently (Table 1) (McMurray and Gottschling 2003). The greater the distance between a marker and its centromere, the higher the probability that the intervening DNA will be damaged. LOH can result from either reciprocal or nonreciprocal repair pathways. In reciprocal recombination (crossing over), heterozygous cells produce a mother cell homozygous for one allele, and a daughter cell homozygous for the other allele (Fig. 1). In nonreciprocal events, either the mother or the daughter loses heterozygosity, while the other remains heterozygous (Fig. 1). We found in young cells spontaneous LOH at the *MET15* locus occurred primarily via crossing over (McMurray and Gottschling 2003). In contrast, the age-induced LOH events occurred predominantly by a nonreciprocal pathway (McMurray and Gottschling 2003). These results identify a mechanistic difference between the pathway of mitotic recombination normally responsible for LOH in young cells and the pathway resulting in increased levels of LOH as cells age.

These results suggested that age-induced LOH events may arise through break-induced replication (BIR) or terminal chromosome deletions. To further test this notion and distinguish between these two mechanisms, we examined the copy number of alleles present at five additional loci on the right arm of chromosome XII in clones in which age-induced *MET15* LOH had occurred. In the vast majority of events (>99%), LOH at *MET15* was accompanied by LOH at all centromere-distal loci (McMurray and Gottschling 2003), consistent with our findings that centromere-distant loci are more susceptible to age-induced LOH (Table 1). Furthermore, in >95% of *MET15* LOH events, the loss of one allele at a locus was accompanied by duplication of the remaining allele (homozygosity) (McMurray and Gottschling 2003), eliminating terminal deletion as a major cause of age-induced LOH. We conclude that BIR is the predominant pathway

Table 1. The Frequency of Age-induced Loss of Heterozygosity (LOH) Correlates with the Centromere-Marker Distance

Marked locus	Distance from centromere (kb)	Number of marker loss events detected
MET15	2200*	167
SAM2	1000	71
GDH1	220	8

The number of age-induced LOH events (daughter colonies with an eighth sector or larger of homozygous cells) observed by pedigree analysis of 79 mother cells is given for each locus examined, together with the centromere-marker interval, in kilobasepairs (kb).
*The size of this interval depends upon the number of rDNA repeats, which was estimated as 150.

through which age-induced LOH events occur. Thus, aging alters both the rate and the mechanism of LOH events.

Mother Knows Best: The Asymmetry of Age-induced LOH

As described above, BIR is an inherently nonreciprocal mechanism of LOH, resulting in homozygosity of one cell after mitosis, but preserving both alleles in the other. Based on the presumed mechanics of BIR (Kraus et al. 2001), we predicted that these events would occur with equal likelihood in mother and daughter cells. Surprisingly, however, there was an ~20-fold daughter bias for age-induced LOH, measured as the number of *MET15* or *SAM2* LOH events resulting in a homozygous daughter or mother cell (McMurray and Gottschling 2003). This observation, in combination with the previous analysis of LOH events in the mother cell herself, leads to a surprising conclusion. Although the advanced age of a mother cell drastically alters both the rate and the mechanism of LOH in her progeny, the mother cell herself rarely loses genetic information. Instead, the adverse effects on chromosomal stability induced by aging are primarily manifested in the progeny of old mothers.

Does the DNA Damage Checkpoint Fail in Age-induced LOH?

What leads to daughter-biased, BIR-mediated LOH in yeast cells as they age? We know that BIR is infrequently used to repair double-strand breaks (DSBs) in young wild-type cells (Malkova et al. 1996). Instead, DSBs are usually repaired by local gene conversion without crossing over (Malkova et al. 1996; for review see Paques and Haber 1999). In late S or G_2, the sister chromatid is preferred as the donor for gene conversion (Kadyk and Hartwell 1992); only rare repair events in these phases utilize the homologous chromosome, a small fraction of which will be associated with crossing over (Fig.1). Of those G_2 crossover events that do occur, only half will result in LOH through appropriate sister segregation in the subsequent mitoses (Fig. 1). Thus, broken chromosomes in yeast are normally repaired by non-BIR pathways—using homologous sequences to rejoin the chromosome fragments—that very rarely result in LOH of distal markers.

However, if the two fragments of a broken chromosome are physically separated from one another, the types of available DSB repair pathways become limited (Paulovich et al. 1997). This might occur if cells bypass a G_2/M checkpoint, proceed through mitosis in the presence of an unrepaired chromosome break, and one cell inherits the chromosome fragment containing the centromere (the "centric" fragment) while the other inherits the centromere-less ("acentric") fragment (Fig. 3). In the following G_1 or S phase, when neither the sister chromatid nor the other portion of the broken chromosome is available, the best template for repair is provided by the homologous chromosome (Fig. 3). BIR events initiating from the centric fragment of the broken chromosome will duplicate sequences from the homologous chromosome all the way to the telomere and result in LOH of all markers distal to the break. Thus, a G_2/M checkpoint defect could create situations favoring LOH through BIR by separating the two chromosome fragments created by a DSB, In this scenario, only one of the cells—either the daughter or the mother—would undergo LOH after mitosis, while the other remained heterozygous (Fig. 3).

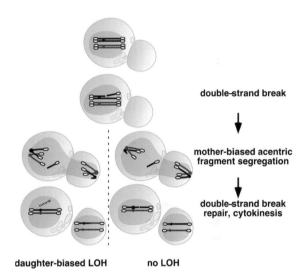

Figure 3. A model for asymmetric age-induced LOH. A diploid mother yeast cell is depicted with two homologous chromosomes (*red* and *black*; centromeres are *filled circles*), contained within the nucleus (*blue*). The cell wall is shown in *green* and bud scars are depicted. *Top*, a mother cell after DNA replication, with duplicated chromosomes. A double-strand break (DSB) in one sister chromatid of the black chromosome is followed by mitosis without repair, resulting in two potential outcomes: On the *left*, the broken centromere-containing chromosome fragment segregates to the daughter; on the *right*, it segregates to the mother. In both cases, the acentric chromosome fragment remains in the mother cell after cytokinesis. On the *left*, the two fragments of the broken chromosome are separated by mitosis, and repair of the broken centromere-containing fragment occurs by break-induced relication (BIR), resulting in duplication from the homologous chromosome of all sequences centromere-distal to the break. The acentric fragment remaining in the mother cell is shown to be degraded (*dashed line*), but could have other fates. On the *right*, where the mother inherits both fragments, DSB repair by nonhomologous end joining or local gene conversion without crossing over preserves both alleles at distal loci (no loss of heterozygosity [LOH]). Note that if, before DNA replication, crossing over did accompany gene conversion, LOH would not occur.

In agreement with such a model, pronounced cell cycle arrests—while easily recognized during pedigree analysis—were essentially never detected in the cell divisions preceding the appearance of a homozygous mother or daughter cell (see Fig. 2B) (McMurray and Gottschling 2003). Thus, in the progeny of aging cells, BIR events causing LOH were not associated with DNA damage checkpoint-mediated cell cycle delay. Interestingly, BIR in young cells may result from G_2/M checkpoint failure (Fasullo et al. 1998; Galgoczy and Toczyski 2001), offering further support for the involvement of a checkpoint defect in age-induced LOH.

Missegregation of Chromosome Fragments and Daughter-biased LOH

How can we explain the ~20-fold daughter bias of age-induced LOH? The answer may lie in the segregation bias of the acentric fragment of a broken chromosome. Acentric circular plasmids display a strong (~20-fold) mother segregation bias (Murray and Szostak 1983); indeed, this is the basis of the accumulation of ERCs in aging mother cells (Sinclair and Guarente 1997). If linear chromosome fragments behave similarly to these circular molecules, then there is an inherent mother bias to the segregation of acentric chromosome fragments. As a result of random segregation of centromere-containing molecules, mother cells will often inherit both fragments of the broken chromosome after mitosis, allowing for standard mechanisms of DSB repair (Fig. 3) (Malkova et al. 1996). On the other hand, the daughter of an old mother will rarely inherit both chromosome fragments, because of the biased retention of the acentric fragment in the mother. This will result in increased rates of LOH by BIR in the following G_1 of the daughter (Fig. 3). Thus, mother-biased segregation of acentric chromosome fragments may underlie the asymmetry of age-induced LOH by allowing mothers to undergo repair that conserves heterozygosity, while forcing daughters into BIR (Fig. 3).

According to our model of daughter-biased LOH, an old mother cell should often inherit the "extra" chromosome fragment that failed to segregate to the bud. What is the fate of these acentric fragments? Others have shown that BIR forks initiating from acentric molecules do not appear to replicate past the centromeres of intact chromosomes to create aneuploidy (Morrow et al. 1997). These acentric fragments may eventually be degraded in the mother during subsequent S phases, as previously observed for a plasmid with a single DSB (Raghuraman et al. 1994).

HERITABILITY OF THE HYPERRECOMBINATIONAL STATE

Sectored Daughter Colonies Suggest Semistable Inheritance

The detailed pedigree analysis we performed led to the discovery that old mother cells rarely experience LOH, yet frequently produce daughter cells that do. However, the majority of age-induced LOH events were observed as sectored daughter colonies (Fig. 2B), suggesting they occurred in cell divisions after the daughter cells were removed from the mother. LOH occurring after one, two, or three generations following separation of the daughter from the mother will result in half-, quarter-, or eighth-sectored daughter colonies, respectively (Fig. 4A). If the hyperrecombinational state is caused by a genetic event—such as mutation of a gene normally required for genome stability—it should be inherited equally by all the progeny of an old mother cell, affecting a great-great-great-granddaughter to the same extent as the mother cell herself.

By examining the frequency of half-, quarter-, and eighth-sectored daughter colonies, we asked whether a constant rate of genome instability is maintained through multiple cell divisions. Specifically, the classes of sectored colonies should be distributed predictably if the rate of LOH is equal for every cell. For example, the frequency of quarter-sectored colonies should be twice that of half-sectored colonies, because there are twice as many opportunities for a quarter-sectored colony to arise (Fig. 4A,B). Similarly, the frequency of eighth-sectored colonies should be four times that of half-sectored colonies.

We found the relative numbers of each class of sectored colonies among spontaneous *MET15* LOH events in young cells approximated the expected ratios (Fig. 4B). This trend was not observed among the daughter colonies produced by pedigree analysis of old mothers. Instead, approximately the same number of eighth-sectored and half-sectored colonies were detected (Fig. 4B). This could indicate that the rate of LOH in every cell decreases as the cells continue to divide. Alternatively, the rate may remain high in one or a few cells for multiple generations, while all other cells have a low rate of LOH. Given the similarity in the frequency of half- and eighth-sectored colonies, it is tempting to speculate that the daughter of an old mother, but not the granddaughters or more distant progeny, maintains a high rate of LOH for at least three cell divisions (Fig. 4A). We interpret these data as evidence that the age-induced hyperrecombinational state is inherited in a semistable manner, and that the "switch" to hyper-LOH is eventually "reset" in the distant descendants of old mothers.

Resetting the Hyperrecombinational Switch

It is worth noting that the analysis described above assumes that all cells in the early divisions of daughter colony formation enjoy full viability and similar cell-doubling times. In principle, frequent cell death or cell cycle arrests could alter the appearance of sectored daughter colonies. Because our pedigree analysis did not preclude these possibilities, we addressed the heritability of the age-induced hyperrecombinational state in another way. If the state of hyperrecombination is "diluted" through the daughter lineage, then the rate of LOH in young cells many generations removed from an old mother should return to the rate characteristic of cultured cells. Indeed, we found that when cells taken from the daughter colonies of "postswitch" mothers (after the first

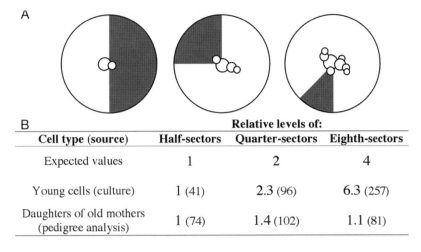

Figure 4. The age-induced hyperrecombinational state is not stably inherited. (*A*) The results of loss of heterozygosity (LOH) events in various cell divisions during the growth of a colony from a single cell are depicted in the *gray* color within the *large circles*, which are superimposed upon illustrations of the relevant progenitor cell(s). Note throughout this illustration that LOH in any single cell would generate the indicated class of sectored colony; only a single possibility is shown. (*B*) The expected proportions of each class of sectored colony are given relative to the number of half sectors, arbitrarily chosen as 1 ("Expected values"). Below, the proportions of each indicated class of sectored colony observed experimentally are presented for the indicated source of cells. Values in parentheses give the actual number of sectored colonies scored. (*C*) The rate of *MET15* LOH was measured by the frequency of brown/white half-sectored colonies for cells of a *MET15/met15Δ* strain. As indicated, cells were either plated directly from culture or taken from the white portions of brown/white sectored daughter colonies produced by old mothers. The old mothers had previously produced multiple other daughter colonies sectored with *MET15* LOH. Values represent the means of measurements from at least three independent cultures/colonies and are accompanied by the 95% confidence interval of the Poisson distribution for the number of half-sectors observed.

LOH event in the pedigree) were assayed at loci that were still heterozygous, the rate of spontaneous LOH was indistinguishable from that of cultured young cells of the same genotype (Fig. 4C; and data not shown). Because a daughter colony is composed of approximately 10^7 cells, the majority of the cells assayed were approximately 25 generations removed from the original old mother. Taken together, these results suggest that the hyperrecombinational state acquired by an old mother cell and transmitted to her immediate progeny is eventually lost through the daughter lineage after many successive asymmetric cell divisions.

A "SENESCENCE FACTOR" FOR GENOME STABILITY?

The observations presented above describe a new phenotype associated with aged mother cells: chromosomal instability leading to daughter-biased loss of heterozygosity. This trait appears to be heritable, but only transiently, as it is eventually lost in the distant descendents of old cells. A similar "dilution process" affects the factor(s) causing the end of a yeast cell's life span, manifested in the gradual reestablishment of full replicative capacity in the great-great-granddaughters of old mothers (Kennedy et al. 1994). As described earlier, additional aging phenotypes are also transiently inherited by the progeny of old mothers. To explain these phenomena, others have invoked a "senescence factor," which accumulates in aging mother cells but does not affect their offspring until a critical concentration is reached, at which point the daughters of old mothers inherit senescence factor and display aging phenotypes (Egilmez and Jazwinski 1989; Kennedy et al. 1994). Accumulation with age depends on mother-biased inheritance of the factor at each cell division (Fig. 5); this asymmetry breaks down in old cells. Upon their discovery as a cause of aging, ERCs were proposed to be the senescence factor (Sinclair and Guarente 1997), and their behavior is indeed consistent with some of the aging phenomenology. However, we showed that when the formation of ERCs is suppressed, the kinetics and magnitude of age-induced LOH is unaffected (McMurray and Gottschling 2003); ERCs are not responsible for age-induced LOH.

What could be the genome stability senescence factor? A class of candidates emerges from the recent discovery that oxidatively damaged proteins are preferentially segregated to a yeast mother cell at cytokinesis (Aguilaniu et al. 2003). Because of this asymmetric segregation, aging mother cells accumulate high levels of oxidized proteins, whereas levels in young cells are relatively low (Aguilaniu et al. 2003; Reverter-Branchat et al. 2004). However,

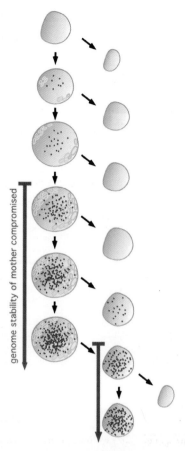

Figure 5. Model for the accumulation and inheritance of a senescence factor that affects genome stability as yeast age. On the *left*, a mother yeast cell undergoes replicative aging, producing a virgin daughter cell at each division, shown on the *right* of the mother cell. A hypothetical senescence factor (*red dots*) accumulates in the mother at each division, reaching high concentrations after many successive divisions. Past a certain threshold (*red arrows*), the concentration of the senescence factor begins to negatively impact the genome stability of the cell; the daughter of an old mother may be affected and begins accumulating more senescence factor once it becomes a mother. However, the granddaughter of an old cell (*far right*) does not inherit the senescence factor and retains normal genome integrity.

after the mother reaches a certain age, damaged proteins are equally distributed between her and her subsequent daughters (Aguilaniu et al. 2003). The behavior of oxidatively damaged proteins in aging cells thus fulfills the phenomenological requirements for a senescence factor.

We suggest that damaged versions of proteins critical for maintaining chromosome integrity accumulate in old cells. Once enough of the protein is in a damaged form, it acts as a "dominant negative" (Herskowitz 1987) and interferes with normal function, initiating an increased frequency of LOH (Fig. 5). As damaged protein continues to accumulate with increasing age of the mother cell, the damaged protein spills over into the daughter cell, and she too begins to display increased LOH. However, the amount of damaged protein is diluted away in further progeny (granddaughters, great-granddaughters, etc.), because the damaged protein is preferentially maintained in mother cells (for reasons that are unknown; Aguilaniu et al. 2003) and is replaced by new protein synthesis in daughter cells.

CONCLUSIONS

We have shown that aging yeast cells—a prospective model for human stem cells—switch to a state characterized by high rates of loss of heterozygosity. The switch does not occur stochastically during the life span, instead affecting mother cells only after approximately 25 successive cell divisions. Thus, age itself influences the rate of new genetic events. Furthermore, this apparently genome-wide effect on chromosome integrity is not itself the result of a primary genetic change: The hyperrecombinational state is inherited by the progeny of an old mother, but is eventually lost in successive cell divisions through the daughter cell lineage.

We believe these findings may be relevant as a model of human cancer. For instance, in order to explain the number of mutations, rearrangements, and other genetic events in cancer cells, a "mutator phenotype" has been postulated to represent an early alteration in tumor progression that facilitates the subsequent changes required for tumorigenesis (Loeb 1991; Loeb et al. 2003). This theory holds that, in the absence of mutator cells, a human lifetime is too short to allow the accumulation of the number of genetic events observed in cancer cells, given the low rate of events per cell division in normal cells (Loeb 1991; Loeb et al. 2003). We suggest that an epigenetic mechanism akin to what we have observed in aging yeast mother cells could provide the increase in genomic instability that is ultimately manifested in tumors, particularly in those that might originate in the stem cell compartments of older individuals.

Increased genomic instability is generally considered to be detrimental to cell vitality. If this is so, why does it occur in aging yeast cells? On one hand, this may be explained by a variation of the classical argument that natural selection cannot act upon traits manifested past the age of reproduction (Medawar 1952). For the unicellular organism *S. cerevisiae*, an old mother cell (age > 25) makes an infinitesimal ($1/2^{26}$) contribution to the fitness of a clone. Thus, decreased fitness in old mothers and their progeny due to age-induced LOH would likely be beyond the reach of natural selection. On the other hand, age-induced genetic instability in microbial systems can also be viewed in a beneficial light. The rarity of progeny of old mother cells in an exponentially growing population may afford a degree of genotypic flexibility to the population without a significant detriment to fitness. Rare cells that have undergone genetic changes may have decreased fitness under normal conditions, but vastly increased fitness in a changing environment. In fact, such a scenario not only applies to the selective pressures incurred by a microbial system, but is also relevant when considering cells that ultimately escape normal cellular control and evolve to become cancer cells.

ACKNOWLEDGMENTS

We thank Laurie Carr and Derek Lindstrom for their comments on the manuscript. This work was supported by a National Institutes of Health grant (AG23779) and an Ellison Medical Foundation Senior Scholar Award to D.E.G.

REFERENCES

ACS (American Cancer Society). 2004. *Cancer facts & figures 2004.* American Cancer Society, Atlanta.

Aguilaniu H., Gustafsson L., Rigoulet M., and Nystrom T. 2003. Asymmetric inheritance of oxidatively damaged proteins during cytokinesis. *Science* **299:** 1751.

Armitage P. and Doll R. 1954. The age distribution of cancer and a multi-stage theory of carcinogenesis. *Br. J. Cancer* **8:** 1.

———. 1957. A two-stage theory of carcinogenesis in relation to the age distribution of human cancer. *Br. J. Cancer* **11:** 161.

Bandyopadhyay D. and Medrano E.E. 2003. The emerging role of epigenetics in cellular and organismal aging. *Exp. Gerontol.* **38:** 1299.

Bitterman K.J., Medvedik O., and Sinclair D.A. 2003. Longevity regulation in *Saccharomyces cerevisiae:* Linking metabolism, genome stability, and heterochromatin. *Microbiol. Mol. Biol. Rev.* **67:** 376.

Bohr V.A. 2002. Repair of oxidative DNA damage in nuclear and mitochondrial DNA, and some changes with aging in mammalian cells. *Free Radic. Biol. Med.* **32:** 804.

Cost G.J. and Boeke J.D. 1996. A useful colony colour phenotype associated with the yeast selectable/counter-selectable marker *MET15*. *Yeast* **12:** 939.

Defossez P.A., Prusty R., Kaeberlein M., Lin S.J., Ferrigno P., Silver P.A., Keil R.L., and Guarente L. 1999. Elimination of replication block protein Fob1 extends the life span of yeast mother cells. *Mol. Cell* **3:** 447.

DePinho R.A. 2000. The age of cancer. *Nature* **408:** 248.

Egilmez N.K. and Jazwinski S.M. 1989. Evidence for the involvement of a cytoplasmic factor in the aging of the yeast *Saccharomyces cerevisiae*. *J. Bacteriol.* **171:** 37.

Fasullo M., Bennett T., Ahching P., and Koudelik J. 1998. The *Saccharomyces cerevisiae RAD9* checkpoint reduces the DNA damage-associated stimulation of directed translocations. *Mol. Cell. Biol.* **18:** 1190.

Frank S.A. 2004. Age-specific acceleration of cancer. *Curr. Biol.* **14:** 242.

Galgoczy D.J. and Toczyski D.P. 2001. Checkpoint adaptation precedes spontaneous and damage-induced genomic instability in yeast. *Mol. Cell. Biol.* **21:** 1710.

Hanahan D. and Weinberg R.A. 2000. The hallmarks of cancer. *Cell* **100:** 57.

Herskowitz I. 1987. Functional inactivation of genes by dominant negative mutations. *Nature* **329:** 219.

Jackson A.L. and Loeb L.A. 2001. The contribution of endogenous sources of DNA damage to the multiple mutations in cancer. *Mutat. Res.* **477:** 7.

Jazwinski S.M. 2001. New clues to old yeast. *Mech. Ageing Dev.* **122:** 865.

Jones P.A. and Laird P.W. 1999. Cancer epigenetics comes of age. *Nat. Genet.* **21:** 163.

Kadyk L.C. and Hartwell L.H. 1992. Sister chromatids are preferred over homologs as substrates for recombinational repair in *Saccharomyces cerevisiae*. *Genetics* **132:** 387.

Kaeberlein M., Kirkland K.T., Fields S., and Kennedy B.K. 2004. Sir2-independent life span extension by calorie restriction in yeast. *PLoS Biol.* **2:** E296.

Kennedy B.K., Austriaco N.R., Jr., and Guarente L. 1994. Daughter cells of *Saccharomyces cerevisiae* from old mothers display a reduced life span. *J. Cell Biol.* **127:** 1985.

Kraus E., Leung W.-Y., and Haber J.E. 2001. Break-induced replication: A review and an example in budding yeast. *Proc. Natl. Acad. Sci.* **98:** 8255.

Lengauer C., Kinzler K.W., and Vogelstein B. 1998. Genetic instabilities in human cancers. *Nature* **396:** 643.

Loeb L.A. 1991. Mutator phenotype may be required for multistage carcinogenesis. *Cancer Res.* **51:** 3075.

Loeb L.A., Loeb K.R., and Anderson J.P. 2003. Multiple mutations and cancer. *Proc. Natl. Acad. Sci.* **100:** 776.

Malkova A., Ivanov E.L., and Haber J.E. 1996. Double-strand break repair in the absence of *RAD51* in yeast: A possible role for break-induced DNA replication. *Proc. Natl. Acad. Sci.* **93:** 7131.

McMurray M.A. and Gottschling D.E. 2003. An age-induced switch to a hyper-recombinational state. *Science* **301:** 1908.

Medawar P.B. 1952. *An unsolved problem of biology* (an inaugural lecture delivered at University College, London, December 6, 1951). H.K. Lewis, London.

Morrow D.M., Connelly C., and Hieter P. 1997. "Break copy" duplication: A model for chromosome fragment formation in *Saccharomyces cerevisiae*. *Genetics* **147:** 371.

Mortimer R. and Johnston J. 1959. Lifespan of individual yeast cells. *Nature* **183:** 1751.

Murray A.W. and Szostak J.W. 1983. Pedigree analysis of plasmid segregation in yeast. *Cell* **34:** 961.

Paques F. and Haber J.E. 1999. Multiple pathways of recombination induced by double-strand breaks in *Saccharomyces cerevisiae*. *Microbiol. Mol. Biol. Rev.* **63:** 349.

Paulovich A.G., Toczyski D.P., and Hartwell L.H. 1997. When checkpoints fail. *Cell* **88:** 315.

Raghuraman M.K., Brewer B.J., and Fangman W.L. 1994. Activation of a yeast replication origin near a double-stranded DNA break. *Genes Dev.* **8:** 554.

Reverter-Branchat G., Cabiscol E., Tamarit J., and Ros J. 2004. Oxidative damage to specific proteins in replicative and chronological-aged *Saccharomyces cerevisiae:* Common targets and prevention by calorie restriction. *J. Biol. Chem.* **279:** 31983.

Roman H. 1957. Studies of gene mutation in *Saccharomyces*. *Cold Spring Harbor Symp. Quant. Biol.* **21:** 175.

Sinclair D.A. and Guarente L. 1997. Extrachromosomal rDNA circles—A cause of aging in yeast. *Cell* **91:** 1033.

Sinclair D.A., Mills K., and Guarente L. 1998. Molecular mechanisms of yeast aging. *Trends Biochem. Sci.* **23:** 131.

Restructuring the Genome in Response to Adaptive Challenge: McClintock's Bold Conjecture Revisited

R.A. JORGENSEN
Department of Plant Sciences, University of Arizona, Tucson, Arizona 85721-0036

I believe there is little reason to question the presence of innate systems that are able to restructure a genome. It is now necessary to learn of these systems and to determine why many of them are quiescent and remain so over very long periods of time only to be triggered into action by forms of stress, the consequences of which vary according to the nature of the challenge to be met.

—Barbara McClintock, 1978

The prevailing view in evolutionary biology has long been that natural selection acts on genetic variants that arise by accident in a manner unrelated to adaptive challenges. Barbara McClintock's revolutionary discovery that genome structure is dynamic and responsive to conditions of stress led to a revised view incorporating the possibility that mutations can arise at widely varying frequencies, even in "bursts," in response to a variety of influences, some of which may represent adaptive challenges to the organism (McDonald 1983; Walbot and Cullis 1985). This new view of the origin of mutations retains the assumption that the mutations induced by any particular adaptive challenge are no more likely to address that challenge than any other.

McClintock, however, citing the sophistication with which organisms perceive and respond to the environment, boldly advocated a more radical position: that organisms respond to challenges by inducing mutations likely to have some adaptive value toward the particular challenge experienced. Expanding on this theme in her Nobel lecture, she challenged biologists "to determine the extent of knowledge the cell has of itself, and how it utilizes this knowledge in a 'thoughtful' manner when challenged" (McClintock 1984).

This provoked one prominent critic to suggest that McClintock's speculations "would indeed seem to verge on mysticism" and to ask (almost incredulously): "[D]oes the organism ... have foresight, conjuring up just the kind of restructuring that the occasion demands?" (Fincham 1992). It is significant, however, that McClintock placed "thoughtful" in quotation marks; never did she claim any conscious effort on the part of the plant. As Shapiro (1992) explains, McClintock often spoke in her later years of "the concept of 'smart cells', a phrase she slipped in humorously at the end of her lectures" that was used to express her deep awareness of the sophistication and complexities of cells' abilities to sense, evaluate, and respond. Thus, McClintock's phrase, "a 'thoughtful' manner," might best be interpreted as referring to *a complex process that integrates information and responds according to the nature of that information.*

At the time she made this conjecture, understanding of the nucleus and the genome was too limited to propose mechanisms that might underlie it. Advances in molecular biology, cell biology, and genomics over the past 25 years have shed much light on the dynamic behavior of the nucleus and its genome, making it attractive to revisit McClintock's suggestion. The purpose of any such exercise is not to show how things *are*, but simply to illustrate how they *might be*, and to help identify questions that will need to be explored experimentally before reaching any conclusions about the validity of McClintock's proposal. Here I explore how *paramutation*, an epigenetic gene-silencing phenomenon discovered by another pioneering maize geneticist, R. Alexander Brink, might act as both a challenge and a genomic imprinting process that could generate novel variants more likely to address this challenge than randomly generated mutations would be.

CONNECTING A CHALLENGE TO A LOCUS

Any mechanism capable of generating a biased set of mutations that are potentially adaptive in the face of the particular challenge inducing them would seem to require two principal elements: (a) information must flow preferentially to those genetic loci that are affected by the challenge, and (b) mutagenesis must occur preferentially at such loci. The suggestion has often been made that chromosomally based epigenetic states could provide a medium for guiding DNA restructuring machinery to "challenged" genes (Cullis 1987; Jablonka and Lamb 1989; Maynard Smith 1990; Monk 1990; Jorgensen 1993; Shapiro 1993). The DNA of eukaryotic genes is enveloped in chromatin, which varies in structural and functional organization and composition according to the physiological and developmental (i.e., epigenetic) information to which it has been exposed. It seems reasonable to consider that the chromatin configuration at or near a locus can influence not only gene expression, but also the accessibility of DNA restructuring enzymes to that region (Jablonka and Lamb 1995). The immune system provides an example of such a mechanism in that epigenetic modification guides the choice of one allele for rearrangement in V(D)J recombination (Rada and Ferguson-Smith 2002).

DIRECTED GENOME RESTRUCTURING IN RESPONSE TO PARAMUTATION?

McClintock (1978) regarded sexual hybridization between taxonomically distinct species to be a prime example of the kind of "frequent accident" she thought likely to activate genome-restructuring mechanisms. Natural hybridization plays an important role in the evolution of plants through two creative outcomes: (a) the origin or transfer of new traits via introgression from one species to another and (b) the origin of entirely new species (Grant 1981; Arnold 2004). How might natural hybridization create an adaptive challenge that could trigger the production of a biased set of mutations in gene(s) whose alteration might be adaptive in the face of that particular challenge? The possibility explored here is that, in flowering plants, paramutation could constitute such a challenge by virtue of the fact that it "labels" the genes it impairs.

Paramutation is an interaction between two types of alleles, one that is "paramutagenic" and another that is "paramutable" (Brink 1960). In the presence of a paramutagenic allele, a paramutable allele is altered to become a new, "paramutant" allele, which is somatically and germinally heritable and can remain paramutant even after loss or segregation of the paramutagenic allele. Many, though not all, paramutants are metastable, exhibiting erratic expression and variable inheritance. Paramutation does not result in DNA sequence alterations, so far as is known, but is instead thought to involve changes in chromatin organization (Chandler et al. 2000). Paramutation-like interactions are not limited to alleles, but also may occur between unlinked genes, especially between unlinked transgenes as well as between transgenes and unlinked, homologous, endogenous genes, suggesting that any gene might be subject to paramutation under the right circumstances.

A favored explanation for how a paramutagenic allele (or locus) can alter a paramutable allele (or locus) invokes double-stranded RNA (dsRNA) produced by the paramutagenic gene and homologous to sequences in or adjacent to the transcriptional control elements of the paramutable gene (Finnegan and Matzke 2003). dsRNA induces changes in DNA methylation and/or chromatin organization patterns in regions of the target gene that play roles in transcription initiation.

A viable alternative explanation invokes a "homology sensing" process by which homologous sequences briefly pair and chromatin-based information is transferred from one gene to the other (Hagemann 1969; Jorgensen 1992; Matzke and Matzke 1993; Patterson et al. 1993). The existence of a homology-sensing process in somatic cells that brings homologous sequences, whether allelic or ectopic, into brief association is thought to be widespread in eukaryotes and to be capable of scanning the entire genome in somatic cells (Tartof and Henikoff 1991; Kleckner and Weiner 1993).

The hypothesis here suggests that species whose genomes are sufficiently diverged become "paramutationally incompatible" such that natural hybridization between them results in new epigenetic variants (paramutants) with aberrant chromatin organization at some loci, and that aberrant chromatin not only disrupts the expression of paramutant genes, but also makes them preferentially susceptible to DNA sequence rearrangements. The hypersusceptibility of such a locus would persist as long as its aberrant chromatin does, i.e., until the DNA is restructured in such a way that the new allele can be organized into "normal" chromatin. Those rearrangements that also confer an adaptive expression pattern on the new allele would survive in nature and would constitute adaptive mutations, having been produced preferentially at affected loci in response to a particular challenge.

Because plants do not sequester the germ line during early development, substantial opportunities exist for sexual transmission of somatically arising mutations and epigenetic states. Plant development subsequent to embryogenesis is based largely on apical and axillary meristems, groups of relatively undifferentiated cells that reiteratively produce vegetative structures, sometimes for many years, until they perceive a developmental or physiological signal directing them to produce reproductive structures (Walbot 1985; Klekowski 1988; Jablonka and Lamb 1995). As a consequence of this flexibility, a large developmental window (potentially lasting hundreds of years in some species) exists, during which challenges could trigger mutations in meristematic cells that will later give rise to germ cells capable of transmitting new genetic variants to sexual progeny. Somatic mutants can be subjected to selection prior to sexual transmission, and so the opportunities for somatic restructuring of genes with aberrant chromatin and incorporating selectively advantageous derivatives into the germ line are significant in plants.

A ROLE FOR TRANSPOSABLE ELEMENTS IN DIRECTED GENOME RESTRUCTURING?

The possibility that transposable elements (TEs) could play a causal role in paramutation has been discussed often (see, e.g., Krebbers et al. 1987; Martienssen 1996; Matzke et al. 1996). Here I also consider the reciprocal possibility: that certain TEs mediate rearrangements that can "repair" alleles exhibiting aberrant behavior.

Plant genomes evolve via many types of small-scale DNA sequence rearrangements, and though the linear order of genes on a chromosome tends to be conserved, individual genes are embedded in a complex and diverse matrix of sequence elements that is rapidly evolving and is largely comprised of TEs of various types (Bennetzen 2000; Feschotte et al. 2002). TE insertions are known to have created evolutionarily significant gene regulatory mutations (McDonald 1995; Kidwell and Lisch 1997), as was proposed by McClintock. TEs can also generate a wide variety of secondary rearrangements and transpositions in the locale of a TE (Lönnig and Saedler 2002), and so it is attractive to propose TEs as candidates for a DNA restructuring system that can preferentially alter aberrant loci and generate new variants.

Obviously, the appropriate cellular machinery would need to be available for restructuring the DNA sequences that underlie aberrant chromatin states, and yet eukaryotic TEs are normally found to be in a quiescent state, often associated with repressive chromatin. However, a variety of "stresses" are known to activate silent TEs, particularly interspecific hybridization (McClintock 1978, 1984; Wessler 1996). Interestingly, an interspecific mammalian hybrid exhibits genome-wide activation of retroelement movement and genome-wide restructuring, as well as undermethylation of DNA (O'Neill et al. 1998). In plants, newly formed polyploids often exhibit rapid, large-scale genome-wide changes in a significant fraction of sequences, up to several percent of tested sequences in just a few generations (Rieseberg 2001). Genome restructuring directed by chromatin states has been proposed to explain rapid genomic evolution in newly synthesized allopolyploids (Comai et al. 2003). A role for TEs in such restructuring has been inferred, but this still requires further investigation.

Importantly, TEs vary widely in target site specificity (Bennetzen 2000; Lönnig and Saedler 2002). Class 2 elements (DNA elements), such as McClintock's *Activator* (*Ac*) and *Dissociation* (*Ds*) elements, typically have high target specificity for genic regions in plants, including and perhaps especially the transcriptional control regions that lie 5′ to the transcribed regions of genes. *P* elements in *Drosophila* have a strong insertional bias, with insertions generally occurring in 5′ gene regulatory regions at the expense of coding sequences. Although the majority of class 1 elements (retrotransposons) prefer to insert into nongenic regions, the rice element *Tos17* strongly prefers to insert into genic regions (Miyao et al. 2003). Particularly interesting is the yeast long terminal repeat (LTR) retrotransposon *Ty1*, which preferentially inserts adjacent to genes transcribed by RNA polymerase III. A hot spot for *Ty1* insertions that occur in vivo is not a hot spot in vitro, suggesting that, in addition to a DNA sequence preference, *Ty1* also targets some aspect of chromatin organization or nuclear environment. Chromatin state also determines the target site preference of the *Ty5* element, again suggesting that retrotransposons can recognize specific chromatin domains (Zou and Voytas 1997).

If paramutation is mediated by dsRNA molecules produced by paramutagenic genes, it is easy to see how TEs could also play a significant role in the origin of (some) paramutagenic genes. Many DNA elements tend to transpose locally and even insert at adjacent sites or within themselves, often in inverse orientation (see, e.g., Jiang and Wessler 2001). They can also create inverse repeats of sequences neighboring the insertion site, some of which cause semidominant mutations (see, e.g., Coen and Carpenter 1988). Clearly, transcriptional readthrough of such inverse repeats could result in dsRNA transcript production. In fact, TEs are known to produce dsRNA and siRNA molecules that can block gene expression (Mette et al. 2002; Sijen and Plasterk 2003; Slotkin et al. 2003). Furthermore, the effects of siRNA can spread over time to adjacent sequences (Slotkin et al. 2003).

Some inverse repeats might be expected to negatively affect the expression of adjacent or unlinked genes by dsRNA- and siRNA-mediated DNA methylation of homologous sequences. These would be eliminated by natural selection, of course, whereas other siRNA-producing DNA rearrangements might be selectively neutral and so would persist until lost by mutation. When natural hybridization brings neutral variants into contact with a diverged genome, however, the possibility exists that some loci in the latter genome may have undergone independent changes that happen to make them susceptible to siRNA-producing loci in the other genome (i.e., paramutable). TEs are known to contribute sequences to promoters that are essential for normal promoter function; thus, the more divergent two populations are, the more likely a variant will arise in one that could be susceptible to a new variant in the other, which, though it may be selectively neutral in its own population, behaves as paramutagenic locus in the hybrid genotype.

Corresponding arguments could be put forward for DNA:DNA interaction–mediated paramutation in which changes in chromatin organization arise as a consequence of "ectopic pairing-like" interactions. These too could be mediated or influenced by differences in TE organization patterns in or near transcriptional control regions. Thus, the hypothesis here does not have to be limited to dsRNA-mediated mechanisms of paramutation.

AN "ADAPTATION DOMAIN" FOR RESTRUCTURING THE DNA THAT UNDERLIES ABERRANT CHROMATIN?

An intriguing further possibility is that metastable loci might reside in a replication domain distinct from the domains in which active and inactive genes are typically found, and that this domain might be preferentially targeted for DNA restructuring. In higher eukaryotes, nuclear genes are replicated and expressed in a variety of distinct temporal and functional domains (Spector 1993, 2003). In general, active genes replicate early in the cell cycle and inactive genes replicate late. Interestingly, these temporal domains can be distinguished by two distinct types of replication foci—those possessing maintenance DNA methyltransferase and those lacking it (Leonhardt et al. 1992)—and imposition of epigenetic states can require S phase and replication (Fox et al. 1997).

It is interesting to speculate that aberrant chromatin might comprise a novel "adaptation" domain, which is targeted by the molecular machinery for generating rearrangements of the DNA underlying the aberrant chromatin residing there. To illustrate the potential for such a mechanism, it is interesting to consider that in the immune system epigenetic modifications have been implicated in the choice of an allele for early replication and subsequent rearrangement (Rada and Ferguson-Smith 2002). Thus, an allele's epigenetic state can determine its replication domain, and this domain can be preferentially targeted for rearrangement. Changes in chromatin structure can even be associated with altered positioning of chromosome territories (Spector 2003). In parental imprinting in mammals, allele-specific epigenetic modifications determine differential replication timing and subnuclear localization (Gribnau et al. 2003), a basis for suggesting a distinct physical location for an adaptation domain within the nucleus.

Because the establishment of a new epigenetic state can require passage through S phase and can occur at replication foci, rearrangement of DNA underlying aberrant chromatin could even be accompanied by establishment of a new chromatin state, perhaps as the DNA leaves the replication focus. If the new chromatin state is a normal, active state, the gene would be returned to an early replicating, actively expressed domain. If not, the gene would remain in the adaptation domain and continue to be a target for DNA rearrangement.

In fragile X syndrome, mutant alleles with an expanded CGG repeat exhibit delayed replication timing not only locally, but also over a large region of at least 180 kb surrounding the repeat array (Hansen et al. 1993). This illustrates how a relatively small change in DNA sequence organization (an increase of only several hundred base pairs) can affect the replication timing of a region nearly three orders of magnitude larger. Thus, rearrangements occurring at a considerable physical distance from transcribed sequences could alter the replication domain of a gene.

CONCLUSIONS

The hypothesis presented here suggests that, in an interspecific plant hybrid, (a) some gene pairs will interact in such a way that a metastable, paramutant state arises in one or both, (b) activation of certain families of TEs is induced as another consequence of hybridization, and (c) aberrant chromatin states of paramutant genes allow preferential integration at or near these genes, as well as associated or subsequent adjacent rearrangements.

Once initiated, restructuring events would continue to arise locally as long as the chromatin remains aberrant and the restructuring machinery remains active and available. The process of "repairing" a paramutant might take several to many generations, depending on the likelihood that a new mutation alters the locus in such a way that it will move from the proposed aberrant chromatin domain to a normal chromatin domain. Natural selection, of course, will act on all new variants, favoring those that are beneficially expressed in the hybrid organism and its progeny. Obviously, recombination and segregation could also remove alleles exhibiting aberrant chromatin, but the fact that extensive, rapid restructuring does in fact occur in hybrids and allopolyploids that derive from "wide crosses" (Rieseberg 2001; Comai et al. 2003) indicates that new variants do arise quickly and abundantly enough to support such a hypothesis.

The principal objection raised to McClintock's conjecture of adaptive genome restructuring has been that it is difficult to conceive of any mechanism behind it. A counterargument suggested that natural selection ought to favor the evolution of any system able to respond more efficiently to the challenges it faces by connecting a particular challenge to the generation of a potentially "useful" spectrum of mutations to meet that challenge and so it is worthwhile looking for evidence of such systems (Shapiro 1993). To bridge this gap in perceptions, it will be necessary to vastly improve our understanding of the processes underlying genome evolution by openly considering realistic molecular possibilities and developing testable hypotheses.

The hypothesis presented here will be useful if it encourages both discussion and experimental investigation of the possibilities and questions raised by it, such as whether paramutation-like events occur commonly in sexual hybrids between substantially diverged plant taxa, whether some loci exhibiting aberrant epigenetic states inhabit a distinct temporal or functional domain in the nucleus, and whether aberrant chromatin is a preferred target for certain TEs and/or other DNA restructuring systems. Also, by first addressing McClintock's conjecture in the narrow sense of paramutation, it is hoped that readers may more easily see the larger possibilities that could derive from the hypothesis, especially that of adaptive restructuring in response to environmental challenge.

Beyond Paramutation

Paramutation and transposition are generally regarded as aberrant events resulting from the breakdown of normal cellular processes and parasitism of selfish DNA elements. McClintock, on the contrary, believed that the most biologically important role of transposable elements is of a higher order, as components of sophisticated genome restructuring systems. Similarly, Brink argued that although paramutation sensu stricto is observed as an aberration, it is perhaps best viewed more broadly as a reflection of normal gene regulatory mechanisms. To explain eukaryotic gene regulation in 1960, he proposed a "paragenetic" function for chromosomes that is superimposed on their primary, genetic function of ensuring stable transmission of the genetic material. This paragenetic function would have a dual purpose: (a) to control gene expression and (b) to receive and record information about the cellular environment and transmit it mitotically throughout growth and development (Brink 1960). It is now clearly established that many gene expression states in plants that are developmentally or environmentally determined are based on chromatin that can be reprogrammed (Goodrich and Tweedie 2002), i.e., they are effectively paragenetic states as defined by Brink (Jorgensen 1994).

Later, McClintock (1967) embraced a similar view to Brink's after observing the "setting" and "erasure" of gene expression states in maize:

> [A] locus is in no manner permanently modified by the events responsible for setting and erasure. It can undergo repeated cycles of this type of programming of action, ... a type of regulation of gene action that may have general significance. Action of genes could be programmed at one stage of development in a manner that would regulate their expressions at a later stage. An imposed program could be erased subsequently and the locus again readied for future programming.

Thus, stepping only slightly beyond the hypothesis described here, it is very tempting to speculate that environmental factors that normally modulate gene expression via paragenetic states might, under "stressful" conditions, produce aberrant chromatin states that are preferentially targeted for DNA restructuring by TE-mediated processes, thereby generating a biased spectrum of mutations potentially adaptive to the particular environmental challenge. Taking together Brink's paragenetic view of gene regulation with McClintock's systems view of transposable elements in adaptive evolution, it is not difficult to envision a sophisticated information processing system whose function is to employ aberrant paragenetic states arising under adverse circumstances to improve the evolutionary odds that an organism will discover useful genetic solutions to some of the unanticipated challenges it encounters—perhaps not a "thoughtful" process per se, but certainly an attentive one. To paraphrase McClintock (1978), it is time to explore the nature and evolutionary significance of these attentive systems for adaptive genome restructuring in response to stress, *"the consequences of which vary according to the nature of the challenge to be met."*

ACKNOWLEDGMENTS

The author's research in epigenetics is supported by grants from the National Science Foundation under Grant No. 9975930 in the Plant Genome Research Program and the Department of Energy's Office of Basic Energy Sciences under Grant No. DE-FG03-98ER20308. I have generally cited reviews in order to keep the list of references to a reasonable length, and so I must apologize to those whose important primary research has not been cited here. Readers may access the primary literature via these reviews.

REFERENCES

Arnold M.L. 2004. Transfer and origin of adaptations through natural hybridization: Were Anderson and Stebbins right? *Plant Cell* **16**: 562.

Bennetzen J.L. 2000. Transposable element contributions to plant gene and genome evolution. *Plant Mol. Biol.* **42**: 251.

Brink R.A. 1960. Paramutation and chromosome organization. *Q. Rev. Biol.* **35**: 120.

Chandler V.L., Eggleston W.B., and Dorweiler J.E. 2000. Paramutation in maize. *Plant Mol. Biol.* **43**: 121.

Coen E.S. and Carpenter R. 1988. A semi-dominant allele, *niv-525*, acts *in trans* to inhibit expression of its wild-type homologue in *Antirrhinum majus*. *EMBO J.* **7**: 877.

Comai L., Madlung A., Josefsson C., and Tyagi A. 2003. Do the different parental 'heteromes' cause genomic shock in newly formed allopolyploids? *Philos. Trans. R. Soc. Lond. B Biol. Sci.* **358**: 1149.

Cullis C.A. 1987. The generation of somatic and heritable variation in response to stress. *Am. Nat.* (suppl.) **103**: S62.

Feshotte C., Jiang N., and Wessler S.R. 2002. Plant transposable elements: Where genetics meets genomics. *Nat. Rev. Genet.* **3**: 329.

Fincham J.R.S. 1992. Book review: "The dynamic genome: Barbara McClintock's ideas in the century of genetics." *Nature* **358**: 631.

Finnegan E.J. and Matzke M.A. 2003. The small RNA world. *J. Cell Sci.* **116**: 4689.

Fox C.A., Ehrenhofer-Murray A.E., Loo S., and Rine J. 1997. The origin recognition complex, *SIR1*, and the S phase requirement for silencing. *Science* **276**: 1547.

Goodrich J. and Tweedie S. 2002. Remembrance of things past: Chromatin remodeling in plant development. *Annu. Rev. Cell Dev. Biol.* **18**: 707.

Grant V. 1981. The species situation in plants. In *Plant speciation*, 2nd edition, p. 70. Columbia University Press, New York, New York.

Gribnau J., Hochedlinger K., Hata K., Li E., and Jaenisch R. 2003. Asynchronous replication timing of imprinted loci is independent of DNA methylation, but consistent with differential subnuclear localization. *Genes Dev.* **17**: 759.

Hagemann R. 1969. Somatic conversion (paramutation) at the Sulfurea locus of Lycopersicon esculentum Mill. III. Studies with trisomics. *Can. J. Genet. Cytol.* **11**: 346.

Hansen R.S., Canfield T.K., Lamb M.M., Gartler S.M., and Laird C.D. 1993. Association of Fragile X Syndrome with delayed replication of the *FMR1* gene. *Cell* **73**: 1403.

Jablonka E. and Lamb M.J. 1989. The inheritance of acquired epigenetic variations. *J. Theor. Biol.* **139**: 69.

———. 1995. *Epigenetic inheritance and evolution*. Oxford University Press, Oxford, United Kingdom.

Jiang N. and Wessler S.R. 2001. Insertion preference of maize and rice miniature inverted repeat transposable elements as revealed by the analysis of nested elements. *Plant Cell* **13**: 2553.

Jorgensen R. 1992. Silencing of plant genes by homologous transgenes. *Agbiotech News Info.* **4**: 265N.

———. 1993. The germinal inheritance of epigenetic information in plants. *Philos. Trans. R. Soc. Lond. B Biol. Sci.* **339**: 173.

———. 1994. Developmental significance of epigenetic impositions on the plant genome: A paragenetic function for chromosomes. *Dev. Genet.* **15**: 523.

Kidwell M.G. and Lisch D. 1997. Transposable elements as sources of variation in animals and plants. *Proc. Natl. Acad. Sci.* **94**: 7704.

Kleckner N. and Weiner B.M. 1993. Potential advantages of unstable interactions for pairing of chromosomes in meiotic, somatic, and premeiotic cells. *Cold Spring Harbor Symp. Quant. Biol.* **58**: 553.

Klekowski E.J., Jr. 1988. *Mutation, developmental selection, and plant evolution*. Columbia University Press, New York.

Krebbers E., Hehl R., Piotrowiak R., Lönnig W.E., Sommer H., and Saedler H. 1987. Molecular analysis of paramutant plants of *Antirrhinum majus* and the involvement of transposable elements. *Mol. Gen. Genet.* **209**: 499.

Leonhardt H., Page A.W., Weier H., and Bestor T.H. 1992. A targeting sequence directs DNA methyltransferase to sites of DNA replication in mammalian nuclei. *Cell* **69**: 865.

Lönnig W.-E. and Saedler H. 2002. Chromosome rearrangements and transposable elements. *Annu. Rev. Genet.* **36**: 389.

Martienssen R. 1996. Epigenetic phenomena: Paramutation and gene silencing in plants. *Curr. Biol.* **6**: 810.

Matzke M.A. and Matzke A.J.M. 1993. Genomic imprinting in plants: Parental effects and *trans*-inactivation phenomena. *Annu. Rev. Plant Physiol. Plant Mol. Biol.* **44**: 53.

Matzke M.A., Matzke A.J.M., and Eggleston W.B. 1996. Paramutation and transgene silencing: A common response to invasive DNA? *Trends Plant Sci.* **1**: 382.

Maynard Smith J. 1990. Models of a dual inheritance system. *J. Theor. Biol.* **143**: 41.

McClintock B. 1967. Genetic systems regulating gene expression during development. *Dev. Biol. Suppl.* **1**: 84.

———. 1978. Mechanisms that rapidly reorganize the genome. *Stadler Genet. Symp.* **10**: 25.

———. 1984. The significance of responses of the genome to challenge. *Science* **226**: 792.

McDonald J.F. 1983. The molecular basis of adaptation: A critical review of relevant ideas and observations. *Annu. Rev. Ecol. Syst.* **14**: 77.

———. 1995. Transposable elements: Possible catalysts of organismic evolution. *Trends Ecol. Evol.* **10**: 123.

Mette M.F., van der Winden J., Matzke M., and Matzke A.J.M. 2002. Short RNAs can identify new candidate transposable element families in *Arabidopsis*. *Plant Physiol.* **130:** 6.

Miyao A., Tanaka K., Murata K., Sawaki H., Takeda S., Abe K., Shinozuka Y., Onosato K., and Hirochika H. 2003. Target site specificity of the Tos17 retrotransposon shows a preference for insertion within genes and against insertion in retrotransposon-rich regions of the genome. *Plant Cell* **15:** 1771.

Monk M. 1990. Variation in epigenetic inheritance. *Trends Genet.* **6:** 110.

O'Neill B.J., O'Neill M.J., and Graves J.A. 1998. Undermethylation associated with retroelement activation and chromosome remodelling in an interspecific mammalian hybrid. *Nature* **393:** 68.

Patterson G.I., Thorpe C.J., and Chandler V.L. 1993. Paramutation, an allelic interaction, is associated with a stable and heritable reduction of transcription of the maize *b* regulatory gene. *Genetics* **135:** 881.

Rada R. and Ferguson-Smith A.C. 2002. Epigenetics: Monoallelic expression in the immune system. *Curr. Biol.* **12:** R108.

Rieseberg L.H. 2001. Polyploid evolution: Keeping the peace at genomic reunions. *Curr. Biol.* **11:** R925.

Shapiro J.A. 1992. Barbara McClintock, 1902-1992. *Bioessays* **14:** 791.

———. 1993. Genome organization, natural genetic engineering, and adaptive mutation. *Trends Genet.* **13:** 98.

Sijen T. and Plasterk R.H.A. 2003. Transposon silencing in the *Caenorhabditis elegans* germ line by natural RNAi. *Nature* **426:** 310.

Slotkin R.K., Freeling M., and Lisch D. 2003. *Mu killer* causes the heritable inactivation of the *Mutator* family of transposable elements in *Zea mays*. *Genetics* **165:** 781.

Spector D.L. 1993. Macromolecular domains within the cell nucleus. *Annu. Rev. Cell Biol.* **9:** 265.

———. 2003. The dynamics of chromosome organization and gene regulation. *Annu. Rev. Biochem.* **72:** 573.

Tartof K.D. and Henikoff S. 1991. *Trans*-sensing effects from *Drosophila* to humans. *Cell* **65:** 201.

Walbot V. 1985. On the life strategies of plants and animals. *Trends Genet.* **1:** 165.

Walbot V. and Cullis C.A. 1985. Rapid genomic change in higher plants. *Annu. Rev. Plant Physiol.* **36:** 367.

Wessler S.R. 1996. Plant retrotransposons: Turned on by stress. *Curr. Biol.* **6:** 959.

Zou S. and Voytas D.F. 1997. Silent chromatin determines the target preference of the *Saccharomyces* retrotransposon *Ty5*. *Proc. Natl. Acad. Sci.* **94:** 7412.

Poetry of *b1* Paramutation: *cis*- and *trans*-Chromatin Communication

V.L. CHANDLER
Department of Plant Sciences, University of Arizona, Tucson, Arizona 85721

The term paramutation was coined by Alexander Brink in the 1950s to describe a meiotically heritable change in the expression of one allele when it was combined with another specific allele. In Brink's classic work he proposed that chromosomes possess a paragenetic function in addition to their genetic one (Brink 1960). Brink distinguished mutations, which arise as discrete events in undirected fashion, are stably transmitted, and are very rarely reversible, from "paramutations," which occur in directed fashion under specific conditions and are transmitted, but are often unstable, progressively changing back to the original phenotype. Brink was using the term "paramutation" in its broadest sense, equivalent to the term "epimutation" commonly used today. In addition to the classic examples of interactions between alleles, paramutation-like interactions have also been described between homologous transgenes or between transgenes and homologous endogenous genes at allelic or nonallelic positions in a number of species, including plants, fungi, animals, and humans. Thus, a more general definition of paramutation would be "*trans* interactions between homologous sequences that set up distinct epigenetic states that are heritable through meiosis."

In all the systems described to date, only specific alleles undergo paramutation. Alleles sensitive to altered expression are termed paramutable, and alleles inducing the change, paramutagenic. Following paramutation, previously sensitive alleles are termed paramutant (or paramutated). Most alleles do not participate in paramutation; these alleles are referred to as either neutral or nonparamutagenic. Two assays are routinely used to monitor paramutation: (1) the ability of a paramutagenic allele to cause a heritable change in the paramutable allele (referred to as establishment of paramutation) and (2) the heritable alteration of the paramutant allele into a paramutagenic allele. In a few examples, such as the *b1* system described in this review, these two phenotypes always occur simultaneously and completely. In contrast, at most other loci, the extent of paramutagenicity acquired by a paramutant allele depends on the circumstances of the cross, if it occurs at all (for summary, see Chandler and Stam 2004).

The degree of penetrance with which paramutation occurs and the subsequent stability of the altered state can vary among different phenomena. The most common situation is illustrated by the classic system studied by Brink, *r1*, in which the paramutant state, although heritable, is frequently unstable. In contrast, the other classic maize system studied by Coe, *b1*, shows 100% penetrance with the paramutated allele always becoming fully paramutagenic and the resulting paramutagenic state being extremely stable. A summary of all known paramutation phenomena and a comparison of their penetrance and stability has recently been published (Chandler and Stam 2004).

The most thoroughly analyzed examples of paramutation are in maize, involving four different genes that encode transcription factors that activate the colored flavonoid pigment biosynthetic pathways (for review, see Chandler et al. 2000, 2002). The fact that the amount of pigment observed reflects the levels of the regulatory proteins (Patterson et al. 1993; Hollick et al. 2000) and that even subtle changes in expression (two- to fourfold) are visually distinguishable likely explains why paramutation was discovered at these genes.

This review summarizes key experiments involving the *b1* locus in maize, including experiments that have revealed the key sequences required for paramutation. The nature of these sequences combined with the isolation of mutations that prevent paramutation suggest models to explain both the *cis*- and *trans*-communication.

b1 PARAMUTATION PHENOTYPES

Expression of the *b1* gene, which encodes a transcription factor required to activate the anthocyanin biosynthetic pathway (Chandler et al. 1989), results in purple pigment. The two alleles involved in *b1* paramutation, *B'* and *B-I*, are expressed in the seedling and most of the mature plant tissues. The highly expressed paramutable *B-I* allele is unstable, it spontaneously gives rise to *B'* at frequencies between ~0.1% and 10% (Coe 1966). In contrast to *B-I*, the *B'* allele is extremely stable, no changes to *B-I* have been observed with more than 100,000 plants examined (Coe 1966; Patterson and Chandler 1995; V. Chandler, unpubl.). When *B'* and *B-I* are crossed together, paramutation always occurs; *B-I* is always changed into *B'* (Fig. 1). The chromosome that enters the cross carrying *B-I* exits the cross carrying *B'*; thus, the presence of *B'* always changes the unstable *B-I* allele into *B'*. The new *B'* allele, symbolized as *B'** is just as paramutagenic as the parental *B'*, as demonstrated in Figure 1.

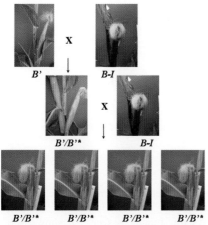

Figure 1. Paramutation at *b1*. Paramutable *B-I* (*dark purple* plant) is always changed to *B′* (*light purple* plant) in *B′/B-I* heterozygotes. Newly paramutated *B′*, *B′**, is indistinguishable from *B′*; it always paramutates naïve *B-I* in subsequent crosses.

Most *b1* alleles do not participate in paramutation. The assay to test whether an allele participates in paramutation is diagrammed in Figure 2, with the results typically seen with *b1* alleles.

Paramutation Causes Reduced Transcription But Does Not Require a Functional *b1* Protein of *B-I*

Once the *b* gene was cloned it became possible to address the molecular basis for altered pigment in *B′* and *B-I*. Comparison of *b* mRNA levels and relative transcription rates in nuclei isolated from *B′* and *B-I* plants (Patterson et al. 1993) indicate that *B′* has a 10–20-fold reduction in transcription relative to *B-I*. No change in the size or 5′ ends of the mRNA was observed. The proteins encoded by the *B′* and *B-I* alleles are equally capable of

Figure 2. Paramutation test result with neutral alleles. Neutral alleles can be recessive or dominant to *B′* with respect to pigment levels. The *B-N* example shown here is more pigmented than *B′* and is dominant such that the F1 looks like the *B-N* allele. The test of whether the allele can become paramutagenic is to cross the F1 plant with *B-I* as shown. If the allele is neutral, it does not become paramutagenic and cannot paramutate *B-I*, which will segregate in 50% of the progeny as shown.

Figure 3. Paramutation test result with paramutagenic alleles. The paramutagenic allele shown is *b-v*, which has a small deletion in the *b*-coding region causing a frameshift and no functional B protein to be made. This allele becomes paramutagenic when crossed to *B′*, which is demonstrated by crossing the F1 plant to *B-I*; all the progeny in the next generation are *B′* because both *b-v* and *B′* are paramutagenic, symbolized by the ′*.

activating the anthocyanin pathway (Patterson et al. 1993). Extensive examination of DNA methylation differences between *B′* and *B-I* failed to show any differences in DNA methylation in the promoter proximal or coding regions (Patterson et al. 1993).

To date, all alleles that participate in *b1* paramutation are transcribed, but some, such as *b-v*, are nonfunctional. The *b-v* allele contains a frameshift and does not produce a protein capable of activating the anthocyanin pathway (M. Beaudet et al., unpubl.). When crossed to *B′*, the *b-v* allele becomes *b′-v* (Coe 1966), capable of paramutating *B-I* (Fig. 3). This result demonstrates that the B protein is not required for paramutation and this experiment illustrates how an allele that is not pigmented can be monitored for paramutation activity. Because the *b-v* allele results in a slightly different transcript, it was also used to demonstrate that reduced transcription is observed at both alleles in *B′/B-I* plants (G. Patterson and V. Chandler, unpubl.).

GENETIC DISSECTION OF *b1* PARAMUTATION

The unusual features of *b1* paramutation, i.e., its full penetrance and extreme stability, enabled a genetic dissection, which has delimited the key sequences required for paramutation and identified multiple genes that, when mutant, prevent paramutation.

Identification of the Key Sequences Required for *b1* Paramutation

Crosses with neutral alleles demonstrated that the ability to participate in paramutation is linked to *B′* and *B-I* (Coe 1966; Patterson and Chandler 1995). Intergenic and intragenic recombination mapping indicated that sequences required for paramutation are tightly linked to *b*, and located in the upstream 5′-flanking region (Patterson and Chandler 1995). Restriction mapping (~100 sites) within the coding and flanking regions of *B-I* and *B'* detected no differences within ~15 kb including the 4.5-kb transcribed region (Patterson et al. 1993). Our recent fine-structure recombination mapping experiments demonstrated that the sequences required for both paramutation and high transcription (enhancer activity) are within a 6-kb region located 100 kb upstream of the *B-I* transcription start site (Stam et al. 2002a,b). Within this region *B'* and *B-I* have seven tandem repeats of an 853-bp sequence otherwise unique in the genome; neutral alleles have one. Examination of recombinant alleles with different numbers of tandem repeats (one, three, and five) demonstrated that the repeats are required for both paramutation and enhancer function (Stam et al. 2002a). Experiments are in progress to determine if the repeats are sufficient for paramutation and enhancer activities.

The 6-kb region is identical in *B-I* and *B'*, demonstrating that epigenetic mechanisms mediate the stable silencing associated with *B'* paramutation. The tandem repeat sequences are differentially methylated in *B-I* relative to *B'*, but the change in repeat methylation follows establishment of the *B'* epigenetic state (Stam et al. 2002a). In the repeats, *B-I* has a more open chromatin structure relative to *B'* and DNaseI hypersensitivity differences developmentally precede transcription at *b*, suggesting the repeat chromatin structure could be the heritable imprint distinguishing the two transcription states (Stam et al. 2002a).

Mutations That Prevent *b1* Paramutation

Forward genetic screens using the *b1* paramutation system identified three genes required for paramutation (Dorweiler et al. 2000; V. Chandler, unpubl.). The phenotypes of each mutation in the presence of *B'* are illustrated in Figure 4. The full penetrance of paramutation within the F_1 and transmission of only *B'* to the F_2 generation have enabled screens at both the F_1 and F_2 generations for mutations that affect paramutation. All F_1 progeny should be lightly pigmented because paramutation occurs with 100% penetrance; thus rare dark F_1 plants are candidates for dominant mutations that prevent paramutation and rare families, segregating one-quarter dark plants in the F_2 generation, are candidates for carrying a recessive mutation that increases the expression of the *B'* allele. The first mutation isolated was *mop1-1* (*mediator of paramutation1*), which was isolated by crossing *B-I* plants that contained active *Mutator* transposable elements to *B'* plants, self-pollinating the resulting F_1 plants and screening for recessive mutations that resulted in dark purple seedlings (Dorweiler et al. 2000).

Two other mutations were isolated from a chemical mutagenesis experiment: EMS-mutagenized *B'* pollen was crossed onto ears of *B-I* plants to generate F_1 *B'/B-I* plants. The F_1 plants were screened for rare dark plants; one dark plant was identified. Subsequent crosses demonstrated that the mutation, designated *mop2-1*, was dominant and it prevented the establishment of paramutation (K. Kubo et al., unpubl.). Further crosses also revealed that when the *mop2-1* allele is homozygous it causes a modest increase in the level of pigment that accumulates in *B'* plants, suggesting it is in fact semidominant (J. Dorweiler and V. Chandler, unpubl.).

All other F_1 plants were light, and these were self-pollinated to generate F_2 progeny to screen for recessive mutations. Screens of F_2 progeny for segregation of darkly pigmented seedlings and plants identified a second recessive mutation, *mop3-1* (J. Dorweiler et al., unpubl.). Subsequent experiments demonstrated that both *mop1-1* and *mop3-1* increased the transcription of *B'*, making plants homozygous for the mutations look like *B-I*, but the mutations did not heritably change *B'* into *B-I* (Dorweiler et al. 2000; J. Dorweiler and V. Chandler, unpubl.). When *B' mop1-1* or *B' mop3-1* homozygotes are outcrossed to a recessive *b* allele, all of the resulting progeny have the *B'* phenotype, indicating that the mutations must be present to increase *B'* pigment levels and the *B'* allele was not heritably changed to *B-I*. Importantly, like *mop2-1*, both *mop1-1* and *mop3-1* prevent the establishment of paramutation. Examples of crosses illustrating how prevention of paramutation is examined can be found in Dorweiler et al. (2000).

A similar screen was done using paramutation at the *pl1* gene, which encodes a myb-related transcription factor required for pigment in seedlings and many mature plant tissues. The initial screens also identified another allele of *mop1* (*mop1-2*) and two additional genes, referred to as *rmr1* (*required for maintenance of repression1*) and *rmr2* (Hollick and Chandler 2001). Subsequent experiments have demonstrated that mutations in all five of these genes affect paramutation at both *pl1* and *b1* (Dorweiler et al. 2000; V. Chandler et al., unpubl.). In addition, the *mop1* mutation has also been shown to prevent paramutation at the *r1* locus (Dorweiler et al. 2000), which is intriguing as *r1* paramutation appears quite distinct from *b1* and *pl1* paramutation (for discussion, see Chandler et al. 2000; Chandler and Stam 2004). The effect of the other mutations on *r1* paramutation is under investigation.

In addition to affecting paramutation at the three maize loci, the *mop1* gene is also required for maintaining the extensive DNA methylation associated with silencing of *Mutator* transposons (Lisch et al. 2002). However, *mop1-1* is distinct from the *ddm1* mutation, which affects transposon DNA methylation and silencing in *Arabidopsis* (Miura et al. 2001; Singer et al. 2001), as *mop1-1* does not have a global effect on the methylation of repeated sequences such as centromeres and rDNA (Dorweiler et

B' *mop1-1*   B' *mop2-1*   B' *mop3-1*

Figure 4. Phenotypes of *trans*-acting mutants affecting paramutation. Plants that are homozygous for *B'* and for each of the three mutations are shown. Plants that are homozygous for *B'* and heterozygous for each of the mutations look like *B'* (not shown).

2000). Intriguingly, within the first generation of introducing homozygous recessive *mop1* mutant alleles, *Mutator* methylation is decreased, but it takes several more generations in the presence of the *mop1* mutation before transposition of the element is reactivated (Lisch et al. 2002). In unpublished studies, all five paramutation mutations have been shown to reactivate some, but not all, of the transcriptionally silent transgenes tested (V. Chandler, unpubl.).

Potential Function of *mop* Genes

In previous publications we have speculated that the *mop* and *rmr* mutations are involved in some aspect of chromatin remodeling (Dorweiler et al. 2000; Hollick et al. 2000). The molecular nature of the *mop* or *rmr* genes is not yet known, as cloning is still underway. However, a candidate gene approach has eliminated approximately 140 genes that were theoretical candidates for *mop1* and *mop2*, based on their role in chromatin and epigenetic silencing in other species. Using molecular markers, the *mop1* and *mop2* genes have been mapped to a 3-cM and 1-cM interval in maize, respectively (Y. Cai et al., unpubl.). The map positions of these two genes have been compared to those of a number of maize candidates for genes involved in chromatin and epigenetic silencing. The candidate genes that have been excluded based on their map position include many genes highly homologous to genes previously shown to be involved in chromatin and epigenetic gene regulation: the 6 known DNA methyltransferase genes in maize (Cao et al. 2000; Papa et al. 2001), the 2 maize genes homologous to *ddm1* (Jeddeloh et al. 1998), 13 genes homologous to other putative *snf2*-related chromatin remodeling components (Verbsky and Richards 2001), the maize gene most homologous to *mom1* (Amedeo et al. 2000), 25 genes homologous to putative histone methyl transferases (Springer et al. 2003), and 8 and 7 genes homologous to histone acetyltransferases and histone deacetyltransferases, respectively (Pandey et al. 2002; data generated by K. Cone, University of Missouri and posted on The Plant Chromatin Database; www.chromDB.org). Experiments are in progress to map an additional approximately 100 maize genes.

MODELS FOR *b1* PARAMUTATION

Models for *b1* paramutation need to account for the following features. The high expressing *B-I* allele can be spontaneously silenced, while *trans*-silencing occurs at a much higher frequency than the spontaneous changes. Both spontaneous and directed paramutation are established in somatic cells and the states are heritably transmitted through mitosis and meiosis. The sequences required for paramutation are also required for the high transcription state associated with *B-I*. Tandem repeats are required for *b1* paramutation.

The observation that *mop1* affects multiple epigenetic phenomena suggests that paramutation is likely to share mechanistic components with those phenomena, i.e., transposon and transgene silencing. It is tempting to hypothesize a role for RNA interference (RNAi) in paramutation, as RNAi has been shown to contribute to both transposon and transcriptional transgene silencing in several species (Ketting et al. 1999; Tabara et al. 1999; Pal-Bhadra et al. 2002; Lippman et al. 2003), features shared with *mop1*. It is also likely that that multiple mechanisms are involved in different examples of paramutation, as *mop1* does not affect all aspects of paramutation at each locus. There is precedent for multiple mechanisms mediating an epigenetic phenomenon, as recent studies of transposon reactivation in *Arabidopsis thaliana* revealed distinct pathways are involved in transposon silencing (Lippman et al. 2003).

Below separate models are discussed for the *cis* interactions (long-distance enhancement of *B-I* expression and the spontaneous change from *B-I* to *B'*), and the *trans* communication that occurs between *B'* and *B-I*.

Model for *cis*-interactions

Paradigms for the long-distance *cis*-interactions between the tandem repeats and the *b1* promoter proximal region derive from studies on the globin loci and imprinted genes in mammals, particularly results from the *Igf2* and *H19* loci (Chandler et al. 2002). Figure 5A presents our favorite model, "boundary element gone awry."

Another possibility is that protein factors that bind to the *b1* enhancer sequences in the *B-I* state mediate sequestration into compartments compatible with expression. When paramutation occurs those proteins would be unable to bind to the sequences when they are in the *B'* state, resulting in an altered chromatin state that localizes the adjacent gene to a nonpermission chromatin environment and reduced expression. The repeats associated with *b1* may then be able to communicate with other homologous sequences *in trans*, either through pairing or *trans*-RNA mechanisms, causing those sequences to localize to a nonpermissive environment for transcription as well. Further dissection of the *b1* sequences required for enhancer and paramutation activities using transgenic approaches should address whether paramutation and enhancer activities are distinct or overlap.

Model for *trans*-interactions

Two major models for *trans*-communication are RNA-mediated chromatin changes and direct interactions between chromatin complexes (Fig. 5B). Our current understanding of *b1* paramutation is consistent with either or both models being involved in establishment or maintenance of *b1* paramutation. A more detailed discussion of these models is in Chandler and Stam (2004).

Pairing Model. There are several examples of studies from *Drosophila* and yeast in which tandem repeats can mediate physical interactions *in trans*. Tandem repeat arrays of *Drosophila* transgenes can mediate silencing in a copy-number-dependent way, demonstrating long-range interactions with other sequences *in cis* and *in trans* (Henikoff 1998). In yeast, tandem repeat arrays of the

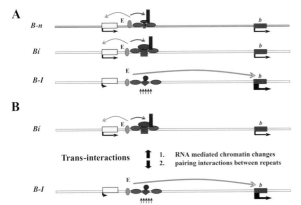

Figure 5. Model for *b1* paramutation. (*A*) Model for *cis*-communication by the tandem repeats. Two genes exist within the 120-kb region diagrammed, the *b1* gene (*purple box*) and an unknown gene identified as an expressed sequence tag (*yellow box*); the direction of transcription is indicated by the *arrows* under the boxes. The region containing the sequences required for paramutation is indicated in *red*: the neutral allele (*B-n*), which only contains a single copy of the sequences that are tandemly repeated in *B'* and *B-I*, is symbolized by a *small rectangle*; the *large rectangle* and *diamond* in *B'* and *B-I* symbolize distinct chromatin structures associated with the tandem repeats. The enhancer is indicated as a *green oval*; the exact location of the enhancer has not been determined. The neutral allele and the repeat region in the *B'* allele are both postulated to form a functional boundary element (symbolized by a *brown rectangle*) that prevents the enhancer from communicating with the *b* gene located 100-kb downstream. Instead the enhancer is envisioned to communicate with a different gene located upstream (*green arrows*, communication; *red arrows*, no communication). In the *B-I* allele the boundary element is postulated not to form because of differences in chromatin structure and/or distinct proteins bound. The *vertical black arrows* symbolize the DNaseI hypersensitive sites seen in *B-I*. The different sized and colored *ovals* indicate distinct protein complexes bound to the different alleles. (*B*) *trans*-communication could occur through direct interactions between the repeats (pairing model) or via a RNA-based communication (*trans*-RNA model); see the text for discussion of these models.

lacO or tetO operators associate in nonmeiotic cells irrespective of their genomic location, with associations occurring with similar frequencies between allelic and ectopic arrays (Aragon-Alcaide and Strunnikov 2000; Fuchs et al. 2002).

One frequent test for the involvement of pairing is to ask if translocations, which should disrupt pairing, disrupt the phenomenon under investigation. Several different translocations have been examined for both *b1* and *r1* paramutation, and none to date disrupt paramutation (for review, see Coe 1966; Brink 1973; Chandler et al. 2000). However, in all examples cited the translocation breakpoints were not adjacent to the loci, leaving open the possibility that pairing could still occur. In addition, once the epigenetic state is established it is heritable in the absence of the inducing allele; thus, brief physical interactions rather than stable synapsis may be all that is required to establish paramutation.

One possibility is that arrays of repeat sequences increase the local concentration of bound proteins, and the higher concentration confers distinct chromatin properties that can mediate pairing of homologous sequences or association of regions with similar heterochromatic properties (Sage and Csink 2003). Such pairing can result in subnuclear localization to heterochromatin domains (Csink and Henikoff 1996; Dernburg et al. 1996). Pairing would then enable the transfer of chromatin complexes from one sequence to another, similar to what has been hypothesized for the communication between complexes at enhancers and RNA polymerase initiation sites resulting in transvection (Morris et al. 1999).

Another possibility is that pairing is mediated by proteins with properties similar to the Polycomb complex (PcG), which binds to Polycomb response elements (PREs). In *Drosophila*, silenced states initiated at PREs can affect the expression of genes over considerable distances (Orlando and Paro 1993), and PRE-containing transposons inserted on different chromosomes can physically interact, resulting in enhanced silencing (Sigrist and Pirrotta 1997). Biochemical studies have also shown that PcG complexes can mediate template bridging; proteins bound to one polynucleosome template can recruit a second template from solution (Lavigne et al. 2004).

trans-*RNA* Model. There are examples of transgene arrays that mediate transcriptional (for review, see Vaucheret and Fagard 2001) and posttranscriptional silencing (for review, see Vaucheret et al. 2001; Matzke et al. 2002) through production of double-stranded RNA homologous to the promoter or transcribed regions, which then triggers RNA interference (RNAi). In many of the reported instances the transgene arrays contain inverted repeats; thus it is easy to envision how dsRNA is produced. However, there are also examples of tandemly repeated transgenes that are efficiently silenced in a variety of species (Assaad et al. 1993; Cogoni et al. 1996; Ketting and Plasterk 2000; Mourrain et al. 2000; Ma and Mitra 2002; Pal-Bhadra et al. 2004). Tandem repeats are involved in maintaining the proper chromatin structure necessary for centromere function (Dawe 2003), and in *Schizosaccharomyces pombe* components of the RNAi machinery have been shown to process transcripts derived from these repeats and mediate the formation and maintenance of silent chromatin, essential for centromere function (Volpe et al. 2002, 2003). The RNAi machinery is also required for heterochromatin formation and silencing of tandemly repeated transgenes in *Drosophila* (Pal-Bhadra et al. 2004).

Results from studies of the *b1* system demonstrated that seven or five copies of the tandem repeat were highly paramutagenic, while one copy was completely inert (neutral allele). Interestingly, an allele with three copies of the repeat was paramutagenic, but unstable and less penetrant (Stam et al. 2002a). These results suggest a cellular mechanism for sensing the number of repeats. A model for efficient maintenance of silencing within the tandem repeats at centromeres invokes that siRNAs are regenerated efficiently from tandem array transcripts by RNA-dependent RNA polymerase (RdRP) and Dicer, while siRNAs from single-copy sequences or dispersed repeats would eventually become depleted through subsequent rounds of priming by RdRP with further downstream primers (Martienssen 2003). Thus, one speculation is that an increased number of tandem repeats results in a higher abundance of siRNAs, required to stably maintain the silencing associated with paramutation.

A second possibility is that longer RNAs are mediating *trans* interactions that induce chromatin structural changes. X-inactivation in mammals is mediated through a *cis*-acting RNA, *Xist*, which confers a chromatin-based mechanism of inactivation on adjacent sequences (for review, see Chow and Brown 2003). *Xist* action is repressed by the antisense gene, *Tsix*, whose full-length RNA product is complementary to *Xist* RNA in mice (Shibata and Lee 2003). Thus, a model for *b1* paramutation is that expression of noncoding RNAs (sense and/or antisense) could mediate altered chromatin formation, and the RNAs could function both *in cis* and *in trans*. It is also possible both mechanisms are employed; the chromatin states established *in cis* through an RNA mechanism might be communicated to the other allele via pairing interactions.

POTENTIAL CONSEQUENCES OF PARAMUTATION

Numerous organisms employ mechanisms for sensing and counting homologous sequences. Paramutation may represent a defense against invasive transposons and viruses, postulated from studies of transgene silencing (Matzke et al. 2002). Another possibility is that paramutation represents the aberrant function of machinery involved in identifying and maintaining chromatin boundaries between genes and nearby repetitive sequences. Independent of the function of paramutation, there are significant evolutionary implications, such as providing an adaptive mechanism for transferring favorable expression states to progeny and a mechanism for establishing functional homozygosity in polyploids. Paramutation-like phenomena may also contribute to the low penetrance and other aspects of non-Mendelian inheritance frequently observed for genes involved in complex human diseases and the segregation of quantitative characters. It is also possible that paramutation contributes to inbreeding depression and hybrid vigor. These possibilities are discussed further in Chandler and Stam (2004).

ACKNOWLEDGMENTS

It has been a terrific pleasure working with the following postdoctoral fellows and graduate and undergraduate students, who have made the work described herein possible: Christiane Belele, Yu Cai, Jane Dorweiler, Jay Hollick, Ken Kubo, Yan Lin, Damon Lisch, Garth Patterson, Lyudmila Sidorenko, Maike Stam, and Josh White. This work was supported by grants MCB-0235329, MCB-9982447, and MCB-9603638 from the National Science Foundation and GM35971 from the National Institutes of Health.

REFERENCES

Amedeo P., Habu Y., Afsar K., Scheid O.M., and Paszkowski J. 2000. Disruption of the plant gene MOM releases transcriptional silencing of methylated genes. *Nature* **405**: 203.

Aragon-Alcaide L. and Strunnikov A.V. 2000. Functional dissection of in vivo interchromosome association in *Saccharomyces cerevisiae*. *Nat. Cell Biol.* **2**: 812.

Assaad F.F., Tucker K.L., and Signer E.R. 1993. Epigenetic repeat-induced gene silencing (RIGS) in *Arabidopsis*. *Plant Mol. Biol.* **22**: 1067.

Brink R.A. 1960. Paramutation and chromosome organization. *Q. Rev. Biol.* **35**: 120.

———. 1973. Paramutation. *Annu. Rev. Genet.* **7**: 129.

Cao X., Springer N.M., Muszynski M.G., Phillips R.L., Kaeppler S., and Jacobsen S.E. 2000. Conserved plant genes with similarity to mammalian de novo DNA methyltransferases. *Proc. Natl. Acad. Sci.* **97**: 4979.

Chandler V.L. and Stam M. 2004. Chromatin conversations: Mechanisms and implications of paramutation. *Nat. Rev. Genet.* **5**: 532.

Chandler V.L., Eggleston W.B., and Dorweiler J.E. 2000. Paramutation in maize. *Plant Mol. Biol.* **43**: 121.

Chandler V.L., Stam M., and Sidorenko L.V. 2002. Long-distance *cis* and *trans* interactions mediate paramutation. *Adv. Genet.* **46**: 215.

Chandler V.L., Radicella J.P., Robbins T.P., Chen J., and Turks D. 1989. Two regulatory genes of the maize anthocyanin pathway are homologous: Isolation of *B* utilizing *R* genomic sequences. *Plant Cell* **1**: 1175.

Chow J.C. and Brown C.J. 2003. Forming facultative heterochromatin: Silencing of an X chromosome in mammalian females. *Cell Mol. Life Sci.* **60**: 2586.

Coe E.H. 1966. The properties, origin and mechanism of conversion-type inheritance at the *b* locus in maize. *Genetics* **53**: 1035.

Cogoni C., Irelan J.T., Schumacher M., Schmidhauser T.J., Selker E.U., and Macino G. 1996. Transgene silencing of the *al-1* gene in vegetative cells of *Neurospora* is mediated by a cytoplasmic effector and does not depend on DNA-DNA interactions or DNA methylation. *EMBO J.* **15**: 3153.

Csink A.K. and Henikoff S. 1996. Genetic modification of heterochromatic association and nuclear organization in *Drosophila*. *Nature* **381**: 529.

Dawe R.K. 2003. RNA interference, transposons, and the centromere. *Plant Cell* **15**: 297.

Dernburg A.F., Broman K.W., Fung J.C., Marshall W.F., Phillips J., Agard D.A., and Sedat J.W. 1996. Perturbation of nuclear architecture by long-distance chromosome interactions. *Cell* **85**: 745.

Dorweiler J.E., Carey C.C., Kubo K.M., Hollick J.B., Kermicle J.L., and Chandler V.L. 2000. *Mediator of paramutation1* is required for establishment and maintenance of paramutation at multiple maize loci. *Plant Cell* **12**: 2101.

Fuchs J., Lorenz A., and Loidl J. 2002. Chromosome associations in budding yeast caused by integrated tandemly repeated transgenes. *J. Cell Sci.* **115**: 1213.

Henikoff S. 1998. Conspiracy of silence among repeated transgenes. *Bioessays* **20**: 532.

Hollick J.B. and Chandler V.L. 2001. Genetic factors required to maintain repression of a paramutagenic maize *pl1* allele. *Genetics* **157**: 369.

Hollick J.B., Patterson G.I., Asmundsson I.M., and Chandler V.L. 2000. Paramutation alters regulatory control of the maize *pl* locus. *Genetics* **154**: 1827.

Jeddeloh J.A., Bender J., and Richards E.J. 1998. The DNA methylation locus DDM1 is required for maintenance of gene silencing in *Arabidopsis*. *Genes Dev.* **12**: 1714.

Ketting R.F. and Plasterk R.H. 2000. A genetic link between cosuppression and RNA interference in *C. elegans*. *Nature* **404**: 296.

Ketting R.F., Haverkamp T.H., van Luenen H.G., and Plasterk R.H. 1999. Mut-7 of *C. elegans*, required for transposon silencing and RNA interference, is a homolog of Werner syndrome helicase and RNaseD. *Cell* **99**: 133.

Lavigne M., Francis N.J., King I.F., and Kingston R.E. 2004. Propagation of silencing; recruitment and repression of naive chromatin in trans by polycomb repressed chromatin. *Mol. Cell* **13**: 415.

Lippman Z., May B., Yordan C., Singer T., and Martienssen R. 2003. Distinct mechanisms determine transposon inheritance and methylation via small interfering RNA and histone modification. *PLoS Biol.* **1:** E67.

Lisch D., Carey C.C., Dorweiler J.E., and Chandler V.L. 2002. A mutation that prevents paramutation in maize also reverses *Mutator* transposon methylation and silencing. *Proc. Natl. Acad. Sci.* **99:** 6130.

Ma C. and Mitra A. 2002. Intrinsic direct repeats generate consistent post-transcriptional gene silencing in tobacco. *Plant J.* **31:** 37.

Martienssen R.A. 2003. Maintenance of heterochromatin by RNA interference of tandem repeats. *Nat. Genet.* **35:** 213.

Matzke M.A., Aufsatz W., Kanno T., Mette M.F., and Matzke A.J. 2002. Homology-dependent gene silencing and host defense in plants. *Adv. Genet.* **46:** 235.

Miura A., Yonebayashi S., Watanabe K., Toyama T., Shimada H., and Kakutani T. 2001. Mobilization of transposons by a mutation abolishing full DNA methylation in *Arabidopsis*. *Nature* **411:** 212.

Morris J.R., Geyer P.K., and Wu C.T. 1999. Core promoter elements can regulate transcription on a separate chromosome in trans. *Genes Dev.* **13:** 253.

Mourrain P., Beclin C., Elmayan T., Feuerbach F., Godon C., Morel J.B., Jouette D., Lacombe A.M., Nikic S., Picault N., Remoue K., Sanial M., Vo T.A., and Vaucheret H. 2000. *Arabidopsis* SGS2 and SGS3 genes are required for posttranscriptional gene silencing and natural virus resistance. *Cell* **101:** 533.

Orlando V. and Paro R. 1993. Mapping Polycomb-repressed domains in the bithorax complex using in vivo formaldehyde cross-linked chromatin. *Cell* **75:** 1187.

Pal-Bhadra M., Bhadra U., and Birchler J.A. 2002. RNAi related mechanisms affect both transcriptional and posttranscriptional transgene silencing in *Drosophila*. *Mol. Cell* **9:** 315.

Pal-Bhadra M., Leibovitch B.A., Gandhi S.G., Rao M., Bhadra U., Birchler J.A., and Elgin S.C. 2004. Heterochromatic silencing and HP1 localization in *Drosophila* are dependent on the RNAi machinery. *Science* **303:** 669.

Pandey R., Muller A., Napoli C.A., Selinger D.A., Pikaard C.S., Richards E.J., Bender J., Mount D.W., and Jorgensen R.A. 2002. Analysis of histone acetyltransferase and histone deacetylase families of *Arabidopsis thaliana* suggests functional diversification of chromatin modification among multicellular eukaryotes. *Nucleic Acids Res.* **30:** 5036.

Papa C.M., Springer N.M., Muszynski M.G., Meeley R., and Kaeppler S.M. 2001. Maize chromomethylase Zea methyltransferase2 is required for CpNpG methylation. *Plant Cell* **13:** 1919.

Patterson G.I. and Chandler V.L. 1995. Paramutation in maize and related allelic interactions. *Curr. Top. Microbiol. Immunol.* **197:** 121.

Patterson G.I., Thorpe C.J., and Chandler V.L. 1993. Paramutation, an allelic interaction, is associated with a stable and heritable reduction of transcription of the maize *b* regulatory gene. *Genetics* **135:** 881.

Sage B.T. and Csink A.K. 2003. Heterochromatic self-association, a determinant of nuclear organization, does not require sequence homology in *Drosophila*. *Genetics* **165:** 1183.

Shibata S. and Lee J.T. 2003. Characterization and quantitation of differential Tsix transcripts: Implications for Tsix function. *Hum. Mol. Genet.* **12:** 125.

Sigrist C.J. and Pirrotta V. 1997. Chromatin insulator elements block the silencing of a target gene by the *Drosophila* polycomb response element (PRE) but allow trans interactions between PREs on different chromosomes. *Genetics* **147:** 209.

Singer T., Yordan C., and Martienssen R.A. 2001. *Robertson's Mutator* transposons in *A. thaliana* are regulated by the chromatin-remodeling gene *Decrease in DNA Methylation* (*DDM1*). *Genes Dev.* **15:** 591.

Springer N.M., Napoli C.A., Selinger D.A., Pandey R., Cone K.C., Chandler V.L., Kaeppler H.F., and Kaeppler S.M. 2003. Comparative analysis of SET domain proteins in maize and *Arabidopsis* reveals multiple duplications preceding the divergence of monocots and dicots. *Plant Physiol.* **132:** 907.

Stam M., Belele C., Dorweiler J.E., and Chandler V.L. 2002a. Differential chromatin structure within a tandem array 100 kb upstream of the maize *b1* locus is associated with paramutation. *Genes Dev.* **16:** 1906.

Stam M., Belele C., Ramakrishna W., Dorweiler J.E., Bennetzen J.L., and Chandler V.L. 2002b. The regulatory regions required for *B'* paramutation and expression are located far upstream of the maize *b1* transcribed sequences. *Genetics* **162:** 917.

Tabara H., Sarkissian M., Kelly W.G., Fleenor J., Grishok A., Timmons L., Fire A., and Mello C.C. 1999. The *rde-1* gene, RNA interference, and transposon silencing in *C. elegans*. *Cell* **99:** 123.

Vaucheret H. and Fagard M. 2001. Transcriptional gene silencing in plants: Targets, inducers and regulators. *Trends Genet.* **17:** 29.

Vaucheret H., Beclin C., and Fagard M. 2001. Post-transcriptional gene silencing in plants. *J. Cell Sci.* **114:** 3083.

Verbsky M.L. and Richards E.J. 2001. Chromatin remodeling in plants. *Curr. Opin. Plant Biol.* **4:** 494.

Volpe T.A., Kidner C., Hall I.M., Teng G., Grewal S.I., and Martienssen R.A. 2002. Regulation of heterochromatic silencing and histone H3 lysine-9 methylation by RNAi. *Science* **297:** 1833.

Volpe T., Schramke V., Hamilton G.L., White S.A., Teng G., Martienssen R.A., and Allshire R.C. 2003. RNA interference is required for normal centromere function in fission yeast. *Chromosome Res.* **11:** 137.

RNA Silencing Pathways in Plants

A.J. Herr and D.C. Baulcombe
The Sainsbury Laboratory, Norwich NR4 7UH, United Kingdom

In one of the first described examples of RNA silencing there was coordinate suppression (cosuppression) of a transgene and an endogenous gene. Petunia plants with a flower pigment transgene (chalcone synthase [CS]) had white flowers because the transgene and the endogenous gene were silenced (Napoli et al. 1990; van der Krol et al. 1990). There was no reduction in the transcription of the endogenous gene and it seemed likely that an RNA turnover mechanism was specifically targeted at CS RNA (van Blokland et al. 1994). In this and in other cosuppression systems, the target RNA had a nucleotide sequence that was identical to the transgene. When these findings were first described, it was not understood how such cosuppression could have such sequence specificity. However, based on discoveries in animal and plant systems, it is now apparent that cosuppression is an RNA silencing process (Baulcombe 2004) in which RNA and DNA sequences are targeted specifically depending on the nucleotide sequence.

The defining feature of RNA silencing is the involvement of long double-stranded (ds) RNA, which is processed by an RNAse III-type enzyme (Dicer) into short (21–25-nucleotide) interfering (si) RNAs that initially exist in a ds form with two nucleotide overhangs at the 3′ ends. These short RNAs are separated into a single-stranded (ss) form that guides effector complexes of silencing to a target RNA by Watson–Crick base pairing. Proteins of the Argonaute class are key components of these effector complexes. They bind siRNAs (Song et al. 2003; Yan et al. 2003; Ma et al. 2004) and are the "Slicer" ribonuclease that degrades the target RNA (Liu et al. 2004; Meister et al. 2004; Rand et al. 2004).

In the petunia cosuppression system, the dsRNA and siRNA were derived from the transgene and the target RNAs were the CS RNAs from either the transgene or the endogenous gene (Stam et al. 2000). A similar process is also involved in silencing of endogenous RNAs. In this natural mechanism Dicer cleaves noncoding RNAs with partially ds regions into short (21–24-nucleotide) regulatory RNAs known as microRNAs (miRNAs). In plants the miRNAs are targeted to mRNAs of proteins affecting development and responses to environmental stimuli (Bartel 2004). Transcription factors are heavily represented in the subset of miRNAs that are conserved in distantly related plants (Jones-Rhoades and Bartel 2004). However, less is known about more recently evolved miRNAs and their targets (Allen et al. 2004).

This basic Dicer-Slicer pathway accounts for the sequence specificity of RNA silencing, but it does not explain why silencing of CS was not uniform in certain petunia plants with cosuppression. Occasionally these plants produced colored flowers with spots and streaks of CS silencing (Jorgensen et al. 1995). Moreover, the basic pathway does not explain the late stages of RNA virus-induced transgene silencing. The early stages of this process can be understood in terms of the basic pathway: A virus vector with its transgene-specific insert is a source of dsRNA that is processed by Dicer into siRNA and, within two or three weeks, the virus and transgene RNA are reduced to undetectable levels by the silencing mechanism (Ruiz et al. 1998). However, in the later stages, the transgene itself becomes the source of the siRNAs and the silencing persists throughout the life of the plant, although not into the next generation (Ruiz et al. 1998). There must be epigenetic and other mechanisms that could account for the instability of cosuppression and the persistence of virus-induced transgene RNA silencing.

In this paper, we describe variations on the basic silencing pathway and their implications for understanding of the mechanisms and natural roles of RNA silencing. The emerging picture is of a network of RNA silencing pathways. Some of these pathways involve DNA and chromatin modification. There may also be feedback loops and positive or negative interactions between different pathways.

MOLECULAR ANALYSIS OF POSTTRANSCRIPTIONAL GENE SILENCING

Our laboratory originally set up two different transgene silencing systems that were intended to mimic cosuppression in petunia. The first system was in *Nicotiana benthamiana* and we infiltrated the lower leaves of these plants with an *Agrobacterium* culture to initiate silencing. These bacterial cells had a silencer GFP transgene in their Ti plasmid that would have been transferred into plant cells in the infiltrated leaves (Voinnet and Baulcombe 1997). A GFP transgene in the genome of the *N. benthamiana* served as a reporter of the silencer activity.

The silencer and reporter GFP transgenes were cosuppressed, as expected, in the infiltrated leaves of these plants. However, there was also silencing of the GFP reporter in a ring of cells around the infiltrated region and in upper parts of the plant (Fig. 1) (Voinnet and Baulcombe 1997; Himber et al. 2003). It seems, as found independently by Vaucheret and colleagues (Palauqui et al. 1997), that a mobile silencing signal moves between cells and through the vascular system of the plant. Systemic silencing is nucleotide sequence-specific and it is

Figure 1. Systemic silencing in *Nicotiana benthamiana*. This upper leaf was imaged under UV light 11 days after silencing had been initiated by transient expression of a GFP transgene on a lower leaf. The *green* areas indicate parts of the leaf where GFP is expressed and the *red* areas indicate loss of GFP due to spread of a silencing signal out of the veins.

likely that RNA intermediates—dsRNA or siRNA—are the signal, either as naked RNA or in association with proteins. The spots and streaks of CS cosuppression in transgenic petunias may be due to movement of a silencing signal (Jorgensen et al. 1995).

RNA silencing also spreads systemically in *Caenorhabditis elegans* but it is unlikely that the transport mechanism is the same as in plants because the signal crosses a membrane when it enters the animal and requires a putative RNA transporter protein (Feinberg and Hunter 2003). This type of protein would not be required in plants because most of the cells, including the phloem cells of the vascular system, are interconnected through plasmodesmatal channels that are a continuation of the endoplasmic reticulum (Haywood et al. 2002).

The *N. benthamiana* system also allowed us to investigate the specificity determinant of RNA silencing. It seemed likely, to explain the nucleotide sequence specificity and posttranscriptional basis of the silencing process, that this determinant would be an antisense RNA and that it would be present only after induction of silencing by transient expression of the silencer GFP transgene. Andrew Hamilton had first seen evidence for such antisense RNAs in transgenic tomatoes that were undergoing cosuppression of the polygalacturonase fruit ripening gene: They were the short RNAs that are now referred to as siRNAs. However, his finding that GFP siRNAs correlated with systemic GFP silencing in *N. benthamiana* provided some of the first confirmation that they were functionally important (Fig. 2) (Hamilton and Baulcombe 1999; Hamilton et al. 2002).

GENETIC ANALYSIS OF POSTTRANSCRIPTIONAL GENE SILENCING

Our second silencing system also involved a GFP reporter and was designed so that we could screen for loss of silencing mutants in *Arabidopsis* (Dalmay et al. 2000).

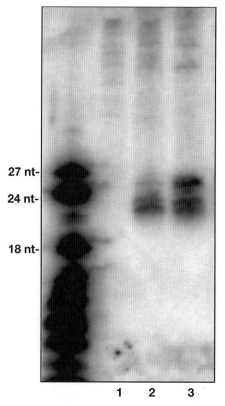

Figure 2. siRNA. The autoradiograph shows a northern blot of GFP siRNA extracted from GFP transgenic *N. benthamiana* that was either not silenced (lane *1*) or silenced (lanes *2* and *3*). The siRNAs migrate in 24-nucleotide and 21/22-nucleotide size classes.

The first characterized locus, formerly *SDE1/SGS2*, is now referred to as *RDR6*, because it encodes one of a family of RNA-dependent RNA polymerase (RDR) homologs. These RDRs are required for RNA silencing in *C. elegans* (Smardon et al. 2000; Sijen et al. 2001) and fungi (Cogoni and Macino 1999; Volpe et al. 2002), as well as in *Arabidopsis* (Dalmay et al. 2000; Mourrain et al. 2000; Xie et al. 2004). They share a common sequence motif that is distantly related to the catalytic domain of DNA-dependent RNA polymerases (Iyer et al. 2003) and it is therefore likely that they are an ancient group of proteins although they do not have homologs in the genomes of mammals and *Drosophila*. Other mutants in our screen implicated a putative RNA helicase SDE3 and a protein of unknown function (SGS2) that had previously been described (Mourrain et al. 2000). The *sde5* locus (T. Dalmay and D.C. Baulcombe, unpubl.) also encodes a protein of unknown function and *sde4* is discussed below.

A likely role of the RDR proteins in RNA silencing is in the synthesis of a dsRNA from an ss template (Fig. 3A). It seems that the RDR is able to use certain RNAs as a template because it somehow recognizes them as being "aberrant." Consistent with this idea, in vitro assays with *Neurospora crassa* (Makeyev and Bamford 2002) and tomato enzymes (Schiebel et al. 1993) indicate de novo synthesis of dsRNA on a ssRNA template that may recapitulate this role of RDR proteins. Similarly, in wheat germ extracts, an ssRNA can be copied into complemen-

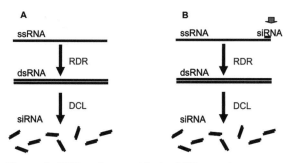

Figure 3. RDR pathways. (*A*) An RDR protein generates dsRNA on a ssRNA template and DCL Dicer produces primary siRNA. (*B*) The RDR proteins may also use the primary siRNA as a primer so that dsRNA and siRNA production is accelerated.

tary RNA by an unidentified enzyme that, presumably, is an RDR (Tang et al. 2002). An indication of how an RNA might be recognized as "aberrant" is from the finding that RNA without a 5′ cap gives enhanced RDR-dependent silencing (Gazzani et al. 2004). Factors that bind the 5′ cap stimulate ribosome loading in concert with factors bound to the polyA tail (Sachs and Varani 2000). In the absence of the interaction between the 5′ and 3′ ends, the RNA would be depleted of ribosomes and perhaps more accessible to the RDR.

The RDR proteins may also have a second mode of action in which RDR-mediated synthesis of dsRNA is primed by an siRNA (Fig. 3B). The resulting dsRNA would then be processed by Dicer into secondary siRNAs that, in turn, are primers for additional cycles of siRNA synthesis. Consistent with this mechanism, the QDE1 RDR protein from *N. crassa* incorporates a labeled 20-nucleotide antisense RNA into the complementary strand of an ssRNA in vitro (Makeyev and Bamford 2002). There is also indirect genetic evidence for primed synthesis from silencing systems in *C. elegans* and *Arabidopsis thaliana* that are initiated by primary siRNAs corresponding to part of the target RNA. The secondary siRNAs that eventually accumulate in these systems do not correspond only to the primary siRNA region; they also correspond to the adjacent regions of the target RNA (Sijen et al. 2001; Vaistij et al. 2002). Production of these secondary siRNAs is dependent on RDR proteins and, in *C. elegans*, they are from the 5′ side of the primary siRNA (relative to the template RNA), as would be predicted by the primer model. In *A. thaliana* and *N. benthamiana* it is likely that the primary siRNAs use both sense and antisense RNAs as a template because the secondary siRNAs are from both sides (Voinnet et al. 1998; Vaistij et al. 2002). There are many antisense transcripts from the *A. thaliana* genome (Yamada et al. 2003) and so there would be ample opportunity for this bidirectional synthesis of secondary siRNAs.

RNA SILENCING AND SUSCEPTIBILITY TO PLANT VIRUSES

Some of the first detailed analyses of RNA silencing were with plants that were carrying viral transgenes (Lindbo et al. 1993; Longstaff et al. 1993). There was co-suppression in these plants of the transgene and the corresponding viral gene so that the plants were resistant against the transgene-specific virus but not against related strains. In some instances, this resistance was so strong that the plants were completely immune and it seemed possible that the transgenic plants were recapitulating a natural resistance mechanism that was activated in virus-infected plants. Subsequently, a series of analyses confirmed this interpretation through the findings that viruses in nontransgenic plants can activate an RNA-mediated and sequence-specific resistance mechanism that was associated with virus-specific siRNAs (Ratcliff et al. 1997; Hamilton and Baulcombe 1999). Also consistent with this finding was the discovery that viruses encode suppressor proteins of the RNA silencing pathway (Anandalakshmi et al. 1998; Brigneti et al. 1998; Kasschau and Carrington 1998). These proteins represent the counterdefense system that would inevitably evolve in response to the defense role of silencing. The link of RNA silencing and virus resistance was further reinforced by the finding that loss of function in RDR1 and RDR6 results in hypersusceptibility to virus infection (Mourrain et al. 2000; Xie et al. 2001).

Our analysis of an RNA virus (potato virus X-PVX) indicates a link between the virus defense role of RNA silencing, the silencing signal, and the role of RDR6. We found that suppression of silencing by the PVX p25 is necessary for cell to cell movement of the virus (Voinnet et al. 2000; E. Bayne and D.C. Baulcombe, unpubl.) and that RDR6 is necessary to prevent invasion of the meristem by the virus (F. Schwach and D.C. Baulcombe, unpubl.). To explain these findings we propose that the RNA signal of silencing can spread through the plant either with or ahead of the virus and that, directly or indirectly, it can silence the viral RNA. In a wild-type plant with a wild-type virus this signal would be normally suppressed but not completely eliminated by the p25 silencing suppressor. The virus would move freely out of the initially infected cell because this low level of signal would be less abundant than the viral RNA. If the p25 suppressor is deleted, however, the signal would be much more abundant and, as observed, the virus would be unable to spread out of the initially infected cells.

Our model further proposes that, in the growing point of the plant including the meristem, the signal blocks virus accumulation through a mechanism that is dependent on RDR6. A plausible scenario for this process requires that the systemic silencing signal is an siRNA. It would enter the meristematic cells with the viral RNA and, using this RNA as template, would prime RDR6-mediated synthesis of secondary siRNAs as described above (Fig. 3B). This secondary siRNA would have an antiviral effect in the meristem because, after several rounds of RDR6-mediated amplification, it would be more abundant than the viral RNA.

A similar model may also account for the observations of S.A. Wingard at the Virginia Experimental Station in 1928. He described infected tobacco plants in which only the initially infected leaves were necrotic and diseased due to tobacco ringspot virus (Wingard 1928). The upper leaves had somehow become specifically immune to the virus and, consequently, were asymptomatic and resistant to secondary infection. We now know that recovery from virus disease involves RNA silencing that is targeted specifically at the viral RNA (Covey et al. 1997; Ratcliff et al. 1997) and, as the immunity is effective in newly developing leaves, it is likely that it is established in the meristem.

In animals the NS1 and E3L proteins of influenza and vaccinia viruses (Li et al. 2004) and the B2 protein of flock house mosaic virus (Li et al. 2002) have silencing suppressor activity. In addition there are five different miRNAs in Epstein–Barr-virus-infected mammalian cells corresponding to inverted repeat regions in the viral genome (Pfeffer et al. 2004). However, this is the only report of siRNA or miRNAs corresponding to mammalian viruses and the silencing suppressor activity of the influenza and vaccinia virus proteins could be a side effect of their dsRNA binding activity (Lichner et al. 2003). Thus, based on current evidence, it seems unlikely that RNA silencing in mammals is a general defense against viruses as it is in plants. Perhaps it is effective against a subset of mammalian viruses or is an antiviral defense in embryonic or other cells in which the systems of innate and humoral immunity are not effective.

RNA-MEDIATED TRANSCRIPTIONAL SILENCING

RNA silencing in plants is often associated with changes to the methylation of histones and DNA at the target gene locus (Mathieu and Bender 2004). The basic mechanism of RNA silencing, as described above, does not explain a link with chromatin/DNA modification. However, it has been known since 1994 that there is the potential for such a link because DNA methylation in plants can be directed by RNA (Wassenegger et al. 1994). The first described example of RNA-directed DNA methylation (RdDM) involved plants carrying a transgene derived from the genome of a noncoding RNA pathogen—a viroid. The transgene DNA was not methylated in noninfected plants but, after viroid inoculation, it was methylated specifically in the regions corresponding to the viroid RNA (Wassenegger et al. 1994).

The link of RdDM and RNA silencing pathway was subsequently confirmed from the demonstration that transcribed inverted repeat transgenes could target DNA methylation and transcriptional silencing if they corresponded to a promoter sequence (Mette et al. 2000; Aufsatz et al. 2002a). The link with RNA silencing correlated with the presence of siRNA derived by Dicer-mediated processing of the foldback transcripts of the inverted repeat RNA. Similarly, when an RNA virus was engineered to carry a transgene promoter sequence as RNA, the transgene was silenced in the infected plant and the promoter was methylated. In this example, the promoter methylation and silencing persisted in the progeny of the infected plant, although the virus was not transmitted between generations (Jones et al. 1999).

Further evidence for a link of RdDM and RNA silenc-

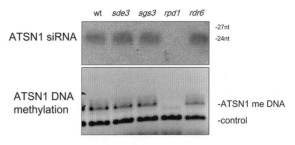

Figure 4. The effect of mutations on endogenous siRNA AtSN1 and DNA methylation. (*Upper panel*) An siRNA northern blot probed for AtSN1. The *nrpd1a(sde4)* mutant lacks this siRNA. (*Lower panel*) Polymerase chain reaction (PCR) analysis of AtSN1 DNA methylation in which the slower migrating PCR product corresponds to the methylated DNA that is absent in the *rpd1(sde4)* mutant.

ing was from analysis of an *Arabidopsis* mutant (*nrpd1a(sde4)*) with a partial loss of transgene RNA silencing phenotype. This mutant also lacked a 24-nucleotide class of endogenous siRNAs including those corresponding to a short interspersed nucleotide element (SINE) AtSN1 (Fig. 4) and an intergenic region of the 5S rDNA repeat (1003) (Hamilton et al. 2002; Herr et al. 2005). Corresponding to this absence of siRNAs, the AtSN1 DNA and the 5S rDNA were hypomethylated in the mutant (Hamilton et al. 2002; Herr et al. 2005) and, at the AtSN1 locus, there is H3K4 rather than H3K9 histone methylation (Zilberman et al. 2004). Other mutants affecting RdDM and siRNA production at AtSN1 include an RNA-dependent RNA polymerase RDR2, a Dicer DCL3, an Argonaute protein AGO4, and a DNA methyltransferase MET1 (Zilberman et al. 2003; Lippman et al. 2004; Xie et al. 2004).

The product of *NRPD1a* is a subunit of putative RNA polymerase (Pol) IV. Pol IV is structurally distinct from the conventional DNA-dependent RNA polymerases I–III and we have proposed that it is a silencing-specific polymerase (Herr et al. 2005). A likely scenario is that Pol IV transcripts are converted into a dsRNA form by RDR2 and that the dsRNA is then cleaved into 24-nucleotide siRNA by Dicer DCL3. By analogy with a similar chromatin silencing mechanism in fission yeast, the role of AGO4 is a likely component of an effector complex that is guided to AtSN1 and other target loci by the 24-nucleotide siRNAs. This effector complex would then mediate the DNA and histone H3K9 methylation. However, AtSN1 RNA levels are affected by *ago4* mutations (Zilberman et al. 2003) that would be expected to be downstream of siRNA production and by *met1* (Lippman et al. 2004). To explain these findings it seems that, as has been proposed for RNA-mediated chromatin silencing in *Saccharomyces pombe* (Sugiyama et al. 2005), there is a self-reinforcing silencing pathway in which pol IV would be dependent on AGO4 and MET1. Correspondingly, AGO4 and MET1 would be dependent on pol IV (Fig. 5A).

At present it is not clear whether this Pol IV/RDR2/DCL3 pathway is the only mechanism for RdDM in plants. Mutant screens for loss of transgene

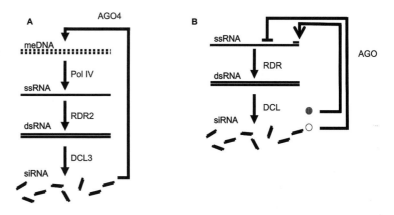

Figure 5. Interactions in silencing pathways. (*A*) A self-reinforcing circular pathway of chromatin silencing in which AGO4 is required for a positive feedback loop that reinforces the basic silencing pathway involving pol IV/RDR2/DCL3-mediated siRNA synthesis. (*B*) An RNA silencing cycle in which the siRNA participates in both negative and positive feedback. The siRNA in a Slicer complex (AGO) degrades the ssRNA and slows down the cycle (O). Alternatively, the siRNA can accelerate production of secondary RNA by priming RDR-mediated synthesis of dsRNA (●).

RdDM have revealed the involvement of a chromatin remodeling protein and a histone deacetylase that are not known to affect AtSN1 and other endogenous siRNAs (Aufsatz et al. 2002b; Kanno et al. 2004). Moreover, these screens have not so far produced mutants in any of the pol IV/RDR2/DCL3/AGO4 proteins described above. The only overlap with the endogenous siRNA-mediated chromatin pathway is DNA methyltransferases (Cao et al. 2003; Aufsatz et al. 2004). It is possible, therefore, that DNA methyltransferases are shared components in several RdDM pathways.

SILENCING PATHWAYS

There are at least three different RNA silencing pathways in plants in which dsRNA is processed into siRNA or miRNA. There are the miRNA and viral RNA pathways in which the targets are RNA that is silenced by turnover and/or an effect at the translational level. The third pathway is targeted to DNA and the end result is DNA methylation and heterochromatin formation (Baulcombe 2004). Equivalent pathways may operate in animals and fungi indicating that the diversity of silencing pathways may predate the evolutionary divergence of plants and animals.

In plants the diversity of silencing pathways is reflected in the existence of multigene families for silencing genes. For example, the Dicer DCL1 is required for miRNA silencing whereas DCL3 is implicated in the nuclear pathway (Xie et al. 2004). Similarly the RNA-dependent RNA polymerases RDR1 and RDR6 influence viral RNA silencing, whereas RDR2 is involved in the chromatin silencing pathway. There are ten AGO homologs encoded in the *A. thaliana* genome and at least some of them are indicators of diversity in silencing pathways. The function of AGO7 in *A. thaliana*, for example, is associated with RDR6 (Hunter et al. 2003), whereas AGO4 and AGO1 are involved in chromatin silencing and miRNA pathways (Zilberman et al. 2003; Vaucheret et al. 2004). It seems likely, by extrapolation, that at least some of the currently unassigned AGO functions could be associated with effector complexes of RNA silencing that are adapted to silence in specialized cells or at particular developmental stages.

These silencing pathways, particularly those with an RDR protein, may form self-reinforcing loops. The transient presence of even a small amount of siRNA in such a pathway could initiate a series of cycles leading eventually to an abundance of siRNAs and strong silencing of the target DNA or RNA (Fig. 3B). In effect, there would be positive feedback that, given the potential for systemic signaling, could lead to a gene being silenced throughout a plant as a result of a localized initiation event.

More complex feedback mechanisms might also operate because ssRNAs are both precursors and targets of siRNA. High levels of the ssRNA, for example, would initially provide an abundant RDR substrate and siRNA levels would increase (Figs. 2 and 4B). The ssRNA would then be targeted by Slicer (Fig. 4B) and its levels would decline. However, as the level of ssRNA decreases, there would be a knockon effect on RDR-mediated siRNA production. Eventually the Slicer activity against the ssRNA would be weakened. The ssRNA levels would then increase and the cycle would start again.

Feedback mechanisms also operate in the miRNA pathways because the mRNA transcripts encoding the Dicer enzyme DCL1 (Xie et al. 2003) and the AGO1 Slicer (Vaucheret et al. 2004) are themselves targets of miRNAs (miR162 and miR168, respectively). The resulting feedback mechanisms might explain the otherwise paradoxical increase in miRNAs in the presence of viral suppressors of silencing (Mallory et al. 2002; Chapman et al. 2004; Dunoyer et al. 2004): The suppression of silencing would uncouple the feedback loop so that the abundance of AGO1, DCL1, and the associated miRNAs would be unchecked by the normal mechanisms.

Further complexity may result from the potential of silencing pathways to interact. In some instances, as illustrated by a GFP transgene silencing phenotype in *Arabidopsis* inflorescences, the interaction is cooperative: GFP silencing requires the combined action of the RDR6 and the pol IV/RDR2 chromatin silencing pathways (Herr et al. 2005). There may also be an interaction between the miRNA and chromatin silencing pathways because the methylation of *phabulosa* DNA is reduced if there is a mutation in a miRNA target sequence (miR167) (Bao et al. 2004).

There is also the potential for antagonism between silencing pathways. It is possible, for example, to ablate the posttranscriptional silencing capability of a gene by targeting transcriptional silencing to a promoter (Hamilton and Baulcombe 1999). This type of antagonism between transcriptional and posttranscriptional RNA silencing may explain the lateral shoots with pigmented flowers that sometimes appear on CS-cosuppressed plants (Jorgensen et al. 1995): It could be that the CS silencer transgene has become transcriptionally silenced and so has lost the capability to produce siRNAs that target the endogenous CS RNA.

The biological consequence of these various interactions will vary depending on the nature of the silencing pathway and possibly on the target sequence. An RNA or DNA target that is more accessible to an siRNA or miRNA, for example, will contribute to feedback loops more actively than an inaccessible target. To unravel the significance of the silencing pathway interactions requires better understanding of RNA silencing mechanisms and of their biological roles. Until now, the only work on the biological role of silencing pathways in plants has involved virus susceptibility, miRNAs affecting gene expression in growth and development (Bartel 2004), and some recent analyses of *trans*-acting siRNAs (Peragine et al. 2004; Vazquez et al. 2004).

ACKNOWLEDGMENTS

We are grateful to the Gatsby Charitable Foundation and the Biotechnology and Biological Sciences Research Council for supporting research in my laboratory. A.J.H. acknowledges a fellowship from the Burroughs Wellcome Fund.

REFERENCES

Allen E., Xie Z., Gustafson A.M., Sung G.-H., Spatafora J.W., and Carrington J.C. 2004. Evolution of microRNA genes by inverted duplication of target gene sequences in *Arabidopsis thaliana*. *Nat. Genet.* **36:** 1282.

Anandalakshmi R., Pruss G.J., Ge X., Marathe R., Smith T.H., and Vance V.B. 1998. A viral suppressor of gene silencing in plants. *Proc. Natl. Acad. Sci.* **95:** 13079.

Aufsatz W., Mette M.F., Matzke A.J.M., and Matzke M. 2004. The role of MET1 in RNA-directed de novo and maintenance methylation of CG dinucleotides. *Plant Mol. Biol.* **54:** 793.

Aufsatz W., Mette M.F., van der Winden J., Matzke A.J.M., and Matzke M. 2002a. RNA-directed DNA methylation in *Arabidopsis*. *Proc. Natl. Acad. Sci.* **99:** 16499.

Aufsatz W., Mette M.F., van der Winden J., Matzke M., and Matzke A.J.M. 2002b. HDA6, a putative histone deacetylase needed to enhance DNA methylation induced by double-stranded RNA. *EMBO J.* **21:** 6832.

Bao N., Lye K.-W., and Barton M.K. 2004. MicroRNA binding sites in *Arabidopsis* class III HD-ZIP mRNAs are required for methylation of the template chromosome. *Dev. Cell* **7:** 653.

Bartel D.P. 2004. MicroRNAs: Genomics, biogenesis, mechanism, and function. *Cell* **116:** 281.

Baulcombe D. 2004. RNA silencing in plants. *Nature* **431:** 356.

Brigneti G., Voinnet O., Li W.X., Ji L.H., Ding S.W., and Baulcombe D.C. 1998. Viral pathogenicity determinants are suppressors of transgene silencing in *Nicotiana benthamiana*. *EMBO J.* **17:** 6739.

Cao X., Aufsatz W., Zilberman D., Mette M.F., Huang M.S., Matzke M., and Jacobsen S.E. 2003. Role of the *DRM* and *CMT3* methyltransferases in RNA-directed DNA methylation. *Curr. Biol.* **13:** 2212.

Chapman E.J., Prokhnevsky A.I., Gopinath K., Dolja V., and Carrington J.C. 2004. Viral RNA silencing suppressors inhibit the microRNA pathway at an intermediate step. *Genes Dev.* **18:** 1179.

Cogoni C. and Macino G. 1999. Gene silencing in *Neurospora crassa* requires a protein homologous to RNA-dependent RNA polymerase. *Nature* **399:** 166.

Covey S.N., Al-Kaff N.S., Langara A., and Turner D.S. 1997. Plants combat infection by gene silencing. *Nature* **385:** 781.

Dalmay T., Hamilton A.J., Rudd S., Angell S., and Baulcombe D.C. 2000. An RNA-dependent RNA polymerase gene in *Arabidopsis* is required for posttranscriptional gene silencing mediated by a transgene but not by a virus. *Cell* **101:** 543.

Dunoyer P., Lecellier C.H., Parizotto E.A., Himber C., and Voinnet O. 2004. Probing the microRNA and small interfering RNA pathways with virus-encoded suppressors of RNA silencing. *Plant Cell* **16:** 1235.

Feinberg E.H. and Hunter C.P. 2003. Transport of dsRNA into cells by the transmembrane protein SID-1. *Science* **301:** 1545.

Gazzani S., Lawrenson T., Woodward C., Headon D., and Sablowski R. 2004. A link between mRNA turnover and RNA interference in *Arabidopsis*. *Science* **306:** 1046.

Hamilton A.J. and Baulcombe D.C. 1999. A species of small antisense RNA in post-transcriptional gene silencing in plants. *Science* **286:** 950.

Hamilton A.J., Voinnet O., Chappell L., and Baulcombe D.C. 2002. Two classes of short interfering RNA in RNA silencing. *EMBO J.* **21:** 4671.

Haywood V., Kragler F., and Lucas W.J. 2002. Plasmodesmata: Pathways for protein and ribonucleoprotein signaling. *Plant Cell* **14:** S303.

Herr A.J., Jensen M.B., Dalmay T., and Baulcombe D.C. 2005. RNA polymerase IV directs silencing of endogenous DNA. *Science* (in press).

Himber C., Dunoyer P., Moissiard G., Ritzenthaler C., and Voinnet O. 2003. Transitivity-dependent and -independent cell-to-cell movement of RNA silencing. *EMBO J.* **22:** 4523.

Hunter C., Sun H., and Poethig R.S. 2003. The *Arabidopsis* heterochronic gene ZIPPY is an *ARGONAUTE* family member. *Curr. Biol.* **13:** 1734.

Iyer L.M., Koonin E.V., and Aravind L. 2003. Evolutionary connection between the catalytic subunits of DNA-dependent RNA polymerases and eukaryotic RNA-dependent RNA polymerases and the origin of RNA polymerases. *BMC Struct. Biol.* **3:** 1. (http://www.biomedcentral.com/1472-6807/3/1)

Jones L., Hamilton A.J., Voinnet O., Thomas C.L., Maule A.J., and Baulcombe D.C. 1999. RNA-DNA interactions and DNA methylation in post-transcriptional gene silencing. *Plant Cell* **11:** 2291.

Jones-Rhoades M.W. and Bartel D.P. 2004. Computational identification of plant MicroRNAs and their targets, including a stress-induced miRNA. *Mol. Cell* **14:** 787.

Jorgensen R.A., Que Q.D., English J.J., Cluster P., and Napoli C. 1995. Sense-suppression of flower color genes as a sensitive reporter of epigenetic states of gene-expression in plant development. *Plant Physiol.* **108:** 14.

Kanno T., Mette F., Kreil D.P., Aufsatz W., Matzke A.J.M., and Matzke M. 2004. Involvement of putative SNF2 chromatin remodelling protein DRD1 in RNA-directred DNA methylation. *Curr. Biol.* **14:** 801.

Kasschau K.D. and Carrington J.C. 1998. A counterdefensive strategy of plant viruses: Suppression of post-transcriptional gene silencing. *Cell* **95:** 461.

Li H., Li W.X., and Ding S.W. 2002. Induction and suppression of RNA silencing by an animal virus. *Science* **296:** 1319.

Li W.-X., Li H., Lu R., Li F., Dus M., Atkinson P., Brydon E.W.A., Johnson K.L., Garcia-Sastre A., Ball L.A., Palease P., and Ding S.-W. 2004. Interferon antagonist proteins of influenza and vaccinia virus are suppressors of RNA silencing. *Proc. Natl. Acad. Sci.* **101:** 1350.

Lichner Z., Silhavy D., and Burgyan J. 2003. Double-stranded

RNA-binding proteins could suppress RNA interference-mediated antiviral defences. *J. Gen. Virol.* **84:** 975.

Lindbo J.A., Silva-Rosales L., Proebsting W.M., and Dougherty W.G. 1993. Induction of a highly specific antiviral state in transgenic plants: Implications for regulation of gene expression and virus resistance. *Plant Cell* **5:** 1749.

Lippman Z., May B., Yordan C., Singer T., and Martienssen R. 2004. Distinct mechanisms determine transposon inheritance and methylation via small interfering RNA and histone modification. *PLoS Biol.* **1:** 420.

Liu J., Carmell M.A., Rivas F.V., Marsden C.G., Thomson M., Song J.J., Hammond S.M., Joshua-Tor L., and Hannon G.J. 2004. Argonaute2 is the catalytic engine of mammalian RNAi. *Science* **305:** 1437.

Longstaff M., Brigneti G., Boccard F., Chapman S.N., and Baulcombe D.C. 1993. Extreme resistance to potato virus X infection in plants expressing a modified component of the putative viral replicase. *EMBO J.* **12:** 379.

Ma J.B., Ye K.Q., and Patel D.J. 2004. Structural basis for overhang-specific small interfering RNA recognition by the PAZ domain. *Nature* **429:** 318.

Makeyev E.V. and Bamford D.H. 2002. Cellular RNA-dependent RNA polymerase involved in posttranscriptional gene silencing has two distinct activity modes. *Mol. Cell* **10:** 1417.

Mallory A.C., Reinhart B.J., Bartel D., Vance V.B., and Bowman L.H. 2002. A viral suppressor of RNA silencing differentially regulates the accumulation of short interfering RNAs and microRNAs in tobacco. *Proc. Natl. Acad. Sci.* **99:** 15228.

Mathieu O. and Bender J. 2004. RNA-directed DNA methylation. *J. Cell Sci.* **117:** 4881.

Meister G., Landthaler M., Patkaniowska A., Dorsett Y., Teng G., and Tuschl T. 2004. Human Argonaute2 mediates RNA cleavage targeted by miRNAs and siRNAs. *Mol. Cell* **15:** 185.

Mette M.F., Aufsatz W., van der Winden J., Matzke M.A., and Matzke A.J.M. 2000. Transcriptional silencing and promoter methylation triggered by double-stranded RNA. *EMBO J.* **19:** 5194.

Mourrain P., Beclin C., Elmayan T., Feuerbach F., Godon C., Morel J.-B., Jouette D., Lacombe A.M., Nikic S., Picault N., Remoue K., Sanial M., Vo T.A., and Vaucheret H. 2000. *Arabidopsis SGS2* and *SGS3* genes are required for posttranscriptional gene silencing and natural virus resistance. *Cell* **101:** 533.

Napoli C., Lemieux C., and Jorgensen R.A. 1990. Introduction of a chimeric chalcone synthase gene into petunia results in reversible co-suppression of homologous genes *in trans*. *Plant Cell* **2:** 279.

Palauqui J.C., Elmayan T., Pollien J.M., and Vaucheret H. 1997. Systemic acquired silencing: Transgene-specific post-transcriptional silencing is transmitted by grafting from silenced stocks to non-silenced scions. *EMBO J.* **16:** 4738.

Peragine A., Yoshikawa M., Wu G., Albrecht H.L., and Poethig R.S. 2004. SGS3 and SGS2/SDE1/RDR6 are required for juvenile development and the production of trans-acting siRNAs in *Arabidopsis*. *Genes Dev.* **18:** 2368.

Pfeffer S., Zavolan M., Grasser F.A., Chien M., Russo J.J., Ju J., John B., Enright A.J., Marks D.S., Sander C., and Tuschl T. 2004. Identification of virus-encoded MicroRNAs. *Science* **304:** 734.

Rand T.A., Ginalski K., Grishin N.V., and Wang X. 2004. Biochemical identification of Argonaute 2 as the sole protein required for RNA-induced silencing complex activity. *Proc. Natl. Acad. Sci.* **101:** 14385.

Ratcliff F., Harrison B.D., and Baulcombe D.C. 1997. A similarity between viral defense and gene silencing in plants. *Science* **276:** 1558.

Ruiz M.T., Voinnet O., and Baulcombe D.C. 1998. Initiation and maintenance of virus-induced gene silencing. *Plant Cell* **10:** 937.

Sachs A.B. and Varani G. 2000. Eukaryotic translation initiation: There are (at least) two sides to every story. *Nat. Struct. Biol.* **7:** 356.

Schiebel W., Haas B., Marinkovic S., Klanner A., and Sanger H.L. 1993. RNA-directed RNA polymerase from tomato leaves. II. Catalytic *in vitro* properties. *J. Biol. Chem.* **268:** 11858.

Sijen T., Fleenor J., Simmer F., Thijssen K.L., Parrish S., Timmons L., Plasterk R.H.A., and Fire A. 2001. On the role of RNA amplification in dsRNA-triggered gene silencing. *Cell* **107:** 465.

Smardon A., Spoerke J.M., Stacey S.C., Klein M.E., Mackin N., and Maine E.M. 2000. EGO-1 is related to RNA-directed RNA polymerase and functions in germ-line development and RNA interference in *C. elegans*. *Curr. Biol.* **10:** 169.

Song J.J., Liu J., Tolia N.H., Schneiderman J., Smith S.K., Martienssen R.A., Hannon G.J., and Joshua-Tor L. 2003. The crystal structure of the Argonaute2 PAZ domain reveals an RNA binding motif in RNAi effector complexes. *Nat. Struct. Biol.* **10:** 1026.

Stam M., de Bruin R., van Blokland R., van der Hoorn R.A.L., Mol J.N.M., and Kooter J.M. 2000. Distinct features of posttranscriptional gene silencing by antisense transgenes in single copy and inverted T-DNA repeat loci. *Plant J.* **21:** 27.

Sugiyama T., Cam H., Verdel A., Moazed D., and Grewal S.I.S. 2005. RNA-dependent RNA polymerase is an essential component of a self-enforcing loop coupling heterochromatin assembly to siRNA production. *Proc. Natl. Acad. Sci.* **102:** 152.

Tang G., Reinhart B.J., Bartel D., and Zamore P.D. 2002. A biochemical framework for RNA silencing in plants. *Genes Dev.* **17:** 49.

Vaistij F.E., Jones L., and Baulcombe D.C. 2002. Spreading of RNA targeting and DNA methylation in RNA silencing requires transcription of the target gene and a putative RNA-dependent RNA polymerase. *Plant Cell* **14:** 857.

van Blokland R., Van der Geest N., Mol J.N.M., and Kooter J.M. 1994. Transgene-mediated suppression of chalcone synthase expression in *Petunia hybrida* results from an increase in RNA turnover. *Plant J.* **6:** 861.

van der Krol A.R., Mur L.A., Beld M., Mol J.N.M., and Stuitje A.R. 1990. Flavonoid genes in petunia: Addition of a limited number of gene copies may lead to a suppression of gene expression. *Plant Cell* **2:** 291.

Vaucheret H., Vazquez F., Crete P., and Bartel D.P. 2004. The action of ARGONAUTE1 in the miRNA pathway and its regulation by the miRNA pathway are crucial for plant development. *Genes Dev.* **18:** 1187.

Vazquez F., Vaucheret H., Rajagopalan R., Lepers C., Gasciolli V., Mallory A.C., Hilbert J.-L., Bartel D.P., and Crete P. 2004. Endogenous *trans*-acting siRNAs regulate the accumulation of *Arabidopsis* mRNAs. *Mol. Cell* **16:** 69.

Voinnet O. and Baulcombe D.C. 1997. Systemic signalling in gene silencing. *Nature* **389:** 553.

Voinnet O., Lederer C., and Baulcombe D.C. 2000. A viral movement protein prevents spread of the gene silencing signal in *Nicotiana benthamiana*. *Cell* **103:** 157.

Voinnet O., Vain P., Angell S., and Baulcombe D.C. 1998. Systemic spread of sequence-specific transgene RNA degradation is initiated by localised introduction of ectopic promoterless DNA. *Cell* **95:** 177.

Volpe T., Kidner C., Hall I.M., Teng G., Grewal S.I.S., and Martienssen R. 2002. Regulation of heterochromatic silencing and histone H3 lysine-9 methylation by RNAi. *Science* **297:** 1833.

Wassenegger M., Heimes S., Riedel L., and Sanger H.L. 1994. RNA-directed de novo methylation of genomic sequences in plants. *Cell* **76:** 567.

Wingard S.A. 1928. Hosts and symptoms of ring spot, a virus disease of plants. *J. Agric. Res.* **37:** 127.

Xie Z., Kasschau K.D., and Carrington J.C. 2003. Negative feedback regulation of *Dicer-Like1* in *Arabidopsis* by microRNA-guided mRNA degradation. *Curr. Biol.* **13:** 784.

Xie Z., Fan B., Chen C.H., and Chen Z. 2001. An important role of an inducible RNA-dependent RNA polymerase in plant antiviral defense. *Proc. Natl. Acad. Sci.* **98:** 6516.

Xie Z., Johansen L.K., Gustafson A.M., Kasschau K.D., Lellis A.D., Zilberman D., Jacobsen S.E., and Carrington J.C. 2004. Genetic and functional diversification of small RNA pathways in plants. *PLoS Biol.* **2:** E104.

Yamada K., Lim J., Dale J.M., Chen H., Shinn P., Palm C.J.,

Southwick A.M., Wu H.C., Kim C., Nguyen M., Pham P., Cheuk R., Karlin-Newmann G., Liu S.X., Lam B., Sakano H., Wu T., Yu G., Miranda M., Quach H.L., Tripp M., Chang C.H., Lee J.M., Toriumi M., Chan M.M., et al. 2003. Empirical analysis of transcriptional activity in the *Arabidopsis* genome. *Science* **302:** 842.

Yan K.S., Yan S., Farooq A., Han A., Zeng L., and Zhou M.M. 2003. Structure and conserved RNA binding of the PAZ domain. *Nature* **426:** 468.

Zilberman D., Cao X., and Jacobsen S.E. 2003. *ARGONAUTE4* control of locus specific siRNA accumulation and DNA and histone methylation. *Science* **299:** 716.

Zilberman D., Cao X., Johansen L.K., Xie Z., Carrington J.C., and Jacobsen S.E. 2004. Role of *Arabidopsis ARGONAUTE4* in RNA-directed DNA methylation triggered by inverted repeats. *Curr. Biol.* **14:** 1214.

Transposons, Tandem Repeats, and the Silencing of Imprinted Genes

R. Martienssen,* Z. Lippman,* B. May, M. Ronemus, and M. Vaughn

*Cold Spring Harbor Laboratory and *Watson School of Biological Sciences, Cold Spring Harbor, New York 11724*

Most eukaryotic cells cleave double-stranded RNA (dsRNA) into small 21–24-bp RNA fragments via the RNAse III helicase Dicer (Hannon 2002). These small interfering RNA (siRNA) are unwound and loaded onto the PAZ domain of Argonaute proteins, where they guide the RNAseH-related PIWI domain to cleave single-stranded RNA complementary to the small RNA (Song et al. 2003, 2004; Liu et al. 2004). Cleavage sites can initiate second-strand RNA synthesis via RNA-dependent RNA polymerase (RdRP) (Makeyev and Bamford 2002). Further processing of this dsRNA by Dicer leads to further (secondary) siRNA, which can reinitiate RdRP synthesis and thus amplify the cycle. RdRP can also incorporate siRNA directly into dsRNA by primer extension (Makeyev and Bamford 2002).

In the fission yeast *Saccharomyces pombe*, centromeric repeats are transcribed into dsRNA, which is subject to cleavage into siRNA and resynthesis, by Dicer, Argonaute, and RdRP, respectively (Volpe et al. 2002). However, while one strand of the outer repeats is continually transcribed, the other strand is transcriptionally silenced, as are reporter genes integrated within the repeats. This transcriptional silencing depends on the HP-1 homolog Swi6, which is recruited to the repeats by histone H3 lysine-9 dimethylation (H3mK9) via a chromodomain (Bannister et al. 2001). Histone modification of the repeats themselves depends in part on Dicer (Dcr1[+]) and on RdRP (Rdp[+]), and to a lesser extent on Argonaute (Ago1[+]), while histone modification of the reporter genes depends strongly on all three genes (Volpe et al. 2002). The accumulation of siRNA also depends on the histone lysine 9 methyltransferase Clr4[+] (Schramke and Allshire 2003), indicating a "reverse pathway" from histone modification to RNAi. Ago1[+] is found in the RITS (RNA-induced initiation of transcriptional gene silencing) complex, along with Chp1[+]+ (a chromodomain protein) and Tas3[+], as well as siRNA from heterochromatic repeats (Verdel et al. 2004). This complex was initially proposed to mediate initiation of transcriptional silencing via RNA interference (RNAi) (hence its name) but recent evidence suggests it may have a prominent role in spreading of silencing to reporter genes integrated nearby (Sigova et al. 2004), consistent with the substantial loss of H3mK9 from the reporter genes (Volpe et al. 2002).

Argonaute is conserved in the higher plant, *Arabidopsis*, but the other members of the RITS complex, Chp1[+] and Tas3[+], have no close homologs in plants or generally in eukaryotes, respectively. siRNA, on the other hand, has been found corresponding to *Arabidopsis* centromeric repeats and transposons (Llave et al. 2002; Xie et al. 2004), both of which are also associated with histone H3 lysine 9 dimethylation (Gendrel et al. 2002; Johnson et al. 2002; Soppe et al. 2002; Lippman et al. 2004; Tariq and Paszkowski 2004). Unlike in yeast, cytosines are methylated in *Arabidopsis* transposons, both in CG and CXG di- and trinucleotides, as well as other noncanonical contexts. This methylation may mask some of the effects of RNAi on transcriptional silencing. Further, Dicer, Argonaute, and RdRP are encoded by multiple *Arabidopsis* genes, as are SET (Suppressor of variegation, Enhancer of Zeste, Trithorax) domain proteins (Lippman and Martienssen 2004). Even so, histone and DNA methylation are lost from a handful of transposons in *ago1* and *ago4*, in which transposon siRNA is reduced, and in *dcl3* and *rdr2*, in which siRNA is absent (Lippman et al. 2003; Zilberman et al. 2003; Xie et al. 2004).

Unlike in fission yeast, *Arabidopsis* siRNA are maintained in the absence of histone H3 lysine 9 dimethylation mediated by the methyltransferase KYP/SUVH4 (Lippman et al. 2003), reflecting either redundancy with other methyltransferases or the absence of RITS. However, some siRNA are lost in *met1*, which encodes a homolog of the mammalian CG methyltransferase dnmt1, and in *ddm1*, which encodes a homolog of the mammalian SWI/SNF ATPase Lsh1 (Jeddeloh et al. 1999), both of which also lose histone H3mK9 (Gendrel et al. 2002; Johnson et al. 2002; Soppe et al. 2002; Lippman et al. 2003; Tariq et al. 2003). These results are consistent with the presence of a complex containing MET1, DDM1, and siRNA (Lippman et al. 2003). Some siRNA are lost in the dnmt3 double mutants *drm1 drm2*, indicating that multiple pathways may include transposon siRNA (Lippman and Martienssen 2004; Matzke et al. 2004).

Here we review the role of siRNA and heterochromatin in transcriptional silencing of transposons and nearby genes in *Arabidopsis*. Transposons containing tandem repeats have more prevalent siRNA than other transposons and may contribute to silencing that depends on such repeats. The imprinted genes *MEDEA* and *FWA* are flanked by transposons and tandem repeats, and corresponding siRNA may be involved in silencing. A similar role for tandem repeats in mammalian cells could contribute to imprinting as well as silencing mediated by triplet repeats.

SILENCING OF ROBERTSON'S MUTATOR TRANSPOSONS IN *ARABIDOPSIS*

Most transposable elements (TEs) in *Arabidopsis* remain silent and methylated in the RNA interference mutants *dcl3*, *rdr2*, *ago1*, and *ago4*, as well as in the histone H3 lysine-9 methyltransferase mutant *kyp*, indicating that silencing can be maintained in the absence of RNAi (Lippman and Martienssen 2004; Xie et al. 2004). An exception is AtSN1, a SINE element that has three- to eightfold elevated expression levels and loses ~50% of associated histone H3 methylated on lysine-9 in mutants in RNAi (Xie et al. 2004). However, most TEs do not exhibit even these modest changes. The Robertson's Mutator (*Mu*) elements are found in *Arabidopsis* as they are in maize, and full-length elements, capable of autonomous transposition, have been found at low copy numbers in many different plant genomes. An *Arabidopsis Mu*, *AtMu1*, is located on chromosome 4L, just distal to the pericentromeric heterochromatin, where it is associated with both histone H3 K9 methylation and histone H3 K4 methylation (Fig. 1). H3K9 methylation is lost in *met1* and *ddm1* mutants, while H3K4 methylation is enriched. These histone modifications remain unchanged in the chromodomain methyltransferase mutant *cmt3*, which impacts CXG and non-CG methylation (Fig. 1) (Lippman et al. 2003). DNA methylation at *AtMu1* was measured using semiquantitative polymerase chain reaction (PCR) following digestion with the methylation-dependent restriction enzyme mcrBC, and was reduced in each of the mutants (Fig. 1B).

AtMu1 germinal transposition frequencies were elevated more than 20-fold in *ddm1*, and transposed elements were found all over the genome (Singer et al. 2001). In contrast, transposition frequencies were elevated only two- to threefold in *met1* (data not shown) despite loss of histone H3 lysine 9 methylation and CG DNA methylation, and a substantial increase in transposase gene expression (Fig. 1B). In *cmt3 met1* double mutants, however, transposition could be readily detected on Southern blots (Fig. 1C). Transposition of the CACTA elements *CAC1* and *CAC2* is similarly regulated (Kato et al. 2003). One explanation is that *met1* and *cmt3* act in distinct pathways to regulate transposons, perhaps reflecting distinct methylation-sensitive DNA-binding sites in *AtMu1* for transposase on the one hand and the transcriptional machinery on the other. In this way MET1 might regulate silencing, while CMT3 regulates transposition. Transcripts are only slightly elevated in *cmt3* relative to *met1*, supporting this idea (Lippman et al. 2003).

Levels of *AtMu1* RNA are also slightly elevated in *ago1*, which encodes one of ten argonaute homologs in *Arabidopsis* and is likely to bind siRNA (Song et al. 2003). We therefore examined levels of siRNA corresponding to *AtMu1* in wild type (WT) as well as in mutant strains defective in RNAi, histone, and DNA modification (Lippman et al. 2003). We found low but detectable levels of siRNA in WT L*er* strains, as well as in *ago1*, the histone deacetylase *sil1*, *cmt3*, and *kyp* (Lippman et al. 2003), but they failed to accumulate in the dicer mutant *dcl3*, as well as in the RdRP mutant *rdr2* (Fig. 1D). These mutants fail to accumulate siRNA from most

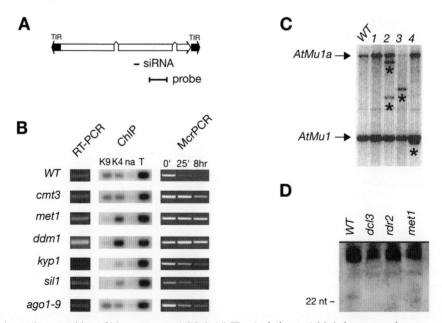

Figure 1. Silencing and transposition of Mu transposon AtMu1. (*A*) The *Arabidopsis AtMu1* element on chromosome 4 has long terminal inverted repeats (TIRs), but cloned small interfering RNA (siRNA) corresponds to the transposase gene itself. Adjacent probes (shown) were used to detect siRNA on northern blots. (*B*) AtMu1 transcripts (RT-PCR [reverse transcriptase polymerase chain reaction]), associated histone modifications (ChIP [chromatin immunoprecipitation]), and DNA methylation (McrPCR) were detected by PCR amplification using primers from the region shown in *A*. A time course of predigestion with mcrBC is shown. (*C*) Transposed AtMu1 elements could be detected only at low frequency in wild type (WT) (1%), *met1* (6%), and *cmt3* (2%), but in high frequency (25%) in *cmt3 met1* double mutants (lanes *1–4*). Loss of a band (lane *3*) most likely represents excision. (*D*) siRNA from AtMu1 could be detected with the probe shown in *A* in WT, but not in *dcl3*, *rdr2*, or *met1* mutant plants.

if not all other transposons, indicating they are specific for 24-nt heterochromatic siRNA (Xie et al. 2004). Interestingly, *AtMu1* siRNA also failed to accumulate in *met1* and *ddm1* both in Columbia (Fig. 1) and in the Landsberg background (Lippman et al. 2003). Many, though not all heterochromatic siRNA, are lost in *met1* (Lippman et al. 2004). One possibility is that RNAi of these sequences depends on heterochromatic modification (Lippman and Martienssen 2004). Alternatively, MET1 might form a complex with some siRNA to guide it to specific sequences (Lippman et al. 2003).

No changes in histone modification were detectable in *ago1* using probes from the terminal inverted repeat (TIR) (Lippman et al. 2003). However, DNA methylation was detectably reduced in these same repeats (Fig. 1) indicating a weak role for RNAi in their modification (Lippman et al. 2003). Comparable levels of demethylation (loss of one methylated cytosine out of three tested) has been reported for *AtMu1* TIR in *ago4* (Zilberman et al. 2003), raising the possibility that AGO1 and AGO4 might have partially overlapping functions.

Two closely related Mutator elements, *AtMu1a* and *AtMu1b*, are found on chromosomes 5L and 1L, respectively, in the Columbia background (Singer et al. 2001). Both elements are heavily methylated and do not transpose. In Landsberg *erecta*, *AtMu1a* has transposed to the short arm of chromosome 1 and is transcribed but transposes at a very low frequency (data not shown).

TANDEM REPEATS

In addition to transposons, heterochromatic tandem repeats in *Arabidopsis* are found in pericentromeric domains, nucleolar organizers, and in the interstitial knobs on chromosomes 4 and 5 (*Arabidopsis* Genome Initiative 2000; CSHL/WUGSC/PEB *Arabidopsis* Sequencing Consortium 2000; Tabata et al. 2000). The interstitial knob repeats on chromosomes 4 and 5 are derived from CACTA and Mutator class transposons, respectively, indicating that the arrangement of the repeats, in addition to their sequence, gives rise to heterochromatic modification.

We examined the correlation between tandem repeats, transposons, and small RNAs in the *Arabidopsis* genome (Fig. 2a). Tandem repeats were discovered using "Tandem Repeats Finder" (Benson 1999) and transposon-like repeats were annotated using CENSOR/Repbase (Jurka et al. 1996). Sequence homologies to siRNA were determined by BLASTN analysis of *Arabidopsis* Small RNA Project database sequences cloned from WT plants (Xie et al. 2004). We excluded microRNAs, which are derived

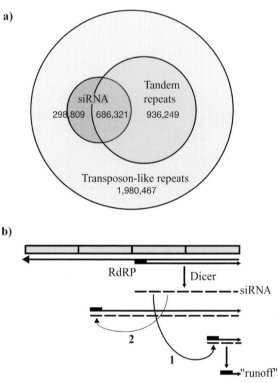

Figure 2. Tandem repeats and small interfering RNA (siRNA). (*a*) Tandem repeats significantly influence the abundance of siRNA in *Arabidopsis*. Approximately one third of transposon-like repeat sequences contained tandem repeats, and one third had siRNA homology by BLASTN. Of the sequences matching siRNA, two thirds were found in conjunction with tandem repeats, suggesting that the presence of tandem repeats within a given transposon enhances the generation of small RNAs. (*b*) Long tandem repeats could maintain populations of small RNA if they are transcribed from one strand. This is because siRNA can initiate multiple rounds of second-strand synthesis both upstream and downstream of their origin in the sequence (*arrow 2*). Short tandem arrays also generate siRNA, but these lead to longer "runoff" products (*arrow 1*) when they initiate second-strand synthesis downstream that are too short for Dicer digestion (Martienssen 2003). Initiation of second-strand synthesis can occur either by primer extension or by cleavage by Argonaute, followed by initiation at the cleavage site. Both depend on siRNA.

from hairpin precursor genes (Hannon 2002). Each transposon-like repeat in a 42-Mb sample of euchromatic sequence was assayed by bioinformatic means for presence of siRNA homology and correspondence to annotated tandem repeats. Results were tabulated and expressed as total base pair lengths per logical category. Transposon-like repeats accounted for 2.92 Mb of the sampled sequence (7%). Of this transposon-like sequence, almost a third (0.94 Mb) included tandem repeats. Randomly sequenced siRNA corresponded to 73% of transposon sequences containing tandem repeats, but only 15% of transposon sequences lacking them. This strongly implicates tandem repeats in influencing the abundance of siRNA.

When repeats are arranged in tandem, secondary siRNA generated by Dicer can match repeats upstream as well as downstream of the primary siRNA used to initiate second-strand synthesis by RdRP (Fig. 2b). In this way, iterative rounds of siRNA-dependent RdRP synthesis can be maintained along with the population of siRNA (Martienssen 2003). Dispersed repeats that match an siRNA will not have this property. Instead, with each round of RdRP and Dicer activity, secondary siRNA, derived from dsRNA, will always lie downstream of the primary siRNA in each round of second-strand synthesis. The RdRP-generated transcripts will become shorter and shorter. Finally, RdRP will generate products that are no longer processed by Dicer, and these will accumulate instead of the siRNA (Martienssen 2003). The ability of a tandemly repeated array to reiterate second-strand synthesis in this way is expected to depend on the number of repeats, so that longer arrays will be more stably silenced than shorter ones (Fig. 2B).

SILENCING IMPRINTED GENES BY TRANSPOSONS AND TANDEM REPEATS

Chromosomal imprints are imposed early in development, but inherited through multiple mitotic divisions long after the imprinting signal has gone (Martienssen 1998a; Alleman and Doctor 2000). When imprinting occurs in one germ line but not the other, monoallelic gene expression results following fertilization. Imprinting occurs in seed plants and placental mammals and is thought to mediate conflicting parental interests in viviparous embryonic growth, as well as proliferation of extraembryonic tissues—namely, the placenta and the endosperm (Alleman and Doctor 2000; Sleutels and Barlow 2002). Chromosomal imprinting involves epigenetic modifications, such as histone H3 lysine-9 methylation, as well as DNA methylation in plants and mammals (Adams et al. 2000; Sleutels and Barlow 2002). However, imprinting was first discovered in insects (Mohan et al. 2002), and related phenomena occur in *Drosophila* (Joanis and Lloyd 2002), which do not bear live offspring in this way, implicating a more ancient function.

In plants, imprinting is mostly, or entirely, limited to the endosperm, which is a terminally differentiated tissue, and so (unlike in animals) imprinting does not need to be reset during meiosis (Alleman and Doctor 2000). One example is the *FWA* gene on chromosome 4. This gene encodes a homeodomain protein that is mono-allelically expressed in the endosperm from the maternal chromosome (Kinoshita et al. 2004). In the body of the plant, *FWA* is silenced by MET1 and DDM1, and silencing is associated with both DNA methylation and histone H3 lysine 9 methylation (Soppe et al. 2000; Johnson et al. 2002; Lippman et al. 2004). In *met1* mutants, ectopic expression of *FWA* delays flowering, and late flowering epialleles of *FWA* are subsequently inherited independently of *met1* (Soppe et al. 2000).

Reannotation of the *FWA* tandem repeats has revealed that they are part of a short interspersed nucleotide element (SINE) and were likely generated by reiterated integration of the transposon (Lippman et al. 2004). The SINE provides the *FWA* promoter and is therefore responsible for silencing. siRNA corresponding to the SINE could be detected in WT and *ddm1*, but not in *met1*. This could account for the high frequency of heritable epialleles of *FWA* that arise in *met1*, if silencing cannot be reestablished in the absence of siRNA. Consistent with these results, transgenic *FWA* is rapidly silenced in WT plants, but not in the RNAi mutants *dcl3*, *rdr2*, or *ago4*, resulting in late flowering phenotypes (Chan et al. 2004; Xie et al.. 2004). Presumably, these mutants have lost the siRNA.

The *MEDEA* locus on chromosome 1 is also imprinted in the endosperm, such that expression is limited to the central cell in the maternal gametophyte (before fertilization) and to the maternal allele in the endosperm in the first few days after fertilization (Grossniklaus et al. 1998; Luo et al. 2000). *mea* gametophytes give rise to aborted embryos if fertilized (Grossniklaus et al. 1998) and to proliferation of the endosperm if left unfertilized (Luo et al. 2000). DNA methylation influences imprinting in both plants and animals (Adams et al. 2000), but in plants the influence seems to be indirect (Vielle-Calzada et al. 1999; Luo et al. 2000). The DNA glycosylase DEMETER is required to activate maternal transcription of *MEA* in WT, but not in *met1*, suggesting that DNA methylation is responsible for silencing (Choi et al. 2002; Xiao et al. 2003). MET1-dependent DNA methylation has been detected in two or three regions upstream of *MEA*, depending on genetic background, as well as in the 3´ region (Xiao et al. 2003; Zilberman et al. 2003). A 4.5-kb-upstream sequence is sufficient to restrict reporter gene expression to the maternal allele (Xiao et al. 2003).

Reannotation of the *MEA* locus reveals that the 5´ and 3´ methylated regions correspond to two clusters of unrelated tandem repeats, ~4 kb upstream and 3 kb downstream, respectively (Fig. 3). The upstream repeat (330 bp) is part of a 554-bp helitron transposon (*ATREP2*) while the downstream repeat (1 kb) is of unknown origin. siRNA corresponding to both repeats was detected in WT plants and was greatly reduced in *met1* and *ddm1* (Fig. 4). Interestingly, probes from the upstream repeat detected both 24-nucleotide siRNA and longer small RNA >40 nucleotides. These longer RNA did not change in *met1*, but were absent in *ddm1*. Longer RNA may correspond to the terminal "runoff" products of second-strand synthesis by RdRP, which are predicted to accumulate in short tandem arrays (Fig. 2). This is because extension by RdRP near

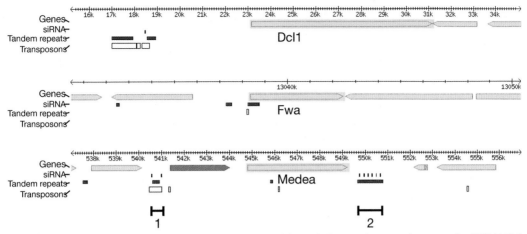

Figure 3. Tandem repeats and transposons surrounding imprinted loci in *Arabidopsis*. The genomic context for *SIN1/DCL1* and for the imprinted genes, *FWA* and *MEDEA*, are shown along with flanking repeats, transposons, and cloned siRNA. Probes for small RNA blots shown in Fig. 4 are indicated for *MEDEA*. Tandem repeats are found in each of the upstream regions required for imprinted expression and are generated by transposable elements.

the edges of the array will result in dsRNA products that are too short to be processed further by DICER (Martienssen 2003). In long tandem arrays, such as the one downstream of *MEA*, small RNA from central repeats are continually regenerated following extension (Fig. 2), so that longer products accumulate to a lesser degree (Fig. 4). This downstream repeat (also known as *MEA-ISR*) is heavily methylated, and non-CG methylation depends on *AGO4*, although this is not the case at *FWA* (Zilberman et al. 2003).

SILENCING MECHANISMS

Silencing of transposons and transgenes in *Arabidopsis* depends on at least three hierarchical mechanisms, one involving MET1, DDM1, and the histone deacetylase SIL1, another involving histone methylation and the chromodomain methyltransferase CMT3, and a third that utilizes the dnmt3 homologs DRM1 and DRM2 (Cao et al. 2003; Lippman et al. 2003; Matzke et al. 2004). RNAi has been implicated in all three mechanisms (Lippman and Martienssen 2004) and each complex has partially overlapping roles. For example, the chromomethylase CMT3 plays a role in transcriptional silencing via the histone modification pathway, consistent with recruitment to methylated histones via the chromodomain (Bartee et al. 2001; Jackson et al. 2002; Johnson et al. 2002). However, CMT3 also has a role in transposition of *AtMu1* and CACTA-class DNA transposons (Kato et al. 2003). In contrast, MET1 alone has a major role in transcriptional silencing, without effecting transposition, while DDM1 affects both (Miura et al. 2001; Singer et al. 2001).

AtMu1 siRNA, whether detected by northern analysis or as cloned fragments, is derived from the transposase gene (Fig. 1). In this respect it differs from transposon *Tc1* siRNA in *Caenorhabditis elegans*, which corresponds to

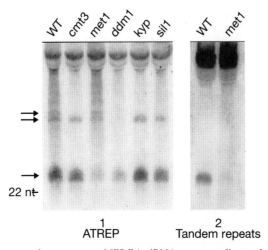

Figure 4. Small RNA corresponding to tandem repeats at *MEDEA*. siRNA corresponding to the upstream *ATREP* helitron transposon (*1*) and downstream *MEA-ISR* (*2*) repeats can be detected in WT, in the histone K9 methyltransferase mutant *kyp*, in the histone deacetylase mutant *sil1*, and in the chromomethylase mutant *cmt3*. They are substantially reduced in *met1* and *ddm1*. Longer short RNA (*arrows*) is found only in the upstream repeats and depends on *DDM1*.

long TIRs. It has been proposed that *Tc1* siRNA is derived from dsRNA by read-through transcription starting outside the element and foldback of the resulting transcript (Sijen and Plasterk 2003). Another origin might be found in extrachromosomal circles, which are found in both maize and *C. elegans* (Ruan and Emmons 1984; Sundaresan and Freeling 1987). Transcription of these circles would generate hairpin RNA from the TIR, which could act as a potential source of siRNA. Retrotransposons generate siRNA corresponding to the LTR repeats (Lippman et al. 2003). This is important, because transposon promoters are encoded within these repeats and can therefore be subject to transcriptional silencing when targeted by siRNA. Silencing targeted to the coding region of *AtMu1*, by contrast, would have to spread to the promoter to fully silence transcription, accounting perhaps for the low mRNA levels observed in WT plants (Fig. 1B). Robertson's Mutator transposons in maize also correspond to siRNA, although its role in silencing is unclear (Rudenko et al. 2003).

In addition to providing promoters for themselves, transposons provide promoters for neighboring genes. For example, SINEs provide promoters and exons for neighboring genes in mammalian genomes (Allen et al. 2004) as well as in *Arabidopsis* (Lippman et al. 2004). Both DNA-class and retrotransposable elements provide promoters for neighboring genes in animals, plants, yeast, and filamentous fungi, as well as in prokaryotes (Martienssen 1998b; Martienssen and Colot 2001). This may be a fundamental property of transposons, in accordance with their role as controlling elements first envisioned by McClintock (1957). Transposon promoters are regulated by transposon proteins, including transposase itself. In *Arabidopsis*, the regulatory proteins FAR1 and FHY3 each closely resemble the maize Mutator transposase *Jittery* and presumably regulate genes via fragments of defective elements integrated in their promoters (Hudson et al. 2003). Unlike other Mutator elements, this family of related genes is largely unmethylated (R.A. Martienssen et al., unpubl.).

Tandem repeats are a common constituent of most eukaryotic genomes. In the human genome, they are highly polymorphic and are often generated by insertion of SINE elements, much like the repeats at *FWA* (Bois et al. 1998). In general, such polymorphisms are neutral, but occasionally short tandem repeats arise in genes. At *FMR1*, a gene responsible for Fragile X syndrome, expansion of a triplet CGG tandem repeat from 20 to 60 or more copies in the untranslated leader leads to DNA and histone H3 lysine-9 methylation, gene silencing, and sex-linked disease (Jin and Warren 2000; Coffee et al. 2002). It is possible that silencing in this case results from the mechanism described here, given that siRNA induces transcriptional silencing in humans as well as plants (Morris et al. 2004). FMR repeats are transcribed, and siRNA could be maintained if sufficient copies of the repeat were present to permit reiterative rounds of second-strand synthesis and degradation. Sixty copies of a triplet repeat would correspond to seven sequential 24-nucleotide repeats in three different registers, providing a rich source of siRNA. Conversely, loss of tandem repeats would be expected to result in loss of gene silencing, ectopic expression, and autosomal dominant disease.

Tandem repeat and transposon-mediated silencing are attractive mechanisms for imprinting. This is because transcription of the repeats in the germ line would establish silencing in the next generation, which could then be maintained in *cis* by epigenetic modifications. Expression of tandem repeats in the female germ line would result in silencing of the maternal allele, while expression in the male germ line would result in paternal chromosome silencing. Tandem repeats have been found associated with many imprinting control regions in the mouse, although only some of them have been implicated in silencing (Lewis et al. 2004). They have also been found associated with imprinted and paramutable genes in maize (Alleman and Doctor 2000; Stam et al. 2002). In the case of *FWA*, transposon-derived tandem repeats regulate silencing and correspond to siRNA (Lippman et al. 2004). RNAi is required for establishment of silencing when mutant plants are transformed, but it is not required to maintain silencing, as *FWA* remains silent in *dcl3* and *rdr2*. This is consistent with a role for RNAi in establishing imprints in the germ line (Chan et al. 2004). Alternatively, RNAi could contribute to silencing throughout the plant and might be countered in the endosperm only by other mechanisms (Xiao et al. 2003).

In addition to 24-nucleotide siRNA, longer 40–80-nucleotide products accumulate from the *MEA* upstream tandem repeats (2.5 copies), and these depend on *DDM1* but not *MET1* (Fig. 4). Longer products are predicted to accumulate from shorter tandem repeat arrays as "runoff" products of second-strand synthesis (Fig. 2). It is therefore tempting to speculate that DDM1 may play a role in second-strand RNA synthesis and only indirectly in the synthesis of siRNA. As well as having chromatin remodeling activity (Brzeski and Jerzmanowski 2003), DDM1 is a DEAD box helicase (Jeddeloh et al. 1999) and may have a role in RNA metabolism.

In *Arabidopsis*, DNA methylation has a major role in silencing of the imprinted genes *MEDEA* and *FWA*. The DNA glycosylase DEMETER acts in opposition to MET1 to activate *MEA* and *FWA* in the central cells of the female gametophyte, which go on to form the extraembryonic endosperm (Xiao et al. 2003; Kinoshita et al. 2004). Activation is inherited for a few cell divisions following fertilization, so that the maternal allele is expressed in the early endosperm. Maternal *mea* seed abortion cannot be rescued when *met1* is transmitted through the pollen, consistent with a role for *MEA* and *MET1* in pre- but not postfertilization (Xiao et al. 2003). DDM1 does not effect imprinting directly (Vielle-Calzada et al. 1999; Xiao et al. 2003) and may not be functional in the gametophyte, unlike MET1 (Vongs et al. 1993; Saze et al. 2003).

In plants, DNA methylation is targeted to transposable elements (Rabinowicz et al. 2003), which implicates transposons in imprinting, as well as other silencing mechanisms (Martienssen 1998a,b). Imprinted genes were first identified in maize, and at least some imprinted alleles have transposon sequences within their promoters (Alleman and Doctor 2000). As well as *FWA* and *MEA*, methylated repeats are found upstream of a third gene, *SHORT INTEGUMENTS1 (SIN1)* which has a tandem repeat 3.5 kb upstream (Fig. 3). This upstream region directs maternal monoallelic reporter gene expression in transgenic plants (Golden et al. 2002). *SIN1* encodes the dicer-like gene *DCL1*. The methylated repeats upstream of *FWA*, *MEA*, and *DCL1* are derived from a 422-bp SINE element, a 554-bp helitron, and a 1186-bp helitron, respectively, each of which corresponds to siRNA (Fig. 3). Loss of SINE and helitron siRNA in *met1* implicates RNAi in silencing and indicates transposons may regulate imprinted genes (Lippman et al. 2004). Some retrotransposons in mammals are also imprinted (Peaston et al. 2004) and can influence gene expression depending on parent of origin (Morgan et al. 1999). The finding of noncoding RNA at differentially methylated regions of imprinted loci in the mouse suggests RNA-mediated mechanisms may play a role (Sleutels et al. 2002).

If siRNA mediates transposon silencing, why do most transposons remain silent in mutants in RNAi? In *Arabidopsis*, redundancy among RNAi genes may explain this, but retrotransposons also remain largely silent in fission yeast RNAi mutants (Hansen et al. 2004), even though some lose H3mK9 (Schramke and Allshire 2003). One explanation is that RNAi is required only for initiation of silencing and plays no further role once silencing is established. Although this is not the case at the centromere (Volpe et al. 2002), it could be the case at the mating-type locus (Hall et al. 2002). However, recent studies indicate this is not the entire explanation. First, maintenance of silencing at the mating-type locus does depend on RNAi after all (as it does at the centromere), but only in the absence of certain transcription factors (Jia et al. 2004). Second, the RITS complex remains bound to the mating type locus in $dcr1^-$, but not in $clr4^-$ (Sigova et al. 2004), indicating H3mK9 is required for binding but siRNA is not. Instead of initiation, RITS might be involved in spreading of heterochromatic silencing, as loss of H3mK9 from centromeric reporter genes is more severe in RITS mutants than loss from neighboring heterochromatic repeats (Volpe et al. 2002; Sigova et al. 2004).

Nonetheless, RNAi is required to initially silence *FWA* transgenes in *Arabidopsis* following infection of the female gametophyte with *Agrobacterium* (see above), while it is not required to maintain silencing of the endogenous *FWA* gene (Chan et al. 2004). Similarly, epialleles of *FWA* are inherited from *met1*, which loses siRNA, but not from *ddm1*, which does not, implicating siRNA in transposon-mediated silencing in *cis* (Lippman et al. 2004). Further genetic analysis should clarify this issue in plants as well as in fission yeast.

CONCLUSION

Heterochromatin is composed primarily of transposons and related repeats. In fission yeast and plants, at least, these repetitive sequences correspond to siRNA, which helps to guide heterochromatic modifications to these regions of the chromosome. Tandem repeats can maintain large populations of siRNA and are frequently associated with epigenetic regulation, such as imprinting, paramutation, and centromere activity. Imprinted genes in *Arabidopsis*, such as *FWA* and *MEDEA*, are associated with tandem repeats, transposable elements, and siRNA, which are coregulated by the DNA methyltransferase MET1. Transposons and siRNA are thus implicated in imprinting and developmental control.

ACKNOWLEDGMENTS

We would like to thank Bob Fischer, Jim Carrington, Steve Jacobsen, and Greg Hannon for helpful discussions, as well as our colleagues Anne-Valerie Gendrel, Tom Volpe, and especially Vincent Colot for sharing their data and their advice. M.V. is a National Science Foundation bioinformatics postdoctoral fellow and Z.L. is an Arnold and Mabel Backman graduate fellow in the Watson School of Biological Sciences. M.R. has a postdoctoral fellowship from NIH. This work was supported by NSF grant DBI 000074, and NIH grant R01-067014 to R.M.

REFERENCES

Adams S., Vinkenoog R., Spielman M., Dickinson H.G., and Scott R.J. 2000. Parent-of-origin effects on seed development in *Arabidopsis thaliana* require DNA methylation. *Development* **127:** 2493.

Alleman M. and Doctor J. 2000. Genomic imprinting in plants: Observations and evolutionary implications. *Plant Mol. Biol.* **43:** 147.

Allen T.A., Von Kaenel S., Goodrich J.A., and Kugel J.F. 2004. The SINE-encoded mouse B2 RNA represses mRNA transcription in response to heat shock. *Nat. Struct. Mol. Biol.* **11:** 816.

Arabidopsis Genome Initiative. 2000. Analysis of the genome sequence of the flowering plant *Arabidopsis thaliana*. *Nature* **408:** 796.

Bannister A.J., Zegerman P., Partridge J.F., Miska E.A., Thomas J.O., Allshire R.C., and Kouzarides T. 2001. Selective recognition of methylated lysine 9 on histone H3 by the HP1 chromo domain. *Nature* **410:** 120.

Bartee L., Malagnac F., and Bender J. 2001. *Arabidopsis* cmt3 chromomethylase mutations block non-CG methylation and silencing of an endogenous gene. *Genes Dev.* **15:** 1753.

Benson G. 1999. Tandem repeats finder: A program to analyze DNA sequences. *Nucleic Acids Res.* **27:** 573.

Bois P., Williamson J., Brown J., Dubrova Y.E., and Jeffreys A.J. 1998. A novel unstable mouse VNTR family expanded from SINE B1 elements. *Genomics* **49:** 122.

Brzeski J. and Jerzmanowski A. 2003. Deficient in DNA methylation 1 (DDM1) defines a novel family of chromatin-remodeling factors. *J. Biol. Chem.* **278:** 823.

Cao X., Aufsatz W., Zilberman D., Mette M.F., Huang M.S., Matzke M., and Jacobsen S.E. 2003. Role of the DRM and CMT3 methyltransferases in RNA-directed DNA methylation. *Curr. Biol.* **13:** 2212.

Chan S.W., Zilberman D., Xie Z., Johansen L.K., Carrington J.C., and Jacobsen S.E. 2004. RNA silencing genes control de novo DNA methylation. *Science* **303:** 1336.

Choi Y., Gehring M., Johnson L., Hannon M., Harada J.J., Goldberg R.B., Jacobsen S.E., and Fischer R.L. 2002. DEMETER, a DNA glycosylase domain protein, is required for endosperm gene imprinting and seed viability in *Arabidopsis*. *Cell* **110:** 33.

Coffee B., Zhang F., Ceman S., Warren S.T., and Reines D. 2002. Histone modifications depict an aberrantly heterochromatinized FMR1 gene in fragile x syndrome. *Am. J. Hum. Genet.* **71:** 923.

CSHL/WUGSC/PEB *Arabidopsis* Sequencing Consortium. 2000. The complete sequence of a heterochromatic island from a higher eukaryote. *Cell* **100:** 377.

Gendrel A.V., Lippman Z., Yordan C., Colot V., and Martienssen R.A. 2002. Dependence of heterochromatic histone H3 methylation patterns on the *Arabidopsis* gene DDM1. *Science* **297:** 1871.

Golden T.A., Schauer S.E., Lang J.D., Pien S., Mushegian A.R., Grossniklaus U., Meinke D.W., and Ray A. 2002. SHORT INTEGUMENTS1/SUSPENSOR1/CARPEL FACTORY, a Dicer homolog, is a maternal effect gene required for embryo development in *Arabidopsis*. *Plant Physiol.* **130:** 808.

Grossniklaus U., Vielle-Calzada J.P., Hoeppner M.A., and Gagliano W.B. 1998. Maternal control of embryogenesis by MEDEA, a polycomb group gene in *Arabidopsis*. *Science* **280:** 446.

Hall I.M., Shankaranarayana G.D., Noma K., Ayoub N., Cohen A., and Grewal S.I. 2002. Establishment and maintenance of a heterochromatin domain. *Science* **297:** 2232.

Hannon G.J. 2002. RNA interference. *Nature* **418:** 244.

Hansen K.R., Burns G., Mata J., Volpe T.A., Martienssen R.A., Bahler J., and Thon G. 2005. Global effects on gene expression in fission yeast by silencing and RNA interference machineries. *Mol. Cell. Biol.* **25:** 590.

Hudson M.E., Lisch D.R., and Quail P.H. 2003. The FHY3 and FAR1 genes encode transposase-related proteins involved in regulation of gene expression by the phytochrome A-signaling pathway. *Plant J.* **34:** 453.

Jackson J.P., Lindroth A.M., Cao X., and Jacobsen S. 2002. Control of CpNpG DNA methylation by the KRYPTONITE histone H3 methyltransferase. *Nature* **416:** 556.

Jeddeloh J.A., Stokes T.L., and Richards E.J. 1999. Maintenance of genomic methylation requires a SWI2/SNF2-like protein. *Nat. Genet.* **22:** 94.

Jia S., Noma K., and Grewal S.I. 2004. RNAi-independent heterochromatin nucleation by the stress-activated ATF/CREB family proteins. *Science* **304:** 1971.

Jin P. and Warren S.T. 2000. Understanding the molecular basis of fragile X syndrome. *Hum. Mol. Genet.* **9:** 901.

Joanis V. and Lloyd V.K. 2002. Genomic imprinting in *Drosophila* is maintained by the products of Suppressor of variegation and trithorax group, but not Polycomb group, genes. *Mol. Genet. Genomics* **268:** 103.

Johnson L., Cao X., and Jacobsen S. 2002. Interplay between two epigenetic marks. DNA methylation and histone H3 lysine 9 methylation. *Curr. Biol.* **12:** 1360.

Jurka J., Klonowski P., Dagman V., and Pelton P. 1996. CENSOR: A program for identification and elimination of repetitive elements from DNA sequences. *Comput. Chem.* **20:** 119.

Kato M., Miura A., Bender J., Jacobsen S.E., and Kakutani T. 2003. Role of CG and non-CG methylation in immobilization of transposons in *Arabidopsis*. *Curr. Biol.* **13:** 421.

Kinoshita T., Miura A., Choi Y., Kinoshita Y., Cao X., Jacobsen S.E., Fischer R.L., and Kakutani T. 2004. One-way control of FWA imprinting in *Arabidopsis* endosperm by DNA methylation. *Science* **303:** 521.

Lewis A., Mitsuya K., Constancia M., and Reik W. 2004. Tandem repeat hypothesis in imprinting: Deletion of a conserved direct repeat element upstream of H19 has no effect on imprinting in the Igf2-H19 region. *Mol. Cell. Biol.* **24:** 5650.

Lippman Z. and Martienssen R. 2004. The role of RNA interference in heterochromatic silencing. *Nature* **431:** 364.

Lippman Z., May B., Yordan C., Singer T., and Martienssen R. 2003. Distinct mechanisms determine transposon inheritance and methylation via small interfering RNA and histone modification. *PLoS Biol.* **1:** E67.

Lippman Z.L., Gendrel A.V., Black M., Vaughn M.W., Dedhia D., McCombie W.R., Lavine K., Mittal V., May B., Kasschau K.D., Carrington J.C., Doerge R.W., Colot V., and Martienssen R.A. 2004. Role of transposable elements in heterochromatin and epigenetic control. *Nature* **430:** 471.

Liu J., Carmell M.A., Rivas F.V., Marsden C.G., Thomson J.M., Song J.J., Hammond S.M., Joshua-Tor L., and Hannon G.J. 2004. Argonaute2 is the catalytic engine of mammalian RNAi. *Science* **305:** 1437.

Llave C., Kasschau K.D., Rector M.A., and Carrington J.C. 2002. Endogenous and silencing-associated small RNAs in plants. *Plant Cell* **14:** 1605.

Luo M., Bilodeau P., Dennis E.S., Peacock W.J., and Chaudhury A. 2000. Expression and parent-of-origin effects for FIS2, MEA, and FIE in the endosperm and embryo of developing *Arabidopsis* seeds. *Proc. Natl. Acad. Sci.* **97:** 10637.

Makeyev E.V. and Bamford D.H. 2002. Cellular RNA-dependent RNA polymerase involved in posttranscriptional gene silencing has two distinct activity modes. *Mol. Cell* **10:** 1417.

Martienssen R. 1998a. Chromosomal imprinting in plants. *Curr. Opin. Genet. Dev.* **8:** 240.

―――. 1998b. Transposons, DNA methylation and gene control. *Trends Genet.* **14:** 263.

―――. 2003. Maintenance of heterochromatin by RNA interference of tandem repeats. *Nat. Genet.* **35:** 213.

Martienssen R.A. and Colot V. 2001. DNA methylation and epigenetic inheritance in plants and filamentous fungi. *Science* **293:** 1070.

Matzke M., Aufsatz W., Kanno T., Daxinger L., Papp I., Mette M.F., and Matzke A.J. 2004. Genetic analysis of RNA-mediated transcriptional gene silencing. *Biochim. Biophys. Acta* **1677:** 129.

McClintock B. 1957. Controlling elements and the gene. *Cold Spring Harbor Symp. Quant. Biol.* **21:** 197.

Miura A., Yonebayashi S., Watanabe K., Toyama T., Shimada H., and Kakutani T. 2001. Mobilization of transposons by a mutation abolishing full DNA methylation in *Arabidopsis*. *Nature* **411:** 212.

Mohan K.N., Ray P., and Chandra H.S. 2002. Characterization of the genome of the mealybug *Planococcus lilacinus*, a model organism for studying whole-chromosome imprinting and inactivation. *Genet. Res.* **79:** 111.

Morgan H.D., Sutherland H.G., Martin D.I., and Whitelaw E. 1999. Epigenetic inheritance at the agouti locus in the mouse. *Nat. Genet.* **23:** 314.

Morris K.V., Chan S.W., Jacobsen S.E., and Looney D.J. 2004. Small interfering RNA-induced transcriptional gene silencing in human cells. *Science* **305:** 1289.

Peaston A.E., Evsikov A.V., Graber J.H., de Vries W.N., Holbrook A.E., Solter D., and Knowles B.B. 2004. Retrotransposons regulate host genes in mouse oocytes and preimplantation embryos. *Dev. Cell* **7:** 597.

Rabinowicz P.D., Palmer L.E., May B.P., Hemann M.T., Lowe S.W., McCombie W.R., and Martienssen R.A. 2003. Genes and transposons are differentially methylated in plants, but not in mammals. *Genome Res.* **13:** 2658.

Ruan K. and Emmons S.W. 1984. Extrachromosomal copies of transposon Tc1 in the nematode *Caenorhabditis elegans*. *Proc. Natl. Acad. Sci.* **81:** 4018.

Rudenko G.N., Ono A., and Walbot V. 2003. Initiation of silencing of maize MuDR/Mu transposable elements. *Plant J.* **33:** 1013.

Saze H., Scheid O.M., and Paszkowski J. 2003. Maintenance of CpG methylation is essential for epigenetic inheritance during plant gametogenesis. *Nat. Genet.* **34:** 65.

Schramke V. and Allshire R. 2003. Hairpin RNAs and retrotransposon LTRs effect RNAi and chromatin-based gene silencing. *Science* **301:** 1069.

Sigova A., Rhind N., and Zamore P.D. 2004. A single Argonaute protein mediates both transcriptional and posttranscriptional silencing in *Schizosaccharomyces pombe*. *Genes Dev.* **18:** 2359.

Sijen T. and Plasterk R.H. 2003. Transposon silencing in the *Caenorhabditis elegans* germ line by natural RNAi. *Nature* **426:** 310.

Singer T., Yordan C., and Martienssen R.A. 2001. Robertson's Mutator transposons in *A. thaliana* are regulated by the chromatin-remodeling gene Decrease in DNA Methylation (DDM1). *Genes Dev.* **15:** 591.

Sleutels F. and Barlow D.P. 2002. The origins of genomic imprinting in mammals. *Adv Genet.* **46:** 119.

Sleutels F., Zwart R., and Barlow D.P. 2002. The non-coding Air RNA is required for silencing autosomal imprinted genes. *Nature* **415:** 810.

Song J.J., Smith S.K., Hannon G.J., and Joshua-Tor L. 2004. Crystal structure of Argonaute and its implications for RISC slicer activity. *Science* **305:** 1434.

Song J.J., Liu J., Tolia N.H., Schneiderman J., Smith S.K., Martienssen R.A., Hannon G.J., and Joshua-Tor L. 2003. The crystal structure of the Argonaute2 PAZ domain reveals an RNA binding motif in RNAi effector complexes. *Nat. Struct. Biol.* **10:** 1026.

Soppe W.J., Jacobsen S.E., Alonso-Blanco C., Jackson J.P., Kakutani T., Koornneef M., and Peeters A.J. 2000. The late flowering phenotype of fwa mutants is caused by gain-of-function epigenetic alleles of a homeodomain gene. *Mol. Cell* **6:** 791.

Soppe W.J., Jasencakova Z., Houben A., Kakutani T., Meister A., Huang M.S., Jacobsen S.E., Schubert I., and Fransz P.F. 2002. DNA methylation controls histone H3 lysine 9 methylation and heterochromatin assembly in *Arabidopsis*. *EMBO J.* **21:** 6549.

Stam M., Belele C., Dorweiler J.E., and Chandler V.L. 2002. Differential chromatin structure within a tandem array 100 kb upstream of the maize b1 locus is associated with paramutation. *Genes Dev.* **16:** 1906.

Sundaresan V. and Freeling M. 1987. An extrachromosomal form of the Mu transposons of maize. *Proc. Natl. Acad. Sci.* **84:** 4924.

Tabata S., Kaneko T., Nakamura Y., Kotani H., Kato T., Asamizu E., Miyajima N., Sasamoto S., Kimura T., Hosouchi T., Kawashima K., Kohara M., Matsumoto M., Matsuno A., Muraki A., Nakayama S., Nakazaki N., Naruo K., Okumura S., Shinpo S., Takeuchi C., Wada T., Watanabe A., Yamada M., Yasuda M., et al. 2000. Sequence and analysis of chromosome 5 of the plant *Arabidopsis thaliana*. *Nature* **408:** 823.

Tariq M. and Paszkowski J. 2004. DNA and histone methylation in plants. *Trends Genet.* **20:** 244.

Tariq M., Saze H., Probst A.V., Lichota J., Habu Y., and Paszkowski J. 2003. Erasure of CpG methylation in *Arabidopsis* alters patterns of histone H3 methylation in heterochromatin. *Proc. Natl. Acad. Sci.* **100:** 8823.

Verdel A., Jia S., Gerber S., Sugiyama T., Gygi S., Grewal S.I., and Moazed D. 2004. RNAi-mediated targeting of heterochromatin by the RITS complex. *Science* **303:** 672.

Vielle-Calzada J.P., Thomas J., Spillane C., Coluccio A., Hoeppner M.A., and Grossniklaus U. 1999. Maintenance of genomic imprinting at the *Arabidopsis* medea locus requires zygotic DDM1 activity. *Genes Dev.* **13:** 2971.

Volpe T.A., Kidner C., Hall I.M., Teng G., Grewal S.I., and Martienssen R.A. 2002. Regulation of heterochromatic silencing and histone H3 lysine-9 methylation by RNAi. *Science* **297:** 1833.

Vongs A., Kakutani T., Martienssen R.A., and Richards E.J. 1993. *Arabidopsis thaliana* DNA methylation mutants. *Science* **260:** 1926.

Xiao W., Gehring M., Choi Y., Margossian L., Pu H., Harada J.J., Goldberg R.B., Pennell R.I., and Fischer R.L. 2003. Imprinting of the MEA Polycomb gene is controlled by antagonism between MET1 methyltransferase and DME glycosylase. *Dev. Cell* **5:** 891.

Xie Z., Johansen L.K., Gustafson A.M., Kasschau K.D., Lellis A.D., Zilberman D., Jacobsen S.E., and Carrington J.C. 2004. Genetic and functional diversification of small RNA pathways in plants. *PLoS Biol.* **2:** E104.

Zilberman D., Cao X., and Jacobsen S.E. 2003. ARGONAUTE4 control of locus-specific siRNA accumulation and DNA and histone methylation. *Science* **299:** 716.

Transposon Silencing and Imprint Establishment in Mammalian Germ Cells

T.H. BESTOR AND D. BOURC'HIS

Department of Genetics and Development, College of Physicians and Surgeons of Columbia University, New York, New York 10032

Activators and repressors cannot explain the regulation of all genes. Some mammalian genes are not expressed even in the presence of all the required transcription factors, and the behavior of these genes also fails to conform to Mendelian laws of transmission genetics. Such genes are said to be under epigenetic control or to be subject to gene silencing. Mammalian gene silencing is especially conspicuous at genes subject to X-chromosome inactivation, at imprinted genes, and at the large number of transposable elements that contain promoter sequences (Bestor 2003). Genes on the inactive X chromosome are clearly in the presence of all factors required for their expression, as shown by the activity of the homologous alleles on the active X, but they remain silent for very long periods of time. The same is true of imprinted genes. The state of activity of a gene subject to X-inactivation or genomic imprinting can be predicted only if the history of the gene is known; in the case of X-inactivation, the critical event occurs in somatic cells soon after implantation, while the state of activity of imprinted genes is determined during spermatogenesis and oogenesis in the previous generation. The promoters of retroposons are silenced in premeiotic prospermatogonia in males and meiotic dictyate oocytes in females. Once established, the silent state can persist for the life of the organism, which in humans can exceed 100 years, and can only be reset in the next reproductive cycle.

Epigenetic gene silencing is unlikely to depend on inheritance of patterns of histone modifications. Histone modifications have not been shown to be heritable and there is no plausible mechanism that might allow heritability. As Henikoff et al. (2004) have noted, histone replacement (especially replacement of histone H3 by H3.3 and of histone H2A by H2A.Z) can provide some stability to chromatin states, but dilution by S phase histones in cycling cells will obviate true heritability. Ongoing transcription is required to maintain high levels of H3.3 and H2A.Z in dividing cells. This clearly cannot explain genomic imprinting at the *H19* locus, where a decision is taken to irreversibly inactivate *H19* within a few days of birth and the restricted expression potential maintained for decades in the absence of H19 expression. Allele-specific *H19* expression is only realized in offspring that can be separated by decades and by more than 50 mitotic divisions from the event that restricted the potential of the *H19* gene. No plausible mechanism by which histone modification could mediate such an effect has been put forward (for review, see Goll and Bestor 2002), even though it is widely assumed that histone modifications mediate all epigenetic effects.

The heritability of cytosine methylation and the inactivation of promoters by methylation have been confirmed in many experiments over the last 25 years (for an especially convincing recent example, see Lorincz et al. 2002). Nearly all available data indicate that epigenetic effects in mammals depend largely on heritable genomic methylation patterns. Demethylation of the genome causes the loss of most genomic imprinting, the inactivation of all X chromosomes by reactivation of *Xist*, and the fulminating expression of endogenous retroviruses (for review, see Bestor 2000). However, almost nothing is known of the mechanisms that target specific sequences for de novo methylation. This is true in large part because most de novo methylation takes place in germ cells and early embryos, which have been much less studied than adult tissues, where genomic methylation patterns are more static. DNA methyltransferase 3-like (Dnmt3L; Aapola et al. 2000) is a noncatalytic regulatory factor expressed specifically in germ cells, and genetic studies of this factor have begun to reveal the roles of genomic methylation patterns in the germ lines of both males and females.

THE DNA METHYLTRANSFERASES OF MAMMALS

There are five mammalian proteins that share sequence relatedness with the DNA cytosine-5 methyltransferases of other organisms (Fig. 1), but only three (Dnmt1, Dnmt3A, and Dnmt3B) have been shown to be active transmethylases in both biochemical and genetic tests; Dnmt3L has no in vitro activity but by genetic tests is required for de novo methylation specifically in germ cells, as will be described later. Dnmt2 has shown no evidence of DNA methyltransferase activity in biochemical or genetic assays.

None of the active DNA methyltransferases has inherent sequence specificity beyond the CpG dinucleotide and the mechanisms that guide methylation to specific sequences are unknown. The Dnmt1, Dnmt2, and Dnmt3 families diverged prior to the separation of the plant and animal kingdoms and are as distantly related to each other as they are to bacterial restriction methyltransferases that produce 5-methylcytosine (m^5C). A brief introduction to the mammalian DNA methyltransferase families follows.

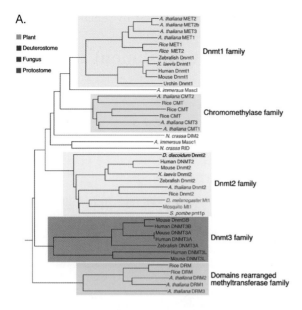

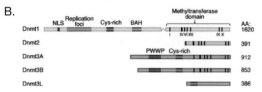

Figure 1. Sequence relationships among the catalytic domains of eukaryotic DNA cytosine methyltransferase. (*A*) ClustalW analysis of relationship of DNA methyltransferases from plants, deuterostomes, fungi, and protostomes. Note that the mammalian Dnmt1, Dnmt2, and Dnmt3 families all have homologs in plants. The most widely distributed DNA methyltransferase homolog is Dnmt2, whose function is unknown. Note deep branching of Dnmt3L from Dnmt3 branch. (*B*) Domain and motif organization within the mammalian DNA methyltransferase homologs. Dnmt3L lacks the Dnmt signature motifs that mediate methyl transfer (*dark vertical bars marked with Roman numerals* for Dnmt1) but is clearly related to Dnmt3A and Dnmt3B in framework regions.

Dnmt1

The first mammalian DNA methyltransferase to be identified was cloned by chromatographic purification of the protein to homogeneity, determination of the sequence of cyanogen bromide peptides by vapor phase Edman degradation, and the preparation of degenerate oligonucleotide probes corresponding to the peptide sequences (Bestor et al. 1988). The enzyme cloned in this way was named DNA methyltransferase 1 (Dnmt1). Dnmt1 contains a carboxy-terminal domain of about 500 amino acids that is closely related to bacterial cytosine-5 restriction methyltransferases and a large (~1000 amino acid) amino-terminal domain that has multiple regulatory functions, which include coordination of DNA replication and maintenance methylation, import of Dnmt1 protein into nuclei (Leonhardt et al. 1992), suppression of de novo methylation (Bestor 1992), and regulation of cell-cycle-dependent protein degradation (Ding and Chaillet 2002).

The Dnmt1 gene was disrupted by means of homologous recombination in embryonic stem (ES) cells (Li et al. 1992). Dnmt1-deficient mouse embryos had severely demethylated genomes and died at the headfold stage, while mutant ES cells grew normally but died when induced to differentiate in vitro or in vivo (Li et al. 1992). The lethal differentiation phenotype is unique to the Dnmt1-deficient genotype. Further work showed that Dnmt1-deficient mouse embryos lost monoallelic expression at most imprinted loci (Li et al. 1993) and showed inactivation of all X chromosomes as a result of *Xist* reactivation (Panning and Jaenisch 1996). It was also shown that retroposons of the intracisternal A particle (IAP) class were transcribed at very high levels in Dnmt1 mutants; controls showed little or no expression (Walsh et al. 1998). Loss of methylation also destabilized the genome (Chen et al. 1998). These findings confirmed the essential role of genomic methylation patterns in mammalian development, something that had been in doubt because of the lack of cytosine methylation in popular model organisms such as *Drosophila melanogaster* and *Caenorhabditis elegans*. Cytosine methylation is now known to have essential roles in genomic imprinting, X-chromosome inactivation, host defense against transposons, and genome stability (for review, see Bestor 2003). Incipient developmental abnormalities and ectopic or precocious activation of tissue-specific genes have not been seen in embryos with demethylated genomes; these and other lines of evidence are incompatible with the well-accepted (but hardly well-established) view that changes in methylation patterns regulate gene expression during development (Walsh and Bestor 1999).

Dnmt1 has been assigned the role of maintenance methyltransferase (that is, able to methylate only the hemimethylated DNA produced by semiconservative DNA replication) in order to satisfy predictions of distinct maintenance and de novo methyltransferases published almost 30 years ago (Holliday and Pugh 1975; Riggs 1975). However, Dnmt1 is more abundant and has a much higher specific activity on unmethylated DNA than does any other mammalian DNA methyltransferase (Yoder et al. 1997). Groudine and colleagues have shown that cells that lack both Dnmt3A and Dnmt3B (which have been held to be the sole de novo DNA methyltransferases [Okano et al. 1999]) remain capable of de novo methylation under certain conditions (Lorincz et al. 2002). While there is evidence that Dnmt1 is predominantly a maintenance DNA methyltransferase and is dispensable for imprint establishment in oogenesis (Howell et al. 2001), the exact functions of Dnmt1 cannot be delimited from the available data and a role in de novo methylation remains possible.

Dnmt2

Dnmt2 was the first DNA methyltransferase homolog to be identified by searches of EST libraries (Yoder and Bestor 1998). Dnmt2 has all ten of the sequence motifs usually diagnostic of DNA cytosine-5 methyltransferases of both prokaryotes and eukaryotes, although the variable amino-terminal extensions that characterize the mammalian enzymes are absent from Dnmt2. Crystallographic studies have shown that every catalytic side chain is in the correct conformation to mediate the transmethylation reaction (Dong et al. 2001), but the protein has not displayed the expected activity in biochemical tests (Yoder and Bestor 1998; Dong et al. 2001) and ES cells and mice that lack Dnmt2 are viable and have no discernible defects in genomic methylation patterns or other discernible phenotypes (Okano et al. 1998; M.G. Goll and T.H. Bestor, unpubl.). Where Dnmt2 homologs are found they are very well conserved, and the phylogenetic distribution of Dnmt2 is the widest but most variable of any DNA cytosine-5 methyltransferase homolog: It is present in the fission yeast *Schizosaccharomyces pombe* (but not in *Saccharomyces cerevisiae* or any other fungus whose genome has been sequenced) and in *D. melanogaster* and other insects but not in *C. elegans*. It is found in all protozoa, vertebrates, and plants (including diatoms, ferns, and mosses) tested to date and is present in two species of one prokaryotic genus (the sulfur-reducing genus *Geobacter*). We and our collaborators have constructed strains of *S. pombe*, *D. melanogaster*, *A. thaliana*, and mice that lack Dnmt2. None have shown any detectable phenotype after generations of propagation in the homozygous state. Dnmt2 is the only DNA methyltransferase homolog dispensable for survival under laboratory conditions. The biological function of Dnmt2 remains enigmatic.

Dnmt3A AND Dnmt3B

Dnmt1 and Dnmt2 are singleton proteins in mammals, but the Dnmt3 family contains three members, Dnmt3A, Dnmt3B, and Dnmt3L. Dnmt3A and Dnmt3B are closely related and have low but approximately equivalent enzymatic activities on unmethylated and hemimethylated substrates; they have been referred to as the "long sought" de novo DNA methyltransferases (Okano et al. 1998), again to satisfy predictions of 1975 (Holliday and Pugh 1975; Riggs 1975). Deletion of Dnmt3A does not cause detectable alteration of genomic methylation patterns in somatic cells of homozygous mice, although mutant male mice lack germ cells and both sexes die of a condition similar to aganglionic megacolon (Okano et al. 1999). Mice that lack Dnmt3B die as embryos with demethylation of minor satellite DNA but normally methylated euchromatic DNA; the Dnmt3A-Dnmt3B double mutant dies very early with demethylation of all genomic sequences in a manner similar to that of Dnmt1 null mutants (Okano et al. 1999). We showed that the rare human genetic disorder ICF syndrome (immunodeficiency, centromere instability, and facial anomalies) is due to recessive loss of function mutations in the *DNMT3B* gene (Xu et al. 1999). Patients with ICF syndrome fail to methylate classical satellite (also known as satellite 2 and 3) sequences on the juxtacentromeric long arms of chromosomes 1, 9, and 16; these demethylated chromosomes gain and lose long arms at a very high rate to produce the multiradiate pinwheel chromosomes unique to this disorder.

Dnmt3A and Dnmt3B are clearly required for the establishment of genomic methylation patterns, but neither enzyme has inherent sequence specificity. The factors that designate specific sequences for de novo methylation are not known. Perhaps the outstanding problem in the mammalian DNA methylation field is the source of the sequence specificity for de novo methylation.

FUNCTIONS OF Dnmt3L IN OOGENESIS

As diagrammed in Figure 1B, Dnmt3L lacks the conserved motifs that mediate transmethylation but is related to Dnmt3A and Dnmt3B in framework regions (Aapola et al. 2000). Dnmt3L also fails to methylate DNA in biochemical tests. As shown in the diagram of Figure 2, expression of full-length Dnmt3L mRNA is confined to germ cells. (Sterile transcripts that initiate at a promoter located between exons 9 and 10 are expressed in spermatids. These truncated transcripts account for the large number of SAGE and EST hits in somatic cells [T. Shovlin et al., unpubl.].) Dnmt3L was of special interest because it is the only DNA methyltransferase homolog whose expression is confined to germ cells (Bourc'his et al. 2001). Disruption of the Dnmt3L gene by gene targeting in ES cells and insertion of a promoterless β-geo marker into the locus showed that Dnmt3L is expressed in growing oocytes (Bourc'his et al. 2004), the stage at which maternal genomic imprints are established (Kono et al. 1996). Mice homozygous for the disrupted *Dnmt3L* gene were viable and without overt phenotype, although both sexes were sterile. Males were azoospermic (the origin of this defect will be described in the following section), but oogenesis and early development of heterozygous embryos derived from homozygous oocytes was normal; the lethal phenotype was only manifested at e9. Such embryos showed signs of nutritional deprivation, and further analysis revealed a failure of chorioallantoic fusion and other dysmorphia of extraembryonic structures (Bourc'his et al. 2001). Analysis of expression of imprinted genes showed a complete lack of imprinting at maternally silenced loci and a lack of methylation of maternally methylated differentially methylated regions (DMRs). Bisulfite genomic sequencing showed that the imprinting defect was due to a failure to establish genomic imprints in the oocyte, and the normal imprinting of paternally silenced genes in heterozygous offspring of homozygous Dnmt3L-deficient females showed that imprint maintenance in the embryo was normal (Bourc'his et al. 2001). This contrasted with the situation in mice that lack Dnmt1o (an oocyte-specific isoform of Dnmt1) in which imprint establishment was normal but maintenance in preimplantation embryos was deficient (Howell et al. 2001). Methylation of sequences other than imprinted regions was normal in heterozygous embryos derived from homozygous Dnmt3L mutant oocytes (Bourc'his et al. 2001).

Dnmt3L behaves as a maternal-effect factor that is required only for imprint establishment in oocytes; as will described in the next section, the functions of Dnmt3L in male germ cells are completely different.

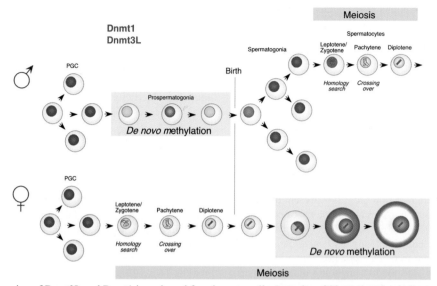

Figure 2. Expression of Dnmt3L and Dnmt1 in male and female germ cells. Intensity of *blue* coloration indicates relative levels of Dnmt3L; *red* indicates levels of Dnmt1. Dnmt3L is expressed in premeiotic male germ cells but only in growing oocytes in females. The full-length somatic form of Dnmt1 is expressed in proliferating primordial germ cells in both sexes and in proliferating spermatogonia in males; the truncated but active Dnmt1o isoform accumulates to very high levels in the cytoplasm of growing oocytes and is required for imprint maintenance specifically at the eight-cell stage (Howell et al. 2001). Dnmt3L is present only at the stages where genomic imprints are established and where transposons undergo de novo DNA methylation. (Data from Mertineit et al. 1998; Bourc'his et al. 2001; and La Salle et al. 2004.)

FUNCTIONS OF Dnmt3L IN SPERMATOGENESIS

In male mice Dnmt3L is expressed at significant levels only in perinatal prospermatogonia, the stage at which paternal genomic imprints are established (Davis et al. 1999) and transposons undergo de novo methylation (Walsh et al. 1998). Male mice that lack Dnmt3L are outwardly normal except for hypogonadism as adults (Bourc'his et al. 2001). The germ cell population is normal at birth, but only the first cohort of germ cells begins meiosis, and none reach the pachytene stage. All meiotic cells show extreme abnormalities of synapsis; grossly abnormal concentrations of synaptonemal complex proteins and nonhomologous synapsis (Fig. 3) are obvious in nearly all leptotene and zygotene spermatocytes. Adult males are devoid of all germ cells. This is in striking contrast to Dnmt3L-deficient females, where oogenesis is outwardly normal and a phenotype is only apparent in heterozygous offspring of homozygous females (Bourc'his et al. 2001).

The fact that Dnmt3L-deficient male germ cells show a phenotype only after the stage at which Dnmt3L protein is no longer expressed suggested an epigenetic or gene silencing defect. Homozygous mutant male germ cells were purified by flow sorting after staining with an antibody against germ cell nuclear antigen (GCNA; Enders and May 1994; the kind gift of G.C. Enders) and inspected for abnormalities of genomic methylation patterns. There is global genome demethylation in mutant spermatogonia and early spermatocytes as shown by increased sensitivity to methylation-sensitive restriction endonucleases (Fig. 4) (Bourc'his and Bestor 2004).

Transposons contain the large majority of m^5C present in the mammalian genome (Yoder et al. 1997), and demethylation of the major transposon classes (IAP elements and LINE-1 elements) was observed in Dnmt3L-deficient male germ cells. However, there was little or no demethylation of major or minor satellite DNA when compared with controls. This indicates that the methylation of heterochromatic satellite DNA is controlled by mechanisms distinct from those that control the methylation of euchromatic sequences. Other data support this conclusion; mutations in the *DNMT3B* gene in humans cause demethylation only of classical satellite (which is analogous to mouse major satellite) in ICF syndrome patients (Xu et al. 1999), and the methylation status of ma-

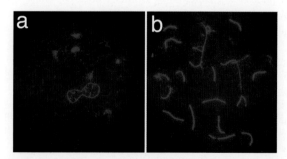

Figure 3. Meiotic catastrophe in spermatocytes derived from Dnmt3L-deficient prospermatogonia. (*a*) Formation of highly aberrant complexes of synaptonemal proteins in the form of interlocked rings and complex three-dimensional structures in spermatocytes that had lacked Dnmt3L in the earlier prospermatogonial stage. (*b*) Branching and anastomosing synaptonemal complexes in Dnmt3L-deficient spermatocytes. Many chromosomes are unpaired or engaged in nonhomologous synapsis; such nonhomologous pairing is not seen in control specimens. Similar staining patterns were seen after labeling with antibodies to Scp3 or combination of Scp1 and Scp3 antibodies.

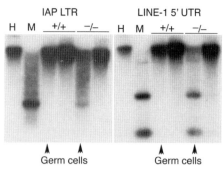
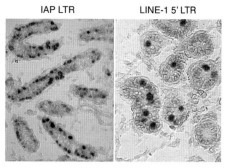

Figure 4. Transposon demethylation in Dnmt3L-deficient male germ cells. Germ cells from 17 dpp testes were purified by fluorescence activated cell sorting after staining for GCNA1 (Enders and May 1994). DNA was digested with the methylation-sensitive restriction endonuclease *HpaII* prior to DNA blot analysis with the indicated probes, except in lanes headed M, which contained DNA that had been cleaved with *MspI*, a methylation-insensitive isoschizomer of *HpaII*. Full methylation of IAP LTR and LINE-1 5′ UTR is visible in the wild-type lanes; note substantial demethylation in Dnmt3L-deficient male germ cells. (*Arrowheads*) Wild type (*left* in each blot) and Dnmt3L$^{-/-}$ (*right* in each blot).

Figure 5. Transcriptional reactivation of retrotransposons in Dnmt3L-deficient germ cells. In situ hybridization against sections of testes from mice at 2 dpp showed expression of high levels of LINE-1 and IAP transcripts in Dnmt3L-deficient prospermatogonia. Further analysis showed expression of LINE-1 and IAP transposons in dividing spermatogonia and spermatocytes as well.

jor satellite (but not of other sequences) is affected by loss of the histone methyltransferases Suv39h1 and Suv39h2 (Lehnertz et al. 2003).

The host defense hypothesis predicts that demethylation of transposons will cause their transcriptional activation (Yoder et al. 1997; Bestor 2003). As shown in Figure 5, there is massive reactivation of LINE-1 and IAP transcripts observed by in situ hybridization. Dnmt3L is therefore the first gene shown to be required for the silencing of transposons in germ cells of any organism. It is notable that homozygous loss of function mutations in Dnmt1 causes reaction of IAP transcription in somatic cells (Walsh et al. 1998), but LINE-1 elements are not reactivated (Bourc'his and Bestor 2004). LINE-1 elements are believed to be the source of reverse transcriptase for most retroposons, and the coexpression of IAP elements and LINE-1 elements suggests that active transposition of multiple retroposon classes will occur in Dnmt3L-deficient germ cells.

Kaneda et al. (2004) reported complete demethylation of the *H19* DMR in Dnmt3L-deficient male germ cells. Our analysis shows only partial (50%) demethylation in a larger set of data (Fig. 6). The recovery of only a single sequence by Kaneda et al. suggests that they may have sequenced the PCR products arising from a single demethylated DNA molecule. Our bisulfite sequencing results, which show no evidence of clonality, indicate that that removal of Dnmt3L has a much smaller effect on the establishment of methylation imprints at the *H19* DMR than was reported by Kaneda et al.

As shown in Figures 1 and 7, Dnmt3L is evolving at a much higher rate than any other mammalian DNA methyltransferase-related protein; mouse and human Dnmt3L proteins are <60% identical, while Dnmt1, Dnmt2, Dnmt3A, and Dnmt3B are all >80% identical. The rapid evolution of Dnmt3L is likely to reflect the involvement of the protein in transposon control. Host defense measures place transposons under selective pressure to evade these measures, which in turn pressures the host to evolve new countermeasures. This evolutionary chase is expected to manifest as rapid evolution of the regulatory factor in the host defense system; analysis of rates of sequence divergence indicate that Dnmt3L is under positive selection for rapid evolution and is likely to play a regulatory role, while the almost perfect conservation of mouse and human Dnmt3A suggests that this protein is under strong negative selection and has an essential catalytic function.

SEXUAL DIMORPHISM IN GENOMIC IMPRINTING

The sexual dimorphism in Dnmt3L phenotypes is striking and without precedent. The loss of Dnmt3L from male germ cells has a far smaller effect on imprint estab-

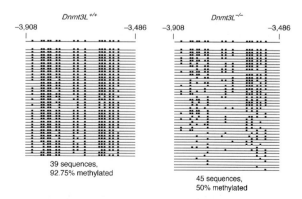

Figure 6. Partial imprint establishment at H19 in Dnmt3L-deficient male germ cells. Germ cells were isolated by flow sorting after staining with antibodies to GCNA1 (Enders and May 1994) and DNA subjected to methylation analysis by bisulfite genomic sequencing (Bourc'his et al. 2001). Establishment of imprints at H19 is only partially dependent on Dnmt3L

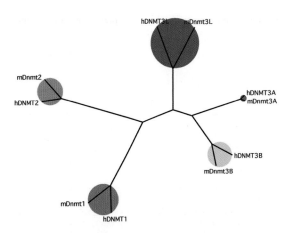

Figure 7. Rapid divergence of Dnmt3L protein sequence as compared to other mammalian DNA methyltransferase family members. ClustalW analysis of full-length sequences of mouse and human DNA methyltransferase-related sequences is shown. Dnmt3L can be seen to be diverging much more rapidly than other Dnmt pairs; sequence identity is <60% for mouse and human Dnmt3L.

lishment than in female germ cells. Imprints at *H19* are only slightly reduced in Dnmt3L-deficient male germ cells, and there is no indication of activation of expression of the *H19* gene in these cells (Bourc'his and Bestor 2004). Dnmt3L is required for imprint establishment in female germ cells but not for normal meiosis or for global genome methylation, but in male germ cells is dispensable for imprint establishment but is required for meiosis and global genome methylation. Imprinting is therefore strongly dimorphic both in terms of the categories of genes that are imprinted in male and female germ cells (Reik and Walter 2001) and in the mechanisms that mediate promoter silencing in the two sexes (Bourc'his et al. 2001, and data shown here).

DEDUCTION OF THE BIOLOGICAL EFFECT OF GENOMIC IMPRINTING FROM THE PHENOTYPE OF IMPRINT-FREE MOUSE EMBRYOS

The nature of the biological function or functions of genomic imprinting has been controversial in part because phenotypic analysis has been restricted to animals that lack one or a few imprinted genes or that have only maternal or paternal imprints (Barton et al. 1984; McGrath and Solter 1984). The true nature of genomic imprinting could be revealed most clearly by analysis of development of mice that lack imprints altogether. This is now possible: Parthenogenetic activation of homozygous Dnmt3L$^-$ oocytes with restoration of diploidy by inhibition of second polar body extrusion can be used to derive mouse embryos that lack both maternal and paternal imprints without global genome demethylation, as would occur if conditional alleles of Dnmt1 were to be employed. While there are recent indications that paternal imprints may be partially or completely dispensable for normal development under some conditions (Kono et al. 2004), it should be noted that the phenotypic differences between embryos derived by parthenogenetic activation of homozygous mutant Dnmt3L oocytes and fertilized Dnmt3L mutant oocytes will provide a direct measurement of the importance of and the biological effects of paternal imprints.

ACKNOWLEDGMENTS

Supported by grants from the National Institutes of Health and by a fellowship from the Rett Syndrome Research Foundation. We thank members of the laboratory for stimulating discussions.

REFERENCES

Aapola U., Kawasaki K., Scott H.S., Ollila J., Vihinen M., Heino M., Shintani A., Kawasaki K., Minoshima S., Krohn K., Antonarakis S.E., Shimizu N., Kudoh J., and Peterson P. 2000. Isolation and initial characterization of a novel zinc finger gene, DNMT3L, on 21q22.3, related to the cytosine-5-methyltransferase 3 gene family. *Genomics* **65**: 293.

Barton S.C., Surani M.A., and Norris M.L. 1984. Role of paternal and maternal genomes in mouse development. *Nature* **311**: 374.

Bestor T.H. 1992. Activation of mammalian DNA methyltransferase by cleavage of a Zn-binding regulatory domain. *EMBO J.* **11**: 2611.

———. 2000. The DNA methyltransferases of mammals. *Hum. Mol. Genet.* **9**: 2395.

———. 2003. Cytosine methylation mediates sexual conflict. *Trends Genet.* **19**: 185.

Bestor T.H., Laudano A., Mattaliano R., and Ingram V. 1988. Cloning and sequencing of a cDNA encoding DNA methyltransferase of mouse cells. The carboxyl-terminal domain of the mammalian enzyme is related to bacterial restriction methyltransferases. *J. Mol. Biol.* **203**: 971.

Bourc'his D. and Bestor T.H. 2004. Meiotic catastrophe and retrotransposon reactivation in male germ cells lacking Dnmt3L. *Nature* **431**: 96.

Bourc'his D., Xu G.L., Lin C.S., Bollman B., and Bestor T.H. 2001. Dnmt3L and the establishment of maternal genomic imprints. *Science* **294**: 2536.

Chen R.Z., Pettersson U., Beard C., Jackson-Grusby L., and Jaenisch R. 1998. DNA hypomethylation leads to elevated mutation rates. *Nature* **395**: 89.

Davis T.L., Trasler J.M., Moss S.B., Yang G.J., and Bartolomei M.S. 1999. Acquisition of the H19 methylation imprint occurs differentially on the parental alleles during spermatogenesis. *Genomics* **58**: 18-28.

Ding F. and Chaillet J.R. 2002. In vivo stabilization of the Dnmt1 (cytosine-5)-methyltransferase protein. *Proc. Natl. Acad. Sci.* **99**: 14861.

Dong A., Yoder J.A., Zhang X., Zhou L., Bestor T.H., and Cheng X. 2001. Structure of human DNMT2, an enigmatic DNA methyltransferase homologue that displays denaturant-resistant binding to DNA. *Nucleic Acids Res.* **29**: 439.

Enders G.C. and May J.J., II. 1994. Developmentally regulated expression of a mouse germ cell nuclear antigen examined from embryonic day 11 to adult in male and female mice. *Dev. Biol.* **163**: 331.

Goll M.G. and Bestor T.H. 2002. Histone modification and replacement in chromatin activation. *Genes Dev.* **16**: 1739.

Henikoff S., Furuyama T., and Ahmad K. 2004. Histone variants, nucleosome assembly and epigenetic inheritance. *Trends Genet.* **7**: 320.

Holliday R. and Pugh J.E. 1975. DNA modification mechanisms and gene activity during development. *Science* **187**: 226.

Howell C.Y., Bestor T.H., Ding F., Latham K.E., Mertineit C., Trasler J.M., and Chaillet J.R. 2001. Genomic imprinting disrupted by a maternal-effect mutation in the Dnmt1 gene. *Cell* **104:** 829.

Kaneda M., Okano M., Hata K., Sado T., Tsujimoto N., Li E., and Sasaki H. 2004. Essential role for de novo DNA methyltransferase 3a in paternal and maternal imprinting. *Nature* **429:** 900.

Kono T., Obata Y., Yoshimzu T., Nakahara T., and Carroll J. 1996. Epigenetic modifications during oocyte growth correlates with extended parthenogenetic development in the mouse. *Nat. Genet.* **13:** 91.

Kono T., Obata Y., Wu Q., Niwa K., Ono Y., Yamamoto Y., Park E.S., Seo J.S., and Ogawa H. 2004. Birth of parthenogenetic mice that can develop to adulthood. *Nature* **428:** 860.

La Salle S., Mertineit C., Taketo T., Moens P.B., Bestor T.H., and Trasler J.M. 2004. Windows for sex-specific methylation marked by DNA methyltransferase expression profiles in mouse germ cells. *Dev. Biol.* **268:** 403.

Lehnertz B., Ueda Y., Derijck A.A., Braunschweig U., Perez-Burgos L., Kubicek S., Chen T., Li E., Jenuwein T., and Peters A.H. 2003. Suv39h-mediated histone H3 lysine 9 methylation directs DNA methylation to major satellite repeats at pericentric heterochromatin. *Curr. Biol.* **13:** 1192.

Leonhardt H., Page A.W., Weier H.-U., and Bestor T.H. 1992. A targeting sequence directs DNA methyltransferase to sites of DNA replication in mammalian nuclei. *Cell* **71:** 865.

Li E., Beard C., and Jaenisch R. 1993. Role for DNA methylation in genomic imprinting. *Nature* **366:** 362.

Li E., Bestor T.H., and Jaenisch R. 1992. Targeted mutation of the DNA methyltransferase gene results in embryonic lethality. *Cell* **69:** 915.

Lorincz M.C., Schubeler D., Hutchinson S.R., Dickerson D.R., and Groudine M. 2002. DNA methylation density influences the stability of an epigenetic imprint and Dnmt3a/b-independent de novo methylation. *Mol. Cell. Biol.* **22:** 7572.

McGrath J. and Solter D. 1984. Completion of mouse embryogenesis requires both the maternal and paternal genomes. *Cell* **37:** 179.

Mertineit C., Yoder J.A., Taketo T., Laird D.W., Trasler J.M., and Bestor T.H. 1998. Sex-specific exons control DNA methyltransferase in mammalian germ cells. *Development* **125:** 889.

Okano M., Xie S., and Li E. 1998. Dnmt2 is not required for de novo and maintenance methylation of viral DNA in embryonic stem cells. *Nucleic Acids Res.* **26:** 2536.

Okano M., Bell D.W., Haber D.A., and Li E. 1999. DNA methyltransferases Dnmt3a and Dnmt3b are essential for de novo methylation and mammalian development. *Cell* **99:** 247.

Panning B. and Jaenisch R. 1996. DNA hypomethylation can activate Xist expression and silence X-linked genes. *Genes Dev.* **15:** 1991.

Reik W. and Walter J. 2001. Genomic imprinting: Parental influence on the genome. *Nat. Rev. Genet.* **2:** 21.

Riggs A.D. 1975. X inactivation, differentiation, and DNA methylation. *Cytogenet. Cell Genet.* **14:** 9.

Walsh C.P. and Bestor T.H. 1999. Cytosine methylation and mammalian development. *Genes Dev.* **13:** 26.

Walsh C.P., Chaillet J.R., and Bestor T.H. 1998. Transcription of IAP endogenous retroviruses is constrained by cytosine methylation. *Nat. Genet.* **20:** 116.

Xu G.-L., Bestor T.H., Bourc'his D., Hsieh C.-L., Tommerup N., Bugge G., Hulten M., Qu X., Russo J.J., and Viegas-Péquignot E. 1999. Chromosome instability and immunodeficiency syndrome caused by mutations in a DNA methyltransferase gene. *Nature* **402:** 187.

Yoder J.A. and Bestor T.H. 1998. A candidate mammalian DNA methyltransferase related to pmt1p of fission yeast. *Hum. Mol. Genet.* **7:** 279.

Yoder J.A., Walsh C.P., and Bestor T.H. 1997. Cytosine methylation and the ecology of intragenomic parasites. *Trends Genet.* **13:** 335.

RNA Interference, Heterochromatin, and Centromere Function

R.C. ALLSHIRE

Wellcome Trust Centre for Cell Biology, University of Edinburgh, The King's Buildings, Edinburgh EH9 3JR, Scotland, United Kingdom

The centromere regions of most eukaryotic metaphase chromosomes are distinctive. The narrower, more compact appearance of these regions reflects two important related features: centromeric cohesion and extensive blocks of heterochromatin. First, sister centromeres remain associated because the cohesin complex remains intact at centromeres until anaphase. Second, the chromatin coating these regions is very different since they remain condensed throughout the cell cycle and are clearly visible as the most concentrated foci of DNA in interphase cells; this has been termed heterochromatin (Bernard and Allshire 2002).

It has been known for many years that marker genes placed close to or within such centromeric heterochromatin regions are transcriptionally silenced, often resulting in a variegated phenotype (PEV) reflecting variability in the ability to silence marker genes. Genetic analyses in *Drosophila* allowed the identification of two important conserved proteins involved in the assembly of this silent heterochromatin. The original mutants were termed Su(var) (suppressor of variegation): The *su(var)3-9* and *su(var)2-5* genes were found to encode Suv39 (histone H3 lysine 9 methyl transferase) and HP1 (Heterochromatin protein 1), respectively (Reuter and Spierer 1992; Dillon and Festenstein 2002; Richards and Elgin 2002; Schotta et al. 2003). Both Suv39 and HP1 proteins associate with each other and are concentrated in the large blocks of heterochromatin in pericentromeric regions in both fly and mammalian cells (James et al. 1989; Wreggett et al. 1994; Aagaard et al. 1999; Minc et al. 1999).

Fission yeast centromere regions bear some resemblance to the large repetitive structures found at fly, plant, and mammalian centromeres (Takahashi et al. 1992; Steiner et al. 1993). Large inverted repetitive structures surround a central core domain (Fig. 1). The central region and the inner repeats have a relatively high A+T content and are packaged in unusual chromatin that generates a diffuse smear, rather than a canonical ladder pattern, upon limited digestion with micrococcal nuclease (Polizzi and Clarke 1991; Takahashi et al. 1992). This probably reflects the fact that most H3 is replaced by the conserved kinetochore specific histone H3-like protein CENP-A[cnp1] in this central domain (B. Mellone and R.C. Allshire, unpubl.). The outer repeats are coated in nucleosomes which are generally underacetylated on lysines in the amino-terminal tails of histones H3 and H4. This hypoacetylated state is important for centromere integrity and function (Ekwall et al. 1997).

The formation of a functional centromere and kinetochore assembly requires at least one outer repeat and a central core region (Takahashi et al. 1992; Marschall and Clarke 1995; Ngan and Clarke 1997). The central core alone is unable to be packaged in unusual chromatin when replicated on an extrachromosomal plasmid or other sites in the genome or when replicated in *Saccharomyces cerevisiae* (Polizzi and Clarke 1991; Takahashi et al. 1992). Thus, this "unusual" CENP-A chromatin is not specified by the underlying DNA sequence within this region alone, but it must be induced by the context in which these sequences are placed. The outer repeats must provide this contextual specificity (Fig. 1).

VARIABLE ESTABLISHMENT OF FUNCTIONAL CENTROMERES

It is apparent that epigenetic processes act to govern the assembly of a functional centromere in fission yeast. First, erasing the hypoacetylated state of the outer repeats by transiently blocking histone deacetylation (and presumably histone H3 lysine 9 methylation) generates a defective, hypoacetylated centromere state that can be propagated through several divisions and meiosis (Ekwall et al. 1997). Second, truncated centromeres on episomal plasmids in the fission yeast nucleus can provide centromere function resulting in reasonable mitotic segregation (Takahashi et al. 1992; Baum et al. 1994). However, the introduction of minimal constructs with part of an outer repeat plus a central domain frequently appear nonfunctional. Nevertheless, in a variable proportion of these transformants functional centromeres are established, imparting mitotic stability on the episome (Steiner and Clarke 1994; Ngan and Clarke 1997). Once established this functional state is propagated through mitotic and meiotic divisions. It seems likely that these minimal constructs struggle to set up the contextual information required to trigger CENP-A and kinetochore assembly over the adjacent central domain; however, this has not been directly assessed. Consistent with this, only constructs that establish the functional state display "unusual" central core/kinetochore chromatin, but it is not known how this correlates with CENP-A incorporation (Marschall and Clarke 1995). Even so, the removal of outer repeat hete-

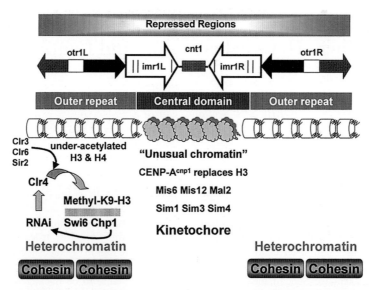

Figure 1. Schematic diagram of fission yeast centromere 1. Outer repeats (otr1) along with innermost repeats (imr1) form a long, almost perfect 18-kb inverted repeat around the central core (Cnt1). Vertical lines represent tRNA genes. The imr and cnt1 make up the central domain. Insertion of the *ura4* gene anywhere within the outer repeats or central domain results in its transcriptional repression. Different factors are required to mediate silencing within the central domain and the outer repeats. An "unusual" chromatin structure is associated with the central domain where H3 is replaced by the kinetochore-specific histone H3-like protein CENP-A. Mutations of the gene encoding CENP-A^{cnp1} (sim2), Sim1, Sim3, Sim4, Mis6, Mis12, or Mal2 all specifically affect central core silencing. CENP-A, Sim4, Mis6, Mis12, and Mal2 are kinetochore-specific proteins and only associate with the central domain. Where examined, most mutants that affect kinetochore function disrupt the "unusual" central core chromatin structure. Nucleosomes with hypoacetylated amino-terminal tails on H3 and H4 coat the outer repeats, thus the action of HDACs allows dimethylation of lysine 9 on H3 by the histone methyltransferase Clr4 and binding of Swi6 and Chp1 via their chromodomains. This results in the recruitment of a high density of cohesin over the outer repeats and this mediates sister centromere cohesion counteracting the pulling forces upon bilateral spindle attachment. The RNAi components Ago1, Dcr1, Rdp1, and Tas3 are required to process outer repeat transcripts and bring about methylation of histone H3 on lysine 9. Chp1 is also required for methylation of lysine 9 H3. Methylation of lysine 9 still occurs in the absence of Swi6.

rochromatin from established functional centromeres (e.g., by deletion of RNAi components, *clr4+*, or *swi6+*) leaves centromeres that retain a reasonably functional kinetochore. In contrast, DNA constructs lacking outer repeat DNA are unable to establish the functional state. This implies that the heterochromatic outer repeats are required to provide a favorable environment for kinetochore assembly but that once established the kinetochore no longer requires flanking heterochromatin for its propagation.

PROPAGATION OF THE FUNCTIONAL CENTROMERE STATE

Because CENP-A chromatin must be pivotal in inducing kinetochore assembly over the central region, there must be specific factors that ensure that CENP-A is only assembled into chromatin in this central region and not elsewhere on chromosomes. Mutations in several kinetochore factors affect the "unusual" kinetochore specific chromatin structure and CENP-A accumulation at centromeres (Fig. 1) (Saitoh et al. 1997; Goshima et al. 1999; Takahashi et al. 2000; Jin et al. 2002; Pidoux et al. 2003). Analyses in other organisms indicate that sites of kinetochore assembly are plastic and epigenetically regulated (Karpen and Allshire 1997; Sullivan et al. 2001). Why might this be? It is well known that there must be one and only one centromere per chromosome: Dicentric chromosomes are unstable and this genome instability leads to loss of genetic information. Consequently it would seem sensible to have the process of active centromere duplication and propagation coupled in some manner to the machinery that habitually senses the formation of functional kinetochores. It is well documented that eukaryotes possess a metaphase–anaphase checkpoint that operates by sensing microtubule attachment and thus tension generated between sister centromeres upon capturing microtubules rooted in opposite poles (Millband et al. 2002; Musacchio and Hardwick 2002).

An attractive idea is that the tension and associated conformational changes generated by correct biorientation of sister kinetochores and progression into anaphase induces a "mark" that is subsequently interpreted to allow CENP-A incorporation (Fig. 2) (Mellone and Allshire 2003). This mark could be read by either a replication-coupled or a replication-independent mechanism of CENP-A deposition and chromatin assembly. Such a device would allow fully active functional centromeres to be propagated at the same site through multiple cell divisions whereas defective centromeres might wither after several divisions. In addition, if the correct circumstances contrive to generate a favorable context elsewhere on a chromosome, CENP-A incorporation might allow the establishment of a new site of kinetochore assembly—a neocentromere (Karpen and Allshire 1997; Sullivan et al. 2001; Amor and Choo 2002)—which, if it biorients on a regular basis, will be propagated through multiple divisions.

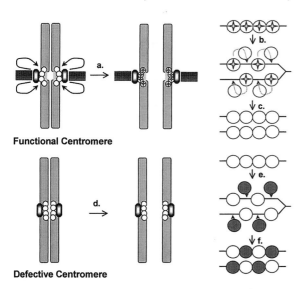

Figure 2. Model for the propagation and loss of a site of centromere activity. Active centromeres allow the assembly of sister kinetochores that capture microtubules emanating from opposite spindle poles during mitosis, resulting in biorientation, chromatin stretching, and the generation of tension between sister kinetochores. Microtubule attachment and tension are recognized as signals for progression into anaphase (*a*) when sister chromatids separate. We propose that the resulting tension of a correctly bioriented kinetochore also marks the underlying active kinetochore chromatin in some manner (directly or indirectly) (✧) so that perhaps during the next S phase this mark is recognized at the replication fork and ensures that CENP-A nucleosomes rather than H3 nucleosomes are assembled in its wake (*b*). Completion of S phase results in two sister centromeres with a full complement of CENP-A and thus a robust kinetochore (*c*). Alternatively, replication-independent assembly mechanisms may recognize the marked CENP-A chromatin and strip out any potentially hindering H3, replacing it with CENP-A after replication is completed. If a centromere fails to biorient, no tension is generated so that the mark is not placed at the centromere at metaphase/anaphase. Thus in the following S phase (*e*), there is no signal to direct CENP-A chromatin assembly and therefore normal H3-containing nucleosomes are incorporated and remain at this defective centromere (*f*). If this centromere fails again to biorient it will again lose more CENP-A, and thus it will be more likely to lose function completely.

OUTER REPEAT HETEROCHROMATIN

The outer repeats of fission yeast centromeres resemble to some extent the arrays of repeats associated with metazoan centromere regions. Because no steady-state transcripts could be detected, it was suggested that they may be heterochromatic (Fishel et al. 1988). It has also been noted that these tandem outer repeats resemble transposable elements (Halverson et al. 1997). Mobile elements of various types are associated with centromere regions of many species and may contribute to the heterochromatic nature of these regions. The high density of various repeats can be detected as heterochromatin coincident with the large foci of DAPI staining in interphase nuclei. In general, genes that become embedded in this heterochromatin are transcriptionally silenced (Reuter and Spierer 1992; Dillon and Festenstein 2002; Richards and Elgin 2002; Schotta et al. 2003). The insertion of the *ura4*⁺ marker gene at any site within the centromeric outer repeats, inner repeats, or central core results in repression of its transcription (Allshire et al. 1994, 1995). The quality of silencing associated with the central region (imr + central core) and outer repeats differs and presumably reflects the distinctly different chromatin associated with these regions. Silencing within the central kinetochore region is unstable so that expressing and nonexpressing colonies are generated at high frequencies. Marker genes with compromised promoters allow more robust phenotypic silencing when inserted in the central core. This has allowed genetic screens for factors affecting repression within the central domain (Pidoux et al. 2003). Such screens have identified CENP-A itself, novel kinetochore proteins, and other proteins that are required for the association of normal levels of CENP-A with the central region (Fig. 1).

In contrast, silencing of markers placed in the outer repeats is much tighter with very few expressing colonies detected. The outer repeats are packaged in hypoacetylated nucleosomes (Ekwall et al. 1997). This state is mediated by the action of three histone deacetylases, Clr3, Clr6, and Sir2 (Nakayama et al. 2001; Bjerling et al. 2002; Shankaranarayana et al. 2003). Clr3 activity is mainly directed toward deacetylation of histone H3 on lysine 14; Clr6 has a more broad affect on deacetylation of several lysine residues in H3 and H4; and Sir2 participates in deacetylating lysine 9 in the process of spreading the assembly of this silent chromatin outward from a nucleation site along the chromatin fiber.

The action of these histone deacetylases paves the way for the activity of the histone methyltransferase Clr4 (orthologous to Suv39) (Rea et al. 2000). Clr4 methylates histone H3 on lysine 9 forming a binding site for the key heterochromatin protein Swi6 (counterpart of HP1) to form transcriptionally repressive chromatin at centromeres (Allshire et al. 1995; Bannister et al. 2001; Lachner et al. 2001; Nakayama et al. 2001).

Small fragments (~1 kb) taken from the centromeric outer repeats can mediate silent chromatin formation when placed at an ectopic euchromatic locus (Partridge et al. 2002). What directs this silent chromatin assembly? Such experiments might suggest that DNA sequence specific binding proteins simply directly recruit chromatin modifying and remodeling complexes that mediate heterochromatin formation. However, the astonishing finding that the RNA interference pathway is required for the formation of this outer repeat heterochromatin made a seminal link between heterochromatin and RNAi that is likely conserved (Fig. 1) (Volpe et al. 2002).

Deletion of genes encoding three components of the RNAi pathway, Argonaute (Ago1), Dicer (Dcr1), or RNA-dependent RNA polymerase (Rdp1), leads to loss of methylation of histone H3 on lysine 9 of, and Swi6 binding to, outer repeat chromatin coupled with alleviation of marker gene silencing within this centromeric heterochromatin but not in the central domain occupied by the kinetochore (Volpe et al. 2002). Noncoding transcripts of specific sizes from the outer repeats can be detected in these cells that lack active RNAi but not in wild-type cells. Transcripts from top and bottom outer repeat strands, which overlap in sequence, are produced. Thus this potentially allows the formation of double-stranded RNA, which then serves as a substrate for Dicer and the generation of homologous siRNAs (Reinhart and Bartel 2002; Volpe et al. 2002). The incorporation of these siRNAs into a RNA-induced silencing complex (RISC)-like complex (e.g., RNA-induced initiation of transcriptional gene silencing [RITS]) allows homologous loci to be targeted and degraded (Verdel et al. 2004).

The chromodomain protein Chp1 is required for methylation of histone H3 on lysine 9, binds dimethylated lysine 9 H3 peptide in vitro, and is dependent on the activity of the Clr4 lysine 9 H3 HMTase for recruitment to outer repeats in vivo (Partridge et al. 2002). Chp1 has been shown to interact with Ago1 in the RITS complex making a link between the RNAi machinery and silent chromatin (see Fig. 1) (Verdel et al. 2004).

HOW DO siRNAs EFFECT CHROMATIN MODIFICATION ON HOMOLOGOUS DNA?

Our view is that siRNAs act on nascent transcripts and that the chromatin modifying activities, a lysine 9 histone H3 deacetylase (Clr3/Clr6) and histone methyltransferase (Clr4), piggyback on the RNAi machinery to modify the juxtaposed chromatin (Fig. 3). An alternative model is suggested by experiments in plants that demonstrated that

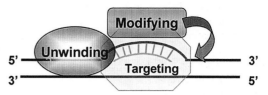

Figure 4. siRNA–DNA interactions in chromatin modification and silencing. The hybridization of siRNA with homologous DNA may induce the recruitment of DNA/chromatin-modifying activities that promote silent chromatin assembly. siRNA may gain access passively during replication or alternatively DNA unwinding activities may act in a complex with siRNAs to unwind DNA and allow access.

siRNAs homologous to a transgene promoter can also bring about RNA-dependent target modification and robust silencing (Fig. 4). Because it is expected that the promoter itself is not transcribed, the prevailing view is that siRNA–DNA interactions could mediate silencing of these transgenes (Jones et al. 1999; Mette et al. 2000).

However, our analyses in fission yeast suggest that transcription of the homologous target sequences is required (R.C. Allshire and V. Schramke, unpubl.). This suggests that siRNA may home in on nascent transcripts and in so doing recruit various histone-modifying activities (HMAs: HDACs and HMTase) that mediate histone modification and heterochromatin assembly. This is somewhat analogous to the association of other histone-modifying activities with elongating RNA polymerase II; for example, the SET1 and SET2 H3 lysine 4 and lysine 36 HMT´ases are recruited via the CTD of RNApolII (Hampsey and Reinberg 2003). It thus seems plausible that siRNAs and associated proteins may also allow the recruitment of histone deacetylases and Clr4 histone metyltranferase to a homologous nascent transcript in the act of being produced by elongating RNA polymerase.

RETROTRANSPOSONS SHOW A SIMILAR REGULATION TO CENTROMERE REPEATS

If the repetitive regions associated with centromeres are indeed derived from centromeres, one might expect that retrotransposons and remnant, solo long terminal repeats (LTRs) may also be assembled in RNAi-dependent heterochromatin. Our analyses demonstrated that fission yeast TF1 and TF2 solo LTRs are packaged in chromatin that is methylated on lysine 9 of histone H3 and, just like at centromeres, this heterochromatin is lost when $ago1^+$, $dcr1^+$, or $rdp1^+$ are deleted, resulting in the detection of transcripts homologous to both template strands (Schramke and Allshire 2003). Further analyses indicated that the repression of nearby genes is enforced by these LTRs.

This raises the possibility that the generation of double-stranded RNA alone is sufficient to bring about the assembly of heterochromatin on homologous DNA. To test this we expressed a hairpin RNA homologous to the middle of the $ura4^+$ transcription unit (Fig. 3). This resulted in loss of $ura4^+$ transcripts, methylation of the $ura4^+$ gene, recruitment of Swi6, and Rad21 cohesin

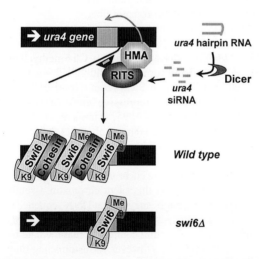

Figure 3. Hairpin RNAs induce heterochromatin assembly on a euchromatic gene. Expression of a 280-bp region as an inverted hairpin, homologous to a central portion of the $ura4^+$ gene, results in transcriptional silencing of the normally euchromatic $ura4^+$ gene. The resulting siRNAs generated by the action of Dicer are taken up by a RISC-like complex (e.g., RITS) and home in on homologous nascent ura4 transcripts at the $ura4^+$ locus in the nucleus. Association with nascent transcripts allows the recruitment of histone-modifying activities (HMAs: HDACs and Clr4 HMTase) that mediate methylation of histone H3 in the vicinity, allowing recruitment of Swi6 and even Rad21-cohesin. This silent chromatin spreads approximately 750 bp into the $ura4^+$ promoter in wild-type cells. In cells lacking Swi6, methylation is still detected over the region of $ura4^+$ homologous to the 280-bp hairpin but it no longer spreads to the promoter. This is consistent with Swi6 binding allowing further Clr4 recruitment so that silent chromatin extends along the chromatin fiber from the nucleation site (Shankaranarayana et al. 2003).

(Schramke and Allshire 2003). Therefore, it appears that no specific DNA sequence is required for the assembly of heterochromatin. Thus silent chromatin assembly may be inextricably linked to the generation of homologous double-stranded RNA and may form part of an endogenous surveillance mechanism that acts to clear such aberrant dsRNAs from the cell.

WHAT IS THE ROLE OF HETEROCHROMATIN AT CENTROMERES?

Heterochromatin regions are known to be "sticky" and sister chromatids remain associated in heterochromatic regions (Bernard and Allshire 2002). In the absence of RNAi, Clr4, or Swi6, cells display elevated rates of chromosome loss and a specific defect in mitosis, lagging chromosomes on late anaphase spindles (Ekwall et al. 1995; 1996; Hall et al. 2003; Volpe et al. 2003). It is now clear that mutants that lack Swi6 at centromeres also display a defect in cohesion at centromeres but not along chromosome arms (Bernard et al. 2001; Nonaka et al. 2002). Thus, centromeric Swi6 heterochromatin is required to mediate strong physical cohesion between sister centromeres and thereby ensure normal chromosome segregation. The defect in segregation must be due to aberrant kinetochore–spindle microtubule interactions. One possibility is that centromeres lacking heterochromatin are disorganized so that the multiple microtubule (MT) binding sites at a stable kinetochore are not rigidly oriented toward one pole but can connect with MTs rooted in opposite poles. This configuration is known as merotelic orientation and is very prevalent in somatic vertebrate cells (Cimini et al. 2001). We propose that the heterochromatin that flanks the kinetochore is required to present the kinetochore in a rigid structure so that the multiple MT binding sites all face the same direction and only connect with MTs emanating from the same pole (Fig. 5).

CONCLUSION

The fission yeast genome is borne on three chromosomes and fission yeast centromeres are 150 times larger than the best characterized "point" centromeres in budding yeast. The main difference is that fission yeast centromeres are subdivided into kinetochore chromatin and flanking repetitive heterochromatin, an organization that resembles that of regional metazoan centromeres. Surprisingly, budding yeast lacks centromeric heterochromatin, RNAi components, histone H3 methylated on lysine 9, and a Swi6/HP1 homolog. One of the main differences is that fission yeast centromeres associate with multiple microtubules during mitosis whereas one microtuble contacts each *S. cerevisiae* centromere (Ding et al. 1993; O'Toole et al. 1999). It seems that one role for this heterochromatin is to mediate robust physical cohesion between sister centromeres and thus presumably to counteract the pulling forces created by biorientation. However, this heterochromatin also appears to play an architectural role preventing merotelic association of multiple microtubule binding sites at a single kinetochore from capturing spindle fibers rooted in opposite poles. Another possible role for this heterochromatin is that it may provide a favorable environment for the assembly of kinetochore-specific CENP-A^{cnp1} chromatin leading to kinetochore formation.

The fact that similar RNAi-dependent Swi6-silent chromatin is formed over retrotransposon LTRs and centromere outer repeats suggests that these outer repeats may be derived from ancient mobile elements, which now take on a key role in centromere structure and function. The fact that a synthetic hairpin RNA can also elicit heterochromatin assembly on homologous genes indicates that no "magic" *cis* acting sequence is necessary for heterochromatin assembly. The siRNAs generated by the double-stranded RNA are sufficient to trigger the required chromatin modification events at the homologous locus.

ACKNOWLEDGMENTS

R.C.A. thanks past and present members of the lab for discussions and input, especially Alison Pidoux and Vera Schramke. Thanks also to Elaine Dunleavy and Sharon White for reading and making comments on the manuscript. R.C.A. is a Wellcome Trust Principal Research Fellow.

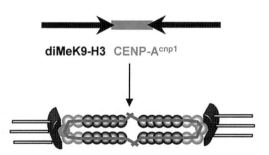

Figure 5. A structural role for heterochromain at centromeres. RNAi, Clr4, Swi6, and cohesin mutants all display a high frequency of lagging chromosomes on late anaphase spindles. Centromeres in metazoans are known to be modular and other observations indicate that stretches of CENP-A and H3 chromatin are intermingled along the chromatin fiber. This has prompted the suggestion that these fibers are coiled or looped to present the CENP-A chromatin on the outer surface of the chromosome so that it induces kinetochore assembly on opposite sides ensuring that sisters interact with microtubules rooted in opposite poles (Sullivan et al. 2001). The left and right sides of fission yeast centromeres may fold on each other, forming a superstructure that holds the central CENP-A domains in a rigid back-to-back orientation to promote biorientation (see Fishel et al. 1988; Takahashi et al. 1992). Loss of heterochromatin disrupts this superstructure, allowing perhaps merotelic orientation of single kinetochores and thus lagging chromosomes during late anaphase.

REFERENCES

Aagaard L., Laible G., Selenko P., Schmid M., Dorn R., Schotta G., Kuhfittig S., Wolf A., Lebersorger A., Singh P.B., Reuter G., and Jenuwein T. 1999. Functional mammalian homologues of the *Drosophila* PEV-modifier Su(var)3-9 encode centromere-associated proteins which complex with the heterochromatin component M31. *EMBO J.* **18:** 1923.

Allshire R.C., Javerzat J.P., Redhead N.J., and Cranston G. 1994. Position effect variegation at fission yeast centromeres. *Cell* **76:** 157.

Allshire R.C., Nimmo E.R., Ekwall K., Javerzat J.P., and Cranston G. 1995. Mutations derepressing silent centromeric domains in fission yeast disrupt chromosome segregation. *Genes Dev.* **9:** 218.

Amor D.J. and Choo K.H. 2002. Neocentromeres: Role in human disease, evolution, and centromere study. *Am. J. Hum. Genet.* **71:** 695.

Bannister A.J., Zegerman P., Partridge J.F., Miska E.A., Thomas J.O., Allshire R.C., and Kouzarides T. 2001. Selective recognition of methylated lysine 9 on histone H3 by the HP1 chromo domain. *Nature* **410:** 120.

Baum M., Ngan V.K., and Clarke L. 1994. The centromeric K-type repeat and the central core are together sufficient to establish a functional *Schizosaccharomyces pombe* centromere. *Mol. Biol. Cell* **5:** 747.

Bernard P. and Allshire R. 2002. Centromeres become unstuck without heterochromatin. *Trends Cell Biol.* **12:** 419.

Bernard P., Maure J.F., Partridge J.F., Genier S., Javerzat J.P., and Allshire R.C. 2001. Requirement of heterochromatin for cohesion at centromeres. *Science* **294:** 2539.

Bjerling P., Silverstein R.A., Thon G., Caudy A., Grewal S., and Ekwall K. 2002. Functional divergence between histone deacetylases in fission yeast by distinct cellular localization and in vivo specificity. *Mol. Cell. Biol.* **22:** 2170.

Cimini D., Howell B., Maddox P., Khodjakov A., Degrassi F., and Salmon E.D. 2001. Merotelic kinetochore orientation is a major mechanism of aneuploidy in mitotic mammalian tissue cells. *J. Cell Biol.* **153:** 517.

Dillon N. and Festenstein R. 2002. Unravelling heterochromatin: Competition between positive and negative factors regulates accessibility. *Trends Genet.* **18:** 252.

Ding R., McDonald K.L., and McIntosh J.R. 1993. Three-dimensional reconstruction and analysis of mitotic spindles from the yeast, *Schizosaccharomyces pombe*. *J. Cell Biol.* **120:** 141.

Ekwall K., Olsson T., Turner B.M., Cranston G., and Allshire R.C. 1997. Transient inhibition of histone deacetylation alters the structural and functional imprint at fission yeast centromeres. *Cell* **91:** 1021.

Ekwall K., Javerzat J.P., Lorentz A., Schmidt H., Cranston G., and Allshire R. 1995. The chromodomain protein Swi6: A key component at fission yeast centromeres. *Science* **269:** 1429.

Ekwall K., Nimmo E.R., Javerzat J.P., Borgstrom B., Egel R., Cranston G., and Allshire R. 1996. Mutations in the fission yeast silencing factors clr4+ and rik1+ disrupt the localisation of the chromo domain protein Swi6p and impair centromere function. *J. Cell Sci.* **109:** 2637.

Fishel B., Amstutz H., Baum M., Carbon J., and Clarke L. 1988. Structural organization and functional analysis of centromeric DNA in the fission yeast *Schizosaccharomyces pombe*. *Mol. Cell. Biol.* **8:** 754.

Goshima G., Saitoh S., and Yanagida M. 1999. Proper metaphase spindle length is determined by centromere proteins Mis12 and Mis6 required for faithful chromosome segregation. *Genes Dev.* **13:** 1664.

Hall I.M., Noma K., and Grewal S.I. 2003. RNA interference machinery regulates chromosome dynamics during mitosis and meiosis in fission yeast. *Proc. Natl. Acad. Sci.* **100:** 193.

Halverson D., Baum M., Stryker J., Carbon J., and Clarke L. 1997. A centromere DNA-binding protein from fission yeast affects chromosome segregation and has homology to human CENP-B. *J. Cell Biol.* **136:** 487.

Hampsey M. and Reinberg D. 2003. Tails of intrigue: Phosphorylation of RNA polymerase II mediates histone methylation. *Cell* **113:** 429.

James T.C., Eissenberg J.C., Craig C., Dietrich V., Hobson A., and Elgin S.C. 1989. Distribution patterns of HP1, a heterochromatin-associated nonhistone chromosomal protein of *Drosophila*. *Eur. J. Cell Biol.* **50:** 170.

Jin Q.W., Pidoux A.L., Decker C., Allshire R.C., and Fleig U. 2002. The mal2p protein is an essential component of the fission yeast centromere. *Mol. Cell. Biol.* **22:** 7168.

Jones L., Hamilton A.J., Voinnet O., Thomas C.L., Maule A.J., and Baulcombe D.C. 1999. RNA–DNA interactions and DNA methylation in post-transcriptional gene silencing. *Plant Cell* **11:** 2291.

Karpen G.H. and Allshire R.C. 1997. The case for epigenetic effects on centromere identity and function. *Trends Genet.* **13:** 489.

Lachner M., O'Carroll D., Rea S., Mechtler K., and Jenuwein T. 2001. Methylation of histone H3 lysine 9 creates a binding site for HP1 proteins. *Nature* **410:** 116.

Marschall L.G. and Clarke L. 1995. A novel cis-acting centromeric DNA element affects *S. pombe* centromeric chromatin structure at a distance. *J. Cell Biol.* **128:** 445.

Mellone B.G. and Allshire R.C. 2003. Stretching it: putting the CEN(P-A) in centromere. *Curr. Opin. Genet. Dev.* **13:** 191.

Mette M.F., Aufsatz W., van der Winden J., Matzke M.A., and Matzke A.J. 2000. Transcriptional silencing and promoter methylation triggered by double-stranded RNA. *EMBO J.* **19:** 5194.

Millband D.N., Campbell L., and Hardwick K.G. 2002. The awesome power of multiple model systems: Interpreting the complex nature of spindle checkpoint signaling. *Trends Cell Biol.* **12:** 205.

Minc E., Allory Y., Worman H.J., Courvalin J.C., and Buendia B. 1999. Localization and phosphorylation of HP1 proteins during the cell cycle in mammalian cells. *Chromosoma* **108:** 220.

Musacchio A. and Hardwick K.G. 2002. The spindle checkpoint: Structural insights into dynamic signalling. *Nat. Rev. Mol. Cell Biol.* **3:** 731.

Nakayama J., Rice J.C., Strahl B.D., Allis C.D., and Grewal S.I. 2001. Role of histone H3 lysine 9 methylation in epigenetic control of heterochromatin assembly. *Science* **292:** 110.

Ngan V.K. and Clarke L. 1997. The centromere enhancer mediates centromere activation in *Schizosaccharomyces pombe*. *Mol. Cell. Biol.* **17:** 3305.

Nonaka N., Kitajima T., Yokobayashi S., Xiao G., Yamamoto M., Grewal S.I., and Watanabe Y. 2002. Recruitment of cohesin to heterochromatic regions by Swi6/HP1 in fission yeast. *Nat. Cell Biol.* **4:** 89.

O'Toole E.T., Winey M., and McIntosh J.R. 1999. High-voltage electron tomography of spindle pole bodies and early mitotic spindles in the yeast *Saccharomyces cerevisiae*. *Mol. Biol. Cell* **10:** 2017.

Partridge J.F., Scott K.S., Bannister A.J., Kouzarides T., and Allshire R.C. 2002. cis-acting DNA from fission yeast centromeres mediates histone H3 methylation and recruitment of silencing factors and cohesin to an ectopic site. *Curr. Biol.* **12:** 1652.

Pidoux A.L., Richardson W., and Allshire R.C. 2003. Sim4: A novel fission yeast kinetochore protein required for centromeric silencing and chromosome segregation. *J. Cell Biol.* **161:** 295.

Polizzi C. and Clarke L. 1991. The chromatin structure of centromeres from fission yeast: Differentiation of the central core that correlates with function. *J. Cell Biol.* **112:** 191.

Rea S., Eisenhaber F., O'Carroll D., Strahl B.D., Sun Z.W., Schmid M., Opravil S., Mechtler K., Ponting C.P., Allis C.D., and Jenuwein T. 2000. Regulation of chromatin structure by site-specific histone H3 methyltransferases. *Nature* **406:** 593.

Reinhart B.J. and Bartel D.P. 2002. Small RNAs correspond to centromere heterochromatic repeats. *Science* **297:** 1831.

Reuter G. and Spierer P. 1992. Position effect variegation and chromatin proteins. *Bioessays* **14:** 605.

Richards E.J. and Elgin S.C. 2002. Epigenetic codes for heterochromatin formation and silencing: Rounding up the usual suspects. *Cell* **108:** 489.

Saitoh S., Takahashi K., and Yanagida M. 1997. Mis6, a fission yeast inner centromere protein, acts during G1/S and forms specialized chromatin required for equal segregation. *Cell* **90:** 131.

Schotta G., Ebert A., Dorn R., and Reuter G. 2003. Position-effect variegation and the genetic dissection of chromatin regulation in *Drosophila*. *Semin. Cell Dev. Biol.* **14:** 67.

Schramke V. and Allshire R. 2003. Hairpin RNAs and retrotransposon LTRs effect RNAi and chromatin-based gene silencing. *Science* **301:** 1069.

Shankaranarayana G.D., Motamedi M.R., Moazed D., and Grewal S.I. 2003. Sir2 regulates histone H3 lysine 9 methylation and heterochromatin assembly in fission yeast. *Curr. Biol.* **13:** 1240.

Steiner N.C. and Clarke L. 1994. A novel epigenetic effect can alter centromere function in fission yeast. *Cell* **79:** 865.

Steiner N.C., Hahnenberger K.M., and Clarke L. 1993. Centromeres of the fission yeast *Schizosaccharomyces pombe* are highly variable genetic loci. *Mol. Cell. Biol.* **13:** 4578.

Sullivan B.A., Blower M.D., and Karpen G.H. 2001. Determining centromere identity: Cyclical stories and forking paths. *Nat. Rev. Genet.* **2:** 584.

Takahashi K., Chen E.S., and Yanagida M. 2000. Requirement of Mis6 centromere connector for localizing a CENP-A-like protein in fission yeast. *Science* **288:** 2215.

Takahashi K., Murakami S., Chikashige Y., Funabiki H., Niwa O., and Yanagida M. 1992. A low copy number central sequence with strict symmetry and unusual chromatin structure in fission yeast centromere. *Mol. Biol. Cell* **3:** 819.

Verdel A., Jia S., Gerber S., Sugiyama T., Gygi S., Grewal S.I., and Moazed D. 2004. RNAi-mediated targeting of heterochromatin by the RITS complex. *Science* **303:** 672.

Volpe T.A., Kidner C., Hall I.M., Teng G., Grewal S.I., and Martienssen R.A. 2002. Regulation of heterochromatic silencing and histone H3 lysine-9 methylation by RNAi. *Science* **297:** 1833.

Volpe T., Schramke V., Hamilton G.L., White S.A., Teng G., Martienssen R.A., and Allshire R.C. 2003. RNA interference is required for normal centromere function in fission yeast. *Chromosome Res.* **11:** 137.

Wreggett K.A., Hill F., James P.S., Hutchings A., Butcher G.W., and Singh P.B. 1994. A mammalian homologue of *Drosophila* heterochromatin protein 1 (HP1) is a component of constitutive heterochromatin. *Cytogenet. Cell Genet.* **66:** 99.

RNA Interference, Transposon Silencing, and Cosuppression in the *Caenorhabditis elegans* Germ Line: Similarities and Differences

V.J.P. ROBERT, N.L. VASTENHOUW, AND R.H.A. PLASTERK

Hubrecht Laboratory, 3584 CT Utrecht, The Netherlands

Like all genomes analyzed to date, the *Caenorhabditis elegans* genome contains numerous transposable elements (or remnants of them); in *C. elegans*, only the DNA transposons have remained active. Jumping of these elements can be frequently detected in the soma but not in the germ line of the most studied lab strain Bristol N2 (Emmons and Yesner 1984). Nevertheless, in some natural isolates, such as Bergerac, transposition also occurs in the germ line (Moerman and Waterston 1984; Eide and Anderson 1985; Collins et al. 1987). In addition, it is possible after mutagenesis to recover lines from Bristol N2 in which germ line transposition occurs (Ketting et al. 1999; Tabara et al. 1999). Taken together, these observations suggest that, in the *C. elegans* germ line, an active system of regulation exists to silence transposition.

Besides these naturally occurring repeated sequences, it is also possible to create artificial repetitive sequences, by microinjecting DNA into the germ line (Stinchcomb et al. 1985). Upon injection, DNA is rearranged to form extrachromosomal transgenic arrays, carrying hundreds of copies of the injected sequence. The sequences present in the array are rapidly silenced in the germ line but not in the soma. Moreover, if the injected DNA (1) shares sequences with an endogenous germ-line-expressed gene and (2) is under the control of an active promoter, the expression of the endogenous locus can also be silenced. This *trans*-silencing, named "cosuppression," phenocopies a loss-of-function allele of the targeted gene (Dernburg 2000; Ketting and Plasterk et al. 2000).

Here, we describe how the mechanisms of transposon silencing and cosuppression are related to each other and to experimental dsRNA-induced posttranscriptional gene silencing, known as RNA interference (RNAi) (Fire et al. 1998; Dernburg et al. 2000; Ketting and Plasterk 2000). Moreover, we present recent data that address the relation between transposon silencing and RNAi. Finally, we present the phenotypic analysis of *ppw-2*, a gene found to be involved in transposon silencing and cosuppression.

COMMON GENES INVOLVED IN RNAi, TRANSPOSON SILENCING, COSUPPRESSION, AND TRANSGENE SILENCING

Forward genetic screens have been performed in Craig Mello's group (Tabara et al. 1999) and in ours (Ketting et al. 1999) to identify *RNAi-de*ficient (*rde*) and *mut*ator (*mut*) genes, which are involved in RNAi and transposon silencing, respectively. Surprisingly, it turned out that the results from these screens were partially overlapping, since genes such as *mut-7, mut-15* (also known as *rde-5*; R. Ketting et al., unpubl.), *mut-16*, and *rde-2* (also known as *mut-8*; B. Tops et al., unpubl.) were identified in both screens (Fig. 1). It is interesting to note that silencing of all active types of transposon is deregulated in these mutants, suggesting that these genes are involved in a general mechanism regulating all transposable elements. Further investigation demonstrated that these same genes are also involved in cosuppression (Ketting and Plasterk 2000), thus these genes required for RNAi, transposon silencing, and cosuppression can be referred to as *mut/rde/cde* (*cde*=*c*osuppression *de*fective) genes.

For the first time, these results established the existence of a mechanistic link between RNAi, transposon silencing, and cosuppression, but molecular characterization of the different steps and intermediary products involved in those silencing processes was necessary to establish the nature of this link.

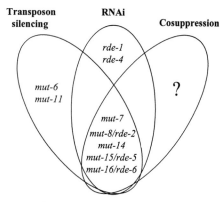

Figure 1. RNAi, transposon silencing, and cosuppression are related but distinct mechanisms. Two independent forward screens were performed to identify *rde* and *mut* genes. The results partially overlap, suggesting that RNAi and transposon silencing are related processes. Specific *rde* and *mut* genes have also been identified in those screens, showing that these mechanisms are nevertheless distinct. Interestingly, the *rde*/*mut* genes are also required for cosuppression.

RNAi IS A POSTTRANSCRIPTIONAL GENE-SILENCING PHENOMENON INDUCED BY A dsRNA MOLECULE AND MEDIATED BY siRNAs

RNAi is the posttranscriptional experimental silencing of genes following the introduction of double-stranded RNA (dsRNA) molecules that share sequence similarity with the targeted gene(s). RNAi was discovered by Andrew Fire, Craig Mello, and their collaborators in *C. elegans* (Fire et al. 1998). Since then, this mechanism has been shown to be conserved in numerous organisms such as protozoa (Ngo et al. 1998), arthropods (Kennerdell and Carthew 1998), mammals (Wianny and Zernicka-Goetz 2000), and plants (Waterhouse et al. 1998). Its mechanism has been extensively studied and our understanding has grown so fast that it is now, only a few years after its discovery, possible to present a model of how this process works (Fig. 2).

In *C. elegans*, dsRNA can be introduced into the animal by feeding, soaking, or injection. To become an active interfering agent, the dsRNA needs to be cleaved by the RNase III-related enzyme Dicer (Bernstein et al. 2001; Grishok et al. 2001; Ketting et al. 2001; Knight and Bass 2001) into short interfering agents (or siRNA) (Zamore et al. 2000), which are RNA duplexes of 21–23 nucleotides (nt) in length with characteristic 2-nt 3′-hydroxyl overhanging ends and 5′-phosphate termini (Elbashir et al. 2001). They are the active agents inducing the degradation of cognate mRNAs. It has been suggested that siRNAs directly produced from the dsRNA trigger can trigger RNA-dependent RNA polymerases (RdRPs) to transform the mRNA into a new dsRNA molecule (Sijen et al. 2001), which can subsequently be processed by dsRNA-specific nucleases. Genes encoding the RdRPs involved in this pathway have been isolated (Table 1). They are *ego-1* (Smardon et al. 2000) and *rrf-1* (Sijen et al. 2001), involved in RNAi in the germ line and the soma, respectively. Secondary siRNAs are potentially produced by this pathway and can be incorporated into the general pool of siRNAs. Alternatively, siRNAs are also proposed to be assembled into a ribonucleoprotein complex, known as RISC (RNA-induced silencing complex) (Hammond et al. 2000), which mediates the endonucleolytic cleavage of the cognate mRNA, as an initial event for further exonucleolytic degradation. In *Drosophila* and human cells (see, e.g., Hutvagner and Zamore 2002; Martinez et al. 2002), such a complex has been isolated and it has recently been reported to have a 5′-phosphomonoester-producing RNA endonuclease activity (Martinez and Tuschl 2004; Schwarz et al. 2004). In *C. elegans*, a RISC-like structure is thought to exist because most of the *Drosophila* and human components of RISC are conserved (Caudy et al. 2003), but it has not been isolated yet. At this point, it is difficult to figure out how RdRPs and RISC are connected. In our working model (Fig. 2), all the siRNA are assembled in RISC and those complexes mediate direct or indirect (via RdRPs) mRNA degradation. Further biochemical studies are necessary to test this model.

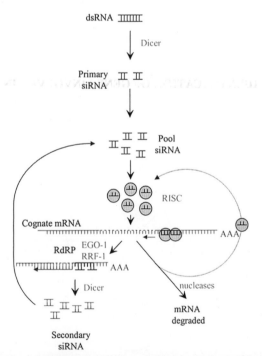

Figure 2. Model for the RNAi mechanism in *C. elegans*. RNAi is a posttranscriptional gene silencing process triggered by the experimental introduction of dsRNA. The trigger is processed by the RNase III-related enzyme Dicer into short interfering RNA (siRNAs). These interfering agents induce degradation of cognate mRNAs. Several factors involved in this degradation have been identified but it is still unknown whether they defined a single pathway or several ones. Our working model is that mRNA degradation is mediated by the RISC either directly or following amplification by RdRPs.

RNAi-LIKE PTGS IS INVOLVED IN TRANSPOSON SILENCING

The "cut-and-paste" transposition mechanism (van Luenen et al. 1994) has been extensively studied for *Tc1* (Emmons et al. 1983) and *Tc3* (Collins et al. 1989), the most active transposable elements in *C. elegans* genome. They both belong to the Tc1/mariner superfamily and contain a single gene (encoding a transposase enzyme) flanked by terminal inverted repeats (TIRs). Their transposition is mediated by self-encoded transposases that specifically recognize their TIRs and induce a double-strand break of the genomic DNA necessary for the excision/reintegration of the transposon (Vos et al. 1996). Their transposition is tightly regulated in the germ line, and the first chemically induced genetic mutants in which increased germ line transposition can be detected were reported more than 15 years ago (Collins et al. 1987). Our understanding of this regulatory mechanism has significantly improved only recently, after the establishment of its link with RNAi mechanism (Ketting et al. 1999; Tabara et al. 1999).

By analogy with what is known for RNAi, dsRNA molecules expressed from transposons have been searched for in *C. elegans* strains. DsRNA molecules have been detected for Tc1, Tc3, and Tc5 (Sijen and Plas-

Table 1. Examples of Genes Involved in Silencing Processes in *C. elegans*

Class	Gene name	Brief description	Identification[a]	mut	rde	cde
mut/rde/cde	mut-7	Nuclease	EMS screen[1]	+	+[b]	+[14,15]
	mut-14	RNA helicase	EMS screen[1,2]	+	+	+
	mut-16	PolyQ	EMS screen[1]/RNAi screen[3]	+	+	+
	ppw-2	Argonaute	RNAi screen[4]	+	+,−	+
	mut-2	Not cloned yet	EMS screen[5]	+	+	+[14]
	rde-2/mut-8	B. Tops et al., unpubl.	EMS screen[1,6]	+	+[a]	+[14,15]
	mut-15	R. Ketting et al., unpubl.	EMS screen[1]	+	+	+
rde	rde-4	DsRNA binding	EMS screen[6,7]	−	+	−
	rde-1	Argonaute	EMS screen[6]	−	+	−[14,15]
	ppw-1	Argonaute	Natural isolate[8]	−	+[b]	−
mut	mut-6	Not cloned yet	Spontaneous[16]	+	−	−[14]
	mut-11	Not cloned yet	EMS screen[1]	+	−	−
still not classified	dcr-1	Nuclease	Reverse genetics[9–11]	?	+	?
	rde-3	Not cloned yet	EMS screen[6]	+	+	?
	ego-1	RdRP	EMS screen[12]	?	+[b]	?
	rrf-1	RdRP	Reverse genetics[13]	−	+[c]	?

+ and − indicate that the gene is involved in the considered silencing process.
[a]References. [1]Ketting et al. 1999; [2]Tijsterman et al. 2002b; [3]Vastenhouw et al. 2003; [4]this study; [5]Collins et al. 1987; [6]Tabara et al. 1999; [7]Tabara et al. 2002; [8]Tijsterman et al. 2002a; [9]Ketting et al. 2001; [10]Knight and Bass 2001; [11]Grishok et al. 2001; [12]Smardon et al. 2000; [13]Sijen et al. 2001; [14]Ketting and Plasterk 2000; [15]Dernburg et al. 2000; [16]Mori et al. 1988.
[b]Germ line specific
[c]Soma specific

terk 2003). They are derived from the TIR sequences. This finding implies that transposon transcripts form dsRNA intramolecularly by snapping back into a panhandle structure. A low amount of Tc1-siRNAs-like molecules, 20–27 nt in length, were detected in wild-type (WT) and *rde* (non *mut*) background animals. They are (almost) absent in *mut/rde/cde* and *mut* backgrounds, are mostly derived from the TIR sequences, and are assumed to be produced by a Dicer-mediated degradation of the TIR dsRNA molecules.

To test whether the detected Tc1-siRNA molecules could induce silencing, TIR sequences were fused to the *gfp* reporter gene in a construct designed to be expressed in the germ line. This chimaeric construct introduced in low copy number in a WT background is indeed silenced. This silencing is directly dependent on TIR sequences, since similar transgenes are normally expressed when unrelated sequences replace the Tc1 TIR sequences. In addition, Tc1 TIR sequences induce silencing only if they are present in the transcribed sequences, strongly suggesting that the observed silencing occurs at the posttranscriptional level. Finally, the silencing induced by Tc1 TIR sequences is dependent on *mut/rde/cde* genes. Comparison of the ratio spliced-to-unspliced *gfp*-Tc1 transcripts in context where these transgenes are silenced or not, confirms that, as previously demonstrated, a posttranscriptional component is involved in their silencing. The presence of Tc1 dsRNA and Tc1 siRNA as well as the observation that transgenes carrying Tc1 sequences are posttranscriptionally silenced in a *mut/rde/cde* genes-dependent way strongly suggest that transposons are silenced at the posttranscriptional level and that Tc1-siRNAs mediate this silencing.

Similar studies aimed at cosuppression have been difficult, as loss of germ-line-expressed genes usually strongly affects the fertility or viability of *C. elegans*. To circumvent this problem, we developed fertile strains in which a *gfp* reporter gene expressed in the germ line is co-suppressed by the presence of multiple copies of *gfp*. These lines are now analyzed to establish cosuppression requirements and identify molecular intermediaries and genes involved in cosuppression (V. Robert et al., unpubl.).

IDENTIFICATION OF GENES INVOLVED IN SILENCING PROCESSES IN *C. ELEGANS*

Forward and genetics screens have been performed to identify genes involved in silencing processes in *C. elegans* (Collins et al. 1987; Ketting et al. 1999; Tabara et al. 1999; Dudley et al. 2002; Vastenhouw et al. 2003). They established the existence of *mut/rde/cde* genes as well as the existence of specific *mut* and *rde* genes. In addition, we carried out an Ethyl Methane Sulfonate (EMS) mutagenesis for *cde* genes (Fig. 3) (V. Robert and R. Plasterk, unpubl.). We have recovered mutants of *mut/rde/cde* genes, but also found mutants affecting some, but not all, phenomena. A short list of some of the genes is given in Table 1. To analyze the relationship between the different silencing processes, we classified them according to their involvement in RNAi, transposon silencing, and cosuppression.

Three *mut/rde/cde* genes have been identified at the molecular level so far (Table 1): *mut-7* encodes a RNaseD domain-containing protein (Ketting et al. 1999); *mut-14*, a putative RNA helicase (Tijsterman et al. 2002b); and *mut-16*, a protein with a polyglutamine domain (Vastenhouw et al. 2003). How could these factors act in the silencing process? One possibility is that they are required for a step common to all silencing processes. Alternatively, the *mut/rde/cde* genes could encode components of the backbone of complex(es) common to all silencing

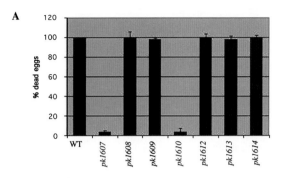

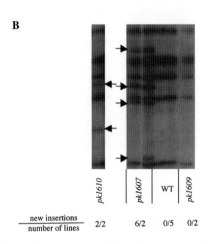

Figure 3. An EMS mutagenesis revealed the existence of specific *cde* genes. Analysis of seven *cde* mutants isolated in an EMS screen were studied for their RNAi competence by their capacity to resist targeting of *pos-1*, a gene required for embryonic viability (*A*). *Pk1607* and *pk1610* are also affected in RNAi. In addition, Tc1 silencing deregulation was observed by transposon insertion display in *pk1607* and *pk1610* genetic backgrounds. (*B*) These alleles belong to *mut/rde/cde* complementation groups. In contrast, Tc1 is still silenced in a *pk1609* background and this allele is still RNAi sensitive. Thus *pk1609* is an allele of a not-yet-cloned specific *cde* gene.

an essential role in stem cell fate (Cox et al. 1998). In addition, they have been implicated in a variety of homology-dependent mechanisms that involve small RNAs and induce transcriptional (see, e.g., Pal-Bhadra et al. 2002; Zilberman et al. 2003; Verdel et al. 2004), posttranscriptional (see, e.g., Tabara et al. 1999; Tijsterman et al. 2002a), or translational (see, e.g., Grishok et al. 2001; Vaucheret et al. 2004) gene silencing. Moreover, the *Schizosaccharomyces pombe* (Hall et al. 2003) and *Trypanosoma brucei* (Durand-Dubief and Bastin 2003) Argonaute proteins are required for chromosome segregation. The connection between these mechanisms and the maintenance of stem cells is still not clearly established.

Three-dimensional nuclear magnetic resonance (Lingel et al. 2003) and crystal (Song et al. 2003; Yan et al. 2003; Ma et al. 2004) structures of PAZ domains have been reported. They suggest that PAZ is a conserved RNA-binding domain, which might play as a siRNA-end-binding module, and that the Argonaute proteins could be shuttles, which drive siRNAs between the different ribonucleoprotein complexes involved in silencing processes. In agreement with this model, Argonaute proteins have been found in siRNAs-containing ribonucleoprotein complexes in *Drosophila* (Hammond et al. 2001), *Neurospora crassa* (Catalanotto et al. 2002), *S. pombe* (Verdel et al. 2004), and human (Martinez et al. 2002).

THE Argonaute FAMILY AND SILENCING PROCESSES IN *C. ELEGANS*

Rde-1 is the first Argonaute gene to have been implicated in RNAi (Tabara et al. 1999). It is encoded, together with 29 other Argonaute proteins, by the *C. elegans* genome. Its function seems to be restricted to RNAi. RDE-1 has been found to interact with RDE-4, a dsRNA binding protein, Dicer, and the RNA helicase DRH-1 (Tabara et al. 2002). It has been proposed that this complex could have a role in selecting and processing double-stranded RNA molecules for the RNAi pathway.

Additional Argonaute genes have been found to be involved in silencing processes in *C. elegans*. First, *alg-1* and *alg-2* (Grishok et al. 2001), together with *dcr-1*, are both required for maturation of the small temporal RNA *lin-4*, and *alg-2*, at least, is required for maturation of the small temporal RNA *let-7*. Second, *ppw-1* (Tijsterman et al. 2002a) is required for RNAi in the germ line but not in the soma, while transposon silencing is still efficient in a genetic background where *ppw-1* is deleted (Tijsterman et al. 2002a; V. Robert and R. Plasterk, unpubl.). Third, RNAi screens performed to identify genes involved in transposon silencing (Vastenhouw et al. 2003) and cosuppression (V. Robert et al, unpubl.) showed that *ppw-2* is required for both transposon silencing and cosuppression and that two other members of the Argonaute family are implicated in cosuppression.

To further investigate the involvement of *ppw-2* in silencing processes, we isolated and analyzed the deletion *pk1673* covering a large region of *ppw-2*, including the PAZ and the PIWI domains. *Ppw-2(pk1673)* homozygous animals exhibit a *cde/mut* phenotype and a partial

processes, whose specificity would be defined by their association with specific RDE, MUT, or CDE factors or siRNAs. To test this, biochemical characterization of (ribonucleo)protein complexes involved in silencing processes (in *C. elegans*) is now necessary.

THE Argonaute GENES ARE WIDELY CONSERVED AND INVOLVED IN SILENCING PROCESSES IN MANY ORGANISMS

The Argonaute proteins are defined by the presence of two conserved domains named PAZ and PIWI (Cerutti et al. 2000). For example, these proteins have been found in *Arabidopsis thaliana*, *C. elegans*, *Drosophila melanogaster*, and mammals (for review, see Carmell et al. 2002). Some of these proteins (the piwi subfamily as defined by Carmell et al. [2002]) have been shown to play

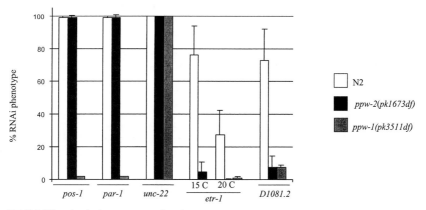

Figure 4. *ppw-2* and RNAi. N2 animals, homozygotes for *pk1673(ppw-2)* and homozygotes for *pk3511(ppw-1)* were assayed for their sensitivity to RNAi. Animals were fed for 48 hr at 15°C with food expressing dsRNA targeting specific *C. elegans* genes. They were then singled onto fresh targeting foods for 24 hr at 20°C. Adults were eliminated and progeny was allowed to develop at 20°C. Embryonic and postembryonic phenotypes were scored after 24 and 48 hr. The targeted gene is indicated below the graph. Scored phenotypes: for *pos-1* and *par-1*, dead embryos; for *unc-22*, twitching animals; and for *etr-1* and *D1081.2*, body abnormalities. For *etr-1*, two independent assays, respectively carried out at 15°C and 20°C, are shown. Between two and ten animals were scored in each assay. Standard deviation is indicated.

rde phenotype (Fig. 4): Depending on the gene targeted by RNAi, it appears that *ppw-2(pk1673)* homozygotes behave as wild type (Fig. 4, RNAi targeting genes *pos-1*, *par-1*, and *unc-22*) or as a mutant (Fig. 4, targeting genes *etr-1* and *D1081.2*). Strikingly, RNAi known to induce a complete loss of function of the targeted gene (for example, against *pos-1*, *par-1*, and *unc-22*) seems to remain efficient in the absence of PPW-2, whereas RNAi known to induce partial loss of function of the targeted gene (e.g., targeting *etr-1* and *D1081.2*) seems to require PPW-2. In this latter case, we assume that less siRNA are available and we propose that PPW-2 is necessary to make them active (for instance, by processing them or by targeting them to an RDE complex). In transposon silencing, which seems to be mediated by Tc1 siRNA produced from endogenous Tc1 dsRNA molecules, PPW-2 could play a similar role and be required, possibly because the Tc1 siRNA are not abundant. In agreement with this hypothesis, Tc1 siRNAs can still be detected in *ppw-2(pk1673)* animals (data not shown), suggesting that *ppw-2* acts downstream of the production of Tc1 siRNA.

CONCLUSIONS

Many factors involved in silencing processes have now been identified. Genetic analysis indicates that on the one hand the three silencing phenomena (RNAi, transposon silencing, and cosuppression) have much in common, while on the other hand some mutations affect one process but not the others. This may indicate that specialized complexes carry out each of these silencing phenomena.

ACKNOWLEDGMENTS

We thank Dr. René Ketting and Dr. Robin May for critical reading of the manuscript.

REFERENCES

Bernstein E., Caudy A.A., Hammond S.M., and Hannon G.J. 2001. Role for a bidentate ribonuclease in the initiation step of RNA interference. *Nature* **409:** 363.

Carmell M.A., Xuan Z., Zhang M.Q., and Hannon G.J. 2002. The Argonaute family: Tentacles that reach into RNAi, developmental control, stem cell maintenance, and tumorigenesis. *Genes Dev.* **16:** 2733.

Catalanotto C., Azzalin G., Macino G., and Cogoni C. 2002. Involvement of small RNAs and role of the qde genes in the gene silencing pathway in *Neurospora*. *Genes Dev.* **16:** 790.

Caudy A.A., Ketting R.F., Hammond S.M., Denli A.M., Bathoorn A.M., Tops B.B., Silva J.M., Myers M.M., Hannon G.J., and Plasterk R.H. 2003. A micrococcal nuclease homologue in RNAi effector complexes. *Nature* **425:** 411.

Cerutti L., Mian N., and Bateman A. 2000. Domains in gene silencing and cell differentiation proteins: The novel PAZ domain and redefinition of the Piwi domain. *Trends Biochem. Sci.* **25:** 481.

Collins J., Forbes E., and Anderson P. 1989. The Tc3 family of transposable genetic elements in *Caenorhabditis elegans*. *Genetics* **121:** 47.

Collins J., Saari B., and Anderson P. 1987. Activation of a transposable element in the germ line but not the soma of *Caenorhabditis elegans*. *Nature* **328:** 726.

Cox D.N., Chao A., Baker J., Chang L., Qiao D., and Lin H. 1998. A novel class of evolutionarily conserved genes defined by piwi are essential for stem cell self-renewal. *Genes Dev.* **12:** 3715.

Dernburg A.F., Zalevsky J., Colaiacovo M.P., and Villeneuve A.M. 2000. Transgene-mediated cosuppression in the *C. elegans* germ line. *Genes Dev.* **14:** 1578.

Dudley N.R., Labbe J.C., and Goldstein B. 2002. Using RNA interference to identify genes required for RNA interference. *Proc. Natl. Acad. Sci.* **99:** 4191.

Durand-Dubief M. and Bastin P. 2003. TbAGO1, an Argonaute protein required for RNA interference, is involved in mitosis and chromosome segregation in *Trypanosoma brucei*. *BMC Biol.* **1:** 2.

Eide D. and Anderson P. 1985. Transposition of Tc1 in the nematode *Caenorhabditis elegans*. *Proc. Natl. Acad. Sci.* **82:** 1756.

Elbashir S.M., Lendeckel W., and Tuschl T. 2001. RNA interference is mediated by 21- and 22-nucleotide RNAs. *Genes Dev.* **15:** 188.

Emmons S.W. and Yesner L. 1984 High-frequency excision of transposable element Tc 1 in the nematode *Caenorhabditis elegans* is limited to somatic cells. *Cell* **36:** 599.

Emmons S.W., Yesner L., Ruan K.S., and Katzenberg D. 1983. Evidence for a transposon in *Caenorhabditis elegans*. *Cell* **32:** 55.

Fire A., Xu S., Montgomery M.K., Kostas S.A., Driver S.E., and Mello C.C. 1998. Potent and specific genetic interference by double-stranded RNA in *Caenorhabditis elegans*. *Nature* **391:** 806.

Grishok A., Pasquinelli A.E., Conte D., Li N., Parrish S., Ha I., Baillie D.L., Fire A., Ruvkun G., and Mello C.C. 2001. Genes and mechanisms related to RNA interference regulate expression of the small temporal RNAs that control *C. elegans* developmental timing. *Cell* **106:** 23.

Hall I.M., Noma K., and Grewal S.I. 2003. RNA interference machinery regulates chromosome dynamics during mitosis and meiosis in fission yeast. *Proc. Natl. Acad. Sci.* **100:** 193.

Hammond S.M., Bernstein E., Beach D., and Hannon G.J. 2000. An RNA-directed nuclease mediates post-transcriptional gene silencing in *Drosophila* cells. *Nature* **404:** 293.

Hammond S.M., Boettcher S., Caudy A.A., Kobayashi R., and Hannon G.J. 2001. Argonaute2, a link between genetic and biochemical analyses of RNAi. *Science* **293:** 1146.

Hutvagner G. and Zamore P.D. 2002. A microRNA in a multiple-turnover RNAi enzyme complex. *Science* **297:** 2056.

Kennerdell J.R. and Carthew R.W. 1998. Use of dsRNA-mediated genetic interference to demonstrate that frizzled and frizzled 2 act in the wingless pathway. *Cell* **95:** 1017.

Ketting R.F. and Plasterk R.H. 2000. A genetic link between co-suppression and RNA interference in *C. elegans*. *Nature* **404:** 296.

Ketting R.F., Haverkamp T.H., van Luenen H.G., and Plasterk R.H. 1999. Mut-7 of *C. elegans*, required for transposon silencing and RNA interference, is a homolog of Werner syndrome helicase and RNaseD. *Cell* **99:** 133.

Ketting R.F., Fischer S.E., Bernstein E., Sijen T., Hannon G.J., and Plasterk R.H. 2001. Dicer functions in RNA interference and in synthesis of small RNA involved in developmental timing in *C. elegans*. *Genes Dev.* **15:** 2654.

Knight S.W. and Bass B.L. 2001. A role for the RNase III enzyme DCR-1 in RNA interference and germ line development in *C. elegans*. *Science* **2:** 2.

Lingel A., Simon B., Izaurralde E., and Sattler M. 2003. Structure and nucleic-acid binding of the *Drosophila* Argonaute 2 PAZ domain. *Nature* **426:** 465.

Ma J.B., Ye K., and Patel D.J. 2004. Structural basis for overhang-specific small interfering RNA recognition by the PAZ domain. *Nature* **429:** 318.

Martinez J. and Tuschl T. 2004. RISC is a 5′ phosphomonoester-producing RNA endonuclease. *Genes Dev.* **18:** 975.

Martinez J., Patkaniowska A., Urlaub H., Luhrmann R., and Tuschl T. 2002. Single-stranded antisense siRNAs guide target RNA cleavage in RNAi. *Cell* **110:** 563.

Moerman D.G. and Waterston R.H. 1984. Spontaneous unstable unc-22 IV mutations in *C. elegans* var. Bergerac. *Genetics* **108:** 859.

Mori I., Moerman D.G., and Waterston R.H. 1988 Analysis of a mutator activity necessary for germline transposition and excision of Tc1 transposable elements in *Caenorhabditis elegans*. *Genetics* **120:** 397.

Ngo H., Tschudi C., Gull K., and Ullu E. 1998. Double-stranded RNA induces mRNA degradation in *Trypanosoma brucei*. *Proc. Natl. Acad. Sci.* **95:** 14687.

Pal-Bhadra M., Bhadra U., and Birchler J.A. 2002. RNAi related mechanisms affect both transcriptional and posttranscriptional transgene silencing in *Drosophila*. *Mol. Cell* **9:** 315.

Schwarz D.S., Tomari Y., and Zamore P.D. 2004. The RNA-induced silencing complex is a Mg2+-dependent endonuclease. *Curr. Biol.* **14:** 787.

Sijen T. and Plasterk R.H. 2003. Transposon silencing in the *Caenorhabditis elegans* germ line by natural RNAi. *Nature* **426:** 310.

Sijen T., Fleenor J., Simmer F., Thijssen K.L., Parrish S., Timmons L., Plasterk R.H., and Fire A. 2001. On the role of RNA amplification in dsRNA-triggered gene silencing. *Cell* **107:** 465.

Smardon A., Spoerke J.M., Stacey S.C., Klein M.E., Mackin N., and Maine E.M. 2000. EGO-1 is related to RNA-directed RNA polymerase and functions in germ-line development and RNA interference in *C. elegans*. *Curr. Biol.* **10:** 169.

Song J.J., Liu J., Tolia N.H., Schneiderman J., Smith S.K., Martienssen R.A., Hannon G.J., and Joshua-Tor L. 2003. The crystal structure of the Argonaute2 PAZ domain reveals an RNA binding motif in RNAi effector complexes. *Nat. Struct. Biol.* **10:** 1026.

Stinchcomb D.T., Shaw J.E., Carr S.H., and Hirsh D. 1985. Extrachromosomal DNA transformation of *Caenorhabditis elegans*. *Mol. Cell. Biol.* **5:** 3484.

Tabara H., Yigit E., Siomi H., and Mello C.C. 2002. The dsRNA binding protein RDE-4 interacts with RDE-1, DCR-1, and a DExH-box helicase to direct RNAi in *C. elegans*. *Cell* **109:** 861.

Tabara H., Sarkissian M., Kelly W.G., Fleenor J., Grishok A., Timmons L., Fire A., and Mello C.C. 1999. The rde-1 gene, RNA interference, and transposon silencing in *C. elegans*. *Cell* **99:** 123.

Tijsterman M., Okihara K.L., Thijssen K., and Plasterk R.H. 2002a. PPW-1, a PAZ/PIWI protein required for efficient germline RNAi, is defective in a natural isolate of *C. elegans*. *Curr. Biol.* **12:** 1535.

Tijsterman M., Ketting R.F., Okihara K.L., Sijen T., and Plasterk R.H. 2002b. RNA helicase MUT-14-dependent gene silencing triggered in *C. elegans* by short antisense RNAs. *Science* **295:** 694.

van Luenen H.G., Colloms S.D., and Plasterk R.H. 1994. The mechanism of transposition of Tc3 in *C. elegans*. *Cell* **79:** 293.

Vastenhouw N.L., Fischer S.E., Robert V.J., Thijssen K.L., Fraser A.G., Kamath R.S., Ahringer J., and Plasterk R.H. 2003. A genome-wide screen identifies 27 genes involved in transposon silencing in *C. elegans*. *Curr. Biol.* **13:** 1311.

Vaucheret H., Vazquez F., Crete P., and Bartel D.P. 2004. The action of ARGONAUTE1 in the miRNA pathway and its regulation by the miRNA pathway are crucial for plant development. *Genes Dev.* **18:** 1187.

Verdel A., Jia S., Gerber S., Sugiyama T., Gygi S., Grewal S.I., and Moazed D. 2004. RNAi-mediated targeting of heterochromatin by the RITS complex. *Science* **303:** 672.

Vos J.C., De Baere I., and Plasterk R.H. 1996. Transposase is the only nematode protein required for in vitro transposition of Tc1. *Genes Dev.* **10:** 755.

Waterhouse P.M., Graham M.W., and Wang M.B. 1998. Virus resistance and gene silencing in plants can be induced by simultaneous expression of sense and antisense RNA. *Proc. Natl. Acad. Sci.* **95:** 13959.

Wianny F. and Zernicka-Goetz M. 2000. Specific interference with gene function by double-stranded RNA in early mouse development. *Nat. Cell Biol.* **2:** 70.

Yan K.S., Yan S., Farooq A., Han A., Zeng L., and Zhou M.M. 2003. Structure and conserved RNA binding of the PAZ domain. *Nature* **426:** 468

Zamore P.D., Tuschl T., Sharp P.A., and Bartel D.P. 2000. RNAi: Double-stranded RNA directs the ATP-dependent cleavage of mRNA at 21 to 23 nucleotide intervals. *Cell* **101:** 25.

Zilberman D., Cao X., and Jacobsen S.E. 2003. ARGONAUTE4 control of locus-specific siRNA accumulation and DNA and histone methylation. *Science* **299:** 716.

Plant RNA Interference in Vitro

C. MATRANGA AND P.D. ZAMORE*
Department of Biochemistry and Molecular Pharmacology, University of Massachusetts Medical School, Worcester, Massachusetts 01605

In the RNA interference (RNAi) pathway, small interfering RNAs (siRNAs) direct the sequence-specific silencing of complementary RNA. RNAi and similar posttranscriptional gene silencing (PTGS) phenomena are found in other eukaryotes, including animals, fungi, protozoa, and plants (Cogoni et al. 1996; Fire et al. 1998; Kennerdell and Carthew 1998; Ngo et al. 1998; Timmons and Fire 1998; Vaucheret et al. 1998; Waterhouse et al. 1998; Lohmann et al. 1999; Sánchez-Alvarado and Newmark 1999; Wianny and Zernicka-Goetz 2000; Caplen et al. 2001; Elbashir et al. 2001c; Volpe et al. 2002; Schramke and Allshire 2003). The RNase III endonuclease, Dicer, initiates RNAi by converting long, double-stranded RNA (dsRNA) into siRNAs (Zamore et al. 2000; Bernstein et al. 2001; Billy et al. 2001), ~22-nucleotide guides that direct mRNA cleavage as components of a protein-RNA complex, the RNA-induced silencing complex (RISC) (Hamilton and Baulcombe 1999; Hammond et al. 2000, 2001; Elbashir et al. 2001a). In *Drosophila melanogaster*, another protein-RNA complex, the RISC loading complex (RLC), assembles one of the two strands of an siRNA into the RISC. The RLC comprises double-stranded siRNA, a heterodimer of the RNase III endonuclease Dicer-2 (Dcr-2) and the siRNA-binding protein R2D2, and other yet identified components (Liu et al. 2003; Pham et al. 2004; Tomari et al. 2004b). Assembly of functional RISC also requires Armitage, a homolog of the *Arabidopsis thaliana* protein SDE3, a putative helicase required for RNA silencing triggered by sense RNA-expressing transgenes (Dalmay et al. 2001). Armitage is required for oogenesis and RNA interference in vivo (Cook et al. 2004); biochemical evidence suggests that Armitage acts in RISC assembly after the RLC (Tomari et al. 2004b).

Argonaute family proteins form the catalytic core of the RISC (Hammond et al. 2001; Caudy et al. 2002; Hutvágner and Zamore 2002; Martinez et al. 2002; Mourelatos et al. 2002; Tabara et al. 2002; Hutvágner et al. 2004; Liu et al. 2004; Martinez and Tuschl 2004; Meister et al. 2004; Parker et al. 2004; Rand et al. 2004; Song et al. 2004) and are thus required genetically for RNAi (Tabara et al. 1999; Fagard et al. 2000; Grishok et al. 2000; Catalanotto et al. 2002; Caudy et al. 2002; Morel et al. 2002; Pal-Bhadra et al. 2002; Williams and Rubin 2002; Doi et al. 2003). Transfer of the siRNA guide strand, the strand complementary to the target RNA, from the RLC to the RISC requires Argonaute 2, consistent with its proposed role as an acceptor of unwound siRNA guide strand (Okamura et al. 2004; Tomari et al. 2004a).

Extracts from *Drosophila* and mammalian cells are important tools for dissecting the mechanism of RNAi in vitro, yet no comparable in vitro system is available for plant RNAi initiated by siRNAs. In plants, wheat germ extract is an important tool for studying RNA silencing, but wheat germ extracts cannot be programmed with synthetic duplex siRNAs. Wheat extracts recapitulate some RNA silencing activities, including RNA-dependent RNA polymerase activity and small RNA production from long dsRNA ("dicing") (Tang et al. 2003). Wheat germ extract also contains functional RISCs programmed with microRNAs (miRNAs), small, endogenous RNA guides that control the expression of messenger RNA (mRNA) targets, typically by cleaving them, suggesting that the extracts are specifically defective in RISC assembly (Tang et al. 2003; Mallory et al. 2004). Exogenous siRNAs trigger target cleavage activity in *Drosophila* embryo lysates and extracts from cultured mammalian cells (Boutla et al. 2001; Elbashir et al. 2001a,c; Nykänen et al. 2001), but not wheat germ extracts. The siRNAs produced in *Drosophila* lysate are both double- and single-stranded (Nykänen et al. 2001; Tang et al. 2003), but those produced in wheat germ extract remain double-stranded (Tang et al. 2003), suggesting that synthetic siRNA duplexes do not trigger RNAi in wheat germ extract because they are not unwound.

Here, we demonstrate that wheat germ extract cannot be programmed with exogenous siRNA duplexes because it does not unwind siRNA. In contrast, single-stranded siRNAs directed endonucleolytic cleavage of a corresponding target RNA in wheat germ extract. Wild-type *Drosophila* embryo and ovary lysate complemented the defect in siRNA unwinding in wheat germ extract. *Drosophila* ovary lysate defective in RNAi but not siRNA unwinding also rescued wheat germ extract, whereas mutant lysate defective for siRNA unwinding did not. Biochemical complementation of wheat germ RNAi requires both components of the RLC and core components of mature *Drosophila* RISC.

RESULTS

Wheat Germ Extract Cannot Unwind siRNA

Double-stranded siRNAs direct target cleavage in *Drosophila* embryo lysate (Boutla et al. 2001; Elbashir et

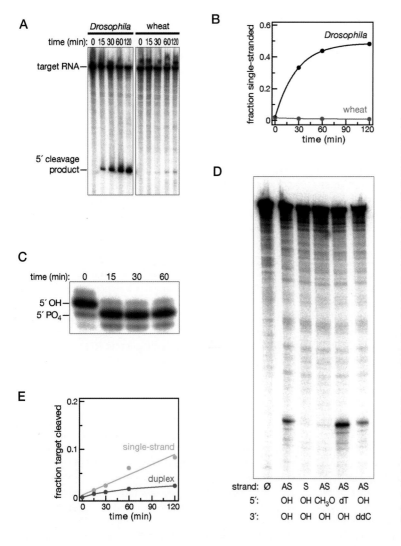

Figure 1. Single-stranded but not double-stranded siRNA triggers RNAi in wheat germ extract. (*A*) Double-stranded siRNAs direct target cleavage in *Drosophila* embryo lysate, but triggered only low levels of target cleavage in wheat germ extract. (*B*) In the presence of ATP, *Drosophila* embryo lysate (*black*), but not wheat germ extract (*red*), unwound an siRNA duplex. (*C*) A 3′-^{32}P-radiolabeled siRNA duplex was rapidly phosphorylated upon incubation in wheat germ extract. (*D*) Single-stranded siRNAs containing 5′ and 3′ hydroxy (OH) termini directed efficient target cleavage in wheat germ extract. The siRNA was rapidly 5′ phosphorylated in the wheat germ extract, as shown in *C*. A 5′ methoxy (CH$_3$O) modification, which blocks 5′ phosphorylation, inhibited target cleavage in wheat germ lysate. In order to introduce the 5′ methoxy modification, the first nucleotide of the siRNA was changed from riboU to deoxyT. This change, in the absence of the methoxy modification, did not inhibit target cleavage, as shown by the 5′ dT siRNA. A 2′,3′ dideoxy (ddC) modification of the 3′ end of the siRNA similarly had no significant effect on cleavage. The unmodified sense (S) siRNA triggered no target cleavage and is presented as a negative control. (*E*) Single-stranded (*green*) but not double-stranded (*blue*) siRNAs triggered efficient RNAi in wheat germ extracts. The appearance of the ~70-nucleotide 5′ cleavage product was monitored over time.

al. 2001b,c; Nykänen et al. 2001), but triggered only low levels of target cleavage in wheat germ extract (Fig. 1A). We asked if the failure of double-stranded siRNAs to trigger efficient target cleavage in wheat germ extract might reflect a defect in unwinding siRNA duplexes or in assembling the unwound strands into RISC. We monitored siRNA unwinding in wheat germ extract in the presence of ATP; in *Drosophila* embryo lysate, siRNA unwinding requires ATP (Nykänen et al. 2001). Wheat germ extract, unlike *Drosophila* embryo lysate, did not unwind double-stranded siRNAs: no single-stranded siRNA accumulated (Fig. 1B) nor did the double-stranded siRNA decrease with time (data not shown). We conclude that wheat germ extract lacks a robust siRNA unwinding activity.

siRNA must contain a 5′ phosphate group to function in RNAi; blocking siRNA 5′ phosphorylation inactivates the siRNA (Nykänen et al. 2001; Chiu and Rana 2002; Schwarz et al. 2002). We examined whether the lack of siRNA unwinding in wheat germ extract was caused by the absence of an siRNA kinase. We incubated a 3′ ^{32}P-radiolabeled siRNA bearing a 5′ hydroxyl group in wheat germ extract and monitored its phosphorylation. The siRNA was rapidly phosphorylated in the wheat germ extract, as evidenced by its faster electrophoretic mobility (Fig. 1C). Thus, wheat germ extract contains an siRNA kinase.

If siRNA unwinding, rather than siRNA phosphorylation, is defective in wheat germ extract, then single-stranded, but not double-stranded, siRNAs might direct target mRNA cleavage in the extract. In vitro, in both *Drosophila* embryo lysates and extracts of cultured mammalian cells, and ex vivo, in cultured mammalian cells, single-stranded siRNAs act as guides for endonucleolytic cleavage, albeit with reduced efficiency (Schwarz et al. 2002). Figures 1D and 1E show that single-stranded siRNAs also direct target cleavage in wheat germ extracts. Single-stranded siRNAs functioned in wheat germ extract only when they contained a 5′ phosphate; the siRNAs did not cleave the target when the 5′ end was blocked by a methoxy group (Fig. 1D), consistent with their guiding target cleavage as a component of RISC. A 5′ methoxy modification blocks RNAi in *Drosophila* embryo lysates and cultured human cells (Schwarz et al. 2002), because it blocks the assembly of the RLC (Tomari et al. 2004b). Thus, a 5′ phosphate is an essential feature of functional siRNAs in both plants and animals.

In contrast, a 2′,3′ dideoxy-modified, single-stranded siRNA guided target cleavage in wheat germ extract, excluding a role for the siRNA as a primer of an RNA-dependent RNA polymerase (RdRP) acting to convert the target RNA into dsRNA that is subsequently destroyed by dicing. Over time, only an ~70-nucleotide product accumulated (Fig. 1E), consistent with the single-stranded siRNA acting directly as a guide without involvement of an RdRP. In *Arabidopsis*, RdRP proteins are required for PTGS initiated by transgenes overproducing single-strand RNA (Dalmay et al. 2000; Mourrain et al. 2000), but not for target destruction initiated by dsRNA (Waterhouse et al. 1998).

Biochemical Complementation of Wheat Germ RNAi

RNA silencing pathways are conserved between plants and animals. Therefore we asked if *Drosophila* proteins might complement the defect in siRNA unwinding in wheat germ extract. Wheat germ extract was supplemented with serial dilutions of *Drosophila* embryo lysate and incubated with double-stranded siRNA and an RNA target (Figs. 2A and 2B). After incubation for 60 minutes, we assayed target cleavage (Fig. 2A). Supplementing the wheat germ extract with as little as 1 part *Drosophila* embryo lysate per 200 activated the wheat germ extract for double-stranded siRNA-directed target cleavage. This amount of *Drosophila* embryo lysate on its own did not support RNAi, demonstrating that our assay measures biochemical complementation of wheat germ extract by *Drosophila* proteins. Thus, *Drosophila* embryo lysate complements wheat germ RNAi activity, even though the two organisms diverged during evolution over one billion years ago (Hedges 2002).

Mutant *Drosophila* Ovary Lysates Rescue the Wheat Defect

Because wild-type *Drosophila* embryo lysate rescued wheat germ, we asked if ovary lysates from *Drosophila* mutants that are defective in RNAi could rescue wheat germ RNAi (Tomari et al. 2004a,b). Wild-type *Drosophila* ovary complemented the wheat germ extract for RNAi (Fig. 3A). Lysate from *dicer-2*, *r2d2*, and *armitage* mutant ovaries are all defective in loading RISC with siRNA. We therefore determined if ovary lysate from these mutants can complement wheat germ extract, allowing it to support target cleavage triggered by double-stranded siRNA. We supplemented wheat germ extract with ovary lysate from mutant flies defective for RNAi in vivo and in vitro: dcr-$2^{L811fsX}$ and dcr-2^{G31R}, $r2d2$, and $armi^{72.1}$. $armi^{72.1}$ lysate rescued the wheat germ defect (Fig. 3B), whereas ovary lysate from $r2d2$ (Fig. 3C) and $dcr2^{L811fsX}$ (Fig. 3D) flies did not complement wheat germ RNAi. Dicer-2 and R2D2 are required for siRNA-mediated RNAi (Liu et al. 2003; Lee et al. 2004; Pham et al. 2004), because they are core components of the RLC (Tomari et al. 2004b), which initiates siRNA unwinding (Tomari et al. 2004a). In contrast, dcr-2^{G31R} mu-

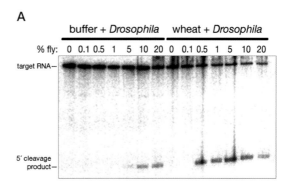

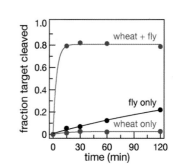

Figure 2. *Drosophila* RNAi components complement the defect in wheat germ extract, allowing double-stranded siRNA to trigger target cleavage. (*A*) When wheat germ extract was supplemented with as little as 0.5% *Drosophila* embryo lysate, by volume, double-stranded siRNA directed target RNA cleavage. On its own, this amount of *Drosophila* embryo lysate was insufficient to trigger target cleavage, demonstrating that the *Drosophila* components complement the defect in wheat germ extract. (*B*) When the wheat germ was supplemented with *Drosophila* embryo lysate (*red*; 5% of the wheat germ reaction, by volume) nearly all the target RNA was converted to 5′ cleavage product by 15 min. By contrast, little or no target cleavage was observed for the wheat germ extract alone (*blue*) at 2 hr. At that time, the diluted embryo lysate (*black*) had converted only ~20% of the target into 5′ cleavage product.

tant lysate rescued the defect in wheat germ at a concentration insufficient to support RNAi on its own (Fig. 3E). The dcr-2^{G31R} mutation prevents *Drosophila* Dcr-2 protein from dicing long dsRNA into siRNAs, but preserves its function in unwinding siRNA duplexes and loading one of the two siRNA strands into RISC (Lee et al. 2004; Pham et al. 2004; Tomari et al. 2004a). Moreover, recombinant Dicer2/R2D2 heterodimer, which can dice long dsRNA in vitro, did not rescue wheat germ RNAi (Fig. 3F). Together these data suggest that multiple RLC components are missing from wheat germ extract and that these components are essential for the transfer of one of the siRNA strands from the double-stranded siRNA into the single-strand-containing, active RISC.

Next, we tested if *Drosophila* Argonaute2 (Ago2) is required to complement wheat germ extract. Ago2 is required in flies for siRNA unwinding (Okamura et al. 2004), because the RLC will not initiate siRNA unwinding in the absence of Ago2 (Tomari et al. 2004a). Ago2 is a core component of RISC and is the endonuclease that

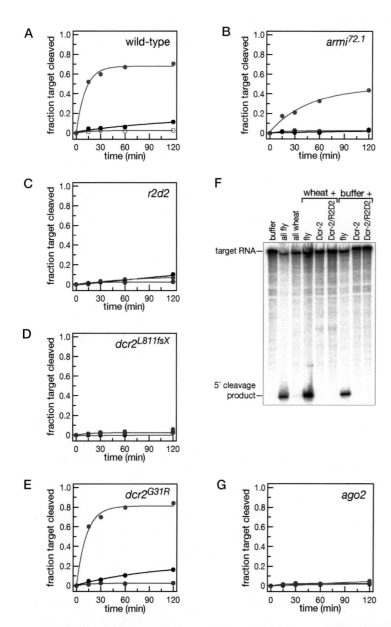

Figure 3. Biochemical complementation of target cleavage (RNAi) directed by double-stranded siRNA in wheat germ extract by *Drosophila* ovary lysates. Panels *A–E* and *G* each show a time course of target RNA cleavage directed by a double-stranded siRNA in wheat germ extract alone (*blue*), *Drosophila* ovary lysate alone (*black*), and wheat germ extract supplemented with *Drosophila* ovary lysate (*red*). Wild-type (*A*) and *armitage* (*armi^{72.1}*) (*B*) ovary lysate complemented the wheat germ lysate. In contrast, ovary lysate from $dcr\text{-}2^{L811fsX}$ (*C*) or *r2d2* (*D*) did not. Ovary lysate from the $dcr\text{-}2^{G31R}$ allele (*E*), which is defective for dicing long dsRNA into siRNAs, but not for unwinding siRNA duplexes, complemented wheat germ extract. (*F*) Recombinant Dcr-2 protein or recombinant Dcr-2/R2D2 heterodimer were not sufficient to rescue wheat germ, when used at a physiologically relevant concentration. (*G*) Ovary lysates from flies mutant for *ago2* did not rescue the RNAi in the wheat germ extract.

"slices" the target RNA (Liu et al. 2004; Meister et al. 2004; Parker et al. 2004; Song et al. 2004). However, it is not known if the endonuclease activity of Ago2 is required for its role in RISC assembly. We prepared ovary lysates from $ago2^{16608}$ flies, a strong *ago2* allele. Intriguingly, $ago2^{16608}$ lysates, unlike those from $armi^{72.1}$ mutant flies, did not rescue the wheat germ RNAi defect (Fig. 3G), suggesting that wheat germ extract requires the function of the RLC and that the wheat germ extract lacks the RISC assembly machinery.

Conclusions

Here, we show that single-stranded siRNAs direct target RNA cleavage in wheat germ extracts. Like RNAi in animals, a 5′ phosphate is required for this siRNA function. Reconstitution of immunopurified human Argonaute2 similarly requires a 5′ phosphate (Liu et al. 2004), and structural and kinetic studies argue that Argonaute proteins in general contain a pocket that binds the siRNA 5′ end, perhaps via phosphate interactions (Haley and Zamore 2004; Parker et al. 2004). Our data are consistent with this phosphate-binding pocket also being conserved in plant Argonaute proteins.

Although single-stranded siRNAs can trigger RNAi in wheat germ extract, these extracts cannot be programmed with double-stranded siRNA. Our data suggest that siRNA duplexes do not enter the RNAi pathway in wheat germ extract because they are not unwound. Biochemical complementation experiments with wild-type and mutant *Drosophila* ovary lysates show that wheat germ extract lacks multiple components of the RISC-loading machinery. We do not know if this reflects the absence of these components from the extract or if they are not present in intact wheat embryos. In contrast, wheat embryos clearly contain miRNA-programmed RISCs, since many functional miRNAs are present in wheat germ extract (G. Tang and P.D. Zamore, unpubl.). Wheat embryos may contain a functional RISC-assembly machinery dedicated strictly to the production of miRNA-containing RISC, but may lack the comparable assembly machinery for siRNAs. Alternatively, RISC may be loaded with miRNAs during wheat germ cell development, and mature wheat embryos may lack some or all RISC-assembly components.

Ovary lysates lacking the *Drosophila* protein *armitage*, which is required for RISC assembly and RNAi in vitro and in vivo (Cook et al. 2004; Tomari et al. 2004b) complement wheat germ extracts for RNAi triggered by siRNA duplexes. The Armitage protein is the animal homolog of the plant protein SDE3, which may be present in plant cell extracts. In contrast, $ago2^{16608}$, $r2d2$, and $dicer2^{L811fsX\ mutant}$ lysates, all of which are defective in siRNA unwinding and RISC assembly, do not complement wheat germ extract. Thus, at least three different *Drosophila* RISC assembly proteins are required to rescue the defect in wheat germ extract.

MATERIALS AND METHODS

General Methods

Wheat germ extract preparation (Tang et al. 2003), target cleavage assays (Haley et al. 2003), RNAi triggered with single-stranded siRNAs (Schwarz et al. 2002), siRNA phosphorylation and unwinding assays (Nykänen et al. 2001), and *Drosophila* ovary lysate preparation (Tomari et al. 2004b) were as described previously.

siRNAs

siRNAs were prepared by standard synthesis (Dharmacon Research). Antisense siRNAs targeting firefly luciferase mRNA were 5′-HO-UCG AAG UAU UCC GCG UAC GUG-3′(5′ OH, riboU); 5′-CH$_3$O-dTCG AAG UAU UCC GCG UAC GUG-3′ (5′ CH$_3$O, dT); 5′-HO-UCG AAG UAU UCC GCG UAC GUddC (2′,3′dideoxyC). *let-7* siRNAs contained a guide strand with the sequence 5′-HO-UGA GGU AGU AGG UUG UAU AGU-3′. Sense strands used were 5′-HO-CGU ACG CGG AAU ACU UCG AAA-3′ for *Pp*-luc and 5′-HO-GCU ACA ACC UAC UAC CUC CUU-3′ for the *let-7* siRNA. Sense and antisense strands were annealed as described (Elbashir et al. 2001c). siRNAs were labeled using polynucleotide kinase (New England Biolabs) and γ-^{32}P-ATP (NEN) or poly(A) polymerase (Life Technologies) and α-^{32}P-cordycepin-5′-triphosphate (NEN) as described (Haley et al. 2003).

ACKNOWLEDGMENTS

We thank Guiliang Tang, Tingting Du, and Yukihide Tomari for help preparing wheat germ extract, ovary lysate, and recombinant proteins; members of the Zamore laboratory for advice and support; and Qinghua Liu, Dean Smith, Richard Carthew, Erik Sontheimer, and Makiko and Haruhiko Siomi for *Drosophila* stocks. P.D.Z. is a W.M. Keck Foundation Young Scholar in Medical Research. Supported in part by grants from the National Institutes of Health to P.D.Z. (GM62862-01 and GM65236-01).

REFERENCES

Bernstein E., Caudy A.A., Hammond S.M., and Hannon G.J. 2001. Role for a bidentate ribonuclease in the initiation step of RNA interference. *Nature* **409**: 363.

Billy E., Brondani V., Zhang H., Muller U., and Filipowicz W. 2001. Specific interference with gene expression induced by long, double-stranded RNA in mouse embryonal teratocarcinoma cell lines. *Proc. Natl. Acad. Sci.* **98**: 14428.

Boutla A., Delidakis C., Livadaras I., Tsagris M., and Tabler M. 2001. Short 5′-phosphorylated double-stranded RNAs induce RNA interference in *Drosophila*. *Curr. Biol.* **11**: 1776.

Caplen N.J., Parrish S., Imani F., Fire A., and Morgan R.A. 2001. Specific inhibition of gene expression by small double-stranded RNAs in invertebrate and vertebrate systems. *Proc. Natl. Acad. Sci.* **98**: 9742.

Catalanotto C., Azzalin G., Macino G., and Cogoni C. 2002. Involvement of small RNAs and role of the qde genes in the gene silencing pathway in *Neurospora*. *Genes Dev.* **16**: 790.

Caudy A.A., Myers M., Hannon G.J., and Hammond S.M. 2002. Fragile X-related protein and VIG associate with the RNA interference machinery. *Genes Dev.* **16**: 2491.

Chiu Y.-L. and Rana T.M. 2002. RNAi in human cells: Basic structural and functional features of small interfering RNA. *Mol. Cell* **10**: 549.

Cogoni C., Irelan J.T., Schumacher M., Schmidhauser T.J., Selker E.U., and Macino G. 1996. Transgene silencing of the al-1 gene in vegetative cells of *Neurospora* is mediated by a cytoplasmic effector and does not depend on DNA–DNA interactions or DNA methylation. *EMBO J.* **15**: 3153.

Cook H.A., Koppetsch B.S., Wu J., and Theurkauf W.E. 2004. The *Drosophila* SDE3 homolog armitage is required for oskar mRNA silencing and embryonic axis specification. *Cell* **116**: 817.

Dalmay T., Horsefield R., Braunstein T.H., and Baulcombe D.C. 2001. SDE3 encodes an RNA helicase required for post-transcriptional gene silencing in *Arabidopsis*. *EMBO J.* **20**: 2069.

Dalmay T., Hamilton A., Rudd S., Angell S., and Baulcombe D.C. 2000. An RNA-dependent RNA polymerase gene in *Arabidopsis* is required for posttranscriptional gene silencing mediated by a transgene but not by a virus. *Cell* **101**: 543.

Doi N., Zenno S., Ueda R., Ohki-Hamazaki H., Ui-Tei K., and Saigo K. 2003. Short-interfering-RNA-mediated gene silencing in mammalian cells requires Dicer and eIF2C translation initiation factors. *Curr. Biol.* **13**: 41.

Elbashir S.M., Lendeckel W., and Tuschl T. 2001a. RNA interference is mediated by 21- and 22-nucleotide RNAs. *Genes Dev.* **15**: 188.

Elbashir S.M., Martinez J., Patkaniowska A., Lendeckel W., and Tuschl T. 2001b. Functional anatomy of siRNAs for mediating efficient RNAi in *Drosophila melanogaster* embryo lysate. *EMBO J.* **20**: 6877.

Elbashir S.M., Harborth J., Lendeckel W., Yalcin A., Weber K., and Tuschl T. 2001c. Duplexes of 21-nucleotide RNAs mediate RNA interference in cultured mammalian cells. *Nature* **411**: 494.

Fagard M., Boutet S., Morel J.-B., Bellini C., and Vaucheret H. 2000. AGO1, QDE-2, and RDE-1 are related proteins required for post-transcriptional gene silencing in plants, quelling in fungi, and RNA interference in animals. *Proc. Natl. Acad. Sci.* **97**: 11650.

Fire A., Xu S., Montgomery M.K., Kostas S.A., Driver S.E., and Mello C.C. 1998. Potent and specific genetic interference by double-stranded RNA in *Caenorhabditis elegans*. *Nature* **391**: 806.

Grishok A., Tabara H., and Mello C. 2000. Genetic requirements for inheritance of RNAi in *C. elegans*. *Science* **287**: 2494.

Haley B. and Zamore P.D. 2004. Kinetic analysis of the RNAi enzyme complex. *Nat. Struct. Mol. Biol.* **11:** 599.

Haley B., Tang G., and Zamore P.D. 2003. In vitro analysis of RNA interference in *Drosophila melanogaster*. *Methods* **30:** 330.

Hamilton A.J. and Baulcombe D.C. 1999. A species of small antisense RNA in posttranscriptional gene silencing in plants. *Science* **286:** 950.

Hammond S.M., Bernstein E., Beach D., and Hannon G.J. 2000. An RNA-directed nuclease mediates post-transcriptional gene silencing in *Drosophila* cells. *Nature* **404:** 293.

Hammond S.M., Boettcher S., Caudy A.A., Kobayashi R., and Hannon G.J. 2001. Argonaute2, a link between genetic and biochemical analyses of RNAi. *Science* **293:** 1146.

Hedges S.B. 2002. The origin and evolution of model organisms. *Nat. Rev. Genet.* **3:** 838.

Hutvágner G. and Zamore P.D. 2002. A microRNA in a multiple-turnover RNAi enzyme complex. *Science* **297:** 2056.

Hutvágner G., Simard M.J., Mello C.C., and Zamore P.D. 2004. Sequence-specific inhibition of small RNA function. *PLoS Biol.* **2:** 465.

Kennerdell J.R. and Carthew R.W. 1998. Use of dsRNA-mediated genetic interference to demonstrate that *frizzled* and *frizzled 2* act in the wingless pathway. *Cell* **95:** 1017.

Lee Y.S., Nakahara K., Pham J.W., Kim K., He Z., Sontheimer E.J., and Carthew R.W. 2004. Distinct roles for *Drosophila* Dicer-1 and Dicer-2 in the siRNA/miRNA silencing pathways. *Cell* **117:** 69.

Liu J., Carmell M.A., Rivas F.V., Marsden C.G., Thomson J.M., Song J.J., Hammond S.M., Joshua-Tor L., and Hannon G.J. 2004. Argonaute2 is the catalytic engine of mammalian RNAi. *Science* **305:** 1437.

Liu Q., Rand T.A., Kalidas S., Du F., Kim H.E., Smith D.P., and Wang X. 2003. R2D2, a bridge between the initiation and effector steps of the *Drosophila* RNAi pathway. *Science* **301:** 1921.

Lohmann J.U., Endl I., and Bosch T.C. 1999. Silencing of developmental genes in Hydra. *Dev. Biol.* **214:** 211.

Mallory A., Reinhart B., Jones-Rhoades M., Tang G., Zamore P., Barton M., and Bartel D. 2004. MicroRNA control of PHABULOSA in leaf development: Importance of pairing to the microRNA 5′ region. *EMBO J.* **23:** 3356.

Martinez J. and Tuschl T. 2004. RISC is a 5′ phosphomonoester-producing RNA endonuclease. *Genes Dev.* **18:** 975.

Martinez J., Patkaniowska A., Urlaub H., Lührmann R., and Tuschl T. 2002. Single-stranded antisense siRNA guide target RNA cleavage in RNAi. *Cell* **110:** 563.

Meister G., Landthaler M., Patkaniowska A., Dorsett Y., Teng G., and Tuschl T. 2004. Human Argonaute2 mediates RNA cleavage targeted by miRNAs and siRNAs. *Mol. Cell* **15:** 185.

Morel J.B., Godon C., Mourrain P., Beclin C., Boutet S., Feuerbach F., Proux F., and Vaucheret H. 2002. Fertile hypomorphic ARGONAUTE (ago1) mutants impaired in post-transcriptional gene silencing and virus resistance. *Plant Cell* **14:** 629.

Mourelatos Z., Dostie J., Paushkin S., Sharma A.K., Charroux B., Abel L., Rappsilber J., Mann M., and Dreyfuss G. 2002. miRNPs: A novel class of ribonucleoproteins containing numerous microRNAs. *Genes Dev.* **16:** 720.

Mourrain P., Beclin C., Elmayan T., Feuerbach F., Godon C., Morel J.B., Jouette D., Lacombe A.M., Nikic S., Picault N., Remoue K., Sanial M., Vo T.A., and Vaucheret H. 2000. *Arabidopsis* SGS2 and SGS3 genes are required for posttranscriptional gene silencing and natural virus resistance. *Cell* **101:** 533.

Ngo H., Tschudi C., Gull K., and Ullu E. 1998. Double-stranded RNA induces mRNA degradation in *Trypanosoma brucei*. *Proc. Natl. Acad. Sci.* **95:** 14687.

Nykänen A., Haley B., and Zamore P.D. 2001. ATP requirements and small interfering RNA structure in the RNA interference pathway. *Cell* **107:** 309.

Okamura K., Ishizuka A., Siomi H., and Siomi M.C. 2004. Distinct roles for Argonaute proteins in small RNA-directed RNA cleavage pathways. *Genes Dev.* **18:** 1655.

Pal-Bhadra M., Bhadra U., and Birchler J.A. 2002. RNAi related mechanisms affect both transcriptional and posttranscriptional transgene silencing in *Drosophila*. *Mol. Cell* **9:** 315.

Parker J.S., Roe S.M., and Barford D. 2004. Crystal structure of a PIWI protein suggests mechanisms for siRNA recognition and slicer activity. *EMBO J.* **23:** 4727.

Pham J.W., Pellino J.L., Lee Y.S., Carthew R.W., and Sontheimer E.J. 2004. A Dicer-2-dependent 80s complex cleaves targeted mRNAs during RNAi in *Drosophila*. *Cell* **117:** 83.

Rand T.A., Ginalski K., Grishin N.V., and Wang X. 2004. Biochemical identification of Argonaute 2 as the sole protein required for RNA-induced silencing complex activity. *Proc. Natl. Acad. Sci.* **101:** 14385.

Sánchez-Alvarado A. and Newmark P.A. 1999. Double-stranded RNA specifically disrupts gene expression during planarian regeneration. *Proc. Natl. Acad. Sci.* **96:** 5049.

Schramke V. and Allshire R. 2003. Hairpin RNAs and retrotransposon LTRs effect RNAi and chromatin-based gene silencing. *Science* **301:** 1069.

Schwarz D.S., Hutvágner G., Haley B., and Zamore P.D. 2002. Evidence that siRNAs function as guides, not primers, in the *Drosophila* and human RNAi pathways. *Mol. Cell* **10:** 537.

Song J.J., Smith S.K., Hannon G.J., and Joshua-Tor L. 2004. Crystal structure of Argonaute and its implications for RISC slicer activity. *Science* **305:** 1434.

Tabara H., Yigit E., Siomi H., and Mello C.C. 2002. The dsRNA binding protein RDE-4 interacts with RDE-1, DCR-1, and a DexH-box helicase to direct RNAi in *C. elegans*. *Cell* **109:** 861.

Tabara H., Sarkissian M., Kelly W.G., Fleenor J., Grishok A., Timmons L., Fire A., and Mello C.C. 1999. The rde-1 gene, RNA interference, and transposon silencing in *C. elegans*. *Cell* **99:** 123.

Tang G., Reinhart B.J., Bartel D.P., and Zamore P.D. 2003. A biochemical framework for RNA silencing in plants. *Genes Dev.* **17:** 49.

Timmons L. and Fire A. 1998. Specific interference by ingested dsRNA. *Nature* **395:** 854.

Tomari Y., Matranga C., Haley B., Martinez N., and Zamore P.D. 2004a. A protein sensor for siRNA asymmetry. *Science* **306:** 1377.

Tomari Y., Du T., Haley B., Schwarz D.S., Bennett R., Cook H.A., Koppetsch B.S., Theurkauf W.E., and Zamore P.D. 2004b. RISC assembly defects in the *Drosophila* RNAi mutant armitage. *Cell* **116:** 831.

Vaucheret H., Beclin C., Elmayan T., Feuerbach F., Godon C., Morel J.B., Mourrain P., Palauqui J.C., and Vernhettes S. 1998. Transgene-induced gene silencing in plants. *Plant J.* **16:** 651.

Volpe T.A., Kidner C., Hall I.M., Teng G., Grewal S.I.S., and Martienssen R.A. 2002. Regulation of heterochromatic silencing and histone H3 lysine-9 methylation by RNAi. *Science* **297:** 1833.

Waterhouse P.M., Graham M.W., and Wang M.B. 1998. Virus resistance and gene silencing in plants can be induced by simultaneous expression of sense and antisense RNA. *Proc. Natl. Acad. Sci.* **95:** 13959.

Wianny F. and Zernicka-Goetz M. 2000. Specific interference with gene function by double-stranded RNA in early mouse development. *Nat. Cell Biol.* **2:** 70.

Williams R.W. and Rubin G.M. 2002. ARGONAUTE1 is required for efficient RNA interference in *Drosophila* embryos. *Proc. Natl. Acad. Sci.* **99:** 6889.

Zamore P.D., Tuschl T., Sharp P.A., and Bartel D.P. 2000. RNAi: Double-stranded RNA directs the ATP-dependent cleavage of mRNA at 21 to 23 nucleotide intervals. *Cell* **101:** 25.

A Conserved microRNA Signal Specifies Leaf Polarity

M.C.P. TIMMERMANS,* M.T. JUAREZ,*,† AND T.L. PHELPS-DURR*
*Cold Spring Harbor Laboratory, Cold Spring Harbor, New York 11724; †Graduate Program in Genetics,
Stony Brook University, Stony Brook, New York 11794

Plant shoots are characterized by indeterminate growth resulting from the action of a population of stem cells in the central zone of the shoot apical meristem (SAM) (Fig. 1). These stem cells give rise to peripheral derivatives from which lateral organs, such as leaves and flowers, arise. Leaves of higher plants exhibit a varying degree of asymmetry along their adaxial/abaxial (upper/lower) axis. This asymmetry is thought to reflect inherent positional differences in the developing organ relative to the SAM (Wardlaw 1949). The adaxial/dorsal side of the leaf develops in close proximity to the stem cells in the SAM, whereas the abaxial/ventral side develops at a greater distance from the apex. Evidence that the meristem is required for pattern formation within lateral organs came from early surgical experiments in potato (Sussex 1951, 1955). Incisions that separate the incipient primordium from the central zone of the SAM resulted in formation of a radially symmetric abaxialized leaf, suggesting that a signal from the SAM is required to specify adaxial cell fate. The exact nature of this adaxializing signal remains elusive despite the recent identification of several gene families required for the determination of adaxial or abaxial identity. These recent studies have, however, identified a microRNA (miRNA) signal that originates from below the incipient leaf and sets up the abaxial domain. Adaxial/abaxial leaf polarity may thus be established by two opposing signals that both originate outside the incipient primordium; the classical adaxializing signal from the tip of the SAM and the miRNA signal from a potential signaling center below the incipient leaf.

ESTABLISHMENT OF POLARITY IN DEVELOPING LEAVES

In *Arabidopsis*, PHABULOSA (PHB), PHAVOLUTA (PHV), and REVOLUTA (REV), members of the class III homeodomain-leucine zipper (HD-ZIPIII) family of proteins, promote adaxial identity in developing lateral organs (McConnell et al. 2001; Otsuga et al. 2001; Emery et al. 2003). By contrast, the *KANADI* (*KAN*) genes, which encode transcriptional regulators of the GARP family, are required for abaxial cell fate determination (Eshed et al. 2001, 2004; Kerstetter et al. 2001). Both the *HD-ZIPIII* and the *KAN* genes are expressed throughout the incipient leaf primordium, but shortly after the primordium emerges from the SAM, their expression becomes restricted to mutually exclusive domains on the adaxial and abaxial side of the developing organ, respectively. Once this adaxial/abaxial polarity is established, it is interpreted by other downstream genes that lead to the differentiation of adaxial and abaxial specific cell types and to mediolateral blade outgrowth at the adaxial/abaxial boundary (Waites and Hudson 1995). Among these are the *YABBY* genes, a third family of transcriptional regulators, that act at least in part downstream of the *HD-ZIPIII* and *KAN* genes and are required for abaxial cell fate and lamina expansion (Sawa et al. 1999; Siegfried et al. 1999; Eshed et al. 2001, 2004; Kumaran et al. 2002).

The mechanisms by which the *HD-ZIPIII* and *KAN* expression domains become restricted to opposing sides of the primordium upon its emergence from the SAM is unclear. Persistent uniform expression of *KAN1* leads to the abaxialization of lateral organs (Eshed et al. 2001; Kerstetter et al. 2001). Thus, specification of adaxial identity requires the suppression of abaxial determinants. The phenotype resulting from loss of *PHB*, *PHV*, and *REV* function resembles the phenotype caused by ectopic *KAN* expression, suggesting that the *HD-ZIPIII* genes act at least in part to spatially restrict the *KAN* expression domain (Emery et al. 2003). HD-ZIPIII proteins contain a

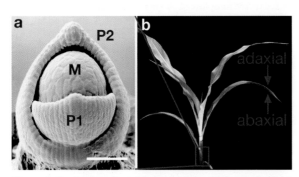

Figure 1. Leaf primordia arise on the flank of the shoot apical meristem (SAM) and become patterned along the adaxial/abaxial axis in response to signals from the SAM. (*a*) Scanning electron micrograph of a maize shoot apex. The meristem (M) contains a population of stem cells that allows the continued initiation of leaf primordia. The youngest leaf primordium is indicated as P1, the second youngest as P2, etc. (*b*) 14-day-old maize seedling. The *red box* marks the approximate position of the SAM within the surrounding older leaves that have distinct adaxial/upper and abaxial/lower surfaces.

START lipid-sterol binding-like domain that is required for protein function (Otsuga et al. 2001). Therefore, PHB, PHV, and REV may become activated upon interaction with a ligand, perhaps the meristem-borne signal proposed by the surgical experiments (McConnell et al. 2001; Juarez et al. 2004a). Such activation could direct the down-regulation of *KAN* expression, thus suppressing abaxial identity and leading to the specification of adaxial cell fate. However, what suppresses the expression of *HD-ZIPIII* genes on the abaxial side? This spatial restriction of *PHB*, *PHV*, and *REV* expression is likely to involve an RNA-interference (RNAi)-like mechanism. Transcripts from all three genes contain a complementary site for miRNA 165 (miR165) and miR166, which can direct their cleavage in vitro (Reinhart et al. 2002; Rhoades et al. 2002; Tang et al. 2003; see below). Disruption of the miR165/166 complementary site, as in gain-of-function *phb-d* and *phv-d* alleles, prevents this miRNA-directed cleavage. Such mutations lead to ectopic abaxial expression of mutant transcripts and adaxialization of leaves and other lateral organs (McConnell et al. 2001). Similar dominant mutations in *REV* are less severe, but cause adaxial/abaxial patterning defects in the normally polarized vascular bundles (Emery et al. 2003; Zhong and Ye 2004).

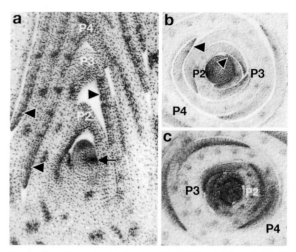

Figure 2. *rld1* expression is altered in *Rld1* leaf primordia. (*a,b*) Longitudinal (*a*) and transverse (*b*) sections through wild-type apices showing *rld1* expression in the SAM, vasculature, and on the adaxial side of P1 and older leaf priomordia (*arrowheads*). The *arrow* marks *rld1* expression near the site of leaf initiation. (*c*) Transverse section through a *Rld1* mutant apex showing misexpression of *rld1* on the abaxial side of young leaf primordia. Leaf primordia P2–P4 are indicated. (Reprinted, with permission, from Juarez et al. 2004b [©Nature Publishing Group; http://www.nature.com].)

SPECIFICATION OF ADAXIAL/ABAXIAL POLARITY DURING MAIZE LEAF DEVELOPMENT

Adaxial/abaxial asymmetry in maize leaves similarly depends on the polarized expression of members of the *hd-zipIII* gene family (Juarez et al. 2004a,b). *rolled leaf1* (*rld1*), which encodes a close homolog of REV, is normally expressed in the presumptive central zone of the SAM and in a stripe of cells from the center of the SAM to the site of leaf initiation (Fig. 2a,b). In the P1 primordium, *rld1* is expressed along the adaxial domain as well as in the midvein region. *rld1* expression persists during primordium development in the vasculature and on the adaxial side near the margins. The expression pattern of a maize *PHB* homolog is comparable to that of *rld1* although the level of *phb* is relatively more abundant in the developing vasculature. These expression patterns resemble those of the *HD-ZIPIII* genes in *Arabidopsis* and are consistent with a conserved role for the maize *hd-zipIII* genes in meristem function and adaxial fate determination. However, the overlapping expression patterns of *rld1* and *phb* suggest these genes probably act redundantly in these processes.

The adaxial-specific expression of *rld1*, *phb*, and potential other *hd-zipIII* family members is in part mediated by *leafbladeless1* (*lbl1*) (Juarez et al. 2004a; M. Juarez and M. Timmermans, unpubl.). Recessive mutations in *lbl1* lead to the formation of abaxialized leaves. The most severely affected *lbl1* mutant leaves are radially symmetric and resemble the abaxialized leaves that arise following the surgical separation of leaf initials from the central region of the SAM (Timmermans et al. 1998). The levels of *rld1* and *phb* transcripts are dramatically reduced or lost in *lbl1* leaves, suggesting that these genes act downstream of *lbl1* in the pathway leading to adaxial identity.

Adaxial-specific expression of *rld1* may also depend on the action of miR165 or miR166. The miR165/166 complementary site is conserved between *rld1* and the *Arabidopsis HD-ZIPIII* genes. Moreover, four dominant mutant alleles of *rld1* were shown to result from single nucleotide substitutions in the 5′ end of the miR165/166 complementary site (Fig. 3a) (Juarez et al. 2004b). Such dominant *Rld1* mutations have no effect on the meristematic expression pattern of *rld1*, but lead to misexpression of *rld1* on the abaxial side of incipient and P1 leaf primordia (Fig. 2c). As in wild type, the *rld1* expression domain in older *Rld1* primordia becomes increasingly more confined to the nondetermined cells near the margins. However, the domain of strong *rld1* expression at the margins is broader and includes the abaxial side. Moreover, weak *rld1* expression persists on both the adaxial and abaxial site in the central region of P2 and older *Rld1* primordia. These latter changes in *rld1* expression are associated with an upward curling of the *Rld1* leaf blade caused by adaxialization or partial reversal of adaxial/abaxial leaf polarity (Fig. 3c) (Nelson et al. 2002; Juarez et al. 2004a). The ligule normally forms on the adaxial side of the leaf, and the wild-type adaxial epidermis has distinctive hairs and strengthening cells (Fig. 3b). Adaxial/abaxial polarity is also reflected in the patterning of the vasculature, in that xylem tissue differentiates toward the adaxial side whereas phloem forms on the abaxial side of the leaf. In *Rld1* leaves, epidermal cell types are displaced from the adaxial to the abaxial leaf surface. In addition, adaxialized sectors often arise on either side of the midvein. Such sectors lack minor vascular bundles and their associated photosynthetic cell types and develop an ectopic ligule on the abaxial side of the leaf (Fig. 3d).

Characterization of the dominant *Rld1*, *phb-d*, *phv-d* and *rev-d* alleles thus implicates an miRNA in the establishment of adaxial/abaxial polarity in both maize and *Arabidopsis*. The spatial regulation of *hd-zipIII* genes by miR165/166 may even date back to the mosses, which last shared a common ancestor with *Arabidopsis* and maize over 400 million years ago (Floyd and Bowman 2004). Other miRNAs with complementarity to developmentally important regulatory genes have been identified in plants as well as animals. The facts that miRNAs are present in such diverse species and control a broad range of targets imply an important role for this recently discovered gene regulatory mechanism in development and suggest that miRNAs may constitute a new class of developmental signaling molecules.

miRNA BIOGENESIS AND FUNCTION IN GENE REGULATION

miRNAs are endogenous small (~22-mer) noncoding RNAs that mediate the cleavage or translational repression of target transcripts containing a complementary sequence. miRNAs were first identified in *Caenorhabditis elegans* (Lee et al. 1993; Reinhart et al. 2000), but have since been found in organisms as evolutionary distinct as plants and humans (Lagos-Quintana et al. 2001; Lau et al. 2001; Lee and Ambros 2001; Llave et al. 2002a; Park et al. 2002; Reinhart et al. 2002). To date, the cloning of small RNAs from *Arabidopsis* has lead to the identification of more than 40 distinct miRNAs that can be grouped into 22 miRNA families based on sequence similarity and target specificity (Jones-Rhoades and Bartel 2004). Most miRNAs are encoded by multigene families, such that the *Arabidopsis* genome includes 92 potential miRNA loci. However, this number is almost certainly an underestimate as additional miRNA families have been predicted using various computational approaches, but these await experimental verification (Bonnet et al. 2004; Jones-Rhoades and Bartel 2004). miRNAs are frequently conserved among distantly related animal species (Pasquinelli et al. 2000; see Bartel 2004). Similarly, 18 of the 22 *Arabidopsis* miRNA families have homologs, or predicted homologs, in rice (Reinhart et al. 2002; Bonnet et al. 2004; Jones-Rhoades and Bartel 2004; Wang et al. 2004). Such conservation suggests an important role for these miRNAs and the posttranscriptional regulation of their target genes.

miRNAs are initially transcribed as long primary transcripts called pri-miRNAs. These transcripts are processed into ~70–300-nt stem-loop intermediates, known as the miRNA precursor or pre-miRNA. pre-miRNAs are subsequently cleaved to yield an imperfect duplex comprising the mature miRNA and the so-called miRNA*, which is derived from the opposite arm of the pre-miRNA stem. Production of the miRNA:miRNA* duplex from the pri-miRNA involves slightly different processes in animals and plants. In animals, the initial processing step is performed by the nuclear RNase III endonuclease, Drosha (Lee et al. 2003); whereas pre-miRNAs are cleaved in the cytoplasm by Dicer (Grishok et al. 2001; Hutvagner et al. 2001; Ketting et al. 2001), the RNase III enzyme that is also required for the production of siRNAs from double-stranded RNAs during RNAi (Bernstein et al. 2001). Drosha homologs are absent from plants, and both processing steps seem to be executed in the nucleus by the Dicer homolog, DICERLIKE1 (DCL1) (Park et al. 2002; Reinhart et al. 2002; Papp et al. 2003; Xie et al. 2003). Accordingly, pre-miRNAs do not accumulate to easily detectable levels in plants, as they do in animals.

The miRNA strand of the duplex becomes incorporated into a RNA-induced silencing complex (RISC), whereas the miRNA* gets degraded. The mechanism of

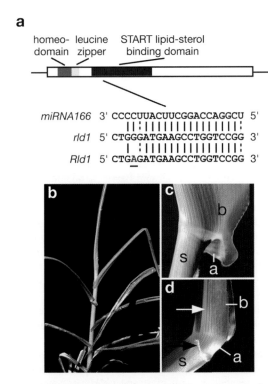

Figure 3. Dominant *Rld1* alleles result from a point mutation in the miR166 complementary site and affect adaxial/abaxial polarity in the leaf. (*a*) Organization of RLD1, a member of the HD-ZIPIII family of proteins. *rld1* encodes an 840-amino-acid protein containing a homeodomain (*green*), leucine-zipper domain (*yellow*), and START lipid-sterol binding domain (*blue*). The approximate position of the miR166 complementary site and an alignment of miR166 with the target sites in *rld1* and the mutant *Rld1* alleles are shown. *Solid lines*, Watson–Crick base pairs; *dotted lines*, RNA base pairs. The nucleotide substitution present in the four dominant *Rld1* alleles is highlighted in *red*. (*b*–*d*) Effects of *Rld1* on leaf polarity. (*b*) Mature *Rld1* plant illustrating the tightly upward curled leaves. (*c*) Abaxial surface of a wild-type adult maize leaf, which comprises sheath (s) and blade (b) tissues separated by the auricle (a). (*d*) Abaxial surface of a partially adaxialized *Rld1* leaf. *Rld1* leaves frequently develop an ectopic abaxial ligule (*black arrow*) and clear sectors with fewer minor veins and no photosynthetic cell types (*white arrow*). (Reprinted, with permission, from Juarez et al. 2004b [©Nature Publishing Group; http://www.nature.com].)

RISC assembly is largely unknown, but is thought to involve a helicase activity that unwinds the miRNA:miRNA* duplex and selects, based on the thermodynamics of the duplex, the strand to enter the RISC (Khvorova et al. 2003; Schwarz et al. 2003). RISC loading in plants may also involve the double-stranded RNA (dsRNA) binding proteins HEN1 and HYPONASTIC LEAVES1 (HYL1), which affect the accumulation of miRNAs (Park et al. 2002; Han et al. 2004; Vazquez et al. 2004). Similar proteins from C. elegans and Drosophila facilitate the transfer of siRNAs to RISC (Tabara et al. 2002; Liu et al. 2003). Interestingly, both HYL1 and HEN1 are nuclear localized suggesting that, unlike in animal systems, miRNA production and RISC assembly in plants may occur entirely in the nucleus.

The miRNA-RISC complex can mediate mRNA cleavage or stall translation. The exact mechanism of repression is dictated in part by the level of complementarity between the miRNA and the target transcript (Hutvagner and Zamore 2002; Doench et al. 2003; Zeng et al. 2003). In animals, this complementarity is relatively limited and predominantly located within the 5′ end of the miRNA. As a result, most animal miRNAs act at the translational level. In contrast, plant miRNAs and their targets frequently possess near-perfect complementarity throughout. Most plant miRNA–mRNA interactions lead to cleavage of the target transcript at a position corresponding to nucleotides 10 and 11 of the miRNA (Llave et al. 2002b; Kasschau et al. 2003). However, the level of complementarity is not the sole factor that distinguishes between RNA cleavage and translational repression, at least not in plants. miR172 has near perfect complementarity to *APETALA2* (*AP2*) but mainly affects AP2 translation (Aukerman and Sakai 2003; Chen 2004). The mode of miRNA action may also depend on the protein composition of RISC. ARGONAUTE (AGO) proteins form an integral component of all RISCs (Hammond et al. 2001; Mourelatos et al. 2002; Caudy et al. 2003), but most organisms encode multiple AGO family members that seem to mediate distinct RISC functions. For instance, the *Arabidopsis zippy/ago7* mutants display developmental phenotypes consistent with a defect in miRNA-mediated gene regulation, but AGO4 is required for transposon silencing, and AGO1 plays a role in both posttranscriptional and miRNA-mediated silencing (Fagard et al. 2000; Hunter et al. 2003; Zilberman et al. 2003; Vaucheret et al. 2004).

miRNAs AS DEVELOPMENTAL SIGNALING MOLECULES

The observation that the first plant miRNAs had near-perfect complementarity to their target mRNAs has enabled the identification of many target genes using computational approaches (Rhoades et al. 2002; Bonnet et al. 2004; Jones-Rhoades and Bartel 2004). The *Arabidopsis* miRNAs identified so far show a strong propensity to target transcription factor families or other genes controlling development. miRNAs that regulate other aspects of plant biology, such as basic metabolism and the response to particular pathogens or environmental stresses, have also been identified (Bonnet et al. 2004; Jones-Rhoades and Bartel 2004). Consistent with a role for miRNAs in a variety of developmental processes, mutations affecting their biogenesis are pleiotropic. For instance, *dcl1*, *hen1*, and *hyl1* mutants exhibit overlapping defects in vegetative, floral, and reproductive development (Jacobsen et al. 1999; Chen et al. 2002; Han et al. 2004; Vazquez et al. 2004). Remarkably similar developmental defects are also observed in plants expressing viral suppressors of RNA silencing that interfere with miRNA function, such as P1/HC-Pro of Turnip mosaic virus, P19 of Tomato bushy stunt virus, or the P15 protein from Peanut clump virus (Kasschau et al. 2003; Dunoyer et al. 2004). However, not all mutations affecting miRNA production or function are as pleiotropic (e.g., *zippy/ago7* and *hasty* [Bollman et al. 2003; Hunter et al. 2003]), which could indicate partial redundancy or branching in the miRNA pathway. Interestingly, mutations in *AGO1* and the closely related *PINHEAD/ZWILLE* (*PNH/ZLL*) gene affect meristem function and adaxial/abaxial polarity (Bohmert et al. 1998; Moussian et al. 1998; Lynn et al. 1999; Kidner and Martienssen 2004), supporting a role for miRNA-mediated cleavage in the spatial regulation of *hd-zipIII* transcripts.

miRNA-directed cleavage products are relatively stable and numerous miRNA targets have been verified through detection of the predicted transcript fragments (see, e.g., Jones-Rhoades and Bartel 2004). Other targets have been confirmed by mutational analysis. These include the above-mentioned *hd-zipIII* family members, which are required for meristem function and adaxial/abaxial patterning (McConnell et al. 2001; Emery et al. 2003; Juarez et al. 2004b; McHale and Koning 2004; Zhong and Ye 2004). Dominant gain-of-function alleles of miR-JAW were identified in a collection of activation-tagged lines (Palatnik et al. 2003). miR-JAW regulates a subset of *TCP* genes required for the proper temporal transition from cell division to differentiation (Nath et al. 2003). This transition is delayed in the *jaw-D* mutants, which alters leaf shape. Plants expressing a miR-JAW-insensitive allele of *TCP4* appear to have the opposite phenotype as they arrest early in development (Palatnik et al. 2003). Similar types of mutations also confirmed a role for miR164 in organ boundary formation by defining the expression domains of *CUP-SHAPED COTYLEDON1* (*CUC1*) and *CUC2* (Laufs et al. 2004; Mallory et al. 2004) and for miR159 and miR172 in floral initiation and floral organ development. miR159 regulates the accumulation of *AtMYB33*, a component of the gibberellin response pathway and activator of the floral inducer *LEAFY* (Palatnik et al. 2003; Achard et al. 2004), whereas miR172 targets AP2 and several AP2-like transcription factors (Aukerman and Sakai 2003; Chen 2004). Interestingly, *DCL1* and *AGO1* are also targets for miRNA mediated posttranscriptional regulation, which suggests that miRNA biogenesis and function is controlled by a negative feedback mechanism (Xie et al. 2003; Vaucheret et al. 2004).

ADAXIAL/ABAXIAL AXIS SPECIFICATION BY miR166

One possible reason why so many miRNAs target transcription factors or other regulators of plant development is that the active degradation of such transcripts may help facilitate changes in cell fate (Rhoades et al. 2002). For example, the down-regulation of *hd-zipIII* expression required for abaxial cell fate specification upon primordium emergence could be achieved by suppressing transcription of the *hd-zipIII* genes in cells on the abaxial side. However, abaxial-specific expression of miRNAs would enable a more rapid switch in cell fate by actively eliminating *hd-zipIII* transcripts inherited from the incipient primordium. Although this scenario is supported by characterization of the dominant *hd-zipIII* mutations, a comparative analysis of the *hd-zipIII* and miR165 or miR166 expression patterns at the cellular level would provide direct evidence that miRNA expression patterns can establish patterns of tissue organization during development. The relatively large size of the maize apex might simplify such a detailed comparison. We therefore took advantage of sequence conservation between the *Arabidopsis* and rice *MIR166* genes and cloned fragments of several miR166 precursors from maize, which were used to determine the precise miR166 expression pattern by in situ hybridization (Juarez et al. 2004b).

miR165 and miR166 differ by a single C-U transition at position 17. Even though this nucleotide difference slightly decreases the complementarity of miR166 to the *HD-ZIPIII* transcripts, by substituting a G:C base pair with a G:U wobble, it does not affect the in vitro cleavage of *PHV* transcripts (Tang et al. 2003). *Arabidopsis* contains two *MIR165* and seven *MIR166* loci (Reinhart et al. 2002). Most pri-miRNA genes within a miRNA family show obvious sequence conservation in the miRNA*. Sequences outside this duplex usually exhibit little primary sequence similarity but maintain a propensity to form a dsRNA structure. Surprisingly, pair-wise comparisons between the *MIR165* and *MIR166* family members revealed extensive sequence conservation surrounding the predicted miRNA:miRNA* duplexes (Juarez et al. 2004b). The two *MIR165* loci share six conserved domains, including the miRNA and miRNA* sequences, over an interval of ~850 nucleotides (nt). The pri-miR166 loci, *MIR166a* and *MIR166b*, share nine conserved sequence motifs over a region of ~1.5 kb. Similarly, *MIR166c* contains six sequence motifs that are conserved in *MIR166d* and four sequence motifs that are conserved in *MIR166e*. *MIR166d* also shows sequence homology to *MIR166g* in six domains over an ~1 kb region. Most of these conserved sequence motifs are 30–50 nt in length, but some extend over 100 bases.

Because of the very efficient processing of pri-miRNAs into mature miRNAs, the *MIR166* precursor transcripts are not detectable by northern blot analysis (Reinhart et al. 2002). However, primers derived from each of the nine conserved sequence motifs in *MIR166a* and *MIR166b* did allow the amplification of overlapping cDNA fragments from seedling RNA (T. Phelps-Durr and M. Timmermans, unpubl.). These conserved sequence motifs thus constitute part of the same pri-miRNA, which spans at least 1.4 kb. Some pri-miRNA transcripts are processed by splicing (Aukerman and Sakai 2003), but these conserved sequences do not correspond to exons. They may, however, contain elements important for miRNA regulation or processing.

The rice genomic sequence contains at least six loci with the potential to produce miR166 homologs although *MIR165* loci have not been identified (Reinhart et al. 2002). Like the *Arabidopsis MIR165* and *MIR166* family members, several of the potential *MIR166* loci from rice contain sequence homology outside the predicted miRNA and miRNA* sequences. *OsMIR166c* and *OsMIR166f* share five regions of homology. *OsMIR166b* and *OsMIR166d* share only three conserved sequence motifs and these include the miR166 and miRNA*. However, the sequence and arrangement of these three motifs is also conserved in *Arabidopsis MIR166a* and *MIR166b* (Juarez et al. 2004b). Irrespective of whether this third sequence motif is also part of the pri-miRNA in rice, its conservation between *Arabidopsis* and rice suggests it is likely conserved in maize as well.

Degenerate primers derived from these three conserved sequence motifs did indeed allow the amplification of several partial *mir166* cDNA clones from maize vegetative apex and inflorescence tissues (Juarez et al. 2004b). Six distinct reverse-transcriptase (RT)-dependent polymerase chain reaction (PCR) products were amplified from immature tassel RNA, but fewer *mir166* genes appear to be expressed in vegetative apices. Sequence analysis suggests at least four of the maize *mir166* genes (*mir166a-mir166d*) contain the three sequence motifs found to be conserved between *Arabidopsis* and rice. No other primary sequence similarities exist between these four maize genes and between the maize, rice, and *Arabidopsis* genes. Consistent with the amplification of fewer PCR products from vegetative apices, *mir166a* and *mir166b* are expressed in both vegetative and inflorescence tissues, whereas expression of *mir166c* and *mir166d* is limited to the inflorescence. As predicted by their regulatory role in development, miRNAs in both animals and plants accumulate in a temporal or tissue-specific pattern. However, these RT-PCR results suggest that individual pri-miRNA genes can display distinct tissue specificities. This could potentially be very significant, as this would reduce the level of redundancy among the pri-miRNA genes.

To examine directly whether miR166 defines the abaxial domain by restricting *hd-zipIII* expression to the adaxial side, we used in situ hybridization to compare the *rld1* and miR166 expression patterns at the cellular level (Juarez et al. 2004b). Approximately 60 rounds of amplification were required to detect *mir166a* pri-miRNA or pre-miRNA transcripts by RT-PCR. Accordingly, no hybridization signal was detected when vegetative apex sections were hybridized with a probe specific for the pri-miRNA. However, in situ hybridizations using a slightly larger fragment of *mir166a* that includes an antisense copy of miR166 revealed expression in developing leaf primordia (Fig. 4). miR166 accumulates on the abaxial side of the P1 leaf. In older leaf primordia, miR166 accumulates in a progressively broader domain extending laterally and adaxially. Only near the margins does miR166 expression remain limited to the abaxial side (Fig. 4a). *rld1* and miR166 thus exhibit complementary expression patterns in developing leaf primordia, consistent with a role for miR166 in defining the domain of *rld1* expression. In contrast, the *rld1* expression domain in meristematic tissue is mainly controlled at the transcriptional level. *rld1* is expressed in the central zone of the SAM and in a stripe of cells that includes the incipient leaf primordium (Fig. 4b), whereas miR166 accumulates in a group of cells immediately below the incipient leaf opposite the P1 primordium. Interestingly, a gradient of weaker miR166 expression extends into the abaxial side of the P0 leaf (Fig. 4c) and a gradual decline in miR166 expression may also exist in older leaf primordia. As expected, *rld1* expression is ectopically expressed in the dominant *Rld1* mutant and includes the miR166 expression domain below the incipient leaf and on the abaxial side of the P1 primordium (Fig. 4d), but *rld1* expression in the *Rld1* SAM is unaffected. *hd-zipIII* genes also play a role in the adaxial/abaxial patterning of vascular bundles (Zhong and Ye 1999; Ratcliffe et al. 2000). *rld1* and *phb* are expressed in provascular strands and expression becomes localized to adaxial pro-xylem cells when distinct phloem and xylem poles become apparent (Fig. 4e). At that time, miR166 accumulates in the abaxial phloem tissue (Fig. 4f).

DISCUSSION

Despite the differences in monocot and dicot leaf development, establishment of adaxial/abaxial polarity in both classes of angiosperms requires the spatial restriction of *hd-zipIII* expression to the adaxial side. In both *Arabidopsis* and maize this is mediated by the complementary, abaxial expression of miR166 and miR165 (Juarez et al. 2004b; Kidner and Martienssen 2004; T. Phelps-Durr and M. Timmermans, unpubl.). The miR165/166-directed cleavage of *hd-zipIII* transcripts is conserved even in the basal lineages of land plants and dates back more than 400 million years ago to the last common ancestor of the mosses and seed plants (Floyd and Bowman 2004). Conservation of *let-7* across the animal phylogeny suggests a similar early origin for miRNA-mediated gene regulation in the metazoans (Pasquinelli et al. 2000). The differences in biogenesis, target gene complementarity, and usual mode of gene regulation between plant and animal miRNAs, together with the fact that no miRNA has thus far been identified that is conserved between animals and plants, suggest nonetheless that miRNAs may have arisen independently in each lineage (Bartel 2004). Alternatively, the regulation of genes by small RNAs may have arisen only once but the divergence of target genes may have lead to the fixation of distinct miRNA families in plants and animals.

The regulation of *hd-zipIII* gene expression by miR165/166 thus predates the origin of angiosperm leaves. Because *hd-zipIII* expression both on the adaxial side of lateral organs and in the adaxial pro-xylem cells is defined by the pattern of miR166 accumulation, these genes may have had an ancestral role in establishing polarity in vascular tissue of nonleafy plants, which was later co-opted in the adaxial/abaxial patterning of leaves (Juarez et al. 2004a). The role of the *KAN* genes in abaxial fate determination in developing leaves may similarly be derived from an ancestral function in vascular patterning (Emery et al. 2003). In *Arabidopsis*, *KAN* expression is limited to the abaxial or peripheral phloem cells, and mutational analysis suggests the *KAN* and *HD-ZIPIII* genes act antagonistically during both primordium and vascular development (Kerstetter et al. 2001; Emery et al. 2003).

Certain aspects of the pathway leading to adaxial/abaxial polarity have evolved between monocots and dicots. Maize *yabby* genes, in contrast to those of *Arabidopsis*, are expressed on the adaxial side of incipient and young leaf primordia, suggesting divergence in their regulation and function (Juarez et al. 2004a). Whereas the *Arabidopsis HD-ZIPIII* genes suppress *YABBY* expression (Siegfried et al. 1999; Eshed et al. 2001), *yabby* genes in maize are positively regulated by *rld1*. The maize and *Arabidopsis yabby* genes may share a role in mediating lateral outgrowth along the adaxial/abaxial boundary (Eshed et al. 2004; Juarez et al. 2004a). However, the *Ara-*

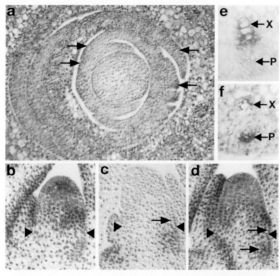

Figure 4. The *hd-zipIII* genes and miR166 have complementary expression patterns in developing leaf primordia. (*a*) Transverse section through a wild-type apex showing miR166 expression in leaf primordia but not in the SAM. *Arrows* mark regions with only abaxial miRNA166 expression. (*b,c*) Longitudinal sections through wild-type apices showing *rld1* expression (*b*) at the site of leaf initiation and miR166 expression (*c*) on the abaxial side (*arrow*) and below the incipient leaf. *Arrowheads* mark the base of the incipient leaf. (*d*) Longitudinal sections through a *Rld1* apex showing *rld1* misexpression below and at the base of the incipient leaf (*arrows*). (*e, f*) Transverse sections through wild-type vascular bundles showing *phb* expression (*e*) in adaxial pro-xylem cells and miR166 expression (*f*) in abaxial phloem tissue. X, xylem; P, phloem. (Reprinted, with permission, from Juarez et al. 2004b [©Nature Publishing Group; http://www.nature.com].)

bidopsis *YABBY* genes also specify abaxial fate, and this function is not conserved in maize (Sawa et al. 1999; Siegfried et al. 1999; Kumaran et al. 2002). Moreover, the miRNA loci involved in adaxial/abaxial patterning have diverged between *Arabidopsis*, rice, and maize. *MIR165* loci have not been identified in the rice genome sequence (Reinhart et al. 2002), and no miR165 expression was observed in maize vegetative apices (T. Phelps-Durr and M. Timmermans, unpubl.). The number of *MIR166* genes also varies between these species, and their sequences exhibit very limited primary sequence similarity outside the miR166 and miRNA*. Moreover, preliminary data suggest that the pattern of miR166 accumulation in *Arabidopsis* is not directly comparable to that in maize, although miR166 shows a dynamic pattern of expression throughout leaf development in both species (T. Phelps-Durr and M. Timmermans, unpubl.).

miR165/166 thus constitutes an important highly conserved polarizing signal. miR166 initially accumulates immediately below the incipient leaf but subsequently in a progressively broader domain including the adaxial side. This dynamic expression pattern and the gradient of miR166 expression into the incipient primordium is reminiscent of a movable signal, suggesting that expression of miR166 may be under such control just as expression of miR159 is regulated by the phytohormone gibberillin (Achard et al. 2004). Alternatively, miR166 may itself move between cells to set up a gradient of expression. This could imply a more active signaling role for miRNAs in development similar to that of peptide ligands or hormones. Although this hypothesis is currently unproven, the accumulation of miR166 in the phloem could be consistent with miR166 movement from the site of *mir166* expression. Similarly suggestive, expression of a viral movement protein in the SAM that disrupts RNA trafficking causes formation of radial adaxialized leaves (Foster et al. 2002). The site of *mir166* expression is unknown but may coincide with the initial strong miR166 accumulation below the incipient primordium. If so, the *mir166* expression domain is likely established independently of abaxial determinants that function in the incipient primordium. *mir166* expression is also established independently of *lbl1* and miR166 acts upstream of *rld1* (M. Juarez and M. Timmermans, unpubl.).

Classical surgical experiments indicated that specification of adaxial cell fate requires a signal from the meristem (Sussex 1951, 1955). This signal could act via RLD1 and other HD-ZIPIII family members as they contain a START lipid-sterol binding domain. If so, RLD1 and other HD-ZIPIII proteins may specify adaxial/abaxial polarity in developing leaves by incorporating positional information established by two opposing signals that originate outside the incipient primordium: the adaxializing signal from the SAM and the miR166 signal from a potential signaling center below the incipient leaf (Fig. 5).

ACKNOWLEDGMENTS

The authors thank Catherine Kidner for helpful discussions and comments on the manuscript. We also thank Julie

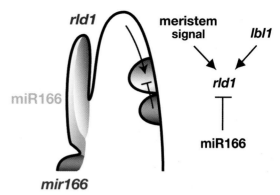

Figure 5. Adaxial/abaxial polarity in the maize leaf may be established by two distinct signals that originate outside the incipient leaf and act upon *rld1*. Expression of *rld1* in the incipient leaf depends on *lbl1* and miR166. *lbl1* positively affects the accumulation of *rld1* transcripts, whereas miR166 directs their cleavage. The miR166 signal initially accumulates immediately below the incipient leaf but gradually spreads into the abaxial domain, thus restricting *rld1* expression to the adaxial side. Specification of adaxial fate also requires a signal from the meristem, which may exert its effects via the START lipid-sterol binding domain of RLD1.

Thomas, Francois Bolduc, and Pawel Mazur for technical assistance; David Jackson for the SEM image of the maize SAM; and Tim Mulligan for plant care. This work was supported by grants from the NSF to M.C.P.T., and M.T.J. was in part funded by a W. Burghardt Turner fellowship.

REFERENCES

Achard P., Herr A., Baulcombe D.P., and Harberd N.P. 2004. Modulation of floral development by a gibberellin-regulated microRNA. *Development* **131:** 3357.

Aukerman M.J. and Sakai H. 2003. Regulation of flowering time and floral organ identity by a microRNA and its *APETALA2*-like target genes. *Plant Cell* **15:** 2730.

Bartel D.P. 2004. MicroRNAs: Genomics, biogenesis, mechanism, and function. *Cell* **116:** 281.

Bernstein E., Caudy A.A., Hammond S.M., and Hannon G.J. 2001. Role for a bidentate ribonuclease in the initiation step of RNA interference. *Nature* **409:** 363.

Bohmert K., Camus I., Bellini C., Bouchez D., Caboche M., and Benning C. 1998. *AGO1* defines a novel locus of *Arabidopsis* controlling leaf development. *EMBO J.* **17:** 170.

Bollman K.M., Aukerman M.J., Park M.Y., Hunter C., Berardini T.Z., and Poethig R.S. 2003. HASTY, the *Arabidopsis* ortholog of exportin 5/MSN5, regulates phase change and morphogenesis. *Development* **130:** 1493.

Bonnet E., Wuyts J., Rouze P., and Van de Peer Y. 2004. Detection of 91 potential conserved plant microRNAs in *Arabidopsis thaliana* and *Oryza sativa* identifies important target genes. *Proc. Natl. Acad. Sci.* **101:** 11511.

Caudy A.A., Ketting R.F., Hammond S.M., Denli A.M., Bathoorn A.M., Tops B.B., Silva J.M., Myers M.M., Hannon G.J., and Plasterk R.H. 2003. A micrococcal nuclease homologue in RNAi effector complexes. *Nature* **425:** 411.

Chen X. 2004. A microRNA as a translational repressor of *APETALA2* in *Arabidopsis* flower development. *Science* **303:** 2022.

Chen X., Liu J., Cheng Y., and Jia D. 2002. *HEN1* functions pleiotropically in *Arabidopsis* development and acts in C function in the flower. *Development* **129:** 1085.

Doench J.G., Petersen C.P., and Sharp P.A. 2003. siRNAs can function as miRNAs. *Genes Dev.* **17:** 438.

Dunoyer P., Lecellier C.H., Parizotto E.A., Himber C., and

Voinnet O. 2004. Probing the microRNA and small interfering RNA pathways with virus-encoded suppressors of RNA silencing. *Plant Cell* **16:** 1235.

Emery J.F., Floyd S.K., Alvarez J., Eshed Y., Hawker N.P., Izhaki A., Baum S.F., and Bowman J.L. 2003. Radial patterning of *Arabidopsis* shoots by class III *HD-ZIP* and *KANADI* Genes. *Curr. Biol.* **13:** 1768.

Eshed Y., Baum S.F., Perea J.V., and Bowman J.L. 2001. Establishment of polarity in lateral organs of plants. *Curr. Biol.* **11:** 1251.

Eshed Y., Izhaki A., Baum S.F., Floyd S.K., and Bowman J.L. 2004. Asymmetric leaf development and blade expansion in *Arabidopsis* are mediated by *KANADI* and *YABBY* activities. *Development* **131:** 2997.

Fagard M., Boutet S., Morel J.B., Bellini C., and Vaucheret H. 2000. AGO1, QDE-2, and RDE-1 are related proteins required for post-transcriptional gene silencing in plants, quelling in fungi, and RNA interference in animals. *Proc. Natl. Acad. Sci.* **97:** 11650.

Floyd S.K. and Bowman J.L. 2004. Gene regulation: Ancient microRNA target sequences in plants. *Nature* **428:** 485.

Foster T.M., Lough T.J., Emerson S.J., Lee R.H., Bowman J.L., Forster R.L., and Lucas W.J. 2002. A surveillance system regulates selective entry of RNA into the shoot apex. *Plant Cell* **14:** 1497.

Grishok A., Pasquinelli A.E., Conte D., Li N., Parrish S., Ha I., Baillie D.L., Fire A., Ruvkun G., and Mello C.C. 2001. Genes and mechanisms related to RNA interference regulate expression of the small temporal RNAs that control *C. elegans* developmental timing. *Cell* **106:** 23.

Hammond S.M., Boettcher S., Caudy A.A., Kobayashi R., and Hannon G.J. 2001. Argonaute2, a link between genetic and biochemical analyses of RNAi. *Science* **293:** 1146.

Han M.H., Goud S., Song L., and Fedoroff N. 2004. The *Arabidopsis* double-stranded RNA-binding protein HYL1 plays a role in microRNA-mediated gene regulation. *Proc. Natl. Acad. Sci.* **101:** 1093.

Hunter C., Sun H., and Poethig R.S. 2003. The *Arabidopsis* heterochronic gene *ZIPPY* is an *ARGONAUTE* family member. *Curr. Biol.* **13:** 1734.

Hutvagner G. and Zamore P.D. 2002. A microRNA in a multiple-turnover RNAi enzyme complex. *Science* **297:** 2056.

Hutvagner G., McLachlan J., Pasquinelli A.E., Balint E., Tuschl T., and Zamore P.D. 2001. A cellular function for the RNA-interference enzyme Dicer in the maturation of the *let-7* small temporal RNA. *Science* **293:** 834.

Jacobsen S.E., Running M.P., and Meyerowitz E.M. 1999. Disruption of an *RNA helicase/RNAse III* gene in *Arabidopsis* causes unregulated cell division in floral meristems. *Development* **126:** 5231.

Jones-Rhoades M.W. and Bartel D.P. 2004. Computational identification of plant microRNAs and their targets, including a stress-induced miRNA. *Mol. Cell* **14:** 787.

Juarez M.T., Twigg R.W., and Timmermans M.C.P. 2004a. Specification of adaxial cell fate during maize leaf development. *Development* **131:** 4533.

Juarez M.T., Kui J.S., Thomas J., Heller B.A., and Timmermans M.C.P. 2004b. microRNA-mediated repression of *rolled leaf1* specifies maize leaf polarity. *Nature* **428:** 84.

Kasschau K.D., Xie Z., Allen E., Llave C., Chapman E.J., Krizan K.A., and Carrington J.C. 2003. P1/HC-Pro, a viral suppressor of RNA silencing, interferes with *Arabidopsis* development and miRNA function. *Dev. Cell* **4:** 205.

Kerstetter R.A., Bollman K., Taylor R.A., Bomblies K., and Poethig R.S. 2001. *KANADI* regulates organ polarity in *Arabidopsis*. *Nature* **411:** 706.

Ketting R.F., Fischer S.E., Bernstein E., Sijen T., Hannon G.J., and Plasterk R.H. 2001. Dicer functions in RNA interference and in synthesis of small RNA involved in developmental timing in *C. elegans*. *Genes Dev.* **15:** 2654.

Khvorova A., Reynolds A., and Jayasena S.D. 2003. Functional siRNAs and miRNAs exhibit strand bias. *Cell* **115:** 209.

Kidner C.A. and Martienssen R.A. 2004. Spatially restricted microRNA directs leaf polarity through *ARGONAUTE1*. *Nature* **428:** 81.

Kumaran M.K., Bowman J.L., and Sundaresan V. 2002. *YABBY* polarity genes mediate the repression of *KNOX* homeobox genes in *Arabidopsis*. *Plant Cell* **14:** 2761.

Lagos-Quintana M., Rauhut R., Lendeckel W., and Tuschl T. 2001. Identification of novel genes coding for small expressed RNAs. *Science* **294:** 853.

Lau N.C., Lim L.P., Weinstein E.G., and Bartel D.P. 2001. An abundant class of tiny RNAs with probable regulatory roles in *Caenorhabditis elegans*. *Science* **294:** 858.

Laufs P., Peaucelle A., Morin H., and Traas J. 2004. MicroRNA regulation of the *CUC* genes is required for boundary size control in *Arabidopsis* meristems. *Development* **131:** 4311.

Lee R.C. and Ambros V. 2001. An extensive class of small RNAs in *Caenorhabditis elegans*. *Science* **294:** 862.

Lee R.C., Feinbaum R.L., and Ambros V. 1993. The *C. elegans* heterochronic gene *lin-4* encodes small RNAs with antisense complementarity to *lin-14*. *Cell* **75:** 843.

Lee Y., Ahn C., Han J., Choi H., Kim J., Yim J., Lee J., Provost P., Radmark O., Kim S., and Kim V.N. 2003. The nuclear RNase III Drosha initiates microRNA processing. *Nature* **425:** 415.

Liu Q., Rand T.A., Kalidas S., Du F., Kim H.E., Smith D.P., and Wang X. 2003. R2D2, a bridge between the initiation and effector steps of the *Drosophila* RNAi pathway. *Science* **301:** 1921.

Llave C., Kasschau K.D., Rector M.A., and Carrington J.C. 2002a. Endogenous and silencing-associated small RNAs in plants. *Plant Cell* **14:** 1605.

Llave C., Xie Z., Kasschau K.D., and Carrington J.C. 2002b. Cleavage of *Scarecrow-like* mRNA targets directed by a class of *Arabidopsis* miRNA. *Science* **297:** 2053.

Lynn K., Fernandez A., Aida M., Sedbrook J., Tasaka M., Masson P., and Barton M.K. 1999. The *PINHEAD/ZWILLE* gene acts pleiotropically in *Arabidopsis* development and has overlapping functions with the *ARGONAUTE1* gene. *Development* **126:** 469.

Mallory A.C., Dugas D.V., Bartel D.P., and Bartel B. 2004. MicroRNA regulation of NAC-domain targets is required for proper formation and separation of adjacent embryonic, vegetative, and floral organs. *Curr. Biol.* **14:** 1035.

McConnell J.R., Emery J., Eshed Y., Bao N., Bowman J., and Barton M.K. 2001. Role of *PHABULOSA* and *PHAVOLUTA* in determining radial patterning in shoots. *Nature* **411:** 709.

McHale N.A. and Koning R.E. 2004. MicroRNA-directed cleavage of *Nicotiana sylvestris PHAVOLUTA* mRNA regulates the vascular cambium and structure of the apical meristem. *Plant Cell* **16:** 1730.

Mourelatos Z., Dostie J., Paushkin S., Sharma A., Charroux B., Abel L., Rappsilber J., Mann M., and Dreyfuss G. 2002. miRNPs: A novel class of ribonucleoproteins containing numerous microRNAs. *Genes Dev.* **16:** 720.

Moussian B., Schoof H., Haecker A., Jurgens G., and Laux T. 1998. Role of the *ZWILLE* gene in the regulation of central shoot meristem cell fate during *Arabidopsis* embryogenesis. *EMBO J.* **17:** 1799.

Nath U., Crawford B.C., Carpenter R., and Coen E. 2003. Genetic control of surface curvature. *Science* **299:** 1404.

Nelson J.M., Lane B., and Freeling M. 2002. Expression of a mutant maize gene in the ventral leaf epidermis is sufficient to signal a switch of the leaf's dorsoventral axis. *Development* **129:** 4581.

Otsuga D., DeGuzman B., Prigge M.J., Drews G.N., and Clark S.E. 2001. *REVOLUTA* regulates meristem initiation at lateral positions. *Plant J.* **25:** 223.

Palatnik J.F., Allen E., Wu X., Schommer C., Schwab R., Carrington J.C., and Weigel D. 2003. Control of leaf morphogenesis by microRNAs. *Nature* **425:** 257.

Papp I., Mette M.F., Aufsatz W., Daxinger L., Schauer S.E., Ray A., van der Winden J., Matzke M., and Matzke A.J. 2003. Evidence for nuclear processing of plant micro RNA and short interfering RNA precursors. *Plant Physiol.* **132:** 1382.

Park W., Li J., Song R., Messing J., and Chen X. 2002. CARPEL FACTORY, a Dicer homolog, and HEN1, a novel protein, act in microRNA metabolism in *Arabidopsis thaliana*. *Curr. Biol.* **12**: 1484.

Pasquinelli A.E., Reinhart B.J., Slack F., Martindale M.Q., Kuroda M.I., Maller B., Hayward D.C., Ball E.E., Degnan B., Muller P., Spring J., Srinivasan A., Fishman M., Finnerty J., Corbo J., Levine M., Leahy P., Davidson E., and Ruvkun G. 2000. Conservation of the sequence and temporal expression of *let-7* heterochronic regulatory RNA. *Nature* **408**: 86.

Ratcliffe O.J., Riechmann J.L., and Zhang J.Z. 2000. *INTERFASCICULAR FIBERLESS1* is the same gene as *REVOLUTA*. *Plant Cell* **12**: 315.

Reinhart B.J., Weinstein E.G., Rhoades M.W., Bartel B., and Bartel D.P. 2002. MicroRNAs in plants. *Genes Dev.* **16**: 1616.

Reinhart B.J., Slack F.J., Basson M., Pasquinelli A.E., Bettinger J.C., Rougvie A.E., Horvitz H.R., and Ruvkun G. 2000. The 21-nucleotide *let-7* RNA regulates developmental timing in *Caenorhabditis elegans*. *Nature* **403**: 901.

Rhoades M.W., Reinhart B.J., Lim L.P., Burge C.B., Bartel B., and Bartel D.P. 2002. Prediction of plant microRNA targets. *Cell* **110**: 513.

Sawa S., Watanabe K., Goto K., Liu Y.G., Shibata D., Kanaya E., Morita E.H., and Okada K. 1999. *FILAMENTOUS FLOWER*, a meristem and organ identity gene of *Arabidopsis*, encodes a protein with a zinc finger and HMG-related domains. *Genes Dev.* **13**: 1079.

Schwarz D.S., Hutvagner G., Du T., Xu Z., Aronin N., and Zamore P.D. 2003. Asymmetry in the assembly of the RNAi enzyme complex. *Cell* **115**: 199.

Siegfried K.R., Eshed Y., Baum S.F., Otsuga D., Drews G.N., and Bowman J.L. 1999. Members of the *YABBY* gene family specify abaxial cell fate in *Arabidopsis*. *Development* **126**: 4117.

Sussex I.M. 1951. Experiments on the cause of dorsiventrality in leaves. *Nature* **167**: 651.

———. 1955. Morphogenesis in *Solanum tuberosum* L.: Experimental investigation of leaf dorsiventrality and orientation in the juvenile shoot. *Phytomorphology* **5**: 286.

Tabara H., Yigit E., Siomi H., and Mello C.C. 2002. The dsRNA binding protein RDE-4 interacts with RDE-1, DCR-1, and a DExH-box helicase to direct RNAi in *C. elegans*. *Cell* **109**: 861.

Tang G., Reinhart B.J., Bartel D.P., and Zamore P.D. 2003. A biochemical framework for RNA silencing in plants. *Genes Dev.* **17**: 49.

Timmermans M.C.P., Schultes N.P., Jankovsky J.P., and Nelson T. 1998. *Leafbladeless1* is required for dorsoventrality of lateral organs in maize. *Development* **125**: 2813.

Vaucheret H., Vazquez F., Crete P., and Bartel D.P. 2004. The action of *ARGONAUTE1* in the miRNA pathway and its regulation by the miRNA pathway are crucial for plant development. *Genes Dev.* **18**: 1187.

Vazquez F., Gasciolli V., Crete P., and Vaucheret H. 2004. The nuclear dsRNA binding protein HYL1 is required for microRNA accumulation and plant development, but not posttranscriptional transgene silencing. *Curr. Biol.* **14**: 346.

Waites R. and Hudson A. 1995. *phantastica*: A gene required for dorsoventrality of leaves in *Antirrhinum majus*. *Development* **121**: 2143.

Wang J.F., Zhou H., Chen Y.Q., Luo Q.J., and Qu L.H. 2004. Identification of 20 microRNAs from *Oryza sativa*. *Nucleic Acids Res.* **32**: 1688.

Wardlaw C.W. 1949. Experiments on organogenesis in ferns. *Growth* (suppl.) **13**: 93.

Xie Z., Kasschau K.D., and Carrington J.C. 2003. Negative feedback regulation of *DICER-LIKE1* in *Arabidopsis* by microRNA-guided mRNA degradation. *Curr. Biol.* **13**: 784.

Zeng Y., Yi R., and Cullen B.R. 2003. MicroRNAs and small interfering RNAs can inhibit mRNA expression by similar mechanisms. *Proc. Natl. Acad. Sci.* **100**: 9779.

Zhong R. and Ye Z.H. 1999. *IFL1*, a gene regulating interfascicular fiber differentiation in *Arabidopsis*, encodes a homeodomain-leucine zipper protein. *Plant Cell* **11**: 2139.

———. 2004. *Amphivasal vascular bundle1*, a gain-of-function mutation of the *IFL1/REV* gene, is associated with alterations in the polarity of leaves, stems and carpels. *Plant Cell Physiol.* **45**: 369.

Zilberman D., Cao X., and Jacobsen S.E. 2003. ARGONAUTE4 control of locus-specific siRNA accumulation and DNA and histone methylation. *Science* **299**: 716.

RNA Interference and Epigenetic Control of Heterochromatin Assembly in Fission Yeast

H. Cam and S.I.S. Grewal

Laboratory of Molecular Cell Biology, National Cancer Institute, National Institutes of Health, Bethesda, Maryland 20892

Large tracts of eukaryotic chromosomes contain no or relatively few genes and are often replete with repetitive elements. Many of these regions are heterochromatic in nature, being highly condensed and inaccessible to a variety of DNA-modifying factors, and therefore are generally thought to be synonymous with gene silencing. Importantly, a phenomenon of position effect variegation (PEV) first observed in flies, in which a euchromatic gene when juxtaposed to heterochromatin becomes repressed in a stochastic fashion but once silenced can be clonally inherited, suggests that cellular factors intimately associated with heterochromatin can stably maintain epigenetic gene silencing through multiple cell divisions without altering the underlying genetic sequence (Grewal and Elgin 2002).

PEV has also been observed in the fission yeast *Schizosaccharomyces pombe* whose genome contains large stretches of heterochromatin with properties similar to heterochromatin present in higher eukaryotes (Grewal 2000; Hall and Grewal 2003). Because of its rather compact genome and tractable genetic system, heterochromatin-mediated gene silencing is arguably best understood in this organism. In the last few years, great strides have been made in uncovering the molecular constituents of heterochromatin and how they effect epigenetic gene silencing. In particular, recent works from our lab and others have elucidated a number of molecules involved in heterochromatin formation, with the surprising discovery of the critical role of the RNA interference (RNAi) pathway in the targeting and assembly of heterochromatin at specific loci of the *S. pombe* genome. A basic picture of how heterochromatin assembles, propagates, and is stably maintained through mitosis and meiosis has begun to emerge.

HETEROCHROMATIC REGIONS IN *S. POMBE*

The three chromosomes of fission yeast contain several silenced domains that are constitutively associated with heterochromatin, namely the telomeres, centromeres, the mating-type (*mat*) region, and the ribosomal DNA (rDNA) tandem loci (Grewal 2000; Thon and Verhein-Hansen 2000). Unlike the small centromeres of the budding yeast *Saccharomyces cerevisiae*, which can comprise only a few hundreds base pairs (bp) of DNA, centromeres in *S. pombe* are relatively large complex structures (35–110 kilobases [kb]) that resemble higher eukaryotic centromeres in structure and organization, having partitions of large inverted-repeat elements flanking the central core (*cnt*), a site of kinetochore assembly and microtubule attachment (Fig. 1A). Clusters of tRNA genes pepper within the inner (*imr*) domains, while tandem but variable copies of the *dg* and *dh* repeats, whose sequences are remarkably conserved among themselves, orient in an inverted symmetry within the centromeric outer (*otr*) domains surrounding both sides of the central core of each chromosome (Takahashi et al. 1992). Reporter genes ectopically placed at the *otr*, *imr*, and *cnt* domains are transcriptionally silenced (Allshire et al. 1994).

Although less is known about the nature of telomeres and rDNA loci, evidence of PEV, suppression of DNA recombination, and common heterochromatin-associated factors localized to these regions as to the centromeres indicate similar mechanisms of heterochromatin formation operating at these loci as well. In contrast, the mating-type region of fission yeast located on the right side of the centromere of chromosome II is a well-studied locus containing several *cis*-acting elements that are known to contribute to transcriptional silencing at the *mat* locus (Fig. 1B). Heterochromatin spanning across the *K* region, a 20-kb interval between *mat2* and *mat3* loci of the mating-type region, has been shown to be important for maintaining gene silencing and suppression of interchromosomal recombination. Interestingly, heterochromatin at the mating-type region has been implicated in the control of directionality of mating-type switching, in which mating-type information present at the upstream *mat1* locus is unidirectionally replaced with the opposite mating-type allele copied from donor cassettes present downstream at *mat2* and *mat3* loci. In fact, genetic screens looking for mating-type switching mutants identified factors that also participate in heterochromatin assembly (Egel et al. 1984; Grewal and Klar 1997).

HETEROCHROMATIN ASSEMBLY FACTORS IN FISSION YEAST

Several *trans*-acting factors involved in effecting gene silencing at heterochromatic loci within the *S. pombe* genome have been identified (Table 1). These factors directly or indirectly affect heterochromatin assembly at these loci. Importantly, many of these factors have been shown to have homologs in higher eukaryotes that par-

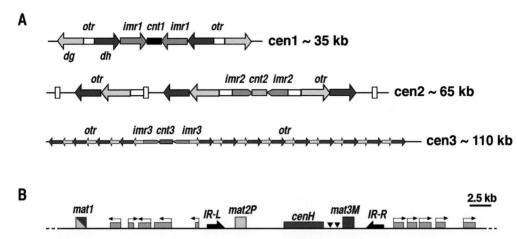

Figure 1. Diagrammatic representation of the *S. pombe* centromeres and the mating-type region. (*A*) The central domain (*cnt*) at each centromere is flanked by a series of inner (*imr*) and outer (*otr*) repeat elements. The *otr* regions contain *dg* and *dh* repeats that are highly conserved among themselves and double-stranded RNA (dsRNA) transcripts have been detected from these regions. (*B*) The mating-type region contains three loci involved in mating-type switching—*mat1*, *mat2*, and *mat3*—and a number of silent *cis*-acting elements, among the most dominant is *cenH*, which lies within the 20-kb *mat2/3* interval, a heterochromatic region exhibiting both suppression of transcription and recombination. *cenH* possesses *dg* and *dh* repeats that are highly homologous to centromeric *dg* and *dh* repeats. dsRNA transcripts whose regulation is RNAi dependent have also been detected at *cenH*. The *double black triangles* between *cenH* and *mat3* indicate Atf1/Pcr1-binding sites.

take in heterochromatin formation as well. spSir2, Clr3, and Clr6 are histone deacetylases (HDACs), enzymes that remove acetyl group from histones' tails (Grewal et al. 1998; Bjerling et al. 2002; Nakayama et al. 2003; Shankaranarayana et al. 2003). spSir2 belongs to a special class of HDACs, present also in budding yeast and higher eukaryotes, that requires the cofactor nicotinamide adenine (NAD) for its catalytic activity (Shankaranarayana et al. 2003). Clr3 and Clr6 are homologous to mammalian HDACs class 1 (HDAC1 and HDAC2) and class 2 (HDAC2 and HDAC5), respectively. HDAC activity has been shown in many systems, including *S. cerevisiae*, to be a prerequisite for establishing a repressive chromatin state, and genome-wide analyses have indicated that heterochromatic regions tend to exhibit overall hypoacetylation relative to euchromatic regions (Kurdistani and Grunstein 2003; Schubeler et al. 2004).

Clr4 is homologous to Su(var)3-9 in *Drosophila* and SUV39H1 in mammals. Both loss of Clr4 and Su(var)3-9 suppress PEV in *S. pombe* and *Drosophila*, respectively (Grewal and Elgin 2002). These proteins possess a conserved amino-terminal chromodomain and a carboxy-terminal SET domain. Clr4 has been shown to preferentially methylate lysine 9 residue on the amino tail of histone H3 (H3-K9) (Rea et al. 2000; Nakayama et al. 2001b). While the SET domain confers histone methyltransferase (HMT) activity on Clr4, the chromodomain is thought to be important for the localization of Clr4 to silenced loci (Nakayama et al. 2001b). Swi6 functions as a chromatin modifier and is structurally similar to heterochromatin protein 1 (HP1) of *Drosophila* and mammals (Lorentz et al. 1994). Swi6/HP1 contains a chromodomain at its amino terminus that has been shown to bind to H3-K9-methyl residues created by Clr4/Suv39H1, a variable hinge region, and the chromoshadow domain at its carboxyl terminus. The chromoshadow domain of Swi6/HP1 promotes Swi6/HP1 oligomerization, which is thought to be important for chromatin compaction and heterochromatin spreading (Brasher et al. 2000; Cowieson et al. 2000; Wang et al. 2000). Chp1 and Chp2

Table 1. Factors Involved in Heterochromatin-mediated Silencing

Locus	Function	Motifs/similarity
spSir2	NAD-dependent histone deacetylase	*S. cerevisiae*, *Drosophila*, human Sir2
clr1	Putative DNA-binding protein	Three zinc fingers
clr2	Unknown	No similarities in database
clr3	Histone deacetylase	Human HDAC4 and HDAC5, *S. cerevisiae* Hda1
clr4	Histone methyltransferase	SET domains and chromodomains/*Drosophila* Su(var)3-9 and Polycomb, human SUV39H1
clr6	Histone deacetylase	Human HDAC1 and HDAC2, *S. cerevisiae* Rpd3
swi6	Chromatin modifier	Chromodomains and shadow domains/*Drosophila* and human HP-1 and Polycomb
chp1	Chromatin modifier	Chromodomain, homologous to Swi6
chp2	Chromatin modifier	Chromodomains and shadow domains, homologous to Swi6
rik1	Putative RNA/DNA-binding protein	11 WD40-like repeats

are also chromodomain proteins. Chp2 is similar to Swi6 in structure possessing a chromodomain as well as a chromoshadow domain. Loss of Chp1, Chp2, or Swi6 results in derepression at a number of heterochromatic regions including at the centromeres and the rDNA tandem repeat loci (Thon and Verhein-Hansen 2000). Rik1, which shares homology to Ddb1, a group of evolutionarily conserved proteins involved in cullin-mediated ubiquitination and nucleotide excision repair (Tuzon et al. 2004), has been shown to be essential for Clr4 methylation of H3-K9 at heterochromatic loci (Nakayama et al. 2001b).

HISTONE CODE AND HETEROCHROMATIN ASSEMBLY

The basic unit of eukaryotic chromatin is nucleosome, an octamer of histone proteins made up of two molecules each of H2A, H2B, H3, and H4. A stretch of 146 bp of genomic DNA coils tightly around the carboxy-terminal two-third of the core histones, leaving the amino-terminal tails of histones to freely interact with other chromosomal factors. Certain key amino acid residues on the histone tails are targeted by a number of enzymes to a variety of posttranslational modifications such as acetylation, methylation, phosphorylation, ADP-ribosylation, and ubiquitylation. These modified residues have been shown to serve as markers that can recruit chromatin modifier proteins. Accumulating evidence suggests these covalent modifications on histone tails together may constitute a "histone code" that is read by nonhistone chromosomal proteins to effect changes ranging from a few nucleosomes at a locus to large-scale global organization of an entire chromosome (Strahl and Allis 2000; Jenuwein and Allis 2001). Recent findings in fission yeast and higher eukaryotes have led to a model of heterochromatin assembly based on a sequential series of histone modifications involving concerted hypoacetylation by HDACs and methylation by Clr4/Suv39H1 and other HMT proteins.

An important advance in which a molecular marker could be identified with heterochromatin occurred when work performed in our lab by Jun-ichi Nakayama, in collaboration with David Allis's group, provided the first in vivo evidence of H3-K9 methylation associated with heterochromatin (Nakayama et al. 2001b). Nakayama et al. detected H3-K9 methylation at the fission yeast's centromeres and the mating-type region, and this methylation pattern depends on the presence of Clr4: Loss of Clr4 completely abolishes H3-K9 methylation at these loci (Nakayama et al. 2001b). Moreover, point mutations in the SET domain as well as the carboxy-terminal post-SET motif severely hinder the ability of Clr4 to methylate H3-K9 (Nakayama et al. 2001b). The discovery of stable methylation marks on heterochromatin-associated histone H3-K9 reveals that this may be part of the histone code, signaling to recruit heterochromatin assembly factors. Indeed, one protein that is capable of binding to methyl H3-K9 marks is Swi6/HP1. Swi6/HP1 can bind to methylated H3-K9 in vitro, but with preference for dimethyl and trimethyl H3-K9, which is thought to correspond to the degree of gene silencing (Bannister et al.

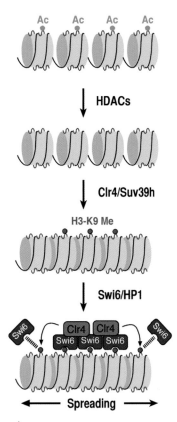

Figure 2. Stepwise model of heterochromatin assembly. Heterochromatin assembly is thought to require concerted deacetylation and methylation on the histones' tails. Deacetylation by the HDACs' enzymes creates a repressive chromatin state and allows histone methyltransferases such as Clr4/Suv39h1 to methylate lysine 9 on histone H3 (H3-K9 Me). Methylation of H3-K9 provides the binding sites for Swi6/HP1 proteins, which, once bound to chromatin, can facilitate heterochromatin spreading to adjacent sequences.

2001; Lachner et al. 2001; Jacobs and Khorasanizadeh 2002; Nielsen et al. 2002). In fission yeast, localization of Swi6 to centromeric and the mating-type loci is entirely dependent on the presence of H3-K9 methylation (Nakayama et al. 2001b). Deletion of Clr4 or loss of Clr4 HMT activity delocalizes Swi6 from heterchromatic loci suggesting that the binding of Swi6 to chromatin is primarily dependent on Clr4 to methylate H3-K9 (Nakayama et al. 2001b). Interestingly, loss of Swi6 prevents the spreading of H3-K9 methylation throughout the *mat* locus while mutations in Clr4 attenuate Swi6 association with heterochromatin (Nakayama et al. 2001b; Hall et al. 2002). Together, these data support a model of heterochromatin assembly in which Swi6 initially localizes to chromatin via its binding to methylated H3-K9 established by Clr4 (Fig. 2). Once bound to methylated H3-K9, Swi6 can stimulate self-oligomerization through its chromoshadow domain and the recruitment of additional Clr4 proteins to methylate H3-K9 residues on adjacent nucleosomes (G. Xiao and S.I.S. Grewal, unpubl.), thereby creating more binding sites for Swi6. The result is a feedback loop that allows heterochromatin to spread to neighboring sequences (Fig. 2).

Although methylation of H3-K9 by Clr4 seems to be the critical step in the assembly of heterochromatin, it is likely to be preceded by a series of deacetylation steps, by Clr3, spSir2, and Clr6 HDACs, because loss of any of these proteins either reduces or abolishes H3-K9 methylation altogether (Nakayama et al. 2001b, 2003; Shankaranarayana et al. 2003).

RNAi-MEDIATED HETEROCHROMATIN FORMATION

Repetitive DNA elements have long been known to play an important role in inducing heterochromatin (Selker 1999; Hsieh and Fire 2000; Hall and Grewal 2003). In fission yeast, a *cenH* element homologous to centromeric *dg* and *dh* repeats present within the *mat2/3* interval contributes to epigenetic silencing of the mating-type region (Grewal and Klar 1996). Mutant strain carrying a deletion of *cenH* fails to efficiently establish heterochromatin at the *mat* locus (Grewal and Klar 1996, 1997). Insertion of *cenH* confers repression on a reporter gene and induces heterochromatin at an ectopic euchromatic site (Ayoub et al. 2000). Recent findings suggest *cenH* and similar repetitive DNA elements collaborate with the RNAi machinery to induce heterochromatin formation at loci containing these repetitive elements (Hall et al. 2002; Hall and Grewal 2003).

The fission yeast contains three evolutionarily conserved components of the RNAi machinery—namely, Ago1 (a PAZ/Piwi family Argonaute), Dcr1 (a RNaseIII-like enzyme Dicer), and Rdp1 (an RNA-dependent RNA polymerase)—and all have been found to be necessary for heterochromatin formation in *S. pombe* (Hall et al. 2002; Volpe et al. 2002). In RNAi mutants lacking either Ago1, Dcr1, or Rdp1, there is a complete loss of H3-K9 methylation and Swi6 localization at centromeric repeats and at *cenH* ectopic site (Hall et al. 2002; Volpe et al. 2002). Interestingly, H3-K9 methylation and Swi6 localization are apparently normal at the mating-type region in these RNAi mutants (Hall et al. 2002). However, in contrast to wild-type strains, RNAi mutants grown in the presence of a deacetylase inhibitor trichostatin A (TSA), which has been shown to remove epigenetic imprint of transcriptional silencing at the *mat* locus, fail to efficiently reestablish heterochromatin upon the removal of TSA (Hall et al. 2002). These results distinguish heterochromatin formation into two distinct phases: the initial targeting and nucleation of heterochromatin, and the spreading and maintenance of heterochromatin. Importantly, the results support the critical role for RNAi in the initial phase of heterochromatin formation, while at the mating-type region RNAi becomes dispensable at the subsequent phase of heterochromatin maintenance.

A multiprotein RNAi effector complex called RISC (RNA-induced silencing complex) has been biochemically characterized in *Drosophila* (Hammond et al. 2001; Murchison and Hannon 2004). RISC contains Argonaute and small interfering RNA (siRNA) molecules of 21–24 nucleotides (nt) cleaved by the enzyme Dicer from longer double-stranded RNA (dsRNA) transcripts (Bernstein et al. 2001). These siRNAs are thought to provide the specificity for RISC-targeted degradation or translational inhibition of mRNA and transcriptional repression. In fission yeast, siRNAs corresponding to centromeric *dg* and *dh* repeats have also been detected in wild-type cells (Reinhart and Bartel 2002). Interestingly, dsRNA transcripts derived from centromeric *otr* repeats and from *cenH* of the mating-type region, while not present in wild-type strains, could be detected in RNAi mutants, suggesting that in wild-type cells, dsRNA transcripts are actively being processed into siRNAs (Volpe et al. 2002; K. Noma and S. Grewal, unpubl.). However, the connection between siRNAs and heterochromatin formation was not readily understood until the recent purification of RITS (RNA-induced initiation of transcriptional gene silencing) in fission yeast (Verdel et al. 2004).

RITS is an RNAi effector complex containing Chp1, a heterochromatin-associated chromodomain protein that is required for H3-K9 methylation and Swi6 localization at the centromeres (Partridge et al. 2002), a novel protein Tas3, and importantly Ago1 and siRNAs (Verdel et al. 2004). RITS can localize to centromeric *otr* repeats in an RNAi-dependent manner. In the absence of Dcr1, RITS can still assemble but without siRNAs and can no longer localize to the centromeres, suggesting RITS-associated siRNAs might provide the locus-specific sequence that can guide RITS to designated genomic site. Loss of any of the RITS components abolishes H3-K9 methylation and Swi6 localization at the centromeres. These results

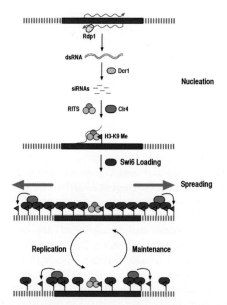

Figure 3. Model of RNAi-mediated heterochromatin formation in *S. pombe*. Double-stranded RNAs (dsRNAs) produced (possibly aided by Rdp1) from regions containing repetitive elements are recognized and converted into siRNAs by Dcr1. siRNAs contain the locus-specific sequence and, in complex with RITS or yet unknown protein(s), may facilitate the recruitment of heterochromatin assembly factors such as Clr4 to site with homologous sequence. Once H3-K9 has been methylated by Clr4, Swi6 can localize to the target site and stimulate the spreading and establishment of heterochromatin.

implicate RITS as the complex that links the RNAi pathway to heterochromatin assembly, providing a cellular mechanism for how RNAi can induce heterochromatin formation at specific genomic loci. A basic picture of how the RNAi machinery can cooperate with chromatin modifier factors to assemble heterochromatin at regions producing dsRNAs can now be delineated as follows (Fig. 3).

The first step in RNAi-mediated heterochromatin assembly involves the synthesis of double-stranded transcripts from genomic regions containing highly repetitive elements. These transcripts are recognized and converted by Dcr1 into siRNAs. siRNAs in association with RITS facilitate in guiding RITS to genomic loci with homologous sequences. Once bound to chromatin, RITS can mediate the recruitment of histone modifier proteins including Clr4. Alternatively, RITS presumably recruits the RNAi machinery to the target site to promote the generation of siRNAs that in turn stimulates the recruitment of heterochromatin assembly factors. However, once Clr4 is recruited and the initial methylation of H3-K9 is established, Swi6 can abet in the spreading of heterochromatin in *cis* into neighboring sequences. The persistent presence of RITS on chromatin once heterochromatin has already been assembled suggests a posttranscriptional role for RITS (K. Noma and S. Grewal, unpubl.). Perhaps RITS participates in posttranscriptional repression and, conceivably in collaboration with Rdp1 and Dcr1, processes dsRNA into siRNAs, similar to the function of RISC seen in other systems (Murchison and Hannon 2004).

RNAi-INDEPENDENT HETEROCHROMATIN FORMATION AT THE MATING-TYPE REGION

While RNAi mutants display a complete loss of heterochromatin at the centromeres, heterochromatin and gene silencing are not affected at the mating-type region (Hall et al. 2002). This rather paradoxical result suggests that other *cis*-acting silent elements besides *cenH* could compensate for the loss of RNAi to maintain heterochromatin at the *mat* locus. Indeed, since RNAi and *cenH* act in the same pathway and *cenH*-deleted mutant strains containing a reporter gene inserted at the *mat2/3* interval could still acquire an epigenetic *off* state, these observations suggest the existence of RNAi-independent factors that can initiate heterochromatin formation at the mating-type region in the absence of RNAi (Grewal and Klar 1997; Hall et al. 2002). Songtao Jia from our lab analyzed the *mat2/3* interval and identified two heptamer binding sites for the transcription factor Atf1/Pcr1 located near the *mat3* locus (Jia et al. 2004). Whereas RNAi mutants retain normal epigenetic silencing at the mating-type region, RNAi mutants lacking either Atf1 or Pcr1 exhibit a dramatic reduction in silencing, accompanied by significant decreases in H3-K9 methylation and Swi6 level at the *mat* locus (Jia et al. 2004). Chromatin immunoprecipitation (ChIP) assay confirms Atf1 and Pcr1 localization to *mat3* region containing the Atf1/Pcr1-binding sites. Deletion of those sites negates the binding of Atf1 and Pcr1 to *mat3* locus. Interestingly, Atf1/Pcr1 localization to the mating-type region is not affected in the absence of Swi6. Taken together, these results suggest that Atf1/Pcr1 can bind directly to the *mat3* region, and it acts probably upstream of Swi6 and is capable of recruiting heterochromatin assembly factors such as HDACs and Clr4 to nucleate heterochromatin formation independent of the RNAi pathway (Jia et al. 2004).

PROPAGATION AND INHERITANCE OF HETEROCHROMATIC STRUCTURES

Results from our studies of the fission yeast have revealed multiple parallel pathways that could lead to heterochromatin formation. Once heterochromatin is established, its propagation and stability need to be maintained with fidelity through multiple cell divisions without being "diluted" by macrocellular processes such as DNA replication and chromosome condensation and segregation. For example, during DNA replication, epigenetic imprint from one DNA molecule has to be transferred to the newly synthesized molecule. Since heterochromatin propagates in *cis* (Nakayama et al. 2000), this probably requires heterochromatin assembly factors to collaborate with the DNA replication machinery to reestablish epigenetic marks (imprinting) on the new DNA molecule as it is being synthesized. Alternatively, components of the DNA replication machinery such as proliferating cell nuclear antigen (PCNA) and chromatin assembly factor-1 (CAF-1) can recognize epigenetic imprints at heterochromatic loci and direct heterochromatin assembly factors to newly duplicated regions marked for heterochromatin reassembly (Shibahara and Stillman 1999; Zhang et al. 2000).

We have proposed a chromatin-replication model in which modified histones and histone-associated proteins segregate semiconservatively to each DNA strand during DNA replication that could promote self-reassembly of heterochromatin on both duplicated chromosomes (Grewal and Klar 1996; Hall and Grewal 2003). Swi6 is stably associated with heterochromatin throughout the mitotic cell cycle and the entire period of meiosis (Nakayama et al. 2000). Also, Swi6 can associate with DNA polymerase α, a component of the DNA replication machinery (Ahmed et al. 2001; Nakayama et al. 2001a), and mouse HP1 has been shown to interact with CAF-1 (Murzina et al. 1999). These findings suggest that components of heterochromatin such as Swi6/HP1 can collaborate with the DNA replication apparatus to mediate heterochromatin reassembly during DNA replication. The simplest scenario of the chromatin-assembly model entails methylated histone H3-K9 being stochastically redistributed to each of the two DNA strands as the replication fork passes through a heterochromatic site. Next, Swi6 can bind to the "diluted" methylated histone H3-K9 on the newly duplicated DNA molecules and helps to recruit HDACs and HMT proteins such as Clr4 to fully restore the epigenetic marks and reestablish heterochromatin to the original locus on both duplicated chromosomes. Alternatively, Swi6 and methylated histone H3-K9 are randomly transferred to the newly duplicated DNA molecules together as a unit. Interestingly, a number of studies have used fluorescence recovery after photobleaching (FRAP) analysis to closely monitor the kinetics of Swi6 and HP1 association with heterochromatin in real time, and all have found that the turnover rates for Swi6 and HP1 proteins are quite rapid, suggesting that the binding of Swi6/HP1 proteins to methylated H3-K9 is a dynamic event and that under conditions that could temporarily disrupt heterochromatin Swi6/HP1 can rapidly reassociate with H3-K9 methyl marks to reestablish heterochromatin (Cheutin et al. 2003, 2004; Festenstein et al. 2003).

In fission yeast, RNAi mutants are unable to retain heterochromatin at the centromeres, which implies the requirement for RNAi in stably maintaining heterochromatin. Indeed, our recent data suggest that stable association of RITS with chromatin is critical for the generation of RITS-associated siRNAs and the maintenance of heterochromatin at centromeres (K. Noma and S. Grewal, unpubl.). Therefore, even though heterochromatin once established is relatively stable, it tends to attenuate over time because of macrocellular processes such as DNA replication and chromosome condensation that could disrupt long-lived chromosomal structures such as heterochromatin. Consequently, there would be a continual need for heterochromatin nucleation factors, such as the RNAi machinery and/or Atf1/Pcr1, to recruit heterochromatin assembly factors, such as Clr4 and Swi6, to maintain heterochromatin so as to faithfully preserve epigenetic imprints across multiple cell divisions.

BOUNDARY DETERMINANTS OF HETEROCHROMATIN

The existence of constitutive heterochromatin domains in fission yeast and higher eukaryotes suggests the need for robust cellular mechanisms that can demarcate clear boundaries partitioning heterochromatic regions from euchromatic regions. In addition, chromatin domains have to be organized in a way that gives deference to a chromosome's structural constraints imposed by diverse cellular processes impinging upon the entire chromosome. For instance, chromatin surrounding the central core (*cnt*) and inner repeats (*imr*) of *S. pombe* centromeres differ in structure and organization from the *outer* (*otr*) repeats regions (Partridge et al. 2000). These two chromatin domains are epigenetically silenced and are enriched with Mis6 and CENP-A centromeric proteins, which are essential for proper biorientation of sister centromeres during mitotic metaphase (Saitoh et al. 1997; Partridge et al. 2000; Takahashi et al. 2000). Surprisingly, Swi6 is virtually excluded from the *cnt* and most of the *imr* regions and instead is confined to the outer repeats (*otr*) regions (Partridge et al. 2000). Evidently, a cluster of tRNA genes situated at the exterior region of *imr* domains act as barriers separating one silenced domain from another silenced domain with disparate functions (Partridge et al. 2000). Interestingly, tRNA genes that act as boundary elements as well have also been demonstrated in *S. cerevisiae* (Donze et al. 1999).

The embedded heterochromatic *K*-region of the *mat* locus within a larger euchromatic domain also implies the existence of boundary elements. Whereas marker genes placed within the *mat2/3* interval or *K*-region are subject to epigenetic silencing, markers genes inserted outside the *K*-region are not repressed, suggesting elements present on both sides of the *K*-region could act to insulate the heterochromatic *mat2/3* interval from the surrounding euchromatin. In fact, two 2-kb identical inverted repeats termed IR-L and IR-R flank the left and right sides of the *K*-region, respectively. While H3-K9 methylation is associated with transcriptional silenced domains, H3-K4 methylation correlates with active transcribed regions (Strahl et al. 1999; Noma et al. 2001). Ken-ichi Noma in our lab performed a high-resolution mapping of H3-K9

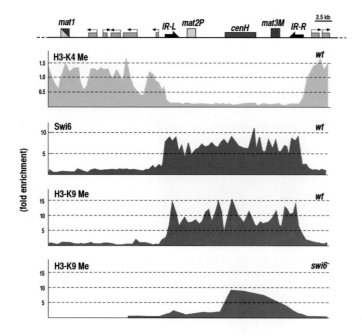

Figure 4. Distinct and mutually exclusive histone H3-K4 and H3-K9 methylation patterns respectively delimit euchromatic and heterochromatic domains at the *mat* locus. Distribution of H3-K4 and H3-K4 methylation patterns and Swi6 levels at the mating-type region shown in alignment with a physical map of the *mat* locus (*top*). IR-L and IR-R flanking the left and right side of the *mat2/3* interval, respectively, act as boundary elements partitioning heterochromatin spanning the *mat2/3* interval from surrounding euchromatin regions. Loss of Swi6 prevents the spreading of H3-K9 methylation beyond the left side of *cenH* (*bottom*).

and H3-K4 methylation across the entire 47 kb *mat* locus (Noma et al. 2001). Consistent with evidence of the *K*-region being silenced, there was considerable enrichment of H3-K9 methylation throughout the *mat2/3* interval while H3-K4 methylation was virtually absent within this region (Fig. 4). Noma et al. also mapped the level of Swi6 proteins across the *mat* locus and found its localization pattern identically paralleled H3-K9 methylation pattern throughout the mating-type region, confirming the idea that H3-K9-methyl marks act as binding sites for Swi6 proteins (Noma et al. 2001). Unexpectedly, there was a sudden sharp concomitant decline of H3-K9 methylation and Swi6 levels on both sides of the *K*-region where the two IR elements reside (Noma et al. 2001). Outside of the *K*-region, there was a gradual increase in H3-K4 methylation on both sides where a number of transcriptionally active loci lie. These results indicate that methylation of H3-K4 or H3-K9 is associated in a mutually exclusive manner with euchromatin and heterochromatin, respectively (Noma et al. 2001). Intriguingly, ChIP mapping across the 53-kb chicken β-globin locus also found similar patterns of H3-K4 methylation and H3 acetylation that are negatively correlated with the pattern of H3-K9 methylation (Litt et al. 2001).

The high level of methylated H3-K9 seen at the two IR elements on the side of the *K*-region while its level precipitously dropped at the immediate outside region of the *mat2/3* interval suggest IR-L and IR-R may act as boundaries for the heterochromatic *mat2/3* interval. Indeed, removal of either IR element allows heterochromatin to spread outside of the *mat2/3* interval and weakens heterochromatin at the *K*-region (Noma et al. 2001; Thon et al. 2002).

CONCLUDING REMARKS

While it has become increasingly clear that cells utilize heterochromatin to impose transcriptional silencing constitutively to a particular genomic region, the larger roles that heterochromatin plays in other cellular processes—such as chromosome segregation, suppression of recombination, and maintenance of genomic integrity—have only begun to be appreciated. For instance, the condensed structures of heterochromatin in *S. pombe* have been shown to be important for proper functioning of the centromeres and telomeres. Heterochromatin loss at the centromeres affects proper orientation and assembly of the kinetochore (Pidoux and Allshire 2000; Sullivan et al. 2001). In addition, components of heterochromatin such as Swi6 have been reported to recruit cohesin proteins to the centromeres as well as other heterochromatic regions along the chromosome arms, including the *mat2/3* interval (Hall and Grewal 2003). In RNAi mutants in which heterochromatin is absent at the centromeres, centromeric cohesion was compromised, accompanied by severe mitotic and meotic segregation defects (Hall et al. 2003). Last, at the telomeres, telomere-specific silencing factors, such as Taz1 and Rap1, and RNAi components contribute to heterochromatin formation. The presence of heterochromatin at the telomeres is important for the maintenance of proper telomere length and the clustering of telomeres at the spindle pole body during meiosis (Cooper et al. 1998; Nimmo et al. 1998; Hall et al. 2003).

The many commonalities shared between heterochromatic regions of the *S. pombe* genome and genomes of other higher eukaryotes—from homologous molecular components of heterochromatin to similar processes of heterochromatin assembly and, even still, to similar higher-order chromatin organization of the genome, such as the structures of centromeres and telomeres—suggest that insights gained from the studies of heterochromatin in *S. pombe* will be equally applicable to the studies of heterochromatin in organisms with larger and more complex genomes. Particularly, in view of the fact that a relatively small portion of mammalian genomes consists of euchromatin, with euchromatic "islands" often scattered within larger heterochromatic domains, the role of heterochromatin in these systems will be vital for the normal functioning of the cell and the biology of the organism as a whole.

ACKNOWLEDGMENTS

We thank members of the Grewal laboratory for helpful discussions and Ken-ichi Noma for critical reading of the manuscript and help with figures. Research in our laboratory is supported by the National Cancer Institute.

REFERENCES

Ahmed S., Saini S., Arora S., and Singh J. 2001. Chromodomain protein Swi6-mediated role of DNA polymerase alpha in establishment of silencing in fission yeast. *J. Biol. Chem.* **276**: 47814.

Allshire R.C., Javerzat J.P., Redhead N.J., and Cranston G. 1994. Position effect variegation at fission yeast centromeres. *Cell* **76**: 157.

Ayoub N., Goldshmidt I., Lyakhovetsky R., and Cohen A. 2000. A fission yeast repression element cooperates with centromere-like sequences and defines a mat silent domain boundary. *Genetics* **156**: 983.

Bannister A.J., Zegerman P., Partridge J.F., Miska E.A., Thomas J.O., Allshire R.C., and Kouzarides T. 2001. Selective recognition of methylated lysine 9 on histone H3 by the HP1 chromo domain. *Nature* **410**: 120.

Bernstein E., Caudy A.A., Hammond S.M., and Hannon G.J. 2001. Role for a bidentate ribonuclease in the initiation step of RNA interference. *Nature* **409**: 363.

Bjerling P., Silverstein R.A., Thon G., Caudy A., Grewal S.I.S., and Ekwall K. 2002. Functional divergence between histone deacetylases in fission yeast by distinct cellular localization and in vivo specificity. *Mol. Cell. Biol.* **22**: 2170.

Brasher S.V., Smith B.O., Fogh R.H., Nietlispach D., Thiru A., Nielsen P.R., Broadhurst R.W., Ball L.J., Murzina N.V., and Laue E.D. 2000. The structure of mouse HP1 suggests a unique mode of single peptide recognition by the shadow chromo domain dimer. *EMBO J.* **19**: 1587.

Cheutin T., Gorski S.A., May K.M., Singh P.B., and Misteli T. 2004. In vivo dynamics of Swi6 in yeast: Evidence for a stochastic model of heterochromatin. *Mol. Cell. Biol.* **24**: 3157.

Cheutin T., McNairn A.J., Jenuwein T., Gilbert D.M., Singh P.B., and Misteli T. 2003. Maintenance of stable heterochromatin domains by dynamic HP1 binding. *Science* **299**: 721.

Cooper J.P., Watanabe Y., and Nurse P. 1998. Fission yeast Taz1 protein is required for meiotic telomere clustering and recombination. *Nature* **392**: 828.

Cowieson N.P., Partridge J.F., Allshire R.C., and McLaughlin P.J. 2000. Dimerisation of a chromo shadow domain and distinctions from the chromodomain as revealed by structural analysis. *Curr. Biol.* **10:** 517.

Donze D., Adams C.R., Rine J., and Kamakaka R.T. 1999. The boundaries of the silenced HMR domain in *Saccharomyces cerevisiae*. *Genes Dev.* **13:** 698.

Egel R., Beach D.H., and Klar A.J. 1984. Genes required for initiation and resolution steps of mating-type switching in fission yeast. *Proc. Natl. Acad. Sci.* **81:** 3481.

Festenstein R., Pagakis S.N., Hiragami K., Lyon D., Verreault A., Sekkali B., and Kioussis D. 2003. Modulation of heterochromatin protein 1 dynamics in primary mammalian cells. *Science* **299:** 719.

Grewal S.I.S. 2000. Transcriptional silencing in fission yeast. *J. Cell. Physiol.* **184:** 311.

Grewal S.I. and Elgin S.C. 2002. Heterochromatin: New possibilities for the inheritance of structure. *Curr. Opin. Genet. Dev.* **12:** 178.

Grewal S.I.S. and Klar A.J. 1996. Chromosomal inheritance of epigenetic states in fission yeast during mitosis and meiosis. *Cell* **86:** 95.

———. 1997. A recombinationally repressed region between mat2 and mat3 loci shares homology to centromeric repeats and regulates directionality of mating-type switching in fission yeast. *Genetics* **146:** 1221.

Grewal S.I.S., Bonaduce M.J., and Klar A.J. 1998. Histone deacetylase homologs regulate epigenetic inheritance of transcriptional silencing and chromosome segregation in fission yeast. *Genetics* **150:** 563.

Hall I.M. and Grewal S.I.S. 2003. Structure and function of heterochromatin: Implications for epigenetic gene silencing and genome organization. In *RNAi: A guide to gene silencing* (ed. G. Hannon), p. 205. Cold Spring Harbor Laboratory Press, Cold Spring Harbor, New York.

Hall I.M., Noma K., and Grewal S.I.S. 2003. RNA interference machinery regulates chromosome dynamics during mitosis and meiosis in fission yeast. *Proc. Natl. Acad. Sci.* **100:** 193.

Hall I.M., Shankaranarayana G.D., Noma K., Ayoub N., Cohen A., and Grewal S.I.S. 2002. Establishment and maintenance of a heterochromatin domain. *Science* **297:** 2232.

Hammond S.M., Boettcher S., Caudy A.A., Kobayashi R., and Hannon G.J. 2001. Argonaute2, a link between genetic and biochemical analyses of RNAi. *Science* **293:** 1146.

Hsieh J. and Fire A. 2000. Recognition and silencing of repeated DNA. *Annu. Rev. Genet.* **34:** 187.

Jacobs S.A. and Khorasanizadeh S. 2002. Structure of HP1 chromodomain bound to a lysine 9-methylated histone H3 tail. *Science* **295:** 2080.

Jenuwein T. and Allis C.D. 2001. Translating the histone code. *Science* **293:** 1074.

Jia S., Noma K., and Grewal S.I.S. 2004. RNAi-independent heterochromatin nucleation by the stress-activated ATF/CREB family proteins. *Science* **304:** 1971.

Kurdistani S.K. and Grunstein M. 2003. Histone acetylation and deacetylation in yeast. *Nat. Rev. Mol. Cell Biol.* **4:** 276.

Lachner M., O'Carroll D., Rea S., Mechtler K., and Jenuwein T. 2001. Methylation of histone H3 lysine 9 creates a binding site for HP1 proteins. *Nature* **410:** 116.

Litt M.D., Simpson M., Gaszner M., Allis C.D., and Felsenfeld G. 2001. Correlation between histone lysine methylation and developmental changes at the chicken beta-globin locus. *Science* **293:** 2453.

Lorentz A., Ostermann K., Fleck O., and Schmidt H. 1994. Switching gene swi6, involved in repression of silent mating-type loci in fission yeast, encodes a homologue of chromatin-associated proteins from *Drosophila* and mammals. *Gene* **143:** 139.

Murchison E.P. and Hannon G.J. 2004. miRNAs on the move: miRNA biogenesis and the RNAi machinery. *Curr. Opin. Cell Biol.* **16:** 223.

Murzina N., Verreault A., Laue E., and Stillman B. 1999. Heterochromatin dynamics in mouse cells: Interaction between chromatin assembly factor 1 and HP1 proteins. *Mol. Cell* **4:** 529.

Nakayama J., Klar A.J., and Grewal S.I.S. 2000. A chromo-domain protein, Swi6, performs imprinting functions in fission yeast during mitosis and meiosis. *Cell* **101:** 307.

Nakayama J., Allshire R.C., Klar A.J., and Grewal S.I.S. 2001a. A role for DNA polymerase alpha in epigenetic control of transcriptional silencing in fission yeast. *EMBO J.* **20:** 2857.

Nakayama J., Rice J.C., Strahl B.D., Allis C.D., and Grewal S.I.S. 2001b. Role of histone H3 lysine 9 methylation in epigenetic control of heterochromatin assembly. *Science* **292:** 110.

Nakayama J., Xiao G., Noma K., Malikzay A., Bjerling P., Ekwall K., Kobayashi R., and Grewal S.I.S. 2003. Alp13, an MRG family protein, is a component of fission yeast Clr6 histone deacetylase required for genomic integrity. *EMBO J.* **22:** 2776.

Nielsen P.R., Nietlispach D., Mott H.R., Callaghan J., Bannister A., Kouzarides T., Murzin A.G., Murzina N.V., and Laue E.D. 2002. Structure of the HP1 chromodomain bound to histone H3 methylated at lysine 9. *Nature* **416:** 103.

Nimmo E.R., Pidoux A.L., Perry P.E., and Allshire R.C. 1998. Defective meiosis in telomere-silencing mutants of *Schizosaccharomyces pombe* (comments). *Nature* **392:** 825.

Noma K., Allis C.D., and Grewal S.I.S. 2001. Transitions in distinct histone H3 methylation patterns at the heterochromatin domain boundaries. *Science* **293:** 1150.

Partridge J.F., Borgstrom B., and Allshire R.C. 2000. Distinct protein interaction domains and protein spreading in a complex centromere. *Genes Dev.* **14:** 783.

Partridge J.F., Scott K.S., Bannister A.J., Kouzarides T., and Allshire R.C. 2002. cis-acting DNA from fission yeast centromeres mediates histone H3 methylation and recruitment of silencing factors and cohesin to an ectopic site. *Curr. Biol.* **12:** 1652.

Pidoux A.L. and Allshire R.C. 2000. Centromeres: Getting a grip of chromosomes. *Curr. Opin. Cell Biol.* **12:** 308.

Rea S., Eisenhaber F., O'Carroll D., Strahl B.D., Sun Z.W., Schmid M., Opravil S., Mechtler K., Ponting C.P., Allis C.D., and Jenuwein T. 2000. Regulation of chromatin structure by site-specific histone H3 methyltransferases. *Nature* **406:** 593.

Reinhart B.J. and Bartel D.P. 2002. Small RNAs correspond to centromere heterochromatic repeats. *Science* **297:** 1831.

Saitoh S., Takahashi K., and Yanagida M. 1997. Mis6, a fission yeast inner centromere protein, acts during G1/S and forms specialized chromatin required for equal segregation. *Cell* **90:** 131.

Schubeler D., MacAlpine D.M., Scalzo D., Wirbelauer C., Kooperberg C., van Leeuwen F., Gottschling D.E., O'Neill L.P., Turner B.M., Delrow J., Bell S.P., and Groudine M. 2004. The histone modification pattern of active genes revealed through genome-wide chromatin analysis of a higher eukaryote. *Genes Dev.* **18:** 1263.

Selker E.U. 1999. Gene silencing repeats that count. *Cell* **97:** 157.

Shankaranarayana G.D., Motamedi M.R., Moazed D., and Grewal S.I.S. 2003. Sir2 regulates histone H3 lysine 9 methylation and heterochromatin assembly in fission yeast. *Curr. Biol.* **13:** 1240.

Shibahara K. and Stillman B. 1999. Replication-dependent marking of DNA by PCNA facilitates CAF-1-coupled inheritance of chromatin. *Cell* **96:** 575.

Strahl B.D. and Allis C.D. 2000. The language of covalent histone modifications. *Nature* **403:** 41.

Strahl B.D., Ohba R., Cook R.G., and Allis C.D. 1999. Methylation of histone H3 at lysine 4 is highly conserved and correlates with transcriptionally active nuclei in *Tetrahymena*. *Proc. Natl. Acad. Sci.* **96:** 14967.

Sullivan B.A., Blower M.D., and Karpen G.H. 2001. Determining centromere identity: Cyclical stories and forking paths. *Nat. Rev. Genet.* **2:** 584.

Takahashi K., Chen E.S., and Yanagida M. 2000. Requirement of Mis6 centromere connector for localizing a CENP-A-like protein in fission yeast. *Science* **288:** 2215.

Takahashi K., Murakami S., Chikashige Y., Funabiki H., Niwa O., and Yanagida M. 1992. A low copy number central sequence with strict symmetry and unusual chromatin structure in fission yeast centromere. *Mol. Biol. Cell* **3:** 819.

Thon G. and Verhein-Hansen J. 2000. Four chromo-domain proteins of *Schizosaccharomyces pombe* differentially repress transcription at various chromosomal locations. *Genetics* **155:** 551.

Thon G., Bjerling P., Bunner C.M., and Verhein-Hansen J. 2002. Expression-state boundaries in the mating-type region of fission yeast. *Genetics* **161:** 611.

Tuzon C.T., Borgstrom B., Weilguny D., Egel R., Cooper J.P., and Nielsen O. 2004. The fission yeast heterochromatin protein Rik1 is required for telomere clustering during meiosis. *J. Cell Biol.* **165:** 759.

Verdel A., Jia S., Gerber S., Sugiyama T., Gygi S., Grewal S.I.S., and Moazed D. 2004. RNAi-mediated targeting of heterochromatin by the RITS complex. *Science* **303:** 672.

Volpe T.A., Kidner C., Hall I.M., Teng G., Grewal S.I.S., and Martienssen R.A. 2002. Regulation of heterochromatic silencing and histone H3 lysine-9 methylation by RNAi. *Science* **297:** 1833.

Wang G., Ma A., Chow C.M., Horsley D., Brown N.R., Cowell I.G., and Singh P.B. 2000. Conservation of heterochromatin protein 1 function. *Mol. Cell. Biol.* **20:** 6970.

Zhang Z., Shibahara K., and Stillman B. 2000. PCNA connects DNA replication to epigenetic inheritance in yeast. *Nature* **408:** 221.

Regulation of *Caenorhabditis elegans* RNA Interference by the *daf-2* Insulin Stress and Longevity Signaling Pathway

D. WANG AND G. RUVKUN

Department of Molecular Biology, Massachusetts General Hospital and Department of Genetics, Harvard Medical School, Boston, Massachusetts 02114

A conserved insulin-like signaling pathway regulates metabolism, development, stress resistance, and life span in *Caenorhabditis elegans* (Lee et al. 2003). This pathway involves an insulin-like receptor DAF-2, a phosphatidylinositol-3-OH kinase (PI3K) AGE-1, and the kinases AKT-1, AKT-2 (Paradis and Ruvkun 1998), and PDK-1 (Paradis et al. 1999), as well as the fork head transcription factor DAF-16 (Ogg et al. 1997; Lee et al. 2003). Loss-of-function mutations in *daf-2* or *age-1* cause the worm to arrest at the dauer diapause stage rather than proceeding to reproductive development (Ogg et al. 1997). In addition, the *daf-2* and *age-1* mutants have increased life span and stress resistance, compared with wild type (Ogg et al. 1997; Honda and Honda 2002; Lee et al. 2003). Both the dauer-constitutive and life span phenotypes of the *daf-2* or *age-1* mutants are suppressed by loss-of-function mutation in the downstream *daf-16* gene (Ogg et al. 1997).

RNA interference, or RNAi, is well conserved across phylogeny and protects the genome from invasive genetic elements such as transposons and viruses (Denli and Hannon 2003). Exogenous double-stranded RNA (dsRNA) is cleaved by the RNase III Dicer into 22-nucleotide short interfering RNA (siRNA) (Bernstein et al. 2001). siRNA is incorporated into a protein complex called RNA-induced silencing complex (RISC), which recognizes and destroys the target mRNA based on the sequence homology between the target mRNA and the trigger dsRNA (Song et al. 2004).

Here we show that *C. elegans* mutants with decreased insulin-like signaling have a more intense RNAi response than wild type. Such regulation of RNAi by this stress and longevity signaling pathway suggests a role in response to pathogens such as viruses.

SECTION THEMES

Decreased *age-1*/PI3-Kinase Signaling Enhances RNAi

Mutants lacking *age-1* activity show enhanced response to RNAi. We used dsRNAs, which cause less severe or penetrant phenotypes than the corresponding loss-of-function mutations, to test the insulin signaling–defective mutants. *age-1(mg305)* is a temperature-sensitive allele that causes dauer arrest at 25°C or higher temperatures. *age-1(mg305)* animals display a temperature-sensitive enhanced RNAi response at the nonpermissive temperature of 25°C. For instance, loss-of-function mutation in the *lin-1* ETS transcription factor gene causes production of multiple vulvae (Muv). RNAi of *lin-1* causes multiple vulvae in 96% of the *age-1(mg305)* mutants at 25°C, but 0% at 20°C, and 0% of wild-type animals at both temperatures. Similarly, mutations in the *daf-2*/insulin-like receptor gene show enhanced response to RNAi (Table 1). The increase in longevity, stress resistance, and fat storage caused by *daf-2* or *age-1* mutations is strongly suppressed by mutations in *daf-16*, a fork head transcription factor coding gene. The enhanced RNAi phenotype is also mediated by signaling via DAF-16: the *age-1;daf-16* double mutant shows wild-type response to *lin-1* dsRNA (Table 2). These data suggest that insulin-like signaling normally inhibits RNAi via the DAF-16 transcriptional cascade.

age-1 Mediates RNAi Response via mRNA Abundance

The phenotypic difference between *age-1* and wild type in response to dsRNA correlates with the different change of the target mRNA level. RNAi of the histone gene *his-44* gene causes 100% early larval arrest in *age-1(mg305)*, but does not cause this phenotype in wild type. Similarly, other RNAi of a wide range of histone genes is enhanced in *age-1(mg305)*. Consistent with the enhanced RNAi phenotype, northern analysis shows that after feeding *his-44* dsRNA, *his-44* mRNA level is significantly decreased in *age-1(mg305)*, whereas no change is observed in wild type (Fig. 1). This decrease in *his-44* mRNA level was similar to that induced by the previously known RNAi enhancer mutant *rrf-3(pk1426)* (Simmer et al. 2002), which also enhances the lethality induced by *his-44* RNAi. Therefore, the insulin-like pathway enhances RNAi by facilitating the degradation of target mRNA.

The Dauer-Regulatory Pathways Overlap with the RNAi Pathway

Three parallel signaling cascades regulate dauer arrest in *C. elegans*: the *daf-2*/insulin-like, *daf-7*/TGF-β-like, and *daf-11*/cyclic GMP pathways (Li et al. 2003). Mutants lacking activity of either pathway arrest at the dauer stage; however, only mutations in the *daf-2* insulin-like pathway

Table 1. The Dauer-Constitutive Mutants Show Different Sensitivity to RNAi

Strains	Feeding RNAi (phenotype)			
	his-44 (% L1/L2 arrest)	*lin-1* (% Muv)	*hmr-1* (% embryonic lethal)	*col-183* (dumpy)
age-1(mg305ts)	75 ± 2	96 ± 7	81 ± 8	+
daf-2(e1370ts)	91 ± 6	65 ± 6	79 ± 5	+
daf-7(mg1372ts)	7 ± 3	22 ± 5	41 ± 9	−
daf-11(mg295ts)	9 ± 6	0 ± 0	35 ± 7	−
rrf-3(pk1426)	86 ± 11	77 ± 10	96 ± 2	++
Wild type (N2)	12 ± 9	0 ± 0	28 ± 6	−

Animals were fed bacteria expressing dsRNA targeting *his-44*, *lin-1*, or *col-183* from the L1 stage at 25°C. For the *lin-1* and *col-183* experiment, dauers from 25°C were recovered at 20°C and allowed to proceed to adulthood when the multiple vulval (Muv) and dumpy body shape phenotypes were scored. For the *col-183* experiment, −, +, and ++ indicate no (−), dumpy (+), and extremely dumpy (++) body shape. For the *hmr-1* experiment, animals at young adult stage were fed *hmr-1* dsRNA and the percentage of embryonic lethality among their progeny was scored.

increase longevity and stress resistance of reproductively growing animals (Tissenbaum and Ruvkun 1998; Wolkow et al. 2000). To decide whether the RNAi-enhanced phenotype is caused by some general dauer-inducing signal shared by all three pathways, we tested the RNAi response of dauer-constitutive mutants in the other two pathways, *daf-7/TGF-β* and *daf-11/guanylate cyclase*. *daf-11(mg295ts)* is no more sensitive to RNAi than wild type. *daf-7(mg1372)* is slightly more sensitive than wild type to *lin-1* (22% Muv vs. 0% in wild type) and *hmr-1* (41% embryonic lethal vs. 28% in wild type) dsRNAs, but no different than wild type in response to *his-44* dsRNA (7% embryonic lethal vs. 12% in wild type) or *col-183* dsRNA (not dumpy) (Table 1). The weak RNAi enhancement of a *daf-7* mutant may reflect the known crosstalk between the insulin and TGF-β pathways (Lee et al. 2001; Li et al. 2003). We conclude that RNAi, like stress resistance and longevity, is most affected by mutations in the insulin-like pathway.

CONCLUSIONS

Mutations in the insulin-like pathway enhance RNAi response and this enhancement is dependent on the DAF-16 fork head transcription factor. The insulin-like metabolic and longevity signaling is transduced by the fork head transcription factor DAF-16 (Ogg et al. 1997).

The nuclear localization of DAF-16 is regulated by insulin-like signaling (Lee et al. 2001). One model for how insulin-like signaling affects RNAi is that components that positively regulate RNAi, like *dcr-1* and *rde-1*, are positively regulated by DAF-16; or components that negatively regulate RNAi, like *eri-1* (Kennedy et al. 2004) and *rrf-3* (Simmer et al. 2002), are negatively regulated by DAF-16. The DAF-16 binding site has been determined and 947 *C. elegans* genes were identified to bear at least one DAF-16 binding site within 1 kilobase (kb) upstream of their predicted transcriptional start sites. Among these genes, 17 genes are orthologous between *Drosophila* and *C. elegans*, highlighting the response pathways that may be conserved (Lee et al. 2003). None of those conserved genes with consensus DAF-16 binding sites correspond to known RNAi factors, but many regulatory steps in RNAi have yet to be identified.

Recent study shows that a conserved and pancreatic islet-specific microRNA, miR375, regulates the insulin signaling pathway in mammals (Poy et al. 2004). Overexpression of miR375 suppresses glucose-induced insulin secretion and, conversely, inhibition of endogenous *miR-375* function enhances insulin secretion (Poy et al. 2004). MicroRNA processing shares the same machinery with RNAi, which includes the RNase III Dicer and the *Argonaute* proteins ALG-1 and ALG-2 (Grishok et al. 2001). The fact that mutation in the insulin-like pathway enhances RNAi suggests a possible negative feedback between insulin signaling and miRNA/siRNA process-

Table 2. The *age-1* Mutant Has Enhanced Response to *lin-1* dsRNA

Strains	% of Muv animals	
	20°C	25°C
age-1(mg305ts)	0 ± 0	96 ± 7
daf-16(mgDf47)	1 ± 3	5 ± 5
age-1(mg305ts); daf-16(mgDf47)	1 ± 2	7 ± 3
rrf-3(pk1426)	64 ± 6	77 ± 4
Wild type (N2)	0 ± 0	0 ± 0

Animals of the indicated genotypes were fed bacteria expressing *lin-1* dsRNA at 20°C or 25°C since the L1 larval stage. The percentage of animals with multiple vulval structures (Muv) was scored. For the experiment with *age-1(mg305)* at 25°C, animals arrested as dauers and were then recovered at 20°C to proceed to adulthood when the Muv phenotype was scored.

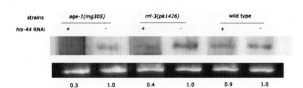

Figure 1. Mutation in *age-1* enhances the suppression of the target mRNA of RNAi. *age-1(mg305)*, *rrf-3(pk1426)*, and wild type were fed bacteria expressing *his-44* dsRNA (+) or a control vector L4440 (−) for 12 hr at 25°C. The change of mRNA was determined by northern blot.

ing. Such a feedback loop could control the amount of insulin signaling via regulation of microRNAs that in turn regulate the secretion of insulins.

Alternatively, the interaction between the insulin and RNAi pathways may occur at the chromatin level. In *C. elegans*, dauer larvae express histone H1 at a higher level than nondauer animals (Jones et al. 2001). The *C. elegans* homolog of SIR-2, which mediates chromatin silencing in yeast, functions in the insulin-like signaling pathway. Overexpression of the *sir-2* gene mimics the loss-of-function mutation in *daf-2* or *age-1* to extend the life span and favors the dauer entry, and both phenotypes are suppressed by a mutation in the downstream *daf-16* gene (Tissenbaum and Guarente 2001). A distinct chromatin packaging in dauers and animals deficient in insulin signaling may enhance the action of siRNAs in chromatin remodeling to suppress the transcription of RNAi target genes. On the other hand, components of the RNAi machinery, including Dicer and *Argonaute*, are required for the assembly and localization of heterochromatin in flies (Pal-Bhadra et al. 2004) and fission yeast (Verdel et al. 2004). There may be a competition between heterochromatin formation and RNAi. It is possible that the change in chromatin structure in the *daf-2* mutants releases RNAi components from heterochromatin to now allow a more robust silencing of mRNAs.

Mutants lacking the insulin-like signaling, such as *daf-2* and *age-1*, are more resistant than wild type to environmental stresses (Honda and Honda 2002) as well as pathogens (Garsin et al. 2003). RNAi components also mediate silencing of transposons in *C. elegans* (Sijen and Plasterk 2003) as well as resistance to virus infection in *Arabidopsis* (Mourrain et al. 2000). Thus the coupling of RNAi responses to the stress resistance pathway of *daf-2*/insulin-like signaling makes biological sense—as part of a stress resistance pathway, RNAi responses may be enhanced. It is intriguing to contemplate that the aging regulation by the *daf-2*/insulin signaling pathway could also depend in part on RNAi pathways.

ACKNOWLEDGMENTS

We thank W. Li, S. Kennedy, C. Wolkow, H.A. Tissenbaum, A. Hart, and J. Kaplan for insightful discussion.

REFERENCES

Bernstein E., Caudy A.A., Hammond S.M., and Hannon G.J. 2001. Role for a bidentate ribonuclease in the initiation step of RNA interference. *Nature* **409**: 363.
Denli A.M. and Hannon G.J. 2003. RNAi: An ever-growing puzzle. *Trends Biochem. Sci.* **28**: 196.
Garsin D.A., Villanueva J.M., Begun J., Kim D.H., Sifri C.D., Calderwood S.B., Ruvkun G., and Ausubel F.M. 2003. Long-lived *C. elegans* daf-2 mutants are resistant to bacterial pathogens. *Science* **300**: 1921.
Grishok A., Pasquinelli A.E., Conte D., Li N., Parrish S., Ha I., Baillie D.L., Fire A., Ruvkun G., and Mello C.C. 2001. Genes and mechanisms related to RNA interference regulate expression of the small temporal RNAs that control *C. elegans* developmental timing. *Cell* **106**: 23.
Honda Y. and Honda S. 2002. Oxidative stress and life span determination in the nematode *Caenorhabditis elegans*. *Ann. N.Y. Acad. Sci.* **959**: 466.
Jones S.J., Riddle D.L., Pouzyrev A.T., Velculescu V.E., Hillier L., Eddy S.R., Stricklin S.L., Baillie D.L., Waterston R., and Marra M.A. 2001. Changes in gene expression associated with developmental arrest and longevity in *Caenorhabditis elegans*. *Genome Res.* **11**: 1346.
Kennedy S., Wang D., and Ruvkun G. 2004. A conserved siRNA-degrading RNase negatively regulates RNA interference in *C. elegans*. *Nature* **427**: 645.
Lee R.Y., Hench J., and Ruvkun G. 2001. Regulation of *C. elegans* DAF-16 and its human ortholog FKHRL1 by the daf-2 insulin-like signaling pathway. *Curr. Biol.* **11**: 1950.
Lee S.S., Kennedy S., Tolonen A.C., and Ruvkun G. 2003. DAF-16 target genes that control *C. elegans* life-span and metabolism. *Science* **300**: 644.
Li W., Kennedy S.G., and Ruvkun G. 2003. daf-28 encodes a *C. elegans* insulin superfamily member that is regulated by environmental cues and acts in the DAF-2 signaling pathway. *Genes Dev.* **17**: 844.
Mourrain P., Beclin C., Elmayan T., Feuerbach F., Godon C., Morel J.B., Jouette D., Lacombe A.M., Nikic S., Picault N., Remoue K., Sanial M., Vo T.A., and Vaucheret H. 2000. *Arabidopsis* SGS2 and SGS3 genes are required for posttranscriptional gene silencing and natural virus resistance. *Cell* **101**: 533.
Ogg S., Paradis S., Gottlieb S., Patterson G.I., Lee L., Tissenbaum H.A., and Ruvkun G. 1997. The Fork head transcription factor DAF-16 transduces insulin-like metabolic and longevity signals in *C. elegans*. *Nature* **389**: 994.
Pal-Bhadra M., Leibovitch B.A., Gandhi S.G., Rao M., Bhadra U., Birchler J.A., and Elgin S.C. 2004. Heterochromatic silencing and HP1 localization in *Drosophila* are dependent on the RNAi machinery. *Science* **303**: 669.
Paradis S. and Ruvkun G. 1998. *Caenorhabditis elegans* Akt/PKB transduces insulin receptor-like signals from AGE-1 PI3 kinase to the DAF-16 transcription factor. *Genes Dev.* **12**: 2488.
Paradis S., Ailion M., Toker A., Thomas J.H., and Ruvkun G. 1999. A PDK1 homolog is necessary and sufficient to transduce AGE-1 PI3 kinase signals that regulate diapause in *Caenorhabditis elegans*. *Genes Dev.* **13**: 1438.
Poy M.N., Eliasson L., Krutzfeldt J., Kuwajima S., Ma X., Macdonald P.E., Pfeffer S., Tuschl T., Rajewsky N., Rorsman P., and Stoffel M. 2004. A pancreatic islet-specific microRNA regulates insulin secretion. *Nature* **432**: 226.
Sijen T. and Plasterk R.H. 2003. Transposon silencing in the *Caenorhabditis elegans* germ line by natural RNAi. *Nature* **426**: 310.
Simmer F., Tijsterman M., Parrish S., Koushika S.P., Nonet M.L., Fire A., Ahringer J., and Plasterk R.H. 2002. Loss of the putative RNA-directed RNA polymerase RRF-3 makes *C. elegans* hypersensitive to RNAi. *Curr. Biol.* **12**: 1317.
Song J.J., Smith S.K., Hannon G.J., and Joshua-Tor L. 2004. Crystal structure of Argonaute and its implications for RISC slicer activity. *Science* **305**: 1434.
Tissenbaum H.A. and Guarente L. 2001. Increased dosage of a sir-2 gene extends lifespan in *Caenorhabditis elegans*. *Nature* **410**: 227.
Tissenbaum H.A. and Ruvkun G. 1998. An insulin-like signaling pathway affects both longevity and reproduction in *Caenorhabditis elegans*. *Genetics* **148**: 703.
Verdel A., Jia S., Gerber S., Sugiyama T., Gygi S., Grewal S.I., and Moazed D. 2004. RNAi-mediated targeting of heterochromatin by the RITS complex. *Science* **303**: 672.
Wolkow C.A., Kimura K.D., Lee M.S., and Ruvkun G. 2000. Regulation of *C. elegans* life-span by insulinlike signaling in the nervous system. *Science* **290**: 147.

Interrelationship of RNA Interference and Transcriptional Gene Silencing in *Drosophila*

M. PAL-BHADRA,*‡ U. BHADRA,†‡ AND J.A. BIRCHLER
Division of Biological Sciences, University of Missouri—Columbia, Columbia, Missouri 65211

Gene silencing processes that operate within the nucleus first came to light with studies of transgene silencing in plants (Matzke et al. 1989; Napoli et al. 1990; van der Krol et al. 1990). This body of work suggested the involvement of double-stranded RNA molecules as the catalyst for posttranscriptional silencing (Metzlaff et al. 1997). Studies of plant viruses, which are targets of posttranscriptional silencing (Goodwin et al. 1996), resulted in the finding of RNA-dependent DNA methylation in the nucleus (Wassenegger et al. 1994; Mette et al. 1999, 2000; Jones et al. 2001). When the finding was made that the experimental procedure of RNAi likely worked through a double-stranded RNA process (Fire et al. 1998), biochemical studies were inspired that led to the discovery of what is referred to as the "RNAi machinery" (Tuschl et al. 1999; Hammond et al. 2000, 2001; Zamore et al. 2000; Bernstein et al. 2001). This machinery not only works to destroy homologous messenger RNAs in a posttranscriptional manner (Montgomery et al. 1998), using siRNAs as a guide (Hamilton and Baulcombe 1999), but also is involved with processing of micro-RNAs that play a role in normal development of plants and animals (Grishok et al. 2001; Hutvagner et al. 2001) as well as with targeting sites on the chromosomes for chromatin modifications, which ultimately result in transcriptional silencing (Pal-Bhadra et al. 1997, 2002, 2004; Morel et al. 2000; Sijen et al. 2001; Hall et al. 2002; Volpe et al. 2002; Schramke and Allshire 2003; Verdel et al. 2004). In this paper we will focus on studies in our laboratory concerned with novel aspects of transcriptional gene silencing and its interrelationship with the RNAi machinery.

These studies began with the construction of a promoter–reporter fusion transformant designed to study regulatory factors effective on the *white* eye color gene in *Drosophila* (Rabinow et al. 1991). This plasmid consists of the *white* gene promoter fused to the *Alcohol dehydrogenase* structural gene in the respective 5′ leader sequences. It was transformed into a background carrying an *Adh* allele with very low expression. This *w-Adh* transgene was expressed less when homozygous at any location in the genome (Rabinow et al. 1991) than when only a single copy was present. This behavior was reminiscent of the phenomenon of pairing-sensitive silencing described by Judy Kassis for *engrailed-white* transgenes (Kassis et al. 1991; Kassis 1994; Americo et al. 2002).

A further unusual property of the *w-Adh* transgenes was that they were not additive in expression when multiple copies were present in the cell at dispersed locations (Pal-Bhadra et al. 1997). This type of behavior was analogous to the phenomenon of cosuppression that had been described in plant species for numerous transgenes. This silencing was modulated by various mutations in the genes encoding the Polycomb complex of repressive chromatin proteins. Indeed, when silencing is operative, the silenced transgenes become ectopically associated with the Polycomb complex. This type of silencing occurs at the transcriptional level (Pal-Bhadra et al. 2002).

Other transgenes, such as a full-length *Adh* construct, exhibit silencing that operates at the posttranscriptional level (Pal-Bhadra et al. 2002). This type of silencing is associated with the production of siRNAs and thus fulfills all the hallmarks of operating via the RNA interference pathway. Mutations in this type of silencing are blocked by mutations in the *piwi* gene (Cox et al. 1998), which is a member of a gene family characterized by the PAZ domain. Members of this family are involved with RNAi functions in many species (Cerutti et al. 2000).

The transcriptional silencing characteristic of an interaction between *w-Adh* and a reciprocal *Adh-white* transgene is also blocked by mutation in the *piwi* gene (Pal-Bhadra et al. 1999, 2002) and to a lesser extent by mutations in another member of the same gene family, *aubergine* (Schupbach and Wieschaus 1991; Schmidt et al. 1999; Birchler et al. 2003). These findings indicated a connection between posttranscriptional and transcriptional silencing in *Drosophila*.

With the finding that heterochromatic silencing in fission yeast was suppressed by mutations in the RNAi machinery (Hall et al. 2002; Volpe et al. 2002), tests were conducted as to whether such mutations in *Drosophila* would suppress various types of heterochromatin-based silencing (Pal-Bhadra et al. 2004). Both *piwi* and *aubergine* suppress *mini-white* insertions in the fourth chromosome heterochromatin. Another gene, *homeless*, which encodes a double-stranded RNA helicase required for RNAi (Aravin et al. 2001; Stapleton et al. 2001; Kennerdell et al. 2002), suppresses these transgene insertions as well as repeat-induced silencing of a tandem array of

---

*Present address: Department of Pharmacology, Indian Institute of Chemical Biology, Hyderabad 500007, India.
†Present address: Functional Genomics and Gene Silencing Group, Centre for Cellular and Molecular Biology, Hyderabad 500007, India.
‡These authors contributed equally.

IS THERE A RELATIONSHIP BETWEEN COSUPPRESSION AND PAIRING-SENSITIVE SILENCING?

The *trans* silencing of *w-Adh* and involvement of the Polycomb complex raises the issue of whether an analogous mechanism operates on the silencing observed with transgenes that carry a Polycomb Response Element (PRE). These transgenes typically exhibit pairing-sensitive silencing, as does *w-Adh*. In those cases examined, increased copy number in the nucleus results in further decline in total expression of the collective homologous transgenes (Hagstrom et al. 1997; Sigrist and Pirrotta 1997; Muller et al. 1999), again in analogy with *w-Adh*.

The *engrailed-white* transgenes carry a PRE and were the prototypical case of pairing-sensitive silencing (Kassis et al. 1991; Kassis 1994; Brown et al. 1998; Americo et al. 2002). We used multiple dispersed copies to test whether dispersed silencing was also operative with this transgene. When a large fragment of *engrailed* (2595 bp) was carried on the transgene adjacent to the *mini-white* reporter insertions at various locations in the genome, they exhibited the normal pairing-sensitive silencing. The general assay for pairing-sensitive silencing is a test of whether two paired copies are expressed less than two copies unpaired in the genome. This is indeed the case for insertions of this class (the G series). An example is shown in Figure 1. However, two unpaired copies are not additive in their expression, and further increase in copy

Figure 2. One to four doses of the *en-w* (series G) transgene shows declining total expression. At the *bottom* is one copy of *en-w*. *Counterclockwise* are two copies, each unpaired; three copies consisting of two copies paired with each other and a third copy unpaired; four copies in which two insert sites are both homozygous. Although not shown, the four copies (two insertion sites homozygous) are expressed less than either paired insert alone.

number causes less total expression (Fig. 2). The pairing effects and *trans*-acting silencing are also reflected in the RNA level of *white* expression (Fig. 3). This departure from additivity illustrates that this class of *en-w* transgene participates in dispersed transgene silencing as well, a phenomenon referred to as cosuppression.

A second class of *en-w* transgenes (the H series) has only 400 bp of *engrailed* adjacent to the *mini-white* reporter. In this case the sequences of *engrailed* thought to be the major determinant of pairing-sensitive silencing are missing. When the transgenes are paired at homologous sites, the expression of the *mini-white* reporter is additive in the homozygotes for the respective insertion sites (Figs. 1 and 4). With respect to dispersed silencing, there is no cosuppression observed with this particular transgene (Figs. 1 and 4).

Last, a transgene containing only the portion of *engrailed* thought to play a major role in attracting the Polycomb complex was tested. This transgene series (TVup) contained only 181 base pairs of the *engrailed* gene. This small segment is sufficient to condition pairing-sensitive silencing (Figs. 1 and 5). Interestingly, however, this transgene construct does not exhibit dispersed transgene silencing. Therefore, although there may be determinants in common between pairing-sensitive silencing and cosuppression, the results with this series suggest that those sequences that can confer pairing-sensitive silencing are insufficient to produce cosuppression or that some sequence in this fragment can interfere with cosuppression.

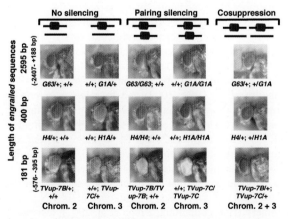

Figure 1. The eye color of multiple *engrailed-white* (*en-w*) transgenes showing a comparison of pairing-sensitive silencing and cosuppression. The length of *engrailed* sequences, chromosomal location, and allelic combinations are noted in the panel. Each *en-w* construct was tested in at least two genomic locations. The *engrailed-white* gene carrying approximately 2595 bp of *engrailed* sequences shows pairing-sensitive silencing. The presence of 400 bp of *engrailed* sequences next to *mini-white* shows a gene dosage effect in paired copies at each location. When only 181 bp upstream to the 400-bp sequence are present, pairing-sensitive silencing occurs, but cosuppression of dispersed copies is not observed.

MUTATIONS IN THE RNAi MACHINERY AFFECT PAIRING-SENSITIVE SILENCING

Given that the Polycomb complex has been implicated in both cosuppression and pairing-sensitive silencing, albeit with mechanistic distinctions, it was of interest to test

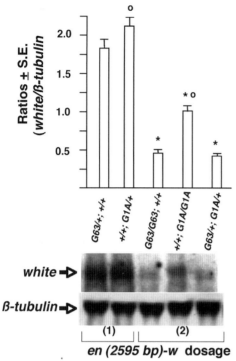

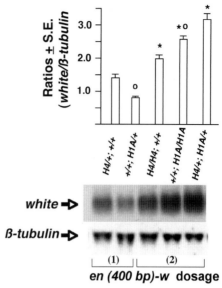

Figure 3. Autoradiograms of Northern blot hybridization from adult female flies carrying 1 and 2 copies of *en-w* constructs demonstrate the trends of interaction at the transcript level. Although only blots using female RNA are shown, males and females follow a similar pattern. The genotypes depicted in the autoradiogram are as follows: (1) *en-w#2/+* (chromosome 2), (2) *+/en-w#3* (chromosome 3), (3) *en-w#2/enw#2*, (4) *en-w#3/en-w#3*, and (5) *en-w#2/en-w#3*. The *bottom* panel represents β-tubulin transcript level after reprobing the same blot, which acts as a gel loading control. The abundance of *white* transcript relative to β-tubulin loading from three independent blots is presented by a bar diagram above the panels. The error bars delimit the 95% confidence intervals. Those values significantly different between the two inserts are designated with an *open circle*. Those significantly different between the paired and unpaired condition are designated with an asterisk. All flies are in a *y w[67c23]* background. The autoradiogram shows a reduction of *white* transcripts for the paired copies relative to the single insert at each location. Note that both pairing silencing and cosuppression are reflected in the level of the *white* transcripts.

Figure 4. The white transcripts of truncated (400-bp) *en-w* inserts are proportionately increased with the copy number, suggesting a gene dosage effect rather than pairing-sensitive silencing and cosuppression. Copy number of the representative transgenes is arranged as described in Fig. 3.

whether pairing-sensitive silencing would be modulated by mutations in the RNAi machinery as is cosuppression. Thus, various insertions of the *engrailed-white* transgene were combined with different alleles of *piwi* and *homeless*.

The insertions A2 and G1A, both of which exhibit pairing-sensitive silencing, were genetically combined with *piwi¹*, *piwi²*, and the heteroallelic combination. The homozygous *piwi* mutations do not have any discernible phenotypic effect on a single unpaired copy of either insertion. However, the paired copies have reduced expression, which is diminished even further in the mutant *piwi* homozygotes. In other words, the silencing is greater in the absence of the functional *piwi* gene product.

Similarly, the combination of *en-w* 1-1B construct, which resides on the X chromosomes and exhibits pairing-sensitive silencing, with *homeless* alleles produced a related result. The alleles of *homeless* tested were *DE8* and *delta125*, as well as their heteroallelic combination. Again, the *homeless* alleles had no apparent phenotypic effect on an unpaired *en-w* transgene when homozygous, but cause a greater silencing of the paired situation. The magnitude of enhancement of pairing-sensitive silencing

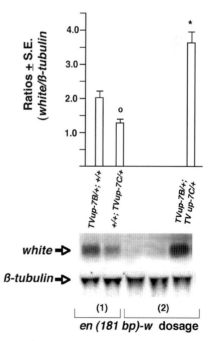

Figure 5. The TVup series exhibits pairing-sensitive silencing, but not cosuppression. Copy number of the representative transgenes is arranged as described in Fig. 3.

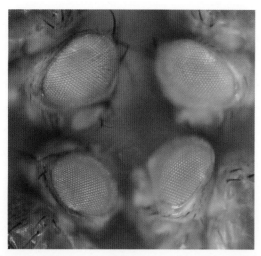

Figure 6. Pairing-sensitive silencing is enhanced by *homeless* mutations. The 1-1B *engrailed-white* construct on the X chromosome was combined with the *homeless* alleles, *delta125* and *DE8*, as well as their heteroallelic combination. Similar results were found in all cases. Depicted are flies segregating for the *delta125* allele. *Bottom right*, a male carrying a single copy of the *en-w* transgene. *Top right*, a female homozygous for this construct, which exhibits pairing-sensitive silencing. *Bottom left*, a male with an unpaired *en-w* insertion and with the *homeless* mutation homozygous. *Top left*, a female homozygous for the transgene and for *homeless*. Note that *homeless* has little effect on the unpaired transgene, but conditions a stronger silencing of *en-w* when it is paired.

by *homeless* was greater than with *piwi*. The effect of *homeless* on pairing-sensitive silencing is shown in Figure 6.

CONCLUSIONS

Several lines of investigations indicate that the RNAi machinery operates via a posttranscriptional mechanism to destroy specific messenger RNAs or viruses. The trigger for these processes appears to be the production of a double-stranded RNA moiety. One natural function is likely to be a defense mechanism against the overexpression of transposable elements (Chabossier et al. 1998; Jensen et al. 1999; Ketting et al. 1999; Tabara et al. 1999) and against viruses. Transposable elements are present at many sites in the chromosomes and can therefore be transcribed in various orientations in such a manner as to contribute to a pool of aberrant RNAs that will trigger such a response. The double-stranded RNAs are cleaved to siRNAs, which are incorporated into the RISC complex to serve as a guide for the enzymatic destruction of the homologous RNAs.

Transcriptional silencing is less well understood but requires functional gene products of many of the same components as posttranscriptional RNA interference (Pal-Bhadra et al. 2002, 2004). Transgene silencing accompanied by the accumulation of the Polycomb complex must involve homology recognition of the related sequences throughout the nucleus for the establishment of this silencing complex. The most likely candidate for this *trans*-acting function are small RNAs, although how they are generated and function at the chromosomal level is still unknown.

Transcriptional silencing of heterochromatic repeats in *Schizosaccharomyces pombe* (Hall et al. 2002; Volpe et al. 2002) and *Drosophila* (Pal-Bhadra et al. 2004) also require the functional gene products of the RNAi machinery. In this case, it is postulated that the repeats generate small RNAs that in turn target chromatin modifications that foster silencing, namely, histone H3 methylated at lysine 9 (H3-mK9). This modified form of histone H3 is bound by Swi6 in fission yeast and by Heterochromatin Protein 1 (HP1) in *Drosophila*. Thus, in this type of transcriptional silencing as well, siRNAs are implicated in guiding the chromatin modification machinery to homologous sequences in the nucleus to silence the various copies of repeated sequences.

An additional type of transcriptional silencing involves the contact or pairing of gene sequences. Transfer of a silenced state from one allele to another was documented during meiosis in *Ascobolus* (Colot et al. 1996). The silencing followed the same parameters as gene conversion, suggesting similarities between homology recognition for silencing and recombination. In *Neurospora*, also during meiosis, silencing of repeated sequences within the nucleus is triggered by the presence of an unpaired sequence (Shiu et al. 2001; Shiu and Metzenberg 2002). Such unpaired DNA will cause the silencing of all homologous copies whether paired or not. This process is referred to as meiotic silencing by unpaired DNA (MSUD). Interestingly, it is suppressed by mutations in the RNAi machinery (Shiu et al. 2001; Shiu and Metzenberg 2002; Lee et al. 2003).

As noted above, a pairing-dependent silencing phenomenon occurs in *Drosophila* and is referred to as pairing-sensitive silencing. It is typically manifested in the behavior of transgenes that express well as one copy, but less well when the transgene is homozygous. In this case, pairing of the transgenes can occur somatically given the property of *Drosophila* homologs to associate in all tissues.

We investigated the relationship of pairing-sensitive silencing to cosuppression and the RNAi machinery in the experiments described above. The larger *engrailed-white* transgenes participate in cosuppression in that they exhibit nonlinearity of expression relative to the copy number present in the nucleus.

Mutations in the RNAi machinery genes, *piwi* and *homeless*, were tested for an impact on pairing-sensitive silencing. Interestingly, they both increase the degree of silencing rather than suppress the loss of expression, raising interesting issues about how the RNAi machinery is involved in a process that appears to involve gene-to-gene contact.

One possibility might be that the paired versus unpaired configuration impacts the nature of gene transcription and modifies the production of messenger RNAs that are abnormal. Abnormal RNAs are thought to enter into the RNAi pathway to eventually generate siRNAs, which may in turn feed back to the gene to modify the local chromatin environment.

A second possibility is simply that the RNAi machinery is intimately associated with the transgenes and acts to inhibit the silencing that might otherwise occur as a result of gene-to-gene contact. When the machinery is lost in the mutants, the silencing can be more severe.

A third possibility is that the gene products of the RNAi machinery are shared with other processes. The RNAi machinery must use the siRNAs as a homology guide for targeting the destruction of mRNAs in the cytoplasm as well as for guiding the sites of chromatin modifications in the nucleus. Pairing-sensitive silencing likely involves homology recognition in some form, followed by a change of chromatin state. It is possible, but not known at present, whether protein components of the RNAi machinery are utilized in other molecular complexes that perform related functions, such as DNA homology recognition.

A fourth possibility is that siRNAs might also be used as a guide to establish chromatin modifications in domains that foster gene activation, as well as those that foster gene silencing. Most research to date implicates siRNAs and the RNAi machinery as the responsible agent for determining the sites of chromatin modifications in the nucleus that result in silencing of all homologous sequences. If, however, the RNAi machinery helps maintain the activity of genes by related means, then the mutational inactivation of the machinery would allow greater silencing to occur.

Polycomb-dependent silencing and heterochromatic silencing are suppressed by mutations in the RNAi machinery. Here we provide evidence that components of this cellular process also affect gene expression as impacted by gene associations. Thus, there is increasing evidence that the RNAi machinery plays an important role in establishing the chromatin configurations along the length of the chromosome.

ACKNOWLEDGMENTS

We thank Judy Kassis for providing *en-w* transgenes and for discussions. Research was supported by a grant from the National Science Foundation MCB 0211376.

REFERENCES

Americo J., Whitely M., Brown J.L., Fujioka M., Jaynes J.B., and Kassis J.A. 2002. A complex array of DNA-binding proteins required for pairing-sensitive silencing by a polycomb group response element from the *Drosophila engrailed* gene. *Genetics* **160:** 1561.

Aravin A.A., Naumova N.M., Tulin A.A., Rozovsky Y.M., and Gvozdev V.A. 2001. Double stranded RNA-mediated silencing of genomic tandem repeats and transposable elements in *Drosophila melanogaster* germline. *Curr. Biol.* **11:** 1017.

Bernstein E., Caudy A.A., Hammond S.M., and Hannon G.J. 2001. Role for a bidentate ribonuclease in the initiation step of RNA interference. *Nature* **409:** 295.

Birchler J.A., Pal-Bhadra M., and Bhadra U. 2003. Transgene cosuppression in animals. In *RNAi: A guide to gene silencing* (ed. G.J. Hannon), p. 23. Cold Spring Harbor Laboratory Press, Cold Spring Harbor, New York.

Brown J.L., Mucci D., Whiteley M., Dirksen M.-L., and Kassis J.A. 1998. The *Drosophila* Polycomb group gene pleiohomeotic encodes a DNA binding protein with homology to the transcription factor YY1. *Mol. Cell* **1:** 1057.

Cerutti L., Mian N., and Bateman A. 2000. Domains in gene silencing and cell differentiation proteins: The novel PAZ domain and redefinition of the *piwi* domain. *Trends Biochem. Sci.* **25:** 481.

Chabossier M.-C., Bucheton A., and Finnegan D.J. 1998. Copy number control of a transposable element, the I factor, a LINE-like element in *Drosophila*. *Proc. Natl. Acad. Sci.* **95:** 11781.

Colot V., Maloisel L., and Rossignol J.L. 1996. Interchromosomal transfer of epigenetic states in Ascobolus: Transfer of DNA methylation is mechanistically related to homologous recombination. *Cell* **86:** 855.

Cox D.N., Chao A., Baker J., Chang L., Qiao D., and Lin H. 1998. A novel class of evolutionary conserved genes defined by *piwi* are essential for stem cell self-renewal. *Genes Dev.* **12:** 3715.

Dorer D.R. and Henikoff S. 1994. Expansions of transgene repeats cause heterochromatin formation and gene silencing in *Drosophila*. *Cell* **77:** 993.

Fire A., Xu S., Montgomery M.K., Kostas S.A., Driver S.E., and Mello C.C. 1998. Potent and specific genetic interference by double-stranded RNA in *Caenorhabditis elegans*. *Nature* **391:** 806.

Goodwin J., Chapman K., Swaney S., Parks T.D., Wernsman E.A., and Doughery W.G. 1996. Genetic and biochemical dissection of transgenic RNA-mediated virus resistance. *Plant Cell* **8:** 95.

Grishok A., Pasquinelli A.E., Conte D., Li N., Parrish S., Ha I., Baillie D.L., Fire A., Ruvkun G., and Mello C.C. 2001. Genes and mechanisms related to RNA interference regulate expression of the small temporal RNAs that control *C. elegans* developmental timing. *Cell* **106:** 23.

Hagstrom K., Muller M., and Schedl P. 1997. A Polycomb and GAGA dependent silencer adjoins the *Fab-7* boundary in the *Drosophila bithorax* complex. *Genetics* **146:** 1365.

Hall I.M., Shankaranarayana G.D., Noma K., Ayoub N., Cohen A., and Grewal S.I. 2002. Establishment and maintenance of a heterochromatin domain. *Science* **297:** 2232.

Hamilton A.J. and Baulcombe D.C. 1999. A species of small antisense RNA in posttranscriptional gene silencing in plants. *Science* **286:** 950.

Hammond S.C., Boettcher S., Caudy A.A., Kobayashi R., and Hannon G.J. 2001. Argonaute2, a link between genetic and biochemical analyses of RNAi. *Science* **293:** 1146.

Hammond S.M., Bernstein E., Beach D., and Hannon G.J. 2000. An RNA-directed nuclease mediates post-transcriptional gene silencing in *Drosophila* cells. *Nature* **404:** 293.

Hutvagner G., McLachlan J., Pasquinelli A.E., Balint E., Tuschl T., and Zamore P.D. 2001. A cellular function for the RNA-interference enzyme Dicer in the maturation of the *let-7* small temporal RNA. *Science* **293:** 834.

Jensen S., Gassama M.-P., and Heidmann T. 1999. Taming of transposable elements by homology dependent gene silencing. *Nat. Genet.* **21:** 209.

Jones L., Ratcliff F., and Baulcombe D.C. 2001. RNA-directed transcriptional gene silencing in plants can be inherited independently of the RNA trigger and requires *Met1* for maintenance. *Curr. Biol.* **11:** 747.

Kassis J.A. 1994. Unusual properties of regulatory DNA from the *Drosophila engrailed* gene: Three "pairing sensitive" sites within a 1.6 kb region. *Genetics* **136:** 1025.

Kassis J.A., VanSickle E.P., and Sensabaugh S.M. 1991. A fragment of *engrailed* regulatory DNA can mediate transvection of the *white* gene in *Drosophila*. *Genetics* **128:** 751.

Kennerdell J.R., Yamaguchi S., and Carthew R.W. 2002. RNAi is activated during *Drosophila* oocyte maturation in a manner dependent on aubergine and spindle-E. *Genes Dev.* **16:** 1884.

Ketting R.F., Haverkamp T.H., van Luenen H.G.A.M., and Plasterk H.A. 1999. *mut-7* of *C. elegans*, required for transposon silencing and RNA interference, is a homolog of Werner syndrome helicase and RNaseD. *Cell* **99:** 133.

Lee D.W., Pratt R.J., McLaughlin M., and Aramayo R. 2003. An argonaute-like protein is required for meiotic silencing. *Genetics* **164:** 821.

Matzke M.A., Primig M., Trnovsky J., and Matzke A.J.M. 1989. Reversible methylation and inactivation of marker genes in sequentially transformed tobacco plants. *EMBO J.* **8:** 643.

Mette M.F., van der Winder J., Matzke M.A., and Matzke A.J.M. 1999. Production of aberrant promoter transcripts contributes to methylation and silencing of unlinked homologous promoters in trans. *EMBO J.* **18:** 241.

Mette M.F., Aufsatz W., van der Winder J., Matzke M.A., and Matzke A.J.M. 2000. Transcriptional silencing and promoter methylation triggered by double-stranded RNA. *EMBO J.* **19:** 5194.

Metzlaff M., O'Dell M., Cluster P.D., and Flavell R.B. 1997. RNA-mediated RNA degradation and chalcone synthase A silencing in petunia. *Cell* **88:** 845.

Montgomery M.K., Xu S., and Fire A. 1998. RNA as a target of dsRNA-mediated genetic interference in *Caenorhabditis elegans. Proc. Natl. Acad. Sci.* **95:** 15502.

Morel J., Mourrain P., Beclin C., and Vaucheret H. 2000. DNA methylation and chromatin structure affect transcriptional and post-transcriptional transgene silencing in *Arabidopsis. Curr. Biol.* **10:** 1591.

Muller M., Hagstrom K., Gyurkovics H., Pirrotta V., and Schedl P. 1999. The *Mcp* element from the *Drosophila melanogaster bithorax* complex mediates long-distance regulatory interactions. *Genetics* **153:** 1333.

Napoli C., Lemieux C., and Jorgenson R. 1990. Introduction of a chimeric chalcone synthase gene in petunia results in reversible co-suppression of homologous genes in trans. *Plant Cell* **2:** 279.

Pal-Bhadra M., Bhadra U., and Birchler J.A. 1997. Cosuppression in *Drosophila:* Gene silencing of *Alcohol dehydrogenase* by *white-Adh* transgenes is Polycomb dependent. *Cell* **90:** 479.

———. 1999. Cosuppression of nonhomologous transgenes in *Drosophila* involves mutually related endogenous sequences. *Cell* **99:** 35.

———. 2002. RNAi related mechanisms affect both transcriptional and posttranscriptional transgene silencing in *Drosophila. Mol. Cell* **9:** 315.

Pal-Bhadra M., Leibovitch B.A., Gandhi S.G., Rao M., Bhadra U., Birchler J.A., and Elgin S.C. 2004. Heterochromatic silencing and HP1 localization in *Drosophila* are dependent on the RNAi machinery. *Science* **303:** 669.

Rabinow L., Nguyen-Huynh A.T., and Birchler J.A. 1991. A trans-acting regulatory gene that inversely affects the expression of the *white, brown* and *scarlet* loci in *Drosophila. Genetics* **129:** 463.

Schmidt A., Palumbo G., Bozzetti M.P., Tritto P., Pimpinelli S., and Schafer U. 1999. Genetic and molecular characterization of *sting*, a gene involved in crystal formation and meiotic drive in the male germ line of *Drosophila melanogaster. Genetics* **151:** 749.

Schramke V. and Allshire R. 2003. Hairpin RNAs and retrotransposon LTRs effect RNAi and chromatin-based gene silencing. *Science* **301:** 1069.

Schupbach T. and Wieschaus E. 1991. Female sterile mutations on the second chromosome of *Drosophila melanogaster. Genetics* **129:** 1119.

Shiu P.K. and Metzenberg R.L. 2002. Meiotic silencing by unpaired DNA: Properties, regulation and suppression. *Genetics* **161:** 1483.

Shiu P.K., Raju N.B., Zickler D., and Metzenberg R.L. 2001. Meiotic silencing by unpaired DNA. *Cell* **107:** 905.

Sigrist C.J.A. and Pirrotta V. 1997. Chromatin insulator elements block the silencing of a target gene by the *Drosophila* polycomb response element (PRE) but allow trans interactions between PREs on different chromosomes. *Genetics* **147:** 209.

Sijen T., Vijn I., Rebocho A., van Blokland R., Roelofs D., Mol J.N., and Kooter J.M. 2001. Transcriptional and posttranscriptional gene silencing are mechanistically related. *Curr. Biol.* **11:** 436.

Stapleton W., Das S., and McKee B.D. 2001. A role of the *Drosophila* homeless gene in repression of Stellate in male meiosis. *Chromosoma* **110:** 228.

Tabara H., Sarkissian M., Kelly W.G., Fleenor J., Grishok A., Timmons L., Fire A., and Mello C.C. 1999. The *rde-1* gene, RNA interference, and transposon silencing in *C. elegans. Cell* **99:** 123.

Tuschl R., Zamore P.D., Lehman R., Bartel D.P., and Sharp P.A. 1999. Targeted mRNA degradation by double-stranded RNA in vitro. *Genes Dev.* **13:** 3191.

van der Krol A.R., Mur L.A., Beld M., Mol J.N.M., and Stuitje A.R. 1990. Flavonoid genes in petunia: Addition of a limited number of gene copies may lead to a suppression of gene expression. *Plant Cell* **2:** 291.

Verdel A., Jia S., Gerber S., Sugiyama T., Gygi S., Grewal S.I.S., and Moazed D. 2004. RNAi-mediated targeting of heterochromatin by the RITS complex. *Science* **303:** 672.

Volpe T.A., Kidner C., Hall I.M., Teng G., Grewal S.I., and Martienssen R.A. 2002 Regulation of heterochromatic silencing and histone H3 lysine-9 mylation by RNAi. *Science* **297:** 1833.

Wassenegger M., Heimes S., Riedel L., and Sanger H.L. 1994. RNA-directed de novo methylation of genomic sequences in plants. *Cell* **76:** 567.

Zamore P.D., Tuschl T., Sharp P.A., and Bartel D.P. 2000. RNAi: Double-stranded RNA directs the ATP-dependent cleavage of mRNA at 21 to 23 nucleotide intervals. *Cell* **101:** 25.

Functional Identification of Cancer-relevant Genes through Large-Scale RNA Interference Screens in Mammalian Cells

T.R. Brummelkamp, K. Berns, E.M. Hijmans, J. Mullenders, A. Fabius, M. Heimerikx,
A. Velds, R.M. Kerkhoven, M. Madiredjo, R. Bernards, and R.L. Beijersbergen

*Division of Molecular Carcinogenesis and Center for Biomedical Genetics, The Netherlands Cancer Institute,
1066 CX Amsterdam, The Netherlands*

The discovery of a cellular response against double-stranded RNA (Fire et al. 1998) has provided one of the most powerful tools to manipulate gene expression and this has revolutionized loss-of-function genetics in *Caenorhabditis elegans* and *Drosophila* (Ashrafi et al. 2003; Kamath et al. 2003; Lum et al. 2003). In most mammalian cells, however, the introduction of long double-stranded RNA provokes an interferon response, leading to a general shutoff of protein synthesis (Stark et al. 1998). This response can be bypassed by using chemically synthesized, 21-base-pair double-stranded short interfering RNA (siRNAs), which can cause strong, but transient, inhibition of gene expression in nearly all mammalian cells (Elbashir et al. 2001). Indeed, small-scale genetic screens with sets of in vitro synthesized siRNAs have recently been performed in mammalian cells to identify modulators of apoptosis (Aza-Blanc et al. 2003). However, the use of these siRNAs is limited by the transient inhibition of gene expression and their high cost. To circumvent these limitations of siRNAs, we and others have developed expression vectors that direct the synthesis of short hairpin RNAs (shRNAs), which are processed in vivo to siRNA-like molecules that can suppress gene expression over prolonged periods of time (Brummelkamp et al. 2002b; Miyagishi and Taira 2002; Paddison et al. 2002; Sui et al. 2002; Yu et al. 2002). We have recently shown the feasibility of using shRNA vectors to identify loss-of-function phenotypes in mammalian cells by creating a set of vectors to suppress nearly all members of the family of de-ubiquitinating enzymes. Using this approach, we identified the cylindromatosis tumor suppressor gene as a key regulator of the transcription factor NF-κB (Brummelkamp et al. 2003). More recently, we constructed a large set of shRNA vectors that together target some 8000 human genes for suppression (Berns et al. 2004). We discuss here several ways in which such shRNA vector libraries can be used to identify novel components of cancer-relevant pathways.

A LIBRARY OF shRNA VECTORS

We generated a shRNA vector library by first selecting nearly 8000 human genes for shRNA-mediated knockdown. The genes were chosen because they were either components of major cellular pathways including cell cycle, transcription regulation, stress signaling, and signal transduction or were involved in important biological processes such as biosynthesis, proteolysis, and metabolism. In addition, a large set of genes implicated in cancer and other diseases was included in the shRNA library. To increase the likelihood of obtaining a functionally significant inhibition of gene expression, we constructed three different shRNA vectors against each gene transcript (Fig. 1A). Thus, the shRNA library consists in total of some 24,000 shRNA vectors. To construct the library, we first selected three 19-mer sequences from each transcript and incorporated each of these 19-mers into two complementary 59-mer DNA oligonucleotides (Fig. 1A). The design of the two 59-mer oligonucleotides was such that, after transcription, the pRETRO-SUPER-encoded RNA transcript will have the form of a short hairpin, in which the two self-complementary stretches in the

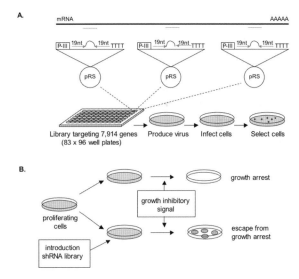

Figure 1. A library of shRNA vectors and its use in mammalian cells. (*A*) Construction of the NKi shRNA library. From each gene transcript three 19-nucleotide (nt) sequences were designed, and 59-mer oligos were cloned into pRETRO-SUPER (pRS). Three vectors targeting one gene were pooled in a single well of a 96-well plate. From each 96-well plate, high-titer polyclonal virus was produced and used to infect BJ-tsLT cells at 32°C. After 2 days cells were shifted to 39°C and after 3 weeks colonies were isolated and analyzed. (*B*) Schematic representation of loss-of-function genetic screen to identify shRNA vectors that overcome a proliferation arrest. The shRNA vectors that induce resistance to the growth inhibitory signal will result in continuous proliferation of those cells. The shRNA vectors can readily be recovered from the positively selected cells.

RNA are separated by a short (9-nucleotide) loop sequence. We have shown previously that this short hairpin precursor transcript is efficiently processed in vivo to yield a gene-specific siRNA-like transcript, causing a gene-specific knockdown of gene expression that lasts as long as the vector is stably expressed (Brummelkamp et al. 2002b). These 59-mers were subsequently annealed and cloned in high throughput fashion into pRETRO-SUPER, a self-inactivating retroviral vector that contains the short hairpin expression cassette (Brummelkamp et al. 2002a). The collection of 24,000 shRNA vectors was cloned in a high-throughput fashion. The bacterial colonies containing one of three shRNA vectors that target the same human transcript for suppression were pooled in a single well of a 96-well plate (Fig. 1A). Using a series of quality control experiments, we estimate that the cloning of the 59-mer oligonucleotides in the retroviral vector was 99% efficient. By testing a number of pools of three knockdown vectors that target the same gene in transient transfection experiments, we estimate that we obtain at least 70% inhibition of gene expression for approximately 70% of the genes in the library (Berns et al. 2004).

shRNA LIBRARY SCREENING STRATEGIES

There are several ways in which shRNA libraries can be screened. The most straightforward approach is to generate high-titer retroviral supernatants of (subsets of) the retroviral shRNA vector library and use these supernatants to infect cells of choice, followed by selection of cells with altered phenotype (Fig. 1A). The simplest iteration of such a loss-of-function genetic screen employs mammalian cell systems in which the knockdown vector confers a selective advantage onto the cells harboring the knockdown vector (Fig. 1B). For example, one could induce cancer cells to undergo proliferation arrest by antiproliferative agents, such as TGF-β or anticancer drugs. Exposure of cells to these agents will induce proliferation arrest, unless the cells have acquired resistance to the antiproliferative agent through expression of a specific shRNA vector. Such shRNA vectors are positively selected by conferring a growth advantage and can be readily recovered from the proliferating cells by a polymerase chain reaction (PCR)-based strategy (Berns et al. 2004). A similar scheme could be used for agents that induce programmed cell death. The advantage of such a genetic screen is that the identified shRNA is causally linked to the pathway of interest. A drawback is that the performance of such screens is time consuming, as the cells from the primary screen often contain multiple shRNA vectors, making a second-round selection mandatory to confirm the identity of the active shRNA in the screen. We have recently performed the first genetic screen of this type in human cells to identify novel components of the p53 tumor suppressor pathway (Berns et al. 2004; and see below). In addition, we describe here the validation of a second way to screen shRNA vector libraries, which we have named "siRNA bar code screening" (Brummelkamp and Bernards 2003). This technology allows the identification of both shRNA vectors that are positively selected and those that are negatively selected under selective conditions.

A SCREEN FOR BYPASS OF p53 PROLIFERATION ARREST

The p53 tumor suppressor pathway is crucial to the maintenance of genome integrity as it transmits both antiproliferative and proapoptotic signals in response to a variety of stress signals (Sherr 1998, 2001). Consequently, the *p53* gene is mutated in more than half of all human cancers. In addition, the other major components of the p53 pathway such as $p19^{ARF}$ in mouse ($p14^{ARF}$ in man) and *MDM2* (*HDM2* in man) are mutated in many forms of cancer. This raises the possibility that additional, yet to be discovered, components of this pathway may also be major players in oncogenesis. We therefore focused on the p53 pathway for the identification of novel components through large-scale loss of function genetic screens using the collection of shRNA vectors described above.

To identify cDNAs that allow bypass of the p53-dependent senescence response, we have previously used a "conditionally immortalized" mouse fibroblast cell line (Shvarts et al. 2002). This cell line harbors a temperature-sensitive allele of SV40 large T antigen (tsLT), which inactivates both pRb and p53 at 32°C, but not at 39°C. As a result, these cells proliferate at 32°C, but enter into a p53-dependent and synchronous proliferation arrest when shifted to 39°C (Shvarts et al. 2002). Because the shRNA library that we constructed targets human genes for suppression, we first generated a human equivalent of these "conditionally immortalized" mouse fibroblasts. In primary human fibroblasts, the ability to proliferate indefinitely is not only determined by the presence of p53, but also by the retinoblastoma protein pRB and by expression of the telomerase catalytic subunit TERT (de Lange 1998; Lundberg et al. 2000). To generate a conditionally immortalized human fibroblast cell line, we used primary foreskin (BJ) fibroblasts (Bodnar et al. 1998). These cells were first modified by retroviral transduction to express the murine ecotropic receptor and the TERT enzyme. We then introduced tsLT into these BJ-TERT fibroblasts (Lee et al. 1995), yielding BJ-TERT-tsLT cells. These cells proliferate when grown at 32°C, but enter into a synchronous proliferation arrest after shift to the nonpermissive temperature (39°C). We found that the growth arrest of these BJ cells is primarily dependent on the p53 tumor suppressor gene, indicating that these cells can be used to identify novel components of the p53 pathway (Berns et al. 2004).

To screen for novel components of the p53 pathway, we isolated polyclonal plasmid DNA from each of the eighty-three 96-well plates of shRNA vectors that together constitute the shRNA library and transfected these pools of vectors into the Phoenix packaging cell line to generate high-titer retroviral supernatants. These retroviruses were then used to infect BJ-TERT-tsLT cells in six-well plates at 32°C. After 2 days, the cells were shifted to 39°C and monitored for the appearance of proliferating colonies over a period of 3 weeks. In a first screen, we identified six genes that allow outgrowth of colonies at 39°C. One of these six genes was *p53* itself, which underscores the quality of the shRNA library. In addition, we found that shRNAs against *RPS6KA6* (ribosomal S6 kinase 4, RSK4), *HTATIP* (histone acetyl transferase TIP60), *HDAC4* (histone deacetylase 4), *KIAA0828* (a putative S-adenosyl-L-homocysteine hydrolase, SAH3), and *CCT2* (T-complex protein 1, b subunit) prevented the p53-dependent growth arrest in the BJ-TERT-tsLT fibroblasts. Subsequent experiments, reported elsewhere (Berns et al. 2004), showed that the identified shRNA constructs also mediated escape from a proliferation arrest induced by the $p19^{ARF}$ tumor suppressor, whose ability to induce cell cycle arrest is strictly p53-dependent (Kamijo et al. 1997). Furthermore, we found that each of the identified siRNAs confers resistance to G1 cell cycle arrest induced by DNA damage (also a p53-dependent process). These data indicate that expression of these shRNA vectors primarily allows bypass of the p53 response and raise the possibility that the targeted genes by these shRNAs are tumor suppressor genes, as their loss abrogates several cancer-relevant aspects of p53 function. Moreover, these data demonstrate the feasibility of the "positive selection" approach (outlined in Fig. 1B) to identify shRNA vectors that confer loss-of-function phenotypes in mammalian cells.

siRNA BAR CODE SCREENS

The shRNA screen described above is time-consuming in that individual colonies of cells must be isolated and hairpin vectors recovered and tested in second-round selection. We have recently proposed an alternative strategy to rapidly screen complex shRNA vector libraries in a polyclonal format to identify individual vectors that either positively or negatively modulate cellular fitness, a technique that we named "siRNA bar code screens" (Brummelkamp and Bernards 2003). This approach was first applied in bacteria and later in yeast (Hensel et al. 1995; Shoemaker et al. 1996; Giaever et al. 1999; Winzeler et al. 1999) and takes advantage of the fact that each hairpin vector contains a unique molecular identifier: The 19-mer sequence in the vector is gene specific. This "molecular bar code" can be used to follow the relative abundance of individual shRNA vectors in a large population of infected cells. The molecular bar codes can be recovered from a cell population by PCR amplification using vector-derived PCR primers that flank the hairpin-encoding DNA sequence. The relative abundance of the shRNA vectors in a cell population (and therefore of their bar code identifier sequences) is influenced by the effect that each knockdown vector has on cellular fitness under the experimental conditions (which will often be a "stress of interest"). Fluorescent labeling of the bar code sequence followed by hybridization to microarrays that contain the gene-specific knockdown bar code oligonucleotides allows one to monitor the relative abundance of each bar-coded DNA fragment in the cell population (Fig. 2) (Brummelkamp and Bernards 2003). Therefore, siRNA bar code screens allow one to detect all genes that have a genetic interaction with a cell population that is exposed to a "stress of interest." As such, this technology allows one to make functional interaction maps of the human genome by being able to efficiently map all genes that modulate a cellular response to a given stress.

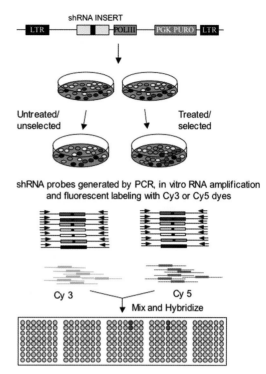

Figure 2. Schematic representation of a bar code screening experiment. Expression of short hairpin RNA (shRNA) molecules in mammalian cells by stably integrated vectors not only creates a gene-specific knockdown phenotype, but also introduces a gene-specific fingerprint (molecular bar code) in cells that express these shRNAs. The 19-base-pair target-gene-specific insert of the shRNA vector is unique in sequence and can be recovered by polymerase chain reaction (PCR) using vector-derived PCR primers that flank the hairpin-encoding DNA sequence. The PCR fragments are amplified by in vitro RNA amplification using the T7 polymerase, whose promoter is present in one of the primer sequences. The abundance of each bar code can be quantified by labeling the RNA fragments with a fluorescent dye, followed by hybridization to a DNA microarray harboring of bar code–complementary DNA fragments. shRNA bar code screens allow the detection of genetic interactions between large sets of genes and almost any biological signal of interest by comparing two cell populations, both of which harbor a collection of shRNA vectors, but only one of which is exposed to a biological signal (e.g., DNA damage, apoptosis-inducing agents, cytotoxic drugs, or inactivation of a tumor suppressor gene). The comparative hybridization of the Cy5- and Cy3-labeled bar-coded RNA fragments allows for the identification of changes (increase or decrease) in the relative abundance of knockdown vectors in response to the stimulus applied.

VALIDATION OF siRNA BAR CODING USING OLIGONUCLEOTIDE MICROARRAYS

The siRNA bar code screening technology provides both opportunities and challenges. One of the challenges is that the 19-mer "molecular bar code" sequence contained within the shRNA vector is part of a hairpin structure that has the ability to fold back on itself (Fig. 3A). This could potentially limit hybridization signals that one obtains when using hairpin probes for hybridization to microarrays. The shRNA vector library was constructed using a set of forty-eight thousand 59-mer oligonucleotides, which include the 19-mer molecular bar code sequences. These 59-mer oligonucleotides provide an attractive source for the generation of siRNA bar code microarrays, as large amounts of these oligonucleotides were left over from the construction of the shRNA library. However, these 59-mers are also self-complementary and might also fold back on themselves rather than hybridize to the Cy-dye-labeled probes to which they are hybridized.

To address whether these hybridization issues pose a problem, we first performed a number of experiments designed to test the relative sensitivity and specificity of hybridization of shRNA probes to DNA microarrays containing the 59-mer oligonucleotide bar code identifiers. Specifically, we asked how efficiently we could identify an arbitrary set of 188 shRNA vectors. To do this, we performed a self–self hybridization with a pool of 188 different 59-mer oligonucleotides used for cloning of the NKi shRNA library. The concentration for each oligonucleotide in the mixture was 200 pg and the mixtures were labeled with either Cy3 or Cy5 followed by hybridization to DNA containing the complementary 59-mer oligonucleotides from the NKi shRNA library. As depicted in Figure 3B, the signal intensities of the spots on the array corresponding to the set of 188 vectors were significantly different from the remainder of the spots (which contained unrelated 59-mer oligonucleotides). The signal intensity from the specific sequences was ~100-fold higher than those from the spots containing noncomplementary 59-mer oligonucleotides. This experiment shows that hairpin probes can be used for hybridization and they do so with good specificity and high signal intensity.

Next, we asked whether the hybridization signals on the array accurately reflect the concentration of the molecular bar code probes in the population. To test this we performed a titration experiment using a pool of 1344 different 59-mer oligonucleotides. In one sample, the amount of the oligonucleotides was held constant at 200 pg; in the other sample the amount of the oligonucleotides was reduced in twofold dilution steps (200 pg, 100 pg, 50 pg, 25 pg, 12.5 pg, 6.25 pg, and 3.12 pg) for groups of 192 oligonucleotides. The samples were labeled with Cy3 or Cy5, respectively, combined, and hybridized to a microarray containing all 23,742 oligonucleotides (representing the complete shRNA library). As can be seen in the representative section of the microarray (Fig. 3C), we

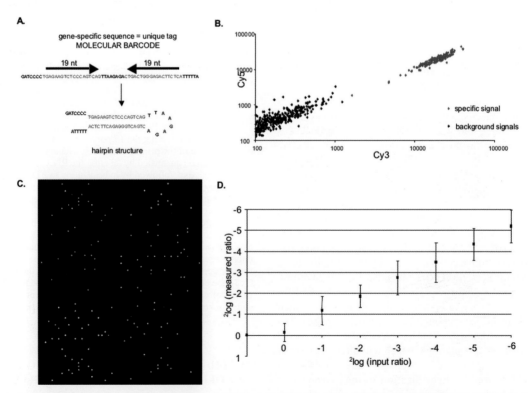

Figure 3. Microarray analysis of shRNA library bar codes. (*A*) Schematic representation of the 59-mer oligonucleotide, including the 19-mer gene-specific sequences, which serves as a molecular bar code sequence. The 59-mer oligonucleotide is part of the shRNA hairpin structure and has the ability to fold back on itself. (*B*) Self–self hybridization of a subset of 59-mer oligonucleotides as depicted in *A* on a microarray containing the complementary 59-mer sequences. A pool of 188 different 59-mer oligonucleotides was used with 200 pg of each oligonucleotide. Depicted in *red* are the hybridization signals obtained from the specific complementary oligonucleotides. Depicted in *black* are 376 nonhomologous oligonucleotides representing nonspecific background hybridization. The average intensity of specific hybridization compared to average background hybridization is more than 100-fold. (*C*) A representative section of a microarray hybridized with a mixture of 1344 different 59-mer oligonucleotides. Hybridizations were performed with two mixtures: a Cy3-labeled mixture in which a constant amount of 200 pg of each oligonucleotide was present and a Cy5-labeled mixture in which the amount was varied (3.1–200 pg) for sets of 192 different oligonucleotides. (*D*) Plot showing the normalized data from the titration study described in *C*. For each set of 192 oligonucleotides the average ratio is calculated plus the standard deviation. Depicted are the ^{2}log from the input ratio (1:1 to 1:64 ratios depicted as 0 to –6) and the ^{2}log from the measured ratio.

obtain specific hybridization with high signal intensities. From the 1344 spots, representing the composition of the probe mixture, we calculated the average ratio and standard deviation for all the different groups of 192 oligonucleotides reflecting a 1:1 to 1:64 dilution series. This ratio was dependent on the relative abundance of the oligonucleotides in the mixture and a twofold change was detected (Fig. 3D). Most importantly, there is a linear correlation between the experimentally determined average ratio of the groups of 192 oligonucleotides compared to the input ratio. The hybridization intensities for the serially diluted oligonucleotides was reflected in linear fashion over the 64-fold dilution range tested (Fig. 3D). This experiment demonstrates the feasibility of using DNA microarrays to determine the relatively abundance of shRNA vectors in a large population and shows that the sensitivity of the approach is sufficient to detect relatively small differences (two- to fourfold) in the presence of shRNA abundance within a complex mixture of shRNAs.

Next we tested whether our experimental approach allowed us to use shRNA probes recovered from cell populations expressing large numbers of shRNA constructs. We generated shRNA cassettes by PCR from cells infected with a pool of 1700 shRNA vectors using vector-derived primers of which one was fused to the T7 promoter sequence. This was used in an in vitro RNA amplification reaction followed by labeling of the in vitro generated RNA with Cy5 dye. The RNA probes were used in a hybridization reaction together with the 1344 59-mer oligonucleotide mixture labeled with Cy3. Figure 4A shows a representative section of a microarray containing all 23,742 oligonucleotides representing the complete NKi shRNA library. The RNA probes generate high signal intensities with virtually no cross-hybridization to noncomplementary spots, similar to what is seen with the 59-mer oligonucleotide probes. In this hybridization experiment, the signal intensity for the RNA probes is more variable than that of the 59-mer oligonucleotides. These differences reflect the variation in abundance of the different shRNA vectors present in the mixture compared to the constant amount of 59-mer oligonucleotides (200 pg each). This experiment demonstrates that shRNA probes generated by PCR followed by linear RNA amplification are suitable for use in microarray hybridization experiments.

Next we tested whether our experimental approach allowed us to use shRNA probes from a large population of shRNA vectors in microarray hybridizations. We generated shRNA probes from two pools of each 5472 shRNA plasmids by PCR followed by in vitro linear RNA amplification. This RNA was subsequently labeled with Cy3 or Cy5 dye and used for hybridization to the DNA microarray. As can be seen in Figure 4, we obtain strong hybridization signals for both Cy3- and Cy5-labeled probe mixtures. The spots harboring oligonucleotides not present in either probe pool hybridize at background levels, emphasizing again the specificity of the hybridization of the bar code oligonucleotides. This experiment demonstrates the feasibility of the use of DNA microarray hy-

Figure 4. Validation of the specific hybridization with RNA bar code probe mixtures. (*A*) A representative fraction of a microarray for the comparison of hybridization with labeled 59-mer oligonucleotides (Cy3, *green*) and labeled amplified RNA probes containing the hairpin sequences (Cy5, *red*). Both probes generate high signal intensities with virtually no cross-hybridization on noncomplementary spots. In this hybridization experiment, the signal intensity for the RNA probes has more variation than the 59-mer oligonucleotides, reflecting the variation in abundance of the different shRNA vectors present compared to the constant amount of 59-mer oligonucleotides (200 pg each). (*B*) Representation of a DNA microarray containing 23,742 oligonucleotides (representing the complete NKi shRNA library) hybridized with two different shRNA probe mixtures containing two different groups of 5000 different shRNA probes labeled Cy3 (*green*) or Cy5 (*red*).

bridizations in conjunction with the full complexity of the shRNA library. It also underscores the sensitivity in this system to detect absolute differences in two large populations of shRNA vectors. This result, in combination with the experiments described above, demonstrates the application of shRNA bar code screens in the detection of small differences in shRNA abundance in parallel screening of large populations of shRNA vectors.

VALIDATION OF BAR CODE SCREENING IN A CELL SYSTEM

As a last validation of the bar code screening concept, we asked whether we could also use the bar coding technology to identify components in the p53 pathway using the conditionally immortalized human fibroblasts (BJ-TERT-tsLT cells) described in more detail by Berns et al. (2004). The design of this experiment is depicted in Figure 5A. We generated a set of 1900 different shRNA vectors targeting 638 different genes. We used this set of vectors to infect the BJ-TERT-tsLT cells, selected for the presence of integrated retroviruses with puromycin. After this, the cells were split at low density and either grown at 32°C (permissive temperature) or at 39°C (restrictive temperature). The cells grown at 32°C continued to proliferate and genomic DNA was isolated when cells reached confluency, after about 8 days. In contrast, the cells cultured at 39°C underwent proliferation arrest and only those cells that harbored a shRNA vector that conferred resistance to the p53-dependent arrest continued to proliferate. As a result, the shRNA vectors associated with the capacity to proliferate at 39°C will be selectively enriched in the population. After 22 days, genomic DNA was isolated from the cultures at 39°C and the relative abundance of shRNA vectors was analyzed on DNA microarrays containing all 1900 shRNA sequences (Fig. 5B). As expected the majority of the shRNA vectors was neither positively or negatively selected (yellow spots) and had a fluorescence intensity ratio (^{2}log ratio) varied from –1 to 1 (Fig. 5C). However, some shRNA vectors were enriched in the population at 39°C (red spots). Upon further analysis (multiple independent temperature-shift experiments and array hybridizations), two shRNA vectors were identified that were consistently enriched in the population at 39°C (Fig. 5C,D). Both shRNA vectors target the cell cycle inhibitor p21^{cip1} (CDKNA1) and were shown previously to suppress p21^{cip1} expression and override the p53-induced cell cycle arrest in the BJ fibroblasts (Berns et al. 2004). The third shRNA vector against p21^{cip1} was not enriched at 39°C and was subsequently

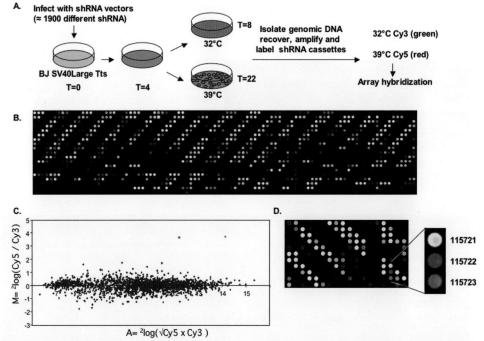

Figure 5. Validation of bar code screening in the BJ-TERT-tsLT system. (*A*) Schematic representation of the experimental setup of the bar code screen in BJ-TERT-tsLT cells. BJ-TERT-tsLT cells were infected with a retroviral supernatant containing 1900 shRNA vectors. Cells were selected for successful retroviral transduction and temperature shifted to induce a senescence-like arrest and used for isolation of the genomic DNA at time (*T*) = 8 days at 32°C or at time (*T*) = 22 days at 39°C. The shRNA cassettes were recovered by PCR, amplified by linear RNA amplification, and labeled with Cy3 (32°C) or Cy5 (39°C). (*B*) A representative section of the microarray hybridization of shRNA fragments isolated from the experiment described in *A*. (*C*) Analysis of the relative abundance of shRNAs in BJ-TERT-tsLT cells recovered from cells grown at 32°C or at 39°C. Data are normalized and depicted as *M*, the ^{2}log (ratio Cy5/Cy3) versus *A*, the ^{2}log ($\sqrt{\text{intensity Cy3} \times \text{Cy5}}$). The data are the average of three independent experiments performed in duplicate and with reversed color. The three *red spots* indicate shRNA vectors targeting the p21 gene. (*D*) Cutout of the section of the DNA microarray containing the three shRNAs for p21. The shRNAs 115722 and 15723 are specifically enriched at 39°C whereas 115721 is not affected. Both shRNAs 115722 and 115723 are active shRNAs against p21, whereas 115721 is inactive.

shown to be inactive in downregulation of $p21^{cip1}$ expression. Together these data show that the siRNA bar code screening can be used to rapidly identify individual shRNA vectors that modulate cellular responses in a large library of vectors. The identification of $p21^{cip1}$ in this screen also validates the application of the bar code screening method in a relevant biological setting in which the relative enrichment of shRNA vectors is influenced by many biological parameters.

CONCLUSIONS

We describe here two approaches to identify shRNA vectors that confer loss-of-function phenotypes in mammalian cells. The siRNA bar code screening technology presented and validated here can be very helpful in the identification of novel anticancer drug targets. For instance, a potentially useful application of siRNA bar code screens is that they facilitate the identification of synthetic lethal interactions (a combination of two nonlethal mutations that together result in cell death). The identification of genes that are specifically toxic in cells having cancer-specific mutations provides an opportunity to make new generations of anticancer drugs that are more tumor-specific and have fewer side effects.

ACKNOWLEDGMENTS

We thank Stephen Friend, Peter Linsley, Guy Cavet, Wei Ge, and Julian Downward for their support of the shRNA library project. This work was supported by grants from the Netherlands Genomics Initiative/Netherlands Organisation for Scientific Research (NWO), Cancer Research UK, the Centre for Biomedical Genetics (CBG), and the Dutch Cancer Society (KWF).

REFERENCES

Ashrafi K., Chang F.Y., Watts J.L., Fraser A.G., Kamath R.S., Ahringer J., and Ruvkun G. 2003. Genome-wide RNAi analysis of *Caenorhabditis elegans* fat regulatory genes. *Nature* **421:** 268.
Aza-Blanc P., Cooper C.L., Wagner K., Batalov S., Deveraux Q.L., and Cooke M.P. 2003. Identification of modulators of TRAIL-induced apoptosis via RNAi-based phenotypic screening. *Mol. Cell* **12:** 627.
Berns K., Hijmans E.M., Mullenders J., Brummelkamp T.R., Velds A., Heimerikx M., Kerkhoven R.M., Madiredjo M., Nijkamp W., Weigelt B., Agami R., Ge W., Cavet G., Linsley P.S., Beijersbergen R.L., and Bernards R. 2004. A large-scale RNAi screen in human cells identifies new components of the p53 pathway. *Nature* **428:** 431.
Bodnar A.G., Ouellette M., Frolkis M., Holt S.E., Chiu C.P., Morin G.B., Harley C.B., Shay J.W., Lichtsteiner S., and Wright W.E. 1998. Extension of life-span by introduction of telomerase into normal human cells. *Science* **279:** 349.
Brummelkamp T.R. and Bernards R. 2003. New tools for functional mammalian cancer genetics. *Nat. Rev. Cancer* **3:** 781.
Brummelkamp T.R., Bernards R., and Agami R. 2002a. Stable suppression of tumorigenicity by virus-mediated RNA interference. *Cancer Cell* **2:** 243.
———. 2002b. A system for stable expression of short interfering RNAs in mammalian cells. *Science* **296:** 550.
Brummelkamp T.R., Nijman S.M., Dirac A.M., and Bernards R. 2003. Loss of the cylindromatosis tumour suppressor inhibits apoptosis by activating NF-kappaB. *Nature* **424:** 797.

de Lange T. 1998. Telomeres and senescence: Ending the debate. *Science* **279:** 334.
Elbashir S.M., Harborth J., Lendeckel W., Yalcin A., Weber K., and Tuschl T. 2001. Duplexes of 21-nucleotide RNAs mediate RNA interference in cultured mammalian cells. *Nature* **411:** 494.
Fire A., Xu S., Montgomery M.K., Kostas S.A., Driver S.E., and Mello C.C. 1998. Potent and specific genetic interference by double-stranded RNA in *Caenorhabditis elegans*. *Nature* **391:** 806.
Giaever G., Shoemaker D.D., Jones T.W., Liang H., Winzeler E.A., Astromoff A., and Davis R.W. 1999. Genomic profiling of drug sensitivities via induced haploinsufficiency. *Nat. Genet.* **21:** 278.
Hensel M., Shea J.E., Gleeson C., Jones M.D., Dalton E., and Holden D.W. 1995. Simultaneous identification of bacterial virulence genes by negative selection. *Science* **269:** 400.
Kamath R.S., Fraser A.G., Dong Y., Poulin G., Durbin R., Gotta M., Kanapin A., Le Bot N., Moreno S., Sohrmann M., Welchman D.P., Zipperlen P., and Ahringer J. 2003. Systematic functional analysis of the *Caenorhabditis elegans* genome using RNAi. *Nature* **421:** 231.
Kamijo T., Zindy F., Roussel M.F., Quelle D.E., Downing J.R., Ashmun R.A., Grosveld G., and Sherr C.J. 1997. Tumor suppression at the mouse INK4a locus mediated by the alternative reading frame product p19ARF. *Cell* **91:** 649.
Lee G.H., Ogawa K., and Drinkwater N.R. 1995. Conditional transformation of mouse liver epithelial cells. An in vitro model for analysis of genetic events in hepatocarcinogenesis. *Am. J. Pathol.* **147:** 1811.
Lum L., Yao S., Mozer B., Rovescalli A., Von Kessler D., Nirenberg M., and Beachy P.A. 2003. Identification of Hedgehog pathway components by RNAi in *Drosophila* cultured cells. *Science* **299:** 2039.
Lundberg A.S., Hahn W.C., Gupta P., and Weinberg R.A. 2000. Genes involved in senescence and immortalization. *Curr. Opin. Cell Biol.* **12:** 705.
Miyagishi M. and Taira K. 2002. U6 promoter-driven siRNAs with four uridine 3´ overhangs efficiently suppress targeted gene expression in mammalian cells. *Nat. Biotechnol.* **20:** 497.
Paddison P.J., Caudy A.A., Bernstein E., Hannon G.J., and Conklin D.S. 2002. Short hairpin RNAs (shRNAs) induce sequence-specific silencing in mammalian cells. *Genes Dev.* **16:** 948.
Sherr C.J. 1998. Tumor surveillance via the ARF-p53 pathway. *Genes Dev.* **12:** 2984.
———. 2001. The ink4a/arf network in tumour suppression. *Nat. Rev. Mol. Cell Biol.* **2:** 731.
Shoemaker D.D., Lashkari D.A., Morris D., Mittmann M., and Davis R.W. 1996. Quantitative phenotypic analysis of yeast deletion mutants using a highly parallel molecular bar-coding strategy. *Nat. Genet.* **14:** 450.
Shvarts A., Brummelkamp T.R., Scheeren F., Koh E.Y., Daley G.Q., Spits H., and Bernards R. 2002. A senescence rescue screen identifies BCL6 as an inhibitor of anti-proliferative p19(ARF)-p53 signaling. *Genes Dev.* **16:** 681.
Stark G.R., Kerr I.M., Williams B.R., Silverman R.H., and Schreiber R.D. 1998. How cells respond to interferons. *Annu. Rev. Biochem.* **67:** 227.
Sui G., Soohoo C., Affar el B., Gay F., Shi Y., and Forrester W.C. 2002. A DNA vector-based RNAi technology to suppress gene expression in mammalian cells. *Proc. Natl. Acad. Sci.* **99:** 5515.
Winzeler E.A., Shoemaker D.D., Astromoff A., Liang H., Anderson K., Andre B., Bangham R., Benito R., Boeke J.D., Bussey H., Chu A.M., Connelly C., Davis K., Dietrich F., Dow S.W., El Bakkoury M., Foury F., Friend S.H., Gentalen E., Giaever G., Hegemann J.H., Jones T., Laub M., Liao H., and Davis R.W. 1999. Functional characterization of the *S. cerevisiae* genome by gene deletion and parallel analysis. *Science* **285:** 901.
Yu J.Y., DeRuiter S.L., and Turner D.L. 2002. RNA interference by expression of short-interfering RNAs and hairpin RNAs in mammalian cells. *Proc. Natl. Acad. Sci.* **99:** 6047.

The New Field Of Epigenomics: Implications for Cancer and Other Common Disease Research

H.T. Bjornsson,[*] H. Cui,[†] D. Gius,[‡] M.D. Fallin,[†] and A.P. Feinberg[†]

[*]Predoctoral Program in Human Genetics and Department of Medicine, and [†]Departments of Molecular Biology & Genetics, and Oncology, Johns Hopkins University School of Medicine, Baltimore, Maryland 21205; [‡]Radiation Oncology Branch, Center for Cancer Research, National Cancer Institute, National Institutes of Health, Bethesda, Maryland 20892

Epigenetic alterations involve information heritable during cell division other than the DNA sequence itself. Epigenetic marks were originally thought to involve only unusual phenomena such as position effect variegation in flies (Tartof and Bremer 1990), telomere silencing (Brachmann et al. 1995), mating-type silencing in yeast (Haber 1998), and transgene-induced gene silencing in plants and animals (Dorer and Henikoff 1994). However, it is increasingly clear that epigenetic inheritance plays a central role in defining cellular growth and differentiation. With exceptions such as antibody gene rearrangements and changes in mitochondrial DNA, there is little evidence to support a role for changes in DNA sequence in development, and thus epigenetics is likely to be at the heart of maintaining the differences between stem cells and somatic cells, one cell type and another, and aged versus younger cells.

Epigenetic inheritance involves three interrelated mechanisms: (1) DNA methylation; (2) posttranslational modifications of histones, including methylation, phosphorylation, acetylation, and sumoylation (Strahl and Allis 2000); and (3) chromatin alterations acting over long distances, such as modifications of specific proteins binding to insulator sequences (Chung et al. 1993). Cytosine DNA methylation is a covalent modification of DNA in which a methyl group is transferred from S-adenosylmethionine to the C-5 position of cytosine by a family of cytosine (DNA-5)-methyltransferases (DNMTs). DNA methylation occurs almost exclusively at CpG nucleotides. The pattern of DNA methylation is transmitted through mitosis and maintained after DNA replication by DNMT1, which has a 100-fold greater affinity for hemimethylated DNA (i.e., parental strand methylated, daughter strand unmethylated) than for unmethylated DNA (Gregory 2001). However, developing cells in the zygote and embryo undergo dramatic shifts in DNA methylation, involving both loss of methylation and de novo methylation.

There are two classes of cytosine DNA methylation in the genome. The first occurs throughout the body of genes that show tissue-specific expression, with methylation generally associated with gene silencing (Riggs 1989; Gregory 2001). The second class involves "CpG islands," or regions rich in CpG dinucleotides (Bird 1986). CpG islands are often described as uniformly unmethylated in normal cells, with the exception of the inactive X chromosome and near imprinted genes (Bird 1986; Riggs and Pfeifer 1992). However, the assumption that autosomal CpG islands (except for imprinted genes) are never methylated is clearly not the case. Strichman-Almashanu demonstrated the presence of normally methylated CpG islands throughout the genome (Strichman-Almashanu et al. 2002). It is also important to note that functionally important methylation information is not always found within CpG islands. For example, the H19 differentially methylated region (DMR) that regulates imprinting of IGF2 and a DMR linked to colorectal cancer risk are not CpG islands (Cui et al. 2002). Thus, epigenome analyses focused solely on these CpG islands would be severely limited in their potential impact.

Epigenetic alteration has two defining characteristics. First is its metastable nature, which involves the capacity for high-frequency alteration, as well as reprogramming in somatic cells and/or specifically through the germ line. The most important consequence of this metastability for human genetics is the apparent high frequency of mutation of affected loci and the ability of large numbers of cells to change the state of their programming, in response to developmental or environmental signals. The second defining characteristic is position effect, in which epigenetic modifications act over a distance along the genome, which in the case of insulators may be in the order of hundreds of kilobases. An important consequence of position effect for human genetics is the ability of regulatory sequences, and epigenetic modifications of them, to act at surprisingly long distances and affect the expression of multiple genes as a group.

Genomic imprinting is a special case of epigenetic modification in which the alteration occurs during germ-line reprogramming, leading to preferential expression (although generally not absolute in humans) of a specific parental allele in somatic cells of the offspring. At least several hundred genes may show imprinting, and imprinted gene expression appears to be important in a number of rare human genetic disorders as well as common cancer. Both mouse and human chromosomes that undergo uniparental disomy (UPD) often show characteristic phenotypic alterations in the offspring (for review, see Cattanach and Beechey 1990; Ledbetter and Engel 1995). These can include overgrowth in the case of paternal

UPD for some chromosomal regions (Cattanach and Beechey 1990; Ledbetter and Engel 1995) and growth retardation in the case of maternal UPD of the same chromosomal regions. There is a strong relationship between imprinted genes and both prenatal and postnatal growth (Moore and Haig 1991). Imprinting is also thought to underlie some quantitative trait loci for growth, with considerable potential commercial application (de Koning et al. 2000). Finally, both imprinted and nonimprinted genes show abnormal expression in animals created by nuclear transfer, and imprinting is thought to be a potential barrier to stem cell transplantation.

Genomics includes a whole-genome approach to genetics. In human disease research, genomics approaches include gene expression arrays, genome scans for sequence variation, and family studies using association tests (transmission test for linkage disequilibrium [Spielman et al. 1993], commonly but erroneously called "transmission disequilibrium test" or TDT).

Epigenomics is defined as a whole-genome approach to epigenetics, similarly advancing epigenetics studies beyond the single-gene level. The field is nascent at present, and efforts to develop it include array-based methylation analysis, array-based hybridization using probes prepared by immunoprecipitation with antibodies to modified histones (so-called "ChIP on chip"), and high-throughput allele-specific expression analysis.

In the following sections, we will describe two efforts to "genomicize" epigenetics. The first is a whole-genome approach to cancer epigenetics, in which array-based gene expression was analyzed after epigenetic modification by combinations of three methods: gene knockout of DNA methyltransferases; treatment with 5-aza-2´-deoxycytidine, an inhibitor of DNA methylation; and treatment with trichostatin A, a histone deacetylase inhibitor. The second approach is an effort to provide a theoretical foundation for a population-based approach to the epigenetic basis of human disease, which we call the "common disease genetic and epigenetic" hypothesis, or CDGE.

CANCER, AN EXAMPLE OF A COMMON DISEASE OF PARTLY EPIGENETIC ORIGIN

Studies of the epigenetics of common human diseases have been generally limited to cancer, and it has not been widely perceived that epigenetics might play a major role in many common complex disease traits. Alterations in DNA methylation were the initial focus (Feinberg and Vogelstein 1983), and epigenetic activation of oncogenes and epigenetic silencing of tumor suppressor genes are both important. Epigenetic activation includes "CT" genes (expressed in cancer and normally only in the testis), e.g., the MAGE gene in melanoma, the PSCA and S100A4 in prostate cancer, and the HPV (human papillomavirus) genome in cervical cancer, to name a few (Feinberg and Tycko 2004). Hypomethylation also leads to chromosomal instability, and this has been shown to promote tumor formation in a mouse model (Gaudet et al. 2003). Hypermethylation is linked to silencing of many tumor suppressor genes, including RB, VHL, and cadherin. There is some controversy whether the methylation changes initiate silencing, but they at least help to maintain it. Methylation changes are ubiquitous in cancer, affecting all known tumor types at nearly universal frequency and are much more common than genetic changes (Feinberg and Tycko 2004).

Genomic imprinting is also important in cancer, first suggested by parent-of-origin specific loss of heterozygosity in several tumor types. Loss of imprinting (LOI) of the autocrine growth factor IGF2, leading to its increased expression, was first observed in Wilms tumor of the kidney and then in many common tumors as well (Rainier et al. 1993; Okamoto et al. 1997). LOI serves as a gatekeeper for some cancers, as methylation changes are found not only in Wilms tumor but in nonneoplastic kidneys surrounding some of the tumors (Moulton et al. 1994; Steenman et al. 1994). Chromatin modifications may also be important in cancer. For example, histone H3 lysine methylation is associated with INK4A tumor suppressor gene silencing (Bachman et al. 2003). More importantly, resilencing of INK4 is established by chromatin modification in methylation-deficient DNMT knockout cell lines, suggesting that chromatin alterations rather than DNA methylation initiate silencing (Bachman et al. 2003). Chromatin modifications at a distance are important in regulating normal imprinting of IGF2, which in fetal development is regulated by a DMR between the IGF2 and H19 and which is methylated on the maternal allele only. The insulator protein CTCF binds to this unmethylated DMR, limiting access to an enhancer shared between H19 and IGF2, causing silencing of the maternal IGF2 allele (Ohlsson et al. 2003). In Wilms tumors, LOI appears to be caused by aberrant methylation of the maternal DMR, blocking CTCF binding and causing activation of the normally silent maternal IGF2 allele (Feinberg et al. 2002). Finally, epigenetic factors may act in *trans* to promote cancer progression. Increased expression of EZH2 is linked to generalized hypermethylation and gene silencing in metastatic prostate cancer (Varambally et al. 2002). EZH2 is an ortholog of the *Drosophila* chromatin repressor protein enhancer of Zeste, and thus overexpression of this gene could cause epigenetic silencing of multiple genes throughout the genome.

While epigenetic mechanisms are generally accepted as important in cancer initiation and progression, the idea that they might play a role in cancer predisposition is relatively new. However, two epigenetic modifications affecting IGF2 imprinting suggest that may be the case. Beckwith–Wiedemann syndrome (BWS), a disorder of prenatal overgrowth, midline abdominal wall defects, and cancer, is caused by several different genetic and epigenetic mechanisms, which are beyond the scope of this paper and discussed elsewhere (DeBaun et al. 2002). Hypermethylation of the H19 DMR occurs in about 15% of BWS patients, and this alteration is specifically associated with cancer risk. Thus, methylation changes in normal tissue serve as a gatekeeper to cancer development. In addition, hypomethylation of a DMR within IGF2 also is linked to LOI in ~10% of the population (Cui et al. 1998, 2002). This LOI appears to be important in colorectal cancer predisposition as it occurs more frequently in patients with a positive history of colon neoplasms or a positive family history of colorectal cancer (Cui et al. 2003; Woodson et al. 2004). Finally, it was found recently that aberrant methylation of H19 is clustered in families, suggesting the existence of epigenetic polymorphisms in the population (Sandovici et al. 2003).

Malignant transformation requires both the inactivation of genomic fidelity surveillance pathways as well as activation of proproliferative/prosurvival signal transduction cascades. The regulation of DNA methylation and the subsequent chromatin structure are significantly altered in tumor cells suggesting a direct role for altered methylation in the process of in vivo cellular transformation. It has also been demonstrated in vitro that c-fos-induced overexpression of DNMT1, or overexpression of DNMT1 alone, transforms immortalized rat fibroblasts in vitro (Bakin and Curran 1999). In addition, this group also demonstrated that the addition of chemical agents that inhibit methyltransferase activity or chromatin compaction significantly reversed the DNMT1-induced transformed phenotype. The results of these experiments raise an intriguing question: Can the silencing of genes alone by methylation induce transformation without the activation of proproliferative/survival pathways or, more interestingly, does hypermethylation activate the expression of genes regulating cellular proproliferative/survival pathways?

AN EPIGENOMIC APPROACH TO CANCER

It now appears clear that methylation plays a central role in transformation, both in vitro and in vivo; however, the mechanism remains to be fully understood. This is due in part to the significant number of genes altered by changes in intracellular methyltransferase activity and the chemical agent used to modulate gene expression, such as 5-aza-CdR, that like all chemicals undoubtedly has nonspecific pharmacological effects. Most previous studies have examined changes at the individual gene level, and a more comprehensive genomic approach would reveal patterns that would otherwise not be apparent. To begin to address these issues we conducted a comprehensive gene expression analysis to reveal the relationship between chemical and genetic manipulation of epigenome. In these studies, HCT116 cells were treated with 5-aza-CdR or TSA followed by microarray analysis to identify changes in gene expression (Gius et al. 2004).

As might be expected, a significant number of genes were increased following exposure to either 5-aza-CdR or TSA, including numerous genes previously shown; however, we also identified multiple genes that are down-regulated following exposure to these agents. A hierarchical cluster analysis identified 231 genes down-regulated at least 1.5-fold and 46 genes at least 2-fold following exposure to 5-aza-CdR and 157 (Table 1) genes down-regulated at least 1.5-fold and 22 genes at least 2-fold following exposure to TSA (Gius et al. 2004). An examination of this microarray analysis demonstrates genes involved in such diverse intracellular processes as cell cycle regulation and growth factors, as well as pro-survival signaling pathways. Interestingly, we observed roughly the same number of genes decreased following

Table 1. Table of Genes Up- and Down-regulated in HCT116 Cells following Treatment with 5-aza-CdR or TSA

Treatment	Increased		Decreased	
	>1.5	>2.0	<1.5	<2.0
5-aza-CdR	280	64	231	46
TSA	143	18	157	22

Total number of genes increased or decreased in HCT116 cells treated with either 5-aza-CdR or TSA. For complete list, see NIH Web page (http://home.ncifcrf.gov/ROSP-Microarray-Lab/DavidGius/ColdSpringHarbor1.html).

chemical exposure as were increased. These preliminary results suggest that in addition to silencing gene expression, hypermethylation is also linked to gene activation. This result would not have been obvious if studies were limited to identifying increases in expression only after demethylation.

To further address this idea, a microarray analysis was done on somatic cell HCT116 knockout lines for DNMT1$^{(-/-)}$, DNMT3B$^{(-/-)}$, and double knockout (DKO) cells. Previous studies had shown that both DNMT1 and DNMT3B cooperatively maintain DNA methylation and gene silencing (Rhee et al. 2002), and genetic disruption of both DNMT1 and DNMT3B significantly inhibited methyltransferase activity and reduced genomic DNA methylation by roughly 95% (Rhee et al. 2000, 2002). Similar to the results with 5-aza-CdR and TSA, there were a significant number of genes decreased in cells lacking methyltransferase activity and this number was roughly equal to the number of genes increased in the DNMT1$^{(-/-)}$, DNMT3B$^{(-/-)}$, and DKO cell lines (Table 2). The genes decreased by chemical or genetic inhibition of methyltransferase activity appear to be involved in a diverse range of critical intracellular processes, including cell cycle regulation, DNA repair, programmed cell death, proliferation, and several signaling cascades.

If genetic inhibition of methyltransferase activity inhibits the activity of specific genes it seems logical that overexpression of DNMT1 would increase expression of genes as well. This idea was recently addressed by the construction of fibroblast cells that constitutively overexpressed DNMT1 and by comparing microarray gene expression patterns between the parent and DNMT1 expression daughter cells. One of the most surprising finds of this study is that following microarray analysis slightly

Table 2. Table of Genes Up- and Down-regulated in Methyltransferase Somatic Cell Knockout Cell Lines versus Control, HCT116 Cells

Cell Type	Gene expression changes			
	Increased		Decreased	
	>1.5	>2.0	<1.5	<2.0
DNMT1$^{(-/-)}$	280	64	231	46
DNMT3B$^{(-/-)}$	143	18	157	22
DKO	143	18	157	22

Total number of genes increased or decreased in DNMT1$^{(-/-)}$, DNMT3B$^{(-/-)}$, or DKO cells as compared to control, parental HCT116 cells. For complete list, see NIH Web page (http://home.ncifcrf.gov/ROSP-Microarray-Lab/DavidGius/ColdSpringHarbor1.html).

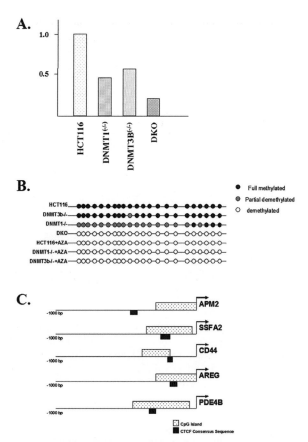

Figure 1. Expression of the APM2 gene is decreased in somatic cell methyltransferase knockout cell lines. (*A*) The total RNA was isolated from control HCT116, as well as dmnt1$^{(-/-)}$, dnmt3B$^{(-/-)}$, and double knockout (DKO) cells and APM2 RNA levels were determined via real-time RT-PCR (reverse transcriptase polymerase chain reaction). Results represent the average of three independent reactions. (*B*) The APM2 promoter CpG island is hypomethylated in somatic cell methyltransferase knockout cell lines. The methylation status of the CpG island in the APM2 upstream promoter region was determined in HCT116, DNMT1$^{(-/-)}$, DNMT3B$^{(-/-)}$, and DKO cells. (*C*) The APM2, SSFA2, CD44, AREG, and PDE4B promoters were examined for the presence of CpG islands and CTCF binding sites using software available at Entrez Genome (http://ncbi.nlm.nih.gov).

more than 350 genes were increased in the CMVdnmt1 cells while roughly 250 genes were decreased, as compared to the parent cells (Ordway et al. 2004). This result is consistent with our finding that the number of genes that appear to be regulated by hypomethylation is as significant as the number of genes that are silenced by hypermethylation. As such, these results suggest, but do not validate, the idea that methylation may increase the expression of specific genes as well as silence gene expression.

If the idea that methylation can activate gene expression is correct, it would be expected that the decrease in the expression of specific genes in the genetic knockout cells would correspond to promoter hypomethylation. To address this, we examined the promoter region of APM2, which was shown to be silenced in the somatic knockout cell lines (Fig. 1A), and found that the degree of transcriptional silencing closely reflected the degree of hypomethylation observed in the APM2 promoter (Fig. 1B). These results suggest a potential direct mechanism for changes in promoter methylation patterns and the regulation of gene expression.

One possible mechanism for explaining how hypomethylation might inhibit gene expression involves DNA-binding factors, such as insulator proteins, that may regulate hypomethylation-mediated gene silencing. For example, CTCF, a chromatin insulator protein, binds to unmethylated GC-rich sequences, causing their silencing by inhibiting enhancer elements (Schoenherr et al. 2003; Fedoriw et al. 2004). To address this idea we conducted a detailed examination of the APM2 gene as well as several other genes that were down-regulated by chemical or genetic inhibition of methyltransferase activity. As can be seen, not only do these genes contain CpG islands in their promoter regions but they also contain consensus CTCF DNA-binding sequence in the upstream regulator regions (Fig. 1C). These results suggest a new paradigm for the role of methylation on gene expression suggesting that, in addition to gene silencing, methylation may activate gene expression as well.

One of the most interesting observations from these microarray data is that several prosurvival or proproliferative factors appear to be regulated by methylation status. This raises an interesting idea: Does overexpression of methyltransferase genes both silence tumor suppressor genes and activate prosurvival or antideath genes?

Finally, we observed that a contiguous domain of metallothionein genes was coordinately regulated in these experiments, and this relationship was highly statistically significant (Gius et al. 2004). Thus, drugs that affect chromatin appear to act in unexpected ways, including similarities in effects between classical "demethylation" and "chromatin decondensation," and to coordinate effects acting over a long distance. These relationships would not be apparent without using an epigenomic approach to their identification.

EPIGENETICS OF RARE HEREDITARY DISORDERS

Two of the known genes involved in DNA methylation have been shown to be involved in rare hereditary diseases. DNA methyltransferase type 3B (DNMT3B) is mutated in the ICF syndrome (immunodeficiency, cytogenetic abnormalities, facial anomalies) (Xu et al. 1999). Centromeric heterochromatin is disrupted in mitogen-treated cells obtained from patients with ICF, suggesting that DNMT3B is involved in this process. However, DNMT3B probably acts more generally in maintaining repressed chromatin, as ICF also involves activation of genes on the inactive X chromosome. Mutations in MECP2 cause Rett syndrome, an X-linked disorder limited to girls (Amir et al. 1999). This syndrome involves normal development in the first year of life, followed by rapid deterioration of higher brain functions, suggesting that methylation is involved in late postnatal brain development.

The number of known mutations in various factors related to chromatin structure and remodeling have increased in number in the last several years and are discussed in several recent reviews (Hendrich and Bickmore 2001; Huang et al. 2003). These include ATRX syndrome, which involves mental retardation and dysmorphic features and is associated with altered DNA methylation and relaxation of repressed chromatin. Rubinstein–Taybi syndrome involves short stature, dysmorphology, and mental retardation and is caused by mutations of CREB binding protein (CBP) (Murata et al. 2001), a histone acetyltransferase. Fascioscapulohumoral dystrophy is associated with contraction of a DNA repeat, which often is associated with hypomethylation of the repeat and expression changes in a gene cluster upstream of the repeat. The epigenetic abnormality is present even in patients with a normal repeat size, indicating a causative role for the epigenetic abnormality (van Overveld et al. 2003).

Several human diseases are known to involve imprinted genes. These include Prader–Willi syndrome and Angelman syndrome, both of which are characterized by short stature, mental retardation, and behavioral disorders, in which the phenotype depends on the parent of origin. Another well known example involving imprinted genes is BWS, a cause of prenatal overgrowth and predisposition to Wilms tumor, hepatoblastoma, and neuroblastoma and of pseudohypoparathyroidism type IA, which involves osteodystrophy and gonadal dysfunction. The common mechanism of BWS is loss of imprinting of the normally silent maternal allele of either insulin-like growth factor II (IGF2) or the antisense RNA LIT1 (DeBaun et al. 2002).

THE COMMON DISEASE GENETIC AND EPIGENETIC HYPOTHESIS (CDGE)

In a recent review we have made the case that epigenetics might play a role in the etiology of common human disease (Bjornsson et al. 2004). The elements of CDGE are as follows.

1. Epigenetic information maintains a developmentally specific pattern, but shows alteration over time. We argue that the stability of epigenetic marks over time is a major factor in the development of common disease. This idea is predicated in part on studies suggesting that an important step in development is the stabilization of patterns of transcriptional activation and silencing by chromatin proteins such as trithorax and polycomb, respectively (Orlando 2003). A recent paper by Ruden has shown a great deal of epigenetic variation in *Drosophila*, which may account for masking of latent mutations, and this masking is environmentally sensitive (e.g., through heat shock [Sollars et al. 2003]). Historically, however, more attention has been focused on the role of DNA methylation in maintaining developmental states. Riggs (1975) and Holliday (Holliday and Pugh 1975) originally suggested that expression patterns were established by DNA methylation. DNA methylation is an attractive candidate as each parental strand remains hemimethylated during replication, thus allowing the methylation to be preserved through replication. This hypothesis has been challenged (Bestor 2000), however, but still remains the dominant hypothesis of the epigenetic literature. Epigenetic modifications are tissue specific, and abnormal epigenetic patterns lead to abnormal development (BWS, Fragile X syndrome, ICF syndrome). Furthermore, the DNA of germ cells goes through extensive demethylation and remethylation stages during gametogenesis and development, thus adding plausability to the idea (Reik et al. 2001).

2. There are several sources of epigenetic variation. These include genetic factors, the individual's environment, the parental environment, the parental genotype and epigenotype, aging, and stochastic events. The fact that epigenetic systems may be influenced by so many events allows a certain developmental "elasticity" for long-lived multicellular organisms. A combined epigenetic and genetic pathogenesis (CDGE model) might provide for a plausible explanation for several presently unexplained features of common complex diseases like late onset, the environmental sensitivity of common human disease, and the apparent tissue specificity of common human disease.

Adding to epigenetic variation may be DNA methylation itself. Genomic DNA methylation is the leading endogenous mutagen accounting for about 30% of germ cell mutations (Krawczak et al. 1998). It is likely that the same phenomenon occurs throughout the somatic life of a cell in that there might be a constant loss of methylated CpG nucleotides, either through point mutation or loss of methyl groups. Many somatic tissues actually have higher total levels of methylation than sperm (Ehrlich et al. 1982) and in those tissues the effect might even be more pronounced than in germ cells. A recent study from mice suggests that the frequency of epimutation might be as much as an order of a magnitude higher than that of genetic mutation (Bennett-Baker et al. 2003).

3. Human disease occurs in part when developmental patterns are lost. Epigenetic modifications could influence disease occurrence either directly or by modifying penetrance. Epigenetic abnormalities predisposing to cancer constitute the best known example of both types. Another example where the epigenetic modification appears to predispose to human disease is that of transient neonatal diabetes, which is a rare disorder with an early diabetes that patients recover from before 18 months of age. This syndrome has recently been shown to be associated with loss of imprinting of two imprinted genes, ZAC and HYMAI (Mackay et al. 2002). Although this is a rare disease, a large proportion of individuals (40%) go on to develop a common adult onset type 2 diabetes mellitus later in life (Mackay et al. 2002). This neonatal form might be an important lead into the pathogenesis of type 2 adult onset diabetes mellitus, a disease that presently is reaching epidemic proportions.

4. The CDGE approach might explain some features of common complex diseases. A combinatorial set of single nucleotide polymorphisms is neither sufficient nor necessary to cause complex diseases, because complex diseases do not occur at birth but usually need a 40-year lag period before the actual disease occurs. To date, the mechanisms through which the environment causes permanent detrimental effects on tissues that act in an additive manner over time have largely remained elusive.

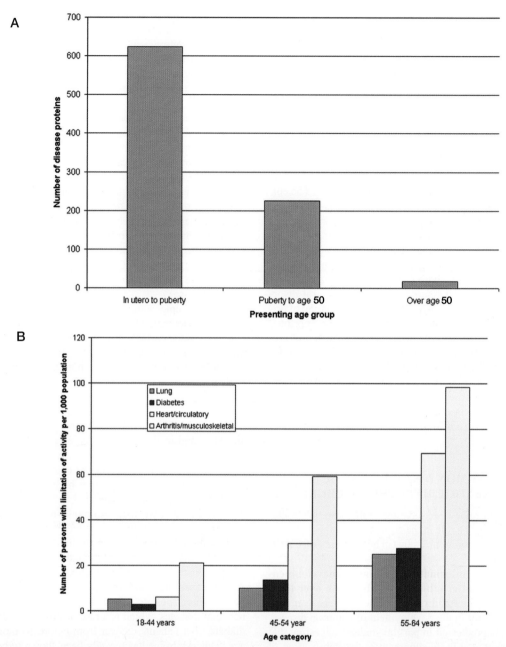

Figure 2. Age of presentation of conventional (*A*) Mendelian disorders and (*B*) common complex traits causing limitation of activity among working-age adults, 1999–2001. Data for panel *A* are from Jimenez-Sanchez et al. (2001), and data for panel *B* are from the U.S. National Center for Health Statistics (Statistics 2003). (Reprinted, with permission, from Bjornsson et al. 2004 [©Elsevier].)

Only two models of adult onset disease and progression provide an obvious explanation for such a phenomenon—a mitochondrial model and an epigenetic model. An epigenetic explanation would be that the normal pattern of epigenetic marks is temporally degraded and when a certain threshold is reached the symptoms of the disease are expressed. Indeed, when a simulation is performed that takes this degeneration of epigenetic patterns into account, the population attributable risk goes up with age, a result that is similar to what is seen for actual common human disease (Figs. 2 and 3) (Bjornsson et al. 2004).

An interesting experimental finding that adds some weight to this idea comes from nuclear transfer experiments. Mice conceived by nuclear transfer develop obesity (Tamashiro et al. 2002), which is not transmitted to their offspring, suggesting an epigenetic role. However, the cloned animals are normal at birth and abnormalities do not present in these mice until 8–10 weeks of age (Tamashiro et al. 2002). The life span of cloned mice that are apparently normal at birth is significantly decreased, but this does not become evident until about 311 days after birth (Ogonuki et al. 2002). Both these cases display a lag time and both are models of epigenetic disease, as the phenotype is not transmitted to offspring.

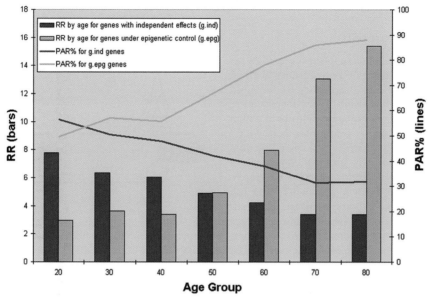

Figure 3. Results from simulations of 40 populations. Simulations were used to create 40 populations containing affected and unaffected individuals for a genetic epidemiologic analysis. Age, environmental status, and genotypes for three different genes were first simulated at random according to specified frequencies. Disease status was then simulated according to CDGE. From these simulated populations, relative risks for g_{ind} and g_{dep} genes were estimated in cross-sectional analysis by each age decade, and averaged over 40 populations (*bars*). Because this risk reflects only the magnitude of a genetic effect, and not the importance of that genotype with respect to all cases in the population, we also estimated population attributable risk percentage (PAR%) at each decade. This reflects the proportion of cases in the population that can be explained by the particular genetic effect. (Reprinted, with permission, from Bjornsson et al. 2004 [©Elsevier].)

Environmental factors play a major role in the causation of complex diseases. For some diseases, such as cardiovascular disease, a partial cause is known. But for most cases of complex disease a mechanism through which an effect of the environment could act is unknown. Perhaps environmental factors interact with the organism at the transcriptional level in a semipermanent manner, possibly through long-term modification of epigenetic marks. A possible reason for a link between permanent changes in gene expression and nutrition could simply be the need for sufficient metabolites for the maintenance of various epigenetic modifications (Cooney 1993). The nutrients needed for maintenance of epigenetic marks are versatile, including methyl donors, many of which are folate dependent for maintenance of methylation of both DNA and histones. Availability of acetyl groups and phosphoryl groups might also in the same way link metabolic intermediates to the control of gene expression and therefore offer a simple explanation of how gene expression is affected by various metabolic states. A recent study found that uremic patients have abnormalities of normal methylation and allele-specific expression patterns that are reversible by folate supplementation (Ingrosso et al. 2003).

5. CDGE provides a theoretical foundation for integrating epigenetic and genetic interaction. While a great deal is now known about chromatin biochemistry, this knowledge has not previously been linked to population genetics. We suggest that epigenetic modifications of some, but certainly not all, genes have an impact on disease penetrance. We refer to such genes as g_{dep}, and the epigenetic modifications themselves may occur directly or may be mediated by variations in the sequence of genes that encode epigenetic modifiers (referred to as g_{epg}). Genes in which epigenetic modification does not play a role in disease are referred to as g_{ind}, even though such genes nevertheless may be epigenetically modified (see Fig. 4).

The implications of this model are that it distinguishes between four possibilities for epigenetic–genetic combinations at a disease-causing locus: (1) At a g_{ind} locus, disease is caused by a genetic variation and is not affected by epigenetics (Fig. 5a), i.e., the special case of CDCV independent of epigenetic variation. Here, the disease phenotype is affected by genetic variation (V) without epigenetic modification playing a role. (2) A locus may exist in which the particular phenotypic contribution at this locus is directly caused by epigenetic variation alone (epg; Fig. 5b). Of course, genetic variation at other loci may still contribute in an additive way to the total disease phenotype, but not at this particular locus. Interestingly, epg can lead to continuously quantitative variation even at a single locus, because epg can be quantitative rather than discrete. (3) A locus with a genetic variant contributing to disease phenotype (like possibility 1 above) may be modified in its penetrance by epigenetic factors (g_{dep}; Fig. 5c).

We think this is likely to be a common mechanism for integrating genetic and epigenetic variation. Finally, (4) genetic variation at a distant locus (g_{epg}) can affect both g_{dep} and epg (Fig. 5d).

CONCLUSION

We chose as our contribution to this meeting some recent efforts of our own group to try to advance the field of epigenomics, because of the history of the Cold Spring Harbor Symposium itself. The symposia were created in an effort to unite disparate fields both within and outside of biology, and also to infuse them with a quantitative approach (http://www.cshl.org/public/history.htm). This series partially spawned contemporary molecular biology. Until recently, epigenetics research, our own included, has been a somewhat eccentric stepchild of this molecular biology. However, recent advances in genomic technology should allow us to approach some of the great questions of human epigenetics, similar to what has been accomplished over the last two decades in human genetics. As occurred in "genomicizing" genetics, the initial stages were method intensive and relatively slow, but that should not deter us in similarly developing human epigenomics technology. Given the vast resources expended toward cancer genomics, the clear importance of epigenomics to cancer research, and the promise of epigenomics for other disorders, we cannot afford to wait.

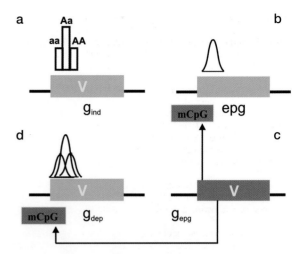

Figure 5. Interaction of genetic and epigenetic variation at a genomic level. (*a*) A gene (*green box*) can contain a sequence variant (V) that contributes to disease. Gene variants that are not influenced by epigenetic modification in their disease contribution (although epigenetics can contribute to normal function) are epigenetic independent (g_{ind}). The distribution of phenotype of a single locus will not be Gaussian in the absence of other factors, e.g., environmental. (*b*) Epigenetic variation can contribute to disease phenotype directly, independent of a genetic variation in the target gene (epg). Epigenetic variation itself is quantitative and thus can impart the quantitative nature to a trait, even at a single locus. (*c*) If the penetrance of a gene sequence variant is affected by epigenetic modification (mCpG), the gene is "epigenetic dependent" (g_{dep}). In this case, the genetic and epigenetic variation together could contribute to a Gaussian distribution, even at a single locus. Note that the epigenetic modification is drawn on the gene but it could be at some distance upstream or downstream from that gene. The epigenetic modification need not be methylation, which is drawn here for convenience. (*d*) A genetic variant that can influence this epigenetic modification (e.g., encoding a chromatin-modifying protein) is referred to as g_{epg}, and its influence is denoted by *arrows*. (Reprinted, with permission, from Bjornsson et al. 2004 [©Elsevier].)

ACKNOWLEDGMENTS

This work was supported by NIH grants HG003233 and CA65145 to A.P.F.

REFERENCES

Amir R.E., Van den Veyver I.B., Wan M., Tran C.Q., Francke U., and Zoghbi H.Y. 1999. Rett syndrome is caused by mutations in X-linked MECP2, encoding methyl-CpG-binding protein 2. *Nat. Genet.* **23**: 185.

Bachman K.E., Park B.H., Rhee I., Rajagopalan H., Herman J.G., Baylin S.B., Kinzler K.W., and Vogelstein B. 2003. Histone modifications and silencing prior to DNA methylation of a tumor suppressor gene. *Cancer Cell* **3**: 89.

Bakin A.V. and Curran T. 1999. Role of DNA 5-methylcytosine transferase in cell transformation by fos. *Science* **283**: 387.

Bennett-Baker P.E., Wilkowski J., and Burke D.T. 2003. Age-associated activation of epigenetically repressed genes in the mouse. *Genetics* **165**: 2055.

Bestor T.H. 2000. The DNA methyltransferases of mammals. *Hum. Mol. Genet.* **9**: 2395.

Bird A.P. 1986. CpG-rich islands and the function of DNA methylation. *Nature* **321**: 209.

Bjornsson H.T., Fallin M.D., and Feinberg A.P. 2004. An integrated epigenetic and genetic approach to common human disease. *Trends Genet.* **20**: 350.

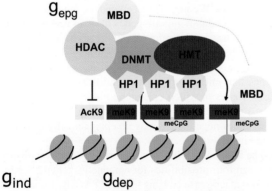

Figure 4. Cooperative and self-reinforcing organization of the chromatin- and DNA-modifying machinery responsible for gene silencing in normal and malignant cells. Histone (H3) modifications include lysine (K) acetylation (Ac) and lysine methylation (Me). Lysines at other positions are also modified. The HP1 protein recognizes MeK9 and, as this protein also binds the histone methyltransferase (HMT), heterochromatin can spread. Histone deacetylases (HDAC) deacetylate lysine residues as a prerequisite for their subsequent methylation. DNA methyltransferases (DNMT) participate in multiprotein complexes that contain HDACs and HMTs, and methyl-C binding proteins (MBD) can be loaded onto methylated DNA through their interactions with both HDACs and HMTs. Much of the evidence comes from studies of constitutive heterochromatin, but recent studies indicate similar interactions of genes silenced de novo in cancer cells. (Reprinted, with permission, from Feinberg and Tycko 2004 [©Nature Publishing Group; http://www.nature.com].)

Brachmann C.B., Sherman J.M., Devine S.E., Cameron E.E., Pillus L., and Boeke J.D. 1995. The SIR2 gene family, conserved from bacteria to humans, functions in silencing, cell cycle progression, and chromosome stability. *Genes Dev.* **9:** 2888.

Cattanach B.M. and Beechey C.V. 1990. Autosomal and X-chromosome imprinting. *Dev. Suppl.* **1990:** 63.

Chung J.H., Whiteley M., and Felsenfeld G. 1993. A 5′ element of the chicken beta-globin domain serves as an insulator in human erythroid cells and protects against position effect in *Drosophila*. *Cell* **74:** 505.

Cooney C.A. 1993. Are somatic cells inherently deficient in methylation metabolism? A proposed mechanism for DNA methylation loss, senescence and aging. *Growth Dev. Aging* **57:** 261.

Cui H., Horon I.L., Ohlsson R., Hamilton S.R., and Feinberg A.P. 1998. Loss of imprinting in normal tissue of colorectal cancer patients with microsatellite instability. *Nat. Med.* **4:** 1276.

Cui H., Onyango P., Brandenburg S., Wu Y., Hsieh C.L., and Feinberg A.P. 2002. Loss of imprinting in colorectal cancer linked to hypomethylation of H19 and IGF2. *Cancer Res.* **62:** 6442.

Cui H., Cruz-Correa M., Giardiello F.M., Hutcheon D.F., Kafonek D.R., Brandenburg S., Wu Y., He X., Powe N.R., and Feinberg A.P. 2003. Loss of IGF2 imprinting: A potential marker of colorectal cancer risk. *Science* **299:** 1753.

DeBaun M.R., Niemitz E.L., McNeil D.E., Brandenburg S.A., Lee M.P., and Feinberg A.P. 2002. Epigenetic alterations of H19 and LIT1 distinguish patients with Beckwith-Wiedemann syndrome with cancer and birth defects. *Am. J. Hum. Genet.* **70:** 604.

de Koning D.J., Rattink A.P., Harlizius B., van Arendonk J.A., Brascamp E.W., and Groenen M.A. 2000. Genome-wide scan for body composition in pigs reveals important role of imprinting. *Proc. Natl. Acad. Sci.* **97:** 7947.

Dorer D.R. and Henikoff S. 1994. Expansions of transgene repeats cause heterochromatin formation and gene silencing in *Drosophila*. *Cell* **77:** 993.

Ehrlich M., Gama-Sosa M.A., Huang L.H., Midgett R.M., Kuo K.C., McCune R.A., and Gehrke C. 1982. Amount and distribution of 5-methylcytosine in human DNA from different types of tissues of cells. *Nucleic Acids Res.* **10:** 2709.

Fedoriw A.M., Stein P., Svoboda P., Schultz R.M., and Bartolomei M.S. 2004. Transgenic RNAi reveals essential function for CTCF in H19 gene imprinting. *Science* **303:** 238.

Feinberg A.P. and Tycko B. 2004. The history of cancer epigenetics. *Nat. Rev. Cancer* **4:** 143.

Feinberg A.P. and Vogelstein B. 1983. Hypomethylation distinguishes genes of some human cancers from their normal counterparts. *Nature* **301:** 89.

Feinberg A.P., Cui H., and Ohlsson R. 2002. DNA methylation and genomic imprinting: Insights from cancer into epigenetic mechanisms. *Semin. Cancer Biol.* **12:** 389.

Gaudet F., Hodgson J.G., Eden A., Jackson-Grusby L., Dausman J., Gray J.W., Leonhardt H., and Jaenisch R. 2003. Induction of tumors in mice by genomic hypomethylation. *Science* **300:** 489.

Gius D., Cui H., Bradbury C.M., Cook J., Smart D.K., Zhao S., Young L., Brandenburg S.A., Hu Y., Bisht K.S., Ho A.S., Mattson D., Sun L., Munson P.J., Chuang E.Y., Mitchell J.B., and Feinberg A.P. 2004. Distinct effects on gene expression of chemical and genetic manipulation of the cancer epigenome revealed by a multimodality approach. *Cancer Cell* **6:** 361.

Gregory P.D. 2001. Transcription and chromatin converge: Lessons from yeast genetics. *Curr. Opin. Genet. Dev.* **11:** 142.

Haber J.E. 1998. Mating-type gene switching in *Saccharomyces cerevisiae*. *Annu. Rev. Genet.* **32:** 561.

Hendrich B. and Bickmore W. 2001. Human diseases with underlying defects in chromatin structure and modification. *Hum. Mol. Genet.* **10:** 2233.

Holliday R. and Pugh J.E. 1975. DNA modification mechanisms and gene activity during development. *Science* **187:** 226.

Huang C., Sloan E.A., and Boerkoel C.F. 2003. Chromatin remodeling and human disease. *Curr. Opin. Genet. Dev.* **13:** 246.

Ingrosso D., Cimmino A., Perna A.F., Masella L., De Santo N.G., De Bonis M.L., Vacca M., D'Esposito M., D'Urso M., Galletti P., and Zappia V. 2003. Folate treatment and unbalanced methylation and changes of allelic expression induced by hyperhomocysteinaemia in patients with uraemia. *Lancet* **361:** 1693.

Jimenez-Sanchez G., Childs B., and Valle D. 2001. Human disease genes. *Nature* **409:** 853.

Krawczak M., Ball E.V., and Cooper D.N. 1998. Neighboring-nucleotide effects on the rates of germ-line single-base-pair substitution in human genes. *Am. J. Hum. Genet.* **63:** 474.

Ledbetter D.H. and Engel E. 1995. Uniparental disomy in humans: Development of an imprinting map and its implications for prenatal diagnosis. *Hum. Mol. Genet.* (Spec No.) **4:** 1757.

Mackay D.J., Coupe A.M., Shield J.P., Storr J.N., Temple I.K., and Robinson D.O. 2002. Relaxation of imprinted expression of ZAC and HYMAI in a patient with transient neonatal diabetes mellitus. *Hum. Genet.* **110:** 139.

Moore T. and Haig D. 1991. Genomic imprinting in mammalian development: A parental tug-of-war. *Trends Genet.* **7:** 45.

Moulton T., Crenshaw T., Hao Y., Moosikasuwan J., Lin N., Dembitzer F., Hensle T., Weiss L., McMorrow L., Loew T., et al. 1994. Epigenetic lesions at the H19 locus in Wilms' tumour patients. *Nat. Genet.* **7:** 440.

Murata T., Kurokawa R., Krones A., Tatsumi K., Ishii M., Taki T., Masuno M., Ohashi H., Yanagisawa M., Rosenfeld M.G., Glass C.K., and Hayashi Y. 2001. Defect of histone acetyltransferase activity of the nuclear transcriptional coactivator CBP in Rubinstein-Taybi syndrome. *Hum. Mol. Genet.* **10:** 1071.

Ogonuki N., Inoue K., Yamamoto Y., Noguchi Y., Tanemura K., Suzuki O., Nakayama H., Doi K., Ohtomo Y., Satoh M., Nishida A., and Ogura A. 2002. Early death of mice cloned from somatic cells. *Nat. Genet.* **30:** 253.

Ohlsson R., Kanduri C., Whitehead J., Pfeifer S., Lobanenkov V., and Feinberg A.P. 2003. Epigenetic variability and the evolution of human cancer. *Adv. Cancer Res.* **88:** 145.

Okamoto K., Morison I.M., Taniguchi T., and Reeve A.E. 1997. Epigenetic changes at the insulin-like growth factor II/H19 locus in developing kidney is an early event in Wilms tumorigenesis. *Proc. Natl. Acad. Sci.* **94:** 5367.

Ordway J.M., Williams K., and Curran T. 2004. Transcription repression in oncogenic transformation: Common targets of epigenetic repression in cells transformed by Fos, Ras or Dnmt1. *Oncogene* **23:** 3737.

Orlando V. 2003. Polycomb, epigenomes, and control of cell identity. *Cell* **112:** 599.

Rainier S., Johnson L.A., Dobry C.J., Ping A.J., Grundy P.E., and Feinberg A.P. 1993. Relaxation of imprinted genes in human cancer. *Nature* **362:** 747.

Reik W., Dean W., and Walter J. 2001. Epigenetic reprogramming in mammalian development. *Science* **293:** 1089.

Rhee I., Jair K.W., Yen R.W., Lengauer C., Herman J.G., Kinzler K.W., Vogelstein B., Baylin S.B., and Schuebel K.E. 2000. CpG methylation is maintained in human cancer cells lacking DNMT1. *Nature* **404:** 1003.

Rhee I., Bachman K.E., Park B.H., Jair K.W., Yen R.W., Schuebel K.E., Cui H., Feinberg A.P., Lengauer C., Kinzler K.W., Baylin S.B., and Vogelstein B. 2002. DNMT1 and DNMT3b cooperate to silence genes in human cancer cells. *Nature* **416:** 552.

Riggs A.D. 1975. X inactivation, differentiation, and DNA methylation. *Cytogenet. Cell Genet.* **14:** 9.

———. 1989. DNA methylation and cell memory. *Cell Biophys.* **15:** 1.

Riggs A.D. and Pfeifer G.P. 1992. X-chromosome inactivation and cell memory. *Trends Genet.* **8:** 169.

Sandovici I., Leppert M., Hawk P.R., Suarez A., Linares Y., and Sapienza C. 2003. Familial aggregation of abnormal methylation of parental alleles at the IGF2/H19 and IGF2R differentially methylated regions. *Hum. Mol. Genet.* **12:** 1569.

Schoenherr C.J., Levorse J.M., and Tilghman S.M. 2003. CTCF maintains differential methylation at the Igf2/H19 locus. *Nat. Genet.* **33:** 66.

Sollars V., Lu X., Xiao L., Wang X., Garfinkel M.D., and Ruden D.M. 2003. Evidence for an epigenetic mechanism by which Hsp90 acts as a capacitor for morphological evolution. *Nat. Genet.* **33:** 70.

Spielman R.S., McGinnis R.E., and Ewens W.J. 1993. Transmission test for linkage disequilibrium: The insulin gene region and insulin-dependent diabetes mellitus (IDDM). *Am. J. Hum. Genet.* **52:** 506.

Statistics (National Center for Health Statistics). 2003. Health, United States, 2003. http://www.cdc.gov/nchs/hushtm.

Steenman M.J., Rainier S., Dobry C.J., Grundy P., Horon I.L., and Feinberg A.P. 1994. Loss of imprinting of IGF2 is linked to reduced expression and abnormal methylation of H19 in Wilms' tumour. *Nat. Genet.* **7:** 433.

Strahl B.D. and Allis C.D. 2000. The language of covalent histone modifications. *Nature* **403:** 41.

Strichman-Almashanu L.Z., Lee R.S., Onyango P.O., Perlman E., Flam F., Frieman M.B., and Feinberg A.P. 2002. A genome-wide screen for normally methylated human CpG islands that can identify novel imprinted genes. *Genome Res.* **12:** 543.

Tamashiro K.L., Wakayama T., Akutsu H., Yamazaki Y., Lachey J.L., Wortman M.D., Seeley R.J., D'Alessio D.A., Woods S.C., Yanagimachi R., and Sakai R.R. 2002. Cloned mice have an obese phenotype not transmitted to their offspring. *Nat. Med.* **8:** 262.

Tartof K.D. and Bremer M. 1990. Mechanisms for the construction and developmental control of heterochromatin formation and imprinted chromosome domains. *Dev. Suppl.* **1990:** 35-45.

van Overveld P.G., Lemmers R.J., Sandkuijl L.A., Enthoven L., Winokur S.T., Bakels F., Padberg G.W., van Ommen G.J., Frants R.R., and van der Maarel S.M. 2003. Hypomethylation of D4Z4 in 4q-linked and non-4q-linked facioscapulohumeral muscular dystrophy. *Nat. Genet.* **35:** 315.

Varambally S., Dhanasekaran S.M., Zhou M., Barrette T.R., Kumar-Sinha C., Sanda M.G., Ghosh D., Pienta K.J., Sewalt R.G., Otte A.P., Rubin M.A., and Chinnaiyan A.M. 2002. The polycomb group protein EZH2 is involved in progression of prostate cancer. *Nature* **419:** 624.

Woodson K., Flood A., Green L., Tangrea J.A., Hanson J., Cash B., Schatzkin A., and Schoenfeld P. 2004. Loss of insulin-like growth factor-II imprinting and the presence of screen-detected colorectal adenomas in women. *J. Natl. Cancer Inst.* **96:** 407.

Xu G.L., Bestor T.H., Bourc'his D., Hsieh C.L., Tommerup N., Bugge M., Hulten M., Qu X., Russo J.J., and Viegas-Pequignot E. 1999. Chromosome instability and immunodeficiency syndrome caused by mutations in a DNA methyltransferase gene. *Nature* **402:** 187.

Epigenetic Regulation in the Control of Flowering

J. MYLNE, T. GREB, C. LISTER, AND C. DEAN

Department of Cell and Developmental Biology, John Innes Centre, Norwich NR4 7UH, United Kingdom

The timing of flowering has significant consequences for the reproductive success of plants. Plants need to gauge when both environmental and endogenous cues are optimal before undergoing the switch from vegetative to reproductive development. To achieve this, a complex regulatory network has evolved consisting of multiple pathways that quantitatively regulate a set of genes—the floral pathway integrators (Simpson and Dean 2002). The activity of these genes causes the transition of the shoot apical meristem to reproductive development and the production of flowers rather than leaves. The major environmental cues that regulate flowering are day length (photoperiod), light quality, and temperature.

Temperature plays different roles in the flowering process. First, in all plants, ambient temperature during the growing season affects growth rate and time to flower (Blázquez et al. 2003). Second, many plant species, particularly those growing in high latitudes, require a prolonged period (weeks to several months) of cold temperature (2–10ºC) before they will flower. This process, called vernalization, is an adaptation to ensure that plants overwinter before flowering, thus aligning flowering and seed production with the favorable environmental conditions of spring. A requirement for vernalization has also been extensively selected for in many crop plants, including wheat, *Brassica* species (Fig. 1A), barley (Fig. 1B), and sugar beet (Fig. 1C). Indeed, breeders have been so successful at introducing a strong vernalization requirement into crops that many accessions of wheat and oilseed rape will not flower without vernalization and so, if sown in spring, would stay vegetative through the whole summer, only flowering the following year after winter. The small garden weed *Arabidopsis thaliana* is an ideal model to study the molecular basis of this process as many *Arabidopsis thaliana* ecotypes also have a vernalization requirement and these have a very different growth form without vernalization to the classic rapid-cycling ecotypes used in the lab (Fig. 2C). The physiological properties of vernalization in *Arabidopsis* match those of all other plants: It is a quantitative response, with increasing weeks of cold progressively accelerating flowering (Figs. 1E, 2A); it is reversible if vernalized seeds are subsequently subjected to a brief period of heat stress; and it is not graft-transmissible (unlike the photoperiodic flowering signal). Importantly, the vernalization response is mitotically stable; the prolonged cold stimulus happens at one stage of development with flowering often occurring many months later. Also, cuttings from vernalized plants "remember" that they have experienced winter and flower at the appropriately early time. However, the process is reset at some point during meiosis or seed development so that seedlings need to be vernalized each generation to align flowering with spring.

Figure 1. Vernalization requirement and response in crops and *Arabidopsis thaliana*. (*A*) Vernalization requirement is a trait bred in *Brassicas*. Pictured are cabbage (*Brassica oleracea ssp. capitata*) plants that were either vernalized as young plants (*left*) or not (*right*). (*B*) Barley varieties with a vernalization requirement (winter varieties, *left*) and a spring variety that has no vernalization requirement (*right*), all grown without vernalization. (*C*) Sugar beet plants require vernalization to flower, so are sown in late spring in order that they remain vegetative. An increasing problem in sugar beet fields is weed beet (the flowering plant in the picture), plants that have broken the vernalization requirement and flowered. (*D*) Both of the *Arabidopsis thaliana* laboratory strains Columbia or Landsberg *erecta* (pictured) do not have a strong vernalization requirement. Nonvernalized Landsberg *erecta* (*left*, non-vern) flower early with a similar number of leaves (vegetative phase) as vernalized L*er* (*right*, 6 weeks vernalization). (*E*) An active copy of the gene *FRIGIDA* (*FRI*) confers a strong vernalization requirement. *Arabidopsis thaliana* laboratory strains Columbia or Landsberg *erecta* (L*er*) have mutations in their *FRIGIDA* genes. Nonvernalized L*er* plants containing active *FRIGIDA* (*left*) are late flowering but this late flowering may be suppressed if vernalized (*right*). (*D,E*: Reprinted, with permission, from Henderson et al. 2003 [© Annual Reviews: www.annualreviews.org].)

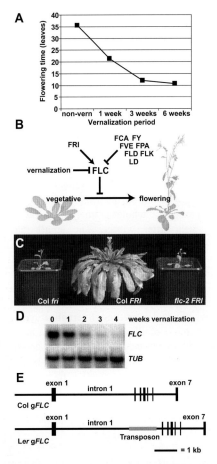

Figure 2. The role of FLC in vernalization requirement and response. (*A*) A response curve for *FRI*-containing Landsberg *erecta* (L*er*) showing the effect of increasing vernalization on flowering time. In all cases, SE was less than one leaf. (*B*) The pathways controlling expression of the MADS-box floral repressor FLC. The vernalization requirement conferred by mutations in *FCA, FY, FVE, FPA, FLC, LD, FLK*, or active *FRI* genes is by elevation of *FLC* levels that are suppressed by vernalization. (*C*) Wild-type *Arabidopsis thaliana* ecotype Columbia with inactive *fri* (*left*), Columbia containing an active copy of *FRI* (*middle*), Columbia containing active *FRIGIDA* but with a fast-neutron deletion of *FLC* (*flc-2* mutation) (*right*). (*D*) Northern blot showing *FLC* levels in *fca-1* declines with increasing vernalization periods. *fca-1* seeds were nonvernalized or vernalized for 1–4 weeks and then harvested after 14 days growth. The blot was reprobed with β-TUBULIN as a loading control (TUB). (*E*) The structure of *FLC* differs in Columbia and L*er*; a transposon in intron 1 of L*er* FLC reduces the steady-state *FLC* RNA levels in L*er* vs. Col backgrounds but does not affect the response to vernalization.

et al. 2004), and evidence that this occurs directly has been found for SOC1/AGL20 (Hepworth et al. 2002). Pathways that confer a vernalization requirement increase levels of *FLC* expression (Fig. 2B). This is the case for *FRIGIDA* (*FRI*), a major determinant of flowering-time variation and vernalization requirement in natural *Arabidopsis thaliana* ecotypes (Fig. 2C) (Johanson et al. 2000) and for a series of mutations whose late-flowering phenotype can be corrected by vernalization (e.g., *fca, fy*) (Koornneef et al. 1991). Vernalization antagonizes the activities of these pathways by decreasing levels of *FLC* expression (Fig. 2D). An in vivo analysis of various *FLC* transgenes carrying deletions has identified *cis*-elements in the *FLC* promoter and first intron required to mediate regulation by these different pathways (Sheldon et al. 2002). Different *Arabidopsis* ecotypes used in the molecular analysis show molecular variation at or near these *cis*-elements that influences *FLC* regulation. For example, compared to the Columbia allele (the accession for which the entire genomic sequence is available), C24 (used in the analyses by Sheldon et al. 2002) is missing one-half of a 30-bp tandem repeat in intron 1 (Gazzani et al. 2003) and Landsberg *erecta* (the background in which most of the original late-flowering mutants were identified) has a Mutator-like transposable element at the 3′ end of intron 1 (Fig. 2E) (Gazzani et al. 2003; Michaels et al. 2003). The presence of the Mutator-element restrains the upregulation of *FLC* levels in response to FRIGIDA or mutations such as *fca, fy* but does not affect the decrease in flowering time in response to vernalization.

vrn MUTANTS IDENTIFY GENES REQUIRED TO MAINTAIN *FLC* REPRESSION

In order to define the molecular events occurring during the downregulation of *FLC* levels during the cold and the subsequent maintenance of *FLC* repression, a series of mutants were defined that have a reduced response to vernalization based on a late flowering after vernalization phenotype (*vrn1* to *vrn7* for reduced *vern*alization) (Chandler et al. 1996). Another group isolated the *vin3* mutant (*v*ernalization *in*sensitive) (Sung and Amasino 2004), which is allelic to *vrn7*. Phenotypically *vrn* mutants look quite normal apart from delayed flowering (Fig. 3B), and they are all reduced in response to vernalization (Fig. 3C). Analysis of *FLC* levels in *vrn* mutants showed that the decrease in *FLC* after prolonged cold is almost as in wild type. However, unlike wild type, *FLC* levels increase again during subsequent growth in all the *vrn* mutants (Fig. 3D). This result demonstrates that the major function of the *VRN* genes is in the maintenance of the repressed *FLC* state.

VRN GENES IMPLICATE CHROMATIN REMODELING IN *FLC* REGULATION

The first *VRN* gene to be cloned, *VRN2*, encodes a protein with homology to Su(z)12, a zinc finger protein that is a component of the E(z)/ESC Polycomb complex in *Drosophila* and humans (Fig. 4A) (Gendall et al.

FLC IS A CENTRAL PLAYER IN THE VERNALIZATION PROCESS

Analysis of the genetic basis of vernalization requirement in natural accessions revealed that *FLC*, which encodes a MADS box transcriptional regulator, plays a central role in vernalization (Michaels and Amasino 1999; Sheldon et al. 1999). Increasing levels of FLC progressively delay flowering in a "rheostat"-like mechanism. FLC is thought to antagonize the activation of the floral pathway integrator genes (Simpson and Dean 2002; Boss

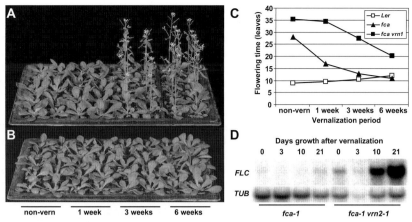

Figure 3. *VRN* genes are required to maintain suppression of *FLC* after vernalization. (*A*) Mutations of the *FCA* gene in L*er* (*fca-1* pictured) cause a strong vernalization requirement. Nonvernalized *fca-1* seeds (*left*, non-vern) produce late-flowering plants with a long vegetative phase. However as *fca-1* seeds are vernalized for progressively longer times (to *right*) their vegetative phase becomes much shorter and if vernalized long enough they produce similar numbers of leaves to wild-type L*er* (*right*, 6 weeks vernalization, see L*er* in Fig. 1*D*). (*B*) Vernalization response of *vrn1 fca-1* double mutants nonvernalized or vernalized as seeds for 1, 3, or 6 weeks. *vrn* mutants typically have a reduced response to vernalization and no pleiotropic phenotypes. (*C*) Flowering time is easily assayed as leaf number. L*er*, *fca-1*, and *vrn1 fca-1* mutants were nonvernalized or vernalized for 1, 3, or 6 weeks and their flowering time measured by counting the total number of leaves. In all cases, SE was less than two leaves. (*D*) The reduced vernalization response of *vrn* mutants is due to a failure to maintain suppression of *FLC* after vernalization during subsequent growth in long days. Seeds from *fca-1* and *vrn2 fca-1* were vernalized for 8 weeks and harvested immediately after the cold (0 days) and at different stages of subsequent growth. *FLC* levels in *fca-1* remain stably repressed after vernalization, but in *vrn2 fca-1 FLC* expression is initially repressed after vernalization but returns during subsequent growth. (*A,B*: Reprinted, with permission, from Henderson et al. 2003 [©Annual Reviews: www.annualreviews.org].)

2001; Czermin et al. 2002; Müller et al. 2002). The Polycomb complex has been shown in *Drosophila* to repress gene expression, including the HOX genes, by maintaining silent chromatin states. This appears to be analogous to the role VRN2 has in the stable repression of *FLC*.

VRN1 encodes a protein with two plant-specific DNA-binding domains that bind a range of DNA sequences in vitro (Levy et al. 2002). *VIN3* encodes a protein with a PHD (plant homeodomain) and a fibronectin III domain (Sung and Amasino 2004). PHD proteins have been found in chromatin complexes associated with histone deacety-

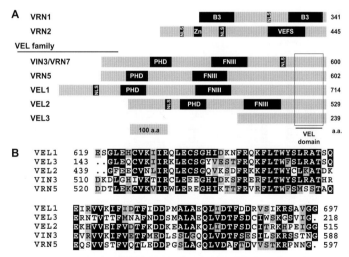

Figure 4. VRN proteins and the VEL (*VERNALIZATION5/VIN3*-Like) family. (*A*) Overview of genes involved in vernalization. Mutants in *VRN1*, *VRN2*, *VRN5*, and *VIN3/VRN7* were isolated in screens for mutants with reduced vernalization responses. VRN1 protein has strong in vitro DNA-binding properties and contains a putative nuclear localization sequence (NLS) and two plant-specific DNA-binding B3 domains. VRN2 shares domains with Su(z)12 and other PcG transcriptional regulators including a zinc finger (Zn), two NLSs, and a region highly conserved in VRN2, EMF2, FIS2, and SU(Z)12 (VEFS). *VRN5* and *VIN3/VRN7* are members of the VEL (*VERNALIZATION5/VIN3-Like*) gene family consisting of five members in *Arabidopsis*, which also includes *VEL1*, *VEL2*, and *VEL3*. With the exception of *VEL3*, all *VEL* genes carry a PHD (plant homeodomain) and a fibronectin III domain (FNIII). *VEL3* is much shorter than the other members but like all VEL proteins carries the highly conserved carboxyl terminus unique to the VEL family. (*B*) Pileup of the VEL domain unique to members of the VEL family. The gene numbers are At4G30200 for *VEL1*, At2G18870 for *VEL3*, At2G18880 for *VEL2*, At5g57380 for *VIN3*, and At3g24440 for *VRN5*.

lation and recently some PHD fingers have been found to function as nuclear phosphoinositide receptors, opening up the possibility that phosphoinositide signaling may play a role during the prolonged cold phase of vernalization (Gozani et al. 2003). During our screens for mutants with reduced vernalization responses we isolated a mutant *vrn5*, which has lesions in a *VIN3* homolog (T. Greb et al., unpubl.). *VRN5* and *VIN3* define a small gene family consisting of five members in *Arabidopsis* that we have called the *VEL* (VE*RNALIZATION5/VIN3-Like*) gene family (Fig. 4A). This opens up the question whether other *VEL* family members are involved in vernalization. With the exception of *VEL3* all members carry a PHD finger as mentioned above and a fibronectin III domain that has been characterized mainly in animals as a protein–protein interaction domain (Potts and Campbell 1996). At their carboxyl terminus all *VEL* genes carry a highly conserved domain that shows no similarities to any described protein motif from plants or animals (Fig. 4B). Therefore we term the *VEL*-specific carboxy-terminal domain the "VEL domain." According to the available annotation, *VEL3* is much smaller and only contains the VEL domain. Furthermore, no EST or cDNA clone is reported for *VEL3*. It is yet to be determined whether *VEL3* is a functional gene. Lesions in *VRN5* and *VIN3* have shown that at least two members of the *VEL* family are involved in the repression of *FLC*. Further experiments will clarify the role of other *VEL* genes in the vernalization response. A functional molecular analysis of the newly defined *VEL* domain will be especially interesting.

USE OF AN *FLC:luciferase* TRANSLATIONAL FUSION TO SELECT SPECIFICALLY FOR MUTANTS IN *FLC* REGULATION

In order to saturate for mutations disrupting the maintenance of *FLC* repression and to target genes required early in the cold-induced repression, we have established a mutagenesis strategy based on *FLC* expression rather than flowering time. The firefly luciferase coding sequence was cloned into exon 6 of an *FLC* clone from Columbia DNA resulting in a translational fusion (Fig. 5A). The *FLC:luciferase* fusion was transformed into *Arabidopsis thaliana* Landsberg *erecta*, which had previously been transformed with a functional *FRIGIDA* allele. Transformants were screened by genetic segregation and Southern blots for lines containing one simple T-DNA insert (to avoid complications induced by silencing phenomena associated with complex T-DNA integration). Of those, a line expressing a high level of *FLC:luciferase* in the presence of *FRI* was selected and the repression of *FLC:luciferase* expression after vernalization was confirmed by luciferase assays (Fig. 6C–E) and northern analysis (Fig. 6F).

In order to determine the functionality of the *FLC:luciferase* translational fusion it was also transformed into *flc-3 FRI*, a Columbia null *FLC* mutant with active *FRIGIDA* introgressed from a late-flowering *Arabidopsis* ecotype (Michaels and Amasino 1999). The flowering time of these transformants was compared to *flc-3 FRI* and wild-type Columbia containing active *FRI* (Fig.

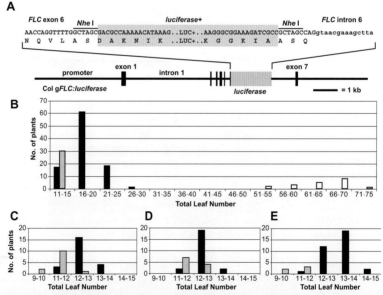

Figure 5. Structure of the genomic *FLC:luciferase* protein fusion construct and its functionality in planta. (*A*) The firefly luciferase (codon modified Luc+, Promega) coding region was inserted in-frame at the *Nhe*I site at the end of Columbia *FLC* exon 6 of a 12.1-kb *Sac*I *FLC* clone from Columbia DNA and cloned into an *Agrobacterium* binary vector pSLJ755I6 (Jones et al. 1992) and transformed into *Arabidopsis thaliana* Landsberg *erecta*, which had previously been transformed with a functional *FRIGIDA* allele. (*B*) The FLC:luciferase fusion has very little effect on flowering time. The flowering time of 100 transgenic *flc-3 FRI* lines containing the FLC:luciferase fusion (*black bars*) were slightly later than *flc-3 FRI* (*gray bars*), but far from the flowering time of *FLC FRI* (*open bars*). (*C–E*) Subsequent analysis of the next generation of three lines that flowered with 18, 23, and 25 leaves in the T1 (*C–E*, respectively) showed that *flc-3 FRI* containing the *FLC:luciferase* transgene (*black bars*) flowered at the same time or, at most, four leaves later than nontransgenic *flc-3 FRI* segregants (*gray bars*), indicating the FLC:luciferase fusion has very weak or negligible effect on flowering time.

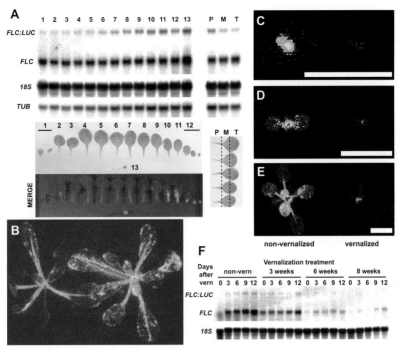

Figure 6. Analysis of *FLC:luciferase* mRNA and FLC:luciferase expression pattern in the L*er FRIGIDA FLC:luciferase* line. (*A*) Northern blot analysis of *FLC* and *FLC:luciferase* mRNA in leaves of different ages. Leaves from ten nonvernalized plants were grouped based on age (*1*, cotyledons; *2–12*, oldest to youngest leaves; and *13*, remaining plant center, root, and apical meristem) as well as into petioles, middle, and tips (P, M, T) and the RNA from these was probed with *FLC*, before stripping and reprobing with β-*TUBULIN* (*TUB*) and then *18S rDNA* probes. (*B*) L*er FRIGIDA FLC:luciferase* (*left*) and the same line containing a *vrn2* mutation (*right*) after a nonsaturating 4-week vernalization and 4 weeks of growth. (*C*) During vernalization, seeds germinate so the L*er FRIGIDA FLC:luciferase* seedling on the *right* was vernalized for 6 weeks and luciferase levels were imaged simultaneously with a nonvernalized seedling at a similar stage (3 days growth). Scale bar (for *C–E*), 1 cm. (*D*) Nonvernalized seedling (10 days, *left*) compared to the same line 7 days after vernalization (*right*). (*E*) Nonvernalized seedling (19 days, *left*) compared to the same line 16 days after vernalization (*right*). (*F*) Northern analysis to test the response of *FLC* and *FLC:luciferase* in the screen conditions. Seedlings were grown for 6 days before different vernalization treatments (non-vern, 3, 6, 8 weeks) and then harvested at different days of subsequent growth in long days.

5B–E). The flowering time of 100 different T1 transgenics suggested that the *FLC:luciferase* fusion is only weakly functional in delaying flowering time (Fig. 5B). Reexamination of the transgenic and nontransgenic segregants in the T2 generation of three individuals (Fig. 5C–E) revealed that, at most, the *FLC:luciferase* fusion may delay flowering by one or two leaves—thus the *FLC:luciferase* fusion mimics the endogenous *FLC* expression without significantly delaying flowering.

The FLC:luciferase signal was strongest in the apical meristem, stronger in younger leaves than older leaves, and stronger in petioles (leaf stems) than leaf blades (Fig. 6A,B,E). To determine if this was an artifact of the reporter or a consequence of differences in spatiotemporal expression of *FLC*, leaves of ten nonvernalized *FLC:luciferase FRI* plants were separated into age and tissue classes and RNA was isolated. Northern blots showed that both endogenous *FLC* and *FLC:luciferase* mRNA expression do not change dramatically with age or across the leaf blade (Fig. 6A). The most likely explanation for the difference in mRNA levels and strength of luciferase reporter signal is cell-size variation in the different tissues. For instance, in older leaves cells are larger and so if the amount of protein per cell is the same in young and old leaves, the reporter signal will be weaker in old leaves because the FLC:luciferase protein is spread over a wider area.

To obtain mutants defective in *FLC* suppression, L*er FLC:luciferase FRI* seeds were mutagenized with 0.3% ethyl methyl sulfonate (EMS) for 9 hours and M2 seeds were harvested from 25–35 M1 plants and pooled. M2 seeds were sterilized and sown on media lacking glucose. Plates were chilled for 2 days at 4°C to synchronize germination and then grown for 6 days at 20°C (16-hr photoperiod). After 6 days of growth, the seeds had germinated, the cotyledons had fully expanded, and the first true leaf pair was forming. The plates were transferred to a cabinet for 6 weeks of vernalization (4°C, 8-hr photoperiod) and were imaged for luciferase activity immediately after the cold to isolate mutants defective in the downregulation of FLC:luciferase expression during vernalization. The remaining plants were grown for a further 2 weeks at 20°C (16-hr photoperiod) to isolate mutants defective in the maintenance of low FLC:luciferase levels after vernalization.

The first mutants that were investigated were those which, in addition to their inability to suppress FLC:luciferase subsequent to vernalization, were late flowering, indicating that the suppression of endogenous *FLC* was also defective. Using this screen we isolated three new noncomplementing mutations we named *vrn7*. Subsequent mapping of this gene reduced it to a 420-kb region containing 56 genes, including the recently published *VIN3* gene (Sung and Amasino 2004). Sequencing confirmed that *VRN7* was *VIN3* (J. Mylne and C. Dean, unpubl.). In addition to the *vrn7/vin3* alleles, we also isolated six alleles of *vrn5*.

HISTONE MODIFICATIONS ASSOCIATED WITH EPIGENETIC SILENCING OF *FLC*

The maintenance of *FLC* repression following vernalization suggested that this gene may be epigenetically silenced. Epigenetic silencing of genes is mediated by numerous covalent modifications of both DNA and histones (Bird 2002; Fischle et al. 2003). Previous work on the epigenetic regulation of vernalization focused on the role of DNA cytosine methylation in control of vernalization (Sheldon et al. 1999), but to date there is no evidence of a direct link between DNA methylation and vernalization. However, recent data have demonstrated a more important role for histone modifications at the *FLC* locus during vernalization (Bastow et al. 2004; Sung and Amasino 2004). Specific residues of histone H3 tails can be modified by acetylation and methylation and changes in these modifications serve as part of a "histone code" specifying active or repressed gene activity states (Fischle et al. 2003). Vernalization was found to increase histone H3 deacetylation in the 5´-region of *FLC*, a modification typically associated with gene repression (Fischle et al. 2003; Sung and Amasino 2004). Vernalization also induced increases in methylation of histone H3 lysine residues 9 and 27, modifications associated with repressed gene states (Bastow et al. 2004; Sung and Amasino 2004). These marks are bound by further mediators of gene silencing, which include Heterochromatin protein 1 in animals (Orlando 2003); however, what these components are in plants is as yet unclear. Interestingly, the histone modifications that are observed at *FLC* are localized to specific regions of the gene (Bastow et al. 2004), colocalizing with sequences shown to be involved in the regulation of *FLC* by vernalization (Sheldon et al. 2002).

CHANGES IN THE HISTONE MODIFICATION IN *vrn/vin* MUTANTS SUGGEST A SEQUENCE OF EVENTS

The vernalization-mediated repression of *FLC* and coincident decrease in histone acetylation and increase in H3 K9 and K27 methylation observed in wild-type *Arabidopsis* are not found in *vin3* mutants (Sung and Amasino 2004). Similarly, in *vrn2* mutants increases in methylation of histone H3 at lysine residues 27 and 9 do not occur at the *FLC* locus during vernalization (Fig. 7) (Bastow et al. 2004; Sung and Amasino 2004). Unlike

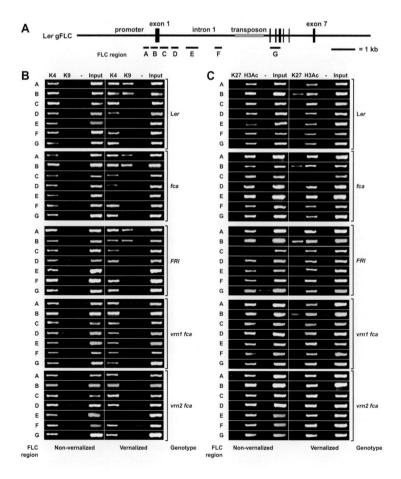

Figure 7. Histone modifications at *FLC* chromatin: PCR analysis of chromatin immunoprecipitates obtained with antibodies specific for H3 dimethyl K4, H3 dimethyl K9, H3 dimethyl K27, and acetylated H3 (Upstate). (*A*) Genomic structure of L*er FLC* and the regions tested (*A–G*) in ChIP assays. (*B*) H3 dimethyl K4 and H3 dimethyl K9 histone modifications at *FLC* are associated with vernalization and are dependent on the *VRN* genes. In ChIP analyses of *FLC* in L*er*, *fca-1*, L*er FRI*, *vrn1 fca-1*, and *vrn2 fca-1*, the vernalization-induced changes in H3 dimethyl K9 and H3 dimethyl K4, were lost in *vrn1 fca-1*; regions A and B did not show increased H3 dimethyl K9; and region E was precipitated by anti-H3 dimethyl K4. In vernalized *vrn2 fca-1*, no increase of H3 dimethyl K9 in regions A and B was found; however, the vernalization-induced loss of H3 dimethyl K4 from region E was maintained or restored only weakly. (*C*) The H3 dimethyl K27 histone modification at *FLC* is associated with vernalization and is dependent on VRN2, but not VRN1. The vernalization-induced H3 dimethyl K27 in region B was retained in *vrn1 fca-1*. In vernalized *vrn2 fca-1*, no increase of H3 dimethyl K27 in region B was found. Precipitation using the antibody to acetylated H3 is also shown. The reason for the low representation of region C only in nonvernalized L*er FRI* is not known. (*B*,*C*: Adapted from supplementary Figures 1 and 2 in Bastow et al. 2004.)

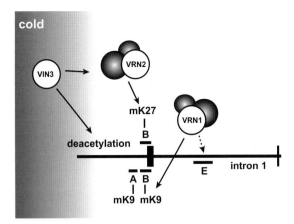

Figure 8. Potential sequence of events at the *FLC* locus during vernalization. Prolonged cold induces a series of events including *VIN3* expression and these result in *FLC* histone deacetylation. This is followed by VRN2 activity causing H3 dimethylation of lysine 27 (and possibly lysine 9) near the 5′ end of *FLC*, followed by, or in parallel with, VRN1 activity causing H3 dimethylation of lysine 9.

VRN2, VRN1 is required only for increases in histone H3 lysine 9 methylation in response to vernalization and not methylation of lysine 27 (Fig. 7) (Bastow et al. 2004; Sung and Amasino 2004). This suggests that it may function downstream or independently of VRN2 during *FLC* repression. Intriguingly, *VIN3* expression increases with cold and only significantly accumulates after a period of cold effective for vernalization (Sung and Amasino 2004). Therefore the sequence of events may entail cold activation of VIN3/VRN7 function resulting in histone deacetylation at *FLC*, followed by VRN2 activity causing H3 dimethylation of lysine 27 (and possibly lysine 9) in specific domains, followed by, or in parallel with, VRN1 activity causing H3 dimethylation of lysine 9 (Fig. 8). The double mutant phenotype of *vrn1* and *vrn2* is much stronger than either single mutant phenotype supporting parallel activities or multiple targets (T. Gendall and C. Dean, unpubl.).

Understanding how prolonged cold induces expression of *VIN3/VRN7* is a key question. The presence of the PHD finger, shown recently to act as a nuclear phosphoinositide receptor, also suggests a role for cold-induced signaling in the induction of VIN3/VRN7 activity. This is consistent with overexpression of *VIN3* being able to rescue the *vin3* mutation but not being able to lead to vernalization-independent early flowering (Sung and Amasino 2004). Constitutive expression of *VRN2* did not lead to pleiotropy or vernalization-independent flowering (T. Gendall and C. Dean, unpubl.); however, overexpression of VRN1 protein revealed a vernalization-independent function for VRN1 (Levy et al. 2002). When constitutively expressed, VRN1 caused *FLC*-independent early flowering and pleiotropic phenotypes, predominantly through a dramatic elevation of levels of the floral pathway integrator *FT*, and demonstrated that VRN1 requires vernalization-specific factors to target *FLC* (Levy et al. 2002).

INTEGRATION OF DIFFERENT FLOWERING PATHWAYS

The vernalization pathway is only one of many that regulate *FLC* levels. In turn, *FLC* is only one of many regulators of the floral pathway integrators, key regulators in the decision to flower. Therefore, an important aspect for future research will be establishing how the predominance of these different pathways is established and what changes these interactions during development or on the sensing of changed environmental cues. We are investigating this by analyzing how FCA/FY and FRIGIDA interact with the vernalization pathway (Fig. 2B). FCA and FY are two interacting proteins that promote flowering and encode an RNA-binding protein and a polyadenylation factor, respectively (Macknight et al. 1997; Simpson et al. 2003). The role of these posttranscriptional regulators in *FLC* expression raises some interesting questions. They may function to repress *FLC* directly or indirectly. However, they may also be components of a posttranscriptional mechanism that feeds back to regulate *FLC* transcriptionally, perhaps via chromatin regulation. There are now many examples where noncoding RNAs act to direct chromatin modifications (Verdel et al. 2004). Whatever their activities, both FCA/FY and vernalization function antagonistically with FRIGIDA, a novel protein that strongly upregulates *FLC* RNA levels (Johanson et al. 2000). Experiments are under way to address whether FRIGIDA also regulates *FLC* chromatin structure. The ability to screen for mutants in order to identify novel genes involved in flowering, together with the large resource of natural variants of *Arabidopsis thaliana* collected from a wide range of environments, will enable a detailed dissection of the molecular interactions of all these pathways.

ACKNOWLEDGMENTS

The authors would like to thank Tony Gendall for Northern blots images used in Figures 2D and 3D. Thanks to Judith Irwin for the image in Figure 1A, David Laurie for the image in Figure 1B, and Andrew Davis for the image in Figure 1C. Thanks also to Catherine Baxter for helpful comments on the manuscript. C.D. and C.L. are funded by a Biotechnology and Biological Sciences Research Council (BBSRC) core strategic grant to the John Innes Centre; J.M. is funded by European Commission grant QLK5-CT-2001-01412 and T.G. is funded by BBSRC grant 208/G15996.

REFERENCES

Bastow R., Mylne J.S., Lister C., Lippman Z., Martienssen R.A., and Dean C. 2004. Vernalization requires epigenetic silencing of *FLC* by histone methylation. *Nature* **427**: 164.

Bird A. 2002. DNA methylation patterns and epigenetic memory. *Genes Dev.* **16**: 6.

Blázquez M., Ahn J.H., and Weigel D. 2003. A thermosensory pathway controlling flowering time in *Arabidopsis thaliana*. *Nat. Genet.* **33**: 168.

Boss P.K., Bastow R.M., Mylne J.S., and Dean C. 2004. Multiple pathways in the decision to flower: Enabling, promoting, and resetting. *Plant Cell* (suppl.) **16**: S18.

Chandler J., Wilson A., and Dean C. 1996. *Arabidopsis* mutants showing an altered response to vernalization. *Plant J.* **10:** 637.

Czermin B., Melfi R., McCabe D., Seitz V., Imhof A., and Pirrotta V. 2002. *Drosophila* enhancer of Zeste/ESC complexes have a histone H3 methyltransferase activity that marks chromosomal Polycomb sites. *Cell* **111:** 185.

Fischle W., Wang Y., and Allis C.D. 2003. Histone and chromatin cross-talk. *Curr. Opin. Cell Biol.* **15:** 172.

Gazzani S., Gendall A.R., Lister C., and Dean C. 2003. Analysis of the molecular basis of flowering time variation in *Arabidopsis* accessions. *Plant Physiol.* **132:** 1107.

Gendall A.R., Levy Y.Y., Wilson A., and Dean C. 2001. The *VERNALIZATION 2* gene mediates the epigenetic regulation of vernalization in *Arabidopsis*. *Cell* **107:** 525.

Gozani O., Karuman P., Jones D.R., Ivanov D., Cha J., Lugovskoy A.A., Baird C.L., Zhu H., Field S.J., Lessnick S.L., Villasenor J., Mehrotra B., Chen J., Rao V.R., Brugge J.S., Ferguson C.G., Payrastre B., Myszka D.G., Cantley L.C., Wagner G., Divecha N., Prestwich G.D., and Yuan J. 2003. The PHD finger of the chromatin-associated protein ING2 functions as a nuclear phosphoinositide receptor. *Cell* **114:** 99.

Henderson I.R., Shindo C., and Dean C. 2003. The need for winter in the switch to flowering. *Annu. Rev. Genet.* **37:** 371.

Hepworth S.R., Valverde F., Ravenscroft D., Mouradov A., and Coupland G. 2002. Antagonistic regulation of flowering-time gene *SOC1* by CONSTANS and FLC via separate promoter motifs. *EMBO J.* **21:** 4327.

Johanson U., West J., Lister C., Michaels S., Amasino R., and Dean C. 2000. Molecular analysis of *FRIGIDA*, a major determinant of natural variation in *Arabidopsis* flowering time. *Science* **290:** 344.

Jones J.D.G., Shlumukov L., Carland F., English J., Scofield S.R., Bishop G.J., and Harrison K. 1992. Effective vectors for transformation, expression of heterologous genes, and assaying transposon excision in transgenic plants. *Transgenic Res.* **1:** 285.

Koornneef M., Hanhart C.J., and Van der Veen J.H. 1991. A genetic and physiological analysis of late flowering mutants in *Arabidopsis thaliana*. *Mol. Gen. Genet.* **229:** 57.

Levy Y.Y., Mesnage S., Mylne J.S., Gendall A.R., and Dean C. 2002. Multiple roles of *Arabidopsis VRN1* in vernalization and flowering time control. *Science* **297:** 243.

Macknight R., Bancroft I., Page T., Lister C., Schmidt R., Love K., Westphal L., Murphy G., Sherson S., Cobbett C., and Dean C. 1997. *FCA*, a gene controlling flowering time in *Arabidopsis*, encodes a protein containing RNA-binding domains. *Cell* **89:** 737.

Michaels S.D. and Amasino R.M. 1999. *FLOWERING LOCUS C* encodes a novel MADS domain protein that acts as a repressor of flowering. *Plant Cell* **11:** 949.

Michaels S.D., He Y., Scortecci K.C., and Amasino R.M. 2003. Attenuation of FLOWERING LOCUS C activity as a mechanism for the evolution of a summer-annual flowering behavior in *Arabidopsis*. *Proc. Natl. Acad. Sci.* **100:** 10102.

Müller J., Hart C.M., Francis N.J., Vargas M.L., Sengupta A., Wild B., Miller E.L., O'Connor M.B., Kingston R.E., and Simon J.A. 2002. Histone methyltransferase activity of a *Drosophila* Polycomb group repressor complex. *Cell* **111:** 197.

Orlando V. 2003. Polycomb, epigenomes, and control of cell identity. *Cell* **112:** 599.

Potts J.R. and Campbell I.D. 1996. Structure and function of fibronectin modules. *Matrix Biol.* **15:** 313.

Sheldon C.C., Conn A.B., Dennis E.S., and Peacock W.J. 2002. Different regulatory regions are required for the vernalization-induced repression of *FLOWERING LOCUS C* and for the epigenetic maintenance of repression. *Plant Cell* **14:** 2527.

Sheldon C.C., Burn J.E., Perez P.P., Metzger J., Edwards J.A., Peacock W.J., and Dennis E.S. 1999. The *FLF* MADS box gene: A repressor of flowering in *Arabidopsis* regulated by vernalization and methylation. *Plant Cell* **11:** 445.

Simpson G.G. and Dean C. 2002. *Arabidopsis*, the Rosetta stone of flowering time? *Science* **296:** 285.

Simpson G.G., Dijkwel P.P., Quesada V., Henderson I., and Dean C. 2003. FY is an RNA 3' end-processing factor that interacts with FCA to control the *Arabidopsis* floral transition. *Cell* **113:** 777.

Sung S. and Amasino R.M. 2004. Vernalization in *Arabidopsis thaliana* is mediated by the PHD finger protein VIN3. *Nature* **427:** 159.

Verdel A., Jia S., Gerber S., Sugiyama T., Gygi S., Grewal S.I., and Moazed D. 2004. RNAi-mediated targeting of heterochromatin by the RITS complex. *Science* **303:** 672.

Transposons and Tandem Repeats Are Not Involved in the Control of Genomic Imprinting at the *MEDEA* Locus in *Arabidopsis*

C. SPILLANE,*† C. BAROUX,* J.-M. ESCOBAR-RESTREPO,* D.R. PAGE,*
S. LAOUEILLE,† AND U. GROSSNIKLAUS*

Institute of Plant Biology and Zürich-Basel Plant Science Center, University of Zürich, 8008 Zürich, Switzerland; †Plant Molecular Genetics, Department of Biochemistry, University College Cork, Cork, Ireland

Transposons are selfish mobile DNA elements that can insert into nonhomologous target sites, thereby amplifying their copy number in the genome. Yet, transposition is considered tightly controlled, because unregulated amplification could have severe consequences for the fitness of the host organism. Nonetheless, transposons constitute ~45% of the human genome and ~80% of the maize genome. Transposable elements can act as both "attractors" and "mediators" of epigenetic regulation across the genome. The potential for transposons to show epigenetic activity leading to effects on phenotypic variation was first recognized by McClintock (see McClintock 1984) as "changes in phase." Some years ago it was proposed that transposons may be involved in a variety of epigenetic phenomena such as gene silencing, paramutation, and genomic imprinting (Martienssen 1996; Matzke et al. 1996). Indeed, there is growing evidence that transposons can act as epigenetic mediators of phenotypic variation. Here, we briefly review the role of transposons and repeated sequences in epigenetic gene regulation and investigate their potential role in controlling genomic imprinting at the *MEDEA* locus of *Arabidopsis*.

TRANSPOSONS AND REPETITIVE SEQUENCES AS EPIGENETIC ATTRACTORS AND MEDIATORS

Transposons are well known to be evolutionary drivers of chromosomal repatterning by reorganizing genome structure through transposition and by causing chromosomal rearrangements such as deletions, inversions, and translocations. But they can also modulate the transcription patterns of genes adjacent to the sites of transposon insertions (for review, see Kazazian 2004). Over evolutionary time most transposons have accumulated mutations that render them incapable of transposition, but many of their promoters remain active (Whitelaw and Martin 2001). Retrotransposons often have strong constitutive promoters that can affect the transcription of adjacent genes. Indeed, transposons can serve as alternative promoters for many mammalian genes (van de Lagemaat et al. 2003). For instance, transposon insertions proximal to genes can lead to overexpression causing hypermorphic alleles or to the production of chimeric transcripts that encode proteins with anti- or neomorphic activity. Such effects have been demonstrated for the *agouti* (Michaud et al. 1994; Argeson et al. 1996) and the mouse intracysternal A-type particle (IAP)-promoted *Mipp* gene (Chang-Yeh et al. 1993).

The effects described above may be viewed as a consequence of genetic changes since they rely on a read-through transcription from a transposon promoter and thus a restructured genome. However, transposons may also provide a link between genetic and epigenetic processes, via their activities as both transcribed genes and *cis*-acting repeats. In many organisms, duplicated or repetitive elements including transposons act as epigenetic "attractors" of mechanisms that lead to their inactivation or reduce their copy number (Matzke et al. 1996; Yoder et al. 1997; Whitelaw and Martin 2001; Lyon 2003). The mechanisms differ between repeat systems and organisms, but involve many of the classical epigenetic regulatory systems, such as DNA methylation, chromatin modification, and transcriptional interference.

DNA Methylation and Histone Modifications Regulate Transposon Activity

It has been proposed that cytosine DNA methylation acts primarily to suppress transcription from "intragenomic" parasitic elements (e.g., transposons) across the genomes of higher eukaryotes (Matzke et al. 1996; Yoder et al. 1997; Bestor 2003). DNA methylation can suppress transposition by making the transposon inaccessible to its transposase. In addition, transposon promoters can be inactivated by methylation either epigenetically or genetically, because of the increased frequency of $C \rightarrow T$ base transitions at methylated sites. Studies in both mammals and plants have demonstrated that demethylation of the genome can trigger remobilization of epigenetically silenced transposons (Walsh et al. 1998; Hirochika et al. 2000; Kato et al. 2003). It is likely that the majority of cytosine methylation found in eukaryotic genomes is associated with suppression of multicopy transposons and centromeric satellite DNA (both enriched for CpG content). This seems to be the case for plants and fungi where methylation is mainly associated with transposons and repetitive DNA, whereas in mammals coding regions also can be methylated (Martienssen and Colot 2001; Lippman et al. 2004). Transposon insertions may also affect adjacent genes via spreading of CpG methylation into "innocent bystander" genes, leading to their silencing. Yates and colleagues (1999) demonstrated such an effect for tandem B1 repetitive elements on the silencing of the adjacent adenine phosphoribosyltransferase *Aprt* gene in mice.

In recent years it has become clear that histone modifications play an important role in the epigenetic regulation of gene activity (for review, see Imhof 2003; Wang et al. 2004). There are mechanistic links between DNA methylation and histone modifications affecting chromatin structure (for review, see Tariq and Paszkowski 2004), and transposons can also be suppressed by targeting them for heterochromatin formation, which would act to suppress their transcription, mobility, and recombinational activity. Indeed, studies of transposons have demonstrated that transposon loci are subject to histone methylation (Rea et al. 2000; Gendrel et al. 2002). For instance, *Arabidopsis* mutants affecting epigenetic regulation were investigated for effects on the activity and inheritance of six transposon classes (Lippman et al. 2003). It was found that two distinct epigenetic mechanisms silence transposons and that transposon silencing complexes interact via histone modifications and RNA interference (RNAi). There is mounting evidence for a role for RNAi in chromatin modifications that regulate transposable element activity at centromeric heterochromatin (Volpe et al. 2002; Dawe 2003). In *Caenorhabditis elegans*, RNAi-deficient strains exhibit mobilization of endogenous transposons indicating that the RNAi machinery is involved in suppression of transposon activity (Tabara et al. 1999). The emerging picture is that heterochromatic regions can generate small RNAs that direct an RNAi-based modification of the chromatin in heterochromatic repeats and transposable elements (Bender 2004; Lippman and Martienssen 2004).

Transposons and Classical Epigenetic Phenomena

Metastable epialleles are alleles where the epigenetic state can switch and be mitotically inherited, yet the establishment of the epigenetic state is a probabilistic event (Rakyan et al. 2002). All metastable alleles that have been investigated at the molecular level have been shown to be associated with a transposon insertion (Rakyan et al. 2002). For instance, Rakyan and coworkers (2003) have demonstrated a role for retrotransposon-based regulation of the classical metastable mutant allele Axin-fused ($Axin^{Fu}$). The presence or absence of the $Axin^{Fu}$ phenotype, a kinked tail, correlated with differential DNA methylation of a retrotransposon within $Axin^{Fu}$. Affected transcripts arising adjacent to the retrotransposon long terminal repeat (LTR), usually containing a promoter, are considered as likely causes of the phenotype. A similar case was described for an *agouti* allele, where the insertion of an IAP retrotransposon into the upstream region caused a range of phenotypes that showed partial epigenetic maternal inheritance due to incomplete erasure of the epigenetic modification at *agouti* (Morgan et al. 1999).

Transposons are also proposed to play a role in mammalian X-chromosome inactivation. Interspersed repeats, in particular long interspersed nucleotide elements (LINEs), have been suggested as features that act as attractors of the X-inactivation machinery (Lyon 2003). In support of this model is the observation that L1 LINE content is lower in regions of the X chromosome containing genes that escape inactivation (Bailey et al. 2000).

Several years ago it was proposed that transposons and repetitive elements may be mechanistically linked to the phenomenon of paramutation (Martienssen 1996; Matzke et al. 1996; Della Vedova and Cone 2004). Paramutation is an allelic interaction that results in meiotically heritable changes in gene expression (Brink 1973). By analyzing the physical structure of 28 haplotypes at the *red1* (*r1*) locus of maize, a strict correlation of paramutability (the ability to become silenced) and structural features could be established (Walker and Panavas 2001). The *r1* locus is complex, often containing several *r1* gene copies encoding helix-loop-helix transcription factors. All paramutable alleles contain an *S* subcomplex that includes two *S* genes (*r1* homologs) forming a head-to-head inverted repeat and a *q* gene fragment (homologous to the *r1* promoter). These elements of the *S* subcomplex usually contain sequences derived from a *Doppia* transposable element. The paramutagenicity (the ability to cause silencing) of *r1* haplotypes, on the other hand, does not correlate with structural features but paramutagenic alleles show consistently higher levels of DNA methylation (Walker and Panavas 2001). At the *booster1* (*b1*) locus, which is also subject to paramutation, Stam and coworkers (2002) have shown that tandem repeats of an 853-bp sequence located ~100 kb upstream of the *b1* gene are required for paramutagenicity. A further link between paramutation and transposons is illustrated by the fact that paramutation at three different loci in maize and silencing of *Mutator* transposable elements are coordinately affected in certain inbred backgrounds (Walbot 2001) and by mutations at the *modifier of paramutation 1* (*mop1*) locus (Dorweiler et al. 2000; Lisch et al. 2002).

EPIGENETIC REGULATION OF TRANSPOSONS AND IMPRINTED GENES

Genomic imprinting refers to an epigenetic phenomenon where paternally and maternally inherited alleles are expressed differentially after fertilization. Most imprinted genes in mammals display parent-of-origin-specific methylation patterns. Compelling evidence that transposons are not neutral genomic parasites but actively influence epigenetic gene regulation poses the question whether they play a role in genomic imprinting as well. Indeed, the epigenetic regulation of some transposons is analogous to that of imprinted genes whereby an autosomal locus can be differentially expressed depending on the sex of the parent from which it was inherited. In mammals, L1 elements and IAP retroviruses are methylated when inherited paternally, but not methylated when inherited maternally (Sanford et al. 1987). The opposite situation occurs for mammalian Alu elements (Rubin et al.

1994). Early during mouse embryogenesis, genome-wide DNA demethylation occurs, followed by de novo remethylation. For most imprinted genes, the unmethylated allele escapes postimplantation de novo methylation; differentially methylated transposons, however, do not (Yoder et al. 1997; Walsh et al. 1998).

A survey of more than 30 imprinted genes brought the first correlative evidence for a possible involvement of transposons in genomic imprinting. Neumann et al. (1995) highlighted that one of the characteristics of known imprinted genes was that they tended to be enriched in short direct repeats. In mammals, the accumulation of short interspersed nucleotide elements (SINEs) is constrained in promoter regions of imprinted genes, whereas L1 LINE transposons preferentially accumulate in the vicinity of paternally expressed imprinted genes (Greally 2002; Fazzari and Greally 2004). Furthermore, this dual feature of imprinted regions points toward a mechanistic role of the transposons, where paucity in one type (the SINEs) would ensure that imprinted regions are isolated in a distinct genomic compartment, potentially enabling distinct regulatory mechanisms, and the other type (the L1 LINEs), being asymmetrically distributed, would provide a genomic signature to undergo preferential maternal or paternal silencing in the gametes. This remains a postulate, which needs to be experimentally verified, but it opens the field of investigation toward elucidating the mechanisms of genomic imprinting.

GENOMIC IMPRINTING IN PLANTS

Long before genomic imprinting was studied in mammals, Kermicle (1970) demonstrated that specific alleles of the maize *r1* locus are regulated by genomic imprinting (for review, see Kermicle 1994; Baroux et al. 2002). Maternal and paternal alleles of several other maize and, more recently, *Arabidopsis* genes were shown to be differentially expressed during seed development. However, of all plant genes suggested to be regulated by genomic imprinting, the maternally inherited allele is active, and most of these genes are already expressed prior to fertilization. Thus, although unlikely, the differential steady-state levels of maternally and paternally derived transcripts might be due to expression of the maternal allele prior to fertilization and not due to active expression postfertilization. In such cases, a clear demonstration of genomic imprinting requires not only the detection of differential expression levels of maternally and paternally derived transcripts but also an assay showing that the maternal allele is actively transcribed in at least one of the products of double fertilization (embryo and endosperm).

Regulation by genomic imprinting was proposed for several maize genes that are active in the endosperm, including specific alleles of *r1* (Kermicle 1970), the *delta zein regulator1* gene (*dzr1*; Chaudhuri and Messing 1994), and specific α-tubulin and *zein* genes (Lund et al. 1995a,b). Except for *dzr1*, which is not yet cloned, the high expression levels of maternal genes correlated with hypomethylation of the maternally inherited alleles (for review, see Alleman and Doctor 2000; Baroux et al. 2002). For *r1*, genomic imprinting could unambiguously be demonstrated in an elegant genetic analysis that excluded a prefertilization component (for instance, long-lived, stored transcripts [Kermicle 1970]). For the other potentially imprinted loci, early expression in ovules was not analyzed; therefore, a prefertilization cause for differential transcript levels cannot unambiguously be excluded. Recently, three additional endosperm-specific maize genes have been described that show differential expression levels during seed development but are not expressed at all prior to fertilization. Thus, *ZmFie1* (Springer et al. 2002; Danilevskaya et al. 2003), *no apical meristem related protein1* (*nrp1*; Guo et al. 2003), and *maternally expressed gene1* (*meg1*; Gutiérrez-Marcos et al. 2004) are clearly regulated by genomic imprinting.

The *FIS*-Class Genes in *Arabidopsis*

In *Arabidopsis*, the *MEDEA* (*MEA*) gene, which was isolated in a screen for gametophytic maternal effect mutations (Grossniklaus et al. 1998), was shown to be regulated by genomic imprinting. Differential expression levels of transcripts derived from the two parental alleles were demonstrated by allele-specific reverse transcriptase polymerase chain reaction (RT-PCR) (Kinoshita et al. 1999; Vielle-Calzada et al. 1999) and active transcription after fertilization was shown using an in situ hybridization method analogous to RNA-FISH (fluorescence in situ hybridization) (Vielle-Calzada et al. 1999). This analysis showed that only two of the three *MEA* copies present in the endosperm are actively transcribed. Although the latter method could be applied only to endosperm nuclei, allele-specific quantitative PCR analyses showed that paternally derived transcripts were not detectable at any stage of seed development (up to 10 days after pollination; Page 2004). Thus, the *MEA* gene, which is expressed in both embryo and endosperm (Vielle-Calzada et al. 1999), is likely regulated by genomic imprinting in both fertilization products.

MEA encodes a *Polycomb* group (PcG) protein with high similarity to *Enhancer of zeste* from *Drosophila* (Grossniklaus et al. 1998). Several other independent screens identified additional loci with similar parent-of-origin-dependent phenotypes (Ohad et al. 1996; Chaudhury et al. 1997; Guitton et al. 2004). This class of mutations is referred to as the *fis* class (Grossniklaus et al. 2001) and includes the *MEA* (Grossniklaus et al. 1998), *FERTILIZATION-INDEPENDENT ENDOSPERM* (*FIE;* Ohad et al. 1999), *FERTILIZATION-INDEPENDENT SEED2* (*FIS2;* Luo et al. 1999), and *MSI1* (Köhler et al. 2003) genes. These *FIS* proteins form a multiprotein complex that is analogous to the E(z)-Esc complex of *Drosophila* and the Enx-Eed complex of mammals (Köhler et al. 2003; for review, see Reyes and Grossniklaus 2003). Differential expression of maternally and paternally inherited alleles has been described for the *FIE* and *FIS2* genes (Luo et al. 2000; Yadegari et al. 2000). However, since both of these genes are also expressed prior to fertilization (Luo et al. 2000: Spillane et al. 2000; Yadegari et al. 2000) and active transcription at postfertilization stages has not been investigated, regulation by genomic imprinting has not been demonstrated unambiguously.

Transposons and Genomic Imprinting at the FWA Locus

Recently, genomic imprinting was reported for an additional *Arabidopsis* gene: the *FWA* locus (Kinoshita et al. 2004), which was originally identified as a late flowering mutant (Koornneef et al. 1991). Late flowering in the *fwa* epimutant is caused by ectopic expression of the *FWA* gene due to hypomethylation of repeats upstream of the transcriptional start site (Soppe et al. 2000). Kinoshita et al. (2004) found that *FWA* is expressed in the endosperm and could detect transcripts derived only from the maternally inherited allele, suggesting the gene may be regulated by genomic imprinting. However, as *FWA* is also expressed prior to fertilization and a conclusive test for active transcription at postfertilization stages has not been performed, the definitive proof is missing. Nevertheless, it is highly likely that *FWA* is regulated by genomic imprinting because it also shares upstream regulators with *MEA*. On the one hand, the maternal activity of both *MEA* and *FWA* depends on *DEMETER* (*DME*), a gene that was identified based on its phenotype that is similar to that of *mea* (Choi et al. 2002; Guitton et al. 2004). *DME* encodes a DNA-glycosylase homolog whose activity is required for an active maternal *MEA* allele (Choi et al. 2002, 2004). On the other hand, the DNA methyltransferase *MET1* is a regulator of both genes. However, while *MET1* was proposed to act antagonistically to *DME* on the maternal *MEA* allele (Xiao et al. 2003), it repressed the paternal *FWA* allele (Kinoshita et al. 2004), while *MET1* does not seem to affect the paternal *MEA* allele (Luo et al. 2000).

Interestingly, at the Symposium, Robert Martienssen (see Martienssen et al., this volume) reported a link between the imprinted control of *FWA* expression and transposons. His group investigated McClintock's hypothesis (1952) that transposons ("controlling elements") might reside in heterochromatic regions (for instance, heterochromatic knobs), but also exercise regulatory functions across the genome. In support of this hypothesis, Lippman and coworkers (2004) have demonstrated that heterochromatin in *Arabidopsis* is determined by transposons and related tandem repeats, which are epigenetically regulated by the chromatin remodeling ATPase DDM1. It was further shown that transposons can exercise epigenetic regulation of adjacent genes and that this was the likely explanation for the epigenetic inheritance patterns observed at the imprinted *FWA* gene. In addition, small interfering RNAs (siRNAs) associated with the epigenetically regulated transposon type were found in both heterochromatin and the promoter of the *FWA* locus. It was proposed that the transposon brings the *FWA* locus under the control of *DDM1* and is responsible for its epigenetic regulation (Lippman et al. 2004).

Over the last few years, we have investigated the potential role of transposable elements and repeated sequences in the regulation of the imprinted *MEA* locus. Unlike for *FWA*, no evidence for a role of repeats was found, suggesting that different mechanisms are responsible for the regulation of *MEA* and *FWA* by genomic imprinting.

EXPERIMENTAL PROCEDURES

Plant Material and Growth Conditions

The *mea-1* mutant line in Ler-0 genetic background used was previously described (Grossniklaus et al. 1998). All *Arabidopsis* ecotypes were obtained from the *Arabidopsis* Biological Resource Center (ABRC) at Ohio State University. Seeds were surface sterilized using 2% sodium hypochlorite and allowed to germinate on Murashige and Skoog (MS) medium (Duchefa) supplemented with 10 g/l of sucrose, 8 g/l of agar prior to transfer of seedlings to soil. To generate interecotype F1 hybrids between the *mea-1* mutant and the ecotypes Ler-0, Yo-0, and Kb-0, each ecotype was crossed with pollen from a *mea-1* plant and the hybrid F1 progeny seeds were selected on MS medium containing 50 mg/L of kanamycin (Sigma) as described above. Interecotype F1 *mea-1* seedlings displaying kanamycin resistance were chosen for further analysis. Seedlings of both ecotypes and F1 interecotype progeny were transplanted to "ED73 mit Bims" soil (Tränkle Einheiteserde) and transferred to a growth chamber with 70% humidity and a day/night cycle of 16 hours light at 21°C and 8 hours dark at 18°C.

A set of 50 evolutionarily divergent ecotypes (comprising a core collection kindly provided by Tom Mitchell-Olds, MPI Jena) was chosen for genetic and molecular analyses. These ecotypes were (stock center accession numbers in parentheses) Ler-0 (CS20), Mh-0 (CS904), Blh-1 (CS1030), Cit-0 (CS1080), Co-1 (CS1084), Col-0 (CS1092), Cvi-0 (CS1096), Di-0 (CS1106), Ei-2 (CS1124), Est-0 (CS1148), Fe-1 (CS1154), Ga-0 (CS1180), Gr-1 (CS1198), Gu-0 (CS1212), Ha-0 (CS1218), Ita-0 (CS1244), Kas-1 (CS1264), Kb-0 (CS1268), Kil-0 (CS1270), Kin-0 (CS1272), Le-0 (CS1308), L1-0 (CS1338), Lo-1 (CS1346), Lz-0 (CS1354), Me-0 (CS1364), Nd-0 (CS1390), Np-0 (CS1396), Nok-0 (CS1398), Pa-1 (CS1438), Pla-0 (CS1458), Pog-0 (CS1476), Rsch-0 (CS1490), Ru-0 (CS1496), Sah-0 (CS1500), Ta-0 (CS1548), Tu-1 (CS1568), Uk-1 (CS1574), Ws-0 (CS1602), Wt-1 (CS1604), Yo-0 (CS1622), Wl-0 (CS1630), Wei-0 (CS3110), RLD1 (CS913), XX-0 (N1618), Mt-0 (N1380), Ko-2 (N1288), C24 (N906), CS22491, CS22495, CS22484, CS22493, and Hodja.

Molecular Biology

Genomic DNA template for PCR and sequence analysis of each of the ecotypes was extracted as described (Edwards et al. 1991). To determine whether the *MEA-AtREP2* helitron was present in each ecotype, a PCR assay was developed using two primers spanning the *MEA-AtREP2* insertion site. The primers used were MEAP RAD S1: 5′-GATATGTTGG GTCCGTCGG-3′ and MEAP RAD AS1: 5′-CTATGCT CGTCTAGCTAC-3′.

For the PCR analysis spanning the *MEA-AtREP2* helitron region, the PCR conditions consisted of annealing temperature of 55°C (15 sec) and extension time of 30 seconds for 30 cycles.

A series of four different combinations of primer pairs were used for PCR spanning the MEA-ISR region. Primer pair 1 consisted of MEA S40: 5′-GCTATGGACCAGAACATGC-3′ and MEA AS42: 5′-AGGGTTTGCTCTTGAAGTCAG-3′. Primer pair 2 consisted of MEA3′REP1: 5′-GTGGCTGTAGCTTACGAAAGG-3′ and MEA AS42: 5′-AGGGTTTGCTCTTGAAGTCAG-3′. Primer pair 3 consisted of MEA 3′REP1F: 5′-GTGGCTGTAGCTTACGAAAGG-3′ and MEA 3′REP2R: 5′-GTTTGGATTCGTGATATACACC-3′. Primer pair 4 consisted of MEA S40: 5′-GCTATGGACCAGAACATGC-3′ and MEA3′REP2R: 5′-GTTTGGATTCGTGATATACACC-3′. For the PCR analysis spanning the MEA-ISR, the PCR conditions were annealing temperature of 50°C (15 sec) and extension time of 90 seconds for 30 cycles. *Bam*HI restriction analysis of PCR products was conducted using standard protocols on MEA-ISR PCR products obtained from the following ecotypes: Bla-1, Bla-14, Blh-1, C24, Cit-0, Co-1, Co-2, Col-0, Col-1, Cs22493, Cs22495, Ct-1, Cvi-0, Di-0, Ei-2, Est-0, Estland, Fr-2, Ga-0, Gr-1, Gu-0, Ha-0, Hodja, Kas-1, Kb-0, Kil-0, Kin-0, KN-0, Le-0, L*er*-0, Lo-1, Me-0, Ms-0, Mt-0, Nd-0, No-0, Nok-0, Np-0, Pa-1, Pog-0, RLD1, Rsch-0, Ru-0, Sah-0, Sf-1, Te-0, Tsu-0, Tu-1, UK-1, Wl-0, Wil-1, Ws, Ws-0, Wt-1, XX-0, and Yo-0. All of these ecotypes produced a PCR product with at least one primer pair combination spanning the MEA-ISR.

Bioinformatics

The *MEA-AtREP2* helitron element was initially identified using the repeat element mapping program Repeat View (http://www.itb.cnr.it/webgene/). The CpG islands were identifed using the CpG islands prediction program available on the Webgene Web site (http://www.itb.cnr.it/webgene/). The *MEA* and *FWA* tandem repeats were identified by the Tandem Repeats Finder program (http://c3.biomath.mssm.edu/trf.basic.submit.html). The large tandem duplication in the *MEA* upstream region was identified from restriction enzyme profiles of the *MEA* locus that exhibited similar restriction patterns indicative of a tandem duplication.

RESULTS AND DISCUSSION

Transposons and Tandem Repeats at the *MEA* and *FWA* Loci

To determine whether the two imprinted loci *MEA* and *FWA* in *Arabidopsis* contained any common structural features that could be associated with epigenetic regulation (i.e., genomic imprinting) we used a range of bioinformatics tools for comparative purposes. The sequences analyzed were generated by the *Arabidopsis* Genome Initiative (2000) and represent the Colombia (Col-0) accession.

In mammals, the regulatory regions of imprinted genes frequently contain a combination of features including tandem repeats associated with differentially methylated CpG islands (Moore 2001). Transposable elements are found in the upstream regions of both *FWA* and *MEA*, although each imprinted gene is proximal to a different type of transposable element in their upstream regions. While *FWA* contains an *AtSINE3* element 980 bp upstream of its start codon, the closest transposable element to the imprinted *MEA* locus is an *AtREP2* helitron found 4363 bp upstream of the *MEA* start codon (Fig. 1A).

The imprinted *MEA* and *FWA* loci also contain tandem repeats (Fig. 1). The *MEA* upstream region contains two tandemly duplicated segments (~1450 bp and ~1690 bp) spanning ~3140 bp. However, no analogous large tandem

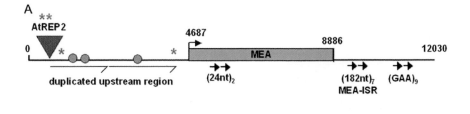

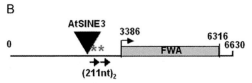

Figure 1. *cis*-elements in known imprinted *Arabidopsis* genes as candidate imprinting control elements (ICEs). (*A*) *MEA* locus (Col-0 accession) with *AtREP2* helitron transposon (*triangle*) and CpG islands (*circles*). Methylated regions upstream of *MEA* as previously reported for stamen and entire seeds (Xiao et al. 2003) are represented as *asterisks*. MEA-ISR refers to MEA intergenic subtelomeric repeat region consisting of eight tandem repeats of 182 nucleotides in Col-0. Other repeats at the MEA locus are a tandem 24-nucleotide repeat and a trinucleotide GAA repeated nine times. (*B*) *FWA* locus (Col-0 accession) with a *AtSINE3* transposon (*triangle*) and tandem 211-nucleotide repeat in the upstream region. Methylated regions as previously reported for embryo, seed coat, and vegetative tissues (Soppe et al. 2000; Kinoshita et al. 2004) are represented by *asterisks*.

duplications are found at the imprinted *FWA* locus. The *MEA* locus contains three different types of tandem repeats: (i) a 24-nucleotide tandem doublet in the third exon of the *MEA* ORF, (ii) a downstream tandem repeat region of seven 182 nucleotide repeats (182nt)$_7$, and (iii) a downstream GAA trinucleotide (GAA)$_9$. The *FWA* locus contains less extensive tandem duplications than the *MEA* locus and contains simply a 211-nucleotide tandem doublet 977 bp upstream of the *FWA* start codon. We also determined whether each of the two imprinted loci contain CpG islands and found that the imprinted *MEA* locus contains 3 CpG islands 3331 bp, 3028 bp, and 1215 bp upstream of its start codon, but the imprinted *FWA* locus contains no CpG islands that we could detect using the GpC island prediction program.

The *AtREP2* Helitron Is Not Required for Imprinting at the *MEA* Locus

The 5′ upstream region of *MEA* contains a nonautonomous *AtREP2*-type *Helitron* transposable element (Kapitonov and Jurka 2001). Helitrons are a novel class of eukaryotic DNA transposons that can transpose by rolling circle replication (Kapitonov and Jurka 2001). Both autonomous and nonautonomous helitrons can be found in eukaryotic genomes. The autonomous rolling-circle (RC) helitrons encode a 5′-to-3′ DNA helicase and nuclease/ligase similar to those encoded by known RC replicons. In addition, numerous nonautonomous RC helitron derivatives can be found throughout some eukaryotic genomes. In *C. elegans*, helitrons (autonomous and nonautonomous) can constitute ~2% of the genome (Kapitonov and Jurka 2001). The *MEA-AtREP2* element is a nonautonomous helitron.

We investigated whether presence of the *AtREP2* transposable element 4363 bp upstream of the *MEA* start codon (Fig. 1A) in the Col-0 and Ler-0 ecotypes is correlated with genomic imprinting at the *MEA* locus taking advantage of the natural variation resources available in *Arabidopsis*. It is known that transposons can accumulate to varying extents between *Arabidopsis* accessions. For instance, a few classes of DNA transposons have been found to be completely absent from some *Arabidopsis* accessions, yet they are prevalent in others. Examples include the low-frequency *Tag1* element absent from Colombia and WS (Frank et al. 1998) and the CACTA family transposons absent from C24 (Kato et al. 2003). However, it is not known whether the RC helitrons display any significant polymorphism between *Arabidopsis* ecotypes.

To identify ecotypes lacking the *MEA-AtREP2* helitron insertion, we used a PCR-based strategy to amplify across the *MEA-AtREP2* insertion site in 33 evolutionarily divergent *Arabidopsis* accessions, four of which are shown in Figure 2. Out of 33 ecotypes screened from an *Arabidopsis* core collection, we identified nine ecotypes where an ~250-bp PCR product was observed (see example in Fig. 2B for Yo-0 and Kb-0). As this size is smaller than the expected ~850 bp, these were candidates where the *MEA-AtREP2* insertion may be absent. To confirm that the *At-REP2* helitron was indeed absent in these accessions, we sequenced of the PCR product spanning the *MEA-AtREP2* insertion site (Fig. 2C). This analysis showed that several accessions, including Yo-0 and Kb-

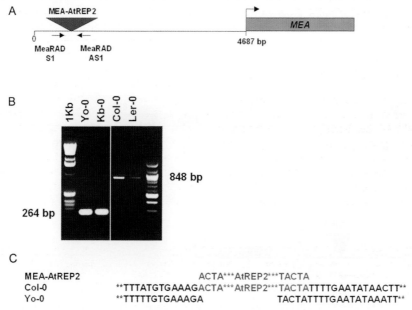

Figure 2. *MEA-AtREP2* helitron is absent in accessions where *MEA* remains imprinted. (*A*) Schematic representation of *MEA* locus indicating *MEA-AtREP2* helitron transposon upstream of *MEA*. Primers (MeaRAD S1 and Mea RAD AS1) spanning the *MEA-AtREP2* insertion are represented by *arrows*. (*B*) Gel electrophoresis of PCR products of four different ecotypes, indicating that this *MEA-AtREP2* is not present in the accessions Yo-0 and Kb-0. (*C*) Sequence analysis across the *MEA-AtREP2* insertion site was used to confirm that the *MEA-AtREP2* helitron was absent in accessions such as Yo-0.

0, completely lack this *AtREP2* helitron.

To link these data with genomic imprinting we investigated whether the accessions lacking the *AtREP2* transposons still contain an imprinted *MEA* locus. A genetic study performed in our laboratory aimed at the identification of modifiers of genomic imprinting at the *MEA* locus within the *Arabidopsis* gene pool (C. Spillane and U. Grossniklaus, unpubl.). To determine whether an accession contained any *cis*- or *trans*-acting modifiers of *mea*-related seed abortion we crossed each accession with a *mea/MEA* pollen donor, and selected F1 hybrids with the genotype *mea/MEA*. Analysis of the F2 seed (aborted: normal) from the selfed *mea/MEA* F1 hybrids allowed us to test whether any genetic modifiers of *mea*-related seed abortion were present. For instance, under the assumption that paternally supplied *MEA* activity can rescue *mea* maternal effect seed abortion, the survival of a seed inheriting a mutant *mea* allele from the mother would indicate that the paternally inherited *MEA* allele was active unlike in the L*er* accession. In general, any modifier leading to the survival of seeds that inherited a mutant maternal *mea* allele, or to the abortion of seeds that inherited a wild-type maternal *MEA* allele, distort the 50% seed abortion ratio observed in heterozygous *mea/MEA* plants (Grossniklaus et al. 1998). Thus, if we observed a F2 seed abortion ratio of 50% from a selfed *mea/MEA* F1 hybrid, the accession contains no modifier, whether paternally, maternally, or zygotically acting. This reasoning applies irrespective of whether the imprint corresponds to the maternal activated state, the paternal silent state, or both. Our analysis showed that the accessions Yo-0 and Kb-0 do not contain any genetic modifiers of genomic imprinting, because all seeds inheriting a mutant maternal *mea* allele abort (Table 1). As both of the ecotypes Yo-0 and Kb-0 lack the *MEA-AtREP2* insertion upstream of the *MEA* locus, this strongly suggests that the *MEA-AtREP2* helitron is not involved in imprinting at the *MEA* locus.

The Repetitive MEA-ISR Tandem Repeats Are Not Required for Imprinting at the *MEA* Locus

Direct tandem repeats have been found proximal to several imprinted genes in mice and humans. The "tandem repeat hypothesis" has been proposed, suggesting that repeats may be important in targeting methylation to differentially methylated regions (DMRs) (Neumann et al. 1995; Lewis et al. 2004). In mammals, evidence for a causal role for tandem repeats in imprinting regulation remains inconclusive (Lewis et al. 2004). In plants, it has been proposed that tandem repeats associated with a SINE transposable element insertion and associated tandem repeats adjacent to the *FWA* gene are the likely cause of imprinting at the *FWA* locus (Lippman et al. 2004).

Downstream from the *MEA* gene we found a conspicuous cluster of short repeats. These 182 nucleotide repeats are also found in 12 other genomic locations in the *Arabidopsis* genome, all of which are also subtelomeric. Hence, Cao and Jacobsen (2002) named the $(182)_7$ repeat region MEA-ISR for intergenic subtelomeric repeat region. They showed that this region attracts high levels of DNA methylation in wild-type strains, namely, 87% at CpG, 47% at CpNpG, and 18% at asymmetric sites. All asymmetric and CpNpG methylation was abolished at the MEA-ISR (and also at the *FWA* 211nt direct repeats) in *drm1*, *drm2* double-mutant and *drm1*, *drm2*; *cmt3-7* triple-mutant backgrounds, while CpG methylation levels remained similar to the wild type (Cao and Jacobsen 2002). Subsequently, Zilberman et al. (2003) demonstrated that the *ARGONAUTE4* (*AGO4*) gene involved in RNA-mediated silencing was also necessary for asymmetric and CpNpG, but not CpG, methylation at the MEA-ISR. In contrast, loss of *AGO4* activity had no effect on asymmetric, CpNpG, or CpG methylation at the *FWA* 211-nucleotide direct repeats.

To determine whether the MEA-ISR region is involved in genomic imprinting at the *MEA* locus, we used a combination of PCR and restriction enzyme-based assays. We screened evolutionarily divergent accessions from an *Arabidopsis* core collection to determine whether any accession lacked the MEA-ISR. For the 56 ecotypes for which we obtained PCR products, our results indicate that the MEA-ISR region has undergone substantial expansions and contractions between accessions, likely because of differences in the number of the 182-nucleotide repeats in each accession (Fig. 3).

In the Col-0 genome, each of the 182-nucleotide repeat regions in the MEA-ISR contains a *Bam*HI (B) restriction site (Fig. 3A). This allowed us to develop a simple assay to test whether each PCR product obtained from primers (e.g., MEA S40 and MEA AS42) spanning the MEA-ISR contained at least one *Bam*HI site. We were interested to identify MEA-ISR-derived PCR products containing no *Bam*HI sites, as these are candidates where the 182-nucleotide repeat region may be absent. *Bam*HI restriction analysis of PCR products from 19 different ecotypes indicated that one or more *Bam*HI sites were present in the MEA-ISR region in all 19 ecotypes tested (results not

Table 1. The Presence of Upstream *MEA-ATREP2* Helitron and Downstream MEA-ISR 182 bp Is Not Necessary for *MEA* Imprinting

Arabidopsis accession	Candidate *cis*-acting element	Aborted F2 seeds	Normal F2 seeds	Total F2 seeds	% aborted F2 seeds	$\chi^{2\,a}$	Best-fit model[b]
L*er*-0	*MEA-AtREP2* and MEA-ISR $(182nt)_7$	215	233	448	47.99	0.362	no modifier
Yo-0	*MEA-AtREP2* absent	1192	1315	2507	47.55	3.019	no modifier
Kb-0	*MEA-AtREP2* absent	277	270	547	50.64	0.045	no modifier
Pa-1	MEA-ISR $(182nt)_1$	1274	1158	2432	52.38	2.768	no modifier

[a]Test of contingency with the expected values for a 1:1 aborted :normal seed ratio corresponding to a "no modifier" model.
[b]For $\alpha < 0.05$ ($\chi^2 = 3.84$, df = 1).

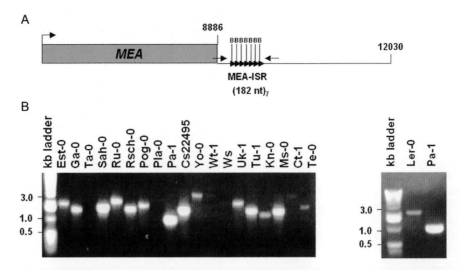

Figure 3. Expansion and contraction of the repetitive MEA-ISR tandem repeats downstream of the imprinted *MEA* locus. (*A*) Schematic representation of MEA-ISR region, where *Bam*HI sites (*B*) are contained in each 182-nucleotide repeat unit. (*B*) Gel electrophoresis of PCR products using primers spanning the MEA-ISR region indicate that the MEA-ISR region undergoes expansions and contractions because of differing numbers of repeat units.

shown). Sequencing of the smallest PCR products allowed the identification of an ecotype (Pa-1) containing only a single MEA-ISR repeat copy (i.e., no repeats of the 182 nucleotides). As the MEA locus remains imprinted in the Pa-1 accession, which contains no modifiers of *MEA* imprinting (Table 1), we conclude that tandem repeats in the MEA-ISR region do not constitute an essential *cis*-acting imprinting control region for the *MEA* locus.

Identification of a Promoter Region Sufficient for Parent-of-Origin-dependent Expression Rules Out a Role for Potential Epigenetic Attractors

To test independently for regions required for imprinted expression of *MEA*, we investigated whether a promoter fragment driving a reporter gene can reproduce the imprinted expression pattern of *MEA*. Transgenic experiments provided evidence that *cis*-acting elements required for imprinting are present in the proximal upstream region of the *MEA* gene. A truncated promoter fragment missing these candidate sequences, but spanning the CpG islands, and comprising the first intron of the *MEA* open reading frame (Fig. 4A), is able to confer imprinted expression on the bacterial *uidA* reporter gene encoding β-glucuronidase (GUS). Reciprocal crosses between transgenic *Arabidopsis* lines carrying the *MEAp:GUS* construct and wild-type plants demonstrated that this promoter fragment is maternally active but paternally inactive in the embryo and endosperm, thereby recapitulating the imprinted expression profile of the endogenous *MEA* gene (Fig. 4B). In agreement with our natural variation studies, these results strongly suggest that the potential epigenetic attractor sites are not required for the parent-of-origin-dependent expression of *MEA*. Neither the *AtREP2* helitron found in the upstream region nor the direct repeats located in 3′ region of the *MEA* gene (both the 187-nucleotide repeats and the GAA trinucleotide repeats) are involved in genomic imprinting at the *MEA* locus.

CONCLUSIONS

In this report we have reviewed the potential role of transposons and repeated sequences in epigenetic gene regulation. There is accumulating evidence that such elements can serve as attractors of epigenetic regulation and that they are involved in gene silencing, paramutation, and genomic imprinting. Recently, a role for the *AtSINE3* transposon in genomic imprinting at the *FWA* locus in *Arabidopsis* was reported (Lippman et al. 2004). In contrast, we could not find evidence for the involvement in genomic imprinting of a transposon upstream, or direct repeat seqences (MEA-ISR and GAA trinucleotide repeats) downstream, of the *MEA* locus. This difference suggests that distinct molecular mechanisms are involved in epigenetic gene regulation by genomic imprinting in plants.

At the *FWA* locus, the *AtSINE3* transposon inserted close to the gene and, in fact, contributes the first two exons to the *FWA* gene (Lippman et al. 2004). Thus, a genetic change led to the formation of a chimeric gene consisting partly of a transposon, which in turn attracts epigenetic modifications. How the transposon is affected differentially in male versus female germ cells, as are several transposons in mammals, is an open question that will attract much attention in the future. At present, very little is known about the function of epigenetic mechanisms during gametogenesis in plants. Of the many *Arabidopsis* mutants affecting epigenetic processes, effects in the gametes have been reported only for mutations in *MET1* (Saze et al. 2003). Our study on the *MEA* locus,

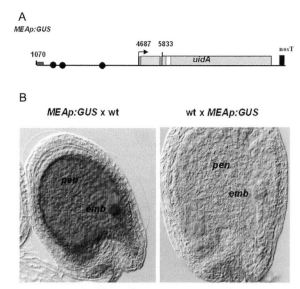

Figure 4. The *AtREP2* helitron and the direct repeats are not required for imprinting. (*A*) The *MEAp:GUS* transgene comprises the *uidA* reporter gene (*purple*) under the transcriptional control of a 4.8-kb *medea* genomic fragment. The fragment spans 3.8-kb promoter sequences lacking the *AtREP2* transposon, except for its distal 70 bp, and incorporates 1 kb upstream open reading frame (*green*; intron in *pale green*), but lacks the direct repeats from the 3′ UTR. (*B*) Imprinted activity of the truncated promoter in developing seeds. *Arabidopsis* lines carrying the *MEAp:GUS* transgene are crossed as female (*left panel*) or male (*right panel*) to wild-type plants (wt). Transgene expression is monitored by the histochemical localization of the GUS enzyme (*blue*) encoded by the *uidA* gene. Abbreviations: pen, peripheral endosperm; emb, embryo at a globular stage.

which ruled out an essential role in imprinting for conspicuous elements such as transposons and repeats, shows that other mechanisms exist as well. The further dissection of *cis*-acting elements required for imprinted expression and of *trans*-acting factors regulating maternal and paternal *MEA* alleles promises to shed light on these alternative mechanisms.

ACKNOWLEDGMENTS

We thank Tom Mitchell-Olds for providing us with a list of an *Arabidopsis* core collection that samples the most divergent accessions and was the basis for our approach relying on natural variation. Our gratitude goes to Shane McManus, who helped with the analysis of helitron variation while he was an exchange student in our lab in summer 2001 and to Andrea Steimer for helpful comments on the manuscript. This project was supported by the University of Zürich, Roche Research Foundation Fellowships to C.B. and D.R.P., as well as grant 31-64061.00 of the Swiss National Science Foundation and a Searle Scholarship to U.G.

REFERENCES

Alleman M. and Doctor J. 2000. Genomic imprinting in plants: Observations and evolutionary implications. *Plant Mol. Biol.* **43:** 147.

Arabidopsis Genome Initiative. 2000. Analysis of the genome sequence of the flowering plant *Arabidopsis thaliana*. *Nature* **408:** 796.

Argeson A.C., Nelson K.K., and Siracusa L.D. 1996. Molecular basis of the pleiotropic phenotype of mice carrying the hypervariable yellow (Ahvy) mutation at the agouti locus. *Genetics* **142:** 557.

Bailey J.A., Carrel L., Chakravarti A., and Eichler E.E. 2000. Molecular evidence for a relationship between LINE-1 elements and X chromosome inactivation: The Lyon repeat hypothesis. *Proc. Natl. Acad. Sci.* **97:** 6634.

Baroux C., Spillane C., and Grossniklaus U. 2002. Evolutionary origins of the endosperm in flowering plants. *Genome Biol.* **3:** 1026.1.

Bender J. 2004. Chromatin-based silencing mechanisms. *Curr. Opin. Plant Biol.* **7:** 521.

Bestor T.H. 2003. Cytosine methylation mediates sexual conflict. *Trends Genet.* **19:** 185.

Brink R.A. 1973. Paramutation. *Annu. Rev. Genet.* **7:** 129.

Cao X. and Jacobsen S.E. 2002. Locus-specific control of asymmetric and CpNpG methylation by the DRM and CMT3 methyltransferase genes. *Proc. Natl. Acad. Sci.* **99:** 16491.

Chang-Yeh A., Mold D.E., Brilliant M.H., and Huang R.C. 1993. The mouse intracisternal A particle-promoted placental gene retrotransposition is mouse-strain-specific. *Proc. Natl. Acad. Sci.* **90:** 292.

Chaudhuri S. and Messing J. 1994. Allele-specific parental imprinting of *dzr1*, a posttranscriptional regulator of zein accumulation. *Proc. Natl. Acad. Sci.* **91:** 4867.

Chaudhury A.M., Ming L., Miller C., Craig S., Dennis E.S., and Peacock W.J. 1997. Fertilization-independent seed development in *Arabidopsis thaliana*. *Proc. Natl. Acad. Sci.* **94:** 4223.

Choi Y., Harada J.J., Goldberg R.B., and Fischer R.L. 2004. An invariant aspartic acid in the DNA glycosylase domain of DEMETER is necessary for transcriptional activation of the imprinted *MEDEA* gene. *Proc. Natl. Acad. Sci.* **101:** 7481.

Choi Y., Gehring M., Johnson L., Hannon M., Harada J.J., Goldberg R.B., Jacobsen S.E., and Fischer R.L. 2002. DEMETER, a DNA glycosylase domain protein, is required for endosperm gene imprinting and seed viability in *Arabidopsis*. *Cell* **110:** 33.

Danilevskaya O.N., Hermon P., Hantke S., Muszynski M.G., Kollipara K., and Ananiev E.V. 2003. Duplicated *fie* genes in maize: Expression pattern and imprinting suggest distinct functions. *Plant Cell* **15:** 425.

Dawe R.K. 2003. RNA interference, transposons, and the centromere. *Plant Cell* **15:** 297.

Della Vedova C.B. and Cone K.C. 2004. Paramutation: The chromatin connection. *Plant Cell* **16:** 1358.

Dorweiler J.E., Carey C.C., Kubo K.M., Hollick J.B., Kermicle J.L., and Chandler V.L. 2000. *mediator of paramutation 1* is required for establishment and maintenance of paramutation at multiple maize loci. *Plant Cell* **12:** 2101.

Edwards K., Johnstone C., and Thompson C. 1991. A simple and rapid method for the preparation of plant genomic DNA for PCR analysis. *Nucleic Acids Res.* **19:** 1349.

Fazzari M.J. and Greally J.M. 2004. Epigenomics: Beyond CpG islands. *Nat. Rev. Genet.* **5:** 446.

Frank M.J., Preuss D., Mack A., Kuhlmann T.C., and Crawford N.M. 1998. The *Arabidopsis* transposable element *Tag1* is widely distributed among *Arabidopsis* ecotypes. *Mol. Gen. Genet.* **257:** 478.

Gendrel A.V., Lippman Z., Yordan C., Colot V., and Martienssen R.A. 2002. Dependence of heterochromatic histone H3 methylation patterns on the *Arabidopsis* gene *DDM1*. *Science* **297:** 1871.

Greally J.M. 2002. Short interspersed transposable elements (SINEs) are excluded from imprinted regions in the human genome. *Proc. Natl. Acad. Sci.* **99:** 327.

Grossniklaus U., Spillane C., Page D.R., and Köhler C. 2001. Genomic imprinting and seed development: Endosperm formation with and without sex. *Curr. Opin. Plant Biol.* **4:** 21.

Grossniklaus U., Vielle-Calzada J.P., Hoeppner M.A., and Gagliano W.B. 1998. Maternal control of embryogenesis by

MEDEA, a *Polycomb* group gene in *Arabidopsis. Science* **280**: 446.

Guitton A.-E., Page D.R., Chambrier P., Lionnet C., Faure J.-E., Grossniklaus U., and Berger F. 2004. Identification of new members of *Fertilisation Independent Seed Polycomb* Group pathway involved in the control of seed development in *Arabidopsis thaliana. Development* **131**: 2971.

Guo M., Rupe M.A., Danilevskaya O.N., Yang X., and Hu Z. 2003. Genome-wide mRNA profiling reveals heterochronic allelic variation and a new imprinted gene in hybrid maize endosperm. *Plant J.* **36**: 30.

Gutiérrez-Marcos J.F., Costa L.M., Biderre-Petit C., Khbaya B., O'Sullivan D.M., Wormald M., Perez P., and Dickinson H.G. 2004. *Maternally expressed gene1* is a novel maize endosperm transfer cell-specific gene with a maternal parent-of-origin pattern of expression. *Plant Cell* **16**: 1288.

Hirochika H., Okamoto H., and Kakutani T. 2000. Silencing of retrotransposons in *Arabidopsis* and reactivation by the *ddm1* mutation. *Plant Cell* **12**: 357.

Imhof A. 2003. Histone modifications: An assembly line for active chromatin? *Curr. Biol.* **13**: R22.

Kapitonov V.V. and Jurka J. 2001. Rolling-circle transposons in eukaryotes. *Proc. Natl. Acad. Sci.* **98**: 8714.

Kato M., Miura A., Bender J., Jacobsen S.E., and Kakutani T. 2003. Role of CG and non-CG methylation in immobilization of transposons in *Arabidopsis. Curr. Biol.* **13**: 421.

Kazazian H.H., Jr. 2004. Mobile elements: Drivers of genome evolution. *Science* **303**: 1626.

Kermicle J.L. 1970. Dependence of the *R*-mottled aleurone phenotype in maize on mode of sexual transmission. *Genetics* **66**: 69.

———. 1994. Epigenetic silencing and activation of a maize *r* gene. In *Epigenetic mechanisms of gene regulation* (ed. V.E.A. Russo et al.), p. 267. Cold Spring Harbor Laboratory Press, Cold Spring Harbor, New York.

Kinoshita T., Yadegari R., Harada J.J., Goldberg R.B., and Fischer R.L. 1999. Imprinting of the *MEDEA Polycomb* gene in the *Arabidopsis* endosperm. *Plant Cell* **11**: 1945.

Kinoshita T., Miura A., Choi Y., Kinoshita Y., Cao X., Jacobsen S.E., Fischer R.L., and Kakutani T. 2004. One-way control of *FWA* imprinting in *Arabidopsis* endosperm by DNA methylation. *Science* **303**: 521.

Köhler C., Hennig L., Bouveret R., Gheyselinck J., Grossniklaus U., and Gruissem W. 2003. *Arabidopsis* MSI1 is a component of the MEA/FIE *Polycomb* group complex and required for seed development. *EMBO J.* **22**: 4804.

Koornneef M., Hanhart C.J., and van der Veen J.H. 1991. A genetic and physiological analysis of late flowering mutants in *Arabidopsis thaliana. Mol. Gen. Genet.* **229**: 57.

Lewis A., Mitsuya K., Constancia M., and Reik W. 2004. Tandem repeat hypothesis in imprinting: Deletion of a conserved direct repeat element upstream of H19 has no effect on imprinting in the Igf2-H19 region. *Mol. Cell. Biol.* **24**: 5650.

Lippman Z. and Martienssen R. 2004. The role of RNA interference in heterochromatin silencing. *Nature* **431**: 364.

Lippman Z., May B., Yordan C., Singer T., and Martienssen R. 2003. Distinct mechanisms determine transposon inheritance and methylation via small interfering RNA and histone modification. *PLoS Biol.* **1**: E67.

Lippman Z., Gendrel A.V., Black M., Vaughn M.W., Dedhia N., McCombie W.R., Lavine K., Mittal V., May B., Kasschau K.D., Carrington J.C., Doerge R.W., Colot V., and Martienssen R. 2004. Role of transposable elements in heterochromatin and epigenetic control. *Nature* **430**: 471.

Lisch D., Carey C.C., Dorweiler J.E., and Chandler V.L. 2002. A mutation that prevents paramutation in maize also reverses *Mutator* transposon methylation and silencing. *Proc. Natl. Acad. Sci.* **99**: 6130.

Lund G., Ciceri P., and Viotti A. 1995a. Maternal-specific demethylation and expression of specific alleles of *zein* genes in the endosperm of *Zea mays* L. *Plant J.* **8**: 571.

Lund G., Messing J., and Viotti A. 1995b. Endosperm-specific demethylation and activation of specific alleles of *alpha-tubulin* genes of *Zea mays* L. *Mol. Gen. Genet.* **246**: 716.

Luo M., Bilodeau P., Dennis E.S., Peacock W.J., and Chaudhury A. 2000. Expression and parent-of-origin effects for *FIS2*, *MEA*, and *FIE* in the endosperm and embryo of developing *Arabidopsis* seeds. *Proc. Natl. Acad. Sci.* **97**: 10637.

Luo M., Bilodeau P., Koltunow A., Dennis E.S., Peacock W.J., and Chaudhury A.M. 1999. Genes controlling fertilization-independent seed development in *Arabidopsis thaliana. Proc. Natl. Acad. Sci.* **96**: 296.

Lyon M.F. 2003. The Lyon and the LINE hypothesis. *Semin. Cell Dev. Biol.* **14**: 313.

Martienssen R.A. 1996. Epigenetic phenomena: Paramutation and gene silencing in plants. *Curr. Biol.* **6**: 810.

Martienssen R.A. and Colot V. 2001. DNA methylation and epigenetic inheritance in plants and filamentous fungi. *Science* **293**: 1070.

Matzke M.A., Matzke A.J.M., and Eggleston W.B. 1996. Paramutation and transgene silencing: A common response to invasive DNA? *Trends Plant Sci.* **1**: 382.

McClintock B. 1952. Chromosome organization and genic expression. *Cold Spring Harbor Symp. Quant. Biol.* **16**: 13.

———. 1984. The significance of responses of the genome to challenge (Nobel Lecture). *Science* **226**: 792.

Michaud E.J., van Vugt M.J., Bultman S.J., Sweet H.O., Davisson M.T., and Woychik R.P. 1994. Differential expression of a new dominant *agouti* allele (Aiapy) is correlated with methylation state and is influenced by parental lineage. *Genes Dev.* **8**: 1463.

Moore T. 2001. Genetic conflict, genomic imprinting and establishment of the epigenotype in relation to growth. *Reproduction* **122**: 185.

Morgan H.D., Sutherland H.G., Martin D.I., and Whitelaw E. 1999. Epigenetic inheritance at the *agouti* locus in the mouse. *Nat. Genet.* **23**: 314.

Neumann B., Kubicka P., and Barlow D.P. 1995. Characteristics of imprinted genes. *Nat. Genet.* **9**: 12.

Ohad N., Margossian L., Hsu Y.C., Williams C., Repetti P., and Fischer R.L. 1996. A mutation that allows endosperm development without fertilization. *Proc. Natl. Acad. Sci.* **93**: 5319.

Ohad N., Yadegari R., Margossian L., Hannon M., Michaeli D., Harada J.J., Goldberg R.B., and Fischer R.L. 1999. Mutations in FIE, a WD polycomb group gene, allow endosperm development without fertilization. *Plant Cell* **11**: 407.

Page D.R. 2004. "Maternal effects during seed development in *Arabidopsis thaliana:* An expression analysis of the *FIS* class gene *MEDEA* and a search for new *FIS* class mutants." Ph.D. thesis, University of Zürich, Zürich, Switzerland.

Rakyan V.K., Blewitt M.E., Druker R., Preis J.I., and Whitelaw E. 2002. Metastable epialleles in mammals. *Trends Genet.* **18**: 348.

Rakyan V.K., Chong S., Champ M.E., Cuthbert P.C., Morgan H.D., Luu K.V., and Whitelaw E. 2003. Transgenerational inheritance of epigenetic states at the murine Axin(Fu) allele occurs after maternal and paternal transmission. *Proc. Natl. Acad. Sci.* **100**: 2538.

Rea S., Eisenhaber F., O'Carroll D., Strahl B.D., Sun Z.W., Schmid M., Opravil S., Mechtler K., Ponting C.P., Allis C.D., and Jenuwein T. 2000. Regulation of chromatin structure by site-specific histone H3 methyltransferases. *Nature* **406**: 593.

Reyes J.C. and Grossniklaus U. 2003. Diverse functions of Polycomb group proteins during plant development. *Semin. Cell Dev. Biol.* **14**: 77.

Rubin C.M., VandeVoort C.A., Teplitz R.L., and Schmid C.W. 1994. Alu repeated DNAs are differentially methylated in primate germ cells. *Nucleic Acids Res.* **22**: 5121.

Sanford J.P., Clark H.J., Chapman V.M., and Rossant J. 1987. Differences in DNA methylation during oogenesis and spermatogenesis and their persistence during early embryogenesis in the mouse. *Genes Dev.* **1**: 1039.

Saze H., Mittelsten Scheid O., and Paszkowski J. 2003. Maintenance of CpG methylation is essential for epigenetic inheritance during plant gametogenesis. *Nat. Genet.* **34**: 65.

Soppe W.J., Jacobsen S.E., Alonso-Blanco C., Jackson J.P.,

Kakutani T., Koornneef M., and Peeters A.J. 2000. The late flowering phenotype of *fwa* mutants is caused by gain-of-function epigenetic alleles of a homeodomain gene. *Mol. Cell* **6:** 791.

Spillane C., MacDougall C., Stock C., Kohler C., Vielle-Calzada J.P., Nunes S.M., Grossniklaus U., and Goodrich J. 2000. Interaction of the *Arabidopsis Polycomb* group proteins FIE and MEA mediates their common phenotypes. *Curr. Biol.* **10:** 1535.

Springer N.M., Danilevskaya O.N., Hermon P., Helentjaris T.G., Phillips R.L., Kaeppler H.F., and Kaeppler S.M. 2002. Sequence relationships, conserved domains, and expression patterns for maize homologs of the *Polycomb* group genes *E(z)*, *esc*, and *E(Pc)*. *Plant Physiol.* **128:** 1332.

Stam M., Belele C., Dorweiler J.E., and Chandler V.L. 2002. Differential chromatin structure within a tandem array 100 kb upstream of the maize *b1* locus is associated with paramutation. *Genes Dev.* **16:** 1906.

Tabara H., Sarkissian M., Kelly W.G., Fleenor J., Grishok A., Timmons L., Fire A., and Mello C.C. 1999. The rde-1 gene, RNA interference, and transposon silencing in *C. elegans*. *Cell* **99:** 123.

Tariq M. and Paszkowski J. 2004. DNA and histone methylation in plants. *Trends Genet.* **20:** 244.

van de Lagemaat L.N., Landry J.R., Mager D.L., and Medstrand P. 2003. Transposable elements in mammals promote regulatory variation and diversification of genes with specialized functions. *Trends Genet.* **19:** 530.

Vielle-Calzada J.P., Thomas J., Spillane C., Coluccio A., Hoeppner M.A., and Grossniklaus U. 1999. Maintenance of genomic imprinting at the *Arabidopsis medea* locus requires zygotic *DDM1* activity. *Genes Dev.* **13:** 2971.

Volpe T.A., Kidner C., Hall I.M., Teng G., Grewal S.I., and Martienssen R.A. 2002. Regulation of heterochromatic silencing and histone H3 lysine-9 methylation by RNAi. *Science* **297:** 1833.

Walbot V. 2001. Imprinting of *R-r*, paramutation of *B-I* and *Pl*, and epigenetic silencing of *MuDR/Mu* transposons in *Zea mays* L. are coordinately affected by inbred background. *Genet. Res.* **77:** 219.

Walker E.L. and Panavas T. 2001. Structural features and methylation patterns associated with paramutation at the *r1* locus of *Zea mays*. *Genetics* **159:** 1201.

Walsh C.P., Chaillet J.R., and Bestor T.H. 1998. Transcription of IAP endogenous retroviruses is constrained by cytosine methylation. *Nat. Genet.* **20:** 116.

Wang Y., Fischle W., Cheung W., Jacobs S., Khorasanizadeh S., and Allis C.D. 2004. Beyond the double helix: Writing and reading the histone code. *Novartis Found. Symp.* **259:** 3.

Whitelaw E. and Martin D.I. 2001. Retrotransposons as epigenetic mediators of phenotypic variation in mammals. *Nat. Genet.* **27:** 361.

Xiao W., Gehring M., Choi Y., Margossian L., Pu H., Harada J.J., Goldberg R.B., Pennell R.I., and Fischer R.L. Imprinting of the *MEA Polycomb* gene is controlled by antagonism between *MET1* methyltransferase and *DME* glycosylase. *Dev. Cell* **5:** 891.

Yadegari R., Kinoshita T., Lotan O., Cohen G., Katz A., Choi Y., Nakashima K., Harada J.J., Goldberg R.B., Fischer R.L., and Ohad N. 2000. Mutations in the *FIE* and *MEA* genes that encode interacting *Polycomb* proteins cause parent-of-origin effects on seed development by distinct mechanisms. *Plant Cell* **12:** 2367.

Yates P.A., Burman R.W., Mummaneni P., Krussel S., and Turker M.S. 1999. Tandem B1 elements located in a mouse methylation center provide a target for de novo DNA methylation. *J. Biol. Chem.* **274:** 36357.

Yoder J.A., Walsh C.P., and Bestor T.H. 1997. Cytosine methylation and the ecology of intragenomic parasites. *Trends Genet.* **13:** 335.

Zilberman D., Cao X., and Jacobsen S.E. 2003. *ARGONAUTE4* control of locus-specific siRNA accumulation and DNA and histone methylation. *Science* **299:** 716.

Toward Molecular Understanding of Polar Overdominance at the Ovine Callipyge Locus

M. Georges,[*] C. Charlier,[*] M. Smit,[†] E. Davis,[*] T. Shay,[†] X. Tordoir,[*] H. Takeda,[*] F. Caiment,[*] and N. Cockett[†]

[*]*Department of Genetics, Faculty of Veterinary Medicine, University of Liege (B43), 4000 Liege, Belgium;*
[†]*Department of Animal, Dairy and Veterinary Sciences, College of Agriculture, Utah State University, Logan, Utah 84322-4700*

The callipyge phenotype (Gk *calli-* beautiful + *-pyge* buttocks) is a generalized muscular hypertrophy described in sheep. It is due to an increase in the size and proportion of fast twitch muscle fibers. It manifests itself only after birth at ~1 month of age. It exhibits a rostrocaudal gradient being more pronounced in the muscle of the pelvic limb and torso, hence its name. It is accompanied by a decrease in all measures of fatness. Affected animals are characterized by an improved feed efficiency and dressing percentage (for review, see Cockett et al. 2001). Quite logically, the callipyge phenotype initially caught the attention of animal breeders because of its potential agronomic value. Ensuing studies, however, would reveal some remarkable features of the callipyge phenotype, especially its non-Mendelian mode of inheritance. These would quickly attract more attention among the scientific community than its potential economic value, especially since the quality of callipyge meat appeared to be mediocre!

The aim of this paper is to update and complement a recent review describing the present understanding of the genetics and epigenetics of the callipyge phenomenon (Georges et al. 2003).

POLAR OVERDOMINANCE AT THE OVINE *CLPG* LOCUS

The callipyge phenotype was first reported in the 1980s, showing in ~15% of offspring of a Dorset ram called "Solid Gold." When mated to wild-type ewes, callipyge rams descending from Solid Gold produced 50% callipyge offspring, irrespective of sex. This Mendelian segregation ratio suggested that the callipyge phenotype results from an autosomal, nonrecessive mutation referred to as "*CLPG*" (Cockett et al. 1994). This monogenic hypothesis was confirmed when the *CLPG* locus was mapped to a 4.5-cM marker interval on distal chromosome 18 (Cockett et al. 1994; Freking et al. 1998; Shay et al. 2001).

Unexpectedly, crosses involving callipyge ewes and wild-type rams did not produce any callipyge offspring, despite the transmission of the *CLPG*-carrying chromosome from the mothers to half their offspring. This nonequivalence of reciprocal crosses suggested the involvement of a gene undergoing parental imprinting that would only be expressed from the paternal allele. This hypothesis was supported by the observation that nonexpressing $CLPG^{Mat}/+^{Pat}$ rams would transmit the callipyge phenotype to their $CLPG^{Pat}$-bearing offspring when mated to +/+ wild-type ewes. Parent-of-origin effects associated with uniparental disomies (UPDs) of the orthologous 14q32 region in man (pUPD14; MIM #608149) and distal 12 region in the mouse (Georgiades et al. 2000) pointed toward the possible sharing of an imprinted locus in the three species.

However, matings performed subsequently between callipyge ewes and rams, each known to be of $+^{Mat}/CLPG^{Pat}$ genotype, would not yield 50% callipyge offspring as expected in case of parental imprinting, but rather the unusual 75% wild-type versus 25% callipyge phenotypic ratio. Marker analysis indicated that only the $+^{Mat}/CLPG^{Pat}$ genotype was associated with the callipyge phenotype, the $CLPG^{Mat}/CLPG^{Pat}$ offspring being wild type although carrying the *CLPG* mutation on their paternal chromosome. This non-Mendelian inheritance pattern was referred to as "polar overdominance" (Cockett et al. 1996). It was postulated to result either from a mutation that would switch the imprinting of the *CLPG* gene from paternal to maternal expression, or would simultaneously knock out an imprinted paternally expressed *trans*-acting repressor and its target *CLPG* gene. In the former case, the callipyge phenotype would result from the illegitimate absence of the *CLPG* gene product, in the latter from its illegitimate presence. The latter model turns out to share many features with reality as we understand it today.

THE *CLPG* MUTATION MAPS TO THE *DLK1-GTL2*-IMPRINTED DOMAIN

A BAC contig spanning the *CLPG* locus was constructed (Segers et al. 2000; Berghmans et al. 2001; Shay et al. 2001), and ~500 contiguous kilobases predicted to contain the mutation were sequenced (Charlier et al. 2001b and unpubl.). In silico annotation of this sequence showed that the *CLPG* mutation mapped to the newly described *DLK1-GTL2*-imprinted domain (Fig. 1).

This evolutionary conserved domain, which spans ~1 Mb, harbors at least four protein-encoding genes with preferential expression from the paternal allele (*BEGAIN*,

DLK1, PEG11, and *DIO3*), as well as multiple "long" (*GTL2, anti-PEG11, MEG8,* and *MIRG*) and "small" (C/D snoRNAs and miRNAs) noncoding RNA genes (ncRNA) with preferential expression from the maternal allele (Kobayashi et al. 2000; Miyoshi et al. 2000; Schmidt et al. 2000; Takada et al. 2000; Wylie et al. 2000; Charlier et al. 2001b; Paulsen et al. 2001; Cavaillé et al. 2002; Hernandez et al. 2002; Tsai et al. 2002; Yevtodiyenko et al. 2002; Seitz et al. 2003, 2004; M.A. Smit et al., in prep.).

BEGAIN (*b*rain-*e*nriched *g*uanylate kinase-*a*ssociated prote*in*) encodes a protein that binds to the guanylate kinase domain of PSD-95/SAP90, a scaffolding protein at the postsynaptic cell membrane (Deguchi et al. 1998). It is widely expressed, producing multiple transcripts—as a result of alternative promoter usage and splicing—that exhibit paternal or biallelic expression in a tissue- and promoter-specific manner (M.A. Smit et al., in prep.). *DLK1* is a member of the *EGF* domain containing *Notch/Delta/Serrate* protein family whose function remains poorly understood. It has been implicated in adipogenesis, hematopoiesis, lymphopoiesis, and neuroendocrine differentiation, as well as tumorigenesis (Laborda 2000). *Dlk1* null mice display accelerated adiposity, as well as symptoms shared with mUPD12 mice and mUPD14 humans: growth retardation, blepharophimosis, and skeletal abnormalities (Moon et al. 2002). A possible involvement of *Dlk1* in myogenesis is suggested by the myofiber hypertrophy and delayed maturation observed in murine pUPD12 fetuses expressing a double dose of *Dlk1* (Georgiadis et al. 2000), and by the known involvement of *Notch* signaling in myogenesis (see, e.g., Hirsinger et al. 2001; Conboy et al. 2003). *PEG11* corresponds to a long (~1300 residues), uninterrupted open reading frame (ORF) that has the potential to code for a protein with a central portion that is highly similar to the gag and pol polyproteins of gypsy-like long terminal repeat (LTR) retrotransposons. It was shown to be hypermethylated in skeletal muscle as expected for a retroelement, yet is not flanked by sequences matching LTRs expected for such an element. The conservation of such a long ORF across mammals suggests that it fulfills an important function, yet, to the best of our knowledge, there is no evidence for its translation in any tissue so far. *DIO3* codes for a type 3 iodothyronine deiodinase that degrades both T3 and T4 by catalyzing 5-deiodination of the inner ring, thereby contributing to the regulation of thyroid hormone levels in several tissues (St. Germain and Galton 1997).

The organization of the maternally expressed ncRNA genes remains more blurry. The first to be identified in a gene trap screen was *GTL2*, also known as *MEG3* (Schuster-Gossler et al. 1996, 1998; Miyoshi et al. 2000). *GTL2* is characterized by at least 12 exons and produces multiple transcripts by means of alternative splicing. The exon–intron organization is fairly well conserved across mammals, yet there is no evidence for a conserved ORF, suggesting that it is noncoding (see, e.g., Charlier et al. 2001b). *MEG8* and *MIRG* were subsequently identified and characterized (Charlier et al. 2001b; Cavaillé et al. 2002; Seitz et al. 2004). They were shown to share a conserved exon–intron organization, extensive alternative splicing, and lack of conserved ORF with *GTL2*. Antisense *PEG11* transcripts expressed exclusively from the maternal allele were discovered as well (Charlier et al. 2001b). As all these supposedly ncRNAs are contiguous, expressed from the same strand and from the maternal allele, they might represent a single very long transcriptional unit akin to *AIR* (Lyle et al. 2000).

Interestingly, these "long" ncRNA genes are hosts for a multititude of "small" ncRNAs. Cavaillé et al. (2002) first identified tandem clusters of C/D snoRNA located in the introns of *MEG8* (also referred to as *Rian*), and—as their host gene—expressed exclusively from the maternal allele. It subsequently appeared that the same "long" ncRNAs might be hosting multiple miRNAs. By May 2004, 11 of the approximately 150 miRNAs known in human (e.g., the microRNA registry: http://www.sanger.ac.uk/Software/Rfam/mirna/index.shtml) mapped to the *DLK1-GTL2* domain. Most of these would map to the introns of *MIRG* or in the vicinity of *PEG11*. As a matter of fact, two, *miR127* and *miR136,* mapped within the boundaries of the paternally expressed *PEG11*, making them the first miRNA genes in mammals with perfect complementarity to a potential target gene and raising the intriguing possibility of a mechanical link between imprinting and RNAi (Seitz et al. 2003). More recently, approximately 40 miRNAs have been identified in the *DLK1-GTL2* domain, most of them being part of tandem clusters located within the introns of *MIRG* (Seitz et al. 2004).

In the mouse, expression of the ncRNA genes and repression of the mRNA genes from the maternal allele has been shown to require a functional intergenic (located between *DLK1* and *GTL2*), germ-line-derived differentially methylated region (*IG-DMR*), as transmission of the Δ-*IG-DMR* deletion causes a paternal epigenotype and modus operandi, while paternal transmission has no effect (Lin et al. 2003).

THE *CLPG* MUTATION IS A POINT MUTATION AFFECTING A MUSCLE-SPECIFIC, LONG-RANGE *cis*-ACTING CONTROL ELEMENT

We monitored the expression of the genes in the *DLK1-GTL2* domain in a range of tissues and throughout development for sheep representing the four possible *CLPG* genotypes. In none of the examined tissues and developmental stages was imprinting of any of the studied genes (*DLK1, PEG11, BEGAIN, GTL2, antiPEG11, MEG8,* and *MIRG*) affected by *CLPG* genotype: mRNAs were always preferentially expressed from the paternal allele, while ncRNA genes were always preferentially expressed from the maternal allele (Charlier et al. 2001a; M.A. Smit et al., in prep. and unpubl.). Thus, this allowed us to exclude the first model for polar overdominance assuming a switch in imprinting of the *CLPG* gene from paternal to maternal expression.

However, the expression levels in skeletal muscle of a cluster of centrally positioned genes (*DLK1, PEG11, GTL2, antiPEG11, MEG8,* and *MIRG*, but not *BEGAIN*

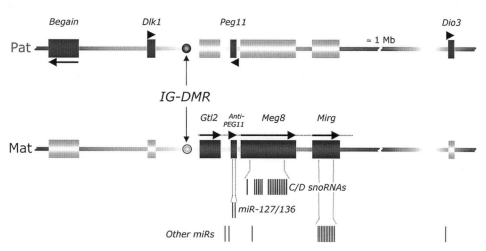

Figure 1. Schematic representation of the imprinted *DLK1-GTL2* domain. The genes shown in *red* are protein-encoding genes that are preferentially transcribed from the paternal allele (Pat) in the directions indicated by the *arrows*. The genes shown in *blue* are non-coding RNA genes that are preferentially transcribed from the maternal allele (Mat) in the direction shown by the *arrow*. The C/D snoRNAs and miRNAs hosted, respectively, by *MEG8*, *antiPEG11*, and *MIRG* are shown underneath. The IG-DMR is an imprinting control element that is methylated (*open circle*) and inactive on the paternal allele, while unmethylated (*closed circle*) and active on the maternal chromosome, thereby leading to the expression of the ncRNA genes and silencing of the protein-encoding genes.

and *DIO3*) were clearly influenced by the *CLPG* genotype (Charlier et al. 2001a; M.A. Smit et al., in prep. and unpubl.). More specifically, the concentrations of mRNA (*DLK1*, *PEG11*) were markedly increased in the two genotypes sharing the *CLPG* mutation on their paternally inherited chromosome ($+^{Mat}/CLPG^{Pat}$ and $CLPG^{Mat}/CLPG^{Pat}$), while the concentrations of ncRNA (*GTL2*, *antiPEG11*, *MEG8*, and *MIRG*) were markedly increased in the two genotypes sharing the *CLPG* mutation on their maternally inherited chromosome ($CLPG^{Mat}/+^{Pat}$ and $+^{Mat}/+^{Pat}$). The effect was limited to skeletal muscle. The easiest interpretation of these findings is that the *CLPG* mutation affects a muscle-specific *cis*-acting regulatory element that controls the expression level of a subset of genes in the *DLK1-GTL2* imprinted domain.

It is worthwhile noting that the *DLK1* and especially *PEG11* mRNA levels were higher in skeletal muscle of $+^{Mat}/CLPG^{Pat}$ when compared to $CLPG^{Mat}/CLPG^{Pat}$ individuals, while the ncRNA levels were higher in skeletal muscle of $CLPG^{Mat}/CLPG^{Pat}$ when compared to $CLPG^{Mat}/+^{Pat}$ individuals (Charlier et al. 2001a).

Having demonstrated the physical integrity of the locus for the *CLPG* allele by chromosome combing (absence of gross deletions or inversions; C. Charlier et al., unpubl.), we and others resequenced >200 Kb encompassing the *DLK1*, *GTL2*, *PEG11*, *antiPEG11*, and *MEG8* genes in order to identify the predicted *cis*-acting element and *CLPG* mutation in it. To minimize confusion from neutral background mutations, both groups selected a wild-type and a *CLPG* chromosome coalescing very recently for resequencing. This was achieved by selecting callipyge animals (therefore heterozygous $+^{Mat}/CLPG^{Pat}$) that were homozygous for all known polymorphisms within and in the vicinity of the *CLPG* locus. Three hundred and twenty polymorphisms were identified differentiating the resequenced + and *CLPG* alleles from the reference BAC sequence. The resequenced alleles, on the contrary, were identical over their entire length with the exception of a single A (+ allele) to G (*CLPG* allele) transition located between the *DLK1* and *GTL2* genes, at 32.8 Kb from the closest genes, i.e., *GTL2* (Freking et al. 2002; Smit et al. 2003).

Was this the *CLPG* mutation? A number of arguments strongly support this conjecture. First, the A to G transition affects the third base pair of a dodecamer motif that is perfectly conserved in the 13 mammalian species in which it has been sequenced. The extensive conservation of this motif strongly suggests that it fulfills an important function. The dodecamer motif is itself embedded in a 2-Kb segment exhibiting >70% similarity between human and sheep. Second, the G allele was only encountered in the callipyge flock, i.e., descendents of Solid Gold, despite the screening of more than 270 animals representing 13 distinct breeds, including wild-type Dorset. Third, and most convincingly, Solid Gold was shown to be mosaic (A/A+G), the G alleles representing ~20% of the residues in its leucocyte DNA (Smit et al. 2003). The observation that only 15% of the approximately 150 offspring sired by Solid Gold were callipyge is in agreement with the ram being germ-line mosaic as well. This strongly suggests that the A to G transition occurred during Solid Gold's early development and virtually proves that it is the causative mutation. To provide final proof of its causality, we are generating transgenic mice with the corresponding nucleotide substitution by gene targeting and hope to recapitulate the callipyge phenotype and mode of inheritance in this more tractable model.

How might this putative *cis*-acting element operate? So far, classical approaches for the functional analysis of regulatory elements have provided relatively little insight. An ~500-bp fragment encompassing the wild-type or mutant dodecamer motif in its center does not seem to affect the expression level of a luciferase reporter gene driven either by the *CMV* or *IGF2* P3 promotor in C2C12 myoblast cells (C. Charlier et al., unpubl.). Freking et al. (2002) indicated that the *CLPG* mutation lies at the end of a putative 10-bp *MyoD* binding motif and presented experimental evidence supporting binding to the *MyoD/E47* transcription factor. However, the 10-bp motif is not well conserved and the *CLPG* mutation does not seem to affect binding. We have obtained preliminary evidence from gel retardation experiments that an oligonucleotide encompassing the dodecamer motif binds a nuclear factor present in skeletal muscle, and that the *CLPG* mutation decreases the affinity for this factor (M.A. Smit et al., unpubl.). DNase hypersensitivity experiments are also underway (H. Takeda et al., unpubl.).

Intriguingly, Freking et al. (2002) identified a rare, unspliced transcript ("*CLPG1*") by random oligoprimed reverse transcriptase polymerase chain reaction (RT-PCR) from fetal skeletal muscle RNA, suggesting that the region encompassing the *CLPG* mutation might be transcribed. We have now extended this work and show that in skeletal muscle of 8-week-old animals, transcripts encompassing the mutation are produced from both strands, albeit exclusively from the *CLPG* allele (H. Takeda et al., unpubl.). It is unknown at this point whether this is a consequence or a cause of the more open chromatin configuration in the region. Note that a region encompassing the mutation and including 11 CpG dinucleotides was shown by bisulfite sequencing to be generally more methylated on the + than on the *CLPG* allele, but that there was no simple correlation between allelic state and methylation for any of the examined sites (M.A. Smit et al., unpubl.).

ECTOPIC EXPRESSION OF DLK1 PROTEIN IN SKELETAL MUSCLE OF $+^{Mat}/CLPG^{Pat}$ INDIVIDUALS CAUSES THE CALLIPYGE PHENOTYPE

The observed effect of the *CLPG* mutation on transcript levels of neighboring genes in skeletal muscle did not satisfactorily explain why only $+^{Mat}/CLPG^{Pat}$ individuals express the callipyge phenotype. Indeed, $+^{Mat}/CLPG^{Pat}$ individuals share the overexpression of the *DLK1* and *PEG11* transcripts with $CLPG^{Mat}/CLPG^{Pat}$ individuals and the lack of overexpression of *GTL2*, *antiPEG11*, *MEG8*, and *MIRG* with $+^{Mat}/+^{Pat}$ individuals, neither of which exhibits callipyge features. It is thus the combination of mRNA overexpression and lack of ncRNA overexpression that seems unique.

To gain a better understanding of why this might be, we monitored the expression of DLK1 at the protein level by immunochemistry (Davis et al. 2004). DLK1 protein could not be detected in skeletal muscle of $+^{Mat}/+^{Pat}$, $CLPG^{Mat}/+^{Pat}$, and $CLPG^{Mat}/CLPG^{Pat}$ individuals, irrespective of muscle group and developmental stage. Remarkably, in $+^{Mat}/CLPG^{Pat}$ individuals, DLK1 protein was found abundantly, albeit exclusively, in skeletal muscle exhibiting the muscular hypertrophy, e.g., *Longissimus dorsi* after 1 month of age. At earlier developmental stages or in nonhypertrophied muscle groups, DLK1 protein could not be detected in $+^{Mat}/CLPG^{Pat}$ individuals either. We thus observed a perfect association between the expression of DLK1 protein in skeletal muscle and their hypertrophy, both when differentiating individuals by *CLPG* genotype and when distinguishing muscle groups within $+^{Mat}/CLPG^{Pat}$ individuals.

To test whether this association might be causal, we generated transgenic mice expressing the membrane-bound form of the DLK1 protein (as we demonstrated that the corresponding *DLK1* transcripts are by far the most abundant in skeletal muscle of callipyge animals) in skeletal muscle under the dependence of a myosin light chain promoter and enhancer (Davis et al. 2004). The two transgenic lines that were produced indeed exhibited a muscular hypertrophy as a result of an increase in myofiber diameter. This strongly suggests that the ectopic expression of DLK1 protein that is observed in hypertrophied muscle of callipyge sheep causes this hypertrophy or at least contributes to it. We are presently using the same transgenic based approach to test the putative effect on muscle mass of ectopic expression of *PEG11*.

POLAR OVERDOMINANCE SUPPORTS THE *trans*-INTERACTION BETWEEN THE PRODUCTS OF RECIPROCALLY IMPRINTED GENES

The question remains why DLK1 protein is detected in skeletal muscle of $+^{Mat}/CLPG^{Pat}$ individuals but not in that of $CLPG^{Mat}/CLPG^{Pat}$ animals, while *DLK1* mRNA is present at comparable concentrations in skeletal muscle of both genotypes, transcribed from the paternal *CLPG* allele. Obviously the difference between the two genotypes is the maternal allele that is wild type and essentially silent in skeletal muscle of $+^{Mat}/CLPG^{Pat}$ individuals while being *CLPG* and thus producing ncRNA in skeletal muscle of $CLPG^{Mat}/CLPG^{Pat}$ animals. This thus suggests that in $CLPG^{Mat}/CLPG^{Pat}$ animals, the ncRNAs are blocking the translation of the *DLK1* mRNAs in *trans* explaining the absence of DLK1 protein (Fig. 2).

It is particularly interesting in this regard that the *DLK1-GTL2* locus is remarkably rich in miRNA genes (cf. above). Could it be that one or several of these maternally expressed miRNAs target *DLK1* transcripts, thus mediating the translational *trans* inhibition postulated to underlie polar overdominance? To test this hypothesis we compared the affinity of 43 miRNAs predicted by Mirscan (Lim et al. 2003) in the *DLK1-GTL2* domain using the human, murine, and rat sequence for the corresponding 3´UTR of *DLK1* (X. Tordoir et al., unpubl.). Targetscan was used to quantify the affinity of the miRNAs for their target (Lewis et al. 2003), and affinities were summed across species. As the boundaries of the miRNAs cannot be determined unambiguously, multiple candidate miRNAs were actually tested for each pri-miRNA. The affinity of the predicted miRNAs for the 3´UTR of *DLK1* was compared with their affinity for a set of 676 size-matched control 3´UTRs. The affinity of the miRNAs was either tested as a group (combinatorial rheostat hypothesis; Bartel 2004) or individually (individual rheostat hypothesis). We found no evidence supporting the fact that, as a group, the miRNAs predicted in the *DLK1-GTL2* have a higher affinity for the 3´UTR of *DLK1* than for the 3´UTRs of a random set of control

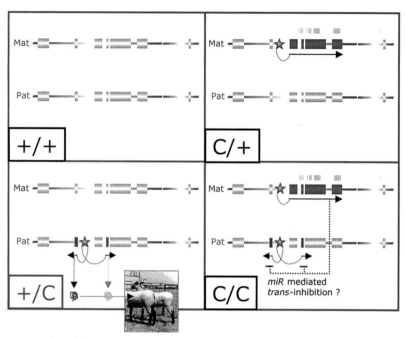

Figure 2. Schematic representation of the expression status of the genes in the *DLK1-GTL2* imprinted domain in skeletal muscle of 8-week-old sheep representing the four possible callipyge genotypes: +/+, C/+, +/C, and C/C (maternal allele/paternal allele). Only the +/C animals express the phenotype and are therefore labeled *red*. The star represents the CLPG mutation, which enhances the expression of a core group of genes (*DLK1, GTL2, PEG11, antiPEG11, MEG8, MIRG, C/D* snoRNAs, miRNAs) in *cis* (*dotted lines* and *arrows*) without altering their imprinting status. Only in callipyge (+/C) animals is DLK1 (and possibly PEG11) protein detected in skeletal muscle causing the phenotype. The absence of DLK1 (and possibly PEG11) protein in C/C animals is postulated to result from a *trans*-inhibition mediated by the noncoding RNA genes expressed from the maternal allele. At present, the best candidate mediators of this *trans* effect are the miRNA genes present in the *DLK1-GTLK2* domain.

genes ($p = 0.49$). When tested individually, however, we found one miRNA that had an affinity for *DLK1* superior to 90% of the best affinities found with the miRNA set for the 3′UTR of the control genes. Noteworthy, this miRNA also obtained the highest Mirscan score (19). The potential interaction between this miRNA and *DLK1* transcripts is being tested experimentally.

It is noteworthy that there is strong evidence for a *trans* interaction between the maternally expressed *miR127* and *miR136* miRNA genes embedded in *antiPEG11* and the perfectly complementary paternally *PEG11* transcripts. First, Lin et al. (2003) provided indirect evidence by showing that concentrations of *PEG11* transcripts were sixfold rather than twofold higher in mice inheriting the Δ-IG-DMR deletion on their maternal chromosome when compared to wild-type controls. This is postulated to be due to the absence of *miR127* and *miR136* expression in the former. More directly, we have recently cloned *PEG11* cleavage products from skeletal muscle of $CLPG^{Mat}/CLPG^{Pat}$ sheep predicted to result from the action of *miR127* (E. Davis et al., in prep.). This observation is in perfect agreement with the lower levels of *PEG11* transcripts observed in $CLPG^{Mat}/CLPG^{Pat}$ when compared to $+^{Mat}/CLPG^{Pat}$ animals.

Although miRNAs are the most attractive candidate mediators of the postulated *trans* effect, alternative hypotheses should not be overlooked. It is intriguing in this regard that noncoding *H19* transcripts might bind to the same *IMP* (*IGF2* mRNA binding protein) postulated to regulate translatability of some *IGF2* mRNAs (Runge et al. 2000). Might this point to a translational *trans* regulation of *IGF2* mediated by H19 ncRNA? Might similar mechanisms operate at the *DLK1-GTL2* domain?

CONCLUSIONS

The study of the callipyge phenomenon has provided some unique opportunities to probe the novel epigenetic mechanisms that underlie its unusual mode of inheritance—polar overdominance. More than being just an ovine idiosyncrasy, the study of polar overdominance is likely to shed light on molecular mechanisms that might be involved in the inheritance of other phenotypes, including complex inherited disorders in the human. The study of the callipyge phenotype once again illustrates the potential value of domestic animal biodiversity in unraveling fundamental biological processes (Andersson and Georges 2004).

ACKNOWLEDGMENTS

This work was supported by grants from (i) the FRFC (no. 2.4525.96), (ii) Crédit aux Chercheurs (no. 1.5.134.00) from the FNRS, (iii) Crédit à la Recherche from the ULg, (iv) the SSTC (no. 0135), (v) the European Union (Callimir), (vi) the Utah Center of Excellence Program, (vii) the USDA/NRICGP (Grants #94-04358, #96-35205, and #98-03455), and (viii) the Utah Agricultural

Experiment Station, USU. Erica Davis is a fellow of the Belgian American Educational Foundation. Carole Charlier is Chercheur Qualifié from the FNRS. We are grateful to Anne Ferguson-Smith and Jérôme Cavaillé for sharing some of their results prior to publication.

REFERENCES

Andersson L. and Georges M. 2004. Domestic animal genomics: Deciphering the genetics of complex traits. *Nat. Rev. Genet.* **5:** 202.

Bartel D.P. 2004. MicroRNAs: Genomics, biogenesis, mechanism, and function. *Cell* **116:** 281.

Berghmans S., Segers K., Shay T., Georges M., Cockett N.E., and Charlier C. 2000. Breakpoint mapping positions the callipyge gene within a 285 kilobase chromosome segment containing the Gtl-2 gene. *Mamm. Genome* **12:** 183.

Cavaillé J., Seitz H., Paulsen M., Ferguson-Smith A.C., and Bachellerie J.-P. 2002. Identification of tandemly-repeated C/D snoRNA genes at the imprinted human 14q32 domain reminiscent of those at the Prader-Willi/Angelman syndrome region. *Hum. Mol. Genet.* **11:** 1527.

Charlier C., Segers K., Karim L., Shay T., Gyapay G., Cockett N., and Georges M. 2001a. The callipyge (CLPG) mutation enhances the expression of the coregulated *DLK1, GTL2, PEG11* and *MEG8* genes in *cis* without affecting their imprinting status. *Nat. Genet.* **27:** 367.

Charlier C., Segers K., Wagenaar D., Karim L., Berghmans S., Jaillon O., Shay T., Weissenbach J., Cockett N., Gyapay G., and Georges M. 2001b. Human-ovine comparative sequencing of a 250 kilobase imprinted domain encompassing the callipyge (*clpg*) gene and identification of six imprinted transcripts: *DLK1, DAT, GTL2, PEG11, antiPEG11* and *MEG8. Genome Res.* **11:** 850.

Cockett N.E., Shay T.L., and Smit M. 2001. Analysis of the sheep genome. *Physiol. Genomics* **7:** 69.

Cockett N.E., Jackson S.P., Shay T.L., Nielsen D., Green R.D., and Georges M. 1994. Chromosomal localisation of the callipyge gene in sheep (Ovis aries) using bovine DNA markers. *Proc. Natl. Acad. Sci.* **91:** 3019.

Cockett N.Y., Jackson S.P., Shay T.L., Famir F., Berghmans S., Snowder G.D., Nielsen D.M., and Georges M. 1996. Polar overdominance at the ovine callipyge locus. *Science* **273:** 236.

Conboy I.M., Conboy M.J., Smythe G.M., and Rando T.A. 2003. Notch-mediated restoration of regenerative potential to aged muscle. *Science* **302:** 1575.

Davis E., Jensen C.H., Schroder H.D., Farnir F., Shay-Hadfield T., Kliem A., Cockett N., Georges M., and Charlier C. 2004. Ectopic expression of DLK1 protein in skeletal muscle of padumnal heterozygotes causes the callipyge phenotype. *Curr. Biol.* **14:** 1858.

Deguchi M., Hata Y., Takeuchi M., Ide N., Hirao K., Yao I., Irie M., Toyoda A., and Takai Y. 1998. BEGAIN (brain-enriched gyanylate kinase-associated protein), a novel neuronal PSD-95/SAP90-binding protein. *J. Biol. Chem.* **273:** 26269.

Freking B.A., Keele J.W., Beattie C.W., Kappes S.M., Smith T.P., Sonstegard T.S., Nielsen M.K., and Leymaster K.A. 1998. Evaluation of the ovine callipyge locus. I. Relative chromosomal position and gene action. *J. Anim. Sci.* **76:** 2062.

Freking B.A., Murphy S.K., Wylie A.A., Rhodes S.J., Keele J.W., Leymaster K.A., Jirtle R.L., and Smith T.P. 2002. Identification of the single base change causing the *callipyge* muscle hypertrophy phenotype, the only known example of polar overdominance in mammals. *Genome Res.* **12:** 1496.

Georges M., Charlier C., and Cockett N. 2003. The callipyge locus: Evidence for the *trans* interaction of reciprocally imprinted genes. *Trends Genet.* **19:** 248.

Georgiades P., Watkins M., Surani M.A., and Ferguson-Smith A.C. 2000. Parental origin-specific developmental defects in mice with uniparental disomy for chromosome 12. *Development* **127:** 4719.

Hernandez A., Fiering S., Martinez E., Galton V.A., and St. Germain D. 2002. The gene locus encoding iodothyronine deiodinase type 3 (*Dio3*) is imprinted in the fetus and expresses antisense transcripts. *Endocrinology* **143:** 4483.

Hirsinger E., Malapert P., Dubrulle J., Delfini M.C., Duprez D., Henrique D., Ish-Horowicz D., and Pourquie O. 2001. Notch signalling acts in postmitotic avian myogenic cells to control MyoD activation. *Development* **128:** 107.

Kobayashi S., Wagatsuma H., Ono R., Ichikawa H., Yamazaki M., Tashiro H., Aisaka K., Miyoshi N., Kohda T., Ogura A., Ohki M., Kaneko-Ishino T., and Ishino F. 2000. Mouse *Peg9/Dlk1* and human *PEG9/DLK1* are paternally expressed imprinted genes closely located to the maternally expressed imprinted genes: Mouse *Meg3/Gtl2* and human *MEG3. Genes Cells* **5:** 1029.

Laborda J. 2000. The role of epidermal growth factor-like protein DLK in cell differentiation. *Histol. Histopathol.* **15:** 119.

Lewis B.P., Shih I., Jones-Rhoades M.W., and Bartel D.P. 2003. Prediction of mammalian microRNA targets. *Cell* **115:** 787.

Lim L.P., Lau N.C., Weinstein E.G., Abdelhakim A., Yekta S., Rhoades M.W., Burge C.B., and Bartel D.P. 2003. The microRNAs of Caenorhabditis elegans. *Genes Dev.* **17:** 991.

Lin S.-P., Youngson N., Takada S., Seitz H., Reik W., Paulsen M., Cavaillé J., and Ferguson-Smith A.C. 2003. Asymmetric regulation of imprinting on the maternal and paternal chromosomes at the *Dlk1-Gtl2* imprinted cluster on mouse chromosome 12. *Nat. Genet.* **35:** 97.

Lyle R., Watanabe D., te Vruchte D., Lerchner W., Smrzka O.W., Wutz A., Schageman J., Hahner L., Davies C., and Barlow D.P. 2000. The imprinted antisense RNA at the *Igf2r* locus overlaps but does not imprint *Mas1. Nat. Genet.* **25:** 19.

Miyoshi N., Wagatsuma H., Wakana S., Shiroishi T., Nomura M., Aisaka K., Kohda T., Surani M.A., Kaneko-Ishino T., and Ishino F. 2000. Identification of an imprinted gene, *Meg3/Gtl2* and its human homologue *MEG3*, first mapped on mouse distal chromosome 12 and human chromosome 14q. *Genes Cells* **5:** 211.

Moon Y.S., Smas C.M., Lee K., Villena J.A., Kim K.H., Yun E.J., and Sul H.S. 2002. Mice lacking paternally expressed Pref-1/Dlk1 display growth retardation and accelerated adiposity. *Mol. Cell. Biol.* **22:** 5585.

Paulsen M., Takada S., Youngson N.A., Benchaib M., Charlier C., Segers K., Georges M., and Ferguson-Smith A. 2001. Detailed sequence analysis of the imprinted *Dlk1-Gtl2* locus in three mammalian species identifies highly conserved genomic elements and a domain structure different from the *Igf2-H19* region. *Genome Res.* **11:** 2085.

Runge S., Nielsen F.C., Nielsen J., Lykke-Andersen J., Weweri U.M., and Christiansen J. 2000. H19 RNA binds four molecules of insulin-like growth factor II mRNA-binding protein. *J. Biol. Chem.* **275:** 29562.

Schmidt J.V., Matteson P.G., Jones B.K., Guan X.-J., and Tilghman S.M. 2000. The *Dlk1* and *Gtl2* genes are linked and reciprocally imprinted. *Genes Dev.* **14:** 1997.

Schuster-Gossler K., Bilinski P., Sado T., Ferguson-Smith A., and Gossler A. 1998. The mouse *GTL2* gene is differentially expressed during embryonic development, encodes multiple alternatively spliced transcripts, and may act as an RNA. *Dev. Dyn.* **212:** 214.

Schuster-Gossler K., Simon D., Guénet J.-L., Zachgo J., and Gossler A. 1996. GTL2lacz, an insertional mutation on mouse chromosome 12 with parental origin dependent phenotype. *Mamm. Genome* **7:** 20.

Segers K., Vaiman D., Berghmans S., Shay T., Beever J., Cockett N., Georges M., and Charlier C. 2000. Construction and characterization of an ovine BAC contig spanning the callipyge locus. *Anim. Genet.* **31:** 352.

Seitz H., Royo H., Bortolin M.-L., Lin S.-P., Ferguson-Smith A.C., and Cavaillé J. 2004. A large imprinted microRNA gene cluster at the mouse Dlk1-Gtl2 domain. *Genome Res.* **14:** 1741.

Seitz H., Youngson N., Lin S.-P., Dalbert S., Paulsen M., Bachellerie J.P., Ferguson-Smith A.C., and Cavaillé J. 2003.

Imprinted microRNA genes transcribed antisense to a reciprocally imprinted retrotransposon-like gene. *Nat. Genet.* **34:** 261.

Shay T.L., Berghmans S., Segers K., Meyers S., Beever J.E., Womack J.E., Georges M., Charlier C., and Cockett N.E. 2001. Fine mapping and construction of a bovine contig spanning the ovine callipyge locus. *Mamm. Genome* **12:** 141.

Smit M., Segers K., Shay T., Baraldi F., Gyapay G., Snowder G., Georges M., Cockett N., and Charlier C. 2003. Mosaicism of Solid Gold supports the causality of a non-coding A to G transition in the determinism of the callipyge phenotype. *Genetics* **163:** 453.

St. Germain D.L. and Galton V.A. 1997. The deiodinase family of selenoproteins. *Thyroid* **7:** 655.

Takada S., Tevendale M., Baker J., Georgiades P., Campbell E., Freeman T., Johnson M.H., Paulsen M., and Ferguson-Smith A.C. 2000. *Delta-like* and *Gtl2* are reciprocally expressed, differentially methylated linked imprinted genes on mouse chromosome 12. *Curr. Biol.* **10:** 1135.

Tsai C.-E., Lin S.-P., Ito M., Takagi N., Takada S., and Ferguson-Smith A.C. 2002. Genomic imprinting contributes to thyroid hormone metabolism in the mouse embryo. *Curr. Biol.* **12:** 1221.

Wylie A.A., Murphy S.K., Orton T.C., and Jirtle R.L. 2000. Novel imprinted *DLK1/GTL2* domain on human chromosome 14 contains motifs that mimic those implicated in *IGF2/H19* regulation. *Genome Res.* **10:** 1711.

Yevtodiyenko A., Carr M.S., Patel N., and Schmidt J.V. 2002. Analysis of candidate imprinted genes linked to Dlk1-Gtl2 using a congenic mouse line. *Mamm. Genome* **13:** 633.

Dscam-mediated Self- versus Non-Self-Recognition by Individual Neurons

G. NEVES AND A. CHESS

Center for Human Genetic Research, Department of Medicine, Massachusetts General Hospital and Harvard Medical School, and Whitehead Institute for Biomedical Research, Cambridge, Massachusetts 02142

Clearly, gene regulation (the turning on and off of the multitudinous genes in the genome) allows the specification of different parts of the nervous system as it allows specification of all parts of the developing animal. Even within a group of neurons, distinguishing similar neurons also can be accomplished by the specific turning on and off of various genes via differential activities of various transcription factors. Moreover, differences in levels of expression of different genes and both regulated and stochastic aspects of posttranscriptional and posttranslational events are all important.

Alternative splicing represents another important mechanism that can render distinct cell populations different from one another. The mechanisms regulating the generation of alternative mRNA transcripts are known to have tissue specificity, and elucidation of this type of regulation is an important area of exploration. The generally accepted concept is that tissue-specific accessory splicing factors bind to *cis*-acting elements in the primary transcript and thus regulate the relative levels of different possible alternative splicing choices.

The alternative splicing of the Down syndrome cell adhesion molecule (Dscam) gene in *Drosophila* represents a particularly striking example of extensive alternative splicing with 38,016 possible splice forms (Schmucker et al. 2000): exon 4 has 12 possible forms; exon 6 has 48 possible forms; and exon 9 has 33 possible forms. Each of these three alternative exons specifies a portion of an immunoglobulin domain in the extracellular region of the protein. The variable exons are quite divergent (as low as 22% identity at the amino acid level) so that the encoded immunoglobulin domains are predicted to have very different three-dimensional structures. The mammalian Dscam has been shown to have homophilic interactions in vitro (Agarwala et al. 2000); this had earlier led to the idea that the different Dscam isoforms will differ in their interactions (Brummendorf and Lemmon 2001). In fact, recent elegant biochemical experiments of Zipursky and colleagues (Wojtowicz et al. 2004) have suggested that each one of the 19,008 different extracellular domains of Dscam may be able to recognize each other specifically. This represents a specificity of interaction reminiscent of the exquisite specificity of the antibody–antigen interaction with the added interesting twist that the recognition molecule recognizes itself. Prior to our work, while the exact specificity of individual Dscam isoforms had not been studied, prevailing hypotheses suggested that the extraordinary diversity would be used to distinguish different types of neurons from one another. Having each distinct neuronal cell type express a restricted portion of the repertoire would contribute to the different characteristic properties of the different cell types. This type of model emanated from the prevailing thinking in the field regarding the specification of cellular properties through the transcriptional regulation of the genome. As detailed below, our experiments lead us to an unanticipated and striking finding that Dscam alternative splicing allows a stochastic aspect to the generation of diversity of neurons. We find that while different populations of neurons have few limitations on choice and are therefore probably not distinguished by Dscam, at the level of individual neurons things become more interesting: Each individual neuron appears to be different from its neighboring cells by virtue of expression of very different Dscam isoforms. This allows each neuron to know when a portion of its membrane is contacting itself as opposed to contacting an adjacent neuron (that could be identical except for the Dscam difference). This striking self- versus non-self- awareness, at the single-cell level, allows the possibility that two adjacent cells, with theoretically identical levels of transcription of every gene in the genome, are distinct from one another.

EXPERIMENTS ESTABLISHING THE DISPOSITION OF Dscam SPLICE FORMS

When we began our studies, the existence of 38,016 possible alternative splice forms of Dscam posed the question of whether Dscam expression can provide a molecular signature of a distinct subpopulation of cells. The inordinately large number of possible alternative splice forms made it technically difficult to use standard methods of analyzing expression patterns such as in situ hybridization, immunohistochemistry, and transgenic animals. Therefore, to circumvent these technical obstacles, we developed an approach combining a customized oligonucleotide microarray with a sensitive reverse transcriptase polymerase chain reaction (RT-PCR) assay that allowed us to examine not only distinct populations of cells, but also single cells. The ability to examine single cells turned out to be critical to discovering the way in which Dscam diversity is used in neuronal specification.

The experiments we carried out on individual cells have been published elsewhere (Neves et al. 2004) and will be summarized here. We designed a microarray containing probes for all 93 alternative exons in the three clusters that encode the variable extracellular immunoglobulin domains (exons 4, 6, and 9). Using the microarray, we first analyzed whole flies at various developmental time points. These initial experiments showed that splicing of Dscam is developmentally regulated, particularly with respect to exon 9. For example, 5 of the 33 exon 9 isoforms are expressed at significantly higher levels in embryos than at the larval stages. Our results also confirm previous observations that the relative representation of exon 4.2 increases from embryos to adults (Celloto and Graveley 2001).

We then turned to the question of how the Dscam diversity is used, asking whether subsets of Dscam isoforms define different cell types. This was an intriguing possibility regarding the regulation of alternative splicing of the Dscam gene: that particular cell populations would each express a specific small subset of all the available isoforms. We analyzed a variety of cell types to search for significant differences in use of the Dscam repertoire by a number of different types of neuronal and nonneuronal cells. For example, we used the GAL4-UAS system (Brand and Perrimon 1993) to label specific populations of cells in the developing eye, and showed that both R3/R4 and R7 populations of photoreceptors express a large number of different isoforms at all three exon clusters. Each type of photoreceptor still expressed a repertoire of Dscam forms that was calculated to have a complexity well over 10,000 different forms (out of ~19,000 possible combinations for the extracellular diverse exons 4, 6, and 9). These results ruled out the model in which a specific class of photoreceptor that projects to a particular brain region expresses one or a small number of distinct Dscam isoforms. The observed broad spectra notwithstanding, clustering analysis indicates that the profiles of Dscam isoforms expressed by the two populations are significantly different from each other and from the entire eye-antennal imaginal disk. Thus, each specific cell type expresses a broad yet distinctive spectrum of Dscam isoforms. The differences in repertoire usage in the distinct cell populations may be functionally important, but more likely are not the primary reason for having such distinct repertoires. The likely reason for the enormous repertoire of Dscam emerged only from our single-cell analyses (see below).

The use of the microarray also allowed us to examine Dscam expression in S2 cell lines and subclones thereof. A very striking profile of Dscam isoforms is expressed in S2 cell lines. While the usage of different isoforms of exons 4 and 6 is roughly equal, for exon 9 apparently only 5 out of the 33 isoforms are used. Thus, there is a somewhat restricted repertoire of Dscam forms present within the population of S2 cells. Similar profiles were also observed with two other cell lines that have, like S2 cells, been suggested to share similar properties with hemocytes. Hemocytes isolated by fluorescence-activated cell sorter (FACS) from third-instar larvae that express green fluorescent protein (GFP) under the control of the promoter region of the hemocyte marker hemolectin (Goto et al. 2001) also show a similar profile. Also note that the same exon 9 isoforms are overrepresented in both whole embryos, larval eye-antennal disks, and larval hemocytes. Whether the use of these particular isoforms is functionally relevant, or whether they represent a "default" splicing mode that can be overwritten in more differentiated cells, remains an open question.

SINGLE-CELL ANALYSES OF ALTERNATIVE SPLICING OF Dscam

We developed and applied a sensitive and accurate RT-PCR approach to determining the number of splice forms made by individual cells (Neves et al. 2004). The splitting of each single-cell RT reaction into multiple tubes allowed us to assess how robust the single-cell analysis was for each cell and to then employ statistical tests to arrive at firm conclusions regarding the disposition of the various splice forms in individual cells. Our first set of analyses included analyses of photoreceptor cells from the developing eye, and we also analyzed individual hemocytes. A number of important points emerged from our analyses. First, while a number of distinct Dscam transcripts were detected in each cell, there were not an extremely large number of transcripts detected. The range varied with the different cell types but allowed us to estimate that individual cells express in the range of 10 to 50 distinct mRNAs for Dscam at a given time. A second important point that emerged was the apparent randomness of the actual splice variants present in a given cell. These analyses indicate that a given cell's Dscam mRNA profile arises from a series of stochastic alternative splicing events for each Dscam transcript. The probability of selecting each individual alternative exon will be a function of the splicing factors expressed by each cell type. The choice of a small set of Dscam transcripts (~50) from among many thousands of possible choices ensures that each individual neuron will be different from its neighbors, in terms of the Dscam repertoire that it expresses at its surface.

SELF- VERSUS NON-SELF-DISCRIMINATION AT THE LEVEL OF INDIVIDUAL CELLS

Our results suggest that each transcribed mRNA molecule will contain a novel combination of alternatively spliced exons, chosen from the broad spectrum of splice variants expressed by a given cell type. Since each cell expresses far fewer transcripts (~50) than the number of possible splice variants (~19,000 extracellular forms), this results in every cell containing a Dscam repertoire that is different from those of its neighbors. The generation of a unique profile of cell surface protein isoforms in each cell through alternative splicing is an intriguing potential mechanism of specifying cell identity in the nervous system. It is conceivable that each cell could know itself as distinct from any other contacted cell solely by "reading" the contacted cell's pattern of Dscam isoforms in the context of its own Dscam repertoire. It is also possible that diverse cell compartments are generated using localized protein synthesis and mRNA localization mechanisms.

This concept is a type of self- versus non-self-distinction that has not been previously appreciated in the nervous system. Of course, in the immune system there are the well known examples where the expression of distinct T cell receptors by individual T cells allows each T cell a unique identity. Similarly, immunoglobulins on the surface of B cells afford unique identities to these cells. The unique identities of T and B cells are critical for the ability of these immune system cells to communicate with each other and coordinated the immune response. The concept of self versus non-self in the case of the immune system is referring to the complex ability of an "educated" T cell to distinguish cells from its own body as different from transplanted or infected cells. The unique identity given to each T cell by its T-cell receptor is central to the proper functioning of the immune system. It is interesting to note that Dscam alternative splicing provides an even greater difference between otherwise similar cells because, after cell division, while two daughter T cells will now share the same T-cell receptor, two daughter Dscam-expressing cells will over time elaborate distinct repertoires of Dscam alternative splice forms.

An interesting emerging model for how self- versus non-self-distinction is used by neurons invokes a dual function for Dscam. In this model—which is supported by studies of mushroom body neurons we published as part of a collaborative effort with Zipursky and colleagues that included genetic data (Zhan et al. 2004), as well as genetic studies of T. Lee and colleagues (Wang et al. 2004)—the first function is a homophilic adhesivity based on the overall structure common to all Dscam isoforms. In other words, irrespective of the exact isoform, any two molecules of Dscam whose extracellular domains encounter each other (in the correct orientation relative to each other) will have an adhesive interaction. This function is similar to the function ascribed to the mammalian Dscams (Barlow et al. 2002). The key to the model is the second function of Dscam. The second function is an ability to transduce a signal that leads to repulsion if two identical Dscam isoforms interact. It is possible that the triggering of the repulsive interaction could be accomplished by the strength of the adhesion or by some other signal transduction that can occur only when two identical Dscam molecules interact. This model is in certain ways reminiscent of antigen recognition by T cells especially during thymic education; the use of the strongest of interactions to mediate a negative signal has been observed in T-cell receptor interactions with peptide antigens presented in the context of the major histocompatibility complex (MHC).

Let us now consider how this model would play out when two portions of membrane touch one another. In the normal situation, two portions of membrane from two distinct axons would contact one another; Dscam molecules on each membrane would sense the presence of Dscam on the other membrane; and, then, through a combination of simple adhesion and perhaps a positive signal activating other adhesivity molecules, the Dscam interaction would promote fasciculation of the two axons.

Importantly, the sensing of the Dscam in this model would include a determination, as it were, of the fact that the two Dscam molecules are different from one another. In the unlikely event that the exact matching Dscam were to be present on the two interacting axons, then a negative signal would be generated, thus promoting repulsion.

Now consider the expectation under various abnormal settings. In the absence of Dscam the expectation is relatively straightforward: Without the Dscam interaction, the portions of membrane will not have the Dscam-mediated adhesive interaction (direct or indirect) and thus the axons will not fasciculate. In another scenario wherein the axons encountering each other have the same Dscam isoform, the second function of Dscam that can mediate a repulsive interaction wins the day and the axons will not fasciculate. To date, the available genetic studies (Wang et al. 2004; Zhan et al. 2004) are consistent with this model. The most striking data indicate that a Dscam null mutation can be rescued by a single form in a single-cell clone, but not in a larger clone. A single-cell clone expressing a single form of Dscam will be different from any cell it encounters and will therefore still be able to determine that it is different from its neighboring cells.

IS SELF- VERSUS NON-SELF-DISCRIMINATION A MORE GENERAL PRINCIPLE?

Extensive alternative splicing of Dscam orthologs is conserved in disparate insect species, and, while Dscam does not have extensive alternative splicing in mammals (Barlow et al. 2002), other cell surface protein families involved in neuronal development such as neurexins (Tabuchi and Sudhof 2002) and protocadherins (Wu et al. 2001) have numerous differentially spliced forms. The neurexins in mammals, for example, have well over 1000 possible forms and play roles in cell adhesion and synapse formation. Neurexins in flies and worms are much simpler and have little or no alternative splicing (Tabuchi and Sudhof 2002). The protocadherins also have an interesting complexity in mammals that is not found in fly homologs. Each protocadherin gene has a single constant region with a multitude of variable regions upstream of it. It appears that differential promoter usage leads to the observed expression of distinct combinations of isoforms in individual neurons. Whether the promoters are regulated like classical promoters, or through some alternative, stochastic mechanism remains to be determined. If stochastic regulation leads to the observed expression patterns, one would expect to observe the independent regulation of the two chromosomes leading to instances of monoallelic expression and instances of biallelic expression. Single-cell RT-PCR and microarray analyses of alternatively spliced isoforms could allow determination of the distribution of alternative isoforms for these other genes and their possible roles in specifying neuronal identity.

Each of these gene families that undergo extensive alternative splicing may be accomplishing two distinct functions. A "baseline" function would be whatever the gene is doing in the species in which it is not alternatively spliced. This baseline function would presumably need to be maintained even in the context of a second function of generating diversity at the cell surface of each individual neuron. This second function, a requirement for the generation of diversity, has been accomplished by increasing the splicing-mediated diversity of different genes in different types of organisms.

Any individual gene has to be looked at as one of many genes that influence any given behavior of a type of cell. The properties of neurons are especially complex. The significance of studying the alternative splicing of Dscam emanates from the extraordinary diversity of splice forms that are possible. Also adding to the significance is the pattern of usage of different splice forms by different cells that is beginning to emerge from our studies. Taken together with the fact that other genes such as neurexins may play similar roles in mammals, these studies will be an important component of an increasing understanding of the generation of an enormous amount of diversity in neuronal type in complex nervous systems. Upon the emergence of a more complete picture of the patterns of different splice forms displayed by different cells, we will be in a position to begin to examine the ways in which the enormous diversity is interplaying with other gene products in mediating specific properties of classes of neurons and individual neurons.

ACKNOWLEDGMENTS

We thank colleagues in the Chess laboratory for stimulating discussions. We are thankful for support of this work from the NIH (NINDS) to A.C.

REFERENCES

Agarwala K.L., Nakamura S., Tsutsumi Y., and Yamakawa K. 2000. Down syndrome cell adhesion molecule Dscam mediates homophilic intercellular adhesion. *Brain Res. Mol. Brain Res.* **79:** 118.

Barlow G.M., Micales B., Chen X.N., Lyons G.E., and Korenberg J.R. 2002. Mammalian Dscams: Roles in the development of the spinal cord, cortex, and cerebellum? *Biochem. Biophys. Res. Commun.* **293:** 881.

Brand A.H. and Perrimon N. 1993. Targeted gene expression as a means of altering cell fates and generating dominant phenotypes. *Development* **118:** 401.

Brummendorf T. and Lemmon V. 2001. Immunoglobulin superfamily receptors: *cis*-interactions, intracellular adapters and alternative splicing regulate adhesion. *Curr. Opin. Cell Biol.* **13:** 611.

Celotto A.M. and Graveley B.R. 2001. Alternative splicing of the *Drosophila* Dscam pre-mRNA is both temporally and spatially regulated. *Genetics* **159:** 599.

Goto A., Kumagai T., Kumagai C., Hirose J., Narita H., Mori H., Kadowaki T., Beck K., and Kitagawa Y.A. 2001. *Drosophila* haemocyte-specific protein, hemolectin, similar to human von Willebrand factor. *Biochem. J.* **359:** 99.

Neves G., Zucker J., Daly M., and Chess A. 2004. Stochastic yet biased expression of multiple Dscam splice variants by individual cells. *Nat. Genet.* **36:** 240.

Schmucker D., Clemens J.C., Shu H., Worby C.A., Xiao J., Muda M., Dixon J.E., and Zipursky S.L. 2000. *Drosophila* Dscam is an axon guidance receptor exhibiting extraordinary molecular diversity. *Cell* **101:** 671.

Tabuchi K. and Sudhof T.C. 2002. Structure and evolution of neurexin genes: Insight into the mechanism of alternative splicing. *Genomics* **79:** 849.

Wang J., Ma X., Yang J.S., Zheng X., Zugates C.T., Lee C.-H.J., and Lee T. 2004. Transmembrane/juxtamembrane domain-dependent Dscam distribution and function during mushroom body neuronal morphogenesis. *Neuron* **43:** 663.

Wojtowicz W.M., Flanagan J.J., Millard S.S., Zipursky S.L., and Clemens J.C. 2004. Alternative splicing of *Drosophila* Dscam generates axon guidance receptors that exhibit isoform-specific homophilic binding. *Cell* **118:** 619.

Wu Q., Zhang T., Cheng J.F., Kim Y., Grimwood J., Schmutz J., Dickson M., Noonan J.P., Zhang M.Q., Myers R.M., and Maniatis T. 2001. Comparative DNA sequence analysis of mouse and human protocadherin gene clusters. *Genome Res.* **11:** 389.

Zhan X.L., Clemens J.C., Neves G., Hattori D., Flanagan J.J., Hummel T., Vasconcelos M.L., Chess A., and Zipursky S.L. 2004. Analysis of Dscam diversity in regulating axon guidance in *Drosophila* mushroom bodies. *Neuron* **43:** 673.

Prions of Yeast Are Genes Made of Protein: Amyloids and Enzymes

R.B. WICKNER, H.K. EDSKES, E.D. ROSS, M.M. PIERCE, F. SHEWMAKER, U. BAXA, AND A. BRACHMANN

Laboratory of Biochemistry and Genetics, National Institute of Diabetes, Digestive and Kidney Diseases, National Institutes of Health, Bethesda, Maryland 20892-0830

In 1994 we described two infectious proteins (prions) of *Saccharomyces cerevisiae*, showing that the nonchromosomal genes [URE3] and [PSI] are inactive, self-propagating forms of Ure2p and Sup35p, respectively (Wickner 1994). Since then, [Het-s], a prion of *Podospora anserina* (Coustou et al. 1997), and [PIN+], another *S. cerevisiae* prion (Derkatch et al. 1997, 2001; Sondheimer and Lindquist 2000), have been found. All of the above prions are based on self-propagating amyloids. Recently, we described another prion, called [β], which is based on the *trans* self-activation by the yeast vacuolar protease B (Roberts and Wickner 2003). Evidence has appeared suggesting that C, the non-Mendelian gene of *Podospora anserina* determining Crippled Growth, is also a self-activating prion of a MAP kinase kinase kinase (Kicka and Silar 2004). Study of these yeast and fungal phenomena has established that proteins can be genes and infectious entities and has revealed many details of what makes a protein infectious.

"EPIGENETICS" AND "PRIONS": WHAT DOES IT ALL MEAN?

"Epigenetic" has been defined as ". . . heritable changes in gene function that cannot be explained by changes in DNA sequence" (Riggs and Porter 1997). The prions (infectious proteins) alter the expression of genes without changes in DNA sequence, but do so by altering the gene product, without necessarily altering the function of the gene itself. However, "epigenetic" is more broadly used to mean any heritable change not due to a change in nucleic acid sequence.

The notion of infectious proteins originated in attempts to explain the relative radiation resistance of the scrapie agent (Alper et al. 1967; Griffith 1967). The term "prion," which we now use to mean "infectious protein without a needed nucleic acid" (the protein–only model), was later coined as a synonym for the agent causing transmissible spongiform encephalopathies (TSEs), without requiring that this be its mechanism, but including this possibility (Prusiner 1982). In spite of two Nobel prizes, it seems that there is still doubt in some quarters about whether the TSEs are infectious proteins (Farquhar et al. 1998; Chesebro 2003; Priola et al. 2003). Our view is that there is considerable evidence pointing in this direction (see Wickner et al. 2004), but we remain agnostic about whether this is enough. Two potential "final proofs" have famously not yet worked: (a) amyloid formed in vitro from recombinant PrP is not infectious for normal mice, and (b) brain extracts of transgenic mice dying with a scrapie-like syndrome as a result of carrying PrP genes with mutations that cause inherited Creutzfeldt–Jakob disease (CJD) in humans are again not infectious for normal mice.

FOUR TYPES OF PRION

There are four known types of prion, depending on whether the altered form of the protein is an amyloid or not, and whether the prion is toxic to the individual, is simply inactive for its normal function, or positively helps the cell (Fig. 1). In the TSEs, assuming they are infectious proteins, the amyloid form of PrP is cytotoxic. The [URE3] and [PSI+] prions of yeast are simply inac-

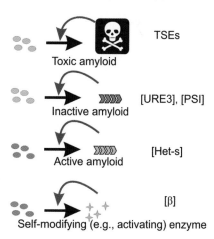

Figure 1. Four prion types. The known prions differ in whether they are based on amyloid or not, and what effect they have on the organism. The TSEs are lethal because of a novel harmful effect of the amyloid form of PrP. Cells with [URE3] and [PSI] partially lose the normal function of Ure2p and Sup35p, respectively. In [Het-s] strains, the HETs protein carries out a function for the cell only in the amyloid form (and is generally found in this form in wild isolates). The [β] prion is simply the active form of the vacuolar protease B, has no relation to amyloid, and functions for the cell in starvation and meiosis.

tive (or partially inactive) amyloid forms of yeast Ure2p and Sup35p, respectively. [Het-s] of *Podospora anserina* is an amyloid-based prion that performs a function for the cell, helping in the heterokaryon incompatibility that this organism uses to avoid harmful infectious agents (such as viruses and mitochondrial senility factors) that spread by fusion of cellular processes (hyphal anastomosis). The [β] prion of *S. cerevisiae* is not amyloid-based but is simply an enzyme whose active form is necessary *in trans* for activation of its own inactive precursor. [β] functions for the cell in promoting survival under starvation conditons and allowing meiosis and spore formation.

GENETIC CRITERIA FOR PRIONS

The first yeast prions were discovered based on their paradoxical genetic properties, properties that are unknown (and unexpected) for nucleic acid replicons, but are expected for prions in which the phenotype produced is due to simple inactivation of the prion protein's normal function (Fig. 2) (Wickner 1994). These properties were emphatically not the observed properties of the TSEs, both because they were designed for a prion of a microorganism and because TSEs cannot be a consequence of simple inactivation of PrP, since deletion of the *PRNP* gene produces essentially no phenotype (Bueler et al. 1992).

Reversible Curing

If a prion can be cured, then it should be possible for it to arise again in the cured strain, because the protein with the potential to become the prion is still present. Reversible curing is to be distinguished from revertible mutation. Curing is the efficient elimination of a nonchromosomal gene by some treatment or condition that, for a nucleic acid replicon, would mean elimination of the replicon—generally an irreversible event. Mutation is generally not so efficient and can revert.

Overproduction of the Protein Increases Frequency of Prion Generation

More of the protein with the potential to become a prion should result in a higher frequency of de novo prion generation, regardless of the mechanism.

Gene Required for Prion Propagation and Phenotype Relationship

The same gene whose overexpression increases prion generation should be essential for prion propagation. Furthermore, for prions of the "inactive amyloid" type (see Fig. 1), the phenotype of the presence of the prion should resemble the phenotype of a recessive mutation in the gene, since in each case there is a deficiency of the normal form.

All three genetic criteria apply strictly to the prions of the inactive amyloid type, indeed to any prion whose phenotype is due to self-propagating inactivation of the nor-

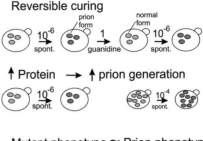
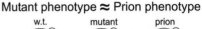

Figure 2. Genetic criteria for a prion. These three properties are expected of a nonchromosomal gene if it is a prion, but are not expected if it is a nucleic acid replicon (Wickner 1994).

mal form (whether by amyloid formation or not). The final genetic feature only half applies to other prions; the gene encoding the protein must be necessary for the propagation of the prion, but the phenotype relation will not be as described (Wickner 1994).

[URE3] AND [PSI+] ARE PRIONS OF Ure2p AND Sup35p, RESPECTIVELY

The non-Mendelian gene [URE3] and the chromosomal *ure2* mutant were discovered by Lacroute in the same genetic selection for strains that could take up ureidosuccinate to replace the absence of that intermediate in a *ura2* (aspartate transcarbamylase) mutant (USA+) (Lacroute 1971; Drillien and Lacroute 1972). The phenotypes of [URE3] and *ure2* strains are thus similar, and, remarkably, *ure2* mutants cannot propagate [URE3] (Drillien and Lacroute 1972; Aigle and Lacroute 1975; Wickner 1994). [URE3] can be cured by growth of cells in the presence of millimolar guanidine (Aigle and Lacroute 1975), but from the cured clones can again be isolated [URE3] subclones (Wickner 1994). Finally, overproduction of Ure2p (*URE2* mRNA does not suffice) induces a 100-fold increase in the frequency with which [URE3] arises de novo (Wickner 1994). This showed that [URE3] is a prion of Ure2p. When yeast is presented with a good nitrogen source, such as ammonia, glutamine, or glutamate, it represses the production of the enzymes and transporters needed to utilize poor nitrogen sources. Ure2p is a central player in this process, binding the Gln3p transcription factor and keeping it in the cytoplasm under repressing conditions (for review, see Cooper 2002).

[PSI+] is a nonchromosomal gene that increases the efficiency of weak non-sense suppressor tRNAs (Cox 1965; Cox et al. 1988). Results similar to those for [URE3] implied that [PSI+] is a prion of the Sup35 protein (Wickner 1994). [PSI+] is efficiently cured by growth in hyperosmotic media (Singh et al. 1979), but

[PSI+] derivatives can again be isolated from cured cells (Lund and Cox 1981). Overproduction of Sup35p increases the frequency with which [PSI+] arises (Chernoff et al. 1993; Derkatch et al. 1996). Finally, the phenotype of the presence of [PSI] resembles that of recessive *sup35* mutants, and [PSI] propagation depends on the *SUP35* gene (Doel et al. 1994; Ter-Avanesyan et al. 1994). Sup35p is a subunit of the translation termination factor (Frolova et al. 1994; Stansfield et al. 1995), explaining these effects on suppression of non-sense mutations.

These results showed that [URE3] and [PSI+] are prions of Ure2p and Sup35p (Wickner 1994), respectively; these conclusions have been amply confirmed by numerous later studies. For [URE3], it was shown that Ure2p becomes protease-resistant in [URE3] strains (Masison and Wickner 1995), and that this protease resistance was not a consequence of the altered nitrogen regulation (Masison et al. 1997). It is the Ure2 protein (not the mRNA or DNA gene) whose overproduction induces the de novo appearance of [URE3] (Masison and Wickner 1995; Masison et al. 1997), a result that may well be sufficient to prove a prion. Sup35p is aggregated in [PSI+] cells (Patino et al. 1996; Paushkin et al. 1996) and it is the protein, not the RNA or gene, that induces [PSI+] to arise de novo (Derkatch et al. 1996).

THE AMINO TERMINUS OF Ure2p CONSTITUTES A PRION DOMAIN

Deletion analysis of Ure2p showed that the amino-terminal 65 residues were sufficient to induce the appearance of [URE3] (Masison and Wickner 1995). This region was also sufficient to propagate [URE3] in the complete absence of the carboxy-terminal part of the molecule (Masison et al. 1997). In contrast, the carboxy-terminal residues 66–354 (Coschigano and Magasanik 1991; Masison and Wickner 1995) or even residues 90–354 (Maddelein and Wickner 1999) are sufficient to carry out the nitrogen regulation function of Ure2p (Fig. 3). These results indicated that the interaction between Ure2p molecules that propagated the [URE3] prion were between one amino-terminal prion domain and another, and that the carboxy-terminal part of the molecule was only the indicator of the prion change, but not its mediator.

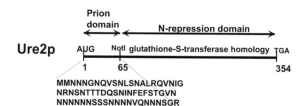

Figure 3. Prion domain of Ure2p. Deletion analysis showed that residues 1–65 are sufficient to induce [URE3] prion formation and necessary for Ure2p to participate in the prion change (Masison and Wickner 1995; Masison et al. 1997). This region can also propagate [URE3] in the absence of the carboxy-terminal domain. Residues 66–354 can carry out the nitrogen regulation function of Ure2p (Masison and Wickner 1995).

THE [URE3] PRION IS A SELF-PROPAGATING AMYLOID FORM OF Ure2p

The protease-resistance of Ure2p in [URE3] strains (Masison and Wickner 1995), and the appearance of Ure2-GFP fusion protein in [URE3] cells (Edskes et al. 1999a) suggested that the [URE3] change of Ure2p involved aggregation. Chemically synthesized Ure2p$^{1-65}$, the prion domain peptide, spontaneously forms amyloid in vitro, suggesting a role for this abnormal protein form in the [URE3] prion (Taylor et al. 1999). Just as overproduced Ure2p$^{1-65}$ induces the full-length protein to undergo the prion change in vivo, the same prion domain peptide induces amyloid formation by the otherwise stably soluble full length Ure2p in vitro (Taylor et al. 1999). This amyloid formation could be further propagated by addition of more native Ure2p. The pattern of protease-resistant bands of Ure2p seen in extracts of [URE3] cells (Masison and Wickner 1995) is identical to that seen for amyloid of Ure2p formed in vitro (Taylor et al. 1999). In each case, the most protease-resistant part of the molecule is the prion domain itself (Masison and Wickner 1995; Taylor et al. 1999; Baxa et al. 2003). It is possible to detect networks of filaments specifically in [URE3]-carrying cells (Speransky et al. 2001). Using antibody to Ure2p linked to gold particles, it was shown that these filaments contained Ure2p.

ARCHITECTURE OF AMYLOID FILAMENT FORMED FROM Ure2p

The fact that the prion domain could propagate the [URE3] prion in the complete absence of the carboxy-terminal domain (Masison et al. 1997) implies that the primary interaction between molecules in the prion (amyloid) form of Ure2p is between one amino-terminal prion domain and another. This conclusion was confirmed by showing that protease digestion of filaments formed in vitro or Ure2p in extracts of [URE3] cells leaves a protease-resistant core consisting of the prion domain. The filaments become more narrow on digestion and are similar to those formed from the synthetic prion domain peptide (Taylor et al. 1999; Baxa et al. 2003).

The carboxy-terminal functional domain is only inactivated by prion formation if it is covalently attached to the prion domain (Masison et al. 1997). This could mean that the carboxy-terminal domain is conformationally changed by association with the amyloid-forming prion domain, or that amyloid formation makes the carboxy-terminal domain sterically unable to interact with Gln3p, its normal function. A series of fusion proteins were prepared consisting of the amino-terminal prion domain linked to enzymes whose structures are known and whose substrates are too small to be subject to steric hindrance by amyloid formation. Each of these fusion proteins readily formed amyloid filaments and in each case the enzyme was fully active in the amyloid filaments. This surprising result is particularly significant in the case of the Ure2p$^{1-65}$-glutathione-S-transferase fusion protein because the carboxy-terminal part of Ure2p is homologous to GSTs in both sequence and structure (Coschigano and Magasanik 1991; Bousset et al. 2001; Umland et

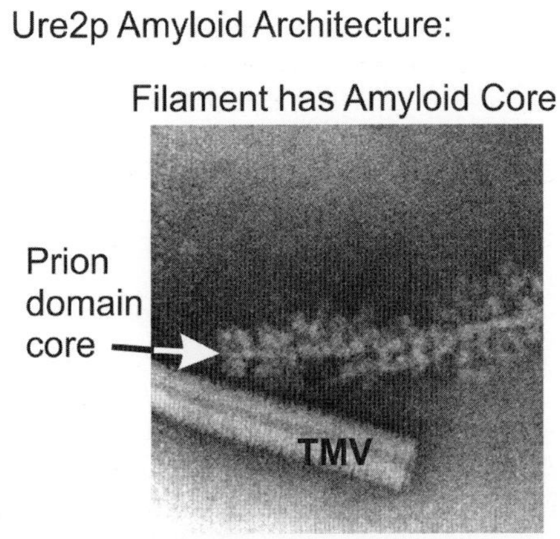

Ure2p Amyloid Architecture:

Filament has Amyloid Core

Figure 4. Architecture of amyloid filaments of Ure2p. Scanning transmission electron micrograph of a Ure2p amyloid filament showing the central core of amyloid, composed of the prion domain, and the appended carboxy-terminal domain (Baxa et al. 2003).

2001). This indicates that the carboxy-terminal domain of Ure2p is not conformationally changed in the prion (amyloid) form, but is unable to interact with Gln3p nonetheless. One intriguing possible explanation (Baxa et al. 2002) is that Gln3p association with Ure2p is a diffusion-limited reaction that is inefficient when Ure2p is restricted to a small part of the cytoplasm in the network of filaments (Speransky et al. 2001).

Scanning transmission electron microscopic images of Ure2p amyloid filaments show that each has a central core around which is arrayed larger masses attached rather loosely to the central core (Fig. 4) (Baxa et al. 2003). The central core of these filaments was shown by Western blot and mass spectrometry to be comprised of Ure2p residues 1–65 (with some fraying in some molecules) (Baxa et al. 2003). The masses arrayed around the core are the GST-like carboxy-terminal domains, consisting of residues 95–354. The carboxy-terminal domains are connected to the amyloid cores by a tether comprising roughly residues 70–95. This region remains highly protease-sensitive in the amyloid form, when both the prion domain and carboxy-terminal domain are structured (Baxa et al. 2003).

The structure of the amyloid core remains to be determined. The recent demonstration by Tycko and coworkers of a parallel β-sheet structure of the Aβ peptide (Petkova et al. 2002; Tycko 2003) suggests that a similar structure may be found for the Ure2p prion domain amyloid. All Ure2p amyloid filaments have one monomer per 0.45 nm, consistent with this type of structure (Baxa et al. 2003).

AMINO ACID COMPOSITION IS THE MAIN DETERMINANT OF Ure2p PRION FORMATION

The N/Q-rich nature of the Ure2p and Sup35p prion domains (Ter-Avanesyan et al. 1994; Masison and Wickner 1995) and the demonstrated importance of Q residues in Sup35p (DePace et al. 1998) and runs of N in Ure2p (Maddelein and Wickner 1999) implicate at least this feature as important in prion formation. However, there are over 100 yeast proteins with N/Q-rich regions, and, aside from Ure2p and Sup35p, only Rnq1p has been shown to be a prion (Derkatch et al. 1997, 2001; Sondheimer and Lindquist 2000). To determine whether there were sequence requirements for prion (or amyloid) formation by Ure2p, we constructed a series of derivatives in which the amino acids of the prion domain of Ure2p had been randomly shuffled without changing the overall composition or codon usage (Fig. 5). We found that all five such random sequences were capable of becoming prions in vivo and the purified proteins were able to form amyloid in vitro (Ross et al. 2004). This shows that, in the case of Ure2p, the major factor determining prion (and amyloid) formation is amino acid composition. There were differences in how robust were the prions formed by the various shuffled derivatives, showing that sequence is indeed a lesser factor but a factor nonetheless (Ross et al. 2004). Future studies will determine whether other prion domains can be shuffled without loss of activity, what aspects of composition (aside from N/Q richness) are important, and what are the sequence determinants of prion robustness.

Shuffled prion domains:

	Prion	Amyloid
Wild-type Ure2: MMNNNGNQVSNLSNALRQVNIGNRNSNTTTDQSNINFEFSTGVNNNNNNN SSSNNNNVQNNNSGRNGSQNNDNENNIKNTLEQHRQQQQ	+	+
Scrambled sequence #1: MVDGNQMNNKSRRNSSQRGNSNQRVNNQNENNFNGLAQSSNNNNSITTT FTNNNQINSQLNGINNNVNQTDQNVQNHGNSNENNSENL	+	+
Scrambled sequence #2: MQSHQAESNSSQNGDQNGTNNLQNNRSNGINNFGNNNRNQNNLESQRVNN TINNNKLNQFNGNNEVNNVQNQSSDNTNNNMSIVTTRNS	+	+
Scrambled sequence #3: MNIRNQNQSTAVLNVNQQSNNGTSNSVNNLNFNNSGMQNHGRNFNQSTRN NNTNEKGGNNILNSNDERINNQQNQENNNTVDNSQNNSS	+	+
Scrambled sequence #4: MMQRNGQQEGTNNNHHSNINTQRNVFNNSANNNRNNNEGLNNNNSNFNNLV SNNQQVNVSSNSNINNQDNNKSILSGTSNDTTENRGQQQ	+/–	+
Scrambled sequence #5: MNTNNSQGSFVDENQNRSIVKSRTVNMSQNNNTGNNNNAQLNNILNNTDS GHVSNNENRLGRQNNDFNQNSSQTNNGNNQQQSNNNNNI	+	+

Conclusion: For the Ure2 prion domain, prion-forming ability is determined by amino acid content, not by amino acid sequence.

Figure 5. Shuffled prion domains can be prions and form amyloid. Residues 1–89 of Ure2p were randomly ordered, and replaced the normal sequence at the normal *URE2* chromosomal locus. The amino acid content was unchanged, but the sequence was completely random. Ability of each of these five randomized proteins to form prions in vivo and amyloid in vitro was tested and each was capable of both (Ross et al. 2004).

CHAPERONES INVOLVED IN [URE3] PROPAGATION

That [PSI+] requires the disaggregating chaperone Hsp104 for its propagation and the curing of [PSI+] by overproduction of Hsp104 were crucial findings, opening up a vital new area in prion (and amyloid) studies (Chernoff et al. 1995). In addition, guanidine curing of yeast prions has been shown to occur by inhibition of Hsp104 (Ferreira et al. 2001; Jung and Masison 2001; Jung et al. 2002). Guanidine treatment was used as a tool to study the mechanism of Hsp104's contribution to [PSI+] propagation. It was found that guanidine treatment had no immediate effect on the incorporation of newly synthesized Sup35p into aggregates, but instead blocked the formation of new prion seeds (Ness et al. 2002; Tuite and Cox 2003). Hsp70s have also been implicated in prion propagation (Chernoff et al. 1999; Newnam et al. 1999; Jung et al. 2000; Jones and Masison 2003). Likewise for the Hsp40 family and various cochaperones (Kushnirov et al. 2000; Jones et al. 2004).

We find that propagation of [URE3] requires Hsp104, but that, unlike [PSI+], overproduction of Hsp104 does not cure [URE3] (Moriyama et al. 2000). However, overproduction of Ydj1p, an Hsp40 family member, does cure [URE3] (Moriyama et al. 2000). Mutation of *SSA2*, encoding the most abundant of the cytoplasmic Hsp70s, also results in loss of [URE3] (Roberts et al. 2004). Interestingly, overexpression of Ssa1p eliminates [URE3], although identical overexpression of Ssa2p does not (Schwimmer and Masison 2002).

GENETIC CONTROL OF [URE3] GENERATION: Mks1p AND [PIN+]

Our finding that overproduction of Mks1p negatively affects the ability of Ure2p to regulate nitrogen catabolism (Edskes et al. 1999b) suggested the possibility that Mks1p might likewise affect prion generation or propagation. We found that *mks1Δ* strains were nearly unable to generate [URE3] derivatives, even when Ure2p or one of its highly prion-inducing derivatives was overexpressed (Edskes and Wickner 2000). Modest overproduction of Mks1p resulted in a substantial increase in [URE3] prion generation. The same *mks1Δ* strains were fully competent to propagate [URE3] and to show the USA+ phenotype typical of [URE3] strains (Edskes and Wickner 2000). This shows that the effect of *mks1Δ* on prion formation was an effect on prion generation, not an effect on propagation or detection of the [URE3] prion.

MKS1 was first defined as a gene whose overexpression slows growth, an effect antagonized by the *RAS*-cAMP pathway (Matsuura and Anraku 1993). We found that expression of the constitutive ("oncogenic") Ras^{val19} mutant dramatically inhibited [URE3] generation, as one would expect if it inhibited Mks1p action (Edskes and Wickner 2000).

The [PIN+] nonchromosomal genetic element was discovered as a factor necessary for [PSI+] inducibility on overproduction of Sup35p (Derkatch et al. 1997). Subsequent genetic analysis showed that [PIN+] is a prion of the Rnq1 protein, a nonessential protein rich in asparagine and glutamine residues throughout most of its length (Derkatch et al. 2000, 2001; Sondheimer and Lindquist 2000). The [PIN+] effect may be a result of cross-seeding between Q/N-rich amyloid formed by Rnq1 and the prion domain of Sup35p. A similar, but less dramatic, effect is seen with [URE3]. [PIN+] strains are 10- to 100-fold more frequently induced to form [URE3] than are isogenic [pin–] strains (Bradley et al. 2002).

Ure2 HOMOLOGS IN PATHOGENIC YEASTS

Although Ure2p is weakly homologous to many GSTs, and it has not been found to have GST activity in vitro, it is reported to confer a low level of metal resistance (Rai et al. 2003). However, close relatives of Ure2p are found in a series of yeasts, including other species of *Saccharomyces*, *Schizosaccharomyces pombe*, *Candida albicans*, *Candida lypolytica*, *Candida maltosa*, *Ashbya gossypii*, and others (Edskes and Wickner 2002). Most of these foreign *URE2* genes were able to complement an *S. cerevisiae ure2Δ* mutant, reflecting the 80–100% identity in their carboxy-terminal domains (Edskes and Wickner 2002). Their amino-terminal domains show far lower similarity, and, in the case of *Saccharomyces pombe*, no amino-terminal domain at all. Those amino-terminal extensions are all Q/N-rich, suggesting some utility for the prion domain. Indeed, expression of only the carboxy-terminal domain gives a leaky phenotype, with some expression of normally nitrogen-repressed genes (Masison and Wickner 1995). Since [URE3] and *ure2* mutant cells grow slowly on most media, it is likely that the evolutionarily selected function of the Ure2p prion domain is not for the rare prion formation but rather for the promotion of the normal function of Ure2p. However, there is at least one case of a prion that is functional for its host, the [Het-s] prion of *Podospora*.

[Het-s], AN ADAPTIVE AMYLOID PRION OF *PODOSPORA ANSERINA*

When two colonies of a filamentous fungus meet, they fuse their cellular processes (hyphae) perhaps to share nutrients, but only if they are genetically identical. Genetic identity is tested by checking identity at around a dozen polymorphic loci scattered around the genome. These are call *het* loci in *Podospora* and include *het-s*, whose alleles are *het-s* and *het-S*. When colonies meet, a few hyphae fuse in a sort of trial. If they differ in alleles at even a single *het* locus, these fused hyphae degenerate and die and set up a barrier to further fusions. This reaction is called "heterokaryon incompatibility." The *het-s* locus encodes a 289-amino-acid protein differing at 13 residues between the *het-s* and *het-S* alleles.

Studies by Saupe and colleagues at Bordeaux have shown that only if the *het-s* protein is in a prion form is the proper incompatibility reaction observed (Coustou et al. 1997; Saupe et al. 2000; Maddelein et al. 2002). This prion is called [Het-s] and is based on a self-propagating amyloid form of the protein encoded by the *het-s* allele. Since heterokaryon incompatibility is a normal fungal function, possibly designed to prevent the spread of debilitating fungal viruses and plasmids between unrelated strains, it is reasonable to assume that this prion is carrying out a function for the cell. This hypothesis is supported by the fact that essentially all isolates of *Podospora* with the *het-s* allele carry the prion.

All yeast and fungal prions are nonchromosomal genes that spread via cell–cell fusion such as occurs in sexual matings or in nonsexual matings such as the hyphal anastomosis discussed above for *Podospora*. This infectious mode of spread means that these elements should gradually invade the population unless they are a disadvantage to their hosts. Thus, finding a prion in the wild does not imply it is an advantage, but failure to find it in the wild is a strong indication that the prion is a disadvantage to

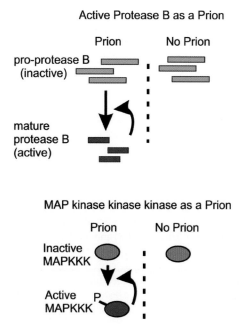

Figure 6. A new type of prion: an enzyme needed for its own activation. Vacuolar protease B of *S. cerevisiae* can be a prion in the absence of protease A, which normally catalyzes its activation (Roberts and Wickner 2003). The properties of the "C" nonchromosomal gene of the filamentous fungus *Podospora anserina* suggest that it too may be a prion of this type, based on self-activation of a MAP kinase kinase kinase (Kicka and Silar 2004). It is possible that the MAP KKK activation is done by one of the kinases further down in the cascade, MAP KK or MAP K.

the host. For example, two efforts to find the [PSI+] prion in wild strains have been unsuccessful (Chernoff et al. 2000; Resende et al. 2002), indicating that this prion is, on the net, a disadvantage to the host.

[β], A DIFFERENT KIND OF PRION

The word prion means "infectious protein," and, as first discussed by Griffith (Griffith 1967), need not involve amyloid. If the active (or, more generally, modified) form of an enzyme is necessary for its own activation (or modification), then a cell that starts out without the active (modified) form will never develop it unless it comes from outside the cell. A cell with the active (modified) form will continue to activate (modify) newly synthesized protein molecules and should pass the trait "has active enzyme" on to its offspring. Transmission of the active protein to a cell lacking it should likewise transmit the self-propagating trait as well (Roberts and Wickner 2003).

Based on extensive work by Beth Jones (Carnegie Mellon University; Zubenko and Jones 1982) and Deiter Wolf (University of Stuttgart; Teichert et al. 1989), we have identified such a prion—namely, the active form of the vacuolar protease B of *S. cerevisiae* encoded by the *PRB1* gene (Fig. 6) (Roberts and Wickner 2003). Protease B, encoded by *PRB1*, is made as an inactive precursor, whose activation by deletion of amino-terminal and carboxy-terminal extensions is carried out in normal cells by combined action of itself and protease A (for review, see Jones 1991). However, in mutants deleted for the protease A structural gene (*pep4Δ*), protease B self-activation can continue indefinitely as cells grow (Roberts and Wickner 2003). The active state can be transmitted by cytoplasmic mixing from a cell with active protease B to one lacking it, showing that this trait is a nonchromosomal gene. The state in which cells have protease B activity is called [β] and its absence [β-o].

[β] has the characteristics expected of a prion for which the prion form is the active form of the protein. Cells that start out with protease B activity give rise to offspring, nearly all of which have active enzyme. Such a strain is efficiently cured of protease B activity by growth on rich dextrose media, a condition that partially represses *PRB1* transcription, leading to the self-activation cycle rapidly dying out. But of these cured cells, about 1 in 10^5 descendants has spontaneously developed protease B activity (reversible curing). Overexpression of *PRB1* increases this frequency to several percent (genetic criterion 2 for a prion) (Roberts and Wickner 2003). [β] depends on *PRB1* for its propagation, but, of course, the phenotype of [β] is not the same as a *prb1* mutant because the prion form is the active form of the protein.

[β] is advantageous to the cell, since [β] cells can undergo meiosis, while [β-o] cells cannot, and cells with [β] survive starvation better than do those without (Roberts and Wickner 2003).

The importance of the [β] prion lies not in its biochemical mechanism, which has been recognized for decades, but in the facts that (1) a protease can be a gene and (2) there are many potentially self-modifying enzymes (protein kinases, protein acetylases, protein methylases, etc.) that, under some circumstances, might become prions. Indeed, recent work on the Crippled Growth phenomenon of *Podospora anserina* appears to be just such a new prion (Kicka and Silar 2004).

CRIPPLED GROWTH AS A NEW PRION OF *PODOSPORA ANSERINA*

Crippled growth (CG) is a condition that can develop in *Podospora* which is characterized by increased pigmentation and slowed growth (Silar et al. 1999). CG is the result of carrying the nonchromosomal gene "C." If a piece of a colony with CG is placed on a non-CG colony, heterokaryon formation leads to cytoplasmic mixing and the non-CG colony acquires the CG trait and the C genetic element.

Recent work has shown that the propagation of C (not just the phenotype, but the cytoplasmic gene) requires a MAP kinase kinase kinase similar to that encoded by *BCK1* of *S. cerevisiae*, and called PaASK1 because of its homology to human ASK1 (Kicka and Silar 2004). Overexpression of PaASK1 results in generation of the C element under conditions where its formation would be rare. C can be cured, but will arise anew in the cured strains. The phenotype of C is different from that of mutants in PaASK1, indicating that it is not inactivity of PaASK1 that produces C (Kicka and Silar 2004). While evidence of altered activity of the kinase cascade in C-carrying strains will be needed, it seems likely that C is a new case of a prion of the same type as [β] (Fig. 6). As previously suggested by Silar (Silar and Daboussi 1999), filamentous fungi may prove to be a rich source of prions. Their syncytial structure allows newly arisen prions to spread throughout the colony, and heterokaryon formation provides a rapid route for infection of neighbors.

CONCLUSIONS

The discovery of yeast prions has ended the debate about whether there can be such a thing as a prion, although it has not ended the debate about whether TSEs are such entities. The ease of genetic manipulation of yeast has led to identification of a number of cellular components affecting the generation and propagation of prions, and produced a great deal of insight into prion phenomena, long pioneered in the more difficult animal systems. The new class of nonamyloid prions (now that there are apparently two of them, we are justified in calling it a "class") expands the possibilities for "infectious proteins" to a host of new mechanisms. It is expected that yeast and other fungi will continue to be useful in revealing the range of prion phenomena and their mechanisms.

REFERENCES

Aigle M. and Lacroute F. 1975. Genetical aspects of [URE3], a non-Mendelian, cytoplasmically inherited mutation in yeast. *Mol. Gen. Genet.* **136:** 327.

Alper T., Cramp W.A., Haig D.A., and Clarke M.C. 1967. Does the agent of scrapie replicate without nucleic acid? *Nature* **214:** 764.

Baxa U., Speransky V., Steven A.C., and Wickner R.B. 2002. Mechanism of inactivation on prion conversion of the *Saccharomyces cerevisiae* Ure2 protein. *Proc. Natl. Acad. Sci.* **99:** 5253.

Baxa U., Taylor K.L., Wall J.S., Simon M.N., Cheng N., Wickner R.B., and Steven A. 2003. Architecture of Ure2p prion filaments: The N-terminal domain forms a central core fiber. *J. Biol. Chem.* **278:** 43717.

Bousset L., Beirhali H., Janin J., Melki R., and Morera S. 2001. Structure of the globular region of the prion protein Ure2 from the yeast *Saccharomyces cerevisiae*. *Structure* **9:** 39.

Bradley M.E., Edskes H.K., Hong J.Y., Wickner R.B., and Liebman S.W. 2002. Interactions among prions and prion "strains" in yeast. *Proc. Natl. Acad. Sci.* (suppl. 4) **99:** 16392.

Bueler H., Fischer M., Lang Y., Bluethmann H., Lipp H.P., DeArmond S.J., Prusiner S.B., Aguet M., and Weissmann C. 1992. Normal development and behavior of mice lacking the neuronal cell-surface PrP protein. *Nature* **356:** 577.

Chernoff Y.O., Derkach I.L., and Inge-Vechtomov S.G. 1993. Multicopy SUP35 gene induces de-novo appearance of psi-like factors in the yeast *Saccharomyces cerevisiae*. *Curr. Genet.* **24:** 268.

Chernoff Y.O., Lindquist S.L., Ono B.-I., Inge-Vechtomov S.G., and Liebman S.W. 1995. Role of the chaperone protein Hsp104 in propagation of the yeast prion-like factor [psi$^+$]. *Science* **268:** 880.

Chernoff Y.O., Newnam G.P., Kumar J., Allen K., and Zink A.D. 1999. Evidence for a protein mutator in yeast: Role of the Hsp70-related chaperone Ssb in formation, stability and toxicity of the [PSI+] prion. *Mol. Cell. Biol.* **19:** 8103.

Chernoff Y.O., Galkin A.P., Lewitin E., Chernova T.A., Newnam G.P., and Belenkly S.M. 2000. Evolutionary conservation of prion-forming abilities of the yeast Sup35 protein. *Mol. Microbiol.* **35:** 865.

Chesebro B. 2003. Introduction to the transmissible spongiform encephalopathies or prion diseases. *Br. Med. Bull.* **66:** 1.

Cooper T.G. 2002. Transmitting the signal of excess nitrogen in *Saccharomyces cerevisiae* from the Tor proteins to the GATA factors: Connecting the dots. *FEMS Microbiol. Rev.* **26:** 223.

Coschigano P.W. and Magasanik B. 1991. The *URE2* gene product of *Saccharomyces cerevisiae* plays an important role in the cellular response to the nitrogen source and has homology to glutathione S-transferases. *Mol. Cell. Biol.* **11:** 822.

Coustou V., Deleu C., Saupe S., and Begueret J. 1997. The protein product of the *het-s* heterokaryon incompatibility gene of the fungus *Podospora anserina* behaves as a prion analog. *Proc. Natl. Acad. Sci.* **94:** 9773.

Cox B.S. 1965. PSI, a cytoplasmic suppressor of super-suppressor in yeast. *Heredity* **20:** 505.

Cox B.S., Tuite M.F., and McLaughlin C.S. 1988. The Psi factor of yeast: A problem in inheritance. *Yeast* **4:** 159.

DePace A.H., Santoso A., Hillner P., and Weissman J.S. 1998. A critical role for amino-terminal glutamine/asparagine repeats in the formation and propagation of a yeast prion. *Cell* **93:** 1241.

Derkatch I.L., Bradley M.E., Hong J.Y., and Liebman S.W. 2001. Prions affect the appearance of other prions: The story of *[PIN]*. *Cell* **106:** 171.

Derkatch I.L., Bradley M.E., Zhou P., Chernoff Y.O., and Liebman S.W. 1997. Genetic and environmental factors affecting the de novo appearance of the [PSI+] prion in *Saccharomyces cerevisiae*. *Genetics* **147:** 507.

Derkatch I.L., Chernoff Y.O., Kushnirov V.V., Inge-Vechtomov S.G., and Liebman S.W. 1996. Genesis and variability of [PSI] prion factors in *Saccharomyces cerevisiae*. *Genetics* **144:** 1375.

Derkatch I.L., Bradley M.E., Masse S.V., Zadorsky S.P., Polozkov G.V., Inge-Vechtomov S.G., and Liebman S.W. 2000. Dependence and independence of [PSI^+] and [PIN^+]: A two-prion system in yeast? *EMBO J.* **19:** 1942.

Doel S.M., McCready S.J., Nierras C.R., and Cox B.S. 1994. The dominant $PNM2^-$ mutation which eliminates the [PSI] factor of *Saccharomyces cerevisiae* is the result of a missense mutation in the *SUP35* gene. *Genetics* **137:** 659.

Drillien R. and Lacroute F. 1972. Ureidosuccinic acid uptake in yeast and some aspects of its regulation. *J. Bacteriol.* **109:** 203.

Edskes H.K. and Wickner R.B. 2000. A protein required for prion generation: [URE3] induction requires the Ras-regulated Mks1 protein. *Proc. Natl. Acad. Sci.* **97:** 6625.

———. 2002. Conservation of a portion of the *S. cerevisiae* Ure2p prion domain that interacts with the full-length protein. *Proc. Natl. Acad. Sci.* (suppl. 4) **99:** 16384.

Edskes H.K., Gray V.T., and Wickner R.B. 1999a. The [URE3] prion is an aggregated form of Ure2p that can be cured by overexpression of Ure2p fragments. *Proc. Natl. Acad. Sci.* **96:** 1498.

Edskes H.K., Hanover J.A., and Wickner R.B. 1999b. Mks1p is a regulator of nitrogen catabolism upstream of Ure2p in *Saccharomyces cerevisiae*. *Genetics* **153:** 585.

Farquhar C.F., Somerville R.A., and Bruce M.E. 1998. Straining the prion hypothesis. *Nature* **391:** 345.

Ferreira P.C., Ness F., Edwards S.R., Cox B.S., and Tuite M.F. 2001. The elimination of the yeast [PSI+] prion by guanidine hydrochloride is the result of Hsp104 inactivation. *Mol. Microbiol.* **40:** 1357.

Frolova L., LeGoff X., Rasmussen H.H., Cheperegin S., Drugeon G., Kress M., Arman I., Haenni A.-L., Celis J.E., Philippe M., Justesen J., and Kisselev L. 1994. A highly conserved eukaryotic protein family possessing properties of polypeptide chain release factor. *Nature* **372:** 701.

Griffith J.S. 1967. Self-replication and scrapie. *Nature* **215:** 1043.

Jones E.W. 1991. Three proteolytic systems in the yeast *Saccharomyces cerevisiae*. *J. Biol. Chem.* **266:** 7963.

Jones G.W. and Masison D.C. 2003. *Saccharomyces cerevisiae* Hsp70 mutations affect [PSI(+)] prion propagation and cell growth differently and implicate Hsp40 and tetratricopeptide repeat cochaperones in impairment of [PSI(+)]. *Genetics* **163:** 495.

Jones G., Song Y., Chung S., and Masison D.C. 2004. Propagation of yeast [PSI+] prion impaired by factors that regulate Hsp70 substrate binding. *Mol. Cell. Biol.* **24:** 3928.

Jung G. and Masison D.C. 2001. Guanidine hydrochloride inhibits Hsp104 activity *in vivo*: A possible explanation for its effect in curing yeast prions. *Curr. Microbiol.* **43:** 7.

Jung G., Jones G., and Masison D.C. 2002. Amino acid residue 184 of yeast Hsp104 chaperone is critical for prion-curing by guanidine, prion propagation, and thermotolerance. *Proc. Natl. Acad. Sci.* **99:** 9936.

Jung G., Jones G., Wegrzyn R.D., and Masison D.C. 2000. A role for cytosolic Hsp70 in yeast [PSI+] prion propagation and [PSI+] as a cellular stress. *Genetics* **156:** 559.

Kicka S. and Silar P. 2004. PaASK1, a mitogen-activated protein kinase kinase kinase that controls cell degeneration and cell differentiation in *Podospora anserina. Genetics* **166:** 1241.

Kushnirov V.V., Kryndushkin D.S., Boguta M., Smirnov V.N., and Ter-Avanesyan M.D. 2000. Chaperones that cure yeast artificial [PSI+] and their prion-specific effects. *Curr. Biol.* **10:** 1443.

Lacroute F. 1971. Non-Mendelian mutation allowing ureidosuccinic acid uptake in yeast. *J. Bacteriol.* **106:** 519.

Lund P.M. and Cox B.S. 1981. Reversion analysis of [psi-] mutations in *Saccharomyces cerevisiae. Genet. Res.* **37:** 173.

Maddelein M.-L. and Wickner R.B. 1999. Two prion-inducing regions of Ure2p are non-overlapping. *Mol. Cell. Biol.* **19:** 4516.

Maddelein M.-L., Dos Reis S., Duvezin-Caubet S., Coulary-Salin B., and Saupe S.J. 2002. Amyloid aggregates of the HET-s prion protein are infectious. *Proc. Natl. Acad. Sci.* **99:** 7402.

Masison D.C. and Wickner R.B. 1995. Prion-inducing domain of yeast Ure2p and protease resistance of Ure2p in prion-containing cells. *Science* **270:** 93.

Masison D.C., Maddelein M.-L., and Wickner R.B. 1997. The prion model for [URE3] of yeast: Spontaneous generation and requirements for propagation. *Proc. Natl. Acad. Sci.* **94:** 12503.

Matsuura A. and Anraku Y. 1993. Characterization of the *MKS1* gene, a new negative regulator of the ras-cyclic AMP pathway in *Saccharomyces cerevisiae. Mol. Gen. Genet.* **238:** 6.

Moriyama H., Edskes H.K., and Wickner R.B. 2000. [URE3] prion propagation in *Saccharomyces cerevisiae:* Requirement for chaperone Hsp104 and curing by overexpressed chaperone Ydj1p. *Mol. Cell. Biol.* **20:** 8916.

Ness F., Ferreira P., Cox B.S., and Tuite M.F. 2002. Guanidine hydrochloride inhibits the generation of prion "seeds" but not prion protein aggregation in yeast. *Mol. Cell. Biol.* **22:** 5593.

Newnam G.P., Wegrzyn R.D., Lindquist S.L., and Chernoff Y.O. 1999. Antagonistic interactions between yeast chaperones Hsp104 and Hsp70 in prion curing. *Mol. Cell. Biol.* **19:** 1325.

Patino M.M., Liu J.-J., Glover J.R., and Lindquist S. 1996. Support for the prion hypothesis for inheritance of a phenotypic trait in yeast. *Science* **273:** 622.

Paushkin S.V., Kushnirov V.V., Smirnov V.N., and Ter-Avanesyan M.D. 1996. Propagation of the yeast prion-like [*psi*$^+$] determinant is mediated by oligomerization of the *SUP35*-encoded polypeptide chain release factor. *EMBO J.* **15:** 3127.

Petkova A.T., Ishii Y., Balbach J.J., Antzutkin O.N., Leapman R.D., Delaglio F., and Tycko R. 2002. A structural model for Alzheimer's beta-amyloid fibrils based on experimental constraints from solid state NMR. *Proc. Natl. Acad. Sci.* **99:** 16742.

Priola S.A., Chesebro B., and Caughey B. 2003. Biomedicine. A view from the top—Prion diseases from 10,000 feet. *Science* **300:** 917.

Prusiner S.B. 1982. Novel proteinaceous infectious particles cause scrapie. *Science* **216:** 136.

Rai R., Tate J.J., and Cooper T.G. 2003. Ure2, a prion precursor with homology to glutathione S-transferase, protects *Saccharomyces cerevisiae* cells from heavy metal ion and oxidant toxicity. *J. Biol. Chem.* **278:** 12826.

Resende C., Parham S.N., Tinsley C., Ferreira P.C., Duarte J.A.B., and Tuite M.F. 2002. The *Candida albicans* Sup35p protein (CaSup35p): Function, prion-like behavior and an associated polyglutamine length polymorphism. *Microbiology* **148:** 1049.

Riggs A.D. and Porter T.N. 1997. Overview of epigenetic mechanisms. In *Epigenetic mechanisms of gene regulation* (ed. V.E.A. Russo et al.), p. 29. Cold Spring Harbor Laboratory Press, Cold Spring Harbor, New York.

Roberts B.T., Moriyama H., and Wickner R.B. 2003. [URE3] prion propagation is abolished by a mutation of the primary cytosolic Hsp70 of budding yeast. *Yeast* **21:** 107.

Roberts B.T. and Wickner R.B. 2004. A class of prions that propagate via covalent auto-activation. *Genes Dev.* **17:** 2083.

Ross E.D., Baxa U., and Wickner R.B. 2004. Scrambled prion domains form prions and amyloid. *Mol. Cell. Biol.* **24:** 7206.

Saupe S.J., Clave C., and Begueret J. 2000. Vegetative incompatibility in filamentous fungi: *Podospora* and *Neurospora* provide some clues. *Curr. Opin. Microbiol.* **3:** 608.

Schwimmer C. and Masison D.C. 2002. Antagonistic interactions between yeast [PSI+] and [URE3] prions and curing of [URE3] by Hsp70 protein chaperone Ssa1p but not by Ssa2p. *Mol. Cell. Biol.* **22:** 3590.

Silar P. and Daboussi M.J. 1999. Non-conventional infectious elements in filamentous fungi. *Trends Genet.* **15:** 141.

Silar P., Haedens V., Rossingnol M., and Lalucque H. 1999. Propagation of a novel cytoplasmic, infectious and deleterious determinant is controlled by translational accuracy in *Podospora anserina. Genetics* **151:** 87.

Singh A.C., Helms C., and Sherman F. 1979. Mutation of the non-Mendelian suppressor ψ^+ in yeast by hypertonic media. *Proc. Natl. Acad. Sci.* **76:** 1952.

Sondheimer N. and Lindquist S. 2000. Rnq1: An epigenetic modifier of protein function in yeast. *Mol. Cell* **5:** 163.

Speransky V., Taylor K.L., Edskes H.K., Wickner R.B., and Steven A. 2001. Prion filament networks in [URE3] cells of *Saccharomyces cerevisiae. J. Cell Biol.* **153:** 1327.

Stansfield I., Jones K.M., Kushnirov V.V., Dagkesamanskaya A.R., Poznyakovski A.I., Paushkin S.V., Nierras C.R., Cox B.S., Ter-Avanesyan M.D., and Tuite M.F. 1995. The products of the *SUP45* (eRF1) and *SUP35* genes interact to mediate translation termination in *Saccharomyces cerevisiae. EMBO J.* **14:** 4365.

Taylor K.L., Cheng N., Williams R.W., Steven A.C., and Wickner R.B. 1999. Prion domain initiation of amyloid formation *in vitro* from native Ure2p. *Science* **283:** 1339.

Teichert U., Mechler B., Muller H., and Wolf D.H. 1989. Lysosomal (vacuolar) proteinases of yeast are essential catalysts for protein degradation, differentiation, and cell survival. *J. Biol. Chem.* **264:** 16037.

Ter-Avanesyan A., Dagkesamanskaya A.R., Kushnirov V.V., and Smirnov V.N. 1994. The *SUP35* omnipotent suppressor gene is involved in the maintenance of the non-Mendelian determinant [psi+] in the yeast *Saccharomyces cerevisiae. Genetics* **137:** 671.

Tuite M.F. and Cox B.S. 2003. Propagation of yeast prions. *Nat. Rev. Mol. Cell Biol.* **4:** 878.

Tycko R. 2003. Insights into the amyloid folding problem from solid-state NMR. *Biochemistry* **42:** 3151.

Umland T.C., Taylor K.L., Rhee S., Wickner R.B., and Davies D.R. 2001. The crystal structure of the nitrogen catabolite regulatory fragment of the yeast prion protein Ure2p. *Proc. Natl. Acad. Sci.* **98:** 1459.

Wickner R.B. 1994. Evidence for a prion analog in *S. cerevisiae*: The [URE3] non-Mendelian genetic element as an altered *URE2* protein. *Science* **264:** 566.

Wickner R.B., Edskes H.K., Ross E.D., Pierce M.M., Baxa U., Brachmann A., and Shewmaker F. 2004. Prion genetics: New rules for a new kind of gene. *Annu. Rev. Genet.* **38:** 681.

Zubenko G.S., Park F.J., and Jones E.W. 1982. Genetic properties of mutations at the *PEP4* locus in *Saccharomyces cerevisiae. Genetics* **102:** 679.

A Possible Epigenetic Mechanism for the Persistence of Memory

K. Si,* S. Lindquist,† and E. Kandel*

*Columbia University, Center for Neurobiology, New York, New York 10032;
†Whitehead Institute, Cambridge, Massachusetts 02142

Synaptic plasticity, the change in synaptic efficacy in response to external cues, has been shown to be one of the critical cellular mechanisms of memory storage (Bliss and Collingridge 1993; Lisman and McIntyre 2001; Morris 2003). Like behavioral memory, synaptic plasticity has at least two temporally distinct forms: a short-term form lasting minutes and a long-term form lasting days. At the molecular level long-term synaptic plasticity differs from short-term synaptic plasticity in that it requires the synthesis of new mRNA and protein (Steward and Schuman 2001; Malinow and Malenka 2002; Ehlers 2003) and is accompanied by the structural alteration of preexisting synapses and growth of new synapses (Bailey and Kandel 1993; Yuste and Bonhoeffer 2001). Even though it requires the synthesis of new mRNA in the nucleus and the nucleus is shared by all the synapses of a given cell, long-term synaptic plasticity can still be synapse specific. How could a genetic program that is cell wide give rise to synapse specificity? Work in rodents and in the marine snail *Aplysia* has given rise to the idea that the synapse could be selectively marked by synaptic stimulation. The synaptic marking results in either selective transport or selective utilization of plasticity-related molecules only at the marked synapse (Martin et al. 1997; Frey and Morris 1998). An early attempt to characterize the molecular nature of the synaptic mark revealed that there is a rapamycin-sensitive, local protein synthesis–dependent component that is needed for the long-term maintenance of synaptic facilitation (Casadio et al. 1999).

To begin to address how synaptic change can be maintained over a long time, we focused on this rapamycin-sensitive, local protein synthesis–dependent component of the synaptic mark. Since mRNAs are made in the cell body and presumably can be transported to all the synapses of a neuron, the need for the local translation of some mRNAs may reflect a requirement that these mRNAs may be dormant until they reach the activated synapse. If this were true, one way of activating synaptic protein synthesis would be to recruit a translational regulator capable of activating translationally dormant mRNAs. In search of such a molecule we focused on cytoplasmic polyadenylation element binding protein (CPEB), a molecule that activates dormant mRNAs through the elongation of their polyA tail (Mendez and Richter 2001). CPEB was first identified in oocytes and subsequently in hippocampal neurons (Hake and Richter 1994; Wu et al. 1998). However, there is a novel, neuron-specific isoform of CPEB active in the processes of marine snail *Aplysia* sensory neurons (Liu and Schwartz 2003; Si et al. 2003b). Stimulation with serotonin, a neurotransmitter released during learning in *Aplysia*, increases the amount of CPEB protein in the synapse (Si et al. 2003b). The increase in the amount of CPEB coincides with the activation of mRNAs encoding synaptic structural molecules such as actin and tubulin. This CPEB isoform is not unique to *Aplysia*. A similar neuronal CPEB is also present in *Drosophila*, mice, and humans (Theis et al. 2003). In *Aplysia* the local activity of CPEB is needed not for the initiation or early expression of long-term facilitation, but for its long-term maintenance (Si et al. 2003b). These data suggest that the maintenance, but not the initiation, of long-term synaptic plasticity requires a new set of molecules in the synapse and that some of these new molecules are made by CPEB-dependent translational activation. These results also support the idea that there are separate mechanisms for the initiation of long-term synaptic plasticity and its maintenance.

The observation that the activity of CPEB, a translation regulator, is needed not for the initiation but rather for the maintenance of long-term synaptic facilitation is quite intriguing. It raises some fundamental questions about the molecular basis of long-term synapse-specific changes: (1) Is there a continuous need for the local synthesis of a set of molecules to maintain the learning-related synaptic changes over long periods of time? (2) Is the experience-dependent molecular change of the synapse indeed maintained for a long time? (3) If so, since biological molecules (such as CPEB) have a relatively short half-life (hours to days) compared with the duration of memories (years), how is the altered molecular composition of a synapse maintained for such a long time?

A plausible answer to how unstable molecules can produce a stable change in the synaptic strength came from the subsequent finding that the neuronal isoform of CPEB shares properties with prion-like proteins (Si et al. 2003a). Prions are proteins that can assume at least two stable conformational states (Prusiner 1994; Wickner et al. 1999; Uptain and Lindquist 2002). Usually one of these conformational states is active while the other is inactive. Furthermore, one of the conformational states, the prion state, is self-perpetuating, promoting the conformational conversion of other proteins of the same type. We have found that *Aplysia* neuronal CPEB exists in two stable, physical states that are functionally distinct. As with other prions, one of these states has the ability to self-perpetuate in a dominant epigenetic fashion. This dominant form is the active form of the protein capable of activating translationally dormant mRNAs.

Based on these properties of *Aplysia* neuronal CPEB, we propose a model for the perpetuation of long-term synapse-specific changes. We hypothesize that CPEB in the neuron has at least two conformational states: one is inactive, or acts as a repressor, while the other is active. In a naive synapse, the basal level of CPEB is low, but unlike conventional prions the protein in this state is in its inactive or repressive state. An increase in the amount of neuronal CPEB by serotonin (or other neurotransmitters), either by itself or in conjunction with other signals, triggers the conversion of CPEB to the prion-like, self-perpetuating state. The prion-like state is either more active or devoid of the inhibitory function of the basal state. Once the prion state is established in an activated synapse, dormant mRNAs that are made in the cell body and distributed globally to all synapses are activated locally through the activated CPEB. Because the activated CPEB can be self-perpetuating, it could contribute to a self-sustaining, synapse-specific, long-term molecular change and provide a mechanism for the persistence of memory.

REFERENCES

Bailey C.H. and Kandel E.R. 1993. Structural changes accompanying memory storage. *Annu. Rev. Physiol.* **55**: 397.

Bliss T.V. and Collingridge G.L. 1993. A synaptic model of memory: Long-term potentiation in the hippocampus. *Nature* **361**: 31.

Casadio A., Martin K.C., Giustetto M., Zhu H., Chen M., Bartsch D., Bailey C.H., and Kandel E.R. 1999. A transient, neuron-wide form of CREB-mediated long-term facilitation can be stabilized at specific synapses by local protein synthesis. *Cell* **99**: 221.

Ehlers M.D. 2003. Activity level controls postsynaptic composition and signaling via the ubiquitin-proteasome system. *Nat. Neurosci.* **6**: 231.

Frey U. and Morris R.G. 1998. Synaptic tagging: Implications for late maintenance of hippocampal long-term potentiation. *Trends Neurosci.* **21**: 181.

Hake L.E. and Richter J.D. 1994. CPEB is a specificity factor that mediates cytoplasmic polyadenylation during *Xenopus* oocyte maturation. *Cell* **79**: 617.

Lisman J.E. and McIntyre C.C. 2001. Synaptic plasticity: A molecular memory switch. *Curr. Biol.* **11**: R788.

Liu J. and Schwartz J.H. 2003. The cytoplasmic polyadenylation element binding protein and polyadenylation of messenger RNA in *Aplysia* neurons. *Brain Res.* **959**: 68.

Malinow R. and Malenka R.C. 2002. AMPA receptor trafficking and synaptic plasticity. *Annu. Rev. Neurosci.* **25**: 103.

Martin K.C., Casadio A., Zhu H., Yaping E., Rose J.C., Chen M., Bailey C.H., and Kandel E.R. 1997. Synapse-specific, long-term facilitation of aplysia sensory to motor synapses: A function for local protein synthesis in memory storage. *Cell* **91**: 927.

Mendez R. and Richter J.D. 2001. Translational control by CPEB: A means to the end. *Nat. Rev. Mol. Cell Biol.* **2**: 521.

Morris R.G. 2003. Long-term potentiation and memory. *Philos. Trans. R. Soc. Lond. B Biol. Sci.* **358**: 643.

Prusiner S.B. 1994. Biology and genetics of prion diseases. *Annu. Rev. Microbiol.* **48**: 655.

Si K., Lindquist S., and Kandel E.R. 2003a. A neuronal isoform of the aplysia CPEB has prion-like properties. *Cell* **115**: 879.

Si K., Giustetto M., Etkin A., Hsu R., Janisiewicz A.M., Miniaci M.C., Kim J.H., Zhu H., and Kandel E.R. 2003b. A neuronal isoform of CPEB regulates local protein synthesis and stabilizes synapse-specific long-term facilitation in aplysia. *Cell* **115**: 893.

Steward O. and Schuman E.M. 2001. Protein synthesis at synaptic sites on dendrites. *Annu. Rev. Neurosci.* **24**: 299.

Theis M., Si K., and Kandel E.R. 2003. Two previously undescribed members of the mouse CPEB family of genes and their inducible expression in the principal cell layers of the hippocampus. *Proc. Natl. Acad. Sci.* **100**: 9602.

Uptain S.M. and Lindquist S. 2002. Prions as protein-based genetic elements. *Annu. Rev. Microbiol.* **56**: 703.

Wickner R.B., Edskes H.K., Maddelein M.L., Taylor K.L., and Moriyama H. 1999. Prions of yeast and fungi. Proteins as genetic material. *J. Biol. Chem.* **274**: 555.

Wu L., Wells D., Tay J., Mendis D., Abbott M.A., Barnitt A., Quinlan E., Heynen A., Fallon J.R., and Richter J.D. 1998. CPEB-mediated cytoplasmic polyadenylation and the regulation of experience-dependent translation of alpha-CaMKII mRNA at synapses. *Neuron* **21**: 1129.

Yuste R. and Bonhoeffer T. 2001. Morphological changes in dendritic spines associated with long-term synaptic plasticity. *Annu. Rev. Neurosci.* **24**: 1071.

An Epigenetic Hypothesis for Human Brain Laterality, Handedness, and Psychosis Development

A.J.S. KLAR

Gene Regulation and Chromosome Biology Laboratory, National Cancer Institute at Frederick, Frederick, Maryland 21702-1201

Mendelian genetics deals with heredity of variation through generations, and it does not directly concern itself with cellular differentiation required for development in higher multicellular organisms. Indeed, in the 1940s, developmental biologists, believing that genes were not important for development, went their separate ways from geneticists. How could the multiple arrays of cell types be generated when all cells of an organism possess the same set of genes questioning the importance of genetics to development? They thought something other than genes must be important for development. Only after the discovery of the phenomenon of gene regulation in *Escherichia coli* (Jacob and Monod 1961) did these fields begin to converge. Questions still remain unanswered about how development occurs. What controls the expressions of genes that regulate development with such remarkable precision such that different organs develop only at specific locations on the body? Thus far, the major paradigm for developmental gene regulation comprises the "morphogen-gradient model" (Brown and Wolpert 1990) that remains to be experimentally verified. For example, it is not known what controls the development of human brain laterality whereby left and right hemispheres structurally and functionally differ from one another. Because there is an unexplained partial association of human left- versus right-hand-use preference with the hemisphere of the brain that develops language, handedness and brain laterality development must be somehow related. Furthermore, an excess of left-handedness is associated with psychiatric diseases of schizophrenia and bipolar disorders; anomalies of brain laterality development may be relevant for disease causation. In this broad overview, we will discuss possible connections between the seemingly unrelated fields of cell-type determination in fission yeast and brain development in humans. We will summarize results of studies that suggest that the traits of brain laterality, handedness, and psychosis development result from an embryonic asymmetric cell division. A mitotic cell genetic ("mitogenetics") model postulates that a Watson versus Crick DNA strand-specific imprinting of a chromosome and its patterned segregation produces differentiated daughter cells and that psychosis results from genetics, without a conventional Mendelian gene mutation in families carrying chromosome 11 rearrangements.

COMPLEX CORRELATION OF BRAIN HEMISPHERIC LATERALITY WITH HAND-USE PREFERENCE

The brain contains left and right hemispheres, and often they are functionally nonequivalent. The one that processes the language, logic, and some math functions is called the "dominant" hemisphere, while the one that deals with things such as spatial perception, intuition, and creativity is called the "automatic" hemisphere (Klar 1999). The functional specialization is associated with development of structural differences in specific regions of the brain, such as the Broca`s area that processes language cognition. Compared to the left/right-axis specification of visceral organs such as heart, lungs, liver, and stomach in vertebrates (Klar 1994), human brain hemispheric structural differences are relatively subtle, although prominent differences lie with the distribution of cognitive functions. What is fascinating is that 97% of individuals who prefer to use their right hand (right-handers, RH) for unimanual tasks, such as handwriting, develop a dominant hemisphere in the left side and the balance process language in the right or both hemispheres (Coren 1992). Thus, RH preference is highly correlated with brain laterality. The situation with left-handers and ambidextrous individuals, collectively termed non-right-handers (NRH), is rather unclear but more interesting because in them the handedness trait and brain laterality are not correlated. Thus, there is a very clear association between these traits in most RH, but not in NRH. The reason and the significance of this complex association have remained a key unanswered question in the fields of brain development and psychology. Because of such association, psychologists have conducted all studies for defining the etiology of handedness. As nearly all RH develop a lateralized brain such that the dominant hemisphere develops on the left side of the brain, clearly brain laterality must be biologically/genetically specified. Had it not been genetically determined, there should have been random left- versus right-sided distribution of specialized hemispheres, as may be the case with NRH (Klar 1996). The mechanism for developing differentiated hemispheres remains unknown, but the basis for the complex association between these traits is becoming clearer. As a way to directly define the biological basis of brain laterality development, many studies in the past have been designed to define the etiology of its indirect measure, i.e., of handedness preference.

HANDEDNESS ETIOLOGY

The causes of hand-use preference have been debated for centuries. Often this preference is considered to be a complex trait where several genes contribute, along with the culturally taught behavior, to affect one's hand practice. Three main arguments have been advanced against a strictly genetic etiology. First, nearly one-half of the children of NRH x NRH cross are RH (Rife 1940). Second, in certain cultures the left-hand use is discouraged, so the hand preference can be culturally influenced. Third, and the pastiest problem, is that 18% of monozygotic twins, each co-twin possessing identical genetic constitutions, are discordant—one "lefty" and one "righty"—for handedness (Rife 1940). Because of these observations, hand preference is proposed in many studies to be a culturally taught behavior. Others posit brain damage during birth or a sudden surge of testosterone levels during pregnancy to influence one's handedness (for review, see Coren 1992). Such psychology models are difficult to test as they fail to make verifiable predictions. Instead, three one-locus models proposed random (50% RH: 50% NRH) handedness of individuals carrying the nonfunctional allele. Two of these models propose additive effects of two alleles of a locus (Annett 1985; McManus 1991), while the more recent "random-recessive model" proposes a fully penetrant and dominant *RGHT1* (for right-handedness) allele and a recessive *r* (for *r*andom) allele (Klar 1996). Results of several relatively recent tests of the random-recessive model are summarized below.

First, the value of 7.6% NRH born to RH x RH parents (Rife 1940) is predicted from the calculated 61% *RGHT1*: 39% *r* allele frequency of RH individuals. This allele frequency was derived from the result of 19.6% NRH born to RH x NRH parents (Klar 1996). Second, both RH and NRH born to NRH x NRH are expected from the random handedness of *r/r* homozygotes (Annett 1985; McManus 1991; Klar 1996). Third, RH (whose both parents were NRH) x standard RH produce increased proportion of NRH progeny, a result similar to the RH x NRH cross and different from that of the standard RH x RH cross (Klar 1996). Fourth, RH (twin of a discordant pair) x RH crosses produced a higher proportion of NRH children, similar to the NRH twin and the standard NRH when they are married with standard, unselected RH (Klar 2003). Thus, the genotype of discordant twins must be similar to that of NRH. Fifth, and the most decisive, finding comprises the recently discovered partial association of handedness with the direction of perital hair-whorl rotation found on the top of human head. Over 95% of individuals support a single whorl that rotates clockwise or counterclockwise. Moreover, a majority (91.6%) of individuals in the general public carry clockwise hair whorls. Clearly, the direction of hair-whorl rotation is genetically determined. The most telling result is that NRH develop random hair-whorl orientation. Confirming this finding, individuals chosen only because of their counterclockwise swirls were equally divided between RH and NRH. Thus, the genetics that cause one to become RH must also cause the development of a clockwise hair whorl, and the genetics that cause NRH lead to randomness in hair-whorl orientation. Therefore, because of the association of handedness to brain laterality, we conclude that the *RGHT1* allele causes the coupled development of dominant hemisphere on the left side of the brain, RH preference, and clockwise hair rotation. Those with the *r/r* constitution develop these traits, but the traits are uncoupled from each other and are distributed randomly to the left versus right side of the body. We therefore propose that the counterclockwise whorl orientation signifies the *r/r* genotype. This phenotype is much easier and definitive to score than handedness, and can be exploited to study the contribution of genetics of brain laterality development to other behavioral traits. For example, a 3.6-fold excess of counterclockwise whorl rotation is found in homosexual men (A.J.S. Klar, unpubl.). In the general public, only 9–11% of persons are NRH (Rife 1940). Curiously, the value of NRH goes up ~3-fold in patients suffering from major psychiatric diseases of schizophrenia and bipolar-affective disorders (Boklage 1977). Explaining this association is the main reason for deciphering the handedness etiology that in turn may shed light on psychosis etiology.

PSYCHOSES ETIOLOGY REMAINS ENIGMATIC

Schizophrenia is characterized by the inability of patients to differentiate between real and imaginary images and voices, suffering delusions and hallucinations, and the inability to function socially. Bipolar disease patients fluctuate between bouts of manic depression and high activity called mania. There is, however, considerable overlap in the symptoms of both diseases, including suicidal behavior, so that several studies have suggested a common etiology for both diseases. Both disorders are highly debilitating and remain incurable. Each disease affects about 1% of the population worldwide. Family, adoption, and twins studies point to genetic factors but, thus far, unambiguous identification of the important susceptibility genes has not been accomplished (for review, see Kennedy et al. 2003). A large number of studies of genome-wide scans produced significant evidence for linkage to many regions of the genome, but no result has been convincingly replicated. The lack of replication has been frequently attributed to the small number of patients investigated. However, a recent pooling of the data of large pedigrees from several studies failed to find significant linkage to chromosome regions for both disorders (Lewis et al. 2003; Segurado et al. 2003). Despite that, consensus of the field is that these are multifactor or "complex" traits and that they result from the combined effects of multiple genetic and environmental factors that interact with each other (Kennedy et al. 2003). As the consensus is based on negative results of not finding major disease-causing loci, the scientific basis for the consensus was recently questioned (Klar 2002). The best guess at present is that diseases result from anomalies of brain development as patients have less lateralized brain hemispheres, compared with healthy controls (Crow 1990; DeLisi et al. 1997). Significantly, NRH single borns and handedness-discordant twins also exhibit reduced brain hemisphere laterality, although they are not diseased (DeLisi et al. 1997; Geschwind et al. 2002). As noted above, psychotic patients are three times more likely to be NRH as compared with controls; it has been suggested that the genetics causing NRH may be a disease-predisposing factor (Boklage 1977; Annett 1985; Crow 1990; McManus 1991; Klar 1999). However, the exact cause is unknown; genetic factors have not been identified; and at present the cause of psychosis remains one of the most challenging problems in all of biology. Because of its importance for human health, it is an area of research vigorously pursued by a large number of investigators. Finding the cause of handedness may shed light on the cause of psychosis. Even so, identification of the handedness-determining gene is unlikely to be the end of the story, since the next great challenge facing us will be to determine how would the *RGHT1* gene specify hemispheric laterality. Thinking ahead, we can imagine that the *RGHT1* gene may control brain laterality development by promoting an asymmetric cell division during embryogenesis. The primary basis of asymmetric cell division at the single cell level is best understood by studies of a model system of the fission yeast. Could such a mechanism, combined with other speculations concerning diploid genome, explain handedness, brain laterality and psychosis development?

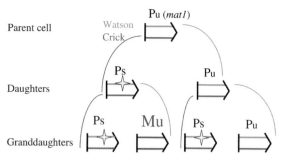

Figure 1. The strand-segregation mechanism explains the switching pattern of fission yeast cells. The Pu ("u" for unswitchable) parent cell mostly produces one Ps ("s" for switchable), and the other is always a Pu daughter cell. This is because the Ps daughter inherits the parental-Watson-strand-containing chromatid (indicated in *green*) that is imprinted (imprint indicated by star on the upper strand), and Pu inherits a nonimprinted parental-Crick-strand-containing chromatid (in *red*). The imprint is installed at *mat1* during DNA replication by a strand-, site-, and sequence-specific DNA alteration. In this mechanism, only the arbitrarily named Watson strand is imprinted. Next, when the imprinted strand is used as a template for replication (i.e., in the Ps cell), a transient double-stranded chromatid break results that initiates recombination interaction with *mat2-P* or *mat3-M* donor loci, resulting in *mat1* switching. The sister of the switched cell is most often switching-competent as indicated with a Ps designation. Thus, two consecutive asymmetric cell divisions result in switching of only one-in-four granddaughters of a Pu cell. The DNA sequence of the *mat1-M* allele is distinguished in *blue*, as it differs from that of the *mat1-P* allele. The newly switched cell is of Mu type and its progeny will also switch by the one-in-four pattern.

"DIFFERENTIATED" WATSON VERSUS CRICK CHAIN INHERITANCE OF THE PARENTAL CHROMOSOME DIFFERENTIATES FISSION YEAST CELLS

Schizosaccharomyces pombe is a haploid yeast whose rod-shaped cells grow by equational division. Its cells exist in two cell/sex types, called "P" (for "plus") and "M" (for "minus"), respectively controlled by the alternate alleles of the mating-type locus (*mat1*). An ~40-kb mating-type region located in chromosome 2 also contains closely linked silent copies of the mating-type information in *mat2-P* and *mat3-M* "donor loci." The P and M cell types interconvert by a transposition/substitution reaction in which a copy of either the *mat2* or *mat3* locus is unidirectionally transferred to the expressed *mat1* locus (for review, see Klar 2001). The *mat1* switching is highly efficient; however, all cells do not exhibit a uniform ability to switch. A remarkable pattern of switching is observed in mitotic pedigrees such that only one of four granddaughter cells of a newly switched cell switches mating type (Miyata and Miyata 1981). As shown in Figure 1, the Pu ("u" for unswitchable) cell nearly always produces one daughter (Pu) like itself, while the other daughter is advanced in its developmental program as it acquires switching competence (Ps, "s" for switchable). The Ps cell divides to produce one of the switched daughters to M type. This pattern is called the "one-in-four" or the "granddaughter" switching rule. Thus, only one (Ps in Fig. 1) of the two daughter cells of an unswitchable cell (Pu) is competent to produce a single switched daughter cell in the following generation.

Further pedigree analysis produced another "consecutive" switching rule: The sister of the recently switched cell is switching-competent (Egel and Eie 1987; Klar 1990). The M cells likewise switch to P type by following the same rules. Notably, this switching process follows the stem cell-like pattern found in some systems of cellular differentiation in eukaryotes. The key observation is that each cell division produces developmentally nonequivalent daughter cells. It is remarkable that this fungus follows a precise developmental program that takes two generations to complete even though it is a single-celled organism! Are we witnessing an evolution of a biological process that may be crucial for evolving multicellular organisms? Understanding the mechanism of an asymmetric cell division in any system may help us explain cellular differentiation required for development in higher eukaryotes.

The asymmetric cell division is explained with the strand-segregation model (Fig. 1) (Klar 1987, 1990). According to the model, sister cells differ simply because one inherits the "imprinted" Watson strand from the parental chromosome while the sister inherits the nonimprintable Crick strand. "Imprint" is defined as an epigenetically inherited, reversible and nonmutational alteration of DNA. Attached proteins, modified bases including ribose nucleotides, or a nick in one strand of the DNA were hypothesized for the biochemical nature of the imprint (Klar 1987, 1990). Replication of the imprinted strand results in a transient double-stranded break (DSB) of the chromatid, and the break initiates recombination of *mat1* with donor loci by the DSB repair mechanism. Meiotic *mat1* gene conversion analysis showed that the imprint is chromosomally borne, linked to *mat1*, and installed in one generation, but it is used in the next one for *mat1* conversion (Klar and Bonaduce 1993). Two other tests of the model predicted that cells engineered to carry inverted *mat1* duplication should produce developmentally equivalent daughter cells since both daughters will inherit identical sequences, one through the parental Watson strand and the other through the Crick strand. Consequently, each daughter should inherit one (never both) imprinted, switching-competent cassette that should exhibit DSB; and, second, both daughters should produce one switched and one unchanged daughter of their own. Thus, only one cassette should show DSB break in a cell; second, unlike the one-in-four rule followed by wild-type cells, two (cousins)-in-four granddaughters should switch. Both predictions were experimentally verified, establishing the strand-segregation model (Klar 1987, 1990). Recent studies have further supported the details of the model. For example, the imprint is installed only on the strand that is replicated by the lagging-strand replication complex (Dalgaard and Klar 1999) and it may be a site-specific nick (Arcangioli 1998) or an alkali-labile modified base (Dalgaard and Klar 1999), such as a ribose nucleotide moiety (Vengrova and Dalgaard 2004). The key lessons learned from yeast studies is that developmental asymmetry of sister cells derives from Watson and Crick DNA chains being complementary and not identical (Watson and Crick 1953) and that differentiated chromatids result from DNA replication because of strand-specific imprinting. Since the model is well established by several genetic and molecular studies, hereafter it will be referred to as the strand-segregation mechanism. Another important lesson we learned is that epigenetically established states of gene expression causing silencing of *mat2* and *mat3* are passed on to progeny as conventional Mendelian/chromosomal markers, both in mitosis and in meiosis (Grewal and Klar 1996). It is clear that sometimes the Mendelian gene is composed of DNA plus an epigenetic moiety, and both of them are faithfully replicated during chromosome replication as Mendelian variations (Klar 1998). In principle, such mechanisms for asymmetric cell division and stable states of gene expression discovered in yeast can be imagined to control the development of human brain hemispheric laterality.

THE STRAND-SPECIFIC IMPRINTING AND SEGREGATION (SSIS) MODEL FOR BRAIN LATERALITY AND PSYCHOSIS DEVELOPMENT

The yeast cells are haploid, but human cells are diploid and therefore contain two homologs of each chromosome. To apply the strand-segregation mechanism to humans, one has to invoke coordination of imprinting and patterned segregation of differentiated chromatids (hence parental strands) of both copies of a chromosome to daughter cells during mitosis to affect an asymmetric cell division. Such a model was proposed recently (Klar 1999, 2002, 2004). Genetic evidence supporting it is covered in primary papers to which the readers should refer for details. Only salient features will be highlighted here with the aim to place those studies in a wider context.

The key proposal of the SSIS model is to make differentiated sister chromatids by expressing a hypothetical gene named *DOH1* in one, say, parental Watson strand-containing chromatid, but epigenetically silencing it in the other chromatid (Fig. 2). This occurs during DNA replication of both copies of a specific chromosome in a certain cell division when brain laterality development is initiated. In addition, another hypothetical *SEG* site must exist elsewhere in the chromosome used to nonrandomly place sister chromatids at the metaphase plate such that a parental Watson with Watson and Crick with Crick pattern of segregation will result. The *trans*-acting factor encoded by the *RGHT1* gene may directly or indirectly interact with the *SEG* site to effect patterned segregation of DNA chains to progeny cells. Thus, one daughter cell will inherit *DOH1* "epialleles" that are transcriptionally active (*ON*), while the other daughter inherits unexpressed (*OFF*) epialleles. Such a single asymmetric cell division in embryogenesis is hypothesized to the development of brain hemispheres in which language is processed in the left hemisphere and the right one becomes an automatic hemisphere. It should be noted that epigenetic controls for gene regulation come in many forms: The epigenetic control may exhibit a parent-of-origin effect; one of the two homologs of a chromosome may be stably inactivated such as with the X chromosome; the genes, such as *mat*-donor loci in fission yeast, and genomic parasites, such as retroelements, are permanently silenced. The SSIS model adds to this list of epigenetic controls by postulating the evolution of somatic strand-specific imprinting to cause asymmetric cell division. It is simply a developmental biology model postulating strand-specific imprinting and patterned-segregation mechanism in embryogenesis for regulating the expression of a developmentally important gene. The model implies that perhaps an important function of the epigenetic phenomenon concerns somatic cell differentiation. The model does not specify the molecular nature of the epigenetic event, but its nature is likely to involve an assembly of heterochromatin structure analogous to the silencing mechanism of fission yeast *mat* loci (Grewal and Klar 1996; Klar 1998).

As no other hypothesis explains the biology behind brain laterality development, we are forced to consider new ideas. The SSIS model remains formally an abstract model, as we do not know which cell division and which chromosome are involved; whether the *DOH1* gene actually exists, what function it performs, and the mechanisms of strand-specific imprinting and segregation remain unknown. Despite these unknowns, a test of the model is provided by studies of rearrangements of the relevant chromosome where the *SEG* site is dissociated

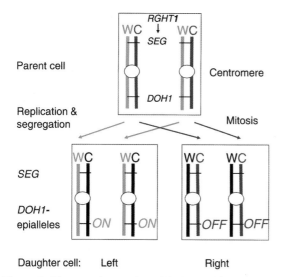

Figure 2. The somatic strand-specific imprinting/segregation model—the SSIS model—advanced to produce nonequivalent daughter cells in mitosis during human embryonic development. The model postulates production of developmentally nonequivalent sister chromatids at a specific cell division. The hypothetical *DOH1* (for Dominant hemisphere-specifying) gene is transcriptionally active (*ON*) in parental chromosome-derived Watson (W) strand-containing chromatid, and it remains silenced (*OFF*) by an epigenetic mechanism in the parental Crick (C) strand-containing sister chromatid. Second, a linked hypothetical *SEG* (segregation) site resides elsewhere in the chromosome to effect patterned distribution of differentiated chromatids to specific ("leftward" versus "rightward" placed with respect to dorsal/ventral axis of the embryo) daughter cells in the embryo. The *SEG* element may or may not coincide with the centromere. Third, *RGHT1* (for right-handedness) gene-encoded factor, acts on the *SEG* site directly, or indirectly, for patterned distribution of sister chromatids of both homologs to specific daughter cells. The chains are color-coded to indicate their distribution to specific daughter cells. An asymmetric segregation occurs so that both parental W chains (*green*) go to one daughter cell, and both Cs (*red*) to the other; the newly synthesized chains are indicated in *black*. In summary, differentiated daughter cells result by inheriting *DOH1* epialleles and, after subsequent growth, result in development of differentiated brain hemispheres.

from the *DOH1* gene owing to a chromosome translocation in one of the homologs. In this situation, the *DOH1* epialleles in the translocation chromosome will be randomly distributed, as the chromosome lacks an *SEG* site, but the epialleles in the wild-type chromosome will be distributed in a patterned fashion. Consequently, one-half of the translocation heterozygotes should produce a healthy, lateralized brain (Fig. 2). In the remaining one-half, both daughters will be become equivalent, as both will inherit *ON/OFF DOH1* epialleles, resulting in development of symmetrical hemispheres and possibly causing psychosis. This explanation was recently advanced to explain the result of 18 diseased among 36 heterozygous translocation carriers of a large Scottish pedigree with the t(1q42;11q14) translocation (Evans et al. 2001). Interestingly, of the 18, nine were diseased with schizophrenia and the remainder with the bipolar disorder. Thus, both of these disorders can be considered as manifestations of the same genetic etiology. However, results of a single translocation do not rule out the conventional explanation in which the translocation produces a dominant disease-causing mutation. Indeed, two overlapping genes on chromosome 1, named *DISC* (for "disrupted in schizophrenia") *1* and *2*, carry mutations in the translocation. According to the conventional explanation, incomplete penetrance of these mutations has been postulated to explain disease occurrence in some and not in other translocation heterozygotes (Evans et al. 2001).

To identify the relevant chromosome according to the SSIS model, be it 1 or 11, and to test whether a specific gene is mutated by the t(1q42;11q14), studies of other translocations may be useful. A recent study searched the Medline database with the query "psychosis and translocation." In addition to studies describing the aforementioned translocation, the search found several papers describing other translocations that also *partially* cosegregate with psychosis (Klar 2004): t(17q24;11q23), only one case found (Hoshi 1999); t(6q14;11q25), two diseased and two healthy (Holland and Gosden 1990); and t(9p24;11q23), six diseased and five healthy (Baysal et al. 1998). Based on these findings, the following conclusions were reached (Klar 2004): First, only chromosome 11 is relevant to psychosis development, as it is a common participant in all the translocations; second, as only one-half of translocation carriers are diseased, the feature of patterned segregation of chromosome 11 strands is supported; and, third, the translocations lie at three far apart regions spanning ~40% of the chromosome, so it is unlikely that a single gene is mutated by disparate translocations. Also, three groups working on these translocations failed to find linkage of the translocation junction region with inheritance of the disease in other families without chromosomal rearrangements (summarized in Klar 2004). In one case the nearest gene is located 299 kb away from both junction regions. Moreover, the linkage to the chromosome 1q arm where *DISC1* and *2* reside is also not supported by a multicenter study of a large number of patients (Levinson et al. 2002). Therefore, we assume that genetic alterations at the translocation junction regions are not causing disease. Together, these findings suggest that disease occurs in translocation carriers by disruption of patterned distribution of *DOH1* epialleles, that there is no conventional Mendelian gene mutation, and that it is clearly a genetically caused disease, as it is partially cosegregating with translocations. This analysis does not imply that all diseased individuals should carry chromosome 11 rearrangements. As noted above, the *RGHT1* gene controls the development of brain laterality, presumably through its interaction with the *SEG* site to effect patterned segregation mechanism (Fig. 2); and combined with the observation that psychotic patients are three times more likely to be NRH as compared with healthy controls, nearly all psychotically diseased persons in the general public may be that way because of their *r/r* genetic constitution. Clearly, only a minority of NRH is diseased; it remains to be tested whether *r/r* genetic constitution alone or in combination with other variations causes the disease. It should be stated that at present there is no experimental evidence demonstrating the hypothesized r/r genotype as the disease-predisposing factor. At present, we should only consider results with translocations as the best evidence favoring genetic etiology.

PATTERNED DNA CHAIN SEGREGATION PROBABLY OCCURS IN MICE

For the phenomenon of strand-specific imprinting in a diploid organism to be biologically useful, the SSIS model also requires evolution of the process of patterned chromatid segregation. As yeast is a haploid and single-celled organism, no need can be perceived for the existence of patterned segregation process. After it was discovered that sister chromatids of fission yeast are nonequivalent simply because of inheriting parental Watson and Crick chains (Klar 1987, 1990), the process of patterned segregation had to be invoked if such a mechanism were to evolve for evolving multicellular eukaryotes. It is generally assumed that identical daughter chromosomes are produced by DNA replication; therefore, experimentally testing the patterned segregation possibility seems unwarranted. Would some unrelated study fortuitously shed light on this possibility?

Liu et al. (2002) investigated whether the *Cre-loxP* recombination system could be used for genetic alterations of mouse chromosomes. They placed the recombination cassettes at allelic sites near the centromeres of two chromosomes, and then they induced recombination by introducing the plasmid expressing the recombinase enzyme in embryonic stem (ES) cells. Site-specific recombination was observed for both chromosomes. A marker distal to the crossover point was maintained in the heterozygous constitution in some and homozygosis occurred in other recombinants of a chromosome. This outcome is expected from random segregation of DNA chains after recombination in the G_2 phase of the cell cycle. Curiously, all 432 recombinants exhibited homozygosis of a distal marker located near the tip of chromosome 7. To explain this unusual result, the authors suggested that the process of recombination might have so placed recombinant chromatids at the metaphase plate, causing them to segregate away from each other. An alternate possibility was suggested where patterned segregation of Watson with Watson and Crick with Crick occurs (Klar 2004), as indicated by the SSIS model (Fig. 2). Following this suggestion, one mouse chromosome undergoes random segregation but chromosome 7 follows the patterned chain distribution. It should be pointed out that if all chromosomes were to undergo patterned segregation at each cell division, the process would not be useful as a gene regulation control for development. For it to be useful, different sets of chromosomes should follow patterned distribution in different cell types. It was then noted that mouse chromosome 7, which presumably undergoes patterned segregation, displays a 36% sequence synteny with human chromosome 11 implicated in psychosis (Klar 2004). It would be interesting to determine the strand segregation pattern of human chromosome 11 as well.

To facilitate the patterned segregation process, it would

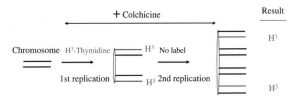

Figure 3. Nonrandom distribution of newly synthesized DNA chains to daughter chromatids in diplochromosomes of endo-reduplicated human blood cells (Schwarzacher and Schnedl 1966). The cells were labeled with H^3-thymidine in the first of the two DNA replication cycles. Because of the colchicine treatment, a second round of replication occurs without cell division to produce diplochromosomes because daughter chromatids fail to separate at mitosis. Autoradiograph of the metaphase arrested diplochromosomes showed label restricted to the outside chromatids of each chromosome.

be helpful if the location of DNA chains at the centromere were fixed. Sister chromatids sometimes fail to separate at mitosis when cells are treated with the colchicine drug. Following another round of replication, four chromatids (i.e., eight DNA chains) remain paired in a structure called "diplochromosome." When human blood cultures were labeled with H^3-thymidine in the first replication cycle, followed by another without the label, remarkably, the diplochromosomes had the outer chromatids labeled (Fig. 3), and not the inner ones (Schwarzacher and Schnedl 1966). Thus, DNA chains must hold fixed positions on the centromere and the newly synthesized chains are segregated to the "outer" chromatids. The importance of this result should now be viewed in the context of the SSIS model. This experiment also showed that mitotic recombination is rare, as most chromosomes did not show crossover events. Clearly, much remains to be done to test details of the SSIS model. Unfortunately, testing details of the SSIS model in humans is nearly impossible. Perhaps a useful approach will be to study hypothesized processes in experimentally amenable model systems, such as yeast, mice, and *Drosophila*. Interestingly, the fly has been recently shown to develop brain hemispheric asymmetry and it is implicated in long-term memory (Pascual et al. 2004).

CONCLUSIONS

The most fundamental unanswered question in developmental biology is how the diversity of cell types develops from a single cell, such as a fertilized egg. It is often thought that cellular differentiation must be controlled by cytoplasm factors that are unevenly expressed or distributed to daughter cells. Instead, nonequivalent daughter chromatids are produced in yeast, causing daughter cells to become different by the strand-segregation mechanism. It is taken for granted that two brain hemispheres develop differently from each other in healthy individuals, but the biology behind that development remains unknown and unapproachable. The SSIS model suggests a mechanism to effect differential hemisphere-specific gene regulation. It is suggested that brain laterality results from Watson and Crick chains being inherently different in acquiring an epigenetic event that regulates developmentally important gene(s). Accordingly, psychosis results when the proposed normally nonrandom segregation of DNA chains is randomized by chromosome 11 translocations. The most novel aspect of this interpretation is that the disorder is not due to conventional gene mutations or malregulation of genes due to epigenetic alterations at the breakpoint region, yet it is clearly a genetically caused disease. The model neatly explains a perplexing problem of how each of three different translocations causes disease only in one-half of the heterozygous rearrangement carriers, thus explaining the genetic oddity as they cause both dominant and recessive mutations. The view of genetic cause without Mendelian mutation is not at variance with the Watson–Crick base pairing or Mendel's rules. Mendelian genetics apply to inheritance through meiosis, but mitotic epigenetic control mechanisms strictly concern the use of genetic material in mitosis for accomplishing development even though all cells of the organism contain the same set of Mendel's genes. This mitogenetic concept implies that Watson and Crick chains can carry additional heritable (epi)genetic information to be used for somatic cell differentiation and that somatic interchromosomal and sister chromatid recombination would be deleterious as it would disrupt distribution of epialleles. For example, such unwanted mitotic recombinant events may lead to loss of heterozygosity for imprinted genes causing cancer development. Such a requirement may be one of the reasons for mitotic recombination rates to be low as compared with meiotic rates. Furthermore, the SSIS model also implies that for regulating developmentally important genes such as *Hox* genes, their linkage and order will be conserved during speciation. Overall, this view may provide a new paradigm for deciphering the cause of highly debilitating diseases and for explaining development in general. The brain tissue and hair whorls are derived from an ectoderm layer of cells where this mechanism may have evolved for brain laterality development. The same mechanism for visceral left/right axis determination in vertebrates has been also proposed for organs derived from mesoderm and endoderm tissues (Klar 1994). It is possible that a single chromosome dictates left/right-axis specification of neuronal and visceral organs by functioning in different layers of embryonic cells. At present, the prominent paradigm for development comprises the morphogen-gradient model, but it remains nontestable in most biological systems because the model fails to make refutable predictions. Cell-lineage-based strand-segregation mechanisms, those exploiting somatically imposed epigenetic mechanisms of gene controls, should be considered as another way to accomplish development.

ACKNOWLEDGMENTS

The author thanks colleagues of the Gene Regulation and Chromosome Biology Laboratory for discussions over many years that helped sharpen ideas presented here. The National Cancer Institute funded the work.

REFERENCES

Annett M. 1985. *Left, right, hand and brain: The right shift theory*. Lawrence Erlbaum, London.
Arcangioli B. 1998. A site- and strand-specific DNA break confers asymmetric switching potential in fission yeast. *EMBO J.* **17:** 4503.
Baysal B.E., Potkin S.G., Farr J.E., Higgins M.J., Korcz J., Gollin S.M., James M.R., Evans G.A., and Richard C.W., III. 1998. Bipolar affective disorder partially cosegregates with a balanced t(9;11)(p24;q23.1) chromosomal translocation in a small pedigree. *Am. J. Med. Genet.* **81:** 81.
Boklage C.E. 1977. Schizophrenia, brain asymmetry development, and twinning: Cellular relationship with etiological and possible prognosis implications. *Biol. Psychol.* **12:** 19.
Brown N.A. and Wolpert L. 1990. The development of handedness in left/right asymmetry. *Development* **109:** 1.
Coren S. 1992. *Left-hander syndrome—The causes and consequences of left-handedness*. Macmillan, New York.
Crow T.J. 1990. The continuum of psychosis and its genetic origins. The sixty-fifth Maudsley Lecture. *Br. J. Psychiatry* **156:** 788.
Dalgaard J.Z. and Klar A.J. 1999. Orientation of DNA replication establishes mating-type switching pattern in *S. pombe*. *Nature* **400:** 181.
DeLisi L.E., Sakuma M., Kushner M., Finer D.L., Hoff A.L., and Crow T.J. 1997. Anomalous cerebral asymmetry and language processing in schizophrenia. *Schizophr. Bull.* **23:** 255.
Egel R. and Eie E. 1987. Cell lineage asymmetry for *Schizosaccharomyces pombe:* Unilateral transmission of a high-frequency state of mating-type switching in diploid pedigrees. *Curr. Genet.* **3:** 5.
Evans K.L., Muir W.J., Blackwood D.H., and Porteous D.J. 2001. Nuts and bolts of psychiatric genetics: Building on the Human Genome Project. *Trends Genet.* **17:** 35.
Geschwind D.H., Miller B.L., DeCarli C., and Carmelli D. 2002. Heritability of lobar brain volumes in twins supports genetic models of cerebral laterality and handedness. *Proc. Natl. Acad. Sci.* **99:** 3176.
Grewal S.I. and Klar A.J. 1996. Chromosomal inheritance of epigenetic states in fission yeast during mitosis and meiosis. *Cell* **86:** 95.
Holland T. and Gosden C. 1990. A balanced chromosomal translocation partially co-segregating with psychotic illness in a family. *Psychiatry Res.* **32:** 1.
Hoshi S. 1999. Acute promyelocytic leukemia with t(11;17)(q23;q21). *Rinsho Ketsueki* **40:** 119.
Jacob F. and Monod J. 1961. Genetic regulatory mechanisms in the synthesis of proteins. *J. Mol. Biol.* **3:** 318.
Kennedy J.L., Farrer L.A., Andreasen N.C., Mayeux R., and St. George-Hyslop P. 2003. The genetics of adult-onset neuropsychiatric disease: Complexities and conundra? *Science* **302:** 822.
Klar A.J.S. 1987. Differentiated parental DNA strands confer developmental asymmetry on daughter cells in fission yeast. *Nature* **326:** 466.
———. 1990. The developmental fate of fission yeast cells is determined by the pattern of inheritance of parental and grandparental DNA strands. *EMBO J.* **9:** 1407.
———. 1994. A model for specification of the left-right axis in vertebrates. *Trends Genet.* **10:** 391.
———. 1996. A single locus, RGHT, specifies preference for hand utilization in humans. *Cold Spring Harbor Symp. Quant. Biol.* **61:** 59.
———. 1998. Propagating epigenetic states through meiosis: Where Mendel's gene is more than a DNA moiety. *Trends Genet.* **14:** 299.
———. 1999. Genetic models for handedness, brain lateralization, schizophrenia, and manic depression. *Schizophr. Res.* **39:** 207.
———. 2001. Differentiated parental DNA chain causes stem cell pattern of cell-type switching in *Schizosaccharomyces pombe*. In *Stem cell biology* (ed. D.R. Marshak et al.), p. 17. Cold Spring Harbor Laboratory Press, Cold Spring Harbor, New York.

———. 2002. The chromosome 1;11 translocation provides the best evidence supporting genetic etiology for schizophrenia and bipolar affective disorders. *Genetics* **160:** 1745.

———. 2003. Human handedness and scalp hair-whorl direction develop from a common genetic mechanism. *Genetics* **165:** 269.

———. 2004. A genetic mechanism implicates chromosome 11 in schizophrenia and bipolar diseases. *Genetics* **167:** 1833.

Klar A.J. and Bonaduce M.J. 1993. The mechanism of fission yeast mating-type interconversion: Evidence for two types of epigenetically inherited chromosomal imprinted events. *Cold Spring Harbor Symp. Quant. Biol.* **58:** 457.

Levinson D.F., Holmans P.A., Laurent C., Riley B., Pulver A.E., Gejman P.V., Schwab S.G., Williams N.M., Owen M.J., Wildenauer D.B., et al. 2002. No major schizophrenia locus detected on chromosome 1q in a large multicenter sample. *Science* **296:** 739.

Lewis C.M., Levinson D.F., Wise L.H., DeLisi L.E., Straub R.E., Hovatta I., Williams N.M., Schwab S.G., Pulver A.E., Faraone S.V., et al. 2003. Genome scan meta-analysis of schizophrenia and bipolar disorder. II. Schizophrenia. *Am. J. Hum. Genet.* **73:** 34.

Liu P., Jenkins N.A., and Copeland N.G. 2002. Efficient Cre-loxP-induced mitotic recombination in mouse embryonic stem cells. *Nat. Genet.* **30:** 66.

McManus I.C. 1991. The inheritance of left-handedness. In *Biological asymmetry and handedness* (ed. J. Marsh), p. 251. Wiley, Chichester, United Kingdom.

Miyata H. and Miyata M. 1981. Mode of conjugation in homothallic cells of *Schizosaccharomyces pombe*. *J. Gen. Appl. Microbiol.* **27:** 365.

Pascual A., Huang K.L., Neveu J., and Preat T. 2004. Neuroanatomy: Brain asymmetry and long-term memory. *Nature* **427:** 605.

Rife D.C. 1940. Handedness, with special reference to twins. *Genetics* **28:** 178.

Schwarzacher H.G. and Schnedl W. 1966. Position of labelled chromatids in diplochromosomes of endo-reduplicated cells after uptake of tritiated thymidine. *Nature* **209:** 107.

Segurado R., Detera-Wadleigh S.D., Levinson D.F., Lewis C.M., Gill J.I., Nurnberger J.I., Jr., Craddock N., DePaulo J.R., Baron M., Gershon E.S., et al. 2003. Genome scan meta-analysis of schizophrenia and bipolar disorder. III. Bipolar disorder. *Am. J. Hum. Genet.* **73:** 49.

Vengrova S. and Dalgaard J.Z. 2004. RNase-sensitive DNA modification(s) initiates *S. pombe* mating-type switching. *Genes Dev.* **18:** 794.

Watson J.D. and Crick F.H.C. 1953. Molecular structure of nucleic acids. *Nature* **171:** 737.

Summary: Epigenetics—from Phenomenon to Field

D.E. GOTTSCHLING

Fred Hutchinson Cancer Research Center, Seattle, Washington 98109

When Bruce Stillman made his opening remarks at the 69th Cold Spring Harbor Symposium, one of the things he said he hoped to learn was a way to easily explain what "epigenetics" meant to his wife Grace. After a week of discussions, it became clear that such a request was akin to asking someone to define "family values"—everyone knew what it meant, but it had a different meaning for each person. Part of the reason for the range of opinions may be understood from the etymology of "epigenetics" as explained by David Haig: The word had two distinct origins in the biological literature in the past century and the meaning has continued to evolve. Waddington first coined the term for the study of "causal mechanisms" by which "the genes of the genotype bring about phenotypic effects" (see Haig). Later, Nanney used it to explain his realization that cells with the same genotype could have different phenotypes that persisted for many generations. I define an epigenetic phenomenon as a change in phenotype that is heritable but does not involve DNA mutation. Furthermore, the change in phenotype must be switch-like, ON or OFF, rather than a graded response, and it must be heritable even if the initial conditions that caused the switch disappear. Thus, I consider epigenetic phenomena to include the lambda bacteriophage switch between lysis and lysogeny (Ptashne 2004), pili switching in uropathogenic *Escherichia coli* (Hernday et al. 2003), position-effect variegation in *Drosophila* (Haynes et al.), heritable changes in cortical patterning of *Tetrahymena* (Frankel 1990), prion diseases (Wickner et al.), and X-chromosome inactivation (Huynh and Lee; Heard et al.).

This Symposium comes on the 100th anniversary of genetics as a field of study at Cold Spring Harbor Laboratory, making it a very timely occasion to consider epigenetics. Given this historical context, I thought it appropriate to provide an examination of epigenetics through the portal of previous Cold Spring Harbor Symposia. While this is the first Symposium dedicated to the topic, epigenetic phenomena and their study have been presented throughout the history of this distinguished series. The history I present is narrowed further by my limitations and likings. For a more complete and scholarly portrayal, I can recommend the more than 1000 reviews on epigenetics that have been written in the four years leading up to the 69th Symposium.

In presenting this chronological account, I hope to convey a sense of how a collection of apparently disparate phenomenon coalesced into a field of study that impacts all areas of biology.

All authors cited here without dates refer to papers in this volume.

A HISTORY OF EPIGENETICS AT COLD SPRING HARBOR SYMPOSIA

1941

In the 9th Symposium, the great *Drosophila* geneticist H.J. Muller described developments on his original "eversporting displacement," in which gross chromosomal rearrangements resulted in the mutant mosaic expression of genes near the breakpoint (Muller 1941). By the time of this meeting, he referred to it as "position effect variegation." It was well established that the affected genes had been transferred "into the neighborhood of a heterochromatic region," that the transferred euchromatic regions had been "partly, but variably, transformed into a heterochromatic condition— 'heterochromatized'," and that *addition* of extra copies of heterochromatic chromosomes "allowed the affected gene to become more normal in its functioning." This latter observation was an unexpected quandary at the time, which we now know to be the result of a titration of limiting heterochromatin components.

1951

In the 16th Symposium, a detailed understanding of the gene was of high priority. This may explain why little progress had been made on understanding position-effect variegation (PEV), though more examples were being discovered. However, the opening speaker noted that PEV would be an exciting area for future research (Goldschmidt 1952). Barbara McClintock noted that chromosomal position effects were the basis of differences in "mutable loci" of maize, and she speculated that the variation of mutability she observed likely had its roots in the same mechanisms underlying PEV in *Drosophila* (McClintock 1952).

1956

By the time of the 21st Symposium, McClintock's ideas about "controlling elements" had developed (McClintock 1957). Two were particularly relevant with regard to epigenetics. In the *Spm* controlling element system, she had uncovered variants that allowed her to distinguish between *trans*-acting factors that could "suppress" a gene (reduce or eliminate its phenotypic expression), and those factors that could mutate it. She also noted that some controlling elements could suppress gene action, not only at the locus where they had inserted, but also at loci that were located some distance on either side of them. Others were discovering this "spreading effect" as well. J. Schultz presented a biochemical and physical characterization of whole *Drosophila* that contained different amounts of heterochromatin (Schultz 1957). While the work was quite primitive and the conclusions drawn were limited, the work represented early attempts to dissect the structure of heterochromatin and demonstrated just how difficult the problem would be.

1958

Two talks at the 23rd Symposium were landmarks with respect to our present day Symposium. First, R.A. Brink described his stunning observations of "paramutation" at the *R* locus in maize. If two alleles (R^{st} and R^r) with distinct phenotypes as homozygotes are combined to form a heterozygote, and this R^{st}/R^r plant is in turn crossed again, the resulting progeny that contain the R^r allele will *always* have an R^{st} phenotype, even though the R^{st} is no longer present (Brink 1959). However, this phenotype is metastable; in subsequent crosses the phenotype reverts to the normal R^r phenotype. He meant for the word paramutation "to be applied in this context in its literal sense, as referring to a phenomenon distinct from, but not wholly unlike, mutation." Second, D.L. Nanney went to great lengths to articulate "conceptual and operational distinctions between genetic and epigenetic systems" (Nanney 1959). In essence, he defined "epigenetics" differently from how it had been originally intended by Waddington (for details, see Haig). He found it necessary to do so to describe phenomenon he observed in *Tetrahymena*. He found evidence that the cytoplasmic history of conjugating parental cells influenced the mating type determination of resulting progeny. His definition encompassed observations made by others as well, including Brink's work on the *R* locus and McClintock's work noted in the 21st Symposium.

1964

Mary Lyon's recently proposed hypothesis of X-chromosome inactivation in female mammals (Lyon 1961) was of considerable interest at the 29th Symposium. S. Gartler, E. Beutler, and W.E. Nance presented further experimental evidence in support of it (Beutler 1965; Gartler and Linder 1965; Nance 1965). Beutler reviewed multiple examples of mosaic expression of X-linked genes in women, supporting the random nature of X-inactivation. From careful quantitative analysis of an X-linked gene product, Nance deduced that X-inactivation occurred before the 32-cell stage of the embryo.

1973

The 38th Symposium on Chromosome Structure and Function represented a return to the examination of eukaryotic chromosomes. Significant progress had been made studying prokaryotic and phage systems and consequently bacterial gene expression had dominated much of the thinking in the burgeoning field of molecular biology. An appreciation for chromatin (DNA with histones and nonhistone proteins) in eukaryotes was building, but it was unclear whether it played a role in chromosome structure or function or both (Swift 1974). Nevertheless, several groups began to speculate that posttranslational modification of chromatin proteins, including histones, was associated with gene transcription or overall chromosome structure (Allfrey et al. 1974; Louie et al. 1974; Weintraub 1974). There was only a hint of epigenetic phenomena in the air. It had been hypothesized that repetitive DNA regulated most genes in eukaryotes, partly based on the fact that McClintock's "controlling elements" were repeated in the genome. However, it was reported that most repeated DNA sequences were unlinked to genes (Peacock et al. 1974; Rudkin and Tartof 1974). From these observations, the idea that repeated elements regulated gene expression lost significant support from those in attendance. More importantly though, these same studies discovered that most of the repetitive DNA was located in heterochromatin.

1977

The 42nd Symposium demonstrated that in four years an amazing number of technical and intellectual advances had transformed the study of eukaryotic chromosomes (Chambon 1978). This included the use of DNA restriction enzymes, development of recombinant DNA technology, routine separation of proteins and nucleic acids, the ability to perform Southern and Northern analysis, rapid DNA and RNA sequencing, and immunofluorescent visualization of chromosomes. The nucleosome hypothesis had been introduced and mRNA splicing had been discovered. Biochemical and cytological differences in chromatin structure, especially between actively transcribed and inactive genes, was the primary interest at this meeting. However, most relevant to epigenetics, Hal Weintraub and colleagues presented ideas about how chromatin could impart variegated gene expression to an organism (Weintraub et al. 1978).

1980

The 45th Symposium was a celebration of Barbara McClintock's discoveries—Movable Genetic Elements (Yarmolinsky 1981). Mechanistic studies of bacterial transposition had made enormous progress and justifiably represented about half the presentations, while others presented evidence that transposition and regulated genomic reorganization occurred not only in maize, but also in other eukaryotes, including flies, snapdragons, *Trypanosomes, Ascobolus*, and budding yeast. In the context of this meeting, all observed variegated expression events were ascribed to transposition. Moreover, there was a reticence to seriously consider that "controlling elements" were responsible for most gene regulation (Campbell 1981), which led some to suggest that "the sole function of these elements is to promote genetic variability." In essence, the idea that heterochromatin was responsible for the regulated expression in PEV was called into question. With respect to future epigenetic studies, perhaps the most noteworthy discussion was the firm establishment of "silent mating cassettes" in *Saccharomyces cerevisiae* (Nasmyth et al. 1981; Rine et al. 1981).

1982

Leading up to the 47th Symposium, a general correlation had been established in vertebrate systems that the overall level of cytosine methylation in CpG DNA sequences was lower for genes that were transcribed than for those that were not. However, there were exceptions to this generalization, and more detailed analysis was presented that methylation of specific areas of a gene's promoter was most important (Cedar et al. 1983; Doerfler et al. 1983; La Volpe et al. 1983). Based on the restriction/modification systems of bacteria, it was thought that DNA methylation prevented binding of key regulatory proteins. Furthermore, it had been shown that DNA methylation patterns could be mitotically inherited in vertebrates, which led to the hypothesis that DNA methylation could serve as a means of transcriptional "memory" as cells divided through development (Shapiro and Mohandas 1983). Another major epigenetic-related finding was the identification of DNA sequences on either side of the "silent mating cassettes" in budding yeast that were responsible for transcriptional repression of genes within the cassettes—these defined the first DNA sequences required for chromosomal position effects (Abraham et al. 1983).

1985

The Molecular Biology of Development was the topic for the 50th Symposium and it too encompassed a number of important advances. Perhaps one of the most exciting developments was the overall awareness that fundamental molecular properties were conserved throughout evolution—e.g., human RAS functioned in budding yeast and homeobox proteins were conserved between flies and humans (Rubin 1985). New efforts to understand chromosome imprinting began with the development of nuclear transfer in mice (Solter et al. 1985). These studies revealed that parent-of-origin information was stored within the paternal and maternal genomes of a new zygote; it was not just the DNA that was important, but the chromosomes contained additional information about which parent they had passed through, and the information was required for successful development of an embryo. Part of the answer was thought to lie in the fact that differential gene expression was dependent on the parental origin of a chromosome (Cattanach and Kirk 1985).

There were a number of studies aimed at understanding the regulation of the bithorax complex. Notably, E.B. Lewis made special mention of the curious nature of known *trans* regulators of the locus; nearly all were repressors of the locus (Lewis 1985). Thus, the importance of maintaining a gene in a silenced state for many cell doublings was imperative for normal development. This contrasted with much of the thinking at the time—i.e., that gene activation/induction was where the critical regulatory decisions of development would be.

DNA transformation and insertional mutagenesis techniques had recently been achieved for a number of organisms. One particularly creative and epigenetic-related use of this technology came in *Drosophila*. A P-element transposon with the *white* eye color gene on it was created and "hopped" throughout the genome (Rubin et al. 1985). This provided a means to map sites throughout the *Drosophila* genome where PEV could occur.

This meeting also highlighted the first genetic approaches to dissecting sex determination and sex chromosome dosage compensation—in *Drosophila* (Belote et al. 1985; Maine et al. 1985) and *Caenorhabditis elegans* (Hodgkin et al. 1985; Wood et al. 1985).

1993

The 58th Symposium highlighted the celebration of the 40th anniversary of Watson and Crick's discovery. Part of the celebration was a coming out party for epigenetic phenomenon: New phenomena had been identified, molecular analysis of other phenomena had begun, and sufficient progress had been made in a number of systems to propose hypotheses and to test them.

In *Trypanosomes*, the family of variable surface antigen genes (VSGs) located near telomeres are largely silenced, with only one VSG expressed at a time. While this organism does not appear to contain methylated DNA, it was reported that the silenced VSGs contained a novel minor base: β-D-glucosylhydroxymethyluracil (Borst et al. 1993). This base appeared to be in place of thymidine in the DNA. Parallels between this base and cytosine methylation in other organisms were easy to draw; the modifications were important for maintaining a silenced gene. But how the base was introduced into the DNA or how it imparted such a function was unclear.

Progress had also been made in vertebrate epigenetic phenomena, including chromosomal imprinting and X-inactivation (Ariel et al. 1993; Li et al. 1993; Tilghman et al. 1993; Willard et al. 1993). It had become clear by this time that numerous loci were subject to imprinting in mammals—only one allele was expressed in diploid cells and expression was dependent on parental origin. The *Igf2-H19* locus was of particular interest, primarily because it contained two nearby genes that were regulated in opposing fashion. *Igf2* is expressed from the paternal chromosome while the maternal copy is repressed, whereas the paternal allele of *H19* is repressed and its maternal allele is expressed. Interestingly, methylated CpG was observed just upstream of both genes on the paternal chromosome. It was proposed that the differential methylation regulated access of the two genes to a nearby enhancer element; the enhancer was closer to, and just downstream of, *H19* (Tilghman et al. 1993). A mutually exclusive competition between the two genes for the enhancer was envisioned; when the *H19* gene was methylated, the enhancer was free to activate the more distant *Igf2* gene. Support for the idea that DNA methylation played a regulatory role in this process came from mouse studies. Mutation of the first vertebrate gene encoding a 5-methyl-cytosine DNA methyltransferase in embryonic stem cells showed that, as embryos developed, the paternal copy of H19 became hypomethylated and the gene became transcriptionally active (Li et al. 1993).

An important step in how ^{5Me}CpG mediated its effects came from the purification of the first ^{5Me}CpG DNA-binding complex (MeCP1) (Bird 1993). Not only did it bind DNA, but when tethered upstream of a reporter gene, MeCP1 caused the gene to be repressed. While this did not explain regulation at the *Igf2-H19* locus, it did provide a potential mechanism to explain the general correlation between DNA methylation and gene repression.

Genetic mapping over a number of years had identified a portion of the human X chromosome as being critical for X-inactivation. Molecular cloning studies of this X-inactivation center led to the discovery of the *Xist* gene (Willard et al. 1993), an ~17-kb noncoding RNA that was expressed only on the inactive X chromosome. The mouse version of *Xist* was surprisingly homologous in structure and sequence and held the promise of being an excellent model system to dissect how this RNA functioned to repress most of the X chromosome.

Two notable findings were described in *Neurospora* (Selker et al. 1993): first, cytosine DNA methylation was not limited to CpG dinucleotides, but could occur in seemingly any DNA context; and, second, the amazing phenomenon of repeat-induced point mutation (RIP). Sequences become "RIP'd" when there is a sequence duplication (linked or unlinked) in a haploid genome and the genome is put through the sexual cycle via conjugation. Two events occur: Both copies of the duplicated DNA pick up G:C → A:T mutations, and DNA within a few hundred base pairs of the RIP'd sequences becomes methylated. This double attack on the genome is quite efficient—50% of unlinked loci succumb to RIP, while tightly linked loci approach 100%—and readily abolishes gene function.

The *brown* gene in *Drosophila*, when translocated near heterochromatin, displays dominant PEV; the translocated copy can cause repression of the wild-type copy. In searching for enhancers and suppressors of this *trans*-inactivation phenomenon, Henikoff discovered that duplication of the gene located near heterochromatin *increased* the level of repression on the normal copy (Martin-Morris et al., 1993). While the mechanism underlying this event remained mysterious, it was postulated that the phenomenon might be similar to RIP in *Neurospora*, though it had to occur in the absence of DNA methylation, which does not take place in *Drosophila*.

Paul Schedl elucidated the concept of chromosomal "boundary elements" (Vazquez et al. 1993). The first were located on either side of the "puff" region at a heat shock locus in *Drosophila* and were defined by their unusual chromatin structure—an ~300-bp nuclease-resistant core bordered by nuclease-hypersensitive sites. It was postulated that such elements separated chromatin domains along the chromosome. Two in vivo assays supported this hypothesis: (1) When bordering either side of a reported gene, boundary elements effectively eliminated chromosomal position effects when the construct was inserted randomly throughout the genome; and (2) the boundary element was also defined by its ability to block enhancer function. When inserted between a gene promoter and its enhancer, the boundary element blocked the gene's expression. While not as well defined, the concept of boundary elements was also developing in other organisms, especially at the globin locus in mammals (Clark et al. 1993).

Budding yeast shined the light on a mechanistic inroad to chromatin-related epigenetic phenomena. It had already been established that the silencers at the silent mating type loci were sites for several DNA-binding proteins. Their binding appeared to be context dependent, as exemplified by the Rap1 protein, which not only was important in silencing, but also bound upstream of a number of genes to activate transcription (for review, see Laurenson and Rine 1992).

Over the years, numerous links had been made between DNA replication and transcriptionally quiescent regions of the genome. The inactive X chromosome, heterochromatin and silenced imprinted loci had all been reported to replicate late in S phase relative to transcriptionally active regions of the genome. In addition, it had been shown that the establishment of silencing at the silent mating type loci required passage through S phase, suggesting that silent chromatin had to be built on newly replicated DNA. Thus, there was great interest when it was reported that one of the silencers was found to be an origin of DNA replication and that its origin activity could not be separated from silencing function (Fox et al. 1993). Furthermore, mutants in the recently identified origin recognition complex (ORC) were found to cripple silencing (Bell et al. 1993; Fox et al. 1993).

The discovery that telomeres in *S. cerevisiae*, like those in *Drosophila*, exerted PEV opened another avenue for dissecting heterochromatic structure and its influence on gene expression. Reporter genes inserted near telomeres give variegated expression in a colony. The repressed state of the genes is dependent on many of the same gene products (*SIR2*, *SIR3*, and *SIR4*) as those required for silencing at the silent mating type loci. Several key aspects of the silent chromatin structure and the regulation of the variegated expression were described. It is worth noting that heterochromatin is defined cytologically as condensed chromatin, but silent chromatin in *S. cerevisiae* has never been visualized in this way. Nevertheless, because of similarities to PEV in *Drosophila*, there was enthusiasm to consider silent chromatin in yeast to be a functional equivalent of heterochromatin (described in Weintraub 1993).

A number of fundamental concepts came to light from the yeast studies. First, the importance of histone H3 and H4 became evident. In particular, the NH$_2$-terminal tail of histone H3 and H4 tails appeared to be directly involved in the formation of silent heterochromatin (Thompson et al. 1993). Specific mutations in the tails of these histones alleviated or crippled silencing and led to the notion that both the net charge of the residues on the tails and specific residues within the tails contributed to silencing. In addition, these early days of chromatin immunoprecipitation (ChIP) demonstrated that the lysines in the NH$_2$-terminal tail of histone H4 were hypoacetylated in regions of silent chromatin relative to the rest of the genome. Moreover, one of the histone mutants identified K16 of histone H4, which could be acetylated, as critical for forming silent chromatin.

Telomeres appeared to provide the simplest system in which to develop an understanding of how Sir proteins mediated silencing. The concept of recruiting silencing proteins was being developed. Briefly, the telomeric DNA-binding protein, Rap1p, was found to interact with Sir3p and Sir4p by two-hybrid methods (described in Palladino et al. 1993). Thus, Rap1 could "recruit" these Sir proteins to telomeric region of the genome. There was evidence that Sir3p and Sir4p could bind to one another and, most importantly, Sir3p and perhaps Sir4p interacted with the tails of histone H3 and H4 (Thompson et al. 1993). Furthermore, overexpression for Sir3p caused it to "spread" inward along the chromatin fiber from the telomere, suggesting that it was a limiting component of silent chromatin and could "polymerize" along the chromatin (Renauld et al. 1993). Taken together, there appeared to be a large interaction network important for silencing; The Sir proteins initiated assembly at telomeric DNA, because of their interaction with Rap1p, and then "polymerized" from the telomere along the chromatin fiber, presumably by binding to the tails of histones H3 and H4.

Switching between transcriptional states in variegated telomeric expression appeared to be the result of a competition between silent and active gene expression (Aparicio and Gottschling 1994; described in Weintraub 1993). If the transcriptional activator for a telomeric gene was deleted, the gene's basal transcriptional machinery was insufficient for expression and the gene was constitutively silenced. Conversely, overexpression of the activator caused the telomeric gene to be expressed continuously; the gene was never silenced. In the absence of *SIR3* (or *SIR2* or *SIR4*) basal gene expression was sufficient, whereas increased dosage of *SIR3* increased the fraction of cells that were silenced. While a transcriptional activator could overcome silencing throughout the cell cycle, it was most effective when cells were arrested in S phase, presumably when chromatin was being replicated and hence most susceptible to competition. Somewhat surprisingly, cells arrested in G_2/M also could be easily switched, suggesting that silent chromatin had not yet been fully assembled by this time.

Silent chromatin in yeast was shown to be recalcitrant to nucleases and DNA-modification enzymes, suggesting that the underlying DNA was much less accessible relative to most of the genome (described in Thompson et al. 1993).

It also appeared that there was a hierarchy of silencing within the yeast genome: The telomeres were the most sensitive to perturbation, *HML* was next, and *HMR* was the least sensitive. In fact, when the *SIR1* gene was mutated, the normally completely silenced *HM* loci displayed variegated expression (Pillus and Rine 1989).

Lastly, Sir3p and Sir4p were localized to the nuclear periphery, as were the telomeres. It was proposed that the nucleus was organized such that the nuclear envelope provided a special environment for silencing (Palladino et al. 1993).

Schizosaccharomyces pombe also has silent mating cassettes that were suspected to behave similarly to the case in *S. cerevisiae*. However, in *S. pombe* there was an added twist to the story of mating type switching. In an elegant set of experiments, Amar Klar proposed how a "mark" is imprinted on one strand of DNA in a cell (Klar and Bonaduce 1993). The mark is manifested, after two cell divisions in one of the four granddaughter cells, as a double-stranded break that facilitates mating type switching. This yeast does not have any known DNA modifications (methylation, etc.); hence, a different type of mark was postulated to be left on the DNA strand.

1994

The 59th Symposium was on The Molecular Genetics of Cancer. The concept of epigenetic regulation in oncogenesis had begun to develop after the idea of tumor suppressor genes became established. While there had been a couple of studies supporting such a notion, an interesting twist to the story came in studies of Beckwith–Wiedemann syndrome and Wilms' tumor patients. Mutations in both types of patients had been mapped to a locus that included the imprinted *H19-IGF2* genes. Feinberg et al. discovered "loss of imprinting" (LOI) for these genes in affected patients: The maternal locus lost its imprint, *H19* was repressed, and *IGF2* was expressed (Feinberg et al. 1994). Thus, LOI, which in principle could occur elsewhere in the genome, could cause either biallelic expression and/or extinction of genes critical in oncogenesis.

1998

In the couple of years leading up to the 63rd Symposium on Mechanisms of Transcription, several important developments occurred that would impact the molecular understanding of several epigenetic phenomena. Histone-modifying enzymes were identified—specifically histone acetylases and deacetylases. Some of these enzymes proved to play critical roles in regulating gene expression and provided an entry into gene products that directly affected PEV and silencing. The tip of this iceberg was presented at the symposium (see Losick 1998). Molecular dissection of the Sir3p- and Sir4p-silencing proteins in yeast revealed the polyvalent nature of their interactions and how the network of interactions between all the Sir proteins, the histones, and various DNA-binding factors set up silent chromatin, as well as the molecular details of how various loci (telomeres, the rDNA, *HM* loci, and double-stranded breaks) could compete for the limited supply of Sir proteins. By crippling the ability of a specific locus to recruit silencing factors, Sir protein levels were increased at the other loci (Cockell et al. 1998). This provided direct evidence that principles of mass action were at work and explained how silencing at one locus could impact the epigenetic silencing at other loci—an idea originally put forth in studies on PEV in *Drosophila* (Locke et al. 1988).

Another finding explained how DNA methylation could regulate gene expression through chromatin. This came with the identification of protein complexes composed of MeCP2, which bind both methylated DNA and histone deacetylases (Wade et al. 1998). Methylated

DNA could serve as a point of recruiting deacetylases to a locus and thus facilitate silencing of nearby genes.

The concept of boundary elements was extended from *Drosophila* to mammals, with clear evidence provided at the β-globin locus, thus indicating that chromatin boundaries were indeed likely conserved in metazoans and perhaps all eukaryotes (Bell et al. 1998).

1999

The 64th Symposium on Signaling and Gene Expression in the Immune System provided evidence about how monoallelic expression arose and that it might be more widespread than previously thought. Monoallelic expression at the immunoglobulin loci had been obvious in lymphocytes for some time; it guaranteed the production of a single receptor type per lymphoid cell (Mostoslavsky et al. 1999). The allele to be expressed was chosen early in development, apparently at random; both alleles began in a repressed state, but over time one became demethylated. It was unclear how a single allele was chosen, but the phenomenon appeared at other loci, too, where the necessity of monoallelism was not obvious. For instance, only one allele of genes encoding the cytokines IL-2 and IL-4 was expressed (Pannetier et al. 1999).

2000

The most significant epigenetic-related talk at the 65th Symposium was the discovery that the Sir2 protein was a histone deacetylase (Imai et al. 2000). This was the only Sir protein that had clear homologs in all other eukaryotes and that regulated PEV. It seemed to be the enzyme primarily responsible for removing acetyl moieties from histones in silent chromatin. Furthermore, because it was an NAD-dependent enzyme, it linked the regulation of silencing (heterochromatin) to cellular physiology.

2003

The 68th Symposium on The Genome of *Homo sapiens* was an important landmark in genetics and, while there is still much genetic work to be done, the complete sequencing of this and other genomes signified that it was time to move "above genetics"—a literal meaning of epigenetics.

This historical account highlights several themes shared with many other areas of research. First, it demonstrates the episodic nature of advances in epigenetics. Second, as molecular mechanisms underlying epigenetic phenomenon began to be understood, it became easier to connect epigenetics to biological regulation in general. Third, it showed that people we now consider scientific luminaries had made these connections early on—it just took a while for most others to "see" the obvious.

THE 69TH SYMPOSIUM

Over the years a few general principles common to all epigenetic phenomena have been identified, and they serve to guide experimental approaches in the search for a mechanistic understanding. First, the differences between the two phenotypic states (OFF and ON) always have a corresponding difference in structure at a key regulatory point—i.e., form translates into function. Hence, identifying the two distinct structures, the components that compose them, and the compositional differences between them has been a primary task. Second, the distinct structures must have the ability to be maintained and perpetuated in a milieu of competing factors and entropic forces. Thus, each structure requires self-reinforcement or positive-feedback loops that ensure that it is maintained and propagated over many cellular divisions. In some cases, such as X-chromosome inactivation, this appears to be on the order of a lifetime.

Many of the mechanistic principles defined in the earlier symposia continued to be refined in the 69th Symposium, but there were also new developments. To put these new developments in context, it is important to note that two other discoveries had a major impact on epigenetics. One was the discovery of RNA interference and related RNA-based mechanisms of regulation. The other was the discovery of mechanisms underlying the prion hypothesis. Both of these fields advanced rapidly in the past decade, with some of the studies contributing to knowledge about chromatin-based epigenetics and others providing new perspectives about heritable transmission of phenotypes.

Below, I highlight but a fraction of the accomplishments from the Symposium that are advances I was able to appreciate. At the conclusion of this review, I will try to distill the most important concepts I took away from the meeting.

Propagating a Chromatin Mark

DNA methylation provides an easy-to-understand mechanism for propagating a phenotypic state as cells divide, but organisms such as yeast, flies, and worms propagate phenotypic states, even through meiosis, without DNA methylases. In these organisms, chromatin seems to be the likely structure that must be heritably propagated. Consequently, a mark on histones that persists and is duplicated in chromatin has become an attractive model for imparting an epigenetic mark (see Smith et al. 2002). While there have been a number of hypotheses put forth to explain how marked nucleosomes could be perpetuated, there have been no detailed mechanistic demonstrations of how this is accomplished.

The major issue hangs on the question: How is a mark in a parental nucleosome propagated to both sister chromatids following DNA replication? Histones H3 and H4 readily form a stable tetramer in vitro (Ruiz-Carrillo et al. 1979), and, in genetic analyses, modifications of these two histones play a critical role in silencing (Johnson et al. 1990; Thompson et al. 1994). The inherent structure of the nucleosome suggests that in contrast to the two heterodimers of histone H2A and H2B on either end of the nucleosome, the H3/H4 tetramer is likely to remain as a single unit (Luger et al. 1997), and in vivo isotope labeling experiments confirm the overall stability of the tetramer (Jackson 1988). Thus, during DNA replication it seems that only one of the two sister chromatids will receive the histone H3/H4 "mark" of the parental nucleosome, while the other sister chromatid will receive new, naive histone H3/H4.

Data presented at the Symposium called these ideas into question and offered a model by which marks could be passed on to both sister chromatids (Nakatani et al.). Biochemical analysis of chromatin assembly factors that are known to be important for chromatin silencing in vivo revealed that these factors could bind to heterodimers, rather than the expected tetramers, of histone H3/H4 and assemble the dimers into mixed tetramers of H3/H4 in vitro. If this kind of mixing turns out to be true in vivo, then it provides a mechanism for passing equal amounts of marked parental histones to both sister chromatids. Further, it is implied in such a model that these mixed tetramers would employ a positive reinforcement mechanism by which the new histones would be rapidly modified to be identical to the parental histones.

Histone Modifications

Since 1995–1996, when the first histone-modifying enzyme genes were identified, there has been a wonderfully productive effort to identify enzymes responsible for adding or removing the myriad of posttranslational modifications observed on histones (see many of the papers and poster abstracts from the meeting). But just as we think we are approaching the end of this race, we are informed that the finish line has been moved—more than 20 new modification sites have been identified on histones in the past year (Zhang et al. 2003).

Nevertheless, some fundamental principles are understood. As observed in other multisubunit complexes, the modifications modulate interactions of the histones with other proteins, sometimes increasing the affinity of a protein for the nucleosome (methylated K9-histone H3 with HP1), or at other times diminishing an interaction (acetylation of histone H3 and H4 tails with Sir3 protein). In addition, there can be interdependence among modifications. For instance, methylation of K4 or K79 on histone H3 depends on the monoubiquitination of histone H2B (Osley 2004).

Another issue focuses on the number of methyl moieties on any given histone residue; primary amines of lysine can be mono-, di-, or trimethylated. In cases where the amine of lysine is required for binding with another factor, modification by even a single methyl group could interfere with achieving an interaction, as would the di- and trimethyl forms. However, if methylation is important to create an interaction, then only one methylated state may be relevant. With regard to this latter situation, R. Paro (pers. comm.) and Ringrose et al. (2004) provided beautiful evidence by in situ peptide competition that only trimethylated forms of K9 or K27 of histone H3 bound efficiently to Polycomb proteins in *Drosophila*. Furthermore, T. Jenuwein (Lachner et al.) provided evidence that the intermediate methylated states were also important in providing information on chromatin. For instance, using specific antibodies, he showed that heterochromatin in mouse cells reacted best with monomethyl K27-H3 and trimethyl K9-H3 and K20-H4. It will be worth following these findings with genetic and in situ competition experiments to determine whether all these specific marks are critical for heterochromatin formation and, if so, how the marks are made and how the marks mediate formation. Does each site contribute to the increased affinity of one protein (e.g., HP1) or do some of these marks help to lower the affinity for competing factors (e.g., euchromatic proteins in heterochromatin)?

The Histone Code Hypothesis

In considering histone modifications and their potential information content, there were many discussions about the "histone code hypothesis" (Jenuwein and Allis 2001). Most of those that I participated in, or overheard, were informal and rather lively. The proponents of the "code" cite examples such as trimethylation of K9 histone H3 and its greater affinity for the HP1 class of heterochromatin proteins (Jenuwein and Allis 2001), while those on the other side cite biochemical and genetic evidence that the net charge on the NH_2-terminal tail of histone H4, irrespective of the position of the charge, has dramatic effects on DNA binding or phenotype (Megee et al. 1995; Zheng and Hayes 2003).

Grunstein presented data that included genome-wide analysis of histone acetylation modifications and chromatin associated proteins using specific antibodies and ChIP-Chip in *S. cerevisiae* (Millar et al.). His focus was on the epigenetic switch associated with K16 acetylation for binding, or not binding, particular chromatin proteins—thus supporting the histone code hypothesis. Though not discussed, some of his data appeared to support reports from others that for much of the genome, there is no correlation between specific histone modifications and gene expression (i.e., all active genes have the same marks, and these marks are not present on inactive genes) (Schubeler et al. 2004; Dion et al. 2005). Taking all the results together, I suspect that both specific modifications *and* general net charge effects are used as mechanisms for regulating chromatin structure and gene expression.

Dynamic Silent Chromatin

I must confess that, based on static images of heterochromatin and the refractory nature of silent chromatin, I was convinced that, once established, a heterochromatic state was as solid as granite. Only when it was time for DNA replication would the impervious structure become relaxed. In thinking this way, I had foolishly ignored principles of equilibrium dynamics I learned in undergraduate chemistry. But these lessons were brought home again by studies of silent chromatin and heterochromatin, where it was shown that silencing proteins of yeast (Sir3) and heterochromatin proteins in mammalian cells (HP1) were in a dynamic equilibrium—proteins were rapidly exchanged between heterochromatin and the soluble compartment—even when the chromatin was in its most impervious state (Cheng and Gartenberg 2000; Cheutin et al. 2003). The realization of its dynamic qualities forced a different view of how an epigenetic chromatin state is maintained and propagated. It suggests that, in some systems, the epigenetic state can be reversed at any time, not just during DNA replication. Hence, we can infer that mechanisms of reinforcement and propagation for silenced chromatin must function constantly.

Methylation of histones was widely held to be the modification that would indeed impart a "permanent" mark on the chromatin (for review, see Kubicek and Jenuwein 2004). In contrast to all other histone modification (e.g., phosphorylation, acetylation, and ubiquitination), there were no enzymes known that could reversibly remove a methyl group from the amine of lysine or arginine. Furthermore, removing the methyl group under physiological conditions by simple hydrolysis was considered thermodynamically disfavored and thus unlikely to occur spontaneously.

Those thinking that methylation marks were permanent had their belief system shaken a bit by several reports. First, it was shown that a nuclear peptidylarginine deiminase (PAD4) could eliminate monomethylarginine from histone H3 (Wang et al.). While this methyl removal process results in the arginine residue being converted into citrulline, and hence is not a true reversal of the modification, it nevertheless provided a mechanism for eliminating a "permanent" methyl mark.

Robin Allshire provided a tantalizing genetic argument that the tis2 gene from S. pombe reversed K9 dimethylation on histone H3 (R. Allshire, per comm.). He may have been on the right track because a few months after the meeting, the unrelated LSD1 enzyme from mammals was shown to specifically demethylate di- and monomethyl K4 on histone H3 (Shi et al. 2004), thus reversing an "active" chromatin mark. Quite interestingly, LSD1 did not work on trimethylated K4; thus, methylation could be reversed during the "marking" process, but reversal was not possible once the mark was fully matured.

However, Steve Henikoff presented a way by which a "permanent" trimethyl lysine mark could be eliminated. He showed that the variant histone H3.3 could replace canonical histone H3 in a replication-independent transcription-coupled manner (Henikoff et al.). In essence, a histone that contained methyl marks for silencing could be removed and replaced with one that was more conducive to transcription. When total chromatin was isolated, histone H3.3 had many more "active" chromatin methylation marks (e.g., MeK79) on it than canonical histone H3 did.

In considering all these results, it seems that there may not be a simple molecular modification within histones that serves as a memory mark for propagating the silent chromatin state through cell division. Rather, there must be a more tenuous set of interactions that increase the probability that a silent state will be maintained, though they do not guarantee it.

Nuclear Organization

Correlations between nuclear location and gene expression have been made for many years (Mirkovitch et al. 1987). These observations began to drive the notion that there were special compartments within the cell where gene expression or silencing were restricted. It was argued that this organization was necessary to keep the complexity of the genome and its regulation in a workable order. This idea was supported by studies in S. cerevisiae, where telomeres are preferentially located at the nuclear periphery, as are key components of the silencing complex, such as Sir4 (Palladino et al. 1993). Mutations that released the telomeres or Sir4 from the nuclear periphery resulted in a loss of telomeric silencing (Andrulis et al. 2002; Laroche et al. 1998). Furthermore, artificially tethering a partially silenced gene to the periphery caused it to become fully silenced (Andrulis et al. 1998).

But in an insightful experiment, Gasser showed that if both the telomeres and the silencing complex were released from the periphery and free to move throughout the nucleus, telomeric silencing was readily established (Gasser et al.). Thus, there does not appear to be a special need for localizing loci to a compartment. This is more consistent with the findings that rapid movement of chromatin proteins on and off chromosomes can still mediate effective regulation such as silencing. Perhaps some of the localization is necessary to maintain high local concentrations of relevant factors under special—perhaps stressful—conditions.

Double-stranded RNA Mechanisms

Since the discovery that double-stranded RNAs could regulate gene expression and DNA transposition in plants, fungi, and C. elegans, some of the fundamentals have been worked out and follow a basic process (for review, see Tomari and Zamore 2005). First, double-stranded RNA (dsRNA) must be generated. In some cases it is introduced by humans (injected dsRNA or as transcribed inverted hairpin sequences). It could also be produced as the result of transcription of inverted repeats from viruses or transposons, from normal endogenous RNA with an inherent hairpin structure, or it may be generated by RNA-dependent RNA polymerases (RdRP) that reverse-transcribe "aberrant" RNAs within the cell (what constitutes "aberrant" is a matter for further research). Next, the dsRNA is cleaved into small (21–27-bp) RNA fragments by the RNase III enzyme Dicer. These small RNAs, which include small interfering RNA (siRNA) and microRNA (miRNA), are unwound and loaded into the RNA-induced silencing complex (RISC) or RNA-induced initiation of transcriptional gene silencing (RITS). At the heart of these complexes is the Argonaute protein, which binds the single-stranded RNA and uses it as a guide to direct the complex to RNA with the complementary sequence. Then, depending on the complex, the larger complementary mRNA may have its translation inhibited, it may be cleaved by an RNase H-related PIWI domain within Argonaute, or it may direct chromatin-mediated silencing of the gene producing the RNA. In systems where the RNA is cleaved, RdRP can initiate second-strand RNA synthesis. These new dsRNA molecules are processed further by Dicer to generate secondary siRNA, which can reinitiate RdRP synthesis and thus amplify the original signal in a cycle. However, it appears that some organisms do not utilize all the regulatory options outlined. For example, no RdRP homologs have been identified in *Drosophila* or mammals, where RNAi has a relatively short duration in dividing cells. Needless to say, there are still many details to be worked out in all these systems.

Many of the Symposium presentations addressed whether dsRNA-mediated pathways were involved in different epigenetic events. For instance, mutants in the RNAi pathways were tested to see if they affected PEV, cosuppression, or other phenomena. (Pal-Bhadra et al.; Robert et al.; Martienssen et al.).

In *S. pombe*, some of the differences were being assessed at the mechanistic level. While silencing at the mating type loci and heterochromatin within the centromere uses virtually all the same histone modifications and chromatin-silencing proteins, RNAi machinery does not appear to affect the mating loci, whereas the RNAi pathway is required for silencing within the centromeric heterochromatin (Allshire; Cam and Grewal). This appears to be the result of having two pathways for recruiting these chromatin-based silencing proteins. At the centromere, the silencing proteins are primarily recruited via the RNAi pathway (apparently via the RITS complex), while at the mating type loci, these proteins appear to be recruited via DNA-binding proteins, as occurs in *S. cerevisiae* (Cam and Grewal).

S. pombe also provided the best evidence that RNA, as it is being transcribed, serves as the recruitment site for RNAi-mediated chromatin silencing (Allshire). When a 280-nucleotide sequence was transcribed in vivo as an inverted repeat to form a 280-bp hairpin of RNA, this served to silence—through Swi6 and Me-K9 histone H3-based mechanisms—a second gene containing the 280 nucleotides at the 3′ end of its coding sequence. However, when a transcriptional terminator was inserted between the 5′ end of the coding sequence and the 280-base sequence, no silencing or chromatin-related proteins were detected at the gene. Thus production of a complementary RNA *in cis* appears to be necessary to recruit silencing machinery. This result leads to a bigger question. If the silencing machinery has been recruited, why is the message still being expressed?

A potential answer to this paradox may be provided by the discovery of DNA-dependent RNA polymerase IV, which is required for RNA-directed silencing via chromatin/DNA modification (Herr and Baulcombe). RNA Pol IV is distinct from its better known cousins—the DNA-dependent RNA polymerases I, II, and III, which transcribe ribosomal RNA genes, messenger RNA genes, and tRNA genes, respectively. RNA Pol IV may transcribe genes that are silenced via RNA-directed pathways to provide a cycle of reinforcement for the process. Whether it can transcribe within silenced chromatin remains to be determined, but it is worth noting that there are genes that can be expressed only when they are in heterochromatin (e.g., the *light* gene in *Drosophila*; Hearn et al. 1991).

Protecting the Genome

For some time, it has been thought that several of the epigenetic processes discussed at the Symposium (DNA methylation, heterochromatic repression, and RNAi machinery) were systems that were used to defend the cellular genome from foreign invaders such as viruses and transposons (Bestor and Tycko 1996). For host defense, transposition must be inhibited to protect the rest of the genome. While "spreading" of transposons may occur in the soma and even have phenotypic consequences, transposition in the soma of a metazoan is a "dead end." Successful reproduction for the transposon requires that it be passed on through the host's germ line. Consequently, transposition in germ cells is the only place where the transposon will be ultimately successful. If such a premise is correct, then the host must have a germ-line-specific defense system—but evidence for such a system has been lacking. This changed with the characterization of mammalian *DNMT3L* (Bestor and Bourc'his). This member of the DNA methyltransferase genes appears to be expressed in only germ cells of both male and female mice. Homozygous mutants of this gene are sterile and show rampant expression of two retrotransposons, LINE-1 and IAP, in germ cells. Intriguingly, meiotic catastrophe is observed in spermatocytes in the mutant, which may be the result of promiscuous transposition by these and other elements. The *DNMT3L* gene appears to be evolving rapidly in mammals, as if it may be "chasing" an ever-changing collection of transposons in the ultimate genetic "arms race."

Prions

Wickner provided an overview and criteria for defining a prion, and from his description it is clear they are part of the epigenetic landscape (Wickner et al.). In the simplest molecular sense, prions are proteins that can cause heritable phenotypic changes, by acting on and altering their cognate gene product. No DNA sequence changes occur; rather the prion typically confers a structural change in its substrate. The best-studied and -understood class of prions cause soluble forms of a protein to transition into amyloid fibers. In many cases, the amyloid form reduces or abolishes normal activity of the protein, thus producing a change in phenotype. Wickner defined another class of prions that do not form amyloid filaments. These are enzymes that require activation by their own enzymatic activity. If a cell should have only inactive forms of the enzyme, then an external source of the active enzyme is required to start what would then become a self-propagating trait, as long as at least one active molecule was passed on to each cell. He provided two examples and the expectation that this class of proteins will define a new set of epigenetic mechanisms to pursue.

Si presented preliminary evidence that a prion model may explain learned memory in *Aplysia* (Si et al.). Protein translation of a number of stored mRNAs in neuronal cells is important for the maintenance of short-term memory in this snail. He found that a regulator of protein translation,

CPEB (cytoplasmic polyadenylation element binding protein), can exist in two forms, and that the activated form of CPEB acts dominantly to perpetuate itself. It is still early days in testing this idea, but it offers an exciting new way to consider the mechanism by which memory in the brain occurs.

New Phenomenon

The description of a new and unexpected phenomenon always holds our imagination. One presentation in particular held my thoughts for weeks after the Symposium. Standard genetic analysis of mutant alleles of the *HOTHEAD* gene, which regulates organ fusion in *Arabidopsis*, revealed that normal rules of Mendelian genetics were not being followed (R. Pruitt, pers comm.). It was discovered that if heterozygous *HOTHEAD/hothead* plants self-fertilized and produced a homozygous *hothead/hothead* plant, and then this homozygous *hothead/hothead* plant was allowed to self-fertilize, the progeny from this homozygous parent reverted to a *HOTHEAD/hothead* genotype at a frequency of up to 15%. This stunning level of wild-type reversion produced an exact duplicate, at the nucleotide level, of the wild-type gene seen in the earlier generations. This reversion was not limited to the *HOTHEAD* locus—several other loci had similar frequencies of reversion to wild-type alleles. However, all the reversions required that the parent be homozygous *hothead/hothead*. The gene product of *HOTHEAD* did not offer an obvious explanation as to how this could occur, but discussions certainly suggested that an archival copy of the wild-type gene was transmitted, perhaps via RNA, through successive generations. While it could be argued that this phenomenon is outside the purview of "epigenetics" because of the change in DNA sequence, the heritable transmission of the putative archived copy does not follow normal genetic rules. Nevertheless, this phenomenon has enormous implications for the field of genetics, especially in evolutionary thinking.

Further Implications of Epigenetics in Human Health

There are abundant examples of how inappropriate gene regulation via epigenetic mechanisms can lead to human disease. This has led to a broader awareness about the field of epigenetics. For instance, it has resulted in pharmaceutical development of drugs that selectively inhibit many of the enzymes discussed at this Symposium (Curtin and Glaser 2003). It has also compelled some to develop a set of guidelines to aid health workers in assessing whether a disease has an epigenetic basis (Bjornsson et al.).

However, the basic science studies have led even further. I found it particularly exciting that A. Bird's continued work on MeCP2 mutants has led to a tangible explanation for Rett Syndrome—pointing straight to a metabolic defect in the brain (Bird and Macleod; A. Bird, pers. comm.). It was also a treat to have A. Klar close the Symposium by applying his creative energy to explain human psychosis based on epigenetic principles he originally developed by studying *S. pombe* (Klar). While their ideas wait further testing, I hope they are right.

There was an important presentation by R. Jaenisch, in which he described the successful nuclear transfer from a melanoma cancer cell into embryonic stem cells that ultimately led to the generation of mice (Jaenisch et al.). While the cells of these mice had the aberrant chromosomes of the original tumor and developed melanomas with a high penetrance and frequency, they also developed a wide spectrum of other types of tumors, though fibroblasts did not form tumors. These results set the stage to dissect the genetic and epigenetic events that are necessary for cancer to develop in different tissues. It holds great promise as a new tool for understanding oncogenesis.

CLOSING THOUGHTS

So what more needs to be done to understand epigenetic mechanisms? For the most part, we are still collecting (discovering) the components. Just as the full sequence of a genome has greatly facilitated progress in genetics, a clearer understanding for epigenetics will likely come when all the parts are known. It is encouraging to see the great strides that have been made in the last decade.

I confess that I cannot discern whether we are close to, or far away from, having an accurate mechanistic understanding about how epigenetic states are maintained and propagated. The prion-based phenomenon may be the first to be understood, but those that are chromatin-based seem the farthest off. The polyvalent nature of interactions that seem to be required to establish a silenced state on a chromosome increases the complexity of the problem. This is further compounded by the dynamic nature of silent chromatin. The ability to know more about movement of components in and out of chromatin structures requires application of enhanced or new methods for an eventual understanding. While ChIP has been important in establishing which components reside in a structure, it has temporarily blinded us to the dynamics.

I suspect that given the complexity, simply measuring binding and equilibrium constants between all the components and trying to derive a set of differential equations to simulate epigenetic switches may not be an effective use of resources, nor will it necessarily result in better comprehension. Rather, I speculate that a new type of mathematical approach will need to be developed and combined with new experimental measuring methods to eventually understand epigenetic events. Part of this may require development of in vitro systems that faithfully recapitulate an epigenetic switch between states. The in vitro system presented by Kingston (Fan et al.; R. Kingston, pers. comm.) provides the first steps toward this enormous task.

The idea of competition between two states in most epigenetic phenomena likely reflects an "arms race" that is happening at many levels in the cell, followed by attempts to rectify "collateral damage." For instance, silencing proteins may have evolved to protect the genome from transposons. However, because silencing proteins work through the ubiquitous nucleosomes, some critical genes become repressed. To overcome this, histone modifications (e.g., methylation of K4 and K79 of H3) and variant replacement histones (H2A.Z) evolved to prevent silencing proteins from binding to critical genes. Depending on subsequent events, these changes may be co-opted for other processes—e.g., repression of some of the genes by the silencing proteins may have become useful (silent mating loci). The silencing mechanisms may have been co-opted for other functions as well, such as promoting chromosome segregation. And so it goes.

I look forward to having the genomes of more organisms sequenced, because this might lead us to understand an order of events through evolution that set up the epigenetic processes we see today. For instance, *S. cerevisiae* does not have RNAi machinery, but many other fungi do. By filling in some of the phylogenetic gaps between species, we may discover what events led to *S. cerevisiae* no longer "needing" the RNAi system.

Perhaps more than any other field of biological research, the study of epigenetics is founded on the attempt to understand unexpected observations, ranging from H.J. Muller's position-effect variegation (Muller 1941), to polar overdominance in the *callipyge* phenotype (Georges et al.). The hope of understanding something unusual serves as the bait to draw us in, but we soon become entranced by the cleverness of the mechanisms employed. This may explain why this field has drawn more than its share of lighthearted and clever minds. I suspect it will continue to do so as we develop a deeper understanding of the cleverness and as new and unexpected epigenetic phenomena are discovered.

In closing, I thank Bruce Stillman and David Stewart for organizing the Symposium, and all the faculty and staff at Cold Spring Harbor Laboratory for permitting epigenetic researchers a chance to share ideas and take a few more steps toward understanding.

ACKNOWLEDGMENTS

I thank my colleagues at the Hutch for making it such a wonderful place to study epigenetic phenomena and the National Institutes of Health for support.

REFERENCES

Abraham J., Feldman J., Nasmyth K.A., Strathern J.N., Klar A.J., Broach J.R., and Hicks J.B. 1983. Sites required for position-effect regulation of mating-type information in yeast. *Cold Spring Harbor Symp. Quant. Biol.* **47**: 989.

Allfrey V.G., Inoue A., Karn J., Johnson E.M., and Vidali G. 1974. Phosphorylation of DNA-binding nuclear acidic proteins and gene activation in the HeLa cell cycle. *Cold Spring Harbor Symp. Quant. Biol.* **38**: 785.

Andrulis E.D., Neiman A.M., Zappulla D.C., and Sternglanz R. 1998. Perinuclear localization of chromatin facilitates transcriptional silencing. *Nature* **394**: 592.

Andrulis E.D., Zappulla D.C., Ansari A., Perrod S., Laiosa C.V., Gartenberg M.R., and Sternglanz R. 2002. Esc1, a nuclear periphery protein required for Sir4-based plasmid anchoring and partitioning. *Mol. Cell. Biol.* **22**: 8292.

Aparicio O.M. and Gottschling D.E. 1994. Overcoming telomeric silencing: A trans-activator competes to establish gene expression in a cell cycle-dependent way. *Genes Dev.* **8**: 1133.

Ariel M., Selig S., Brandeis M., Kitsberg D., Kafri T., Weiss A., Keshet I., Razin A., and Cedar H. 1993. Allele-specific structures in the mouse Igf2-H19 domain. *Cold Spring Harbor Symp. Quant. Biol.* **58**: 307.

Bell A., Boyes J., Chung J., Pikaart M., Prioleau M.N., Recillas F., Saitoh N., and Felsenfeld G. 1998. The establishment of active chromatin domains. *Cold Spring Harbor Symp. Quant. Biol.* **63**: 509.

Bell S.P., Marahrens Y., Rao H., and Stillman B. 1993. The replicon model and eukaryotic chromosomes. Cold Spring Harbor Symp. Quant. Biol. 58: 435.

Belote J.M., McKeown M.B., Andrew D.J., Scott T.N., Wolfner M.F., and Baker B.S. 1985. Control of sexual differentiation in *Drosophila melanogaster*. *Cold Spring Harbor Symp. Quant. Biol.* **50**: 605.

Bestor T.H. and Tycko B. 1996. Creation of genomic methylation patterns. *Nat. Genet.* **12**: 363.

Beutler E. 1965. Gene inactivation: The distribution of gene products among populations of cells in heterozygous humans. *Cold Spring Harbor Symp. Quant. Biol.* **29**: 261.

Bird A.P. 1993. Functions for DNA methylation in vertebrates. *Cold Spring Harbor Symp. Quant. Biol.* **58**: 281.

Borst P., Gommers-Ampt J.H., Ligtenberg M.J., Rudenko G., Kieft R., Taylor M.C., Blundell P.A., and van Leeuwen F. 1993. Control of antigenic variation in African trypanosomes. *Cold Spring Harbor Symp. Quant. Biol.* **58**: 105.

Brink R.A. 1959. Paramutation at the R locus in maize. *Cold Spring Harbor Symp. Quant. Biol.* **23**: 379.

Campbell A. 1981. Some general questions about movable elements and their implications. *Cold Spring Harbor Symp. Quant. Biol.* **45**: 1.

Cattanach B.M. and Kirk M. 1985. Differential activity of maternally and paternally derived chromosome regions in mice. *Nature* **315**: 496.

Cedar H., Stein R., Gruenbaum Y., Naveh-Many T., Sciaky-Gallili N., and Razin A. 1983. Effect of DNA methylation on gene expression. *Cold Spring Harbor Symp. Quant. Biol.* **47**: 605.

Chambon P. 1978. Summary: The molecular biology of the eukaryotic genome is coming of age. *Cold Spring Harbor Symp. Quant. Biol.* **42**: 1209.

Cheng T.H. and Gartenberg M.R. 2000. Yeast heterochromatin is a dynamic structure that requires silencers continuously. *Genes Dev.* **14**: 452.

Cheutin T., McNairn A.J., Jenuwein T., Gilbert D.M., Singh P.B., and Misteli T. 2003. Maintenance of stable heterochromatin domains by dynamic HP1 binding. *Science* **299**: 721.

Clark D., Reitman M., Studitsky V., Chung J., Westphal H., Lee E., and Felsenfeld G. 1993. Chromatin structure of transcriptionally active genes. *Cold Spring Harbor Symp. Quant. Biol.* **58**: 1.

Cockell M., Gotta M., Palladino F., Martin S.G., and Gasser S.M. 1998. Targeting Sir proteins to sites of action: A general mechanism for regulated repression. *Cold Spring Harbor Symp. Quant. Biol.* **63**: 401.

Curtin M. and Glaser K. 2003. Histone deacetylase inhibitors: The Abbott experience. *Curr. Med. Chem.* **10**: 2373.

Dion M.F., Altschuler S.J., Wu L.F., and Rando O.J. 2005. Genomic characterization reveals a simple histone H4 acetylation code. *Proc. Natl. Acad. Sci.* **102**: 5501.

Doerfler W., Kruczek I., Eick D., Vardimon L., and Kron B. 1983. DNA methylation and gene activity: The adenovirus system as a model. *Cold Spring Harbor Symp. Quant. Biol.* **47**: 593.

Feinberg A.P., Kalikin L.M., Johnson L.A., and Thompson J.S. 1994. Loss of imprinting in human cancer. *Cold Spring Harbor Symp. Quant. Biol.* **59**: 357.

Fox C.A., Loo S., Rivier D.H., Foss M.A., and Rine J. 1993. A transcriptional silencer as a specialized origin of replication that establishes functional domains of chromatin. *Cold Spring Harbor Symp. Quant. Biol.* **58**: 443.

Frankel J. 1990. Positional order and cellular handedness. *J. Cell Sci.* **97**: 205.

Gartler S.M. and Linder D. 1965. Selection in mammalian mosaic cell populations. *Cold Spring Harbor Symp. Quant. Biol.* **29:** 253.

Goldschmidt R.B. 1952. The theory of the gene: Chromosomes and genes. *Cold Spring Harbor Symp. Quant. Biol.* **16:** 1.

Hearn M.G., Hedrick A., Grigliatti T.A., and Wakimoto B.T. 1991. The effect of modifiers of position-effect variegation on the variegation of heterochromatic genes of *Drosophila melanogaster*. *Genetics* **128:** 785.

Hernday A.D., Braaten B.A., and Low D.A. 2003. The mechanism by which DNA adenine methylase and PapI activate the pap epigenetic switch. *Mol. Cell* **12:** 947.

Hodgkin J., Doniach T., and Shen M. 1985. The sex determination pathway in the nematode *Caenorhabditis elegans:* Variations on a theme. *Cold Spring Harbor Symp. Quant. Biol.* **50:** 585.

Imai S., Johnson F.B., Marciniak R.A., McVey M., Park P.U., and Guarente L. 2000. Sir2: An NAD-dependent histone deacetylase that connects chromatin silencing, metabolism, and aging. *Cold Spring Harbor Symp. Quant. Biol.* **65:** 297.

Jackson V. 1988. Deposition of newly synthesized histones: Hybrid nucleosomes are not tandemly arranged on daughter DNA strands. *Biochemistry* **27:** 2109.

Jenuwein T. and Allis C.D. 2001. Translating the histone code. *Science* **293:** 1074.

Johnson L.M., Kayne P.S., Kahn E.S., and Grunstein M. 1990. Genetic evidence for an interaction between SIR3 and histone H4 in the repression of the silent mating loci in *Saccharomyces cerevisiae*. *Proc. Natl. Acad. Sci.* **87:** 6286.

Klar A.J. and Bonaduce M.J. 1993. The mechanism of fission yeast mating-type interconversion: Evidence for two types of epigenetically inherited chromosomal imprinted events. *Cold Spring Harbor Symp. Quant. Biol.* **58:** 457.

Kubicek S. and Jenuwein T. 2004. A crack in histone lysine methylation. *Cell* **119:** 903.

La Volpe A., Taggart M., Macleod D., and Bird A. 1983. Coupled demethylation of sites in a conserved sequence of *Xenopus* ribosomal DNA. *Cold Spring Harbor Symp. Quant. Biol.* **47:** 585.

Laroche T., Martin S.G., Gotta M., Gorham H.C., Pryde F.E., Louis E.J., and Gasser S.M. 1998. Mutation of yeast Ku genes disrupts the subnuclear organization of telomeres. *Curr. Biol.* **8:** 653.

Laurenson P. and Rine J. 1992. Silencers, silencing, and heritable transcriptional states. *Microbiol. Rev.* **56:** 543.

Lewis E.B. 1985. Regulation of the genes of the bithorax complex in *Drosophila*. *Cold Spring Harbor Symp. Quant. Biol.* **50:** 155.

Li E., Beard C., Forster A.C., Bestor T.H., and Jaenisch R. 1993. DNA methylation, genomic imprinting, and mammalian development. *Cold Spring Harbor Symp. Quant. Biol.* **58:** 297.

Locke J., Kotarski M.A., and Tartof K.D. 1988. Dosage-dependent modifiers of position effect variegation in *Drosophila* and a mass action model that explains their effect. *Genetics* **120:** 181.

Losick R. 1998. Summary: Three decades after sigma. *Cold Spring Harbor Symp. Quant. Biol.* **63:** 653.

Louie A.J., Candido E.P., and Dixon G.H. 1974. Enzymatic modifications and their possible roles in regulating the binding of basic proteins to DNA and in controlling chromosomal structure. *Cold Spring Harbor Symp. Quant. Biol.* **38:** 803.

Luger K., Mader A.W., Richmond R.K., Sargent D.F., and Richmond T.J. 1997. Crystal structure of the nucleosome core particle at 2.8 Å resolution. *Nature* **389:** 251.

Lyon M.F. 1961. Gene action in the X-chromosome of the mouse (*Mus musculus* L.). *Nature* **190:** 372.

Maine E.M., Salz H.K., Schedl P., and Cline T.W. 1985. Sex-lethal, a link between sex determination and sexual differentiation in *Drosophila melanogaster*. *Cold Spring Harbor Symp. Quant. Biol.* **50:** 595.

Martin-Morris L.E., Loughney K., Kershisnik E.O., Poortinga G., and Henikoff S. 1993. Characterization of sequences responsible for trans-inactivation of the *Drosophila* brown gene. *Cold Spring Harbor Symp. Quant. Biol.* **58:** 577.

McClintock B. 1952.. Chromosome organization and genic expression. *Cold Spring Harbor Symp. Quant. Biol.* **16:** 13.

———. 1957. Controlling elements and the gene. *Cold Spring Harbor Symp. Quant. Biol.* **21:** 197.

Megee P.C., Morgan B.A., and Smith M.M. 1995. Histone H4 and the maintenance of genome integrity. *Genes Dev.* **9:** 1716.

Mirkovitch J., Gasser S.M., and Laemmli U.K. 1987. Relation of chromosome structure and gene expression. *Philos. Trans. R. Soc. Lond. B Biol. Sci.* **317:** 563.

Mostoslavsky R., Kirillov A., Ji Y.H., Goldmit M., Holzmann M., Wirth T., Cedar H., and Bergman Y. 1999. Demethylation and the establishment of κ allelic exclusion. *Cold Spring Harbor Symp. Quant. Biol.* **64:** 197.

Muller H.J. 1941. Induced mutations in *Drosophila*. *Cold Spring Harbor Symp. Quant. Biol.* **9:** 151.

Nance W.E. 1965. Genetic tests with a sex-linked marker: Glucose-6-phosphate dehydrogenase. *Cold Spring Harbor Symp. Quant. Biol.* **29:** 415.

Nanney D.L. 1959. Epigenetic factors affecting mating type expression in certain ciliates. *Cold Spring Harbor Symp. Quant. Biol.* **23:** 327.

Nasmyth K.A., Tatchell K., Hall B.D., Astell C., and Smith M. 1981. Physical analysis of mating-type loci in *Saccharomyces cerevisiae*. *Cold Spring Harbor Symp. Quant. Biol.* **45:** 961.

Osley M.A. 2004. H2B ubiquitylation: The end is in sight. *Biochim. Biophys. Acta* **1677:** 74.

Palladino F., Laroche T., Gilson E., Pillus L., and Gasser S.M. 1993. The positioning of yeast telomeres depends on SIR3, SIR4, and the integrity of the nuclear membrane. *Cold Spring Harbor Symp. Quant. Biol.* **58:** 733.

Pannetier C., Hu-Li J., and Paul W.E. 1999. Bias in the expression of IL-4 alleles: The use of T cells from a GFP knock-in mouse. *Cold Spring Harbor Symp. Quant. Biol.* **64:** 599.

Peacock W.J., Brutlag D., Goldring E., Appels R., Hinton C.W., and Lindsley D.L. 1974. The organization of highly repeated DNA sequences in *Drosophila melanogaster* chromosomes. *Cold Spring Harbor Symp. Quant. Biol.* **38:** 405.

Pillus L. and Rine J. 1989. Epigenetic inheritance of transcriptional states in *S. cerevisiae*. *Cell* **59:** 637.

Ptashne M. 2004. *A genetic switch: Phage lambda revisited*, 3rd edition, Cold Spring Harbor Laboratory Press, Cold Spring Harbor, New York.

Renauld H., Aparicio O.M., Zierath P.D., Billington B.L., Chhablani S.K., and Gottschling D.E. 1993. Silent domains are assembled continuously from the telomere and are defined by promoter distance and strength, and by SIR3 dosage. *Genes Dev.* **7:** 1133.

Rine J., Jensen R., Hagen D., Blair L., and Herskowitz I. 1981. Pattern of switching and fate of the replaced cassette in yeast mating-type interconversion. *Cold Spring Harbor Symp. Quant. Biol.* **45:** 951.

Ringrose L., Ehret H., and Paro R. 2004. Distinct contributions of histone H3 lysine 9 and 27 methylation to locus-specific stability of polycomb complexes. *Mol. Cell* **16:** 641.

Rubin G.M. 1985. Summary. *Cold Spring Harbor Symp. Quant. Biol.* **50:** 905.

Rubin G.M., Hazelrigg T., Karess R.E., Laski F.A., Laverty T., Levis R., Rio D.C., Spencer F.A., and Zuker C.S. 1985. Germ line specificity of P-element transposition and some novel patterns of expression of transduced copies of the white gene. *Cold Spring Harbor Symp. Quant. Biol.* **50:** 329.

Rudkin G.T. and Tartof K.D. 1974. Repetitive DNA in polytene chromosomes of *Drosophila melanogaster*. *Cold Spring Harbor Symp. Quant. Biol.* **38:** 397.

Ruiz-Carrillo A., Jorcano J.L., Eder G., and Lurz R. 1979. In vitro core particle and nucleosome assembly at physiological ionic strength. *Proc. Natl. Acad. Sci.* **76:** 3284.

Schubeler D., MacAlpine D.M., Scalzo D., Wirbelauer C., Kooperberg C., van Leeuwen F., Gottschling D.E., O'Neill L.P., Turner B.M., Delrow J., Bell S.P., and Groudine M. 2004. The histone modification pattern of active genes revealed through genome-wide chromatin analysis of a higher eukaryote. *Genes Dev.* **18:** 1263.

Schultz J. 1957. The relation of the heterochromatic chromosome regions to the nucleic acids of the cell. *Cold Spring Harbor Symp. Quant. Biol.* **21**: 307.

Selker E.U., Richardson G.A., Garrett-Engele P.W., Singer M.J., and Miao V. 1993. Dissection of the signal for DNA methylation in the zeta-eta region of *Neurospora*. *Cold Spring Harbor Symp. Quant. Biol.* **58**: 323.

Shapiro L.J. and Mohandas T. 1983. DNA methylation and the control of gene expression on the human X chromosome. *Cold Spring Harbor Symp. Quant. Biol.* **47**: 631.

Shi Y., Lan F., Matson C., Mulligan P., Whetstine J.R., Cole P.A., and Casero R.A. 2004. Histone demethylation mediated by the nuclear amine oxidase homolog LSD1. *Cell* **119**: 941.

Smith C.M., Haimberger Z.W., Johnson C.O., Wolf A.J., Gafken P.R., Zhang Z., Parthun M.R., and Gottschling D.E. 2002. Heritable chromatin structure: Mapping "memory" in histones H3 and H4. *Proc. Natl. Acad. Sci.* (suppl. 4) **99**: 16454.

Solter D., Aronson J., Gilbert S.F., and McGrath J. 1985. Nuclear transfer in mouse embryos: Activation of the embryonic genome. *Cold Spring Harbor Symp. Quant. Biol.* **50**: 45.

Swift H. 1974. The organization of genetic material in eukaryotes: Progress and prospects. *Cold Spring Harbor Symp. Quant. Biol.* **38**: 963.

Thompson J.S., Hecht A., and Grunstein M. 1993. Histones and the regulation of heterochromatin in yeast. *Cold Spring Harbor Symp. Quant. Biol.* **58**: 247.

Thompson J.S., Ling X., and Grunstein M. 1994. Histone H3 amino terminus is required for telomeric and silent mating locus repression in yeast. *Nature* **369**: 245.

Tilghman S.M., Bartolomei M.S., Webber A.L., Brunkow M.E., Saam J., Leighton P.A., Pfeifer K., and Zemel S. 1993. Parental imprinting of the H19 and Igf2 genes in the mouse. *Cold Spring Harbor Symp. Quant. Biol.* **58**: 287.

Tomari Y. and Zamore P.D. 2005. Perspective: Machines for RNAi. *Genes Dev.* **19**: 517.

Vazquez J., Farkas G., Gaszner M., Udvardy A., Muller M., Hagstrom K., Gyurkovics H., Sipos L., Gausz J., Galloni M., Hogga I., Karch F., and Schedl P. 1993. Genetic and molecular analysis of chromatin domains. *Cold Spring Harbor Symp. Quant. Biol.* **58**: 45.

Wade P.A., Jones P.L., Vermaak D., Veenstra G.J., Imhof A., Sera T., Tse C., Ge H., Shi Y.B., Hansen J.C., and Wolffe A.P. 1998. Histone deacetylase directs the dominant silencing of transcription in chromatin: Association with MeCP2 and the Mi-2 chromodomain SWI/SNF ATPase. *Cold Spring Harbor Symp. Quant. Biol.* **63**: 435.

Weintraub H. 1974. The assembly of newly replicated DNA into chromatin. *Cold Spring Harbor Symp. Quant. Biol.* **38**: 247.

———. 1993. Summary: Genetic tinkering—Local problems, local solutions. *Cold Spring Harbor Symp. Quant. Biol.* **58**: 819.

Weintraub H., Flint S.J., Leffak I.M., Groudine M., and Grainger R.M. 1978. The generation and propagation of variegated chromosome structures. *Cold Spring Harbor Symp. Quant. Biol.* **42**: 401.

Willard H.F., Brown C.J., Carrel L., Hendrich B., and Miller A.P. 1993. Epigenetic and chromosomal control of gene expression: Molecular and genetic analysis of X chromosome inactivation. *Cold Spring Harbor Symp. Quant. Biol.* **58**: 315.

Wood W.B., Meneely P., Schedin P., and Donahue L. 1985. Aspects of dosage compensation and sex determination in *Caenorhabditis elegans*. *Cold Spring Harbor Symp. Quant. Biol.* **50**: 575.

Yarmolinsky M.B. 1981. Summary. *Cold Spring Harbor Symp. Quant. Biol.* **45**: 1009.

Zhang L., Eugeni E.E., Parthun M.R., and Freitas M.A. 2003. Identification of novel histone post-translational modifications by peptide mass fingerprinting. *Chromosoma* **112**: 77.

Zheng C. and Hayes J.J. 2003. Structures and interactions of the core histone tail domains. *Biopolymers* **68**: 539.

Author Index

A
Abate-Shen, C., 171
Ahmad, K., 235
Allis, C.D., 161
Allshire, R.C., 389
Almouzni, G., 273
Ancelin, K., 1

B
Bai, X., 81
Baldwin, K., 19
Bao, Y., 227
Barlow, D.P., 55
Baroux, C., 465
Bartolomei, M.S., 39
Baubec, T., 55
Baulcombe, D.C., 363
Baxa, U., 489
Beijersbergen, R.L., 439
Bender, J., 145
Bernards, R., 439
Berns, K., 439
Bestor, T.H., 381
Bhadra, U., 433
Bickmore, W.A., 251
Birchler, J.A., 433
Bird, A., 113
Bjornsson, H.T., 447
Blelloch, R., 19
Bone, J.R., 81
Bourc'his, D., 381
Boutsma, E., 319
Brachmann, A., 489
Braidotti, G., 55
Bruggeman, S.W.M., 319
Brummelkamp, T.R., 439
Burgess-Beusse, B., 245

C
Caiment, F., 477
Cam, H., 419
Cao, R., 309
Cedar, H., 131
Chakravarthy, S., 227
Chambeyron, S., 251
Chandler, V.L., 355
Charlier, C., 477
Chaumeil, J., 89
Chess, A., 485
Chuikov, S., 171
Cockett, N., 477
Coonrod, S.A., 161
Craig, C., 267
Csankovszki, G., 71
Cui, H., 447

D
Davis, E., 477
Dean, C., 457
Dean, W., 29
Dellino, G.I., 301
De Vries, W.N., 11

E
Edskes, H.K., 489
Eggan, K., 19
Elgin, S.C.R., 267
Engel, N.I., 39
Escobar-Restrepo, J.-M., 465
Evsikov, A.V., 11

F
Fabius, A., 439
Fallin, M.D., 447
Fan, H.-Y., 183
Farnham, P., 171
Farrell, C., 245
Fedoriw, A.M., 39
Feil, R., 29
Feinberg, A.P., 447
Felsenfeld, G., 245
Franco, A.A., 201
Fry, C.J., 219

G
Gartenberg, M.R., 327
Gasser, S.M., 327
Georges, M., 477
Ghirlando, R., 245
Gius, D., 447
Glaszner, M., 245
Goren, A., 131
Gottschling, D.E., 339, 507
Greb, T., 457
Grewal, S.I.S., 419
Grossniklaus, U., 465
Grunstein, M., 193

H
Haig, D., 67
Hajkova, P., 1
Hashimshony, T., 131
Hata, K., 125
Haynes, K.A., 267
Heard, E., 89
Hediger, F., 327
Heimerikx, M., 439
Henikoff, S., 235
Hernández-Muñoz, I., 319
Herr, A.J., 363
Higgins, M., 29
Hiiragi, T., 11
Hijmans, E.M., 439
Hochedlinger, K., 19
Huang, S., 245
Hulsman, D., 319
Huynh, K.D., 103

J
Jaenisch, R., 19
Jenuwein, T., 209
Jin, C., 245
Jones, R.S., 309
Jorgensen, R.A., 349
Juarez, M.T., 409

K
Kahn, T.G., 301
Kakutani, T., 139
Kandel, E., 497
Kaneda, M., 125
Karachentsev, D., 171
Kato, M., 139
Kaufman, P.D., 201
Kerkhoven, R.M., 439
Keshet, I., 131
Kingston, R.E., 183
Kinoshita, T., 139
Kirmizis, A., 171
Klar, A.J.S., 499
Knowles, B.B., 11
Kurdistani, S.K., 193
Kuroda, M.I., 81
Kuzmichev, A., 171

L
Lachner, M., 209
Lande-Diner, L., 131
Lange, U.C., 1
Laoueille, B., 465
Lee, J.T., 103
Leibovitch, B.A., 267
Leonelli, L., 161
Lewis, A., 29
Li, E., 125
Lindquist, S., 497
Lippman, Z., 371
Lister, C., 457
Litt, M., 245
Luger, K., 227
Lund, A.H., 319

M
Macleod, D., 113
Madiredjo, M., 439
Magdinier, F., 245
Mahy, N.L., 251
Margueron, R., 171
Martienssen, R., 371
Masui, O., 89
Matranga, C., 403
May, B., 371
McDonel, P., 71
McKittrick, E., 235
McMurray, M.A., 339
Meyer, B.J., 71
Millar, C.B., 193
Mitsuya, K., 29
Miura, A., 139
Moyer, J., 11
Mullenders, J., 439
Murrell, A., 29

Mutskov, V., 245
Muyrers-Chen, I., 319
Mylne, J., 457

N
Nakatani, Y., 245, 273
Narlikar, G.J., 183
Neumann, F.R., 327
Neves, G., 485
Nishioka, K., 171
Noback, S., 319

O
Oh, H., 81
Okamoto, I., 89
Okano, M., 125

P
Page, D.R., 465
Pal-Bhadra, M., 433
Park, Y., 81
Paro, R., 47
Pauler, F., 55
Payer, B., 1
Peaston, A.E., 11
Perlin, J.R., 161
Peterson, C.L., 219
Phelps-Durr, T.L., 409
Pierce, M.M., 489
Pillus, L., 259
Pirrotta, V., 301
Plasterk, R.H.A., 397
Preissner, T.S., 171

Q
Quivy, J.-P., 273

R
Ralston, E., 71
Rangwala, S.H., 267
Ray-Gallet, D., 273
Reik, W., 29
Reinberg, D., 171
Richards, E.J., 155
Rine, J., 259
Robert, V.J.P., 397
Roberts, V.A., 227
Ronemus, M., 371
Ross, E.D., 489
Ruvkun, G., 429

S
Sado, T., 125
Saitou, M., 1
Sarma, K., 171
Sasaki, H., 125
Schoenfelder, S., 47
Schotta, G., 209

Schwartz, Y.B., 301
Seidl, C., 55
Selker, E.U., 119
Sengupta, R., 209
Shay, T., 477
Shewmaker, F., 489
Shogren-Knaak, M.A., 219
Si, K., 497
Smit, M., 477
Smrzka, O., 55
Solter, D., 11
Spillane, C., 465
Steward, R., 171
Stokes, T.L., 155
Stricker, S., 55
Surani, M.A., 1

T

Taddei, A., 327
Tagami, H., 245, 273
Taghavi, P., 319
Takeda, H., 477
Tanger, E., 319
Theunissen, H., 319
Timmermans, M.C.P., 409
Tolhuis, B., 319
Tordoir, X., 477
Tremethick, D., 227
Tsujimoto, N., 125

U

Umlauf, D., 29

V

Valk-Lingbeek, M.E., 319
van der Stoop, P., 319
van Lohuizen, M., 319
Vaquero, A., 171
Vastenhouw, N.L., 397
Vaughn, M., 371
Velds, A., 439
Verhoeven, E., 319

W

Wang, D., 429
Wang, L., 309
Wang, Y., 161
West, A., 245

Western, P., 1
Wickner, R.B., 489
Woo, H.-R., 155
Wysocka, J., 161

Y

Yamada, Y., 19
Yi, H., 155
Yotova, I., 55
Yusufzai, T., 245

Z

Zamore, P.D., 403
Zhang, J., 131
Zhang, Y., 309

Subject Index

A

acetylation of histone H3. *See* histone H3 phosphorylation/acetylation
acetylation of lysine residues. *See* histone H4-K16 acetylation in yeast
ACF, 281
ACF1, 284, 285
adaptive challenge. *See* paramutation
age-1, 429
AG gene, 140
Agrobacterium, 363
Air noncoding RNA
    characteristics of mouse, 59–60
    as a *cis*-acting silencer, 58–59
    cluster dissociation from chromosome, 62
    cluster formation phases, 63
    DNA methylation and, 57–58
    expression of *Igf2r*, 61, 61f
    genomic imprinting and X-inactivation similarities, 63–64
    question of a human "H" AIR, 59
    replication asynchrony and, 58
    restriction to paternal chromosome, 61
    RNA FISH results, 60, 61f
    sequence and function, 56–57
    specificity of *Igf2r* RNA FISH signals, 60–61
    weak detection in undifferentiated ES cells, 62–63
alg-1, 400
amyloid fibers, 491–492. *See also* Ure2p and prions
Angelman syndrome, 451
animal-vegetal (A-V) axis, 15
anthocyanin biosynthetic pathway, 355
Aplysia neuronal CPEB, 497–498
Arabidopsis Mu (*AtMu1*), 372
Arabidopsis thaliana
    CAF-1 and, 203
    chromatin marks in, 215
    DNA methylation and
        ddm1-induced developmental abnormalities, 140
        developmental abnormalities due to mutations, 139
        epigenetic variation study (*see* epigenetic variation)
        mobilization of *CACTA* due to loss, 140–142
        transposon distribution within the genome, 142
    FWA inheritability, 139–140
    PAI genes and (*see PAI* genes and DNA methylation)
    silencing in (*see* silencing of imprinted genes in *Arabidopsis*)
    vernalization requirement (*see* flowering and epigenetic regulation)
Argonaute, 371, 391
    involvement in silencing, 400–401
    RISC and, 403
Arg residues methylation, 162, 163f, 165
Asf1, 204, 205, 274, 277
AsSN1, 366
At4g16890, 156–157
ATPase
    domain identity related to remodeling capability, 188–190
    domain role in defining remodeling action, 190–191
ATPase ISWI, 283
AtREP2 helitron and imprinting at *MEA* locus, 470–471
ATRX syndrome, 451
AtSN1, 372
A-V (animal-vegetal) axis, 15

B

b1 paramutation. *See also* paramutation
    function of *mop* genes, 358
    key sequences required, 356–357
    models for
        cis-interactions, 358
        trans-interactions pairing, 358–359
        trans-RNA, 359–360
    mutations preventing, 357–358
    phenotypes, 355–356
    potential consequences of paramutation, 360
    reduced transcription due to, 356
bal, 140, 156–157
BDF1, 197
Beckwith–Wiedemann syndrome (BWS), 448, 451
BEGAIN, 478
Berry, R.J., 68
[β] prion, 494
beta globin expression, 31
bipolar disease, 500
BIR (break-induced replication), 342–343
Blimp-1 and PGC specification, 5–6
Bmi1, 322, 323, 324
bone morphogenetic protein 4 (BMP4), 3
Boris, 42, 43, 52
brain development
    asymmetric cell division and gene expression, 501–502
    correlation of hemispheric laterality with hand-use preference, 499–500
    epigenetics applied to, 504–505
    handedness etiology and random-recessive model, 500
    patterned chromatin segregation, 504
    psychoses etiology and handedness, 500–501
    strand-specific imprinting and segregation model, 502–504
break-induced replication (BIR), 342–343
BRG1
    BRG1 vs. SNF2h ability to create accessible DNA sites, 187–188
    remodeled products produced by, 185–187
Brink, Alexander, 355
Bristol N2, 397
BWS (Beckwith–Wiedemann syndrome), 448, 451

C

CACTA, 140–142, 372, 373
Caenorhabditis elegans
    epigenetic reprogramming, 6
    formation of germ cell precursors, 2–3
    modes of germ cell specification, 1
    RNAi regulation (*see* RNAi regulation in *C. elegans*)
    RNA silencing pathways in, 364
    sex and X-chromosome-wide repression (*see* dosage compensation in *C. elegans*)
    silencing process in (*see* silencing process in *C. elegans*)
    somatic program repression in, 7
CAF-1 (Chromatin Assembly Factor-1), 203, 205, 274, 275–276
Cahn, M.B., 69
Cahn, R.D., 69
callipyge phenotype
    CLPG mutation identified as a point mutation, 478–480
    CLPG mutation mapped to *DLK1-GTL2*-imprinted domain, 477–478
    ectopic expression of DLK1 protein, 480
    polar overdominance at the *CLPG* locus, 477, 480–481
cancer
    epigenomics and, 448, 449–450
    Ezh2/Eed expression and, 178–180
    functional identification of relevant genes (*see* short hairpin RNAs)
    incidence increase with age, 339, 346
    PcG regulation and, 323–324
    reprogramming of nuclei by nuclear transfer, 24–25
carboxy-terminal domain (CTD), 2
CDGE (common disease genetic and epigenetic hypothesis), 451–454
Cellular Memory Modules (CMMs), 319
cenH element, 422, 423
CENP-A, 238, 389. *See also* centromere function
centromere function
    heterochromatin's role at centromeres, 393
    outer repeat heterochromatin, 391–392
    propagation of functional centromere state, 390, 391f
    retrotransposons and centromere repeats, 392–393
    siRNA's effect on chromatin modification, 392

Page numbers followed by f indicate a figure and by t indicate a table.

SUBJECT INDEX

centromere function (*continued*)
  variable establishment of, 389–390
cerebellar granule neuron precursor cells (CGNPs), 323
chicken beta-globin insulator properties study. *See* chromatin boundaries and domains
Chp1, 422–423
chromatin
  accessibility complex (*see* Chromatin Accessibility Complex (CHRAC))
  boundaries and domains (*see* chromatin boundaries and domains)
  current research/discovery, 512, 513–514
  described, 201
  epigenetic mechanisms and, 11–12
  functions of, 281
  histone modifications and, 165
  Hox gene regulation (*see Hox* gene regulation)
  mark propagation, 512
  marks in *Arabidopsis thaliana*, 215
  noncovalent modifications (*see* chromatin noncovalent modification)
Chromatin Accessibility Complex (CHRAC)
  ATP-dependent nucleosome remodeling activities, 281–282
  catalysis of nucleosome sliding, 282–283
  chromatin function, 281
  mechanistic considerations, 283–284
  molecular anatomy, 283
  nucleosome remodeling function, 285–286
  regulation of activity, 284–285
  regulation of ISWI activity by subunits, 284
Chromatin Assembly Factor-1 (CAF-1), 203, 205, 274
chromatin boundaries and domains
  condensed chromatin region and, 248
  enhancer-blocking insulator activity
    biological role of insulators, 246, 247f
    CTCF identification, 245–246
    enhancer mechanisms, 246–247
    factors leading to interactions, 246
  functions of, 247
  histone modifications and, 248
  insulator properties, 245
Chromatin Conformation Capture (3C), 31, 320
chromatin noncovalent modification
  ATPase domain identity related to remodeling capability, 188–190
  ATPase domain role in defining remodeling action, 190–191
  behavior of SWI/SNG and ISWI family complexes, 184–185
  BRG1 vs. SNF2h ability to create accessible DNA sites, 187–188
  mechanisms involved in remodeling reaction, 190–191
  range of chromatin structure, 183
  remodeled products produced by BRG1 and SNF2h, 185–187
  role in the genome, 183–184
  structural alterations during remodeling, 184
  SWI/SFN formation of stably remodeled structures, 185
chromodomain adaptor in histone lysine methylation, 211
chromosome 4 *R*-gene cluster, 156–157
chromosome looping, 31–32
CID, 238
Circe effect, 335
citrullination, 162, 165
cloning, nuclear. *See* nuclear transfer
cloning and epigenetic information transfer, 21–22, 162
CLPG. *See* callipyge phenotype
CLR4, 171
CMMs (Cellular Memory Modules), 319
CMT3 DMTase, 150
CNG methylation, 150
Coffin–Lowry Syndrome (CLS), 220
Cold Spring Harbor Symposium on Quantitative Biology, 68
Columbia (Col), 146
common disease genetic and epigenetic hypothesis (CDGE), 451–454
condensin complex, 72, 73f
cosuppression in *C. elegans*, 397. *See also* silencing process in *C. elegans*
CPEB (cytoplasmic polyadenylation element binding), 13, 497–498
CpG, 42, 43, 114, 447
crippled growth in *Podospora anserina*, 494–495
CTCF, 30, 31
  identification, 245–246
  interaction with imprinting control region, 47–48
  methylation of DMRs and, 40–41
  Su(Hw) compared to, 52
CTD (carboxy-terminal domain), 2
CULLIN3, 322
CUP-SHAPED COTYLEDON1 (CUC1), 412
cytoplasmic polyadenylation element binding (CPEB), 13, 497–498
cytosine methylation in *Arabidopsis thaliana*, 157, 158

D

daf-2, 429
Dam identification (DamID), 320
ddm1
  DNA hypomethylation mutation, 139, 140
  induced alterations in *Arabidopsis thaliana*, 156–157
de novo DNA methyltransferases
  comprehensive view of role, 128–129
  imprinting of autosomal genes, parental line, 125–127
  imprinting of *Mash2*, maternal line, 127
  initiation of imprinted X-chromosome inactivation, 127–128
  initiation of random X-chromosome inactivation, 127
deubiquitylating enzymes (DUBs), 291

DHS (DNaseI hypersensitive site), 84
Dicer, 371, 391, 403
Dicer1, 14
Dicer-like (DCL) proteins, 148–149
Dicer-Slicer pathway, 363
differentially methylated region (DMR), 30
DLK1-GTL2-imprinted domain, 477–478
DNA
  demethylation, 12
  epigenetic mechanisms and, 11–12
  methylation (*see* DNA methylation)
  methyltransferases in germ cells
    Dnmt1, 382
    Dnmt2, 382–383
    Dnmt3A and Dnmt3B, 383
    lack of inherent sequence specificity, 381
  postmitotic neurons from cloning, 23
  replication of (*see* histone deposition proteins)
DNA methylation
  Air noncoding RNA and, 57–58
  contribution to gene repression (*see* gene repression)
  development control in *Arabidopsis thaliana*
    ddm1-induced developmental abnormalities, 140
    developmental abnormalities due to mutations, 139
    epigenetic variation study (*see* epigenetic variation)
    mobilization of *CACTA* due to loss, 140–142
    transposon distribution within the genome, 142
  epigenetic differences during cloning and, 21–22
  genome defense in *Neurospora* (*see* genome defense and DNA methylation in *Neurospora*)
  histone methylation and, 163, 164
  importance in imprinting, 39
  methyltransferases role (*see* de novo DNA methyltransferases)
  PAI genes and (*see PAI* genes and DNA methylation)
  signal reading (*see* DNA methylation signal reading)
  transposon activity regulation, 465–466
DNA methylation signal reading
  biological consequences of DNA methylation, 114
  MeCP and disease, 117
  MeCP2 target genes search, 115–117
  models for interpretation of histone modifications, 114
  potential for gene marking, 113
  transcription repression by CpG, 114
  transcription repression mediation by MBD proteins, 114–115
DNaseI hypersensitive site (DHS), 84
DNMT1, 449
Dnmt family, 39, 47
  germ cells and (*see* germ cells, mammalian)
  imprinting of autosomal genes and, 125
  methyltransferases in germ cells and, 382–383
DOH1 gene, 502–503

SUBJECT INDEX

domains-rearranged methytransferase (DRM), 150
dosage compensation in *C. elegans*
  machinery composition, 72–73
  molecular identification of X-recognition elements, 77–78
  process of, 71–72
  recruitment and spreading along X-recognition elements, 75–77
  recruitment for gene-specific vs. chromosome-wide repression, 75
  sex-specific targeting of the complex, 74–75
dosage compensation in *Drosophila*. *See* Male-Specific Lethal complex (MSL)
dosage compensation in mice
  discrepancies in classical model, 104–105
  forms of X-chromosome inactivation, 103–104
  nature and purpose of, 103
  relationship between imprinted and random XCI, 104
  silent XP in preimplantation embryos
    de novo inactivation hypothesis, 107
    preinactivation from an evolutionary perspective, 108–109
    preinactivation hypothesis, 105–107
    preinactivation vs. de novo X-inactivation, 107–108, 109f
  summary, 109–110
double-stranded RNA (dsRNA), 350, 363
Down syndrome cell adhesion molecule (Dscam). *See* Dscam isoform
DPY-27, 73
DRM (domains-rearranged methyltransferase), 150
Drosophila melanogaster
  alternative splicing and (*see* Dscam isoform)
  Asf1 isolation, 204
  biochemical complementation of wheat germ RNAi, 405
  chromatin marks in, 215
  dosage compensation in (*see* Male-Specific Lethal complex)
  epigenetic marks and, 163–164
  epigenetic reprogramming, 6
  genome partitioning in, 210
  heterochromatin formation (*see* heterochromatin formation in *Drosophila*)
  histone H3 variants (*see* histone H3 variants and PEV)
  loss of *pgc* in, 3
  modes of germ cell specification, 1
  polycomb silencing mechanisms (*see* Polycomb group (PcG) silencing mechanisms)
  PR-Set7 expression and, 173
  RNAi and transcriptional gene silencing, 433–437
  silencer elements in genetic imprinting, 48
  somatic program repression in, 1–3, 7
Dscam isoform
  alternative splicing and, 485
  cellular uses of Dscam diversity, 486
  disposition of splice forms, 485–486
  self- vs. non-self-discrimination as a general principle, 487–488
  self- vs. non-self-discrimination at cell level, 486–487
  single-cell analyses of alternative splicing, 486
dsRNA (double-stranded RNA), 350, 363, 514–515
DUBs (deubiquitylating enzymes), 291

E

Edman degradation analysis, 240
Eed/Ezh2. *See* Ezh2/Eed complex
EGL (external granular layer), 323
EPA-1, 328
Ephrussi, Boris, 68, 69
epigenetic mechanisms in early mammalian development
  cytoplasmic localization and polarity, 14–16
  DNA, chromatin, and imprinting, 11–12
  mRNA translational control, 12–13, 14t
  retrotransposons and microRNA, 14
  transcription control, 12
epigenetics
  alteration characteristics, 447
  application of the term, 67–68
  brain development and (*see* brain development)
  covalent histone modifications and (*see* histone modifications)
  current research/discovery overview
    chromatin mark propagation, 512
    dsRNA, 514–515
    dynamic silent chromatin, 513–514
    genome protection, 515
    histone code hypothesis, 513
    histone modifications, 512–513
    HOTHEAD/hothead, 515
    nuclear organization, 514
    prions, 515
  definition, 67, 507
  future outlook, 516
  genetic vs. epigenetic, 68, 69
  historical overview of past symposia, 507–512
  implications for human health, 516
  inheritance mechanism, 447, 507
  Lederberg's concept of epinucleic information, 69
  mammalian development and (*see* epigenetic mechanisms in early mammalian development)
  Nanney's use of the term, 68
  origins of term, 67, 507
  regulation of vernalization requirement (*see* flowering and epigenetic regulation)
  reprogramming regulation (*see* germ cell specification; nuclear transfer)
  tumorigenesis and, 166–167
epigenetic variation
  among natural accessions, 157–158
  chromosome 4 *R*-gene cluster and, 156–157
  cytosine methylation and, 157, 158
  ddm1-induced alterations, 156–157
  inherited, 158
  view of an evolutionary change role, 155
epigenomics
  cancer and, 448, 449–450
  CDGE elements, 451–454
  classes of cytosine DNA methylation in the genome, 447
  defined, 448
  epigenetic alteration characteristics, 447
  epigenetic inheritance mechanism, 447
  genomic imprinting, 447–448
  rare hereditary disorders and, 450–451
ESC-E(Z) complex, 314
ESC protein, 175
ethyl methane sulfonate (EMS)-induced mutations, 150
Evx1, 4
external granular layer (EGL), 323
extrachromosomal rDNA circles (ERCs), 339
E(Z)/ESC complex, 302, 305
Ezh2/Eed complex, 6, 7, 34
  cancer and, 178–180
  histone lysine methylation and, 213
  inheritance of repressive marks in histones
    associated gene repression, 174
    identification of PRC-regulated genes, 177–178
    mouse model for prostate cancer, 180, 181f
    overexpression in cancer cells, 178–180
    PRC2 and PRC3 analysis, 176
    PRC2 protein complex analysis, 175–176
    PRC4 identification, 177
    proteins' importance in eliciting activity, 174–175
    SirT1 association with PRC4, 177
  purification and characterization of, 310–311

F

Facilitates Chromatin Transcription/Transactions (FACT), 204–205
FIS-class genes in *Arabidopsis*, 467–468
fission yeast. *See Schizosaccharomyces pombe*
FLC
  FLC:luciferase fusion used to select for mutants, 460–462
  histone modifications associated with *FLC* silencing, 462
  role in vernalization process, 458
  vrn mutants maintenance of repressed *FLC* state, 458
FLO genes, 328
flowering and epigenetic regulation
  FLC:luciferase fusion used to select for mutants, 460–462
  FLC role in vernalization process, 458
  histone modifications associated with *FLC* silencing, 462
  histone modifications changes in *vrn/vin* mutants, 462–463
  integration of flowering pathways, 463

flowering and epigenetic regulation (*continued*)
    vernalization requirement in plants, 457
    VIN3/VRN7 expression induction by cold, 463
    VRN genes and chromatin remodeling, 458–460
    vrn mutants maintenance of repressed *FLC* state, 458
fragilis, 3, 5
FRIGIDA (FRI), 458
FWA
    gene, 374
    inheritability in *Arabidopsis thaliana*, 139–140
    locus, 468, 469–470

G

G9a, 5
GAGA factor, 303, 314
gametogenesis, 20, 161
gcl (germ cell-less), 2
Gcn5, 171
gene repression
    correlation between expression patterns and gene structure, 133
    effect of DNA methylation on chromatin structure, 131–132
    effect of removing DNA methylation, 136
    gene silencing due to DNA modification, 133
    Group I–IV tissue-specific genes and, 133–134
    histone code hypothesis, 136–137, 513
    by late replication, 135
    replication timing in S phase, 134–135
    stepwise gene activation, 137
genetic instability in aging yeast
    asymmetry of age-induced LOH, 343
    break-induced replication and LOH, 342–343
    DNA damage checkpoint and, 343–344
    heritability of hyperrecombinational state, 344–345, 346
    inheriting of LOH instability, 340–341
    LOH analysis, 340
    proposed cause of yeast aging, 339
    segregation bias in broken chromosome, 344
    senescence factor, 345–346
    two-state switch and, 341–342
genome defense and DNA methylation in *Neurospora*
    control of DNA methylation, 121
    de novo and maintenance methylation, 120
    forward genetics approach, 121
    investigation of HP1 and RNAi involvement, 121–122
    methylation characterization, 119
    purpose of control of DNA methylation, 122
    repeat-induced point mutation function, 119–120
    specificity of methylation, 120–121
    study areas, 122
genomic imprinting
    Ctcf binding during oogenesis and, 43–44
    described, 447–448
    DMDs division into maternally or paternally hypermethylated domains, 39
    DNA methylation and (*see* de novo DNA methyltransferases)
    DNA methylation's importance, 39
    establishment in mammalian germ cells (*see* germ cells, mammalian)
    at *FWA* locus, 468
    H19 and *Igf2* and, 39–40
    H19 DMD maternal identity establishment, 41–42
    H19 DMD paternal identity establishment, 42–43
    methylation of DMRs in gametes, 40–41
    models for imprint establishment, 41
    in plants, 467
    possibility that hypermethylated state is actively determined, 43
    role in mammalian development, 125
    silencer elements, 48
    similarities with X-inactivation, 63–64
    transposons and, 466–467
    variety of mechanisms DMRs employ, 44
genomics, 448
germ cell-less (gcl), 2
germ cells, mammalian
    DNA methyltransferases, 382–383
        lack of inherent sequence specificity, 381
    Dnmt3L functions in oogenesis, 382f, 383
    Dnmt3L functions in spermatogenesis, 384–385
    sexual dimorphism in imprinting, 385–386
    study of biological effect of imprinting, 386
germ cell specification
    epigenetic reprogramming, 6
    Ezh2/Eed complex role, 6
    maternal inheritance of epigenetic modifiers, 6–7
    modes of, 1
    primordial germ cells, 1
    regulation by Blimp-1, 5–6
    reverting to embryonic germ cells, 7
    somatic program repression in mice
        acquisition of germ cell competence, 3–4
        process of specification of PGCs, 3, 7
    somatic program repression in nonmammals, 1–3, 7
    summary of PGC specification events, 4–5

H

Haig, D., 67
handedness. *See* brain development
Harris, M., 69
HATs (histone acetyltransferases), 193
HDACs (histone deacetylases), 193, 420
HD-AIPIII (homeodomain-leucine zipper), 409. *See also* leaf polarity
hemisphere, brain, 499–500
her-1, 75
hereditary disorders and epigenomics, 450–451
Herring, S.W., 68
[Het-s] in *Podospora anserina*, 493–494
heterochromatin and centromere function. *See* centromere function
heterochromatin assembly in *S. pombe*
    assembly factors, 419–420
    boundary determinants of heterochromatin, 424–425
    heterochromatic regions, 419
    histone code and, 421–422
    propagation and inheritance of structures, 423–424
    RNAi-independent formation, 423
    RNAi-mediated formation, 422–423
heterochromatin formation in *Drosophila*
    chromosome 4 characteristics, 267
    heterochromatin's ability to spread in *cis*, 267
    link between gene silencing and nuclear location, 269
    maintenance and spread, 267–268
    RNAi machinery dependencies, 270
    RNAi targeting, 270–271
    support for a model of interspersed domains, 268–269
Heterochromatin Protein 1 (HP1), 267
heterokaryon incompatibility, 493
high performance liquid chromatography (HPLC), 240
HIRA, 275
HIR proteins, 204
histone acetyltransferases (HATs), 193
histone deacetylases (HDACs), 193, 420
histone deposition proteins
    active coupling of deposition and DNA replication, 205
    chaperone function during DNA replication, 203–204
    deposition during DNA synthesis, 202–203
    DNA replication fundamentals, 201–202
    functional overlap in, 204–205
    nucleosome architecture and assembly, 201
histone H2A variants
    described, 227–228
    evolutionary targets, 231–232, 233f
    histone features that allow variants, 228–229
    localization patterns, 228
    structural characteristics
        H2A.Bbd, 231
        H2A.X, 229
        H2A.Z, 229–230
        macroH2A, 230–231
histone H2B ubiquitylation
    deubiquitylation by Ubp8 and transcription, 291–293
    mechanisms linking methylation, 291
    roles, sites, and cross-talk, 290–291
    roles of ubiquitylation and deubiquitylation, 295–296
    silencing and Ubp10, 293–294
    ubiquitylation described, 289–290
    Ubp8 vs. Ubp10 processes, 295
histone H3, 248, 249f
histone H3-K9 (histone H3/lysine 9), 11. *See also* histone lysine methylation states
histone H3-K27

SUBJECT INDEX

methylation states (*see* histone lysine methylation states)
Polycomb group silencing and, 311, 313
histone H3 Lys methylation variations, 91–92, 162, 164
histone H3meK4, 3
histone H3meK9, 5, 239, 372
histone H3meK27, 6
histone H3 phosphorylation/ acetylation coupling mechanism, 221–223
    decreasing of binding affinity of transcriptional repressors, 224–225
    evidence for coupled marks, 220–221
    evidence of a transcription role, 219–220
    modulation of chromatin structure, 223–224
histone H3 variants and PEV
    activation dependency on transcription factor abundance, 236–237
    alternative nucleosome assembly pathways, 238
    H3.3 enrichment in active lysine modifications, 240–241
    H3.3 replacement of H3K9me at activated loci, 239
    heterochromatin during *Drosophila* development, 236
    inheritance of chromatin states, 237
    PEV described, 235
    RI nucleosome assembly and chromatin inheritance, 241
    substrate requirements for alternate pathways, 238–239
histone H3.1 and H3.3 complexes
    de novo deposition, 276–277
    deposition through distinct nucleosomal assembly pathways, 276, 277f
    existence as a dimer, 275
    HIRA- and CAF-1-dependent assembly pathways, 275–276
    nucleosome deposition pathways, 273–275
    potential sources of dimers, 277–279
histone H4-K16 acetylation in yeast
    acetylation and transcriptional activity, 195–196
    acetylation/deacetylation as a protein binding switch, 197–198
    BDF1 binding and the H4 tail, 197
    gene clusters with signature patterns, 196
    global nature of, 194
    nonredundancy of sites, 193
    patterns of, 194–195
    regulation during gene activity, 196–197
    telomeric heterochromatin and, 193–194
histone H4-K20 methylation states. *See* histone lysine methylation states; PR-SET7 catalysis of H4 lysine-20 methylation
histone *H19* and *Igf2*
    Ctcf binding during oogenesis and, 43–44
    genomic imprinting and, 39–40
histone H19 DMR, 448
histone *H19* gene, 246

histone *H19* imprinting control region, 385, 386
    experimental procedures, 48
    methylation-regulated insulator, 47–48
    silencer elements and, 48
    Su(Hw) as a suppressor of silencing, 49
    Su(Hw) binding of silencer in vitro, 50–51
    Su(Hw) binding of transgene in vivo, 49–50
    Su(Hw)-binding region deletion results, 51
    Su(Hw)-binding site role as chromatin insulator, 52
    Su(Hw) compared to CTCF, 52
    Su(Hw) role in gene silencing, 52–53
histone lysine methylation states
    areas for study, 216
    epigenetic control of eukaryotic genomes, 210–211
    epigenetic mechanisms overview, 209
    evolutionary conservation of, 214–215
    functional roles of methylated positions, 211–212
    genome vs. epigenomes, 209f
    HMTase networks indexing function, 213
    in mouse vs. *S. pombe*, 214
    pattern contribution to distinct subdomains, 212–213
    positions in the H3 and H4 tails, 210
    roles of the distinct marks, 216
    systems components, 211
histone lysine methyltransferase (HKMT), 171
histone modifications
    Arg methylation marks, 165
    associated with *FLC* silencing, 462
    changes in *vrn/vin* mutants, 462–463
    chromatin boundaries and domains and, 248
    cross-talk of DNA and Lys methylation, 162–163
    current research/discovery overview, 512–513
    differentiation and a chromatin state, 165
    epigenetic marks and establishing the mark, 163–164
        inheritance of (*see* inheritance of repressive methyl-lysine marks in histones)
        maintaining the mark, 164
        resetting in germ cells and embryos, 164
    epigenetic mechanisms underlying differentiation, 161–162
    epigenetic tumorigenesis and, 166–167
    H2B ubiquitylation (*see* histone H2B ubiquitylation)
    heterochromatin assembly and (*see* heterochromatin assembly in *S. pombe*)
    methylation of Lys and Arg, 162, 163f
    role in epigenetic information mechanisms, 162
    transposon activity regulation, 465–466
    types of, 162
HKMT (histone lysine methyltransferase), 171
HMGB1, 285
HML and *HMR*, 328–330, 332–333

Holliday, R., 69
homeobox genes, 4
homeodomain-leucine zipper (HD-AIPIII). *See also* leaf polarity
HOTHEAD/hothead, 515
Hox gene regulation, 5
    activation ex vivo, 251, 252f
    chromatin decondensation and histone acetylation, 254
    chromatin decondensation at *HoxB*, 253
    chromatin decondensation correlation with altered nuclear organization, 253–254
    chromatin decondensation trigger mechanism, 255–256
    expression pattern models, 254–255
    HoxB temporal program of gene expression, 252–253
HP1 (Heterochromatin Protein 1), 267
hPC2, 322
HPLC (high performance liquid chromatography), 240
HSE *(Xenopus* egg extract system), 275
hsp-10-white, 268
hsp26 promoter, 306
Hsp104, 492–493
Hunchback, 301–302
Huxley, J., 68
hydroxyurea (HU), 205

I

IAP elements, 384, 385
IC1 domain. *See IGF2-H19* locus
IC2 domain, 33–35
ICF syndrome, 450
Iftm gene, 3
IGF2-H19 locus (IC1 domain)
    chromosome loop model test, 31–32
    epigenetic switch observation, 32
    models for imprinting, 30–31
    structure of, 30
Igf2r
    Air noncoding RNA and, 56, 58, 60, 61f
    expression of, 246
imaginal disc enhancers, 301–302
imprinted genes
    clustering of, 56
    epigenetic mechanisms and, 12
    features of mammalian, 55, 56t
    genomic (*see* genomic imprinting)
    silencing in *Arabidopsis* (*see* silencing of imprinted genes in *Arabidopsis*)
imprinting regulation
    clustering of imprinted genes, 29
    H19 imprinting control region (*see* histone *H19* ICR)
    IGF2-H19 locus
        chromosome loop model test, 31–32
        epigenetic switch observation, 32
        models for imprinting, 30–31
        structure of, 30
    Kcnq1ot1 locus
        DMRs related to imprinting, 33
        histone methylation imprint and, 33–35
        insulator-loop model, 35
        noncoding RNA model, 35
        structure of, 33
    sharing of regulatory elements, 29

INCENP, 230
infectious proteins. *See* prions
inheritance of chromatin states. *See*
    histone H3 variants and
    PEV
inheritance of repressive methyl-lysine
    marks in histones
  chromatin structures and histone tails,
    171
  connection between chromatin and
    gene expression regulation,
    171
  Ezh2/Eed complex role
    associated gene repression, 174
    identification of PRC-regulated
      genes, 177–178
    mouse model for prostate cancer,
      180, 181f
    overexpression in cancer cells,
      178–180
    PRC2 and PRC3 analysis, 176
    PRC2 protein complex analysis,
      175–176
    PRC4 identification, 177
    proteins importance in eliciting
      activity, 174–175
    SirT1 association with PRC4, 177
  PR-SET7 catalysis of H4 lysine-20
    methylation
    basis for propagation of mark
      through cell divisions, 173,
      174f
    evidence of a repressive marker,
      172
    functional role of PR-Set7, 173
    mediation of H4-K20 di- and
      trimethylation, 172–173
    PEV suppression, 173
    PR-Set7 expression during mitosis,
      172
  stability of lysine-methyl bond, 171
Ink4a/Arf tumor-suppressor locus, 323,
    324
INK4A tumor suppressor gene, 448
insulator-loop model for imprinting, 35
insulators, 31
insulators and chromatin
  biological role of insulators, 246, 247f
  CTCF identification, 245–246
  enhancer mechanisms, 246–247
  factors leading to interactions, 246
  insulator properties, 245
insulin-like growth factor 2 gene (*Igf2*),
    30. *See also* IGF2-H19
    locus
ISWI-family complexes, 183, 184–185,
    198, 284

J

JJAZ1 protein, 175

K

K9/K27 histone, 34
Kaiso, 115, 116
Kcnq1ot1 locus (IC2 domain), 33–35
KRYPTONITE, 163

L

L1 repeat elements (LINEs), 90, 384, 385
lactacystin, 13
leafbladeless1 (lbl1), 410

leaf polarity
  establishment in developing leaves,
    409–410
  evolution of polarity pathway,
    414–415
  miR166 adaxial/abaxial axis
    specification, 413–414
  miRNA biogenesis and function,
    411–412
  miRNAs as developmental signaling
    molecules, 412
  regulation of *hd-zipIII* gene
    expression by miR165/166,
    414
  specification of adaxial/abaxial
    polarity, 410–411
Lederberg, J., 69
leukemia, 323
LINEs (L1 repeat elements), 90, 384, 385
LOI (loss of imprinting), 448
long-term memory, 497–498
loss of heterozygosity (LOH), 340. *See*
    also genetic instability in
    aging yeast
Lys residues methylation, 162, 163f. *See*
    also histone lysine
    methylation states

M

macroH2A, 230–231, 322
maize
  b1 paramutation in (*see b1*
    paramutation)
  specification of adaxial/abaxial
    polarity, 410–411
Male-Specific Lethal complex (MSL), 164
  bases for redundancy of *roX1* and
    roX2, 85–86
  binding and spreading model, 85
  cis-acting elements, 86–87
  local spreading from *roX* genes, 83–84
  preference for X-chromosome
    targeting, 81–83
  targeting through high-affinity sites on
    X, 84
  twofold upregulation of gene
    expression, 86
Markert, C.L., 69
marks, epigenetic
  chromatin mark propagation, 512
  histone deposition and, 202–203
  histone modifications and, 163–164
  inheritance of (*see* inheritance of
    repressive methyl-lysine
    marks in histones)
  PR-SET and, 172, 173, 174f
Mash2, 127
mat locus, 423, 501
McClintock, Barbara, 349
MEDEA gene, 467–468, 469–472
meiotic sex chromosome inactivation
    (MSCI), 107
memory, long-term, 497–498
merotelic orientation, 393
MES 2/3/6 complex, 6
metastable epialleles, 466
methylamine, 165
Methyl-CpG-binding Protein (MeCP),
    114, 120. *See also* DNA
    methylation signal reading
mice/mouse
  Air noncoding RNA, 59–60
  cloning and, 23–24

  dosage compensation in (*see* dosage
    compensation in mice)
  genome partitioning in, 210
  germ cell specification (*see* germ cell
    specification)
  histone lysine methylation states, 214
  prostate cancer model, 180, 181f
  somatic program repression in, 3–4
microR166, 413–414, 415
microR-JAW, 412
microRNAs (miRNAs), 14, 363
  biogenesis and function, 411–412
  as developmental signaling molecules,
    412
MIX-1, 72–73
Mks1p, 493
MMU11 chromosome territory, 253
mnDp1, 76
Mod(mdg4), 52
Moller, F., 69
mop1/mop2/mop3 alleles, 357–358
morphogen-gradient model, 499
mouse. *See* mice/mouse
mRNA translational control, 12–13, 14t
MSCI (meiotic sex chromosome
    inactivation), 107
MSL. *See* Male-Specific Lethal complex
    (MSL)
multiple microtubule (MT) binding, 393
Mus musculus. See mice/mouse
mutator genes (*mut*), 397
mut/rde/cde gene complex, 397, 399–400.
    See also silencing process
    in *C. elegans*
Myosin-like proteins (Mlp), 327, 329f. *See*
    also telomere clustering in
    yeast
MYST family, 261–262

N

Nanney, D.L., 68
nanos, 2, 6
ncRNA. *See* noncoding RNA genes
NE (nuclear envelope), 327
neurexins, 487
Neurospora crassa
  chromatin marks in, 215
  genome defense in (*see* genome
    defense and DNA
    methylation in *Neurospora*)
  RNA silencing pathways and, 364–365
neviparine, 14
Nicotiana benthamiana, 363
noncoding RNA genes (ncRNA), 478
  features of mammalian, 55–56
  imprinting model, 35
  mouse chromosome 17 (*see Air*
    noncoding RNA)
non-right handers (NRH), 499. *See also*
    brain development
NORs (nucleolus organizer regions),
    157–158
nuclear envelope (NE), 327
nuclear transfer
  biological barriers to safe cloning, 21
  cloned animals from terminally
    differentiated cells
    obtaining cloned mice from
      olfactory neurons, 23–24
    obtaining monoclonal mice from
      immune cells, 23
    potential of somatic vs. ES cell
      use, 22–23

current research/discovery, 514
epigenetic reprogramming
  during cleavage, 20
  during gametogenesis, 20
  postimplantation, 20–21
  postnatal, 21
  problems with inadequate reprogramming, 19–20
question of normalcy of clones, 21
reprogramming of cancer nuclei, 24–25
therapeutic cloning outlook, 25–26
totipotency of neuronal nuclei, 24
nucleolus organizer regions (NORs), 157–158
Nucleosome Remodeling Factor (NURF), 281, 283

O

Oct-4-like genes, 20, 21, 25
Okazaki fragment, 202
olfactory neurons and cloning, 23–24
Origin Recognition Complex (ORC), 261, 262

P

p53 tumor suppressor gene, 290, 440–441
PAD (peptidylarginine deiminase), 165
Paf1 complex, 291
PAI genes and DNA methylation
  evidence supporting a *PAI*-specific methylation signal, 147
  heterochromatin and, 145
  non-CG methylation maintenance, 150–151
  normal gene arrangement, 146
  PAI methylation establishment, 151
  transcription of methylated *PAI* genes, 147–149
  WS-*PAI* arrangement, 146–147
  WS-*PAI* genes as reporters, 149–150
paramutation
  aberrant chromatin organization, 350
  adaptation domain existence possibility, 351–352
  adaptive challenge connected to a locus, 349
  b1 locus in maize (*see b1* paramutation)
  description and functioning, 350
  mutation vs., 355
  natural hybridization and adaptive challenge, 350
  objections to idea of an adaptive genome, 352
  paramutable and paramutagenic alleles, 355
  potential consequences of, 360
  role of, 352–353
  role of TEs in directed genome restructuring, 350–351
Pax5, 5
pBluescript, 50
PcG (Polycomb group), 6, 174, 183
PCNA (Proliferating-Cell Nuclear Antigen), 202, 205
peptidylarginine deiminase (PAD), 165
petunia plants and RNA silencing. *See* RNA silencing pathways in plants
PEV (position effect variegation), 173, 174, 235, 267

pgc (*polar granule component*), 2
PGCs (primordial germ cells), 1, 3, 4–5, 7
phb, 409, 410
PHD (plant homeodomain), 459
PHO (Pleiohomeotic), 303, 314
Pho-like (Phol), 314
phosphorylation of histone H3. *See* histone H3 phosphorylation/acetylation
PIE-1, 3
PIN (prostate intraepithelial neoplasia), 180, 181f
[PIN+], 493
PINHEAD/ZWILLE (PNH/ZLL), 412
Pipsqueak, 303, 314
plant homeodomain (PHD), 459
plant RNAi in vitro
  biochemical complementation of, 405
  failure of ds siRNAs to trigger target cleavage, 403–405, 406–407
  materials and methods, 407
  ovary lysate rescuing of wheat defect, 405–406
  wheat germ extract as a study tool, 403
Pleiohomeotic (PHO), 303, 314
Podospora anserina
  crippled growth in, 494–495
  [Het-s] in, 493–494
polar granule component (*pgc*) gene, 2
polar overdominance. *See* callipyge phenotype
pole cells, 2, 3
Pol IV/RDR2/DCL3, 366–367
Polycomb (PcG) genes, 6
Polycomb group (PcG), 174, 183
  b1 paramutation and, 359
  histone lysine methylation and, 211–212, 213
  proteins, 92–93
Polycomb group (PcG) and trithorax group (trxG)
  connections with stem cells and cancer formation, 323–324
  enzymatic activity of, 321
  interaction with CIS elements concerning gene activity, 319
  mammalian PREs/CMMs search, 320–321
  PRC2 complex and, 319
  silencing models, 319–320
  X-inactivation and multilayer silencing, 321–323
Polycomb group (PcG) silencing mechanisms
  binding specificity of PC chromodomain to methyl-K27, 311–313
  conservation of PRE sequences, 303–304
  DNA-binding domain targeting of complexes, 304–305
  in *Drosophila* and mammals, 309–310
  evolutionary conservation of, 315
  GAGA factor, 303
  H3-K27 methylation contribution to recruitment, 313
  H3-K27 methylation requirement and *Ubx* gene silencing, 311
  hierarchical recruitment of complexes, 314–315

histone methylation role in the repressed *Ubx* gene, 305–306
multiple DNA-binding recruiters, 303
PHO factor, 303
PRE maintenance of the epigenetic state, 301–302
protein complex assembly, 302
purification and characterization of EED-EZH2 HMTase complex, 310–311
recruiter and effector role, 315–316
recruitment evidence, 302–303
relationship to HMTase activity, 311
silencing mechanisms and condensation, 306–307
trxG and transcription activity, 309
X-inactivation and (*see* Polycomb group (PcG) and trithorax group (trxG))
Polycomb Repressive Complex 2 (PRC2), 319
Polycomb Response Element (PRE), 301–302, 311, 434
polymerase (Pol alpha and delta), 202
position effect variegation (PEV), 173, 174, 235, 267
posttranscriptional gene silencing (PTGS), 403
posttranscriptional silencing. *See* RNA silencing pathways in plants
ppw-1 involvement in silencing, 400
Prader–Willi syndrome, 451
PRC2 (Polycomb Repressive Complex 2), 319
PRE (Polycomb Response Element), 301–302, 311, 434
preformation, 1
primordial germ cells (PGCs), 1, 3, 4–5, 7
prions
  crippled growth in *Podospora anserina*, 494–495
  current research/discovery, 515
  genetic criteria for, 490
  [Het-s] in *Podospora anserina*, 493–494
  neuronal CPEB, 497–498
  notion of infectious proteins, 489
  protease B activity, 494
  types of, 489–490
  [URE3] and [PSI+], 490–491
  Ure2p (*see* Ure2p and prions)
Proliferating-Cell Nuclear Antigen (PCNA), 202, 205
prostate intraepithelial neoplasia (PIN), 180, 181f
protocadherins, 487
PR-SET7 catalysis of H4 lysine-20 methylation
  basis for propagation of mark through cell divisions, 173, 174f
  evidence of a repressive marker, 172
  functional role of PR-Set7, 173
  mediation of H4-K20 di- and trimethylation, 172–173
  PEV suppression, 173
  PR-Set7 expression during mitosis, 172
psychoses etiology and handedness, 500–501, 503
PTGS (posttranscriptional gene silencing), 403
pumilio, 2

R

Rad6/Ubc2 E2 conjugase, 290
Rad53, 205
RdDM (RNA-directed DNA methylation), 366
rde (RNAi-deficient gene), 397, 400–401
Rdp1 (RNA-dependent RNA polymerase), 391, 405
RdRP, 371
RDR proteins and RNA silencing, 364–365
repeat-induced point (RIP) mutation, 119–120
replication-independent assembly (RI), 238
required for maintenance of repression (rmr), 357–358
retrotransposons and microRNA, 14
Rett Syndrome, 116, 117, 450
reversible curing, 490
RI (replication-independent assembly), 238
right-handers (RH), 499. *See also* brain development
Ring1B/Rnf2, 322
RIP (repeat-induced point mutation), 119–120
RISC. *See* RNA-induced silencing complex
RITS (RNA-induced initiation of transcriptional gene silencing), 371, 422–423
rmr (required for maintenance of repression), 357–358
RNA-dependent RNA polymerase (Rdp1), 391, 405
RNA-directed DNA methylation (RdDM), 366
RNAi-deficient gene (*rde*), 397
RNAi machinery
  b1 paramutation and, 359
  heterochromatin formation and, 422–423
  involvement in leaf polarity, 410 (*see also* RNA silencing pathways in plants)
  involvement in silencing (*see* RNA silencing pathways in plants; silencing process in *C. elegans*)
  outer repeat heterochromatin and, 391
  regulation in *C. elegans* (*see* RNAi regulation in *C. elegans*)
  transcriptional gene silencing and cosuppression related to pairing-sensitive silencing, 434
  function as a defense against overexpression of TEs, 436
  gene sequences pairing and increase in degree of silencing, 436–437
  mutations affecting pairing-sensitive silencing, 434–436
  w-Adh transgenes expression and, 433
RNA-induced initiation of transcriptional gene silencing (RITS), 371, 422–423

RNA-induced silencing complex (RISC), 392
  in *C. elegans* (*see* silencing process in *C. elegans*)
  miRNA and, 411–412
  in plants (*see* plant RNAi in vitro)
RNA interference pathway. *See* RNAi machinery
RNAi regulation in *C. elegans*
  age-1 mediation of responses, 429
  dauer-regulatory pathways overlap, 429–430
  enhancement by decreased *age-1* signaling, 429
  interaction of insulin and RNAi pathways, 430–431
RNA polymerase II CTD, 2
RNA polymerase II (RNAPII), 291
RNA silencing pathways in plants
  diversity of silencing pathways, 367–368
  evidence for a link between RdDM and, 366–367
  genetic analysis, 364–365
  involvement of dsRNA, 363
  molecular analysis, 363–364
  potential of silencing pathways to interact, 367
  RNA-directed DNA methylation, 366
  susceptibility to viruses and, 365–366
RNA TRAP, 320
Robertson's mutator (*Mu*), 372–373
rolled leaf1 (*rld1*), 410–411
roX genes
  bases for redundancy of *roX1* and *roX2*, 85–86
  MSL complex spreading from, 83–84
  MSL complex targeting through high-affinity sites on X and, 84
RPA, 201–202
Rpd3, 171
Rubinstein–Taybi syndrome, 451

S

Saccharomyces cerevisiae
  chromatin marks in, 214
  DNA replication and, 203
  genetic instability in aging (*see* genetic instability in aging yeast)
  genome partitioning in, 210, 211
  histone acetylation study (*see* histone H4-K16 acetylation in yeast)
  histone ubiquitylation (*see* histone H2B ubiquitylation)
  SIR1 and transcription silencing
    epigenetic function, 263
    HP1 as an analog of *SIR1*, 262–263
    mutant alleles and, 259–260
    mutant enhancers and chromatin connections, 261–262
    SIR1 mechanical roles, 261
    suppression of recombination, 260–261
    telomeric position effects, 260
  telomere clustering in (*see* telomere clustering in yeast)

yeast prions (*see* prions)
SAGA complex, 222, 290, 292–293
SAGA/Gcn5-dependent genes, 291–293
SAM (shoot apical meristem), 409. *See also* leaf polarity
SANT domain, 284
schizophrenia, 500
Schizosaccharomyces pombe
  asymmetric cell division, 501–502
  chromatin marks in, 214–215
  genome partitioning in, 210
  heterochromatin and (*see* heterochromatin assembly in *S. pombe*)
  histone methylation and, 210
SDC-2, 74
sdc genes, 72
Searle, A.G., 68
SEG site, 502–503
senescence factor, 345–346
SET domain, 5, 6, 211, 214. *See also* PR-SET7 catalysis of H4 lysine-20 methylation
sex and X-chromosome-wide repression. *See* dosage compensation in mice
Sgf11, 292–293
sheep. *See* callipyge phenotype
shoot apical meristem (SAM), 409. *See also* leaf polarity
short hairpin RNAs (shRNAs)
  library screening strategies, 440
  screen for bypass of p53 proliferation arrest, 440–441
  siRNA bar code screens, 441
  validation of cell system bar code screening, 444–445
  validation of siRNA bar coding, 441–444
  vector library construction, 439–440
short interfering RNA. *See* siRNAs
short interspersed nucleotide element (SINE), 366
shRNAs. *See* short hairpin RNAs
silencing and RNAi machinery
  cosuppression related to pairing-sensitive silencing, 434
  function as a defense against overexpression of TEs, 436
  gene sequences pairing and increase in degree of silencing, 436–437
  mutations affecting pairing-sensitive silencing, 434–436
  w-Adh transgenes expression and, 433
silencing and X-chromosome. *See* dosage compensation in mice
silencing mechanisms in PcG. *See* Polycomb group (PcG) and trithorax group (trxG)
silencing of imprinted genes in *Arabidopsis*
  correlation between tandem repeats, transposons, and small RNAs, 373–374
  mechanisms for silencing of transposons and transgenes, 375–377
  overview of siRNA and heterochromatin roles, 371

Robertson's mutator transposons, 372–373
role of transposons and tandem repeats, 374–375
silencing pathways in plants. *See* RNA silencing pathways in plants
silencing process in *C. elegans*
    Argonaute gene involvement in, 400–401
    commonality of genes involved, 397
    cosuppression, 399
    identification of involved genes, 399–400
    RNAi mechanism, 398
    Tc1-siRNA involvement, 398–399
silent information regulator (SIR), 194
SINE (short interspersed nucleotide element), 366
SIR1 and transcription silencing
    epigenetic function, 263
    HP1 as an analog of *SIR1*, 262–263
    mutant alleles and, 259–260
    mutant enhancers and chromatin connections, 261–262
    SIR1 mechanical roles, 261
    suppression of recombination, 260–261
    telomeric position effects, 260
siRNAs (short interfering RNA), 363, 365
    effect on chromatin modification, 392
    in plants (*see* RNA silencing pathways in plants)
    silencing in *Arabidopsis* (*see* silencing of imprinted genes in *Arabidopsis*)
    silencing in *C. elegans* (*see* silencing process in *C. elegans*)
Slc22a2/3, 57, 58. *See also Air* noncoding RNA
SLIDE domain, 284
Smad1 and *Smad4*, 4
snail (*Aplysia*), 497
SNF2h
    BRG1 vs. SNF2h ability to create accessible DNA sites, 187–188
    remodeled products produced by, 185–187
Sp1, 41
S phase, 203, 205
SPOP protein, 322
stDp2, 76
stella, 4, 5
strand-specific imprinting and segregation (SSIS) model, 502–504
Su(Hw)
    binding of silencer in vitro, 50–51
    binding of transgene in vivo, 49–50
    binding region deletion results, 51
    binding site role as chromatin insulator, 52
    compared to CTCF, 52
    role in gene silencing, 52–53
    as a suppressor of silencing, 49
sumoylation, 162
Sup35 protein, 490–491. *See also* prions
SUP gene, 140

Suv39H1, 171, 212–213
Su(var) (suppressor of variegation), 389
SUVH4, 150
SUZ12, 11
Swi6, 393, 421. *See also* heterochromatin assembly in *S. pombe*
SWI/SNG. *See* chromatin noncovalent modification
synaptic plasticity, 497–498

T

tandem repeat hypothesis, 471
telomere clustering in yeast
    biological functions of, 327
    Circe effect, 335
    creation of repressive subcompartments, 335
    experimental procedures, 335–336
    genetic diversity through variegation, 327–328
    localization along the nuclear membrane, 327, 328f
    model for chromatin anchoring role in silencing, 334f
    perinuclear anchorage role in repression, 331–334
    perinuclear anchorage types, 330–331
    silent chromatin anchoring and, 328–330
    spontaneous and self-propagating Sir pools, 334–335
TEs (transposable elements), 350–351. *See also* paramutation
therapeutic cloning, 25–26
3C (Chromatin Conformation Capture), 31, 320
transcription control
    mammal gametes, 12
    silencing in yeast (*see Saccharomyces cerevisiae*)
transdifferentiation, 25
transmissible spongiform encephalopathies (TSEs), 489. *See also* prions
transposable elements (TEs), 350–351. *See also* paramutation
transposons and repetitive sequences
    activity regulation, 465–466
    epigenetic attractors and mediators role, 465
    epigenetic regulation of, 466–467
    evidence of a epigenetic regulation attractor role, 472–473
    experimental procedures, 468–469
    FIS-class genes in *Arabidopsis*, 467–468
    genomic imprinting at *FWA* locus, 468
    genomic imprinting in plants, 467
    results
        AtREP2 helitron and imprinting at *MEA* locus, 470–471
        MEA and *FWA* loci features, 469–470
        MEA-ISR and imprinting at *MEA* locus, 471–472
        test for regions required for imprinting, 471–472
transposon silencing in *C. elegans*, 397. *See also* silencing process in *C. elegans*
Trichostatin A, 248

trithorax-group (trxG), 183. *See also* Polycomb group (PcG) and trithorax group (trxG)
Trithorax system, 302
TSEs (transmissible spongiform encephalopathies), 489. *See also* prions
tumorigenesis and epigenetics, 166–167. *See also* cancer

U

UASGFP gene, 236
ub carboxy-terminal hydrolases (UCHs), 291
ubiquitylation. *See* histone H2B ubiquitylation
Ubp8-medicate H2B deubiquitylation, 291–293
Ubp10, 293–294, 295
ub-specific proteases (UBPs), 291–293
UCHs (ub carboxy-terminal hydrolases), 291
Ultrabithorax (*Ubx*) gene, 301, 305–306, 311, 314
Ure2p and prions
    amino acid composition and formation of, 492
    amino terminus of, 491
    amyloid filament architecture formed from, 491–492
    chaperones involved in [URE3] propagation, 492–493
    genetic control of [URE3] generation, 493
    homologs in pathogenic yeasts, 493
    [URE3] as a self-propagating amyloid form of, 491

V

Vernalization5/VIN3-like (VEL), 460
vernalization requirement. *See* flowering and epigenetic regulation
vrn mutants, 458–460

W

WAC homology, 283
Waddington, C.H., 67, 68
Wassileqskija (WS), 146. *See also PAI* genes and DNA methylation
wheat germ extract. *See* plant RNAi in vitro
Wilms tumor, 448

X

X-chromosome inactivation (XCI)
    asynchronous replication timing, 93
    b1 paramutation and, 360
    DNA methylation and (*see* de novo DNA methyltransferases)
    dosage compensation and (*see* dosage compensation in mice)
    evolutionary considerations, 93–95
    inactive state spreading via Xist RNA, 91–92
    initiation and the Xic, 89–91
    kinetics of, 91f, 96–97

X-chromosome inactivation (XCI) (*continued*)
    macro-H2A.1 and, 93
    maintenance of inactive site, 92–93
    nature of the imprints, 95–96
    nuclear organization context, 98, 99f
    polycomb group silencing role (*see* Polycomb group (PcG) and trithorax group (trxG))
    relative importance of epigenetic marks, 93
    reprogramming of inactive X chromosome, 97–98
    role in mammalian development, 125
    similarities with genomic imprinting, 63–64
X-chromosome recognition elements, 75–78
Xenopus, 204, 275
Xenopus egg extract system (HSE), 275
Xist (X-inactive-specific-transcript), 89, 127, 321–323
Xist noncoding RNA, 63–64
xol-1, 74–75

Y

yeast. See *Saccharomyces cerevisiae*; *Schizosaccharomyces pombe*
yeast prions. *See* prions
yKu (end-binding complex), 327. *See also* telomere clustering in yeast

Z

Zeste, 303

WITHDRAWN